Student Organizer

Course Information

Course Number: _____ Name: _____

Location: _____ Days/Time: _____

Contact Information

Contact	Name	Email	Phone	Office Hours	Location
Instructor					
Tutor					
Math Lab					
Classmate					
Classmate					

Semester Organizer

Week	Homework	Quizzes and Tests	Other

At a Glance: Introductory and Intermediate Algebra

Operations with Real Numbers

$-18 + 3 = -15$ $-9 \cdot 6 = -54$
$-6 + (-4) = -10$ $-5 \cdot (-3) = 15$
$9 - 12 = -3$ $18 \div (-3) = -6$
$-7 - (-10) = 3$ $-10 \div (-2) = 5$

Absolute value: $|-4| = 4$

The opposite of $-\frac{3}{7}$ is $\frac{3}{7}$.

The reciprocal of $-\frac{2}{9}$ is $-\frac{9}{2}$.

Order of Operations

1. Do all calculations within grouping symbols before operations outside.
2. Evaluate all exponential expressions.
3. Do all multiplications and divisions in order from left to right.
4. Do all additions and subtractions in order from left to right.

Exponents

$x^0 = 1$; $x^1 = x$; $x^{-3} = \dfrac{1}{x^3}$;

$x^2 \cdot x^5 = x^7$; $\dfrac{x^5}{x^2} = x^3$; $(x^2)^5 = x^{10}$

Polynomials

Multiplying:
$(y - 4)(3y + 5) = 3y^2 - 7y - 20$
$(q - 5)(q + 5) = q^2 - 25$
$(2a - 3)^2 = 4a^2 - 12a + 9$

Factoring:
$2x^2 - 5x - 12 = (2x + 3)(x - 4)$
$25x^2 - 4 = (5x - 2)(5x + 2)$
$9x^2 + 6x + 1 = (3x + 1)^2$
$x^3 + 64 = (x + 4)(x^2 - 4x + 16)$
$x^3 - 1000 = (x - 10)(x^2 + 10x + 100)$

Set-Builder Notation and Interval Notation

$\{x \mid x \text{ is a real number}\} = (-\infty, \infty)$
$\{x \mid x < 3\} = (-\infty, 3)$
$\{x \mid -3 \le x < 3\} = [-3, 3)$
$\{x \mid x \ge 3\} = [3, \infty)$

Linear Function and Slope

$Ax + By = C$: $2x - 3y = 6$;

$y = mx + b$: $y = \dfrac{2}{3}x - 2$;

$f(x) = mx + b$: $f(x) = \dfrac{2}{3}x - 2$

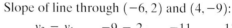

Slope $(m) = \dfrac{2}{3}$

y-intercept $(0, b) = (0, -2)$

Slope of line through $(-6, 2)$ and $(4, -9)$:

$m = \dfrac{y_2 - y_1}{x_2 - x_1} = \dfrac{-9 - 2}{4 - (-6)} = \dfrac{-11}{10} = -\dfrac{11}{10}$

The slope of a horizontal line is 0.
The slope of a vertical line is not defined.

Quadratic Functions

$f(x) = ax^2 + bx + c$
$f(x) = x^2 - x - 6$
$\quad = (x + 2)(x - 3)$

Function values:
$f(0) = -6$, $f(1) = -6$,
$f(-2) = 0$, $f(3) = 0$,
$f(-1) = -4$, $f(2) = -4$

x-intercepts: $(-2, 0)$ and $(3, 0)$

Vertex: $\left(-\dfrac{b}{2a}, f\left(-\dfrac{b}{2a}\right)\right) = \left(\dfrac{1}{2}, -6\dfrac{1}{4}\right)$

Axis of symmetry: $x = \dfrac{1}{2}$

Domain: $(-\infty, \infty)$

Range: $\left[-6\dfrac{1}{4}, \infty\right)$

Parallel Lines and Perpendicular Lines

Two lines are parallel if they have the same slope and different y-intercepts;

$y = 2x - 3$ and $y = 2x + 4$ are parallel.

Two nonvertical lines are perpendicular if the product of their slopes is -1: $m_1 \cdot m_2 = -1$;

$y = \dfrac{1}{2}x + 3$ and $y = -2x - 7$ are perpendicular.

Pythagorean Theorem

$$a^2 + b^2 = c^2$$

Solving Equations

Using the Principle of Zero Products

$$x^2 + 3x = 54$$
$$x^2 + 3x - 54 = 0$$
$$(x + 9)(x - 6) = 0$$
$$x + 9 = 0 \quad or \quad x - 6 = 0$$
$$x = -9 \quad or \quad x = 6$$

The solutions are -9 and 6.

Using the Quadratic Formula

Quadratic Formula: $x = \dfrac{-b \pm \sqrt{b^2 - 4ac}}{2a}$

$x^2 - 6x + 2 = 0; \, a = 1, b = -6, c = 2$

$$x = \dfrac{-(-6) \pm \sqrt{(-6)^2 - 4 \cdot 1 \cdot 2}}{2 \cdot 1} = \dfrac{6 \pm \sqrt{28}}{2}$$
$$= \dfrac{6 \pm 2\sqrt{7}}{2} = 3 \pm \sqrt{7}$$

The solutions are $3 + \sqrt{7}$ and $3 - \sqrt{7}$, or $3 \pm \sqrt{7}$.

Containing Absolute Value

$$|x - 2| = 5$$
$$x - 2 = -5 \quad or \quad x - 2 = 5$$
$$x = -3 \quad or \quad x = 7$$

The solutions are -3 and 7.

Multiplying by the LCM

$$\dfrac{5}{4x} + \dfrac{1}{x} = 2$$
$$4x \cdot \left(\dfrac{5}{4x} + \dfrac{1}{x}\right) = 4x \cdot 2$$
$$5 + 4 = 8x$$
$$9 = 8x$$
$$\dfrac{9}{8} = x$$

The solution is $\dfrac{9}{8}$.

Using the Principle of Powers

$$\sqrt{x - 1} - 3 = 9$$
$$\sqrt{x - 1} = 12$$
$$(\sqrt{x - 1})^2 = 12^2$$
$$x - 1 = 144$$
$$x = 145$$

The solution is 145.

Solving Systems of Equations Using the Elimination Method

$$x - 3y = -7 \longrightarrow -2x + 6y = 14$$
$$2x + 5y = -3 \longrightarrow \underline{2x + 5y = -3}$$
$$11y = 11$$
$$y = 1$$

Substitute 1 for y in either equation and solve for x:

$$2x + 5 \cdot 1 = -3$$
$$2x = -8$$
$$x = -4.$$

The solution is $(-4, 1)$.

Solving Inequalities

Using the Addition Principle and the Multiplication Principle

$$-5x + 2 \leq -78$$
$$-5x \leq -80$$
$$x \geq 16$$

The solution set is $\{x | x \geq 16\}$, or $[16, \infty)$.

Containing Absolute Value

$$|x - 2| \leq 5$$
$$-5 \leq x - 2 \leq 5$$
$$-3 \leq x \leq 7$$

The solution set is $\{x | -3 \leq x \leq 7\}$, or $[-3, 7]$.

$$|x - 2| > 5$$
$$x - 2 < -5 \quad or \quad x - 2 > 5$$
$$x < -3 \quad or \quad x > 7$$

The solution set is $\{x | x < -3 \, or \, x > 7\}$, or $(-\infty, -3) \cup (7, \infty)$.

Variation

Direct: $y = kx; \, y = 6x$

Inverse: $y = \dfrac{k}{x}; \, y = \dfrac{2}{x}$

Joint: $y = kxz; \, y = 9xz$

Complex Numbers

$i = \sqrt{-1}; \, i^2 = -1$

$(2 - 3i) + (6 + 2i) = 8 - i$

$\sqrt{-4} \cdot \sqrt{-15} = 2i \cdot \sqrt{15}i = 2\sqrt{15}i^2 = -2\sqrt{15}$

$\dfrac{-3 + 4i}{1 - 6i} = \dfrac{-3 + 4i}{1 - 6i} \cdot \dfrac{1 + 6i}{1 + 6i} = \dfrac{-27 - 14i}{1 - 36i^2} = -\dfrac{27}{37} - \dfrac{14}{37}i$

Properties of Logarithms

Product Rule: $\log_a(M \cdot N) = \log_a M + \log_a N$

Power Rule: $\log_a M^k = k \cdot \log_a M$

Quotient Rule: $\log_a \dfrac{M}{N} = \log_a M - \log_a N$

Your Guide to Success in Math

Complete Step 0 as soon as you begin your math course.

STEP 0: Plan Your Semester
- ☐ Register for the online part of the course (if there is one) as soon as possible.
- ☐ Fill in your Course and Contact information on this pull-out card.
- ☐ Write important dates from your syllabus in the Semester Organizer section on this pull-out card.

Follow Steps 1–3 during your course. Your instructor will tell you which resources to use—and when—in the textbook or eText, *MyMathGuide* workbook, videos, and MyLab Math. Use these resources for extra help and practice.

STEP 1: PREPARE: Studying the Concepts
- ☐ Do the Just-in-Time Review in the text and/or watch the videos in MyLab Math.
- ☐ Read the textbook or eText, listen to your instructor's lecture, and/or watch the section videos. You can work in *MyMathGuide* as you do this, saving all your work to review throughout the course.
- ☐ Work the Skill Review exercises and/or watch the videos in MyLab Math in each section.
- ☐ Stop and do the Margin Exercises, including the Guided Solution Exercises, as directed.

STEP 2: PARTICIPATE: Making Connections through Active Exploration
- ☐ Explore the concepts using the Animations in MyLab Math.
- ☐ Work the Visualizing for Success or Translating for Success exercises in the text and/or in MyLab Math.
- ☐ Answer the Check Your Understanding exercises in the Section Exercises in the textbook and/or in MyLab Math.

STEP 3: PRACTICE: Reinforcing Understanding
- ☐ Complete your assigned homework from the textbook and/or in MyLab Math.
  - ☐ When doing homework from the textbook, use the answer section to check your work.
  - ☐ When doing homework in MyLab Math, use the Learning Aids, such as Help Me Solve This and View an Example, as needed, working toward being able to complete exercises without the aids.
- ☐ Work the exercises in the Mid-Chapter Review.
- ☐ Read the Study Guide and work the Review Exercises in the Chapter Summary and Review.
- ☐ Take the Chapter Test as a practice exam. To watch an instructor solve each problem, go to the Chapter Test Prep Videos in MyLab Math.

Use the ***Studying for Success*** tips in the text and the ***MyLab Math Study Skills*** modules (with videos) to help you develop effective time-management, note-taking, test-prep, and other skills.

EDITION 6

Introductory and Intermediate Algebra

Marvin L. Bittinger
Indiana University Purdue University Indianapolis

Judith A. Beecher

Barbara L. Johnson
Ivy Tech Community College of Indiana

Director, Courseware Portfolio Management:	Michael Hirsch
Courseware Portfolio Manager:	Cathy Cantin
Courseware Portfolio Management Assistants:	Shannon Bushee; Shannon Slocum
Managing Producer:	Karen Wernholm
Content Producer:	Ron Hampton
Producer:	Erin Carreiro
Manager, Courseware QA:	Mary Durnwald
Manager, Content Development:	Eric Gregg
Field Marketing Managers:	Jennifer Crum; Lauren Schur
Product Marketing Manager:	Kyle DiGiannantonio
Product Marketing Assistant:	Brooke Imbornone
Senior Author Support/ Technology Specialist:	Joe Vetere
Manager, Rights and Permissions:	Gina Cheselka
Manufacturing Buyer:	Carol Melville, LSC Communications
Associate Director of Design:	Blair Brown
Program Design Lead:	Barbara T. Atkinson
Text Design:	Geri Davis/The Davis Group, Inc.
Editorial and Production Service:	Jane Hoover/Lifland et al., Bookmakers
Composition:	Cenveo® Publisher Services
Illustration:	Network Graphics; William Melvin
Cover Design:	Cenveo® Publisher Services
Cover Image:	Nick Veasey/Untitled X-Ray/Getty Images

Library of Congress Cataloging-in-Publication Data is on file with the publisher.

Copyright © 2019, 2015, 2011 by Pearson Education, Inc. All Rights Reserved. Printed in the United States of America. This publication is protected by copyright, and permission should be obtained from the publisher prior to any prohibited reproduction, storage in a retrieval system, or transmission in any form or by any means, electronic, mechanical, photocopying, recording, or otherwise. For information regarding permissions, request forms and the appropriate contacts within the Pearson Education Global Rights & Permissions department, please visit www.pearsoned.com/permissions.

Attributions of third-party content appear on page G-6, which constitutes an extension of this copyright page.

PEARSON, ALWAYS LEARNING, and MYLAB MATH are exclusive trademarks owned by Pearson Education, Inc. of its affiliates in the U.S. and/or other countries.

Unless otherwise indicated herein, any third-party trademarks that may appear in this work are the property of their respective owners and any reference to third-party trademarks, logos or other trade dress are for demonstrative or descriptive purposes only. Such references are not intended to imply any sponsorship, endorsement, authorization, or promotion of Pearson's products by the owners of such marks, or any relationship between the owner and Pearson Education, Inc., or its affiliates, authors, licensees, or distributors.

ISBN 13: 978-0-13-468648-6
ISBN 10: 0-13-468648-9

Contents

Index of Activities vii
Index of Animations viii
Preface x
Index of Applications xvi

JUST-IN-TIME REVIEW 1

1. All Factors of a Number 2
2. Prime Factorizations 3
3. Greatest Common Factor 4
4. Least Common Multiple 5
5. Equivalent Expressions and Fraction Notation 7
6. Mixed Numerals 8
7. Simplify Fraction Notation 9
8. Multiply and Divide Fraction Notation 10
9. Add and Subtract Fraction Notation 12
10. Convert from Decimal Notation to Fraction Notation 14
11. Add and Subtract Decimal Notation 15
12. Multiply and Divide Decimal Notation 16
13. Convert from Fraction Notation to Decimal Notation 17
14. Rounding with Decimal Notation 18
15. Convert between Percent Notation and Decimal Notation 19
16. Convert between Percent Notation and Fraction Notation 21
17. Exponential Notation 23
18. Order of Operations 24

1 INTRODUCTION TO REAL NUMBERS AND ALGEBRAIC EXPRESSIONS 27

1.1 Introduction to Algebra 28
1.2 The Real Numbers 35
1.3 Addition of Real Numbers 46
1.4 Subtracton of Real Numbers 54

Mid-Chapter Review 61

1.5 Multiplication of Real Numbers 63
1.6 Division of Real Numbers 70
1.7 Properties of Real Numbers 79
1.8 Simplifying Expressions; Order of Operations 92

Summary and Review 101
Test 107

2 SOLVING EQUATIONS AND INEQUALITIES 109

2.1 Solving Equations: The Addition Principle 110
2.2 Solving Equations: The Multiplication Principle 116
2.3 Using the Principles Together 122
2.4 Formulas 133

Mid-Chapter Review 141

2.5 Applications of Percent 143
2.6 Applications and Problem Solving 151

Translating for Success 162

2.7 Solving Inequalities 168
2.8 Applications and Problem Solving with Inequalities 180

Summary and Review 188
Test 193
Cumulative Review 195

3 GRAPHS OF LINEAR EQUATIONS 197

3.1 Introduction to Graphing 198
3.2 Graphing Linear Equations 205
3.3 More with Graphing and Intercepts 216

Visualizing for Success   221
Mid-Chapter Review   227

3.4   Slope and Applications   229

Summary and Review   240
Test   246
Cumulative Review   249

4 POLYNOMIALS: OPERATIONS   251

4.1   Integers as Exponents   252
4.2   Exponents and Scientific Notation   262
4.3   Introduction to Polynomials   274
4.4   Addition and Subtraction of Polynomials   287

Mid-Chapter Review   295

4.5   Multiplication of Polynomials   297
4.6   Special Products   304

Visualizing for Success   310

4.7   Operations with Polynomials in Several Variables   315
4.8   Division of Polynomials   324

Summary and Review   331
Test   337
Cumulative Review   339

5 POLYNOMIALS: FACTORING   341

5.1   Introduction to Factoring   342
5.2   Factoring Trinomials of the Type $x^2 + bx + c$ 350
5.3   Factoring $ax^2 + bx + c, a \neq 1$: The FOIL Method   360
5.4   Factoring $ax^2 + bx + c, a \neq 1$: The ac-Method   368

Mid-Chapter Review   374

5.5   Factoring Trinomial Squares and Differences of Squares   376
5.6   Factoring Sums or Differences of Cubes   386
5.7   Factoring: A General Strategy   391
5.8   Solving Quadratic Equations by Factoring   399
5.9   Applications of Quadratic Equations   407

Translating for Success   413
Summary and Review   419
Test   425
Cumulative Review   427

6 RATIONAL EXPRESSIONS AND EQUATIONS   429

6.1   Multiplying and Simplifying Rational Expressions   430
6.2   Division and Reciprocals   440
6.3   Least Common Multiples and Denominators   445
6.4   Adding Rational Expressions   449
6.5   Subtracting Rational Expressions   457

Mid-Chapter Review   465

6.6   Complex Rational Expressions   467
6.7   Solving Rational Equations   473
6.8   Applications Using Rational Equations and Proportions   481

Translating for Success   488
6.9   Variation and Applications   495

Summary and Review   506
Test   513
Cumulative Review   515

7 GRAPHS, FUNCTIONS, AND APPLICATIONS   517

7.1   Functions and Graphs   518
7.2   Finding Domain and Range   532

Mid-Chapter Review   537

7.3   Linear Functions: Graphs and Slope   539
7.4   More on Graphing Linear Equations   550

Visualizing for Success   557

7.5   Finding Equations of Lines; Applications   562

Summary and Review   573
Test   582
Cumulative Review   585

8 SYSTEMS OF EQUATIONS 587

- 8.1 Systems of Equations in Two Variables 588
- 8.2 Solving by Substitution 597
- 8.3 Solving by Elimination 603
- 8.4 Solving Applied Problems: Two Equations 612

Translating for Success 620
Mid-Chapter Review 625

- 8.5 Systems of Equations in Three Variables 627
- 8.6 Solving Applied Problems: Three Equations 634

Summary and Review 641
Test 647
Cumulative Review 649

9 MORE ON INEQUALITIES 651

- 9.1 Sets, Inequalities, and Interval Notation 652

Translating for Success 661

- 9.2 Intersections, Unions, and Compound Inequalities 668

Mid-Chapter Review 680

- 9.3 Absolute-Value Equations and Inequalities 682
- 9.4 Systems of Inequalities in Two Variables 693

Visualizing for Success 702
Summary and Review 707
Test 713
Cumulative Review 715

10 RADICAL EXPRESSIONS, EQUATIONS, AND FUNCTIONS 717

- 10.1 Radical Expressions and Functions 718
- 10.2 Rational Numbers as Exponents 729
- 10.3 Simplifying Radical Expressions 736
- 10.4 Addition, Subtraction, and More Multiplication 745

Mid-Chapter Review 751

- 10.5 More on Division of Radical Expressions 753
- 10.6 Solving Radical Equations 758
- 10.7 Applications Involving Powers and Roots 769

Translating for Success 771

- 10.8 The Complex Numbers 776

Summary and Review 787
Test 793
Cumulative Review 795

11 QUADRATIC EQUATIONS AND FUNCTIONS 797

- 11.1 The Basics of Solving Quadratic Equations 798
- 11.2 The Quadratic Formula 812
- 11.3 Applications Involving Quadratic Equations 819

Translating for Success 825

- 11.4 More on Quadratic Equations 831

Mid-Chapter Review 840

- 11.5 Graphing $f(x) = a(x - h)^2 + k$ 842
- 11.6 Graphing $f(x) = ax^2 + bx + c$ 851

Visualizing for Success 856

- 11.7 Mathematical Modeling with Quadratic Functions 860
- 11.8 Polynomial Inequalities and Rational Inequalities 871

Summary and Review 879
Test 885
Cumulative Review 887

12 EXPONENTIAL FUNCTIONS AND LOGARITHMIC FUNCTIONS 889

- 12.1 Exponential Functions 890
- 12.2 Composite Functions and Inverse Functions 904
- 12.3 Logarithmic Functions 921
- 12.4 Properties of Logarithmic Functions 932

Mid-Chapter Review 938	**APPENDIXES** 987
12.5 Natural Logarithmic Functions 940	A Introductory Algebra Review 988
Visualizing for Success 945	B Mean, Median, and Mode 990
12.6 Solving Exponential Equations and Logarithmic Equations 949	C Synthetic Division 993
	D Determinants and Cramer's Rule 996
12.7 Mathematical Modeling with Exponential Functions and Logarithmic Functions 956	E Elimination Using Matrices 1001
	F The Algebra of Functions 1005
Translating for Success 964	G Distance, Midpoints, and Circles 1008
Summary and Review 970	**Answers** A-1
Test 978	**Guided Solutions** A-47
Cumulative Review 981	**Glossary** G-1
	Index I-1

Index of Activities

Chapter	Title
1	American Football
2	Pets in the United States
3	Going Beyond High School
4	Finding the Magic Number
5	Visualizing Factoring
6	Data and Downloading
7	Going Beyond High School
8	Fruit Juice Consumption
	Waterfalls
	Construction
9	Pets in the United States
10	Firefighting Formulas
11	Music Downloads
	Let's Go to the Movies
12	Earthquake Magnitude
Appendix F	Mobile Data

Index of Animations

Section	Title
1.2d	Order on the Number Line
2.7b	Graphing Inequalities
3.2a	Graphing Linear Inequalities
3.4a	Slope
3.4a	Slope of a Line
4.1f	Negative Exponents
4.6c	Special Products
5.8b	Intercepts and Solutions
6.8a	Motion Problems
7.1c	Graphing Functions
7.2a	Domain and Range of a Function
7.3b	Slope
7.3b	Slope of a Line
7.3b	Slope-Intercept Form
7.3b	Equations of Lines: Slope-Intercept Form
7.5b	Equations of Lines: Point-Slope Form
8.4a	Mixture Problems
9.1b	Graphing Inequalities

9.3e	Absolute-Value Equations and Inequalities
9.4b	Linear Inequalities in Two Variables
10.1a	Graphs of Radical Functions
11.3a	Motion Problems
11.5c	Quadratic Functions and Their Graphs
11.5c	Graphs of Quadratic Functions
11.6b	Intercepts and Solutions
11.7a	Application: Height of a Baseball
11.8a	Polynomial and Rational Inequalities
12.1a	Graphs of Exponential Functions
12.2d	Graphing Functions and Their Inverses
12.2d	Graphs of Inverse Functions
12.3a	Graphs of Logarithmic Functions
Appendix F	Sum and Difference of Two Functions
	Product and Quotient of Two Functions

Preface

Math doesn't change, but students' needs—and the way students learn—do.

With this in mind, *Introductory and Intermediate Algebra,* 6th edition, continues the Bittinger tradition of objective-based, guided learning, while integrating many updates with the proven pedagogy. These updates are motivated by feedback that we received from students and instructors, as well as our own experience in the classroom. In this edition, our focus is on guided learning and retention: helping each student (and instructor) get the most out of all the available program resources—wherever and whenever they engage with the math.

We believe that student success in math hinges on four key areas: **Foundation, Engagement, Application,** and **Retention.** In the 6th edition, we have added key new program features (highlighted below, for quick reference) in each area to make it easier for each student to personalize his or her learning experience. In addition, you will recognize many proven features and presentations from the previous edition of the program.

FOUNDATION

Studying the Concepts

Students can learn the math concepts by reading the textbook or the eText, participating in class, watching the videos, working in the *MyMathGuide* workbook—or using whatever combination of these course resources works best for them.

> In order to understand new math concepts, students must recall and use skills and concepts previously studied. To support student learning, we have integrated two important new features throughout the 6th edition program:
>
> ☐ *New!* Just-in-Time Review at the beginning of the text and the eText is a set of quick reviews of the key topics from previous courses that are prerequisites for the new material in this course. A note on each Chapter Opener in Chapters 1–6 alerts students to the topics they should review for that chapter. In MyLab Math, students will find a concise presentation of each topic in the **Just-in-Time Review Videos.**
>
> ☐ *New!* Skill Review, in nearly every section of the text and the eText, reviews a previously presented skill at the objective level where it is key to learning the new material. This feature offers students two practice exercises with answers. In MyLab Math, new **Skill Review Videos,** created by the Bittinger author team, offer a concise, step-by-step solution for each Skill Review exercise.

Margin Exercises with Guided Solutions, with fill-in blanks at key steps in the problem-solving process, appear in nearly every text section and can be assigned in MyLab Math.

Algebraic–Graphical Connections in the text draw explicit connections between the algebra and the corresponding graphical visualization.

Introductory and Intermediate Algebra **Video Program,** our comprehensive program of objective-based, interactive videos, can be used hand-in-hand with our *MyMathGuide* workbook. **Interactive Your Turn exercises** in the videos prompt students to solve problems and receive instant feedback. These videos can be accessed at the section, objective, and example levels.

MyMathGuide offers students a guided, hands-on learning experience. This objective-based workbook (available in print and in MyLab Math) includes vocabulary, skill, and concept review—as well as problem-solving practice with space for students to fill in the answers and stepped-out solutions to problems, to show (and keep) their work, and to write notes. Students can use *MyMathGuide* while watching the videos, listening to the instructor's lecture, or reading the text or the eText in order to reinforce and self-assess their learning.

Studying for Success sections are checklists of study skills designed to ensure that students develop the skills they need to succeed in math, school, and life. They are available at the beginning of selected sections.

ENGAGEMENT
Making Connections through Active Exploration

Since understanding the big picture is key to student success, we offer many active learning opportunities for the practice, review, and reinforcement of important concepts and skills.

> - *New!* **Chapter Opener Applications** with graphics use current data and applications to present the math in context. Each application is related to exercises in the text to help students model, visualize, learn, and retain the math.
> - *New!* **Student Activities,** included with each chapter, have been developed as multistep, data-based activities for students to apply the math in the context of an authentic application. Student Activities are available in *MyMathGuide* and in MyLab Math. (See the Index of Activities on p. vii.)
> - *New!* **Interactive Animations** can be manipulated by students in MyLab Math through guided and open-ended exploration to further solidify their understanding of important concepts. (See the Index of Animations on p. viii.)

Translating for Success offers extra practice with the important first step of the process for solving applied problems. **Visualizing for Success** asks students to match an equation or an inequality with its graph by focusing on characteristics of the equation or the inequality and the corresponding attributes of the graph. Both of these activities are available in the text and in MyLab Math.

Calculator Corner is an optional feature in each chapter that helps students use a calculator to perform calculations and to visualize concepts.

Learning Catalytics uses students' mobile devices for an engagement, assessment, and classroom intelligence system that gives instructors real-time feedback on student learning.

APPLICATION
Reinforcing Understanding

As students explore the math, they have frequent opportunities to apply new concepts, practice, self-assess, and reinforce their understanding.

Margin Exercises, labeled "Do Exercise . . . ," give students frequent opportunities to apply concepts just discussed by solving problems that parallel text examples.

Exercise Sets in each section offer abundant opportunity for practice and review in the text and in MyLab Math. The Section Exercises are grouped by objective for ease of use, and each set includes the following special exercise types:

- ☐ *New!* **Check Your Understanding,** with **Reading Check** and **Concept Check** exercises, at the beginning of each exercise set gives students the opportunity to assess their grasp of the skills and concepts before moving on to the objective-based section exercises. In MyLab Math, many of these exercises use drag & drop functionality.
- ☐ **Skill Maintenance Exercises** offer a thorough review of the math in the preceding sections of the text.
- ☐ **Synthesis Exercises** help students develop critical-thinking skills by requiring them to use what they know in combination with content from the current and previous sections.

RETENTION
Carrying Success Forward

Because continual practice and review is so important to retention, we have integrated both throughout the program in the text and in MyLab Math.

- ☐ *New!* **Skill Builder** adaptive practice, available in MyLab Math, offers each student a personalized learning experience. When a student struggles with the assigned homework, Skill Builder exercises offer just-in-time additional adaptive practice. The adaptive engine tracks student performance and delivers to each individual questions that are appropriate for his or her level of understanding. When the system has determined that the student has a high probability of successfully completing the assigned exercise, it suggests that the student return to the assigned homework.

Mid-Chapter Review offers an opportunity for active review midway through each chapter. This review offers four types of practice problems:

Concept Reinforcement, Guided Solutions, Mixed Review, and Understanding Through Discussion and Writing

Summary and Review is a comprehensive learning and review section at the end of each chapter. Each of the five sections—**Vocabulary Reinforcement** (fill-in-the-blank), **Concept Reinforcement** (true/false), **Study Guide** (examples with stepped-out solutions paired with similar practice problems), **Review Exercises,** and **Understanding Through Discussion and Writing**—includes references to the section in which the material was covered to facilitate review.

Chapter Test offers students the opportunity for comprehensive review and reinforcement prior to taking their instructor's exam. **Chapter Test Prep Videos** in MyLab Math show step-by-step solutions to the questions on the chapter test.

Cumulative Review follows each chapter beginning with Chapter 2. These reviews revisit skills and concepts from all preceding chapters to help students retain previously presented material.

Resources for Success

MyLab Math Online Course for Bittinger, Beecher, and Johnson, *Introductory and Intermediate Algebra*, 6th edition
(access code required)

MyLab™ Math is available to accompany Pearson's market-leading text offerings. To give students a consistent tone, voice, and teaching method, the pedagogical approach of the text is tightly integrated throughout the accompanying MyLab Math course, making learning the material as seamless as possible.

UPDATED! Learning Path

Structured, yet flexible, the updated Learning Path highlights author-created, faculty-vetted content—giving students what they need exactly when they need it. The Learning Path directs students to resources such as two new types of video: **Just-in-Time Review** (concise presentations of key topics from previous courses) and **Skill Review** (author-created exercises with step-by-step solutions that reinforce previously presented skills), both available in the Multimedia Library and assignable in MyLab Math.

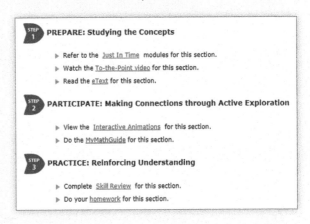

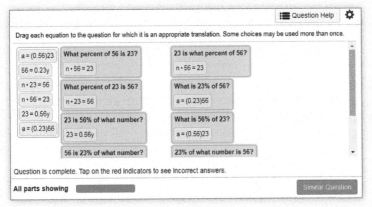

NEW! Drag-and-Drop Exercises

Drag-and-drop exercises are now available in MyLab Math. This new assignment type allows students to drag answers and values within a problem, providing a new and engaging way to test students' concept knowledge.

NEW and UPDATED! Animations

New animations encourage students to learn key concepts through guided and open-ended exploration. Animations are available through the Learning Path and Multimedia Library, and they can be assigned within MyLab Math.

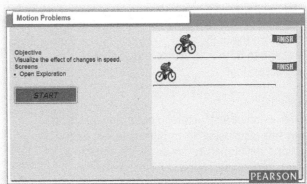

pearson.com/mylab/math

Resources for Success

Instructor Resources

Additional resources can be downloaded from **www.pearsonhighered.com** or hardcopy resources can be ordered from your sales representative.

Annotated Instructor's Edition
ISBN: 0134707435
- Answers to all text exercises.
- Helpful teaching tips, including suggestions for incorporating Student Activities in the course.

Instructor's Resource Manual with Tests and Minilectures
(download only)
ISBN: 0134707397
- Resources designed to help both new and experienced instructors with course preparation and class management.
- Chapter teaching tips and support for media supplements.
- Multiple versions of multiple-choice and free-response chapter tests, as well as final exams.

Instructor's Solutions Manual
(download only)
By Judy Penna
ISBN: 0134707494

The *Instructor's Solutions Manual* includes brief solutions for the even-numbered exercises in the exercise sets and fully worked-out annotated solutions for all the exercises in the Mid-Chapter Reviews, the Summary and Reviews, the Chapter Tests, and the Cumulative Reviews.

PowerPoint® Lecture Slides
(download only)
- Editable slides present key concepts and definitions from the text.
- Available to both instructors and students.
- Fully accessible.

TestGen®
TestGen enables instructors to build, edit, print, and administer tests using a computerized test bank of questions developed to cover all the objectives of the text. (www.pearsoned.com/testgen)

Student Resources

Introductory and Intermediate Algebra Lecture Videos
- Concise, interactive, and objective-based videos.
- View a whole section, choose an objective, or go straight to an example.

Chapter Test Prep Videos
- Step-by-step solutions for every problem in the chapter tests.

Just-in-Time Review Videos
- One video per review topic in the Just-in-Time Review at the beginning of the text.
- View examples and worked-out solutions that parallel the concepts reviewed in each review topic.

Skill Review Videos
Students can review previously presented skills at the objective level with two practice exercises before moving forward in the content. Videos include a step-by-step solution for each exercise.

MyMathGuide: Notes, Practice, and Video Path
ISBN: 0134707400
- Guided, hands-on learning in a workbook format with space for students to show their work and record their notes and questions.
- Highlights key concepts, skills, and definitions; offers quick reviews of key vocabulary terms with practice problems, examples with guided solutions, similar Your Turn exercises, and practice exercises with readiness checks.
- Includes student activities utilizing real data.
- Available in MyLab Math and as a printed manual.

Student's Solutions Manual
ISBN: 0134707451
By Judy Penna
- Includes completely worked-out annotated solutions for odd-numbered exercises in the text, as well as all the exercises in the Mid-Chapter Reviews, the Summary and Reviews, the Chapter Tests, and the Cumulative Reviews.
- Available in MyLab Math and as a printed manual.

pearson.com/mylab/math

Acknowledgments

Our deepest appreciation to all the instructors and students who helped to shape this revision of our program by reviewing our texts and courses, providing feedback, and sharing their experiences with us at conferences and on campus. In particular, we would like to thank the following for reviewing the titles in our worktext program for this revision:

> Amanda L. Blaker, *Gallatin College*
> Jessica Bosworth, *Nassau Community College*
> Judy G. Burn, *Trident Technical College*
> Abushieba A. Ibrahim, *Nova Southeastern University*
> Laura P. Kyser, *Savannah Technical College*
> David Mandelbaum, *Nova Southeastern University*

An outstanding team of professionals was involved in the production of this text. We want to thank Judy Penna for creating the new Skill Review videos and for writing the *Student's Solutions Manual* and the *Instructor's Solutions Manual*. We also thank Laurie Hurley for preparing *MyMathGuide*, Robin Rufatto for creating the new Just-in-Time videos, and Tom Atwater for supporting and overseeing the new videos. Accuracy checkers Judy Penna, Laurie Hurley, and Susan Meshulam contributed immeasurably to the quality of the text.

Jane Hoover, of Lifland et al., Bookmakers, provided editorial and production services of the highest quality, and Geri Davis, of The Davis Group, performed superb work as designer, art editor, and photo researcher. Their countless hours of work and consistent dedication have led to products of which we are immensely proud.

In addition, a number of people at Pearson, including the Developmental Math Team, have contributed in special ways to the development and production of our program. Special thanks are due to Cathy Cantin, Courseware Portfolio Manager, for her visionary leadership and development support. In addition, Ron Hampton, Content Producer, contributed invaluable coordination for all aspects of the project. We also thank Erin Carreiro, Producer, and Kyle DiGiannantonio, Product Marketing Manager, for their exceptional support.

Our goal in writing this textbook was to make mathematics accessible to every student. We want you to be successful in this course and in the mathematics courses you take in the future. Realizing that your time is both valuable and limited, and that you learn in a uniquely individual way, we employ a variety of pedagogical and visual approaches to help you learn in the best and most efficient way possible. We wish you a positive and successful learning experience.

Marv Bittinger
Judy Beecher
Barbara Johnson

Index of Applications

Agriculture
Composting, 868
Corral design, 867
Feed lot, 826
Fenced-in land, 867
Filling a grain bin, 796
Flower bed, 26
Flower bulbs, 622
Garden design, 868
Gardening, 490
Grain flow, 964
Grass seed, 492
Harvesting, 490, 796
Livestock feed, 622
Mixing fertilizers, 616–617
Mulching flowerbeds, 514
Vegetable seeds, 622

Astronomy
Distance from the sun to Earth, 251, 267
Earth vs. Jupiter, 251, 271
Earth vs. Saturn, 337
Space travel, 272
Stars in the known universe, 271
Stars in the Milky Way galaxy, 264
Surface temperature on a planet, 56, 60
Weight on Mars, 503

Automotive
Automobile pricing, 638
Automotive maintenance, 623, 624
Automotive repair, 791–792
Car assembly line, 234
Classic cars, 969
Daytime accidents, 869
Electric vehicles, 150
Fuel economy, 624
Gas mileage, 485, 492
Insurance-covered repairs, 186
Median age of cars, 582
Nighttime accidents, 869
Speed of a skidding car, 767
Stopping distance of a car, 504, 868
Vehicle production, 549

Biology
Animal speeds, 489
Bacteria, 991
Bees and honey, 497
Blue whales, 511, 965
Butterfly wings, 408
DNA, 268
Elephant measurements, 315
Endangered species, 27, 28–29, 75, 493, 902
Fish population, 270, 484, 492, 494
Frog population, 492
Gray whale calves, 193
Gray wolves, 192
Growth of bacteria, 903, 977
Hawaiian goose population, 890, 902
Honey bees, 493
Length of an *E. coli* bacterium, 264
Mass of water in the human body, 503
Number of humpback whales, 494
Number of manatees, 27, 28–29, 75
Otter population, 966
Sharks' teeth, 423
Speed of a black racer snake, 491
Speed of sea animals, 481–482
Zebra population, 514

Business
Book sale, 638
Bookstores, 149
Catering, 622
Clients, 661
Container sizes, 692
Copy machine rental, 194
Deli trays, 491
Delivering leaflets, 768
Delivery service, 666
Delivery truck rental, 155–156
Disney revenue, 571
eBook revenue, 148
Fabric manufacturing, 498
Fruit quality, 485–486
Gift card sales, 524–525, 579
Gold production in China, 527
Holiday sales, 245, 546
Home listing price, 163
Homespun Jellies, 581
Lens production, 639
Manufacturing computers, 990
Markup, 339
Maximizing profit, 868
Minimizing cost, 867
Newspaper circulation, 547
Number of eating and drinking places, 549
Office budget, 488
Office expense, 620

Office supplies, 271
Online sales growth, 884, 968
Paperback book revenue, 149
Quality control, 492, 511, 514, 586, 650
Radio advertising, 650
Renting office space, 666, 675
Sales, 613, 621, 645, 648, 864, 968
Sales meeting attendance, 413
Selling a condominium, 161
Selling a house, 160–161
Shipping, 502
Sports sponsorship, 150
Super Bowl ad spending, 549, 623
Tea Mug Collective, 904
Telemarketing, 639
Ticket profits, 866
Total profit, 878
Warehouse storing nuts, 261
Worldwide smartphone sales by vendor, 717

Chemistry
Ammonia solutions, 985
Carbon dating, 963, 967, 977, 980, 985
Chemical reaction, 66, 68, 78, 108
Chlorine for a pool, 486
Gold temperatures, 184
Half-life, 968
Hydrogen ion concentration, 957, 965
Ink remover, 622
Mixing solutions, 626, 648, 650, 839
Orange drink mixtures, 645
Oxygen dissolved in water, 480
pH of liquids, 957, 965
pH of tomatoes, 980
Temperature conversions, 528
Temperatures of liquids, 678
Temperatures of solids, 666
Zinc and copper in pennies, 493

Construction
Architecture, 43, 417, 599–600, 867, 888
Beam weight, 504
Beipanjiang River Bridge in China, 810
Blueprints, 184
Board cutting, 163, 194
Box construction, 827
Bridges needing repair, 572
Burj Khalifa in Dubai, 810
Clifton Suspension Bridge in Bristol, UK, 865–866
Concrete work, 514
Constructing stairs, 238
Covered bridges, 968
Diagonal braces in a lookout tower, 416
Fencing, 661
Flipping houses, 53
Footing of a house, 770
Gateway Arch, 810
Grade of a stairway, 549
Highway work, 511
Home construction, 968
Jackhammer noise, 965
Kitchen design, 243
Kitchen island, 407

Ladders, 411, 412, 472, 820, 821
Masonry, 234, 426
Molding plastics, 867
Observatory paint costs, 323
Patio design, 867
Pipe cutting, 163
Pitch of a roof, 237, 548
Plank height, 771
Plumbing, 491, 774, 968
Rafters of a house, 487
Rain gutter design, 418
Roofing, 157–158, 418, 586, 775
Square footage, 515
Stained-glass window design, 861
Two-by-four, 165
Washington Monument, 810
Wire cutting, 162, 196, 250, 413, 428, 825, 964
Wiring, 491
Wood scaffold, 411

Consumer
Aluminum usage, 503
Average price, 167
Babysitting costs, 34
Banquet costs, 184
Bottled water consumption, 239
Catering costs, 180–181
Cheddar cheese prices, 992
Coffee consumption, 148
Consumption of potatoes, 547
Copy center account, 52
Copy machine rental, 194
Cost of a necklace, 567–568
Cost of a service call, 568
Cost of an entertainment center, 191
Cost of operating a microwave oven, 133
Electrician visits, 187
Energy use, 19
Entertainment expenditures, 621
Fitness club costs, 571
Free ornamental tree offer, 611
Furnace repairs, 185
Gasoline prices, 825
Home ownership, 210
Incorrect bill, 192
Insurance benefits, 666
Insurance claims, 666
Juice consumption, 69
Laptop on sale, 14
Moving costs, 711, 713
Orange juice consumption, 238
Parking costs, 166, 185, 187
Phone rates, 185
Pizza consumption, 335
Pizza prices, 886
Popcorn prices, 646
Price before sale, 192, 428
Price reduction, 34, 196
Prices, 165
Repair rates, 648
Store credit for a return, 192
Taxi fares, 166
Tipping, 149, 167

Index of Applications **xvii**

TV prices, 902
U.S. home ownership rates, 797, 870
Utility cost, 650
Van rental, 166
Wedding costs, 666

Economics
Median price of a home, 530
Revenue from gas taxes, 572
Sales tax, 162, 413, 586
Salvage value, 902, 966
Stimulating the hometown economy, 608
Stock market changes, 107
Stock prices, 37, 53, 62, 69, 105, 166, 868
Supply and demand, 667, 869
U.S. public debt, 549
U.S. tax code, 967
Worldwide demand for oil, 981

Education
Advanced placement exams, 239, 637
Change in class size, 50
Code-school graduates, 966
College acceptance, 493
College course load, 185
College enrollment, 138, 163
College tuition, 185
Community college credits, 611
Distance education, 150
Enrollment costs, 32
Exam scores, 681
Grade average, 196
Grades, 665
High school dropout rate, 215
High school graduation rate, 869
Quiz scores, 330, 661
Reading assignment, 661
SAT scores, 620
School enrollment, 588–589
School fund-raiser, 571
School photos, 491
Student loans, 149, 197, 198, 209–210, 615, 620, 622, 623, 639, 868
Test questions, 167
Test scores, 159, 167, 182, 184, 192, 444, 512, 990
Tuition and fees at two-year colleges, 667
Valley Community College, 868, 869

Engineering
Beam load, 985
Bridge expansion, 774
Cell-phone tower, 413
Current and resistance, 503
Design, 602
Design of a TI-84 Plus graphing calculator, 415
Design of a window panel, 417
Distance over water, 774
Electrical power, 138
Electrical resistance, 501
Furnace output, 137
Guy wire, 417, 773, 825
Height of a tower, 767
Landscape design, 820–821, 860–861
Microchips, 270

Nuclear energy, 969
Ohm's Law, 496
Power generated from renewable sources, 889
Power of electric current, 512
Road design, 415, 416
Road grade, 233, 237, 245, 545
Rocket liftoff, 37
Solar capacity, 283
Solar power, 961–962
Town planning, 820
Wind energy, 238, 409
Wind power, 505

Environment
Banda Cliffs in Australia, 810
Coral reefs, 271
Dewpoint spread, 667
Distance from lightning, 133
Elevations, 36, 59, 60, 108
Glaciers, 638
Hurricanes, 869, 992
Low points on continents, 60
Melting snow, 496, 528
Niagara Falls water flow, 267
Ocean depth, 59
Pond depth, 187
Precipitation in Sonoma, California, 864
Record temperature drop, 215
Recycling aluminum cans, 903
River depth, 870
River discharge, 271
Slope of a river, 238, 548
Slope of Longs Peak, 238
Tallest mountain, 52
Temperature, 37, 43, 53, 56, 60, 62, 69
Tornadoes, 992
Tree supports, 423
Waste generation, 716
Water contamination, 271
Water level, 36
Waterfalls, 706
Wind chill temperature, 775

Finance
Bank account balance, 53, 69
Banking transactions, 50
Borrowing money, 428, 515, 771
Budgeting, 194
Charitable contributions, 964
Checking accounts, 60, 61
Compound interest, 323, 896, 897, 898, 901, 902, 939, 940, 958, 982
Credit cards, 53, 60, 166
Debt, 215
Decreasing number of banks, 568–569
Interest compounded continuously, 959–960, 967, 977
Investment, 196, 323, 615, 622, 623, 626, 636, 639, 648, 666, 681, 711, 896, 940, 977
Loan interest, 166
Making change, 623
Money remaining, 34, 115
Savings account, 194
Savings interest, 166
Simple interest, 32, 160

Small business loans, 643
Total assets, 105

Geometry
Angle measures, 162, 640
Angles of a triangle, 158, 166, 191, 261, 340, 373, 407, 413, 635, 637, 646, 650, 771
Area of a circle, 253, 259, 340, 505
Area of a parallelogram, 33
Area of a rectangle, 29, 186, 261, 330, 340, 414
Area of a square, 32, 259, 515
Area of a triangle, 32, 33, 121, 186, 774, 796
Bookcase width, 792
Complementary angles, 602, 610
Diagonal of a rectangle, 770, 773, 774
Diagonal of a square, 771, 775
Dimensions of a rectangular region, 165, 336, 407, 413, 414, 415, 417, 424, 426, 428, 488, 601, 620, 626, 648, 650, 661, 825, 826, 827, 828, 883, 964
Dimensions of a sail, 408, 415
Dimensions of an open box, 418
Distance between the parallel sides of an octagon, 775
Golden rectangle, 830
Height of a parallelogram, 586
Lawn area, 290
Length of a side of a square, 423, 488, 549, 792, 794
Length of a side of a triangle, 549, 773, 775, 825, 826, 827, 828
Lengths of a rectangle, 194, 771
Maximum area, 860–861, 867, 884, 885, 888, 985
Perimeter of a hexagon, 888
Perimeter of a rectangular region, 156–157, 162, 186, 194, 413, 472, 771
Perimeter of a square, 888
Perimeter of a triangle, 186
Perimeters of polygons, 985
Quilt dimensions, 827
Radius of a circle, 428
Radius of a sphere, 373
Right triangles, 412, 416, 417, 418
Room perimeter, 340
Supplementary angles, 602, 610, 620
Surface area, 830
Surface area of a cube, 137
Surface area of a right circular cylinder, 320
Surface area of a silo, 320
Triangle dimensions, 413, 415, 417, 426, 428, 444, 586, 620, 744, 825, 826, 827, 964
Volume of a box, 338
Volume of a cube, 303
Volume of a tree, 500–501
Volume of carpeting, 390
Width of a pool sidewalk, 418
Width of a rectangle, 162, 192, 413
Width of the margins in a book, 424

Health/Medicine
Acidosis, 957
Alkalosis, 965
Audiology, 957
Blood types, 250
Body mass index, 665, 679
Body surface area, 726
Body temperatures, 184
Calories, 137, 320, 859
Cancer center advertising, 546
Cholesterol levels, 638, 668
Drug overdose, 966
Exercise, 624, 678, 706
Fat content of fast food, 637
Fat intake, 503
Federal government hospitals, 146
Final adult height, 150
Food science, 623
Health insurance cost, 248
Hemoglobin, 492
Impulses in nerve fibers, 484
Influenza virus, 272
Lung capacity, 319
Medicaid enrollment, 984
Medicine dosage, 276, 679, 859
Milk alternatives, 164
Muscle weight, 515
Nutrition, 181–182, 185, 639
Patients with the flu, 330
Pharmaceutical marketing, 488
Prescription coverage, 666
Prescription drug costs, 214
Protein needs, 492
Reduced fat foods, 186, 187
Staphylococcus bacterium, 272
Weight gain, 187
Weight loss, 69

Labor
Amazon employees, 967
Bonuses, 608
Commission, 191, 661
Earnings, 495
Employment outlook, 146
Median income, 146
Newsroom employment, 150
Part-time job, 34
Salary, 34, 192, 196, 651, 660, 661, 666, 991
Vacation days, 234–235
Work rate, 503
Work recycling, 484
Work time, 250
Working alone, 768, 830
Working together, 483–484, 490, 491, 498, 511, 514, 771, 985

Miscellaneous
Apartment numbers, 164
Apples, 167
Archaeology, 417, 601
Blending granola, 622
Blending spices, 613–614
Book pages, 330
Books in libraries, 215
Brussels sprouts, 992
Butter temperatures, 181
Cake servings, 870
Central Park, 773
City park space, 149
Cleaning bleachers, 498
Coffee, 492, 614, 638, 839, 957
Coin mixture, 167, 771, 825, 964
Converting dress sizes, 919
Cost of raising a child, 194

Cutting a submarine sandwich, 261
Digits, 640
Dishwasher noise, 965
Easter Island, 826
Eggs, 638
Elevators, 248
Envelope size, 185
Filling a tank, 716, 888
Filling time for a pool, 488
First class mail, 192
Flag dimensions, 826, 827
Furniture polish, 622
Gold leaf, 272
Gourmet sandwiches, 154
Hands on a clock, 494
Hat size as a function of head circumference, 569
Height of a flagpole, 487
Height of a telephone pole, 424
Jewelry design, 634–635
Junk mail, 149
Knitted scarf, 153–154
Landscaping, 819, 825, 827
Limited-edition prints, 154–155
Locker numbers, 162, 825
Lunch orders, 612–613
Medals of Honor, 164
Memorizing words, 283
Mine rescue, 66
Mirror framing, 827
Mixing cleaning solutions, 617
Mystery numbers, 620
National park visitation, 105
Nontoxic floor wax, 622
Nontoxic scouring powder, 602
Package sizes, 185
Page dimensions, 828
Page numbers, 340, 410, 416
Pet ownership, 638
Picture matting and framing, 828, 883
Pieces of mail, 78
Pizza, 511, 830
Post office box numbers, 164
Pulitzer Prize books and authors, 526
Pumping rate, 503
Pumping time, 512
Pyramide du Louvre, 773
Raffle tickets, 165, 194, 640, 827
Raking, 490
Reducing a drawing on a copier, 196
Shoveling snow, 490
Shoveling time, 771
Sighting to the horizon, 767, 794
Snow removal, 245
Socks from cotton, 134
Sodding a yard, 429, 483–484
Sound of an alarm clock, 965
Stacking spheres, 282
Supreme Court justices and appointing president, 519
Tattoos, 149
Typing, 623
Uniform numbers, 964
Value of a lawn mower, 571
Value of a rare stamp, 939

Warning dye, 796
Wireless internet sign, 32
Yield sign, 32

Physics
Acoustics, 957
Altitude of a launched object, 319
Apparent size, 650
Atmospheric drag, 505
Centripetal force, 985
Combined gas law, 505
Falling distance, 501, 823–824
Free-falling objects, 810, 885
Height of a rocket, 417
Height of a thrown object, 878
Hooke's Law, 503
Intensity of light, 501, 504
Law of gravity, 499
Motion of a spring, 692
Musical pitch, 504
Pendulums, 744, 768, 822–823
Pressure at sea depth, 528, 678, 713
Sound levels, 956–957, 965, 976
Speed of light, 491
Speed of sound, 764
Temperature and the speed of sound, 768
Temperature as a function of depth, 528
Volume and pressure, 504
Water flow and nozzle pressure, 505, 768
Wavelength and frequency, 503
Wavelength of a musical note, 137
Wavelength of light, 270
Weight of a sphere, 505

Social Science
Adoption, 150
Age, 165, 192, 640
Fraternity or sorority membership, 162
Handshakes, 416
Ramps for the disabled, 774
Siblings, 624
Snapchat, 902
Social networking, 267–268
Spoken languages, 621
Spread of a rumor, 980
Volunteer work, 187

Sports/Entertainment
Amusement park visitors, 109, 145
Baseball diamond, 770, 774
Baseball throw, 774
Basketball scoring, 611, 639
Batting average, 484
Beach volleyball, 819
Bicycling, 488, 828, 829
Book club, 984
Boston Marathon, 149
Bungee jumping, 830
Capitol concert, 764
Contemporary art, 969
Cross-country skiing, 490
Cycling in Vietnam, 152
Diver's position, 69

Dubai ski run, 233
Earned run average, 505
Fastest roller coasters, 158–159
500 Festival Mini-Marathon, 164
Football yardage, 37, 53, 105
Games in a sports league, 275, 415
Golf, 639, 887
Gondola aerial lift, 245
Grade of a treadmill, 233, 546, 548
Hang time, 806–807, 809, 823, 883, 885
High school basketball court, 157
Hockey, 602, 716
Kingda Ka roller coaster, 810
Lacrosse, 156–157
Magazine reading, 150
Major League baseball attendance, 359
Marathon times, 810
Marching band performance, 764
Media usage, 272
Movie screens in China, 964
Movie theatre tickets, 194
Movies released, 990
Pool dimensions, 413
Race numbers, 409–410
Rollerblade costs, 196
Run differential, 60
Running, 152, 184, 448, 549, 620
SCAD diving, 282
Size of a league schedule, 137
Skiing, 187, 828
Skydiving, 282
Soccer field, 601
Super Bowl TV viewership, 78
Television viewership of NASCAR, 235
Tennis court, 648
Ticket revenue, 646
Ticket sales, 868
Vertical leap, 807
Walking, 490, 492, 966
Zipline, 423

Statistics/Demographics
Ages 65 and older, 1, 20
Ages 0–14 in Kenya, 22
Areas of Texas and Alaska, 146
Average number of motorcyclists killed, 864
Dimensions of Wyoming, 191
Haitian population ages 0–14, 145
Island population, 162
Life expectancy, 572, 864
Median age of men at first marriage, 584
Millennials living with parents, 515
Population, 863, 864
Population decrease, 69, 108, 239, 661, 968
Population growth, 239, 959, 967, 969, 980, 986
Population of the United States, 270
Senior population, 194, 283
Urban population, 571
World population, 148, 214, 967

Technology
Apps in the iTunes store, 78, 409
Computer pricing, 638
Computer repair, 581
Computer screens, 769
Copying time, 514
Digital media usage, 341, 359
Global mobile data traffic, 962
Information technology, 272
Internet traffic, 977
Length of a smartphone screen, 773
Memory board, 885
Office copiers, 491
Office printers, 491
Relative aperture, 503
Value of a computer, 549, 571
Video game delivery format, 587
Video game production, 621
Web developers, 651, 659

Transportation
Air travel, 619, 624, 645, 648, 828, 985
Airline tickets, 635–636
Airplane seating, 602
Airplane speed, 511
Airport control tower, 164
Auto travel, 618
Bicycle speed, 490
Bicycle travel, 768
Boat speed, 490
Boat travel, 771
Boating, 624, 626, 811
Bus travel, 497
Canoeing, 624, 828, 870
Car speed, 489, 490, 829, 841
Car travel, 515, 623, 825, 964
Car trips, 828
Chartering a bus, 977
Commuting, 30, 32, 775
Cycling distance, 488
Distance traveled, 32, 34, 115, 121, 137
Driving speed, 482, 490, 494, 514
Escape ramp on an airliner, 546
Grade of a transit system, 238
Height of an airplane, 767, 794
International travelers to the United States, 527
Interstate mile markers, 155, 191
Marine travel, 619, 775, 822, 825, 885, 888
Minimizing tolls, 678
Motorcycle travel, 646, 821–822, 883
Navigation, 247, 829
Parking lots, 726, 826, 827
Passports, 78
Point of no return, 624
Rate of travel, 504
Road-pavement messages, 770, 773
Sailing, 826, 827
Scheduled airline passengers, 531
Shipwreck, 43
Sightseeing boat, 620
States with the most miles of tollways, 517
Submarine, 37
Tractor speed, 490
Train speed, 489, 511, 771
Train travel, 248, 514, 618, 623, 645, 964, 985
Trucking speed, 490
U.S. transcontinental railroad, 637

Just-in-Time Review

The U.S. population ages 65 and older has continually increased since the baby boomers (those who were born between 1946 and 1964) began turning 65. These increases have both social and economic implications, most notably for Social Security and Medicare. By 2050, the population ages 65 and older is expected to be double that of 2012. Analyzing changes in the percentage of the population in all age groups is important when creating new programs. The graph at left shows percentages for two age groups in selected countries.

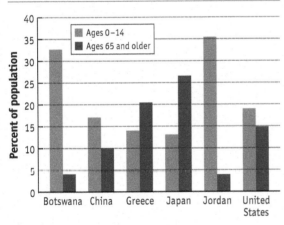

Population Ages 0–14 and Ages 65 and Older

DATA: *The CIA World Factbook 2017*

Data: U.S. Census Bureau, "An Aging Nation: The Older Population in the United States," by Jennifer M. Ortman, Victoria A. Vilkoff, and Howard Hogan

1. All Factors of a Number
2. Prime Factorizations
3. Greatest Common Factor
4. Least Common Multiple
5. Equivalent Expressions and Fraction Notation
6. Mixed Numerals
7. Simplify Fraction Notation
8. Multiply and Divide Fraction Notation
9. Add and Subtract Fraction Notation
10. Convert from Decimal Notation to Fraction Notation
11. Add and Subtract Decimal Notation
12. Multiply and Divide Decimal Notation
13. Convert from Fraction Notation to Decimal Notation
14. Rounding with Decimal Notation
15. Convert between Percent Notation and Decimal Notation
16. Convert between Percent Notation and Fraction Notation
17. Exponential Notation
18. Order of Operations

In Example 3 of Just-in-Time 15, we will express as a percentage the portion of the U.S. population ages 65 and older projected for 2060.

Just-in-Time Review

1 ALL FACTORS OF A NUMBER

Factoring is a necessary skill for addition and subtraction with fraction notation. Factoring is also an important skill in algebra. The numbers we will be factoring are **natural numbers**:

$$1, 2, 3, 4, 5, \text{ and so on.}$$

To **factor** a number means to express the number as a product. Consider the product $12 = 3 \cdot 4$. We say that 3 and 4 are **factors** of 12 and that $3 \cdot 4$ is a **factorization** of 12. Since $12 = 1 \cdot 12$ and $12 = 2 \cdot 6$, we also know that 1, 12, 2, and 6 are factors of 12 and that $1 \cdot 12$ and $2 \cdot 6$ are factorizations of 12.

EXAMPLE 1 Find all the factors of 77.

We first find some factorizations:

$$77 = 1 \cdot 77,$$
$$77 = 7 \cdot 11.$$

The factors of 77 are 1, 7, 11, and 77.

EXAMPLE 2 Find all the factors of 150.

We first find some factorizations:

$$150 = 1 \cdot 150,$$
$$150 = 2 \cdot 75,$$
$$150 = 3 \cdot 50,$$
$$150 = 5 \cdot 30,$$
$$150 = 6 \cdot 25,$$
$$150 = 10 \cdot 15.$$

The factors of 150 are 1, 2, 3, 5, 6, 10, 15, 25, 30, 50, 75, and 150.

Do Exercises 1–8.

EXERCISES

Find all factors of the given number.

1. 20
2. 39
3. 81
4. 47
5. 160
6. 45
7. 28
8. 36

2 PRIME FACTORIZATIONS

A natural number that has *exactly two different factors*, itself and 1, is called a **prime number**.

EXAMPLE 1 Which of these numbers are prime? 7, 4, 11, 18, 1

7 is prime. It has exactly two different factors, 1 and 7.
4 is *not* prime. It has three different factors, 1, 2, and 4.
11 is prime. It has exactly two different factors, 1 and 11.
18 is *not* prime. It has factors 1, 2, 3, 6, 9, and 18.
1 is *not* prime. It does not have two *different* factors.

In the margin at right is a table of the prime numbers from 2 to 157. These prime numbers will be helpful to you in this text.

If a natural number, other than 1, is not prime, we call it **composite**. Every composite number can be factored into a product of prime numbers. Such a factorization is called a **prime factorization**.

EXAMPLE 2 Find the prime factorization of 36.

We begin by factoring 36 any way we can. One way is like this:

$36 = 4 \cdot 9$ The factors 4 and 9 are not prime, so we factor them.
$= 2 \cdot 2 \cdot 3 \cdot 3$.

The factors in the last factorization are all prime, so we now have the *prime factorization* of 36. Note that 1 is *not* part of this factorization because it is not prime.

Another way to find the prime factorization of 36 is like this:

$36 = 2 \cdot 18 = 2 \cdot 3 \cdot 6 = 2 \cdot 3 \cdot 2 \cdot 3$.

EXAMPLE 3 Find the prime factorization of 80.

One way to factor 80 is $8 \cdot 10$. Here we use a factor tree to find the prime factors of 80.

Each factor in $2 \cdot 2 \cdot 2 \cdot 2 \cdot 5$ is prime. This is the prime factorization.

Do Exercises 1–8. ▶

EXERCISES

Find the prime factorization of the given number.

1. 33
2. 121
3. 18
4. 56
5. 120
6. 90
7. 210
8. 91

A TABLE OF PRIMES
2, 3, 5, 7, 11, 13, 17, 19, 23, 29, 31, 37, 41, 43, 47, 53, 59, 61, 67, 71, 73, 79, 83, 89, 97, 101, 103, 107, 109, 113, 127, 131, 137, 139, 149, 151, 157

Just-in-Time Review

3 GREATEST COMMON FACTOR

The numbers 20 and 30 have several factors in common, among them 2 and 5. The greatest of the common factors is called the **greatest common factor, GCF**. One way to find the GCF is by making a list of factors of each number.

List all the factors of 20: <u>1</u>, <u>2</u>, 4, <u>5</u>, <u>10</u>, and 20.
List all the factors of 30: <u>1</u>, <u>2</u>, 3, <u>5</u>, 6, <u>10</u>, 15, and 30.

We underline the common factors. The greatest common factor, the GCF, is 10.

The preceding procedure gives meaning to the notion of a GCF, but the following method, using prime factorizations, is generally faster.

EXAMPLE 1 Find the GCF of 20 and 30.

We find the prime factorization of each number. Then we draw lines between the common factors.

$$20 = 2 \cdot 2 \cdot 5$$
$$30 = 2 \cdot 3 \cdot 5$$

The GCF $= 2 \cdot 5 = 10$.

EXAMPLE 2 Find the GCF of 54, 90, and 252.

We find the prime factorization of each number. Then we draw lines between the common factors.

$$54 = 2 \cdot 3 \cdot 3 \cdot 3$$
$$90 = 2 \cdot 3 \cdot 3 \cdot 5$$
$$252 = 2 \cdot 2 \cdot 3 \cdot 3 \cdot 7$$

The GCF $= 2 \cdot 3 \cdot 3 = 18$.

EXAMPLE 3 Find the GCF of 30 and 77.

We find the prime factorization of each number.

$$30 = 2 \cdot 3 \cdot 5$$
$$77 = 7 \cdot 11$$

Since there is no common prime factor, the GCF is 1.

Do Exercises 1–10.

EXERCISES

Find the GCF of the given numbers.

1. 36, 48
2. 13, 52
3. 27, 40
4. 54, 180
5. 18, 66
6. 30, 135
7. 40, 220
8. 14, 42
9. 15, 40, 60
10. 70, 105, 350

4 LEAST COMMON MULTIPLE

Two or more numbers always have many multiples in common. From lists of multiples, we can find common multiples. To find the common multiples of 2 and 3, we circle the multiples that appear in both lists of multiples:

2, 4, ⑥, 8, 10, ⑫, 14, 16, ⑱, 20, 22, ㉔, 26, 28, ㉚, . . . ;
3, ⑥, 9, ⑫, 15, ⑱, 21, ㉔, 27, ㉚, ...

The common multiples of 2 and 3 are 6, 12, 18, 24, 30. . . .

The *least*, or smallest, of those common multiples is 6. We abbreviate **least common multiple** as **LCM**.

EXAMPLE 1 Find the LCM of 9 and 15.

We first look at the factorizations of 9 and 15:

$$9 = 3 \cdot 3, \quad 15 = 3 \cdot 5.$$

Any multiple of 9 must have *two* 3's as factors. Any multiple of 15 must have *one* 3 and *one* 5 as factors. The smallest multiple of 9 and 15 is

$$3 \cdot 3 \cdot 5 = 45.$$

— Two 3's; 9 is a factor
— One 3, one 5; 15 is a factor

The LCM must have all the factors of 9 and all the factors of 15, *but the factors are not repeated when they are common to both numbers.*

To find the LCM of several numbers using prime factorizations:
a) Write the prime factorization of each number.
b) Form the LCM by writing the product of the different factors from step (a), using each factor the greatest number of times that it occurs in any *one* of the factorizations.

EXAMPLE 2 Find the LCM of 40 and 100.
a) We first find the prime factorizations:

$$40 = 2 \cdot 2 \cdot 2 \cdot 5,$$
$$100 = 2 \cdot 2 \cdot 5 \cdot 5.$$

b) The different prime factors are 2 and 5. We write 2 as a factor three times (the greatest number of times that it occurs in any *one* factorization). We write 5 as a factor two times (the greatest number of times that it occurs in any *one* factorization).

The LCM is $2 \cdot 2 \cdot 2 \cdot 5 \cdot 5$, or 200.

(continued)

Just-in-Time Review

4 LEAST COMMON MULTIPLE (continued)

EXAMPLE 3 Find the LCM of 27, 90, and 84.

a) We first find the prime factorizations:

$27 = 3 \cdot 3 \cdot 3,$
$90 = 2 \cdot 3 \cdot 3 \cdot 5,$
$84 = 2 \cdot 2 \cdot 3 \cdot 7.$

b) We write 2 as a factor two times, 3 three times, 5 one time, and 7 one time.

The LCM is $2 \cdot 2 \cdot 3 \cdot 3 \cdot 3 \cdot 5 \cdot 7$, or 3780.

EXAMPLE 4 Find the LCM of 7 and 21.

Since 7 is prime, it has no prime factorization. It still, however, must be a factor of the LCM:

$7 = 7,$
$21 = 3 \cdot 7.$

The LCM is $7 \cdot 3$, or 21.

> If one number is a factor of another, then the LCM is the larger of the two numbers.

EXAMPLE 5 Find the LCM of 8 and 9.

We have

$8 = 2 \cdot 2 \cdot 2,$
$9 = 3 \cdot 3.$

The LCM is $2 \cdot 2 \cdot 2 \cdot 3 \cdot 3$, or 72.

> If two or more numbers have no common prime factor, then the LCM is the product of the numbers.

Do Exercises 1–10. ▶

EXERCISES

Find the LCM of the given numbers.

1. 24, 27
2. 3, 15
3. 50, 60
4. 13, 23
5. 45, 72
6. 30, 36
7. 12, 28
8. 8, 16, 22
9. 5, 12, 15
10. 24, 35, 45

5 EQUIVALENT EXPRESSIONS AND FRACTION NOTATION

An example of **fraction notation** for a number is

$\dfrac{2}{3}$ ← Numerator
← Denominator

The **whole numbers** consist of the natural numbers and 0:

0, 1, 2, 3, 4, 5, ...

The **arithmetic numbers**, also called the **nonnegative rational numbers**, consist of the whole numbers and the fractions, such as 8, $\frac{2}{3}$, and $\frac{9}{5}$. All these numbers can be named with fraction notation $\frac{a}{b}$, where a and b are whole numbers and $b \neq 0$.

Note that all whole numbers can be named with fraction notation. For example, we can name the whole number 8 as $\frac{8}{1}$. We call 8 and $\frac{8}{1}$ **equivalent expressions**. Two simple but powerful properties of numbers that allow us to find equivalent expressions are the identity properties of 0 and 1.

THE IDENTITY PROPERTY OF 0 (ADDITIVE IDENTITY)
For any number a,

$$a + 0 = a.$$

(Adding 0 to any number gives that same number—for example, $12 + 0 = 12$.)

THE IDENTITY PROPERTY OF 1 (MULTIPLICATIVE IDENTITY)
For any number a,

$$a \cdot 1 = a.$$

(Multiplying any number by 1 gives that same number—for example, $\frac{3}{5} \cdot 1 = \frac{3}{5}$.)

EQUIVALENT EXPRESSIONS FOR 1
For any number a, $a \neq 0$,

$$\dfrac{a}{a} = 1.$$

(For example, some ways to name the number 1 are $\frac{5}{5}$, $\frac{3}{3}$, and $\frac{26}{26}$.)

EXERCISES

Write an equivalent expression with the given denominator.

1. $\dfrac{7}{8}$ (Denominator: 24)

2. $\dfrac{5}{6}$ (Denominator: 48)

3. $\dfrac{5}{4}$ (Denominator: 16)

4. $\dfrac{2}{9}$ (Denominator: 54)

5. $\dfrac{3}{11}$ (Denominator: 77)

6. $\dfrac{13}{16}$ (Denominator: 80)

(continued)

Just-in-Time Review

5   EQUIVALENT EXPRESSIONS AND FRACTION NOTATION (continued)

EXAMPLE 1   Write a fraction expression equivalent to $\frac{2}{3}$ with a denominator of 15.

$$\frac{2}{3} = \frac{2}{3} \cdot 1 \qquad \text{Using the identity property of 1}$$

$$= \frac{2}{3} \cdot \frac{5}{5} \qquad \text{Using } \frac{5}{5} \text{ for 1}$$

$$= \frac{10}{15}. \qquad \text{Multiplying numerators and denominators}$$

Do Exercises 1–6 on the preceding page. ▶

6   MIXED NUMERALS

A mixed numeral like $2\frac{3}{8}$ represents a sum: $2 + \frac{3}{8}$.

Whole number ⎯⎯⎯⎦   ⎣⎯⎯⎯ Fraction less than 1.

To convert $2\frac{3}{8}$ from a mixed numeral to fraction notation:

(a) Multiply the whole number 2 by the denominator 8: $2 \cdot 8 = 16$.
(b) Add the result 16 to the numerator 3: $16 + 3 = 19$.
(c) Keep the denominator 8.

$$2\frac{3}{8} = \frac{19}{8}$$

EXAMPLE 1   Convert to fraction notation.

$$6\frac{2}{3} = \frac{20}{3} \qquad 6 \cdot 3 = 18,\ 18 + 2 = 20$$

To convert from fraction notation to a mixed numeral, divide.

$$\frac{13}{5} \qquad 5\overline{)13} \qquad 2\frac{3}{5}$$

- The quotient: 2
- The divisor: 5
- The remainder: 3

EXAMPLE 2   Convert to a mixed numeral.

$$\frac{69}{10} \qquad 10\overline{)69} \qquad \frac{69}{10} = 6\frac{9}{10}$$

Do Exercises 1–10. ▶

MyLab Math   VIDEO

EXERCISES

Convert to fraction notation.

1. $5\frac{2}{3}$

2. $9\frac{1}{10}$

3. $30\frac{4}{5}$

4. $1\frac{5}{8}$

5. $66\frac{2}{3}$

Convert to a mixed numeral.

6. $\frac{18}{5}$

7. $\frac{29}{6}$

8. $\frac{57}{10}$

9. $\frac{40}{3}$

10. $\frac{757}{100}$

8

7   SIMPLIFY FRACTION NOTATION

We know that $\frac{1}{2}, \frac{2}{4}, \frac{4}{8}$, and so on, all name the same number. Any arithmetic number can be named in many ways. The **simplest fraction notation** is the notation that has the smallest numerator and denominator. We call the process of finding the simplest fraction notation **simplifying**. When simplifying fractions, we remove factors of 1.

EXAMPLE 1   Simplify: $\frac{10}{15}$.

$$\frac{10}{15} = \frac{2 \cdot 5}{3 \cdot 5}$$ Factoring the numerator and the denominator. In this case, each is the prime factorization.

$$= \frac{2}{3} \cdot \frac{5}{5}$$ Factoring the fraction expression

$$= \frac{2}{3} \cdot 1 \quad \frac{5}{5} = 1$$

$$= \frac{2}{3}$$ Using the identity property of 1 (removing a factor of 1)

EXAMPLE 2   Simplify: $\frac{36}{24}$.

$$\frac{36}{24} = \frac{2 \cdot 3 \cdot 2 \cdot 3}{2 \cdot 2 \cdot 3 \cdot 2} = \frac{2 \cdot 3 \cdot 2}{2 \cdot 3 \cdot 2} \cdot \frac{3}{2} = 1 \cdot \frac{3}{2} = \frac{3}{2}$$

Canceling

Canceling is a shortcut that you may have used to remove a factor of 1 when working with fraction notation. With *great* concern, we mention it as a possible way to speed up your work. You should use canceling only when removing common factors in numerators and denominators. Each common factor allows us to remove a factor of 1 in a product. **Canceling** *cannot* **be done when adding.** Example 2 might have been done faster as follows:

$$\frac{36}{24} = \frac{\cancel{2} \cdot \cancel{3} \cdot \cancel{2} \cdot 3}{\cancel{2} \cdot \cancel{2} \cdot \cancel{3} \cdot 2} = \frac{3}{2}, \quad \text{or} \quad \frac{36}{24} = \frac{3 \cdot \cancel{12}}{2 \cdot \cancel{12}} = \frac{3}{2}, \quad \text{or} \quad \frac{\overset{3}{\cancel{\underset{\cancel{12}}{\underset{2}{36}}}}}{\underset{2}{\cancel{24}}} = \frac{3}{2}.$$

EXAMPLE 3   Simplify: $\frac{18}{72}$.

$$\frac{18}{72} = \frac{2 \cdot 9}{8 \cdot 9} = \frac{2}{8} = \frac{2 \cdot 1}{2 \cdot 4} = \frac{1}{4}, \quad \text{or} \quad \frac{18}{72} = \frac{1 \cdot 18}{4 \cdot 18} = \frac{1}{4}$$

Do Exercises 1–12. ▶

EXERCISES

Simplify.

1. $\dfrac{18}{27}$

2. $\dfrac{49}{56}$

3. $\dfrac{30}{72}$

4. $\dfrac{240}{600}$

5. $\dfrac{56}{14}$

6. $\dfrac{45}{60}$

7. $\dfrac{216}{18}$

8. $\dfrac{6}{42}$

9. $\dfrac{33}{81}$

10. $\dfrac{2600}{1400}$

11. $\dfrac{84}{126}$

12. $\dfrac{325}{625}$

Just-in-Time Review

8 — MULTIPLY AND DIVIDE FRACTION NOTATION

MULTIPLYING FRACTIONS

To multiply fractions, multiply the numerators and multiply the denominators:

$$\frac{a}{b} \cdot \frac{c}{d} = \frac{a \cdot c}{b \cdot d}.$$

EXAMPLE 1 Multiply and simplify: $\frac{5}{6} \cdot \frac{9}{25}$.

$$\frac{5}{6} \cdot \frac{9}{25} = \frac{5 \cdot 9}{6 \cdot 25} \quad \text{Multiplying numerators and multiplying denominators}$$

$$= \frac{5 \cdot 3 \cdot 3}{2 \cdot 3 \cdot 5 \cdot 5} \quad \text{Factoring the numerator and the denominator}$$

$$= \frac{\cancel{5} \cdot \cancel{3} \cdot 3}{2 \cdot \cancel{3} \cdot \cancel{5} \cdot 5} \quad \text{Removing a factor of 1: } \frac{3 \cdot 5}{3 \cdot 5} = 1$$

$$= \frac{3}{10} \quad \text{Simplifying}$$

Two numbers whose product is 1 are called **reciprocals**, or **multiplicative inverses**. All the arithmetic numbers, except zero, have reciprocals.

EXAMPLES

2. The reciprocal of $\frac{2}{3}$ is $\frac{3}{2}$ because $\frac{2}{3} \cdot \frac{3}{2} = \frac{6}{6} = 1$.
3. The reciprocal of 9 is $\frac{1}{9}$ because $9 \cdot \frac{1}{9} = \frac{9}{1} \cdot \frac{1}{9} = \frac{9}{9} = 1$.
4. The reciprocal of $\frac{1}{4}$ is 4 because $\frac{1}{4} \cdot 4 = \frac{4}{4} = 1$.

Reciprocals and the number 1 can be used to justify a quick way to divide arithmetic numbers. We multiply by 1, carefully choosing the expression for 1.

This is a symbol for 1.

$$\frac{2}{3} \div \frac{7}{5} = \frac{\frac{2}{3}}{\frac{7}{5}} = \frac{\frac{2}{3}}{\frac{7}{5}} \cdot \frac{\frac{5}{7}}{\frac{5}{7}} \quad \text{Multiplying by } \frac{\frac{5}{7}}{\frac{5}{7}}. \text{ We use } \frac{5}{7} \text{ because it is the reciprocal of the divisor, } \frac{7}{5}.$$

$$= \frac{\frac{2}{3} \cdot \frac{5}{7}}{\frac{7}{5} \cdot \frac{5}{7}} = \frac{\frac{10}{21}}{\frac{35}{35}} = \frac{\frac{10}{21}}{1} = \frac{10}{21}$$

This is the same result that we would have found if we had multiplied $\frac{2}{3}$ by the reciprocal of the divisor:

$$\frac{2}{3} \cdot \frac{5}{7} = \frac{2 \cdot 5}{3 \cdot 7} = \frac{10}{21}.$$

(continued)

8 MULTIPLY AND DIVIDE FRACTION NOTATION (continued)

DIVIDING FRACTIONS
To divide fractions, multiply by the reciprocal of the divisor:

$$\frac{a}{b} \div \frac{c}{d} = \frac{a}{b} \cdot \frac{d}{c}.$$

EXAMPLE 5 Divide by multiplying by the reciprocal of the divisor: $\frac{1}{2} \div \frac{3}{5}$.

$$\frac{1}{2} \div \frac{3}{5} = \frac{1}{2} \cdot \frac{5}{3} \qquad \text{$\frac{5}{3}$ is the reciprocal of $\frac{3}{5}$}$$
$$= \frac{5}{6} \qquad \text{Multiplying}$$

EXAMPLE 6 Divide and simplify: $\frac{2}{3} \div \frac{4}{9}$.

$$\frac{2}{3} \div \frac{4}{9} = \frac{2}{3} \cdot \frac{9}{4} \qquad \text{$\frac{9}{4}$ is the reciprocal of $\frac{4}{9}$}$$
$$= \frac{2 \cdot 9}{3 \cdot 4} \qquad \text{Multiplying numerators and denominators}$$
$$= \frac{2 \cdot 3 \cdot 3}{3 \cdot 2 \cdot 2} \qquad \text{Removing a factor of 1: } \frac{2 \cdot 3}{2 \cdot 3} = 1$$
$$= \frac{3}{2}$$

EXAMPLE 7 Divide and simplify: $\frac{5}{6} \div 30$.

$$\frac{5}{6} \div 30 = \frac{5}{6} \div \frac{30}{1} = \frac{5}{6} \cdot \frac{1}{30} = \frac{5 \cdot 1}{6 \cdot 30} = \frac{5 \cdot 1}{6 \cdot 5 \cdot 6} = \frac{1}{6 \cdot 6} = \frac{1}{36}$$

Removing a factor of 1: $\frac{5}{5} = 1$

EXAMPLE 8 Divide and simplify: $24 \div \frac{3}{8}$.

$$24 \div \frac{3}{8} = \frac{24}{1} \div \frac{3}{8} = \frac{24}{1} \cdot \frac{8}{3} = \frac{24 \cdot 8}{1 \cdot 3} = \frac{3 \cdot 8 \cdot 8}{1 \cdot 3} = \frac{8 \cdot 8}{1} = 64$$

Removing a factor of 1: $\frac{3}{3} = 1$

Do Exercises 1–10.

EXERCISES
Compute and simplify.

1. $\dfrac{3}{7} \cdot \dfrac{9}{10}$

2. $5 \cdot \dfrac{2}{3}$

3. $\dfrac{3}{4} \div \dfrac{3}{7}$

4. $\dfrac{15}{16} \cdot \dfrac{8}{5}$

5. $\dfrac{2}{5} \div \dfrac{7}{3}$

6. $\dfrac{8}{9} \div \dfrac{4}{15}$

7. $\dfrac{1}{20} \div \dfrac{1}{5}$

8. $\dfrac{22}{35} \cdot \dfrac{5}{11}$

9. $\dfrac{10}{11} \cdot \dfrac{11}{10}$

10. $\dfrac{3}{4} \div 8$

Just-in-Time Review

9 ADD AND SUBTRACT FRACTION NOTATION

MyLab Math
VIDEO

ADDING OR SUBTRACTING FRACTIONS WITH LIKE DENOMINATORS
To add or subtract fractions when denominators are the same, add the numerators and keep the same denominator:

$$\frac{a}{c} + \frac{b}{c} = \frac{a+b}{c}; \quad \frac{a}{c} - \frac{b}{c} = \frac{a-b}{c}.$$

EXAMPLES

1. $\dfrac{4}{8} + \dfrac{5}{8} = \dfrac{4+5}{8} = \dfrac{9}{8}$

2. $\dfrac{13}{5} - \dfrac{6}{5} = \dfrac{13-6}{5} = \dfrac{7}{5}$

ADDING OR SUBTRACTING FRACTIONS WITH DIFFERENT DENOMINATORS
To add or subtract fractions when denominators are different:
a) Find the least common multiple of the denominators. That number is the least common denominator, LCD.
b) Multiply by 1, using the appropriate notation n/n for each fraction to express fractions in terms of the LCD.
c) Add or subtract the numerators, keeping the same denominator.
d) Simplify, if possible.

EXAMPLE 3 Add and simplify: $\dfrac{3}{8} + \dfrac{5}{12}$.

The LCM of the denominators, 8 and 12, is 24. Thus the LCD is 24. We multiply each fraction by 1 to obtain the LCD:

$\dfrac{3}{8} + \dfrac{5}{12} = \dfrac{3}{8} \cdot \dfrac{3}{3} + \dfrac{5}{12} \cdot \dfrac{2}{2}$ Multiplying by 1. Since $3 \cdot 8 = 24$, we multiply the first number by $\frac{3}{3}$. Since $2 \cdot 12 = 24$, we multiply the second number by $\frac{2}{2}$.

$= \dfrac{9}{24} + \dfrac{10}{24}$

$= \dfrac{9+10}{24}$ Adding the numerators and keeping the same denominator

$= \dfrac{19}{24}.$ $\frac{19}{24}$ is in simplest form.

CALCULATOR CORNER

Operations on Fractions We can perform operations on fractions on a graphing calculator. Selecting the ▶FRAC option from the MATH menu causes the result to be expressed in fraction form. The calculator display is shown below.

    3/4+1/2▶Frac
                    5/4

EXERCISES: Perform each calculation. Give the answer in fraction notation.

1. $\dfrac{5}{6} + \dfrac{7}{8}$

2. $\dfrac{13}{16} - \dfrac{4}{7}$

3. $\dfrac{15}{4} \cdot \dfrac{7}{12}$

4. $\dfrac{1}{5} \div \dfrac{3}{10}$

(continued)

9 ADD AND SUBTRACT FRACTION NOTATION (continued)

EXAMPLE 4 Add and simplify: $\dfrac{11}{30} + \dfrac{5}{18}$.

We first look for the LCM of 30 and 18. That number is then the LCD. We find the prime factorization of each denominator:

$$\dfrac{11}{30} + \dfrac{5}{18} = \dfrac{11}{5 \cdot 2 \cdot 3} + \dfrac{5}{2 \cdot 3 \cdot 3}.$$

The LCD is $5 \cdot 2 \cdot 3 \cdot 3$, or 90. To get the LCD in the first denominator, we need a factor of 3. To get the LCD in the second denominator, we need a factor of 5. We get these numbers by multiplying by 1:

$\dfrac{11}{30} + \dfrac{5}{18} = \dfrac{11}{5 \cdot 2 \cdot 3} \cdot \dfrac{3}{3} + \dfrac{5}{2 \cdot 3 \cdot 3} \cdot \dfrac{5}{5}$ Multiplying by 1

$= \dfrac{33}{5 \cdot 2 \cdot 3 \cdot 3} + \dfrac{25}{2 \cdot 3 \cdot 3 \cdot 5}$ The denominators are now the LCD.

$= \dfrac{58}{5 \cdot 2 \cdot 3 \cdot 3}$ Adding the numerators and keeping the LCD

$= \dfrac{2 \cdot 29}{5 \cdot 2 \cdot 3 \cdot 3}$ Factoring the numerator and removing a factor of 1

$= \dfrac{29}{45}.$ Simplifying

EXAMPLE 5 Subtract and simplify: $\dfrac{9}{8} - \dfrac{4}{5}$.

$\dfrac{9}{8} - \dfrac{4}{5} = \dfrac{9}{8} \cdot \dfrac{5}{5} - \dfrac{4}{5} \cdot \dfrac{8}{8}$ The LCD is 40.

$= \dfrac{45}{40} - \dfrac{32}{40}$

$= \dfrac{45 - 32}{40}$ Subtracting the numerators and keeping the same denominator

$= \dfrac{13}{40}$ $\dfrac{13}{40}$ is in simplest form.

EXAMPLE 6 Subtract and simplify: $\dfrac{7}{10} - \dfrac{1}{5}$.

$\dfrac{7}{10} - \dfrac{1}{5} = \dfrac{7}{10} - \dfrac{1}{5} \cdot \dfrac{2}{2}$ The LCD is 10; $\dfrac{7}{10}$ already has the LCD.

$= \dfrac{7}{10} - \dfrac{2}{10} = \dfrac{7-2}{10}$

$= \dfrac{5}{10}$

$= \dfrac{1 \cdot \cancel{5}}{2 \cdot \cancel{5}} = \dfrac{1}{2}$ Removing a factor of 1: $\dfrac{5}{5} = 1$

Do Exercises 1–12. ▶

EXERCISES
Compute and simplify.

1. $\dfrac{5}{11} + \dfrac{3}{11}$

2. $\dfrac{12}{5} - \dfrac{2}{5}$

3. $\dfrac{11}{12} - \dfrac{3}{8}$

4. $\dfrac{4}{9} + \dfrac{13}{18}$

5. $\dfrac{3}{10} + \dfrac{8}{15}$

6. $\dfrac{3}{16} - \dfrac{1}{18}$

7. $\dfrac{7}{30} + \dfrac{5}{12}$

8. $\dfrac{15}{16} - \dfrac{5}{12}$

9. $\dfrac{11}{12} - \dfrac{2}{5}$

10. $\dfrac{1}{4} + \dfrac{1}{3}$

11. $\dfrac{9}{8} + \dfrac{7}{12}$

12. $\dfrac{147}{50} - 2$

13

Just-in-Time Review

10 ▸ CONVERT FROM DECIMAL NOTATION TO FRACTION NOTATION

A laptop is on sale for $1576.98. This amount is given in **decimal notation**. The following place-value chart shows the place value of each digit in 1576.98.

PLACE-VALUE CHART								
Ten Thousands	Thousands	Hundreds	Tens	Ones	Tenths	Hundredths	Thousandths	Ten-Thousandths
10,000	1000	100	10	1	$\frac{1}{10}$	$\frac{1}{100}$	$\frac{1}{1000}$	$\frac{1}{10,000}$
	1	5	7	6 .	9	8		

Look for a pattern in the following products:

$$0.6875 = 0.6875 \times 1 = 0.6875 \times \frac{10,000}{10,000} = \frac{0.6875 \times 10,000}{10,000} = \frac{6875}{10,000};$$

$$53.47 = 53.47 \times 1 = 53.47 \times \frac{100}{100} = \frac{53.47 \times 100}{100} = \frac{5347}{100}.$$

To convert from decimal notation to fraction notation:
a) Count the number of decimal places.   4.98   2 places
b) Move the decimal point that many places to the right.   4.98.   Move 2 places.
c) Write the result over a denominator with a 1 followed by that number of zeros. $\frac{498}{100}$ 2 zeros

EXAMPLES Convert to fraction notation. Do not simplify.

1. 0.876   0.876. $0.876 = \frac{876}{1000}$
    3 places   3 zeros

2. 1.5018   1.5018. $1.5018 = \frac{15,018}{10,000}$
    4 places   4 zeros

Do Exercises 1–8. ▸

EXERCISES

Convert to fraction notation. Do not simplify.

1. 5.3
2. 0.67
3. 4.0008
4. 1122.3
5. 14.703
6. 0.9
7. 183.42
8. 0.006

11 ADD AND SUBTRACT DECIMAL NOTATION

Adding with decimal notation is similar to adding whole numbers. First we line up the decimal points. Then we add the digits with the same place value going from right to left, carrying if necessary.

EXAMPLE 1 Add: $74 + 26.46 + 0.998$.

$$\begin{array}{r} \overset{1\ 1\ 1}{}\\ 7\,4.\\ 2\,6.4\,6\\ +0.9\,9\,8\\ \hline 1\,0\,1.4\,5\,8 \end{array}$$

You can place extra zeros to the right of any decimal point so that there are the same number of decimal places in all the addends, but this is not necessary. If you did so, the preceding problem would look like this:

$$\begin{array}{r} \overset{1\ 1\ 1}{}\\ 7\,4.0\,0\,0\\ 2\,6.4\,6\,0\\ +0.9\,9\,8\\ \hline 1\,0\,1.4\,5\,8 \end{array}$$

  Adding zeros to 74
  Adding a zero to 26.46

Subtracting with decimal notation is similar to subtracting whole numbers. First we line up the decimal points. Then we subtract the digits with the same place value going from right to left, borrowing if necessary. Extra zeros can be added if needed.

EXAMPLES

2. Subtract: $76.14 - 18.953$.

$$\begin{array}{r} 7\,6.1\,4\,0\\ -\,1\,8.9\,5\,3\\ \hline 5\,7.1\,8\,7 \end{array}$$

3. Subtract: $200 - 0.68$.

$$\begin{array}{r} 2\,0\,0.0\,0\\ -0.6\,8\\ \hline 1\,9\,9.3\,2 \end{array}$$

Do Exercises 1–8. ▶

EXERCISES

Add.

1. $4\,1\,5.7\,8$
   $+2\,9.1\,6$

2. $3\,5.$
   $7.2\,1\,4$
   $+1\,2\,8.6\,3$

3. $17.95 + 16.99 + 28.85$

4. $0.6 + 2000.43 + 7.213$

Subtract.

5. $7\,8.1\,1\,0$
   $-\,4\,5.8\,7\,6$

6. $3\,8.7$
   $-\,1\,1.8\,6\,5$

7. $2.6 - 1.08$

8. $3 - 1.0807$

CALCULATOR CORNER

Operations with Decimal Notation We can perform operations with decimals on a graphing calculator. The following calculator display illustrates $62.043 - 48.915$ and 6.73×2.18. Note that the subtraction operation key ⊖ must be used rather than the opposite key ⊝ when subtracting. We will discuss the use of the ⊝ key in Chapter 1.

```
62.043−48.915
                13.128
6.73∗2.18
                14.6714
```

EXERCISES: Use a calculator to perform each operation.

1. $26 + 13.47 + 0.95$
2. $9.03 - 5.7$
3. 0.159×4.36
4. $135.66 \div 57$

Just-in-Time Review

12 MULTIPLY AND DIVIDE DECIMAL NOTATION

Look at this product.

$$5.14 \times 0.8 = \frac{514}{100} \times \frac{8}{10} = \frac{514 \times 8}{100 \times 10} = \frac{4112}{1000} = 4.112$$

2 places, 1 place, 3 places

We can also do this calculation more quickly by first ignoring the decimal points and multiplying the whole numbers. Then we can determine the position of the decimal point by adding the number of decimal places in the original factors.

EXAMPLE 1 Multiply: 5.14×0.8.

$$\begin{array}{r} 5.1\,4 \leftarrow \text{2 decimal places} \\ \times \quad 0.8 \leftarrow \text{1 decimal place} \\ \hline 4.1\,1\,2 \end{array}$$

— 3 decimal places

When dividing with decimal notation when the divisor is a whole number, we place the decimal point in the quotient directly above the decimal point in the dividend. Then we divide as we do with whole numbers.

EXAMPLE 2 Divide: $216.75 \div 25$.

$$25\overline{)216.75}$$
Place the decimal point.

$$\begin{array}{r} 8.6\,7 \\ 25\overline{)216.75} \\ \underline{200} \\ 167 \\ \underline{150} \\ 175 \\ \underline{175} \\ 0 \end{array}$$

Divide as though dividing whole numbers.

EXAMPLE 3 Divide: $54 \div 8$.

$$\begin{array}{r} 6.7\,5 \\ 8\overline{)54.00} \\ \underline{48} \\ 60 \\ \underline{56} \\ 40 \\ \underline{40} \\ 0 \end{array}$$

Extra zeros are written to the right of the decimal point as needed.

EXERCISES

Multiply.

1. $\begin{array}{r} 7.3\,4 \\ \times \quad 1.8 \\ \hline \end{array}$

2. $\begin{array}{r} 0.8\,6 \\ \times 0.9\,3 \\ \hline \end{array}$

3. $\begin{array}{r} 0.0\,0\,2\,4 \\ \times \quad 0.0\,1\,5 \\ \hline \end{array}$

4. $\begin{array}{r} 0.4\,5\,7 \\ \times \quad 3.0\,8 \\ \hline \end{array}$

Divide.

5. $7.8\overline{)72.54}$

6. $72\overline{)165.6}$

7. $1.05\overline{)693}$

8. $0.47\overline{)0.1222}$

(continued)

12 MULTIPLY AND DIVIDE DECIMAL NOTATION (continued)

When dividing with decimal notation when the divisor is not a whole number, we move the decimal point in the divisor as many places to the right as it takes to make it a whole number. Next, we move the decimal point in the dividend the same number of places to the right and place the decimal point above it in the quotient. Then we divide as we would with whole numbers, inserting zeros if necessary.

EXAMPLE 4 Divide: $83.79 \div 0.098$.

$$
0.098.\overline{)83.790.}
$$

$$
\begin{array}{r}
855. \\
0.098\overline{)83.790} \\
\underline{784} \\
539 \\
\underline{490} \\
490 \\
\underline{490} \\
0
\end{array}
$$

Do Exercises 1–8 on the preceding page. ▶

13 CONVERT FROM FRACTION NOTATION TO DECIMAL NOTATION

MyLab Math
VIDEO

To convert from fraction notation to decimal notation when the denominator is not a number like 10, 100, or 1000, we divide the numerator by the denominator.

EXAMPLE 1

Convert to decimal notation: $\frac{5}{16}$.

$$
\begin{array}{r}
0.3125 \\
16\overline{)5.0000} \\
\underline{48} \\
20 \\
\underline{16} \\
40 \\
\underline{32} \\
80 \\
\underline{80} \\
0
\end{array}
$$

If we get a remainder of 0, we say that the decimal *terminates*. Thus, $\frac{5}{16} = 0.3125$.

EXAMPLE 2

Convert to decimal notation: $\frac{7}{12}$.

$$
\begin{array}{r}
0.5833 \\
12\overline{)7.0000} \\
\underline{60} \\
100 \\
\underline{96} \\
40 \\
\underline{36} \\
40 \\
\underline{36} \\
4
\end{array}
$$

The number 4 repeats as a remainder, so the digit 3 will repeat in the quotient. Thus, $\frac{7}{12} = 0.583333\ldots = 0.58\overline{3}$. We can write a bar over the repeating digit(s).

Do Exercises 1–8. ▶

EXERCISES

Convert to decimal notation.

1. $\frac{11}{32}$

2. $\frac{7}{8}$

3. $\frac{13}{11}$

4. $\frac{17}{12}$

5. $\frac{5}{9}$

6. $\frac{5}{6}$

7. $\frac{19}{9}$

8. $\frac{9}{11}$

Just-in-Time Review

14 ROUNDING WITH DECIMAL NOTATION

When working with decimal notation in real-life situations, we often shorten notation by **rounding**. Although there are many rules for rounding, we will use the rules listed below.

To round decimal notation to a certain place:
a) Locate the digit in that place.
b) Consider the digit to its right.
c) If the digit to the right is 5 or higher, round up. If the digit to the right is less than 5, round down.

EXAMPLE 1 Round 3872.2459 to the nearest tenth.

a) We locate the digit in the tenths place, 2.

   3 8 7 2 . 2 4 5 9

b) Then we consider the next digit to the right, 4.

   3 8 7 2 . 2 4 5 9

c) Since that digit, 4, is less than 5, we round down.

   3 8 7 2 . 2 ← This is the answer.

EXAMPLE 2 Round 3872.2459 to the nearest thousandth, hundredth, tenth, one, ten, hundred, and thousand.

thousandth:	3872.246
hundredth:	3872.25
tenth:	3872.2
one:	3872
ten:	3870
hundred:	3900
thousand:	4000

.... **Caution!** ....

Each time you round, use the original number.

In rounding, we sometimes use the symbol ≈, which means "is approximately equal to." Thus, $46.124 \approx 46.1$.

EXAMPLE 3 Divide and round $\frac{2}{7}$ to the nearest ten-thousandth, thousandth, hundredth, tenth, and one.

Dividing, we have $\frac{2}{7} = 0.\overline{285714}$. Thus we have

ten-thousandth:	0.2857
thousandth:	0.286
hundredth:	0.29
tenth:	0.3
one:	0

Do Exercises 1–6. ▶

EXERCISES

Round to the nearest hundredth, tenth, one, ten, and hundred.

1. 745.06534

2. 6780.50568

Round to the nearest cent (nearest hundredth) and to the nearest dollar (nearest one).

3. $17.988

4. $20.492

Divide and round to the nearest ten-thousandth, thousandth, hundredth, tenth, and one.

5. $\dfrac{5}{12}$

6. $\dfrac{1000}{81}$

15 CONVERT BETWEEN PERCENT NOTATION AND DECIMAL NOTATION

On average, 43% of residential energy use is for heating and cooling. This means that of every 100 units of energy used, 43 units are used for heating and cooling. Thus, 43% is a ratio of 43 to 100.

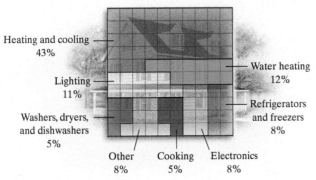

DATA: U.S. Department of Energy

The percent symbol % means "per hundred." We can regard the percent symbol as a part of a name for a number. For example,

28% is defined to mean

28×0.01, or   Replacing n% with $n \times 0.01$

$28 \times \dfrac{1}{100}$, or   Replacing n% with $n \times \dfrac{1}{100}$

$\dfrac{28}{100}$.   Replacing n% with $\dfrac{n}{100}$

NOTATION FOR n%

n% means $n \times 0.01$, or $n \times \dfrac{1}{100}$, or $\dfrac{n}{100}$.

EXAMPLE 1 *Energy Use.* The U.S. Department of Energy has determined that, on average, 8% of residential energy use is for electronics. Convert 8% to decimal notation.

$8\% = 8 \times 0.01$ Replacing % with $\times 0.01$
$ = 0.08$

(continued)

Just-in-Time Review

15 — CONVERT BETWEEN PERCENT NOTATION AND DECIMAL NOTATION (continued)

FROM PERCENT NOTATION TO DECIMAL NOTATION
To convert from percent notation to decimal notation, move the decimal point *two* places to the *left* and drop the percent symbol.

EXAMPLE 2 Convert 43.67% to decimal notation.

$$43.67\% \qquad 0.43.67 \qquad 43.67\% = 0.4367$$

Move the decimal point two places to the left.

By applying the definition of percent in reverse, we can convert from decimal notation to percent notation. We multiply by 1, expressing it as 100×0.01 and replacing $\times 0.01$ with %.

EXAMPLE 3 *Ages 65 and Older.* By 2060, it is projected that 0.236 of the total U.S. population will be ages 65 and older. Convert 0.236 to percent notation.
Data: Decennial Censuses and Population Projections Program, U.S. Census Bureau, U.S. Department of Commerce

$$\begin{aligned}
0.236 &= 0.236 \times 1 & &\text{Identity property of 1} \\
&= 0.236 \times (100 \times 0.01) & &\text{Expressing 1 as } 100 \times 0.01 \\
&= (0.236 \times 100) \times 0.01 & & \\
&= 23.6 \times 0.01 & & \\
&= 23.6\% & &\text{Replacing } \times 0.01 \text{ with \%}
\end{aligned}$$

FROM DECIMAL NOTATION TO PERCENT NOTATION
To convert from decimal notation to percent notation, move the decimal point *two* places to the *right* and write the percent symbol.

EXAMPLE 4 Convert 0.082 to percent notation.

$$0.082 \qquad 0.08.2 \qquad 0.082 = 8.2\%$$

Move the decimal point two places to the right.

Do Exercises 1–12. ▶

EXERCISES

Convert to decimal notation.

1. 63%
2. 94.1%
3. 240%
4. 0.81%
5. 2.3%
6. 100%

Convert to percent notation.

7. 0.76
8. 5
9. 0.093
10. 0.0047
11. 0.675
12. 1.34

16 CONVERT BETWEEN PERCENT NOTATION AND FRACTION NOTATION

We can convert from percent notation to fraction notation by replacing % with $\times \frac{1}{100}$ and then multiplying.

EXAMPLE 1 Convert 88% to fraction notation.

$88\% = 88 \times \frac{1}{100}$ Replacing % with $\times \frac{1}{100}$

$= \frac{88}{100}$ Multiplying. You need not simplify.

EXAMPLE 2 Convert 34.8% to fraction notation.

$34.8\% = 34.8 \times \frac{1}{100}$ Replacing % with $\times \frac{1}{100}$

$= \frac{34.8}{100}$

$= \frac{34.8}{100} \cdot \frac{10}{10}$ Multiplying by 1 to get a whole number in the numerator

$= \frac{348}{1000}$ You need not simplify.

We can convert from fraction notation to percent notation by first finding decimal notation for the fraction. Then we move the decimal point *two* places to the *right* and write the percent symbol.

EXAMPLE 3 Convert $\frac{5}{8}$ to percent notation.

We first divide to find decimal notation for $\frac{5}{8}$.

```
    0.6 2 5
8 ) 5.0 0 0
    4 8
    ─────
      2 0
      1 6
      ───
        4 0
        4 0
        ───
          0
```

Thus, $\frac{5}{8} = 0.625$.

Next, we convert the decimal notation to percent notation.

0.62.5 $\frac{5}{8} = 62.5\%$, or $62\frac{1}{2}\%$ $0.5 = \frac{5}{10} = \frac{1}{2}$

(continued)

Just-in-Time Review

16 CONVERT BETWEEN PERCENT NOTATION AND FRACTION NOTATION (continued)

EXAMPLE 4 Convert $\frac{227}{150}$ to percent notation.

We first divide to find decimal notation for $\frac{227}{150}$.

$$
\begin{array}{r}
1.5\,1\,3\,3\ldots \\
150\,\overline{)\,2\,2\,7.0\,0\,0\,0} \\
\underline{1\,5\,0} \\
7\,7\,0 \\
\underline{7\,5\,0} \\
2\,0\,0 \\
\underline{1\,5\,0} \\
5\,0\,0 \\
\underline{4\,5\,0} \\
5\,0
\end{array}
$$

We get a repeating decimal: $1.51\overline{3}$.

Next, we convert the decimal notation to percent notation by moving the decimal point *two* places to the *right* and writing the percent symbol.

$1.51\overline{3}\frac{227}{150} = 151.\overline{3}\%,\text{or}151\frac{1}{3}\%0.\overline{3} = \frac{1}{3}$

EXAMPLE 5 *Ages 0–14 in Kenya.* As of July 2015, $\frac{21}{50}$ of the population of Kenya was 0–14 years old. Convert $\frac{21}{50}$ to percent notation.

Data: *The CIA World Factbook 2017*

KENYA • Nairobi
AFRICA

We can use division. Or, since $2 \cdot 50 = 100$, we can multiply by a form of 1 in order to obtain 100 in the denominator:

$\frac{21}{50} = \frac{21}{50} \cdot \frac{2}{2} = \frac{42}{100} = 42\%.$

Do Exercises 1–12.

EXERCISES

Convert to fraction notation. Do not simplify.

1. 60%
2. 28.9%
3. 110%
4. 0.042%
5. 320%
6. 3.47%

Convert to percent notation.

7. $\frac{7}{10}$
8. $\frac{14}{25}$
9. $\frac{3.17}{100}$
10. $\frac{17}{50}$
11. $\frac{3}{8}$
12. $\frac{1}{6}$

17 EXPONENTIAL NOTATION

Exponents provide a shorter way of writing products. An abbreviation for a product in which the factors are the same is called a **power**. An expression for a power is called **exponential notation**. For

$$\underbrace{10 \cdot 10 \cdot 10}_{\text{3 factors of 10}}, \text{ we write } 10^3.$$

This is read "ten to the third power." We call the number 3 an **exponent** and we say that 10 is the **base**. For example,

$$a \cdot a \cdot a \cdot a = a^4.$$

← This is the exponent.
← This is the base.

An exponent of 2 or greater tells how many times the base is used as a factor.

EXPONENTIAL NOTATION
For any natural number n greater than or equal to 2,

$$b^n = \overbrace{b \cdot b \cdot b \cdot b \cdots b}^{n \text{ factors}}.$$

EXAMPLE 1 Write exponential notation for $10 \cdot 10 \cdot 10 \cdot 10 \cdot 10$.

$$10 \cdot 10 \cdot 10 \cdot 10 \cdot 10 = 10^5$$

EXAMPLE 2 Evaluate: 3^4.

$$3^4 = 3 \cdot 3 \cdot 3 \cdot 3 = 9 \cdot 9 = 81.$$

Do Exercises 1–10. ▶

EXERCISES

Write exponential notation.

1. $5 \times 5 \times 5 \times 5$

2. $3 \times 3 \times 3 \times 3 \times 3$

3. $4.2 \times 4.2 \times 4.2$

4. $9 \cdot 9$

5. $\dfrac{2}{11} \cdot \dfrac{2}{11} \cdot \dfrac{2}{11}$

Evaluate.

6. 4^3

7. 1^7

8. $(2.5)^2$

9. 10^6

10. $\left(\dfrac{3}{2}\right)^3$

CALCULATOR CORNER

Exponents and Powers We use the ⌃ key to evaluate exponential notation on a graphing calculator. Here we see three examples. Note that parentheses are needed when evaluating $\left(\dfrac{5}{8}\right)^3$.

The calculator has a special x^2 key that can be used to raise a number to the second power.

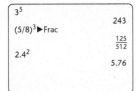

```
3^5
                    243
(5/8)^3▶Frac
                    125
                    ---
                    512
2.4^2
                    5.76
```

EXERCISES: Evaluate.

1. 7^9
2. 1.8^4
3. 23.4^3
4. $\left(\dfrac{2}{3}\right)^6$

Just-in-Time Review

18 ORDER OF OPERATIONS

What does $4 + 5 \times 2$ mean? If we add 4 and 5 and multiply the result by 2, we get 18. If we multiply 5 and 2 and add 4 to the result, we get 14. Since the results are different, we see that the order in which we carry out operations is important. To indicate which operation is to be done first, we use grouping symbols such as parentheses ( ), or brackets [ ], or braces { }. For example, $(3 \times 5) + 6 = 15 + 6 = 21$, but $3 \times (5 + 6) = 3 \times 11 = 33$.

Grouping symbols tell us what to do first. If there are no grouping symbols, there is a set of rules for the order in which operations should be done.

RULES FOR ORDER OF OPERATIONS
1. Do all calculations within grouping symbols before operations outside.
2. Evaluate all exponential expressions.
3. Do all multiplications and divisions in order from left to right.
4. Do all additions and subtractions in order from left to right.

EXAMPLE 1 Calculate: $15 - 2 \times 5 + 3$.

$$15 - 2 \times 5 + 3 = 15 - 10 + 3 \quad \text{Multiplying}$$
$$= 5 + 3 \quad \text{Subtracting}$$
$$= 8 \quad \text{Adding}$$

Always calculate within parentheses first. When there are exponents and no parentheses, simplify powers first.

EXAMPLE 2 Calculate: $(3 \times 4)^2$.

$$(3 \times 4)^2 = (12)^2 \quad \text{Working within parentheses first}$$
$$= 144 \quad \text{Evaluating the exponential expression}$$

EXAMPLE 3 Calculate: 3×4^2.

$$3 \times 4^2 = 3 \times 16 \quad \text{Evaluating the exponential expression}$$
$$= 48 \quad \text{Multiplying}$$

Note that Examples 2 and 3 show that $(3 \times 4)^2 \neq 3 \times 4^2$.

MyLab Math VIDEO

CALCULATOR CORNER

Order of Operations Computations are generally entered on a graphing calculator as they are written. We enter grouping symbols (parentheses, brackets, and braces) using the (and) keys. We indicate that a fraction bar acts as a grouping symbol by enclosing both the numerator and the denominator in parentheses. To calculate $\dfrac{38 + 142}{47 - 2}$, for example, we rewrite it with grouping symbols as $(38 + 142) \div (47 - 2)$.

```
3+4*2
              11
7(13-2)-40
              37
(38+142)/(47-2)
              4
```

EXERCISES: Calculate.
1. $68 - 8 \div 4 + 3 \cdot 5$
2. $\dfrac{311 - 17^2}{13 - 2}$
3. $(15 + 3)^3 + 4(12 - 7)^2$
4. $3.2 + 4.7[159.3 - 2.1(60.3 - 59.4)]$
5. $785 - \dfrac{5^4 - 285}{17 + 3 \cdot 51}$
6. $12^5 - 12^4 + 11^5 \div 11^3 - 10.2^2$

(continued)

18 ORDER OF OPERATIONS (continued)

EXAMPLE 4 Calculate: $7 + 3 \times 29 - 4^2$.

$7 + 3 \times 29 - 4^2 = 7 + 3 \times 29 - 16$ There are no parentheses, so we find 4^2 first.

$ = 7 + 87 - 16$ Multiplying

$ = 94 - 16$ Adding

$ = 78$ Subtracting

EXAMPLE 5 Calculate: $100 \div 20 \div 2$.

$100 \div 20 \div 2 = 5 \div 2$ Doing the divisions in order from left to right

$ = \dfrac{5}{2}$, or 2.5   Doing the second division

EXAMPLE 6 Calculate: $1000 \div \tfrac{1}{10} \cdot \tfrac{4}{5}$.

$1000 \div \dfrac{1}{10} \cdot \dfrac{4}{5} = (1000 \cdot 10) \cdot \dfrac{4}{5}$ Doing the division first, multiplying by the reciprocal of the divisor

$\phantom{1000 \div \dfrac{1}{10} \cdot \dfrac{4}{5}} = 10{,}000 \cdot \dfrac{4}{5}$ Multiplying inside the parentheses

$\phantom{1000 \div \dfrac{1}{10} \cdot \dfrac{4}{5}} = 8000$ Multiplying

Sometimes combinations of grouping symbols are used. The rules for order of operations still apply. We begin with the innermost grouping symbols and work to the outside.

EXAMPLE 7 Calculate: $5[14 - (8 + 2)]$.

$5[14 - (8 + 2)] = 5[14 - 10]$ Adding within the parentheses first

$ = 5[4]$ Subtracting inside the brackets

$ = 20$ Multiplying

EXAMPLE 8 Calculate: $\tfrac{1}{6}[(3^3 - 3) + 12]$.

$\tfrac{1}{6}[(3^3 - 3) + 12] = \tfrac{1}{6}[(27 - 3) + 12]$ Evaluating the exponential expression

$\phantom{\tfrac{1}{6}[(3^3 - 3) + 12]} = \tfrac{1}{6}[24 + 12]$ Subtracting inside the parentheses

$\phantom{\tfrac{1}{6}[(3^3 - 3) + 12]} = \tfrac{1}{6}[36]$ Adding inside the brackets

$\phantom{\tfrac{1}{6}[(3^3 - 3) + 12]} = 6$ Multiplying

Do Exercises 1–12. ▶

EXERCISES

Calculate.

1. $9 + 2 \times 8$
2. $39 - 4 \times 2 + 2$
3. $32 - 8 \div 4 - 2$
4. $3 \cdot 2^3$
5. $4^3 \div 8 - 4$
6. $20 + 4^3 \div 8 - 4$
7. $400 \times 0.64 \div 3.2$
8. $14 - 2 \times 6 + 7$
9. $2000 \div \dfrac{3}{50} \cdot \dfrac{3}{2}$
10. $1000 \div 100 \div 10$
11. $8[11 - (2 + 6)]$
12. $\dfrac{1}{10}[(5^3 - 5) + 30]$

CHAPTER 1

Introduction to Real Numbers and Algebraic Expressions

1.1 Introduction to Algebra
1.2 The Real Numbers
1.3 Addition of Real Numbers
1.4 Subtraction of Real Numbers

Mid-Chapter Review

1.5 Multiplication of Real Numbers
1.6 Division of Real Numbers
1.7 Properties of Real Numbers
1.8 Simplifying Expressions; Order of Operations

Summary and Review

Test

In 1972, the West Indian manatee was placed on the U.S. endangered species list. A large percentage of the number of annual deaths of this species is due to human factors such as collision with watercraft, as illustrated in the accompanying graph. Since 1972, concerted efforts by conservationists have led to a great enough increase in the number of manatees that the species may be reclassified as threatened instead of endangered.

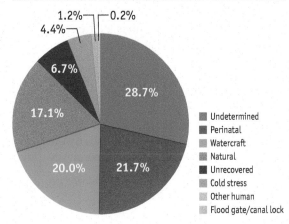

Florida Manatee Mortalities 2016
- 28.7% Undetermined
- 21.7% Perinatal
- 20.0% Watercraft
- 17.1% Natural
- 6.7% Unrecovered
- 4.4% Cold stress
- 1.2% Other human
- 0.2% Flood gate/canal lock

DATA: Florida Fish and Wildlife Conservation Commission

JUST IN TIME Review topics 8, 9, 11, 12, 13, 15, and 18 in the **Just In Time** section at the front of the text. This provides excellent prerequisite skill review for this chapter.

MyLab Math VIDEO

We will calculate a percent decrease in the number of manatees in Example 24 of Section 1.6.

STUDYING FOR SUCCESS Getting Off to a Good Start

- ☐ Your syllabus for this course is extremely important. Read it carefully, noting required texts and materials.
- ☐ If you have an online component in your course, register for it as soon as possible.
- ☐ At the front of the text, you will find a Student Organizer card. This pullout card will help you keep track of important dates and useful contact information.

Introduction to Algebra

1.1

OBJECTIVES

a Evaluate algebraic expressions by substitution.

b Translate phrases to algebraic expressions.

The study of algebra involves the use of equations to solve problems. Equations are constructed from algebraic expressions.

a EVALUATING ALGEBRAIC EXPRESSIONS

SKILL REVIEW *Simplify fraction notation.* [J7]
Simplify.

1. $\dfrac{100}{20}$ 2. $\dfrac{78}{3}$

Answers: 1. 5 2. 26

MyLab Math VIDEO

In arithmetic, you have worked with expressions such as

$$49 + 75, \quad 8 \times 6.07, \quad 29 - 14, \quad \text{and} \quad \dfrac{5}{6}.$$

In algebra, we can use letters to represent numbers and work with *algebraic expressions* such as

$$x + 75, \quad 8 \times y, \quad 29 - t, \quad \text{and} \quad \dfrac{a}{b}.$$

Sometimes a letter can represent various numbers. In that case, we call the letter a **variable**. Let a = your age. Then a is a variable because a changes from year to year. Sometimes a letter can stand for just one number. In that case, we call the letter a **constant**. Let b = your year of birth. Then b is a constant.

We often use algebraic expressions when we are solving applied problems. For example, data from aerial surveys of manatees in Florida are shown in the bar graph at left. Suppose that we want to know how many more manatees were counted in 2016 than in 1991. If we represent the increase in the number of manatees by the letter n, we can describe the number of manatees in 2016 as the number in 1991 plus the increase, or

$$1267 + n. \quad \text{This is an } \textit{algebraic expression.}$$

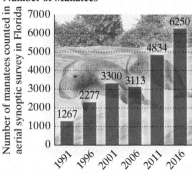

Number of Manatees

DATA: Florida Fish and Wildlife Conservation Commission

28   CHAPTER 1   Introduction to Real Numbers and Algebraic Expressions

Since we know that the number of manatees in 2016 was 6250, we can make a statement of equality:

$$1267 + n = 6250.$$ This is an *algebraic equation*.

We could solve this equation to determine how many more manatees were counted in 2016 than in 1991.

Do Exercise 1.

An **algebraic expression** consists of variables, constants, numerals, operation signs, and/or grouping symbols. When we replace a variable with a number, we say that we are **substituting** for the variable. When we replace all of the variables in an expression with numbers and carry out the operations in the expression, we are **evaluating the expression**.

EXAMPLE 1 Evaluate $x + y$ when $x = 37$ and $y = 29$.

We substitute 37 for x and 29 for y and carry out the addition:

$$x + y = 37 + 29 = 66.$$

The number 66 is called the **value** of the expression when $x = 37$ and $y = 29$.

Algebraic expressions involving multiplication can be written in several ways. For example, "8 times a" can be written as

$$8 \times a, \quad 8 \cdot a, \quad 8(a), \quad \text{or simply} \quad 8a.$$

Two letters written together without an operation symbol, such as ab, also indicate a multiplication.

Do Exercises 2–4.

EXAMPLE 2 *Area of a Rectangle.* The area A of a rectangle of length l and width w is given by the formula $A = lw$. Find the area when l is 24.5 in. and w is 16 in.

We substitute 24.5 in. for l and 16 in. for w and carry out the multiplication:

$$A = lw = (24.5 \text{ in.})(16 \text{ in.})$$
$$= (24.5)(16)(\text{in.})(\text{in.})$$
$$= 392 \text{ in}^2, \text{ or } 392 \text{ square inches.}$$

Do Exercise 5.

Algebraic expressions involving division can also be written in several ways. For example, "8 divided by t" can be written as

$$8 \div t, \quad \frac{8}{t}, \quad 8/t, \quad \text{or} \quad 8 \cdot \frac{1}{t},$$

where the fraction bar is a division symbol.

EXAMPLE 3 Evaluate $\dfrac{a}{b}$ when $a = 63$ and $b = 9$.

We substitute 63 for a and 9 for b and carry out the division:

$$\frac{a}{b} = \frac{63}{9} = 7.$$

1. Translate this problem to an equation.

   Population Estimates. The numbers of manatees in Florida counted in aerial surveys for various years between 1991 and 2016 are shown in the bar graph on the preceding page. How many more manatees were counted in 2011 than in 1996? Let x represent the increase in the number of manatees counted.

2. Evaluate $a + b$ when $a = 38$ and $b = 26$.

3. Evaluate $x - y$ when $x = 57$ and $y = 29$.

4. Evaluate $4t$ when $t = 15$.

GS 5. Find the area of a rectangle when l is 24 ft and w is 8 ft.

$A = lw$
$A = (24 \text{ ft})()$
$= (24)()(\text{ft})(\text{ft})$
$= 192 $, or
192 square feet

Answers
1. $2277 + x = 4834$ 2. 64   3. 28
4. 60   5. 192 ft^2

Guided Solution:
5. 8 ft; 8; ft^2

SECTION 1.1   Introduction to Algebra   **29**

6. Evaluate a/b when $a = 200$ and $b = 8$.

7. Evaluate $10p/q$ when $p = 40$ and $q = 25$.

EXAMPLE 4 Evaluate $\dfrac{12m}{n}$ when $m = 8$ and $n = 16$.

$$\frac{12m}{n} = \frac{12 \cdot 8}{16} = \frac{96}{16} = 6$$

◀ Do Exercises 6 and 7.

EXAMPLE 5 *Commuting via Bicycle.* Commuting to work via bicycle has increased in popularity with the emerging concept of sharing bicycles. Bikes are picked up and returned at docking stations. The payment is approximately $1.50 per 30 min. Richard bicycles 18 mi to work. The time t, in hours, that it takes to bike 18 mi is given by

$$t = \frac{18}{r},$$

where r is the speed. Find the time that it takes Richard to commute to work if his speed is 15 mph.

We substitute 15 for r and carry out the division:

$$t = \frac{18}{r} = \frac{18}{15} = 1.2 \text{ hr.}$$

8. *Commuting via Bicycle.* Find the time that it takes to bike 22 mi if the speed is 16 mph.

◀ Do Exercise 8.

b TRANSLATING TO ALGEBRAIC EXPRESSIONS

We translate problems to equations. The different parts of an equation are translations of word phrases to algebraic expressions. It is easier to translate if we know that certain words often translate to certain operation symbols.

Key Words, Phrases, and Concepts

ADDITION (+)	SUBTRACTION (−)	MULTIPLICATION (·)	DIVISION (÷)
add	subtract	multiply	divide
added to	subtracted from	multiplied by	divided by
sum	difference	product	quotient
total	minus	times	per
plus	less than	of	
more than	decreased by		
increased by	take away		

EXAMPLE 6 Translate to an algebraic expression:

Twice (or two times) some number.

Think of some number, say, 8. We can write 2 times 8 as 2×8, or $2 \cdot 8$. We multiplied by 2. Do the same thing using a variable. We can use any variable we wish, such as x, y, m, or n. Let's use y to represent some number. If we multiply by 2, we get an expression

$$y \times 2, \quad 2 \times y, \quad 2 \cdot y, \quad \text{or} \quad 2y.$$

Answers
6. 25   7. 16   8. 1.375 hr

EXAMPLE 7 Translate to an algebraic expression:

Thirty-eight percent of some number.

We let n = the number. The word "of" translates to a multiplication symbol, so we could write any of the following expressions as a translation:

$38\% \cdot n$, $0.38 \times n$,   or $0.38n$.

EXAMPLE 8 Translate to an algebraic expression:

Seven less than some number.

We let x represent the number. If the number were 10, then 7 less than 10 is $10 - 7$, or 3. If we knew the number to be 34, then 7 less than the number would be $34 - 7$. Thus if the number is x, then the translation is

$x - 7$.

EXAMPLE 9 Translate to an algebraic expression:

Eighteen more than a number.

We let t = the number. If the number were 6, then the translation would be $6 + 18$, or $18 + 6$. If we knew the number to be 17, then the translation would be $17 + 18$, or $18 + 17$. Thus if the number is t, then the translation is

$t + 18$,   or $18 + t$.

EXAMPLE 10 Translate to an algebraic expression:

A number divided by 5.

We let m = the number. If the number were 7, then the translation would be $7 \div 5$, or $7/5$, or $\frac{7}{5}$. If the number were 21, then the translation would be $21 \div 5$, or $21/5$, or $\frac{21}{5}$. If the number is m, then the translation is

$m \div 5$, $m/5$,   or $\frac{m}{5}$.

EXAMPLE 11 Translate each phrase to an algebraic expression.

PHRASE	ALGEBRAIC EXPRESSION
Five more than some number	$n + 5$, or $5 + n$
Half of a number	$\frac{1}{2}t, \frac{t}{2}$, or $t/2$
Five more than three times some number	$3p + 5$, or $5 + 3p$
The difference of two numbers	$x - y$
Six less than the product of two numbers	$mn - 6$
Seventy-six percent of some number	$76\%z$, or $0.76z$
Four less than twice some number	$2x - 4$

Do Exercises 9–17.

Caution!

Note that $7 - x$ is *not* a correct translation of the expression in Example 8. The expression $7 - x$ is a translation of "seven minus some number" or "some number less than seven."

Translate each phrase to an algebraic expression.

9. Eight less than some number

10. Eight more than some number

11. Four less than some number

12. One-third of some number

13. Six more than eight times some number

14. The difference of two numbers

15. Fifty-nine percent of some number

16. Two hundred less than the product of two numbers

17. The sum of two numbers

Answers

9. $x - 8$ **10.** $y + 8$, or $8 + y$
11. $m - 4$ **12.** $\frac{1}{3} \cdot p$, or $\frac{p}{3}$ **13.** $8x + 6$, or $6 + 8x$ **14.** $a - b$ **15.** $59\%x$, or $0.59x$
16. $xy - 200$ **17.** $p + q$

1.1 Exercise Set

✓ Check Your Understanding

Reading Check Determine whether each of the following is an algebraic expression or an algebraic equation.

RC1. $3 + a$ **RC2.** $x + 7 = 16$ **RC3.** $2(a + b) - 10$ **RC4.** $3n = 10$

Concept Check Classify each expression as an algebraic expression involving either multiplication or division.

CC1. $3/q$ **CC2.** $3q$ **CC3.** $3 \cdot q$ **CC4.** $\dfrac{3}{q}$

a Substitute to find values of the expressions in each of the following applied problems.

1. *Commuting Time.* It takes Abigail 24 min less time to commute to work than it does Jayden. Suppose that the variable x stands for the time that it takes Jayden to get to work. Then $x - 24$ stands for the time that it takes Abigail to get to work. How long does it take Abigail to get to work if it takes Jayden 56 min? 93 min? 105 min?

2. *Enrollment Costs.* At Mountain View Community College, it costs $600 to enroll in the 8 A.M. section of Elementary Algebra. Suppose that the variable n stands for the number of students who enroll. Then $600n$ stands for the total amount of tuition collected for this course. How much is collected if 34 students enroll? 78 students? 250 students?

3. *Distance Traveled.* A driver who drives at a constant speed of r miles per hour for t hours will travel a distance of d miles given by $d = rt$ miles. How far will a driver travel at a speed of 65 mph for 4 hr?

4. *Simple Interest.* The simple interest I on a principal of P dollars at interest rate r for time t, in years, is given by $I = Prt$. Find the simple interest on a principal of $4800 at 3% for 2 years.

5. *Wireless Internet Sign.* A highway sign that indicates the availability of wireless internet is in the shape of a square. The sign measures 24 in. on each side. Find its area. The area of a square with side s is given by $A = s \cdot s$.

6. *Yield Sign.* The U.S. Department of Transportation yield sign is in the shape of an equilateral triangle. Each side of the sign measures 30 in., and the height of the triangle is 26 in. Find its area. The area of a triangle with base b and height h is given by $A = \tfrac{1}{2}bh$.

   Data: *Manual on Uniform Traffic Control Devices*, U.S. Department of Transportation, Federal Highway Administration

32 CHAPTER 1 Introduction to Real Numbers and Algebraic Expressions

7. *Area of a Triangle.* The area A of a triangle with base b and height h is given by $A = \frac{1}{2}bh$. Find the area when $b = 45$ m (meters) and $h = 86$ m.

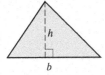

8. *Area of a Parallelogram.* The area A of a parallelogram with base b and height h is given by $A = bh$. Find the area of the parallelogram when the height is 15.4 cm (centimeters) and the base is 6.5 cm.

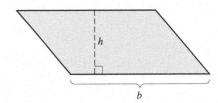

Evaluate.

9. $8x$, when $x = 7$

10. $6y$, when $y = 7$

11. $\dfrac{c}{d}$, when $c = 24$ and $d = 3$

12. $\dfrac{p}{q}$, when $p = 16$ and $q = 2$

13. $\dfrac{3p}{q}$, when $p = 2$ and $q = 6$

14. $\dfrac{5y}{z}$, when $y = 15$ and $z = 25$

15. $\dfrac{x + y}{5}$, when $x = 10$ and $y = 20$

16. $\dfrac{p + q}{2}$, when $p = 2$ and $q = 16$

17. $\dfrac{x - y}{8}$, when $x = 20$ and $y = 4$

18. $\dfrac{m - n}{5}$, when $m = 16$ and $n = 6$

b Translate each phrase to an algebraic expression. Use any letter for the variable(s) unless directed otherwise.

19. Seven more than some number

20. Some number increased by thirteen

21. Twelve less than some number

22. Fourteen less than some number

23. b more than a

24. c more than d

25. x divided by y

26. c divided by h

27. x plus w

28. s added to t

29. m subtracted from n

30. p subtracted from q

31. Twice some number

32. Three times some number

33. Three multiplied by some number

34. The product of eight and some number

35. Six more than four times some number

36. Two more than six times some number

37. Eight less than the product of two numbers

38. The product of two numbers minus seven

39. Five less than twice some number

40. Six less than seven times some number

41. Three times some number plus eleven

42. Some number times 8 plus 5

43. The sum of four times a number plus three times another number

44. Five times a number minus eight times another number

45. Your salary after a 5% salary increase if your salary before the increase was s

46. The price of a chain saw after a 30% reduction if the price before the reduction was P

47. Aubrey drove at a speed of 65 mph for t hours. How far did she travel? (See Exercise 3.)

48. Liam drove his pickup truck at 55 mph for t hours. How far did he travel? (See Exercise 3.)

49. Lisa had $50 before spending x dollars on pizza. How much money remains?

50. Juan has d dollars before spending $820 on four new tires for his truck. How much did Juan have after the purchase?

51. Sid's part-time job pays $12.50 per hour. How much does he earn for working n hours?

52. Meredith pays her babysitter $10 per hour. What does it cost her to hire the sitter for m hours?

Skill Maintenance

This heading indicates that the exercises that follow are Skill Maintenance exercises, which review any skill previously studied in the text. You can expect such exercises in every exercise set. Answers to *all* skill maintenance exercises are found at the back of the book. If you miss an exercise, restudy the objective shown in red.

Find the prime factorization. [J2]

53. 108

54. 192

Add. [J9]

55. $\dfrac{3}{8} + \dfrac{5}{14}$

56. $\dfrac{11}{27} + \dfrac{1}{6}$

Multiply. [J12]

57. 0.05×1.03

58. 43.5×1000

Find the LCM. [J4]

59. 16, 24, 32

60. 18, 36, 44

Synthesis

To the student and the instructor: The Synthesis exercises found at the end of most exercise sets challenge students to combine concepts or skills studied in that section or in preceding parts of the text.

Evaluate.

61. $\dfrac{a - 2b + c}{4b - a}$, when $a = 20$, $b = 10$, and $c = 5$

62. $\dfrac{x}{y} - \dfrac{5}{x} + \dfrac{2}{y}$, when $x = 30$ and $y = 6$

63. $\dfrac{12 - c}{c + 12b}$, when $b = 1$ and $c = 12$

64. $\dfrac{2w - 3z}{7y}$, when $w = 5$, $y = 6$, and $z = 1$

The Real Numbers

1.2

OBJECTIVES

a. State the integer that corresponds to a real-world situation.

b. Graph rational numbers on the number line.

c. Convert from fraction notation for a rational number to decimal notation.

d. Determine which of two real numbers is greater and indicate which, using $<$ or $>$. Given an inequality like $a > b$, write another inequality with the same meaning. Determine whether an inequality like $-3 \leq 5$ is true or false.

e. Find the absolute value of a real number.

A **set** is a collection of objects. For our purposes, we will most often be considering sets of numbers. One way to name a set uses what is called **roster notation**. For example, roster notation for the set containing the numbers 0, 2, and 5 is $\{0, 2, 5\}$.

Sets that are part of other sets are called **subsets**. In this section, we become acquainted with the set of *real numbers* and its various subsets.

Two important subsets of the real numbers are listed below using roster notation.

NATURAL NUMBERS

The set of **natural numbers** = $\{1, 2, 3, \ldots\}$. These are the numbers used for counting.

WHOLE NUMBERS

The set of **whole numbers** = $\{0, 1, 2, 3, \ldots\}$. This is the set of natural numbers and 0.

We can represent these sets on the number line. The natural numbers are to the right of zero. The whole numbers are the natural numbers and zero.

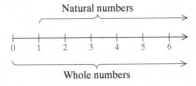

We create a new set, called the *integers*, by starting with the whole numbers, 0, 1, 2, 3, and so on. For each natural number 1, 2, 3, and so on, we obtain a new number to the left of zero on the number line:

For the number 1, there will be an *opposite* number -1 (negative 1).

For the number 2, there will be an *opposite* number -2 (negative 2).

For the number 3, there will be an *opposite* number -3 (negative 3), and so on.

The **integers** consist of the whole numbers and these new numbers.

INTEGERS

The set of **integers** = $\{\ldots, -5, -4, -3, -2, -1, 0, 1, 2, 3, 4, 5, \ldots\}$.

We picture the integers on the number line as follows.

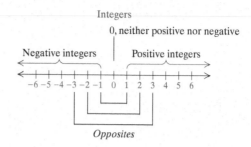

We call the integers to the left of zero **negative integers**. The natural numbers are also called **positive integers**. Zero is neither positive nor negative. We call −1 and 1 **opposites** of each other. Similarly, −2 and 2 are opposites, −3 and 3 are opposites, −100 and 100 are opposites, and 0 is its own opposite. Pairs of opposite numbers like −3 and 3 are the same distance on the number line from zero. The integers extend infinitely on the number line to the left and right of zero.

a INTEGERS AND THE REAL WORLD

Integers correspond to many real-world problems and situations. The following examples will help you get ready to translate problem situations that involve integers to mathematical language.

EXAMPLE 1 Tell which integer corresponds to this situation: Baku, the capital of Azerbaijan, lies on the Caspian Sea. Its elevation is 28 m below sea level.
Data: elevationmap.net

The integer −28 corresponds to the situation. The elevation is −28 m.

EXAMPLE 2 *Water Level.* Tell which integer corresponds to this situation: As the water level of the Mississippi River fell during the drought of 2012, barge traffic was restricted, causing a severe decline in shipping volumes. On August 24, the river level at Greenville, Mississippi, was 10 ft below normal.
Data: Rick Jervis, *USA TODAY*, August 24, 2012

The integer −10 corresponds to the drop in water level.

EXAMPLE 3 *Stock Price Change.* Tell which integers correspond to this situation: Hal owns a stock whose price decreased $16 per share over a recent period. He owns another stock whose price increased $2 per share over the same period.

The integer -16 corresponds to the decrease in the value of the first stock. The integer 2 represents the increase in the value of the second stock.

Do Exercises 1–5.

b THE RATIONAL NUMBERS

We created the set of integers by obtaining a negative number for each natural number and also including 0. To create a larger number system, called the set of **rational numbers**, we consider quotients of integers with nonzero divisors. The following are some examples of rational numbers:

$$\frac{2}{3}, \quad -\frac{2}{3}, \quad \frac{7}{1}, \quad 4, \quad -3, \quad 0, \quad \frac{23}{-8}, \quad 2.4, \quad -0.17, \quad 10\frac{1}{2}.$$

The number $-\frac{2}{3}$ (read "negative two-thirds") can also be named $\frac{-2}{3}$ or $\frac{2}{-3}$; that is,

$$-\frac{a}{b} = \frac{-a}{b} = \frac{a}{-b}.$$

The number 2.4 can be named $\frac{24}{10}$ or $\frac{12}{5}$, and -0.17 can be named $-\frac{17}{100}$. We can describe the set of rational numbers as follows.

RATIONAL NUMBERS

The set of **rational numbers** = the set of numbers $\frac{a}{b}$, where a and b are integers and b is not equal to 0 ($b \neq 0$).

Note that this new set of numbers, the rational numbers, contains the whole numbers, the integers, the arithmetic numbers (also called the nonnegative rational numbers), and the negative rational numbers.

We picture the rational numbers on the number line as follows.

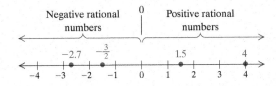

To **graph** a number means to find and mark its point on the number line. Some rational numbers are graphed in the preceding figure.

Tell which integers correspond to each situation.

1. *Temperature High and Low.* The highest recorded temperature in Illinois is 117°F on July 14, 1954, in East St. Louis. The lowest recorded temperature in Illinois is 36°F below zero on January 5, 1999, in Congerville.

   Data: Prairie Research Institute, University of Illinois at Urbana–Champaign

2. *Stock Decrease.* The price of a stock decreased $3 per share over a recent period.

3. At 10 sec before liftoff, ignition occurs. At 148 sec after liftoff, the first stage is detached from the rocket.

4. The halfback gained 8 yd on first down. The quarterback was sacked for a 5-yd loss on second down.

5. A submarine dove 120 ft, rose 50 ft, and then dove 80 ft.

Answers
1. 117; −36 2. −3 3. −10; 148
4. 8; −5 5. −120; 50; −80

Graph each number on the number line.

6. $-\dfrac{7}{2}$

7. 1.4

8. $-\dfrac{11}{4}$

EXAMPLES Graph each number on the number line.

4. -3.2 The graph of -3.2 is $\dfrac{2}{10}$ of the way from -3 to -4.

5. $\dfrac{13}{8}$ The number $\dfrac{13}{8}$ can also be named $1\dfrac{5}{8}$, or 1.625. The graph is $\dfrac{5}{8}$ of the way from 1 to 2.

◀ Do Exercises 6–8.

C NOTATION FOR RATIONAL NUMBERS

SKILL REVIEW *Convert fraction notation to decimal notation.* [J13]
Convert to decimal notation.

1. $\dfrac{17}{8}$ 2. $\dfrac{7}{11}$

Answers: 1. 2.125 2. $0.\overline{63}$

MyLab Math VIDEO

Each rational number can be named using either fraction notation or decimal notation. Decimal notation for rational numbers either *terminates* or *repeats*.

EXAMPLE 6   Convert to decimal notation: $-\dfrac{5}{8}$.

We first find decimal notation for $\dfrac{5}{8}$. Since $\dfrac{5}{8}$ means $5 \div 8$, we divide.

$$
\begin{array}{r}
0.625 \\
8\,\overline{)\,5.000} \\
\underline{4\;8} \\
2\;0 \\
\underline{1\;6} \\
4\;0 \\
\underline{4\;0} \\
0
\end{array}
$$

Thus, $\dfrac{5}{8} = 0.625$, so $-\dfrac{5}{8} = -0.625$. The notation -0.625 is a terminating decimal.

EXAMPLE 7   Convert to decimal notation: $-\dfrac{7}{9}$.

We first find decimal notation for $\dfrac{7}{9}$.

$$
\begin{array}{r}
0.77 \\
9\,\overline{)\,7.00} \\
\underline{6\;3} \\
7\;0 \\
\underline{6\;3} \\
7
\end{array}
$$

Writing a bar over the repeating digit, we see that $\dfrac{7}{9} = 0.\overline{7}$, so $-\dfrac{7}{9} = -0.\overline{7}$. The notation $-0.\overline{7}$ is a repeating decimal.

Answers

6.
7.
8.

Each rational number can be expressed in either terminating decimal notation or repeating decimal notation.

The following are other examples showing how rational numbers can be named using fraction notation or decimal notation:

$$0 = \frac{0}{8}, \quad \frac{27}{100} = 0.27, \quad -8\frac{3}{4} = -8.75, \quad -\frac{13}{6} = -2.1\overline{6}.$$

Do Exercises 9–11.

d THE REAL NUMBERS AND ORDER

Every rational number has a point on the number line. However, there are some points on the line for which there is no rational number. These points correspond to what are called **irrational numbers**.

What kinds of numbers are irrational? One example is the number π, which is used in finding the area and the circumference of a circle: $A = \pi r^2$ and $C = 2\pi r$.

Another example of an irrational number is the square root of 2, named $\sqrt{2}$. It is the length of the diagonal of a square with sides of length 1. It is also the number that when multiplied by itself gives 2—that is, $\sqrt{2} \cdot \sqrt{2} = 2$. There is no rational number that can be multiplied by itself to get 2. But the following are rational *approximations*:

1.4 is an approximation of $\sqrt{2}$ because $(1.4)^2 = 1.96$;

1.41 is a better approximation because $(1.41)^2 = 1.9881$;

1.4142 is an even better approximation because $(1.4142)^2 = 1.99996164$.

We can find rational approximations for square roots using a calculator.

Decimal notation for rational numbers *either* terminates *or* repeats.
Decimal notation for irrational numbers *neither* terminates *nor* repeats.

Some other examples of irrational numbers are $\sqrt{3}$, $-\sqrt{8}$, $\sqrt{11}$, and 0.121221222122221.... Whenever we take the square root of a number that is not a perfect square, we will get an irrational number.

The rational numbers and the irrational numbers together correspond to all the points on the number line and make up what is called the **real-number system**.

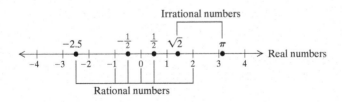

Find decimal notation.

9. $-\dfrac{3}{8}$

10. $-\dfrac{6}{11}$

11. $\dfrac{4}{3}$

MyLab Math
ANIMATION

CALCULATOR CORNER

Approximating Square Roots and π Square roots are found by pressing 2ND √. (√ is the second operation associated with the x^2 key.)

To find an approximation for $\sqrt{48}$, we press 2ND √ 4 8 ENTER.

The number π is used widely enough to have its own key. (π is the second operation associated with the ^ key.) To approximate π, we press 2ND π ENTER.

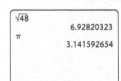

EXERCISES: Approximate.

1. $\sqrt{76}$
2. $\sqrt{317}$
3. $15 \cdot \sqrt{20}$
4. $29 + \sqrt{42}$
5. π
6. $29 \cdot \pi$
7. $\pi \cdot 13^2$
8. $5 \cdot \pi + 8 \cdot \sqrt{237}$

Answers
9. -0.375 10. $-0.\overline{54}$ 11. $1.\overline{3}$

REAL NUMBERS

The set of **real numbers** = The set of all numbers corresponding to points on the number line.

The real numbers consist of the rational numbers and the irrational numbers. The following figure shows the relationships among various kinds of numbers.

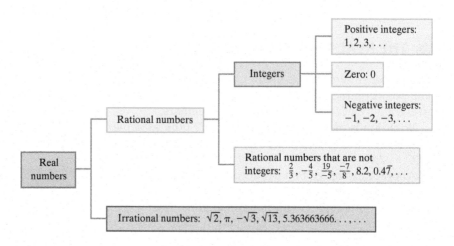

CALCULATOR CORNER

Negative Numbers on a Calculator; Converting to Decimal Notation We use the opposite key (-) to enter negative numbers on a graphing calculator. Note that this is different from the subtraction key, (−).

To convert $-\frac{5}{8}$ to decimal notation, we press (-) 5 ÷ 8 ENTER. The result is −0.625.

```
-5/8
        -.625
```

EXERCISES: Convert to decimal notation.

1. $-\frac{3}{4}$
2. $-\frac{1}{8}$
3. $-\frac{11}{16}$
4. $-\frac{7}{2}$

Order

Real numbers are named in order on the number line, increasing as we move from left to right. For any two numbers on the line, the one on the left is less than the one on the right.

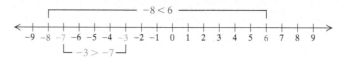

We use the symbol **<** to mean "**is less than.**" The sentence $-8 < 6$ means "−8 is less than 6." The symbol **>** means "**is greater than.**" The sentence $-3 > -7$ means "−3 is greater than −7." The sentences $-8 < 6$ and $-3 > -7$ are **inequalities**.

EXAMPLES Use either < or > for ☐ to write a true sentence.

8. 2 ☐ 9   Since 2 is to the left of 9, 2 is less than 9, so $2 < 9$.
9. −7 ☐ 3   Since −7 is to the left of 3, we have $-7 < 3$.
10. 6 ☐ −12   Since 6 is to the right of −12, then $6 > -12$.
11. −18 ☐ −5   Since −18 is to the left of −5, we have $-18 < -5$.
12. −2.7 ☐ $-\frac{3}{2}$ The answer is $-2.7 < -\frac{3}{2}$.

13. 1.5 ☐ −2.7   The answer is $1.5 > -2.7$.
14. 1.38 ☐ 1.83   The answer is $1.38 < 1.83$.

15. $-3.45 \square 1.32$ The answer is $-3.45 < 1.32$.
16. $-4 \square 0$ The answer is $-4 < 0$.
17. $5.8 \square 0$ The answer is $5.8 > 0$.
18. $\frac{5}{8} \square \frac{7}{11}$ We convert to decimal notation: $\frac{5}{8} = 0.625$ and $\frac{7}{11} = 0.6363\ldots$. Thus, $\frac{5}{8} < \frac{7}{11}$.
19. $-\frac{1}{2} \square -\frac{1}{3}$ The answer is $-\frac{1}{2} < -\frac{1}{3}$.

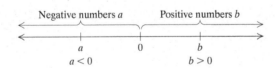

20. $-2\frac{3}{5} \square -\frac{11}{4}$ The answer is $-2\frac{3}{5} > -\frac{11}{4}$.

Do Exercises 12–19. ▶

Note that both $-8 < 6$ and $6 > -8$ are true. Every true inequality yields another true inequality when we interchange the numbers or the variables and reverse the direction of the inequality sign.

ORDER; >, <

$a < b$ also has the meaning $b > a$.

EXAMPLES Write another inequality with the same meaning.

21. $-3 > -8$ The inequality $-8 < -3$ has the same meaning.
22. $a < -5$ The inequality $-5 > a$ has the same meaning.

A helpful mental device is to think of an inequality sign as an "arrow" with the arrowhead pointing to the smaller number.

Do Exercises 20 and 21. ▶

Note that all positive real numbers are greater than zero and all negative real numbers are less than zero.

If b is a positive real number, then $b > 0$.
If a is a negative real number, then $a < 0$.

Use either $<$ or $>$ for $\square$ to write a true sentence.

12. $-3 \square 7$
13. $-8 \square -5$
14. $7 \square -10$
15. $3.1 \square -9.5$
16. $-4.78 \square -5.01$
17. $-\frac{2}{3} \square -\frac{5}{9}$
18. $-\frac{11}{8} \square \frac{23}{15}$
19. $0 \square -9.9$

Write another inequality with the same meaning.

20. $-5 < 7$
21. $x > 4$

Answers
12. <   13. <   14. >   15. >   16. >
17. <   18. <   19. >   20. $7 > -5$
21. $4 < x$

SECTION 1.2   The Real Numbers   **41**

Write true or false for each statement.

22. $-4 \leq -6$

23. $7.8 \geq 7.8$

24. $-2 \leq \dfrac{3}{8}$

CALCULATOR CORNER

Absolute Value Finding absolute value is the first item in the MATH NUM submenu on the TI-84 Plus graphing calculator. To find $|-7|$, we first press **MATH** ▷ **1**. Then we press **(-)** **7** ▷ **ENTER**. The result is 7.
To find $\left|-\dfrac{1}{2}\right|$ and express the result as a fraction, we press **MATH** ▷ **1** **(-)** **1** **÷** **2** ▷ **MATH** **1** **ENTER**. The result is $\dfrac{1}{2}$.

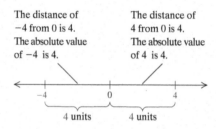

Some older operating systems will display abs(−7) instead of $|-7|$.

EXERCISES: Find the absolute value.

1. $|-5|$ 2. $|17|$
3. $|0|$ 4. $|6.48|$
5. $|-12.7|$ 6. $\left|-\dfrac{5}{7}\right|$

Find the absolute value.

25. $|8|$ 26. $|-9|$

27. $\left|-\dfrac{2}{3}\right|$ 28. $|5.6|$

Answers
22. False 23. True 24. True 25. 8
26. 9 27. $\dfrac{2}{3}$ 28. 5.6

Expressions like $a \leq b$ and $b \geq a$ are also inequalities. We read $a \leq b$ as "*a* **is less than or equal to** *b*." We read $a \geq b$ as "*a* **is greater than or equal to** *b*."

EXAMPLES Write true or false for each statement.

23. $-3 \leq 5.4$ True, since $-3 < 5.4$ is true
24. $-3 \leq -3$ True, since $-3 = -3$ is true
25. $-5 \geq 1\frac{2}{3}$ False, since neither $-5 > 1\frac{2}{3}$ nor $-5 = 1\frac{2}{3}$ is true

◀ Do Exercises 22–24.

e ABSOLUTE VALUE

From the number line, we see that numbers like 4 and −4 are the same distance from zero. We call the distance of a number from zero on the number line the **absolute value** of the number. Because distance is always nonnegative, the absolute value of a number is always nonnegative.

The distance of −4 from 0 is 4. The absolute value of −4 is 4.

The distance of 4 from 0 is 4. The absolute value of 4 is 4.

4 units   4 units

ABSOLUTE VALUE

The **absolute value** of a number is its distance from zero on the number line. We use the symbol $|x|$ to represent the absolute value of a number *x*.

FINDING ABSOLUTE VALUE

a) If a number is negative, its absolute value is its opposite.
b) If a number is positive or zero, its absolute value is the same as the number.

EXAMPLES Find the absolute value.

26. $|-7|$ The distance of −7 from 0 is 7, so $|-7| = 7$.
27. $|12|$ The distance of 12 from 0 is 12, so $|12| = 12$.
28. $|0|$ The distance of 0 from 0 is 0, so $|0| = 0$.
29. $\left|\dfrac{3}{2}\right| = \dfrac{3}{2}$
30. $|-2.73| = 2.73$

◀ Do Exercises 25–28.

1.2 Exercise Set

✓ Check Your Understanding

Reading Check Determine whether each statement is true or false.

RC1. Every integer is a rational number.

RC2. Some numbers are both rational and irrational.

RC3. The absolute value of a number is never negative.

Concept Check Match each number with its graph from the number line below.

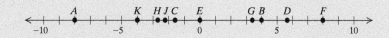

CC1. $-2\frac{5}{7}$ **CC2.** $\left|\frac{0}{-8}\right|$ **CC3.** -2.25 **CC4.** $\frac{17}{3}$ **CC5.** $|-4|$ **CC6.** $3.\overline{4}$

a State the integers that correspond to each situation.

1. On Wednesday, the temperature was 24° above zero. On Thursday, it was 2° below zero.

2. A student deposited her tax refund of $750 in a savings account. Two weeks later, she withdrew $125 to pay technology fees.

3. *Temperature Extremes.* The highest temperature ever created in a lab is 7,200,000,000,000°F. The lowest temperature ever created is approximately 460°F below zero.
Data: Live Science; The Guinness Book of World Records

4. *Extreme Climate.* Verkhoyansk, a river port in northeast Siberia, has the most extreme climate on the planet. Its average monthly winter temperature is 58.5°F below zero, and its average monthly summer temperature is 56.5°F.
Data: The Guinness Book of World Records

5. *Architecture.* The Shanghai Tower in Shanghai, China, has a total height of 2073 ft. The foundation depth is 282 ft below ground level.
Data: travelchinaguide.com

6. *Sunken Ships.* There are numerous sunken ships to explore near Bermuda. One of the most frequently visited sites is the Hermes, a decommissioned freighter that was sunk in 1985 to create an artificial reef. This ship is 80 ft below the surface.
Data: skin-diver.com

b Graph the number on the number line.

7. $\dfrac{10}{3}$

8. $-\dfrac{17}{4}$

9. -5.2

10. 4.78

11. $-4\dfrac{2}{5}$

12. $2\dfrac{6}{11}$

c Convert to decimal notation.

13. $-\dfrac{7}{8}$ 14. $-\dfrac{3}{16}$ 15. $\dfrac{5}{6}$ 16. $\dfrac{5}{3}$ 17. $-\dfrac{7}{6}$

18. $-\dfrac{5}{12}$ 19. $\dfrac{2}{3}$ 20. $-\dfrac{11}{9}$ 21. $\dfrac{1}{10}$ 22. $\dfrac{1}{4}$

23. $-\dfrac{1}{2}$ 24. $\dfrac{9}{8}$ 25. $\dfrac{4}{25}$ 26. $-\dfrac{7}{20}$

d Use either < or > for ☐ to write a true sentence.

27. $8 \,\square\, 0$ 28. $3 \,\square\, 0$ 29. $-8 \,\square\, 3$ 30. $6 \,\square\, -6$

31. $-8 \,\square\, 8$ 32. $0 \,\square\, -9$ 33. $-8 \,\square\, -5$ 34. $-4 \,\square\, -3$

35. $-5 \,\square\, -11$ 36. $-3 \,\square\, -4$ 37. $2.14 \,\square\, 1.24$ 38. $-3.3 \,\square\, -2.2$

39. $-12.88 \;\square\; -6.45$
40. $17.2 \;\square\; -1.67$
41. $-\dfrac{1}{2} \;\square\; -\dfrac{2}{3}$
42. $-\dfrac{5}{4} \;\square\; -\dfrac{3}{4}$

43. $-\dfrac{2}{3} \;\square\; \dfrac{1}{3}$
44. $\dfrac{3}{4} \;\square\; -\dfrac{5}{4}$
45. $\dfrac{5}{12} \;\square\; \dfrac{11}{25}$
46. $-\dfrac{13}{16} \;\square\; -\dfrac{5}{9}$

Write an inequality with the same meaning.

47. $-6 > x$
48. $x < 8$
49. $-10 \le y$
50. $12 \ge t$

Write true or false.

51. $-5 \le -6$
52. $-7 \ge -10$
53. $4 \ge 4$
54. $7 \le 7$

55. $-3 \ge -11$
56. $-1 \le -5$
57. $0 \ge 8$
58. $-5 \le 7$

e Find the absolute value.

59. $|-3|$
60. $|-6|$
61. $|11|$
62. $|0|$

63. $\left|-\dfrac{2}{3}\right|$
64. $|325|$
65. $\left|\dfrac{0}{4}\right|$
66. $|14.8|$

67. $|-2.65|$
68. $\left|-3\dfrac{5}{8}\right|$

Skill Maintenance

Convert to decimal notation. [J15]

69. 110%
70. $23\dfrac{4}{5}\%$

Convert to percent notation. [J16]

71. $\dfrac{13}{25}$
72. $\dfrac{19}{32}$

Evaluate. [J17]

73. 3^4
74. 5^0

Simplify. [J18]

75. $3(7 + 2^3)$
76. $48 \div 8 - 6$

Synthesis

List in order from the least to the greatest.

77. $\dfrac{2}{3}, -\dfrac{1}{7}, \dfrac{1}{3}, -\dfrac{2}{7}, -\dfrac{2}{3}, \dfrac{2}{5}, -\dfrac{1}{3}, \dfrac{2}{5}, \dfrac{9}{8}$

78. $-8\dfrac{7}{8}, 7^1, -5, |-6|, 4, |3|, -8\dfrac{5}{8}, -100, 0, 1^7, \dfrac{7}{2}, -\dfrac{67}{8}$

Given that $0.\overline{3} = \tfrac{1}{3}$ and $0.\overline{6} = \tfrac{2}{3}$, express each of the following as a quotient or a ratio of two integers.

79. $0.\overline{9}$
80. $0.\overline{1}$
81. $5.\overline{5}$

1.3 Addition of Real Numbers

OBJECTIVES

a. Add real numbers without using the number line.

b. Find the opposite, or additive inverse, of a real number.

c. Solve applied problems involving addition of real numbers.

In this section, we consider addition of real numbers. First, to gain an understanding, we add using the number line. Then we consider rules for addition.

> **ADDITION ON THE NUMBER LINE**
>
> To do the addition $a + b$ on the number line, start at 0, move to a, and then move according to b.
> a) If b is positive, move from a to the right.
> b) If b is negative, move from a to the left.
> c) If b is 0, stay at a.

EXAMPLE 1 Add: $3 + (-5)$.

We start at 0 and move to 3. Then we move 5 units left since -5 is negative.

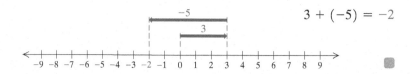

$3 + (-5) = -2$

EXAMPLE 2 Add: $-4 + (-3)$.

We start at 0 and move to -4. Then we move 3 units left since -3 is negative.

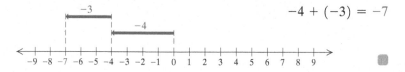

$-4 + (-3) = -7$

EXAMPLE 3 Add: $-4 + 9$.

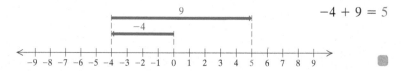

$-4 + 9 = 5$

EXAMPLE 4 Add: $-5.2 + 0$.

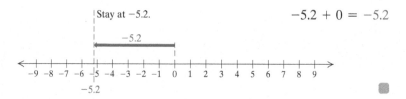

$-5.2 + 0 = -5.2$

Do Exercises 1–6.

a ADDING WITHOUT THE NUMBER LINE

SKILL REVIEW Add using fraction notation. [J9]
Add.

1. $\dfrac{1}{6} + \dfrac{2}{9}$

2. $\dfrac{13}{30} + \dfrac{5}{18}$

Answers: 1. $\dfrac{7}{18}$ 2. $\dfrac{32}{45}$

MyLab Math VIDEO

You may have noticed some patterns in the preceding examples. These lead us to rules for adding without using the number line that are more efficient for adding larger numbers.

> **RULES FOR ADDITION OF REAL NUMBERS**
>
> 1. *Positive numbers*: Add the same as arithmetic numbers. The answer is positive.
> 2. *Negative numbers*: Add absolute values. The answer is negative.
> 3. *A positive number and a negative number*:
>   - If the numbers have the same absolute value, the answer is 0.
>   - If the numbers have different absolute values, subtract the smaller absolute value from the larger. Then:
>   a) If the positive number has the greater absolute value, the answer is positive.
>   b) If the negative number has the greater absolute value, the answer is negative.
> 4. *One number is zero*: The sum is the other number.

Rule 4 is known as the **identity property of 0**. It says that for any real number a, $a + 0 = a$.

EXAMPLES Add without using the number line.

5. $-12 + (-7) = -19$ Two negatives. Add the absolute values: $|-12| + |-7| = 12 + 7 = 19$. Make the answer *negative*: -19.

6. $-1.4 + 8.5 = 7.1$ One negative, one positive. Find the absolute values: $|-1.4| = 1.4$; $|8.5| = 8.5$. Subtract the smaller absolute value from the larger: $8.5 - 1.4 = 7.1$. The *positive* number, 8.5, has the larger absolute value, so the answer is *positive*: 7.1.

7. $-36 + 21 = -15$ One negative, one positive. Find the absolute values: $|-36| = 36$; $|21| = 21$. Subtract the smaller absolute value from the larger: $36 - 21 = 15$. The *negative* number, -36, has the larger absolute value, so the answer is *negative*: -15.

Add using the number line.

1. $0 + (-3)$

2. $1 + (-4)$

3. $-3 + (-2)$

4. $-3 + 7$

5. $-2.4 + 2.4$

6. $-\dfrac{5}{2} + \dfrac{1}{2}$

Answers
1. -3 2. -3 3. -5
4. 4 5. 0 6. -2

SECTION 1.3 Addition of Real Numbers

Add without using the number line.
7. $-5 + (-6)$ 8. $-9 + (-3)$
9. $-4 + 6$ 10. $-7 + 3$
11. $5 + (-7)$ 12. $-20 + 20$
13. $-11 + (-11)$ 14. $10 + (-7)$
15. $-0.17 + 0.7$ 16. $-6.4 + 8.7$
17. $-4.5 + (-3.2)$
18. $-8.6 + 2.4$
19. $\dfrac{5}{9} + \left(-\dfrac{7}{9}\right)$

20. $-\dfrac{1}{5} + \left(-\dfrac{3}{4}\right)$ [GS]

$= -\dfrac{4}{20} + \left(-\dfrac{\boxed{}}{20}\right)$

$= -\dfrac{19}{\boxed{}}$

Add.
21. $(-15) + (-37) + 25 + 42 + (-59) + (-14)$
22. $42 + (-81) + (-28) + 24 + 18 + (-31)$
23. $-2.5 + (-10) + 6 + (-7.5)$
24. $-35 + 17 + 14 + (-27) + 31 + (-12)$

8. $1.5 + (-1.5) = 0$ The numbers have the same absolute value. The sum is 0.

9. $-\dfrac{7}{8} + 0 = -\dfrac{7}{8}$ One number is zero. The sum is $-\dfrac{7}{8}$.

10. $-9.2 + 3.1 = -6.1$

11. $-\dfrac{3}{2} + \dfrac{9}{2} = \dfrac{6}{2} = 3$

12. $-\dfrac{2}{3} + \dfrac{5}{8} = -\dfrac{16}{24} + \dfrac{15}{24} = -\dfrac{1}{24}$

◀ Do Exercises 7–20.

Suppose that we want to add several numbers, some positive and some negative, as follows. How can we proceed?

$$15 + (-2) + 7 + 14 + (-5) + (-12)$$

We can change grouping and order as we please when adding. For instance, we can group the positive numbers together and the negative numbers together and add them separately. Then we add the two results.

EXAMPLE 13 Add: $15 + (-2) + 7 + 14 + (-5) + (-12)$.

a) $15 + 7 + 14 = 36$ Adding the positive numbers
b) $-2 + (-5) + (-12) = -19$ Adding the negative numbers
   $36 + (-19) = 17$ Adding the results in (a) and (b)

We can also add the numbers in any other order we wish—say, from left to right—as follows:

$$\begin{aligned} 15 + (-2) + 7 + 14 + (-5) + (-12) &= 13 + 7 + 14 + (-5) + (-12) \\ &= 20 + 14 + (-5) + (-12) \\ &= 34 + (-5) + (-12) \\ &= 29 + (-12) \\ &= 17 \end{aligned}$$

◀ Do Exercises 21–24.

b OPPOSITES, OR ADDITIVE INVERSES

Suppose that we add two numbers that are **opposites**, such as 6 and -6. The result is 0. When opposites are added, the result is always 0. Opposites are also called **additive inverses**. Every real number has an opposite, or additive inverse.

OPPOSITES, OR ADDITIVE INVERSES

Two numbers whose sum is 0 are called **opposites**, or **additive inverses**, of each other.

Answers
7. -11 8. -12 9. 2 10. -4
11. -2 12. 0 13. -22 14. 3
15. 0.53 16. 2.3 17. -7.7 18. -6.2
19. $-\dfrac{2}{9}$ 20. $-\dfrac{19}{20}$ 21. -58 22. -56
23. -14 24. -12
Guided Solution:
20. 15; 20

EXAMPLES Find the opposite, or additive inverse, of each number.

14. 34   The opposite of 34 is -34 because $34 + (-34) = 0$.
15. -8 The opposite of -8 is 8 because $-8 + 8 = 0$.
16. 0   The opposite of 0 is 0 because $0 + 0 = 0$.
17. $-\dfrac{7}{8}$ The opposite of $-\dfrac{7}{8}$ is $\dfrac{7}{8}$ because $-\dfrac{7}{8} + \dfrac{7}{8} = 0$.

Find the opposite, or additive inverse, of each number.

25. -4 **26.** 8.7

27. -7.74 **28.** $-\dfrac{8}{9}$

29. 0   **30.** 12

Do Exercises 25–30. ▶

To name the opposite, we use the symbol $-$, as follows.

> **SYMBOLIZING OPPOSITES**
>
> The opposite, or additive inverse, of a number a can be named $-a$ (read "the opposite of a," or "the additive inverse of a").

Note that if we take a number, say, 8, and find its opposite, -8, and then find the opposite of the result, we will have the original number, 8, again.

> **THE OPPOSITE OF AN OPPOSITE**
>
> The **opposite of the opposite** of a number is the number itself. (The additive inverse of the additive inverse of a number is the number itself.) That is, for any number a,
>
> $$-(-a) = a.$$

EXAMPLE 18 Evaluate $-x$ and $-(-x)$ when $x = 16$.

If $x = 16$, then $-x = -16$.   The opposite of 16 is -16.
If $x = 16$, then $-(-x) = -(-16) = 16$.   The opposite of the opposite of 16 is 16.

EXAMPLE 19 Evaluate $-x$ and $-(-x)$ when $x = -3$.

If $x = -3$, then $-x = -(-3) = 3$.
If $x = -3$, then $-(-x) = -(-(-3)) = -(3) = -3$.

Note that in Example 19 we used a second set of parentheses to show that we are substituting the negative number -3 for x. Symbolism like $--x$ is not considered meaningful.

Evaluate $-x$ and $-(-x)$ when:

31. $x = 14$.

GS 32. $x = -1.6$.
$-x = -() = 1.6;$
$-(-x) = -(-())$
$ = -() = -1.6$

Do Exercises 31–34. ▶

33. $x = \dfrac{2}{3}$.   **34.** $x = -\dfrac{9}{8}$.

A symbol such as -8 is generally read "negative 8." It could be read "the additive inverse of 8," because the additive inverse of 8 is negative 8. It could also be read "the opposite of 8," because the opposite of 8 is -8. Thus a symbol like -8 can be read in more than one way. It is never correct to read -8 as "minus 8."

.... **Caution!** ....

A symbol like $-x$, which has a variable, should be read "the opposite of x" or "the additive inverse of x" and *not* "negative x," because we do not know whether x represents a positive number, a negative number, or 0. You can check this in Examples 18 and 19.

Answers

25. 4   **26.** -8.7 **27.** 7.74   **28.** $\dfrac{8}{9}$
29. 0   **30.** -12 **31.** $-14; 14$
32. $1.6; -1.6$ **33.** $-\dfrac{2}{3}; \dfrac{2}{3}$ **34.** $\dfrac{9}{8}; -\dfrac{9}{8}$

Guided Solution:
32. $-1.6; -1.6; 1.6$

We can use the symbolism $-a$ to restate the definition of opposite, or additive inverse.

OPPOSITES, OR ADDITIVE INVERSES

For any real number a, the **opposite**, or **additive inverse**, of a, denoted $-a$, is such that
$$a + (-a) = (-a) + a = 0.$$

Signs of Numbers

A negative number is sometimes said to have a "negative sign." A positive number is said to have a "positive sign." When we replace a number with its opposite, we can say that we have "changed its sign."

EXAMPLES Find the opposite. (Change the sign.)

20. -3 $-(-3) = 3$
21. $-\frac{2}{13}$ $-\left(-\frac{2}{13}\right) = \frac{2}{13}$
22. 0 $-(0) = 0$
23. 14 $-(14) = -14$

◀ Do Exercises 35–38.

Find the opposite. (Change the sign.)
35. -4
36. -13.4
37. 0
38. $\frac{1}{4}$

C APPLICATIONS AND PROBLEM SOLVING

Addition of real numbers occurs in many real-world situations.

EXAMPLE 24 *Banking Transactions.* On August 1st, Martias checks his bank account balance on his phone and sees that it is $54. During the next week, the following transactions were recorded: a debit-card purchase of $71, an overdraft fee of $29, a direct deposit of $160, and an ATM withdrawal of $80. What is Martias's balance at the end of the week?

We let $B =$ the ending balance of the bank account. Then the problem translates to the following:

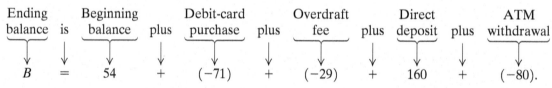

Adding, we have
$B = 54 + (-71) + (-29) + 160 + (-80)$
$= 214 + (-180)$ Adding the positive numbers and adding the negative numbers
$= 34.$

Martias's balance at the end of the week is $34.

◀ Do Exercise 39.

39. *Change in Class Size.* During the first two weeks of the semester in Jim's algebra class, 4 students withdrew, 8 students enrolled late, and 6 students were dropped as "no shows." By how many students had the class size changed at the end of the first two weeks?

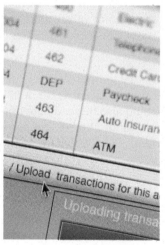

Answers
35. 4   36. 13.4   37. 0   38. $-\frac{1}{4}$
39. -2 students

50   CHAPTER 1   Introduction to Real Numbers and Algebraic Expressions

1.3 Exercise Set

✓ Check Your Understanding

Reading Check Choose the word or words from the list on the right to complete each sentence. Words may be used more than once or not at all.

RC1. To add −3 + (−6), _____ 3 and 6 and make the answer _____.

RC2. To add −11 + 5, _____ 5 from 11 and make the answer _____.

RC3. The sum of two numbers that are _____ is 0.

RC4. The addition −7 + 0 = −7 illustrates the _____ property of 0.

add
subtract
opposites
identity
positive
negative

Concept Check Fill in each blank with either "left" or "right" so that the statements describe the steps when adding numbers with the number line.

CC1. To add 7 + 2, start at 0, move _____ to 7, and then move 2 units _____. The sum is 9.

CC2. To add −3 + (−5), start at 0, move _____ to −3, and then move 5 units _____. The sum is −8.

CC3. To add 4 + (−6), start at 0, move _____ to 4, and then move 6 units _____. The sum is −2.

CC4. To add −8 + 3, start at 0, move _____ to −8, and then move 3 units _____. The sum is −5.

a Add. Do not use the number line except as a check.

1. $2 + (-9)$
2. $-5 + 2$
3. $-11 + 5$
4. $4 + (-3)$
5. $-6 + 6$

6. $8 + (-8)$
7. $-3 + (-5)$
8. $-4 + (-6)$
9. $-7 + 0$
10. $-13 + 0$

11. $0 + (-27)$
12. $0 + (-35)$
13. $17 + (-17)$
14. $-15 + 15$
15. $-17 + (-25)$

16. $-24 + (-17)$
17. $18 + (-18)$
18. $-13 + 13$
19. $-28 + 28$
20. $11 + (-11)$

21. $8 + (-5)$
22. $-7 + 8$
23. $-4 + (-5)$
24. $10 + (-12)$
25. $13 + (-6)$

SECTION 1.3 Addition of Real Numbers

26. $-3 + 14$
27. $-25 + 25$
28. $50 + (-50)$
29. $53 + (-18)$
30. $75 + (-45)$

31. $-8.5 + 4.7$
32. $-4.6 + 1.9$
33. $-2.8 + (-5.3)$
34. $-7.9 + (-6.5)$
35. $-\dfrac{3}{5} + \dfrac{2}{5}$

36. $-\dfrac{4}{3} + \dfrac{2}{3}$
37. $-\dfrac{2}{9} + \left(-\dfrac{5}{9}\right)$
38. $-\dfrac{4}{7} + \left(-\dfrac{6}{7}\right)$
39. $-\dfrac{5}{8} + \dfrac{1}{4}$
40. $-\dfrac{5}{6} + \dfrac{2}{3}$

41. $-\dfrac{5}{8} + \left(-\dfrac{1}{6}\right)$
42. $-\dfrac{5}{6} + \left(-\dfrac{2}{9}\right)$
43. $-\dfrac{3}{8} + \dfrac{5}{12}$
44. $-\dfrac{7}{16} + \dfrac{7}{8}$

45. $-\dfrac{1}{6} + \dfrac{7}{10}$
46. $-\dfrac{11}{18} + \left(-\dfrac{3}{4}\right)$
47. $\dfrac{7}{15} + \left(-\dfrac{1}{9}\right)$
48. $-\dfrac{4}{21} + \dfrac{3}{14}$

49. $76 + (-15) + (-18) + (-6)$

50. $29 + (-45) + 18 + 32 + (-96)$

51. $-44 + \left(-\dfrac{3}{8}\right) + 95 + \left(-\dfrac{5}{8}\right)$

52. $24 + 3.1 + (-44) + (-8.2) + 63$

b Find the opposite, or additive inverse.

53. 24
54. -64
55. -26.9
56. 48.2

Evaluate $-x$ when:

57. $x = 8$.
58. $x = -27$.
59. $x = -\dfrac{13}{8}$.
60. $x = \dfrac{1}{236}$.

Evaluate $-(-x)$ when:

61. $x = -43$.
62. $x = 39$.
63. $x = \dfrac{4}{3}$.
64. $x = -7.1$.

Find the opposite. (Change the sign.)

65. -24
66. -12.3
67. $-\dfrac{3}{8}$
68. 10

c Solve.

69. *Tallest Mountain.* The tallest mountain in the world, when measured from base to peak, is Mauna Kea (White Mountain) in Hawaii. From its base 19,684 ft below sea level in the Hawaiian Trough, it rises 33,480 ft. What is the elevation of the peak above sea level?

    Data: *The Guinness Book of World Records*

70. *Copy Center Account.* Rachel's copy-center bill for July was $327. She made a payment of $200 and then made $48 worth of copies in August. How much did she then owe on her account?

71. *Temperature Changes.* One day, the temperature in Lawrence, Kansas, is 32°F at 6:00 A.M. It rises 15° by noon, but falls 50° by midnight when a cold front moves in. What is the final temperature?

72. *Stock Changes.* On a recent day, the price of a stock opened at a value of $61.38. During the day, it rose $4.75, dropped $7.38, and rose $5.13. Find the value of the stock at the end of the day.

73. *"Flipping" Houses.* Buying run-down houses, fixing them up, and reselling them is referred to as "flipping" houses. Charlie and Sophia bought and sold four houses in a recent year. The profits and losses are shown in the following bar graph. Find the sum of the profits and losses.

Flipping Houses

House 1: $10,500
House 2: −$8300
House 3: −$4100
House 4: $6200

74. *Football Yardage.* In a college football game, the quarterback attempted passes with the following results. Find the total gain or loss.

ATTEMPT	GAIN OR LOSS
1st	13-yd gain
2nd	12-yd loss
3rd	21-yd gain

75. *Credit-Card Bills.* On August 1, Lyle's credit-card bill shows that he owes $470. During the month of August, Lyle makes a payment of $45 to the credit-card company, charges another $160 in merchandise, and then pays off another $500 of his bill. What is the new amount that Lyle owes at the end of August?

76. *Account Balance.* Emma has $460 in a checking account. She uses her debit card for a purchase of $530, makes a deposit of $75, and then writes a check for $90. What is the balance in her account?

Skill Maintenance

77. Evaluate $5a - 2b$ when $a = 9$ and $b = 3$. [1.1a]

78. Write an inequality with the same meaning as the inequality $-3 < y$. [1.2d]

Convert to decimal notation. [1.2c]

79. $-\dfrac{1}{12}$

80. $\dfrac{5}{8}$

Find the absolute value. [1.2e]

81. $|0|$

82. $|-21.4|$

Synthesis

83. For what numbers x is $-x$ negative?

84. For what numbers x is $-x$ positive?

85. If a is positive and b is negative, then $-a + b$ is which of the following?
- **A.** Positive
- **B.** Negative
- **C.** 0
- **D.** Cannot be determined without more information

86. If $a = b$ and a and b are negative, then $-a + (-b)$ is which of the following?
- **A.** Positive
- **B.** Negative
- **C.** 0
- **D.** Cannot be determined without more information

1.4 Subtraction of Real Numbers

OBJECTIVES

a Subtract real numbers and simplify combinations of additions and subtractions.

b Solve applied problems involving subtraction of real numbers.

a SUBTRACTION

We now consider subtraction of real numbers.

> **SUBTRACTION**
>
> The difference $a - b$ is the number c for which $a = b + c$.

Consider, for example, $45 - 17$. *Think*: What number can we add to 17 to get 45? Since $45 = 17 + 28$, we know that $45 - 17 = 28$. Let's consider an example whose answer is a negative number.

EXAMPLE 1 Subtract: $3 - 7$.

Think: What number can we add to 7 to get 3? The number must be negative. Since $7 + (-4) = 3$, we know the number is -4: $3 - 7 = -4$. That is, $3 - 7 = -4$ because $7 + (-4) = 3$.

◀ Do Exercises 1–3.

The definition above does not provide the most efficient way to do subtraction. We can develop a faster way to subtract. As a rationale for the faster way, let's compare $3 + 7$ and $3 - 7$ on the number line.

To find $3 + 7$ on the number line, we start at 0, move to 3, and then move 7 units farther to the right since 7 is positive.

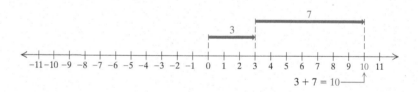

$3 + 7 = 10$

To find $3 - 7$, we do the "opposite" of adding 7: We move 7 units to the *left* to do the subtracting. This is the same as *adding* the opposite of 7, -7, to 3.

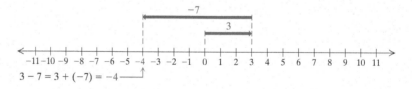

$3 - 7 = 3 + (-7) = -4$

◀ Do Exercises 4–6.

Look for a pattern in the examples shown at right.

SUBTRACTING	ADDING AN OPPOSITE
$5 - 8 = -3$	$5 + (-8) = -3$
$-6 - 4 = -10$	$-6 + (-4) = -10$
$-7 - (-2) = -5$	$-7 + 2 = -5$

Subtract.

1. $-6 - 4$

   Think: What number can be added to 4 to get -6:

   $\square + 4 = -6?$

2. $-7 - (-10)$

   Think: What number can be added to -10 to get -7:

   $\square + (-10) = -7?$

3. $-7 - (-2)$

   Think: What number can be added to -2 to get -7:

   $\square + (-2) = -7?$

Subtract. Use the number line, doing the "opposite" of addition.

4. $5 - 9$

5. $-3 - 2$

6. $-4 - (-3)$

Answers
1. -10 2. 3 3. -5 4. -4
5. -5 6. -1

Do Exercises 7–10.

Perhaps you have noticed that we can subtract by adding the opposite of the number being subtracted. This can always be done.

SUBTRACTING BY ADDING THE OPPOSITE

For any real numbers a and b,
$$a - b = a + (-b).$$
(To subtract, add the opposite, or additive inverse, of the number being subtracted.)

This is the method generally used for quick subtraction of real numbers.

EXAMPLES Subtract.

2. $2 - 6 = 2 + (-6) = -4$ The opposite of 6 is -6. We change the subtraction to addition and add the opposite. Check: $-4 + 6 = 2$.

3. $4 - (-9) = 4 + 9 = 13$ The opposite of -9 is 9. We change the subtraction to addition and add the opposite. Check: $13 + (-9) = 4$.

4. $-4.2 - (-3.6) = -4.2 + 3.6 = -0.6$ Adding the opposite. Check: $-0.6 + (-3.6) = -4.2$.

5. $-\dfrac{1}{2} - \left(-\dfrac{3}{4}\right) = -\dfrac{1}{2} + \dfrac{3}{4}$
$= -\dfrac{2}{4} + \dfrac{3}{4} = \dfrac{1}{4}$
Adding the opposite. Check: $\dfrac{1}{4} + \left(-\dfrac{3}{4}\right) = -\dfrac{1}{2}$.

Do Exercises 11–16.

EXAMPLES Subtract by adding the opposite of the number being subtracted.

6. $3 - 5$ Think: "Three minus five is three plus the opposite of five"
$3 - 5 = 3 + (-5) = -2$

7. $\dfrac{1}{8} - \dfrac{7}{8}$ Think: "One-eighth minus seven-eighths is one-eighth plus the opposite of seven-eighths"
$\dfrac{1}{8} - \dfrac{7}{8} = \dfrac{1}{8} + \left(-\dfrac{7}{8}\right) = -\dfrac{6}{8}$, or $-\dfrac{3}{4}$

8. $-4.6 - (-9.8)$ Think: "Negative four point six minus negative nine point eight is negative four point six plus the opposite of negative nine point eight"
$-4.6 - (-9.8) = -4.6 + 9.8 = 5.2$

9. $-\dfrac{3}{4} - \dfrac{7}{5}$ Think: "Negative three-fourths minus seven-fifths is negative three-fourths plus the opposite of seven-fifths"
$-\dfrac{3}{4} - \dfrac{7}{5} = -\dfrac{3}{4} + \left(-\dfrac{7}{5}\right) = -\dfrac{15}{20} + \left(-\dfrac{28}{20}\right) = -\dfrac{43}{20}$

Do Exercises 17–21.

Complete the addition and compare with the subtraction.

7. $4 - 6 = -2$;
$4 + (-6) =$ _____

8. $-3 - 8 = -11$;
$-3 + (-8) =$ _____

9. $-5 - (-9) = 4$;
$-5 + 9 =$ _____

10. $-5 - (-3) = -2$;
$-5 + 3 =$ _____

Subtract.

GS 11. $2 - 8 = 2 + ($ 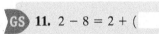 $) = $ _____

12. $-6 - 10$

13. $12.4 - 5.3$

14. $-8 - (-11)$

15. $-8 - (-8)$

16. $\dfrac{2}{3} - \left(-\dfrac{5}{6}\right)$

Subtract by adding the opposite of the number being subtracted.

17. $3 - 11$

18. $12 - 5$

GS 19. $-12 - (-9) = -12 + $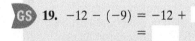
$= $ ____

20. $-12.4 - 10.9$

21. $-\dfrac{4}{5} - \left(-\dfrac{4}{5}\right)$

Answers
7. -2 8. -11 9. 4 10. -2 11. -6
12. -16 13. 7.1 14. 3 15. 0 16. $\dfrac{3}{2}$
17. -8 18. 7 19. -3 20. -23.3 21. 0
Guided Solutions:
11. $-8; -6$ 19. $9; -3$

When several additions and subtractions occur together, we can make them all additions.

EXAMPLES Simplify.

10. $8 - (-4) - 2 - (-4) + 2 = 8 + 4 + (-2) + 4 + 2$ Adding the opposite
$= 16$

11. $8.2 - (-6.1) + 2.3 - (-4) = 8.2 + 6.1 + 2.3 + 4 = 20.6$

12. $\dfrac{3}{4} - \left(-\dfrac{1}{12}\right) - \dfrac{5}{6} - \dfrac{2}{3} = \dfrac{9}{12} + \dfrac{1}{12} + \left(-\dfrac{10}{12}\right) + \left(-\dfrac{8}{12}\right)$

$= \dfrac{9 + 1 + (-10) + (-8)}{12}$

$= \dfrac{-8}{12} = -\dfrac{8}{12} = -\dfrac{2}{3}$

◀ Do Exercises 22–24.

Simplify.

22. $-6 - (-2) - (-4) - 12 + 3$

23. $\dfrac{2}{3} - \dfrac{4}{5} - \left(-\dfrac{11}{15}\right) + \dfrac{7}{10} - \dfrac{5}{2}$

24. $-9.6 + 7.4 - (-3.9) - (-11)$

b APPLICATIONS AND PROBLEM SOLVING

Let's now see how we can use subtraction of real numbers to solve applied problems.

EXAMPLE 13 *Surface Temperatures on Mars.* Surface temperatures on Mars vary from $-128°C$ during polar night to $27°C$ at the equator during midday at the closest point in orbit to the sun. Find the difference between the highest value and the lowest value in this temperature range.
Data: Mars Institute

SKILL REVIEW   *Translate phrases to algebraic expressions.* [1.1b]

Translate to an algebraic expression.
1. The difference of 100 and points missed
2. 10 less than the marked price

Answers: 1. $100 - x$
2. $p - 10$

We let $D = $ the difference in the temperatures. Then the problem translates to the following subtraction:

$$\underbrace{\text{Difference in temperature}}_{D} \; \underbrace{\text{is}}_{=} \; \underbrace{\text{Highest temperature}}_{27} \; \underbrace{\text{minus}}_{-} \; \underbrace{\text{Lowest temperature}}_{(-128)}$$

$D = 27 + 128 = 155.$

The difference in the temperatures is $155°C$.

◀ Do Exercise 25.

25. *Temperature Extremes.* The highest temperature ever recorded in the United States is $134°F$ in Greenland Ranch, California, on July 10, 1913. The lowest temperature ever recorded is $-80°F$ in Prospect Creek, Alaska, on January 23, 1971. How much higher was the temperature in Greenland Ranch than the temperature in Prospect Creek?

Data: National Oceanographic and Atmospheric Administration

Answers

22. -9 **23.** $-\dfrac{6}{5}$ **24.** 12.7 **25.** $214°F$

1.4 Exercise Set

✓ Check Your Understanding

Reading Check Choose the word from the list on the right to complete each sentence. Words may be used more than once or not at all.

RC1. The number 3 is the _____ of −3.

RC2. To subtract, we add the _____ of the number being subtracted.

RC3. The word _____ usually translates to subtraction.

difference
opposite
reciprocal
sum

Concept Check Match the expression with an expression from the column on the right that names the same number.

CC1. $18 - 6$ a) $18 + 6$

CC2. $-18 - (-6)$ b) $-18 + 6$

CC3. $-18 - 6$ c) $18 + (-6)$

CC4. $18 - (-6)$ d) $-18 + (-6)$

a Subtract.

1. $2 - 9$
2. $3 - 8$
3. $-8 - (-2)$
4. $-6 - (-8)$

5. $-11 - (-11)$
6. $-6 - (-6)$
7. $12 - 16$
8. $14 - 19$

9. $20 - 27$
10. $30 - 4$
11. $-9 - (-3)$
12. $-7 - (-9)$

13. $40 - (-40)$
14. $-9 - (-9)$
15. $7 - (-7)$
16. $4 - (-4)$

17. $8 - (-3)$
18. $-7 - 4$
19. $-6 - 8$
20. $6 - (-10)$

21. $-4 - (-9)$
22. $-14 - 2$
23. $-6 - (-5)$
24. $-4 - (-3)$

25. 8 − (−10)   **26.** 5 − (−6)   **27.** −5 − (−2)   **28.** −3 − (−1)

29. −7 − 14   **30.** −9 − 16   **31.** 0 − (−5)   **32.** 0 − (−1)

33. −8 − 0   **34.** −9 − 0   **35.** 7 − (−5)   **36.** 7 − (−4)

37. 2 − 25   **38.** 18 − 63   **39.** −42 − 26   **40.** −18 − 63

41. −71 − 2   **42.** −49 − 3   **43.** 24 − (−92)   **44.** 48 − (−73)

45. −50 − (−50)   **46.** −70 − (−70)   **47.** $-\dfrac{3}{8} - \dfrac{5}{8}$ **48.** $\dfrac{3}{9} - \dfrac{9}{9}$

49. $\dfrac{3}{4} - \dfrac{2}{3}$ **50.** $\dfrac{5}{8} - \dfrac{3}{4}$ **51.** $-\dfrac{3}{4} - \dfrac{2}{3}$ **52.** $-\dfrac{5}{8} - \dfrac{3}{4}$

53. $-\dfrac{5}{8} - \left(-\dfrac{3}{4}\right)$ **54.** $-\dfrac{3}{4} - \left(-\dfrac{2}{3}\right)$ **55.** 6.1 − (−13.8)   **56.** 1.5 − (−3.5)

57. −2.7 − 5.9   **58.** −3.2 − 5.8   **59.** 0.99 − 1   **60.** 0.87 − 1

61. −79 − 114   **62.** −197 − 216   **63.** 0 − (−500)   **64.** 500 − (−1000)

65. −2.8 − 0   **66.** 6.04 − 1.1   **67.** 7 − 10.53   **68.** 8 − (−9.3)

69. $\dfrac{1}{6} - \dfrac{2}{3}$ **70.** $-\dfrac{3}{8} - \left(-\dfrac{1}{2}\right)$ **71.** $-\dfrac{4}{7} - \left(-\dfrac{10}{7}\right)$ **72.** $\dfrac{12}{5} - \dfrac{12}{5}$

73. $-\dfrac{7}{10} - \dfrac{10}{15}$ **74.** $-\dfrac{4}{18} - \left(-\dfrac{2}{9}\right)$ **75.** $\dfrac{1}{5} - \dfrac{1}{3}$ **76.** $-\dfrac{1}{7} - \left(-\dfrac{1}{6}\right)$

77. $\dfrac{5}{12} - \dfrac{7}{16}$ **78.** $-\dfrac{1}{35} - \left(-\dfrac{9}{40}\right)$ **79.** $-\dfrac{2}{15} - \dfrac{7}{12}$ **80.** $\dfrac{2}{21} - \dfrac{9}{14}$

Simplify.

81. $18 - (-15) - 3 - (-5) + 2$

82. $22 - (-18) + 7 + (-42) - 27$

83. $-31 + (-28) - (-14) - 17$

84. $-43 - (-19) - (-21) + 25$

85. $-34 - 28 + (-33) - 44$

86. $39 + (-88) - 29 - (-83)$

87. $-93 - (-84) - 41 - (-56)$

88. $84 + (-99) + 44 - (-18) - 43$

89. $-5.4 - (-30.9) + 30.8 + 40.2 - (-12)$

90. $14.9 - (-50.7) + 20 - (-32.8)$

91. $-\dfrac{7}{12} + \dfrac{3}{4} - \left(-\dfrac{5}{8}\right) - \dfrac{13}{24}$

92. $-\dfrac{11}{16} + \dfrac{5}{32} - \left(-\dfrac{1}{4}\right) + \dfrac{7}{8}$

b Solve.

93. *Elevations in Asia.* The elevation of the highest point in Asia, Mt. Everest, Nepal–Tibet, is 29,035 ft. The lowest elevation, at the Dead Sea, Israel–Jordan, is −1348 ft. What is the difference in the elevations of the two locations?

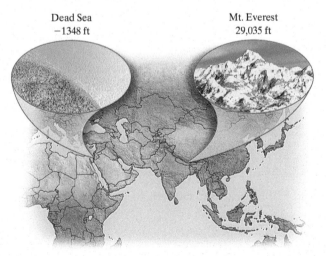

Dead Sea
−1348 ft

Mt. Everest
29,035 ft

94. *Ocean Depth.* The deepest point in the Pacific Ocean is the Marianas Trench, with a depth of 10,924 m. The deepest point in the Atlantic Ocean is the Puerto Rico Trench, with a depth of 8605 m. What is the difference in the elevation of the two trenches?

Data: *The World Almanac and Book of Facts*

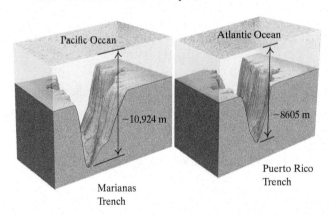

SECTION 1.4 Subtraction of Real Numbers

95. Francisca has a charge of $476.89 on her credit card, but she then returns a sweater that cost $128.95. How much does she now owe on her credit card?

96. Jacob has $825 in a checking account. What is the balance in his account after he has written a check for $920 to pay for a laptop?

97. *Difference in Elevation.* The highest elevation in Japan is 3776 m above sea level at Fujiyama. The lowest elevation in Japan is 4 m below sea level at Hachirogata. Find the difference in the elevations.

Data: *Information Please Almanac*

98. *Difference in Elevation.* The lowest elevation in North America—Death Valley, California—is 282 ft below sea level. The highest elevation in North America—Mount McKinley, Alaska—is 20,320 ft above sea level. Find the difference in elevation between the highest point and the lowest point.

Data: *National Geographic Society*

99. *Low Points on Continents.* The lowest point in Africa is Lake Assal, which is 512 ft below sea level. The lowest point in South America is the Valdes Peninsula, which is 131 ft below sea level. How much lower is Lake Assal than the Valdes Peninsula?

Data: *National Geographic Society*

100. *Temperature Records.* The greatest recorded temperature change in one 24-hr period occurred between January 14 and January 15, 1972, in Loma, Montana, where the temperature rose from −54°F to 49°F. By how much did the temperature rise?

Data: *The Guinness Book of World Records*

101. *Surface Temperature on Mercury.* Surface temperatures on Mercury vary from 840°F on the equator when the planet is closest to the sun to −290°F at night. Find the difference between these two temperatures.

102. *Run Differential.* In baseball, the difference between the number of runs that a team scores and the number of runs that it allows its opponents to score is called the *run differential*. That is,

Run differential = Number of runs scored − Number of runs allowed.

Teams strive for a positive run differential.

Data: *Major League Baseball*

a) In a recent season, the Kansas City Royals scored 676 runs and allowed 746 runs to be scored on them. Find the run differential.
b) In a recent season, the Atlanta Braves scored 700 runs and allowed 600 runs to be scored on them. Find the run differential.

Skill Maintenance

Translate to an algebraic expression. [1.1b]

103. 7 more than y

104. 41 less than t

105. h subtracted from a

106. The product of 6 and c

107. r more than s

108. x less than y

Synthesis

Determine whether each statement is true or false for all integers a and b. If false, give an example to show why. Examples may vary.

109. $a - 0 = 0 - a$

110. $0 - a = a$

111. If $a \neq b$, then $a - b \neq 0$.

112. If $a = -b$, then $a + b = 0$.

113. If $a + b = 0$, then a and b are opposites.

114. If $a - b = 0$, then $a = -b$.

Mid-Chapter Review

Concept Reinforcement

Determine whether each statement is true or false.

_____ 1. All rational numbers can be named using fraction notation. [1.2c]

_____ 2. If $a > b$, then a lies to the left of b on the number line. [1.2d]

_____ 3. The absolute value of a number is always nonnegative. [1.2e]

_____ 4. We can translate "7 less than y" as $7 - y$. [1.1b]

Guided Solutions

 Fill in each blank with the number that creates a correct statement or solution.

5. Evaluate $-x$ and $-(-x)$ when $x = -4$. [1.3b]

   $-x = -(\quad) = \quad$;

   $-(-x) = -(-(\quad)) = -(\quad) = \quad$

Subtract. [1.4a]

6. $5 - 13 = 5 + (\quad) = \quad$

7. $-6 - 7 = -6 + (\quad) = \quad$

Mixed Review

Evaluate. [1.1a]

8. $\dfrac{3m}{n}$, when $m = 8$ and $n = 6$

9. $\dfrac{a + b}{2}$, when $a = 5$ and $b = 17$

Translate each phrase to an algebraic expression. Use any letter for the variable. [1.1b]

10. Three times some number

11. Five less than some number

12. State the integers that correspond to this situation: Jerilyn deposited $450 in her checking account. Later that week, she wrote a check for $79. [1.2a]

13. Graph -3.5 on the number line. [1.2b]

    ⟵+–+–+–+–+–+–+–+–+–+–+–+–⟶
    −6 −5 −4 −3 −2 −1 0 1 2 3 4 5 6

Convert to decimal notation. [1.2c]

14. $-\dfrac{4}{5}$

15. $\dfrac{7}{3}$

Use either $<$ or $>$ for ☐ to write a true sentence. [1.2d]

16. $-5 \,\square\, -3$

17. $-9.9 \,\square\, -10.1$

Mid-Chapter Review: Chapter 1   **61**

Write true or false. [1.2d]

18. $-8 \geq -5$ **19.** $-4 \leq -4$

Write an inequality with the same meaning. [1.2d]

20. $y < 5$ **21.** $-3 \geq t$

Find the absolute value. [1.2e]

22. $|15.6|$ **23.** $|-18|$ **24.** $|0|$ **25.** $\left|-\dfrac{12}{5}\right|$

Find the opposite, or additive inverse, of each number. [1.3b]

26. -5.6 **27.** $\dfrac{7}{4}$ **28.** 0 **29.** -49

30. Evaluate $-x$ when x is -19. [1.3b]

31. Evaluate $-(-x)$ when x is 2.3. [1.3b]

Compute and simplify. [1.3a], [1.4a]

32. $7 + (-9)$ **33.** $-\dfrac{3}{8} + \dfrac{1}{4}$ **34.** $3.6 + (-3.6)$ **35.** $-8 + (-9)$

36. $\dfrac{2}{3} + \left(-\dfrac{9}{8}\right)$ **37.** $-4.2 + (-3.9)$ **38.** $-14 + 5$ **39.** $19 + (-21)$

40. $-4.1 - 6.3$ **41.** $5 - (-11)$ **42.** $-\dfrac{1}{4} - \left(-\dfrac{3}{5}\right)$ **43.** $12 - 24$

44. $-8 - (-4)$ **45.** $-\dfrac{1}{2} - \dfrac{5}{6}$ **46.** $12.3 - 14.1$ **47.** $6 - (-7)$

48. $16 - (-9) - 20 - (-4)$

49. $-4 + (-10) - (-3) - 12$

50. $17 - (-25) + 15 - (-18)$

51. $-9 + (-3) + 16 - (-10)$

Solve. [1.3c], [1.4b]

52. *Temperature Change.* In chemistry lab, Ben works with a substance whose initial temperature is 25°C. During an experiment, the temperature falls to −8°C. Find the difference between the two temperatures.

53. *Stock Price Change.* The price of a stock opened at $56.12. During the day, it dropped $1.18, then rose $1.22, and then dropped $1.36. Find the value of the stock at the end of the day.

Understanding Through Discussion and Writing

54. Give three examples of rational numbers that are not integers. Explain. [1.2b]

55. Give three examples of irrational numbers. Explain the difference between an irrational number and a rational number. [1.2b, d]

56. Explain in your own words why the sum of two negative numbers is always negative. [1.3a]

57. If a negative number is subtracted from a positive number, will the result always be positive? Why or why not? [1.4a]

STUDYING FOR SUCCESS *Using the Textbook*

- ☐ Study the step-by-step solutions to the examples, using the red comments for extra explanations.
- ☐ Stop and do the margin exercises as directed. Answers are at the bottom of the margin, so you can check your work right away.
- ☐ The objective symbols **a**, **b**, **c**, and so on, allow you to refer to the appropriate place in the text whenever you need to review a topic.

Multiplication of Real Numbers

1.5

OBJECTIVES

a Multiply real numbers.

b Solve applied problems involving multiplication of real numbers.

a MULTIPLICATION

SKILL REVIEW

Multiply using fraction notation. [J8]
Multiply and, if possible, simplify.

1. $\dfrac{5}{3} \cdot \dfrac{6}{7}$

2. $\dfrac{1}{3} \cdot 12$

Answers: 1. $\dfrac{10}{7}$ 2. 4

MyLab Math VIDEO

Multiplication of real numbers is very much like multiplication of arithmetic numbers. The only difference is that we must determine whether the product is positive or negative.

Multiplication of a Positive Number and a Negative Number

To see how to multiply a positive number and a negative number, consider the pattern of the following.

This number decreases by 1 each time.
This number decreases by 5 each time.

$3 \cdot 5 = 15$
$2 \cdot 5 = 10$
$1 \cdot 5 = 5$
$0 \cdot 5 = 0$
$-1 \cdot 5 = -5$
$-2 \cdot 5 = -10$
$-3 \cdot 5 = -15$

Do Exercise 1.

As the pattern suggests, the product of a negative number and a positive number is negative.

1. Complete, as in the example.

$3 \cdot 10 = 30$
$2 \cdot 10 = 20$
$1 \cdot 10 =$
$0 \cdot 10 =$
$-1 \cdot 10 =$
$-2 \cdot 10 =$
$-3 \cdot 10 =$

THE PRODUCT OF A POSITIVE NUMBER AND A NEGATIVE NUMBER

To multiply a positive number and a negative number, multiply their absolute values. The product is negative.

Answer
1. 10; 0; −10; −20; −30

Multiply.

2. $-3 \cdot 6$ 3. $20 \cdot (-5)$

4. $-\dfrac{2}{3} \cdot \dfrac{5}{6}$ 5. $-4.23(7.1)$

6. Complete, as in the example.

$3 \cdot (-10) = -30$
$2 \cdot (-10) = -20$
$1 \cdot (-10) = $
$0 \cdot (-10) = $
$-1 \cdot (-10) = $
$-2 \cdot (-10) = $
$-3 \cdot (-10) = $

Multiply.

7. $-9 \cdot (-3)$

8. $-16 \cdot (-2)$

9. $-7 \cdot (-5)$

10. $-\dfrac{4}{7}\left(-\dfrac{5}{9}\right)$

11. $-\dfrac{3}{2}\left(-\dfrac{4}{9}\right)$

12. $-3.25(-4.14)$

EXAMPLES Multiply.

1. $8(-5) = -40$ 2. $-\dfrac{1}{3} \cdot \dfrac{5}{7} = -\dfrac{5}{21}$ 3. $(-7.2)5 = -36$

◀ Do Exercises 2–5.

Multiplication of Two Negative Numbers

How do we multiply two negative numbers? Again, we look for a pattern.

This number decreases by 1 each time.   This number increases by 5 each time.

$3 \cdot (-5) = -15$
$2 \cdot (-5) = -10$
$1 \cdot (-5) = -5$
$0 \cdot (-5) = 0$
$-1 \cdot (-5) = 5$
$-2 \cdot (-5) = 10$
$-3 \cdot (-5) = 15$

◀ Do Exercise 6.

As the pattern suggests, the product of two negative numbers is positive.

THE PRODUCT OF TWO NEGATIVE NUMBERS

To multiply two negative numbers, multiply their absolute values. The product is positive.

◀ Do Exercises 7–12.

The following is another way to consider the rules that we have for multiplication.

To multiply two nonzero real numbers:
a) Multiply the absolute values.
b) If the signs are the same, the product is positive.
c) If the signs are different, the product is negative.

Multiplication by Zero

As with nonnegative numbers, the product of any real number and 0 is 0.

THE MULTIPLICATION PROPERTY OF ZERO

For any real number a,
$$a \cdot 0 = 0 \cdot a = 0.$$

(The product of 0 and any real number is 0.)

Answers

2. -18 3. -100 4. $-\dfrac{5}{9}$ 5. -30.033
6. $-10; 0; 10; 20; 30$ 7. 27 8. 32
9. 35 10. $\dfrac{20}{63}$ 11. $\dfrac{2}{3}$ 12. 13.455

EXAMPLES Multiply.

4. $(-3)(-4) = 12$

5. $-1.6(2) = -3.2$

6. $-193 \cdot 0 = 0$

7. $\left(-\dfrac{5}{6}\right)\left(-\dfrac{1}{9}\right) = \dfrac{5}{54}$

Do Exercises 13–18. ▷

Multiplying More Than Two Numbers

When multiplying more than two real numbers, we can choose order and grouping as we please.

EXAMPLES Multiply.

8. $-8 \cdot 2(-3) = -16(-3)$ Multiplying the first two numbers
$= 48$

9. $-8 \cdot 2(-3) = 24 \cdot 2$ Multiplying the negative numbers. Every pair of negative numbers gives a positive product.
$= 48$

10. $-3(-2)(-5)(4) = 6(-5)(4)$ Multiplying the first two numbers
$= (-30)4$
$= -120$

11. $\left(-\dfrac{1}{2}\right)(8)\left(-\dfrac{2}{3}\right)(-6) = (-4)4$ Multiplying the first two numbers and the last two numbers
$= -16$

12. $-5 \cdot (-2) \cdot (-3) \cdot (-6) = 10 \cdot 18 = 180$

13. $(-3)(-5)(-2)(-3)(-6) = (-30)(18) = -540$

Considering that the product of a pair of negative numbers is positive, we see the following pattern.

> The product of an even number of negative numbers is positive.
> The product of an odd number of negative numbers is negative.

Do Exercises 19–24. ▷

EXAMPLE 14 Evaluate $2x^2$ when $x = 3$ and when $x = -3$.

$2x^2 = 2(3)^2 = 2(9) = 18;$
$2x^2 = 2(-3)^2 = 2(9) = 18$

Let's compare the expressions $(-x)^2$ and $-x^2$.

EXAMPLE 15 Evaluate $(-x)^2$ and $-x^2$ when $x = 5$.

$(-x)^2 = (-5)^2 = (-5)(-5) = 25;$ Substitute 5 for x. Then evaluate the power.

$-x^2 = -(5)^2 = -(25) = -25$ Substitute 5 for x. Evaluate the power. Then find the opposite.

Multiply.

13. $5(-6)$

14. $(-5)(-6)$

15. $(-3.2) \cdot 10$

16. $\left(-\dfrac{4}{5}\right)\left(\dfrac{10}{3}\right)$

17. $0 \cdot (-34.2)$

18. $-\dfrac{5}{7} \cdot 0 \cdot \left(-4\dfrac{2}{3}\right)$

Multiply.

19. $5 \cdot (-3) \cdot 2$

20. $-3 \times (-4.1) \times (-2.5)$

21. $-\dfrac{1}{2} \cdot \left(-\dfrac{4}{3}\right) \cdot \left(-\dfrac{5}{2}\right)$

22. $-2 \cdot (-5) \cdot (-4) \cdot (-3)$

23. $(-4)(-5)(-2)(-3)(-1)$

24. $(-1)(-1)(-2)(-3)(-1)(-1)$

Answers

13. -30 **14.** 30 **15.** -32 **16.** $-\dfrac{8}{3}$
17. 0 **18.** 0 **19.** -30 **20.** -30.75
21. $-\dfrac{5}{3}$ **22.** 120 **23.** -120 **24.** 6

25. Evaluate $3x^2$ when $x = 4$ and when $x = -4$.

26. Evaluate $(-x)^2$ and $-x^2$ when $x = 2$.

27. Evaluate $(-x)^2$ and $-x^2$ when $x = -3$.

In Example 15, we see that the expressions $(-x)^2$ and $-x^2$ are *not* equivalent. That is, they do not have the same value for every allowable replacement of the variable by a real number. To find $(-x)^2$, we take the opposite and then square. To find $-x^2$, we find the square and then take the opposite.

EXAMPLE 16 Evaluate $(-a)^2$ and $-a^2$ when $a = -4$.

$(-a)^2 = [-(-4)]^2 = [4]^2 = 16;$ Using parentheses to substitute
$-a^2 = -(-4)^2 = -(16) = -16$

◀ Do Exercises 25–27.

b APPLICATIONS AND PROBLEM SOLVING

EXAMPLE 17 *Mine Rescue.*
The San Jose copper and gold mine near Copiapó, Chile, collapsed on August 5, 2010, trapping 33 miners. Each miner was safely brought out of the mine with a specially designed capsule that could be lowered into the mine at -137 feet per minute. It took approximately 15 minutes to lower the capsule to the miners' location. Determine how far below the surface of the earth the miners were trapped.
Data: Reuters News

28. *Chemical Reaction.* During a chemical reaction, the temperature in a beaker increased by 3°C every minute until 1:34 P.M. If the temperature was -17°C at 1:10 P.M., when the reaction began, what was the temperature at 1:34 P.M.?

Since the capsule moved -137 feet per minute and it took 15 minutes to reach the miners, we have the depth d given by

$$d = 15 \cdot (-137) = -2055.$$

Thus the miners were trapped at -2055 ft.

◀ Do Exercise 28.

Answers
25. 48; 48   26. 4; −4   27. 9; −9
28. 55°C

1.5 Exercise Set

FOR EXTRA HELP   MyLab Math

✓ Check Your Understanding

Reading Check Fill in the blank with either "positive" or "negative."

RC1. To multiply a positive number and a negative number, multiply their absolute values. The answer is _____.

RC2. To multiply two negative numbers, multiply their absolute values. The answer is _____.

RC3. The product of an even number of negative numbers is _____.

RC4. The product of an odd number of negative numbers is _____.

Concept Check Evaluate.

CC1. -3^2 **CC2.** $(-3)^2$ **CC3.** $-\left(\dfrac{1}{2}\right)^2$ **CC4.** $-\left(-\dfrac{1}{2}\right)^2$

a Multiply.

1. $-4 \cdot 2$
2. $-3 \cdot 5$
3. $8 \cdot (-3)$
4. $9 \cdot (-5)$
5. $-9 \cdot 8$

6. $-10 \cdot 3$
7. $-8 \cdot (-2)$
8. $-2 \cdot (-5)$
9. $-7 \cdot (-6)$
10. $-9 \cdot (-2)$

11. $15 \cdot (-8)$
12. $-12 \cdot (-10)$
13. $-14 \cdot 17$
14. $-13 \cdot (-15)$
15. $-25 \cdot (-48)$

16. $39 \cdot (-43)$
17. $-3.5 \cdot (-28)$
18. $97 \cdot (-2.1)$
19. $9 \cdot (-8)$
20. $7 \cdot (-9)$

21. $4 \cdot (-3.1)$
22. $3 \cdot (-2.2)$
23. $-5 \cdot (-6)$
24. $-6 \cdot (-4)$
25. $-7 \cdot (-3.1)$

26. $-4 \cdot (-3.2)$
27. $\dfrac{2}{3} \cdot \left(-\dfrac{3}{5}\right)$
28. $\dfrac{5}{7} \cdot \left(-\dfrac{2}{3}\right)$
29. $-\dfrac{3}{8} \cdot \left(-\dfrac{2}{9}\right)$

30. $-\dfrac{5}{8} \cdot \left(-\dfrac{2}{5}\right)$
31. -6.3×2.7
32. -4.1×9.5
33. $7 \cdot (-4) \cdot (-3) \cdot 5$

34. $9 \cdot (-2) \cdot (-6) \cdot 7$
35. $-\dfrac{2}{3} \cdot \dfrac{1}{2} \cdot \left(-\dfrac{6}{7}\right)$
36. $-\dfrac{1}{8} \cdot \left(-\dfrac{1}{4}\right) \cdot \left(-\dfrac{3}{5}\right)$
37. $-3 \cdot (-4) \cdot (-5)$

38. $-2 \cdot (-5) \cdot (-7)$
39. $-2 \cdot (-5) \cdot (-3) \cdot (-5)$
40. $-3 \cdot (-5) \cdot (-2) \cdot (-1)$

41. $-4 \cdot (-1.8) \cdot 7$
42. $-8 \cdot (-1.3) \cdot (-5)$
43. $-\dfrac{1}{9}\left(-\dfrac{2}{3}\right)\left(\dfrac{5}{7}\right)$

44. $-\dfrac{7}{2}\left(-\dfrac{5}{7}\right)\left(-\dfrac{2}{5}\right)$
45. $4 \cdot (-4) \cdot (-5) \cdot (-12)$
46. $-2 \cdot (-3) \cdot (-4) \cdot (-5)$

47. $0.07 \cdot (-7) \cdot 6 \cdot (-6)$

48. $80 \cdot (-0.8) \cdot (-90) \cdot (-0.09)$

49. $\left(-\dfrac{5}{6}\right)\left(\dfrac{1}{8}\right)\left(-\dfrac{3}{7}\right)\left(-\dfrac{1}{7}\right)$

50. $\left(\dfrac{4}{5}\right)\left(-\dfrac{2}{3}\right)\left(-\dfrac{15}{7}\right)\left(\dfrac{1}{2}\right)$

51. $(-14) \cdot (-27) \cdot 0$

52. $7 \cdot (-6) \cdot 5 \cdot (-4) \cdot 3 \cdot (-2) \cdot 1 \cdot 0$

53. $(-8)(-9)(-10)$

54. $(-7)(-8)(-9)(-10)$

55. $(-6)(-7)(-8)(-9)(-10)$

56. $(-5)(-6)(-7)(-8)(-9)(-10)$

57. $(-1)^{12}$

58. $(-1)^9$

59. Evaluate $(-x)^2$ and $-x^2$ when $x = 4$ and when $x = -4$.

60. Evaluate $(-x)^2$ and $-x^2$ when $x = 10$ and when $x = -10$.

61. Evaluate $(-y)^2$ and $-y^2$ when $y = \dfrac{2}{5}$ and when $y = -\dfrac{2}{5}$.

62. Evaluate $(-w)^2$ and $-w^2$ when $w = \dfrac{1}{10}$ and when $w = -\dfrac{1}{10}$.

63. Evaluate $-(-t)^2$ and $-t^2$ when $t = 3$ and when $t = -3$.

64. Evaluate $-(-s)^2$ and $-s^2$ when $s = 1$ and when $s = -1$.

65. Evaluate $(-3x)^2$ and $-3x^2$ when $x = 7$ and when $x = -7$.

66. Evaluate $(-2x)^2$ and $-2x^2$ when $x = 3$ and when $x = -3$.

67. Evaluate $5x^2$ when $x = 2$ and when $x = -2$.

68. Evaluate $2x^2$ when $x = 5$ and when $x = -5$.

69. Evaluate $-2x^3$ when $x = 1$ and when $x = -1$.

70. Evaluate $-3x^3$ when $x = 2$ and when $x = -2$.

b Solve.

71. *Chemical Reaction.* The temperature of a chemical compound was 0°C at 11:00 A.M. During a reaction, it dropped 3°C per minute until 11:08 A.M. What was the temperature at 11:08 A.M.?

72. *Chemical Reaction.* The temperature of a chemical compound was −5°C at 3:20 P.M. During a reaction, it increased 2°C per minute until 3:52 P.M. What was the temperature at 3:52 P.M.?

73. *Weight Loss.* Dave lost 2 lb each week for a period of 10 weeks. Express his total weight change as an integer.

74. *Stock Loss.* Each day for a period of 5 days, the value of a stock that Lily owned dropped $3. Express Lily's total loss as an integer.

75. *Stock Price.* The price of a stock began the day at $23.75 per share and dropped $1.38 per hour for 8 hr. What was the price of the stock after 8 hr?

76. *Population Decrease.* The population of Bloomtown was 12,500. It decreased 380 each year for 4 years. What was the population of the town after 4 years?

77. *Diver's Position.* After diving 95 m below sea level, a diver rises at a rate of 7 m/min for 9 min. Where is the diver in relation to the surface at the end of the 9-min period?

78. *Bank Account Balance.* Karen had $68 in her bank account. After she used her debit card to make seven purchases at $13 each, what was the balance in her bank account?

79. *Drop in Temperature.* The temperature in Osgood was 62°F at 2:00 P.M. It dropped 6°F per hour for the next 4 hr. What was the temperature at the end of the 4-hr period?

80. *Juice Consumption.* Oliver bought a 64-oz container of cranberry juice and drank 8 oz per day for a week. How much juice was left in the container at the end of the week?

Skill Maintenance

81. Evaluate $\dfrac{x - 2y}{3}$ when $x = 20$ and $y = 7$. [1.1a]

82. Evaluate $\dfrac{d - e}{3d}$ when $d = 5$ and $e = 1$. [1.1a]

Subtract. [1.4a]

83. $-\dfrac{1}{2} - \left(-\dfrac{1}{6}\right)$

84. $8 - 12.3$

85. $31 - (-13)$

86. $-\dfrac{5}{12} - \left(-\dfrac{1}{3}\right)$

Write true or false. [1.2d]

87. $-10 > -12$

88. $0 \leq -1$

89. $4 < -8$

90. $-7 \geq -6$

Synthesis

91. If a is positive and b is negative, then $-ab$ is which of the following?

   A. Positive
   B. Negative
   C. 0
   D. Cannot be determined without more information

92. If a is positive and b is negative, then $(-a)(-b)$ is which of the following?

   A. Positive
   B. Negative
   C. 0
   D. Cannot be determined without more information

93. Of all possible quotients of the numbers 10, $-\tfrac{1}{2}$, -5, and $\tfrac{1}{5}$, which two produce the largest quotient? Which two produce the smallest quotient?

1.6 Division of Real Numbers

OBJECTIVES

a Divide integers.
b Find the reciprocal of a real number.
c Divide real numbers.
d Solve applied problems involving division of real numbers.

We now consider division of real numbers. The definition of division results in rules for division that are the same as those for multiplication.

a DIVISION OF INTEGERS

DIVISION

The quotient $a \div b$, or $\dfrac{a}{b}$, where $b \neq 0$, is that unique real number c for which $a = b \cdot c$.

Let's use the definition to divide integers.

EXAMPLES Divide, if possible. Check your answer.

1. $14 \div (-7) = -2$ *Think*: What number multiplied by -7 gives 14? That number is -2. *Check*: $(-2)(-7) = 14$.

2. $\dfrac{-32}{-4} = 8$ *Think*: What number multiplied by -4 gives -32? That number is 8. *Check*: $8(-4) = -32$.

3. $\dfrac{-10}{7} = -\dfrac{10}{7}$ *Think*: What number multiplied by 7 gives -10? That number is $-\tfrac{10}{7}$. *Check*: $-\tfrac{10}{7} \cdot 7 = -10$.

4. $\dfrac{-17}{0}$ is **not defined**.   *Think*: What number multiplied by 0 gives -17? There is no such number because the product of 0 and *any* number is 0.

The rules for division are the same as those for multiplication.

To multiply or divide two real numbers (where the divisor is nonzero):
a) Multiply or divide the absolute values.
b) If the signs are the same, the answer is positive.
c) If the signs are different, the answer is negative.

◂ Do Exercises 1–6.

Divide.
1. $6 \div (-3)$
   Think: What number multiplied by -3 gives 6?

2. $\dfrac{-15}{-3}$
   Think: What number multiplied by -3 gives -15?

3. $-24 \div 8$
   Think: What number multiplied by 8 gives -24?

4. $\dfrac{-48}{-6}$ 5. $\dfrac{30}{-5}$

6. $\dfrac{30}{-7}$

Excluding Division by 0

Example 4 shows why we cannot divide -17 by 0. We can use the same argument to show why we cannot divide any nonzero number b by 0. Consider $b \div 0$. We look for a number that when multiplied by 0 gives b. There is no such number because the product of 0 and any number is 0. Thus we cannot divide a nonzero number b by 0.

On the other hand, if we divide 0 by 0, we look for a number c such that $0 \cdot c = 0$. But $0 \cdot c = 0$ for any number c. Thus it appears that $0 \div 0$ could be any number we choose. Getting any answer we want when we divide 0 by 0 would be very confusing. Thus we agree that division by 0 is not defined.

Answers
1. -2 2. 5   3. -3 4. 8   5. -6
6. $-\dfrac{30}{7}$

EXCLUDING DIVISION BY 0

Division by 0 is not defined.

$a \div 0$, or $\dfrac{a}{0}$, is not defined for all real numbers a.

Dividing 0 by Other Numbers

Note that

$0 \div 8 = 0$ because $0 = 0 \cdot 8$; $\dfrac{0}{-5} = 0$ because $0 = 0 \cdot (-5)$.

DIVIDENDS OF 0

Zero divided by any nonzero real number is 0:

$\dfrac{0}{a} = 0;\quad a \neq 0.$

EXAMPLES Divide, if possible.

5. $0 \div (-6) = 0$ **6.** $\dfrac{0}{12} = 0$ **7.** $\dfrac{-3}{0}$ is not defined.

Do Exercises 7 and 8.

Divide, if possible.

7. $\dfrac{-5}{0}$ **8.** $\dfrac{0}{-3}$

b RECIPROCALS

Find the opposite of a real number. [1.3b]
Find the opposite.

1. 37   **2.** $-\dfrac{5}{6}$

Answers: 1. -37 **2.** $\dfrac{5}{6}$

When two numbers like $\tfrac{1}{2}$ and 2 are multiplied, the result is 1. Such numbers are called **reciprocals** of each other. Every nonzero real number has a reciprocal, also called a **multiplicative inverse**.

RECIPROCALS

Two numbers whose product is 1 are called **reciprocals**, or **multiplicative inverses**, of each other.

EXAMPLES Find the reciprocal.

8. $\dfrac{7}{8}$ The reciprocal of $\dfrac{7}{8}$ is $\dfrac{8}{7}$ because $\dfrac{7}{8} \cdot \dfrac{8}{7} = 1$.

9. -5 The reciprocal of -5 is $-\dfrac{1}{5}$ because $-5\left(-\dfrac{1}{5}\right) = 1$.

Answers
7. Not defined   **8.** 0

10. 3.9   The reciprocal of 3.9 is $\frac{1}{3.9}$ because $3.9\left(\frac{1}{3.9}\right) = 1$.

11. $-\frac{1}{2}$ The reciprocal of $-\frac{1}{2}$ is -2 because $\left(-\frac{1}{2}\right)(-2) = 1$.

12. $-\frac{2}{3}$ The reciprocal of $-\frac{2}{3}$ is $-\frac{3}{2}$ because $\left(-\frac{2}{3}\right)\left(-\frac{3}{2}\right) = 1$.

13. $\frac{3y}{8x}$ The reciprocal of $\frac{3y}{8x}$ is $\frac{8x}{3y}$ because $\left(\frac{3y}{8x}\right)\left(\frac{8x}{3y}\right) = 1$.

RECIPROCAL PROPERTIES

For $a \neq 0$, the reciprocal of a can be named $\frac{1}{a}$ and the reciprocal of $\frac{1}{a}$ is a.

The reciprocal of a nonzero number $\frac{a}{b}$ can be named $\frac{b}{a}$.

The number 0 has no reciprocal.

Find the reciprocal.

9. $\frac{2}{3}$ 10. $-\frac{5}{4}$

11. -3 12. $-\frac{1}{5}$

13. 1.3 14. $\frac{a}{6b}$

◀ Do Exercises 9–14.

The reciprocal of a positive number is also a positive number, because the product of the two numbers must be the positive number 1. The reciprocal of a negative number is also a negative number, because the product of the two numbers must be the positive number 1.

THE SIGN OF A RECIPROCAL

The reciprocal of a number has the same sign as the number itself.

⋯⋯⋯⋯⋯⋯⋯⋯⋯⋯⋯⋯⋯⋯⋯⋯ **Caution!** ⋯⋯⋯⋯⋯⋯⋯⋯⋯⋯⋯⋯⋯⋯⋯⋯

It is important *not* to confuse *opposite* with *reciprocal*. The opposite, or additive inverse, of a number is what we add to the number to get 0. The reciprocal, or multiplicative inverse, is what we multiply the number by to get 1.

15. Complete the following table.

NUMBER	OPPOSITE	RECIPROCAL
$\frac{2}{9}$		
$-\frac{7}{4}$		
0		
1		
-8		
-4.7		

Compare the following.

NUMBER	OPPOSITE (Change the sign.)	RECIPROCAL (Invert but do not change the sign.)
$-\frac{3}{8}$	$\frac{3}{8}$	$-\frac{8}{3}$
$\frac{18}{7}$	$-\frac{18}{7}$	$\frac{7}{18}$
-7.9	7.9	$-\frac{1}{7.9}$, or $-\frac{10}{79}$
0	0	None

$\left(-\frac{3}{8}\right)\left(-\frac{8}{3}\right) = 1$

$-\frac{3}{8} + \frac{3}{8} = 0$

◀ Do Exercise 15.

Answers

9. $\frac{3}{2}$ 10. $-\frac{4}{5}$ 11. $-\frac{1}{3}$ 12. -5 13. $\frac{1}{1.3}$, or $\frac{10}{13}$ 14. $\frac{6b}{a}$ 15. $-\frac{2}{9}$ and $\frac{9}{2}$; $\frac{7}{4}$ and $-\frac{4}{7}$; 0 and none; -1 and 1; 8 and $-\frac{1}{8}$; 4.7 and $-\frac{1}{4.7}$, or $-\frac{10}{47}$

72 CHAPTER 1   Introduction to Real Numbers and Algebraic Expressions

c DIVISION OF REAL NUMBERS

We know that we can subtract by adding an opposite. Similarly, we can divide by multiplying by a reciprocal.

> **RECIPROCALS AND DIVISION**
>
> For any real numbers a and b, $b \neq 0$,
> $$a \div b = \frac{a}{b} = a \cdot \frac{1}{b}.$$
>
> (To divide, multiply by the reciprocal of the divisor.)

EXAMPLES Rewrite each division as a multiplication.

14. $-4 \div 3$ $-4 \div 3$ is the same as $-4 \cdot \frac{1}{3}$

15. $\frac{6}{-7}$ $\frac{6}{-7} = 6\left(-\frac{1}{7}\right)$ $\frac{6}{-7}$ means $6 \div (-7)$

16. $\frac{3}{5} \div \left(-\frac{9}{7}\right)$ $\frac{3}{5} \div \left(-\frac{9}{7}\right) = \frac{3}{5}\left(-\frac{7}{9}\right)$

17. $\frac{x+2}{5}$ $\frac{x+2}{5} = (x+2)\frac{1}{5}$ Parentheses are necessary here.

18. $\frac{-17}{1/b}$ $\frac{-17}{1/b} = -17 \cdot b$ $\frac{-17}{1/b}$ means $-17 \div \frac{1}{b}$

Do Exercises 16–20. ▶

When actually doing division calculations, we sometimes multiply by a reciprocal and we sometimes divide directly. With fraction notation, it is usually better to multiply by a reciprocal. With decimal notation, it is usually better to divide directly.

EXAMPLES Divide by multiplying by the reciprocal of the divisor.

19. $\frac{2}{3} \div \left(-\frac{5}{4}\right) = \frac{2}{3} \cdot \left(-\frac{4}{5}\right) = -\frac{8}{15}$

20. $-\frac{5}{6} \div \left(-\frac{3}{4}\right) = -\frac{5}{6} \cdot \left(-\frac{4}{3}\right) = \frac{20}{18} = \frac{10 \cdot 2}{9 \cdot 2} = \frac{10}{9} \cdot \frac{2}{2} = \frac{10}{9}$

↑ Caution!

Be careful *not* to change the sign when taking a reciprocal!

21. $-\frac{3}{4} \div \frac{3}{10} = -\frac{3}{4} \cdot \left(\frac{10}{3}\right) = -\frac{30}{12} = -\frac{5 \cdot 6}{2 \cdot 6} = -\frac{5}{2} \cdot \frac{6}{6} = -\frac{5}{2}$

Do Exercises 21 and 22. ▶

Rewrite each division as a multiplication.

16. $\frac{4}{7} \div \left(-\frac{3}{5}\right)$

17. $\frac{5}{-8}$

18. $\frac{a-b}{7}$

19. $\frac{-23}{1/a}$

20. $-5 \div 7$

Divide by multiplying by the reciprocal of the divisor.

GS 21. $\frac{4}{7} \div \left(-\frac{3}{5}\right)$

$= \frac{4}{7} \cdot \left(-\frac{5}{\square}\right) = \square$

22. $-\frac{12}{7} \div \left(-\frac{3}{4}\right)$

Answers

16. $\frac{4}{7} \cdot \left(-\frac{5}{3}\right)$ 17. $5 \cdot \left(-\frac{1}{8}\right)$

18. $(a-b) \cdot \frac{1}{7}$ 19. $-23 \cdot a$

20. $-5 \cdot \left(\frac{1}{7}\right)$ 21. $-\frac{20}{21}$ 22. $\frac{16}{7}$

Guided Solution:

21. $3; -\frac{20}{21}$

Divide.

23. $21.7 \div (-3.1)$

24. $-20.4 \div (-4)$

CALCULATOR CORNER

Operations on the Real Numbers To perform operations on the real numbers on a graphing calculator, recall that negative numbers are entered using the opposite key, (-), and subtraction is entered using the subtraction operation key, −. Consider the sum $-5 + (-3.8)$. On a graphing calculator, the parentheses around -3.8 are optional. The result is -8.8.

```
-5+-3.8
            -8.8
-5+(-3.8)
            -8.8
```

EXERCISES: Use a calculator to perform each operation.

1. $1.2 - (-1.5)$
2. $-7.6 + (-1.9)$
3. $1.2 \div (-1.5)$
4. $-7.6 \cdot (-1.9)$

Find two equal expressions for each number with negative signs in different places.

25. $\dfrac{-5}{6} = \dfrac{5}{\boxed{}} = -\dfrac{\boxed{}}{6}$ **GS**

26. $-\dfrac{8}{7}$

27. $\dfrac{10}{-3}$

Answers

23. -7 **24.** 5.1 **25.** $\dfrac{5}{-6}; -\dfrac{5}{6}$ **26.** $\dfrac{8}{-7}; \dfrac{-8}{7}$

27. $\dfrac{-10}{3}; -\dfrac{10}{3}$

Guided Solution:
25. $-6; 5$

With decimal notation, it is easier to carry out long division than to multiply by the reciprocal.

EXAMPLES Divide.

22. $-27.9 \div (-3) = \dfrac{-27.9}{-3} = 9.3$ Do the long division $3\overline{)27.9}$. The answer is positive.

23. $-6.3 \div 2.1 = -3$ Do the long division $2.1\overline{)6.3}$. The answer is negative.

◀ Do Exercises 23 and 24.

Consider the following:

1. $\dfrac{2}{3} = \dfrac{2}{3} \cdot 1 = \dfrac{2}{3} \cdot \dfrac{-1}{-1} = \dfrac{2(-1)}{3(-1)} = \dfrac{-2}{-3}$. Thus, $\dfrac{2}{3} = \dfrac{-2}{-3}$.

(A negative number divided by a negative number is positive.)

2. $-\dfrac{2}{3} = -1 \cdot \dfrac{2}{3} = \dfrac{-1}{1} \cdot \dfrac{2}{3} = \dfrac{-1 \cdot 2}{1 \cdot 3} = \dfrac{-2}{3}$. Thus, $-\dfrac{2}{3} = \dfrac{-2}{3}$.

(A negative number divided by a positive number is negative.)

3. $\dfrac{-2}{3} = \dfrac{-2}{3} \cdot 1 = \dfrac{-2}{3} \cdot \dfrac{-1}{-1} = \dfrac{-2(-1)}{3(-1)} = \dfrac{2}{-3}$. Thus, $-\dfrac{2}{3} = \dfrac{2}{-3}$.

(A positive number divided by a negative number is negative.)

We can use the following properties to make sign changes in fraction notation.

SIGN CHANGES IN FRACTION NOTATION

For any numbers a and b, $b \neq 0$:

1. $\dfrac{-a}{-b} = \dfrac{a}{b}$

(The opposite of a number a divided by the opposite of another number b is the same as the quotient of the two numbers a and b.)

2. $\dfrac{-a}{b} = \dfrac{a}{-b} = -\dfrac{a}{b}$

(The opposite of a number a divided by another number b is the same as the number a divided by the opposite of the number b, and both are the same as the opposite of a divided by b.)

◀ Do Exercises 25–27.

d APPLICATIONS AND PROBLEM SOLVING

When we describe a change in quantity, we can discuss either the change in quantity itself or the percent increase or decrease of the change. To determine the change, we subtract the original amount from the new amount. The change is *positive* if there was an increase or *negative* if there was a decrease. To determine the percent increase or decrease, we divide the change by the original amount and convert the decimal answer to percent notation.

EXAMPLE 24 *Endangered Species.* Because of hunting in previous centuries and current coastal development, the Florida manatee is considered an endangered species. Efforts to protect manatees have resulted in an increase in the population. One way that conservationists track the number of manatees is through aerial surveys conducted in the winter when manatees congregate in warmer areas. Although not considered accurate estimates of the manatee population, the surveys do indicate population trends. An aerial survey in 2001 found 3300 manatees and a survey in 2006 found 3113 manatees. What was the percent increase or decrease from 2001 to 2006?

We find the change in population and then the percent increase or decrease.

$$\text{Change in population} = \text{New amount} - \text{original amount}$$
$$= 3113 - 3300 = -187 \quad \text{The count decreased.}$$

$$\text{Percent change} = \frac{\text{Change}}{\text{Original amount}}$$
$$= \frac{-187}{3300} \approx -0.06 = -6\%$$

Since the percent change was negative, there was a percent decrease in the manatee count of 6%.

Do Exercise 28. ▶

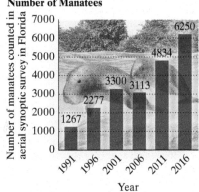

Number of Manatees

DATA: Florida Fish and Wildlife Conservation Commission

28. *Endangered Species.* An aerial survey in 2006 found 3113 manatees in Florida. A survey in 2016 found 6250 manatees. What was the percent increase or decrease?

Answer
28. The percent increase was about 101%.

1.6 Exercise Set

FOR EXTRA HELP   MyLab Math

✓ Check Your Understanding

Reading Check Choose the word or the number below the blank that will make the sentence true.

RC1. The numbers 4 and −4 are called _____ of each other.
     opposites/reciprocals

RC2. The multiplicative inverse, or reciprocal, of a number is what we multiply the number by to get ____.
                                                                                                    0/1

RC3. The additive inverse, or opposite, of a number is what we add to the number to get ____.
                                                                                          0/1

RC4. The numbers $-\frac{9}{4}$ and $-\frac{4}{9}$ are called _____ of each other.
                                                    opposites/reciprocals

Concept Check Fill in the blank with either 0 or 1 to complete each sentence.

CC1. The number ____ has no reciprocal.

CC2. The number ____ is its own reciprocal.

CC3. The number ____ is its own opposite.

CC4. Division by ____ is undefined.

CC5. A nonzero number divided by itself is ____.

a Divide, if possible. Check each answer.

1. $48 \div (-6)$
2. $\dfrac{42}{-7}$
3. $\dfrac{28}{-2}$
4. $24 \div (-12)$

5. $\dfrac{-24}{8}$
6. $-18 \div (-2)$
7. $\dfrac{-36}{-12}$
8. $-72 \div (-9)$

9. $\dfrac{-72}{9}$
10. $\dfrac{-50}{25}$
11. $-100 \div (-50)$
12. $\dfrac{-200}{8}$

13. $-108 \div 9$
14. $\dfrac{-63}{-7}$
15. $\dfrac{200}{-25}$
16. $-300 \div (-16)$

17. $\dfrac{75}{0}$
18. $\dfrac{0}{-5}$
19. $\dfrac{0}{-2.6}$
20. $\dfrac{-23}{0}$

b Find the reciprocal.

21. $\dfrac{15}{7}$
22. $\dfrac{3}{8}$
23. $-\dfrac{47}{13}$
24. $-\dfrac{31}{12}$

25. 13
26. -10
27. -32
28. 15

29. $\dfrac{1}{-7.1}$
30. $\dfrac{1}{-4.9}$
31. $\dfrac{1}{9}$
32. $\dfrac{1}{16}$

33. $\dfrac{1}{4y}$
34. $\dfrac{-1}{8a}$
35. $\dfrac{2a}{3b}$
36. $\dfrac{-4y}{3x}$

C Rewrite each division as a multiplication.

37. $4 \div 17$

38. $5 \div (-8)$

39. $\dfrac{8}{-13}$

40. $-\dfrac{13}{47}$

41. $\dfrac{13.9}{-1.5}$

42. $-\dfrac{47.3}{21.4}$

43. $\dfrac{2}{3} \div \left(-\dfrac{4}{5}\right)$

44. $\dfrac{3}{4} \div \left(-\dfrac{7}{10}\right)$

45. $\dfrac{x}{\frac{1}{y}}$

46. $\dfrac{13}{\frac{1}{x}}$

47. $\dfrac{3x + 4}{5}$

48. $\dfrac{4y - 8}{-7}$

Divide.

49. $\dfrac{3}{4} \div \left(-\dfrac{2}{3}\right)$

50. $\dfrac{7}{8} \div \left(-\dfrac{1}{2}\right)$

51. $-\dfrac{5}{4} \div \left(-\dfrac{3}{4}\right)$

52. $-\dfrac{5}{9} \div \left(-\dfrac{5}{6}\right)$

53. $-\dfrac{2}{7} \div \left(-\dfrac{4}{9}\right)$

54. $-\dfrac{3}{5} \div \left(-\dfrac{5}{8}\right)$

55. $-\dfrac{3}{8} \div \left(-\dfrac{8}{3}\right)$

56. $-\dfrac{5}{8} \div \left(-\dfrac{6}{5}\right)$

57. $-\dfrac{5}{6} \div \dfrac{2}{3}$

58. $-\dfrac{7}{16} \div \dfrac{3}{8}$

59. $-\dfrac{9}{4} \div \dfrac{5}{12}$

60. $-\dfrac{3}{5} \div \dfrac{7}{10}$

61. $\dfrac{-11}{-13}$

62. $\dfrac{-21}{-25}$

63. $-6.6 \div 3.3$

64. $-44.1 \div (-6.3)$

SECTION 1.6 Division of Real Numbers

65. $\dfrac{48.6}{-3}$

66. $\dfrac{-1.9}{20}$

67. $\dfrac{-12.5}{5}$

68. $\dfrac{-17.8}{3.2}$

69. $11.25 \div (-9)$

70. $-9.6 \div (-6.4)$

71. $\dfrac{-9}{17 - 17}$

72. $\dfrac{-8}{-5 + 5}$

d Solve.

73. *Apps.* The number of available apps in the iTunes app store increased from 1.4 million in 2015 to 2.0 million in 2016. What is the percent increase?
Data: statista.com

74. *Passports.* In 2006, approximately 71 million valid passports were in circulation in the United States. This number had increased to approximately 132 million in 2016. What is the percent increase?
Data: U.S. State Department

75. *Super Bowl TV Viewership.* Television viewership of the Super Bowl in the United States decreased from 114.4 million in 2015 to 111.3 million in 2017. What is the percent decrease?
Data: statista.com

76. *Pieces of Mail.* The number of pieces of mail handled by the U.S. Postal Service decreased from 212 billion in 2007 to 154 billion in 2011. What is the percent decrease?
Data: U.S. Postal Service

77. *Chemical Reaction.* During a chemical reaction, the temperature in a beaker decreased every minute by the same number of degrees. The temperature was 56°F at 10:10 A.M. By 10:42 A.M., the temperature had dropped to −12°F. By how many degrees did it change each minute?

78. *Chemical Reaction.* During a chemical reaction, the temperature in a beaker decreased every minute by the same number of degrees. The temperature was 71°F at 2:12 P.M. By 2:37 P.M., the temperature had changed to −14°F. By how many degrees did it change each minute?

Skill Maintenance

Simplify.

79. $\dfrac{1}{4} - \dfrac{1}{2}$ [1.4a]

80. $-9 - 3 + 17$ [1.4a]

81. $35 \cdot (-1.2)$ [1.5a]

82. $4 \cdot (-6) \cdot (-2) \cdot (-1)$ [1.5a]

83. $13.4 + (-4.9)$ [1.3a]

84. $-\dfrac{3}{8} - \left(-\dfrac{1}{4}\right)$ [1.4a]

Convert to decimal notation. [1.2c]

85. $-\dfrac{1}{11}$

86. $\dfrac{11}{12}$

87. $\dfrac{15}{4}$

88. $-\dfrac{10}{3}$

Synthesis

89. Find the reciprocal of −10.5. What happens if you take the reciprocal of the result?

90. Determine those real numbers a for which the opposite of a is the same as the reciprocal of a.

Determine whether each expression represents a positive number or a negative number when a and b are negative.

91. $\dfrac{-a}{b}$

92. $\dfrac{-a}{-b}$

93. $-\left(\dfrac{a}{-b}\right)$

94. $-\left(\dfrac{-a}{b}\right)$

95. $-\left(\dfrac{-a}{-b}\right)$

Properties of Real Numbers

1.7

a EQUIVALENT EXPRESSIONS

In solving equations and doing other kinds of work in algebra, we manipulate expressions in various ways. For example, instead of writing $x + x$, we might write $2x$, knowing that the two expressions represent the same number for any allowable replacement of x. In that sense, the expressions $x + x$ and $2x$ are **equivalent**, as are $\dfrac{3}{x}$ and $\dfrac{3x}{x^2}$, even though 0 is not an allowable replacement because division by 0 is not defined.

OBJECTIVES

a Find equivalent fraction expressions and simplify fraction expressions.

b Use the commutative laws and the associative laws to find equivalent expressions.

c Use the distributive laws to multiply expressions like 8 and $x - y$.

d Use the distributive laws to factor expressions like $4x - 12 + 24y$.

e Collect like terms.

EQUIVALENT EXPRESSIONS

Two expressions that have the same value for all allowable replacements are called **equivalent expressions**.

The expressions $x + 3x$ and $5x$ are *not* equivalent, as we see in Margin Exercise 2.

▶ Do Exercises 1 and 2.

In this section, we will consider several laws of real numbers that will allow us to find equivalent expressions. The first two laws are the *identity properties of 0 and 1*.

THE IDENTITY PROPERTY OF 0

For any real number a,

$$a + 0 = 0 + a = a.$$

(The number 0 is the *additive identity*.)

THE IDENTITY PROPERTY OF 1

For any real number a,

$$a \cdot 1 = 1 \cdot a = a.$$

(The number 1 is the *multiplicative identity*.)

We often refer to the use of the identity property of 1 as "multiplying by 1." We can use this method to find equivalent fraction expressions. Recall from arithmetic that to multiply with fraction notation, we multiply the numerators and multiply the denominators.

EXAMPLE 1 Write a fraction expression equivalent to $\frac{2}{3}$ with a denominator of $3x$:

$$\frac{2}{3} = \frac{\square}{3x}.$$

Complete the table by evaluating each expression for the given values.

1.

VALUE	x + x	2x
$x = 3$		
$x = -6$		
$x = 4.8$		

2.

VALUE	x + 3x	5x
$x = 2$		
$x = -6$		
$x = 4.8$		

Answers
1. 6, 6; −12, −12; 9.6, 9.6   **2.** 8, 10; −24, −30; 19.2, 24

3. Write a fraction expression equivalent to $\frac{3}{4}$ with a denominator of 8:

$$\frac{3}{4} = \frac{3}{4} \cdot 1 = \frac{3}{4} \cdot \frac{\square}{\square} = \frac{\square}{8}.$$

4. Write a fraction expression equivalent to $\frac{3}{4}$ with a denominator of $4t$:

$$\frac{3}{4} = \frac{3}{4} \cdot 1 = \frac{3}{4} \cdot \frac{\square}{\square} = \frac{\square}{4t}.$$

Simplify.

5. $\frac{3y}{4y}$

6. $-\frac{16m}{12m}$

7. $\frac{5xy}{40y}$

8. $\frac{18p}{24pq} = \frac{6p \cdot 3}{6p \cdot \square}$

$= \frac{6p}{6p} \cdot \frac{\square}{4q}$

$= 1 \cdot \frac{3}{4q} = \frac{3}{4q}$

9. Evaluate $x + y$ and $y + x$ when $x = -2$ and $y = 3$.

10. Evaluate xy and yx when $x = -2$ and $y = 5$.

Answers

3. $\frac{6}{8}$ 4. $\frac{3t}{4t}$ 5. $\frac{3}{4}$ 6. $-\frac{4}{3}$ 7. $\frac{x}{8}$ 8. $\frac{3}{4q}$
9. $1; 1$ 10. $-10; -10$

Guided Solutions:

3. $\frac{2}{2}; 6$ 4. $\frac{t}{t}; 3t$ 8. $4q; 3$

Note that $3x = 3 \cdot x$. We want fraction notation for $\frac{2}{3}$ that has a denominator of $3x$, but the denominator 3 is missing a factor of x. Thus we multiply by 1, using x/x as an equivalent expression for 1:

$$\frac{2}{3} = \frac{2}{3} \cdot 1 = \frac{2}{3} \cdot \frac{x}{x} = \frac{2x}{3x}.$$

The expressions $2/3$ and $2x/(3x)$ are equivalent. They have the same value for any allowable replacement. Note that $2x/(3x)$ is not defined for a replacement of 0, but for all nonzero real numbers, the expressions $2/3$ and $2x/(3x)$ have the same value.

◀ Do Exercises 3 and 4.

In algebra, we consider an expression like $2/3$ to be a "simplified" form of $2x/(3x)$. To find such simplified expressions, we use the identity property of 1 to remove a factor of 1.

EXAMPLE 2 Simplify: $-\frac{20x}{12x}$.

$-\frac{20x}{12x} = -\frac{5 \cdot 4x}{3 \cdot 4x}$ We look for the largest factor common to both the numerator and the denominator and factor each.

$= -\frac{5}{3} \cdot \frac{4x}{4x}$ Factoring the fraction expression

$= -\frac{5}{3} \cdot 1$ $\frac{4x}{4x} = 1$

$= -\frac{5}{3}$ Removing a factor of 1 using the identity property of 1

EXAMPLE 3 Simplify: $\frac{14ab}{56a}$.

$$\frac{14ab}{56a} = \frac{14a \cdot b}{14a \cdot 4} = \frac{14a}{14a} \cdot \frac{b}{4} = 1 \cdot \frac{b}{4} = \frac{b}{4}$$

◀ Do Exercises 5–8.

b THE COMMUTATIVE LAWS AND THE ASSOCIATIVE LAWS

The Commutative Laws

Let's examine the expressions $x + y$ and $y + x$, as well as xy and yx.

EXAMPLE 4 Evaluate $x + y$ and $y + x$ when $x = 4$ and $y = 3$.

We substitute 4 for x and 3 for y in both expressions:

$x + y = 4 + 3 = 7;$ $y + x = 3 + 4 = 7.$

EXAMPLE 5 Evaluate xy and yx when $x = 3$ and $y = -12$.

We substitute 3 for x and -12 for y in both expressions:

$xy = 3 \cdot (-12) = -36;$ $yx = (-12) \cdot 3 = -36.$

◀ Do Exercises 9 and 10.

CHAPTER 1 Introduction to Real Numbers and Algebraic Expressions

The expressions $x + y$ and $y + x$ have the same values no matter what the variables stand for. Thus they are equivalent. Therefore, when we add two numbers, the order in which we add does not matter. Similarly, the expressions xy and yx are equivalent. They also have the same values, no matter what the variables stand for. Therefore, when we multiply two numbers, the order in which we multiply does not matter.

The following are examples of general patterns or laws.

> **THE COMMUTATIVE LAWS**
>
> *Addition.* For any numbers a and b,
> $$a + b = b + a.$$
> (We can change the order when adding without affecting the answer.)
>
> *Multiplication.* For any numbers a and b,
> $$ab = ba.$$
> (We can change the order when multiplying without affecting the answer.)

Using a commutative law, we know that $x + 2$ and $2 + x$ are equivalent. Similarly, $3x$ and $x(3)$ are equivalent. Thus, in an algebraic expression, we can replace one with the other and the result will be equivalent to the original expression.

EXAMPLE 6 Use the commutative laws to write an equivalent expression: **(a)** $y + 5$; **(b)** mn; **(c)** $7 + xy$.

a) An expression equivalent to $y + 5$ is $5 + y$ by the commutative law of addition.

b) An expression equivalent to mn is nm by the commutative law of multiplication.

c) An expression equivalent to $7 + xy$ is $xy + 7$ by the commutative law of addition. Another expression equivalent to $7 + xy$ is $7 + yx$ by the commutative law of multiplication. Another equivalent expression is $yx + 7$.

Do Exercises 11–13.

Use the commutative laws to write an equivalent expression.

11. $x + 9$

12. fg

13. $xy + t$

The Associative Laws

Now let's examine the expressions $a + (b + c)$ and $(a + b) + c$. Note that these expressions involve the use of parentheses as *grouping* symbols, and they also involve three numbers. Calculations within parentheses are to be done first.

EXAMPLE 7 Calculate and compare: $3 + (8 + 5)$ and $(3 + 8) + 5$.

$3 + (8 + 5) = 3 + 13$ Calculating within parentheses first; adding 8 and 5

$\qquad\qquad = 16;$

$(3 + 8) + 5 = 11 + 5$ Calculating within parentheses first; adding 3 and 8

$\qquad\qquad = 16$

Answers
11. $9 + x$ **12.** gf
13. $t + xy$, or $yx + t$, or $t + yx$

14. Calculate and compare:

    $8 + (9 + 2)$ and $(8 + 9) + 2$.

15. Calculate and compare:

    $10 \cdot (5 \cdot 3)$ and $(10 \cdot 5) \cdot 3$.

The two expressions in Example 7 name the same number. Moving the parentheses to group the additions differently does not affect the value of the expression.

EXAMPLE 8 Calculate and compare: $3 \cdot (4 \cdot 2)$ and $(3 \cdot 4) \cdot 2$.

$$3 \cdot (4 \cdot 2) = 3 \cdot 8 = 24; \quad (3 \cdot 4) \cdot 2 = 12 \cdot 2 = 24$$

◀ Do Exercises 14 and 15.

You may have noted that when only addition is involved, numbers can be grouped in any way we please without affecting the answer. When only multiplication is involved, numbers can also be grouped in any way we please without affecting the answer.

> **THE ASSOCIATIVE LAWS**
>
> *Addition.* For any numbers a, b, and c,
>
> $$a + (b + c) = (a + b) + c.$$
>
> (Numbers can be grouped in any manner for addition.)
>
> *Multiplication.* For any numbers a, b, and c,
>
> $$a \cdot (b \cdot c) = (a \cdot b) \cdot c.$$
>
> (Numbers can be grouped in any manner for multiplication.)

EXAMPLE 9 Use an associative law to write an equivalent expression: **(a)** $(y + z) + 3$; **(b)** $8(xy)$.

a) An expression equivalent to $(y + z) + 3$ is $y + (z + 3)$ by the associative law of addition.

b) An expression equivalent to $8(xy)$ is $(8x)y$ by the associative law of multiplication.

Use the associative laws to write an equivalent expression.

16. $r + (s + 7)$

17. $9(ab)$

◀ Do Exercises 16 and 17.

The associative laws say that numbers can be grouped in any way we please when only additions or only multiplications are involved. Thus we often omit the parentheses. For example,

$$x + (y + 2) \text{ means } x + y + 2, \quad \text{and} \quad (lw)h \text{ means } lwh.$$

Using the Commutative Laws and the Associative Laws Together

EXAMPLE 10 Use the commutative laws and the associative laws to write at least three expressions equivalent to $(x + 5) + y$.

a) $(x + 5) + y = x + (5 + y)$ Using the associative law first and
   $= x + (y + 5)$ then using the commutative law

b) $(x + 5) + y = y + (x + 5)$ Using the commutative law twice
   $= y + (5 + x)$

c) $(x + 5) + y = (5 + x) + y$ Using the commutative law first and
   $= 5 + (x + y)$ then the associative law

Answers

14. 19; 19   **15.** 150; 150   **16.** $(r + s) + 7$
17. $(9a)b$

EXAMPLE 11 Use the commutative laws and the associative laws to write at least three expressions equivalent to $(3x)y$.

a) $(3x)y = 3(xy)$ Using the associative law first and then using the
   $= 3(yx)$ commutative law

b) $(3x)y = y(3x)$ Using the commutative law twice
   $= y(x \cdot 3)$

c) $(3x)y = (x \cdot 3)y$ Using the commutative law, and then the associative
   $= x(3y)$ law, and then the commutative law again
   $= x(y \cdot 3)$

Do Exercises 18 and 19. ▶

Use the commutative laws and the associative laws to write at least three equivalent expressions.

18. $4(tu)$

19. $r + (2 + s)$

C   THE DISTRIBUTIVE LAWS

The *distributive laws* are the basis of many procedures in both arithmetic and algebra. They are probably the most important laws that we use to manipulate algebraic expressions. The distributive law of multiplication over addition involves two operations: addition and multiplication.

Let's begin by considering a multiplication problem from arithmetic:

$$
\begin{array}{r}
4\,5 \\
\times 7 \\
\hline
3\,5 \\
2\,8\,0 \\
\hline
3\,1\,5
\end{array}
$$

← This is $7 \cdot 5$.
← This is $7 \cdot 40$.
← This is the sum $7 \cdot 5 + 7 \cdot 40$.

To carry out the multiplication, we actually added two products. That is,

$$7 \cdot 45 = 7(5 + 40) = 7 \cdot 5 + 7 \cdot 40.$$

Let's examine this further. If we wish to multiply a sum of several numbers by a factor, we can either add and then multiply, or multiply and then add.

EXAMPLE 12 Compute in two ways: $5 \cdot (4 + 8)$.

a) $5 \cdot (4 + 8)$ Adding within parentheses first, and then multiplying
   $= 5 \cdot 12$
   $= 60$

b) $5 \cdot (4 + 8) = (5 \cdot 4) + (5 \cdot 8)$ Distributing the multiplication to terms within parentheses first and then adding
   $= 20 + 40$
   $= 60$

Do Exercises 20–22. ▶

Compute.

20. a) $7 \cdot (3 + 6)$
     b) $(7 \cdot 3) + (7 \cdot 6)$

21. a) $2 \cdot (10 + 30)$
     b) $(2 \cdot 10) + (2 \cdot 30)$

22. a) $(2 + 5) \cdot 4$
     b) $(2 \cdot 4) + (5 \cdot 4)$

THE DISTRIBUTIVE LAW OF MULTIPLICATION OVER ADDITION

For any numbers a, b, and c,

$$a(b + c) = ab + ac.$$

Answers
18. $(4t)u, (tu)4, t(4u)$; answers may vary
19. $(2 + r) + s, (r + s) + 2, s + (r + 2)$; answers may vary   **20.** (a) $7 \cdot 9 = 63$;
(b) $21 + 42 = 63$ **21.** (a) $2 \cdot 40 = 80$;
(b) $20 + 60 = 80$ **22.** (a) $7 \cdot 4 = 28$;
(b) $8 + 20 = 28$

In the statement of the distributive law, we know that in an expression such as $ab + ac$, the multiplications are to be done first according to the rules for order of operations. So, instead of writing $(4 \cdot 5) + (4 \cdot 7)$, we can write $4 \cdot 5 + 4 \cdot 7$. However, in $a(b + c)$, we cannot simply omit the parentheses. If we did, we would have $ab + c$, which means $(ab) + c$. For example, $3(4 + 2) = 3(6) = 18$, but $3 \cdot 4 + 2 = 12 + 2 = 14$.

Another distributive law relates multiplication and subtraction. This law says that to multiply by a difference, we can either subtract and then multiply, or multiply and then subtract.

> **THE DISTRIBUTIVE LAW OF MULTIPLICATION OVER SUBTRACTION**
>
> For any numbers a, b, and c,
> $$a(b - c) = ab - ac.$$

We often refer to "*the* distributive law" when we mean *either* or *both* of these laws.

◀ Do Exercises 23–25.

What do we mean by the *terms* of an expression? **Terms** are separated by addition signs. If there are subtraction signs, we can find an equivalent expression that uses addition signs.

EXAMPLE 13 What are the terms of $3x - 4y + 2z$?

We have

$3x - 4y + 2z = 3x + (-4y) + 2z.$ Separating parts with + signs

The terms are $3x$, $-4y$, and $2z$.

◀ Do Exercises 26 and 27.

The distributive laws are a basis for **multiplying** algebraic expressions. In an expression like $8(a + 2b - 7)$, we multiply each term inside the parentheses by 8:

$8(a + 2b - 7) = 8 \cdot a + 8 \cdot 2b - 8 \cdot 7 = 8a + 16b - 56.$

EXAMPLES Multiply.

14. $9(x - 5) = 9 \cdot x - 9 \cdot 5$ Using the distributive law of multiplication over subtraction
$ = 9x - 45$

15. $\frac{2}{3}(w + 1) = \frac{2}{3} \cdot w + \frac{2}{3} \cdot 1$ Using the distributive law of multiplication over addition
$\phantom{15.\ \frac{2}{3}(w + 1)} = \frac{2}{3}w + \frac{2}{3}$

16. $\frac{4}{3}(s - t + w) = \frac{4}{3}s - \frac{4}{3}t + \frac{4}{3}w$ Using both distributive laws

◀ Do Exercises 28–30.

Calculate.

23. a) $4(5 - 3)$
  b) $4 \cdot 5 - 4 \cdot 3$

24. a) $-2 \cdot (5 - 3)$
  b) $-2 \cdot 5 - (-2) \cdot 3$

25. a) $5 \cdot (2 - 7)$
  b) $5 \cdot 2 - 5 \cdot 7$

What are the terms of each expression?

26. $5x - 8y + 3$

27. $-4y - 2x + 3z$

Multiply.

28. $3(x - 5)$

29. $5(x + 1)$

30. $\frac{3}{5}(a + b - t)$

Answers

23. (a) $4 \cdot 2 = 8$; (b) $20 - 12 = 8$
24. (a) $-2 \cdot 2 = -4$; (b) $-10 - (-6) = -4$
25. (a) $5(-5) = -25$; (b) $10 - 35 = -25$
26. $5x, -8y, 3$ **27.** $-4y, -2x, 3z$
28. $3x - 15$ **29.** $5x + 5$
30. $\frac{3}{5}a + \frac{3}{5}b - \frac{3}{5}t$

EXAMPLE 17 Multiply: $-4(x - 2y + 3z)$.

$$-4(x - 2y + 3z) = -4 \cdot x - (-4)(2y) + (-4)(3z) \quad \text{Using both distributive laws}$$
$$= -4x - (-8y) + (-12z) \quad \text{Multiplying}$$
$$= -4x + 8y - 12z$$

We can also do this problem by first finding an equivalent expression with all plus signs and then multiplying:

$$-4(x - 2y + 3z) = -4[x + (-2y) + 3z]$$
$$= -4 \cdot x + (-4)(-2y) + (-4)(3z)$$
$$= -4x + 8y - 12z.$$

Do Exercises 31–33. ▶

EXAMPLES Name the property or the law illustrated by each equation.

Equation	Property or Law
18. $5x = x(5)$	Commutative law of multiplication
19. $a + (8.5 + b) = (a + 8.5) + b$	Associative law of addition
20. $0 + 11 = 11$	Identity property of 0
21. $(-5s)t = -5(st)$	Associative law of multiplication
22. $\frac{3}{4} \cdot 1 = \frac{3}{4}$	Identity property of 1
23. $12.5(w - 3) = 12.5w - 12.5(3)$	Distributive law of multiplication over subtraction
24. $y + \frac{1}{2} = \frac{1}{2} + y$	Commutative law of addition

Do Exercises 34–40. ▶

d FACTORING

Find all factors of a number. [J1]
List all factors of each number.
1. 24   **2.** 90

Answers: **1.** 1, 2, 3, 4, 6, 8, 12, 24
2. 1, 2, 3, 5, 6, 9, 10, 15, 18, 30, 45, 90

MyLab Math VIDEO

Factoring is the reverse of multiplying. To factor, we can use the distributive laws in reverse:

$$ab + ac = a(b + c) \quad \text{and} \quad ab - ac = a(b - c).$$

FACTORING

To **factor** an expression is to find an equivalent expression that is a product.

Multiply.

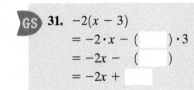

31. $-2(x - 3)$
$= -2 \cdot x - () \cdot 3$
$= -2x - ()$
$= -2x + $

32. $5(x - 2y + 4z)$

33. $-5(x - 2y + 4z)$

Name the property or the law illustrated by each equation.

34. $(-8a)b = -8(ab)$

35. $p \cdot 1 = p$

36. $m + 34 = 34 + m$

37. $2(t + 5) = 2t + 2(5)$

38. $0 + k = k$

39. $-8x = x(-8)$

40. $x + (4.3 + b) = (x + 4.3) + b$

Answers
31. $-2x + 6$ **32.** $5x - 10y + 20z$
33. $-5x + 10y - 20z$ **34.** Associative law of multiplication **35.** Identity property of 1 **36.** Commutative law of addition **37.** Distributive law of multiplication over addition **38.** Identity property of 0 **39.** Commutative law of multiplication **40.** Associative law of addition

Guided Solution:
31. $-2; -6; 6$

To factor $9x - 45$, for example, we find an equivalent expression that is a product: $9(x - 5)$. This reverses the multiplication that we did in Example 14. The expression $9x - 45$ is the difference of $9x$ and 45; the expression $9(x - 5)$ is the product of 9 and $(x - 5)$.

When all the terms of an expression have a factor in common, we can "factor it out" using the distributive laws. Note the following.

$9x$ has the factors $9, -9, 3, -3, 1, -1, x, -x, 3x, -3x, 9x, -9x$;

-45 has the factors $1, -1, 3, -3, 5, -5, 9, -9, 15, -15, 45, -45$

We generally remove the largest common factor. In this case, that factor is 9. Thus,

$$9x - 45 = 9 \cdot x - 9 \cdot 5$$
$$= 9(x - 5).$$

EXAMPLES Factor.

25. $5x - 10 = 5 \cdot x - 5 \cdot 2$ Try to do this step mentally.
$ = 5(x - 2)$ You can check by multiplying.

26. $ax - ay + az = a(x - y + z)$

27. $9x + 27y - 9 = 9 \cdot x + 9 \cdot 3y - 9 \cdot 1 = 9(x + 3y - 1)$

Note in Example 27 that you might, at first, just factor out a 3, as follows:

$$9x + 27y - 9 = 3 \cdot 3x + 3 \cdot 9y - 3 \cdot 3$$
$$= 3(3x + 9y - 3).$$

At this point, the mathematics is correct, but the answer is not because there is another factor of 3 that can be factored out, as follows:

$$3 \cdot 3x + 3 \cdot 9y - 3 \cdot 3 = 3(3x + 9y - 3)$$
$$= 3(3 \cdot x + 3 \cdot 3y - 3 \cdot 1)$$
$$= 3 \cdot 3(x + 3y - 1)$$
$$= 9(x + 3y - 1).$$

We now have a correct answer, but it took more work than we did in Example 27. Thus it is better to look for the *greatest common factor* at the outset.

EXAMPLES Factor. Try to write only the answer, if you can.

28. $5x - 5y = 5(x - y)$

29. $-3x + 6y - 9z = -3(x - 2y + 3z)$

We generally factor out a negative factor when the first term is negative. The way we factor can depend on the situation in which we are working. We might also factor the expression in Example 29 as follows:

$$-3x + 6y - 9z = 3(-x + 2y - 3z).$$

30. $18z - 12x - 24 = 6(3z - 2x - 4)$

31. $\frac{1}{2}x + \frac{3}{2}y - \frac{1}{2} = \frac{1}{2}(x + 3y - 1)$

Remember that you can always check factoring by multiplying. Keep in mind that an expression is factored when it is written as a product.

◀ Do Exercises 41–46.

Factor.

41. $6x - 12$

42. $3x - 6y + 9$

43. $bx + by - bz$

44. $16a - 36b + 42$
$= 2 \cdot 8a - \underline{} \cdot 18b + 2 \cdot 21$
$= \underline{}(8a - 18b + 21)$

45. $\frac{3}{8}x - \frac{5}{8}y + \frac{7}{8}$

46. $-12x + 32y - 16z$

Answers
41. $6(x - 2)$ **42.** $3(x - 2y + 3)$
43. $b(x + y - z)$ **44.** $2(8a - 18b + 21)$
45. $\frac{1}{8}(3x - 5y + 7)$ **46.** $-4(3x - 8y + 4z)$, or $4(-3x + 8y - 4z)$

Guided Solution:
44. 2; 2

e COLLECTING LIKE TERMS

Terms such as $5x$ and $-4x$, whose variable factors are exactly the same, are called **like terms**. Similarly, numbers, such as -7 and 13, are like terms. Also, $3y^2$ and $9y^2$ are like terms because the variables are raised to the same power. Terms such as $4y$ and $5y^2$ are not like terms, and $7x$ and $2y$ are not like terms.

The process of **collecting like terms** is also based on the distributive laws. We can apply a distributive law when a factor is on the right-hand side because of the commutative law of multiplication.

Later in this text, terminology like "collecting like terms" and "combining like terms" will also be referred to as "simplifying."

EXAMPLES Collect like terms. Try to write just the answer, if you can.

32. $4x + 2x = (4 + 2)x = 6x$ Factoring out x using a distributive law

33. $2x + 3y - 5x - 2y = 2x - 5x + 3y - 2y$
$= (2 - 5)x + (3 - 2)y = -3x + 1y = -3x + y$

34. $3x - x = 3x - 1x = (3 - 1)x = 2x$

35. $x - 0.24x = 1 \cdot x - 0.24x = (1 - 0.24)x = 0.76x$

36. $x - 6x = 1 \cdot x - 6 \cdot x = (1 - 6)x = -5x$

37. $4x - 7y + 9x - 5 + 3y - 8 = 13x - 4y - 13$

38. $\frac{2}{3}a - b + \frac{4}{5}a + \frac{1}{4}b - 10 = \frac{2}{3}a - 1 \cdot b + \frac{4}{5}a + \frac{1}{4}b - 10$
$= \left(\frac{2}{3} + \frac{4}{5}\right)a + \left(-1 + \frac{1}{4}\right)b - 10$
$= \left(\frac{10}{15} + \frac{12}{15}\right)a + \left(-\frac{4}{4} + \frac{1}{4}\right)b - 10$
$= \frac{22}{15}a - \frac{3}{4}b - 10$

Do Exercises 47–53.

Collect like terms.

47. $6x - 3x$ 48. $7x - x$

49. $x - 9x$ 50. $x - 0.41x$

51. $5x + 4y - 2x - y$

 52. $3x - 7x - 11 + 8y + 4 - 13y$
$= (3 -)x + (8 - 13)y + (+ 4)$
$= x + ()y + ()$
$= -4x - 5y - 7$

53. $-\frac{2}{3} - \frac{3}{5}x + y + \frac{7}{10}x - \frac{2}{9}y$

Answers
47. $3x$ 48. $6x$ 49. $-8x$ 50. $0.59x$
51. $3x + 3y$ 52. $-4x - 5y - 7$
53. $\frac{1}{10}x + \frac{7}{9}y - \frac{2}{3}$

Guided Solution:
52. $7; -11; -4; -5; -7$

1.7 Exercise Set

FOR EXTRA HELP   MyLab Math

✓ Check Your Understanding

Reading Check Choose from the column on the right an equation that illustrates the given property or law.

RC1. Associative law of multiplication

RC2. Identity property of 1

RC3. Distributive law of multiplication over subtraction

RC4. Commutative law of addition

RC5. Identity property of 0

RC6. Commutative law of multiplication

RC7. Associative law of addition

a) $3 \cdot 5 = 5 \cdot 3$

b) $8 + \left(\frac{1}{2} + 9\right) = \left(8 + \frac{1}{2}\right) + 9$

c) $5(6 + 3) = 5 \cdot 6 + 5 \cdot 3$

d) $3 + 0 = 3$

e) $3 + 5 = 5 + 3$

f) $5(6 - 3) = 5 \cdot 6 - 5 \cdot 3$

g) $8 \cdot \left(\frac{1}{2} \cdot 9\right) = \left(8 \cdot \frac{1}{2}\right) \cdot 9$

h) $\frac{6}{5} \cdot 1 = \frac{6}{5}$

SECTION 1.7 Properties of Real Numbers 87

Concept Check Find the largest common factor of each pair of terms.

CC1. 45, 10x   **CC2.** 7x, 7   **CC3.** 16x, 24y   **CC4.** dy, 4d

a Find an equivalent expression with the given denominator.

1. $\dfrac{3}{5} = \dfrac{\Box}{5y}$
2. $\dfrac{5}{8} = \dfrac{\Box}{8t}$
3. $\dfrac{2}{3} = \dfrac{\Box}{15x}$
4. $\dfrac{6}{7} = \dfrac{\Box}{14y}$
5. $\dfrac{2}{x} = \dfrac{\Box}{x^2}$
6. $\dfrac{4}{9x} = \dfrac{\Box}{9xy}$

Simplify.

7. $-\dfrac{24a}{16a}$
8. $-\dfrac{42t}{18t}$
9. $-\dfrac{42ab}{36ab}$
10. $-\dfrac{64pq}{48pq}$
11. $\dfrac{20st}{15t}$
12. $\dfrac{21w}{7wz}$

b Write an equivalent expression. Use the commutative laws.

13. $y + 8$
14. $x + 3$
15. mn
16. yz

17. $9 + xy$
18. $11 + ab$
19. $ab + c$
20. $rs + t$

Write an equivalent expression. Use the associative laws.

21. $a + (b + 2)$
22. $3(vw)$
23. $(8x)y$
24. $(y + z) + 7$

25. $(a + b) + 3$
26. $(5 + x) + y$
27. $3(ab)$
28. $(6x)y$

Use the commutative laws and the associative laws to write three equivalent expressions.

29. $(a + b) + 2$
30. $(3 + x) + y$
31. $5 + (v + w)$
32. $6 + (x + y)$

33. $(xy)3$
34. $(ab)5$
35. $7(ab)$
36. $5(xy)$

c Multiply.

37. $2(b + 5)$
38. $4(x + 3)$
39. $7(1 + t)$
40. $4(1 + y)$

41. $6(5x + 2)$
42. $9(6m + 7)$
43. $7(x + 4 + 6y)$
44. $4(5x + 8 + 3p)$

45. $7(x - 3)$ **46.** $15(y - 6)$ **47.** $-3(x - 7)$ **48.** $1.2(x - 2.1)$

49. $\frac{2}{3}(b - 6)$ **50.** $\frac{5}{8}(y + 16)$ **51.** $7.3(x - 2)$ **52.** $5.6(x - 8)$

53. $-\frac{3}{5}(x - y + 10)$ **54.** $-\frac{2}{3}(a + b - 12)$ **55.** $-9(-5x - 6y + 8)$ **56.** $-7(-2x - 5y + 9)$

57. $-4(x - 3y - 2z)$ **58.** $8(2x - 5y - 8z)$

59. $3.1(-1.2x + 3.2y - 1.1)$ **60.** $-2.1(-4.2x - 4.3y - 2.2)$

List the terms of each expression.

61. $4x + 3z$ **62.** $8x - 1.4y$ **63.** $7x + 8y - 9z$ **64.** $8a + 10b - 18c$

d Factor. Check by multiplying.

65. $2x + 4$ **66.** $5y + 20$ **67.** $30 + 5y$ **68.** $7x + 28$

69. $14x + 21y$ **70.** $18a + 24b$ **71.** $14t - 7$ **72.** $25m - 5$

73. $8x - 24$ **74.** $10x - 50$ **75.** $18a - 24b$ **76.** $32x - 20y$

77. $-4y + 32$ **78.** $-6m + 24$ **79.** $5x + 10 + 15y$ **80.** $9a + 27b + 81$

81. $16m - 32n + 8$ **82.** $6x + 10y - 2$ **83.** $12a + 4b - 24$ **84.** $8m - 4n + 12$

85. $8x + 10y - 22$ **86.** $9a + 6b - 15$ **87.** $ax - a$ **88.** $by - 9b$

89. $ax - ay - az$ **90.** $cx + cy - cz$ **91.** $-18x + 12y + 6$ **92.** $-14x + 21y + 7$

93. $\frac{2}{3}x - \frac{5}{3}y + \frac{1}{3}$ **94.** $\frac{3}{5}a + \frac{4}{5}b - \frac{1}{5}$ **95.** $36x - 6y + 18z$ **96.** $8a - 4b + 20c$

e   Collect like terms.

97. $9a + 10a$ **98.** $12x + 2x$ **99.** $10a - a$

100. $-16x + x$ **101.** $2x + 9z + 6x$ **102.** $3a - 5b + 7a$

103. $7x + 6y^2 + 9y^2$ **104.** $12m^2 + 6q + 9m^2$ **105.** $41a + 90 - 60a - 2$

106. $42x - 6 - 4x + 2$ **107.** $23 + 5t + 7y - t - y - 27$ **108.** $45 - 90d - 87 - 9d + 3 + 7d$

109. $\frac{1}{2}b + \frac{1}{2}b$ **110.** $\frac{2}{3}x + \frac{1}{3}x$ **111.** $2y + \frac{1}{4}y + y$

112. $\frac{1}{2}a + a + 5a$ **113.** $11x - 3x$ **114.** $9t - 17t$

115. $6n - n$ **116.** $100t - t$ **117.** $y - 17y$

118. $3m - 9m + 4$ **119.** $-8 + 11a - 5b + 6a - 7b + 7$ **120.** $8x - 5x + 6 + 3y - 2y - 4$

121. $9x + 2y - 5x$ **122.** $8y - 3z + 4y$ **123.** $11x + 2y - 4x - y$

124. $13a + 9b - 2a - 4b$ **125.** $2.7x + 2.3y - 1.9x - 1.8y$ **126.** $6.7a + 4.3b - 4.1a - 2.9b$

127. $\frac{13}{2}a + \frac{9}{5}b - \frac{2}{3}a - \frac{3}{10}b - 42$ **128.** $\frac{11}{4}x + \frac{2}{3}y - \frac{4}{5}x - \frac{1}{6}y + 12$

Skill Maintenance

Compute and simplify. [1.3a], [1.4a], [1.5a], [1.6a, c]

129. $18 - (-20)$ **130.** $-3.8 + (-1.1)$ **131.** $-\frac{4}{15} \cdot (-15)$ **132.** $-500 \div (-50)$

133. $\frac{2}{7} \div \left(-\frac{7}{2}\right)$ **134.** $2 \cdot (-53)$ **135.** $-\frac{1}{2} + \frac{3}{2}$ **136.** $-6 - 28$

137. Evaluate $9w$ when $w = 20$. [1.1a] **138.** Find the absolute value: $\left|-\frac{4}{13}\right|$. [1.2e]

Write true or false. [1.2d]

139. $-43 < -40$ **140.** $-3 \geq 0$ **141.** $-6 \leq -6$ **142.** $0 > -4$

Synthesis

Determine whether the expressions are equivalent. Explain why if they are. Give an example if they are not. Examples may vary.

143. $3t + 5$ and $3 \cdot 5 + t$ **144.** $4x$ and $x + 4$

145. $5m + 6$ and $6 + 5m$ **146.** $(x + y) + z$ and $z + (x + y)$

147. Factor: $q + qr + qrs + qrst$. **148.** Collect like terms:
$21x + 44xy + 15y - 16x - 8y - 38xy + 2y + xy$.

1.8 Simplifying Expressions; Order of Operations

OBJECTIVES

a Find an equivalent expression for an opposite without parentheses, where an expression has several terms.

b Simplify expressions by removing parentheses and collecting like terms.

c Simplify expressions with parentheses inside parentheses.

d Simplify expressions using the rules for order of operations.

We now expand our ability to manipulate expressions by first considering opposites of sums and differences. Then we simplify expressions involving parentheses.

a OPPOSITES OF SUMS

What happens when we multiply a real number by -1? Consider the following products:

$$-1(7) = -7, \quad -1(-5) = 5, \quad -1(0) = 0.$$

From these examples, it appears that when we multiply a number by -1, we get the opposite, or additive inverse, of that number.

THE PROPERTY OF -1

For any real number a,

$$-1 \cdot a = -a.$$

(Negative one times a is the opposite, or additive inverse, of a.)

The property of -1 enables us to find expressions equivalent to opposites of sums.

EXAMPLES Find an equivalent expression without parentheses.

1. $-(3 + x) = -1(3 + x)$ Using the property of -1
   $= -1 \cdot 3 + (-1)x$ Using a distributive law, multiplying each term by -1
   $= -3 + (-x)$ Using the property of -1
   $= -3 - x$

2. $-(3x + 2y + 4) = -1(3x + 2y + 4)$ Using the property of -1
   $= -1(3x) + (-1)(2y) + (-1)4$ Using a distributive law
   $= -3x - 2y - 4$ Using the property of -1

◀ Do Exercises 1 and 2.

Suppose that we want to remove parentheses in an expression like

$$-(x - 2y + 5).$$

We can first rewrite any subtractions inside the parentheses as additions. Then we take the opposite of each term:

$$-(x - 2y + 5) = -[x + (-2y) + 5]$$
$$= -x + 2y + (-5) = -x + 2y - 5.$$

The most efficient method for removing parentheses is to replace each term in the parentheses with its opposite ("change the sign of every term"). Doing so for $-(x - 2y + 5)$, we obtain $-x + 2y - 5$ as an equivalent expression.

Find an equivalent expression without parentheses.

1. $-(x + 2)$

2. $-(5x + 2y + 8)$

Answers

1. $-x - 2$ 2. $-5x - 2y - 8$

EXAMPLES Find an equivalent expression without parentheses.
3. $-(5 - y) = -5 + y$ Changing the sign of each term
4. $-(2a - 7b - 6) = -2a + 7b + 6$
5. $-(-3x + 4y + z - 7w - 23) = 3x - 4y - z + 7w + 23$

Do Exercises 3–6.

Find an equivalent expression without parentheses. Try to do this in one step.
3. $-(6 - t)$
4. $-(x - y)$
5. $-(-4a + 3t - 10)$
6. $-(18 - m - 2n + 4z)$

b REMOVING PARENTHESES AND SIMPLIFYING

SKILL REVIEW Use the distributive laws to multiply. [1.7c]
Multiply.
1. $4(x + 5)$ 2. $-7(a + b)$

Answers: 1. $4x + 20$ 2. $-7a - 7b$

When a sum is added to another expression, as in $5x + (2x + 3)$, we can simply remove, or drop, the parentheses and collect like terms because of the associative law of addition: $5x + (2x + 3) = 5x + 2x + 3$.

On the other hand, when a sum is subtracted from another expression, as in $3x - (4x + 2)$, we cannot simply drop the parentheses. However, we can subtract by adding an opposite. We then remove parentheses by changing the sign of each term inside the parentheses and collecting like terms.

EXAMPLE 6 Remove parentheses and simplify.

$$3x - (4x + 2) = 3x + [-(4x + 2)] \quad \text{Adding the opposite of } (4x + 2)$$
$$= 3x + (-4x - 2) \quad \text{Changing the sign of each term inside the parentheses}$$
$$= 3x - 4x - 2$$
$$= -x - 2 \quad \text{Collecting like terms}$$

······················· **Caution!** ·······················

Note that $3x - (4x + 2) \neq 3x - 4x + 2$. You cannot simply drop the parentheses.

Do Exercises 7 and 8.

Remove parentheses and simplify.
7. $5x - (3x + 9)$
8. $5y - 2 - (2y - 4)$

In practice, the first three steps of Example 6 are usually combined by changing the sign of each term in parentheses and then collecting like terms.

EXAMPLES Remove parentheses and simplify.
7. $5y - (3y + 4) = 5y - 3y - 4$ Removing parentheses by changing the sign of every term inside the parentheses
   $= 2y - 4$ Collecting like terms
8. $3x - 2 - (5x - 8) = 3x - 2 - 5x + 8$
   $= -2x + 6$
9. $(3a + 4b - 5) - (2a - 7b + 4c - 8)$
   $= 3a + 4b - 5 - 2a + 7b - 4c + 8$
   $= a + 11b - 4c + 3$

Do Exercises 9–11.

Remove parentheses and simplify.
9. $6x - (4x + 7)$
10. $8y - 3 - (5y - 6)$
11. $(2a + 3b - c) - (4a - 5b + 2c)$

Answers
3. $-6 + t$ 4. $-x + y$ 5. $4a - 3t + 10$
6. $-18 + m + 2n - 4z$ 7. $2x - 9$
8. $3y + 2$ 9. $2x - 7$ 10. $3y + 3$
11. $-2a + 8b - 3c$

Next, consider subtracting an expression consisting of several terms multiplied by a number other than 1 or −1.

EXAMPLE 10 Remove parentheses and simplify.

$$x - 3(x + y) = x + [-3(x + y)] \quad \text{Adding the opposite of } 3(x+y)$$
$$= x + [-3x - 3y] \quad \text{Multiplying } x + y \text{ by } -3$$
$$= x - 3x - 3y$$
$$= -2x - 3y \quad \text{Collecting like terms}$$

Remove parentheses and simplify.

12. $y - 9(x + y)$

13. $5a - 3(7a - 6)$
 $= 5a - \boxed{} + \boxed{}$
 $= \boxed{} + 18$

EXAMPLES Remove parentheses and simplify.

11. $3y - 2(4y - 5) = 3y - 8y + 10$ Multiplying each term in the parentheses by −2
 $= -5y + 10$

12. $(2a + 3b - 7) - 4(-5a - 6b + 12)$
 $= 2a + 3b - 7 + 20a + 24b - 48 = 22a + 27b - 55$

13. $2y - \frac{1}{3}(9y - 12) = 2y - 3y + 4 = -y + 4$

14. $6(5x - 3y) - 2(8x + y) = 30x - 18y - 16x - 2y = 14x - 20y$

◀ Do Exercises 12–16.

14. $4a - b - 6(5a - 7b + 8c)$

15. $5x - \frac{1}{4}(8x + 28)$

16. $4.6(5x - 3y) - 5.2(8x + y)$

c PARENTHESES WITHIN PARENTHESES

In addition to parentheses, some expressions contain other grouping symbols such as brackets [ ] and braces { }.

> When more than one kind of grouping symbol occurs, do the computations in the innermost symbol first. Then work from the inside out.

EXAMPLES Simplify.

15. $2[3 - (7 + 3)] = 2[3 - 10] = 2[-7] = -14$

16. $8 - [9 - (12 + 5)] = 8 - [9 - 17]$ Computing $12 + 5$
 $= 8 - [-8]$ Computing $9 - 17$
 $= 8 + 8 = 16$

Simplify.

17. $12 - (8 + 2)$

18. $9 - [10 - (13 + 6)]$
 $= 9 - [10 - (\boxed{})]$
 $= 9 - [\boxed{}]$
 $= 9 + \boxed{}$
 $= 18$

17. $\left[-4 - 2\left(-\frac{1}{2}\right)\right] \div \frac{1}{4} = [-4 + 1] \div \frac{1}{4}$ Working within parentheses
 $= -3 \div \frac{1}{4}$ Computing $-4 + 1$
 $= -3 \cdot 4 = -12$

18. $4(2 + 3) - \{7 - [4 - (8 + 5)]\}$
 $= 4(5) - \{7 - [4 - 13]\}$ Working with the innermost parentheses first
 $= 4(5) - \{7 - [-9]\}$ Computing $4 - 13$
 $= 4(5) - 16$ Computing $7 - [-9]$
 $= 20 - 16 = 4$

19. $[24 \div (-2)] \div (-2)$

20. $5(3 + 4) - \{8 - [5 - (9 + 6)]\}$

◀ Do Exercises 17–20.

Answers
12. $-9x - 8y$ **13.** $-16a + 18$
14. $-26a + 41b - 48c$ **15.** $3x - 7$
16. $-18.6x - 19y$ **17.** 2   **18.** 18   **19.** 6
20. 17

Guided Solutions:
13. $21a; 18; -16a$ **18.** $19; -9; 9$

EXAMPLE 19 Simplify.

$[5(x + 2) - 3x] - [3(y + 2) - 7(y - 3)]$
$= [5x + 10 - 3x] - [3y + 6 - 7y + 21]$ Working with the innermost parentheses first
$= [2x + 10] - [-4y + 27]$ Collecting like terms within brackets
$= 2x + 10 + 4y - 27$ Removing brackets
$= 2x + 4y - 17$ Collecting like terms

Do Exercise 21.

21. Simplify:
$[3(x + 2) + 2x] - [4(y + 2) - 3(y - 2)]$.

d ORDER OF OPERATIONS

When several operations are to be done in a calculation or a problem, we apply the following.

RULES FOR ORDER OF OPERATIONS

1. Do all calculations within grouping symbols before operations outside.
2. Evaluate all exponential expressions.
3. Do all multiplications and divisions in order from left to right.
4. Do all additions and subtractions in order from left to right.

These rules are consistent with the way in which most computers and scientific calculators perform calculations.

EXAMPLE 20 Simplify: $-34 \cdot 56 - 17$.

There are no parentheses or powers, so we start with the third step.

$-34 \cdot 56 - 17 = -1904 - 17$ Doing all multiplications and divisions in order from left to right
$= -1921$ Doing all additions and subtractions in order from left to right

EXAMPLE 21 Simplify: $25 \div (-5) + 50 \div (-2)$.

There are no calculations inside parentheses and no powers. The parentheses with (-5) and (-2) are used only to represent the negative numbers. We begin by doing all multiplications and divisions.

$\underbrace{25 \div (-5)}_{} + \underbrace{50 \div (-2)}_{}$
$= -5 + (-25)$ Doing all multiplications and divisions in order from left to right
$= -30$ Doing all additions and subtractions in order from left to right

Do Exercises 22-24.

Simplify.
22. $23 - 42 \cdot 30$
23. $32 \div 8 \cdot 2$
24. $-24 \div 3 - 48 \div (-4)$

Answers
21. $5x - y - 8$ **22.** -1237
23. 8 **24.** 4

EXAMPLE 22 Simplify: $-2^4 + 51 \cdot 4 - (37 + 23 \cdot 2)$.

$$-2^4 + 51 \cdot 4 - (37 + 23 \cdot 2)$$
$$= -2^4 + 51 \cdot 4 - (37 + 46)$$ Following the rules for order of operations within the parentheses first
$$= -2^4 + 51 \cdot 4 - 83$$ Completing the addition inside parentheses
$$= -16 + 51 \cdot 4 - 83$$ Evaluating exponential expressions. Note that $-2^4 \neq (-2)^4$.
$$= -16 + 204 - 83$$ Doing all multiplications
$$= 188 - 83$$ Doing all additions and subtractions in order from left to right
$$= 105$$

A fraction bar can play the role of a grouping symbol.

EXAMPLE 23 Simplify: $\dfrac{-64 \div (-16) \div (-2)}{2^3 - 3^2}$.

An equivalent expression with brackets as grouping symbols is

$$[-64 \div (-16) \div (-2)] \div [2^3 - 3^2].$$

This shows, in effect, that we do the calculations in the numerator and then in the denominator, and divide the results:

$$\frac{-64 \div (-16) \div (-2)}{2^3 - 3^2} = \frac{4 \div (-2)}{8 - 9} = \frac{-2}{-1} = 2.$$

◀ Do Exercises 25 and 26.

Simplify.

25. $-4^3 + 5^2 \cdot 5 + 5^3 - (4^2 - 48 \div 4)$ GS
$= \underline{} + 5^2 \cdot 5 + 125 - (\underline{} - 48 \div 4)$
$= -64 + 5^2 \cdot 5 + 125 - (16 - \underline{})$
$= -64 + 5^2 \cdot 5 + 125 - 4$
$= -64 + \underline{} + 125 - 4$
$= \underline{} + 125 - 4$
$= 321 - 4$
$= 317$

26. $\dfrac{5 - 10 - 5 \cdot 2^3}{2^3 + 3^2 - 7}$

Answers
25. 317 **26.** -12
Guided Solution:
25. -64; 16; 12; 260; 196

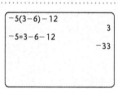

CALCULATOR CORNER

Order of Operations and Grouping Symbols Parentheses are necessary in some calculations. To simplify $-5(3 - 6) - 12$, we must use parentheses. The result is 3. Without parentheses, the computation is $-5 \cdot 3 - 6 - 12$, and the result is -33.

When a negative number is raised to an even power, parentheses also must be used. To find -3 raised to the fourth power, for example, we must use parentheses. The result is 81. Without parentheses, the computation is $-3^4 = -1 \cdot 3^4 = -1 \cdot 81 = -81$. To simplify an expression like $\dfrac{49 - 104}{7 + 4}$, we must enter it as $(49 - 104) \div (7 + 4)$. The result is -5.

EXERCISES: Calculate.
1. $-8 + 4(7 - 9)$
2. $-3[2 + (-5)]$
3. $(-7)^6$
4. $(-17)^5$
5. -7^6
6. -17^5
7. $\dfrac{38 - 178}{5 + 30}$
8. $\dfrac{311 - 17^2}{2 - 13}$

1.8 Exercise Set

✓ Check Your Understanding

Reading Check Choose from the list on the right the most appropriate illustration of each term.

RC1. The opposite of a sum

RC2. Like terms

RC3. Grouping symbols

RC4. Equivalent expressions

a) [ ]
b) $2y, -10y$
c) $-2(x - y), -2x + 2y$
d) $-(3x + 6y + 4z)$

Concept Check In each of the following, name the operation that should be performed first. Do not calculate.

CC1. $10 - 4 \cdot 2 + 5$
CC2. $10 - 4(2 + 5)$
CC3. $(10 - 4) \cdot 2 + 5$

CC4. $5[2(10 \div 5) - 3]$
CC5. $5(10 \div 2 + 5 - 3)$
CC6. $5 \cdot 2 - 4 \cdot 8 \div 2$

a Find an equivalent expression without parentheses.

1. $-(2x + 7)$
2. $-(8x + 4)$
3. $-(8 - x)$
4. $-(a - b)$

5. $-(4a - 3b + 7c)$
6. $-(x - 4y - 3z)$
7. $-(6x - 8y + 5)$
8. $-(4x + 9y + 7)$

9. $-(3x - 5y - 6)$
10. $-(6a - 4b - 7)$
11. $-(-8x - 6y - 43)$
12. $-(-2a + 9b - 5c)$

b Remove parentheses and simplify.

13. $9x - (4x + 3)$
14. $4y - (2y + 5)$
15. $2a - (5a - 9)$

16. $12m - (4m - 6)$
17. $2x + 7x - (4x + 6)$
18. $3a + 2a - (4a + 7)$

19. $2x - 4y - 3(7x - 2y)$
20. $3a - 9b - 1(4a - 8b)$
21. $15x - y - 5(3x - 2y + 5z)$

SECTION 1.8 Simplifying Expressions; Order of Operations

22. $4a - b - 4(5a - 7b + 8c)$

23. $(3x + 2y) - 2(5x - 4y)$

24. $(-6a - b) - 5(2b + a)$

25. $(12a - 3b + 5c) - 5(-5a + 4b - 6c)$

26. $(-8x + 5y - 12) - 6(2x - 4y - 10)$

c Simplify.

27. $9 - 2(5 - 4)$

28. $6 - 5(8 - 4)$

29. $8[7 - 6(4 - 2)]$

30. $10[7 - 4(7 - 5)]$

31. $[4(9 - 6) + 11] - [14 - (6 + 4)]$

32. $[7(8 - 4) + 16] - [15 - (7 + 8)]$

33. $[10(x + 3) - 4] + [2(x - 1) + 6]$

34. $[9(x + 5) - 7] + [4(x - 12) + 9]$

35. $[7(x + 5) - 19] - [4(x - 6) + 10]$

36. $[6(x + 4) - 12] - [5(x - 8) + 14]$

37. $3\{[7(x - 2) + 4] - [2(2x - 5) + 6]\}$

38. $4\{[8(x - 3) + 9] - [4(3x - 2) + 6]\}$

39. $4\{[5(x - 3) + 2] - 3[2(x + 5) - 9]\}$

40. $3\{[6(x - 4) + 5] - 2[5(x + 8) - 3]\}$

d Simplify.

41. $8 - 2 \cdot 3 - 9$

42. $8 - (2 \cdot 3 - 9)$

43. $(8 - 2) \div (3 - 9)$

44. $(8 - 2) \div 3 - 9$

45. $[(-24) \div (-3)] \div \left(-\frac{1}{2}\right)$

46. $[32 \div (-2)] \div \left(-\frac{1}{4}\right)$

47. $16 \cdot (-24) + 50$

48. $10 \cdot 20 - 15 \cdot 24$

49. $2^4 + 2^3 - 10$

50. $40 - 3^2 - 2^3$

51. $5^3 + 26 \cdot 71 - (16 + 25 \cdot 3)$

52. $4^3 + 10 \cdot 20 + 8^2 - 23$

53. $4 \cdot 5 - 2 \cdot 6 + 4$

54. $4 \cdot (6 + 8)/(4 + 3)$

55. $4^3/8$

56. $5^3 - 7^2$

57. $8(-7) + 6(-5)$

58. $10(-5) + 1(-1)$

59. $19 - 5(-3) + 3$

60. $14 - 2(-6) + 7$

61. $9 \div (-3) + 16 \div 8$

62. $-32 - 8 \div 4 - (-2)$

63. $-4^2 + 6$

64. $-5^2 + 7$

65. $-8^2 - 3$

66. $-9^2 - 11$

67. $12 - 20^3$

68. $20 + 4^3 \div (-8)$

69. $2 \cdot 10^3 - 5000$

70. $-7(3^4) + 18$

71. $6[9 - (3 - 4)]$

72. $8[3(6 - 13) - 11]$

73. $-1000 \div (-100) \div 10$

74. $256 \div (-32) \div (-4)$

75. $8 - (7 - 9)$

76. $(16 - 6) \cdot \dfrac{1}{2} + 9$

77. $\dfrac{10 - 6^2}{9^2 + 3^2}$

78. $\dfrac{5^2 - 4^3 - 3}{9^2 - 2^2 - 1^5}$

79. $\dfrac{3(6 - 7) - 5 \cdot 4}{6 \cdot 7 - 8(4 - 1)}$

80. $\dfrac{20(8 - 3) - 4(10 - 3)}{10(2 - 6) - 2(5 + 2)}$

81. $\dfrac{|2^3 - 3^2| + |12 \cdot 5|}{-32 \div (-16) \div (-4)}$

82. $\dfrac{|3 - 5|^2 - |7 - 13|}{|12 - 9| + |11 - 14|}$

Skill Maintenance

Evaluate. [1.1a]

83. $\dfrac{x - y}{y}$, when $x = 38$ and $y = 2$

84. $a - 3b$, when $a = 50$ and $b = 5$

Find the absolute value. [1.2e]

85. $|-0.4|$

86. $\left|\dfrac{15}{2}\right|$

Find the reciprocal. [1.6b]

87. -9

88. $\dfrac{7}{3}$

Subtract. [1.4a]

89. $5 - 30$

90. $-5 - 30$

91. $-5 - (-30)$

92. $5 - (-30)$

Synthesis

Simplify.

93. $x - [f - (f - x)] + [x - f] - 3x$

94. $x - \{x - 1 - [x - 2 - (x - 3 - \{x - 4 - [x - 5 - (x - 6)]\})]\}$

95. Use your calculator to do the following.
  a) Evaluate $x^2 + 3$ when $x = 7$, when $x = -7$, and when $x = -5.013$.
  b) Evaluate $1 - x^2$ when $x = 5$, when $x = -5$, and when $x = -10.455$.

96. Express $3^3 + 3^3 + 3^3$ as a power of 3.

Find the average.

97. $-15, 20, 50, -82, -7, -2$

98. $-1, 1, 2, -2, 3, -8, -10$

CHAPTER 1 Summary and Review

Vocabulary Reinforcement

Complete each statement with the correct term from the column on the right. Some of the choices may not be used.

1. The set of _____ is $\{\ldots, -5, -4, -3, -2, -1, 0, 1, 2, 3, 4, 5, \ldots\}$. [1.2a]

2. Two numbers whose sum is 0 are called _____ of each other. [1.3b]

3. The _____ of addition says that $a + b = b + a$ for any real numbers a and b. [1.7b]

4. The _____ states that for any real number a, $a \cdot 1 = 1 \cdot a = a$. [1.7a]

5. The _____ of multiplication says that $a(bc) = (ab)c$ for any real numbers a, b, and c. [1.7b]

6. Two numbers whose product is 1 are called _____ of each other. [1.6b]

7. The equation $y + 0 = y$ illustrates the _____. [1.7a]

natural numbers
whole numbers
integers
real numbers
multiplicative inverses
additive inverses
commutative law
associative law
distributive law
identity property of 0
identity property of 1
property of -1

Concept Reinforcement

Determine whether each statement is true or false.

_____ 1. Every whole number is also an integer. [1.2a]

_____ 2. The product of an even number of negative numbers is positive. [1.5a]

_____ 3. The product of a number and its multiplicative inverse is -1. [1.6b]

_____ 4. $a < b$ also has the meaning $b \geq a$. [1.2d]

Study Guide

Objective 1.1a Evaluate algebraic expressions by substitution.

Example Evaluate $y - z$ when $y = 5$ and $z = -7$.
$y - z = 5 - (-7) = 5 + 7 = 12$

Practice Exercise
1. Evaluate $2a + b$ when $a = -1$ and $b = 16$.

Objective 1.2d Determine which of two real numbers is greater and indicate which, using $<$ or $>$.

Example Use $<$ or $>$ for $\square$ to write a true sentence:
$-5 \ \square \ -12$.
Since -5 is to the right of -12 on the number line, we have $-5 > -12$.

Practice Exercise
2. Use $<$ or $>$ for $\square$ to write a true sentence:
$-6 \ \square \ -3$.

Objective 1.2e Find the absolute value of a real number.

Example Find the absolute value: **(a)** $|21|$; **(b)** $|-3.2|$; **(c)** $|0|$.

a) The number is positive, so the absolute value is the same as the number.
$$|21| = 21$$

b) The number is negative, so we make it positive.
$$|-3.2| = 3.2$$

c) The number is 0, so the absolute value is the same as the number.
$$|0| = 0$$

Practice Exercise

3. Find the absolute value: $\left|-\dfrac{5}{4}\right|$.

Objective 1.3a Add real numbers without using the number line.

Example Add without using the number line: **(a)** $-13 + 4$; **(b)** $-2 + (-3)$.

a) We have a negative number and a positive number. The absolute values are 13 and 4. The difference is 9. The negative number has the larger absolute value, so the answer is negative.
$$-13 + 4 = -9$$

b) We have two negative numbers. The sum of the absolute values is $2 + 3$, or 5. The answer is negative.
$$-2 + (-3) = -5$$

Practice Exercise

4. Add without using the number line: $-5.6 + (-2.9)$.

Objective 1.4a Subtract real numbers.

Example Subtract: $-4 - (-6)$.
$$-4 - (-6) = -4 + 6 = 2$$

Practice Exercise

5. Subtract: $7 - 9$.

Objective 1.5a Multiply real numbers.

Example Multiply: **(a)** $-1.9(4)$; **(b)** $-7(-6)$.

a) The signs are different, so the answer is negative.
$$-1.9(4) = -7.6$$

b) The signs are the same, so the answer is positive.
$$-7(-6) = 42$$

Practice Exercise

6. Multiply: $-8(-7)$.

Objective 1.6a Divide integers.

Example Divide: **(a)** $15 \div (-3)$; **(b)** $-72 \div (-9)$.

a) The signs are different, so the answer is negative.
$$15 \div (-3) = -5$$

b) The signs are the same, so the answer is positive.
$$-72 \div (-9) = 8$$

Practice Exercise

7. Divide: $-48 \div 6$.

Objective 1.6c Divide real numbers.

Example Divide: **(a)** $-\dfrac{1}{4} \div \dfrac{3}{5}$; **(b)** $-22.4 \div (-4)$.

a) We multiply by the reciprocal of the divisor:
$$-\dfrac{1}{4} \div \dfrac{3}{5} = -\dfrac{1}{4} \cdot \dfrac{5}{3} = -\dfrac{5}{12}.$$

b) We carry out the long division. The answer is positive.

$$\begin{array}{r} 5.6 \\ 4\overline{)22.4} \\ \underline{20} \\ 2\,4 \\ \underline{2\,4} \\ 0 \end{array}$$

Practice Exercise

8. Divide: $-\dfrac{3}{4} \div \left(-\dfrac{5}{3}\right)$.

Objective 1.7a Simplify fraction expressions.

Example Simplify: $-\dfrac{18x}{15x}$.

$$-\dfrac{18x}{15x} = -\dfrac{6 \cdot 3x}{5 \cdot 3x} \quad \text{Factoring the numerator and the denominator}$$

$$= -\dfrac{6}{5} \cdot \dfrac{3x}{3x} \quad \text{Factoring the fraction expression}$$

$$= -\dfrac{6}{5} \cdot 1 \quad \dfrac{3x}{3x} = 1$$

$$= -\dfrac{6}{5} \quad \text{Removing a factor of 1}$$

Practice Exercise

9. Simplify: $\dfrac{45y}{27y}$.

Objective 1.7c Use the distributive laws to multiply expressions like 8 and $x - y$.

Example Multiply: $3(4x - y + 2z)$.
$$3(4x - y + 2z) = 3 \cdot 4x - 3 \cdot y + 3 \cdot 2z$$
$$= 12x - 3y + 6z$$

Practice Exercise

10. Multiply: $5(x + 3y - 4z)$.

Objective 1.7d Use the distributive laws to factor expressions like $4x - 12 + 24y$.

Example Factor: $12a - 8b + 4c$.
$$12a - 8b + 4c = 4 \cdot 3a - 4 \cdot 2b + 4 \cdot c$$
$$= 4(3a - 2b + c)$$

Practice Exercise

11. Factor: $27x + 9y - 36z$.

Objective 1.7e Collect like terms.

Example Collect like terms: $3x - 5y + 8x + y$.
$$3x - 5y + 8x + y = 3x + 8x - 5y + y$$
$$= 3x + 8x - 5y + 1 \cdot y$$
$$= (3 + 8)x + (-5 + 1)y$$
$$= 11x - 4y$$

Practice Exercise

12. Collect like terms: $6a - 4b - a + 2b$.

Objective 1.8b Simplify expressions by removing parentheses and collecting like terms.

Example Remove parentheses and simplify:
$5x - 2(3x - y)$.
$5x - 2(3x - y) = 5x - 6x + 2y = -x + 2y$

Practice Exercise
13. Remove parentheses and simplify:
$8a - b - (4a + 3b)$.

Objective 1.8d Simplify expressions using the rules for order of operations.

Example Simplify: $12 - (7 - 3 \cdot 6)$.
$12 - (7 - 3 \cdot 6) = 12 - (7 - 18)$
$= 12 - (-11)$
$= 12 + 11$
$= 23$

Practice Exercise
14. Simplify: $75 \div (-15) + 24 \div 8$.

Review Exercises

The review exercises that follow are for practice. Answers are at the back of the book. If you miss an exercise, restudy the objective indicated in red after the exercise or the direction line that precedes it.

1. Evaluate $\dfrac{x - y}{3}$ when $x = 17$ and $y = 5$. [1.1a]

2. Translate to an algebraic expression: [1.1b]
   Nineteen percent of some number.

3. Tell which integers correspond to this situation: [1.2a]
   Josh earned $620 for one week's work. While driving to work one day, he received a speeding ticket for $125.

Find the absolute value. [1.2e]

4. $|-38|$ 5. $|126|$

Graph the number on the number line. [1.2b]

6. -2.5 7. $\dfrac{8}{9}$

Use either < or > for ☐ to write a true sentence. [1.2d]

8. -3 ☐ 10 9. -1 ☐ -6

10. 0.126 ☐ -12.6 11. $-\dfrac{2}{3}$ ☐ $-\dfrac{1}{10}$

12. Write another inequality with the same meaning as $-3 < x$. [1.2d]

Write true or false. [1.2d]

13. $-9 \leq 9$ 14. $-11 \geq -3$

Find the opposite. [1.3b]

15. 3.8 16. $-\dfrac{3}{4}$

Find the reciprocal. [1.6b]

17. $\dfrac{3}{8}$ 18. -7

19. Evaluate $-x$ when $x = -34$. [1.3b]

20. Evaluate $-(-x)$ when $x = 5$. [1.3b]

Compute and simplify.

21. $4 + (-7)$ [1.3a]

22. $6 + (-9) + (-8) + 7$ [1.3a]

23. $-3.8 + 5.1 + (-12) + (-4.3) + 10$ [1.3a]

24. $-3 - (-7) + 7 - 10$ [1.4a]

25. $-\dfrac{9}{10} - \dfrac{1}{2}$ [1.4a]

26. $-3.8 - 4.1$ [1.4a]

27. $-9 \cdot (-6)$ [1.5a]

28. $-2.7(3.4)$ [1.5a]

29. $\dfrac{2}{3} \cdot \left(-\dfrac{3}{7}\right)$ [1.5a]

30. $3 \cdot (-7) \cdot (-2) \cdot (-5)$ [1.5a]

31. $35 \div (-5)$ [1.6a]

32. $-5.1 \div 1.7$ [1.6c]

33. $-\dfrac{3}{11} \div \left(-\dfrac{4}{11}\right)$ [1.6c]

Simplify. [1.8d]

34. $2(-3.4 - 12.2) - 8(-7)$

35. $\dfrac{-12(-3) - 2^3 - (-9)(-10)}{3 \cdot 10 + 1}$

36. $-16 \div 4 - 30 \div (-5)$

37. $\dfrac{-4[7 - (10 - 13)]}{|-2(8) - 4|}$

Solve.

38. On the first, second, and third downs, a football team had these gains and losses: 5-yd gain, 12-yd loss, and 15-yd gain, respectively. Find the total gain (or loss). [1.3c]

39. Chang's total assets are $2140. He borrows $2500. What are his total assets now? [1.4b]

40. *Stock Price.* The value of EFX Corp. stock began the day at $17.68 per share and dropped $1.63 per hour for 8 hr. What was the price of the stock after 8 hr? [1.5b]

41. *National Park Visitation.* According to the National Park Service, there were 87,513 visitors to Congaree National Park in 2015 and 143,843 visitors in 2016. What was the percent increase or percent decrease in the number of visitors to the park? [1.6d]

Multiply. [1.7c]

42. $5(3x - 7)$ 　　　　43. $-2(4x - 5)$

44. $10(0.4x + 1.5)$ 　　45. $-8(3 - 6x)$

Factor. [1.7d]

46. $2x - 14$ 　　　　47. $-6x + 6$

48. $5x + 10$ 　　　　49. $-3x + 12y - 12$

Collect like terms. [1.7e]

50. $11a + 2b - 4a - 5b$

51. $7x - 3y - 9x + 8y$

52. $6x + 3y - x - 4y$

53. $-3a + 9b + 2a - b$

Remove parentheses and simplify.

54. $2a - (5a - 9)$ [1.8b]

55. $3(b + 7) - 5b$ [1.8b]

56. $3[11 - 3(4 - 1)]$ [1.8c]

57. $2[6(y - 4) + 7]$ [1.8c]

58. $[8(x + 4) - 10] - [3(x - 2) + 4]$ [1.8c]

59. $5\{[6(x - 1) + 7] - [3(3x - 4) + 8]\}$ [1.8c]

60. Factor out the greatest common factor: $18x - 6y + 30$. [1.7d]
   A. $2(9x - 2y + 15)$ **B.** $3(6x - 2y + 10)$
   C. $6(3x + 5)$ **D.** $6(3x - y + 5)$

61. Which expression is *not* equivalent to $mn + 5$? [1.7b]
   A. $nm + 5$ **B.** $5n + m$
   C. $5 + mn$ **D.** $5 + nm$

Synthesis

Simplify. [1.2e], [1.4a], [1.6a], [1.8d]

62. $-\left|\dfrac{7}{8} - \left(-\dfrac{1}{2}\right) - \dfrac{3}{4}\right|$

63. $(|2.7 - 3| + 3^2 - |-3|) \div (-3)$

64. $2000 - 1990 + 1980 - 1970 + \cdots + 20 - 10$

65. Find a formula for the perimeter of the figure below. [1.7e]

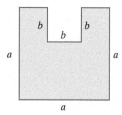

Understanding Through Discussion and Writing

1. Without actually performing the addition, explain why the sum of all integers from -50 to 50 is 0. [1.3b]

2. What rule have we developed that would tell you the sign of $(-7)^8$ and of $(-7)^{11}$ without doing the computations? Explain. [1.5a]

3. Explain how multiplication can be used to justify why a negative number divided by a negative number is positive. [1.6c]

4. Explain how multiplication can be used to justify why a negative number divided by a positive number is negative. [1.6c]

5. The distributive law was introduced before the discussion on collecting like terms. Why do you think this was done? [1.7c, e]

6. Jake keys in $18/2 \cdot 3$ on his calculator and expects the result to be 3. What mistake is he making? [1.8d]

CHAPTER 1 Test

1. Evaluate $\dfrac{3x}{y}$ when $x = 10$ and $y = 5$.

2. Translate to an algebraic expression: Nine less than some number.

Use either < or > for ☐ to write a true sentence.

3. $-3\ \square\ -8$

4. $-\dfrac{1}{2}\ \square\ -\dfrac{1}{8}$

5. $-0.78\ \square\ -0.87$

6. Write an inequality with the same meaning as $x < -2$.

7. Write true or false: $-13 \leq -3$.

Simplify.

8. $|-7|$

9. $\left|\dfrac{9}{4}\right|$

10. $|-2.7|$

Find the opposite.

11. $\dfrac{2}{3}$

12. -1.4

Find the reciprocal.

13. -2

14. $\dfrac{4}{7}$

15. Evaluate $-x$ when $x = -8$.

Compute and simplify.

16. $3.1 - (-4.7)$

17. $-8 + 4 + (-7) + 3$

18. $-\dfrac{1}{5} + \dfrac{3}{8}$

19. $2 - (-8)$

20. $3.2 - 5.7$

21. $\dfrac{1}{8} - \left(-\dfrac{3}{4}\right)$

22. $4 \cdot (-12)$

23. $-\dfrac{1}{2} \cdot \left(-\dfrac{3}{8}\right)$

24. $-45 \div 5$

25. $-\dfrac{3}{5} \div \left(-\dfrac{4}{5}\right)$

26. $4.864 \div (-0.5)$

27. $-2(16) - |2(-8) - 5^3|$

28. $-20 \div (-5) + 36 \div (-4)$

29. Isabella kept track of the changes in the stock market over a period of 5 weeks. By how many points had the market risen or fallen over this time?

WEEK 1	WEEK 2	WEEK 3	WEEK 4	WEEK 5
Down 13 pts	Down 16 pts	Up 36 pts	Down 11 pts	Up 19 pts

30. *Difference in Elevation.* The lowest elevation in Australia, Lake Eyre, is 15 m below sea level. The highest elevation in Australia, Mount Kosciuszko, is 2229 m. Find the difference in elevation between the highest point and the lowest point.

   Data: *The CIA World Factbook 2012*

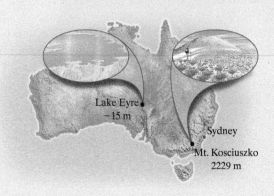

31. *Population Decrease.* The population of Stone City was 18,600. It dropped 420 each year for 6 years. What was the population of the city after 6 years?

32. *Chemical Experiment.* During a chemical reaction, the temperature in a beaker decreased every minute by the same number of degrees. The temperature was 16°C at 11:08 A.M. By 11:52 A.M., the temperature had dropped to −17°C. By how many degrees did it change each minute?

Multiply.

33. $3(6 - x)$

34. $-5(y - 1)$

Factor.

35. $12 - 22x$

36. $7x + 21 + 14y$

Simplify.

37. $6 + 7 - 4 - (-3)$

38. $5x - (3x - 7)$

39. $4(2a - 3b) + a - 7$

40. $4\{3[5(y - 3) + 9] + 2(y + 8)\}$

41. $256 \div (-16) \div 4$

42. $2^3 - 10[4 - 3(-2 + 18)]$

43. Which of the following is *not* a true statement?
   - A. $-5 \leq -5$
   - B. $-5 < -5$
   - C. $-5 \geq -5$
   - D. $-5 = -5$

Synthesis

Simplify.

44. $|-27 - 3(4)| - |-36| + |-12|$

45. $a - \{3a - [4a - (2a - 4a)]\}$

46. Find a formula for the perimeter of the figure shown here.

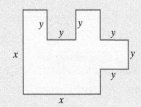

CHAPTER 2

2.1 Solving Equations: The Addition Principle
2.2 Solving Equations: The Multiplication Principle
2.3 Using the Principles Together
2.4 Formulas

Mid-Chapter Review

2.5 Applications of Percent
2.6 Applications and Problem Solving

Translating for Success

2.7 Solving Inequalities
2.8 Applications and Problem Solving with Inequalities

Summary and Review
Test
Cumulative Review

Solving Equations and Inequalities

Roller coasters, water slides, and drop rides help to make amusement parks popular recreation destinations around the world. This is especially true in North America. Although North Americans make up only 5% of the world population, the graph at right indicates that 40% of amusement park visits worldwide are in North America.

Data: worldpopulationreview.com

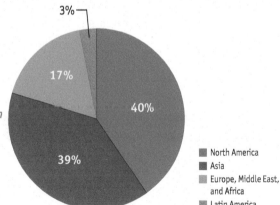

Amusement Park Visitors

- North America
- Asia
- Europe, Middle East, and Africa
- Latin America

DATA: International Association of Amusement Parks and Attractions

JUST IN TIME Review topics 8, 9, 11, 12, 15, 16, and 18 in the **Just In Time** section at the front of the text. This provides excellent prerequisite skill review for this chapter.

MyLab Math VIDEO

We will calculate the number of amusement park visits from North America in Example 7 of Section 2.5 and the speed of the fastest steel roller coaster in Example 7 of Section 2.6.

109

STUDYING FOR SUCCESS   *Learning Resources on Campus*

☐ There may be a learning lab or a tutoring center for drop-in tutoring.
☐ There may be group tutoring sessions for this specific course.
☐ The mathematics department may have a bulletin board or a network for locating private tutors.

2.1 Solving Equations: The Addition Principle

OBJECTIVES

a Determine whether a given number is a solution of a given equation.

b Solve equations using the addition principle.

a EQUATIONS AND SOLUTIONS

SKILL REVIEW Evaluate algebraic expressions by substitution. [1.1a]
1. Evaluate $x - 7$ when $x = 5$.
2. Evaluate $2x + 3$ when $x = -1$.

Answers: 1. -2 2. 1

EQUATION

An **equation** is a number sentence that says that the expressions on either side of the equals sign, =, represent the same number.

Some examples of equations are $14 - 10 = 1 + 3$ and $x + 6 = 13$. The sentence "$14 - 10 = 1 + 3$" asserts that the expressions $14 - 10$ and $1 + 3$ name the same number.

Some equations are true. Some are false. Some are neither true nor false.

EXAMPLES Determine whether each equation is true, false, or neither.

1. $3 + 2 = 5$ The equation is *true*.
2. $7 - 2 = 4$ The equation is *false*.
3. $x + 6 = 13$ The equation is *neither* true nor false, because we do not know what number x represents.

◀ Do Exercises 1–3.

Determine whether each equation is true, false, or neither.
1. $5 - 8 = -4$
2. $12 + 6 = 18$
3. $x + 6 = 7 - x$

SOLUTION OF AN EQUATION

Any replacement for the variable that makes an equation true is called a **solution** of the equation. To solve an equation means to find *all* of its solutions.

One way to determine whether a number is a solution of an equation is to evaluate the expression on each side of the equals sign by substitution. If the values are the same, then the number is a solution.

Answers
1. False 2. True 3. Neither

EXAMPLE 4 Determine whether 7 is a solution of $x + 6 = 13$.

We have

$$\begin{array}{c|c} x + 6 = 13 & \text{Writing the equation} \\ \hline 7 + 6 \;?\; 13 & \text{Substituting 7 for } x \\ 13 & \text{TRUE} \end{array}$$

Since the left side and the right side are the same, 7 is a solution. No other number makes the equation true, so the only solution is the number 7.

EXAMPLE 5 Determine whether 19 is a solution of $7x = 141$.

We have

$$\begin{array}{c|c} 7x = 141 & \text{Writing the equation} \\ \hline 7(19) \;?\; 141 & \text{Substituting 19 for } x \\ 133 & \text{FALSE} \end{array}$$

Since the left side and the right side are not the same, 19 is not a solution of the equation.

Do Exercises 4–7.

Determine whether the given number is a solution of the given equation.

4. $8;\; x + 4 = 12$

5. $0;\; x + 4 = 12$

6. $-3;\; 7 + x = -4$

7. $-\dfrac{3}{5};\; -5x = 3$

b USING THE ADDITION PRINCIPLE

SKILL REVIEW

Add and subtract real numbers. [1.3a], [1.4a]

Add or subtract.

1. $13 + (-8)$ 2. $-7 - 4$ 3. $-\dfrac{1}{2} - \left(-\dfrac{3}{4}\right)$

Answers: **1.** 5 **2.** -11 **3.** $\dfrac{1}{4}$

Consider the equation $x = 7$. We can easily see that the solution of this equation is 7. If we replace x with 7, we get $7 = 7$, which is true.

Now consider the equation of Example 4: $x + 6 = 13$. In Example 4, we discovered that the solution of this equation is also 7, but the fact that 7 is the solution is not obvious. We now begin to consider principles that allow us to begin with an equation like $x + 6 = 13$ and end with an *equivalent equation*, like $x = 7$, in which the variable is alone on one side and for which the solution is easier to find.

EQUIVALENT EQUATIONS

Equations with the same solutions are called **equivalent equations**.

One of the principles that we use in solving equations involves addition. An equation $a = b$ says that a and b represent the same number. Suppose that this is true, and we add a number c to the number a. We get the same answer if we add c to b, because a and b are the same number.

Answers
4. Yes 5. No 6. No 7. Yes

> **THE ADDITION PRINCIPLE FOR EQUATIONS**
>
> For any real numbers a, b, and c,
>
> $$a = b \quad \text{is equivalent to} \quad a + c = b + c.$$

EXAMPLE 6 Solve: $x + 6 = 13$.

We have

$x + 6 = 13$	We want to get x on one side.
$x + 6 + (-6) = 13 + (-6)$	Using the addition principle: adding -6 on both sides because $6 + (-6) = 0$
$x + 0 = 7$	Simplifying
$x = 7.$	Using the identity property of 0: $x + 0 = x$

The solution of $x + 6 = 13$ is 7.

◀ Do Exercise 8.

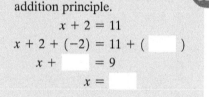

8. Solve $x + 2 = 11$ using the addition principle.
$$x + 2 = 11$$
$$x + 2 + (-2) = 11 + (\,\boxed{}\,)$$
$$x + \boxed{} = 9$$
$$x = \boxed{}$$

When we use the addition principle, we sometimes say that we "add the same number on both sides of the equation." Since

$$a - c = b - c \quad \text{is equivalent to} \quad a + (-c) = b + (-c),$$

the addition principle also tells us that we can "subtract the same number on both sides of the equation."

EXAMPLE 7 Solve: $x + 5 = -7$.

We have

$x + 5 = -7$	
$x + 5 - 5 = -7 - 5$	Using the addition principle: adding -5 on both sides or subtracting 5 on both sides
$x + 0 = -12$	Simplifying
$x = -12.$	Identity property of 0

The solution of the original equation is -12. The equations $x + 5 = -7$ and $x = -12$ are *equivalent*.

◀ Do Exercise 9.

9. Solve using the addition principle, subtracting 5 on both sides:
$$x + 5 = -8.$$

EXAMPLE 8 Solve: $a - 4 = 10$.

We have

$a - 4 = 10$	
$a - 4 + 4 = 10 + 4$	Using the addition principle: adding 4 on both sides
$a + 0 = 14$	Simplifying
$a = 14.$	Identity property of 0

Check:
$$\begin{array}{c|c} a - 4 = 10 \\ \hline 14 - 4 \;?\; 10 \\ 10 \;\bigg|\; \quad \text{TRUE} \end{array}$$

The solution is 14.

◀ Do Exercise 10.

10. Solve: $t - 3 = 19$.

Answers
8. 9   **9.** -13 **10.** 22
Guided Solution:
8. $-2, 0, 9$

EXAMPLE 9 Solve: $-6.5 = y - 8.4$.

We have

$$-6.5 = y - 8.4$$
$$-6.5 + 8.4 = y - 8.4 + 8.4 \quad \text{Using the addition principle: adding 8.4 on both sides to eliminate } -8.4 \text{ on the right}$$
$$1.9 = y.$$

Check:
$$\begin{array}{c|c} -6.5 = y - 8.4 \\ \hline -6.5 \ ? \ 1.9 - 8.4 \\ \ \ | \ -6.5 \quad \text{TRUE} \end{array}$$

The solution is 1.9.

Note that equations are reversible. That is, if $a = b$ is true, then $b = a$ is true. Thus when we solve $-6.5 = y - 8.4$, we can reverse it and solve $y - 8.4 = -6.5$ if we wish.

Do Exercises 11 and 12. ▶

Solve.

11. $8.7 = n - 4.5$

12. $y + 17.4 = 10.9$

EXAMPLE 10 Solve: $-\dfrac{2}{3} + x = \dfrac{5}{2}$.

We have

$$-\dfrac{2}{3} + x = \dfrac{5}{2}$$
$$\dfrac{2}{3} - \dfrac{2}{3} + x = \dfrac{2}{3} + \dfrac{5}{2} \quad \text{Adding } \tfrac{2}{3} \text{ on both sides}$$
$$x = \dfrac{2}{3} + \dfrac{5}{2}$$
$$x = \dfrac{2}{3} \cdot \dfrac{2}{2} + \dfrac{5}{2} \cdot \dfrac{3}{3} \quad \text{Multiplying by 1 to obtain equivalent fraction expressions with the least common denominator 6}$$
$$x = \dfrac{4}{6} + \dfrac{15}{6}$$
$$x = \dfrac{19}{6}.$$

Check:
$$\begin{array}{c|c} -\dfrac{2}{3} + x = \dfrac{5}{2} \\ \hline -\dfrac{2}{3} + \dfrac{19}{6} \ ? \ \dfrac{5}{2} \\ -\dfrac{4}{6} + \dfrac{19}{6} \\ \dfrac{15}{6} \\ \dfrac{5}{2} \quad \text{TRUE} \end{array}$$

The solution is $\dfrac{19}{6}$.

Do Exercises 13 and 14. ▶

Solve.

13. $x + \dfrac{1}{2} = -\dfrac{3}{2}$

14. $t - \dfrac{13}{4} = \dfrac{5}{8}$

Answers

11. 13.2 **12.** -6.5 **13.** -2 **14.** $\dfrac{31}{8}$

SECTION 2.1 Solving Equations: The Addition Principle

2.1 Exercise Set

✓ Check Your Understanding

Reading Check Choose from the column on the right the most appropriate choice for each.

RC1. The equations $x = 3$ and $9x = 27$
RC2. The equation $2 + 3 = 9 - 4$
RC3. The equation $7 - 6 = 6 - 7$
RC4. A replacement that makes an equation true

a) A true equation
b) A false equation
c) Equivalent equations
d) A solution of an equation

Concept Check Choose from the column on the right the most appropriate first step in solving each equation.

CC1. $9 = x - 4$
CC2. $3 + x = -15$
CC3. $x - 3 = 9$
CC4. $x + 4 = 3$

a) Add -4 on both sides.
b) Add 15 on both sides.
c) Subtract 3 on both sides.
d) Subtract 9 on both sides.
e) Add 3 on both sides.
f) Add 4 on both sides.

a Determine whether the given number is a solution of the given equation.

1. 15; $x + 17 = 32$
2. 35; $t + 17 = 53$
3. 21; $x - 7 = 12$
4. 36; $a - 19 = 17$

5. -7; $6x = 54$
6. -9; $8y = -72$
7. 30; $\dfrac{x}{6} = 5$
8. 49; $\dfrac{y}{8} = 6$

9. 20; $5x + 7 = 107$
10. 9; $9x + 5 = 86$
11. -10; $7(y - 1) = 63$
12. -5; $6(y - 2) = 18$

b Solve using the addition principle. Don't forget to check!

13. $x + 2 = 6$
Check: $\dfrac{x + 2 = 6}{?}$

14. $y + 4 = 11$
Check: $\dfrac{y + 4 = 11}{?}$

15. $x + 15 = -5$
Check: $\dfrac{x + 15 = -5}{?}$

16. $t + 10 = 44$
Check: $\dfrac{t + 10 = 44}{?}$

17. $x + 6 = -8$
Check: $\dfrac{x + 6 = -8}{?}$

18. $z + 9 = -14$

19. $x + 16 = -2$

20. $m + 18 = -13$

21. $x - 9 = 6$

22. $x - 11 = 12$

23. $x - 7 = -21$

24. $x - 3 = -14$

25. $5 + t = 7$

26. $8 + y = 12$

27. $-7 + y = 13$

28. $-8 + y = 17$

29. $-3 + t = -9$

30. $-8 + t = -24$

31. $x + \frac{1}{2} = 7$

32. $24 = -\frac{7}{10} + r$

33. $12 = a - 7.9$

34. $2.8 + y = 11$

35. $r + \frac{1}{3} = \frac{8}{3}$

36. $t + \frac{3}{8} = \frac{5}{8}$

37. $m + \frac{5}{6} = -\frac{11}{12}$

38. $x + \frac{2}{3} = -\frac{5}{6}$

39. $x - \frac{5}{6} = \frac{7}{8}$

40. $y - \frac{3}{4} = \frac{5}{6}$

41. $-\frac{1}{5} + z = -\frac{1}{4}$

42. $-\frac{1}{8} + y = -\frac{3}{4}$

43. $7.4 = x + 2.3$

44. $8.4 = 5.7 + y$

45. $7.6 = x - 4.8$

46. $8.6 = x - 7.4$

47. $-9.7 = -4.7 + y$

48. $-7.8 = 2.8 + x$

49. $5\frac{1}{6} + x = 7$

50. $5\frac{1}{4} = 4\frac{2}{3} + x$

51. $q + \frac{1}{3} = -\frac{1}{7}$

52. $52\frac{3}{8} = -84 + x$

Skill Maintenance

53. Divide: $\frac{2}{3} \div \left(-\frac{4}{9}\right)$. [1.6c]

54. Add: $-8.6 + 3.4$. [1.3a]

55. Subtract: $-\frac{2}{3} - \left(-\frac{5}{8}\right)$. [1.4a]

56. Multiply: $(-25.4)(-6.8)$. [1.5a]

Translate to an algebraic expression. [1.1b]

57. Jane had $83 before paying x dollars for a pair of tennis shoes. How much does she have left?

58. Justin drove his pickup truck 65 mph for t hours. How far did he drive?

Synthesis

Solve.

59. $x + \frac{4}{5} = -\frac{2}{3} - \frac{4}{15}$

60. $x + x = x$

61. $16 + x - 22 = -16$

62. $x + 4 = 5 + x$

63. $x + 3 = 3 + x$

64. $|x| + 6 = 19$

2.2 Solving Equations: The Multiplication Principle

OBJECTIVE

a Solve equations using the multiplication principle.

a USING THE MULTIPLICATION PRINCIPLE

SKILL REVIEW

Find the reciprocal of a real number. [1.6b]
Find the reciprocal.

1. 5
2. $-\dfrac{5}{4}$
3. -10

Answers: 1. $\dfrac{1}{5}$ 2. $-\dfrac{4}{5}$ 3. $-\dfrac{1}{10}$

MyLab Math VIDEO

Suppose that $a = b$ is true, and we multiply a by some number c. We get the same number if we multiply b by c, because a and b are the same number.

THE MULTIPLICATION PRINCIPLE FOR EQUATIONS

For any real numbers a, b, and c, $c \neq 0$,

$a = b$ is equivalent to $a \cdot c = b \cdot c$.

When using the multiplication principle, we sometimes say that we "multiply on both sides of the equation by the same number."

EXAMPLE 1 Solve: $5x = 70$.

To get x alone on one side, we multiply by the *multiplicative inverse*, or *reciprocal*, of 5. Then we get the *multiplicative identity* 1 times x, or $1 \cdot x$, which simplifies to x. This allows us to eliminate 5 on the left.

$$5x = 70 \quad \text{The reciprocal of 5 is } \tfrac{1}{5}.$$
$$\tfrac{1}{5} \cdot 5x = \tfrac{1}{5} \cdot 70 \quad \text{Multiplying by } \tfrac{1}{5} \text{ to get } 1 \cdot x \text{ and eliminate 5 on the left}$$
$$1 \cdot x = 14 \quad \text{Simplifying}$$
$$x = 14 \quad \text{Identity property of 1: } 1 \cdot x = x$$

Check: $\dfrac{5x = 70}{5 \cdot 14 \;?\; 70}$
$ 70 \;\;|\;\;$ TRUE

The solution is 14.

The multiplication principle also tells us that we can "divide on both sides of the equation by the same nonzero number." This is because dividing is the same as multiplying by a reciprocal. That is,

$$\dfrac{a}{c} = \dfrac{b}{c} \quad \text{is equivalent to} \quad a \cdot \dfrac{1}{c} = b \cdot \dfrac{1}{c}, \quad \text{when } c \neq 0.$$

In an expression like $5x$ in Example 1, the number 5 is called the **coefficient**. Example 1 could be done as in the next example, dividing by 5, the coefficient of x, on both sides.

EXAMPLE 2 Solve: $5x = 70$.

$$5x = 70$$
$$\frac{5x}{5} = \frac{70}{5} \quad \text{Dividing by 5 on both sides}$$
$$1 \cdot x = 14 \quad \text{Simplifying}$$
$$x = 14 \quad \text{Identity property of 1. The solution is 14.}$$

Do Exercises 1 and 2. ▶

EXAMPLE 3 Solve: $-4x = 92$.

$$-4x = 92$$
$$\frac{-4x}{-4} = \frac{92}{-4} \quad \text{Using the multiplication principle. Dividing by } -4 \text{ on both sides is the same as multiplying by } -\tfrac{1}{4}.$$
$$1 \cdot x = -23 \quad \text{Simplifying}$$
$$x = -23 \quad \text{Identity property of 1}$$

Check:
$$\begin{array}{c|c} -4x = 92 \\ \hline -4(-23) \;?\; 92 \\ 92 \;|\; \text{TRUE} \end{array}$$

The solution is -23.

Do Exercise 3. ▶

EXAMPLE 4 Solve: $-x = 9$.

$$-x = 9$$
$$-1 \cdot x = 9 \quad \text{Using the property of } -1\text{: } -x = -1 \cdot x$$
$$\frac{-1 \cdot x}{-1} = \frac{9}{-1} \quad \text{Dividing by } -1 \text{ on both sides: } -1/(-1) = 1$$
$$1 \cdot x = -9$$
$$x = -9$$

Check:
$$\begin{array}{c|c} -x = 9 \\ \hline -(-9) \;?\; 9 \\ 9 \;|\; \text{TRUE} \end{array}$$

The solution is -9.

Do Exercise 4. ▶

We can also solve the equation $-x = 9$ by multiplying as follows.

EXAMPLE 5 Solve: $-x = 9$.

$$-x = 9$$
$$-1 \cdot (-x) = -1 \cdot 9 \quad \text{Multiplying by } -1 \text{ on both sides}$$
$$-1 \cdot (-1) \cdot x = -9 \quad -x = (-1) \cdot x$$
$$1 \cdot x = -9 \quad -1 \cdot (-1) = 1$$
$$x = -9$$

The solution is -9.

Do Exercise 5. ▶

GS 1. Solve $6x = 90$ by multiplying on both sides.
$$6x = 90$$
$$\frac{1}{6} \cdot 6x = \boxed{} \cdot 90$$
$$1 \cdot x = 15$$
$$\boxed{} = 15$$

Check:
$$\begin{array}{c|c} 6x = 90 \\ \hline 6 \cdot \boxed{} \;?\; 90 \\ 90 \;|\; \text{TRUE} \end{array}$$

2. Solve $4x = -7$ by dividing on both sides.
$$4x = -7$$
$$\frac{4x}{4} = \frac{-7}{\boxed{}}$$
$$1 \cdot x = -\frac{7}{4}$$
$$\boxed{} = -\frac{7}{4}$$

Don't forget to check.

3. Solve: $-6x = 108$.

4. Solve by dividing on both sides.
$$-x = -10$$

5. Solve by multiplying on both sides.
$$-x = -10$$

Answers
1. 15 2. $-\dfrac{7}{4}$ 3. -18 4. 10 5. 10

Guided Solutions:
1. $\dfrac{1}{6}, x, 15$ 2. $4, x$

In practice, it is generally more convenient to divide on both sides of the equation if the coefficient of the variable is in decimal notation or is an integer. If the coefficient is in fraction notation, it is usually more convenient to multiply by a reciprocal.

EXAMPLE 6 Solve: $\dfrac{3}{8} = -\dfrac{5}{4}x$.

$$\dfrac{3}{8} = -\dfrac{5}{4}x$$

The reciprocal of $-\dfrac{5}{4}$ is $-\dfrac{4}{5}$. There is no sign change.

$$-\dfrac{4}{5} \cdot \dfrac{3}{8} = -\dfrac{4}{5} \cdot \left(-\dfrac{5}{4}x\right)$$ Multiplying by $-\dfrac{4}{5}$ to get $1 \cdot x$ and eliminate $-\dfrac{5}{4}$ on the right

$$-\dfrac{12}{40} = 1 \cdot x$$

$$-\dfrac{3}{10} = 1 \cdot x \qquad \text{Simplifying}$$

$$-\dfrac{3}{10} = x \qquad \text{Identity property of 1}$$

Check: $\dfrac{3}{8} = -\dfrac{5}{4}x$

$$\dfrac{3}{8} \;?\; -\dfrac{5}{4}\left(-\dfrac{3}{10}\right)$$
$$\dfrac{3}{8} \qquad \text{TRUE}$$

The solution is $-\dfrac{3}{10}$.

Note that if $a = b$ is true, then $b = a$ is true. Thus we can reverse the equation $\dfrac{3}{8} = -\dfrac{5}{4}x$ and solve $-\dfrac{5}{4}x = \dfrac{3}{8}$ if we wish.

◀ Do Exercise 6.

EXAMPLE 7 Solve: $1.16y = 9744$.

$$1.16y = 9744$$
$$\dfrac{1.16y}{1.16} = \dfrac{9744}{1.16} \qquad \text{Dividing by 1.16 on both sides}$$
$$y = \dfrac{9744}{1.16}$$
$$y = 8400 \qquad \text{Simplifying}$$

Check: $1.16y = 9744$
$$1.16(8400) \;?\; 9744$$
$$9744 \qquad \text{TRUE}$$

The solution is 8400.

◀ Do Exercises 7 and 8.

---

6. Solve: $\dfrac{2}{3} = -\dfrac{5}{6}y$.

$$\dfrac{2}{3} = -\dfrac{5}{6}y$$

$$\square \cdot \dfrac{2}{3} = -\dfrac{6}{5} \cdot \left(-\dfrac{5}{6}y\right)$$

$$-\dfrac{\square}{15} = 1 \cdot y$$

$$-\dfrac{\square}{5} = y$$

Solve.

7. $1.12x = 8736$

8. $6.3 = -2.1y$

Answers

6. $-\dfrac{4}{5}$ 7. 7800   8. -3

Guided Solution:

6. $-\dfrac{6}{5}$, 12, 4

EXAMPLE 8 Solve: $\dfrac{-y}{9} = 14$.

$$\dfrac{-y}{9} = 14$$

$$9 \cdot \dfrac{-y}{9} = 9 \cdot 14 \qquad \text{Multiplying by 9 on both sides}$$

$$-y = 126$$

$$-1 \cdot (-y) = -1 \cdot 126 \qquad \text{Multiplying by } -1 \text{ on both sides}$$

$$y = -126$$

Check:
$$\dfrac{-y}{9} = 14$$
$$\dfrac{-(-126)}{9} \;?\; 14$$
$$\dfrac{126}{9}$$
$$14 \qquad \text{TRUE}$$

The solution is -126.

Another way to solve the equation in Example 8 is by multiplying by -9 on both sides:

$$-9 \cdot \dfrac{-y}{9} = -9 \cdot 14$$

$$\dfrac{9y}{9} = -126$$

$$y = -126.$$

Do Exercise 9.

9. Solve: $-14 = \dfrac{-y}{2}$.

Answer
9. 28

2.2 Exercise Set

FOR EXTRA HELP — MyLab Math

✓ Check Your Understanding

Reading Check Choose from the column on the right the most appropriate term for each.

RC1. For all real numbers $a, b,$ and $c, c \neq 0$, $a = b$ is equivalent to $a \cdot c = b \cdot c$.

RC2. For all real numbers x, $1 \cdot x = x$.

RC3. The number 7 in $7x$

RC4. $\dfrac{2}{3}$ and $\dfrac{3}{2}$

a) Coefficient
b) Reciprocals
c) Identity property of 1
d) Multiplication principle for equations

SECTION 2.2 Solving Equations: The Multiplication Principle

Concept Check Choose from the column on the right the most appropriate first step in solving each equation.

CC1. $3 = -\frac{1}{12}x$

CC2. $-6x = 12$

CC3. $12x = -6$

CC4. $\frac{1}{6}x = 12$

a) Divide by 12 on both sides.
b) Multiply by 6 on both sides.
c) Multiply by 12 on both sides.
d) Divide by -6 on both sides.
e) Divide by 6 on both sides.
f) Multiply by -12 on both sides.

a Solve using the multiplication principle. Don't forget to check!

1. $6x = 36$

Check: $6x = 36$

2. $3x = 51$

Check: $3x = 51$

3. $5y = 45$

Check: $5y = 45$

4. $8y = 72$

Check: $8y = 72$

5. $84 = 7x$

6. $63 = 9x$

7. $-x = 40$

8. $-x = 53$

9. $-1 = -z$

10. $-47 = -t$

11. $7x = -49$

12. $8x = -56$

13. $-12x = 72$

14. $-15x = 105$

15. $-21w = -126$

16. $-13w = -104$

17. $\frac{t}{7} = -9$

18. $\frac{y}{5} = -6$

19. $\frac{n}{-6} = 8$

20. $\frac{y}{-8} = 11$

21. $\frac{3}{4}x = 27$

22. $\frac{4}{5}x = 16$

23. $-\frac{2}{3}x = 6$

24. $-\frac{3}{8}x = 12$

25. $\dfrac{-t}{3} = 7$

26. $\dfrac{-x}{6} = 9$

27. $-\dfrac{m}{3} = \dfrac{1}{5}$

28. $\dfrac{1}{8} = -\dfrac{y}{5}$

29. $-\dfrac{3}{5}r = \dfrac{9}{10}$

30. $-\dfrac{2}{5}y = \dfrac{4}{15}$

31. $-\dfrac{3}{2}r = -\dfrac{27}{4}$

32. $-\dfrac{3}{8}x = -\dfrac{15}{16}$

33. $6.3x = 44.1$

34. $2.7y = 54$

35. $-3.1y = 21.7$

36. $-3.3y = 6.6$

37. $38.7m = 309.6$

38. $29.4m = 235.2$

39. $-\dfrac{2}{3}y = -10.6$

40. $-\dfrac{9}{7}y = 12.06$

41. $\dfrac{-x}{5} = 10$

42. $\dfrac{-x}{8} = -16$

43. $-\dfrac{t}{2} = 7$

44. $\dfrac{m}{-3} = 10$

Skill Maintenance

Collect like terms. [1.7e]

45. $3x + 4x$

46. $6x + 5 - 7x$

47. $-4x + 11 - 6x + 18x$

48. $8y - 16y - 24y$

Remove parentheses and simplify. [1.8b]

49. $3x - (4 + 2x)$

50. $2 - 5(x + 5)$

51. $8y - 6(3y + 7)$

52. $-2a - 4(5a - 1)$

Translate to an algebraic expression. [1.1b]

53. Patty drives her van for 8 hr at a speed of r miles per hour. How far does she drive?

54. A triangle has a height of 10 meters and a base of b meters. What is the area of the triangle?

Synthesis

Solve.

55. $-0.2344m = 2028.732$

56. $0 \cdot x = 0$

57. $0 \cdot x = 9$

58. $4|x| = 48$

59. $2|x| = -12$

Solve for x.

60. $ax = 5a$

61. $3x = \dfrac{b}{a}$

62. $cx = a^2 + 1$

63. $\dfrac{a}{b}x = 4$

64. A student makes a calculation and gets an answer of 22.5. On the last step, she multiplies by 0.3 when she should have divided by 0.3. What is the correct answer?

2.3 Using the Principles Together

OBJECTIVES

a Solve equations using both the addition principle and the multiplication principle.

b Solve equations in which like terms may need to be collected.

c Solve equations by first removing parentheses and collecting like terms; solve equations with an infinite number of solutions and equations with no solutions.

a APPLYING BOTH PRINCIPLES

Consider the equation $3x + 4 = 13$. It is more complicated than those we discussed in the preceding two sections. In order to solve such an equation, we first isolate the x-term, $3x$, using the addition principle. Then we apply the multiplication principle to get x by itself.

EXAMPLE 1 Solve: $3x + 4 = 13$.

$3x + 4 = 13$

$3x + 4 - 4 = 13 - 4$ Using the addition principle: subtracting 4 on both sides

First isolate the x-term. → $3x = 9$ Simplifying

$\dfrac{3x}{3} = \dfrac{9}{3}$ Using the multiplication principle: dividing by 3 on both sides

Then isolate x. → $x = 3$ Simplifying

Check: $\dfrac{3x + 4 = 13}{3 \cdot 3 + 4 \;?\; 13}$
$9 + 4$
$13 \;\;|\;\;$ TRUE

We use the rules for order of operations to carry out the check. We find the product $3 \cdot 3$. Then we add 4.

The solution is 3.

◀ Do Exercise 1.

EXAMPLE 2 Solve: $-5x - 6 = 16$.

$-5x - 6 = 16$

$-5x - 6 + 6 = 16 + 6$ Adding 6 on both sides

$-5x = 22$

$\dfrac{-5x}{-5} = \dfrac{22}{-5}$ Dividing by -5 on both sides

$x = -\dfrac{22}{5},\;\text{or}\;-4\dfrac{2}{5}$ Simplifying

Check: $\dfrac{-5x - 6 = 16}{-5\left(-\dfrac{22}{5}\right) - 6 \;?\; 16}$
$22 - 6$
$16 \;\;|\;\;$ TRUE

The solution is $-\dfrac{22}{5}$.

◀ Do Exercises 2 and 3.

1. Solve: $9x + 6 = 51$.

Solve.
2. $8x - 4 = 28$

3. $-\dfrac{1}{2}x + 3 = 1$

Answers
1. 5 **2.** 4 **3.** 4

EXAMPLE 3 Solve: $45 - t = 13$.

$$45 - t = 13$$
$$-45 + 45 - t = -45 + 13 \quad \text{Adding } -45 \text{ on both sides}$$
$$-t = -32$$
$$-1(-t) = -1(-32) \quad \text{Multiplying by } -1 \text{ on both sides}$$
$$t = 32$$

The number 32 checks and is the solution.

Do Exercise 4.

4. Solve: $-18 - m = -57$.
$$-18 - m = -57$$
$$18 - 18 - m = \boxed{} - 57$$
$$\boxed{} = -39$$
$$\boxed{}(-m) = -1(-39)$$
$$\boxed{} = 39$$

EXAMPLE 4 Solve: $16.3 - 7.2y = -8.18$.

$$16.3 - 7.2y = -8.18$$
$$-16.3 + 16.3 - 7.2y = -16.3 + (-8.18) \quad \text{Adding } -16.3 \text{ on both sides}$$
$$-7.2y = -24.48$$
$$\frac{-7.2y}{-7.2} = \frac{-24.48}{-7.2} \quad \text{Dividing by } -7.2 \text{ on both sides}$$
$$y = 3.4$$

Check:
$$\begin{array}{c|c} 16.3 - 7.2y = -8.18 \\ \hline 16.3 - 7.2(3.4) \; ? \; -8.18 \\ 16.3 - 24.48 \\ -8.18 \; | \; \text{TRUE} \end{array}$$

The solution is 3.4.

Do Exercises 5 and 6.

Solve.
5. $-4 - 8x = 8$
6. $41.68 = 4.7 - 8.6y$

b COLLECTING LIKE TERMS

SKILL REVIEW

Collect like terms. [1.7e]
Collect like terms.
1. $q + 5t - 1 + 5q - t$
2. $7d + 16 - 11w - 2 - 10d$

Answers: **1.** $6q + 4t - 1$
2. $-3d + 14 - 11w$

MyLab Math VIDEO

If there are like terms on one side of the equation, we collect them before using the addition principle or the multiplication principle.

EXAMPLE 5 Solve: $3x + 4x = -14$.

$$3x + 4x = -14$$
$$7x = -14 \quad \text{Collecting like terms}$$
$$\frac{7x}{7} = \frac{-14}{7} \quad \text{Dividing by 7 on both sides}$$
$$x = -2$$

The number -2 checks, so the solution is -2.

Do Exercises 7 and 8.

Solve.
7. $4x + 3x = -21$
8. $x - 0.09x = 728$

Answers
4. 39 **5.** $-\frac{3}{2}$ **6.** -4.3 **7.** -3 **8.** 800

Guided Solution:
4. 18, $-m$, -1, m

If there are like terms on opposite sides of the equation, we get them on the same side by using the addition principle. Then we collect them. In other words, we get all the terms with a variable on one side of the equation and all the terms without a variable on the other side.

EXAMPLE 6 Solve: $2x - 2 = -3x + 3$.

$$2x - 2 = -3x + 3$$
$$2x - 2 + 2 = -3x + 3 + 2 \quad \text{Adding 2}$$
$$2x = -3x + 5 \quad \text{Simplifying}$$
$$2x + 3x = -3x + 3x + 5 \quad \text{Adding } 3x$$
$$5x = 5 \quad \text{Simplifying}$$
$$\frac{5x}{5} = \frac{5}{5} \quad \text{Dividing by 5}$$
$$x = 1 \quad \text{Simplifying}$$

Check:
$$\begin{array}{c|c} 2x - 2 = -3x + 3 \\ \hline 2 \cdot 1 - 2 \ ? \ -3 \cdot 1 + 3 & \text{Substituting in the} \\ 2 - 2 \ | \ -3 + 3 & \text{original equation} \\ 0 \ | \ 0 & \text{TRUE} \end{array}$$

The solution is 1.

◀ **Do Exercises 9 and 10.**

Solve.

9. $7y + 5 = 2y + 10$

10. $5 - 2y = 3y - 5$

In Example 6, we used the addition principle to get all the terms with an x on one side of the equation and all the terms without an x on the other side. Then we collected like terms and proceeded as before. If there are like terms on one side at the outset, they should be collected first.

EXAMPLE 7 Solve: $6x + 5 - 7x = 10 - 4x + 3$.

$$6x + 5 - 7x = 10 - 4x + 3$$
$$-x + 5 = 13 - 4x \quad \text{Collecting like terms}$$
$$4x - x + 5 = 13 - 4x + 4x \quad \text{Adding } 4x \text{ to get all terms with a variable on one side}$$
$$3x + 5 = 13 \quad \text{Simplifying; that is, collecting like terms}$$
$$3x + 5 - 5 = 13 - 5 \quad \text{Subtracting 5}$$
$$3x = 8 \quad \text{Simplifying}$$
$$\frac{3x}{3} = \frac{8}{3} \quad \text{Dividing by 3}$$
$$x = \frac{8}{3} \quad \text{Simplifying}$$

The number $\frac{8}{3}$ checks, so it is the solution.

◀ **Do Exercises 11 and 12.**

Solve.

11. $7x - 17 + 2x = 2 - 8x + 15$

$\square \cdot x - 17 = 17 - 8x$
$8x + 9x - 17 = 17 - 8x + \square$
$\square \cdot x - 17 = 17$
$17x - 17 + 17 = 17 + \square$
$17x = 34$
$\frac{17x}{17} = \frac{34}{17}$
$\square = 2$

12. $3x - 15 = 5x + 2 - 4x$

Clearing Fractions and Decimals

In general, equations are easier to solve if they do not contain fractions or decimals. Consider, for example, the equations

$$\frac{1}{2}x + 5 = \frac{3}{4} \quad \text{and} \quad 2.3x + 7 = 5.4.$$

Answers

9. 1   10. 2   11. 2   12. $\frac{17}{2}$

Guided Solution:
11. 9, 8x, 17, 17, 17, x

If we multiply by 4 on both sides of the first equation and by 10 on both sides of the second equation, we have

$$4\left(\frac{1}{2}x + 5\right) = 4 \cdot \frac{3}{4} \quad \text{and} \quad 10(2.3x + 7) = 10 \cdot 5.4$$

$$4 \cdot \frac{1}{2}x + 4 \cdot 5 = 4 \cdot \frac{3}{4} \quad \text{and} \quad 10 \cdot 2.3x + 10 \cdot 7 = 10 \cdot 5.4$$

$$2x + 20 = 3 \quad \text{and} \quad 23x + 70 = 54.$$

The first equation has been "cleared of fractions" and the second equation has been "cleared of decimals." Both resulting equations are equivalent to the original equations and are easier to solve. *It is your choice* whether to clear fractions or decimals, but doing so often eases computations.

The easiest way to clear an equation of fractions is to multiply *every term on both sides* by the **least common multiple of all the denominators**.

EXAMPLE 8 Solve: $\frac{2}{3}x - \frac{1}{6} + \frac{1}{2}x = \frac{7}{6} + 2x$.

The denominators are 3, 6, and 2. The number 6 is the least common multiple of all the denominators. We multiply by 6 on both sides of the equation.

$6\left(\frac{2}{3}x - \frac{1}{6} + \frac{1}{2}x\right) = 6\left(\frac{7}{6} + 2x\right)$ Multiplying by 6 on both sides

$6 \cdot \frac{2}{3}x - 6 \cdot \frac{1}{6} + 6 \cdot \frac{1}{2}x = 6 \cdot \frac{7}{6} + 6 \cdot 2x$ Using the distributive law (*Caution!* Be sure to multiply *all* the terms by 6.)

$4x - 1 + 3x = 7 + 12x$ Simplifying. Note that the fractions are cleared.

$7x - 1 = 7 + 12x$ Collecting like terms

$7x - 1 - 12x = 7 + 12x - 12x$ Subtracting $12x$

$-5x - 1 = 7$ Collecting like terms

$-5x - 1 + 1 = 7 + 1$ Adding 1

$-5x = 8$ Collecting like terms

$\frac{-5x}{-5} = \frac{8}{-5}$ Dividing by -5

$x = -\frac{8}{5}$

Check:
$$\frac{2}{3}x - \frac{1}{6} + \frac{1}{2}x = \frac{7}{6} + 2x$$

$$\frac{2}{3}\left(-\frac{8}{5}\right) - \frac{1}{6} + \frac{1}{2}\left(-\frac{8}{5}\right) \;?\; \frac{7}{6} + 2\left(-\frac{8}{5}\right)$$

$$-\frac{16}{15} - \frac{1}{6} - \frac{8}{10} \;\bigg|\; \frac{7}{6} - \frac{16}{5}$$

$$-\frac{32}{30} - \frac{5}{30} - \frac{24}{30} \;\bigg|\; \frac{35}{30} - \frac{96}{30}$$

$$-\frac{61}{30} \;\bigg|\; -\frac{61}{30} \quad \text{TRUE}$$

········ **Caution!** ········

Check the possible solution in the *original* equation rather than in the equation that has been cleared of fractions.

The solution is $-\frac{8}{5}$.

CALCULATOR CORNER

Checking Possible Solutions There are several ways to check the possible solutions of an equation on a calculator. One of the most straightforward methods is to substitute and carry out the calculations on each side of the equation just as we do when we check by hand. To check the possible solution, 1, in Example 6, for instance, we first substitute 1 for *x* in the expression on the left side of the equation. We get 0. Next, we substitute 1 for *x* in the expression on the right side of the equation. Again we get 0. Since the two sides of the equation have the same value when *x* is 1, we know that 1 is the solution of the equation.

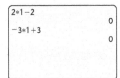

EXERCISE:
1. Use substitution to check the solutions found in Examples 1–5.

13. Solve: $\frac{7}{8}x - \frac{1}{4} + \frac{1}{2}x = \frac{3}{4} + x$.

$\frac{7}{8}x - \frac{1}{4} + \frac{1}{2}x = \frac{3}{4} + x$

$8 \cdot \left(\frac{7}{8}x - \frac{1}{4} + \frac{1}{2}x\right) = \boxed{} \cdot \left(\frac{3}{4} + x\right)$

$8 \cdot \frac{7}{8}x - \boxed{} \cdot \frac{1}{4} + 8 \cdot \frac{1}{2}x$

$ = 8 \cdot \frac{3}{4} + \boxed{} \cdot x$

$\boxed{}\, x - \boxed{} + 4x = 6 + 8x$

$\boxed{}\, x - 2 = 6 + 8x$

$11x - 2 - 8x = 6 + 8x - \boxed{}$

$3x - 2 = \boxed{}$

$3x - 2 + \boxed{} = 6 + 2$

$3x = \boxed{}$

$\frac{3x}{3} = \frac{8}{\boxed{}}$

$x = \frac{8}{3}$

14. Solve: $41.68 = 4.7 - 8.6y$.

◀ Do Exercise 13.

To illustrate clearing decimals, we repeat Example 4, but this time we clear the equation of decimals first. Compare the methods.

To clear an equation of decimals, we count the greatest number of decimal places in any one number. If the greatest number of decimal places is 1, we multiply every term on both sides by 10; if it is 2, we multiply by 100; and so on.

EXAMPLE 9 Solve: $16.3 - 7.2y = -8.18$.

The greatest number of decimal places in any one number is *two*. Multiplying by 100, which has *two* 0's, will clear all decimals.

$100(16.3 - 7.2y) = 100(-8.18)$ Multiplying by 100 on both sides

$100(16.3) - 100(7.2y) = 100(-8.18)$ Using the distributive law

$1630 - 720y = -818$ Simplifying

$1630 - 720y - 1630 = -818 - 1630$ Subtracting 1630

$-720y = -2448$ Collecting like terms

$\frac{-720y}{-720} = \frac{-2448}{-720}$ Dividing by -720

$y = \frac{17}{5}$, or 3.4

The number $\frac{17}{5}$, or 3.4, checks, as shown in Example 4, so it is the solution.

◀ Do Exercise 14.

C   EQUATIONS CONTAINING PARENTHESES

SKILL REVIEW Simplify expressions by removing parentheses. [1.8b]
Simplify.
1. $4(x - 3) - 5$ **2.** $6 - 9(x + 2)$ **3.** $x + 5(6x - 2)$

Answers: **1.** $4x - 17$ **2.** $-9x - 12$ **3.** $31x - 10$

To solve certain kinds of equations that contain parentheses, we first use the distributive laws to remove the parentheses. Then we proceed as before.

EXAMPLE 10 Solve: $8x = 2(12 - 2x)$.

$8x = 2(12 - 2x)$

$8x = 24 - 4x$ Using the distributive law to multiply and remove parentheses

$8x + 4x = 24 - 4x + 4x$ Adding $4x$ to get all the x-terms on one side

$12x = 24$ Collecting like terms

$\frac{12x}{12} = \frac{24}{12}$ Dividing by 12

$x = 2$

The number 2 checks, so the solution is 2.

◀ Do Exercises 15 and 16.

Solve.
15. $2(2y + 3) = 14$

16. $5(3x - 2) = 35$

Answers
13. $\frac{8}{3}$ **14.** $-\frac{43}{10}$, or -4.3 **15.** 2   **16.** 3

Guided Solution:
13. 8, 8, 8, 7, 2, 11, 8x, 6, 2, 8, 3

Here is a procedure for solving the types of equation discussed in this section.

> **AN EQUATION-SOLVING PROCEDURE**
> 1. Multiply on both sides to clear the equation of fractions or decimals. (This is optional, but it can ease computations.)
> 2. If parentheses occur, multiply to remove them using the *distributive laws*.
> 3. Collect like terms on each side, if necessary.
> 4. Get all terms with variables on one side and all numbers (constant terms) on the other side, using the *addition principle*.
> 5. Collect like terms again, if necessary.
> 6. Multiply or divide to solve for the variable, using the *multiplication principle*.
> 7. Check all possible solutions in the original equation.

EXAMPLE 11 Solve: $2 - 5(x + 5) = 3(x - 2) - 1$.

$$2 - 5(x + 5) = 3(x - 2) - 1$$
$$2 - 5x - 25 = 3x - 6 - 1 \quad \text{Using the distributive law to multiply and remove parentheses}$$
$$-5x - 23 = 3x - 7 \quad \text{Collecting like terms}$$
$$-5x - 23 + 5x = 3x - 7 + 5x \quad \text{Adding } 5x$$
$$-23 = 8x - 7 \quad \text{Collecting like terms}$$
$$-23 + 7 = 8x - 7 + 7 \quad \text{Adding } 7$$
$$-16 = 8x \quad \text{Collecting like terms}$$
$$\frac{-16}{8} = \frac{8x}{8} \quad \text{Dividing by 8}$$
$$-2 = x$$

Check:
$$\begin{array}{c|c} 2 - 5(x + 5) = 3(x - 2) - 1 \\ \hline 2 - 5(-2 + 5) \;?\; 3(-2 - 2) - 1 \\ 2 - 5(3) & 3(-4) - 1 \\ 2 - 15 & -12 - 1 \\ -13 & -13 \quad \text{TRUE} \end{array}$$

The solution is -2.

Do Exercises 17 and 18. ▶

Equations with Infinitely Many Solutions

The types of equation that we have considered thus far in Sections 2.1–2.3 have all had exactly one solution. We now look at two other possibilities.
Consider
$$3 + x = x + 3.$$

Let's explore the equation and possible solutions in Margin Exercises 19–22.

Do Exercises 19–22. ▶

Solve.
17. $3(7 + 2x) = 30 + 7(x - 1)$
18. $4(3 + 5x) - 4 = 3 + 2(x - 2)$

Determine whether the given number is a solution of the given equation.
19. 10; $3 + x = x + 3$
20. -7; $3 + x = x + 3$
21. $\frac{1}{2}$; $3 + x = x + 3$
22. 0; $3 + x = x + 3$

Answers
17. -2 18. $-\frac{1}{2}$ 19. Yes 20. Yes
21. Yes 22. Yes

We know by the commutative law of addition that the equation $3 + x = x + 3$ holds for any replacement of x with a real number. (See Section 1.7.) We have confirmed some of these solutions in Margin Exercises 19–22. Suppose that we try to solve this equation using the addition principle:

$$3 + x = x + 3$$
$$-x + 3 + x = -x + x + 3 \quad \text{Adding } -x$$
$$3 = 3. \quad \text{True}$$

We end with a true equation. The original equation holds for all real-number replacements. Every real number is a solution. Thus the number of solutions is **infinite**.

EXAMPLE 12 Solve: $7x - 17 = 4 + 7(x - 3)$.

$$7x - 17 = 4 + 7(x - 3)$$
$$7x - 17 = 4 + 7x - 21 \quad \text{Using the distributive law to multiply and remove parentheses}$$
$$7x - 17 = 7x - 17 \quad \text{Collecting like terms}$$
$$-7x + 7x - 17 = -7x + 7x - 17 \quad \text{Adding } -7x$$
$$-17 = -17 \quad \text{True for all real numbers}$$

Every real number is a solution. There are infinitely many solutions.

Equations with No Solution

Now consider

$$3 + x = x + 8.$$

Let's explore the equation and possible solutions in Margin Exercises 23–26.

◀ Do Exercises 23–26.

None of the replacements in Margin Exercises 23–26 is a solution of the given equation. In fact, there are no solutions. Let's try to solve this equation using the addition principle:

$$3 + x = x + 8$$
$$-x + 3 + x = -x + x + 8 \quad \text{Adding } -x$$
$$3 = 8. \quad \text{False}$$

We end with a false equation. The original equation is false for all real-number replacements. Thus it has **no** solution.

EXAMPLE 13 Solve: $3x + 4(x + 2) = 11 + 7x$.

$$3x + 4(x + 2) = 11 + 7x$$
$$3x + 4x + 8 = 11 + 7x \quad \text{Using the distributive law to multiply and remove parentheses}$$
$$7x + 8 = 11 + 7x \quad \text{Collecting like terms}$$
$$7x + 8 - 7x = 11 + 7x - 7x \quad \text{Subtracting } 7x$$
$$8 = 11 \quad \text{False}$$

There are no solutions.

◀ Do Exercises 27 and 28.

Determine whether the given number is a solution of the given equation.

23. 10; $3 + x = x + 8$

24. -7; $3 + x = x + 8$

25. $\frac{1}{2}$; $3 + x = x + 8$

26. 0; $3 + x = x + 8$

Solve.

27. $30 + 5(x + 3) = -3 + 5x + 48$

28. $2x + 7(x - 4) = 13 + 9x$

When solving an equation, if the result is:

- an equation of the form $x = a$, where a is a real number, then there is one solution, the number a;
- a true equation like $3 = 3$ or $-1 = -1$, then every real number is a solution;
- a false equation like $3 = 8$ or $-4 = 5$, then there is no solution.

Answers
23. No 24. No 25. No 26. No
27. All real numbers 28. No solution

2.3 Exercise Set

FOR EXTRA HELP — MyLab Math

✓ Check Your Understanding

Reading Check Choose from the column on the right the most appropriate word to complete each statement. Not every word will be used.

RC1. When solving equations, we may need to _____ like terms.

RC2. When solving equations, we may wish to _____ fractions.

RC3. We can remove parentheses using the _____ laws.

RC4. We use the _____ principle when dividing both sides of an equation by −1.

clear
addition
distributive
collect
multiplication
commutative

Concept Check Choose from the column on the right the operation that will clear each equation of fractions or decimals.

CC1. $\frac{2}{5}x - 5 + \frac{1}{2}x = \frac{3}{10} + x$

CC2. $0.003y - 0.1 = 0.03 + y$

CC3. $\frac{1}{4} - 8t + \frac{5}{6} = t - \frac{1}{12}$

CC4. $\frac{1}{2}y + \frac{1}{3} = \frac{2}{5}y$

CC5. $\frac{3}{5} - x = \frac{2}{7}x + 4$

a) Multiply by 1000 on both sides.
b) Multiply by 35 on both sides.
c) Multiply by 12 on both sides.
d) Multiply by 10 on both sides.
e) Multiply by 30 on both sides.

a Solve. Don't forget to check!

1. $5x + 6 = 31$
   Check: $5x + 6 = 31$?

2. $7x + 6 = 13$
   Check: $7x + 6 = 13$?

3. $8x + 4 = 68$
   Check: $8x + 4 = 68$?

4. $4y + 10 = 46$
   Check: $4y + 10 = 46$?

5. $4x - 6 = 34$

6. $5y - 2 = 53$

7. $3x - 9 = 33$

8. $4x - 19 = 5$

9. $7x + 2 = -54$

10. $5x + 4 = -41$

11. $-45 = 3 + 6y$

12. $-91 = 9t + 8$

13. $-4x + 7 = 35$

14. $-5x - 7 = 108$

15. $\frac{5}{4}x - 18 = -3$

16. $\frac{3}{2}x - 24 = -36$

SECTION 2.3 Using the Principles Together

b Solve.

17. $5x + 7x = 72$
Check: $\dfrac{5x + 7x = 72}{?}$

18. $8x + 3x = 55$
Check: $\dfrac{8x + 3x = 55}{?}$

19. $8x + 7x = 60$
Check: $\dfrac{8x + 7x = 60}{?}$

20. $8x + 5x = 104$
Check: $\dfrac{8x + 5x = 104}{?}$

21. $4x + 3x = 42$

22. $7x + 18x = 125$

23. $-6y - 3y = 27$

24. $-5y - 7y = 144$

25. $-7y - 8y = -15$

26. $-10y - 3y = -39$

27. $x + \dfrac{1}{3}x = 8$

28. $x + \dfrac{1}{4}x = 10$

29. $10.2y - 7.3y = -58$

30. $6.8y - 2.4y = -88$

31. $8y - 35 = 3y$

32. $4x - 6 = 6x$

33. $8x - 1 = 23 - 4x$

34. $5y - 2 = 28 - y$

35. $2x - 1 = 4 + x$

36. $4 - 3x = 6 - 7x$

37. $6x + 3 = 2x + 11$

38. $14 - 6a = -2a + 3$

39. $5 - 2x = 3x - 7x + 25$

40. $-7z + 2z - 3z - 7 = 17$

41. $4 + 3x - 6 = 3x + 2 - x$

42. $5 + 4x - 7 = 4x - 2 - x$

43. $4y - 4 + y + 24 = 6y + 20 - 4y$

44. $5y - 7 + y = 7y + 21 - 5y$

Solve. Clear fractions or decimals first.

45. $\dfrac{7}{2}x + \dfrac{1}{2}x = 3x + \dfrac{3}{2} + \dfrac{5}{2}x$

46. $\dfrac{7}{8}x - \dfrac{1}{4} + \dfrac{3}{4}x = \dfrac{1}{16} + x$

47. $\dfrac{2}{3} + \dfrac{1}{4}t = \dfrac{1}{3}$

48. $-\dfrac{3}{2} + x = -\dfrac{5}{6} - \dfrac{4}{3}$

49. $\dfrac{2}{3} + 3y = 5y - \dfrac{2}{15}$

50. $\dfrac{1}{2} + 4m = 3m - \dfrac{5}{2}$

51. $\dfrac{5}{3} + \dfrac{2}{3}x = \dfrac{25}{12} + \dfrac{5}{4}x + \dfrac{3}{4}$

52. $1 - \dfrac{2}{3}y = \dfrac{9}{5} - \dfrac{y}{5} + \dfrac{3}{5}$

53. $2.1x + 45.2 = 3.2 - 8.4x$

54. $0.96y - 0.79 = 0.21y + 0.46$

55. $1.03 - 0.62x = 0.71 - 0.22x$

56. $1.7t + 8 - 1.62t = 0.4t - 0.32 + 8$

57. $\dfrac{2}{7}x - \dfrac{1}{2}x = \dfrac{3}{4}x + 1$

58. $\dfrac{5}{16}y + \dfrac{3}{8}y = 2 + \dfrac{1}{4}y$

C Solve.

59. $3(2y - 3) = 27$

60. $8(3x + 2) = 30$

61. $40 = 5(3x + 2)$

62. $9 = 3(5x - 2)$

63. $-23 + y = y + 25$

64. $17 - t = -t + 68$

65. $-23 + x = x - 23$

66. $y - \dfrac{2}{3} = -\dfrac{2}{3} + y$

67. $2(3 + 4m) - 9 = 45$

68. $5x + 5(4x - 1) = 20$

69. $5r - (2r + 8) = 16$

70. $6b - (3b + 8) = 16$

71. $6 - 2(3x - 1) = 2$

72. $10 - 3(2x - 1) = 1$

73. $5(d + 4) = 7(d - 2)$

74. $3(t - 2) = 9(t + 2)$

75. $8(2t + 1) = 4(7t + 7)$

76. $7(5x - 2) = 6(6x - 1)$

77. $5x + 5 - 7x = 15 - 12x + 10x - 10$

78. $3 - 7x + 10x - 14 = 9 - 6x + 9x - 20$

79. $22x - 5 - 15x + 3 = 10x - 4 - 3x + 11$

80. $11x - 6 - 4x + 1 = 9x - 8 - 2x + 12$

SECTION 2.3 Using the Principles Together

81. $3(r - 6) + 2 = 4(r + 2) - 21$

82. $5(t + 3) + 9 = 3(t - 2) + 6$

83. $19 - (2x + 3) = 2(x + 3) + x$

84. $13 - (2c + 2) = 2(c + 2) + 3c$

85. $2[4 - 2(3 - x)] - 1 = 4[2(4x - 3) + 7] - 25$

86. $5[3(7 - t) - 4(8 + 2t)] - 20 = -6[2(6 + 3t) - 4]$

87. $11 - 4(x + 1) - 3 = 11 + 2(4 - 2x) - 16$

88. $6(2x - 1) - 12 = 7 + 12(x - 1)$

89. $22x - 1 - 12x = 5(2x - 1) + 4$

90. $2 + 14x - 9 = 7(2x + 1) - 14$

91. $0.7(3x + 6) = 1.1 - (x + 2)$

92. $0.9(2x + 8) = 20 - (x + 5)$

Skill Maintenance

93. Divide: $-22.1 \div 3.4$. [1.6c]

94. Multiply: $-22.1(3.4)$. [1.5a]

95. Factor: $7x - 21 - 14y$. [1.7d]

96. Factor: $8y - 88x + 8$. [1.7d]

Simplify.

97. $-3 + 2(-5)^2(-3) - 7$ [1.8d]

98. $3x + 2[4 - 5(2x - 1)]$ [1.8c]

99. $23(2x - 4) - 15(10 - 3x)$ [1.8b]

100. $256 \div 64 \div 4^2$ [1.8d]

Synthesis

Solve.

101. $\dfrac{2}{3}\left(\dfrac{7}{8} - 4x\right) - \dfrac{5}{8} = \dfrac{3}{8}$

102. $\dfrac{1}{4}(8y + 4) - 17 = -\dfrac{1}{2}(4y - 8)$

103. $\dfrac{4 - 3x}{7} = \dfrac{2 + 5x}{49} - \dfrac{x}{14}$

104. The width of a rectangle is 5 ft, its length is $(3x + 2)$ ft, and its area is 75 ft². Find x.

Formulas

2.4

a EVALUATING FORMULAS

A **formula** is a "recipe" for doing a certain type of calculation. Formulas are often given as equations. When we replace the variables in an equation with numbers and calculate the result, we are **evaluating** the formula.

Let's consider a formula that has to do with weather. Suppose that you see a flash of lightning during a storm. Then a few seconds later, you hear thunder. Your distance from the place where the lightning struck is given by the formula $M = \frac{1}{5}t$, where t is the number of seconds from the lightning flash to the sound of the thunder and M is in miles.

OBJECTIVES

a Evaluate a formula.

b Solve a formula for a specified letter.

EXAMPLE 1 *Distance from Lightning.* Consider the formula $M = \frac{1}{5}t$. Suppose that it takes 10 sec for the sound of thunder to reach you after you have seen a flash of lightning. How far away did the lightning strike?

We substitute 10 for t and calculate M:

$$M = \tfrac{1}{5}t = \tfrac{1}{5}(10) = 2.$$

The lightning struck 2 mi away.

Do Exercise 1.

1. *Storm Distance.* Refer to Example 1. Suppose that it takes the sound of thunder 14 sec to reach you. How far away is the storm?

EXAMPLE 2 *Cost of Operating a Microwave Oven.* The cost C of operating a microwave oven for 1 year is given by the formula

$$C = \frac{W \times h \times 365}{1000} \cdot k,$$

where W = the wattage, h = the number of hours used per day, and k = the energy cost per kilowatt-hour. Find the cost of operating a 1500-W microwave oven for 0.25 hr per day if the energy cost is $0.13 per kilowatt-hour.

Substituting, we have

$$C = \frac{W \times h \times 365}{1000} \cdot k = \frac{1500 \times 0.25 \times 365}{1000} \cdot \$0.13 \approx \$17.79.$$

The cost for operating a 1500-W microwave oven for 0.25 hr per day for 1 year is about $17.79.

Do Exercise 2.

2. *Microwave Oven.* Refer to Example 2. Determine the cost of operating an 1100-W microwave oven for 0.5 hr per day for 1 year if the energy cost is $0.16 per kilowatt-hour.

Answers

1. 2.8 mi   **2.** $32.12

3. *Socks from Cotton.* Refer to Example 3. Determine the number of socks that can be made from 65 bales of cotton.

EXAMPLE 3 *Socks from Cotton.* Consider the formula $S = 4321x$, where S is the number of socks of average size that can be produced from x bales of cotton. You see a shipment of 300 bales of cotton taken off a ship. How many socks can be made from the cotton?
Data: *Country Woman Magazine*

We substitute 300 for x and calculate S:

$$S = 4321x = 4321(300) = 1{,}296{,}300.$$

Thus, 1,296,300 socks can be made from 300 bales of cotton.

◀ Do Exercise 3.

b SOLVING FORMULAS

SKILL REVIEW   Solve equations.   [2.3a]
Solve.

1. $28 = 7 - 3a$ 2. $\dfrac{1}{2}x - 22 = -20$

Answers: **1.** −7 **2.** 4

MyLab Math
VIDEO

Refer to Example 3. Suppose that a clothing company wants to produce S socks and needs to know how many bales of cotton to order. If this calculation is to be repeated many times, it might be helpful to first solve the formula for x:

$$S = 4321x$$

$$\dfrac{S}{4321} = x. \quad \text{Dividing by 4321}$$

Then we can substitute a number for S and calculate x. For example, if the number of socks S to be produced is 432,100, then

$$x = \dfrac{S}{4321} = \dfrac{432{,}100}{4321} = 100.$$

The company would need to order 100 bales of cotton.

EXAMPLE 4 Solve for z: $H = \tfrac{1}{4}z$.

$H = \tfrac{1}{4}z$ We want this letter alone.
$4 \cdot H = 4 \cdot \tfrac{1}{4}z$ Multiplying by 4 on both sides
$4H = z$

4. Solve for q: $B = \dfrac{1}{3}q$.

EXAMPLE 5 *Distance, Rate, and Time.* Solve for t: $d = rt$.

5. Solve for m: $n = mz$.

$d = rt$ We want this letter alone.
$\dfrac{d}{r} = \dfrac{rt}{r}$ Dividing by r

6. *Electricity.* Solve for I:
$V = IR$. (This formula relates voltage V, current I, and resistance R.)

$\dfrac{d}{r} = \dfrac{r}{r} \cdot t$

$\dfrac{d}{r} = t$ Simplifying

Answers
3. 280,865 socks 4. $q = 3B$
5. $m = \dfrac{n}{z}$ 6. $I = \dfrac{V}{R}$

◀ Do Exercises 4–6.

EXAMPLE 6 Solve for x: $y = x + 3$.

$y = x + 3$ We want this letter alone.
$y - 3 = x + 3 - 3$ Subtracting 3
$y - 3 = x$ Simplifying

EXAMPLE 7 Solve for x: $y = x - a$.

$y = x - a$ We want this letter alone.
$y + a = x - a + a$ Adding a
$y + a = x$ Simplifying

Do Exercises 7–9. ▶

Solve for x.
7. $y = x + 5$
8. $y = x - 7$
9. $y = x - b$

EXAMPLE 8 Solve for y: $6y = 3x$.

$6y = 3x$ We want this letter alone.
$\dfrac{6y}{6} = \dfrac{3x}{6}$ Dividing by 6
$y = \dfrac{x}{2}$, or $\dfrac{1}{2}x$ Simplifying

EXAMPLE 9 Solve for y: $by = ax$.

$by = ax$ We want this letter alone.
$\dfrac{by}{b} = \dfrac{ax}{b}$ Dividing by b
$y = \dfrac{ax}{b}$ Simplifying

Do Exercises 10 and 11. ▶

10. Solve for y: $9y = 5x$.
11. Solve for p: $ap = bt$.

EXAMPLE 10 Solve for x: $ax + b = c$.

$ax + b = c$ We want this letter alone.
$ax + b - b = c - b$ Subtracting b
$ax = c - b$ Simplifying
$\dfrac{ax}{a} = \dfrac{c - b}{a}$ Dividing by a
$x = \dfrac{c - b}{a}$ Simplifying

Do Exercises 12 and 13. ▶

GS 12. Solve for x: $y = mx + b$.
$y = mx + b$
$y - \boxed{} = mx + b - b$
$y - b = \boxed{}$
$\dfrac{y - b}{m} = \dfrac{mx}{\boxed{}}$
$\dfrac{y - b}{m} = \boxed{}$

13. Solve for Q: $tQ - p = a$.

Answers
7. $x = y - 5$ 8. $x = y + 7$
9. $x = y + b$ 10. $y = \dfrac{5x}{9}$, or $\dfrac{5}{9}x$
11. $p = \dfrac{bt}{a}$ 12. $x = \dfrac{y - b}{m}$
13. $Q = \dfrac{a + p}{t}$

Guided Solution:
12. b, mx, m, x

A FORMULA-SOLVING PROCEDURE

To solve a formula for a given letter, identify the letter and:

1. Multiply on both sides to clear fractions or decimals, if that is needed.
2. Collect like terms on each side, if necessary.
3. Get all terms with the letter to be solved for on one side of the equation and all other terms on the other side.
4. Collect like terms again, if necessary.
5. Solve for the letter in question.

EXAMPLE 11 *Circumference.* Solve for r: $C = 2\pi r$. This is a formula for the circumference C of a circle of radius r.

$$C = 2\pi r \quad \text{We want this letter alone.}$$
$$\frac{C}{2\pi} = \frac{2\pi r}{2\pi} \quad \text{Dividing by } 2\pi$$
$$\frac{C}{2\pi} = r$$

14. *Circumference.* Solve for D:
$$C = \pi D.$$
This is a formula for the circumference C of a circle of diameter D.

EXAMPLE 12 *Averages.* Solve for a: $A = \dfrac{a + b + c}{3}$. This is a formula for the average A of three numbers a, b, and c.

$$A = \frac{a + b + c}{3} \quad \text{We want the letter } a \text{ alone.}$$
$$3 \cdot A = 3 \cdot \frac{a + b + c}{3} \quad \text{Multiplying by 3 on both sides}$$
$$3A = a + b + c \quad \text{Simplifying}$$
$$3A - b - c = a \quad \text{Subtracting } b \text{ and } c$$

15. *Averages.* Solve for c:
$$A = \frac{a + b + c + d}{4}.$$

◀ Do Exercises 14 and 15.

Answers

14. $D = \dfrac{C}{\pi}$ **15.** $c = 4A - a - b - d$

2.4 Exercise Set

FOR EXTRA HELP — MyLab Math

✓ Check Your Understanding

Reading Check Determine whether each statement is true or false.

RC1. A formula is often given as an equation.

RC2. Evaluating a formula is the same as solving for a letter.

RC3. If x appears on both sides of the equals sign in a formula, that formula is not solved for x.

Concept Check Determine whether each formula is solved for n.

CC1. $n = \dfrac{p}{f}$ **CC2.** $n = 3xn - y$ **CC3.** $t^2 + 3t + 7 = n$

 Solve.

1. *Wavelength of a Musical Note.* The wavelength w, in meters per cycle, of a musical note is given by
$$w = \frac{r}{f},$$
where r is the speed of the sound, in meters per second, and f is the frequency, in cycles per second. The speed of sound in air is 344 m/sec. What is the wavelength of a note whose frequency in air is 24 cycles per second?

2. *Furnace Output.* Contractors in the Northeast use the formula $B = 30a$ to determine the minimum furnace output B, in British thermal units (Btu's), for a well-insulated house with a square feet of flooring. Determine the minimum furnace output for an 1800-ft^2 house that is well insulated.

Data: U.S. Department of Energy

3. *Calorie Density.* The calorie density D, in calories per ounce, of a food that contains c calories and weighs w ounces is given by
$$D = \frac{c}{w}.$$
Eight ounces of fat-free milk contains 84 calories. Find the calorie density of fat-free milk.

Data: *Nutrition Action Healthletter*, March 2000, p. 9.

4. *Size of a League Schedule.* When all n teams in a league play every other team twice, a total of N games are played, where
$$N = n^2 - n.$$
A soccer league has 7 teams and all teams play each other twice. How many games are played?

5. *Distance, Rate, and Time.* The distance d that a car will travel at a rate, or speed, r in time t is given by
$$d = rt.$$
a) A car travels at 75 miles per hour (mph) for 4.5 hr. How far will it travel?
b) Solve the formula for t.

6. *Surface Area of a Cube.* The surface area A of a cube with side s is given by
$$A = 6s^2.$$

a) Find the surface area of a cube with sides of 3 in.
b) Solve the formula for s^2.

SECTION 2.4 Formulas

7. *College Enrollment.* At many colleges, the number of "full-time-equivalent" students f is given by

$$f = \frac{n}{15},$$

where n is the total number of credits for which students have enrolled in a given semester.
a) Determine the number of full-time-equivalent students on a campus in which students registered for a total of 21,345 credits.
b) Solve the formula for n.

8. *Electrical Power.* The power rating P, in watts, of an electrical appliance is determined by

$$P = I \cdot V,$$

where I is the current, in amperes, and V is measured in volts.
a) A microwave oven requires 12 amps of current and the voltage in the house is 115 volts. What is the wattage of the microwave?
b) Solve the formula for I; for V.

b Solve for the indicated letter.

9. $y = 5x$, for x

10. $d = 55t$, for t

11. $a = bc$, for c

12. $y = mx$, for x

13. $n = m + 11$, for m

14. $z = t + 21$, for t

15. $y = x - \frac{3}{5}$, for x

16. $y = x - \frac{2}{3}$, for x

17. $y = 13 + x$, for x

18. $t = 6 + s$, for s

19. $y = x + b$, for x

20. $y = x + A$, for x

21. $y = 5 - x$, for x

22. $y = 10 - x$, for x

23. $y = a - x$, for x

24. $y = q - x$, for x

25. $8y = 5x$, for y

26. $10y = -5x$, for y

27. $By = Ax$, for x

28. $By = Ax$, for y

29. $W = mt + b$, for t

30. $W = mt - b$, for t

31. $y = bx + c$, for x

32. $y = bx - c$, for x

33. *Area of a Parallelogram:*
$A = bh$, for h
(Area A, base b, height h)

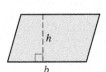

34. *Distance, Rate, Time:*
$d = rt$, for r
(Distance d, speed r, time t)

35. *Perimeter of a Rectangle:*
$P = 2l + 2w$, for w
(Perimeter P, length l, width w)

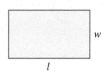

36. *Area of a Circle:*
$A = \pi r^2$, for r^2
(Area A, radius r)

37. *Average of Two Numbers:*
$A = \dfrac{a + b}{2}$, for a

38. *Area of a Triangle:*
$A = \dfrac{1}{2}bh$, for b

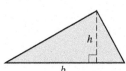

39. $A = \dfrac{a + b + c}{3}$, for b

40. $A = \dfrac{a + b + c}{3}$, for c

41. $A = at + b$, for t

42. $S = rx + s$, for x

43. $Ax + By = c$, for x

44. $Q = \dfrac{p - q}{2}$, for p

45. *Force:*
$F = ma$, for a
(Force F, mass m, acceleration a)

46. *Simple Interest:*
$I = Prt$, for P
(Interest I, principal P, interest rate r, time t)

47. *Relativity:*
$E = mc^2$, for c^2
(Energy E, mass m, speed of light c)

48. $Ax + By = c$, for y

49. $v = \dfrac{3k}{t}$, for t

50. $P = \dfrac{ab}{c}$, for c

Skill Maintenance

51. Evaluate $\dfrac{3x - 2y}{y}$ when $x = 6$ and $y = 2$. [1.1a]

52. Remove parentheses and simplify:
$4a - 8b - 5(5a - 4b)$. [1.8b]

Subtract. [1.4a]

53. $-45.8 - (-32.6)$

54. $-\dfrac{2}{3} - \dfrac{5}{6}$

55. $87\dfrac{1}{2} - 123$

Add. [1.3a]

56. $-\dfrac{5}{12} + \dfrac{1}{4}$

57. $0.082 + (-9.407)$

58. $-2\dfrac{1}{2} + 6\dfrac{1}{4}$

Solve.

59. $2y - 3 + y = 8 - 5y$ [2.3b]

60. $10x + 4 = 3x - 2 + x$ [2.3b]

61. $2(5x + 6) = x - 15$ [2.3c]

62. $5a = 3(6 - 3a)$ [2.3c]

Synthesis

Solve.

63. $H = \dfrac{2}{a - b}$, for b; for a

64. $P = 4m + 7mn$, for m

65. In $A = lw$, if l and w both double, what is the effect on A?

66. In $P = 2a + 2b$, if P doubles, do a and b necessarily both double?

67. In $A = \tfrac{1}{2}bh$, if b increases by 4 units and h does not change, what happens to A?

68. Solve for F: $D = \dfrac{1}{E + F}$.

Mid-Chapter Review

Concept Reinforcement

Determine whether each statement is true or false.

_____ 1. $3 - x = 4x$ and $5x = -3$ are equivalent equations. [2.1b]

_____ 2. For any real numbers a, b, and c, $a = b$ is equivalent to $a + c = b + c$. [2.1b]

_____ 3. We can use the multiplication principle to divide on both sides of an equation by the same nonzero number. [2.2a]

_____ 4. Every equation has at least one solution. [2.3c]

Guided Solutions

GS Fill in each blank with the number, variable, or expression that creates a correct statement or solution.

Solve. [2.1b], [2.2a]

5. $x + 5 = -3$
    $x + 5 - 5 = -3 - \boxed{}$
    $x + \boxed{} = -8$
    $x = \boxed{}$

6. $-6x = 42$
    $\dfrac{-6x}{-6} = \dfrac{42}{\boxed{}}$
    $\boxed{} \cdot x = -7$
    $x = \boxed{}$

7. Solve for y: $5y + z = t$. [2.4b]
    $5y + z = t$
    $5y + z - z = t - \boxed{}$
    $5y = \boxed{}$
    $\dfrac{5y}{5} = \dfrac{t - z}{\boxed{}}$
    $y = \dfrac{\boxed{}}{5}$

Mixed Review

Solve. [2.1b], [2.2a], [2.3a, b, c]

8. $x + 5 = 11$

9. $x + 9 = -3$

10. $8 = t + 1$

11. $-7 = y + 3$

12. $x - 6 = 14$

13. $y - 7 = -2$

14. $-\dfrac{3}{2} + z = -\dfrac{3}{4}$

15. $-3.3 = -1.9 + t$

16. $7x = 42$

17. $17 = -t$

18. $6x = -54$

19. $-5y = -85$

20. $\dfrac{x}{7} = 3$

21. $\dfrac{2}{3}x = 12$

22. $-\dfrac{t}{5} = 3$

23. $\dfrac{3}{4}x = -\dfrac{9}{8}$

24. $3x + 2 = 5$

25. $5x + 4 = -11$

26. $6x - 7 = 2$

27. $-4x - 9 = -5$

28. $6x + 5x = 33$

29. $-3y - 4y = 49$

30. $3x - 4 = 12 - x$

31. $5 - 6x = 9 - 8x$

32. $4y - \dfrac{3}{2} = \dfrac{3}{4} + 2y$

33. $\dfrac{4}{5} + \dfrac{1}{6}t = \dfrac{1}{10}$

34. $0.21n - 1.05 = 2.1 - 0.14n$

35. $5(3y - 1) = -35$

36. $7 - 2(5x + 3) = 1$

37. $-8 + t = t - 8$

38. $z + 12 = -12 + z$

39. $4(3x + 2) = 5(2x - 1)$

40. $8x - 6 - 2x = 3(2x - 4) + 6$

Solve for the indicated letter. [2.4b]

41. $A = 4b$, for b

42. $y = x - 1.5$, for x

43. $n = s - m$, for m

44. $4t = 9w$, for t

45. $B = at - c$, for t

46. $M = \dfrac{x + y + z}{2}$, for y

Understanding Through Discussion and Writing

47. Explain the difference between equivalent expressions and equivalent equations. [1.7a], [2.1b]

48. Are the equations $x = 5$ and $x^2 = 25$ equivalent? Why or why not? [2.1b]

49. When solving an equation using the addition principle, how do you determine which number to add or subtract on both sides of the equation? [2.1b]

50. Explain the following mistake made by a fellow student. [2.1b]

$$x + \dfrac{1}{3} = -\dfrac{5}{3}$$

$$x = -\dfrac{4}{3}$$

51. When solving an equation using the multiplication principle, how do you determine by what number to multiply or divide on both sides of the equation? [2.2a]

52. Devise an application in which it would be useful to solve the equation $d = rt$ for r. [2.4b]

STUDYING FOR SUCCESS   *A Valuable Resource—Your Instructor*

- ☐ Don't be afraid to ask questions in class. Other students probably have the same questions you do.
- ☐ Visit your instructor during office hours if you need additional help.
- ☐ Many instructors welcome e-mails with questions from students.

2.5 Applications of Percent

a TRANSLATING AND SOLVING

OBJECTIVE

a Solve applied problems involving percent.

SKILL REVIEW
Convert between percent notation and decimal notation.   [J15]
1. Convert to decimal notation: 16.7%.
2. Convert to percent notation: 0.0007.

Answers: 1. 0.167 **2.** 0.07%

MyLab Math VIDEO

In solving percent problems, we first *translate* the problem to an equation. The key words in the translation are as follows.

KEY WORDS IN PERCENT TRANSLATIONS

"**Of**" translates to "·" or "×".

"**Is**" translates to "=".

"**What number**" or "**what percent**" translates to any letter.

"**%**" translates to "× $\frac{1}{100}$" or "× 0.01".

EXAMPLE 1   Translate:

28% of 5 is what number?
↓   ↓ ↓ ↓   ↓
28% · 5 = a This is a percent equation.

EXAMPLE 2   Translate:

45% of what number is 28?
↓   ↓   ↓   ↓ ↓
45% × b = 28

EXAMPLE 3   Translate:

What percent of 90 is 7?
↓   ↓ ↓ ↓
n · 90 = 7

Do Exercises 1–6. ▶

Translate to an equation. Do not solve.

1. 13% of 80 is what number?

2. What number is 60% of 70?

3. 43 is 20% of what number?

GS 4. 110% of what number is 30?

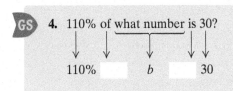

5. 16 is what percent of 80?

6. What percent of 94 is 10.5?

Answers
1. 13% · 80 = a **2.** a = 60% · 70
3. 43 = 20% · b **4.** 110% · b = 30
5. 16 = n · 80   **6.** n · 94 = 10.5
Guided Solution:
4. ·, =

SECTION 2.5   Applications of Percent   **143**

Percent problems are actually of three different types. Although the method we present does *not* require that you be able to identify which type we are studying, it is helpful to know them. Let's begin by using a specific example to find a standard form for a percent problem. We know that

$$15 \text{ is } 25\% \text{ of } 60, \quad \text{or} \quad 15 = 25\% \times 60.$$

We can think of this as:

> Amount = Percent number × Base.

Each of the three types of percent problem depends on which of the three pieces of information is missing in the statement

$$\text{Amount} = \text{Percent number} \times \text{Base}.$$

1. Finding the *amount* (the result of taking the percent)

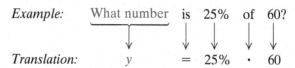

2. Finding the *base* (the number you are taking the percent of)

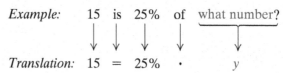

3. Finding the *percent number* (the percent itself)

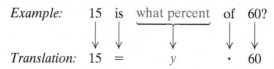

Finding the Amount

EXAMPLE 4 What number is 11% of 49?

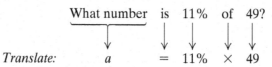

Solve: The letter is by itself. To solve the equation, we need only convert 11% to decimal notation and multiply:

$$a = 11\% \times 49 = 0.11 \times 49 = 5.39.$$

Thus, 5.39 is 11% of 49. The answer is 5.39.

◀ Do Exercise 7.

7. What number is 2.4% of 80?

Finding the Base

EXAMPLE 5 3 is 16% of what number?

Translate: 3 = 16% × b

3 = 0.16 × b Converting 16% to decimal notation

Answer
7. 1.92

Solve: In this case, the letter is not by itself. To solve the equation, we divide by 0.16 on both sides:

$$3 = 0.16 \times b$$
$$\frac{3}{0.16} = \frac{0.16 \times b}{0.16} \quad \text{Dividing by 0.16}$$
$$18.75 = b. \quad \text{Simplifying}$$

The answer is 18.75.

Do Exercise 8.

8. 25.3 is 22% of what number?

$$25.3 \quad\quad 22\% \quad\quad x$$
$$25.3 = \square \cdot x$$
$$\frac{25.3}{\square} = \frac{0.22x}{0.22}$$
$$\square = x$$

Finding the Percent Number

In solving these problems, you *must* remember to convert to percent notation after you have solved the equation.

EXAMPLE 6 $32 is what percent of $50?

$$\text{Translate:} \quad 32 = p \times 50$$

Solve: To solve the equation, we divide by 50 on both sides and convert the answer to percent notation:

$$32 = p \times 50$$
$$\frac{32}{50} = \frac{p \times 50}{50} \quad \text{Dividing by 50}$$
$$0.64 = p$$
$$64\% = p. \quad \text{Converting to percent notation}$$

Thus, $32 is 64% of $50. The answer is 64%.

Do Exercise 9.

9. What percent of $50 is $18?

10. *Haitian Population Ages 0–14.* The population of Haiti is approximately 10,980,000. Of this number, 33.74% are ages 0–14. How many Haitians are ages 14 and younger? Round to the nearest 1000.

Data: worldpopulationreview.com; World Bank

EXAMPLE 7 *Amusement Parks.* Worldwide, amusement parks welcome approximately 940 million guests each year. About 40% of these guests visit amusement parks in North America. How many guests do North American amusement parks welcome each year?
Data: International Association of Amusement Parks and Attractions

To solve this problem, we first reword and then translate. We let n = the annual number of guests visiting North American amusement parks, and we write 940 million as 940,000,000.

Rewording: What number is 40% of 940 million?
Translating: $n = 40\% \times 940{,}000{,}000$

Solve: The letter is by itself. To solve the equation, we need only convert 40% to decimal notation and multiply:

$$n = 40\% \times 940{,}000{,}000$$
$$= 0.40 \times 940{,}000{,}000 = 376{,}000{,}000.$$

Thus North American amusement parks welcome approximately 376 million guests each year.

Do Exercise 10.

Number of Amusement Park Visitors Worldwide

North America 40%
Asia 39%
Europe, Middle East, and Africa 17%
Latin America 3%

DATA: International Association of Amusement Parks and Attractions

Answers
8. 115 **9.** 36% **10.** About 3,705,000
Guided Solution:
8. =, ·, 0.22, 0.22, 115

11. *Areas of Texas and Alaska.* The area of the second largest state, Texas, is 268,581 mi². This is about 40.5% of the area of the largest state, Alaska. What is the area of Alaska?

EXAMPLE 8 *Hospitals.* In 2017, there were 212 federal government hospitals in the United States. This was 3.8% of the total number of hospitals registered in the United States. How many hospitals were registered in the United States in 2017?
Data: American Hospital Association

To solve this problem, we first reword and then translate. We let $H =$ the total number of hospitals registered in the United States in 2017.

Rewording: 212 is 3.8% of what number?

Translating: $212 = 3.8\% \times H$

Solve: To solve the equation, we convert 3.8% to decimal notation and divide by 0.038 on both sides:

$$212 = 3.8\% \times H$$
$$212 = 0.038 \times H \quad \text{Converting to decimal notation}$$
$$\frac{212}{0.038} = \frac{0.038 \times H}{0.038} \quad \text{Dividing by 0.038}$$
$$5579 \approx H. \quad \text{Simplifying and rounding to the nearest one}$$

There were about 5579 hospitals registered in the United States in 2017.

◀ Do Exercise 11.

EXAMPLE 9 *Employment Outlook.* Jobs for registered nurses in the United States totaled approximately 2,751,000 in 2014. This number is expected to grow to 3,190,300 by 2024. What is the percent increase?
Data: *Occupational Outlook Handbook*, U.S. Bureau of Labor Statistics

To solve the problem, we must first determine the amount of the increase:

Jobs in 2024 minus Jobs in 2014 = Increase

$$3,190,300 - 2,751,000 = 439,300.$$

Using the job increase of 439,300, we reword and then translate. We let $p =$ the percent increase. We want to know, "what percent of the number of jobs in 2014 is 439,300?"

Rewording: 439,300 is what percent of 2,751,000?

Translating: $439,300 = p \times 2,751,000$

Solve: To solve the equation, we divide by 2,751,000 on both sides and convert the answer to percent notation:

$$439,300 = p \times 2,751,000$$
$$\frac{439,300}{2,751,000} = \frac{p \times 2,751,000}{2,751,000} \quad \text{Dividing by 2,751,000}$$
$$0.160 \approx p \quad \text{Simplifying}$$
$$16.0\% \approx p. \quad \text{Converting to percent notation}$$

The percent increase is about 16.0%.

◀ Do Exercise 12.

12. *Median Income.* The U.S. median family income in 2007, in 2015 adjusted dollars, was $57,423. This number had decreased to $56,516 in 2015. What is the percent decrease?
Data: Federal Reserve Bank of St. Louis

Answers
11. About 663,163 mi² **12.** The percent decrease is about 1.6%.

2.5 Exercise Set

✓ Check Your Understanding

Reading Check Choose from the column on the right the word that best completes each statement. Choices may be used more than once or not at all.

RC1. The symbol % means _____.

RC2. The word "_____" often translates to multiplication.

RC3. In the statement "15 is 30% of 50," the number 50 is the _____.

RC4. When finding the percent number, we must remember to convert to _____ notation.

of
percent
base
is
sum
difference

Concept Check Choose from the column on the right the most appropriate translation of each question.

CC1. 13 is 82% of what number?

CC2. What number is 13% of 82?

CC3. 82 is what percent of 13?

CC4. 82 is 13% of what number?

CC5. 13 is what percent of 82?

CC6. What number is 82% of 13?

a) $82 = 13\% \cdot b$
b) $a = 13\% \cdot 82$
c) $a = 82\% \cdot 13$
d) $13 = 82\% \cdot b$
e) $82 = p \cdot 13$
f) $13 = p \cdot 82$

a Solve.

1. What percent of 180 is 36?

2. What percent of 76 is 19?

3. 45 is 30% of what number?

4. 20.4 is 24% of what number?

5. What number is 65% of 840?

6. What number is 50% of 50?

7. 30 is what percent of 125?

8. 57 is what percent of 300?

9. 12% of what number is 0.3?

10. 7 is 175% of what number?

11. 2 is what percent of 40?

12. 16 is what percent of 40?

13. What percent of 68 is 17?

14. What percent of 150 is 39?

15. What number is 35% of 240?

16. What number is 1% of one million?

17. What percent of 575 is 138?

18. What percent of 60 is 75?

19. What percent of 300 is 48?

20. What percent of 70 is 70?

21. 14 is 30% of what number?

22. 54 is 24% of what number?

23. What number is 2% of 40?

24. What number is 40% of 2?

25. 0.8 is 16% of what number?

26. 40 is 2% of what number?

27. 54 is 135% of what number?

28. 8 is 2% of what number?

World Population by Continent. It has been projected that in 2050, the world population will be 8909 million, or 8.909 billion. The following circle graph shows the breakdown of this total population by continent.

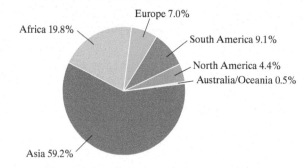

World Population by Continent, 2050

Europe 7.0%
Africa 19.8%
South America 9.1%
North America 4.4%
Australia/Oceania 0.5%
Asia 59.2%

DATA: Central Intelligence Agency

Using the data in the figure, complete the following table of projected populations in 2050. Round to the nearest million.

	Continent	Population		Continent	Population
29.	South America		30.	Europe	
31.	Asia		32.	North America	
33.	Africa		34.	Australia/Oceania	

35. *Coffee Consumption.* Between October 2015 and September 2016, coffee drinkers worldwide consumed 151.3 million 60-kg bags of coffee. Of this, 30.7% was consumed in countries that are exporters of coffee. How many bags of coffee were consumed during this time in countries that are exporters of coffee? Round to the nearest tenth of a million.

Data: International Coffee Organization

36. *eBook Revenue.* In 2016, global revenue from eBook sales totaled approximately $15,870 million. North American revenue accounted for 68.7% of this amount. What was the net revenue from eBook sales in North America in 2016? Round to the nearest million.

Data: statista.com

37. *Student Loans.* To finance her community college education, Sarah takes out a federally subsidized loan for $6500. After a year, Sarah decides to pay off the interest, which is 3.76% of $6500. How much will she pay?

38. *Student Loans.* Paul takes out a PLUS loan for $5000. After a year, Paul decides to pay off the interest, which is 6.31% of $5000. How much will he pay?

39. *Tattoos.* Of the 245,600,000 adults ages 18 and older in the United States, approximately 51,576,000 have at least one tattoo. What percent of adults ages 18 and older have at least one tattoo?

Data: U.S. Census Bureau; Harris Poll of 2016 adults; UPI.com

40. *Boston Marathon.* The 2016 Boston Marathon was the 120th running of the race. Since its first race, the United States has won the men's open division 44 times. What percent of the years did the United States win the men's open?

Data: Boston Athletic Association

41. *Tipping.* William left a $1.50 tip for a meal that cost $12.
   a) What percent of the cost of the meal was the tip?
   b) What was the total cost of the meal including the tip?

42. *Tipping.* Sam, Selena, Rachel, and Clement left a 20% tip for a meal that cost $75.
   a) How much was the tip?
   b) What was the total cost of the meal including the tip?

43. *Tipping.* David left a 15% tip of $4.65 for a meal.
   a) What was the cost of the meal before the tip?
   b) What was the total cost of the meal including the tip?

44. *Tipping.* Addison left an 18% tip of $6.75 for a meal.
   a) What was the cost of the meal before the tip?
   b) What was the total cost of the meal including the tip?

45. *City Park Space.* Portland, Oregon, has 12,959 acres of park space. This is 15.1% of the acreage of the entire city. What is the total acreage of Portland?

Data: Indy Parks and Recreation Master Plan

46. *Junk Mail.* About 46.2 billion pieces of unopened junk mail end up in landfills each year. This is about 44% of all the junk mail that is sent annually. How many pieces of junk mail are sent annually?

Data: Globaljunkmailcrisis.org

47. *Paperback Books.* In the first half of 2016, the revenue from sales of paperback books in the United States was $0.93 billion. This number increased to $1.01 billion in the second half of 2016. What is the percent increase?

Data: ibtimes.com

48. *Bookstores.* Between 2010 and 2016, the number of bookstores in the United States grew from 1651 to 2311. What is the percent increase?

Data: ibtimes.com

49. *Magazine Reading.* In 2010, Americans spent, on average, 24.7 min per day reading magazines. In 2017, they spent, on average, only 16.5 min per day reading magazines. What is the percent decrease?

Data: statista.com

50. *Newsroom Employment.* In 2014, there were 55,000 people employed in newsrooms in the United States. In 2016, this number had decreased to only 32,900. What is the percent decrease?

Data: *State of the News Media 2016*, Pew Research Center

51. *Sports Sponsorship.* In 2014, corporations spent $14.35 billion to sponsor sports in North America. This amount had increased to $15.74 billion in 2016. What is the percent increase?

Data: sponsorship.com

52. *Electric Vehicles.* The range of an electric vehicle is defined as the number of miles that the vehicle can travel between battery charges. The greatest listed range of any electric vehicle in 2014 was 265 mi, and the greatest listed range in 2017 was 315 mi. What is the percent increase?

Data: autotrader.com; caranddriver.com

53. *Adoption.* The number of children in public child welfare who were waiting for adoption declined from 135,000 in 2006 to 112,000 in 2015. What is the percent decrease?

Data: U.S. Department of Health and Human Services

54. *Distance Education.* The number of undergraduate students who were taking at least one online class declined from 7.37 million in 2012 to 5.75 million in 2014. What is the percent decrease?

Data: National Center for Education Statistics

Skill Maintenance

Multiply. [1.7c]

55. $3(4 + q)$

56. $-\dfrac{1}{2}(-10x + 42)$

Simplify. [1.7a]

57. $\dfrac{75yw}{40y}$

58. $-\dfrac{18b}{12b}$

Simplify. [1.8c]

59. $-2[3 - 5(7 - 2)]$

60. $[3(x + 4) - 6] - [8 + 2(x - 5)]$

Synthesis

61. It has been determined that at the age of 15, a boy has reached 96.1% of his final adult height. Jaraan is 6 ft 4 in. at the age of 15. What will his final adult height be?

62. It has been determined that at the age of 10, a girl has reached 84.4% of her final adult height. Dana is 4 ft 8 in. at the age of 10. What will her final adult height be?

Applications and Problem Solving

2.6

OBJECTIVE

a Solve applied problems by translating to equations.

a FIVE STEPS FOR SOLVING PROBLEMS

SKILL REVIEW

Translate phrases to algebraic expressions. [1.1b]
Translate each phrase to an algebraic expression.
1. One-third of a number
2. Two more than a number

Answers: 1. $\frac{1}{3}n$, or $\frac{n}{3}$
2. $x + 2$, or $2 + x$

MyLab Math
VIDEO

We have discussed many new equation-solving tools in this chapter and used them for applications and problem solving. Here we consider a five-step strategy that can be very helpful in solving problems.

FIVE STEPS FOR PROBLEM SOLVING IN ALGEBRA

1. *Familiarize* yourself with the problem situation.
2. *Translate* the problem to an equation.
3. *Solve* the equation.
4. *Check* the answer in the original problem.
5. *State* the answer to the problem clearly.

Of the five steps, the most important is probably the first one: becoming familiar with the problem situation. The following box lists some hints for familiarization.

FAMILIARIZING YOURSELF WITH A PROBLEM

- If a problem is given in words, read it carefully. Reread the problem, perhaps aloud. Try to verbalize the problem as though you were explaining it to someone else.
- Choose a variable (or variables) to represent the unknown(s) and clearly state what each variable represents. Be descriptive! For example, let L = the length, d = the distance, in feet, and so on.
- Make a drawing and label it with known information, using specific units if given. Also, indicate unknown information.
- Find further information. Look up formulas or definitions with which you are not familiar. (Geometric formulas appear on the inside back cover of this text.) Consult the Internet or a reference librarian.
- Create a table that lists all the information you have available. Look for patterns that may help in the translation to an equation.
- Think of a possible answer and check the guess. Note the manner in which the guess is checked.

EXAMPLE 1 *Cycling in Vietnam.* National Highway 1, which runs along the coast of Vietnam, is considered one of the top routes for avid bicyclists. While on sabbatical, a history professor spent six weeks biking 1720 km on National Highway 1 from Hanoi through Ha Tinh to Ho Chi Minh City (commonly known as Saigon). At Ha Tinh, he was four times as far from Ho Chi Minh City as he was from Hanoi. How far had he biked and how far did he still need to bike in order to reach the end?
Data: smh.com; Lonely Planet's Best in 2010

1. **Familiarize.** To become familiar with the problem, we begin by locating a map of Vietnam, such as the one shown at left, and labeling the distance of 1720 km from Hanoi to Ho Chi Minh City. Then we guess a possible distance that the professor is from Hanoi—say, 400 km. Four times 400 km is 1600 km. Since 400 km + 1600 km = 2000 km and 2000 km is greater than 1720 km, we see that our guess is too large. Rather than guess again, let's use the equation-solving skills that we learned in this chapter. We let

   d = the distance, in kilometers, to Hanoi, and

   $4d$ = the distance, in kilometers, to Ho Chi Minh City.

   (We also could let d = the distance to Ho Chi Minh City and $\frac{1}{4}d$ = the distance to Hanoi.) We label these distances on the map.

2. **Translate.** From the map, we see that the lengths of the two parts of the trip must add up to 1720 km. This leads to our translation.

   $$\underbrace{\text{Distance to Hanoi}}_{d} + \underbrace{\text{Distance to Ho Chi Minh City}}_{4d} = 1720$$

3. **Solve.** We solve the equation:

   $$d + 4d = 1720$$
   $$5d = 1720 \quad \text{Collecting like terms}$$
   $$\frac{5d}{5} = \frac{1720}{5} \quad \text{Dividing by 5}$$
   $$d = 344.$$

4. **Check.** As we expected, d is less than 400 km. If $d = 344$ km, then $4d = 1376$ km. Since 344 km + 1376 km = 1720 km, the answer checks.

5. **State.** At Ha Tinh, the professor had biked 344 km from Hanoi and had 1376 km to go to reach Ho Chi Minh City.

◀ Do Exercise 1.

1. *Running.* Yiannis Kouros of Australia holds the record for the greatest distance run in 24 hr by running 188 mi. After 8 hr, he was approximately twice as far from the finish line as he was from the start line. How far had he run?
   Data: Australian Ultra Runners Association

Answer

1. $62\frac{2}{3}$ mi

EXAMPLE 2 *Knitted Scarf.* Lily knitted a scarf with shades of orange and red yarn, starting with an orange section, then a medium-red section, and finally a dark-red section. The medium-red section is one-half the length of the orange section. The dark-red section is one-fourth the length of the orange section. The scarf is 7 ft long. Find the length of each section of the scarf.

1. **Familiarize.** Because the lengths of the medium-red section and the dark-red section are expressed in terms of the length of the orange section, we let

    x = the length of the orange section.

    Then $\frac{1}{2}x$ = the length of the medium-red section

    and $\frac{1}{4}x$ = the length of the dark-red section.

    We make a drawing and label it.

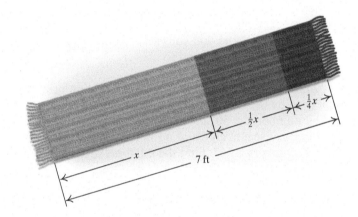

2. **Translate.** From the statement of the problem and the drawing, we know that the lengths add up to 7 ft. This gives us our translation:

    $$\underbrace{x}_{\text{Length of orange section}} + \underbrace{\frac{1}{2}x}_{\text{Length of medium-red section}} + \underbrace{\frac{1}{4}x}_{\text{Length of dark-red section}} = \underbrace{7}_{\text{Total length}}.$$

3. **Solve.** First, we clear fractions and then carry out the solution as follows:

    $$x + \frac{1}{2}x + \frac{1}{4}x = 7 \quad \text{The LCM of the denominators is 4}$$

    $$4\left(x + \frac{1}{2}x + \frac{1}{4}x\right) = 4 \cdot 7 \quad \text{Multiplying by the LCM, 4}$$

    $$4 \cdot x + 4 \cdot \frac{1}{2}x + 4 \cdot \frac{1}{4}x = 4 \cdot 7 \quad \text{Using the distributive law}$$

    $$4x + 2x + x = 28 \quad \text{Simplifying}$$

    $$7x = 28 \quad \text{Collecting like terms}$$

    $$\frac{7x}{7} = \frac{28}{7} \quad \text{Dividing by 7}$$

    $$x = 4.$$

2. *Gourmet Sandwiches.* A sandwich shop specializes in sandwiches prepared in buns of length 18 in. Jenny, Emma, and Sarah buy one of these sandwiches and take it back to their apartment. Since they have different appetites, Jenny cuts the sandwich in such a way that Emma gets one-half of what Jenny gets and Sarah gets three-fourths of what Jenny gets. Find the length of each person's sandwich.

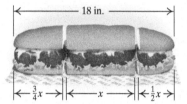

4. **Check.** Do we have an answer to the *original problem*? If the length of the orange section is 4 ft, then the length of the medium-red section is $\frac{1}{2} \cdot 4$ ft, or 2 ft, and the length of the dark-red section is $\frac{1}{4} \cdot 4$ ft, or 1 ft. The sum of these lengths is 7 ft, so the answer checks.

5. **State.** The length of the orange section is 4 ft, the length of the medium-red section is 2 ft, and the length of the dark-red section is 1 ft. (Note that we must include the unit, feet, in the answer.)

◀ Do Exercise 2.

Recall that the set of integers = $\{\ldots, -5, -4, -3, -2, -1, 0, 1, 2, 3, 4, 5, \ldots\}$. Before we solve the next problem, we need to learn some additional terminology regarding integers.

The following are examples of **consecutive integers:** 16, 17, 18, 19, 20; and −31, −30, −29, −28. Note that consecutive integers can be represented in the form $x, x + 1, x + 2$, and so on.

The following are examples of **consecutive even integers:** 16, 18, 20, 22, 24; and −52, −50, −48, −46. Note that consecutive even integers can be represented in the form $x, x + 2, x + 4$, and so on.

The following are examples of **consecutive odd integers:** 21, 23, 25, 27, 29; and −71, −69, −67, −65. Note that consecutive odd integers can be represented in the form $x, x + 2, x + 4$, and so on.

EXAMPLE 3 *Limited-Edition Prints.* A limited-edition print is usually signed and numbered by the artist. For example, a limited edition with only 50 prints would be numbered 1/50, 2/50, 3/50, and so on. An estate donates two prints numbered consecutively from a limited edition with 150 prints. The sum of the two numbers is 263. Find the numbers of the prints.

1. **Familiarize.** The numbers of the prints are consecutive integers. If we let $x =$ the smaller number, then $x + 1 =$ the larger number. Since there are 150 prints in the edition, the first number must be 149 or less. If we guess that $x = 138$, then $x + 1 = 139$. The sum of the numbers is 277. We see that the numbers need to be smaller. We could continue guessing and solve the problem this way, but let's work on developing algebra skills.

Answer

2. Jenny: 8 in.; Emma: 4 in.; Sarah: 6 in.

2. **Translate.** We reword the problem and translate as follows:

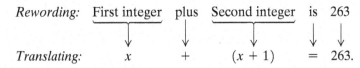

3. **Solve.** We solve the equation:

$$x + (x + 1) = 263$$
$$2x + 1 = 263 \quad \text{Collecting like terms}$$
$$2x + 1 - 1 = 263 - 1 \quad \text{Subtracting 1}$$
$$2x = 262$$
$$\frac{2x}{2} = \frac{262}{2} \quad \text{Dividing by 2}$$
$$x = 131.$$

If $x = 131$, then $x + 1 = 132$.

4. **Check.** Our possible answers are 131 and 132. These are consecutive positive integers and $131 + 132 = 263$, so the answers check.
5. **State.** The print numbers are 131/150 and 132/150.

Do Exercise 3. ▶

EXAMPLE 4 *Delivery Truck Rental.* An appliance business needs to rent a delivery truck for 6 days while one of its trucks is being repaired. The cost of renting a 16-ft truck is $29.95 per day plus $0.29 per mile. If $550 is budgeted for the rental, how many miles can be driven for the budgeted amount?

1. **Familiarize.** Suppose that the truck is driven 1100 mi. The cost is given by the daily charge plus the mileage charge, so we have

which is $498.70. We see that the van can be driven more than 1100 mi on the business' budget of $550. This process familiarizes us with the way in which a calculation is made.

 3. *Interstate Mile Markers.* The sum of two consecutive mile markers on I-90 in upstate New York is 627. (On I-90 in New York, the marker numbers increase from east to west.) Find the numbers on the markers.

Data: New York State Department of Transportation

Let x = the first marker number and $x + 1$ = the second marker number.

Translate and *Solve*:

First marker number + Second marker number = 627

$$\square + (\square) = 627$$
$$\square + 1 = 627$$
$$2x + 1 - 1 = 627 - \square$$
$$2x = \square$$
$$\frac{2x}{\square} = \frac{626}{2}$$
$$x = 313.$$

If $x = 313$, then $x + 1 = \square$. The mile markers are $\square$ and 314.

Answer
3. 313 and 314
Guided Solution:
3. $x, x + 1, 2x, 1, 626, 2, 314, 313$

We let m = the number of miles that can be driven on the budget of $550.

2. **Translate.** We reword the problem and translate as follows:

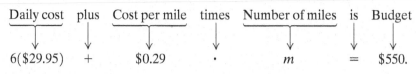

3. **Solve.** We solve the equation:

$$6(29.95) + 0.29m = 550$$
$$179.70 + 0.29m = 550$$
$$0.29m = 370.30 \quad \text{Subtracting 179.70}$$
$$\frac{0.29m}{0.29} = \frac{370.30}{0.29} \quad \text{Dividing by 0.29}$$
$$m \approx 1277. \quad \text{Rounding to the nearest one}$$

4. **Check.** We check our answer in the original problem. The cost for driving 1277 mi is 1277($0.29) = $370.33. The rental for 6 days is 6($29.95) = $179.70. The total cost is then

$$\$370.33 + \$179.70 \approx \$550,$$

which is the $550 budget that was allowed.

5. **State.** The truck can be driven 1277 mi on the truck-rental allotment.

◀ Do Exercise 4.

4. *Delivery Truck Rental.* Refer to Example 4. The business decides to increase its 6-day rental budget to $625. How many miles can be driven for $625?

EXAMPLE 5 *Perimeter of a Lacrosse Field.* The perimeter of a lacrosse field is 340 yd. The length is 50 yd longer than the width. Find the dimensions of the field.
Data: sportsknowhow.com

1. **Familiarize.** We first make a drawing.

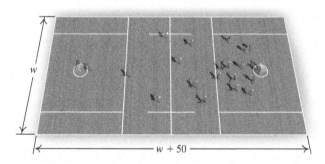

We let w = the width of the rectangle, in yards. Then $w + 50$ = the length. The perimeter P of a rectangle is the distance around the rectangle and is given by the formula $2l + 2w = P$, where

l = the length and w = the width.

Answer
4. 1536 mi

2. **Translate.** To translate the problem, we substitute $w + 50$ for l and 340 for P:

$$2l + 2w = P$$
$$2(w + 50) + 2w = 340.$$

Caution! Parentheses are necessary here.

3. **Solve.** We solve the equation:

$$2(w + 50) + 2w = 340$$
$$2w + 100 + 2w = 340 \quad \text{Using the distributive law}$$
$$4w + 100 = 340 \quad \text{Collecting like terms}$$
$$4w + 100 - 100 = 340 - 100 \quad \text{Subtracting 100}$$
$$4w = 240$$
$$\frac{4w}{4} = \frac{240}{4} \quad \text{Dividing by 4}$$
$$w = 60.$$

Thus the possible dimensions are

$$w = 60 \text{ yd} \quad \text{and} \quad l = w + 50 = 60 + 50, \text{ or } 110 \text{ yd}.$$

4. **Check.** If the width is 60 yd and the length is 110 yd, then the perimeter is $2(60 \text{ yd}) + 2(110 \text{ yd})$, or 340 yd. This checks.

5. **State.** The width is 60 yd and the length is 110 yd.

Do Exercise 5.

5. *Perimeter of High School Basketball Court.* The perimeter of a standard high school basketball court is 268 ft. The length is 34 ft longer than the width. Find the dimensions of the court.

Data: Indiana High School Athletic Association

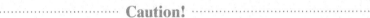

Caution!

Always be sure to answer the original problem completely. For instance, in Example 1, we need to find *two* numbers: the distances from *each* city to the biker. Similarly, in Example 3, we need to find two print numbers, and in Example 5, we need to find two dimensions, not just the width.

EXAMPLE 6 *Roof Gable.* In a triangular gable end of a roof, the angle of the peak is twice as large as the angle of the back side of the house. The measure of the angle on the front side is 20° greater than the angle on the back side. How large are the angles?

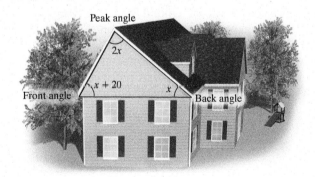

1. **Familiarize.** We first make a drawing as shown above. We let

measure of back angle $= x$.

Then measure of peak angle $= 2x$

and measure of front angle $= x + 20$.

Answer
5. Length: 84 ft; width: 50 ft

2. **Translate.** To translate, we need to know that the sum of the measures of the angles of a triangle is 180°. You might recall this fact from geometry or you can look it up in a geometry book or in the list of formulas inside the back cover of this book. We translate as follows:

$$\underbrace{\text{Measure of back angle}}_{x} + \underbrace{\text{Measure of peak angle}}_{2x} + \underbrace{\text{Measure of front angle}}_{(x+20)} \text{ is } 180°$$

$$x + 2x + (x+20) = 180°.$$

3. **Solve.** We solve the equation:

$$x + 2x + (x + 20) = 180$$
$$4x + 20 = 180$$
$$4x + 20 - 20 = 180 - 20$$
$$4x = 160$$
$$\frac{4x}{4} = \frac{160}{4}$$
$$x = 40.$$

The possible measures for the angles are as follows:

Back angle: $x = 40°$;
Peak angle: $2x = 2(40) = 80°$;
Front angle: $x + 20 = 40 + 20 = 60°$.

4. **Check.** Consider our answers: 40°, 80°, and 60°. The peak is twice the back, and the front is 20° greater than the back. The sum is 180°. The angles check.

5. **State.** The measures of the angles are 40°, 80°, and 60°.

.... **Caution!** ....

Units are important in answers. Remember to include them, where appropriate.

◀ Do Exercise 6.

6. The second angle of a triangle is three times as large as the first. The third angle measures 30° more than the first angle. Find the measures of the angles.

Fastest Roller Coasters

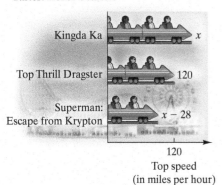

EXAMPLE 7 *Fastest Roller Coasters.* The average top speed of the three fastest steel roller coasters in the United States is 116 mph. The third-fastest roller coaster, Superman: Escape from Krypton (located at Six Flags Magic Mountain, Valencia, California), reaches a top speed of 28 mph less than the fastest roller coaster, Kingda Ka (located at Six Flags Great Adventure, Jackson, New Jersey). The second-fastest roller coaster, Top Thrill Dragster (located at Cedar Point, Sandusky, Ohio), has a top speed of 120 mph. What is the top speed of the fastest steel roller coaster?
Data: Coaster Grotto

Answer
6. First: 30°; second: 90°; third: 60°

1. **Familiarize.** The **average** of a set of numbers is the sum of the numbers divided by the number of addends.

   We are given that the second-fastest speed is 120 mph. Suppose that the three top speeds are 131, 120, and 103. The average is then
   $$\frac{131 + 120 + 103}{3} = \frac{354}{3} = 118,$$
   which is too high. Instead of continuing to guess, let's use the equation-solving skills that we have learned in this chapter. We let $x =$ the top speed of the fastest roller coaster. Then $x - 28 =$ the top speed of the third-fastest roller coaster.

2. **Translate.** We reword the problem and translate as follows:

   $$\frac{\underbrace{\text{Speed of fastest coaster}}_{} + \underbrace{\text{Speed of second-fastest coaster}}_{} + \underbrace{\text{Speed of third-fastest coaster}}_{}}{\text{Number of roller coasters}} = \underbrace{\text{Average speed of three fastest roller coasters}}_{}$$

   $$\frac{x + 120 + (x - 28)}{3} = 116.$$

3. **Solve.** We solve as follows:

   $$\frac{x + 120 + (x - 28)}{3} = 116$$

   $$3 \cdot \frac{x + 120 + (x - 28)}{3} = 3 \cdot 116 \quad \text{Multiplying by 3 on both sides to clear the fraction}$$

   $$x + 120 + (x - 28) = 348$$
   $$2x + 92 = 348 \quad \text{Collecting like terms}$$
   $$2x = 256 \quad \text{Subtracting 92}$$
   $$x = 128. \quad \text{Dividing by 2}$$

4. **Check.** If the top speed of the fastest roller coaster is 128 mph, then the top speed of the third-fastest is $128 - 28$, or 100 mph. The average of the top speeds of the three fastest is
   $$\frac{128 + 120 + 100}{3} = \frac{348}{3} = 116 \text{ mph}.$$
   The answer checks.

5. **State.** The top speed of the fastest steel roller coaster in the United States is 128 mph.

   Do Exercise 7.

7. *Average Test Score.* Sam's average score on his first three math tests is 77. He scored 62 on the first test. On the third test, he scored 9 more than he scored on his second test. What did he score on the second and third tests?

Answer
7. Second: 80; third: 89

EXAMPLE 8 *Simple Interest.* An investment is made at 3% simple interest for 1 year. It grows to $746.75. How much was originally invested (the principal)?

1. **Familiarize.** Suppose that $100 was invested. Recalling the formula for simple interest, $I = Prt$, we know that the interest for 1 year on $100 at 3% simple interest is given by $I = \$100 \cdot 0.03 \cdot 1 = \$3$. Then, at the end of the year, the amount in the account is found by adding the principal and the interest:

   Principal + Interest = Amount
   $100 + $3 = $103.

   In this problem, we are working backward. We are trying to find the principal, which is the original investment. We let $x =$ the principal. Then the interest earned is 3%x.

2. **Translate.** We reword the problem and then translate:

   Principal + Interest = Amount
   x + 3%x = 746.75.   Interest is 3% of the principal.

3. **Solve.** We solve the equation:

   $x + 3\%x = 746.75$
   $x + 0.03x = 746.75$ Converting to decimal notation
   $1x + 0.03x = 746.75$ Identity property of 1
   $(1 + 0.03)x = 746.75$
   $1.03x = 746.75$ Collecting like terms
   $\dfrac{1.03x}{1.03} = \dfrac{746.75}{1.03}$ Dividing by 1.03
   $x = 725.$

4. **Check.** We check by taking 3% of $725 and adding it to $725:

   $3\% \times \$725 = 0.03 \times 725 = \$21.75.$

   Then $725 + $21.75 = $746.75, so $725 checks.

5. **State.** The original investment was $725.

◀ Do Exercise 8.

EXAMPLE 9 *Selling a House.* The Patels are planning to sell their house. If they want to be left with $130,200 after paying 7% of the selling price to a realtor as a commission, for how much must they sell the house?

1. **Familiarize.** Suppose that the Patels sell the house for $138,000. A 7% commission can be determined by finding 7% of $138,000:

   7% of $\$138{,}000 = 0.07(\$138{,}000) = \$9660.$

   Subtracting this commission from $138,000 would leave the Patels with

   $\$138{,}000 - \$9660 = \$128{,}340.$

   This shows that in order for the Patels to clear $130,200, the house must sell for more than $138,000. Our guess shows us how to translate to an equation. We let $x =$ the selling price, in dollars. With a 7% commission, the realtor would receive $0.07x$.

---

8. *Simple Interest.* An investment is made at 5% simple interest for 1 year. It grows to $2520. How much was originally invested (the principal)?

   Let $x =$ the principal. Then the interest earned is 5%x.

   Translate and *Solve*:

   Principal + Interest = Amount
   x + ☐ = 2520
   $x + 0.05x = 2520$
   $(1 + ☐)x = 2520$
   ☐$x = 2520$
   $\dfrac{1.05x}{1.05} = \dfrac{2520}{☐}$
   $x = 2400.$

Answer
8. $2400

Guided Solution:
8. 5%x, 0.05, 1.05, 1.05

2. **Translate.** We reword the problem and translate as follows:

$$\underbrace{\text{Selling price}}_{x} \underbrace{\text{less}}_{-} \underbrace{\text{Commission}}_{0.07x} \underbrace{\text{is}}_{=} \underbrace{\text{Amount remaining}}_{130{,}200.}$$

3. **Solve.** We solve the equation:

   $x - 0.07x = 130{,}200$
   $1x - 0.07x = 130{,}200$
   $(1 - 0.07)x = 130{,}200$
   $0.93x = 130{,}200$ Collecting like terms. Had we noted that after the commission has been paid, 93% remains, we could have begun with this equation.

   $\dfrac{0.93x}{0.93} = \dfrac{130{,}200}{0.93}$ Dividing by 0.93

   $x = 140{,}000.$

4. **Check.** To check, we first find 7% of $140,000:

   7% of $140,000 = 0.07($140,000) = $9800.   This is the commission.

   Next, we subtract the commission from the selling price to find the remaining amount:

   $140,000 − $9800 = $130,200.

   Since, after the commission, the Patels are left with $130,200, our answer checks. Note that the $140,000 selling price is greater than $138,000, as predicted in the *Familiarize* step.

5. **State.** To be left with $130,200, the Patels must sell the house for $140,000.

   Do Exercise 9. ▶

·· **Caution!** ··

The problem in Example 9 is easier to solve with algebra than without algebra. A common error in such a problem is to take 7% of the price after commission and then subtract or add. Note that 7% of the selling price (7% · $140,000 = $9800) is not equal to 7% of the amount that the Patels want to be left with (7% · $130,200 = $9114).

9. *Selling a Condominium.* An investor needs to sell a condominium in New York City. If she wants to be left with $761,400 after paying a 6% commission, for how much must she sell the condominium?

Answer
9. $810,000

Translating for Success

The goal of these matching questions is to practice step (2), Translate, of the five-step problem-solving process. Translate each word problem to an equation and select a correct translation from equations A–O.

1. **Angle Measures.** The measure of the second angle of a triangle is 51° more than that of the first angle. The measure of the third angle is 3° less than twice the first angle. Find the measures of the angles.

2. **Sales Tax.** Tina paid $3976 for a used car. This amount included 5% for sales tax. How much did the car cost before tax?

3. **Perimeter.** The perimeter of a rectangle is 2347 ft. The length is 28 ft greater than the width. Find the length and the width.

4. **Fraternity or Sorority Membership.** At Arches Tech University, 3976 students belong to a fraternity or a sorority. This is 35% of the total enrollment. What is the total enrollment at Arches Tech?

5. **Fraternity or Sorority Membership.** At Moab Tech University, thirty-five percent of the students belong to a fraternity or a sorority. The total enrollment of the university is 11,360 students. How many students belong to either a fraternity or a sorority?

6. **Island Population.** There are 180 thousand people living on a small Caribbean island. The women outnumber the men by 96 thousand. How many men live on the island?

7. **Wire Cutting.** A 384-m wire is cut into three pieces. The second piece is 3 m longer than the first. The third is four-fifths as long as the first. How long is each piece?

8. **Locker Numbers.** The numbers on three adjoining lockers are consecutive integers whose sum is 384. Find the integers.

9. **Fraternity or Sorority Membership.** The total enrollment at Canyonlands Tech University is 11,360 students. Of these, 3976 students belong to a fraternity or a sorority. What percent of the students belong to a fraternity or a sorority?

10. **Width of a Rectangle.** The length of a rectangle is 96 ft. The perimeter of the rectangle is 3976 ft. Find the width.

A. $x + (x - 3) + \frac{4}{5}x = 384$

B. $x + (x + 51) + (2x - 3) = 180$

C. $x + (x + 96) = 180$

D. $2 \cdot 96 + 2x = 3976$

E. $x + (x + 1) + (x + 2) = 384$

F. $3976 = x \cdot 11{,}360$

G. $2x + 2(x + 28) = 2347$

H. $3976 = x + 5\%x$

I. $x + (x + 28) = 2347$

J. $x = 35\% \cdot 11{,}360$

K. $x + 96 = 3976$

L. $x + (x + 3) + \frac{4}{5}x = 384$

M. $x + (x + 2) + (x + 4) = 384$

N. $35\% \cdot x = 3976$

O. $2x + (x + 28) = 2347$

Answers on page A-4

2.6 Exercise Set

✓ Check Your Understanding

Reading Check Choose from the column on the right the word that completes each step in the five steps for problem solving.

RC1. _____ yourself with the problem situation.

RC2. _____ the problem to an equation.

RC3. _____ the equation.

RC4. _____ the answer in the original problem.

RC5. _____ the answer to the problem clearly.

Solve
Familiarize
State
Translate
Check

Concept Check Complete the translation of each statement.

CC1. The sum of two consecutive odd numbers is 32.
If x = the smaller number, then _____ = the larger number.
The translation is x + _____ = _____.

CC2. Including a 6% sales tax, Jaykob paid $36.57 for a sweatshirt.
If x = the marked price, then _____ = the sales tax.
The translation is x + _____ = _____.

a Solve. *Although you might find the answer quickly in some other way, practice using the five-step problem-solving strategy.*

1. *College Enrollment.* In 2014, the U.S. college with the highest enrollment was the University of Phoenix, with 195,059 students. This was 103,880 more students than the number enrolled in the college with the second highest enrollment, Ivy Tech Community College of Indiana. How many students were enrolled in Ivy Tech?

   Data: National Center for Education Statistics

2. *Home Listing Price.* In 2017, the median value of a home in California was $102,100 more than three times the median value of a home in Ohio. The median value of a home in California was $469,300. What was the median value of a home in Ohio?

   Data: Zillow

3. *Pipe Cutting.* A 240-in. pipe is cut into two pieces. One piece is three times as long as the other. Find the lengths of the pieces.

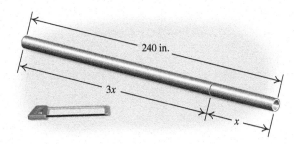

4. *Board Cutting.* A 72-in. board is cut into two pieces. One piece is 2 in. longer than the other. Find the lengths of the pieces.

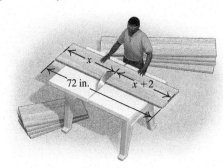

SECTION 2.6 Applications and Problem Solving   163

5. *Medals of Honor.* In 1863, the U.S. Secretary of War presented the first Medals of Honor. The two wars with the most Medals of Honor awarded are the Civil War and World War II. There were 464 recipients of this medal for World War II. This number is 1058 fewer than the number of recipients for the Civil War. How many Medals of Honor were awarded for valor in the Civil War?

Data: U.S. Army Center of Military History; U.S. Department of Defense

6. *Milk Alternatives.* Milk alternatives such as rice, soy, almond, and flax are becoming more available and increasingly popular. A cup of almond milk contains only 60 calories. This number is 89 calories less than the number of calories in a cup of whole milk. How many calories are in a cup of whole milk?

Data: Janet Kinosian, "Nutrition Udder Chaos," *AARP Magazine*, August/September, 2012

7. *500 Festival Mini-Marathon.* On May 6, 2017, 22,752 runners finished the 13.1-mi One America 500 Festival Mini-Marathon. If a runner stopped at a water station that was twice as far from the start as from the finish, how far was the runner from the finish? Round the answer to the nearest hundredth of a mile.

Data: results.xacte.com

8. *Airport Control Tower.* At a height of 385 ft, the FAA airport traffic control tower in Atlanta is the tallest traffic control tower in the United States. Its height is 59 ft greater than the height of the tower at the Memphis airport. How tall is the traffic control tower at the Memphis airport?

Data: Federal Aviation Administration

9. *Consecutive Apartment Numbers.* The apartments in Vincent's apartment building are numbered consecutively on each floor. The sum of his number and his next-door neighbor's number is 2409. What are the two numbers?

10. *Consecutive Post Office Box Numbers.* The sum of the numbers on two consecutive post office boxes is 547. What are the numbers?

11. *Consecutive Ticket Numbers.* The numbers on Sam's three raffle tickets are consecutive integers. The sum of the numbers is 126. What are the numbers?

12. *Consecutive Ages.* The ages of Whitney, Wesley, and Wanda are consecutive integers. The sum of their ages is 108. What are their ages?

13. *Consecutive Odd Integers.* The sum of three consecutive odd integers is 189. What are the integers?

14. *Consecutive Integers.* Three consecutive integers are such that the first plus one-half the second plus seven less than twice the third is 2101. What are the integers?

15. *Photo Size.* A hotel orders a large photo for its newly renovated lobby. The perimeter of the photo is 292 in. The width is 2 in. more than three times the height. Find the dimensions of the photo.

16. *Two-by-Four.* The perimeter of a cross section or end of a "two-by-four" piece of lumber is 10 in. The length is 2 in. more than the width. Find the actual dimensions of the cross section of a two-by-four.

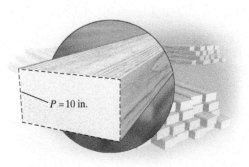

17. *Price of Coffee Beans.* A student-owned and operated coffee shop near a campus purchases gourmet coffee beans from Costa Rica. During a recent 30%-off sale, a 3-lb bag could be purchased for $44.10. What is the regular price of a 3-lb bag?

18. *Price of an iPad Case.* Makayla paid $33.15 for an iPad case during a 15%-off sale. What was the regular price?

19. *Price of a Security Wallet.* Caleb paid $26.70, including a 7% sales tax, for a security wallet. How much did the wallet itself cost?

20. *Price of a Car Battery.* Tyler paid $117.15, including a 6.5% sales tax, for a car battery. How much did the battery itself cost?

21. *Parking Costs.* A hospital parking lot charges $1.50 for the first hour or part thereof, and $1.00 for each additional hour or part thereof. A weekly pass costs $27.00 and allows unlimited parking for 7 days. Suppose that each visit Hailey makes to the hospital lasts $1\frac{1}{2}$ hr. What is the minimum number of times that Hailey would have to visit per week to make it worthwhile for her to buy the pass?

22. *Van Rental.* Value Rent-A-Car rents vans at a daily rate of $84.45 plus 55¢ per mile. Molly rents a van to deliver electrical parts to her customers. She is allotted a daily budget of $250. How many miles can she drive for $250? (*Hint*: 55¢ = $0.55.)

23. *Triangular Field.* The second angle of a triangular field is three times as large as the first angle. The third angle is 40° greater than the first angle. How large are the angles?

24. *Triangular Parking Lot.* The second angle of a triangular parking lot is four times as large as the first angle. The third angle is 45° less than the sum of the other two angles. How large are the angles?

25. *Triangular Backyard.* A home has a triangular backyard. The second angle of the triangle is 5° more than the first angle. The third angle is 10° more than three times the first angle. Find the angles of the triangular yard.

26. *Boarding Stable.* A rancher needs to form a triangular horse pen next to a stable using ropes. The second angle is three times the first angle. The third angle is 15° less than the first angle. Find the angles of the triangular pen.

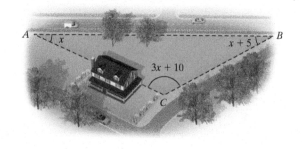

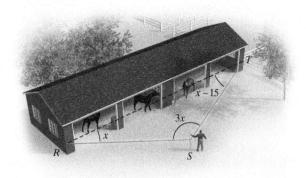

27. *Stock Prices.* Diego's investment in a technology stock grew 28% to $448. How much did he invest?

28. *Savings Interest.* Ella invested money in a savings account at a rate of 6% simple interest. After 1 year, she has $6996 in the account. How much did Ella originally invest?

29. *Credit Cards.* The balance on Will's credit card grew 2%, to $870, in one month. What was his balance at the beginning of the month?

30. *Loan Interest.* Alvin borrowed money from a cousin at a rate of 10% simple interest. After 1 year, $7194 paid off the loan. How much did Alvin borrow?

31. *Taxi Fares.* In New Orleans, Louisiana, taxis charge an initial charge of $3.50 plus $2.00 per mile. How far can one travel for $39.50?
    Data: taxifarefinders.com

32. *Taxi Fares.* In Baltimore, Maryland, taxis charge an initial charge of $1.80 plus $2.20 per mile. How far can one travel for $26?
    Data: taxifarefinders.com

33. *Tipping.* Isabella left a 15% tip for a meal. The total cost of the meal, including the tip, was $44.39. What was the cost of the meal before the tip was added?

34. *Tipping.* Nicolas left a 20% tip for a meal. The total cost of the meal, including the tip, was $24.90. What was the cost of the meal before the tip was added?

35. *Average Test Score.* Mariana averaged 84 on her first three history exams. The first score was 67. The second score was 7 less than the third score. What did she score on the second and third exams?

36. *Average Price.* David paid an average of $34 per shirt for a recent purchase of three shirts. The price of one shirt was twice as much as another, and the remaining shirt cost $27. What were the prices of the other two shirts?

37. If you double a number and then add 16, you get $\frac{2}{3}$ of the original number. What is the original number?

38. If you double a number and then add 85, you get $\frac{3}{4}$ of the original number. What is the original number?

Skill Maintenance

Calculate.

39. $-\frac{4}{5} - \frac{3}{8}$ [1.4a]

40. $-\frac{4}{5} + \frac{3}{8}$ [1.3a]

41. $-\frac{4}{5} \cdot \frac{3}{8}$ [1.5a]

42. $-\frac{4}{5} \div \frac{3}{8}$ [1.6c]

43. $\frac{1}{10} \div \left(-\frac{1}{100}\right)$ [1.6c]

44. $-25.6 \div (-16)$ [1.6c]

45. $-25.6(-16)$ [1.5a]

46. $-25.6 - (-16)$ [1.4a]

47. $-25.6 + (-16)$ [1.3a]

48. $(-0.02) \div (-0.2)$ [1.6c]

49. Use a commutative law to write an equivalent expression for $12 + yz$.   [1.7b]

50. Use an associative law to write an equivalent expression for $(c + 4) + d$.   [1.7b]

Synthesis

51. Apples are collected in a basket for six people. One third, one-fourth, one-eighth, and one-fifth are given to four people, respectively. The fifth person gets ten apples, leaving one apple for the sixth person. Find the original number of apples in the basket.

52. *Test Questions.* A student scored 78 on a test that had 4 seven-point fill-in questions and 24 three-point multiple-choice questions. The student answered one fill-in question incorrectly. How many multiple-choice questions did the student answer correctly?

53. The area of this triangle is 2.9047 in². Find x.

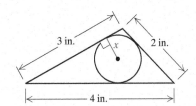

54. Susanne goes to the bank to get $20 in quarters, dimes, and nickels to use to make change at her yard sale. She gets twice as many quarters as dimes and 10 more nickels than dimes. How many of each type of coin does she get?

2.7 Solving Inequalities

OBJECTIVES

a. Determine whether a given number is a solution of an inequality.

b. Graph an inequality on the number line.

c. Solve inequalities using the addition principle.

d. Solve inequalities using the multiplication principle.

e. Solve inequalities using the addition principle and the multiplication principle together.

We now extend our equation-solving principles to the solving of inequalities.

a SOLUTIONS OF INEQUALITIES

SKILL REVIEW *Determine whether an inequality is true or false.* [1.2d]
Write true or false.
1. $-6 \le -8$
2. $1 \ge 1$

Answers: 1. False 2. True

An **inequality** is a number sentence with $>, <, \ge,$ or $\le$ as its verb—for example,

$$-4 > t, \quad x < 3, \quad 2x + 5 \ge 0, \quad \text{and} \quad -3y + 7 \le -8.$$

Some replacements for a variable in an inequality make it true and some make it false. (There are some exceptions to this statement, but we will not consider them here.)

> **SOLUTION OF AN INEQUALITY**
>
> A replacement that makes an inequality true is called a **solution**. The set of all solutions is called the **solution set**. When we have found the set of all solutions of an inequality, we say that we have **solved** the inequality.

EXAMPLES Determine whether each number is a solution of $x < 2$.
1. -2.7 Since $-2.7 < 2$ is true, -2.7 is a solution.
2. 2 Since $2 < 2$ is false, 2 is not a solution.

EXAMPLES Determine whether each number is a solution of $y \ge 6$.
3. 6 Since $6 \ge 6$ is true, 6 is a solution.
4. $-\dfrac{4}{3}$ Since $-\dfrac{4}{3} \ge 6$ is false, $-\dfrac{4}{3}$ is not a solution.

◀ Do Exercises 1 and 2.

b GRAPHS OF INEQUALITIES

Some solutions of $x < 2$ are $-3, 0, 1, 0.45, -8.9, -\pi, \tfrac{5}{8}$, and so on. In fact, there are infinitely many real numbers that are solutions. Because we cannot list them all individually, it is helpful to make a drawing that represents all the solutions.

A **graph** of an inequality is a drawing that represents its solutions. An inequality in one variable can be graphed on the number line. An inequality in two variables can be graphed on the coordinate plane. We will study such graphs in Chapter 9.

Determine whether each number is a solution of the inequality.

1. $x > 3$
   a) 2
   b) 0
   c) -5
   d) 15.4
   e) 3
   f) $-\dfrac{2}{5}$

2. $x \le 6$
   a) 6
   b) 0
   c) -4.3
   d) 25
   e) -6
   f) $\dfrac{5}{8}$

Answers
1. (a) No; (b) no; (c) no; (d) yes; (e) no; (f) no
2. (a) Yes; (b) yes; (c) yes; (d) no; (e) yes; (f) yes

EXAMPLE 5 Graph: $x < 2$.

The solutions of $x < 2$ are all those numbers less than 2. They are shown on the number line by shading all points to the left of 2. The parenthesis at 2 indicates that 2 *is not* part of the graph.

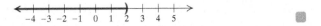

EXAMPLE 6 Graph: $x \geq -3$.

The solutions of $x \geq -3$ are shown on the number line by shading the point for -3 and all points to the right of -3. The bracket at -3 indicates that -3 *is* part of the graph.

EXAMPLE 7 Graph: $-3 \leq x < 2$.

The inequality $-3 \leq x < 2$ is read "-3 is less than or equal to x *and* x is less than 2," or "x is greater than or equal to -3 *and* x is less than 2." In order to be a solution of this inequality, a number must be a solution of both $-3 \leq x$ and $x < 2$. The number 1 is a solution, as are -1.7, 0, 1.5, and $\frac{3}{8}$. We can see from the following graphs that the solution set consists of the numbers that overlap in the two solution sets in Examples 5 and 6.

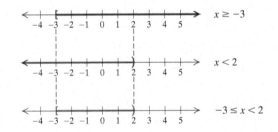

The parenthesis at 2 means that 2 *is not* part of the graph. The bracket at -3 means that -3 *is* part of the graph. The other solutions are shaded.

Do Exercises 3–5.

C SOLVING INEQUALITIES USING THE ADDITION PRINCIPLE

Consider the true inequality $3 < 7$. If we add 2 on both sides, we get another true inequality:

$$3 + 2 < 7 + 2, \quad \text{or} \quad 5 < 9.$$

Similarly, if we add -4 on both sides of $x + 4 < 10$, we get an *equivalent* inequality:

$$x + 4 + (-4) < 10 + (-4),$$
or
$$x < 6.$$

To say that $x + 4 < 10$ and $x < 6$ are **equivalent** is to say that they have the same solution set. For example, the number 3 is a solution of $x + 4 < 10$. It is also a solution of $x < 6$. The number -2 is a solution of $x < 6$. It is also a solution of $x + 4 < 10$. Any solution of one inequality is a solution of the other—they are equivalent.

Graph.

3. $x \leq 4$

4. $x > -2$

5. $-2 < x \leq 4$

Answers

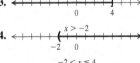

THE ADDITION PRINCIPLE FOR INEQUALITIES

For any real numbers a, b, and c:

$a < b$ is equivalent to $a + c < b + c$;
$a > b$ is equivalent to $a + c > b + c$;
$a \leq b$ is equivalent to $a + c \leq b + c$;
$a \geq b$ is equivalent to $a + c \geq b + c$.

When we add or subtract the same number on both sides of an inequality, the direction of the inequality symbol is not changed.

As with equation solving, when solving inequalities, our goal is to isolate the variable on one side. Then it is easier to determine the solution set.

EXAMPLE 8 Solve: $x + 2 > 8$. Then graph.

We use the addition principle, subtracting 2 on both sides:

$$x + 2 - 2 > 8 - 2$$
$$x > 6.$$

From the inequality $x > 6$, we can determine the solutions directly. Any number greater than 6 makes the last sentence true and is a solution of that sentence. Any such number is also a solution of the original sentence. Thus the inequality is solved. The graph is as follows:

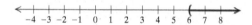

We cannot check all the solutions of an inequality by substitution, as we usually can for an equation, because there are too many of them. A partial check can be done by substituting a number greater than 6—say, 7—into the original inequality:

$$\begin{array}{c|c} x + 2 > 8 \\ \hline 7 + 2 \ ? \ 8 \\ 9 & \text{TRUE} \end{array}$$

Since $9 > 8$ is true, 7 is a solution. This is a partial check that any number greater than 6 is a solution.

EXAMPLE 9 Solve: $3x + 1 \leq 2x - 3$. Then graph.

We have

$$\begin{array}{ll} 3x + 1 \leq 2x - 3 & \\ 3x + 1 - 1 \leq 2x - 3 - 1 & \text{Subtracting 1} \\ 3x \leq 2x - 4 & \text{Simplifying} \\ 3x - 2x \leq 2x - 4 - 2x & \text{Subtracting } 2x \\ x \leq -4. & \text{Simplifying} \end{array}$$

Any number less than or equal to -4 is a solution. The graph is as follows:

In Example 9, any number less than or equal to -4 is a solution. The following are some solutions:

$$-4, \quad -5, \quad -6, \quad -\frac{13}{3}, \quad -204.5, \quad \text{and} \quad -18\pi.$$

Besides drawing a graph, we can also describe all the solutions of an inequality using **set notation**. We could just begin to list them in a set using roster notation (see p. 35), as follows:

$$\left\{-4, -5, -6, -\frac{13}{3}, -204.5, -18\pi, \ldots\right\}.$$

We can never list them all this way, however. Seeing this set without knowing the inequality makes it difficult for us to know what real numbers we are considering. There is, however, another kind of notation that we can use. It is

$$\{x \mid x \leq -4\},$$

which is read

"The set of all x such that x is less than or equal to -4."

This shorter notation for sets is called **set-builder notation**.
From now on, we will use this notation when solving inequalities.

> Do Exercises 6–8. ▶

EXAMPLE 10 Solve: $x + \frac{1}{3} > \frac{5}{4}$.

We have

$$x + \frac{1}{3} > \frac{5}{4}$$
$$x + \frac{1}{3} - \frac{1}{3} > \frac{5}{4} - \frac{1}{3} \qquad \text{Subtracting } \frac{1}{3}$$
$$x > \frac{5}{4} \cdot \frac{3}{3} - \frac{1}{3} \cdot \frac{4}{4} \qquad \text{Multiplying by 1 to obtain a common denominator}$$
$$x > \frac{15}{12} - \frac{4}{12}$$
$$x > \frac{11}{12}.$$

Any number greater than $\frac{11}{12}$ is a solution. The solution set is

$$\left\{x \mid x > \frac{11}{12}\right\},$$

which is read

"The set of all x such that x is greater than $\frac{11}{12}$."

When solving inequalities, you may obtain an answer like $\frac{11}{12} < x$. This has the same meaning as $x > \frac{11}{12}$. Thus the solution set in Example 10 can be described as $\left\{x \mid \frac{11}{12} < x\right\}$ or as $\left\{x \mid x > \frac{11}{12}\right\}$. The latter is used most often.

> Do Exercises 9 and 10. ▶

d SOLVING INEQUALITIES USING THE MULTIPLICATION PRINCIPLE

There is a multiplication principle for inequalities that is similar to that for equations, but it must be modified. When we are multiplying on both sides by a negative number, the direction of the inequality symbol must be changed.

Solve. Then graph.
6. $x + 3 > 5$

$\xleftarrow{}\!\xrightarrow{}$
$\ -5\ -4\ -3\ -2\ -1\ \ 0\ \ 1\ \ 2\ \ 3\ \ 4\ \ 5$

7. $x - 1 \leq 2$

$\xleftarrow{}\!\xrightarrow{}$
$\ -5\ -4\ -3\ -2\ -1\ \ 0\ \ 1\ \ 2\ \ 3\ \ 4\ \ 5$

8. $5x + 1 < 4x - 2$

$\xleftarrow{}\!\xrightarrow{}$
$\ -5\ -4\ -3\ -2\ -1\ \ 0\ \ 1\ \ 2\ \ 3\ \ 4\ \ 5$

Solve.
9. $x + \frac{2}{3} \geq \frac{4}{5}$

GS 10. $5y + 2 \leq -1 + 4y$
$5y + 2 - \boxed{} \leq -1 + 4y - 4y$
$y + 2 \leq -1$
$y + 2 - 2 \leq -1 - \boxed{}$
$y \leq \boxed{}$
The solution set is $\{y \mid y \boxed{} -3\}$.

Answers
6. $\{x \mid x > 2\}$;

$\xleftarrow{}\!\xrightarrow{}$
$\quad\quad 0\quad\ 2$

7. $\{x \mid x \leq 3\}$;

$\xleftarrow{}\!\xrightarrow{}$
$\quad\quad 0\quad\ 3$

8. $\{x \mid x < -3\}$;

$\xleftarrow{}\!\xrightarrow{}$
$\ -3\quad\ 0$

9. $\left\{x \mid x \geq \frac{2}{15}\right\}$ 10. $\{y \mid y \leq -3\}$

Guided Solution:
10. $4y, 2, -3, \leq$

SECTION 2.7   Solving Inequalities   **171**

Consider the true inequality $3 < 7$. If we multiply on both sides by a *positive* number, like 2, we get another true inequality:

$$3 \cdot 2 < 7 \cdot 2, \quad \text{or} \quad 6 < 14. \quad \text{True}$$

If we multiply on both sides by a *negative* number, like -2, and we do not change the direction of the inequality symbol, we get a *false* inequality:

$$3 \cdot (-2) < 7 \cdot (-2), \quad \text{or} \quad -6 < -14. \quad \text{False}$$

The fact that $6 < 14$ is true but $-6 < -14$ is false stems from the fact that the negative numbers, in a sense, mirror the positive numbers. That is, whereas 14 is to the *right* of 6 on the number line, the number -14 is to the *left* of -6. Thus, if we reverse (change the direction of) the inequality symbol, we get a *true* inequality: $-6 > -14$.

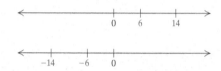

THE MULTIPLICATION PRINCIPLE FOR INEQUALITIES

For any real numbers a and b, and any *positive* number c:

$a < b$ is equivalent to $ac < bc$;
$a > b$ is equivalent to $ac > bc$.

For any real numbers a and b, and any *negative* number c:

$a < b$ is equivalent to $ac > bc$;
$a > b$ is equivalent to $ac < bc$.

Similar statements hold for $\leq$ and $\geq$.

When we multiply or divide by a positive number on both sides of an inequality, the direction of the inequality symbol stays the same. When we multiply or divide by a negative number on both sides of an inequality, the direction of the inequality symbol is reversed.

EXAMPLE 11 Solve: $4x < 28$. Then graph.

We have

$$4x < 28$$
$$\frac{4x}{4} < \frac{28}{4} \quad \text{Dividing by 4}$$

The symbol stays the same.

$$x < 7. \quad \text{Simplifying}$$

The solution set is $\{x \mid x < 7\}$. The graph is as follows:

◀ Do Exercises 11 and 12.

Solve. Then graph.

11. $8x < 64$

12. $5y \geq 160$

Answers

11. $\{x \mid x < 8\}$;

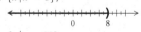

12. $\{y \mid y \geq 32\}$;

EXAMPLE 12 Solve: $-2y < 18$. Then graph.

$$-2y < 18$$
$$\frac{-2y}{-2} > \frac{18}{-2} \quad \text{Dividing by } -2$$
$$\qquad \text{The symbol must be reversed!}$$
$$y > -9. \quad \text{Simplifying}$$

The solution set is $\{y | y > -9\}$. The graph is as follows:

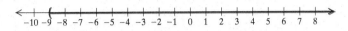

Do Exercises 13 and 14.

Solve.

13. $-4x \leq 24$

14. $-5y > 13$

e USING THE PRINCIPLES TOGETHER

All of the equation-solving techniques used in Sections 2.1–2.3 can be used with inequalities, provided we remember to reverse the inequality symbol when multiplying or dividing on both sides by a negative number.

EXAMPLE 13 Solve: $6 - 5x > 7$.

$$6 - 5x > 7$$
$$-6 + 6 - 5x > -6 + 7 \quad \text{Adding } -6. \text{ The symbol stays the same.}$$
$$-5x > 1 \quad \text{Simplifying}$$
$$\frac{-5x}{-5} < \frac{1}{-5} \quad \text{Dividing by } -5$$
$$\qquad \text{The symbol must be reversed because we are dividing by a } \textit{negative} \text{ number, } -5.$$
$$x < -\frac{1}{5}. \quad \text{Simplifying}$$

The solution set is $\{x | x < -\frac{1}{5}\}$.

Do Exercise 15.

15. Solve: $7 - 4x < 8$.

EXAMPLE 14 Solve: $17 - 5y > 8y - 9$.

$$-17 + 17 - 5y > -17 + 8y - 9 \quad \text{Adding } -17. \text{ The symbol stays the same.}$$
$$-5y > 8y - 26 \quad \text{Simplifying}$$
$$-8y - 5y > -8y + 8y - 26 \quad \text{Adding } -8y$$
$$-13y > -26 \quad \text{Simplifying}$$
$$\frac{-13y}{-13} < \frac{-26}{-13} \quad \text{Dividing by } -13$$
$$\qquad \text{The symbol must be reversed because we are dividing by a } \textit{negative} \text{ number, } -13.$$
$$y < 2$$

The solution set is $\{y | y < 2\}$.

Do Exercise 16.

16. Solve. Begin by subtracting 24 on both sides.
$$24 - 7y \leq 11y - 14$$

Answers

13. $\{x | x \geq -6\}$ **14.** $\{y | y < -\frac{13}{5}\}$
15. $\{x | x > -\frac{1}{4}\}$ **16.** $\{y | y \geq \frac{19}{9}\}$

Typically, we solve an equation or an inequality by isolating the variable on the left side. When we are solving an inequality, however, there are situations in which isolating the variable on the right side will eliminate the need to reverse the inequality symbol. Let's solve the inequality in Example 14 again, but this time we will isolate the variable on the right side.

EXAMPLE 15 Solve: $17 - 5y > 8y - 9$.

Note that if we add $5y$ on both sides, the coefficient of the y-term will be positive after like terms have been collected.

$17 - 5y + 5y > 8y - 9 + 5y$ Adding $5y$
$17 > 13y - 9$ Simplifying
$17 + 9 > 13y - 9 + 9$ Adding 9
$26 > 13y$ Simplifying
$\dfrac{26}{13} > \dfrac{13y}{13}$ Dividing by 13. We leave the inequality symbol the same because we are dividing by a positive number.
$2 > y$

The solution set is $\{y \mid 2 > y\}$, or $\{y \mid y < 2\}$.

◀ Do Exercise 17.

EXAMPLE 16 Solve: $3(x - 2) - 1 < 2 - 5(x + 6)$.

First, we use the distributive law to remove parentheses. Next, we collect like terms and then use the addition and multiplication principles for inequalities to get an equivalent inequality with x alone on one side.

$3(x - 2) - 1 < 2 - 5(x + 6)$
$3x - 6 - 1 < 2 - 5x - 30$ Using the distributive law to multiply and remove parentheses
$3x - 7 < -5x - 28$ Collecting like terms
$3x + 5x < -28 + 7$ Adding $5x$ and 7 to get all x-terms on one side and all other terms on the other side
$8x < -21$ Simplifying
$x < \dfrac{-21}{8}$, or $-\dfrac{21}{8}$.   Dividing by 8

The solution set is $\{x \mid x < -\tfrac{21}{8}\}$.

◀ Do Exercise 18.

17. Solve. Begin by adding $7y$ on both sides.
$24 - 7y \leq 11y - 14$

18. Solve:
$3(7 + 2x) \leq 30 + 7(x - 1)$.
$\underline{} + 6x \leq 30 + 7x - \underline{}$
$21 + 6x \leq \underline{} + 7x$
$21 + 6x - 6x \leq 23 + 7x - \underline{}$
$21 \leq 23 + \underline{}$
$21 - \underline{} \leq 23 + x - 23$
$-2 \leq x$, or
$x \underline{} -2$
The solution set is $\{x \mid x \geq \underline{}\}$.

Answers
17. $\left\{y \mid y \geq \dfrac{19}{9}\right\}$ **18.** $\{x \mid x \geq -2\}$

Guided Solution:
18. 21, 7, 23, 6x, x, 23, ≥, −2

EXAMPLE 17 Solve: $16.3 - 7.2p \leq -8.18$.

The greatest number of decimal places in any one number is *two*. Multiplying by 100, which has two 0's, will clear decimals. Then we proceed as before.

$$16.3 - 7.2p \leq -8.18$$
$$100(16.3 - 7.2p) \leq 100(-8.18) \quad \text{Multiplying by 100}$$
$$100(16.3) - 100(7.2p) \leq 100(-8.18) \quad \text{Using the distributive law}$$
$$1630 - 720p \leq -818 \quad \text{Simplifying}$$
$$1630 - 720p - 1630 \leq -818 - 1630 \quad \text{Subtracting 1630}$$
$$-720p \leq -2448 \quad \text{Simplifying}$$
$$\frac{-720p}{-720} \geq \frac{-2448}{-720} \quad \text{Dividing by } -720$$

The symbol must be reversed.

$$p \geq 3.4$$

The solution set is $\{p \mid p \geq 3.4\}$.

Do Exercise 19. ▶

19. Solve:
$2.1x + 43.2 \geq 1.2 - 8.4x$.

EXAMPLE 18 Solve: $\frac{2}{3}x - \frac{1}{6} + \frac{1}{2}x > \frac{7}{6} + 2x$.

The number 6 is the least common multiple of all the denominators. Thus we first multiply by 6 on both sides to clear fractions.

$$\frac{2}{3}x - \frac{1}{6} + \frac{1}{2}x > \frac{7}{6} + 2x$$
$$6\left(\frac{2}{3}x - \frac{1}{6} + \frac{1}{2}x\right) > 6\left(\frac{7}{6} + 2x\right) \quad \text{Multiplying by 6 on both sides}$$
$$6 \cdot \frac{2}{3}x - 6 \cdot \frac{1}{6} + 6 \cdot \frac{1}{2}x > 6 \cdot \frac{7}{6} + 6 \cdot 2x \quad \text{Using the distributive law}$$
$$4x - 1 + 3x > 7 + 12x \quad \text{Simplifying}$$
$$7x - 1 > 7 + 12x \quad \text{Collecting like terms}$$
$$7x - 1 - 7x > 7 + 12x - 7x \quad \text{Subtracting } 7x. \text{ The coefficient of the } x\text{-term will be positive.}$$
$$-1 > 7 + 5x \quad \text{Simplifying}$$
$$-1 - 7 > 7 + 5x - 7 \quad \text{Subtracting 7}$$
$$-8 > 5x \quad \text{Simplifying}$$
$$\frac{8}{5} > \frac{5x}{5} \quad \text{Dividing by 5}$$
$$-\frac{8}{5} > x$$

The solution set is $\{x \mid -\frac{8}{5} > x\}$, or $\{x \mid x < -\frac{8}{5}\}$.

Do Exercise 20. ▶

20. Solve:
$\frac{3}{4} + x < \frac{7}{8}x - \frac{1}{4} + \frac{1}{2}x$.

Answers

19. $\{x \mid x \geq -4\}$ **20.** $\left\{x \mid x > \frac{8}{3}\right\}$

2.7 Exercise Set

✓ Check Your Understanding

Reading Check Classify each pair of inequalities as either "equivalent" or "not equivalent."

RC1. $x + 10 \geq 12$; $x \leq 2$

RC2. $-3y \leq 30$; $y \leq -10$

RC3. $-y < 8$; $y > -8$

RC4. $2 - t > -3t + 4$; $2t > 2$

Concept Check Insert the symbol $<$, $>$, $\leq$, or $\geq$ to make each pair of inequalities equivalent.

CC1. $y - 6 \geq 3$; $y \,\square\, 9$

CC2. $4x < 20$; $x \,\square\, 5$

CC3. $-5x \leq 50$; $x \,\square\, -10$

CC4. $-\frac{1}{2}n > -5$; $n \,\square\, 10$

a Determine whether each number is a solution of the given inequality.

1. $x > -4$
   a) 4
   b) 0
   c) -4
   d) 6
   e) 5.6

2. $x \leq 5$
   a) 0
   b) 5
   c) -1
   d) -5
   e) $7\frac{1}{4}$

3. $x \geq 6.8$
   a) -6
   b) 0
   c) 6
   d) 8
   e) $-3\frac{1}{2}$

4. $x < 8$
   a) 8
   b) -10
   c) 0
   d) 11
   e) -4.7

b Graph on the number line.

5. $x > 4$

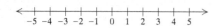

6. $x < 0$

7. $t < -3$

8. $y > 5$

9. $m \geq -1$

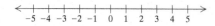

10. $x \leq -2$

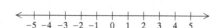

11. $-3 < x \leq 4$

12. $-5 \leq x < 2$

13. $0 < x < 3$

14. $-5 \leq x \leq 0$

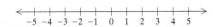

c Solve using the addition principle. Then graph.

15. $x + 7 > 2$

16. $x + 5 > 2$

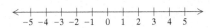

17. $x + 8 \leq -10$

18. $x + 8 \leq -11$

Solve using the addition principle.

19. $y - 7 > -12$

20. $y - 9 > -15$

21. $2x + 3 > x + 5$

22. $2x + 4 > x + 7$

23. $3x + 9 \leq 2x + 6$

24. $3x + 18 \leq 2x + 16$

25. $5x - 6 < 4x - 2$

26. $9x - 8 < 8x - 9$

27. $-9 + t > 5$

28. $-8 + p > 10$

29. $y + \dfrac{1}{4} \leq \dfrac{1}{2}$

30. $x - \dfrac{1}{3} \leq \dfrac{5}{6}$

31. $x - \dfrac{1}{3} > \dfrac{1}{4}$

32. $x + \dfrac{1}{8} > \dfrac{1}{2}$

d Solve using the multiplication principle. Then graph.

33. $5x < 35$

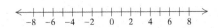

34. $8x \geq 32$

35. $-12x > -36$

36. $-16x > -64$

Solve using the multiplication principle.

37. $5y \geq -2$

38. $3x < -4$

39. $-2x \leq 12$

40. $-3x \leq 15$

SECTION 2.7 Solving Inequalities **177**

41. $-4y \geq -16$ **42.** $-7x < -21$ **43.** $-3x < -17$ **44.** $-5y > -23$

45. $-2y > \dfrac{1}{7}$ **46.** $-4x \leq \dfrac{1}{9}$ **47.** $-\dfrac{6}{5} \leq -4x$ **48.** $-\dfrac{7}{9} > 63x$

e Solve using the addition principle and the multiplication principle.

49. $4 + 3x < 28$ **50.** $3 + 4y < 35$ **51.** $3x - 5 \leq 13$

52. $5y - 9 \leq 21$ **53.** $13x - 7 < -46$ **54.** $8y - 6 < -54$

55. $30 > 3 - 9x$ **56.** $48 > 13 - 7y$ **57.** $4x + 2 - 3x \leq 9$

58. $15x + 5 - 14x \leq 9$ **59.** $-3 < 8x + 7 - 7x$ **60.** $-8 < 9x + 8 - 8x - 3$

61. $6 - 4y > 4 - 3y$ **62.** $9 - 8y > 5 - 7y + 2$ **63.** $5 - 9y \leq 2 - 8y$

64. $6 - 18x \leq 4 - 12x - 5x$ **65.** $19 - 7y - 3y < 39$ **66.** $18 - 6y - 4y < 63 + 5y$

67. $0.9x + 19.3 > 5.3 - 2.6x$ **68.** $0.96y - 0.79 \leq 0.21y + 0.46$ **69.** $\dfrac{x}{3} - 2 \leq 1$

70. $\dfrac{2}{3} + \dfrac{x}{5} < \dfrac{4}{15}$ **71.** $\dfrac{y}{5} + 1 \leq \dfrac{2}{5}$ **72.** $\dfrac{3x}{4} - \dfrac{7}{8} \geq -15$

73. $3(2y - 3) < 27$ **74.** $4(2y - 3) > 28$ **75.** $2(3 + 4m) - 9 \geq 45$

76. $3(5 + 3m) - 8 \leq 88$ **77.** $8(2t + 1) > 4(7t + 7)$ **78.** $7(5y - 2) > 6(6y - 1)$

79. $3(r - 6) + 2 < 4(r + 2) - 21$ **80.** $5(x + 3) + 9 \leq 3(x - 2) + 6$

81. $0.8(3x + 6) \geq 1.1 - (x + 2)$ **82.** $0.4(2x + 8) \geq 20 - (x + 5)$

83. $\dfrac{5}{3} + \dfrac{2}{3}x < \dfrac{25}{12} + \dfrac{5}{4}x + \dfrac{3}{4}$ **84.** $1 - \dfrac{2}{3}y \geq \dfrac{9}{5} - \dfrac{y}{5} + \dfrac{3}{5}$

Skill Maintenance

Add or subtract. [1.3a], [1.4a]

85. $-\dfrac{3}{4} + \dfrac{1}{8}$ **86.** $8.12 - 9.23$

87. $-2.3 - 7.1$ **88.** $-\dfrac{3}{4} - \dfrac{1}{8}$

Simplify.

89. $5 - 3^2 + (8 - 2)^2 \cdot 4$ [1.8d]   **90.** $10 \div 2 \cdot 5 - 3^2 + (-5)^2$ [1.8d]

91. $5(2x - 4) - 3(4x + 1)$ [1.8b]   **92.** $9(3 + 5x) - 4(7 + 2x)$ [1.8b]

Synthesis

93. Determine whether each number is a solution of the inequality $|x| < 3$.
     a) 0   **b)** -2
     c) -3 **d)** 4
     e) 3   **f)** 1.7
     g) -2.8

94. Graph $|x| < 3$ on the number line.

Solve.

95. $x + 3 < 3 + x$ **96.** $x + 4 > 3 + x$

2.8 Applications and Problem Solving with Inequalities

OBJECTIVES

a Translate number sentences to inequalities.

b Solve applied problems using inequalities.

The five steps for problem solving can be used for problems involving inequalities.

a TRANSLATING TO INEQUALITIES

Before solving problems that involve inequalities, we list some important phrases to look for. Sample translations are listed as well.

IMPORTANT WORDS	SAMPLE SENTENCE	TRANSLATION
is at least	Bill is at least 21 years old.	$b \geq 21$
is at most	At most 5 students dropped the course.	$n \leq 5$
cannot exceed	To qualify, earnings cannot exceed $12,000.	$r \leq 12{,}000$
must exceed	The speed must exceed 15 mph.	$s > 15$
is less than	Tucker's weight is less than 50 lb.	$w < 50$
is more than	Nashville is more than 200 mi away.	$d > 200$
is between	The film is between 90 min and 100 min long.	$90 < t < 100$
no more than	Cooper weighs no more than 90 lb.	$w \leq 90$
no less than	Sofia scored no less than 8.3.	$s \geq 8.3$

The following phrases deserve special attention.

> **TRANSLATING "AT LEAST" AND "AT MOST"**
>
> A quantity x is at least some amount q: $x \geq q$.
> (If x is at least q, it cannot be less than q.)
>
> A quantity x is at most some amount q: $x \leq q$.
> (If x is at most q, it cannot be more than q.)

Translate.

1. Sara worked no fewer than 15 hr last week.

2. The price of that Volkswagen Beetle convertible is at most $31,210.

3. The time for the test was between 45 min and 55 min.

4. Camila's weight is less than 110 lb.

5. That number is more than −2.

6. The costs of production of that marketing video cannot exceed $12,500.

7. At most 1250 people attended the concert.

8. Yesterday, at least 23 people got tickets for speeding.

◀ Do Exercises 1–8.

b SOLVING PROBLEMS

EXAMPLE 1 *Catering Costs.* To cater a company's annual lobster-bake cookout, Jayla's Catering charges a $325 setup fee plus $18.50 per person. The cost cannot exceed $3200. How many people can attend the cookout?

1. **Familiarize.** Suppose that 130 people were to attend the cookout. The cost would then be $325 + $18.50(130), or $2730. This shows that more than 130 people could attend the cookout without exceeding $3200. Instead of making another guess, we let n = the number of people in attendance.

Answers
1. $h \geq 15$ 2. $p \leq 31{,}210$ 3. $45 < t < 55$
4. $w < 110$ 5. $n > -2$ 6. $c \leq 12{,}500$
7. $p \leq 1250$ 8. $s \geq 23$

2. **Translate.** Our guess shows us how to translate. The cost of the cookout will be the $325 setup fee plus $18.50 times the number of people attending. We translate to an inequality:

Rewording: The setup fee plus the cost of the meals cannot exceed $3200.

Translating: $325 + 18.50n \leq 3200.$

3. **Solve.** We solve the inequality for n:

$$325 + 18.50n \leq 3200$$
$$325 + 18.50n - 325 \leq 3200 - 325 \quad \text{Subtracting 325}$$
$$18.50n \leq 2875 \quad \text{Simplifying}$$
$$\frac{18.50n}{18.50} \leq \frac{2875}{18.50} \quad \text{Dividing by 18.50}$$
$$n \leq 155.4. \quad \text{Rounding to the nearest tenth}$$

4. **Check.** Although the solution set of the inequality is all numbers less than or equal to about 155.4, since $n =$ the number of people in attendance, we round *down* to 155 people. If 155 people attend, the cost will be $325 + $18.50(155), or $3192.50. If 156 attend, the cost will exceed $3200.

5. **State.** At most, 155 people can attend the lobster-bake cookout.

Do Exercise 9.

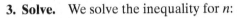   Caution!

Solutions of problems should always be checked using the original wording of the problem. In some cases, answers might need to be whole numbers or integers or rounded off in a particular direction.

EXAMPLE 2 *Nutrition.* The U.S. Department of Agriculture recommends that for a typical 2000-calorie daily diet, no more than 20 g of saturated fat be consumed. In the first three days of a four-day vacation, Ethan consumed 26 g, 17 g, and 22 g of saturated fat. Determine (in terms of an inequality) how many grams of saturated fat Ethan can consume on the fourth day if he is to average no more than 20 g of saturated fat per day.

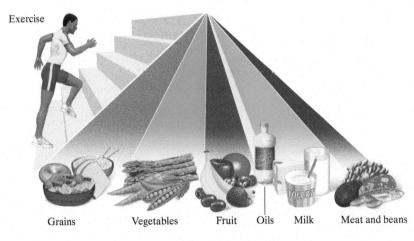

DATA: U.S. Department of Health and Human Services; U.S. Department of Agriculture

GS Translate to an inequality and solve.

9. **Butter Temperatures.** Butter stays solid at Fahrenheit temperatures below 88°. The formula
$$F = \tfrac{9}{5}C + 32$$
can be used to convert Celsius temperatures C to Fahrenheit temperatures F. Determine (in terms of an inequality) those Celsius temperatures for which butter stays solid.

Translate and *Solve*:

$$F < 88$$
$$\tfrac{9}{5}C + 32 < 88$$
$$\tfrac{9}{5}C + 32 - 32 < 88 - \boxed{}$$
$$\tfrac{9}{5}C < 56$$
$$\boxed{} \cdot \tfrac{9}{5}C < \tfrac{5}{9} \cdot 56$$
$$C < \frac{\boxed{}}{9}$$
$$C < 31\tfrac{1}{9}.$$

Butter stays solid at Celsius temperatures less than $31\tfrac{1}{9}°$—that is, $\{C \,|\, C < 31\tfrac{1}{9}°\}$.

Answer
9. $\tfrac{9}{5}C + 32 < 88;\ \{C \,|\, C < 31\tfrac{1}{9}°\}$
Guided Solution:
9. $32, \tfrac{5}{9}, 280$

1. **Familiarize.** Suppose that Ethan consumed 19 g of saturated fat on the fourth day. His daily average for the vacation would then be

$$\frac{26\text{ g} + 17\text{ g} + 22\text{ g} + 19\text{ g}}{4} = \frac{84\text{ g}}{4} = 21\text{ g}.$$

This shows that Ethan cannot consume 19 g of saturated fat on the fourth day, if he is to average no more than 20 g of fat per day. We let $x =$ the number of grams of fat that Ethan can consume on the fourth day.

2. **Translate.** We reword the problem and translate to an inequality:

   Rewording: The average consumption of saturated fat   should be no more than   20 g.

   Translating: $\dfrac{26 + 17 + 22 + x}{4} \qquad \leq \qquad 20.$

3. **Solve.** Because of the fraction expression, it is convenient to use the multiplication principle first to clear the fraction:

$$\frac{26 + 17 + 22 + x}{4} \leq 20$$

$$4\left(\frac{26 + 17 + 22 + x}{4}\right) \leq 4 \cdot 20 \qquad \text{Multiplying by 4}$$

$$26 + 17 + 22 + x \leq 80$$

$$65 + x \leq 80 \qquad \text{Simplifying}$$

$$x \leq 15. \qquad \text{Subtracting 65}$$

4. **Check.** As a partial check, we show that Ethan can consume 15 g of saturated fat on the fourth day and not exceed a 20-g average per day:

$$\frac{26 + 17 + 22 + 15}{4} = \frac{80}{4} = 20.$$

5. **State.** Ethan's average intake of saturated fat for the vacation will not exceed 20 g per day if he consumes no more than 15 g of saturated fat on the fourth day.

◀ **Do Exercise 10.**

Translate to an inequality and solve.

10. *Test Scores.* A pre-med student is taking a chemistry course in which four tests are given. To get an A, she must average at least 90 on the four tests. The student got scores of 91, 86, and 89 on the first three tests. Determine (in terms of an inequality) what scores on the last test will allow her to get an A.

Answer

10. $\dfrac{91 + 86 + 89 + s}{4} \geq 90; \{s | s \geq 94\}$

2.8 Exercise Set

FOR EXTRA HELP 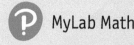 MyLab Math

✓ Check Your Understanding

Reading Check Match each sentence with one of the following.

$q < r \qquad q \leq r \qquad r < q \qquad r \leq q$

RC1. r is at most q.

RC2. q is no more than r.

RC3. r is less than q.

RC4. r is at least q.

RC5. q exceeds r.

RC6. q is no less than r.

182 CHAPTER 2   Solving Equations and Inequalities

Concept Check Determine whether $7 satisfies the requirement in each statement.

CC1. The cost must exceed $7.

CC2. The cost cannot exceed $7.

CC3. The cost is no more than $8.

CC4. The cost is no less than $8.

a   Translate to an inequality.

1. A number is at least 7.

2. A number is greater than or equal to 5.

3. The baby weighs more than 2 kilograms (kg).

4. Between 75 and 100 people attended the concert.

5. The speed of the train was between 90 mph and 110 mph.

6. The attendance was no more than 180.

7. Brianna works no more than 20 hr per week.

8. The amount of acid must exceed 40 liters (L).

9. The cost of gasoline is no less than $3.20 per gallon.

10. The temperature is at most $-2°$.

11. A number is greater than 8.

12. A number is less than 5.

13. A number is less than or equal to -4.

14. A number is greater than or equal to 18.

15. The number of people is at least 1300.

16. The cost is at most $4857.95.

17. The amount of water is not to exceed 500 liters.

18. The cost of ground beef is no less than $3.19 per pound.

19. Two more than three times a number is less than 13.

20. Five less than one-half of a number is greater than 17.

 Solve.

21. *Test Scores.* Xavier is taking a geology course in which four tests are given. To get a B, he must average at least 80 on the four tests. He got scores of 82, 76, and 78 on the first three tests. Determine (in terms of an inequality) what scores on the last test will allow him to get at least a B.

22. *Test Scores.* Chloe is taking a French class in which five quizzes are given. Her first four quiz grades are 73, 75, 89, and 91. Determine (in terms of an inequality) what scores on the last quiz will allow her to get an average quiz grade of at least 85.

23. *Gold Temperatures.* Gold stays solid at Fahrenheit temperatures below 1945.4°. Determine (in terms of an inequality) those Celsius temperatures for which gold stays solid. Use the formula given in Margin Exercise 9.

24. *Body Temperatures.* The human body is considered to be fevered when its temperature is higher than 98.6°F. Using the formula given in Margin Exercise 9, determine (in terms of an inequality) those Celsius temperatures for which the body is fevered.

25. *World Records in the 1500-m Run.* The formula
$$R = -0.075t + 3.85$$
can be used to predict the world record in the 1500-m run t years after 1930. Determine (in terms of an inequality) those years for which the world record will be less than 3.5 min.

26. *World Records in the 200-m Dash.* The formula
$$R = -0.028t + 20.8$$
can be used to predict the world record in the 200-m dash t years after 1920. Determine (in terms of an inequality) those years for which the world record will be less than 19.0 sec.

27. *Blueprints.* To make copies of blueprints, Vantage Reprographics charges a $5 setup fee plus $4 per copy. Myra can spend no more than $65 for copying her blueprints. What numbers of copies will allow her to stay within budget?

28. *Banquet Costs.* The Shepard College women's volleyball team can spend at most $750 for its awards banquet at a local restaurant. If the restaurant charges an $80 setup fee plus $16 per person, at most how many can attend?

29. *Envelope Size.* For a direct-mail campaign, Hollcraft Advertising determines that any envelope with a fixed width of $3\frac{1}{2}$ in. and an area of at least $17\frac{1}{2}$ in^2 can be used. Determine (in terms of an inequality) those lengths that will satisfy the company constraints.

30. *Package Sizes.* Logan Delivery Service accepts packages of up to 165 in. in length and girth combined. (Girth is the distance around the package.) A package has a fixed girth of 53 in. Determine (in terms of an inequality) those lengths for which a package is acceptable.

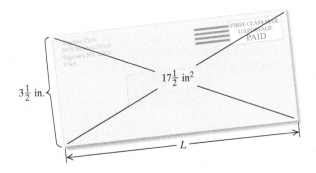

31. *Phone Costs.* Simon claims that it costs him at least $3.00 every time he calls an overseas customer. If his typical call costs 75¢ plus 45¢ for each minute, how long do his calls typically last? (*Hint:* 75¢ = $0.75.)

32. *Parking Costs.* Laura is certain that every time she parks in the municipal garage it costs her at least $6.75. If the garage charges $1.50 plus 75¢ for each half hour, for how long is Laura's car generally parked?

33. *College Tuition.* Angelica's financial aid stipulates that her tuition cannot exceed $1000. If her local community college charges a $35 registration fee plus $375 per course, what is the greatest number of courses for which Angelica can register?

34. *Furnace Repairs.* RJ's Plumbing and Heating charges $45 for a service call plus $30 per hour for emergency service. Gary remembers being billed over $150 for an emergency call. For how long was RJ's there?

35. *Nutrition.* Following the guidelines of the Food and Drug Administration, Dale tries to eat at least 5 servings of fruits or vegetables each day. For the first six days of one week, he had 4, 6, 7, 4, 6, and 4 servings. How many servings of fruits or vegetables should Dale eat on Saturday in order to average at least 5 servings per day for the week?

36. *College Course Load.* To remain on financial aid, Millie needs to complete an average of at least 7 credits per quarter each year. In the first three quarters of 2017, Millie completed 5, 7, and 8 credits. How many credits of course work must Millie complete in the fourth quarter if she is to remain on financial aid?

37. *Perimeter of a Rectangle.* The width of a rectangle is fixed at 8 ft. What lengths will make the perimeter at least 200 ft? at most 200 ft?

38. *Perimeter of a Triangle.* One side of a triangle is 2 cm shorter than the base. The other side is 3 cm longer than the base. What lengths of the base will allow the perimeter to be greater than 19 cm?

39. *Area of a Rectangle.* The width of a rectangle is fixed at 4 cm. For what lengths will the area be less than 86 cm^2?

40. *Area of a Rectangle.* The width of a rectangle is fixed at 16 yd. For what lengths will the area be at least 264 yd^2?

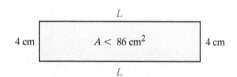

41. *Insurance-Covered Repairs.* Most insurance companies will replace a vehicle if an estimated repair exceeds 80% of the "blue-book" value of the vehicle. Rachel's insurance company paid $8500 for repairs to her Toyota after an accident. What can be concluded about the blue-book value of the car?

42. *Insurance-Covered Repairs.* Following an accident, Jeff's Ford pickup was replaced by his insurance company because the damage was so extensive. Before the damage, the blue-book value of the truck was $21,000. How much would it have cost to repair the truck? (See Exercise 41.)

43. *Reduced-Fat Foods.* In order for a food to be labeled "reduced fat," it must have at least 25% less fat than the regular item. One brand of reduced-fat peanut butter contains 12 g of fat per serving. What can you conclude about the fat content in a serving of the brand's regular peanut butter?

44. *Reduced-Fat Foods.* One brand of reduced-fat chocolate chip cookies contains 5 g of fat per serving. What can you conclude about the fat content of the brand's regular chocolate chip cookies? (See Exercise 43.)

45. *Area of a Triangular Flag.* As part of an outdoor education course at Baxter YMCA, Wendy needs to make a bright-colored triangular flag with an area of at least 3 ft^2. What heights can the triangle be if the base is $1\frac{1}{2}$ ft?

46. *Area of a Triangular Sign.* Zoning laws in Harrington prohibit displaying signs with areas exceeding 12 ft^2. If Flo's Marina is ordering a triangular sign with an 8-ft base, how tall can the sign be?

47. *Pond Depth.* On July 1, Garrett's Pond was 25 ft deep. Since that date, the water level has dropped $\frac{2}{3}$ ft per week. For what dates will the water level not exceed 21 ft?

48. *Weight Gain.* A 3-lb puppy is gaining weight at a rate of $\frac{3}{4}$ lb per week. When will the puppy's weight exceed $22\frac{1}{2}$ lb?

49. *Electrician Visits.* Dot's Electric made 17 customer calls last week and 22 calls this week. How many calls must be made next week in order to maintain a weekly average of at least 20 calls for the three-week period?

50. *Volunteer Work.* George and Joan do volunteer work at a hospital. Joan worked 3 more hr than George, and together they worked more than 27 hr. What possible numbers of hours did each work?

Skill Maintenance

Solve.

51. $-13 + x = 27$ [2.1b]

52. $-6y = 132$ [2.2a]

53. $4a - 3 = 45$ [2.3a]

54. $8x + 3x = 66$ [2.3b]

55. $-\frac{1}{2} + x = x - \frac{1}{2}$ [2.3c]

56. $9x - 1 + 11x - 18 = 3x - 15 + 4 + 17x$ [2.3c]

Solve. [2.5a]

57. What percent of 200 is 15?

58. What is 10% of 310?

59. 25 is 2% of what number?

60. 80 is what percent of 96?

Synthesis

Solve.

61. *Ski Wax.* Green ski wax works best between 5° and 15° Fahrenheit. Determine those Celsius temperatures for which green ski wax works best. Use the formula given in Margin Exercise 9.

62. *Parking Fees.* Mack's Parking Garage charges $4.00 for the first hour and $2.50 for each additional hour. For how long has a car been parked when the charge exceeds $16.50?

63. *Low-Fat Foods.* In order for a food to be labeled "low fat," it must have fewer than 3 g of fat per serving. One brand of reduced-fat tortilla chips contains 60% less fat than regular nacho cheese tortilla chips, but still cannot be labeled low fat. What can you conclude about the fat content of a serving of nacho cheese tortilla chips?

64. *Parking Fees.* When asked how much the parking charge is for a certain car, Mack replies "between 14 and 24 dollars." For how long has the car been parked? (See Exercise 62.)

CHAPTER 2 Summary and Review

Vocabulary Reinforcement

Complete each statement with the correct word or words from the column on the right. Some of the choices may not be used.

addition principle
multiplication principle
solution
equivalent
equation
inequality

1. Any replacement for the variable that makes an equation true is called a(n) _____ of the equation. [2.1a]
2. The _____ for equations states that for any real numbers a, b, and c, $a = b$ is equivalent to $a + c = b + c$. [2.1b]
3. The _____ for equations states that for any real numbers a, b, and c, $a = b$ is equivalent to $a \cdot c = b \cdot c$. [2.2a]
4. An _____ is a number sentence with $<$, $\leq$, $>$, or $\geq$ as its verb. [2.7a]
5. Equations with the same solutions are called _____ equations. [2.1b]

Concept Reinforcement

Determine whether each statement is true or false.

_____ 1. Some equations have no solution. [2.3c]
_____ 2. For any number n, $n \geq n$. [2.7a]
_____ 3. $2x - 7 < 11$ and $x < 2$ are equivalent inequalities. [2.7c]
_____ 4. If $x > y$, then $-x < -y$. [2.7d]

Study Guide

Objective 2.3a Solve equations using both the addition principle and the multiplication principle.
Objective 2.3b Solve equations in which like terms may need to be collected.
Objective 2.3c Solve equations by first removing parentheses and collecting like terms.

Example Solve: $6y - 2(2y - 3) = 12$.

$6y - 2(2y - 3) = 12$
$6y - 4y + 6 = 12$ Removing parentheses
$2y + 6 = 12$ Collecting like terms
$2y + 6 - 6 = 12 - 6$ Subtracting 6
$2y = 6$
$\dfrac{2y}{2} = \dfrac{6}{2}$ Dividing by 2
$y = 3$

The solution is 3.

Practice Exercise

1. Solve: $4(x - 3) = 6(x + 2)$.

188 CHAPTER 2 Solving Equations and Inequalities

Objective 2.3c Solve equations with an infinite number of solutions and equations with no solutions.

Example Solve: $8 + 2x - 4 = 6 + 2(x - 1)$.
$$8 + 2x - 4 = 6 + 2(x - 1)$$
$$8 + 2x - 4 = 6 + 2x - 2$$
$$2x + 4 = 2x + 4$$
$$2x + 4 - 2x = 2x + 4 - 2x$$
$$4 = 4$$

Every real number is a solution of the equation $4 = 4$, so all real numbers are solutions of the original equation. The equation has infinitely many solutions.

Example Solve: $2 + 5(x - 1) = -6 + 5x + 7$.
$$2 + 5(x - 1) = -6 + 5x + 7$$
$$2 + 5x - 5 = -6 + 5x + 7$$
$$5x - 3 = 5x + 1$$
$$5x - 3 - 5x = 5x + 1 - 5x$$
$$-3 = 1$$

This is a false equation, so the original equation has no solution.

Practice Exercises

2. Solve: $4 + 3y - 7 = 3 + 3(y - 2)$.

3. Solve: $4(x - 3) + 7 = -5 + 4x + 10$.

Objective 2.4b Solve a formula for a specified letter.

Example Solve for n: $M = \dfrac{m + n}{5}$.

$$M = \dfrac{m + n}{5}$$
$$5 \cdot M = 5\left(\dfrac{m + n}{5}\right)$$
$$5M = m + n$$
$$5M - m = m + n - m$$
$$5M - m = n$$

Practice Exercise

4. Solve for b: $A = \dfrac{1}{2}bh$.

Objective 2.7b Graph an inequality on the number line.

Example Graph each inequality: **(a)** $x < 2$; **(b)** $x \geq -3$.

a) The solutions of $x < 2$ are all numbers less than 2. We shade all points to the left of 2, and we use a parenthesis at 2 to indicate that 2 *is not* part of the graph.

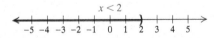

b) The solutions of $x \geq -3$ are all numbers greater than -3 and the number -3 as well. We shade all points to the right of -3, and we use a bracket at -3 to indicate that -3 *is* part of the graph.

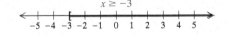

Practice Exercises

5. Graph: $x > 1$.

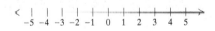

6. Graph: $x \leq -1$.

Objective 2.7e Solve inequalities using the addition principle and the multiplication principle together.

Example Solve: $8y - 7 \leq 5y + 2$.
$$8y - 7 \leq 5y + 2$$
$$8y - 7 - 8y \leq 5y + 2 - 8y$$
$$-7 \leq -3y + 2$$
$$-7 - 2 \leq -3y + 2 - 2$$
$$-9 \leq -3y$$
$$\frac{-9}{-3} \geq \frac{-3y}{-3} \quad \text{Reversing the symbol}$$
$$3 \geq y$$
The solution set is $\{y \mid 3 \geq y\}$, or $\{y \mid y \leq 3\}$.

Practice Exercise

7. Solve: $6y + 5 > 3y - 7$.

Review Exercises

Solve. [2.1b]

1. $x + 5 = -17$
2. $n - 7 = -6$

3. $x - 11 = 14$
4. $y - 0.9 = 9.09$

Solve. [2.2a]

5. $-\frac{2}{3}x = -\frac{1}{6}$
6. $-8x = -56$

7. $-\frac{x}{4} = 48$
8. $15x = -35$

9. $\frac{4}{5}y = -\frac{3}{16}$

Solve. [2.3a]

10. $5 - x = 13$
11. $\frac{1}{4}x - \frac{5}{8} = \frac{3}{8}$

Solve. [2.3b, c]

12. $5t + 9 = 3t - 1$

13. $7x - 6 = 25x$

14. $14y = 23y - 17 - 10$

15. $0.22y - 0.6 = 0.12y + 3 - 0.8y$

16. $\frac{1}{4}x - \frac{1}{8}x = 3 - \frac{1}{16}x$

17. $14y + 17 + 7y = 9 + 21y + 8$

18. $4(x + 3) = 36$

19. $3(5x - 7) = -66$

20. $8(x - 2) - 5(x + 4) = 20 + x$

21. $-5x + 3(x + 8) = 16$

22. $6(x - 2) - 16 = 3(2x - 5) + 11$

Determine whether the given number is a solution of the inequality $x \leq 4$. [2.7a]

23. -3
24. 7
25. 4

Solve. Write set notation for the answers. [2.7c, d, e]

26. $y + \dfrac{2}{3} \geq \dfrac{1}{6}$

27. $9x \geq 63$

28. $2 + 6y > 14$

29. $7 - 3y \geq 27 + 2y$

30. $3x + 5 < 2x - 6$

31. $-4y < 28$

32. $4 - 8x < 13 + 3x$

33. $-4x \leq \dfrac{1}{3}$

Graph on the number line. [2.7b, e]

34. $4x - 6 < x + 3$

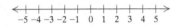

35. $-2 < x \leq 5$

36. $y > 0$

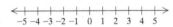

Solve. [2.4b]

37. $C = \pi d$, for d

38. $V = \dfrac{1}{3}Bh$, for B

39. $A = \dfrac{a + b}{2}$, for a

40. $y = mx + b$, for x

Solve. [2.6a]

41. *Dimensions of Wyoming.* The state of Wyoming is roughly in the shape of a rectangle whose perimeter is 1280 mi. The length is 90 mi more than the width. Find the dimensions.

42. *Interstate Mile Markers.* The sum of two consecutive mile markers on I-5 in California is 691. Find the numbers on the markers.

43. An entertainment center sold for $2449 in June. This was $332 more than the cost in February. What was the cost in February?

44. Ty is paid a commission of $4 for each magazine subscription he sells. One week, he received $108 in commissions. How many subscriptions did he sell?

45. The measure of the second angle of a triangle is 50° more than that of the first angle. The measure of the third angle is 10° less than twice the measure of the first angle. Find the measures of the angles.

Solve. [2.5a]

46. What number is 20% of 75?

47. Fifteen is what percent of 80?

48. 18 is 3% of what number?

49. *Gray Wolves.* The number of gray wolves in southwestern New Mexico and southeast Arizona increased from 97 in 2015 to 113 in 2016. What is the percent increase?

Data: seattletimes.com

50. *First-Class Mail.* The volume of first-class mail decreased from 102.4 billion pieces in 2002 to only 62.4 billion pieces in 2015. What is the percent decrease?

Data: United States Postal Service

Solve. [2.6a]

51. After a 30% reduction, a bread maker is on sale for $154. What was the marked price (the price before the reduction)?

52. A restaurant manager's salary is $78,300, which is an 8% increase over the previous year's salary. What was the previous salary?

53. A tax-exempt organization received a bill of $145.90 for janitorial supplies. The bill incorrectly included sales tax of 5%. How much does the organization actually owe?

Solve. [2.8b]

54. *Test Scores.* Noah's test grades are 71, 75, 82, and 86. What is the lowest grade that he can get on the next test and still have an average test score of at least 80?

55. The length of a rectangle is 43 cm. What widths will make the perimeter greater than 120 cm?

56. The solution of the equation
$$4(3x - 5) + 6 = 8 + x$$
is which of the following? [2.3c]
- **A.** Less than −1
- **B.** Between −1 and 1
- **C.** Between 1 and 5
- **D.** Greater than 5

57. Solve for y: $3x + 4y = P$. [2.4b]

- **A.** $y = \dfrac{P - 3x}{4}$
- **B.** $y = \dfrac{P + 3x}{4}$
- **C.** $y = P - \dfrac{3x}{4}$
- **D.** $y = \dfrac{P}{4} - 3x$

Synthesis

Solve.

58. $2|x| + 4 = 50$ [1.2e], [2.3a]

59. $|3x| = 60$ [1.2e], [2.2a]

60. $y = 2a - ab + 3$, for a [2.4b]

Understanding Through Discussion and Writing

1. Would it be better to receive a 5% raise and then an 8% raise or the other way around? Why? [2.5a]

2. Erin returns a tent that she bought during a storewide 25%-off sale that has ended. She is offered store credit for 125% of what she paid (not to be used on sale items). Is this fair to Erin? Why or why not? [2.5a]

3. Are the inequalities $x > -5$ and $-x < 5$ equivalent? Why or why not? [2.7d]

4. Explain in your own words why it is necessary to reverse the inequality symbol when multiplying on both sides of an inequality by a negative number. [2.7d]

5. If f represents Fran's age and t represents Todd's age, write a sentence that would translate to $t + 3 < f$. [2.8a]

6. Explain how the meanings of "Five more than a number" and "Five is more than a number" differ. [2.8a]

CHAPTER 2 Test

Solve.

1. $x + 7 = 15$
2. $t - 9 = 17$
3. $3x = -18$
4. $-\frac{4}{7}x = -28$
5. $3t + 7 = 2t - 5$
6. $\frac{1}{2}x - \frac{3}{5} = \frac{2}{5}$
7. $8 - y = 16$
8. $-\frac{2}{5} + x = -\frac{3}{4}$
9. $3(x + 2) = 27$
10. $-3x - 6(x - 4) = 9$
11. $0.4p + 0.2 = 4.2p - 7.8 - 0.6p$
12. $4(3x - 1) + 11 = 2(6x + 5) - 8$
13. $-2 + 7x + 6 = 5x + 4 + 2x$

Solve. Write set notation for the answers.

14. $x + 6 \leq 2$
15. $14x + 9 > 13x - 4$
16. $12x \leq 60$
17. $-2y \geq 26$
18. $-4y \leq -32$
19. $-5x \geq \frac{1}{4}$
20. $4 - 6x > 40$
21. $5 - 9x \geq 19 + 5x$

Graph on the number line.

22. $y \leq 9$

23. $6x - 3 < x + 2$

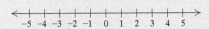

24. $-2 \leq x \leq 2$

Solve.

25. What number is 24% of 75?
26. 15.84 is what percent of 96?
27. 800 is 2% of what number?
28. *Gray Whale Calves.* A rehabilitating gray whale calf weighed 1670 lb at birth. After 4 weeks, the calf weighed 2160 lb. What is the percent increase?

    Data: "Growth of Two Captive Gray Whale Calves," J. L. Sumich, T. Goff, and W. L. Perryman. *Aquatic Mammals*, 2001, **27**(3), 231–233.

29. *Perimeter of a Photograph.* The perimeter of a rectangular photograph is 36 cm. The length is 4 cm greater than the width. Find the width and the length.

30. *Cost of Raising a Child.* It has been estimated that $41,500 will be spent for child care and K–12 education for a child to age 17. This number represents approximately 18% of the total cost of raising a child to age 17. What is the total cost of raising a child to age 17?
Data: U.S. Department of Agriculture

31. *Raffle Tickets.* The numbers on three raffle tickets are consecutive integers whose sum is 7530. Find the integers.

32. *Savings Account.* Money is invested in a savings account at 5% simple interest. After 1 year, there is $924 in the account. How much was originally invested?

33. *Board Cutting.* An 8-m board is cut into two pieces. One piece is 2 m longer than the other. How long are the pieces?

34. *Lengths of a Rectangle.* The width of a rectangle is 96 yd. Find all possible lengths such that the perimeter of the rectangle will be at least 540 yd.

35. *Budgeting.* Jason has budgeted an average of $95 per month for entertainment. For the first five months of the year, he has spent $98, $89, $110, $85, and $83. How much can Jason spend in the sixth month without exceeding his average budget?

36. *Copy Machine Rental.* A catalog publisher needs to lease a copy machine for use during a special project that they anticipate will take 3 months. It costs $225 per month plus 3.2¢ per copy to rent the machine. The company must stay within a budget of $4500 for copies. Determine (in terms of an inequality) the number of copies they can make and still remain within budget.

37. Solve $A = 2\pi rh$ for r.

38. Solve $y = 8x + b$ for x.

39. *Senior Population.* The number of Americans ages 65 and older is projected to grow from 40.4 million to 70.3 million between 2011 and 2030. Find the percent increase.
Data: U.S. Census Bureau
A. 42.5% **B.** 47%
C. 57.5% **D.** 74%

Synthesis

40. Solve $c = \dfrac{1}{a - d}$ for d.

41. Solve: $3|w| - 8 = 37$.

42. A movie theater had a certain number of tickets to give away. Five people got the tickets. The first got one-third of the tickets, the second got one-fourth of the tickets, and the third got one-fifth of the tickets. The fourth person got eight tickets, and there were five tickets left for the fifth person. Find the total number of tickets given away.

Chapters 1–2 Cumulative Review

Evaluate.

1. $\dfrac{y - x}{4}$, when $y = 12$ and $x = 6$

2. $\dfrac{3x}{y}$, when $x = 5$ and $y = 4$

3. $x - 3$, when $x = 3$

4. Translate to an algebraic expression: Four less than twice w.

Use $<$ or $>$ for $\square$ to write a true sentence.

5. $-4 \;\square\; -6$ 6. $0 \;\square\; -5$

7. $-8 \;\square\; 7$

8. Find the opposite and the reciprocal of $\dfrac{2}{5}$.

Find the absolute value.

9. $|3|$ 10. $\left|-\dfrac{3}{4}\right|$ 11. $|0|$

Compute and simplify.

12. $-6.7 + 2.3$ 13. $-\dfrac{1}{6} - \dfrac{7}{3}$

14. $-\dfrac{5}{8}\left(-\dfrac{4}{3}\right)$ 15. $(-7)(5)(-6)(-0.5)$

16. $81 \div (-9)$ 17. $-10.8 \div 3.6$

18. $-\dfrac{4}{5} \div -\dfrac{25}{8}$

Multiply.

19. $5(3x + 5y + 2z)$ 20. $4(-3x - 2)$

21. $-6(2y - 4x)$

Factor.

22. $64 + 18x + 24y$ 23. $16y - 56$

24. $5a - 15b + 25$

Collect like terms.

25. $9b + 18y + 6b + 4y$ 26. $3y + 4 + 6z + 6y$

27. $-4d - 6a + 3a - 5d + 1$

28. $3.2x + 2.9y - 5.8x - 8.1y$

Simplify.

29. $7 - 2x - (-5x) - 8$ 30. $-3x - (-x + y)$

31. $-3(x - 2) - 4x$ 32. $10 - 2(5 - 4x)$

33. $[3(x + 6) - 10] - [5 - 2(x - 8)]$

Solve.

34. $x + 1.75 = 6.25$ 35. $\dfrac{5}{2}y = \dfrac{2}{5}$

36. $-2.6 + x = 8.3$ 37. $4\dfrac{1}{2} + y = 8\dfrac{1}{3}$

38. $-\dfrac{3}{4}x = 36$

39. $\dfrac{2}{5}x = -\dfrac{3}{20}$

40. $5.8x = -35.96$

41. $-4x + 3 = 15$

42. $-3x + 5 = -8x - 7$

43. $4y - 4 + y = 6y + 20 - 4y$

44. $-3(x - 2) = -15$

45. $\dfrac{1}{3}x - \dfrac{5}{6} = \dfrac{1}{2} + 2x$

46. $-3.7x + 6.2 = -7.3x - 5.8$

47. $4(x + 2) = 4(x - 2) + 16$

48. $0(x + 3) + 4 = 0$

49. $3x - 1 < 2x + 1$

50. $3y + 7 > 5y + 13$

51. $5 - y \le 2y - 7$

52. $H = 65 - m$, for m
(To determine the number of heating degree days H for a day with m degrees Fahrenheit as the average temperature)

53. $I = Prt$, for t
(Simple-interest formula, where I is interest, P is principal, r is interest rate, and t is time)

54. What number is 24% of 105?

55. 39.6 is what percent of 88?

56. $163.35 is 45% of what?

Solve.

57. *Price Reduction.* After a 25% reduction, a book is on sale for $18.45. What was the price before reduction?

58. *Rollerblade Costs.* Susan and Melinda purchased rollerblades for a total of $107. Susan paid $17 more for her rollerblades than Melinda did. What did Melinda pay?

59. *Savings Investment.* Money is invested in a savings account at 2% simple interest. After 1 year, there is $1071 in the account. How much was originally invested?

60. *Wire Cutting.* A 143-m wire is cut into three pieces. The second piece is 3 m longer than the first. The third is four-fifths as long as the first. How long is each piece?

61. *Grade Average.* Nadia is taking a mathematics course in which four tests are given. In order to get a B, a student must average at least 80 on the four tests. Nadia scored 82, 76, and 78 on the first three tests. What scores on the last test will earn her at least a B?

62. Simplify: $-125 \div 25 \cdot 625 \div 5$.
   A. $-390{,}625$
   B. -125
   C. -625
   D. 25

Synthesis

63. A technician's salary at the end of a year is $48,418.24. This reflects a 4% salary increase and a later 3% cost-of-living adjustment during the year. What was the salary at the beginning of the year?

64. Ava needs to use a copier to reduce a drawing to fit on a page. The original drawing is 9 in. long and it must fit into a space that is 6.3 in. long. By what percent should she reduce the drawing?

Solve.

65. $4|x| - 13 = 3$

66. $\dfrac{2 + 5x}{4} = \dfrac{11}{28} + \dfrac{8x + 3}{7}$

67. $p = \dfrac{2}{m + Q}$, for Q.

CHAPTER 2 Solving Equations and Inequalities

CHAPTER 3

3.1 Introduction to Graphing
3.2 Graphing Linear Equations
3.3 More with Graphing and Intercepts

Visualizing for Success

Mid-Chapter Review

3.4 Slope and Applications

Summary and Review

Test

Cumulative Review

Graphs of Linear Equations

Approximately 69% of college graduates have some type of student debt, and, on average, each graduate owes about $29,000. In the United States, the outstanding student loan debt is over $1.4 trillion and is increasing at a rate of over $2700 per second. It is not only young people who take out student loans, as illustrated by the circle graph showing the breakdown by age of holders of student loans in 2015. The amount of student loans is growing for every age bracket, but it is growing most quickly for those ages 60 and older.

Data: marketwatch.com; The Institute for College Access and Success, Federal Reserve Bank of New York

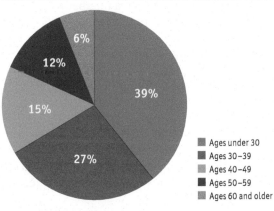

Holders of Student Loans in 2015

- Ages under 30: 39%
- Ages 30–39: 27%
- Ages 40–49: 15%
- Ages 50–59: 12%
- Ages 60 and older: 6%

DATA: Federal Reserve Bank of New York.

JUST IN TIME
Review topics 8, 14, and 18 in the **Just In Time** section at the front of the text. This provides excellent prerequisite skill review for this chapter.

MyLab Math
VIDEO

In Example 6 of Section 3.2, we will evaluate and graph a *linear equation* that models the increase in the number of consumers ages 60 and older with student loan debt.

STUDYING FOR SUCCESS *Preparing for a Test*

- [ ] Make up your own test questions as you study.
- [ ] Do an overall review of the chapter, focusing on the objectives and the examples.
- [ ] Do the exercises in the mid-chapter review and in the summary and review at the end of the chapter.
- [ ] Take the chapter test at the end of the chapter.

3.1 Introduction to Graphing

OBJECTIVES

a Plot points associated with ordered pairs of numbers; determine the quadrant in which a point lies.

b Find the coordinates of a point on a graph.

c Determine whether an ordered pair is a solution of an equation with two variables.

You probably have seen bar graphs like the following in newspapers and magazines. Note that a straight line can be drawn along the tops of the bars. Such a line is a *graph of a linear equation*. In this chapter, we study how to graph linear equations and consider properties such as slope and intercepts. Many applications of these topics will also be considered.

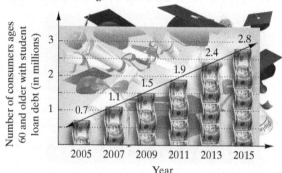

DATA: Federal Reserve Bank of New York

a PLOTTING ORDERED PAIRS

In Chapter 2, we graphed numbers and inequalities in one variable on a line. To enable us to graph an equation that contains two variables, we now learn to graph number pairs on a plane.

On the number line, each point is the graph of a number. On a plane, each point is the graph of a number pair. To form the plane, we use two perpendicular number lines called **axes**. They cross at a point called the **origin**. The arrows show the positive directions.

Consider the **ordered pair** $(3, 4)$. The numbers in an ordered pair are called **coordinates**. In $(3, 4)$, the **first coordinate** (the **abscissa**) is 3 and the **second coordinate** (the **ordinate**) is 4. To plot $(3, 4)$, we start at the origin and move *horizontally* to the 3. Then we move up *vertically* 4 units and make a "dot."

The point $(4, 3)$ is also plotted on the graph at the left. Note that $(3, 4)$ and $(4, 3)$ represent different points. The order of the numbers in the pair is important. We use the term *ordered* pairs because it makes a difference which number comes first. The coordinates of the origin are $(0, 0)$.

EXAMPLE 1 Plot the point $(-5, 2)$.

The first number, -5, is negative. Starting at the origin, we move -5 units in the horizontal direction (5 units to the left). The second number, 2, is positive. We move 2 units in the vertical direction (up).

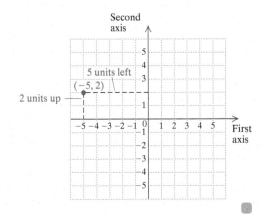

Plot these points on the grid below.

1. $(4, 5)$
2. $(5, 4)$
3. $(-2, 5)$
4. $(-3, -4)$
5. $(5, -3)$
6. $(-2, -1)$
7. $(0, -3)$
8. $(2, 0)$

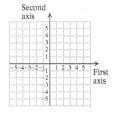

·············· **Caution!** ··············

The *first* coordinate of an ordered pair is always graphed in a *horizontal* direction, and the *second* coordinate is always graphed in a *vertical* direction.

Do Exercises 1–8. ▶

The following figure shows some points and their coordinates. In region I (the *first quadrant*), both coordinates of any point are positive. In region II (the *second quadrant*), the first coordinate is negative and the second positive. In region III (the *third quadrant*), both coordinates are negative. In region IV (the *fourth quadrant*), the first coordinate is positive and the second is negative.

EXAMPLE 2 In which quadrant, if any, are the points $(-4, 5), (5, -5), (2, 4), (-2, -5),$ and $(-5, 0)$ located?

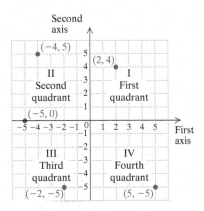

9. What can you say about the coordinates of a point in the third quadrant?

10. What can you say about the coordinates of a point in the fourth quadrant?

In which quadrant, if any, is each point located?

11. $(5, 3)$
12. $(-6, -4)$
13. $(10, -14)$
14. $(-13, 9)$
15. $(0, -3)$
16. $\left(-\dfrac{1}{2}, \dfrac{1}{4}\right)$

The point $(-4, 5)$ is in the second quadrant. The point $(5, -5)$ is in the fourth quadrant. The point $(2, 4)$ is in the first quadrant. The point $(-2, -5)$ is in the third quadrant. The point $(-5, 0)$ is on an axis and is *not in any quadrant*.

Do Exercises 9–16. ▶

Answers

1.–8. [graph showing plotted points]

9. First, negative; second, negative
10. First, positive; second, negative
11. I 12. III 13. IV 14. II 15. On an axis, not in any quadrant 16. II

SECTION 3.1 Introduction to Graphing

b FINDING COORDINATES

To find the coordinates of a point, we see how far to the right or to the left of the origin it is located and how far up or down from the origin.

EXAMPLE 3 Find the coordinates of points A, B, C, D, E, F, and G.

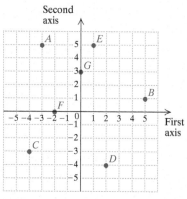

Point A is 3 units to the left (horizontal direction) and 5 units up (vertical direction). Its coordinates are $(-3, 5)$. Point D is 2 units to the right and 4 units down. Its coordinates are $(2, -4)$. The coordinates of the other points are as follows:

B: $(5, 1)$; C: $(-4, -3)$;
E: $(1, 5)$; F: $(-2, 0)$;
G: $(0, 3)$.

◀ Do Exercise 17.

17. Find the coordinates of points A, B, C, D, E, F, and G on the graph below.

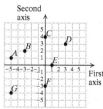

c SOLUTIONS OF EQUATIONS

SKILL REVIEW Determine whether a given number is a solution of a given equation. [2.1a]

Determine whether -3 is a solution of each equation.
1. $8(w - 3) = 0$
2. $15 = -2y + 9$

Answers: **1.** -3 is not a solution. **2.** -3 is a solution.

MyLab Math VIDEO

Now we begin to learn how graphs can be used to represent solutions of equations. When an equation contains two variables, the solutions of the equation are *ordered pairs* in which each number in the pair corresponds to a letter in the equation. Unless stated otherwise, to determine whether a pair is a solution, we use the first number in each pair to replace the variable that occurs first *alphabetically*.

EXAMPLE 4 Determine whether each of the following pairs is a solution of $4q - 3p = 22$: $(2, 7)$ and $(-1, 6)$.

For $(2, 7)$, we substitute 2 for p and 7 for q (using alphabetical order of variables). For $(-1, 6)$, we substitute -1 for p and 6 for q.

$$\begin{array}{c|c} 4q - 3p = 22 & 4q - 3p = 22 \\ \hline 4 \cdot 7 - 3 \cdot 2 \ ? \ 22 & 4 \cdot 6 - 3 \cdot (-1) \ ? \ 22 \\ 28 - 6 & 24 + 3 \\ 22 \ \text{TRUE} & 27 \ \text{FALSE} \end{array}$$

Thus, $(2, 7)$ *is* a solution of the equation, and $(-1, 6)$ *is not* a solution of the equation.

◀ Do Exercises 18 and 19.

18. Determine whether $(2, -4)$ is a solution of $4q - 3p = 22$.

$$4q - 3p = 22$$
$$4 \cdot () - 3 \cdot \ ? \ 22$$
$$-16 - $$
$$ \quad \text{FALSE}$$

Thus, $(2, -4)$ _____ a solution.
    is/is not

19. Determine whether $(2, -4)$ is a solution of $7a + 5b = -6$.

Answers
17. A: $(-5, 1)$; B: $(-3, 2)$; C: $(0, 4)$; D: $(3, 3)$; E: $(1, 0)$; F: $(0, -3)$; G: $(-5, -4)$ **18.** No   **19.** Yes
Guided Solution:
18. $-4, 2, 6, -22$, is not

EXAMPLE 5 Show that the pairs $(3, 7)$, $(0, 1)$, and $(-3, -5)$ are solutions of $y = 2x + 1$. Then graph the three points and use the graph to determine another pair that is a solution.

To show that a pair is a solution, we substitute, replacing x with the first coordinate and y with the second coordinate of each pair:

$$\begin{array}{c|c}
y = 2x + 1 & \\
\hline
7 \overset{?}{\,} 2 \cdot 3 + 1 & \\
6 + 1 & \\
7 & \text{TRUE}
\end{array}
\qquad
\begin{array}{c|c}
y = 2x + 1 & \\
\hline
1 \overset{?}{\,} 2 \cdot 0 + 1 & \\
0 + 1 & \\
1 & \text{TRUE}
\end{array}
\qquad
\begin{array}{c|c}
y = 2x + 1 & \\
\hline
-5 \overset{?}{\,} 2(-3) + 1 & \\
-6 + 1 & \\
-5 & \text{TRUE}
\end{array}$$

In each of the three cases, the substitution results in a true equation. Thus the pairs are all solutions.

We plot the points as shown at right. The order of the points follows the alphabetical order of the variables. That is, x is before y, so x-values are first coordinates and y-values are second coordinates. Similarly, we also label the horizontal axis as the x-axis and the vertical axis as the y-axis.

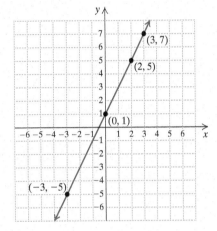

Note that the three points appear to "line up." That is, they appear to be on a straight line. Will other points that line up with these points also represent solutions of $y = 2x + 1$? To find out, we use a straightedge and sketch a line passing through $(3, 7)$, $(0, 1)$, and $(-3, -5)$.

The line appears to pass through $(2, 5)$ as well. Let's see if this pair is a solution of $y = 2x + 1$:

$$\begin{array}{c|c}
y = 2x + 1 & \\
\hline
5 \overset{?}{\,} 2 \cdot 2 + 1 & \\
4 + 1 & \\
5 & \text{TRUE}
\end{array}$$

Thus, $(2, 5)$ is a solution.

Do Exercise 20.

Example 5 leads us to suspect that any point on the line that passes through $(3, 7)$, $(0, 1)$, and $(-3, -5)$ represents a solution of $y = 2x + 1$. In fact, every solution of $y = 2x + 1$ is represented by a point on that line and every point on that line represents a solution. The line is the *graph* of the equation.

> ### GRAPH OF AN EQUATION
>
> The **graph** of an equation is a drawing that represents all of its solutions.

20. Use the graph in Example 5 to find at least two more points that are solutions of $y = 2x + 1$.

Answer
20. $(-2, -3)$, $(1, 3)$; answers may vary

3.1 Exercise Set

FOR EXTRA HELP MyLab Math

✓ Check Your Understanding

Reading Check Determine whether each statement is true or false.

RC1. In the ordered pair $(-6, 2)$, the first coordinate, -6, is also called the abscissa.

RC2. The point $(1, 0)$ is in quadrant I and in quadrant IV.

RC3. The ordered pairs $(4, -7)$ and $(-7, 4)$ name the same point.

RC4. To plot the point $(-3, 5)$, start at the origin and move horizontally to -3. Then move up vertically 5 units and make a "dot."

Concept Check Choose from the column on the right an ordered pair that is a solution of the equation.

CC1. $3p - 2t = 8$

CC2. $2x + y = 0$

CC3. $5d - 4c = 13$

CC4. $a + 4b = -7$

a) $(1, -2)$
b) $(-2, 1)$
c) $(2, -1)$
d) $(-1, 2)$

a

1. Plot these points.
   $(2, 5)$ $(-1, 3)$ $(3, -2)$ $(-2, -4)$
   $(0, 4)$ $(0, -5)$ $(5, 0)$ $(-5, 0)$

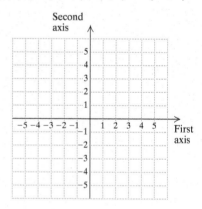

2. Plot these points.
   $(4, 4)$ $(-2, 4)$ $(5, -3)$ $(-5, -5)$
   $(0, 2)$ $(0, -4)$ $(3, 0)$ $(-4, 0)$

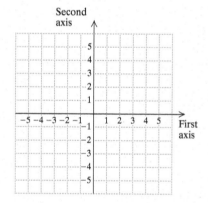

In which quadrant, if any, is each point located?

3. $(-5, 3)$
4. $(1, -12)$
5. $(100, -1)$
6. $(-2.5, 35.6)$

7. $(-6, -29)$
8. $(3.6, 105.9)$
9. $(3.8, 0)$
10. $(0, -492)$

11. $\left(-\dfrac{1}{3}, \dfrac{15}{7}\right)$
12. $\left(-\dfrac{2}{3}, -\dfrac{9}{8}\right)$
13. $\left(12\dfrac{7}{8}, -1\dfrac{1}{2}\right)$
14. $\left(23\dfrac{5}{8}, 81.74\right)$

202 CHAPTER 3 Graphs of Linear Equations

In which quadrant(s) can the point described be located?

15. The first coordinate is negative and the second coordinate is positive.

16. The first and second coordinates are positive.

17. The first coordinate is positive.

18. The second coordinate is negative.

19. The first and second coordinates are equal.

20. The first coordinate is the additive inverse of the second coordinate.

b Find the coordinates of points *A*, *B*, *C*, *D*, and *E*.

21.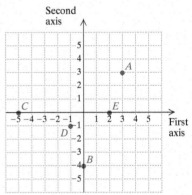

22.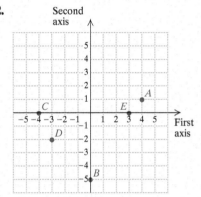

c Determine whether the given ordered pair is a solution of the equation.

23. $(2, 9)$; $y = 3x - 1$

24. $(1, 7)$; $y = 2x + 5$

25. $(4, 2)$; $2x + 3y = 12$

26. $(0, 5)$; $5x - 3y = 15$

27. $(3, -1)$; $3a - 4b = 13$

28. $(-5, 1)$; $2p - 3q = -13$

In each of Exercises 29–34, an equation and two ordered pairs are given. Show that each pair is a solution of the equation. Then use the graph of the equation to determine another solution. Answers may vary.

29. $y = x - 5$; $(4, -1)$ and $(1, -4)$

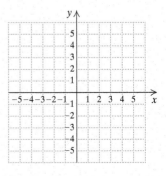

30. $y = x + 3$; $(-1, 2)$ and $(3, 6)$

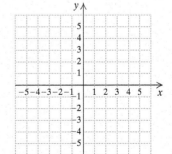

31. $y = \frac{1}{2}x + 3$; $(4, 5)$ and $(-2, 2)$

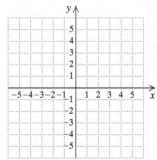

32. $3x + y = 7$; $(2, 1)$ and $(4, -5)$

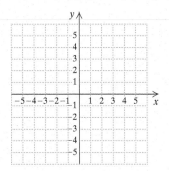

33. $4x - 2y = 10$; $(0, -5)$ and $(4, 3)$

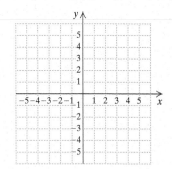

34. $6x - 3y = 3$; $(1, 1)$ and $(-1, -3)$

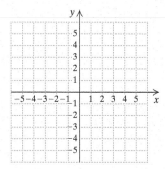

Skill Maintenance

Solve. [2.3c]

35. $6(z - 5) = 2(z + 1)$

36. $8 - 4(q + 2) = -7$

37. $-5 + x = x - 5$

38. $-\dfrac{3}{4} + x = x + \dfrac{1}{4}$

39. $4t - 2 = 3(5 - 2t)$

40. $2b - 5 + b = 6(b - 1)$

Graph on the number line. [2.7b]

41. $-2 < y \leq 1$

42. $p \geq -3$

Solve. [2.5a]

43. What is 15% of $23.80?

44. $7.29 is 15% of what number?

45. 75 is what percent of 500?

Subtract. [1.4a]

46. $\dfrac{1}{4} - \dfrac{2}{5}$

47. $-3.1 - (-3.1)$

48. $-403 - 52$

Synthesis

49. The points $(-1, 1)$, $(4, 1)$, and $(4, -5)$ are three vertices of a rectangle. Find the coordinates of the fourth vertex.

50. Three parallelograms share the vertices $(-2, -3)$, $(-1, 2)$, and $(4, -3)$. Find the fourth vertex of each parallelogram.

51. Graph eight points such that the sum of the coordinates in each pair is 6.

52. Graph eight points such that the first coordinate minus the second coordinate is 1.

53. Find the perimeter of a rectangle whose vertices have coordinates $(5, 3)$, $(5, -2)$, $(-3, -2)$, and $(-3, 3)$.

54. Find the area of a triangle whose vertices have coordinates $(0, 9)$, $(0, -4)$, and $(5, -4)$.

Graphing Linear Equations

3.2

a GRAPHS OF LINEAR EQUATIONS

Equations like $y = 2x + 1$ and $4q - 3p = 22$ are said to be **linear** because the graph of each equation is a straight line. In general, any equation equivalent to one of the form $y = mx + b$ or $Ax + By = C$, where $m, b, A, B,$ and C are constants (not variables) and A and B are not both 0, is linear.

> To graph a linear equation:
> 1. Select a value for one variable and calculate the corresponding value of the other variable. Form an ordered pair using alphabetical order as indicated by the variables.
> 2. Repeat step (1) to obtain at least two other ordered pairs. Two points are necessary in order to determine a straight line. A third point serves as a check.
> 3. Plot the ordered pairs and draw a straight line passing through the points.

OBJECTIVES

a Graph linear equations of the type $y = mx + b$ and $Ax + By = C$, identifying the y-intercept.

b Solve applied problems involving graphs of linear equations.

MyLab Math
ANIMATION

In general, calculating three (or more) ordered pairs is not difficult for equations of the form $y = mx + b$. We simply substitute values for x and calculate the corresponding values for y.

EXAMPLE 1 Graph: $y = 2x$.

First, we find some ordered pairs that are solutions. We choose *any* number for x and then determine y by substitution. Since $y = 2x$, we find y by doubling x. Suppose that we choose 3 for x. Then

$$y = 2x = 2 \cdot 3 = 6.$$

We get a solution: the ordered pair $(3, 6)$.

Suppose that we choose 0 for x. Then

$$y = 2x = 2 \cdot 0 = 0.$$

We get another solution: the ordered pair $(0, 0)$.

For a third point, we make a negative choice for x. If x is -3, we have

$$y = 2x = 2 \cdot (-3) = -6.$$

This gives us the ordered pair $(-3, -6)$.

We now have enough points to plot the line, but if we wish, we can compute more. If a number takes us off the graph paper, we either do not use it or we use larger paper or rescale the axes. Continuing in this manner, we create a table like the one shown on the following page.

CALCULATOR CORNER

Finding Solutions of Equations A table of values representing ordered pairs that are solutions of an equation can be displayed on a graphing calculator. To do this for the equation in Example 1, $y = 2x$, we first access the equation-editor screen. Then we clear any equations that are present. Next, we enter the equation, display the table set-up screen, and set both INDPNT and DEPEND to AUTO.

We will display a table of values that starts with $x = -2$ (TBLSTART) and adds 1 (ΔTBL) to the preceding x-value.

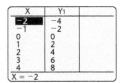

EXERCISE:

1. Create a table of ordered pairs that are solutions of the equations in Examples 2 and 3.

Complete each table and graph.

1. $y = -2x$

x	y	(x, y)
-3	6	(-3, 6)
-1		(-1,)
0	0	(0, 0)
1		(, -2)
3		(3,)

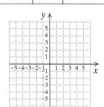

2. $y = \frac{1}{2}x$

x	y	(x, y)
4		
2		
0		
-2		
-4		

Answers

1.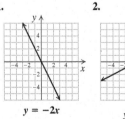

$y = -2x$, $y = \frac{1}{2}x$

Guided Solution:
1. 2, 2, -2, 1, -6, -6

Now we plot these points. Then we draw the line, or graph, with a straight-edge and label it $y = 2x$.

x	y = 2x	(x, y)
3	6	(3, 6)
1	2	(1, 2)
0	0	(0, 0)
-2	-4	(-2, -4)
-3	-6	(-3, -6)

(1) Choose x.
(2) Compute y.
(3) Form the pair (x, y).
(4) Plot the points.

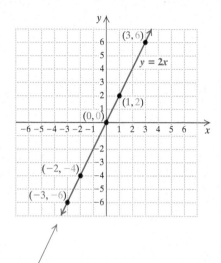

Caution!

Keep in mind that you can choose *any* number for x and then compute y. Our choice of certain numbers in the examples does not dictate which numbers you must choose.

◀ Do Exercises 1 and 2.

EXAMPLE 2 Graph: $y = -3x + 1$.

We select a value for x, compute y, and form an ordered pair. Then we repeat the process for other choices of x.

If $x = 2$, then $y = -3 \cdot 2 + 1 = -5$, and $(2, -5)$ is a solution.
If $x = 0$, then $y = -3 \cdot 0 + 1 = 1$, and $(0, 1)$ is a solution.
If $x = -1$, then $y = -3 \cdot (-1) + 1 = 4$, and $(-1, 4)$ is a solution.

Results are listed in the following table. The points corresponding to each pair are then plotted.

x	y = -3x + 1	(x, y)
2	-5	(2, -5)
0	1	(0, 1)
-1	4	(-1, 4)

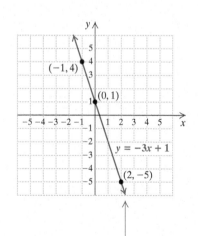

(1) Choose x.
(2) Compute y.
(3) Form the pair (x, y).
(4) Plot the points.

206   CHAPTER 3   Graphs of Linear Equations

Note that all three points line up. If they did not, we would know that we had made a mistake. When only two points are plotted, a mistake is harder to detect. We use a ruler or another straightedge to draw a line through the points. Every point on the line represents a solution of $y = -3x + 1$.

Do Exercises 3 and 4.

In Example 1, we saw that $(0, 0)$ is a solution of $y = 2x$. It is also the point at which the graph crosses the y-axis. Similarly, in Example 2, we saw that $(0, 1)$ is a solution of $y = -3x + 1$. It is also the point at which the graph crosses the y-axis. A generalization can be made: If x is replaced with 0 in the equation $y = mx + b$, then the corresponding y-value is $m \cdot 0 + b$, or b. Thus any equation of the form $y = mx + b$ has a graph that passes through the point $(0, b)$. Since $(0, b)$ is the point at which the graph crosses the y-axis, it is called the **y-intercept**. Sometimes, for convenience, we simply refer to b as the y-intercept.

y-INTERCEPT

The graph of the equation $y = mx + b$ passes through the **y-intercept** $(0, b)$.

EXAMPLE 3 Graph $y = \frac{2}{5}x + 4$ and identify the y-intercept.

We select a value for x, compute y, and form an ordered pair. Then we repeat the process for other choices of x. In this case, using multiples of 5 avoids fractions. We try to avoid graphing ordered pairs with fractions because they are difficult to graph accurately.

If $x = 0$, then $y = \dfrac{2}{5} \cdot 0 + 4 = 4$, and $(0, 4)$ is a solution.

If $x = 5$, then $y = \dfrac{2}{5} \cdot 5 + 4 = 6$, and $(5, 6)$ is a solution.

If $x = -5$, then $y = \dfrac{2}{5} \cdot (-5) + 4 = 2$, and $(-5, 2)$ is a solution.

The following table lists these solutions. Next, we plot the points and see that they form a line. Finally, we draw and label the line.

x	$y = \frac{2}{5}x + 4$	(x, y)
0	4	$(0, 4)$
5	6	$(5, 6)$
-5	2	$(-5, 2)$

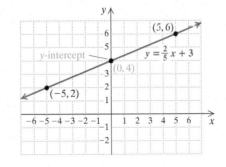

Complete each table and graph.

3. $y = 2x + 3$

x	y	(x, y)

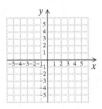

4. $y = -\dfrac{1}{2}x - 3$

x	y	(x, y)

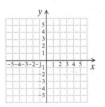

Answers

3.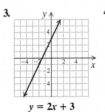
$y = 2x + 3$

4.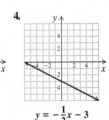
$y = -\dfrac{1}{2}x - 3$

Graph each equation and identify the y-intercept.

5. $y = \dfrac{3}{5}x + 2$

x	y	(x, y)

6. $y = -\dfrac{3}{5}x - 1$

x	y	(x, y)

Graph each equation and identify the y-intercept.

7. $5y + 4x = 0$

x	y
0	

8. $4y = 3x$

x	y
0	

We see that $(0, 4)$ is a solution of $y = \dfrac{2}{5}x + 4$. It is the y-intercept. Because the equation is in the form $y = mx + b$, we can read the y-intercept directly from the equation as follows:

$y = \dfrac{2}{5}x + 4 \quad (0, 4)$ is the y-intercept.

◀ Do Exercises 5 and 6.

Calculating ordered pairs is generally easiest when y is isolated on one side of the equation, as in $y = mx + b$. To graph an equation in which y is not isolated, we can use the addition and multiplication principles to solve for y. (See Sections 2.3 and 2.4.)

EXAMPLE 4 Graph $3y + 5x = 0$ and identify the y-intercept.

To find an equivalent equation in the form $y = mx + b$, we solve for y:

$3y + 5x = 0$

$3y + 5x - 5x = 0 - 5x$ Subtracting $5x$

$3y = -5x$ Collecting like terms

$\dfrac{3y}{3} = \dfrac{-5x}{3}$ Dividing by 3

$y = -\dfrac{5}{3}x.$

Because all the equations above are equivalent, we can use $y = -\dfrac{5}{3}x$ to draw the graph of $3y + 5x = 0$. To graph $y = -\dfrac{5}{3}x$, we select x-values and compute y-values. In this case, if we select multiples of 3, we can avoid fractions for y-values.

If $x = 0$,   then $y = -\dfrac{5}{3} \cdot 0 = 0$.

If $x = 3$,   then $y = -\dfrac{5}{3} \cdot 3 = -5$.

If $x = -3$,   then $y = -\dfrac{5}{3} \cdot (-3) = 5$.

We list these solutions in a table. Next, we plot the points and see that they form a line. Finally, we draw and label the line. The y-intercept is $(0, 0)$.

x	y	
0	0	← y-intercept
3	-5	
-3	5	

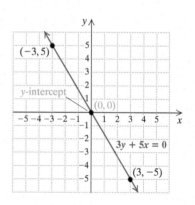

◀ Do Exercises 7 and 8.

Answers

Answers to Margin Exercises 5–8 are on p. 209.

EXAMPLE 5 Graph $4y + 3x = -8$ and identify the y-intercept.

To find an equivalent equation in the form $y = mx + b$, we solve for y:

$$4y + 3x = -8$$
$$4y + 3x - 3x = -8 - 3x \quad \text{Subtracting } 3x$$
$$4y = -3x - 8 \quad \text{Simplifying}$$
$$\frac{1}{4} \cdot 4y = \frac{1}{4} \cdot (-3x - 8) \quad \text{Multiplying by } \tfrac{1}{4} \text{ or dividing by 4}$$
$$y = \frac{1}{4} \cdot (-3x) - \frac{1}{4} \cdot 8 \quad \text{Using the distributive law}$$
$$y = -\frac{3}{4}x - 2. \quad \text{Simplifying}$$

Thus, $4y + 3x = -8$ is equivalent to $y = -\frac{3}{4}x - 2$. The y-intercept is $(0, -2)$. We find two other pairs using multiples of 4 for x to avoid fractions. We then complete and label the graph as shown.

x	y	
0	−2	← y-intercept
4	−5	
−4	1	

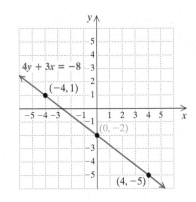

Do Exercises 9 and 10. ▶

Graph each equation and identify the y-intercept.

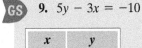

9. $5y - 3x = -10$

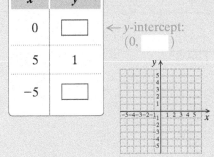

10. $5y + 3x = 20$

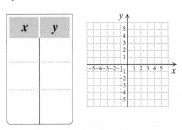

b   APPLICATIONS OF LINEAR EQUATIONS

Mathematical concepts become more understandable through visualization. Throughout this text, you will occasionally see the heading Algebraic–Graphical Connection, as in Example 6, which follows. In this feature, the algebraic approach is enhanced and expanded with a graphical connection. Relating a solution of an equation to a graph can often give added meaning to the algebraic solution.

EXAMPLE 6 *Student Loans.* The number of consumers y, in millions, ages 60 and older with student loan debt can be estimated and projected by the equation

$$y = 0.21t + 0.68,$$

where t is the number of years after 2005. That is, $t = 0$ corresponds to 2005, $t = 3$ corresponds to 2008, and so on.

Data: Federal Reserve Bank of New York

Answers

5. $y = \frac{3}{5}x + 2$

6. $y = -\frac{3}{5}x - 1$

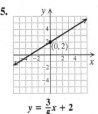

7. $5y + 4x = 0$

8. $4y = 3x$

9. $5y - 3x = -10$

10. $5y + 3x = 20$

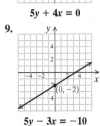 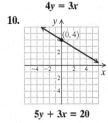

Guided Solution:
9. −2, −5, −2

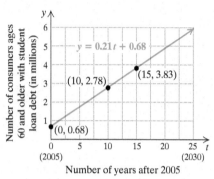

FIGURE 1

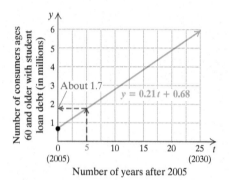

FIGURE 2

11. *Home Ownership.* The home ownership rate r, in percent, in the United States among millennials can be estimated by

$$r = -0.9t + 39.8,$$

where t is the number of years after 2009.

Data: U.S. Census Bureau

a) Find the home ownership rate among millennials in 2009, in 2016, and in 2020.

b) Graph the equation and use the graph to estimate the home ownership rate among millennials in 2013.

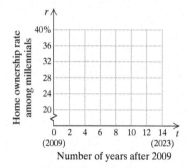

c) In which year will the home ownership rate among millennials be 29%?

a) Estimate the number of consumers ages 60 and older with student loan debt in 2005 and in 2015. Then predict the number in 2020.

b) Graph the equation and then use the graph to estimate the number of consumers ages 60 and older with student loan debt in 2010.

c) In what year could we predict the number of consumers ages 60 and older with student loan debt to be 5.3 million?

a) The years 2005, 2015, and 2020 correspond to $t = 0$, $t = 10$, and $t = 15$, respectively. We substitute 0, 10, and 15 for t and then calculate y:

$y = 0.21(0) + 0.68 = 0 + 0.68 = 0.68;$
$y = 0.21(10) + 0.68 = 2.1 + 0.68 = 2.78;$
$y = 0.21(15) + 0.68 = 3.15 + 0.68 = 3.83.$

The number of consumers ages 60 and older with student loan debt is estimated or projected to be 0.68 million in 2005, 2.78 million in 2015, and 3.83 million in 2020.

ALGEBRAIC ⟩⟨ GRAPHICAL CONNECTION

b) We have three ordered pairs from part (a). We plot these points and see that they line up. Thus our calculations are probably correct. Since we are considering only the number of years after 2005 ($t \geq 0$) and since the number of consumers for those years will be positive ($y > 0$), we need only the first quadrant for the graph. We use the three points we have plotted to draw a straight line. (See Figure 1.)

To use the graph to estimate the number of consumers ages 60 and older with student loan debt in 2010, we note in Figure 2 that this year corresponds to $t = 5$. We need to determine which y-value is paired with $t = 5$. We locate the point on the graph by moving up vertically from $t = 5$ and then finding the value on the y-axis that corresponds to that point. It appears that a good estimate is 1.7.

To check our estimate, we can simply substitute into the equation:

$$y = 0.21(5) + 0.68 = 1.05 + 0.68 = 1.73.$$

This is close to 1.7, so our estimate is good. From the graph, we estimate that about 1.7 million consumers ages 60 and older had student loan debt in 2010.

c) We substitute 5.3 for y and solve for t:

$y = 0.21t + 0.68$
$5.3 = 0.21t + 0.68$
$4.62 = 0.21t$
$22 = t.$

We predict that in 22 years after 2005, or in 2027, 5.3 million consumers ages 60 and older will have student loan debt.

◀ Do Exercise 11 (the answer is on p. 211).

Many equations in two variables have graphs that are not straight lines. Three such nonlinear graphs are shown below. We will cover some such graphs in the optional Calculator Corners throughout the text and in Chapter 11.

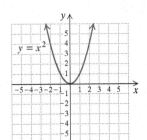
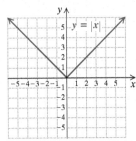
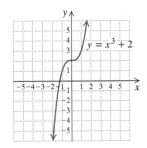

CALCULATOR CORNER

Graphing Equations Equations must be solved for y before they can be graphed on most graphing calculators. Consider the equation $3x + 2y = 6$. Solving for y, we get $y = \dfrac{6 - 3x}{2}$.

We enter this equation as $y_1 = (6 - 3x)/2$ on the equation-editor screen. Then we select the standard viewing window and display the graph.

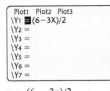
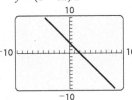

EXERCISES: Graph each equation in the standard viewing window $[-10, 10, -10, 10]$, with $\text{Xscl} = 1$ and $\text{Yscl} = 1$.

1. $y = -5x + 3$
2. $y = 4x - 5$
3. $4x - 5y = -10$
4. $5y + 5 = -3x$

Answer

11. **(a)** 2009: 39.8%; 2016: 33.5%; 2020: 29.9%;
(b) about 36%;

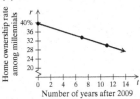

(c) 12 years after 2009, or in 2021

3.2 Exercise Set

FOR EXTRA HELP MyLab Math

✓ Check Your Understanding

Reading Check For each equation, choose from the following list an equivalent equation.

a) $y = \dfrac{5}{2}x + \dfrac{5}{2}$ b) $y = \dfrac{2}{5}x$ c) $y = -\dfrac{5}{2}x - 5$ d) $y = -\dfrac{2}{5}x$

RC1. $2y + 5x = -10$ RC2. $5y - 2x = 0$ RC3. $2x + 5y = 0$ RC4. $5x - 2y = -5$

Concept Check Choose from the following list the y-intercept of the graph of each equation.

a) $\left(\dfrac{5}{2}, 0\right)$ b) $(0, 0)$ c) $\left(0, -\dfrac{5}{2}\right)$ d) $(2, 0)$ e) $(0, -2)$ f) $(0, -5)$

CC1. $2x - 5y = 10$ CC2. $2x = 5y$ CC3. $y = \dfrac{5}{2}x - 5$ CC4. $5x + 2y = -5$

a Graph each equation and identify the *y*-intercept.

1. $y = x + 1$

x	y
−2	
−1	
0	
1	
2	
3	

2. $y = x - 1$

x	y
−2	
−1	
0	
1	
2	
3	

3. $y = x$

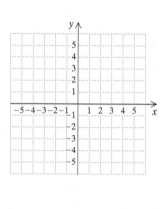

4. $y = -x$

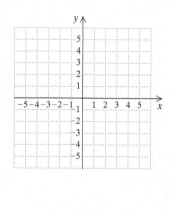

5. $y = \dfrac{1}{2}x$

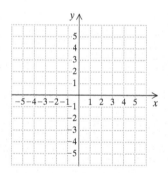

6. $y = \dfrac{1}{3}x$

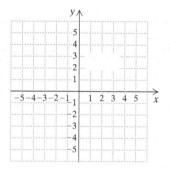

7. $y = x - 3$

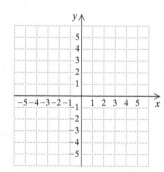

8. $y = x + 3$

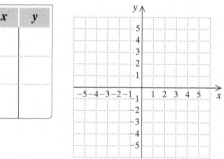

212 CHAPTER 3 Graphs of Linear Equations

9. $y = 3x - 2$

10. $y = 2x + 2$

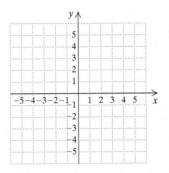

11. $y = \frac{1}{2}x + 1$

12. $y = \frac{1}{3}x - 4$

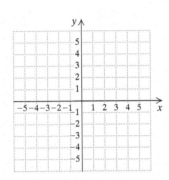

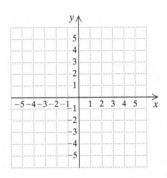

13. $x + y = -5$

14. $x + y = 4$

15. $y = \frac{5}{3}x - 2$

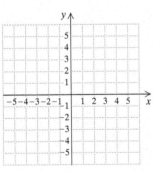
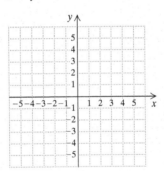
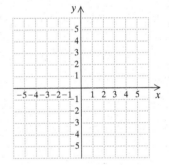

16. $y = \frac{5}{2}x + 3$

17. $x + 2y = 8$

18. $x + 2y = -6$

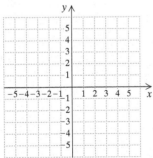
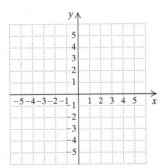
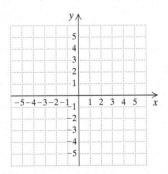

SECTION 3.2 Graphing Linear Equations

19. $y = \dfrac{3}{2}x + 1$

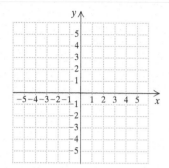

20. $y = -\dfrac{1}{2}x - 3$

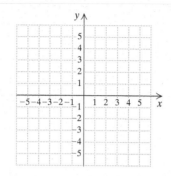

21. $8x - 2y = -10$

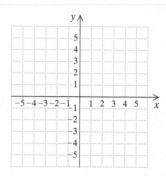

22. $6x - 3y = 9$

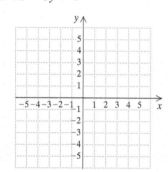

23. $8y + 2x = -4$

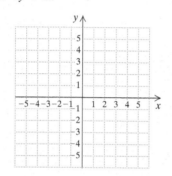

24. $6y + 2x = 8$

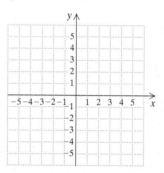

b Solve.

25. *World Population.* The world population P, in billions, can be approximated and projected by the equation
$$P = 0.078t + 6.081,$$
where t is the number of years after 2000.

Data: U.S. Census Bureau

a) Estimate the world population in 2000 and in 2015. Then project the population in 2030.
b) Graph the equation and use the graph to estimate the world population in 2020.

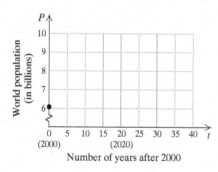

c) In what year does the model project the world population will be 8.811 billion?

26. *Prescription Drug Costs.* The average retail cost c, in dollars, in the United States of a one-year supply of a prescription drug can be approximated and projected by
$$c = 840t + 5050,$$
where t is the number of years after 2006.

Data: AARP Bulletin, April 2016

a) Determine the average retail cost of a one-year supply of a prescription drug in 2006, in 2012, and in 2016.
b) Graph the equation and use the graph to estimate the average retail cost of a one-year supply of a prescription drug in 2018.

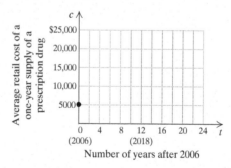

c) In what year will the average retail cost of a one-year supply of a prescription drug be approximately $20,000?

27. *High School Dropout Rate.* The dropout rate d, in percent, of high school students in the United States can be approximated and projected by
$$d = -0.34t + 10.9,$$
where t is the number of years after 2000.

Data: The World Almanac, 2017

a) Find the high school dropout rate in 2000, in 2010, and in 2018.

b) Graph the equation and then use the graph to estimate the high school dropout rate in 2015.

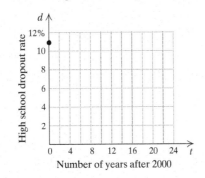

c) At this rate of decline, in what year will the high school dropout rate be 4.1%?

28. *Record Temperature Drop.* On 22 January 1943, the temperature T, in degrees Fahrenheit, in Spearfish, South Dakota, could be approximated by
$$T = -2.15m + 54,$$
where m is the number of minutes after 9:00 that morning.

Data: Information Please Almanac

a) Find the temperature at 9:01 A.M., at 9:08 A.M., and at 9:20 A.M.

b) Graph the equation and use the graph to estimate the temperature at 9:15 A.M.

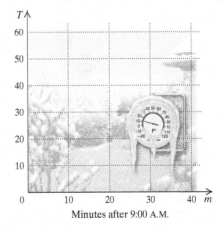

c) The temperature stopped dropping when it reached $-4°F$. At what time did this occur?

Skill Maintenance

Find the absolute value. [1.2e]

29. $|-12|$

30. $|4.89|$

31. $|0|$

32. $\left|-\frac{4}{5}\right|$

Solve. [2.3a]

33. $2x - 14 = 29$

34. $\frac{1}{3}t + 6 = -12$

35. $-10 = 1.2y + 2$

36. $4 - 5w = -16$

Solve. [2.6a]

37. *Books in Libraries.* The Library of Congress houses 33.5 million books. This number is 0.3 million more than twice the number of books in the New York Public Library. How many books are in the New York Public Library?

Data: American Library Association

38. *Debt.* In 2017, the total U.S. student loan debt was $1.4 trillion. This was $160 billion less than twice the total U.S. credit card debt. What was the total U.S. credit card debt in 2017?

Data: studentloanhero.com

Calculate.

39. $-\frac{3}{5} \div 5$ [1.6c]

40. $2.8 - (-0.2)$ [1.4a]

41. $-\frac{9}{16} + \left(-\frac{3}{8}\right)$ [1.3a]

42. $4.2 \times (-100)$ [1.5a]

43. $-\frac{8}{7} \div \left(-\frac{1}{4}\right)$ [1.6c]

44. $23.3 - 32.3$ [1.4a]

3.3 More with Graphing and Intercepts

OBJECTIVES

a Find the intercepts of a linear equation, and graph using intercepts.

b Graph equations equivalent to those of the type $x = a$ and $y = b$.

a GRAPHING USING INTERCEPTS

SKILL REVIEW

Solve equations using both the addition principle and the multiplication principle. [2.3a]

Solve.

1. $5x - 7 = -10$
2. $-20 = \dfrac{7}{4}x + 8$

Answers: 1. $-\dfrac{3}{5}$ 2. -16

In Section 3.2, we graphed linear equations of the form $Ax + By = C$ by first solving for y to find an equivalent equation in the form $y = mx + b$. We did so because it is then easier to calculate the y-value that corresponds to a given x-value. Another convenient way to graph $Ax + By = C$ is to use **intercepts**. Look at the graph of $-2x + y = 4$ shown at left.

The y-intercept is $(0, 4)$. It occurs where the line crosses the y-axis and thus will always have 0 as the first coordinate. The x-intercept is $(-2, 0)$. It occurs where the line crosses the x-axis and thus will always have 0 as the second coordinate.

◀ Do Exercise 1.

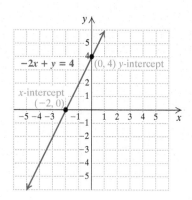

We find intercepts as follows.

INTERCEPTS

The **y-intercept** is $(0, b)$. To find b, let $x = 0$ and solve the equation for y.

The **x-intercept** is $(a, 0)$. To find a, let $y = 0$ and solve the equation for x.

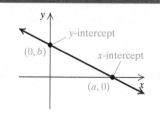

1. Look at the graph shown below.

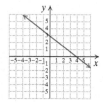

a) Find the coordinates of the y-intercept.

b) Find the coordinates of the x-intercept.

EXAMPLE 1 Consider $4x + 3y = 12$. Find the intercepts. Then graph the equation using the intercepts.

To find the y-intercept, we let $x = 0$. Then we solve for y:

$$4 \cdot 0 + 3y = 12$$
$$3y = 12$$
$$y = 4.$$

Thus, $(0, 4)$ is the y-intercept. Note that finding this intercept involves covering up the x-term and solving the rest of the equation for y.

To find the x-intercept, we let $y = 0$. Then we solve for x:

$$4x + 3 \cdot 0 = 12$$
$$4x = 12$$
$$x = 3.$$

Answer

1. (a) $(0, 3)$; (b) $(4, 0)$

Thus, $(3, 0)$ is the x-intercept. Note that finding this intercept involves covering up the y-term and solving the rest of the equation for x.

We plot these points and draw the line, or graph.

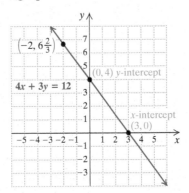

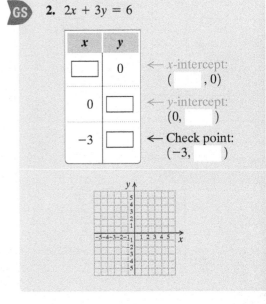

A third point should be used as a check. We substitute any convenient value for x and solve for y. In this case, we choose $x = -2$. Then

$4(-2) + 3y = 12$ Substituting -2 for x
$-8 + 3y = 12$
$3y = 20$ Adding 8 on both sides
$y = \frac{20}{3}$, or $6\frac{2}{3}$. Solving for y

It appears that the point $(-2, 6\frac{2}{3})$ is on the graph, though graphing fraction values can be inexact. The graph is probably correct.

Do Exercises 2 and 3. ▶

Graphs of equations of the type $y = mx$ pass through the origin. Thus the x-intercept and the y-intercept are the same, $(0, 0)$. In such cases, we must calculate another point in order to complete the graph. A third point would also need to be calculated if a check is desired.

EXAMPLE 2 Graph: $y = 3x$.

We know that $(0, 0)$ is both the x-intercept and the y-intercept. We calculate values at two other points and complete the graph, knowing that it passes through the origin $(0, 0)$.

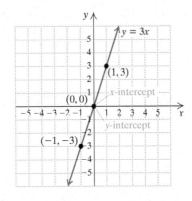

Do Exercises 4 and 5 on the following page. ▶

For each equation, find the intercepts. Then graph the equation using the intercepts.

2. $2x + 3y = 6$

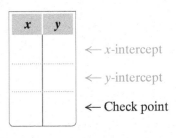

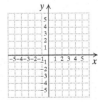

3. $3y - 4x = 12$

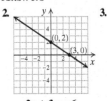

Answers

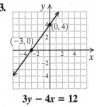

Guided Solution:
2. 3, 2, 4, 3, 2, 4

SECTION 3.3   More with Graphing and Intercepts   **217**

Graph.

4. $y = 2x$

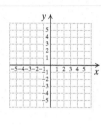

5. $y = -\dfrac{2}{3}x$

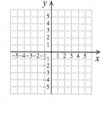

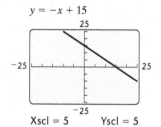

CALCULATOR CORNER

Viewing the Intercepts Knowing the intercepts of a linear equation helps us to determine a good viewing window for the graph of the equation. For example, when we graph the equation $y = -x + 15$ in the standard window, we see only a small portion of the graph in the upper right-hand corner of the screen, as shown on the left below.

Using algebra, as we did in Example 1, we find that the intercepts of the graph of this equation are $(0, 15)$ and $(15, 0)$. This tells us that, if we are to see more of the graph than is shown on the left above, both Xmax and Ymax should be greater than 15. We can try different window settings until we find one that suits us. One good choice is $[-25, 25, -25, 25]$, with Xscl = 5 and Yscl = 5, shown on the right above.

EXERCISES: Find the intercepts of each equation algebraically. Then graph the equation on a graphing calculator, choosing window settings that allow the intercepts to be seen clearly. (Settings may vary.)

1. $y = -7.5x - 15$
2. $y - 2.15x = 43$
3. $6x - 5y = 150$
4. $y = 0.2x - 4$
5. $y = 1.5x - 15$
6. $5x - 4y = 2$

b EQUATIONS WHOSE GRAPHS ARE HORIZONTAL LINES OR VERTICAL LINES

EXAMPLE 3 Graph: $y = 3$.

The equation $y = 3$ tells us that y must be 3, but it doesn't give us any information about x. We can also think of this equation as $0 \cdot x + y = 3$. No matter what number we choose for x, we find that y is 3. We make up a table with all 3's in the y-column.

x	y
	3
	3
	3

Choose any number for $x. \rightarrow$

y must be 3.

x	y
-2	3
0	3	← y-intercept
4	3

Answers

4.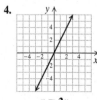
$y = 2x$

5.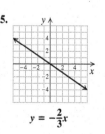
$y = -\dfrac{2}{3}x$

218 CHAPTER 3 Graphs of Linear Equations

When we plot the ordered pairs $(-2, 3)$, $(0, 3)$, and $(4, 3)$ and connect the points, we obtain a horizontal line. Any ordered pair $(x, 3)$ is a solution. So the line is parallel to the x-axis with y-intercept $(0, 3)$.

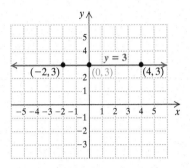

EXAMPLE 4 Graph: $x = -4$.

Consider $x = -4$. We can also think of this equation as $x + 0 \cdot y = -4$. We make up a table with all -4's in the x-column.

x	y
-4	
-4	
-4	
-4	

x must be -4.

x	y
-4	-5
-4	1
-4	3
-4	0

← Choose any number for y.

x-intercept →

When we plot the ordered pairs $(-4, -5)$, $(-4, 1)$, $(-4, 3)$, and $(-4, 0)$ and connect the points, we obtain a vertical line. Any ordered pair $(-4, y)$ is a solution. So the line is parallel to the y-axis with x-intercept $(-4, 0)$.

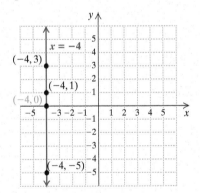

HORIZONTAL LINES AND VERTICAL LINES

The graph of $y = b$ is a **horizontal line**. The y-intercept is $(0, b)$.

The graph of $x = a$ is a **vertical line**. The x-intercept is $(a, 0)$.

Do Exercises 6–9.

Graph.

6. $x = 5$

x	y
5	-4
☐	0
☐	3

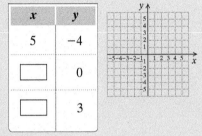

7. $y = -2$

x	y
-1	-2
0	☐
2	☐

8. $x = -3$

x	y

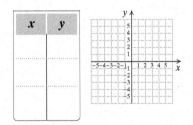

9. $x = 0$

x	y

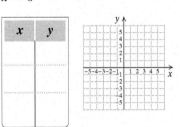

Answers

Answers to Margin Exercises 6–9 and Guided Solutions 6 and 7 are on p. 220.

The following is a general procedure for graphing linear equations.

GRAPHING LINEAR EQUATIONS

1. If the equation is of the type $x = a$ or $y = b$, the graph will be a line parallel to an axis; $x = a$ is vertical and $y = b$ is horizontal.
   Examples.

   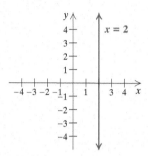 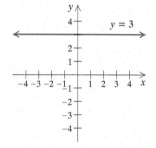

2. If the equation is of the type $y = mx$, both intercepts are the origin, $(0, 0)$. Plot $(0, 0)$ and two other points.
   Example.

   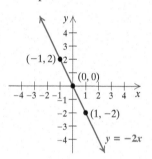

3. If the equation is of the type $y = mx + b$, plot the y-intercept $(0, b)$ and two other points.
   Example.

   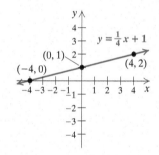

4. If the equation is of the type $Ax + By = C$, but not of the type $x = a$ or $y = b$, then either solve for y and proceed as with the equation $y = mx + b$, or graph using intercepts. If the intercepts are too close together, choose another point or points farther from the origin.
   Examples.

   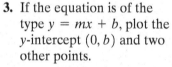

Answers

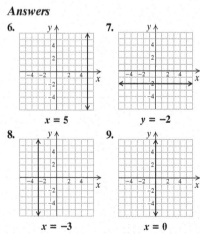

6. y, $x = 5$
7. y, $y = -2$
8. y, $x = -3$
9. y, $x = 0$

Guided Solutions:
6. 5, 5   7. −2, −2

220   CHAPTER 3   Graphs of Linear Equations

Visualizing for Success

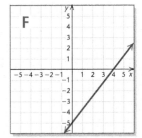

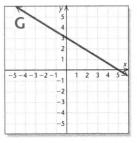

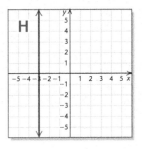

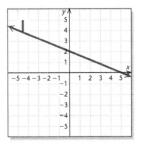

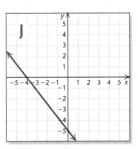

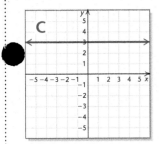

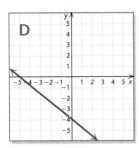

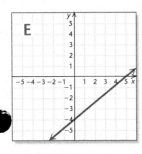

Match each equation with its graph.

1. $5y + 20 = 4x$

2. $y = 3$

3. $3x + 5y = 15$

4. $5y + 4x = 20$

5. $5y = 10 - 2x$

6. $4x + 5y + 20 = 0$

7. $5x - 4y = 20$

8. $4y + 5x + 20 = 0$

9. $5y - 4x = 20$

10. $x = -3$

Answers on page A-7

3.3 Exercise Set

✓ Check Your Understanding

Reading Check Choose from the list on the right the word or the expression that best completes each statement. Not every choice will be used.

RC1. The graph of $y = -3$ is a(n) _____ line with $(0, -3)$ as its _____.

RC2. The x-intercept occurs when a line crosses the _____.

RC3. In the graph of $y = 2x$, the point _____ is both the x-intercept and the y-intercept.

RC4. The graph of $x = 4$ is a(n) _____ line with $(4, 0)$ as its _____.

RC5. The y-intercept occurs when a line crosses the _____.

RC6. Graphs of equations of the type $y = mx$ pass through the _____.

$(0, 2)$
$(0, 0)$
horizontal
vertical
origin
x-intercept
y-intercept
x-axis
y-axis

Concept Check Determine whether each statement is true or false.

CC1. To find the x-intercept of the graph of $2x - 7y = -14$, let $x = 0$.

CC2. The second coordinate of each point of the graph of $x = 5$ is 5.

CC3. The second coordinate of each point of the graph of $y = -5$ is -5.

CC4. To find the y-intercept of the graph of $y = \frac{1}{2}x + 3$, let $x = 0$.

a For each of Exercises 1–4, find (a) the coordinates of the y-intercept and (b) the coordinates of the x-intercept.

1.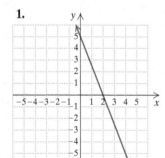
2.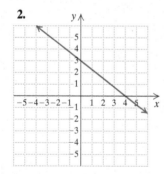
3.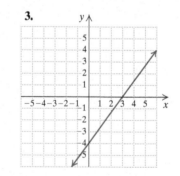
4.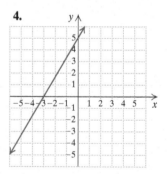

For each of Exercises 5–12, find (a) the coordinates of the y-intercept and (b) the coordinates of the x-intercept. Do not graph.

5. $3x + 5y = 15$
6. $5x + 2y = 20$
7. $7x - 2y = 28$
8. $3x - 4y = 24$

9. $-4x + 3y = 10$
10. $-2x + 3y = 7$
11. $6x - 3 = 9y$
12. $4y - 2 = 6x$

For each equation, find the intercepts. Then use the intercepts to graph the equation.

13. $x + 3y = 6$

x	y
0		← y-intercept
	0	← x-intercept

14. $x + 2y = 2$

x	y
0		← y-intercept
	0	← x-intercept

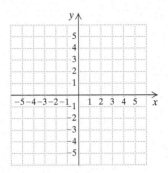

15. $-x + 2y = 4$

x	y
0		← y-intercept
	0	← x-intercept

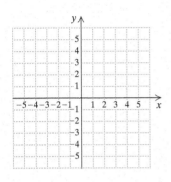

16. $-x + y = 5$

x	y
0		← y-intercept
	0	← x-intercept

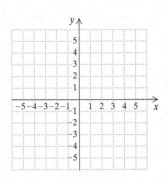

17. $3x + y = 6$

x	y
0		← y-intercept
	0	← x-intercept

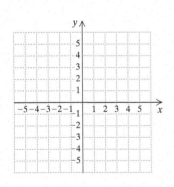

18. $2x + y = 6$

x	y
0		← y-intercept
	0	← x-intercept

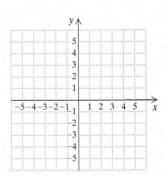

19. $2y - 2 = 6x$

x	y
		← y-intercept
		← x-intercept

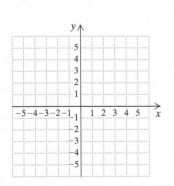

20. $3y - 6 = 9x$

x	y
		← y-intercept
		← x-intercept

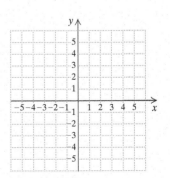

21. $3x - 9 = 3y$

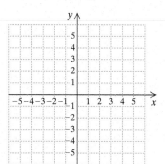

22. $5x - 10 = 5y$

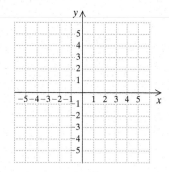

23. $2x - 3y = 6$

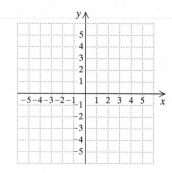

24. $2x - 5y = 10$

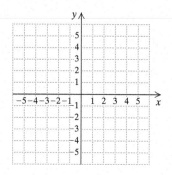

25. $4x + 5y = 20$

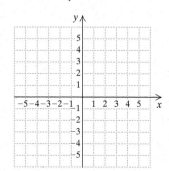

26. $2x + 6y = 12$

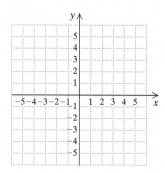

27. $2x + 3y = 8$

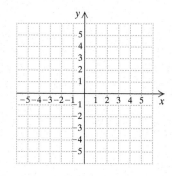

28. $x - 1 = y$

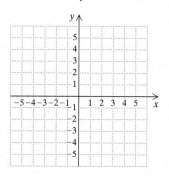

29. $3x + 4y = 5$

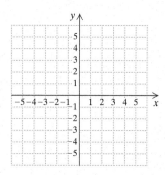

30. $2x - 1 = y$

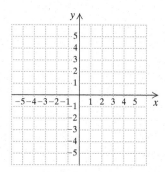

31. $3x - 2 = y$

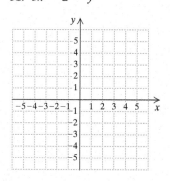

32. $4x - 3y = 12$

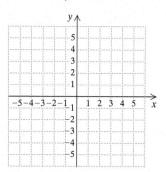

33. $6x - 2y = 12$

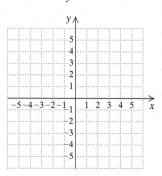

34. $7x + 2y = 6$

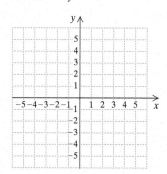

35. $y = -3 - 3x$

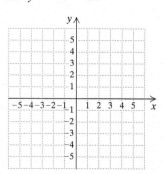

36. $-3x = 6y - 2$

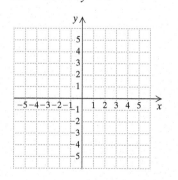

CHAPTER 3 Graphs of Linear Equations

37. $y - 3x = 0$

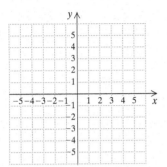

38. $x + 2y = 0$

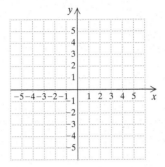

b Graph.

39. $x = -2$

x	y
-2	
-2	
-2	

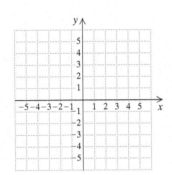

40. $x = 1$

x	y
1	
1	
1	

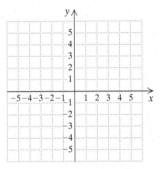

41. $y = 2$

x	y
	2
	2
	2

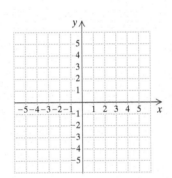

42. $y = -4$

x	y
	-4
	-4
	-4

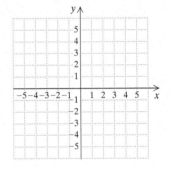

43. $x = 2$

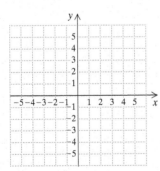

44. $x = 3$

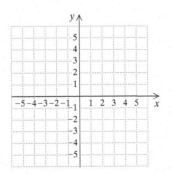

45. $y = 0$

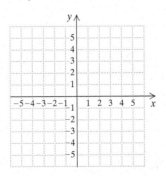

46. $y = -1$

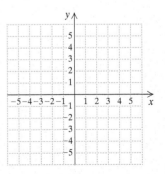

SECTION 3.3 More with Graphing and Intercepts

47. $x = \dfrac{3}{2}$

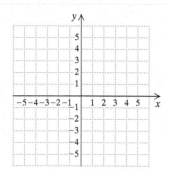

48. $x = -\dfrac{5}{2}$

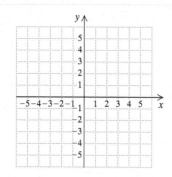

49. $3y = -5$

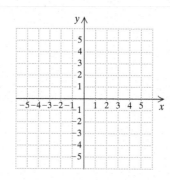

50. $12y = 45$

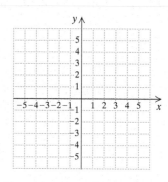

51. $4x + 3 = 0$

52. $-3x + 12 = 0$

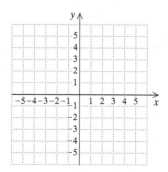

53. $48 - 3y = 0$

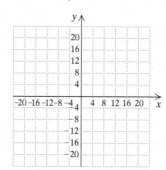

54. $63 + 7y = 0$

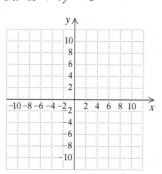

Write an equation for the graph shown.

55.

56.

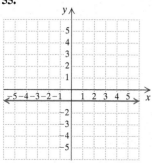

57.

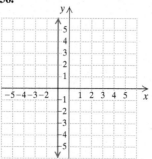

58.

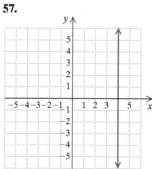

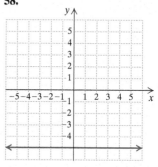

Skill Maintenance

Solve. [2.7e]

59. $x + (x - 1) < (x + 2) - (x + 1)$

60. $6 - 18x \leq 4 - 12x - 5x$

61. $\dfrac{2x}{7} - 4 \leq -2$

62. $\dfrac{1}{4} + \dfrac{x}{3} > \dfrac{7}{12}$

Synthesis

63. Write an equation of a line parallel to the x-axis and passing through $(-3, -4)$.

64. Find the value of m such that the graph of $y = mx + 6$ has an x-intercept of $(2, 0)$.

65. Find the value of k such that the graph of $3x + k = 5y$ has an x-intercept of $(-4, 0)$.

66. Find the value of k such that the graph of $4x = k - 3y$ has a y-intercept of $(0, -8)$.

Mid-Chapter Review

Concept Reinforcement

Determine whether each statement is true or false.

_____ 1. In quadrant II, the first coordinate of all points is less than the second coordinate. [3.1a]

_____ 2. The y-intercept of the graph of $2 - y = 3x$ is $(0, -3)$. [3.3a]

_____ 3. The y-intercept of $Ax + By = C, B \neq 0$, is $\left(0, \dfrac{C}{B}\right)$. [3.3a]

_____ 4. Both coordinates of points in quadrant IV are negative. [3.1a]

Guided Solutions

5. Given the graph of the line below, fill in the letters and numbers that create correct statements. [3.3a]

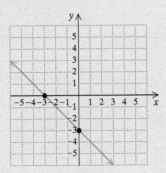

a) The ___-intercept is (___, −3).
b) The ___-intercept is (___, 0).

6. Given the graph of the line below, fill in the letters and numbers that create correct statements. [3.3a]

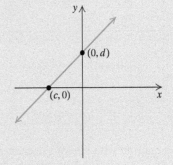

a) The x-intercept is (___, ___).
b) The y-intercept is (___, ___).

Mixed Review

7. Determine the coordinates of points A, B, C, D, and E. [3.1b]

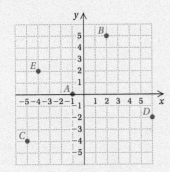

8. Determine the coordinates of points F, G, H, I, and J. [3.1b]

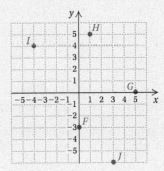

Determine whether the given ordered pair is a solution of the equation. [3.1c]

9. $(8, -5)$; $-2q - 7p = 19$

10. $\left(-1, \dfrac{2}{3}\right)$; $6y = -3x + 1$

Find the coordinates of the x-intercept and the y-intercept. [3.3a]

11. $-3x + 2y = 18$

12. $x - \dfrac{1}{2} = 10y$

13. $x - 40 = 20y$

14. $\dfrac{1}{3}y - \dfrac{5}{6}x = 35$

Graph. [3.2a], [3.3a, b]

15. $-2x + y = -3$

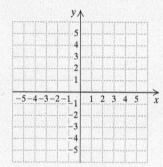

16. $y = -\dfrac{3}{2}$

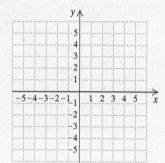

17. $y = -x + 4$

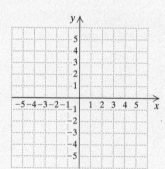

18. $x = 0$

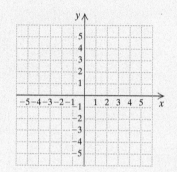

Match each equation with one of the characteristics listed on the right. [3.3a, b]

19. $y = -1$

20. $x = 1$

21. $y = -x - 1$

22. $y = x - 1$

23. $y = x + 1$

A. The y-intercept is $(0, 1)$ and the x-intercept is $(-1, 0)$.
B. The x-intercept is $(-1, 0)$ and the y-intercept is $(0, -1)$.
C. The line is vertical and the x-intercept is $(1, 0)$.
D. The line is horizontal and the y-intercept is $(0, -1)$.
E. The y-intercept is $(0, -1)$ and the x-intercept is $(1, 0)$.

Understanding Through Discussion and Writing

24. Do all graphs of linear equations have y-intercepts? Why or why not? [3.3b]

25. The equations $3x + 4y = 8$ and $y = -\dfrac{3}{4}x + 2$ are equivalent. Which equation is easier to graph and why? [3.2a], [3.3a]

26. If the graph of the equation $Ax + By = C$ is a horizontal line, what can you conclude about A? Why? [3.3b]

27. Explain in your own words why the graph of $x = 7$ is a vertical line. [3.3b]

STUDYING FOR SUCCESS Taking a Test

- ☐ Read each question carefully. Be sure you understand the question before you answer it.
- ☐ Try to answer all the questions first time through, marking those to recheck if you have time.
- ☐ Write your test in a neat and orderly manner.

3.4 Slope and Applications

a SLOPE

We have considered two forms of a linear equation, $Ax + By = C$ and $y = mx + b$. We found that from the form of the equation $y = mx + b$, we know that the y-intercept of the line is $(0, b)$.

$$y = mx + b.$$
$ \uparrow \quad \downarrow$
$\, ? \qquad \text{The } y\text{-intercept is } (0, b).$

What about the constant m? Does it give us information about the line? Look at the graphs in the margin and see if you can make any connection between the constant m and the "slant" of the line.

The graphs of some linear equations slant upward from left to right. Others slant downward. Some are vertical and some are horizontal. Some slant more steeply than others. We now look for a way to describe such possibilities with numbers.

Consider a line with two points marked P and Q. As we move from P to Q, the y-coordinate changes from 1 to 3 and the x-coordinate changes from 2 to 6. The change in y is $3 - 1$, or 2. The change in x is $6 - 2$, or 4.

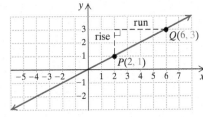

We call the change in y the **rise** and the change in x the **run**. The ratio rise/run is the same for any two points on a line. We call this ratio the **slope** of the line. Slope describes the slant of a line. The slope of the line in the graph above is given by

$$\frac{\text{rise}}{\text{run}} = \frac{\text{the change in } y}{\text{the change in } x}, \text{ or } \frac{2}{4}, \text{ or } \frac{1}{2}.$$

SLOPE

The **slope** of a line containing points (x_1, y_1) and (x_2, y_2) is given by

$$m = \frac{\text{rise}}{\text{run}} = \frac{\text{the change in } y}{\text{the change in } x} = \frac{y_2 - y_1}{x_2 - x_1}.$$

OBJECTIVES

a Given the coordinates of two points on a line, find the slope of the line, if it exists.

b Find the slope of a line from an equation.

c Find the slope, or rate of change, in an applied problem involving slope.

SKILL REVIEW Subtract real numbers. [1.4a]
Subtract.
1. $-4 - 20$
2. $-21 - (-5)$

Answers: 1. -24 2. -16

MyLab Math
ANIMATION

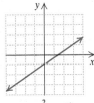
$y = \tfrac{2}{3}x - 1$

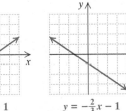
$y = -\tfrac{2}{3}x - 1$

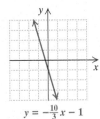
$y = -\tfrac{10}{3}x - 1$

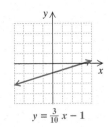
$y = \tfrac{3}{10}x - 1$

SECTION 3.4 Slope and Applications 229

In the preceding definition, (x_1, y_1) and (x_2, y_2) —read "x sub-one, y sub-one and x sub-two, y sub-two"—represent two different points on a line. It does not matter which point is considered (x_1, y_1) and which is considered (x_2, y_2) so long as coordinates are subtracted in the same order in both the numerator and the denominator:

$$\frac{y_2 - y_1}{x_2 - x_1} = \frac{y_1 - y_2}{x_1 - x_2}.$$

EXAMPLE 1 Graph the line containing the points $(-4, 3)$ and $(2, -6)$ and find the slope.

The graph is shown below. We consider (x_1, y_1) to be $(-4, 3)$ and (x_2, y_2) to be $(2, -6)$. From $(-4, 3)$ and $(2, -6)$, we see that the change in y, or the rise, is $-6 - 3$, or -9. The change in x, or the run, is $2 - (-4)$, or 6.

Graph the line containing the points and find the slope in two different ways.

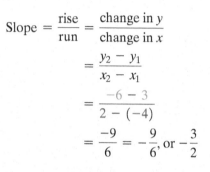

1. $(-2, 3)$ and $(3, 5)$

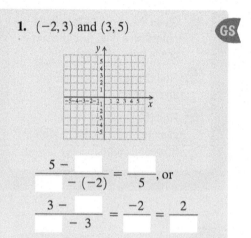

$$\frac{5 - \boxed{}}{\boxed{} - (-2)} = \frac{\boxed{}}{5}, \text{ or}$$

$$\frac{3 - \boxed{}}{\boxed{} - 3} = \frac{-2}{\boxed{}} = \frac{2}{\boxed{}}$$

$$\text{Slope} = \frac{\text{rise}}{\text{run}} = \frac{\text{change in } y}{\text{change in } x}$$
$$= \frac{y_2 - y_1}{x_2 - x_1}$$
$$= \frac{-6 - 3}{2 - (-4)}$$
$$= \frac{-9}{6} = -\frac{9}{6}, \text{ or } -\frac{3}{2}$$

When we use the formula

$$m = \frac{y_2 - y_1}{x_2 - x_1},$$

we must remember to subtract the x-coordinates in the same order in which we subtract the y-coordinates. Let's redo Example 1, where we consider (x_1, y_1) to be $(2, -6)$ and (x_2, y_2) to be $(-4, 3)$:

$$\text{Slope} = \frac{\text{change in } y}{\text{change in } x} = \frac{3 - (-6)}{-4 - 2} = \frac{9}{-6} = -\frac{9}{6} = -\frac{3}{2}.$$

2. $(0, -3)$ and $(-3, 2)$

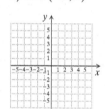

◀ Do Exercises 1 and 2.

The slope of a line tells how it slants. A line with positive slope slants up from left to right. The larger the slope, the steeper the slant. A line with negative slope slants downward from left to right.

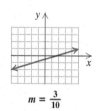 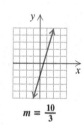 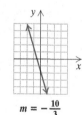 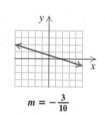

$m = \frac{3}{10}$ $m = \frac{10}{3}$ $m = -\frac{10}{3}$ $m = -\frac{3}{10}$ $m = 0$ m is not defined.

Later in this section, in Examples 7 and 8, we will discuss the slope of a horizontal line and of a vertical line.

Answers
Answers to Margin Exercises 1 and 2 are on p. 231.

b FINDING THE SLOPE FROM AN EQUATION

It is possible to find the slope of a line from its equation. Let's consider the equation $y = 2x + 3$, which is in the form $y = mx + b$. The graph of this equation is shown at right. We can find two points by choosing convenient values for x—say, 0 and 1—and substituting to find the corresponding y-values. We find the two points on the line to be $(0, 3)$ and $(1, 5)$. The slope of the line is found using the definition of slope:

$$m = \frac{\text{change in } y}{\text{change in } x} = \frac{5 - 3}{1 - 0} = \frac{2}{1} = 2.$$

The slope is 2. Note that this is also the coefficient of the x-term in the equation $y = 2x + 3$.

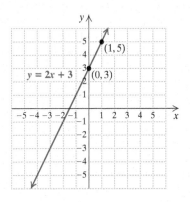

> **DETERMINING SLOPE FROM THE EQUATION $y = mx + b$**
>
> The slope of the line $y = mx + b$ is m. To find the slope of a nonvertical line, solve the linear equation in x and y for y and get the resulting equation in the form $y = mx + b$. The coefficient of the x-term, m, is the slope of the line.

EXAMPLES Find the slope of each line.

2. $y = -3x + \dfrac{2}{9}$
$\quad \longrightarrow m = -3 = \text{Slope}$

3. $y = \dfrac{4}{5}x$
$\quad \longrightarrow m = \dfrac{4}{5} = \text{Slope}$

4. $y = x + 6$
$\quad \longrightarrow m = 1 = \text{Slope}$

5. $y = -0.6x - 3.5$
$\quad \longrightarrow m = -0.6 = \text{Slope}$

Do Exercises 3–6. ▶

To find slope from an equation, we may need to first find an equivalent form of the equation.

EXAMPLE 6 Find the slope of the line $2x + 3y = 7$.

We solve for y to get the equation in the form $y = mx + b$:

$$2x + 3y = 7$$
$$3y = -2x + 7$$
$$y = \frac{1}{3}(-2x + 7)$$
$$y = -\frac{2}{3}x + \frac{7}{3}. \quad \text{This is } y = mx + b.$$

The slope is $-\dfrac{2}{3}$.

Do Exercises 7 and 8. ▶

Find the slope of each line.

3. $y = 4x + 11$

4. $y = -17x + 8$

5. $y = -x + \dfrac{1}{2}$

6. $y = \dfrac{2}{3}x - 1$

Find the slope of each line.

7. $4x + 4y = 7$

GS 8. $5x - 4y = 8$
$$5x = \boxed{} + 8$$
$$5x - \boxed{} = 4y$$
$$\frac{5x - 8}{\boxed{}} = \frac{4y}{4}$$
$$\boxed{} \cdot x - 2 = y, \text{ or}$$
$$y = \boxed{} \cdot x - 2$$
$$\text{Slope is } \boxed{}.$$

Answers

1. $\dfrac{2}{5}$ **2.** $-\dfrac{5}{3}$

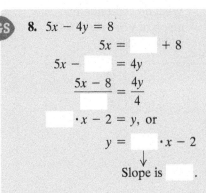

3. 4 **4.** -17 **5.** -1 **6.** $\dfrac{2}{3}$

7. -1 **8.** $\dfrac{5}{4}$

Guided Solutions

1. 3, 3, 2, 5, -2, -5, 5   **8.** $4y, 8, 4, \dfrac{5}{4}, \dfrac{5}{4}, \dfrac{5}{4}$

CALCULATOR CORNER

Visualizing Slope

EXERCISES: Graph each of the following sets of equations using the window settings $[-6, 6, -4, 4]$, with Xscl $= 1$ and Yscl $= 1$.

1. $y = x$, $y = 2x$, $y = 5x$, $y = 10x$
   What do you think the graph of $y = 123x$ will look like?

2. $y = x$, $y = \frac{3}{4}x$, $y = 0.38x$, $y = \frac{5}{32}x$
   What do you think the graph of $y = 0.000043x$ will look like?

3. $y = -x$, $y = -2x$, $y = -5x$, $y = -10x$
   What do you think the graph of $y = -123x$ will look like?

4. $y = -x$, $y = -\frac{3}{4}x$, $y = -0.38x$, $y = -\frac{5}{32}x$
   What do you think the graph of $y = -0.000043x$ will look like?

Find the slope, if it exists, of each line.

9. $x = 7$

10. $y = -5$

What about the slope of a horizontal line or a vertical line?

EXAMPLE 7 Find the slope of the line $y = 5$.

We can think of $y = 5$ as $y = 0x + 5$. Then from this equation, we see that $m = 0$. Consider the points $(-3, 5)$ and $(4, 5)$, which are on the line. The change in $y = 5 - 5$, or 0. The change in $x = -3 - 4$, or -7. We have

$$m = \frac{5 - 5}{-3 - 4}$$
$$= \frac{0}{-7}$$
$$= 0.$$

Any two points on a horizontal line have the same y-coordinate. The change in y is 0. Thus the slope of a horizontal line is 0.

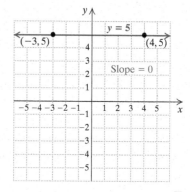

EXAMPLE 8 Find the slope of the line $x = -4$.

Consider the points $(-4, 3)$ and $(-4, -2)$, which are on the line. The change in $y = 3 - (-2)$, or 5. The change in $x = -4 - (-4)$, or 0. We have

$$m = \frac{3 - (-2)}{-4 - (-4)}$$
$$= \frac{5}{0}. \quad \text{Not defined}$$

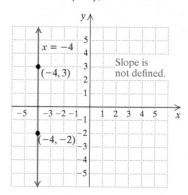

Since division by 0 is not defined, the slope of this line is not defined. The answer in this example is "The slope of this line is not defined."

SLOPE 0; SLOPE NOT DEFINED

The slope of a horizontal line is 0.
The slope of a vertical line is not defined.

◀ Do Exercises 9 and 10.

Answers

9. Not defined   10. 0

C APPLICATIONS OF SLOPE; RATES OF CHANGE

Slope has many real-world applications. For example, numbers like 2%, 3%, and 6% are often used to represent the *grade* of a road, a measure of how steep a road on a hill or a mountain is. For example, a 3% grade ($3\% = \frac{3}{100}$) means that for every horizontal distance of 100 ft, the road rises 3 ft, and a -3% grade means that for every horizontal distance of 100 ft, the road drops 3 ft. (Road signs do not include negative signs.)

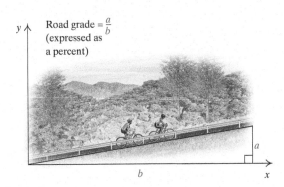

The concept of grade also occurs in skiing or snowboarding, where a 7% grade is considered very tame, but a 70% grade is considered extremely steep.

EXAMPLE 9 *Dubai Ski Run.* Dubai Ski Resort has the fifth longest indoor ski run in the world. It drops 197 ft over a horizontal distance of 1297 ft. Find the grade of the ski run.

The grade of the ski run is its slope, expressed as a percent:

$$m = \frac{197}{1297} \begin{array}{l} \leftarrow \text{Vertical distance} \\ \leftarrow \text{Horizontal distance} \end{array}$$
$$\approx 0.15$$
$$\approx 15\%.$$

Do Exercise 11.

11. *Grade of a Treadmill.* During a stress test, a physician may change the grade, or slope, of a treadmill to measure its effect on heart rate (number of beats per minute). Find the grade, or slope, of the treadmill shown below.

Answer
11. 8%

Slope can also be considered as a **rate of change**.

EXAMPLE 10 *Car Assembly Line.* Cameron, a supervisor in a car assembly plant, prepared the following graph to display data from a recent day's work. Use the graph to determine the slope, or the rate of change of the number of cars that came off an assembly line with respect to time.

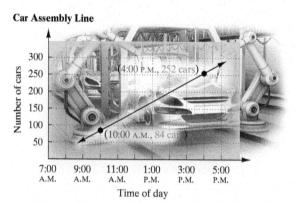

The vertical axis of the graph shows the number of cars, and the horizontal axis shows the time, in units of hours. We can describe the rate of change of the number of cars with respect to time as

$$\frac{\text{Cars}}{\text{Hours}}, \quad \text{or} \quad \text{number of cars per hour.}$$

This value is the slope of the line. We determine two ordered pairs on the graph—in this case,

(10:00 A.M., 84 cars)   and   (4:00 P.M., 252 cars).

This tells us that in the 6 hr between 10:00 A.M. and 4:00 P.M., 252 − 84, or 168, cars came off the assembly line. Thus,

$$\text{Rate of change} = \frac{252 \text{ cars} - 84 \text{ cars}}{4:00 \text{ P.M.} - 10:00 \text{ A.M.}}$$

$$= \frac{168 \text{ cars}}{6 \text{ hours}}$$

$$= 28 \text{ cars per hour.}$$

◀ Do Exercise 12.

12. *Masonry.* Daryl, a mason, graphed data from a recent day's work. Use the following graph to determine the slope, or the rate of change of the number of bricks that he can lay with respect to time.

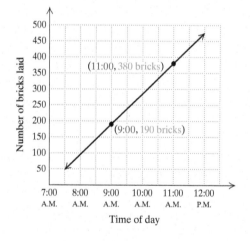

EXAMPLE 11 *Vacation Days.* The average number of vacation days used annually by workers in the United States has been decreasing since 1996. Use the following graph to determine the slope, or rate of change in the average number of vacation days used with respect to time.

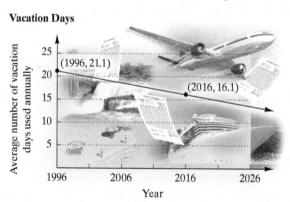

DATA: *USA Today*, 8/19/2016

Answer
12. 95 bricks per hour

The vertical axis of the graph shows the average number of vacation days used annually by workers in the United States, and the horizontal axis shows the years. We can describe the rate of change in the average number of vacation days used with respect to time as

$$\frac{\text{Change in number of annual vacation days used}}{\text{Years}}, \text{ or } \begin{array}{l}\text{change in number of}\\ \text{annual vacation days}\\ \text{used per year.}\end{array}$$

This value is the slope of the line. We determine two ordered pairs on the graph—in this case, (1996, 21.1) and (2016, 16.1). This tells us that in the 20 years from 1996 to 2016, the average number of vacation days used annually by workers in the United States dropped from 21.1 to 16.1. Thus,

$$\text{Rate of change} = \frac{16.1 - 21.1}{2016 - 1996} = \frac{-5}{20} = -0.25 \text{ day per year.}$$

Do Exercise 13.

13. *Television Viewership of NASCAR.* Use the following graph to determine the rate of change in the average television viewership of NASCAR races since 2006.

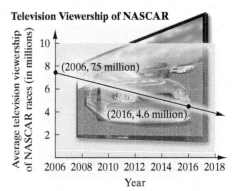

DATA: *The Wall Street Journal*, 2/22/17

Answer
13. −0.29 million viewers per year

3.4 Exercise Set

FOR EXTRA HELP — MyLab Math

✓ Check Your Understanding

Reading Check Match each expression with an appropriate description or value from the column on the right.

RC1. Slope of a horizontal line
RC2. y-intercept of $y = mx + b$
RC3. Change in x
RC4. Slope of a vertical line
RC5. Slope
RC6. Change in y

a) Rise
b) Run
c) Rise/run
d) 0
e) Not defined
f) $(0, b)$

Concept Check Choose from the column on the right the correct description of the slope of the line.

CC1. The line slants upward from left to right.
CC2. The line is horizontal.
CC3. The line is vertical.
CC4. The line slants downward from left to right.

a) Not defined
b) Zero
c) Positive
d) Negative

a Find the slope, if it exists, of each line.

1.
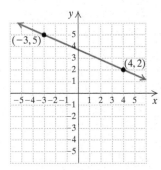

2.
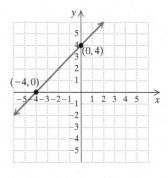

3.
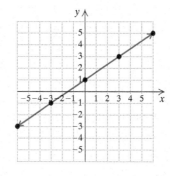

4.
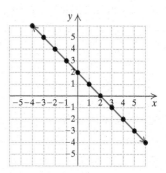

5.
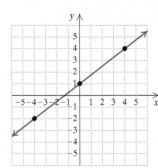

6.
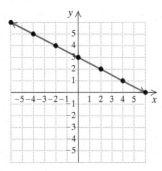

7.
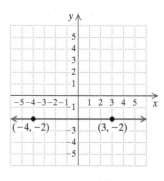

8.
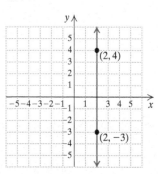

Graph the line containing the given pair of points and find the slope.

9. $(-2, 4), (3, 0)$
10. $(2, -4), (-3, 2)$
11. $(-4, 0), (-5, -3)$
12. $(-3, 0), (-5, -2)$

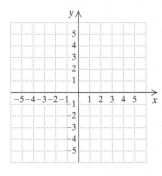
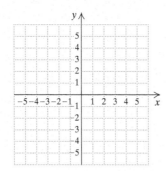
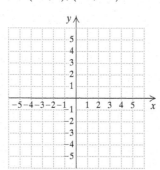
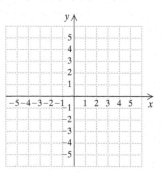

13. $(-4, 1), (2, -3)$
14. $(-3, 5), (4, -3)$
15. $(5, 3), (-3, -4)$
16. $(-4, -3), (2, 5)$

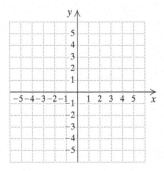
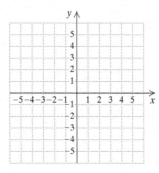
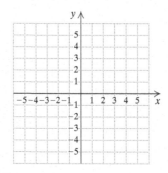
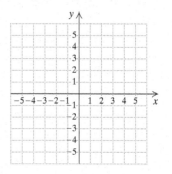

Find the slope, if it exists, of the line containing the given pair of points.

17. $\left(2, -\frac{1}{2}\right), \left(5, \frac{3}{2}\right)$ **18.** $\left(\frac{2}{3}, -1\right), \left(\frac{5}{3}, 2\right)$ **19.** $(4, -2), (4, 3)$ **20.** $(4, -3), (-2, -3)$

21. $(-11, 7), (15, -3)$ **22.** $(-13, 22), (8, -17)$ **23.** $\left(-\frac{1}{2}, \frac{3}{11}\right), \left(\frac{5}{4}, \frac{3}{11}\right)$ **24.** $(0.2, 4), (0.2, -0.04)$

b Find the slope, if it exists, of each line.

25. $y = -10x$ **26.** $y = \frac{10}{3}x$ **27.** $y = 3.78x - 4$ **28.** $y = -\frac{3}{5}x + 28$

29. $3x - y = 4$ **30.** $-2x + y = 8$ **31.** $x + 5y = 10$ **32.** $x - 4y = 8$

33. $3x + 2y = 6$ **34.** $2x - 4y = 8$ **35.** $x = \frac{2}{15}$ **36.** $y = -\frac{1}{3}$

37. $y = 2 - x$ **38.** $y = \frac{3}{4} + x$ **39.** $9x = 3y + 5$ **40.** $4y = 9x - 7$

41. $5x - 4y + 12 = 0$ **42.** $16 + 2x - 8y = 0$ **43.** $y = 4$ **44.** $x = -3$

45. $x = \frac{3}{4}y - 2$ **46.** $3x - \frac{1}{5}y = -4$ **47.** $\frac{2}{3}y = -\frac{7}{4}x$ **48.** $-x = \frac{2}{11}y$

c In each of Exercises 49–52, find the slope (or rate of change).

49. Find the slope (or pitch) of the roof.

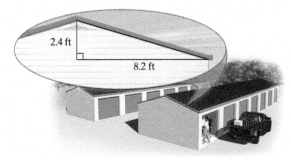

50. Find the slope (or grade) of the road.

51. *Slope of a River.* When a river flows, its strength or force depends on how far it falls vertically compared to how far it flows horizontally. Find the slope of the river shown below.

52. *Constructing Stairs.* Carpenters use slope when designing and building stairs. Public buildings normally include steps with 7-in. risers and 11-in. treads. Find the grade of such a stairway.

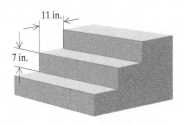

53. *Grade of a Transit System.* The maximum grade allowed between two stations in a rapid-transit rail system is 3.5%. Between station A and station B, which are 280 ft apart, the tracks rise $8\frac{1}{2}$ ft. What is the grade of the tracks between these two stations? Round the answer to the nearest tenth of a percent. Does this grade meet the rapid-transit rail standards?

Data: Brian Burell, *Merriam Webster's Guide to Everyday Math*, Merriam-Webster, Inc., Springfield MA

54. *Slope of Longs Peak.* From a base elevation of 9600 ft, Longs Peak in Colorado rises to a summit elevation of 14,256 ft over a horizontal distance of 15,840 ft. Find the grade of Longs Peak.

In each of Exercises 55–58, use the graph to calculate a rate of change in which the units of the horizontal axis are used in the denominator.

55. *Wind Power Generation.* The power generated by wind, in megawatt-hours (MWh), increased steadily between 2008 and 2016. Use the following graph to find the rate of change in the number of megawatt-hours generated by wind power with respect to time.

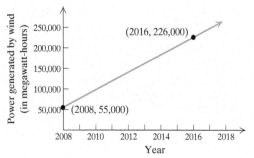

DATA: U.S. Energy Information Administration

56. *Orange Juice Consumption.* In 2002, U.S. per-capita consumption of orange juice was 5.1 gal. Since then, this amount has steadily decreased. Use the following graph to find the rate of change, rounded to the nearest hundredth of a gallon, in the per-capita consumption of orange juice in the United States.

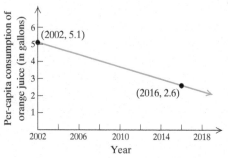

DATA: *The Wall Street Journal*, 2/7/2017

57. *Population Decrease.* The change in the population of Montgomery, Alabama, is illustrated in the following graph. Find the rate of change, to the nearest hundred, in the population with respect to time.

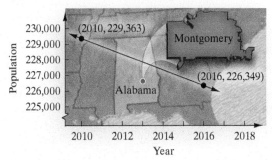

DATA: U.S. Census Bureau

58. *Population Increase.* The change in the population of Austin, Texas, is illustrated in the following graph. Find the rate of change, to the nearest hundred, in the population with respect to time.

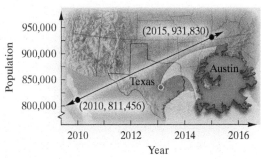

DATA: U.S. Census Bureau

59. *Advanced Placement Exams.* The number of students taking Advanced Placement exams increased from 0.8 million in 2002 to 2.6 million in 2017. Find the rate of change in the number of students taking Advanced Placement exams with respect to time.

Data: The College Board

60. *Bottled Water.* Bottled water consumption per person per year in the United States increased from 16.7 gal in 2000 to 39.3 gal in 2016. Find the rate of change, rounded to the nearest tenth, in the number of gallons of bottled water consumed annually per person per year.

Data: Beverage Marketing Corporation; International Bottled Water Association

Skill Maintenance

Solve. [2.3a]

61. $2x - 11 = 4$

62. $5 - \frac{1}{2}x = 11$

Collect like terms. [1.7e]

63. $\frac{1}{3}p - p$

64. $t - 6 + 4t + 5$

Synthesis

In each of Exercises 65–68, find an equation for the graph shown.

65.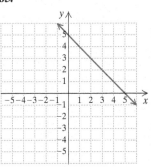

66.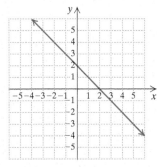

67.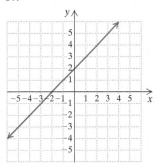

68.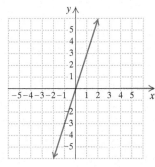

SECTION 3.4 Slope and Applications **239**

CHAPTER 3 Summary and Review

Vocabulary Reinforcement

Complete each statement with the correct term or expression from the list on the right. Some of the choices may not be used.

$(0, b)$
$(b, 0)$
$(a, 0)$
$(0, a)$
0
not defined
x-intercept
y-intercept
vertical
horizontal
coordinates
axes

1. The slope of a vertical line is _____. [3.4b]
2. The graph of $y = b$ is a(n) _____ line. The y-intercept is _____. [3.3b]
3. Consider the ordered pair $(-5, 3)$. The numbers -5 and 3 are called _____. [3.1a]
4. The _____ occurs when a line crosses the y-axis and thus will always have 0 as the first coordinate. [3.3a]
5. The graph of $x = a$ is a(n) _____ line. The x-intercept is _____. [3.3b]
6. The slope of a horizontal line is _____. [3.4b]

Concept Reinforcement

Determine whether each statement is true or false.

_____ 1. The x- and y-intercepts of $y = mx$ are both $(0, 0)$. [3.3a]

_____ 2. A slope of $-\frac{3}{4}$ is steeper than a slope of $-\frac{5}{2}$. [3.4a]

_____ 3. The slope of the line that passes through (a, b) and (c, d) is $\frac{d - b}{c - a}$. [3.4a]

_____ 4. The second coordinate of all points in quadrant III is negative. [3.1a]

_____ 5. The x-intercept of $Ax + By = C$, $C \neq 0$, is $\left(\frac{A}{C}, 0\right)$. [3.3a]

_____ 6. The slope of the line that passes through $(0, t)$ and $(-t, 0)$ is $\frac{1}{t}$. [3.4a]

Study Guide

Objective 3.1b Find the coordinates of a point on a graph.

Example Find the coordinates of points Q, R, and S.

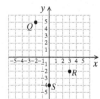

Point Q is 2 units to the left of the origin and 5 units up. Its coordinates are $(-2, 5)$.

Point R is 3 units to the right of the origin and 2 units down. Its coordinates are $(3, -2)$.

Point S is 0 units to the left or right of the origin and 4 units down. Its coordinates are $(0, -4)$.

Practice Exercise

1. Find the coordinates of points F, G, and H.

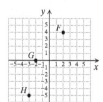

240 CHAPTER 3 Graphs of Linear Equations

Objective 3.2a Graph linear equations of the type $y = mx + b$ and $Ax + By = C$, identifying the y-intercept.

Example Graph $2y + 2 = -3x$ and identify the y-intercept.

To find an equivalent equation in the form $y = mx + b$, we solve for y: $y = -\frac{3}{2}x - 1$. The y-intercept is $(0, -1)$.

We then find two other points using multiples of 2 for x to avoid fractions.

x	y
0	-1	← y-intercept
-2	2
2	-4

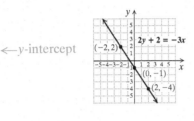

Practice Exercise

2. Graph $x + 2y = 8$ and identify the y-intercept.

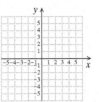

Objective 3.3a Find the intercepts of a linear equation, and graph using intercepts.

Example For $2x - y = -6$, find the intercepts. Then use the intercepts to graph the equation.

To find the y-intercept, we let $x = 0$ and solve for y:
$$2 \cdot 0 - y = -6 \quad \text{and} \quad y = 6.$$
The y-intercept is $(0, 6)$.

To find the x-intercept, we let $y = 0$ and solve for x:
$$2x - 0 = -6 \quad \text{and} \quad x = -3.$$
The x-intercept is $(-3, 0)$.

We find a third point as a check.

x	y
0	6	← y-intercept
-3	0	← x-intercept
-1	4

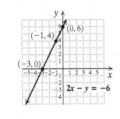

Practice Exercise

3. For $y - 2x = -4$, find the intercepts. Then use the intercepts to graph the equation.

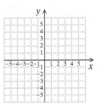

Objective 3.3b Graph equations equivalent to those of the type $x = a$ and $y = b$.

Example Graph. $y = 1$ and $x = -\frac{3}{2}$.

For $y = 1$, no matter what number we choose for x, $y = 1$. The graph is a horizontal line. For $x = -\frac{3}{2}$, no matter what number we choose for y, $x = -\frac{3}{2}$. The graph is a vertical line.

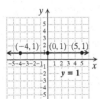 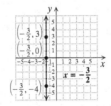

Practice Exercises

Graph.

4. $y = -\frac{5}{2}$

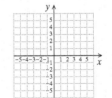

5. $x = 2$

Summary and Review: Chapter 3   241

Objective 3.4a Given the coordinates of two points on a line, find the slope of the line, if it exists.

Example Find the slope, if it exists, of the line containing the given points.

$(-9, 3)$ and $(5, -6)$: $m = \dfrac{-6 - 3}{5 - (-9)} = \dfrac{-9}{14} = -\dfrac{9}{14}$;

$\left(7, \dfrac{1}{2}\right)$ and $\left(-13, \dfrac{1}{2}\right)$: $m = \dfrac{\frac{1}{2} - \frac{1}{2}}{-13 - 7} = \dfrac{0}{-20} = 0$;

$(0.6, 1.5)$ and $(0.6, -1.5)$: $m = \dfrac{-1.5 - 1.5}{0.6 - 0.6} = \dfrac{-3}{0}$, m is not defined.

Practice Exercises

Find the slope, if it exists, of the line containing the given points.

6. $(-8, 20), (-8, 14)$
7. $(2, -1), (16, 20)$
8. $(0.5, 2.8), (1.5, 2.8)$

Objective 3.4b Find the slope of a line from an equation.

Example Find the slope, if it exists, of each line.
a) $5x - 20y = -10$
  We first solve for y: $y = \frac{1}{4}x + \frac{1}{2}$. The slope is $\frac{1}{4}$.
b) $y = -\frac{4}{5}$
  Think: $y = 0 \cdot x - \frac{4}{5}$. This line is horizontal. The slope is 0.
c) $x = 6$
  This line is vertical. The slope is not defined.

Practice Exercises

Find the slope, if it exists, of the line.

9. $x = 0.25$
10. $7y + 14x = -28$
11. $y = -5$

Objective 3.4c Find the slope, or rate of change, in an applied problem involving slope.

Example In 2000, the population of Cincinnati, Ohio, was approximately 331,410. By 2010, the population had decreased to 296,943. Find the rate of change, to the nearest hundred, in the population with respect to time.

Data: U.S. Census Bureau

$$\text{Rate of change} = \dfrac{296{,}943 - 331{,}410}{2010 - 2000}$$

$$= \dfrac{-34{,}467}{10} \approx -3400 \text{ people per year}$$

Practice Example

12. In 2000, the population of Idaho was 1,293,953. By 2010, the population had increased to 1,567,582. Find the rate of change, to the nearest hundred, in the population with respect to time.

Data: U.S. Census Bureau

Review Exercises

Plot each point. [3.1a]

1. $(2, 5)$
2. $(0, -3)$
3. $(-4, -2)$

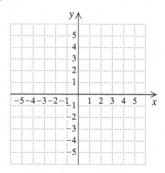

Find the coordinates of each point. [3.1b]

4. A
5. B
6. C

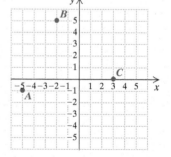

In which quadrant is each point located? [3.1a]

7. (3, −8) **8.** (−20, −14) **9.** (4.9, 1.3)

Determine whether each ordered pair is a solution of $2y - x = 10$. [3.1c]

10. (2, −6) **11.** (0, 5)

12. Show that the ordered pairs (0, −3) and (2, 1) are solutions of the equation $2x - y = 3$. Then use the graph of the equation to determine another solution. Answers may vary. [3.1c]

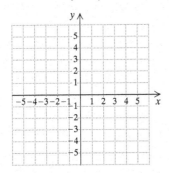

Graph each equation, identifying the y-intercept. [3.2a]

13. $y = 2x - 5$

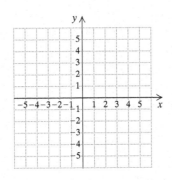

14. $y = -\dfrac{3}{4}x$

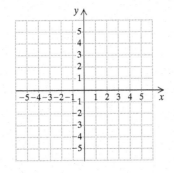

15. $y = -x + 4$

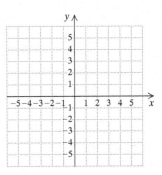

16. $y = 3 - 4x$

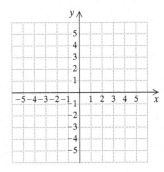

Solve. [3.2b]

17. *Kitchen Design.* Kitchen designers recommend that a refrigerator be selected on the basis of the number of people n in the household. The appropriate size S, in cubic feet, is given by

$$S = \frac{3}{2}n + 13.$$

a) Determine the recommended size of a refrigerator if the number of people is 1, 2, 5, and 10.
b) Graph the equation and use the graph to estimate the recommended size of a refrigerator for 4 people sharing an apartment.

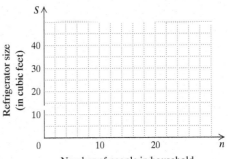

c) A refrigerator is 22 ft³. For how many residents is it the recommended size?

Summary and Review: Chapter 3   **243**

Find the intercepts of each equation. Then graph the equation. [3.3a]

18. $x - 2y = 6$

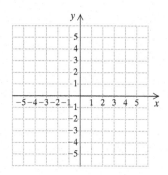

19. $5x - 2y = 10$

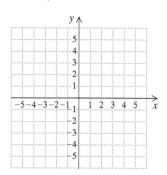

Graph each equation. [3.3b]

20. $y = 3$

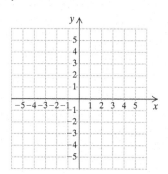

21. $5x - 4 = 0$

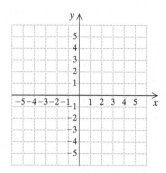

Find the slope. [3.4a]

22.

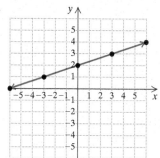

23.

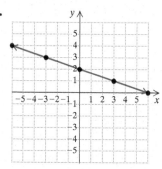

Graph the line containing the given pair of points and find the slope. [3.4a]

24. $(-5, -2), (5, 4)$

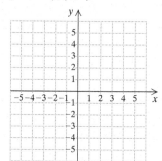

25. $(-5, 5), (4, -4)$

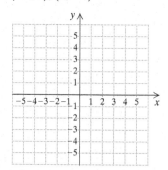

Find the slope, if it exists. [3.4b]

26. $y = -\dfrac{5}{8}x - 3$

27. $2x - 4y = 8$

28. $x = -2$

29. $y = 9$

30. $x - 10y = 20$

31. $6x - 5 = 4y$

244 CHAPTER 3 Graphs of Linear Equations

32. *Snow Removal.* By 3:00 P.M., Erin had plowed 7 driveways, and by 5:30 P.M., she had completed 13 driveways. [3.4c]

   a) Find Erin's plowing rate, in number of driveways per hour.
   b) Find Erin's plowing rate, in number of minutes per driveway.

33. *Road Grade.* Along one stretch, Beartooth Highway in Yellowstone National Park rises 315 ft over a horizontal distance of 4500 ft. Find the slope, or grade, of the road. [3.4c]

34. *Holiday Sales.* Retail sales in the United States during the winter holiday season increased steadily between 2008 and 2016. Use the following graph to determine the slope, or rate of change in holiday retail sales, in billions of dollars, with respect to time. [3.4c]

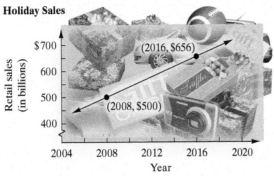

35. Which of the following is the x-intercept of the graph of $5x - y = -15$? [3.3a]
   A. $(0, -3)$ B. $(-15, 0)$
   C. $(0, 15)$ D. $(-3, 0)$

36. What is the slope, if it exists, of the line containing the points $(-8, 8)$ and $(8, -8)$? [3.4a]
   A. 1 B. 0
   C. -1 D. Not defined

Synthesis

37. Find the area and the perimeter of a rectangle for which $(-2, 2)$, $(7, 2)$, and $(7, -3)$ are three of the vertices. [3.1a]

38. *Gondola Aerial Lift.* In Telluride, Colorado, there is a free gondola ride that provides a spectacular view of the town and the surrounding mountains. The gondolas start from the town at an elevation of 8725 ft and travel 5750 ft to Station St. Sophia, whose elevation is 10,550 ft. They then continue 3913 ft to Mountain Village, whose elevation is 9500 ft. [3.4c]

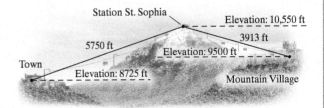

A gondola departs from the town at 11:55 A.M. and with no stop at Station St. Sophia reaches Mountain Village at 12:07 P.M.

   a) Find the gondola's average rate of ascent and descent, in number of feet per minute.
   b) Find the gondola's average rate of ascent and descent, in number of minutes per foot.

Understanding Through Discussion and Writing

1. Explain why the slant of a line with slope $\frac{5}{3}$ is steeper than the slant of a line with slope $\frac{4}{3}$. [3.4a]

2. Do all graphs of linear equations have x-intercepts? Explain. [3.3b]

3. Explain why the first coordinate of the y-intercept is always 0. [3.2a]

4. Explain why the graph of $y = -2$ is a horizontal line. [3.3b]

CHAPTER 3 Test

For Extra Help — For step-by-step test solutions, access the Chapter Test Prep Videos in MyLab Math.

In which quadrant is each point located?

1. $\left(-\frac{1}{2}, 7\right)$

2. $(-5, -6)$

Find the coordinates of each point.

3. A

4. B

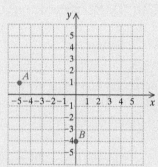

5. Show that the ordered pairs $(-4, -3)$ and $(-1, 3)$ are solutions of the equation $y - 2x = 5$. Then use the graph of the straight line containing the two points to determine another solution. Answers may vary.

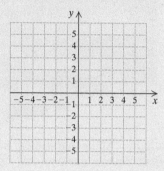

Graph each equation. Identify the y-intercept.

6. $y = 2x - 1$

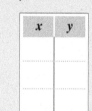 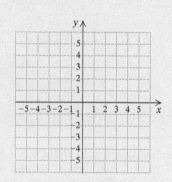

7. $y = -\frac{3}{2}x$

 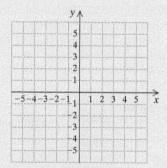

Find the intercepts of each equation. Then graph the equation.

8. $2x - 4y = -8$

← x-intercept
← y-intercept

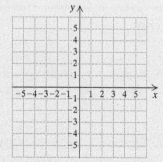

9. $2x - y = 3$

← x-intercept
← y-intercept

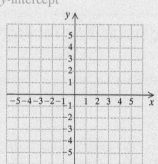

246 CHAPTER 3 Graphs of Linear Equations

Graph each equation.

10. $2x + 8 = 0$

x	y

11. $y = 5$

x	y

12. Find the slope.

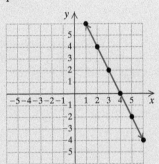

13. Graph the line containing $(-3, 1)$ and $(5, 4)$ and find the slope.

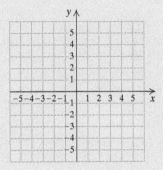

Find the slope, if it exists, of each line.

14. $2x - 5y = 10$

15. $x = -2$

16. $3y = \dfrac{1}{9}$

17. $y = -11x + 6$

18. *Navigation.* Capital Rapids drops 54 ft vertically over a horizontal distance of 1080 ft. What is the slope of the rapids?

19. *Health Insurance Cost.* The total annual cost, employer plus employee, of health insurance can be approximated by
$$C = 606t + 8593,$$
where t is the number of years since 2007. That is, $t = 0$ corresponds to 2007, $t = 3$ corresponds to 2010, and so on.
Data: TW/NBGH Value Purchasing Survey

a) Find the total annual cost of health insurance in 2007, in 2009, and in 2012.
b) Graph the equation and then use the graph to estimate the cost of health insurance in 2016.

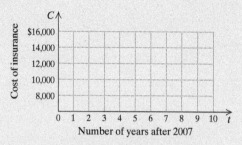

c) Predict the year in which the cost of health insurance will be $15,259.

20. *Elevators.* At 2:38, Serge entered an elevator on the 34th floor of the Regency Hotel. At 2:40, he stepped off at the 5th floor.
a) Find the elevator's average rate of travel, in number of floors per minute.
b) Find the elevator's average rate of travel, in seconds per floor.

21. *Train Travel.* The following graph shows data concerning a recent train ride from Denver to Kansas City. At what rate did the train travel?

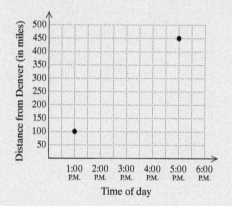

22. Which of the following is the correct description of the line $6x - 1 = 3y + 2$?
A. The slope is -6, and the x-intercept is $\left(-\frac{1}{2}, 0\right)$.
B. The slope is 2, and the y-intercept is $(0, -1)$.
C. The slope is -2, and the x-intercept is $\left(\frac{1}{2}, 0\right)$.
D. The slope is 6, and the y-intercept is $(0, -3)$.

Synthesis

23. Write an equation of a line whose graph is parallel to the x-axis and 3 units above it.

24. A diagonal of a square connects the points $(-3, -1)$ and $(2, 4)$. Find the area and the perimeter of the square.

CHAPTERS 1–3 Cumulative Review

1. Evaluate $\dfrac{2m - n}{5}$ when $m = -1$ and $n = 2$.

2. Multiply: $-\dfrac{2}{3}(x - 6y + 3)$.

3. Factor: $18w - 24 + 9y$.

4. Find decimal notation: $-\dfrac{7}{9}$.

5. Find the absolute value: $\left|-2\dfrac{1}{5}\right|$.

6. Find the opposite of 8.17.

7. Find the reciprocal of $-\dfrac{8}{7}$.

8. Collect like terms: $2x - 5y + (-3x) + 4y$.

Simplify.

9. $-2.6 + (-0.4)$

10. $3 - [81 \div (1 + 2^3)]$

11. $\dfrac{5}{18} \div \left(-\dfrac{5}{12}\right)$

12. $6(x + 4) - 5[x - (2x - 3)]$

13. $\left(-\dfrac{1}{2}\right)(-1.1)(4.8)$

14. $-20 + 30 \div 10 \cdot 6$

Solve.

15. $\dfrac{4}{9}y = -36$

16. $-8 + w = w + 7$

17. $7.5 - 2x = 0.5x$

18. $4(x + 2) = 4(x - 2) + 16$

19. $2(x + 2) \geq 5(2x + 3)$

20. $x - \dfrac{5}{6} = \dfrac{1}{2}$

21. Find the slope, if it exists, of $9x - 12y = -3$.

22. Find the slope, if it exists, of $x = -\dfrac{15}{16}$.

23. Find the slope, if it exists, of $y - 3 = 7$.

24. Solve $7t = sz$ for s.

25. Solve $A = \dfrac{1}{2}h(b + c)$ for h.

26. In which quadrant is the point $(3, -1)$ located?

27. Find the intercepts of $2x - 7y = 21$. Do not graph.

28. Graph on the number line: $-1 < x \leq 2$.

<-+--+--+--+--+--+--+--+--+->
-4 -3 -2 -1 0 1 2 3 4

Graph.

29. $2x + 5y = 10$

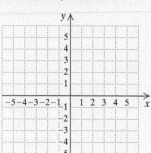

30. $y = -2$

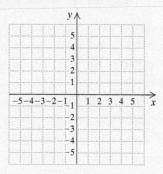

31. $y = -2x + 1$

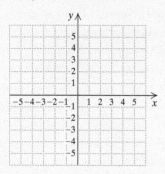

32. $3y + 6 = 2x$

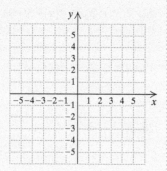

33. $y = -\dfrac{3}{2}x$

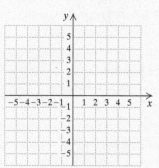

34. $x = 4.5$

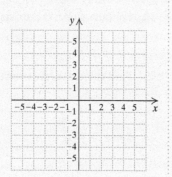

35. Find the slope of the line graphed below.

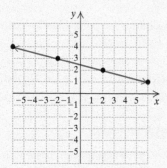

36. *Blood Types.* There are 134.6 million Americans with either O-positive or O-negative blood. Those with O-positive blood outnumber those with O-negative blood by 94.2 million. How many Americans have O-negative blood?

Data: Stanford University School of Medicine

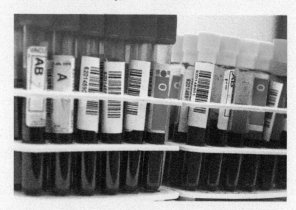

37. *Work Time.* Cory's contract stipulates that he cannot work more than 40 hr per week. For the first four days of one week, he worked 7, 10, 9, and 6 hr. Determine as an inequality the number of hours he can work on the fifth day without violating his contract.

38. *Wire Cutting.* A 143-m wire is cut into three pieces. The second piece is 3 m longer than the first. The third is four-fifths as long as the first. How long is each piece?

39. Compute and simplify: $1000 \div 100 \cdot 10 - 10$.
   A. 90   B. 0
   C. −9   D. −90

40. The slope of the line containing the points $(2, -7)$ and $(-4, 3)$ is which of the following?
   A. $-\dfrac{5}{2}$ B. $-\dfrac{3}{5}$
   C. $-\dfrac{2}{5}$ D. $-\dfrac{5}{3}$

Synthesis

Solve.

41. $4|x| - 13 = 3$

42. $\dfrac{2 + 5x}{4} = \dfrac{11}{28} + \dfrac{8x + 3}{7}$

43. $p = \dfrac{2}{m + Q}$, for Q

250   CHAPTER 3   Graphs of Linear Equations

CHAPTER 4

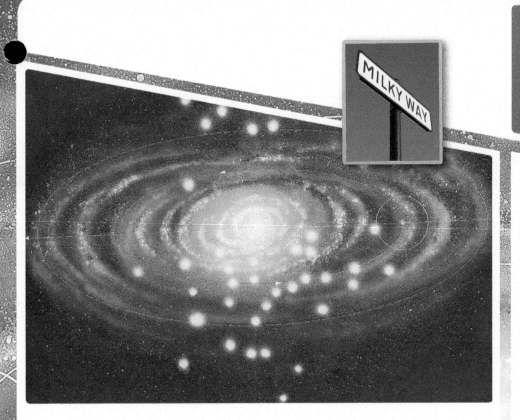

Polynomials: Operations

4.1 Integers as Exponents
4.2 Exponents and Scientific Notation
4.3 Introduction to Polynomials
4.4 Addition and Subtraction of Polynomials

Mid-Chapter Review

4.5 Multiplication of Polynomials
4.6 Special Products

Visualizing for Success

4.7 Operations with Polynomials in Several Variables
4.8 Division of Polynomials

Summary and Review
Test
Cumulative Review

The planet upon which we live, its sun, and, in fact, our entire solar system make up only a tiny part of the universe. The universe is so large that distances within the universe are measured in light-years—one light-year is the distance that light travels in one year and is approximately 5.879 trillion miles. For example, our solar system is 0.0013 light-years in diameter, while the observable universe is about 92 billion light-years in diameter. We often write large numbers such as these using *scientific notation*, as illustrated in the accompanying graph showing the relative size of several galaxies.

Galaxy Diameters

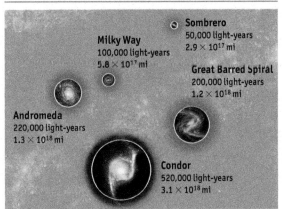

Sombrero
50,000 light-years
2.9×10^{17} mi

Milky Way
100,000 light-years
5.8×10^{17} mi

Great Barred Spiral
200,000 light-years
1.2×10^{18} mi

Andromeda
220,000 light-years
1.3×10^{18} mi

Condor
520,000 light-years
3.1×10^{18} mi

We will calculate the distance from the sun to Earth in Example 23 of Section 4.2 and compare the masses of Earth and Jupiter in Exercise 91 of Exercise Set 4.2.

JUST IN TIME
Review topics 12 and 17 in the **Just In Time** section at the front of the text. This provides excellent prerequisite skill review for this chapter.

MyLab Math
VIDEO

STUDYING FOR SUCCESS Time Management

- ☐ As a rule of thumb, budget two to three hours for homework and study for every hour that you spend in class.
- ☐ Make an hour-by-hour schedule of your week, planning time for leisure as well as work and study.
- ☐ Use your syllabus to help you plan your time. Transfer project deadlines and test dates to your calendar.

4.1 Integers as Exponents

OBJECTIVES

a Tell the meaning of exponential notation.

b Evaluate exponential expressions with exponents of 0 and 1.

c Evaluate algebraic expressions containing exponents.

d Use the product rule to multiply exponential expressions with like bases.

e Use the quotient rule to divide exponential expressions with like bases.

f Express an exponential expression involving negative exponents with positive exponents.

a EXPONENTIAL NOTATION

An exponent of 2 or greater tells how many times the base is used as a factor. For example, $a \cdot a \cdot a \cdot a = a^4$. In this case, the **exponent** is 4 and the **base** is a. An expression for a power is called **exponential notation**.

$$\text{This is the base.} \rightarrow a^n \leftarrow \text{This is the exponent.}$$

EXAMPLE 1 What is the meaning of 3^5? of n^4? of $(2n)^3$? of $50x^2$? of $(-n)^3$? of $-n^3$?

3^5 means $3 \cdot 3 \cdot 3 \cdot 3 \cdot 3$ n^4 means $n \cdot n \cdot n \cdot n$

$(2n)^3$ means $2n \cdot 2n \cdot 2n$ $50x^2$ means $50 \cdot x \cdot x$

$(-n)^3$ means $(-n) \cdot (-n) \cdot (-n)$ $-n^3$ means $-1 \cdot n \cdot n \cdot n$

◀ Do Exercises 1–6.

We read a^n as the **nth power of a**, or simply **a to the nth**, or **a to the n**. We often read x^2 as "**x-squared**" because the area of a square of side x is $x \cdot x$, or x^2. We often read x^3 as "**x-cubed**" because the volume of a cube with length, width, and height x is $x \cdot x \cdot x$, or x^3.

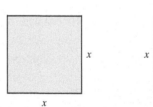
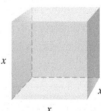

What is the meaning of each of the following?

1. 5^4
2. x^5
3. $(3t)^2$
4. $3t^2$
5. $(-x)^4$
6. $-y^3$

b ONE AND ZERO AS EXPONENTS

Look for a pattern in the following:

On each side, we **divide** by 8 at each step.

$8 \cdot 8 \cdot 8 \cdot 8 = 8^4$
$8 \cdot 8 \cdot 8 = 8^3$
$8 \cdot 8 = 8^2$
$8 = 8^?$
$1 = 8^?$

On this side, the exponents **decrease** by 1 at each step.

To continue the pattern, we would say that $8 = 8^1$ and $1 = 8^0$.

Answers
1. $5 \cdot 5 \cdot 5 \cdot 5$ 2. $x \cdot x \cdot x \cdot x \cdot x$
3. $3t \cdot 3t$ 4. $3 \cdot t \cdot t$
5. $(-x) \cdot (-x) \cdot (-x) \cdot (-x)$ 6. $-1 \cdot y \cdot y \cdot y$

> **EXPONENTS OF 0 AND 1**
>
> $a^1 = a$, for any number a;
> $a^0 = 1$, for any nonzero number a

We consider 0^0 to be not defined. We will explain why later in this section.

EXAMPLE 2 Evaluate 5^1, $(-8)^1$, 3^0, and $(-749.21)^0$.

$5^1 = 5;$ $(-8)^1 = -8;$
$3^0 = 1;$ $(-749.21)^0 = 1$

Do Exercises 7–12.

Evaluate.

7. 6^1 8. 7^0
9. $(8.4)^1$ 10. 8654^0
11. $(-1.4)^1$ 12. 0^1

c EVALUATING ALGEBRAIC EXPRESSIONS

SKILL REVIEW *Simplify expressions using order of operations.* [J18]
Calculate.
1. $3 \cdot 5^2$ 2. $100 - 4^3$ 3. $100 - 5^2 \cdot 2 \div 10$

Answers: 1. 75 **2.** 36 **3.** 95

MyLab Math
VIDEO

We evaluate algebraic expressions by replacing variables with numbers and following the rules for order of operations.

EXAMPLE 3 Evaluate $1000 - x^4$ when $x = 5$.

$1000 - x^4 = 1000 - 5^4$ Substituting
$ = 1000 - 625$ Evaluating 5^4
$ = 375$ Subtracting

EXAMPLE 4 *Area of a Circular Region.* The Richat Structure is a circular eroded geologic dome with a radius of 20 km. Find the area of the structure.

$A = \pi r^2$ Using the formula for the area of a circle
$ = \pi (20 \text{ km})^2$ Substituting
$ = \pi \cdot 20 \text{ km} \cdot 20 \text{ km}$
$ \approx 3.14 \times 400 \text{ km}^2$ Using 3.14 as an approximation for π
$ = 1256 \text{ km}^2$

In Example 4, "km^2" means "square kilometers" and "$\approx$" means "is approximately equal to."

EXAMPLE 5 Evaluate $(5x)^3$ when $x = -2$.

We often use parentheses when substituting a negative number.

$(5x)^3 = [5 \cdot (-2)]^3$ Substituting
$ = [-10]^3$ Multiplying within brackets first
$ = [-10] \cdot [-10] \cdot [-10]$
$ = -1000$ Evaluating the power

Answers
7. 6 **8.** 1 **9.** 8.4 **10.** 1
11. -1.4 **12.** 0

13. Evaluate t^3 when $t = 5$.

14. Evaluate $-5x^5$ when $x = -2$.

15. Find the area of a circle when $r = 32$ cm. Use 3.14 for π.

16. Evaluate $200 - a^4$ when $a = 3$.

17. Evaluate $t^1 - 4$ and $t^0 - 4$ when $t = 7$.

18. **a)** Evaluate $(4t)^2$ when $t = -3$. **GS**
    b) Evaluate $4t^2$ when $t = -3$.
    c) Determine whether $(4t)^2$ and $4t^2$ are equivalent.

   a) $(4t)^2 = [4 \cdot ()]^2$
    $= []^2$
    $=$

   b) $4t^2 = 4 \cdot ()^2$
    $= 4 \cdot ()$
    $=$

   c) Since $144 \neq 36$, the expressions _____ equivalent.
   are/are not

Multiply and simplify.
19. $3^5 \cdot 3^5$

20. $x^4 \cdot x^6$

21. $p^4 p^{12} p^8$

22. $x \cdot x^4$

23. $(a^2 b^3)(a^7 b^5)$

Answers
13. 125 **14.** 160 **15.** 3215.36 cm² **16.** 119
17. 3; −3 **18.** (a) 144; (b) 36; (c) no
19. 3^{10} **20.** x^{10} **21.** p^{24} **22.** x^5 **23.** $a^9 b^8$

Guided Solution:
18. (a) −3, −12, 144; (b) −3, 9, 36; (c) are not

EXAMPLE 6 Evaluate $5x^3$ when $x = -2$.

$5x^3 = 5 \cdot (-2)^3$ Substituting
$= 5 \cdot (-2) \cdot (-2) \cdot (-2)$ Evaluating the power first
$= 5(-8)$ $(-2)(-2)(-2) = -8$
$= -40$

Recall that two expressions are equivalent if they have the same value for all meaningful replacements. Note that Examples 5 and 6 show that $(5x)^3$ and $5x^3$ are *not* equivalent—that is, $(5x)^3 \neq 5x^3$.

◀ Do Exercises 13–18.

d MULTIPLYING POWERS WITH LIKE BASES

We can multiply powers with like bases by adding exponents. For example,

$$a^3 \cdot a^2 = (a \cdot a \cdot a)(a \cdot a) = a \cdot a \cdot a \cdot a \cdot a = a^5.$$

3 factors   2 factors   5 factors

Note that the exponent in a^5 is the sum of those in $a^3 \cdot a^2$. That is, $3 + 2 = 5$. Likewise,

$$b^4 \cdot b^3 = (b \cdot b \cdot b \cdot b)(b \cdot b \cdot b) = b^7, \quad \text{where} \quad 4 + 3 = 7.$$

Adding the exponents gives the correct result.

THE PRODUCT RULE

For any number a and any positive integers m and n,

$$a^m \cdot a^n = a^{m+n}.$$

(When multiplying with exponential notation, if the bases are the same, keep the base and add the exponents.)

EXAMPLES Multiply and simplify.

7. $5^6 \cdot 5^2 = 5^{6+2}$ Adding exponents: $a^m \cdot a^n = a^{m+n}$
   $= 5^8$

8. $m^5 m^{10} m^3 = m^{5+10+3} = m^{18}$

9. $x \cdot x^8 = x^1 \cdot x^8$ Writing x as x^1
   $= x^{1+8}$
   $= x^9$

10. $(a^3 b^2)(a^3 b^5) = (a^3 a^3)(b^2 b^5)$
    $= a^6 b^7$

11. $(4y)^6 (4y)^3 = (4y)^{6+3} = (4y)^9$

◀ Do Exercises 19–23.

e DIVIDING POWERS WITH LIKE BASES

The following suggests a rule for dividing powers with like bases, such as a^5/a^2:

$$\frac{a^5}{a^2} = \frac{a \cdot a \cdot a \cdot a \cdot a}{a \cdot a} = \frac{a \cdot a \cdot a \cdot a \cdot a}{1 \cdot a \cdot a} = \frac{a \cdot a \cdot a}{1} \cdot \frac{a \cdot a}{a \cdot a}$$

$$= \frac{a \cdot a \cdot a}{1} \cdot 1 = a \cdot a \cdot a = a^3.$$

Note that the exponent in a^3 is the difference of those in $a^5 \div a^2$. That is, $5 - 2 = 3$. In a similar way, we have

$$\frac{t^9}{t^4} = \frac{t \cdot t \cdot t \cdot t \cdot t \cdot t \cdot t \cdot t \cdot t}{t \cdot t \cdot t \cdot t} = t^5, \quad \text{where} \quad 9 - 4 = 5.$$

Subtracting exponents gives the correct answer.

THE QUOTIENT RULE

For any nonzero number a and any positive integers m and n,

$$\frac{a^m}{a^n} = a^{m-n}.$$

(When dividing with exponential notation, if the bases are the same, keep the base and subtract the exponent of the denominator from the exponent of the numerator.)

EXAMPLES Divide and simplify.

12. $\dfrac{6^5}{6^3} = 6^{5-3}$ Subtracting exponents

$= 6^2$

13. $\dfrac{x^8}{x} = \dfrac{x^8}{x^1} = x^{8-1}$

$= x^7$

14. $\dfrac{(3t)^{12}}{(3t)^2} = (3t)^{12-2}$

$= (3t)^{10}$

15. $\dfrac{p^5 q^7}{p^2 q^5} = \dfrac{p^5}{p^2} \cdot \dfrac{q^7}{q^5} = p^{5-2} q^{7-5}$

$= p^3 q^2$

The quotient rule can also be used to explain the definition of 0 as an exponent. Consider the expression a^4/a^4, where a is nonzero:

$$\frac{a^4}{a^4} = \frac{a \cdot a \cdot a \cdot a}{a \cdot a \cdot a \cdot a} = 1.$$

This is true because the numerator and the denominator are the same. Now suppose that we apply the rule for dividing powers with the same base:

$$\frac{a^4}{a^4} = a^{4-4} = a^0.$$

Since $a^4/a^4 = 1$ and $a^4/a^4 = a^0$, it follows that $a^0 = 1$, when $a \neq 0$.

We can explain why we do not define 0^0 using the quotient rule. We know that 0^0 is 0^{1-1}. But 0^{1-1} is also equal to $0^1/0^1$, or $0/0$. We have already seen that division by 0 is not defined, so 0^0 is also not defined.

Do Exercises 24–27.

Divide and simplify.

24. $\dfrac{4^5}{4^2}$

25. $\dfrac{y^6}{y^2}$

26. $\dfrac{p^{10}}{p}$

27. $\dfrac{a^7 b^6}{a^3 b^4}$

Answers
24. 4^3 **25.** y^4 **26.** p^9 **27.** $a^4 b^2$

f NEGATIVE INTEGERS AS EXPONENTS

SKILL REVIEW

Subtract real numbers. [1.4a]
Subtract.
1. $3 - 7$ 2. $-3 - 7$ 3. $-3 - (-7)$

Answers: 1. -4 2. -10 3. 4

To develop a definition of exponential notation when the exponent is a negative integer, consider $5^3/5^7$ and first simplify using procedures that we have learned for working with fractions:

$$\frac{5^3}{5^7} = \frac{5 \cdot 5 \cdot 5}{5 \cdot 5 \cdot 5 \cdot 5 \cdot 5 \cdot 5 \cdot 5} = \frac{5 \cdot 5 \cdot 5 \cdot 1}{5 \cdot 5 \cdot 5 \cdot 5 \cdot 5 \cdot 5 \cdot 5}$$

$$= \frac{5 \cdot 5 \cdot 5}{5 \cdot 5 \cdot 5} \cdot \frac{1}{5 \cdot 5 \cdot 5 \cdot 5} = \frac{1}{5^4}.$$

Now we simplify $5^3/5^7$ using the quotient rule:

$$\frac{5^3}{5^7} = 5^{3-7} = 5^{-4}.$$

From these two expressions for $5^3/5^7$, it follows that

$$5^{-4} = \frac{1}{5^4}.$$

This leads to our definition of negative exponents.

NEGATIVE EXPONENT

For any real number a that is nonzero and any integer n,

$$a^{-n} = \frac{1}{a^n}.$$

In fact, the numbers a^n and a^{-n} are reciprocals because

$$a^n \cdot a^{-n} = a^n \cdot \frac{1}{a^n} = \frac{a^n}{a^n} = 1.$$

The following pattern reinforces the definition of negative exponents.

On each side, we **divide** by 5 at each step.

$5 \cdot 5 \cdot 5 = 5^3$
$5 \cdot 5 = 5^2$
$5 = 5^1$
$1 = 5^0$
$\frac{1}{5} = 5^?$
$\frac{1}{25} = 5^?$

On this side, the exponents **decrease** by 1 at each step.

To continue the pattern, it should follow that

$$\frac{1}{5} = \frac{1}{5^1} = 5^{-1} \quad \text{and} \quad \frac{1}{25} = \frac{1}{5^2} = 5^{-2}.$$

EXAMPLES Express using positive exponents. Then simplify.

16. $4^{-2} = \dfrac{1}{4^2} = \dfrac{1}{16}$

17. $(-3)^{-2} = \dfrac{1}{(-3)^2} = \dfrac{1}{(-3)(-3)} = \dfrac{1}{9}$

18. $m^{-3} = \dfrac{1}{m^3}$

19. $ab^{-1} = a\left(\dfrac{1}{b^1}\right) = a\left(\dfrac{1}{b}\right) = \dfrac{a}{b}$

20. $\dfrac{1}{x^{-3}} = x^{-(-3)} = x^3$

21. $3c^{-5} = 3\left(\dfrac{1}{c^5}\right) = \dfrac{3}{c^5}$

Example 20 might also be done as follows:

$$\dfrac{1}{x^{-3}} = \dfrac{1}{\frac{1}{x^3}} = 1 \cdot \dfrac{x^3}{1} = x^3.$$

.... **Caution!** ....

As shown in Examples 16 and 17, a negative exponent does not necessarily mean that an expression is negative.

Do Exercises 28–33. ▶

Express with positive exponents. Then simplify.

28. 4^{-3} 29. 5^{-2}

30. 2^{-4} 31. $(-2)^{-3}$

32. $\dfrac{1}{x^{-2}}$

GS 33. $4p^{-3}$
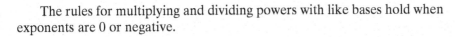

The rules for multiplying and dividing powers with like bases hold when exponents are 0 or negative.

EXAMPLES Simplify. Write the result using positive exponents.

22. $7^{-3} \cdot 7^6 = 7^{-3+6}$ Adding exponents
    $= 7^3$

23. $x^4 \cdot x^{-3} = x^{4+(-3)} = x^1 = x$

24. $\dfrac{5^4}{5^{-2}} = 5^{4-(-2)}$ Subtracting exponents
    $= 5^{4+2} = 5^6$

25. $\dfrac{x}{x^7} = x^{1-7} = x^{-6} = \dfrac{1}{x^6}$

26. $\dfrac{b^{-4}}{b^{-5}} = b^{-4-(-5)}$
    $= b^{-4+5} = b^1 = b$

27. $y^{-4} \cdot y^{-8} = y^{-4+(-8)}$
    $= y^{-12} = \dfrac{1}{y^{12}}$

Simplify.
34. $5^{-2} \cdot 5^4$

35. $x^{-3} \cdot x^{-4}$

36. $\dfrac{7^{-2}}{7^3}$

37. $\dfrac{b^{-2}}{b^{-3}}$

38. $\dfrac{t}{t^{-5}}$

Do Exercises 34–38. ▶

The following is a summary of the definitions and rules for exponents that we have considered in this section.

DEFINITIONS AND RULES FOR EXPONENTS

1 as an exponent:	$a^1 = a$
0 as an exponent:	$a^0 = 1, a \neq 0$
Negative integers as exponents:	$a^{-n} = \dfrac{1}{a^n}, \dfrac{1}{a^{-n}} = a^n; a \neq 0$
Product Rule:	$a^m \cdot a^n = a^{m+n}$
Quotient Rule:	$\dfrac{a^m}{a^n} = a^{m-n}, a \neq 0$

Answers

28. $\dfrac{1}{4^3} = \dfrac{1}{64}$ 29. $\dfrac{1}{5^2} = \dfrac{1}{25}$ 30. $\dfrac{1}{2^4} = \dfrac{1}{16}$
31. $\dfrac{1}{(-2)^3} = -\dfrac{1}{8}$ 32. x^2 33. $\dfrac{4}{p^3}$ 34. $5^2 = 25$
35. $\dfrac{1}{x^7}$ 36. $\dfrac{1}{7^5}$ 37. b 38. t^6

Guided Solution:
33. p^3, p^3

4.1 Exercise Set

✓ Check Your Understanding

Reading Check Choose from the list on the right the word that best completes each sentence.

RC1. In the expression 4^7, the number 4 is the _____.

RC2. We often read s^3 as "s-_____."

RC3. When simplifying $x^3 x^{-10}$, we _____ 3 and -10.

- add
- subtract
- base
- exponent
- squared
- cubed

Concept Check Match each expression with the appropriate value from the column on the right. Choices may be used more than once or not at all.

CC1. ___ y^1

CC2. ___ $y^0, y \neq 0$

CC3. ___ $y^1 \cdot y^1$

CC4. ___ $\dfrac{y^9}{y^8}$

CC5. ___ $\dfrac{y^8}{y^9}$

CC6. ___ $\dfrac{1}{y^{-1}}$

a) 1
b) 0
c) y
d) $\dfrac{1}{y}$
e) y^2

a What is the meaning of each of the following?

1. 3^4
2. 4^3
3. $(-1.1)^5$
4. $(87.2)^6$
5. $\left(\dfrac{2}{3}\right)^4$
6. $\left(-\dfrac{5}{8}\right)^3$
7. $(7p)^2$
8. $(11c)^3$
9. $8k^3$
10. $17x^2$
11. $-6y^4$
12. $-q^5$

b Evaluate.

13. $a^0, a \neq 0$
14. $t^0, t \neq 0$
15. b^1
16. c^1
17. $\left(\dfrac{2}{3}\right)^0$
18. $\left(-\dfrac{5}{8}\right)^0$
19. $(-7.03)^1$
20. $\left(\dfrac{4}{5}\right)^1$
21. 8.38^0
22. 8.38^1
23. $(ab)^1$
24. $(ab)^0, a \neq 0, b \neq 0$
25. $ab^0, b \neq 0$
26. ab^1

258 CHAPTER 4 Polynomials: Operations

c Evaluate.

27. m^3, when $m = 3$
28. x^6, when $x = 2$
29. p^1, when $p = 19$
30. x^{19}, when $x = 0$

31. $-x^4$, when $x = -3$
32. $-2y^7$, when $y = 2$
33. x^4, when $x = 4$
34. y^{15}, when $y = 1$

35. $y^2 - 7$, when $y = -10$
36. $z^5 + 5$, when $z = -2$
37. $161 - b^2$, when $b = 5$
38. $325 - v^3$, when $v = -3$

39. $x^1 + 3$ and $x^0 + 3$, when $x = 7$
40. $y^0 - 8$ and $y^1 - 8$, when $y = -3$

41. Find the area of a circle when $r = 34$ ft. Use 3.14 for π.

42. The area A of a square with sides of length s is given by $A = s^2$. Find the area of a square with sides of length 24 m.

f Express using positive exponents. Then simplify.

43. 3^{-2}
44. 2^{-3}
45. 10^{-3}
46. 5^{-4}
47. a^{-3}

48. x^{-2}
49. $\dfrac{1}{8^{-2}}$
50. $\dfrac{1}{2^{-5}}$
51. $\dfrac{1}{y^{-4}}$
52. $\dfrac{1}{t^{-7}}$

53. $5z^{-4}$
54. $6n^{-5}$
55. xy^{-2}
56. ab^{-3}

Express using negative exponents.

57. $\dfrac{1}{4^3}$
58. $\dfrac{1}{5^2}$
59. $\dfrac{1}{x^3}$
60. $\dfrac{1}{y^2}$
61. $\dfrac{1}{a^5}$
62. $\dfrac{1}{b^7}$

d, f Multiply and simplify.

63. $2^4 \cdot 2^3$
64. $3^5 \cdot 3^2$
65. $9^{17} \cdot 9^{21}$
66. $7^{22} \cdot 7^{15}$

67. $x^4 \cdot x$
68. $y \cdot y^9$
69. $x^{14} \cdot x^3$
70. $x^9 \cdot x^4$

71. $(3y)^4(3y)^8$
72. $(2t)^8(2t)^{17}$
73. $(7y)^1(7y)^{16}$
74. $(8x)^0(8x)^1$

75. $3^{-5} \cdot 3^8$
76. $5^{-8} \cdot 5^9$
77. $x^{-2} \cdot x^2$
78. $x \cdot x^{-1}$

SECTION 4.1 Integers as Exponents 259

79. $x^{-7} \cdot x^{-6}$

80. $y^{-5} \cdot y^{-8}$

81. $a^{11} \cdot a^{-3} \cdot a^{-18}$

82. $a^{-11} \cdot a^{-3} \cdot a^{-7}$

83. $(x^4y^7)(x^2y^8)$

84. $(a^5c^2)(a^3c^9)$

85. $(s^2t^3)(st^4)$

86. $(m^4n)(m^2n^7)$

e, f Divide and simplify. Write the result using positive exponents.

87. $\dfrac{7^5}{7^2}$

88. $\dfrac{5^8}{5^6}$

89. $\dfrac{y^9}{y}$

90. $\dfrac{x^{11}}{x}$

91. $\dfrac{16^2}{16^8}$

92. $\dfrac{7^2}{7^9}$

93. $\dfrac{m^6}{m^{12}}$

94. $\dfrac{a^3}{a^4}$

95. $\dfrac{(8x)^6}{(8x)^{10}}$

96. $\dfrac{(8t)^4}{(8t)^{11}}$

97. $\dfrac{x}{x^{-1}}$

98. $\dfrac{t^8}{t^{-3}}$

99. $\dfrac{z^{-6}}{z^{-2}}$

100. $\dfrac{x^{-9}}{x^{-3}}$

101. $\dfrac{x^{-5}}{x^{-8}}$

102. $\dfrac{y^{-2}}{y^{-9}}$

103. $\dfrac{m^{-9}}{m^{-9}}$

104. $\dfrac{x^{-7}}{x^{-7}}$

105. $\dfrac{a^5b^3}{a^2b}$

106. $\dfrac{s^8t^4}{st^3}$

Matching. In Exercises 107 and 108, match each item in the first column with the appropriate item in the second column by drawing connecting lines. Items in the second column may be used more than once.

107.

5^2	$-\dfrac{1}{10}$
5^{-2}	$\dfrac{1}{10}$
$\left(\dfrac{1}{5}\right)^2$	$-\dfrac{1}{25}$
$\left(\dfrac{1}{5}\right)^{-2}$	10
-5^2	25
$(-5)^2$	-25
$-\left(-\dfrac{1}{5}\right)^2$	$\dfrac{1}{25}$
$\left(-\dfrac{1}{5}\right)^{-2}$	-10

108.

$-\left(\dfrac{1}{8}\right)^2$	16
$\left(\dfrac{1}{8}\right)^{-2}$	-16
8^{-2}	64
8^2	-64
-8^2	$\dfrac{1}{64}$
$(-8)^2$	$-\dfrac{1}{64}$
$\left(-\dfrac{1}{8}\right)^{-2}$	$-\dfrac{1}{16}$
$\left(-\dfrac{1}{8}\right)^2$	$\dfrac{1}{16}$

Skill Maintenance

Solve.

109. A 12-in. submarine sandwich is cut into two pieces. One piece is twice as long as the other. How long are the pieces? [2.6a]

110. The first angle of a triangle is 24° more than the second. The third angle is twice the first. Find the measures of the angles of the triangle. [2.6a]

111. A warehouse stores 1800 lb of peanuts, 1500 lb of cashews, and 700 lb of almonds. What percent of the total is peanuts? cashews? almonds? [2.5a]

112. The width of a rectangle is fixed at 10 ft. For what lengths will the area be less than 25 ft²? [2.8b]

Solve.

113. $2x - 4 - 5x + 8 = x - 3$ [2.3b]

114. $8x + 7 - 9x = 12 - 6x + 5$ [2.3b]

115. $-6(2 - x) + 10(5x - 7) = 10$ [2.3c]

116. $-10(x - 4) = 5(2x + 5) - 7$ [2.3c]

Synthesis

Determine whether each of the following equations is true.

117. $(x + 1)^2 = x^2 + 1$

118. $(x - 1)^2 = x^2 - 2x + 1$

119. $(5x)^0 = 5x^0$

120. $\dfrac{x^3}{x^5} = x^2$

Simplify.

121. $(y^{2x})(y^{3x})$

122. $a^{5k} \div a^{3k}$

123. $\dfrac{a^{6t}(a^{7t})}{a^{9t}}$

124. $\dfrac{\left(\frac{1}{2}\right)^4}{\left(\frac{1}{2}\right)^5}$

125. $\dfrac{(0.8)^5}{(0.8)^3(0.8)^2}$

126. $\dfrac{(x - 3)^5}{x - 3}$

Use >, <, or = for ☐ to write a true sentence.

127. 3^5 ☐ 3^4

128. 4^2 ☐ 4^3

129. 4^3 ☐ 5^3

130. 4^3 ☐ 3^4

Evaluate.

131. $\dfrac{1}{-z^4}$, when $z = -10$

132. $\dfrac{1}{-z^5}$, when $z = -0.1$

133. Determine whether $(a + b)^2$ and $a^2 + b^2$ are equivalent. (*Hint:* Choose values for a and b and evaluate.)

4.2 Exponents and Scientific Notation

OBJECTIVES

a. Use the power rule to raise powers to powers.
b. Raise a product to a power and a quotient to a power.
c. Convert between scientific notation and decimal notation.
d. Multiply and divide using scientific notation.
e. Solve applied problems using scientific notation.

We now consider three rules used to simplify exponential expressions. We then apply our knowledge of exponents to *scientific notation*.

a RAISING POWERS TO POWERS

SKILL REVIEW

Multiply real numbers. [1.5a]
Multiply.
1. $-5 \cdot 8$
2. $(-3)(-5)$

Answers: 1. -40 2. 15

MyLab Math VIDEO

Consider an expression like $(3^2)^4$. We are raising 3^2 to the fourth power:

$$(3^2)^4 = (3^2)(3^2)(3^2)(3^2)$$
$$= (3 \cdot 3)(3 \cdot 3)(3 \cdot 3)(3 \cdot 3)$$
$$= 3 \cdot 3 \cdot 3 \cdot 3 \cdot 3 \cdot 3 \cdot 3 \cdot 3$$
$$= 3^8.$$

Note that in this case we could have multiplied the exponents:

$$(3^2)^4 = 3^{2 \cdot 4} = 3^8.$$

THE POWER RULE

For any real number a and any integers m and n,

$$(a^m)^n = a^{mn}.$$

(To raise a power to a power, multiply the exponents.)

EXAMPLES Simplify. Express the answers using positive exponents.

1. $(3^5)^4 = 3^{5 \cdot 4}$ Multiplying exponents
$= 3^{20}$

2. $(a^{-4})^{-6} = a^{(-4)(-6)} = a^{24}$

3. $(y^{-5})^7 = y^{-5 \cdot 7} = y^{-35} = \dfrac{1}{y^{35}}$

4. $(x^4)^{-2} = x^{4(-2)} = x^{-8} = \dfrac{1}{x^8}$

◀ Do Exercises 1–4.

Simplify. Express the answers using positive exponents.
1. $(3^4)^5$
2. $(x^{-3})^4$
3. $(y^{-5})^{-3}$
4. $(x^4)^{-8}$

b RAISING A PRODUCT OR A QUOTIENT TO A POWER

When an expression inside parentheses is raised to a power, the inside expression is the base. Let's compare $2a^3$ and $(2a)^3$:

$2a^3 = 2 \cdot a \cdot a \cdot a$;   The base is a.

$(2a)^3 = (2a)(2a)(2a)$ The base is $2a$.
$= (2 \cdot 2 \cdot 2)(a \cdot a \cdot a)$ Using the associative and commutative laws of multiplication
$= 2^3 a^3 = 8a^3.$

Answers
1. 3^{20} 2. $\dfrac{1}{x^{12}}$ 3. y^{15} 4. $\dfrac{1}{x^{32}}$

We see that $2a^3$ and $(2a)^3$ are *not* equivalent. We also see that we can evaluate the power $(2a)^3$ by raising each factor to the power 3. This leads us to a rule for raising a product to a power.

RAISING A PRODUCT TO A POWER

For any real numbers a and b and any integer n,
$$(ab)^n = a^n b^n.$$

(To raise a product to the nth power, raise each factor to the nth power.)

EXAMPLES Simplify.

5. $(4x^2)^3 = (4^1 x^2)^3$ $4 = 4^1$
   $= (4^1)^3 \cdot (x^2)^3$ Raising *each* factor to the third power
   $= 4^3 \cdot x^6 = 64x^6$ Using the power rule and simplifying

6. $(-5x^4 y^3)^3 = (-5)^3 (x^4)^3 (y^3)^3$ Raising each factor to the third power
   $= -125 x^{12} y^9$

7. $[(-x)^{25}]^2 = (-x)^{50}$ Using the power rule
   $= (-1 \cdot x)^{50}$ Using the property of -1: $-x = -1 \cdot x$
   $= (-1)^{50} x^{50}$ Raising each factor to the fiftieth power
   $= 1 \cdot x^{50}$ The product of an even number of negative factors is positive.
   $= x^{50}$

8. $(3x^3 y^{-5} z^2)^4 = 3^4 (x^3)^4 (y^{-5})^4 (z^2)^4 = 81 x^{12} y^{-20} z^8 = \dfrac{81 x^{12} z^8}{y^{20}}$

9. $(-x^4)^{-3} = (-1 \cdot x^4)^{-3} = (-1)^{-3} \cdot (x^4)^{-3} = (-1)^{-3} \cdot x^{-12}$
   $= \dfrac{1}{(-1)^3} \cdot \dfrac{1}{x^{12}} = \dfrac{1}{-1} \cdot \dfrac{1}{x^{12}} = -\dfrac{1}{x^{12}}$

10. $(-2x^{-5} y^4)^{-4} = (-2)^{-4} (x^{-5})^{-4} (y^4)^{-4} = \dfrac{1}{(-2)^4} \cdot x^{20} \cdot y^{-16}$
    $= \dfrac{1}{16} \cdot x^{20} \cdot \dfrac{1}{y^{16}} = \dfrac{x^{20}}{16 y^{16}}$

Do Exercises 5–11. ▶

There is a similar rule for raising a quotient to a power.

RAISING A QUOTIENT TO A POWER

For any real numbers a and b, $b \neq 0$, and any integer n,
$$\left(\dfrac{a}{b}\right)^n = \dfrac{a^n}{b^n}.$$

(To raise a quotient to the nth power, raise both the numerator and the denominator to the nth power.)

Simplify.

5. $(2x^5 y^{-3})^4$

6. $(5x^5 y^{-6} z^{-3})^2$

7. $[(-x)^{37}]^2$

8. $(3y^{-2} x^{-5} z^8)^3$

9. $(-y^8)^{-3}$

GS 10. $(-2x^4)^{-2}$
   $= (-2)^{-2} ()^{-2}$
   $= \dfrac{1}{(-2)^{}} \cdot x^{}$
   $= \dfrac{1}{} \cdot \dfrac{1}{x^{}}$
   $= \dfrac{1}{}$

11. $(-3x^2 y^{-5})^{-3}$

Answers

5. $\dfrac{16 x^{20}}{y^{12}}$ 6. $\dfrac{25 x^{10}}{y^{12} z^6}$ 7. x^{74} 8. $\dfrac{27 z^{24}}{y^6 x^{15}}$
9. $-\dfrac{1}{y^{24}}$ 10. $\dfrac{1}{4x^8}$ 11. $-\dfrac{y^{15}}{27 x^6}$

Guided Solution:
10. x^4, 2, -8, 4, 8, $4x^8$

Simplify.

12. $\left(\dfrac{x^6}{5}\right)^2$

13. $\left(\dfrac{2t^5}{w^4}\right)^3$

14. $\left(\dfrac{a^4}{3b^{-2}}\right)^3$

15. $\left(\dfrac{x^4}{3}\right)^{-2}$

Do this two ways.

$\left(\dfrac{x^4}{3}\right)^{-2} = \dfrac{(x^4)^{\square}}{3^{-2}} = \dfrac{x^{\square}}{3^{-2}}$

$= \dfrac{\dfrac{1}{x^{\square}}}{\dfrac{1}{3^2}} = \dfrac{1}{x^8} \div \dfrac{1}{3^2}$

$= \dfrac{1}{x^8} \cdot \dfrac{3^2}{\square} = \dfrac{9}{\square}$

This can be done a second way.

$\left(\dfrac{x^4}{3}\right)^{-2} = \left(\dfrac{3}{x^4}\right)^{\square}$

$= \dfrac{3^2}{(x^4)^{\square}} = \dfrac{9}{\square}$

EXAMPLES Simplify.

11. $\left(\dfrac{x^2}{4}\right)^3 = \dfrac{(x^2)^3}{4^3} = \dfrac{x^6}{64}$ Raising *both* the numerator and the denominator to the third power

12. $\left(\dfrac{3a^4}{b^3}\right)^2 = \dfrac{(3a^4)^2}{(b^3)^2} = \dfrac{3^2(a^4)^2}{b^{3\cdot 2}} = \dfrac{9a^8}{b^6}$

13. $\left(\dfrac{y^2}{2z^{-5}}\right)^4 = \dfrac{(y^2)^4}{(2z^{-5})^4} = \dfrac{(y^2)^4}{2^4(z^{-5})^4} = \dfrac{y^8}{16z^{-20}} = \dfrac{y^8 z^{20}}{16}$

14. $\left(\dfrac{y^3}{5}\right)^{-2} = \dfrac{(y^3)^{-2}}{5^{-2}} = \dfrac{y^{-6}}{5^{-2}} = \dfrac{\dfrac{1}{y^6}}{\dfrac{1}{5^2}} = \dfrac{1}{y^6} \div \dfrac{1}{5^2} = \dfrac{1}{y^6} \cdot \dfrac{5^2}{1} = \dfrac{25}{y^6}$

The following can often be used to simplify a quotient that is raised to a negative power.

For $a \neq 0$ and $b \neq 0$,
$$\left(\dfrac{a}{b}\right)^{-n} = \left(\dfrac{b}{a}\right)^n.$$

Example 14 might also be completed as follows:

$\left(\dfrac{y^3}{5}\right)^{-2} = \left(\dfrac{5}{y^3}\right)^2 = \dfrac{5^2}{(y^3)^2} = \dfrac{25}{y^6}.$

◀ Do Exercises 12–15.

c SCIENTIFIC NOTATION

We can write numbers using different types of notation, such as fraction notation, decimal notation, and percent notation. Another type, **scientific notation**, makes use of exponential notation. Scientific notation is especially useful when calculations involve very large or very small numbers. The following are examples of scientific notation.

The number of stars in the Milky Way galaxy:
$4 \times 10^{11} = 400{,}000{,}000{,}000$

The length of an *E.coli* bacterium:
2×10^{-6} m $= 0.000002$ m

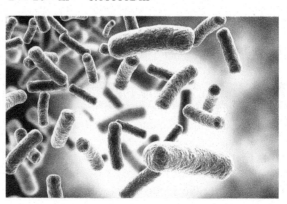

Answers

12. $\dfrac{x^{12}}{25}$ 13. $\dfrac{8t^{15}}{w^{12}}$ 14. $\dfrac{a^{12}b^6}{27}$ 15. $\dfrac{9}{x^8}$

Guided Solution:
15. $-2, -8, 8, 1, x^8; 2, 2, x^8$

SCIENTIFIC NOTATION

Scientific notation for a number is an expression of the type

$$M \times 10^n,$$

where n is an integer, M is greater than or equal to 1 and less than 10 ($1 \leq M < 10$), and M is expressed in decimal notation. 10^n is also considered to be scientific notation when $M = 1$.

Caution!

Each of the following is *not* scientific notation.

$$\underline{12.46} \times 10^7$$

This number is greater than 10.

$$\underline{0.347} \times 10^{-5}$$

This number is less than 1.

You should try to make conversions to scientific notation mentally as often as possible. Here is a handy mental device.

A positive exponent in scientific notation indicates a large number (greater than or equal to 10) and a negative exponent indicates a small number (between 0 and 1).

EXAMPLES Convert to scientific notation.

15. $78{,}000 = 7.8 \times 10^4$

7.8,000.   Large number, so the exponent is positive
4 places

16. $0.0000057 = 5.7 \times 10^{-6}$

0.000005.7   Small number, so the exponent is negative
6 places

Convert to scientific notation.
16. 0.000517

▶ Do Exercises 16 and 17.

17. 523,000,000

EXAMPLES Convert mentally to decimal notation.

17. $7.893 \times 10^5 = 789{,}300$

7.89300.   Positive exponent, so the answer is a large number
5 places

18. $4.7 \times 10^{-8} = 0.000000047$

.00000004.7   Negative exponent, so the answer is a small number
8 places

Convert to decimal notation.
18. 6.893×10^{11}

▶ Do Exercises 18 and 19.

19. 5.67×10^{-5}

d MULTIPLYING AND DIVIDING USING SCIENTIFIC NOTATION

Multiplying

Consider the product

$$400 \cdot 2000 = 800{,}000.$$

In scientific notation, this is

$$(4 \times 10^2) \cdot (2 \times 10^3) = (4 \cdot 2)(10^2 \cdot 10^3) = 8 \times 10^5.$$

Answers
16. 5.17×10^{-4} **17.** 5.23×10^8
18. 689,300,000,000   **19.** 0.0000567

Multiply and write scientific notation for the result.

20. $(1.12 \times 10^{-8})(5 \times 10^{-7})$

21. $(9.1 \times 10^{-17})(8.2 \times 10^{3})$

CALCULATOR CORNER

To find the product in Example 19 and express the result in scientific notation on a graphing calculator, we first set the calculator in Scientific mode using MODE. Then we go to the home screen and enter the computation by pressing 1 . 8 2ND EE 6 × 2 . 3 2ND EE (-) 4 ENTER. (EE is the second operation associated with the , key.) The decimal portion of a number written in scientific notation appears before a small E and the exponent follows the E.

```
1.8E6*2.3E-4
              4.14E2
```

EXERCISE: Multiply or divide and express the answer in scientific notation.

1. $(3.15 \times 10^{7})(4.3 \times 10^{-12})$
2. $(8 \times 10^{9})(4 \times 10^{-5})$
3. $\dfrac{4.5 \times 10^{6}}{1.5 \times 10^{12}}$
4. $\dfrac{4 \times 10^{-9}}{5 \times 10^{16}}$

Divide and write scientific notation for the result.

22. $\dfrac{4.2 \times 10^{5}}{2.1 \times 10^{2}}$

23. $\dfrac{1.1 \times 10^{-4}}{2.0 \times 10^{-7}}$

EXAMPLE 19 Multiply: $(1.8 \times 10^{6}) \cdot (2.3 \times 10^{-4})$.

We apply the commutative and associative laws to get

$$(1.8 \times 10^{6}) \cdot (2.3 \times 10^{-4}) = (1.8 \cdot 2.3) \times (10^{6} \cdot 10^{-4})$$
$$= 4.14 \times 10^{6+(-4)}$$
$$= 4.14 \times 10^{2}.$$

We get 4.14 by multiplying 1.8 and 2.3. We get 10^{2} by adding the exponents 6 and -4.

EXAMPLE 20 Multiply: $(3.1 \times 10^{5}) \cdot (4.5 \times 10^{-3})$.

$$(3.1 \times 10^{5}) \cdot (4.5 \times 10^{-3}) = (3.1 \times 4.5)(10^{5} \cdot 10^{-3})$$
$$= 13.95 \times 10^{2} \quad \text{Not scientific notation; 13.95 is greater than 10.}$$
$$= (1.395 \times 10^{1}) \times 10^{2} \quad \text{Substituting } 1.395 \times 10^{1} \text{ for 13.95}$$
$$= 1.395 \times (10^{1} \times 10^{2}) \quad \text{Associative law}$$
$$= 1.395 \times 10^{3} \quad \text{The answer is now in scientific notation.}$$

◀ Do Exercises 20 and 21.

Dividing

Consider the quotient $800{,}000 \div 400 = 2000$. In scientific notation, this is

$$(8 \times 10^{5}) \div (4 \times 10^{2}) = \dfrac{8 \times 10^{5}}{4 \times 10^{2}} = \dfrac{8}{4} \times \dfrac{10^{5}}{10^{2}} = 2 \times 10^{3}.$$

EXAMPLE 21 Divide: $(3.41 \times 10^{5}) \div (1.1 \times 10^{-3})$.

$$(3.41 \times 10^{5}) \div (1.1 \times 10^{-3}) = \dfrac{3.41 \times 10^{5}}{1.1 \times 10^{-3}} = \dfrac{3.41}{1.1} \times \dfrac{10^{5}}{10^{-3}}$$
$$= 3.1 \times 10^{5-(-3)}$$
$$= 3.1 \times 10^{8}$$

EXAMPLE 22 Divide: $(6.4 \times 10^{-7}) \div (8.0 \times 10^{6})$.

$$(6.4 \times 10^{-7}) \div (8.0 \times 10^{6}) = \dfrac{6.4 \times 10^{-7}}{8.0 \times 10^{6}}$$
$$= \dfrac{6.4}{8.0} \times \dfrac{10^{-7}}{10^{6}}$$
$$= 0.8 \times 10^{-7-6}$$
$$= 0.8 \times 10^{-13} \quad \text{Not scientific notation; 0.8 is less than 1.}$$
$$= (8.0 \times 10^{-1}) \times 10^{-13} \quad \text{Substituting } 8.0 \times 10^{-1} \text{ for 0.8}$$
$$= 8.0 \times (10^{-1} \times 10^{-13}) \quad \text{Associative law}$$
$$= 8.0 \times 10^{-14} \quad \text{Adding exponents}$$

◀ Do Exercises 22 and 23.

Answers

20. 5.6×10^{-15} **21.** 7.462×10^{-13}
22. 2.0×10^{3} **23.** 5.5×10^{2}

e APPLICATIONS WITH SCIENTIFIC NOTATION

EXAMPLE 23 *Distance from the Sun to Earth.* Light from the sun traveling at a rate of 300,000 kilometers per second (km/s) reaches Earth in 499 sec. Find the distance, expressed in scientific notation, from the sun to Earth.

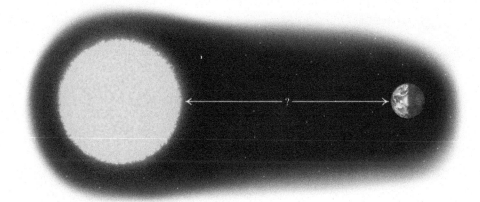

The time t that it takes for light to reach Earth from the sun is 4.99×10^2 sec (s). The speed is 3.0×10^5 km/s. Recall that distance can be expressed in terms of speed and time as

Distance = Speed · Time
$d = rt$.

We substitute 3.0×10^5 for r and 4.99×10^2 for t:

$d = rt$
$= (3.0 \times 10^5)(4.99 \times 10^2)$ Substituting
$= 14.97 \times 10^7$
$= (1.497 \times 10^1) \times 10^7$
$= 1.497 \times (10^1 \times 10^7)$ Converting to scientific notation
$= 1.497 \times 10^8$ km.

Thus, the distance from the sun to Earth is 1.497×10^8 km.

Do Exercise 24.

EXAMPLE 24 *Social Networking.* The social networking site LinkedIn allows registered users to upload information about their professional careers. Users can also verify, or endorse, the skills of other users. In October 2016, the 467 million LinkedIn users had 10 billion endorsements. On average, how many endorsements did each user have?
Data: expandedramblings.com

In order to find the average number of endorsements per LinkedIn user, we divide the total number of endorsements by the number of users. We first write each number using scientific notation:

467 million = 467,000,000 = 4.67×10^8,
10 billion = 10,000,000,000 = 1.0×10^{10}.

---

24. *Niagara Falls Water Flow.* On the Canadian side, the amount of water that spills over Niagara Falls in 1 min during the summer is about
1.3088×10^8 L.
How much water spills over the falls in one day? Express the answer in scientific notation.

Answer
24. 1.884672×10^{11} L

SECTION 4.2 Exponents and Scientific Notation 267

25. DNA. The width of a DNA (deoxyribonucleic acid) double helix is about 2×10^{-9} m. If its length, fully stretched, is 5×10^{-2} m, how many times longer is the helix than it is wide?

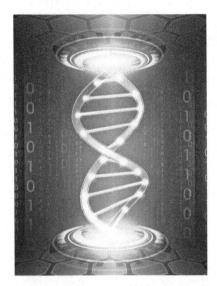

Answer
25. The length of the helix is 2.5×10^7 times its width.

We then divide 1.0×10^{10} by 4.67×10^8:

$$\frac{1.0 \times 10^{10}}{4.67 \times 10^8} = \frac{1.0}{4.67} \times \frac{10^{10}}{10^8}$$

$$\approx 0.21 \times 10^2 = (2.1 \times 10^{-1}) \times 10^2 = 2.1 \times 10.$$

On average, each user has 2.1×10, or 21, endorsements.

◀ **Do Exercise 25.**

The following is a summary of the definitions and rules for exponents that we have considered in this section and the preceding one.

DEFINITIONS AND RULES FOR EXPONENTS

Exponent of 1:	$a^1 = a$
Exponent of 0:	$a^0 = 1, a \neq 0$
Negative exponents:	$a^{-n} = \dfrac{1}{a^n}, \dfrac{1}{a^{-n}} = a^n, a \neq 0$
Product Rule:	$a^m \cdot a^n = a^{m+n}$
Quotient Rule:	$\dfrac{a^m}{a^n} = a^{m-n}, a \neq 0$
Power Rule:	$(a^m)^n = a^{mn}$
Raising a product to a power:	$(ab)^n = a^n b^n$
Raising a quotient to a power:	$\left(\dfrac{a}{b}\right)^n = \dfrac{a^n}{b^n}, b \neq 0;$
	$\left(\dfrac{a}{b}\right)^{-n} = \left(\dfrac{b}{a}\right)^n, b \neq 0, a \neq 0$
Scientific notation:	$M \times 10^n$, where $1 \leq M < 10$

4.2 Exercise Set

FOR EXTRA HELP MyLab Math

✓ Check Your Understanding

Reading Check Choose from the list on the right the appropriate word to complete each statement.

RC1. To raise a power to a power, _____ the exponents.

RC2. To raise a product to the nth power, raise each factor to the _____ power.

RC3. To convert a number less than 1 to scientific notation, move the decimal point to the _____.

RC4. A _____ exponent in scientific notation indicates a number greater than or equal to 10.

add
left
multiply
negative
nth
positive
right

Concept Check State whether scientific notation for each of the following numbers would include a positive power of 10 or a negative power of 10.

CC1. The distance from Earth to the sun, in feet

CC2. The diameter of an atom, in meters

CC3. The time it takes to blink, in hours

CC4. The mass of the Earth, in grams

a, b Simplify.

1. $(2^3)^2$
2. $(5^2)^4$
3. $(5^2)^{-3}$
4. $(7^{-3})^5$

5. $(x^{-3})^{-4}$
6. $(a^{-5})^{-6}$
7. $(a^{-2})^9$
8. $(x^{-5})^6$

9. $(t^{-3})^{-6}$
10. $(a^{-4})^{-7}$
11. $(t^4)^{-3}$
12. $(t^5)^{-2}$

13. $(x^{-2})^{-4}$
14. $(t^{-6})^{-5}$
15. $(ab)^3$
16. $(xy)^2$

17. $(ab)^{-3}$
18. $(xy)^{-6}$
19. $(mn^2)^{-3}$
20. $(x^3y)^{-2}$

21. $(4x^3)^2$
22. $4(x^3)^2$
23. $(3x^{-4})^2$
24. $(2a^{-5})^3$

25. $(x^4y^5)^{-3}$
26. $(t^5x^3)^{-4}$
27. $(x^{-6}y^{-2})^{-4}$
28. $(x^{-2}y^{-7})^{-5}$

29. $(a^{-2}b^7)^{-5}$
30. $(q^5r^{-1})^{-3}$
31. $(5r^{-4}t^3)^2$
32. $(4x^5y^{-6})^3$

33. $(a^{-5}b^7c^{-2})^3$
34. $(x^{-4}y^{-2}z^9)^2$
35. $(3x^3y^{-8}z^{-3})^2$
36. $(2a^2y^{-4}z^{-5})^3$

37. $(-4x^3y^{-2})^2$
38. $(-8x^3y^{-2})^3$
39. $(-a^{-3}b^{-2})^{-4}$
40. $(-p^{-4}q^{-3})^{-2}$

41. $\left(\dfrac{y^3}{2}\right)^2$
42. $\left(\dfrac{a^5}{3}\right)^3$
43. $\left(\dfrac{a^2}{b^3}\right)^4$
44. $\left(\dfrac{x^3}{y^4}\right)^3$

45. $\left(\dfrac{y^2}{2}\right)^{-3}$
46. $\left(\dfrac{a^4}{3}\right)^{-2}$
47. $\left(\dfrac{7}{x^{-3}}\right)^2$
48. $\left(\dfrac{3}{a^{-2}}\right)^3$

49. $\left(\dfrac{x^2y}{z}\right)^3$
50. $\left(\dfrac{m}{n^4p}\right)^3$
51. $\left(\dfrac{a^2b}{cd^3}\right)^{-2}$
52. $\left(\dfrac{2a^2}{3b^4}\right)^{-3}$

c Convert to scientific notation.

53. 28,000,000,000

54. 4,900,000,000,000

55. 907,000,000,000,000

56. 168,000,000,000,000

57. 0.00000304

58. 0.000000000865

59. 0.000000018

60. 0.00000000002

61. 100,000,000,000

62. 0.0000001

63. *Population of the United States.* It is estimated that the population of the United States will be 419,854,000 in 2050. Convert 419,854,000 to scientific notation.
Data: U.S. Census Bureau

64. *Wavelength of Light.* The wavelength of red light is 0.00000068 m. Convert 0.00000068 to scientific notation.

65. *Microchips.* The size of a microchip is, in part, determined by the minimum size of transistors used in the chip. In 2015, IBM built a prototype chip with transistors of width 0.000000007 m. Convert 0.000000007 to scientific notation.
Data: investorplace.com

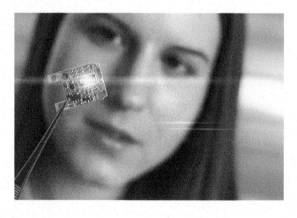

66. *Fish Population.* The biomass of fish living in the ocean layer between depths of 200 m and 1000 m is approximately 10,000,000,000 tons. Convert 10,000,000,000 to scientific notation.
Data: nature.com

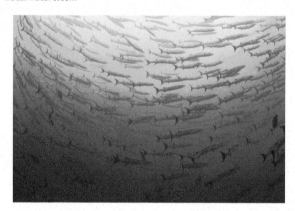

Convert to decimal notation.

67. 8.74×10^7

68. 1.85×10^8

69. 5.704×10^{-8}

70. 8.043×10^{-4}

71. 10^7

72. 10^6

73. 10^{-5}

74. 10^{-8}

d Multiply or divide and write scientific notation for the result.

75. $(3 \times 10^4)(2 \times 10^5)$

76. $(3.9 \times 10^8)(8.4 \times 10^{-3})$

77. $(5.2 \times 10^5)(6.5 \times 10^{-2})$

78. $(7.1 \times 10^{-7})(8.6 \times 10^{-5})$

79. $(9.9 \times 10^{-6})(8.23 \times 10^{-8})$

80. $(1.123 \times 10^4) \times 10^{-9}$

81. $\dfrac{8.5 \times 10^8}{3.4 \times 10^{-5}}$

82. $\dfrac{5.6 \times 10^{-2}}{2.5 \times 10^5}$

83. $(3.0 \times 10^6) \div (6.0 \times 10^9)$

84. $(1.5 \times 10^{-3}) \div (1.6 \times 10^{-6})$

85. $\dfrac{7.5 \times 10^{-9}}{2.5 \times 10^{12}}$

86. $\dfrac{4.0 \times 10^{-3}}{8.0 \times 10^{20}}$

e Solve.

87. *River Discharge.* The average discharge at the mouths of the Amazon River is 4,200,000 cubic feet per second. How much water is discharged from the Amazon River in 1 year? Express the answer in scientific notation.

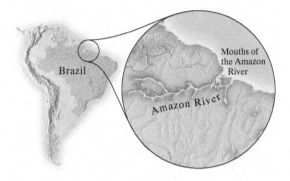

88. *Coral Reefs.* There are 10 million bacteria per square centimeter of coral in a coral reef. The coral reefs near the Hawaiian Islands cover 14,000 km². How many bacteria are there in Hawaii's coral reefs?

Data: livescience.com; U.S. Geological Survey

89. *Stars.* It is estimated that there are 70 billion trillion stars in the known universe. Express the number of stars in scientific notation. (1 billion = 10^9; 1 trillion = 10^{12})

Data: cnn.com

90. *Water Contamination.* Americans who change their own motor oil generate about 150 million gallons of used oil annually. If this oil is not disposed of properly, it can contaminate drinking water and soil. One gallon of used oil can contaminate one million gallons of drinking water. How many gallons of drinking water can 150 million gallons of oil contaminate? Express the answer in scientific notation. (1 million = 10^6)

Data: New Car Buying Guide

91. *Earth vs. Jupiter.* The mass of Earth is about 6×10^{21} metric tons. The mass of Jupiter is about 1.908×10^{24} metric tons. About how many times the mass of Earth is the mass of Jupiter? Express the answer in scientific notation.

92. *Office Supplies.* A ream of copier paper weighs 2.25 kg. How much does a sheet of copier paper weigh?

93. *Media Usage.* Approximately 1.484×10^{11} videos are viewed on YouTube each month by the 1.325×10^9 YouTube users. On average, how many videos does each user view each month?

Data: statisticbrain.com

94. *Information Technology.* In 2014, Danish researchers set a record for data transfer with a single transmitter. Using a laser transmitter, they were able to transfer data at a speed of about 5.4 terabytes per second. If 27 petabytes of information is created daily, how long would it take to transfer one day's information? (*Note:* 1 terabyte = 10^{12} bytes and 1 petabyte = 10^{15} bytes.)

Data: engadget.com

95. *Gold Leaf.* Gold can be milled into a very thin film called gold leaf. This film is so thin that it took only 43 oz of gold to cover the dome of Georgia's state capitol building. The gold leaf used was 5×10^{-6} m thick. In contrast, a U.S. penny is 1.55×10^{-3} m thick. How many sheets of gold leaf are in a stack that is the height of a penny?

Data: georgiaencyclopedia.org

96. *Relative Size.* An influenza virus is about 1.2×10^{-7} m in diameter. A staphylococcus bacterium is about 1.5×10^{-6} m in diameter. How many influenza viruses would it take, laid side by side, to equal the diameter of the bacterium?

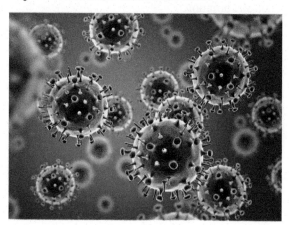

Space Travel. Use the following information for Exercises 97 and 98.

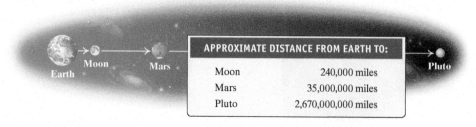

97. *Time to Reach Mars.* Suppose that it takes about 3 days for a space vehicle to travel from Earth to the moon. About how long would it take the same space vehicle traveling at the same speed to reach Mars? Express the answer in scientific notation.

98. *Time to Reach Pluto.* Suppose that it takes about 3 days for a space vehicle to travel from Earth to the moon. About how long would it take the same space vehicle traveling at the same speed to reach the dwarf planet Pluto? Express the answer in scientific notation.

Skill Maintenance

Graph.

99. $y = x - 5$ [3.3a]
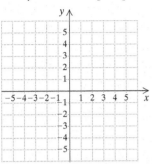

100. $2x + y = 4$ [3.3a]
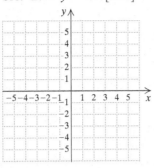

101. $3x - y = 3$ [3.3a]
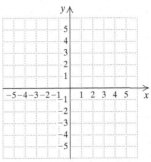

102. $y = -x$ [3.3a]
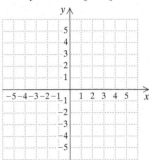

103. $2x = -10$ [3.3b]
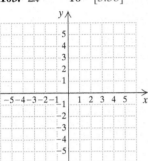

104. $y = -4$ [3.3b]
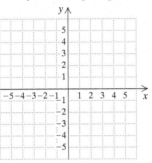

105. $8y - 16 = 0$ [3.3b]
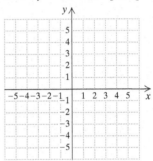

106. $x = 4$ [3.3b]
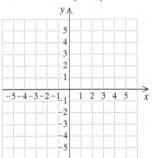

Synthesis

107. 🖩 Carry out the indicated operations. Express the result in scientific notation.

$$\frac{(5.2 \times 10^6)(6.1 \times 10^{-11})}{1.28 \times 10^{-3}}$$

108. Find the reciprocal and express it in scientific notation.

$$6.25 \times 10^{-3}$$

Simplify.

109. $\dfrac{(5^{12})^2}{5^{25}}$

110. $\dfrac{a^{22}}{(a^2)^{11}}$

111. $\dfrac{(3^5)^4}{3^5 \cdot 3^4}$

112. $\left(\dfrac{5x^{-2}}{3y^{-2}z}\right)^0$

113. $\dfrac{49^{18}}{7^{35}}$

114. $\left(\dfrac{1}{a}\right)^{-n}$

115. $\dfrac{(0.4)^5}{[(0.4)^3]^2}$

116. $\left(\dfrac{4a^3 b^{-2}}{5c^{-3}}\right)^1$

Determine whether each equation is true or false for all pairs of integers m and n and all positive numbers x and y.

117. $x^m \cdot y^n = (xy)^{mn}$

118. $x^m \cdot y^m = (xy)^{2m}$

119. $(x - y)^m = x^m - y^m$

120. $-x^m = (-x)^m$

121. $(-x)^{2m} = x^{2m}$

122. $x^{-m} = \dfrac{-1}{x^m}$

4.3 Introduction to Polynomials

OBJECTIVES

a Evaluate a polynomial for a given value of the variable.

b Identify the terms of a polynomial and classify a polynomial by its number of terms.

c Identify the coefficient and the degree of each term of a polynomial and the degree of the polynomial.

d Collect the like terms of a polynomial.

e Arrange a polynomial in descending order, or collect the like terms and then arrange in descending order.

f Identify the missing terms of a polynomial.

We have already learned to evaluate and to manipulate certain kinds of algebraic expressions. We will now consider algebraic expressions called *polynomials*.

The following are examples of *monomials in one variable*:

$$3x^2, \quad 2x, \quad -5, \quad 37p^4, \quad 0.$$

Each expression is a constant or a constant times some variable to a nonnegative integer power.

> **MONOMIAL**
>
> A **monomial** is an expression of the type ax^n, where a is a real-number constant and n is a nonnegative integer.

Algebraic expressions like the following are **polynomials**:

$$\tfrac{3}{4}y^5, \quad -2, \quad 5y + 3, \quad 3x^2 + 2x - 5, \quad -7a^3 + \tfrac{1}{2}a, \quad 6x, \quad 37p^4, \quad x, \quad 0.$$

> **POLYNOMIAL**
>
> A **polynomial** is a monomial or a combination of sums and/or differences of monomials.

The following algebraic expressions are *not* polynomials:

$$(1) \; \frac{x+3}{x-4}, \quad (2) \; 5x^3 - 2x^2 + \frac{1}{x}, \quad (3) \; \frac{1}{x^3 - 2}.$$

Expressions (1) and (3) are not polynomials because they represent quotients, not sums or differences. Expression (2) is not a polynomial because

$$\frac{1}{x} = x^{-1},$$

and x^{-1} is not a monomial because the exponent is negative.

◀ Do Exercise 1.

1. Write three polynomials.

a EVALUATING POLYNOMIALS AND APPLICATIONS

When we replace the variable in a polynomial with a number, the polynomial then represents a number called a **value** of the polynomial. Finding that number, or value, is called **evaluating the polynomial**. We evaluate a polynomial using the rules for order of operations.

EXAMPLE 1 Evaluate each polynomial when $x = 2$.

a) $3x + 5 = 3 \cdot 2 + 5$
$= 6 + 5$
$= 11.$

b) $2x^2 - 7x + 3 = 2 \cdot 2^2 - 7 \cdot 2 + 3$
$= 2 \cdot 4 - 7 \cdot 2 + 3$
$= 8 - 14 + 3$
$= -3$

Answer

1. $4x^2 - 3x + \dfrac{5}{4}$; $15y^3$; $-7x^3 + 1.1$; answers may vary

EXAMPLE 2 Evaluate each polynomial when $x = -4$.

a) $2 - x^3 = 2 - (-4)^3 = 2 - (-64)$
$= 2 + 64 = 66$

b) $-x^2 - 3x + 1 = -(-4)^2 - 3(-4) + 1$
$= -16 + 12 + 1 = -3$

Do Exercises 2–5.

Evaluate each polynomial when $x = 3$.

2. $-4x - 7$

3. $-5x^3 + 7x + 10$

Evaluate each polynomial when $x = -5$.

4. $5x + 7$

ALGEBRAIC ▶◀ GRAPHICAL CONNECTION

An equation like $y = 2x - 2$, which has a polynomial on one side and only y on the other, is called a **polynomial equation**. For such an equation, determining y is the same as evaluating the polynomial. Once the graph of such an equation has been drawn, we can evaluate the polynomial for a given x-value by finding the y-value that is paired with it on the graph.

GS 5. $2x^2 + 5x - 4$
$= 2()^2 + 5() - 4$
$= 2() + () - 4$
$= 50 - - 4$
$= $

EXAMPLE 3 Use *only* the given graph of $y = 2x - 2$ to evaluate the polynomial $2x - 2$ when $x = 3$.

First, we locate 3 on the x-axis. From there we move vertically to the graph of the equation and then horizontally to the y-axis. There we locate the y-value that is paired with 3. It appears that the y-value 4 is paired with 3. Thus, the value of $2x - 2$ is 4 when $x = 3$. We can check this by evaluating $2x - 2$ when $x = 3$.

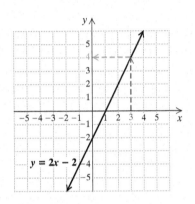

Do Exercise 6.

6. Use *only* the graph shown in Example 3 to evaluate the polynomial $2x - 2$ when $x = 4$ and when $x = -1$.

Polynomial equations can be used to model many real-world situations.

EXAMPLE 4 *Games in a Sports League.* In a sports league of x teams in which each team plays every other team twice, the total number of games N to be played is given by the polynomial equation

$N = x^2 - x.$

A women's slow-pitch softball league has 10 teams and each team plays every other team twice. What is the total number of games to be played?

We evaluate the polynomial when $x = 10$:

$N = x^2 - x = 10^2 - 10 = 100 - 10 = 90.$

The league plays 90 games.

Do Exercise 7.

7. Refer to Example 4. Determine the total number of games to be played in a league of 12 teams in which each team plays every other team twice.

Answers
2. -19 3. -104 4. -18 5. 21
6. 6; -4 7. 132 games
Guided Solution:
5. $-5, -5, 25, -25, 25, 21$

CALCULATOR CORNER

To use a graphing calculator to evaluate the polynomial in Example 2(b), $-x^2 - 3x + 1$, when $x = -4$, we can first graph $y_1 = -x^2 - 3x + 1$ in a window that includes the x-value -4. Then we can use the Value feature from the CALC menu, supplying the desired x-value and pressing ENTER to see $X = -4, Y = -3$ at the bottom of the screen. Thus, when $x = -4$, the value of $-x^2 - 3x + 1$ is -3.

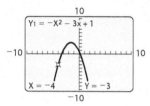

EXERCISES: Use the Value feature to evaluate each polynomial for the given values of x.

1. $-x^2 - 3x + 1$, when $x = -2$, when $x = -0.5$, and when $x = 4$
2. $3x^2 - 5x + 2$, when $x = -3$, when $x = 1$, and when $x = 2.6$

EXAMPLE 5 *Medical Dosage.* The concentration C, in parts per million, of a certain antibiotic in the bloodstream after t hours is given by the polynomial equation

$$C = -0.05t^2 + 2t + 2.$$

Find the concentration after 2 hr.

To find the concentration after 2 hr, we evaluate the polynomial when $t = 2$:

$$\begin{aligned}
C &= -0.05t^2 + 2t + 2 \\
&= -0.05(2)^2 + 2(2) + 2 &&\text{Substituting 2 for } t \\
&= -0.05(4) + 2(2) + 2 &&\text{Carrying out the calculation using the rules for order of operations} \\
&= -0.2 + 4 + 2 \\
&= 3.8 + 2 \\
&= 5.8.
\end{aligned}$$

The concentration after 2 hr is 5.8 parts per million.

ALGEBRAIC ▶◀ GRAPHICAL CONNECTION

The polynomial equation in Example 5 can be graphed if we evaluate the polynomial for several values of t. We list the values in a table and show the graph below. Note that the concentration peaks at the 20-hr mark and after slightly more than 40 hr, the concentration is 0. Since neither time nor concentration can be negative, our graph uses only the first quadrant.

t	C $C = -0.05t^2 + 2t + 2$
0	2
2	5.8 ← Example 5
10	17
20	22
30	17

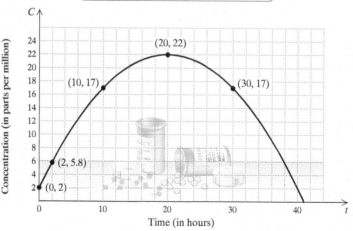

◀ Do Exercises 8 and 9.

---

8. *Medical Dosage.* Refer to Example 5.
   a) Determine the concentration after 3 hr by evaluating the polynomial when $t = 3$.
   b) Use *only* the graph showing medical dosage to check the value found in part (a).

9. *Medical Dosage.* Refer to Example 5. Use *only* the graph showing medical dosage to estimate the value of the polynomial when $t = 26$.

Answers

8. (a) 7.55 parts per million; (b) When $t = 3$, $C \approx 7.5$ so the value found in part (a) appears to be correct. 9. About 20 parts per million

b IDENTIFYING TERMS AND CLASSIFYING POLYNOMIALS

For any polynomial that has some subtractions, we can find an equivalent polynomial using only additions.

EXAMPLES Find an equivalent polynomial using only additions.
6. $-5x^2 - x = -5x^2 + (-x)$
7. $4x^5 - 2x^6 + 4x - 7 = 4x^5 + (-2x^6) + 4x + (-7)$

Do Exercises 10 and 11. ▶

Find an equivalent polynomial using only additions.

10. $-9x^3 - 4x^5$

11. $-2y^3 + 3y^7 - 7y - 9$

When a polynomial is written using only additions, the monomials being added are called **terms**. In Example 6, the terms are $-5x^2$ and $-x$. In Example 7, the terms are $4x^5$, $-2x^6$, $4x$, and -7.

EXAMPLE 8 Identify the terms of the polynomial

$$4x^7 + 3x + 12 + 8x^3 + 5x.$$

Terms: $4x^7, 3x, 12, 8x^3,$ and $5x$.

If there are subtractions, you can *think* of them as additions without rewriting.

EXAMPLE 9 Identify the terms of the polynomial

$$3t^4 - 5t^6 - 4t + 2.$$

Terms: $3t^4, -5t^6, -4t,$ and 2.

Do Exercises 12 and 13. ▶

Identify the terms of each polynomial.

12. $3x^2 + 6x + \dfrac{1}{2}$

13. $-4y^5 + 7y^2 - 3y - 2$

Polynomials with just one term are called **monomials**. Polynomials with just two terms are called **binomials**. Those with just three terms are called **trinomials**. Those with more than three terms are generally not specified with a name.

EXAMPLE 10

MONOMIALS	BINOMIALS	TRINOMIALS	NONE OF THESE
$-4x^2$	$2x + 4$	$3x^5 + 4x^4 + 7x$	$4x^3 - 5x^2 + x - 8$
9	$-9x^7 - 6x$	$4x^2 - 6x - \frac{1}{2}$	$z^5 + 2z^4 - z^3 + 7z + 3$

Do Exercises 14–17. ▶

Classify each polynomial as either a monomial, a binomial, a trinomial, or none of these.

14. $3x^2 + x$

15. $5x^4$

16. $4x^3 - 3x^2 + 4x + 2$

17. $3x^2 + 2x - 4$

c COEFFICIENTS AND DEGREES

The coefficient of the term $5x^3$ is 5. In the following polynomial, the red numbers are the **coefficients**, 3, −2, 5, and 4:

$$3x^5 - 2x^3 + 5x + 4.$$

Answers
10. $-9x^3 + (-4x^5)$
11. $-2y^3 + 3y^7 + (-7y) + (-9)$
12. $3x^2, 6x, \dfrac{1}{2}$ 13. $-4y^5, 7y^2, -3y, -2$
14. Binomial   15. Monomial
16. None of these   17. Trinomial

EXAMPLE 11 Identify the coefficient of each term in the polynomial

$$3x^4 - 4x^3 + \frac{1}{2}x^2 + x - 8.$$

The coefficient of $3x^4$ is 3.
The coefficient of $-4x^3$ is -4.
The coefficient of $\frac{1}{2}x^2$ is $\frac{1}{2}$.
The coefficient of x (or $1x$) is 1.
The coefficient of -8 is -8.

18. Identify the coefficient of each term in the polynomial
$2x^4 - 7x^3 - 8.5x^2 - x - 4.$

◀ Do Exercise 18.

The **degree** of a term is the exponent of the variable. The degree of the term $-5x^3$ is 3.

EXAMPLE 12 Identify the degree of each term of $8x^4 - 3x + 7$.

The degree of $8x^4$ is 4.
The degree of $-3x$ (or $-3x^1$) is 1. $x = x^1$
The degree of 7 (or $7x^0$) is 0. $7 = 7 \cdot 1 = 7 \cdot x^0$, since $x^0 = 1$

Because we can write 1 as x^0, the degree of any constant term (except 0) is 0. The term 0 is a special case. We agree that it has *no* degree because we can express 0 as $0 = 0x^5 = 0x^7$, and so on, using any exponent we wish.

> The degree of any nonzero constant term is 0.

The **degree of a polynomial** is the largest of the degrees of the terms, unless it is the polynomial 0.

EXAMPLE 13 Identify the degree of the polynomial $5x^3 - 6x^4 + 7$.

Identify the degree of each term and the degree of the polynomial.

19. $-6x^4 + 8x^2 - 2x + 9$

20. $4 - x^3 + \frac{1}{2}x^6 - x^5$

$5x^3 - 6x^4 + 7.$ The largest degree is 4.

The degree of the polynomial is 4.

◀ Do Exercises 19 and 20.

Let's summarize the terminology that we have learned, using the polynomial $3x^4 - 8x^3 + x^2 + 7x - 6$.

TERM	COEFFICIENT	DEGREE OF THE TERM	DEGREE OF THE POLYNOMIAL
$3x^4$	3	4	
$-8x^3$	-8	3	
x^2	1	2	4
$7x$	7	1	
-6	-6	0	

Answers
18. $2, -7, -8.5, -1, -4$ **19.** $4, 2, 1, 0; 4$
20. $0, 3, 6, 5; 6$

d COLLECTING LIKE TERMS

SKILL REVIEW

Collect like terms. [1.7e]
Collect like terms.
1. $3x - 4y + 5x + y$
2. $2a - 7b + 6 - 3a - 1$

Answers: 1. $8x - 3y$ 2. $-a - 7b + 5$

When terms have the same variable and the same exponent, we say that they are **like terms**.

EXAMPLES Identify the like terms in each polynomial.

14. $4x^3 + 5x - 4x^2 + 2x^3 + x^2$

Like terms: $4x^3$ and $2x^3$ Same variable and exponent
Like terms: $-4x^2$ and x^2 Same variable and exponent

15. $6 - 3a^2 - 8 - a - 5a$

Like terms: 6 and -8 Constant terms are like terms; note that $6 = 6x^0$ and $-8 = -8x^0$.

Like terms: $-a$ and $-5a$

Do Exercises 21–23.

We can often simplify polynomials by **collecting like terms**, or **combining like terms**. To do this, we use the distributive laws.

EXAMPLES Collect like terms.

16. $2x^3 - 6x^3 = (2 - 6)x^3 = -4x^3$

17. $5x^2 + 7 + 4x^4 + 2x^2 - 11 - 2x^4 = (5 + 2)x^2 + (4 - 2)x^4 + (7 - 11)$
$= 7x^2 + 2x^4 - 4$

EXAMPLE 18 Collect like terms: $3x^5 + 2x^2 - 3x^5 + 8$.

$3x^5 + 2x^2 - 3x^5 + 8 = (3 - 3)x^5 + 2x^2 + 8$
$= 0x^5 + 2x^2 + 8$
$= 2x^2 + 8$

Do Exercises 24–29.

EXAMPLES Collect like terms.

19. $5x^8 - 6x^5 - x^8 = 5x^8 - 6x^5 - 1x^8$ Replacing x^8 with $1x^8$
$= (5 - 1)x^8 - 6x^5$ Using a distributive law
$= 4x^8 - 6x^5$

20. $\frac{2}{3}x^4 - x^3 - \frac{1}{6}x^4 + \frac{2}{5}x^3 - \frac{3}{10}x^3$
$= \left(\frac{2}{3} - \frac{1}{6}\right)x^4 + \left(-1 + \frac{2}{5} - \frac{3}{10}\right)x^3$ $-x^3 = -1 \cdot x^3$
$= \left(\frac{4}{6} - \frac{1}{6}\right)x^4 + \left(-\frac{10}{10} + \frac{4}{10} - \frac{3}{10}\right)x^3$
$= \frac{3}{6}x^4 - \frac{9}{10}x^3 = \frac{1}{2}x^4 - \frac{9}{10}x^3$

Do Exercises 30–32.

Identify the like terms in each polynomial.

21. $4x^3 - x^3 + 2$

22. $4t^4 - 9t^3 - 7t^4 + 10t^3$

23. $5x^2 + 3x - 10 + 7x^2 - 8x + 11$

Collect like terms.

24. $3x^2 + 5x^2$

25. $4x^3 - 2x^3 + 2 + 5$

26. $\frac{1}{2}x^5 - \frac{3}{4}x^5 + 4x^2 - 2x^2$

27. $24 - 4x^3 - 24$

28. $5x^3 - 8x^5 + 8x^5$

GS 29. $-2x^4 + 16 + 2x^4 + 9 - 3x^5$
$= -3x^5 + (-2 +)x^4 + (16 +)$
$= -3x^5 + 0x^4 + $
$= -3x^5 + 25$

Collect like terms.

30. $5x^3 - x^3 + 4$

31. $\frac{3}{4}x^3 + 4x^2 - x^3 + 7$

32. $\frac{4}{5}x^4 - x^4 + x^5 - \frac{1}{5} - \frac{1}{4}x^4 + 10$

Answers
21. $4x^3$ and $-x^3$ **22.** $4t^4$ and $-7t^4$; $-9t^3$ and $10t^3$ **23.** $5x^2$ and $7x^2$; $3x$ and $-8x$; -10 and 11 **24.** $8x^2$
25. $2x^3 + 7$ **26.** $-\frac{1}{4}x^5 + 2x^2$ **27.** $-4x^3$
28. $5x^3$ **29.** $-3x^5 + 25$ **30.** $4x^3 + 4$
31. $-\frac{1}{4}x^3 + 4x^2 + 7$ **32.** $x^5 - \frac{9}{20}x^4 + \frac{49}{5}$

Guided Solution:
29. 2, 9, 25

e DESCENDING ORDER

A polynomial is written in **descending order** when the term with the largest degree is written first, the term with the next largest degree is written next, and so on, in order from left to right.

EXAMPLES Arrange each polynomial in descending order.

21. $6x^5 + 4x^7 + x^2 + 2x^3 = 4x^7 + 6x^5 + 2x^3 + x^2$

22. $\frac{2}{3} + 4x^5 - 8x^2 + 5x - 3x^3 = 4x^5 - 3x^3 - 8x^2 + 5x + \frac{2}{3}$

◀ Do Exercises 33 and 34.

Arrange each polynomial in descending order.

33. $4x^2 - 3 + 7x^5 + 2x^3 - 5x^4$

34. $-14 + 7t^2 - 10t^5 + 14t^7$

EXAMPLE 23 Collect like terms and then arrange in descending order:
$$2x^2 - 4x^3 + 3 - x^2 - 2x^3.$$

We have

$2x^2 - 4x^3 + 3 - x^2 - 2x^3 = x^2 - 6x^3 + 3$ Collecting like terms

$= -6x^3 + x^2 + 3$ Arranging in descending order

◀ Do Exercises 35 and 36.

Collect like terms and then arrange in descending order.

35. $3x^2 - 2x + 3 - 5x^2 - 1 - x$

36. $-x + \frac{1}{2} + 14x^4 - 7x - 1 - 4x^4$

The opposite of descending order is called **ascending order**. Generally, if an exercise is written in a certain order, we give the answer in that same order.

f MISSING TERMS

If a coefficient is 0, we generally do not write the term. If a term with degree less than the degree of the polynomial has a coefficient of 0, we say that we have a **missing term**.

EXAMPLE 24 Identify the missing terms in the polynomial
$$8x^5 - 2x^3 + 5x^2 + 7x + 8.$$

There is no term with x^4. We say that the x^4-term is missing.

◀ Do Exercises 37–39.

Identify the missing term(s) in each polynomial.

37. $2x^3 + 4x^2 - 2$

38. $-3x^4$

39. $x^3 + 1$

We can either write missing terms with zero coefficients or leave space.

EXAMPLE 25 Write the polynomial $x^4 - 6x^3 + 2x - 1$ in two ways: with its missing term and by leaving space for it.

a) $x^4 - 6x^3 + 2x - 1 = x^4 - 6x^3 + 0x^2 + 2x - 1$ Writing with the missing x^2-term

b) $x^4 - 6x^3 + 2x - 1 = x^4 - 6x^3 + 2x - 1$ Leaving space for the missing x^2-term

EXAMPLE 26 Write the polynomial $y^5 - 1$ in two ways: with its missing terms and by leaving space for them.

a) $y^5 - 1 = y^5 + 0y^4 + 0y^3 + 0y^2 + 0y - 1$

b) $y^5 - 1 = y^5 - 1$

◀ Do Exercises 40 and 41.

Write each polynomial in two ways: with its missing term(s) and by leaving space for them.

40. $2x^3 + 4x^2 - 2$

41. $a^4 + 10$

Answers

33. $7x^5 - 5x^4 + 2x^3 + 4x^2 - 3$
34. $14t^7 - 10t^5 + 7t^2 - 14$
35. $-2x^2 - 3x + 2$ **36.** $10x^4 - 8x - \frac{1}{2}$
37. x **38.** x^3, x^2, x, x^0 **39.** x^2, x
40. $2x^3 + 4x^2 + 0x - 2$;
$2x^3 + 4x^2 - 2$
41. $a^4 + 0a^3 + 0a^2 + 0a + 10$;
$a^4 + 10$

4.3 Exercise Set

✓ Check Your Understanding

Reading Check Choose from the column on the right the expression that best fits each description.

RC1. ____ The value of $x^2 - x$ when $x = -1$
RC2. ____ A polynomial written in ascending order
RC3. ____ A coefficient of $5x^4 - 3x + 7$
RC4. ____ A term of $5x^4 - 3x + 7$
RC5. ____ The degree of one of the terms of $5x^4 - 3x + 7$
RC6. ____ An example of a binomial

a) 0
b) 2
c) 5
d) $-3x$
e) $8x - 9$
f) $y + 6y^2 - 2y^8$

Concept Check

CC1. Evaluate $1 - x$ when $x = -1$.

CC2. Evaluate $1 - x^2$ when $x = -1$.

CC3. How many terms are in the polynomial $x^3 + 4x^2 - 7x + 5$?

CC4. What is the coefficient of the term $5x^2$?

CC5. What is the degree of the term $5x^2$?

CC6. Determine whether the polynomial $10 + 8x - 3x^4$ is written in either ascending order or descending order.

CC7. What term is missing in the polynomial $x^2 + 7$?

a Evaluate each polynomial when $x = 4$ and when $x = -1$.

1. $-5x + 2$
2. $-8x + 1$
3. $2x^2 - 5x + 7$
4. $3x^2 + x - 7$
5. $x^3 - 5x^2 + x$
6. $7 - x + 3x^2$

Evaluate each polynomial when $x = -2$ and when $x = 0$.

7. $\frac{1}{3}x + 5$
8. $8 - \frac{1}{4}x$
9. $x^2 - 2x + 1$
10. $5x + 6 - x^2$
11. $-3x^3 + 7x^2 - 3x - 2$
12. $-2x^3 + 5x^2 - 4x + 3$

13. *Skydiving.* During the first 13 sec of a jump, the distance S, in feet, that a skydiver falls in *t* seconds can be approximated by the polynomial equation
$$S = 11.12t^2.$$
In 2009, 108 U.S. skydivers fell headfirst in formation from a height of 18,000 ft. How far had they fallen 10 sec after having jumped from the plane?

Data: www.telegraph.co.uk

14. *Skydiving.* For jumps that exceed 13 sec, the polynomial equation
$$S = 173t - 369$$
can be used to approximate the distance S, in feet, that a skydiver has fallen in *t* seconds. Approximately how far has a skydiver fallen 20 sec after having jumped from a plane?

15. *Stacking Spheres.* In 2004, the journal *Annals of Mathematics* accepted a proof of the so-called Kepler Conjecture: that the most efficient way to pack spheres is in the shape of a square pyramid. The number N of balls in the stack is given by the polynomial equation
$$N = \frac{1}{3}x^3 + \frac{1}{2}x^2 + \frac{1}{6}x,$$
where *x* is the number of layers. A square pyramid with 3 layers is illustrated below. Find the number of oranges in a pyramid with 5 layers.

Data: *The New York Times* 4/6/04

Bottom layer   Second layer   Top layer

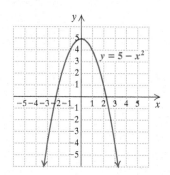

16. *SCAD Diving.* The distance *s*, in feet, traveled by a body falling freely from rest in *t* seconds is approximated by the polynomial equation
$$s = 16t^2.$$
The SCAD thrill ride is a 2.5-sec free fall into a net. How far does the diver fall?

Data: www.scadfreefall.co.uk

17. The graph of the polynomial equation $y = 5 - x^2$ is shown below. Use *only* the graph to estimate the value of the polynomial when $x = -3$, $x = -1$, $x = 0$, $x = 1.5$, and $x = 2$.

18. The graph of the polynomial equation $y = 6x^3 - 6x$ is shown below. Use *only* the graph to estimate the value of the polynomial when $x = -1$, $x = -0.5$, $x = 0.5$, $x = 1$, and $x = 1.1$.

19. *Solar Capacity.* The annual capacity C, in megawatts (MW), of U.S. residential installations generating energy from the sun can be estimated by the polynomial equation
$$C = 400t + 200,$$
where t is the number of years after 2010.

Data: Solar Energy Industries Association

a) Use the equation to estimate the capacity of U.S. residential solar-energy installations in 2018.

b) Check the result of part (a) using the following graph.

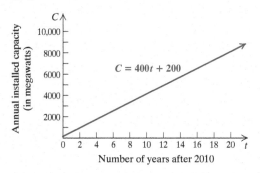

20. *Senior Population.* The number N, in millions, of people in the United States ages 65 and older can be estimated by the polynomial equation
$$N = 1.1t + 28.4,$$
where t is the number of years after 2000.

Data: U.S. Census Bureau

a) Use the equation to estimate the number of people in the United States ages 65 and older in 2030.

b) Check the result of part (a) using the following graph.

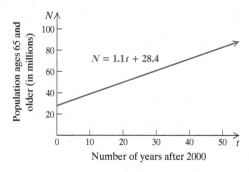

Memorizing Words. Participants in a psychology experiment were able to memorize an average of M words in t minutes, where $M = -0.001t^3 + 0.1t^2$. Use the following graph for Exercises 21–26.

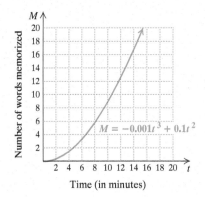

21. Estimate the number of words memorized in 10 min.

22. Estimate the number of words memorized in 14 min.

23. Find the approximate value of M for $t = 8$.

24. Find the approximate value of M for $t = 12$.

25. Estimate the value of M when t is 13.

26. Estimate the value of M when t is 7.

b Identify the terms of each polynomial.

27. $2 - 3x + x^2$

28. $2x^2 + 3x - 4$

29. $-2x^4 + \frac{1}{3}x^3 - x + 3$

30. $-\frac{2}{5}x^5 - x^3 + 6$

Classify each polynomial as either a monomial, a binomial, a trinomial, or none of these.

31. $x^2 - 10x + 25$
32. $-6x^4$
33. $x^3 - 7x^2 + 2x - 4$

34. $x^2 - 9$
35. $4x^2 - 25$
36. $2x^4 - 7x^3 + x^2 + x - 6$

37. $40x$
38. $4x^2 + 12x + 9$

C Identify the coefficient of each term of the polynomial.

39. $-3x + 6$
40. $2x - 4$
41. $5x^2 + \frac{3}{4}x + 3$

42. $\frac{2}{3}x^2 - 5x + 2$
43. $-5x^4 + 6x^3 - 2.7x^2 + x - 2$
44. $7x^3 - x^2 - 4.2x + 5$

Identify the degree of each term of the polynomial and the degree of the polynomial.

45. $2x - 4$
46. $6 - 3x$
47. $3x^2 - 5x + 2$
48. $5x^3 - 2x^2 + 3$

49. $-7x^3 + 6x^2 + \frac{3}{5}x + 7$
50. $5x^4 + \frac{1}{4}x^2 - x + 2$
51. $x^2 - 3x + x^6 - 9x^4$
52. $8x - 3x^2 + 9 - 8x^3$

53. Complete the following table for the polynomial $-7x^4 + 6x^3 - x^2 + 8x - 2$.

TERM	COEFFICIENT	DEGREE OF THE TERM	DEGREE OF THE POLYNOMIAL
$-7x^4$			
$6x^3$	6		
		2	
$8x$		1	
	-2		

54. Complete the following table for the polynomial $3x^2 + x^5 - 46x^3 + 6x - 2.4 - \frac{1}{2}x^4$.

TERM	COEFFICIENT	DEGREE OF THE TERM	DEGREE OF THE POLYNOMIAL
		5	
$-\frac{1}{2}x^4$		4	
	-46		
$3x^2$		2	
	6		
	-2.4		

d Identify the like terms in each polynomial.

55. $5x^3 + 6x^2 - 3x^2$

56. $3x^2 + 4x^3 - 2x^2$

57. $2x^4 + 5x - 7x - 3x^4$

58. $-3t + t^3 - 2t - 5t^3$

59. $3x^5 - 7x + 8 + 14x^5 - 2x - 9$

60. $8x^3 + 7x^2 - 11 - 4x^3 - 8x^2 - 29$

Collect like terms.

61. $2x - 5x$

62. $2x^2 + 8x^2$

63. $x - 9x$

64. $x - 5x$

65. $5x^3 + 6x^3 + 4$

66. $6x^4 - 2x^4 + 5$

67. $5x^3 + 6x - 4x^3 - 7x$

68. $3a^4 - 2a + 2a + a^4$

69. $6b^5 + 3b^2 - 2b^5 - 3b^2$

70. $2x^2 - 6x + 3x + 4x^2$

71. $\frac{1}{4}x^5 - 5 + \frac{1}{2}x^5 - 2x - 37$

72. $\frac{1}{3}x^3 + 2x - \frac{1}{6}x^3 + 4 - 16$

73. $6x^2 + 2x^4 - 2x^2 - x^4 - 4x^2$

74. $8x^2 + 2x^3 - 3x^3 - 4x^2 - 4x^2$

75. $\frac{1}{4}x^3 - x^2 - \frac{1}{6}x^2 + \frac{3}{8}x^3 + \frac{5}{16}x^3$

76. $\frac{1}{5}x^4 + \frac{1}{5} - 2x^2 + \frac{1}{10} - \frac{3}{15}x^4 + 2x^2 - \frac{3}{10}$

e Arrange each polynomial in descending order.

77. $x^5 + x + 6x^3 + 1 + 2x^2$

78. $3 + 2x^2 - 5x^6 - 2x^3 + 3x$

79. $5y^3 + 15y^9 + y - y^2 + 7y^8$

80. $9p - 5 + 6p^3 - 5p^4 + p^5$

Collect like terms and then arrange in descending order.

81. $3x^4 - 5x^6 - 2x^4 + 6x^6$

82. $-1 + 5x^3 - 3 - 7x^3 + x^4 + 5$

83. $-2x + 4x^3 - 7x + 9x^3 + 8$

84. $-6x^2 + x - 5x + 7x^2 + 1$

85. $3x + 3x + 3x - x^2 - 4x^2$

86. $-2x - 2x - 2x + x^3 - 5x^3$

87. $-x + \frac{3}{4} + 15x^4 - x - \frac{1}{2} - 3x^4$

88. $2x - \frac{5}{6} + 4x^3 + x + \frac{1}{3} - 2x$

f Identify the missing terms in each polynomial.

89. $x^3 - 27$

90. $x^5 + x$

91. $x^4 - x$

92. $5x^4 - 7x + 2$

93. $2x^3 - 5x^2 + x - 3$

94. $-6x^3$

Write each polynomial in two ways: with its missing terms and by leaving space for them.

95. $x^3 - 27$

96. $x^5 + x$

97. $x^4 - x$

98. $5x^4 - 7x + 2$

99. $5x^2$

100. $-6x^3$

Skill Maintenance

Perform the indicated operations.

101. $1 + (-20)$ [1.3a]

102. $-\frac{2}{3} + \left(-\frac{1}{3}\right)$ [1.3a]

103. $-4.2 + 1.95$ [1.3a]

104. $5.6 - 8.2$ [1.4a]

105. $\frac{1}{8} - \frac{5}{6}$ [1.4a]

106. $\frac{3}{8} - \left(-\frac{1}{4}\right)$ [1.4a]

107. $\left(-\frac{1}{2}\right)\left(-\frac{2}{3}\right)$ [1.5a]

108. $0.5(-1.2)$ [1.5a]

109. $(-2)(-3)(-4)$ [1.5a]

110. $\frac{4}{5} \div \left(-\frac{1}{2}\right)$ [1.6c]

111. $0 \div (-4)$ [1.6a]

112. $\frac{-6.3}{-5 + 5}$ [1.6c]

Synthesis

Collect like terms.

113. $6x^3 \cdot 7x^2 - (4x^3)^2 + (-3x^3)^2 - (-4x^2)(5x^3) - 10x^5 + 17x^6$

114. $(3x^2)^3 + 4x^2 \cdot 4x^4 - x^4(2x)^2 + ((2x)^2)^3 - 100x^2(x^2)^2$

115. Construct a polynomial in x (meaning that x is the variable) of degree 5 with four terms and coefficients that are integers.

116. What is the degree of $(5m^5)^2$?

Use the VALUE feature from the CALC menu on your graphing calculator to find the values in each of the following. (Refer to the Calculator Corner on p. 276.)

117. Exercise 17

118. Exercise 18

119. Exercise 21

120. Exercise 22

Addition and Subtraction of Polynomials

4.4

OBJECTIVES

a Add polynomials.
b Simplify the opposite of a polynomial.
c Subtract polynomials.
d Use polynomials to represent perimeter and area.

a ADDITION OF POLYNOMIALS

To add two polynomials, we can write a plus sign between them and then collect like terms. In general, if polynomials in an exercise are written in either descending order or ascending order, we write the answer in that same order.

EXAMPLE 1 Add $(-3x^3 + 2x - 4)$ and $(4x^3 + 3x^2 + 2)$.

$(-3x^3 + 2x - 4) + (4x^3 + 3x^2 + 2)$
$= (-3 + 4)x^3 + 3x^2 + 2x + (-4 + 2)$ Collecting like terms
$= x^3 + 3x^2 + 2x - 2$

EXAMPLE 2 Add:

$\left(\frac{2}{3}x^4 + 3x^2 - 2x + \frac{1}{2}\right) + \left(-\frac{1}{3}x^4 + 5x^3 - 3x^2 + 3x - \frac{1}{2}\right)$.

We have

$\left(\frac{2}{3}x^4 + 3x^2 - 2x + \frac{1}{2}\right) + \left(-\frac{1}{3}x^4 + 5x^3 - 3x^2 + 3x - \frac{1}{2}\right)$
$= \left(\frac{2}{3} - \frac{1}{3}\right)x^4 + 5x^3 + (3 - 3)x^2 + (-2 + 3)x + \left(\frac{1}{2} - \frac{1}{2}\right)$ Collecting like terms
$= \frac{1}{3}x^4 + 5x^3 + x$.

We can add polynomials as we do because they represent numbers.

▶ Do Exercises 1–4.

EXAMPLE 3 Add $(3x^2 - 2x + 2)$ and $(5x^3 - 2x^2 + 3x - 4)$.

$(3x^2 - 2x + 2) + (5x^3 - 2x^2 + 3x - 4)$
$= 5x^3 + (3 - 2)x^2 + (-2 + 3)x + (2 - 4)$ You might do this step mentally.
$= 5x^3 + x^2 + x - 2$ Then you would write only this.

▶ Do Exercises 5 and 6.

We can also add polynomials by writing like terms in columns.

EXAMPLE 4 Add $9x^5 - 2x^3 + 6x^2 + 3$ and $5x^4 - 7x^2 + 6$ and $3x^6 - 5x^5 + x^2 + 5$.

We arrange the polynomials with the like terms in columns.

$$\begin{array}{r} 9x^5 - 2x^3 + 6x^2 + 3 \\ 5x^4 - 7x^2 + 6 \\ 3x^6 - 5x^5 + x^2 + 5 \\ \hline 3x^6 + 4x^5 + 5x^4 - 2x^3 + 14 \end{array}$$ We leave spaces for missing terms.   Adding

We write the answer as $3x^6 + 4x^5 + 5x^4 - 2x^3 + 14$ without the space.

▶ Do Exercises 7 and 8 on the following page.

Add.

1. $(3x^2 + 2x - 2) + (-2x^2 + 5x + 5)$

2. $(-4x^5 + x^3 + 4) + (7x^4 + 2x^2)$

3. $(31x^4 + x^2 + 2x - 1) + (-7x^4 + 5x^3 - 2x + 2)$

4. $(17x^3 - x^2 + 3x + 4) + \left(-15x^3 + x^2 - 3x - \frac{2}{3}\right)$

Add mentally. Try to write just the answer.

5. $(4x^2 - 5x + 3) + (-2x^2 + 2x - 4)$

6. $(3x^3 - 4x^2 - 5x + 3) + \left(5x^3 + 2x^2 - 3x - \frac{1}{2}\right)$

Answers

1. $x^2 + 7x + 3$
2. $-4x^5 + 7x^4 + x^3 + 2x^2 + 4$
3. $24x^4 + 5x^3 + x^2 + 1$
4. $2x^3 + \frac{10}{3}$ 5. $2x^2 - 3x - 1$
6. $8x^3 - 2x^2 - 8x + \frac{5}{2}$

SECTION 4.4   Addition and Subtraction of Polynomials   287

Add.

7. $-2x^3 + 5x^2 - 2x + 4$
   $x^4 + 6x^2 + 7x - 10$
   $\underline{-9x^4 + 6x^3 + x^2 - 2}$

8. $-3x^3 + 5x + 2$ and
   $x^3 + x^2 + 5$ and
   $x^3 - 2x - 4$

Simplify.

9. $-(4x^3 - 6x + 3)$

10. $-(-5x^4 + 3x^2 + 7x - 5)$

11. $-(14x^{10} - \frac{1}{2}x^5 + 5x^3 - x^2 + 3x)$

Subtract.

12. $(7x^3 + 2x + 4) - (5x^3 - 4)$

13. $(-3x^2 + 5x - 4) -$
    $(-4x^2 + 11x - 2)$

Answers

7. $-8x^4 + 4x^3 + 12x^2 + 5x - 8$
8. $-x^3 + x^2 + 3x + 3$ 9. $-4x^3 + 6x - 3$
10. $5x^4 - 3x^2 - 7x + 5$
11. $-14x^{10} + \frac{1}{2}x^5 - 5x^3 + x^2 - 3x$
12. $2x^3 + 2x + 8$ 13. $x^2 - 6x - 2$

b OPPOSITES OF POLYNOMIALS

SKILL REVIEW *Remove parentheses and simplify.* [1.8b]
Simplify.
1. $6x - (7x + 4)$
2. $3a + 5a - (2a - 7)$

Answers: 1. $-x - 4$ 2. $6a + 7$

MyLab Math
VIDEO

We can use the property of -1 to write an equivalent expression for an opposite. For example, the opposite of $x - 2y + 5$ can be written as

$$-(x - 2y + 5).$$

We find an equivalent expression by changing the sign of every term:

$$-(x - 2y + 5) = -x + 2y - 5.$$

We use this concept when we subtract polynomials.

OPPOSITES OF POLYNOMIALS

To find an equivalent polynomial for the **opposite**, or **additive inverse**, of a polynomial, change the sign of every term. This is the same as multiplying by -1.

EXAMPLE 5 Simplify: $-(x^2 - 3x + 4)$.

$$-(x^2 - 3x + 4) = -x^2 + 3x - 4$$

EXAMPLE 6 Simplify: $-(-7x^4 - \frac{5}{9}x^3 + 8x^2 - x + 67)$.

$$-(-7x^4 - \tfrac{5}{9}x^3 + 8x^2 - x + 67) = 7x^4 + \tfrac{5}{9}x^3 - 8x^2 + x - 67$$

◀ Do Exercises 9–11.

c SUBTRACTION OF POLYNOMIALS

Recall that we can subtract a real number by adding its opposite, or additive inverse: $a - b = a + (-b)$. This allows us to subtract polynomials.

EXAMPLE 7 Subtract:

$$(9x^5 + x^3 - 2x^2 + 4) - (2x^5 + x^4 - 4x^3 - 3x^2).$$

We have

$(9x^5 + x^3 - 2x^2 + 4) - (2x^5 + x^4 - 4x^3 - 3x^2)$
$= 9x^5 + x^3 - 2x^2 + 4 + [-(2x^5 + x^4 - 4x^3 - 3x^2)]$ Adding the opposite
$= 9x^5 + x^3 - 2x^2 + 4 - 2x^5 - x^4 + 4x^3 + 3x^2$ Finding the opposite by changing the sign of *each* term
$= 7x^5 - x^4 + 5x^3 + x^2 + 4.$ Adding (collecting like terms)

◀ Do Exercises 12 and 13.

We combine steps by changing the sign of each term of the polynomial being subtracted and collecting like terms. Try to do this mentally as much as possible.

EXAMPLE 8 Subtract: $(9x^5 + x^3 - 2x) - (-2x^5 + 5x^3 + 6)$.

$(9x^5 + x^3 - 2x) - (-2x^5 + 5x^3 + 6)$
$= 9x^5 + x^3 - 2x + 2x^5 - 5x^3 - 6$ Finding the opposite by changing the sign of each term
$= 11x^5 - 4x^3 - 2x - 6$ Collecting like terms

Do Exercises 14 and 15.

We can use columns to subtract. We replace coefficients with their opposites, as shown in Example 8.

EXAMPLE 9 Write in columns and subtract:
$(5x^2 - 3x + 6) - (9x^2 - 5x - 3)$.

a) $5x^2 - 3x + 6$ Writing like terms in columns
$\underline{-(9x^2 - 5x - 3)}$

b) $5x^2 - 3x + 6$
$\underline{-9x^2 + 5x + 3}$ Changing signs

c) $5x^2 - 3x + 6$
$\underline{-9x^2 + 5x + 3}$
$-4x^2 + 2x + 9$ Adding

If you can do so without error, you can arrange the polynomials in columns and write just the answer, remembering to change the signs and add.

EXAMPLE 10 Write in columns and subtract:
$(x^3 + x^2 + 2x - 12) - (-2x^3 + x^2 - 3x)$.

$x^3 + x^2 + 2x - 12$
$\underline{-(-2x^3 + x^2 - 3x)}$ Leaving space for the missing term
$3x^3 + 5x - 12$ Changing the signs and adding

Do Exercises 16 and 17.

d POLYNOMIALS AND GEOMETRY

EXAMPLE 11 Find a polynomial for the sum of the areas of these four rectangles.

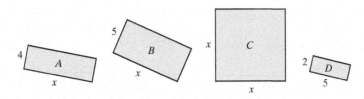

Recall that the area of a rectangle is the product of the length and the width. The sum of the areas is a sum of products. We find these products and then collect like terms.

Subtract.

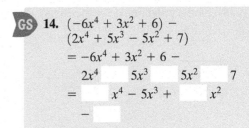

GS **14.** $(-6x^4 + 3x^2 + 6) - (2x^4 + 5x^3 - 5x^2 + 7)$
$= -6x^4 + 3x^2 + 6 \;\square\; 2x^4 \;\square\; 5x^3 \;\square\; 5x^2 \;\square\; 7$
$= \square x^4 - 5x^3 + \square x^2 - \square$

15. $\left(\dfrac{3}{2}x^3 - \dfrac{1}{2}x^2 + 0.3\right) - \left(\dfrac{1}{2}x^3 + \dfrac{1}{2}x^2 + \dfrac{4}{3}x + 1.2\right)$

Write in columns and subtract.
16. $(4x^3 + 2x^2 - 2x - 3) - (2x^3 - 3x^2 + 2)$

17. $(2x^3 + x^2 - 6x + 2) - (-x^5 - 4x^3 - 2x^2 - 4x)$

Answers
14. $-8x^4 - 5x^3 + 8x^2 - 1$
15. $x^3 - x^2 - \dfrac{4}{3}x - 0.9$
16. $2x^3 + 5x^2 - 2x - 5$
17. $x^5 + 6x^3 + 3x^2 - 2x + 2$
Guided Solution:
14. $-, +, -, -8, 8, 1$

18. Find a polynomial for the sums of the perimeters and of the areas of the rectangles.

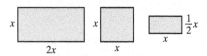

Area of A	plus	Area of B	plus	Area of C	plus	Area of D
$4 \cdot x$	$+$	$5 \cdot x$	$+$	$x \cdot x$	$+$	$2 \cdot 5$

We collect like terms:
$$4x + 5x + x^2 + 10 = x^2 + 9x + 10.$$

◀ Do Exercise 18.

EXAMPLE 12 *Lawn Area.* A new city park is to contain a square grassy area that is x ft on a side. Within that grassy area will be a circular playground, with a radius of 15 ft, that will be mulched. To determine the amount of sod needed, find a polynomial for the grassy area.

We make a drawing, reword the problem, and write the polynomial.

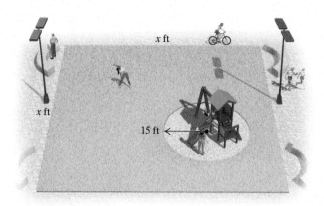

19. *Lawn Area.* An 8-ft by 8-ft shed is placed on a lawn x ft on a side. Find a polynomial for the remaining area.

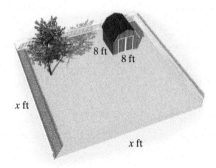

Area of square	$-$	Area of playground	$=$	Area of grass
$x \cdot x$ ft^2	$-$	$\pi \cdot 15^2$ ft^2	$=$	Area of grass

Then $(x^2 - 225\pi)$ ft^2 = Area of grass.

◀ Do Exercise 19.

Answers

18. Sum of perimeters: $13x$; sum of areas: $\frac{7}{2}x^2$
19. $(x^2 - 64)$ ft^2

4.4 Exercise Set

FOR EXTRA HELP MyLab Math

✓ Check Your Understanding

Reading Check Determine whether each statement is true or false.

RC1. To find the opposite of a polynomial, we need change only the sign of the first term.

RC2. We can subtract a polynomial by adding its opposite.

RC3. The sum of two binomials is always a binomial.

RC4. The area of a rectangle is the sum of its length and its width.

Concept Check Simplify.

CC1. $x^2 - (x^2)$ **CC2.** $x^2 - (-x^2)$ **CC3.** $x^2 - (x^2 - x)$

CC4. $x^2 + x - (x^2 + x)$ **CC5.** $x^2 + x - (x^2 - x)$ **CC6.** $x^2 - x - (-x^2 - x)$

a Add.

1. $(3x + 2) + (-4x + 3)$
2. $(6x + 1) + (-7x + 2)$
3. $(-6x + 2) + \left(x^2 + \tfrac{1}{2}x - 3\right)$

4. $\left(x^2 - \tfrac{5}{3}x + 4\right) + (8x - 9)$
5. $(x^2 - 9) + (x^2 + 9)$
6. $(x^3 + x^2) + (2x^3 - 5x^2)$

7. $(3x^2 - 5x + 10) + (2x^2 + 8x - 40)$
8. $(6x^4 + 3x^3 - 1) + (4x^2 - 3x + 3)$

9. $(1.2x^3 + 4.5x^2 - 3.8x) + (-3.4x^3 - 4.7x^2 + 23)$
10. $(0.5x^4 - 0.6x^2 + 0.7) + (2.3x^4 + 1.8x - 3.9)$

11. $(1 + 4x + 6x^2 + 7x^3) + (5 - 4x + 6x^2 - 7x^3)$
12. $(3x^4 - 6x - 5x^2 + 5) + (6x^2 - 4x^3 - 1 + 7x)$

13. $\left(\tfrac{1}{4}x^4 + \tfrac{2}{3}x^3 + \tfrac{5}{8}x^2 + 7\right) + \left(-\tfrac{3}{4}x^4 + \tfrac{3}{8}x^2 - 7\right)$
14. $\left(\tfrac{1}{3}x^9 + \tfrac{1}{5}x^5 - \tfrac{1}{2}x^2 + 7\right) + \left(-\tfrac{1}{5}x^9 + \tfrac{1}{4}x^4 - \tfrac{3}{5}x^5 + \tfrac{3}{4}x^2 + \tfrac{1}{2}\right)$

15. $(0.02x^5 - 0.2x^3 + x + 0.08) + (-0.01x^5 + x^4 - 0.8x - 0.02)$
16. $(0.03x^6 + 0.05x^3 + 0.22x + 0.05) + \left(\tfrac{7}{100}x^6 - \tfrac{3}{100}x^3 + 0.5\right)$

17. $(9x^8 - 7x^4 + 2x^2 + 5) + (8x^7 + 4x^4 - 2x) + (-3x^4 + 6x^2 + 2x - 1)$
18. $(4x^5 - 6x^3 - 9x + 1) + (6x^3 + 9x^2 + 9x) + (-4x^3 + 8x^2 + 3x - 2)$

SECTION 4.4 Addition and Subtraction of Polynomials

19.
$$0.15x^4 + 0.10x^3 - 0.9x^2$$
$$- 0.01x^3 + 0.01x^2 + x$$
$$1.25x^4 + 0.11x^2 + 0.01$$
$$0.27x^3 + 0.99$$
$$-0.35x^4 + 15x^2 - 0.03$$

20.
$$0.05x^4 + 0.12x^3 - 0.5x^2$$
$$- 0.02x^3 + 0.02x^2 + 2x$$
$$1.5x^4 + 0.01x^2 + 0.15$$
$$0.25x^3 + 0.85$$
$$-0.25x^4 + 10x^2 - 0.04$$

b Simplify.

21. $-(-5x)$

22. $-(x^2 - 3x)$

23. $-\left(-x^2 + \frac{3}{2}x - 2\right)$

24. $-\left(-4x^3 - x^2 - \frac{1}{4}x\right)$

25. $-(12x^4 - 3x^3 + 3)$

26. $-(4x^3 - 6x^2 - 8x + 1)$

27. $-(3x - 7)$

28. $-(-2x + 4)$

29. $-(4x^2 - 3x + 2)$

30. $-(-6a^3 + 2a^2 - 9a + 1)$

31. $-\left(-4x^4 + 6x^2 + \frac{3}{4}x - 8\right)$

32. $-(-5x^4 + 4x^3 - x^2 + 0.9)$

c Subtract.

33. $(3x + 2) - (-4x + 3)$

34. $(6x + 1) - (-7x + 2)$

35. $(-6x + 2) - (x^2 + x - 3)$

36. $(x^2 - 5x + 4) - (8x - 9)$

37. $(x^2 - 9) - (x^2 + 9)$

38. $(x^3 + x^2) - (2x^3 - 5x^2)$

39. $(6x^4 + 3x^3 - 1) - (4x^2 - 3x + 3)$

40. $(-4x^2 + 2x) - (3x^3 - 5x^2 + 3)$

41. $(1.2x^3 + 4.5x^2 - 3.8x) - (-3.4x^3 - 4.7x^2 + 23)$

42. $(0.5x^4 - 0.6x^2 + 0.7) - (2.3x^4 + 1.8x - 3.9)$

43. $\left(\frac{5}{8}x^3 - \frac{1}{4}x - \frac{1}{3}\right) - \left(-\frac{1}{8}x^3 + \frac{1}{4}x - \frac{1}{3}\right)$

44. $\left(\frac{1}{5}x^3 + 2x^2 - 0.1\right) - \left(-\frac{2}{5}x^3 + 2x^2 + 0.01\right)$

45. $(0.08x^3 - 0.02x^2 + 0.01x) - (0.02x^3 + 0.03x^2 - 1)$

46. $(0.8x^4 + 0.2x - 1) - \left(\frac{7}{10}x^4 + \frac{1}{5}x - 0.1\right)$

47. $\quad x^2 + 5x + 6$
$\underline{-(x^2 + 2x)}$

48. $\quad x^3 + 1$
$\underline{-(x^3 + x^2)}$

49. $\quad 5x^4 + 6x^3 - 9x^2$
$\underline{-(-6x^4 - 6x^3 + 8x + 9)}$

50. $\quad 5x^4 + 6x^2 - 3x + 6$
$\underline{-(6x^3 + 7x^2 - 8x - 9)}$

51. $\quad x^5 - 1$
$\underline{-(x^5 - x^4 + x^3 - x^2 + x - 1)}$

52. $\quad x^5 + x^4 - x^3 + x^2 - x + 2$
$\underline{-(x^5 - x^4 + x^3 - x^2 - x + 2)}$

d Solve.

Find a polynomial for the perimeter of each figure.

53.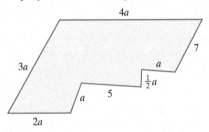

54.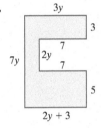

55. Find a polynomial for the sum of the areas of these rectangles.

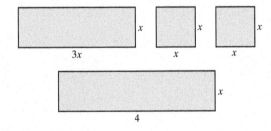

56. Find a polynomial for the sum of the areas of these circles.

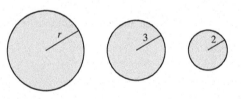

Find two algebraic expressions for the area of each figure. First, regard the figure as one large rectangle, and then regard the figure as a sum of four smaller rectangles.

57.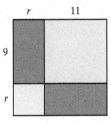

58.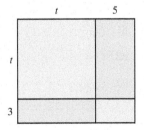

59.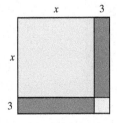

60.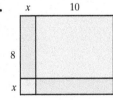

Find a polynomial for the shaded area of each figure.

61.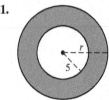

62.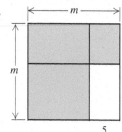

63.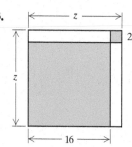

64.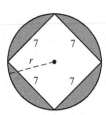

Skill Maintenance

Solve. [2.3b]

65. $8x + 3x = 66$

66. $5x - 7x = 38$

67. $\frac{3}{8}x + \frac{1}{4} - \frac{3}{4}x = \frac{11}{16} + x$

68. $5x - 4 = 26 - x$

69. $1.5x - 2.7x = 22 - 5.6x$

70. $3x - 3 = -4x + 4$

Solve. [2.3c]

71. $6(y - 3) - 8 = 4(y + 2) + 5$

72. $8(5x + 2) = 7(6x - 3)$

Solve. [2.7e]

73. $3x - 7 \leq 5x + 13$

74. $2(x - 4) > 5(x - 3) + 7$

Synthesis

Find a polynomial for the surface area of each right rectangular solid.

75.

76.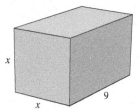

77.

78.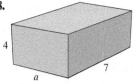

79. Find $(y - 2)^2$ using the four parts of this square.

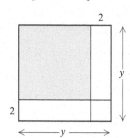

Simplify.

80. $(3x^2 - 4x + 6) - (-2x^2 + 4) + (-5x - 3)$

81. $(7y^2 - 5y + 6) - (3y^2 + 8y - 12) + (8y^2 - 10y + 3)$

82. $(-4 + x^2 + 2x^3) - (-6 - x + 3x^3) - (-x^2 - 5x^3)$

83. $(-y^4 - 7y^3 + y^2) + (-2y^4 + 5y - 2) - (-6y^3 + y^2)$

Mid-Chapter Review

Concept Reinforcement

Determine whether each statement is true or false.

_____ 1. a^n and a^{-n} are reciprocals. [4.1f]

_____ 2. $x^2 \cdot x^3 = x^6$ [4.1d]

_____ 3. $-5y^4$ and $-5y^2$ are like terms. [4.3d]

_____ 4. $4920^0 = 1$ [4.1b]

Guided Solutions

GS Fill in each blank with the expression or operation sign that creates a correct statement or solution.

5. Collect like terms: $4w^3 + 6w - 8w^3 - 3w$. [4.3d]

$4w^3 + 6w - 8w^3 - 3w = (4 - 8)\ \_\_\_ + (6 - 3)\ \_\_\_$
$= \_\_\_ w^3 + \_\_\_ w$

6. Subtract: $(3y^4 - y^2 + 11) - (y^4 - 4y^2 + 5)$. [4.4c]

$(3y^4 - y^2 + 11) - (y^4 - 4y^2 + 5) = 3y^4 - y^2 + 11\ \_\_\_ y^4\ \_\_\_ 4y^2\ \_\_\_ 5$
$= \_\_\_ y^4 + \_\_\_ y^2 + \_\_\_$

Mixed Review

Evaluate. [4.1b, c]

7. z^1

8. 4.56^0

9. a^5, when $a = -2$

10. $-x^3$, when $x = -1$

Multiply and simplify. [4.1d, f]

11. $5^3 \cdot 5^4$

12. $(3a)^2(3a)^7$

13. $x^{-8} \cdot x^5$

14. $t^4 \cdot t^{-4}$

Divide and simplify. [4.1e, f]

15. $\dfrac{7^8}{7^4}$

16. $\dfrac{x}{x^3}$

17. $\dfrac{w^5}{w^{-3}}$

18. $\dfrac{y^{-6}}{y^{-2}}$

Simplify. [4.2a, b]

19. $(3^5)^3$

20. $(x^{-3}y^2)^{-6}$

21. $\left(\dfrac{a^4}{5}\right)^6$

22. $\left(\dfrac{2y^3}{xz^2}\right)^{-2}$

Convert to scientific notation. [4.2c]

23. 25,430,000

24. 0.00012

Convert to decimal notation. [4.2c]

25. 3.6×10^{-5}

26. 1.44×10^8

Multiply or divide and write scientific notation for the result. [4.2d]

27. $(3 \times 10^6)(2 \times 10^{-3})$ **28.** $\dfrac{1.2 \times 10^{-4}}{2.4 \times 10^2}$

Evaluate the polynomial when $x = -3$ and when $x = 2$. [4.3a]

29. $-3x + 7$ **30.** $x^3 - 2x + 5$

Collect like terms and then arrange in descending order. [4.3e]

31. $3x - 2x^5 + x - 5x^2 + 2$

32. $4x^3 - 9x^2 - 2x^3 + x^2 + 8x^6$

Identify the degree of each term of the polynomial and the degree of the polynomial. [4.3c]

33. $5x^3 - x + 4$ **34.** $2x - x^4 + 3x^6$

Classify the polynomial as either a monomial, a binomial, a trinomial, or none of these. [4.3b]

35. $x - 9$ **36.** $x^5 - 2x^3 + 6x^2$

Add or subtract. [4.4a, c]

37. $(3x^2 - 1) + (5x^2 + 6)$

38. $(x^3 + 2x - 5) + (4x^3 - 2x^2 - 6)$

39. $(5x - 8) - (9x + 2)$

40. $(0.1x^2 - 2.4x + 3.6) - (0.5x^2 + x - 5.4)$

41. Find a polynomial for the sum of the areas of these rectangles. [4.4d]

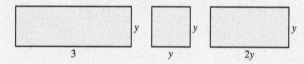

Understanding Through Discussion and Writing

42. Suppose that the length of a side of a square is three times the length of a side of a second square. How do the areas of the squares compare? Why? [4.1d]

43. Suppose that the length of a side of a cube is twice the length of a side of a second cube. How do the volumes of the cubes compare? Why? [4.1d]

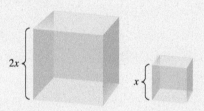

44. Explain in your own words when exponents should be added and when they should be multiplied. [4.1d], [4.2a]

45. Without performing actual computations, explain why 3^{-29} is smaller than 2^{-29}. [4.1f]

46. Is it better to evaluate a polynomial before or after like terms have been collected? Why? [4.3a, d]

47. Is the sum of two binomials ever a trinomial? Why or why not? [4.3b], [4.4a]

STUDYING FOR SUCCESS   *As You Study*
- Find a quiet place to study.
- Be disciplined in your use of electronics such as video games, the Internet, and television.
- Pace yourself. It is usually better to study for shorter periods several times a week than to study in one marathon session each week.

Multiplication of Polynomials

4.5

We now multiply polynomials using techniques based, for the most part, on the distributive law, but also on the associative and commutative laws.

OBJECTIVES

a Multiply monomials.
b Multiply a monomial and any polynomial.
c Multiply two binomials.
d Multiply any two polynomials.

a MULTIPLYING MONOMIALS

Consider $(3x)(4x)$. We multiply as follows:

$(3x)(4x) = 3 \cdot x \cdot 4 \cdot x$ By the associative law of multiplication
$= 3 \cdot 4 \cdot x \cdot x$ By the commutative law of multiplication
$= (3 \cdot 4)(x \cdot x)$ By the associative law
$= 12x^2$.   Using the product rule for exponents

MULTIPLYING MONOMIALS

To find an equivalent expression for the product of two monomials, multiply the coefficients and then multiply the variables using the product rule for exponents.

EXAMPLES Multiply.

1. $5x \cdot 6x = (5 \cdot 6)(x \cdot x)$ By the associative and commutative laws
$= 30x^2$ Multiplying the coefficients and multiplying the variables

2. $(3x)(-x) = (3x)(-1x) = (3)(-1)(x \cdot x) = -3x^2$

3. $(-7y^5)(4y^3) = (-7 \cdot 4)(y^5 \cdot y^3)$
$= -28y^{5+3} = -28y^8$ Adding exponents

Do Exercises 1–8.

Multiply.
1. $(3x)(-5)$ 2. $(-x) \cdot x$
3. $(-x)(-x)$ 4. $(-x^2)(x^3)$
5. $3x^5 \cdot 4x^2$
6. $(4y^5)(-2y^6)$
7. $(-7y^4)(-y)$ 8. $7x^5 \cdot 0$

b MULTIPLYING A MONOMIAL AND ANY POLYNOMIAL

SKILL REVIEW   *Use the distributive law to multiply.* [1.7c]
Multiply.
1. $3(x - 5)$ 2. $2(3y + 4z - 1)$

Answers: 1. $3x - 15$ 2. $6y + 8z - 2$

Answers
1. $-15x$ 2. $-x^2$ 3. x^2 4. $-x^5$
5. $12x^7$ 6. $-8y^{11}$ 7. $7y^5$ 8. 0

To multiply a monomial, such as $2x$, and a binomial, such as $5x + 3$, we use the distributive law and multiply each term of $5x + 3$ by $2x$:

$$2x(5x + 3) = (2x)(5x) + (2x)(3) \quad \text{Using a distributive law}$$
$$= 10x^2 + 6x. \quad \text{Multiplying the monomials}$$

EXAMPLE 4 Multiply: $5x(2x^2 - 3x + 4)$.

$$5x(2x^2 - 3x + 4) = (5x)(2x^2) - (5x)(3x) + (5x)(4)$$
$$= 10x^3 - 15x^2 + 20x$$

MULTIPLYING A MONOMIAL AND A POLYNOMIAL

To multiply a monomial and a polynomial, multiply each term of the polynomial by the monomial.

EXAMPLE 5 Multiply: $-2x^2(x^3 - 7x^2 + 10x - 4)$.

$$-2x^2(x^3 - 7x^2 + 10x - 4)$$
$$= (-2x^2)(x^3) - (-2x^2)(7x^2) + (-2x^2)(10x) - (-2x^2)(4)$$
$$= -2x^5 + 14x^4 - 20x^3 + 8x^2$$

◀ Do Exercises 9–11.

Multiply.

9. $4x(2x + 4)$

10. $3t^2(-5t + 2)$

11. $-5x^3(x^3 + 5x^2 - 6x + 8)$

12. **a)** Multiply: $(y + 2)(y + 7)$.
$(y + 2)(y + 7)$
$= y \cdot (y + 7) + 2 \cdot ()$
$= y \cdot y + y \cdot + 2 \cdot y + 2 \cdot $
$= + 7y + 2y + $
$= y^2 + + 14$

b) Write an algebraic expression that represents the total area of the four smaller rectangles in the following figure.

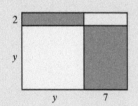

The area is $(y + 7)(y +)$, or, from part (a), $y^2 + + 14$.

C MULTIPLYING TWO BINOMIALS

To find an equivalent expression for the product of two binomials, we use the distributive laws more than once. In Example 6, we use a distributive law three times.

EXAMPLE 6 Multiply: $(x + 5)(x + 4)$.

$$(x + 5)(x + 4) = x(x + 4) + 5(x + 4) \quad \text{Using a distributive law}$$
$$= x \cdot x + x \cdot 4 + 5 \cdot x + 5 \cdot 4 \quad \text{Using a distributive law twice}$$
$$= x^2 + 4x + 5x + 20 \quad \text{Multiplying the monomials}$$
$$= x^2 + 9x + 20 \quad \text{Collecting like terms}$$

To visualize the product in Example 6, consider a rectangle of length $x + 5$ and width $x + 4$. The total area can then be expressed as $(x + 5)(x + 4)$ or, by adding the four smaller areas, $x^2 + 4x + 5x + 20$, or $x^2 + 9x + 20$.

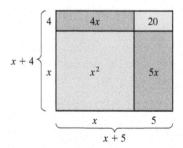

◀ Do Exercises 12–14.

Multiply.

13. $(x + 8)(x + 5)$

14. $(x + 5)(x - 4)$

Answers
9. $8x^2 + 16x$ 10. $-15t^3 + 6t^2$
11. $-5x^6 - 25x^5 + 30x^4 - 40x^3$
12. (a) $y^2 + 9y + 14$; (b) $(y + 2)(y + 7)$, or $y^2 + 2y + 7y + 14$, or $y^2 + 9y + 14$ 13. $x^2 + 13x + 40$
14. $x^2 + x - 20$

Guided Solution:
12. (a) $y + 7, 7, 7, y^2, 14, 9y$; (b) 2, 9y

EXAMPLE 7 Multiply: $(4x + 3)(x - 2)$.

$$(4x + 3)(x - 2) = 4x(x - 2) + 3(x - 2) \quad \text{Using a distributive law}$$
$$= 4x \cdot x - 4x \cdot 2 + 3 \cdot x - 3 \cdot 2 \quad \text{Using a distributive law twice}$$
$$= 4x^2 - 8x + 3x - 6 \quad \text{Multiplying the monomials}$$
$$= 4x^2 - 5x - 6 \quad \text{Collecting like terms}$$

Do Exercises 15 and 16.

Multiply.
15. $(5x + 3)(x - 4)$
16. $(2x - 3)(3x - 5)$

d MULTIPLYING ANY TWO POLYNOMIALS

Let's consider the product of a binomial and a trinomial. We use a distributive law four times. You may see ways to skip some steps and do the work mentally.

EXAMPLE 8 Multiply: $(x^2 + 2x - 3)(x^2 + 4)$.

$$(x^2 + 2x - 3)(x^2 + 4) = x^2(x^2 + 4) + 2x(x^2 + 4) - 3(x^2 + 4)$$
$$= x^2 \cdot x^2 + x^2 \cdot 4 + 2x \cdot x^2 + 2x \cdot 4 - 3 \cdot x^2 - 3 \cdot 4$$
$$= x^4 + 4x^2 + 2x^3 + 8x - 3x^2 - 12$$
$$= x^4 + 2x^3 + x^2 + 8x - 12$$

Do Exercises 17 and 18.

Multiply.
17. $(x^2 + 3x - 4)(x^2 + 5)$

PRODUCT OF TWO POLYNOMIALS

To multiply two polynomials P and Q, select one of the polynomials—say, P. Then multiply each term of P by every term of Q and collect like terms.

GS 18. $(3y^2 - 7)(2y^3 - 2y + 5)$
$= 3y^2(2y^3 - 2y + 5) - \boxed{}(2y^3 - 2y + 5)$
$= 6y^5 - \boxed{} + 15y^2 - 14y^3 + \boxed{} - 35$
$= 6y^5 - \boxed{} + 15y^2 + 14y - 35$

To use columns for long multiplication, we multiply each term in the top row by every term in the bottom row. We write like terms in columns, and then add the results. Such multiplication is like multiplying with whole numbers.

```
    3 2 1              300 + 20 + 1
  ×   1 2            ×         10 + 2
  ─────────          ──────────────────
    6 4 2              600 +  40 + 2      Multiplying the top row by 2
  3 2 1         3000 +  200 + 10          Multiplying the top row by 10
  ─────────    ──────────────────────
  3 8 5 2        3000 + 800 + 50 + 2      Adding
```

EXAMPLE 9 Multiply: $(4x^3 - 2x^2 + 3x)(x^2 + 2x)$.

```
         4x³ − 2x² + 3x
                x² + 2x
       ──────────────────
         8x⁴ − 4x³ + 6x²         Multiplying the top row by 2x
  4x⁵ − 2x⁴ + 3x³                Multiplying the top row by x²
  ──────────────────────
  4x⁵ + 6x⁴ −  x³ + 6x²          Collecting like terms
    ↑     ↑    ↑    ↑ ─────── Line up like terms in columns.
```

Answers
15. $5x^2 - 17x - 12$ 16. $6x^2 - 19x + 15$
17. $x^4 + 3x^3 + x^2 + 15x - 20$
18. $6y^5 - 20y^3 + 15y^2 + 14y - 35$
Guided Solution:
18. $7, 6y^3, 14y, 20y^3$

EXAMPLE 10 Multiply: $(2x^2 + 3x - 4)(2x^2 - x + 3)$.

$$
\begin{array}{r}
2x^2 + 3x - 4 \\
2x^2 - x + 3 \\
\hline
6x^2 + 9x - 12 \\
-2x^3 - 3x^2 + 4x \\
4x^4 + 6x^3 - 8x^2 \\
\hline
4x^4 + 4x^3 - 5x^2 + 13x - 12
\end{array}
\quad
\begin{array}{l}
\text{Multiplying by 3} \\
\text{Multiplying by } -x \\
\text{Multiplying by } 2x^2 \\
\text{Collecting like terms}
\end{array}
$$

◀ Do Exercise 19.

EXAMPLE 11 Multiply: $(5x^3 - 3x + 4)(-2x^2 - 3)$.

If terms are missing, it helps to leave spaces for them and align like terms in columns as we multiply.

$$
\begin{array}{r}
5x^3 - 3x + 4 \\
-2x^2 - 3 \\
\hline
-15x^3 + 9x - 12 \\
-10x^5 + 6x^3 - 8x^2 \\
\hline
-10x^5 - 9x^3 - 8x^2 + 9x - 12
\end{array}
\quad
\begin{array}{l}
\text{Multiplying by } -3 \\
\text{Multiplying by } -2x^2 \\
\text{Collecting like terms}
\end{array}
$$

◀ Do Exercises 20 and 21.

19. Multiply.

$$\begin{array}{r} 3x^2 - 2x - 5 \\ 2x^2 + x - 2 \end{array}$$

Multiply.

20. $3x^2 - 2x + 4$
$ x + 5$

21. $-5x^2 + 4x + 2$
$ -4x^2 - 8$

Answers
19. $6x^4 - x^3 - 18x^2 - x + 10$
20. $3x^3 + 13x^2 - 6x + 20$
21. $20x^4 - 16x^3 + 32x^2 - 32x - 16$

CALCULATOR CORNER

Checking Multiplication of Polynomials A partial check of multiplication of polynomials can be performed graphically. Consider the product $(x + 3)(x - 2) = x^2 + x - 6$. We will use two graph styles to determine whether this product is correct. First, we press **MODE** and select the SEQUENTIAL mode.

Next, on the Y= screen, we enter $y_1 = (x + 3)(x - 2)$ and $y_2 = x^2 + x - 6$. We will select the line-graph style for y_1 and the path style for y_2. To select these graph styles, we use ◁ to position the cursor over the icon to the left of the equation and press **ENTER** repeatedly until the desired style of icon appears, as shown below. Then we graph the equations.

  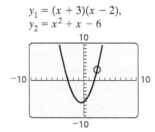

$y_1 = (x + 3)(x - 2),$
$y_2 = x^2 + x - 6$

The graphing calculator will graph y_1 first as a solid line. Then it will graph y_2 as the circular cursor traces the leading edge of the graph, allowing us to determine visually whether the graphs coincide. In this case, the graphs appear to coincide, so the factorization is probably correct.

A table of values can also be used as a check.

EXERCISES: Determine graphically whether each product is correct.
1. $(x + 5)(x + 4) = x^2 + 9x + 20$
2. $(4x + 3)(x - 2) = 4x^2 - 5x - 6$
3. $(5x + 3)(x - 4) = 5x^2 + 17x - 12$
4. $(2x - 3)(3x - 5) = 6x^2 - 19x - 15$

4.5 Exercise Set

✓ Check Your Understanding

Reading Check Determine whether each statement is true or false.

RC1. When multiplying the monomials $4x^2$ and $5x^3$, we multiply the coefficients.

RC2. When multiplying the monomials $4x^2$ and $5x^3$, we multiply the exponents.

RC3. After we have multiplied $(a + b)(c + d)$, there will be four terms in the product.

RC4. The product of two binomials may be a trinomial.

Concept Check Match each expression with an equivalent expression from the column on the right. Choices may be used more than once or not at all.

CC1. $8x \cdot 2x$

CC2. $(-16x)(-x)$

CC3. $2x(8x - 1)$

CC4. $(2x - 1)(8x + 1)$

a) $16x^2$
b) $-16x^2$
c) $16x^2 - 1$
d) $16x^2 - 2x$
e) $16x^2 - 6x - 1$

a Multiply.

1. $(8x^2)(5)$
2. $(4x^2)(-2)$
3. $(-x^2)(-x)$
4. $(-x^3)(x^2)$

5. $(8x^5)(4x^3)$
6. $(10a^2)(2a^2)$
7. $(0.1x^6)(0.3x^5)$
8. $(0.3x^4)(-0.8x^6)$

9. $\left(-\frac{1}{5}x^3\right)\left(-\frac{1}{3}x\right)$
10. $\left(-\frac{1}{4}x^4\right)\left(\frac{1}{5}x^8\right)$
11. $(-4x^2)(0)$
12. $(-4m^5)(-1)$

13. $(3x^2)(-4x^3)(2x^6)$
14. $(-2y^5)(10y^4)(-3y^3)$

b Multiply.

15. $2x(-x + 5)$
16. $3x(4x - 6)$
17. $-5x(x - 1)$
18. $-3x(-x - 1)$

19. $x^2(x^3 + 1)$
20. $-2x^3(x^2 - 1)$
21. $3x(2x^2 - 6x + 1)$
22. $-4x(2x^3 - 6x^2 - 5x + 1)$

23. $(-6x^2)(x^2 + x)$
24. $(-4x^2)(x^2 - x)$
25. $(3y^2)(6y^4 + 8y^3)$
26. $(4y^4)(y^3 - 6y^2)$

c Multiply.

27. $(x+6)(x+3)$ 28. $(x+5)(x+2)$ 29. $(x+5)(x-2)$ 30. $(x+6)(x-2)$

31. $(x-1)(x+4)$ 32. $(x-8)(x+7)$ 33. $(x-4)(x-3)$ 34. $(x-7)(x-3)$

35. $(x+3)(x-3)$ 36. $(x+6)(x-6)$ 37. $(x-4)(x+4)$ 38. $(x-9)(x+9)$

39. $(3x+5)(x+2)$ 40. $(2x+6)(x+3)$ 41. $(5-x)(5-2x)$ 42. $(3-4x)(2-x)$

43. $(2x+5)(2x+5)$ 44. $(3x+4)(3x+4)$ 45. $(x-3)(x-3)$ 46. $(x-6)(x-6)$

47. $\left(x-\tfrac{5}{2}\right)\left(x+\tfrac{2}{5}\right)$ 48. $\left(x+\tfrac{4}{3}\right)\left(x+\tfrac{3}{2}\right)$ 49. $(x-2.3)(x+4.7)$ 50. $(2x+0.13)(2x-0.13)$

Write an algebraic expression that represents the total area of the four smaller rectangles in each figure.

51. 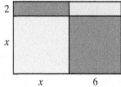   52. 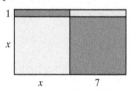   53. 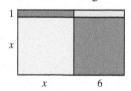   54.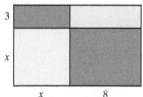

Draw and label rectangles similar to the one following Example 6 to illustrate each product.

55. $x(x+5)$ 56. $x(x+2)$ 57. $(x+1)(x+2)$

58. $(x+3)(x+1)$ 59. $(x+5)(x+3)$ 60. $(x+4)(x+6)$

d Multiply.

61. $(x^2+x+1)(x-1)$ 62. $(x^2+x-2)(x+2)$ 63. $(2x+1)(2x^2+6x+1)$

64. $(3x-1)(4x^2-2x-1)$ 65. $(y^2-3)(3y^2-6y+2)$ 66. $(3y^2-3)(y^2+6y+1)$

67. $(x^3+x^2)(x^3+x^2-x)$ 68. $(x^3-x^2)(x^3-x^2+x)$ 69. $(-5x^3-7x^2+1)(2x^2-x)$

70. $(-4x^3 + 5x^2 - 2)(5x^2 + 1)$ **71.** $(1 + x + x^2)(-1 - x + x^2)$ **72.** $(1 - x + x^2)(1 - x + x^2)$

73. $(2t^2 - t - 4)(3t^2 + 2t - 1)$ **74.** $(3a^2 - 5a + 2)(2a^2 - 3a + 4)$ **75.** $(x - x^3 + x^5)(x^2 - 1 + x^4)$

76. $(x - x^3 + x^5)(3x^2 + 3x^6 + 3x^4)$ **77.** $(x + 1)(x^3 + 7x^2 + 5x + 4)$ **78.** $(x + 2)(x^3 + 5x^2 + 9x + 3)$

79. $\left(x - \tfrac{1}{2}\right)\left(2x^3 - 4x^2 + 3x - \tfrac{2}{5}\right)$ **80.** $\left(x + \tfrac{1}{3}\right)\left(6x^3 - 12x^2 - 5x + \tfrac{1}{2}\right)$

Skill Maintenance

Simplify.
81. $5 - 2[3 - 4(8 - 2)]$ [1.8c]
82. $(10 - 2)(10 + 2)$ [1.8d]

Factor. [1.7d]
83. $16x - 24y + 36$
84. $-9x - 45y + 15$

Synthesis

85. Find a polynomial for the shaded area of the figure.

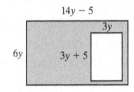

86. Determine what the missing number must be in order for the figure to have the area $x^2 + 7x + 10$.

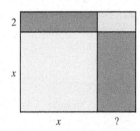

87. Find a polynomial for the volume of the following solid.

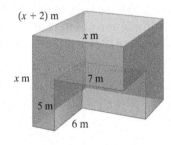

88. An open wooden box is a cube with side length x cm. The box, including its bottom, is made of wood that is 1 cm thick. Find a polynomial for the interior volume of the cube.

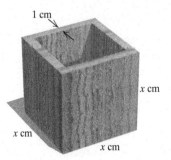

Compute and simplify.
89. $(x - 2)(x - 7) - (x - 7)(x - 2)$

90. $(x + 5)^2 - (x - 3)^2$

91. Extend the pattern and simplify:
$(x - a)(x - b)(x - c)(x - d) \cdots (x - z)$.

92. Use a graphing calculator to check your answers to Exercises 15, 29, and 61. Use graphs, tables, or both, as directed by your instructor.

4.6 Special Products

OBJECTIVES

a. Multiply two binomials mentally using the FOIL method.

b. Multiply the sum and the difference of the same two terms mentally.

c. Square a binomial mentally.

d. Find special products when polynomial products are mixed together.

We encounter certain products so often that it is helpful to have efficient methods of computing them. Such techniques are called *special products*.

a PRODUCTS OF TWO BINOMIALS USING FOIL

SKILL REVIEW

Collect like terms. [1.7e]
Collect like terms.

1. $-3.4x - 3.4x$
2. $-12n + 12n$

Answers: 1. $-6.8x$ 2. 0

MyLab Math
VIDEO

To multiply two binomials, we can select one binomial and multiply each term of that binomial by every term of the other. Then we collect like terms. Consider the product $(x + 3)(x + 7)$:

$$(x + 3)(x + 7) = x(x + 7) + 3(x + 7)$$
$$= x \cdot x + x \cdot 7 + 3 \cdot x + 3 \cdot 7$$
$$= x^2 + 7x + 3x + 21$$
$$= x^2 + 10x + 21.$$

This example illustrates a special technique for finding the product of two binomials:

$$(x + 3)(x + 7) = \underbrace{x \cdot x}_{\text{First terms}} + \underbrace{7 \cdot x}_{\text{Outside terms}} + \underbrace{3 \cdot x}_{\text{Inside terms}} + \underbrace{3 \cdot 7}_{\text{Last terms}}.$$

To remember this method of multiplying, we use the initials **FOIL**.

We can show the FOIL method geometrically as follows. One way to write the area of the large rectangle below is $(A + B)(C + D)$. To find another expression for the area of the large rectangle, we add the areas of the smaller rectangles.

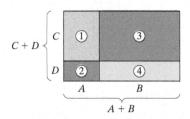

The area of rectangle ① is AC.

The area of rectangle ② is AD.

The area of rectangle ③ is BC.

The area of rectangle ④ is BD.

The area of the large rectangle is the sum of the areas of the smaller rectangles. Thus,

$(A + B)(C + D)$
$= AC + AD + BC + BD$.

THE FOIL METHOD

To multiply two binomials, $A + B$ and $C + D$, multiply the First terms AC, the Outside terms AD, the Inside terms BC, and then the Last terms BD. Then collect like terms, if possible.

$(A + B)(C + D) = AC + AD + BC + BD$

1. Multiply First terms: AC.
2. Multiply Outside terms: AD.
3. Multiply Inside terms: BC.
4. Multiply Last terms: BD.
   ↓
   FOIL

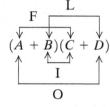

304   CHAPTER 4   Polynomials: Operations

EXAMPLE 1 Multiply: $(x + 8)(x^2 - 5)$.

We have

$$(x + 8)(x^2 - 5) = \overset{F}{x \cdot x^2} + \overset{O}{x \cdot (-5)} + \overset{I}{8 \cdot x^2} + \overset{L}{8(-5)}$$
$$= x^3 - 5x + 8x^2 - 40$$
$$= x^3 + 8x^2 - 5x - 40.$$

Since each of the original binomials is in descending order, we write the product in descending order, as is customary, but this is not a "must."

Often we can collect like terms after we have multiplied.

EXAMPLES Multiply.

2. $(x + 6)(x - 6) = x^2 - 6x + 6x - 36$ Using FOIL
   $ = x^2 - 36$ Collecting like terms

3. $(x + 7)(x + 4) = x^2 + 4x + 7x + 28$
   $ = x^2 + 11x + 28$

4. $(y - 3)(y - 2) = y^2 - 2y - 3y + 6$
   $ = y^2 - 5y + 6$

5. $(x^3 - 1)(x^3 + 5) = x^6 + 5x^3 - x^3 - 5$
   $ = x^6 + 4x^3 - 5$

6. $(4t^3 + 5)(3t^2 - 2) = 12t^5 - 8t^3 + 15t^2 - 10$

Do Exercises 1–8. ▶

EXAMPLES Multiply.

7. $\left(x - \tfrac{2}{3}\right)\left(x + \tfrac{2}{3}\right) = x^2 + \tfrac{2}{3}x - \tfrac{2}{3}x - \tfrac{4}{9}$
   $= x^2 - \tfrac{4}{9}$

8. $(x^2 - 0.3)(x^2 - 0.3) = x^4 - 0.3x^2 - 0.3x^2 + 0.09$
   $= x^4 - 0.6x^2 + 0.09$

9. $(3 - 4x)(7 - 5x^3) = 21 - 15x^3 - 28x + 20x^4$
   $= 21 - 28x - 15x^3 + 20x^4$

(*Note:* If the original polynomials are in ascending order, it is natural to write the product in ascending order, but this is not a "must.")

10. $(5x^4 + 2x^3)(3x^2 - 7x) = 15x^6 - 35x^5 + 6x^5 - 14x^4$
    $= 15x^6 - 29x^5 - 14x^4$

Do Exercises 9–12. ▶

Multiply mentally, if possible. If you need extra steps, be sure to use them.

1. $(x + 3)(x + 4)$
2. $(x + 3)(x - 5)$
3. $(2x - 1)(x - 4)$
4. $(2x^2 - 3)(x^2 - 2)$
5. $(6x^2 + 5)(2x^3 + 1)$
6. $(y^3 + 7)(y^3 - 7)$
7. $(t + 2)(t + 3)$
8. $(2x^4 + x^2)(-x^3 + x)$

Multiply.

9. $\left(x + \dfrac{4}{5}\right)\left(x - \dfrac{4}{5}\right)$
10. $(x^3 - 0.5)(x^2 + 0.5)$
11. $(2 + 3x^2)(4 - 5x^2)$
12. $(6x^3 - 3x^2)(5x^2 - 2x)$

Answers
1. $x^2 + 7x + 12$ 2. $x^2 - 2x - 15$
3. $2x^2 - 9x + 4$ 4. $2x^4 - 7x^2 + 6$
5. $12x^5 + 10x^3 + 6x^2 + 5$
6. $y^6 - 49$ 7. $t^2 + 5t + 6$
8. $-2x^7 + x^5 + x^3$ 9. $x^2 - \dfrac{16}{25}$
10. $x^5 + 0.5x^3 - 0.5x^2 - 0.25$
11. $8 + 2x^2 - 15x^4$
12. $30x^5 - 27x^4 + 6x^3$

SECTION 4.6   Special Products

b   MULTIPLYING SUMS AND DIFFERENCES OF TWO TERMS

Consider the product of the sum and the difference of the same two terms, such as

$$(x + 2)(x - 2).$$

Since this is the product of two binomials, we can use FOIL. This type of product occurs so often, however, that it would be valuable if we could use an even faster method. To find a faster way to compute such a product, look for a pattern in the following:

a) $(x + 2)(x - 2) = x^2 - 2x + 2x - 4$ Using FOIL
$= x^2 - 4;$

b) $(3x - 5)(3x + 5) = 9x^2 + 15x - 15x - 25$
$= 9x^2 - 25.$

◀ Do Exercises 13 and 14.

Perhaps you discovered in each case that when you multiply the two binomials, two terms are opposites, or additive inverses, which add to 0 and "drop out."

> **PRODUCT OF THE SUM AND THE DIFFERENCE OF TWO TERMS**
>
> The product of the sum and the difference of the same two terms is the square of the first term minus the square of the second term:
>
> $$(A + B)(A - B) = A^2 - B^2.$$

It is helpful to memorize this rule in both words and symbols. (If you do forget it, you can, of course, use FOIL.)

EXAMPLES   Multiply. (Carry out the rule and say the words as you go.)

$(A + B)(A - B) = A^2 - B^2$

11. $(x + 4)(x - 4) = x^2 - 4^2$ "The square of the first term, x^2, minus the square of the second, 4^2"
$= x^2 - 16$ Simplifying

12. $(5 + 2w)(5 - 2w) = 5^2 - (2w)^2$
$= 25 - 4w^2$

13. $(3x^2 - 7)(3x^2 + 7) = (3x^2)^2 - 7^2$
$= 9x^4 - 49$

14. $(-4x - 10)(-4x + 10) = (-4x)^2 - 10^2$
$= 16x^2 - 100$

15. $\left(x + \dfrac{3}{8}\right)\left(x - \dfrac{3}{8}\right) = x^2 - \left(\dfrac{3}{8}\right)^2 = x^2 - \dfrac{9}{64}$

◀ Do Exercises 15–19.

Multiply.

13. $(x + 5)(x - 5)$

14. $(2x - 3)(2x + 3)$

Multiply.

15. $(x + 8)(x - 8)$

16. $(x - 7)(x + 7)$

17. $(6 - 4y)(6 + 4y)$ GS
$()^2 - ()^2 = 36 - $

18. $(2x^3 - 1)(2x^3 + 1)$

19. $\left(x - \dfrac{2}{5}\right)\left(x + \dfrac{2}{5}\right)$

Answers
13. $x^2 - 25$ **14.** $4x^2 - 9$ **15.** $x^2 - 64$
16. $x^2 - 49$ **17.** $36 - 16y^2$
18. $4x^6 - 1$ **19.** $x^2 - \dfrac{4}{25}$

Guided Solution:
17. $6, 4y, 16y^2$

c SQUARING BINOMIALS

Consider the square of a binomial, such as $(x + 3)^2$. This can be expressed as $(x + 3)(x + 3)$. Since this is the product of two binomials, we can use FOIL. But again, this type of product occurs so often that we would like to use an even faster method. Look for a pattern in the following.

a) $(x + 3)^2 = (x + 3)(x + 3)$
$= x^2 + 3x + 3x + 9$
$= x^2 + 6x + 9;$

b) $(x - 3)^2 = (x - 3)(x - 3)$
$= x^2 - 3x - 3x + 9$
$= x^2 - 6x + 9$

Do Exercises 20 and 21. ▶

When squaring a binomial, we multiply a binomial by itself. Perhaps you noticed that two terms are the same and when added give twice the product of the terms in the binomial. The other two terms are squares.

Multiply.
20. $(x + 8)(x + 8)$
21. $(x - 5)(x - 5)$

SQUARE OF A BINOMIAL

The square of a sum (or a difference) of two terms is the square of the first term, plus (or minus) twice the product of the two terms, plus the square of the last term:

$$(A + B)^2 = A^2 + 2AB + B^2; \quad (A - B)^2 = A^2 - 2AB + B^2.$$

It is helpful to memorize this rule in both words and symbols.

EXAMPLES Multiply. (Carry out the rule and say the words as you go.)

$(A + B)^2 = A^2 + 2 \cdot A \cdot B + B^2$
16. $(x + 3)^2 = x^2 + 2 \cdot x \cdot 3 + 3^2$ "x^2 plus 2 times x times 3 plus 3^2"
$= x^2 + 6x + 9$

$(A - B)^2 = A^2 - 2 \cdot A \cdot B + B^2$
17. $(t - 5)^2 = t^2 - 2 \cdot t \cdot 5 + 5^2$ "t^2 minus 2 times t times 5 plus 5^2"
$= t^2 - 10t + 25$

18. $(2x + 7)^2 = (2x)^2 + 2 \cdot 2x \cdot 7 + 7^2 = 4x^2 + 28x + 49$

19. $(5x - 3x^2)^2 = (5x)^2 - 2 \cdot 5x \cdot 3x^2 + (3x^2)^2 = 25x^2 - 30x^3 + 9x^4$

20. $(2.3 - 5.4m)^2 = 2.3^2 - 2(2.3)(5.4m) + (5.4m)^2$
$= 5.29 - 24.84m + 29.16m^2$

Do Exercises 22–27. ▶

Multiply.
22. $(x + 2)^2$
23. $(a - 4)^2$
24. $(2x + 5)^2$
25. $(4x^2 - 3x)^2$
26. $(7.8 + 1.2y)(7.8 + 1.2y)$

GS 27. $(3x^2 - 5)(3x^2 - 5)$
$(3x^2)^2 - 2(3x^2)() + 5^2$
$= x^4 - x^2 + 25$

.... **Caution!** ....

Although the square of a product is the product of the squares, the square of a sum is *not* the sum of the squares. That is, $(AB)^2 = A^2B^2$, but

The term $2AB$ is missing.
↓
$(A + B)^2 \neq A^2 + B^2.$

To illustrate this inequality, note, using the rules for order of operations, that $(7 + 5)^2 = 12^2 = 144$, whereas $7^2 + 5^2 = 49 + 25 = 74$, and $74 \neq 144$.

Answers
20. $x^2 + 16x + 64$ **21.** $x^2 - 10x + 25$
22. $x^2 + 4x + 4$ **23.** $a^2 - 8a + 16$
24. $4x^2 + 20x + 25$
25. $16x^4 - 24x^3 + 9x^2$
26. $60.84 + 18.72y + 1.44y^2$
27. $9x^4 - 30x^2 + 25$

Guided Solution:
27. 5, 9, 30

28. In the figure at right, describe in terms of area the sum $A^2 + B^2$. How can the figure be used to verify that $(A + B)^2 \neq A^2 + B^2$?

We can look at the rule for finding $(A + B)^2$ geometrically as follows. The area of the large square is
$$(A + B)(A + B) = (A + B)^2.$$
This is equal to the sum of the areas of the smaller rectangles:
$$A^2 + AB + AB + B^2 = A^2 + 2AB + B^2.$$
Thus, $(A + B)^2 = A^2 + 2AB + B^2$.

◀ Do Exercise 28.

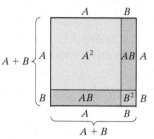

d MULTIPLICATION OF VARIOUS TYPES

Let's now try several types of multiplications mixed together so that we can learn to sort them out. When you multiply, first see what kind of multiplication you have. Then use the best method.

MULTIPLYING TWO POLYNOMIALS

1. Is it the product of a monomial and a polynomial? If so, multiply each term of the polynomial by the monomial.
   Example: $5x(x + 7) = 5x \cdot x + 5x \cdot 7 = 5x^2 + 35x$

2. Is it the product of the sum and the difference of the *same* two terms? If so, use the following:
   $$(A + B)(A - B) = A^2 - B^2.$$
   Example: $(x + 7)(x - 7) = x^2 - 7^2 = x^2 - 49$

3. Is the product the square of a binomial? If so, use the following:
   $$(A + B)(A + B) = (A + B)^2 = A^2 + 2AB + B^2,$$
   or $(A - B)(A - B) = (A - B)^2 = A^2 - 2AB + B^2.$
   Example: $(x + 7)(x + 7) = (x + 7)^2$
   $$= x^2 + 2 \cdot x \cdot 7 + 7^2$$
   $$= x^2 + 14x + 49$$

4. Is it the product of two binomials other than those above? If so, use FOIL.
   Example: $(x + 7)(x - 4) = x^2 - 4x + 7x - 28$
   $$= x^2 + 3x - 28$$

5. Is it the product of two polynomials other than those above? If so, multiply each term of one by every term of the other. Use columns if you wish.
   Example:
   $(x^2 - 3x + 2)(x + 7) = x^2(x + 7) - 3x(x + 7) + 2(x + 7)$
   $$= x^2 \cdot x + x^2 \cdot 7 - 3x \cdot x - 3x \cdot 7$$
   $$+ 2 \cdot x + 2 \cdot 7$$
   $$= x^3 + 7x^2 - 3x^2 - 21x + 2x + 14$$
   $$= x^3 + 4x^2 - 19x + 14$$

Answer

28. $(A + B)^2$ represents the area of the large square. This includes all four sections. $A^2 + B^2$ represents the area of only two of the sections.

Remember that FOIL will *always* work for two binomials. You can use it instead of either of rules (2) and (3), but those rules will make your work go faster.

EXAMPLE 21 Multiply: $(x + 3)(x - 3)$.

$(x + 3)(x - 3) = x^2 - 3^2$
$= x^2 - 9$

This is the product of the sum and the difference of the same two terms. We use $(A + B)(A - B) = A^2 - B^2$.

EXAMPLE 22 Multiply: $(t + 7)(t - 5)$.

$(t + 7)(t - 5) = t^2 + 2t - 35$

This is the product of two binomials, but neither the square of a binomial nor the product of the sum and the difference of two terms. We use FOIL.

EXAMPLE 23 Multiply: $(x + 6)(x + 6)$.

$(x + 6)(x + 6) = x^2 + 2(6)x + 6^2$
$= x^2 + 12x + 36$

This is the square of a binomial. We use $(A + B)(A + B) = A^2 + 2AB + B^2$.

EXAMPLE 24 Multiply: $2x^3(9x^2 + x - 7)$.

$2x^3(9x^2 + x - 7) = 18x^5 + 2x^4 - 14x^3$

This is the product of a monomial and a trinomial. We multiply each term of the trinomial by the monomial.

EXAMPLE 25 Multiply: $(5x^3 - 7x)^2$.

$(5x^3 - 7x)^2 = (5x^3)^2 - 2(5x^3)(7x) + (7x)^2$ $(A - B)^2 = A^2 - 2AB + B^2$
$= 25x^6 - 70x^4 + 49x^2$

EXAMPLE 26 Multiply: $\left(3x + \frac{1}{4}\right)^2$.

$\left(3x + \frac{1}{4}\right)^2 = (3x)^2 + 2(3x)\left(\frac{1}{4}\right) + \left(\frac{1}{4}\right)^2$ $(A + B)^2 = A^2 + 2AB + B^2$
$= 9x^2 + \frac{3}{2}x + \frac{1}{16}$

EXAMPLE 27 Multiply: $\left(4x - \frac{3}{4}\right)\left(4x + \frac{3}{4}\right)$.

$\left(4x - \frac{3}{4}\right)\left(4x + \frac{3}{4}\right) = (4x)^2 - \left(\frac{3}{4}\right)^2$ $(A + B)(A - B) = A^2 - B^2$
$= 16x^2 - \frac{9}{16}$

EXAMPLE 28 Multiply: $(p + 3)(p^2 + 2p - 1)$.

$$\begin{array}{r} p^2 + 2p - 1 \\ p + 3 \\ \hline 3p^2 + 6p - 3 \\ p^3 + 2p^2 - p \\ \hline p^3 + 5p^2 + 5p - 3 \end{array}$$

Finding the product of two polynomials
Multiplying by 3
Multiplying by p

Do Exercises 29–36.

Multiply.

29. $(x + 5)(x + 6)$

30. $(t - 4)(t + 4)$

31. $4x^2(-2x^3 + 5x^2 + 10)$

32. $(9x^2 + 1)^2$

33. $(2a - 5)(2a + 8)$

34. $\left(5x + \frac{1}{2}\right)^2$

35. $\left(2x - \frac{1}{2}\right)^2$

36. $(x^2 - x + 4)(x - 2)$

Answers

29. $x^2 + 11x + 30$ 30. $t^2 - 16$
31. $-8x^5 + 20x^4 + 40x^2$
32. $81x^4 + 18x^2 + 1$ 33. $4a^2 + 6a - 40$
34. $25x^2 + 5x + \frac{1}{4}$ 35. $4x^2 - 2x + \frac{1}{4}$
36. $x^3 - 3x^2 + 6x - 8$

Visualizing for Success

1
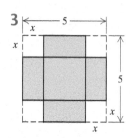

2
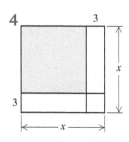

3
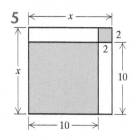

4
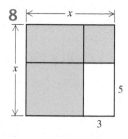

5
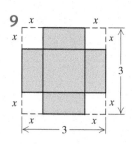

In each of Exercises 1–10, choose from the following list two algebraic expressions for the shaded area of the figure.

A. $9 - 4x^2$

B. $x^2 - (x - 6)^2$

C. $(x + 3)(x - 3)$

D. $10^2 + 2^2$

E. $x^2 + 8x + 15$

F. $(x + 5)(x + 3)$

G. $x^2 - 6x + 9$

H. $(3 - 2x)^2 + 4x(3 - 2x)$

I. $(x + 3)^2$

J. $(5x + 3)^2$

K. $(5 - 2x)^2 + 4x(5 - 2x)$

L. $x^2 - 9$

M. 104

N. $x^2 - 15$

O. $12x - 36$

P. $25x^2 + 30x + 9$

Q. $(x - 5)(x - 3) + 3(x - 5) + 5(x - 3)$

R. $(x - 3)^2$

S. $25 - 4x^2$

T. $x^2 + 6x + 9$

Answers on page A-12

6
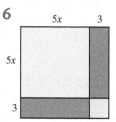

7
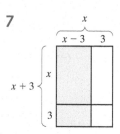

8
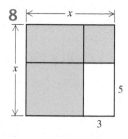

9
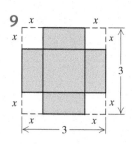

10
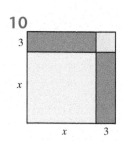

4.6 Exercise Set

✓ Check Your Understanding

Reading Check Choose from the list on the right the appropriate word to complete each statement. A word may be used more than once or not at all.

RC1. For the FOIL multiplication method, the initials F O I L represent the words first, _____, inside, and _____.

RC2. If polynomials being multiplied are written in descending order, we generally write the product in _____ order.

RC3. The expression $(A + B)(A - B)$ is the product of the sum and the _____ of the same two terms.

RC4. The expression $(A + B)^2$ is the _____ of a _____.

RC5. We can find the product of any two _____ using the FOIL method.

RC6. The product of the sum and the difference of the same two terms is the _____ of their squares.

ascending
binomial(s)
descending
difference
last
outside
product
square

Concept Check Choose from the column on the right the best pattern to use for each multiplication.

CC1. $(2x + 3)(2x + 3)$

CC2. $(5x + 7)(5x - 7)$

CC3. $(3x - 4)(3x + 5)$

CC4. $(4x - 1)(4x - 1)$

a) $(A + B)(C + D) = AC + AD + BC + BD$
b) $(A + B)(A - B) = A^2 - B^2$
c) $(A + B)^2 = A^2 + 2AB + B^2$
d) $(A - B)^2 = A^2 - 2AB + B^2$

a Multiply. Try to write only the answer. If you need more steps, be sure to use them.

1. $(x + 1)(x^2 + 3)$
2. $(x^2 - 3)(x - 1)$
3. $(x^3 + 2)(x + 1)$
4. $(x^4 + 2)(x + 10)$

5. $(y + 2)(y - 3)$
6. $(a + 2)(a + 3)$
7. $(3x + 2)(3x + 2)$
8. $(4x + 1)(4x + 1)$

9. $(5x - 6)(x + 2)$
10. $(x - 8)(x + 8)$
11. $(3t - 1)(3t + 1)$
12. $(2m + 3)(2m + 3)$

13. $(4x - 2)(x - 1)$
14. $(2x - 1)(3x + 1)$
15. $(p - \frac{1}{4})(p + \frac{1}{4})$
16. $(q + \frac{3}{4})(q + \frac{3}{4})$

17. $(x - 0.1)(x + 0.1)$
18. $(x + 0.3)(x - 0.4)$
19. $(2x^2 + 6)(x + 1)$
20. $(2x^2 + 3)(2x - 1)$

21. $(-2x + 1)(x + 6)$
22. $(3x + 4)(2x - 4)$
23. $(a + 7)(a + 7)$
24. $(2y + 5)(2y + 5)$

25. $(1 + 2x)(1 - 3x)$
26. $(-3x - 2)(x + 1)$
27. $(\frac{3}{8}y - \frac{5}{6})(\frac{3}{8}y - \frac{5}{6})$
28. $(\frac{1}{5}x - \frac{2}{7})(\frac{1}{5}x + \frac{2}{7})$

29. $(x^2 + 3)(x^3 - 1)$
30. $(x^4 - 3)(2x + 1)$
31. $(3x^2 - 2)(x^4 - 2)$

32. $(x^{10} + 3)(x^{10} - 3)$
33. $(2.8x - 1.5)(4.7x + 9.3)$
34. $(x - \frac{3}{8})(x + \frac{4}{7})$

35. $(3x^5 + 2)(2x^2 + 6)$
36. $(1 - 2x)(1 + 3x^2)$
37. $(4x^2 + 3)(x - 3)$

38. $(7x - 2)(2x - 7)$
39. $(4y^4 + y^2)(y^2 + y)$
40. $(5y^6 + 3y^3)(2y^6 + 2y^3)$

b Multiply mentally, if possible. If you need extra steps, be sure to use them.

41. $(x + 4)(x - 4)$
42. $(x + 1)(x - 1)$
43. $(2x + 1)(2x - 1)$
44. $(x^2 + 1)(x^2 - 1)$

45. $(5m - 2)(5m + 2)$
46. $(3x^4 + 2)(3x^4 - 2)$
47. $(2x^2 + 3)(2x^2 - 3)$
48. $(6x^5 - 5)(6x^5 + 5)$

49. $(3x^4 - 4)(3x^4 + 4)$
50. $(t^2 - 0.2)(t^2 + 0.2)$

51. $(x^6 - x^2)(x^6 + x^2)$
52. $(2x^3 - 0.3)(2x^3 + 0.3)$

53. $(x^4 + 3x)(x^4 - 3x)$
54. $(\frac{3}{4} + 2x^3)(\frac{3}{4} - 2x^3)$
55. $(x^{12} - 3)(x^{12} + 3)$
56. $(12 - 3x^2)(12 + 3x^2)$

57. $(2y^8 + 3)(2y^8 - 3)$
58. $(m - \frac{2}{3})(m + \frac{2}{3})$

59. $(\frac{5}{8}x - 4.3)(\frac{5}{8}x + 4.3)$
60. $(10.7 - x^3)(10.7 + x^3)$

c Multiply mentally, if possible. If you need extra steps, be sure to use them.

61. $(x + 2)^2$ **62.** $(2x - 1)^2$ **63.** $(3x^2 + 1)^2$ **64.** $\left(3x + \frac{3}{4}\right)^2$

65. $\left(a - \frac{1}{2}\right)^2$ **66.** $\left(2a - \frac{1}{5}\right)^2$ **67.** $(3 + x)^2$ **68.** $(x^3 - 1)^2$

69. $(x^2 + 1)^2$ **70.** $(8x - x^2)^2$ **71.** $(2 - 3x^4)^2$ **72.** $(6x^3 - 2)^2$

73. $(5 + 6t^2)^2$ **74.** $(3p^2 - p)^2$ **75.** $\left(x - \frac{5}{8}\right)^2$ **76.** $(0.3y + 2.4)^2$

d Multiply mentally, if possible.

77. $(3 - 2x^3)^2$ **78.** $(x - 4x^3)^2$ **79.** $4x(x^2 + 6x - 3)$ **80.** $8x(-x^5 + 6x^2 + 9)$

81. $\left(2x^2 - \frac{1}{2}\right)\left(2x^2 - \frac{1}{2}\right)$ **82.** $(-x^2 + 1)^2$ **83.** $(-1 + 3p)(1 + 3p)$ **84.** $(-3q + 2)(3q + 2)$

85. $3t^2(5t^3 - t^2 + t)$ **86.** $-6x^2(x^3 + 8x - 9)$ **87.** $(6x^4 + 4)^2$ **88.** $(8a + 5)^2$

89. $(3x + 2)(4x^2 + 5)$ **90.** $(2x^2 - 7)(3x^2 + 9)$ **91.** $(8 - 6x^4)^2$ **92.** $\left(\frac{1}{5}x^2 + 9\right)\left(\frac{3}{5}x^2 - 7\right)$

93. $(t - 1)(t^2 + t + 1)$ **94.** $(y + 5)(y^2 - 5y + 25)$

Compute each of the following and compare.

95. $3^2 + 4^2$; $(3 + 4)^2$ **96.** $6^2 + 7^2$; $(6 + 7)^2$ **97.** $9^2 - 5^2$; $(9 - 5)^2$ **98.** $11^2 - 4^2$; $(11 - 4)^2$

Find the total area of all the shaded rectangles.

99. 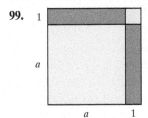   **100.**   **101.** 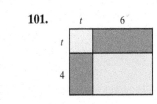   **102.**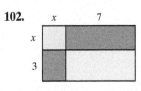

Skill Maintenance

Solve. [2.3c]

103. $3x - 8x = 4(7 - 8x)$

104. $3(x - 2) = 5(2x + 7)$

105. $5(2x - 3) - 2(3x - 4) = 20$

Solve. [2.4b]

106. $3x - 2y = 12$, for y

107. $C = ab - r$, for b

108. $3a - 5d = 4$, for a

Synthesis

Multiply.

109. $5x(3x - 1)(2x + 3)$

110. $[(2x - 3)(2x + 3)](4x^2 + 9)$

111. $[(a - 5)(a + 5)]^2$

112. $(a - 3)^2(a + 3)^2$
(*Hint*: Examine Exercise 111.)

113. $(3t^4 - 2)^2(3t^4 + 2)^2$
(*Hint*: Examine Exercise 111.)

114. $[3a - (2a - 3)][3a + (2a - 3)]$

Solve.

115. $(x + 2)(x - 5) = (x + 1)(x - 3)$

116. $(2x + 5)(x - 4) = (x + 5)(2x - 4)$

117. *Factors and Sums.* To *factor* a number is to express it as a product. Since $12 = 4 \cdot 3$, we say that 12 is *factored* and that 4 and 3 are *factors* of 12. In the following table, the top number has been factored in such a way that the sum of the factors is the bottom number. For example, in the first column, 40 has been factored as $5 \cdot 8$, and $5 + 8 = 13$, the bottom number. Such thinking is important in algebra when we factor trinomials of the type $x^2 + bx + c$. Find the missing numbers in the table.

PRODUCT	40	63	36	72	−140	−96	48	168	110			
FACTOR	5									−9	−24	−3
FACTOR	8									−10	18	
SUM	13	16	−20	−38	−4	4	−14	−29	−21			18

118. Consider the following rectangle.

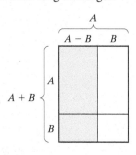

a) Find a polynomial for the area of the entire rectangle.
b) Find a polynomial for the sum of the areas of the two small unshaded rectangles.
c) Find a polynomial for the area in part (a) minus the area in part (b).
d) Find a polynomial for the area of the shaded region and compare this with the polynomial found in part (c).

Use the **TABLE** or **GRAPH** feature to check whether each of the following is correct.

119. $(x - 1)^2 = x^2 - 2x + 1$

120. $(x - 2)^2 = x^2 - 4x - 4$

121. $(x - 3)(x + 3) = x^2 - 6$

122. $(x - 3)(x + 2) = x^2 - x - 6$

Operations with Polynomials in Several Variables

4.7

OBJECTIVES

a Evaluate a polynomial in several variables for given values of the variables.

b Identify the coefficients and the degrees of the terms of a polynomial and the degree of a polynomial.

c Collect like terms of a polynomial in several variables.

d Add polynomials in several variables.

e Subtract polynomials in several variables.

f Multiply polynomials in several variables.

The polynomials that we have been studying have only one variable. A **polynomial in several variables** is an expression like those that you have already seen, but with more than one variable. Here are two examples:

$$3x + xy^2 + 5y + 4, \quad 8xy^2z - 2x^3z - 13x^4y^2 + 15.$$

a EVALUATING POLYNOMIALS

EXAMPLE 1 Evaluate the polynomial

$$4 + 3x + xy^2 + 8x^3y^3$$

when $x = -2$ and $y = 5$.

We replace x with -2 and y with 5:

$$\begin{aligned} 4 + 3x + xy^2 + 8x^3y^3 &= 4 + 3(-2) + (-2) \cdot 5^2 + 8(-2)^3 \cdot 5^3 \\ &= 4 + 3(-2) + (-2) \cdot 25 + 8(-8)(125) \\ &= 4 - 6 - 50 - 8000 \\ &= -8052. \end{aligned}$$

EXAMPLE 2 *Zoology.* The weight, in kilograms, of an elephant with a girth of g centimeters at the heart, a length of l centimeters, and a footpad circumference of f centimeters can be estimated by the polynomial

$$11.5g + 7.55l + 12.5f - 4016.$$

A field zoologist finds that the girth of a 3-year-old female elephant is 231 cm, the length is 135 cm, and the footpad circumference is 86 cm. Approximately how much does the elephant weigh?
Data: "How Much Does That Elephant Weigh?" by Mark MacAllister on fieldtripearth.org

We evaluate the polynomial for $g = 231$, $l = 135$, and $f = 86$:

$$11.5g + 7.55l + 12.5f - 4016 = 11.5(231) + 7.55(135) + 12.5(86) - 4016$$
$$= 734.75.$$

The elephant weighs about 735 kg.

Do Exercises 1–3.

1. Evaluate the polynomial
$$4 + 3x + xy^2 + 8x^3y^3$$
when $x = 2$ and $y = -5$.

2. Evaluate the polynomial
$$8xy^2 - 2x^3z - 13x^4y^2 + 5$$
when $x = -1$, $y = 3$, and $z = 4$.

3. *Zoology.* Refer to Example 2. A 25-year-old female elephant has a girth of 366 cm, a length of 226 cm, and a footpad circumference of 117 cm. How much does the elephant weigh?

Answers
1. -7940 2. -176 3. About 3362 kg

b COEFFICIENTS AND DEGREES

The **degree** of a term is the sum of the exponents of the variables. For example, the degree of $3x^5y^2$ is $5 + 2$, or 7. The **degree of a polynomial** is the degree of the term of highest degree.

EXAMPLE 3 Identify the coefficient and the degree of each term and the degree of the polynomial

$$9x^2y^3 - 14xy^2z^3 + xy + 4y + 5x^2 + 7.$$

TERM	COEFFICIENT	DEGREE	DEGREE OF THE POLYNOMIAL
$9x^2y^3$	9	5	
$-14xy^2z^3$	-14	6	6
xy	1	2	
$4y$	4	1	*Think:* $4y = 4y^1$.
$5x^2$	5	2	
7	7	0	*Think:* $7 = 7x^0$, or $7x^0y^0z^0$.

4. Identify the coefficient of each term:
$-3xy^2 + 3x^2y - 2y^3 + xy + 2.$

5. Identify the degree of each term and the degree of the polynomial
$4xy^2 + 7x^2y^3z^2 - 5x + 2y + 4.$

◀ Do Exercises 4 and 5.

c COLLECTING LIKE TERMS

Like terms have exactly the same variables with exactly the same exponents. For example,

$3x^2y^3$ and $-7x^2y^3$ are like terms;
$9x^4z^7$ and $12x^4z^7$ are like terms.

But

$13xy^5$ and $-2x^2y^5$ are *not* like terms, because the *x*-factors have different exponents;

and

$3xyz^2$ and $4xy$ are *not* like terms, because there is no factor of z^2 in the second expression.

Collecting like terms is based on the distributive laws.

EXAMPLES Collect like terms.

4. $5x^2y + 3xy^2 - 5x^2y - xy^2 = (5 - 5)x^2y + (3 - 1)xy^2 = 2xy^2$

5. $8a^2 - 2ab + 7b^2 + 4a^2 - 9ab - 17b^2 = 12a^2 - 11ab - 10b^2$

6. $7xy - 5xy^2 + 3xy^2 - 7 + 6x^3 + 9xy - 11x^3 + y - 1$
$= 16xy - 2xy^2 - 5x^3 + y - 8$

Collect like terms.
6. $4x^2y + 3xy - 2x^2y$

7. $-3pt - 5ptr^3 - 12 + 8pt + 5ptr^3 + 4$ **GS**

The like terms are $-3pt$ and ____, $-5ptr^3$ and ____, and -12 and ____.

Collecting like terms, we have
$(-3 +)pt + $
$(-5 +)\, ptr^3 + $
$(-12 +)$
$= - 8.$

◀ Do Exercises 6 and 7.

Answers
4. $-3, 3, -2, 1, 2$ **5.** $3, 7, 1, 1, 0; 7$
6. $2x^2y + 3xy$ **7.** $5pt - 8$
Guided Solution:
7. $8pt, 5ptr^3, 4, 8, 5, 4, 5pt$

d ADDITION

We can find the sum of two polynomials in several variables by writing a plus sign between them and then collecting like terms.

EXAMPLE 7 Add: $(-5x^3 + 3y - 5y^2) + (8x^3 + 4x^2 + 7y^2)$.

$$(-5x^3 + 3y - 5y^2) + (8x^3 + 4x^2 + 7y^2)$$
$$= (-5 + 8)x^3 + 4x^2 + 3y + (-5 + 7)y^2$$
$$= 3x^3 + 4x^2 + 3y + 2y^2$$

EXAMPLE 8 Add:

$$(5xy^2 - 4x^2y + 5x^3 + 2) + (3xy^2 - 2x^2y + 3x^3y - 5).$$

We have

$$(5xy^2 - 4x^2y + 5x^3 + 2) + (3xy^2 - 2x^2y + 3x^3y - 5)$$
$$= (5 + 3)xy^2 + (-4 - 2)x^2y + 5x^3 + 3x^3y + (2 - 5)$$
$$= 8xy^2 - 6x^2y + 5x^3 + 3x^3y - 3.$$

Do Exercises 8–10.

Add.
8. $(4x^3 + 4x^2 - 8y - 3) + (-8x^3 - 2x^2 + 4y + 5)$

9. $(13x^3y + 3x^2y - 5y) + (x^3y + 4x^2y - 3xy + 3y)$

10. $(-5p^2q^4 + 2p^2q^2 + 3q) + (6pq^2 + 3p^2q + 5)$

e SUBTRACTION

We subtract a polynomial by adding its opposite, or additive inverse. The opposite of the polynomial $4x^2y - 6x^3y^2 + x^2y^2 - 5y$ is

$$-(4x^2y - 6x^3y^2 + x^2y^2 - 5y) = -4x^2y + 6x^3y^2 - x^2y^2 + 5y.$$

EXAMPLE 9 Subtract:

$(4x^2y + x^3y^2 + 3x^2y^3 + 6y + 10) - (4x^2y - 6x^3y^2 + x^2y^2 - 5y - 8).$

We have

$(4x^2y + x^3y^2 + 3x^2y^3 + 6y + 10) - (4x^2y - 6x^3y^2 + x^2y^2 - 5y - 8)$
$= 4x^2y + x^3y^2 + 3x^2y^3 + 6y + 10 - 4x^2y + 6x^3y^2 - x^2y^2 + 5y + 8$
   Finding the opposite by changing the sign of each term
$= 7x^3y^2 + 3x^2y^3 - x^2y^2 + 11y + 18.$ Collecting like terms. (Try to write just the answer!)

Do Exercises 11 and 12.

Subtract.
11. $(-4s^4t + s^3t^2 + 2s^2t^3) - (4s^4t - 5s^3t^2 + s^2t^2)$

12. $(-5p^4q + 5p^3q^2 - 3p^2q^3 - 7q^4 - 2) - (4p^4q - 4p^3q^2 + p^2q^3 + 2q^4 - 7)$

f MULTIPLICATION

SKILL REVIEW
Use the product rule to multiply exponential expressions with like bases. [4.1d]
Simplify.
1. $(x^2y)(x^3y^4)$
2. $(ab^4c^3)(a^5)$

Answers: **1.** x^5y^5 **2.** $a^6b^4c^3$

MyLab Math
VIDEO

Answers
8. $-4x^3 + 2x^2 - 4y + 2$
9. $14x^3y + 7x^2y - 3xy - 2y$
10. $-5p^2q^4 + 2p^2q^2 + 3p^2q + 6pq^2 + 3q + 5$
11. $-8s^4t + 6s^3t^2 + 2s^2t^3 - s^2t^2$
12. $-9p^4q + 9p^3q^2 - 4p^2q^3 - 9q^4 + 5$

To multiply polynomials in several variables, we can multiply each term of one by every term of the other. We can use columns for long multiplications as with polynomials in one variable. We multiply each term at the top by every term at the bottom. We write like terms in columns, and then we add.

EXAMPLE 10 Multiply: $(3x^2y - 2xy + 3y)(xy + 2y)$.

$$
\begin{array}{r}
3x^2y - 2xy + 3y \\
xy + 2y \\
\hline
6x^2y^2 - 4xy^2 + 6y^2 \\
3x^3y^2 - 2x^2y^2 + 3xy^2 \\
\hline
3x^3y^2 + 4x^2y^2 - xy^2 + 6y^2
\end{array}
$$

Multiplying by $2y$
Multiplying by xy
Adding

◀ Do Exercises 13 and 14.

Multiply.
13. $(x^2y^3 + 2x)(x^3y^2 + 3x)$
14. $(p^4q - 2p^3q^2 + 3q^3)(p + 2q)$

Where appropriate, we use the special products that we have learned.

EXAMPLES Multiply.

11. $(x^2y + 2x)(xy^2 + y^2) = x^3y^3 + x^2y^3 + 2x^2y^2 + 2xy^2$ Using FOIL

12. $(p + 5q)(2p - 3q) = 2p^2 - 3pq + 10pq - 15q^2$ Using FOIL
$= 2p^2 + 7pq - 15q^2$

$(A + B)^2 = A^2 + 2 \cdot A \cdot B + B^2$

13. $(3x + 2y)^2 = (3x)^2 + 2(3x)(2y) + (2y)^2 = 9x^2 + 12xy + 4y^2$

$(A - B)^2 = A^2 - 2 \cdot A \cdot B + B^2$

14. $(2y^2 - 5x^2y)^2 = (2y^2)^2 - 2(2y^2)(5x^2y) + (5x^2y)^2$
$= 4y^4 - 20x^2y^3 + 25x^4y^2$

$(A + B)(A - B) = A^2 - B^2$

Multiply.
15. $(3xy + 2x)(x^2 + 2xy^2)$
16. $(x - 3y)(2x - 5y)$
17. $(4x + 5y)^2$
18. $(3x^2 - 2xy^2)^2$
19. $(2xy^2 + 3x)(2xy^2 - 3x)$
20. $(3xy^2 + 4y)(-3xy^2 + 4y)$
21. $(3y + 4 - 3x)(3y + 4 + 3x)$

15. $(3x^2y + 2y)(3x^2y - 2y) = (3x^2y)^2 - (2y)^2 = 9x^4y^2 - 4y^2$
16. $(-2x^3y^2 + 5t)(2x^3y^2 + 5t) = (5t - 2x^3y^2)(5t + 2x^3y^2)$

The sum and the difference of the same two terms

$= (5t)^2 - (2x^3y^2)^2 = 25t^2 - 4x^6y^4$

$(A - B)(A + B) = A^2 - B^2$

17. $(2x + 3 - 2y)(2x + 3 + 2y) = (2x + 3)^2 - (2y)^2$
$= 4x^2 + 12x + 9 - 4y^2$

22. $(2a + 5b + c)(2a - 5b - c)$
$= [2a + (5b + c)][2a - ()]$
$= (2a)^2 - ()^2$
$= - (25b^2 + 10bc +)$
$= 4a^2 - 25b^2 - 10bc - $

Remember that FOIL will always work when you are multiplying binomials. You can use it instead of the rules for special products, but those rules will make your work go faster.

◀ Do Exercises 15–22.

.... **Caution!** ....

Do not add exponents when collecting like terms—that is,

$7x^3 + 8x^3 \neq 15x^6$; ← Adding exponents is incorrect.
$7x^3 + 8x^3 = 15x^3$. ← Correct

Do add exponents when multiplying terms with like bases—that is,

$(7x^3)(8x^3) = 56x^6$. ← Correct

Answers
13. $x^5y^5 + 2x^4y^2 + 3x^3y^3 + 6x^2$
14. $p^5q - 4p^3q^3 + 3pq^3 + 6q^4$
15. $3x^3y + 6x^2y^3 + 2x^3 + 4x^2y^2$
16. $2x^2 - 11xy + 15y^2$
17. $16x^2 + 40xy + 25y^2$
18. $9x^4 - 12x^3y^2 + 4x^2y^4$
19. $4x^2y^4 - 9x^2$
20. $16y^2 - 9x^2y^4$
21. $9y^2 + 24y + 16 - 9x^2$
22. $4a^2 - 25b^2 - 10bc - c^2$

Guided Solution:
22. $5b + c, 5b + c, 4a^2, c^2, c^2$

4.7 Exercise Set

✓ Check Your Understanding

Reading Check Determine whether each sentence is true or false.

RC1. The variables in the polynomial $8x - xy + t^2 - xy^2$ are $t, x,$ and y.

RC2. The degree of the term $4xy$ is 4.

RC3. The terms $3x^2y$ and $3xy^2$ are like terms.

RC4. When we collect like terms, we add the exponents of the variables.

Concept Check Answer each of the following questions with reference to the polynomial
$$3ax^2 - 2xy^3 - axy + x^3y + 7ax^2.$$

CC1. What are the variables in the polynomial?

CC2. How many terms are in the polynomial?

CC3. Which term has a coefficient of -1?

CC4. Which terms are of degree 3?

CC5. List any pairs of like terms.

a Evaluate the polynomial when $x = 3, y = -2,$ and $z = -5$.

1. $x^2 - y^2 + xy$
2. $x^2 + y^2 - xy$
3. $x^2 - 3y^2 + 2xy$
4. $x^2 - 4xy + 5y^2$
5. $8xyz$
6. $-3xyz^2$
7. $xyz^2 - z$
8. $xy - xz + yz$

9. *Lung Capacity.* The polynomial equation
$$C = 0.041h - 0.018A - 2.69$$
can be used to estimate the lung capacity C, in liters, of a person of height h, in centimeters, and age A, in years. Find the lung capacity of a 20-year-old person who is 165 cm tall.

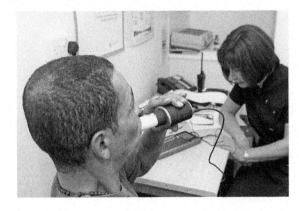

10. *Altitude of a Launched Object.* The altitude h, in meters, of a launched object is given by the polynomial equation
$$h = h_0 + vt - 4.9t^2,$$
where h_0 is the height, in meters, from which the launch occurs, v is the initial upward speed (or velocity), in meters per second (m/s), and t is the number of seconds for which the object is airborne. A rock is thrown upward from the top of the Lands End Arch, near San Lucas, Baja, Mexico, 32 m above the ground. The upward speed is 10 m/s. How high will the rock be 3 sec after it has been thrown?

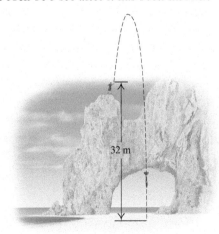

32 m

11. *Male Caloric Needs.* The number of calories needed each day by a moderately active man who weighs w kilograms, is h centimeters tall, and is a years old can be estimated by the polynomial
$$19.18w + 7h - 9.52a + 92.4.$$
Steve is moderately active, weighs 82 kg, is 185 cm tall, and is 67 years old. What is his daily caloric need?

Data: Parker, M., *She Does Math*. Mathematical Association of America

12. *Female Caloric Needs.* The number of calories needed each day by a moderately active woman who weighs w pounds, is h inches tall, and is a years old can be estimated by the polynomial
$$917 + 6w + 6h - 6a.$$
Christine is moderately active, weighs 125 lb, is 64 in. tall, and is 27 years old. What is her daily caloric need?

Data: Parker, M., *She Does Math*. Mathematical Association of America

Surface Area of a Right Circular Cylinder. The surface area S of a right circular cylinder is given by the polynomial equation $S = 2\pi rh + 2\pi r^2$, where h is the height and r is the radius of the base. Use this formula for Exercises 13 and 14.

13. A 12-oz beverage can has a height of 4.7 in. and a radius of 1.2 in. Find the surface area of the can. Use 3.14 for π.

14. A 26-oz coffee can has a height of 6.5 in. and a radius of 2.5 in. Find the surface area of the can. Use 3.14 for π.

Surface Area of a Silo. A silo is a structure that is shaped like a right circular cylinder with a half sphere on top. The surface area S of a silo of height h and radius r (including the area of the base) is given by the polynomial equation $S = 2\pi rh + \pi r^2$. Note that h is the height of the entire silo.

15. A coffee grinder is shaped like a silo, with a height of 7 in. and a radius of $1\frac{1}{2}$ in. Find the surface area of the coffee grinder. Use 3.14 for π.

16. A $1\frac{1}{2}$-oz bottle of roll-on deodorant has a height of 4 in. and a radius of $\frac{3}{4}$ in. Find the surface area of the bottle if the bottle is shaped like a silo. Use 3.14 for π.

b Identify the coefficient and the degree of each term of the polynomial. Then find the degree of the polynomial.

17. $x^3y - 2xy + 3x^2 - 5$

18. $5x^2y^2 - y^2 + 15xy + 1$

19. $17x^2y^3 - 3x^3yz - 7$

20. $6 - xy + 8x^2y^2 - y^5$

c Collect like terms.

21. $a + b - 2a - 3b$

22. $xy^2 - 1 + y - 6 - xy^2$

23. $3x^2y - 2xy^2 + x^2$

24. $m^3 + 2m^2n - 3m^2 + 3mn^2$

25. $6au + 3av + 14au + 7av$

26. $3x^2y - 2z^2y + 3xy^2 + 5z^2y$

27. $2u^2v - 3uv^2 + 6u^2v - 2uv^2$

28. $3x^2 + 6xy + 3y^2 - 5x^2 - 10xy - 5y^2$

d Add.

29. $(2x^2 - xy + y^2) + (-x^2 - 3xy + 2y^2)$

30. $(2zt - z^2 + 5t^2) + (z^2 - 3zt + t^2)$

31. $(r - 2s + 3) + (2r + s) + (s + 4)$

32. $(ab - 2a + 3b) + (5a - 4b) + (3a + 7ab - 8b)$

33. $(b^3a^2 - 2b^2a^3 + 3ba + 4) + (b^2a^3 - 4b^3a^2 + 2ba - 1)$

34. $(2x^2 - 3xy + y^2) + (-4x^2 - 6xy - y^2) + (x^2 + xy - y^2)$

e Subtract.

35. $(a^3 + b^3) - (a^2b - ab^2 + b^3 + a^3)$

36. $(x^3 - y^3) - (-2x^3 + x^2y - xy^2 + 2y^3)$

37. $(xy - ab - 8) - (xy - 3ab - 6)$

38. $(3y^4x^2 + 2y^3x - 3y - 7) - (2y^4x^2 + 2y^3x - 4y - 2x + 5)$

39. $(-2a + 7b - c) - (-3b + 4c - 8d)$

40. Subtract $5a + 2b$ from the sum of $2a + b$ and $3a - b$.

f Multiply.

41. $(3z - u)(2z + 3u)$

42. $(a - b)(a^2 + b^2 + 2ab)$

43. $(a^2b - 2)(a^2b - 5)$

44. $(xy + 7)(xy - 4)$

45. $(a^3 + bc)(a^3 - bc)$

46. $(m^2 + n^2 - mn)(m^2 + mn + n^2)$

47. $(y^4x + y^2 + 1)(y^2 + 1)$

48. $(a - b)(a^2 + ab + b^2)$

49. $(3xy - 1)(4xy + 2)$

50. $(m^3n + 8)(m^3n - 6)$

51. $(3 - c^2d^2)(4 + c^2d^2)$

52. $(6x - 2y)(5x - 3y)$

53. $(m^2 - n^2)(m + n)$

54. $(pq + 0.2) \times (0.4pq - 0.1)$

55. $(xy + x^5y^5) \times (x^4y^4 - xy)$

56. $(x - y^3)(2y^3 + x)$

57. $(x + h)^2$

58. $(y - a)^2$

59. $(3a + 2b)^2$

60. $(2ab - cd)^2$

61. $(r^3t^2 - 4)^2$

62. $(3a^2b - b^2)^2$

63. $(p^4 + m^2n^2)^2$

64. $\left(2a^3 - \frac{1}{2}b^3\right)^2$

65. $3a(a - 2b)^2$

66. $-3x(x + 8y)^2$

67. $(m + n - 3)^2$

68. $(a^2 + b + 2)^2$

69. $(a + b)(a - b)$

70. $(x - y)(x + y)$

71. $(2a - b)(2a + b)$

72. $(w + 3z)(w - 3z)$

73. $(c^2 - d)(c^2 + d)$

74. $(p^3 - 5q)(p^3 + 5q)$

75. $(ab + cd^2) \times (ab - cd^2)$

76. $(xy + pq) \times (xy - pq)$

77. $(x + y - 3)(x + y + 3)$

78. $(p + q + 4)(p + q - 4)$

79. $[x + y + z][x - (y + z)]$

80. $[a + b + c][a - (b + c)]$

81. $(a + b + c)(a + b - c)$

82. $(3x + 2 - 5y)(3x + 2 + 5y)$

83. $(x^2 - 4y + 2)(3x^2 + 5y - 3)$

84. $(2x^2 - 7y + 4)(x^2 + y - 3)$

Skill Maintenance

In which quadrant is each point located? [3.1a]

85. $(2, -5)$ **86.** $(-8, -9)$ **87.** $(16, 23)$ **88.** $(-3, 2)$

89. Find the absolute value: $|-39|$. [1.2e]

90. Convert $\frac{9}{8}$ to decimal notation. [1.2c]

91. Use either $<$ or $>$ for ☐ to write a true sentence: $-17\ \square\ -5$. [1.2d]

92. Evaluate $-(-x)$ when $x = -3$. [1.3b]

Synthesis

Find a polynomial for each shaded area. (Leave results in terms of π where appropriate.)

93.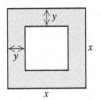

94.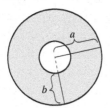

95.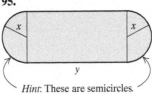
Hint: These are semicircles.

96.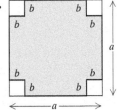

Find a formula for the surface area of each solid object. Leave results in terms of π.

97.

98.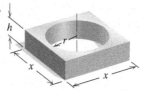

99. *Observatory Paint Costs.* The observatory at Danville University is shaped like a silo that is 40 ft high and 30 ft wide. (See Exercise 15.) The Heavenly Bodies Astronomy Club is to paint the exterior of the observatory using paint that covers 250 ft² per gallon. How many gallons should they purchase?

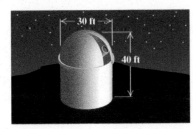

100. *Interest Compounded Annually.* An amount of money P that is invested at the yearly interest rate r grows to the amount
$$P(1 + r)^t$$
after t years. Find a polynomial that can be used to determine the amount to which P will grow after 2 years.

101. Suppose that $10,400 is invested at 3.5%, compounded annually. How much is in the account at the end of 5 years? (See Exercise 100.)

102. Multiply: $(x + a)(x - b)(x - a)(x + b)$.

4.8 Division of Polynomials

OBJECTIVES

a Divide a polynomial by a monomial.

b Divide a polynomial by a divisor that is a binomial.

Division of polynomials is similar to what is done in arithmetic.

a DIVIDING BY A MONOMIAL

SKILL REVIEW *Divide integers.* [1.6a]
Divide.

1. $\dfrac{20}{4}$

2. $\dfrac{-30}{5}$

Answers: 1. 5 2. −6

We first consider division by a monomial. When dividing a monomial by a monomial, we use the quotient rule to subtract exponents when the bases are the same. We also divide the coefficients.

EXAMPLES Divide.

1. $\dfrac{10x^2}{2x} = \dfrac{10}{2} \cdot \dfrac{x^2}{x} = 5x^{2-1} = 5x$

   Caution! The coefficients are divided, but the exponents are subtracted.

2. $\dfrac{x^9}{3x^2} = \dfrac{1x^9}{3x^2} = \dfrac{1}{3} \cdot \dfrac{x^9}{x^2} = \dfrac{1}{3}x^{9-2} = \dfrac{1}{3}x^7$

3. $\dfrac{-18x^{10}}{3x^3} = \dfrac{-18}{3} \cdot \dfrac{x^{10}}{x^3} = -6x^{10-3} = -6x^7$

4. $\dfrac{42a^2b^5}{-3ab^2} = \dfrac{42}{-3} \cdot \dfrac{a^2}{a} \cdot \dfrac{b^5}{b^2} = -14a^{2-1}b^{5-2} = -14ab^3$

◀ Do Exercises 1–4.

Divide.

1. $\dfrac{20x^3}{5x}$

2. $\dfrac{-28x^{14}}{4x^3}$

3. $\dfrac{-56p^5q^7}{2p^2q^6}$

4. $\dfrac{x^5}{4x}$

To divide a polynomial by a monomial, we note that since

$$\dfrac{A}{C} + \dfrac{B}{C} = \dfrac{A+B}{C},$$

it follows that

$$\dfrac{A+B}{C} = \dfrac{A}{C} + \dfrac{B}{C}.$$ Switching the left and right sides of the equation

This is actually the procedure that we use when performing divisions like 86 ÷ 2. Although we might write

$$\dfrac{86}{2} = 43,$$

we could also calculate as follows:

$$\dfrac{86}{2} = \dfrac{80+6}{2} = \dfrac{80}{2} + \dfrac{6}{2} = 40 + 3 = 43.$$

Similarly, to divide a polynomial by a monomial, we divide each term by the monomial.

Answers

1. $4x^2$ 2. $-7x^{11}$ 3. $-28p^3q$ 4. $\dfrac{1}{4}x^4$

EXAMPLE 5 Divide: $(9x^8 + 12x^6) \div (3x^2)$.

We have

$$(9x^8 + 12x^6) \div (3x^2) = \frac{9x^8 + 12x^6}{3x^2}$$

$$= \frac{9x^8}{3x^2} + \frac{12x^6}{3x^2}.$$ To see this, add and get the original expression.

We now perform the separate divisions:

$$\frac{9x^8}{3x^2} + \frac{12x^6}{3x^2} = \frac{9}{3} \cdot \frac{x^8}{x^2} + \frac{12}{3} \cdot \frac{x^6}{x^2}$$

$$= 3x^{8-2} + 4x^{6-2}$$

$$= 3x^6 + 4x^4.$$

Caution!

The coefficients are *divided*, but the exponents are *subtracted*.

To check, we multiply the quotient, $3x^6 + 4x^4$, by the divisor, $3x^2$:

$$3x^2(3x^6 + 4x^4) = (3x^2)(3x^6) + (3x^2)(4x^4) = 9x^8 + 12x^6.$$

This is the polynomial that was being divided, so our answer is $3x^6 + 4x^4$.

Do Exercises 5–7. ▶

EXAMPLE 6 Divide and check: $(10a^5b^4 - 2a^3b^2 + 6a^2b) \div (-2a^2b)$.

$$\frac{10a^5b^4 - 2a^3b^2 + 6a^2b}{-2a^2b} = \frac{10a^5b^4}{-2a^2b} - \frac{2a^3b^2}{-2a^2b} + \frac{6a^2b}{-2a^2b}$$

$$= \frac{10}{-2} \cdot a^{5-2}b^{4-1} - \frac{2}{-2} \cdot a^{3-2}b^{2-1} + \frac{6}{-2}$$

$$= -5a^3b^3 + ab - 3$$

Check: $-2a^2b(-5a^3b^3 + ab - 3) = (-2a^2b)(-5a^3b^3) + (-2a^2b)(ab) - (-2a^2b)(3)$

$$= 10a^5b^4 - 2a^3b^2 + 6a^2b$$

Our answer, $-5a^3b^3 + ab - 3$, checks.

> To divide a polynomial by a monomial, divide each term of the polynomial by the monomial.

Do Exercises 8 and 9. ▶

b DIVIDING BY A BINOMIAL

Let's first consider long division as it is performed in arithmetic. We review this by considering the division $3711 \div 8$.

$$\begin{array}{r} 4 \\ 8 \overline{)3711} \\ \underline{32} \\ 51 \end{array}$$

① Divide: $37 \div 8 \approx 4$.
② Multiply: $4 \times 8 = 32$.
③ Subtract: $37 - 32 = 5$.
④ Bring down the 1.

$$\begin{array}{r} 463 \\ 8 \overline{)3711} \\ \underline{32} \\ 51 \\ \underline{48} \\ 31 \\ \underline{24} \\ 7 \end{array}$$

Divide. Check the result.

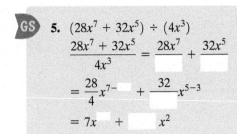

5. $(28x^7 + 32x^5) \div (4x^3)$

$$\frac{28x^7 + 32x^5}{4x^3} = \frac{28x^7}{\square} + \frac{32x^5}{\square}$$

$$= \frac{28}{4}x^{7-\square} + \frac{32}{\square}x^{5-3}$$

$$= 7x^{\square} + \square x^2$$

6. $(2x^3 + 6x^2 + 4x) \div (2x)$

7. $(6x^2 + 3x - 2) \div 3$

Divide and check.

8. $(8x^2 - 3x + 1) \div (-2)$

9. $\dfrac{2x^4y^6 - 3x^3y^4 + 5x^2y^3}{x^2y^2}$

Answers

5. $7x^4 + 8x^2$ 6. $x^2 + 3x + 2$
7. $2x^2 + x - \dfrac{2}{3}$ 8. $-4x^2 + \dfrac{3}{2}x - \dfrac{1}{2}$
9. $2x^2y^4 - 3xy^2 + 5y$

Guided Solution:
5. $4x^3, 4x^3, 3, 4, 4, 8$

To carry out long division:
1. Divide,
2. Multiply,
3. Subtract, and
4. Bring down the next number or term.

To complete the division, we repeat the procedure at left two more times. We obtain the complete division as shown on the preceding page. The quotient is 463. The remainder is 7, expressed as R = 7. We write the answer as

$$463 \text{ R } 7 \quad \text{or} \quad 463 + \frac{7}{8} = 463\frac{7}{8}.$$

We check the answer, 463 R 7, by multiplying the quotient, 463, by the divisor, 8, and adding the remainder, 7, to see if we get the dividend, 3711:

$$8 \cdot 463 + 7 = 3704 + 7 = 3711.$$

Now let's look at long division with polynomials. We use this procedure when the divisor is not a monomial. We write polynomials in descending order and then write in missing terms, if necessary.

EXAMPLE 7 Divide $x^2 + 5x + 6$ by $x + 2$.

$$\begin{array}{r} x \\ x+2\overline{)x^2 + 5x + 6} \\ \underline{x^2 + 2x} \\ 3x \end{array}$$

— Divide the first term by the first term: $x^2/x = x$. Ignore the term 2 for this step.
— Multiply x above by the divisor, $x + 2$.
— Subtract: $(x^2 + 5x) - (x^2 + 2x) = x^2 + 5x - x^2 - 2x = 3x$.

We now "bring down" the next term of the dividend—in this case, 6.

$$\begin{array}{r} x + 3 \\ x+2\overline{)x^2 + 5x + 6} \\ \underline{x^2 + 2x} \\ 3x + 6 \\ \underline{3x + 6} \\ 0 \end{array}$$

— Divide the first term of $3x + 6$ by the first term of the divisor: $3x/x = 3$.
— The 6 has been "brought down."
— Multiply 3 above by the divisor, $x + 2$.
— Subtract: $(3x + 6) - (3x + 6) = 3x + 6 - 3x - 6 = 0$.

The quotient is $x + 3$. The remainder is 0. A remainder of 0 is generally not included in an answer.

To check, we multiply the quotient by the divisor and add the remainder, if any, to see if we get the dividend:

Divisor   Quotient   Remainder   Dividend
$(x + 2) \cdot (x + 3) \; + \quad 0 \quad = x^2 + 5x + 6.$ The division checks.

◀ Do Exercise 10.

10. Divide and check:
$(x^2 + x - 6) \div (x + 3)$.

$$\begin{array}{r} \underline{\quad - \quad} \\ x+3\overline{)x^2 + x - 6} \\ \underline{x^2 + \square} \\ \square - 6 \\ \underline{-2x - \square} \\ \end{array}$$

EXAMPLE 8 Divide and check: $(x^2 + 2x - 12) \div (x - 3)$.

$$\begin{array}{r} x \\ x-3\overline{)x^2 + 2x - 12} \\ \underline{x^2 - 3x} \\ 5x \end{array}$$

— Divide the first term by the first term: $x^2/x = x$.
— Multiply x above by the divisor, $x - 3$.
— Subtract: $(x^2 + 2x) - (x^2 - 3x) = x^2 + 2x - x^2 + 3x = 5x$.

We now "bring down" the next term of the dividend—in this case, -12.

$$\begin{array}{r} x + 5 \\ x-3\overline{)x^2 + 2x - 12} \\ \underline{x^2 - 3x} \\ 5x - 12 \\ \underline{5x - 15} \\ 3 \end{array}$$

— Divide the first term of $5x - 12$ by the first term of the divisor: $5x/x = 5$.
— Bring down the -12.
— Multiply 5 above by the divisor, $x - 3$.
— Subtract: $(5x - 12) - (5x - 15) = 5x - 12 - 5x + 15 = 3$.

Answers
10. $x - 2$
Guided Solution:
10. $\begin{array}{r} x - 2 \\ x+3\overline{)x^2 + x - 6} \\ \underline{x^2 + 3x - 6} \\ -2x - 6 \\ \underline{-2x - 6} \\ 0 \end{array}$

326   CHAPTER 4   Polynomials: Operations

The answer is $x + 5$ with $R = 3$, or

$$\underbrace{x + 5}_{\text{Quotient}} + \underbrace{\dfrac{\overset{\text{Remainder}}{3}}{\underset{\text{Divisor}}{x - 3}}}.$$

(This is the way that answers will be given at the back of the book.)

Check: We can check by multiplying the divisor by the quotient and adding the remainder, as follows:

$$(x - 3)(x + 5) + 3 = x^2 + 2x - 15 + 3$$
$$= x^2 + 2x - 12.$$

When dividing, an answer may "come out even" (that is, have a remainder of 0, as in Example 7), or it may not (as in Example 8). **If a remainder is not 0, we continue dividing until the degree of the remainder is less than the degree of the divisor.**

Do Exercises 11 and 12. ▶

EXAMPLE 9 Divide and check: $(x^3 + 1) \div (x + 1)$.

$$\begin{array}{r}
x^2 - x + 1 \\
x + 1 \overline{\smash{)}\ x^3 + 0x^2 + 0x + 1} \\
\underline{x^3 + x^2 } \\
-x^2 + 0x \\
\underline{-x^2 - x } \\
x + 1 \\
\underline{x + 1} \\
0
\end{array}$$

← Fill in the missing terms. (See Section 4.3.)
— Subtract: $x^3 - (x^3 + x^2) = -x^2$.

— Subtract: $-x^2 - (-x^2 - x) = x$.

— Subtract: $(x + 1) - (x + 1) = 0$.

The answer is $x^2 - x + 1$. The check is left to the student.

EXAMPLE 10 Divide and check: $(9x^4 - 7x^2 - 4x + 13) \div (3x - 1)$.

$$\begin{array}{r}
3x^3 + x^2 - 2x - 2 \\
3x - 1 \overline{\smash{)}\ 9x^4 + 0x^3 - 7x^2 - 4x + 13} \\
\underline{9x^4 - 3x^3 } \\
3x^3 - 7x^2 \\
\underline{3x^3 - x^2 } \\
-6x^2 - 4x \\
\underline{-6x^2 + 2x } \\
-6x + 13 \\
\underline{-6x + 2} \\
11
\end{array}$$

← Fill in the missing term.
— Subtract: $9x^4 - (9x^4 - 3x^3) = 3x^3$.

— Subtract: $(3x^3 - 7x^2) - (3x^3 - x^2) = -6x^2$.

— Subtract: $(-6x^2 - 4x) - (-6x^2 + 2x) = -6x$.

— Subtract: $(-6x + 13) - (-6x + 2) = 11$.

The answer is $3x^3 + x^2 - 2x - 2$ with $R = 11$, or

$$3x^3 + x^2 - 2x - 2 + \dfrac{11}{3x - 1}.$$

Check: $(3x - 1)(3x^3 + x^2 - 2x - 2) + 11$
$= 9x^4 + 3x^3 - 6x^2 - 6x - 3x^3 - x^2 + 2x + 2 + 11$
$= 9x^4 - 7x^2 - 4x + 13$

Do Exercises 13 and 14. ▶

Divide and check.
11. $x - 2 \overline{\smash{)}\ x^2 + 2x - 8}$

12. $x + 3 \overline{\smash{)}\ x^2 + 7x + 10}$

Divide and check.
13. $(x^3 - 1) \div (x - 1)$

14. $(8x^4 + 10x^2 + 2x + 9) \div (4x + 2)$

Answers
11. $x + 4$ 12. $x + 4$ with $R = -2$, or
$x + 4 + \dfrac{-2}{x + 3}$ 13. $x^2 + x + 1$
14. $2x^3 - x^2 + 3x - 1$ with $R = 11$, or
$2x^3 - x^2 + 3x - 1 + \dfrac{11}{4x + 2}$

SECTION 4.8 Division of Polynomials

4.8 Exercise Set

✓ Check Your Understanding

Reading Check Choose from the list on the right the appropriate word(s) to complete each statement. A word may be used more than once.

RC1. When dividing a monomial by a monomial, we _____ exponents and _____ coefficients.

RC2. To divide a polynomial by a monomial, we _____ each term by the monomial.

RC3. To carry out long division, we repeat the following process: divide, _____, _____, and bring down the next term.

RC4. To check division, we _____ the divisor and the quotient, and then _____ the remainder.

add
subtract
multiply
divide

Concept Check Place the dividend and the divisor appropriately, making sure that they are written in the correct form. Do not carry out the division.

CC1. $(x^2 + 5x - 6) \div (x - 1)$

CC2. $\dfrac{x^2 + x + 1}{x - 3}$

CC3. $(x^3 - 4) \div (x - 2)$

a Divide and check.

1. $\dfrac{24x^4}{8}$

2. $\dfrac{-2u^2}{u}$

3. $\dfrac{25x^3}{5x^2}$

4. $\dfrac{16x^7}{-2x^2}$

5. $\dfrac{-54x^{11}}{-3x^8}$

6. $\dfrac{-75a^{10}}{3a^2}$

7. $\dfrac{64a^5b^4}{16a^2b^3}$

8. $\dfrac{-34p^{10}q^{11}}{-17pq^9}$

9. $\dfrac{24x^4 - 4x^3 + x^2 - 16}{8}$

10. $\dfrac{12a^4 - 3a^2 + a - 6}{6}$

11. $\dfrac{u - 2u^2 - u^5}{u}$

12. $\dfrac{50x^5 - 7x^4 + x^2}{x}$

328 CHAPTER 4 Polynomials: Operations

13. $(15t^3 + 24t^2 - 6t) \div (3t)$

14. $(25t^3 + 15t^2 - 30t) \div (5t)$

15. $(20x^6 - 20x^4 - 5x^2) \div (-5x^2)$

16. $(24x^6 + 32x^5 - 8x^2) \div (-8x^2)$

17. $(24x^5 - 40x^4 + 6x^3) \div (4x^3)$

18. $(18x^6 - 27x^5 - 3x^3) \div (9x^3)$

19. $\dfrac{18x^2 - 5x + 2}{2}$

20. $\dfrac{15x^2 - 30x + 6}{3}$

21. $\dfrac{12x^3 + 26x^2 + 8x}{2x}$

22. $\dfrac{2x^4 - 3x^3 + 5x^2}{x^2}$

23. $\dfrac{9r^2s^2 + 3r^2s - 6rs^2}{3rs}$

24. $\dfrac{4x^4y - 8x^6y^2 + 12x^8y^6}{4x^4y}$

b Divide.

25. $(x^2 + 4x + 4) \div (x + 2)$

26. $(x^2 - 6x + 9) \div (x - 3)$

27. $(x^2 - 10x - 25) \div (x - 5)$

28. $(x^2 + 8x - 16) \div (x + 4)$

29. $(x^2 + 4x - 14) \div (x + 6)$

30. $(x^2 + 5x - 9) \div (x - 2)$

31. $\dfrac{x^2 - 9}{x + 3}$

32. $\dfrac{x^2 - 25}{x - 5}$

33. $\dfrac{x^5 + 1}{x + 1}$

34. $\dfrac{x^4 - 81}{x - 3}$

35. $\dfrac{8x^3 - 22x^2 - 5x + 12}{4x + 3}$

36. $\dfrac{2x^3 - 9x^2 + 11x - 3}{2x - 3}$

37. $(x^6 - 13x^3 + 42) \div (x^3 - 7)$

38. $(x^6 + 5x^3 - 24) \div (x^3 - 3)$

39. $(t^3 - t^2 + t - 1) \div (t - 1)$

40. $(y^3 + 3y^2 - 5y - 15) \div (y + 3)$

41. $(y^3 - y^2 - 5y - 3) \div (y + 2)$

42. $(t^3 - t^2 + t - 1) \div (t + 1)$

SECTION 4.8 Division of Polynomials

43. $(15x^3 + 8x^2 + 11x + 12) \div (5x + 1)$

44. $(20x^4 - 2x^3 + 5x + 3) \div (2x - 3)$

45. $(12y^3 + 42y^2 - 10y - 41) \div (2y + 7)$

46. $(15y^3 - 27y^2 - 35y + 60) \div (5y - 9)$

Skill Maintenance

Solve.

47. $-13 = 8d - 5$ [2.3a]

48. $x + \frac{1}{2}x = 5$ [2.3b]

49. $4(x - 3) = 5(2 - 3x) + 1$ [2.3c]

50. $3(r + 1) - 5(r + 2) \geq 15 - (r + 7)$ [2.7e]

51. The number of patients with the flu who were treated at Riverview Clinic increased from 25 one week to 60 the next week. What was the percent increase? [2.5a]

52. Todd's quiz grades are 82, 88, 93, and 92. Determine (in terms of an inequality) what scores on the last quiz will allow him to get an average quiz grade of at least 90.   [2.8b]

53. The perimeter of a rectangle is 640 ft. The length is 15 ft more than the width. Find the area of the rectangle.   [2.6a]

54. *Book Pages.*   The sum of the page numbers on the facing pages of a book is 457. Find the page numbers.   [2.6a]

Synthesis

Divide.

55. $(x^4 + 9x^2 + 20) \div (x^2 + 4)$

56. $(y^4 + a^2) \div (y + a)$

57. $(5a^3 + 8a^2 - 23a - 1) \div (5a^2 - 7a - 2)$

58. $(15y^3 - 30y + 7 - 19y^2) \div (3y^2 - 2 - 5y)$

59. $(6x^5 - 13x^3 + 5x + 3 - 4x^2 + 3x^4) \div (3x^3 - 2x - 1)$

60. $(5x^7 - 3x^4 + 2x^2 - 10x + 2) \div (x^2 - x + 1)$

61. $(a^6 - b^6) \div (a - b)$

62. $(x^5 + y^5) \div (x + y)$

If the remainder is 0 when one polynomial is divided by another, the divisor is a *factor* of the dividend. Find the value(s) of c for which $x - 1$ is a factor of the polynomial.

63. $x^2 + 4x + c$

64. $2x^2 + 3cx - 8$

65. $c^2x^2 - 2cx + 1$

CHAPTER 4 Summary and Review

Vocabulary Reinforcement

Complete each statement with the correct word from the list on the right. Some of the choices may not be used.

1. In the expression 7^5, the number 5 is the _____. [4.1a]
2. The _____ rule asserts that when multiplying with exponential notation, if the bases are the same, we keep the base and add the exponent. [4.1d]
3. An expression of the type ax^n, where a is a real-number constant and n is a nonnegative integer, is a(n) _____. [4.3a, b]
4. A(n) _____ is a polynomial with three terms, such as $5x^4 - 7x^2 + 4$. [4.3b]
5. The _____ rule asserts that when dividing with exponential notation, if the bases are the same, we keep the base and subtract the exponent of the denominator from the exponent of the numerator. [4.1e]
6. If the exponents in a polynomial decrease from left to right, the polynomial is arranged in _____ order. [4.3e]
7. The _____ of a term is the sum of the exponents of the variables. [4.7b]
8. The number 2.3×10^{-5} is written in _____ notation. [4.2c]

ascending
descending
degree
fraction
scientific
base
exponent
product
quotient
monomial
binomial
trinomial

Concept Reinforcement

Determine whether each statement is true or false.

_____ 1. All trinomials are polynomials. [4.3b]

_____ 2. $(x + y)^2 = x^2 + y^2$ [4.6c]

_____ 3. The square of the difference of two expressions is the difference of the squares of the two expressions. [4.6c]

_____ 4. The product of the sum and the difference of the same two expressions is the difference of the squares of the expressions. [4.6b]

Study Guide

Objective 4.1d Use the product rule to multiply exponential expressions with like bases.

Example Multiply and simplify: $x^3 \cdot x^4$.
$x^3 \cdot x^4 = x^{3+4} = x^7$

Practice Exercise
1. Multiply and simplify: $z^5 \cdot z^3$.

Objective 4.1e Use the quotient rule to divide exponential expressions with like bases.

Example Divide and simplify: $\dfrac{x^6 y^5}{xy^3}$.

$$\dfrac{x^6 y^5}{xy^3} = \dfrac{x^6}{x} \cdot \dfrac{y^5}{y^3}$$
$$= x^{6-1} y^{5-3} = x^5 y^2$$

Practice Exercise

2. Divide and simplify: $\dfrac{a^4 b^7}{a^2 b}$.

Objective 4.1f Express an exponential expression involving negative exponents with positive exponents.

Objective 4.2a Use the power rule to raise powers to powers.

Objective 4.2b Raise a product to a power and a quotient to a power.

Example Simplify: $\left(\dfrac{2a^3 b^{-2}}{c^4}\right)^5$.

$$\left(\dfrac{2a^3 b^{-2}}{c^4}\right)^5 = \dfrac{(2a^3 b^{-2})^5}{(c^4)^5}$$
$$= \dfrac{2^5 (a^3)^5 (b^{-2})^5}{(c^4)^5} = \dfrac{32 a^{3 \cdot 5} b^{-2 \cdot 5}}{c^{4 \cdot 5}}$$
$$= \dfrac{32 a^{15} b^{-10}}{c^{20}} = \dfrac{32 a^{15}}{b^{10} c^{20}}$$

Practice Exercise

3. Simplify: $\left(\dfrac{x^{-4} y^2}{3z^3}\right)^3$.

Objective 4.2c Convert between scientific notation and decimal notation.

Example Convert 0.00095 to scientific notation.

0.0009.5
    ↑
    4 places

The number is small, so the exponent is negative.

$0.00095 = 9.5 \times 10^{-4}$

Practice Exercises

4. Convert to scientific notation: 763,000.

Example Convert 3.409×10^6 to decimal notation.

3.409000.
       ↑
       6 places

The exponent is positive, so the number is large.

$3.409 \times 10^6 = 3,409,000$

5. Convert to decimal notation: 3×10^{-4}.

Objective 4.2d Multiply and divide using scientific notation.

Example Multiply and express the result in scientific notation: $(5.3 \times 10^9) \cdot (2.4 \times 10^{-5})$.

$(5.3 \times 10^9) \cdot (2.4 \times 10^{-5}) = (5.3 \cdot 2.4) \times (10^9 \cdot 10^{-5})$
$= 12.72 \times 10^4$

We convert 12.72 to scientific notation and simplify:

$12.72 \times 10^4 = (1.272 \times 10) \times 10^4$
$= 1.272 \times (10 \times 10^4)$
$= 1.272 \times 10^5$.

Practice Exercise

6. Divide and express the result in scientific notation:

$$\dfrac{3.6 \times 10^3}{6.0 \times 10^{-2}}.$$

Objective 4.3d Collect the like terms of a polynomial.

Example Collect like terms:
$4x^3 - 2x^2 + 5 + 3x^2 - 12$.
$4x^3 - 2x^2 + 5 + 3x^2 - 12$
$= 4x^3 + (-2 + 3)x^2 + (5 - 12)$
$= 4x^3 + x^2 - 7$

Practice Exercise

7. Collect like terms: $5x^4 - 6x^2 - 3x^4 + 2x^2 - 3$.

Objective 4.4a Add polynomials.

Example Add: $(4x^3 + x^2 - 8) + (2x^3 - 5x + 1)$.
$(4x^3 + x^2 - 8) + (2x^3 - 5x + 1)$
$= (4 + 2)x^3 + x^2 - 5x + (-8 + 1)$
$= 6x^3 + x^2 - 5x - 7$

Practice Exercise

8. Add: $(3x^4 - 5x^2 - 4) + (x^3 + 3x^2 + 6)$.

Objective 4.5d Multiply any two polynomials.

Example Multiply: $(z^2 - 2z + 3)(z - 1)$.
We use columns. First, we multiply the top row by -1 and then by z, placing like terms of the product in the same column. Finally, we collect like terms.

$$\begin{array}{r} z^2 - 2z + 3 \\ z - 1 \\ \hline -z^2 + 2z - 3 \\ z^3 - 2z^2 + 3z \\ \hline z^3 - 3z^2 + 5z - 3 \end{array}$$

Practice Exercise

9. Multiply: $(x^4 - 3x^2 + 2)(x^2 - 3)$.

Objective 4.6a Multiply two binomials mentally using the FOIL method.

Example Multiply: $(3x + 5)(x - 1)$.
$ \text{F} \text{O} \text{I} \text{L}$
$(3x + 5)(x - 1) = 3x \cdot x + 3x \cdot (-1) + 5 \cdot x + 5 \cdot (-1)$
$= 3x^2 - 3x + 5x - 5$
$= 3x^2 + 2x - 5$

Practice Exercise

10. Multiply: $(y + 4)(2y + 3)$.

Objective 4.6b Multiply the sum and the difference of the same two terms mentally.

Example Multiply: $(3y + 2)(3y - 2)$.
$(3y + 2)(3y - 2) = (3y)^2 - 2^2$
$= 9y^2 - 4$

Practice Exercise

11. Multiply: $(x + 5)(x - 5)$.

Objective 4.6c Square a binomial mentally.

Example Multiply: $(2x - 3)^2$.
$(2x - 3)^2 = (2x)^2 - 2 \cdot 2x \cdot 3 + 3^2$
$= 4x^2 - 12x + 9$

Practice Exercise

12. Multiply: $(3w + 4)^2$.

Objective 4.7e Subtract polynomials in several variables.

Example Subtract:
$(m^4n + 2m^3n^2 - m^2n^3) - (3m^4n + 2m^3n^2 - 4m^2n^2).$
$(m^4n + 2m^3n^2 - m^2n^3) - (3m^4n + 2m^3n^2 - 4m^2n^2)$
$= m^4n + 2m^3n^2 - m^2n^3 - 3m^4n - 2m^3n^2 + 4m^2n^2$
$= -2m^4n - m^2n^3 + 4m^2n^2$

Practice Exercise

13. Subtract:
$(a^3b^2 - 5a^2b + 2ab) - (3a^3b^2 - ab^2 + 4ab).$

Objective 4.8a Divide a polynomial by a monomial.

Example Divide: $(6x^3 - 8x^2 + 15x) \div (3x).$
$$\frac{6x^3 - 8x^2 + 15x}{3x} = \frac{6x^3}{3x} - \frac{8x^2}{3x} + \frac{15x}{3x}$$
$$= \frac{6}{3}x^{3-1} - \frac{8}{3}x^{2-1} + \frac{15}{3}x^{1-1}$$
$$= 2x^2 - \frac{8}{3}x + 5$$

Practice Exercise

14. Divide: $(5y^2 - 20y + 8) \div 5.$

Objective 4.8b Divide a polynomial by a divisor that is a binomial.

Example Divide $x^2 - 3x + 7$ by $x + 1$.

$$\begin{array}{r} x - 4 \phantom{{}+7} \\ x+1{\overline{\smash{\big)}\,x^2 - 3x + 7}} \\ \underline{x^2 + x} \\ -4x + 7 \\ \underline{-4x - 4} \\ 11 \end{array}$$

The answer is $x - 4 + \dfrac{11}{x + 1}$.

Practice Exercise

15. Divide: $(x^2 - 4x + 3) \div (x + 5).$

Review Exercises

Multiply and simplify. [4.1d, f]

1. $7^2 \cdot 7^{-4}$
2. $y^7 \cdot y^3 \cdot y$
3. $(3x)^5 \cdot (3x)^9$
4. $t^8 \cdot t^0$

Divide and simplify. [4.1e, f]

5. $\dfrac{4^5}{4^2}$
6. $\dfrac{a^5}{a^8}$
7. $\dfrac{(7x)^4}{(7x)^4}$

Simplify.

8. $(3t^4)^2$ [4.2a, b]

9. $(2x^3)^2(-3x)^2$ [4.1d], [4.2a, b]

10. $\left(\dfrac{2x}{y}\right)^{-3}$ [4.2b]

11. Express using a negative exponent: $\dfrac{1}{t^5}$. [4.1f]

12. Express using a positive exponent: y^{-4}. [4.1f]

13. Convert to scientific notation: 0.0000328. [4.2c]

14. Convert to decimal notation: 8.3×10^6. [4.2c]

Multiply or divide and write scientific notation for the result. [4.2d]

15. $(3.8 \times 10^4)(5.5 \times 10^{-1})$
16. $\dfrac{1.28 \times 10^{-8}}{2.5 \times 10^{-4}}$

17. *Pizza Consumption.* Each man, woman, and child in the United States eats an average of 46 slices of pizza per year. The U.S. population is projected to be about 340 million in 2020. At this rate, how many slices of pizza would be consumed in 2020? Express the answer in scientific notation. [4.2e]

   Data: Packaged Facts; U.S. Census Bureau

18. Evaluate the polynomial $x^2 - 3x + 6$ when $x = -1$. [4.3a]

19. Identify the terms of the polynomial $-4y^5 + 7y^2 - 3y - 2$. [4.3b]

20. Identify the missing terms in $x^3 + x$. [4.3f]

21. Identify the degree of each term and the degree of the polynomial $4x^3 + 6x^2 - 5x + \frac{5}{3}$. [4.3c]

Classify the polynomial as either a monomial, a binomial, a trinomial, or none of these. [4.3b]

22. $4x^3 - 1$

23. $4 - 9t^3 - 7t^4 + 10t^2$

24. $7y^2$

Collect like terms and then arrange in descending order [4.3e]

25. $3x^2 - 2x + 3 - 5x^2 - 1 - x$

26. $-x + \frac{1}{2} + 14x^4 - 7x^2 - 1 - 4x^4$

Add. [4.4a]

27. $(3x^4 - x^3 + x - 4) + (x^5 + 7x^3 - 3x^2 - 5) + (-5x^4 + 6x^2 - x)$

28. $(3x^5 - 4x^4 + x^3 - 3) + (3x^4 - 5x^3 + 3x^2) + (-5x^5 - 5x^2) + (-5x^4 + 2x^3 + 5)$

Subtract. [4.4c]

29. $(5x^2 - 4x + 1) - (3x^2 + 1)$

30. $(3x^5 - 4x^4 + 3x^2 + 3) - (2x^5 - 4x^4 + 3x^3 + 4x^2 - 5)$

31. Find a polynomial for the perimeter and for the area. [4.4d], [4.5b]

32. Find two algebraic expressions for the area of this figure. First, regard the figure as one large rectangle, and then regard the figure as a sum of four smaller rectangles. [4.4d]

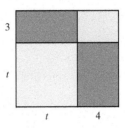

Multiply.

33. $\left(x + \frac{2}{3}\right)\left(x + \frac{1}{2}\right)$ [4.6a]

34. $(7x + 1)^2$ [4.6c]

35. $(4x^2 - 5x + 1)(3x - 2)$ [4.5d]

36. $(3x^2 + 4)(3x^2 - 4)$ [4.6b]

37. $5x^4(3x^3 - 8x^2 + 10x + 2)$ [4.5b]

38. $(x + 4)(x - 7)$ [4.6a]

39. $(3y^2 - 2y)^2$ [4.6c]

40. $(2t^2 + 3)(t^2 - 7)$ [4.6a]

41. Evaluate the polynomial
$$2 - 5xy + y^2 - 4xy^3 + x^6$$
when $x = -1$ and $y = 2$. [4.7a]

42. Identify the coefficient and the degree of each term of the polynomial
$$x^5y - 7xy + 9x^2 - 8.$$
Then find the degree of the polynomial. [4.7b]

Collect like terms. [4.7c]

43. $y + w - 2y + 8w - 5$

44. $m^6 - 2m^2n + m^2n^2 + n^2m - 6m^3 + m^2n^2 + 7n^2m$

45. Add: [4.7d]
$(5x^2 - 7xy + y^2) + (-6x^2 - 3xy - y^2) + (x^2 + xy - 2y^2)$.

46. Subtract: [4.7e]
$(6x^3y^2 - 4x^2y - 6x) - (-5x^3y^2 + 4x^2y + 6x^2 - 6)$.

Multiply. [4.7f]

47. $(p - q)(p^2 + pq + q^2)$

48. $(3a^4 - \frac{1}{3}b^3)^2$

Divide.

49. $(10x^3 - x^2 + 6x) \div (2x)$ [4.8a]

50. $(6x^3 - 5x^2 - 13x + 13) \div (2x + 3)$ [4.8b]

51. The graph of the polynomial equation $y = 10x^3 - 10x$ is shown below. Use *only* the graph to estimate the value of the polynomial when $x = -1$, $x = -0.5$, $x = 0.5$, and $x = 1$. [4.3a]

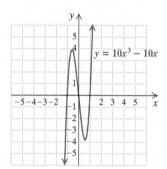

52. Subtract: $(2x^2 - 3x + 4) - (x^2 + 2x)$. [4.4c]

A. $x^2 - 3x - 2$ B. $x^2 - 5x + 4$
C. $x^2 - x + 4$ D. $3x^2 - x + 4$

53. Multiply: $(x - 1)^2$. [4.6c]

A. $x^2 - 1$ B. $x^2 + 1$
C. $x^2 - 2x - 1$ D. $x^2 - 2x + 1$

Synthesis

Find a polynomial for the shaded area in each figure. [4.4d], [4.6b]

54.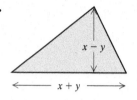

55.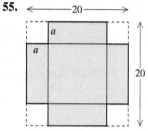

56. Collect like terms: [4.1d], [4.2a], [4.3d]
$-3x^5 \cdot 3x^3 - x^6(2x)^2 + (3x^4)^2 + (2x^2)^4 - 40x^2(x^3)^2$.

57. Solve: [2.3b], [4.6a]
$(x - 7)(x + 10) = (x - 4)(x - 6)$.

58. The product of two polynomials is $x^5 - 1$. One of the polynomials is $x - 1$. Find the other. [4.8b]

59. A rectangular garden is twice as long as it is wide and is surrounded by a sidewalk that is 4 ft wide. The area of the sidewalk is 1024 ft^2. Find the dimensions of the garden. [2.3b], [4.4d], [4.5a], [4.6a]

Understanding Through Discussion and Writing

1. Explain why the expression 578.6×10^{-7} is not written in scientific notation. [4.2c]

2. Explain why an understanding of the rules for order of operations is essential when evaluating polynomials. [4.3a]

3. How can the following figure be used to show that $(x + 3)^2 \neq x^2 + 9$? [4.5c]

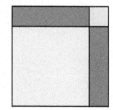

4. On an assignment, Emma *incorrectly* writes
$$\frac{12x^3 - 6x}{3x} = 4x^2 - 6x.$$
What mistake do you think she is making and how might you convince her that a mistake has been made? [4.8a]

5. Can the sum of two trinomials in several variables be a trinomial in one variable? Why or why not? [4.7d]

6. Is it possible for a polynomial in four variables to have a degree less than 4? Why or why not? [4.7b]

CHAPTER 4 Test

Multiply and simplify.

1. $6^{-2} \cdot 6^{-3}$
2. $x^6 \cdot x^2 \cdot x$
3. $(4a)^3 \cdot (4a)^8$

Divide and simplify.

4. $\dfrac{3^5}{3^2}$
5. $\dfrac{x^3}{x^8}$
6. $\dfrac{(2x)^5}{(2x)^5}$

Simplify.

7. $(x^3)^2$
8. $(-3y^2)^3$
9. $(2a^3b)^4$
10. $\left(\dfrac{ab}{c}\right)^3$

11. $(3x^2)^3(-2x^5)^3$
12. $3(x^2)^3(-2x^5)^3$
13. $2x^2(-3x^2)^4$
14. $(2x)^2(-3x^2)^4$

15. Express using a positive exponent: 5^{-3}.

16. Express using a negative exponent: $\dfrac{1}{y^8}$.

17. Convert to scientific notation: 3,900,000,000.

18. Convert to decimal notation: 5×10^{-8}.

Multiply or divide and write scientific notation for the answer.

19. $\dfrac{5.6 \times 10^6}{3.2 \times 10^{-11}}$

20. $(2.4 \times 10^5)(5.4 \times 10^{16})$

21. *Earth vs. Saturn.* The mass of Earth is about 6×10^{21} metric tons. The mass of Saturn is about 5.7×10^{23} metric tons. About how many times the mass of Earth is the mass of Saturn? Express the answer in scientific notation.

22. Evaluate the polynomial $x^5 + 5x - 1$ when $x = -2$.

23. Identify the coefficient of each term of the polynomial $\tfrac{1}{3}x^5 - x + 7$.

24. Identify the degree of each term and the degree of the polynomial $2x^3 - 4 + 5x + 3x^6$.

25. Classify the polynomial $7 - x$ as either a monomial, a binomial, a trinomial, or none of these.

Collect like terms.

26. $4a^2 - 6 + a^2$

27. $y^2 - 3y - y + \dfrac{3}{4}y^2$

28. Collect like terms and then arrange in descending order:
$3 - x^2 + 2x^3 + 5x^2 - 6x - 2x + x^5$.

Add.

29. $(3x^5 + 5x^3 - 5x^2 - 3) +$
$(x^5 + x^4 - 3x^3 - 3x^2 + 2x - 4)$

30. $\left(x^4 + \dfrac{2}{3}x + 5\right) + \left(4x^4 + 5x^2 + \dfrac{1}{3}x\right)$

Subtract.

31. $(2x^4 + x^3 - 8x^2 - 6x - 3) - (6x^4 - 8x^2 + 2x)$

32. $(x^3 - 0.4x^2 - 12) - (x^5 + 0.3x^3 + 0.4x^2 + 9)$

Multiply.

33. $-3x^2(4x^2 - 3x - 5)$

34. $\left(x - \dfrac{1}{3}\right)^2$

35. $(3x + 10)(3x - 10)$

36. $(3b + 5)(b - 3)$

37. $(x^6 - 4)(x^8 + 4)$

38. $(8 - y)(6 + 5y)$

39. $(2x + 1)(3x^2 - 5x - 3)$

40. $(5t + 2)^2$

41. Collect like terms:
$x^3y - y^3 + xy^3 + 8 - 6x^3y - x^2y^2 + 11.$

42. Subtract:
$(8a^2b^2 - ab + b^3) - (-6ab^2 - 7ab - ab^3 + 5b^3).$

43. Multiply: $(3x^5 - 4y^5)(3x^5 + 4y^5).$

Divide.

44. $(12x^4 + 9x^3 - 15x^2) \div (3x^2)$

45. $(6x^3 - 8x^2 - 14x + 13) \div (3x + 2)$

46. The graph of the polynomial equation
$$y = x^3 - 5x - 1$$
is shown at right. Use *only* the graph to estimate the value of the polynomial when $x = -1$, $x = -0.5$, $x = 0.5$, $x = 1$, and $x = 1.1$.

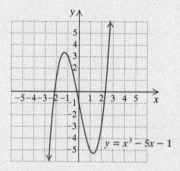

47. Find two algebraic expressions for the area of the following figure. First, regard the figure as one large rectangle, and then regard the figure as a sum of four smaller rectangles.

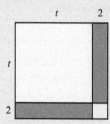

48. Which of the following is a polynomial for the surface area of this right rectangular solid?

A. $28a$
B. $28a + 90$
C. $14a + 45$
D. $45a$

Synthesis

49. The height of a box is 1 less than its length, and the length is 2 more than its width. Find the volume in terms of the length.

50. Solve: $(x - 5)(x + 5) = (x + 6)^2.$

CHAPTERS 1–4 Cumulative Review

1. Evaluate $\dfrac{x}{2y}$ when $x = 10$ and $y = 2$.

2. Evaluate $2x^3 + x^2 - 3$ when $x = -1$.

3. Evaluate $x^3y^2 + xy + 2xy^2$ when $x = -1$ and $y = 2$.

4. Find the absolute value: $|-4|$.

5. Find the reciprocal of 5.

Compute and simplify.

6. $-\dfrac{3}{5} + \dfrac{5}{12}$

7. $3.4 - (-0.8)$

8. $(-2)(-1.4)(2.6)$

9. $\dfrac{3}{8} \div \left(-\dfrac{9}{10}\right)$

10. $(1.1 \times 10^{10})(2 \times 10^{12})$

11. $(3.2 \times 10^{-10}) \div (8 \times 10^{-6})$

Simplify.

12. $\dfrac{-9x}{3x}$

13. $y - (3y + 7)$

14. $3(x - 1) - 2[x - (2x + 7)]$

15. $2 - [32 \div (4 + 2^2)]$

Add.

16. $(x^4 + 3x^3 - x + 7) + (2x^5 - 3x^4 + x - 5)$

17. $(x^2 + 2xy) + (y^2 - xy) + (2x^2 - 3y^2)$

Subtract.

18. $(x^3 + 3x^2 - 4) - (-2x^2 + x + 3)$

19. $\left(\dfrac{1}{3}x^2 - \dfrac{1}{4}x - \dfrac{1}{5}\right) - \left(\dfrac{2}{3}x^2 + \dfrac{1}{2}x - \dfrac{1}{5}\right)$

Multiply.

20. $3(4x - 5y + 7)$

21. $(-2x^3)(-3x^5)$

22. $2x^2(x^3 - 2x^2 + 4x - 5)$

23. $(y^2 - 2)(3y^2 + 5y + 6)$

24. $(2p^3 + p^2q + pq^2)(p - pq + q)$

25. $(2x + 3)(3x + 2)$

26. $(3x^2 + 1)^2$

27. $\left(t + \dfrac{1}{2}\right)\left(t - \dfrac{1}{2}\right)$

28. $(2y^2 + 5)(2y^2 - 5)$

29. $(2x^4 - 3)(2x^2 + 3)$

30. $(t - 2t^2)^2$

31. $(3p + q)(5p - 2q)$

Divide.

32. $(18x^3 + 6x^2 - 9x) \div (3x)$

33. $(3x^3 + 7x^2 - 13x - 21) \div (x + 3)$

Solve.

34. $1.5 = 2.7 + x$

35. $\dfrac{2}{7}x = -6$

36. $5x - 9 = 36$

37. $\dfrac{2}{3} = \dfrac{-m}{10}$

38. $5.4 - 1.9x = 0.8x$

39. $x - \dfrac{7}{8} = \dfrac{3}{4}$

40. $2(2 - 3x) = 3(5x + 7)$

41. $\dfrac{1}{4}x - \dfrac{2}{3} = \dfrac{3}{4} + \dfrac{1}{3}x$

42. $y + 5 - 3y = 5y - 9$

43. $\dfrac{1}{4}x - 7 < 5 - \dfrac{1}{2}x$

44. $2(x + 2) \geq 5(2x + 3)$

45. $A = Qx + P$, for x

Solve.

46. *Markup.* A bookstore sells books at a price that is 80% higher than the price the store pays for the books. A book is priced for sale at $6.30. How much did the store pay for the book?

47. A 6-ft by 3-ft raft is floating in a circular swimming pool of radius r. Find a polynomial for the area of the surface of the pool not covered by the raft.

48. *Consecutive Page Numbers.* The sum of the page numbers on the facing pages of a book is 37. What are the page numbers?

49. *Room Perimeter.* The perimeter of a room is 88 ft. The width is 4 ft less than the length. Find the width and the length.

50. The second angle of a triangle is five times as large as the first. The third angle is twice the sum of the other two angles. Find the measure of the first angle.

Simplify.

51. $y^2 \cdot y^{-6} \cdot y^8$

52. $\dfrac{x^6}{x^7}$

53. $(-3x^3y^{-2})^3$

54. $\dfrac{x^3 x^{-4}}{x^{-5} x}$

55. Find the intercepts of $4x - 5y = 20$ and then graph the equation using the intercepts.

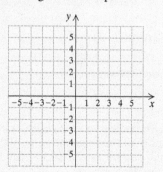

56. *Matching.* Match each item in the first column with the appropriate item in the second column by drawing connecting lines.

3^2 $\dfrac{1}{6}$

3^{-2} $-\dfrac{1}{9}$

$\left(\dfrac{1}{3}\right)^2$ 6

$\left(\dfrac{1}{3}\right)^{-2}$ 9

-3^2 -9

$(-3)^2$ $\dfrac{1}{9}$

$\left(-\dfrac{1}{3}\right)^2$ -6

$\left(-\dfrac{1}{3}\right)^{-2}$ 12

Synthesis

57. A picture frame is x in. square. The picture that it frames is 2 in. shorter than the frame in both length and width. Find a polynomial for the area of the frame.

Add.

58. $[(2x)^2 - (3x)^3 + 2x^2 x^3 + (x^2)^2] + [5x^2(2x^3) - ((2x)^2)^2]$

59. $(x-3)^2 + (2x+1)^2$

Solve.

60. $(x+3)(2x-5) + (x-1)^2 = (3x+1)(x-3)$

61. $(2x^2 + x - 6) \div (2x - 3) = (2x^2 - 9x - 5) \div (x - 5)$

62. $20 - 3|x| = 5$

63. $(x-3)(x+4) = (x^3 - 4x^2 - 17x + 60) \div (x - 5)$

CHAPTER 5

5.1 Introduction to Factoring
5.2 Factoring Trinomials of the Type $x^2 + bx + c$
5.3 Factoring $ax^2 + bx + c$, $a \neq 1$: The FOIL Method
5.4 Factoring $ax^2 + bx + c$, $a \neq 1$: The ac-Method

Mid-Chapter Review

5.5 Factoring Trinomial Squares and Differences of Squares
5.6 Factoring Sums or Differences of Cubes
5.7 Factoring: A General Strategy
5.8 Solving Quadratic Equations by Factoring
5.9 Applications of Quadratic Equations

Translating for Success
Summary and Review
Test
Cumulative Review

Polynomials: Factoring

The total amount of digital media usage time in the United States is growing at a rate of 250 billion minutes per year. As the graph below illustrates, over half of digital media time is spent using mobile apps. Smartphone users download, on average, 3 apps per month, but app use varies by age, with users between the ages of 18 and 24 spending the most time on apps: 125 hours per month, on average. The number of apps available is growing as well. In 2017, there were 5 million apps available, in total, in Google Play and the Apple iTunes Store. These numbers do not include games.

Data: comScore.com; statista.com

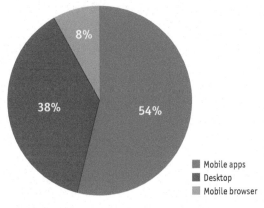

Digital Media Usage Time
- Mobile apps 54%
- Desktop 38%
- Mobile browser 8%

DATA: comscore.com

JUST IN TIME Review topics 1, 2, 3, 8, and 17 in the **Just In Time** section at the front of the text. This provides excellent prerequisite skill review for this chapter.

MyLab Math VIDEO

We will use a *quadratic equation* to estimate the number of apps in the iTunes Store in Example 3 of Section 5.9.

STUDYING FOR SUCCESS Working Exercises

☐ Don't begin solving a homework problem by working backward from the answer given at the back of the text. Remember: Quizzes and tests have no answer section!

☐ Check answers to odd-numbered exercises at the back of the book.

☐ Work some even-numbered exercises as practice doing exercises without answers. Check your answers later with a friend or your instructor.

5.1 Introduction to Factoring

OBJECTIVES

a. Find the greatest common factor, the GCF, of monomials.

b. Factor polynomials when the terms have a common factor, factoring out the greatest common factor.

c. Factor certain expressions with four terms using factoring by grouping.

We introduce factoring with a review of factoring natural numbers. Because we can write 15 as $3 \cdot 5$, we say that 3 and 5 are **factors** of 15 and that $3 \cdot 5$ is a **factorization** of 15.

a FINDING THE GREATEST COMMON FACTOR

The numbers 60 and 100 have several factors in common, among them 2 and 5. The greatest of the common factors is called the **greatest common factor**, GCF. We can find the GCF of a set of numbers using prime factorizations.

EXAMPLE 1 Find the GCF of 60 and 100.

We find the prime factorization of each number. Then we draw lines between the common factors.

$$60 = 2 \cdot 2 \cdot 3 \cdot 5 = 2^2 \cdot 3^1 \cdot 5^1,$$
$$100 = 2 \cdot 2 \cdot 5 \cdot 5 = 2^2 \cdot 5^2$$

The GCF $= 2 \cdot 2 \cdot 5 = 2^2 \cdot 5^1 = 20$. Note that we can use the exponents to determine the GCF. There are 2 lines for the 2's, no line for the 3, and 1 line for the 5.

EXAMPLE 2 Find the GCF of 30 and 77.

We look for factors common to the prime factorizations.

$$30 = 2 \cdot 3 \cdot 5 = 2^1 \cdot 3^1 \cdot 5^1,$$
$$77 = 7 \cdot 11 = 7^1 \cdot 11^1$$

Since there is no common prime factor, the GCF is 1.

EXAMPLE 3 Find the GCF of 54, 90, and 252.

We look for factors common to all three prime factorizations.

$$54 = 2 \cdot 3 \cdot 3 \cdot 3 = 2^1 \cdot 3^3,$$
$$90 = 2 \cdot 3 \cdot 3 \cdot 5 = 2^1 \cdot 3^2 \cdot 5^1,$$
$$252 = 2 \cdot 2 \cdot 3 \cdot 3 \cdot 7 = 2^2 \cdot 3^2 \cdot 7^1$$

The GCF $= 2^1 \cdot 3^2 = 18$.

◀ Do Exercises 1–4.

Find the GCF.

1. 40, 100

2. 7, 21

3. 72, 360, 432

4. 3, 5, 22

Answers
1. 20   2. 7   3. 72   4. 1

Consider the product
$$12x^3(x^2 - 6x + 2) = 12x^5 - 72x^4 + 24x^3.$$
To factor the polynomial on the right side, we reverse the process of multiplication:
$$12x^5 - 72x^4 + 24x^3 = \underbrace{12x^3(x^2 - 6x + 2)}.$$

This is a *factorization*. The *factors* are $(12x^3)$ and $(x^2 - 6x + 2)$.

> **FACTOR; FACTORIZATION**
>
> To **factor** a polynomial is to express it as a product.
>
> A **factor** of a polynomial P is a polynomial that can be used to express P as a product.
>
> A **factorization** of a polynomial is an expression that names that polynomial as a product.

In the factorization
$$12x^5 - 72x^4 + 24x^3 = 12x^3(x^2 - 6x + 2),$$
the monomial $12x^3$ is called the GCF of the terms, $12x^5$, $-72x^4$, and $24x^3$. The first step in factoring polynomials is to find the GCF of the terms. To do this, we find the greatest positive common factor of the coefficients and the greatest common factors of the powers of any variables.

EXAMPLE 4 Find the GCF of $15x^5$, $-12x^4$, $27x^3$, and $-3x^2$.

First, we find a prime factorization of the coefficients, including a factor of -1 for the negative coefficients.

$$15x^5 = 3 \cdot 5 \cdot x^5,$$
$$-12x^4 = -1 \cdot 2 \cdot 2 \cdot 3 \cdot x^4,$$
$$27x^3 = 3 \cdot 3 \cdot 3 \cdot x^3,$$
$$-3x^2 = -1 \cdot 3 \cdot x^2$$

The greatest *positive* common factor of the coefficients is 3.

Next, we find the GCF of the powers of x. That GCF is x^2, because 2 is the smallest exponent of x. Thus the GCF of the set of monomials is $3x^2$.

EXAMPLE 5 Find the GCF of $14p^2y^3$, $-8py^2$, $2py$, and $4p^3$.

We have

$$14p^2y^3 = 2 \cdot 7 \cdot p^2 \cdot y^3,$$
$$-8py^2 = -1 \cdot 2 \cdot 2 \cdot 2 \cdot p \cdot y^2,$$
$$2py = 2 \cdot p \cdot y,$$
$$4p^3 = 2 \cdot 2 \cdot p^3.$$

The greatest positive common factor of the coefficients is 2, the GCF of the powers of p is p, and the GCF of the powers of y is 1 since there is no y-factor in the last monomial. Thus the GCF is $2p$.

Find the GCF.

5. $12x^2$, $-16x^3$

6. $3y^6$, $-5y^3$, $2y^2$

7. $-24m^5n^6$, $12mn^3$, $-16m^2n^2$, $8m^4n^4$

The coefficients are $-24, 12, -16$, and ____.

The greatest positive common factor of the coefficients is ____.

The smallest exponent of the variable m is ____.

The smallest exponent of the variable n is ____.

The GCF $= 4mn$.

8. $-35x^7$, $-49x^6$, $-14x^5$, $-63x^3$

TO FIND THE GCF OF TWO OR MORE MONOMIALS

1. Find the prime factorization of the coefficients, including -1 as a factor if any coefficient is negative.
2. Determine the greatest positive common factor of the coefficients. (If the coefficients have no prime factors in common, the GCF of the coefficients is 1.)
3. Determine the greatest common factor of the powers of any variables. If any variable appears as a factor of all the monomials, include it as a factor, using the smallest exponent of the variable. (If none occurs in all the monomials, the GCF of the variables is 1.)
4. The GCF of the monomials is the product of the results of steps (2) and (3).

◀ Do Exercises 5–8.

b FACTORING WHEN TERMS HAVE A COMMON FACTOR

SKILL REVIEW *Factor expressions using the distributive law.* [1.7d]
Factor.
1. $15m - 10c + 5$
2. $ax - ay + az$

Answers: 1. $5(3m - 2c + 1)$ 2. $a(x - y + z)$

To multiply a monomial and a polynomial with more than one term, we multiply each term of the polynomial by the monomial using the distributive laws:

$$a(b + c) = ab + ac \quad \text{and} \quad a(b - c) = ab - ac.$$

To factor, we express a polynomial as a product using the distributive laws in reverse:

$$ab + ac = a(b + c) \quad \text{and} \quad ab - ac = a(b - c).$$

Compare.

Multiply 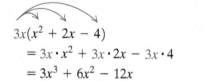 *Factor*

9. a) Multiply: $3(x + 2)$.
   b) Factor: $3x + 6$.

10. a) Multiply: $2x(x^2 + 5x + 4)$.
    b) Factor: $2x^3 + 10x^2 + 8x$.

$3x(x^2 + 2x - 4)$
$= 3x \cdot x^2 + 3x \cdot 2x - 3x \cdot 4$
$= 3x^3 + 6x^2 - 12x$

$3x^3 + 6x^2 - 12x$
$= 3x \cdot x^2 + 3x \cdot 2x - 3x \cdot 4$
$= 3x(x^2 + 2x - 4)$

◀ Do Exercises 9 and 10.

Answers
5. $4x^2$ 6. y^2 7. $4mn^2$ 8. $7x^3$
9. (a) $3x + 6$; (b) $3(x + 2)$
10. (a) $2x^3 + 10x^2 + 8x$; (b) $2x(x^2 + 5x + 4)$

Guided Solution:
7. 8, 4, 1, 2, 2

EXAMPLE 6 Factor: $7x^2 + 14$.

We have
$$7x^2 + 14 = 7 \cdot x^2 + 7 \cdot 2 \quad \text{Factoring each term}$$
$$= 7(x^2 + 2). \quad \text{Factoring out the GCF, 7}$$

Check: We multiply to check:
$$7(x^2 + 2) = 7 \cdot x^2 + 7 \cdot 2 = 7x^2 + 14.$$

EXAMPLE 7 Factor: $16x^3 + 20x^2$.

$$16x^3 + 20x^2 = (4x^2)(4x) + (4x^2)(5) \quad \text{Factoring each term}$$
$$= 4x^2(4x + 5) \quad \text{Factoring out the GCF, } 4x^2$$

Although it is always more efficient to begin by finding the GCF, suppose in Example 7 that you had not recognized the GCF and removed only part of it, as follows:

$$16x^3 + 20x^2 = (2x^2)(8x) + (2x^2)(10)$$
$$= 2x^2(8x + 10).$$

Note that $8x + 10$ still has a common factor of 2. You need not begin again. Just continue factoring out common factors, as follows, until finished:

$$= 2x^2(2 \cdot 4x + 2 \cdot 5)$$
$$= 2x^2[2(4x + 5)]$$
$$= (2x^2 \cdot 2)(4x + 5)$$
$$= 4x^2(4x + 5).$$

EXAMPLE 8 Factor: $15x^5 - 12x^4 + 27x^3 - 3x^2$.

$$15x^5 - 12x^4 + 27x^3 - 3x^2 = (3x^2)(5x^3) - (3x^2)(4x^2) + (3x^2)(9x) - (3x^2)(1)$$
$$= 3x^2(5x^3 - 4x^2 + 9x - 1) \quad \text{Factoring out the GCF, } 3x^2$$

············ **Caution!** ············

Don't forget the term -1.

As you become more familiar with factoring, you will be able to spot the GCF without factoring each term. Then you can write just the answer.

EXAMPLES Factor.

9. $24x^2 + 12x - 36 = 12(2x^2 + x - 3)$
10. $8m^3 - 16m = 8m(m^2 - 2)$
11. $14p^2y^3 - 8py^2 + 2py = 2py(7py^2 - 4y + 1)$
12. $\frac{4}{5}x^2 + \frac{1}{5}x + \frac{2}{5} = \frac{1}{5}(4x^2 + x + 2)$

Do Exercises 11–16. ▶

············ **Caution!** ············

Consider the following:

$$7x^2 + 14 = 7 \cdot x \cdot x + 7 \cdot 2.$$

The terms of the polynomial have been factored, but the polynomial itself has not been factored. This is not what we mean by the factorization of the polynomial. The *factorization* is

$7(x^2 + 2).$ ← A product

The expressions 7 and $x^2 + 2$ are *factors* of $7x^2 + 14$.

Factor. Check by multiplying.
11. $x^2 + 3x$

12. $3y^6 - 5y^3 + 2y^2$

13. $9x^4y^2 - 15x^3y + 3x^2y$

14. $\frac{3}{4}t^3 + \frac{5}{4}t^2 + \frac{7}{4}t + \frac{1}{4}$

15. $35x^7 - 49x^6 + 14x^5 - 63x^3$

16. $84x^2 - 56x + 28$

Answers
11. $x(x + 3)$ 12. $y^2(3y^4 - 5y + 2)$
13. $3x^2y(3x^2y - 5x + 1)$
14. $\frac{1}{4}(3t^3 + 5t^2 + 7t + 1)$
15. $7x^3(5x^4 - 7x^3 + 2x^2 - 9)$
16. $28(3x^2 - 2x + 1)$

Factor.
17. $x^2(x + 7) + 3(x + 7)$

18. $x^3(a + b) - 5(a + b)$

C FACTORING BY GROUPING: FOUR TERMS

Certain polynomials with four terms can be factored using a method called *factoring by grouping*.

EXAMPLE 13 Factor: $x^2(x + 1) + 2(x + 1)$.

The binomial $x + 1$ is a common factor. We factor it out:

$$x^2(x + 1) + 2(x + 1) = (x + 1)(x^2 + 2).$$

The factorization is $(x + 1)(x^2 + 2)$.

◀ Do Exercises 17 and 18.

Consider the four-term polynomial

$$x^3 + x^2 + 2x + 2.$$

There is no factor other than 1 that is common to all the terms. We can, however, factor $x^3 + x^2$ and $2x + 2$ separately:

$x^3 + x^2 = x^2(x + 1);$ Factoring $x^3 + x^2$
$2x + 2 = 2(x + 1).$ Factoring $2x + 2$

When we group the terms as shown above and factor each polynomial separately, we see that $(x + 1)$ appears in *both* factorizations. Thus we can factor out the common binomial factor as in Example 13:

$$x^3 + x^2 + 2x + 2 = (x^3 + x^2) + (2x + 2)$$
$$= x^2(x + 1) + 2(x + 1)$$
$$= (x + 1)(x^2 + 2).$$

This method of factoring is called **factoring by grouping**.

Not all polynomials with four terms can be factored by grouping, but it does give us a method to try.

EXAMPLES Factor by grouping.

14. $6x^3 - 9x^2 + 4x - 6$
    $= (6x^3 - 9x^2) + (4x - 6)$ Grouping the terms
    $= 3x^2(2x - 3) + 2(2x - 3)$ Factoring each binomial
    $= (2x - 3)(3x^2 + 2)$ Factoring out the common factor $2x - 3$

We think through this process as follows:

$$6x^3 - 9x^2 + 4x - 6 = \underline{3x^2(2x - 3)} \;\square\; \underline{(2x - 3)}$$

(1) Factor the first two terms.
(2) The factor $2x - 3$ gives us a hint for factoring the last two terms.
(3) Now we ask ourselves, "What times $2x - 3$ is $4x - 6$?" The answer is $+ 2$.

Answers
17. $(x + 7)(x^2 + 3)$ 18. $(a + b)(x^3 - 5)$

15. $x^3 + x^2 + x + 1 = (x^3 + x^2) + (x + 1)$ — Don't forget the 1.

$= x^2(x + 1) + 1(x + 1)$ Factoring each binomial

$= (x + 1)(x^2 + 1)$ Factoring out the common factor $x + 1$

16. $2x^3 - 6x^2 - x + 3$

$= (2x^3 - 6x^2) + (-x + 3)$ Grouping as two binomials

$= 2x^2(x - 3) - 1(x - 3)$ Check: $-1(x - 3) = -x + 3$.

$= (x - 3)(2x^2 - 1)$ Factoring out the common factor $x - 3$

We can think through this process as follows.

(1) Factor the first two terms: $2x^3 - 6x^2 = 2x^2(x - 3)$.

(2) The factor $x - 3$ gives us a hint for factoring the last two terms:

$2x^3 - 6x^2 - x + 3 = 2x^2(x - 3) \;\square\; (x - 3)$.

(3) We ask, "What times $x - 3$ is $-x + 3$?" The answer is -1.

17. $12x^5 + 20x^2 - 21x^3 - 35 = 4x^2(3x^3 + 5) - 7(3x^3 + 5)$

$= (3x^3 + 5)(4x^2 - 7)$

18. $x^3 + x^2 + 2x - 2 = x^2(x + 1) + 2(x - 1)$

This polynomial is not factorable using factoring by grouping. It may be factorable, but not by methods that we will consider in this text.

Do Exercises 19–24.

There are two important points to keep in mind when factoring.

> **TIPS FOR FACTORING**
>
> - Before doing any other kind of factoring, first try to factor out the GCF.
> - Always check the result of factoring by multiplying.

Factor by grouping.

19. $x^3 + 7x^2 + 3x + 21$

$= x^2(\quad) + 3(\quad)$

$= (\quad)(x^2 + 3)$

20. $8t^3 + 2t^2 + 12t + 3$

21. $3m^5 - 15m^3 + 2m^2 - 10$

22. $3x^3 - 6x^2 - x + 2$

23. $4x^3 - 6x^2 - 6x + 9$

24. $y^4 - 2y^3 - 2y - 10$

Answers

19. $(x + 7)(x^2 + 3)$ 20. $(4t + 1)(2t^2 + 3)$
21. $(m^2 - 5)(3m^3 + 2)$
22. $(x - 2)(3x^2 - 1)$ 23. $(2x - 3)(2x^2 - 3)$
24. Not factorable using factoring by grouping

Guided Solution:
19. $x + 7, x + 7, x + 7$

5.1 Exercise Set

FOR EXTRA HELP   MyLab Math

✓ Check Your Understanding

Reading Check Choose from the column on the right the expression that fits each description.

RC1. _____ A factorization of $36x^2$

RC2. _____ A factorization of $27x$

RC3. _____ The greatest common factor of $36x^2 - 27x$

RC4. _____ A factorization of $36x^2 - 27x$

a) $9x(4x - 3)$
b) $(9x)(4x)$
c) $(9x)(3)$
d) $9x$

Concept Check Determine whether each of the following could be a factorization of a polynomial.

CC1. $x(x-3)$ **CC2.** $x \cdot x - x \cdot 3$ **CC3.** $(x^2 - 2)(x-1)$ **CC4.** $x(x-9) - x$

a Find the GCF.

1. 36, 42
2. 60, 75
3. 48, 72, 120

4. 90, 135, 225
5. 8, 15, 40
6. 12, 20, 75

7. x^2, $-6x$
8. x^2, $5x$
9. $3x^4$, x^2

10. $8x^4$, $-24x^2$
11. $2x^2$, $2x$, -8
12. $8x^2$, $-4x$, -20

13. $-17x^5y^3$, $34x^3y^2$, $51xy$
14. $16p^6q^4$, $32p^3q^3$, $-48pq^2$
15. $-x^2$, $-5x$, $-20x^3$

16. $-x^2$, $-6x$, $-24x^5$
17. x^5y^5, x^4y^3, x^3y^3, $-x^2y^2$
18. $-x^9y^6$, $-x^7y^5$, x^4y^4, x^3y^3

b Factor. Check by multiplying.

19. $x^2 - 6x$
20. $x^2 + 5x$
21. $2x^2 + 6x$

22. $8y^2 - 8y$
23. $x^3 + 6x^2$
24. $3x^4 - x^2$

25. $8x^4 - 24x^2$
26. $5x^5 + 10x^3$
27. $2x^2 + 2x - 8$

28. $8x^2 - 4x - 20$
29. $17x^5y^3 + 34x^3y^2 + 51xy$
30. $16p^6q^4 + 32p^5q^3 - 48pq^2$

31. $6x^4 - 10x^3 + 3x^2$
32. $5x^5 + 10x^2 - 8x$
33. $x^5y^5 + x^4y^3 + x^3y^3 - x^2y^2$

34. $x^9y^6 - x^7y^5 + x^4y^4 + x^3y^3$
35. $2x^7 - 2x^6 - 64x^5 + 4x^3$
36. $8y^3 - 20y^2 + 12y - 16$

37. $1.6x^4 - 2.4x^3 + 3.2x^2 + 6.4x$
38. $2.5x^6 - 0.5x^4 + 5x^3 + 10x^2$

39. $\dfrac{5}{3}x^6 + \dfrac{4}{3}x^5 + \dfrac{1}{3}x^4 + \dfrac{1}{3}x^3$
40. $\dfrac{5}{9}x^7 + \dfrac{2}{9}x^5 - \dfrac{4}{9}x^3 - \dfrac{1}{9}x$

c Factor.

41. $x^2(x + 3) + 2(x + 3)$

42. $y^2(y + 4) + 6(y + 4)$

43. $4z^2(3z - 1) + 7(3z - 1)$

44. $2x^2(4x - 3) + 5(4x - 3)$

45. $2x^2(3x + 2) + (3x + 2)$

46. $3z^2(2z + 7) + (2z + 7)$

47. $5a^3(2a - 7) - (2a - 7)$

48. $m^4(8 - 3m) - 3(8 - 3m)$

Factor by grouping.

49. $x^3 + 3x^2 + 2x + 6$

50. $6z^3 + 3z^2 + 2z + 1$

51. $2x^3 + 6x^2 + x + 3$

52. $3x^3 + 2x^2 + 3x + 2$

53. $8x^3 - 12x^2 + 6x - 9$

54. $10x^3 - 25x^2 + 4x - 10$

55. $12p^3 - 16p^2 + 3p - 4$

56. $18x^3 - 21x^2 + 30x - 35$

57. $5x^3 - 5x^2 - x + 1$

58. $7x^3 - 14x^2 - x + 2$

59. $x^3 + 8x^2 - 3x - 24$

60. $2x^3 + 12x^2 - 5x - 30$

61. $2x^3 - 8x^2 - 9x + 36$

62. $20g^3 - 4g^2 - 25g + 5$

Skill Maintenance

Multiply. [4.5b], [4.6d], [4.7f]

63. $(y + 5)(y + 7)$

64. $(y + 7)^2$

65. $(y + 7)(y - 7)$

66. $(y - 7)^2$

67. $8x(2x^2 - 6x + 1)$

68. $(7w + 6)(4w - 11)$

69. $(7w + 6)^2$

70. $(4w - 11)^2$

71. $(4w - 11)(4w + 11)$

72. $-y(-y^2 + 3y - 5)$

73. $(3x - 5y)(2x + 7y)$

74. $(5x - t)^2$

Synthesis

Factor.

75. $4x^5 + 6x^3 + 6x^2 + 9$

76. $x^6 + x^4 + x^2 + 1$

77. $x^{12} + x^7 + x^5 + 1$

78. $x^3 - x^2 - 2x + 5$

79. $p^3 + p^2 - 3p + 10$

80. $4y^6 + 2y^4 - 12y^3 - 6y$

5.2 Factoring Trinomials of the Type $x^2 + bx + c$

OBJECTIVE

a Factor trinomials of the type $x^2 + bx + c$ by examining the constant term c.

a FACTORING $x^2 + bx + c$

We now begin a study of the factoring of trinomials. We first factor trinomials like

$$x^2 + 5x + 6 \quad \text{and} \quad x^2 + 3x - 10$$

by a refined *trial-and-error process*. In this section, we restrict our attention to trinomials of the type $ax^2 + bx + c$, where the **leading coefficient** a is 1.

Compare the following multiplications:

$$
\begin{aligned}
(x + 2)(x + 5) &= x^2 + \overset{F}{\downarrow}5x + \overset{O}{\downarrow}2x + \overset{I}{\downarrow}2 \cdot \overset{L}{\downarrow}5 \\
&= x^2 + 7x + 10; \\
(x - 2)(x - 5) &= x^2 - 5x - 2x + (-2)(-5) \\
&= x^2 - 7x + 10; \\
(x + 3)(x - 7) &= x^2 - 7x + 3x + 3(-7) \\
&= x^2 - 4x - 21; \\
(x - 3)(x + 7) &= x^2 + 7x - 3x + (-3)7 \\
&= x^2 + 4x - 21.
\end{aligned}
$$

Note that for all four products:

- The product of the two binomials is a trinomial.
- The coefficient of x in the trinomial is the sum of the constant terms in the binomials.
- The constant term in the trinomial is the product of the constant terms in the binomials.

These observations lead to a method for factoring certain trinomials. The first type that we consider has a positive constant term, just as in the first two multiplications above.

Constant Term Positive

To factor $x^2 + 7x + 10$, we think of FOIL in reverse. Since $x \cdot x = x^2$, the first term of each binomial is x.

Next, we look for numbers p and q such that

$$x^2 + 7x + 10 = (x + p)(x + q).$$

To get the middle term and the last term of the trinomial, we look for two numbers p and q whose product is 10 and whose sum is 7. Those numbers are 2 and 5. Thus the factorization is

$$(x + 2)(x + 5).$$

Check: $(x + 2)(x + 5) = x^2 + 5x + 2x + 10$
$= x^2 + 7x + 10.$

SKILL REVIEW

List all factors of a number. [J1]

List all factors of each number.
1. 12
2. 110

Answers:
1. 1, 2, 3, 4, 6, 12
2. 1, 2, 5, 10, 11, 22, 55, 110

EXAMPLE 1 Factor: $x^2 + 5x + 6$.

Think of FOIL in reverse. The first term of each factor is x:

$(x + \Box)(x + \Box)$.

Next, we look for two numbers whose product is 6 and whose sum is 5. All the pairs of factors of 6 are shown in the table on the left below. Since both the product, 6, and the sum, 5, of the pair of numbers must be positive, we need consider only the positive factors, listed in the table on the right.

PAIRS OF FACTORS	SUMS OF FACTORS
1, 6	7
−1, −6	−7
2, 3	5
−2, −3	−5

PAIRS OF FACTORS	SUMS OF FACTORS
1, 6	7
2, 3	**5**

↑ The numbers we need are 2 and 3.

The factorization is $(x + 2)(x + 3)$. We can check by multiplying to see whether we get the original trinomial.

Check: $(x + 2)(x + 3) = x^2 + 3x + 2x + 6 = x^2 + 5x + 6$.

Do Exercises 1 and 2.

Compare these multiplications:

$(x - 2)(x - 5) = x^2 - 5x - 2x + 10 = x^2 - 7x + 10;$
$(x + 2)(x + 5) = x^2 + 5x + 2x + 10 = x^2 + 7x + 10.$

TO FACTOR $x^2 + bx + c$ WHEN c IS POSITIVE

When the constant term of a trinomial is positive, look for two numbers with the same sign. The sign is that of the middle term:

$x^2 - 7x + 10 = (x - 2)(x - 5);$

$x^2 + 7x + 10 = (x + 2)(x + 5).$

EXAMPLE 2 Factor: $y^2 - 8y + 12$.

Since the constant term, 12, is positive and the coefficient of the middle term, −8, is negative, we look for a factorization of 12 in which both factors are negative. Their sum must be −8.

PAIRS OF FACTORS	SUMS OF FACTORS
−1, −12	−13
−2, −6	**−8** ← The numbers we need are −2 and −6.
−3, −4	−7

The factorization is $(y - 2)(y - 6)$. The student should check by multiplying.

Do Exercises 3–5.

Factor.

GS 1. $x^2 + 7x + 12$

Complete the following table.

PAIRS OF FACTORS	SUMS OF FACTORS
1, 12	13
−1, −12	
2, 6	
−2, −6	
3, 4	
−3, −4	

Because both 7 and 12 are positive, we need consider only the ____ factors in the table above.

$x^2 + 7x + 12$
$= (x + 3)()$

2. $x^2 + 13x + 36$

3. Explain why you would *not* consider the pairs of factors listed below in factoring $y^2 - 8y + 12$.

PAIRS OF FACTORS	SUMS OF FACTORS
1, 12	
2, 6	
3, 4	

Factor.
4. $x^2 - 8x + 15$

5. $t^2 - 9t + 20$

Answers
1. $(x + 3)(x + 4)$ **2.** $(x + 4)(x + 9)$
3. The coefficient of the middle term, −8, is negative. **4.** $(x - 5)(x - 3)$
5. $(t - 5)(t - 4)$
Guided Solution:
1. −13, 8, −8, 7, −7; positive; $x + 4$

Constant Term Negative

As we saw in two of the multiplications earlier in this section, the product of two binomials can have a negative constant term:

$$(x + 3)(x - 7) = x^2 - 4x - 21$$

and

$$(x - 3)(x + 7) = x^2 + 4x - 21.$$

Note that when the signs of the constants in the binomials are reversed, only the sign of the middle term in the product changes.

EXAMPLE 3 Factor: $x^2 - 8x - 20$.

The constant term, -20, must be expressed as the product of a negative number and a positive number. Since the sum of these two numbers must be negative (specifically, -8), the negative number must have the greater absolute value.

PAIRS OF FACTORS	SUMS OF FACTORS
1, −20	−19
2, −10	**−8** ←
4, −5	−1
5, −4	1
10, −2	8
20, −1	19

The numbers we need are 2 and −10.

Because these sums are all positive, for this problem all the corresponding pairs can be disregarded. Note that in all three pairs, the positive number has the greater absolute value.

The numbers that we are looking for are 2 and −10. The factorization is $(x + 2)(x - 10)$.

Check: $(x + 2)(x - 10) = x^2 - 10x + 2x - 20$
$= x^2 - 8x - 20.$

> **TO FACTOR $x^2 + bx + c$ WHEN c IS NEGATIVE**
>
> When the constant term of a trinomial is negative, look for two numbers whose product is negative. One must be positive and the other negative:
>
> $$x^2 - 4x - 21 = (x + 3)(x - 7);$$
>
> $$x^2 + 4x - 21 = (x - 3)(x + 7).$$
>
> Consider pairs of numbers for which the number with the larger absolute value has the same sign as b, the coefficient of the middle term.

◀ Do Exercises 6 and 7. (Exercise 7 is on the following page.)

---

6. Consider $x^2 - 5x - 24$.

a) Explain why you would *not* consider the pairs of factors listed below in factoring $x^2 - 5x - 24$.

PAIRS OF FACTORS	SUMS OF FACTORS
−1, 24	
−2, 12	
−3, 8	
−4, 6	

b) Explain why you *would* consider the pairs of factors listed below in factoring $x^2 - 5x - 24$.

PAIRS OF FACTORS	SUMS OF FACTORS
1, −24	
2, −12	
3, −8	
4, −6	

c) Factor: $x^2 - 5x - 24$.

Answer
6. (a) The positive factor has the larger absolute value. (b) The negative factor has the larger absolute value.
(c) $(x + 3)(x - 8)$

EXAMPLE 4 Factor: $t^2 - 24 + 5t$.

We first write the trinomial in descending order: $t^2 + 5t - 24$. Since the constant term, -24, is negative, factorizations of -24 will have one positive factor and one negative factor. The sum of the factors must be 5, so we consider only pairs of factors in which the positive factor has the larger absolute value.

PAIRS OF FACTORS	SUMS OF FACTORS
$-1, 24$	23
$-2, 12$	10
$-3, 8$	5 ← The numbers we need are -3 and 8.
$-4, 6$	2

The factorization is $(t - 3)(t + 8)$. The check is left to the student.

> Do Exercises 8 and 9.

EXAMPLE 5 Factor: $x^4 - x^2 - 110$.

Consider this trinomial as $(x^2)^2 - x^2 - 110$. We look for numbers p and q such that
$$x^4 - x^2 - 110 = (x^2 + p)(x^2 + q).$$
We look for two numbers whose product is -110 and whose sum is -1. The middle-term coefficient, -1, is small compared to -110. This tells us that the desired factors are close to each other in absolute value. The numbers we want are 10 and -11. The factorization is $(x^2 + 10)(x^2 - 11)$.

EXAMPLE 6 Factor: $a^2 + 4ab - 21b^2$.

We consider the trinomial in the equivalent form
$$a^2 + 4ba - 21b^2,$$
and think of $-21b^2$ as the "constant" term and $4b$ as the "coefficient" of the middle term. Then we try to express $-21b^2$ as a product of two factors whose sum is $4b$. Those factors are $-3b$ and $7b$. The factorization is $(a - 3b)(a + 7b)$.

Check: $(a - 3b)(a + 7b) = a^2 + 7ab - 3ba - 21b^2$
$= a^2 + 4ab - 21b^2$.

There are polynomials that are not factorable.

EXAMPLE 7 Factor: $x^2 - x + 5$.

Since 5 has very few factors, we can easily check all possibilities.

PAIRS OF FACTORS	SUMS OF FACTORS
$5, 1$	6
$-5, -1$	-6

There are no factors of 5 whose sum is -1. Thus the polynomial is *not* factorable into factors that are polynomials with rational-number coefficients.

7. Consider $x^2 + 5x - 6$.

 a) Explain why you would *not* consider the pairs of factors listed below in factoring $x^2 + 5x - 6$.

PAIRS OF FACTORS	SUMS OF FACTORS
$1, -6$	
$2, -3$	

 b) Explain why you *would* consider the pairs of factors listed below in factoring $x^2 + 5x - 6$.

PAIRS OF FACTORS	SUMS OF FACTORS
$-1, 6$	
$-2, 3$	

 c) Factor: $x^2 + 5x - 6$.

Factor.

GS 8. $a^2 - 40 + 3a$

First, rewrite in descending order:
$a^2 + 3a - \boxed{}$.

PAIRS OF FACTORS	SUMS OF FACTORS
$-1, 40$	☐
$-2, 20$	☐
$-4, 10$	☐
$-5, 8$	☐

The factorization is $(a - 5)(\boxed{})$.

9. $-18 - 3t + t^2$

Answers
7. (a) The negative factor has the larger absolute value. (b) The positive factor has the larger absolute value. (c) $(x - 1)(x + 6)$
8. $(a - 5)(a + 8)$ **9.** $(t - 6)(t + 3)$
Guided Solution:
8. $40, 39, 18, 6, 3, a + 8$

Factor.

10. $y^2 - 12 - 4y$

11. $t^4 + 5t^2 - 14$

12. $x^2 + 2x + 7$

In this text, a polynomial like $x^2 - x + 5$ that cannot be factored using rational numbers is said to be **prime**. In more advanced courses, polynomials like $x^2 - x + 5$ can be factored and are not considered prime.

◀ Do Exercises 10–12.

Often factoring requires two or more steps. In general, when told to factor, we should be sure to *factor completely*. This means that the final factorization should not contain any factors that can be factored further.

EXAMPLE 8 Factor: $2x^3 - 20x^2 + 50x$.

Always look first for a common factor. This time there is one, $2x$, which we factor out first:

$$2x^3 - 20x^2 + 50x = 2x(x^2 - 10x + 25).$$

Now consider $x^2 - 10x + 25$. Since the constant term is positive and the coefficient of the middle term is negative, we look for a factorization of 25 in which both factors are negative. Their sum must be -10.

PAIRS OF FACTORS	SUMS OF FACTORS
$-25, -1$	-26
$-5, -5$	-10 ← The numbers we need are -5 and -5.

The factorization of $x^2 - 10x + 25$ is $(x - 5)(x - 5)$, or $(x - 5)^2$. The final factorization is $2x(x - 5)^2$. We check by multiplying:

$$2x(x - 5)^2 = 2x(x^2 - 10x + 25)$$
$$= (2x)(x^2) - (2x)(10x) + (2x)(25)$$
$$= 2x^3 - 20x^2 + 50x.$$

Factor.

13. $x^3 + 4x^2 - 12x$

14. $p^2 - pq - 3pq^2$

15. $3x^3 + 24x^2 + 48x$

◀ Do Exercises 13–15.

Once any common factors have been factored out, the following summary can be used to factor $x^2 + bx + c$.

TO FACTOR $x^2 + bx + c$

1. First arrange the polynomial in descending order.
2. Use a trial-and-error process that looks for factors of c whose sum is b.
3. If c is positive, the signs of the factors are the same as the sign of b.
4. If c is negative, one factor is positive and the other is negative. If the sum of two factors is the opposite of b, changing the sign of each factor will give the desired factors whose sum is b.
5. Check by multiplying.

Answers

10. $(y - 6)(y + 2)$ 11. $(t^2 + 7)(t^2 - 2)$
12. Prime 13. $x(x + 6)(x - 2)$
14. $p(p - q - 3q^2)$ 15. $3x(x + 4)^2$

Leading Coefficient −1

EXAMPLE 9 Factor: $10 - 3x - x^2$.

Note that the polynomial is written in ascending order. When we write it in descending order, we get

$$-x^2 - 3x + 10,$$

which has a leading coefficient of −1. Before factoring in such a case, we can factor out a −1, as follows:

$$-x^2 - 3x + 10 = -1 \cdot x^2 + (-1)(3x) + (-1)(-10)$$
$$= -1(x^2 + 3x - 10).$$

Then we proceed to factor $x^2 + 3x - 10$. We get

$$-x^2 - 3x + 10 = -1(x^2 + 3x - 10) = -1(x + 5)(x - 2).$$

We can also express this answer in two other ways by multiplying either binomial by −1. Thus each of the following is a correct answer:

$$10 - 3x - x^2 = -1(x + 5)(x - 2)$$
$$= (-x - 5)(x - 2) \quad \text{Multiplying } x + 5 \text{ by } -1$$
$$= (x + 5)(-x + 2). \quad \text{Multiplying } x - 2 \text{ by } -1$$

Do Exercises 16 and 17.

Factor.
16. $14 + 5x - x^2$
17. $-x^2 + 3x + 18$

Answers
16. $-1(x + 2)(x - 7)$, or $(-x - 2)(x - 7)$, or $(x + 2)(-x + 7)$
17. $-1(x + 3)(x - 6)$, or $(-x - 3)(x - 6)$, or $(x + 3)(-x + 6)$

5.2 Exercise Set

FOR EXTRA HELP — MyLab Math

✓ Check Your Understanding

Reading Check Determine whether each statement is true or false.

RC1. The leading coefficient of $x^2 - 3x - 10$ is 1.

RC2. To factor $x^2 - 3x - 10$, we look for two numbers whose product is −10.

RC3. To factor $x^2 - 3x - 10$, we look for two numbers whose sum is −3.

RC4. The factorization of $x^2 - 3x - 10$ is $(x + 5)(x - 2)$.

Concept Check List all pairs of positive factors of each number.

CC1. 18

CC2. 42

CC3. 96

CC4. 150

a Factor by first filling in each table. Remember that you can check by multiplying.

1. $x^2 + 8x + 15$

PAIRS OF FACTORS	SUMS OF FACTORS

2. $x^2 + 5x + 6$

PAIRS OF FACTORS	SUMS OF FACTORS

3. $x^2 + 7x + 12$

PAIRS OF FACTORS	SUMS OF FACTORS

4. $x^2 + 9x + 8$

PAIRS OF FACTORS	SUMS OF FACTORS

5. $x^2 - 6x + 9$

PAIRS OF FACTORS	SUMS OF FACTORS

6. $y^2 - 11y + 28$

PAIRS OF FACTORS	SUMS OF FACTORS

7. $x^2 - 5x - 14$

PAIRS OF FACTORS	SUMS OF FACTORS

8. $a^2 + 7a - 30$

PAIRS OF FACTORS	SUMS OF FACTORS

9. $b^2 + 5b + 4$

PAIRS OF FACTORS	SUMS OF FACTORS

10. $z^2 - 8z + 7$

PAIRS OF FACTORS	SUMS OF FACTORS

11. $t^2 + 3t - 18$

PAIRS OF FACTORS	SUMS OF FACTORS

12. $t^2 + 8t + 16$

PAIRS OF FACTORS	SUMS OF FACTORS

Factor.

13. $d^2 - 7d + 10$
14. $t^2 - 12t + 35$
15. $y^2 - 11y + 10$
16. $x^2 - 4x - 21$

17. $x^2 + x + 1$
18. $x^2 + 5x + 3$
19. $x^2 - 7x - 18$
20. $y^2 - 3y - 28$

21. $x^3 - 6x^2 - 16x$
22. $x^3 - x^2 - 42x$
23. $y^3 - 4y^2 - 45y$
24. $x^3 - 7x^2 - 60x$

25. $-2x - 99 + x^2$
26. $x^2 - 72 + 6x$
27. $c^4 + c^2 - 56$
28. $b^4 + 5b^2 - 24$

29. $a^4 + 2a^2 - 35$
30. $x^4 - x^2 - 6$
31. $x^2 + x - 42$
32. $x^2 + 2x - 15$

33. $7 - 2p + p^2$
34. $11 - 3w + w^2$
35. $x^2 + 20x + 100$
36. $a^2 + 19a + 88$

37. $2z^3 - 2z^2 - 24z$
38. $5w^4 - 20w^3 - 25w^2$
39. $3t^4 + 3t^3 + 3t^2$
40. $4y^5 - 4y^4 - 4y^3$

41. $x^4 - 21x^3 - 100x^2$
42. $x^4 - 20x^3 + 96x^2$
43. $x^2 - 21x - 72$
44. $4x^2 + 40x + 100$

45. $x^2 - 25x + 144$
46. $y^2 - 21y + 108$
47. $a^2 + a - 132$
48. $a^2 + 9a - 90$

49. $3t^2 + 6t + 3$
50. $2y^2 + 24y + 72$
51. $w^4 - 8w^3 + 16w^2$
52. $z^5 - 6z^4 + 9z^3$

53. $30 + 7x - x^2$
54. $45 + 4x - x^2$
55. $24 - a^2 - 10a$
56. $-z^2 + 36 - 9z$

57. $120 - 23x + x^2$
58. $96 + 22d + d^2$
59. $108 - 3x - x^2$
60. $112 + 9y - y^2$

61. $y^2 - 0.2y - 0.08$
62. $t^2 - 0.3t - 0.10$
63. $p^2 + 3pq - 10q^2$
64. $a^2 + 2ab - 3b^2$

65. $84 - 8t - t^2$
66. $72 - 6m - m^2$
67. $m^2 + 5mn + 4n^2$
68. $x^2 + 11xy + 24y^2$

69. $s^2 - 2st - 15t^2$
70. $p^2 + 5pq - 24q^2$
71. $6a^{10} - 30a^9 - 84a^8$
72. $7x^9 - 28x^8 - 35x^7$

Skill Maintenance

Solve.

73. $-2y + 11y = 108$ [2.3b]

74. $\frac{1}{2}x - \frac{1}{3}x = \frac{2}{3} + \frac{5}{6}x$ [2.3b]

75. $5(t - 1) - 3 = 4t - (7t - 2)$ [2.3c]

76. $10 - (x - 7) = 4x - (1 + 5x)$ [2.3c]

77. $-2x < 48$ [2.7d]

78. $4x - 8x + 16 \geq 6(x - 2)$ [2.7e]

Solve. [2.4b]

79. $A = \dfrac{p + w}{2}$, for p

80. $y = mx + b$, for x

81. *Digital Media Usage.* In 2015, digital media usage in the United States was about 1.4 billion min. Use the information in the following circle graph to estimate how many minutes were spent using mobile apps. [2.5a]

Digital Media Usage

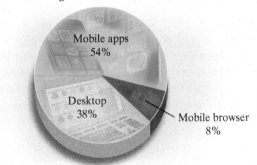

DATA: comscore.com

82. *Major League Baseball Attendance.* Total attendance at Major League baseball games was about 73.2 million in 2016. This was a 0.5% decrease from the attendance in 2015. What was the attendance in 2015? [2.5a]

Data: baseball-reference.com

Synthesis

83. Find all integers m for which $y^2 + my + 50$ can be factored.

84. Find all integers b for which $a^2 + ba - 50$ can be factored.

Factor completely.

85. $x^2 - \dfrac{1}{4}x - \dfrac{1}{8}$

86. $x^2 - \dfrac{2}{5}x + \dfrac{1}{25}$

87. $x^2 + \dfrac{30}{7}x - \dfrac{25}{7}$

88. $\dfrac{1}{3}x^3 + \dfrac{1}{3}x^2 - 2x$

89. $b^{2n} + 7b^n + 10$

90. $a^{2m} - 11a^m + 28$

Find a polynomial in factored form for the shaded area in each figure. (Leave answers in terms of π.)

91.

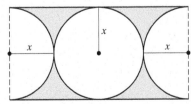

92.

5.3 Factoring $ax^2 + bx + c$, $a \neq 1$: The FOIL Method

OBJECTIVE

a Factor trinomials of the type $ax^2 + bx + c$, $a \neq 1$, using the FOIL method.

a THE FOIL METHOD

In this section, we factor trinomials in which the coefficient of the leading term x^2 is not 1. The procedure we use is a refined trial-and-error method.

We want to factor trinomials of the type $ax^2 + bx + c$. Consider the following multiplication:

$$\begin{array}{c} \text{F}\text{O}\text{I}\text{L} \\ (2x+5)(3x+4) = 6x^2 + 8x + 15x + 20 \\ = 6x^2 + 23x + 20 \end{array}$$

$$\begin{array}{ccc} \text{F} & \text{O} + \text{I} & \text{L} \\ 2 \cdot 3 & 2 \cdot 4 \;\; 5 \cdot 3 & 5 \cdot 4 \end{array}$$

To factor $6x^2 + 23x + 20$, we reverse the above multiplication, using what we might call an "unFOIL" process. We look for two binomials $rx + p$ and $sx + q$ whose product is $(rx + p)(sx + q) = 6x^2 + 23x + 20$. The product of the First terms must be $6x^2$. The product of the Outside terms plus the product of the Inside terms must be $23x$. The product of the Last terms must be 20. We know from the preceding discussion that the answer is $(2x + 5)(3x + 4)$. Generally, however, finding such an answer is a refined trial-and-error process. It turns out that $(-2x - 5)(-3x - 4)$ is also a correct answer, but we usually choose an answer in which the first coefficients are positive.

We will use the following trial-and-error method.

SKILL REVIEW

Multiply binomials using the FOIL method. [4.6a]

Multiply.

1. $(2x + 3)(x + 1)$
2. $(3x + 4)(2x - 1)$

Answers: 1. $2x^2 + 5x + 3$ 2. $6x^2 + 5x - 4$

MyLab Math VIDEO

The ac-method in Section 5.4

To the student: In Section 5.4, we will consider an alternative method for the same kind of factoring. It involves factoring by grouping and is called the ac-method.

To the instructor: We present two ways to factor general trinomials in Sections 5.3 and 5.4: the FOIL method in Section 5.3 and the ac-method in Section 5.4. You can teach both methods and let the student use the one that he or she prefers or you can select just one.

THE FOIL METHOD

To factor $ax^2 + bx + c$, $a \neq 1$, using the FOIL method:

1. Factor out the largest common factor, if one exists.
2. Find two First terms whose product is ax^2.

   $(\Box x +)(\Box x +) = ax^2 + bx + c$.

   FOIL

3. Find two Last terms whose product is c:

   $(x + \Box)(x + \Box) = ax^2 + bx + c$.

   FOIL

4. Look for Outer and Inner products resulting from steps (2) and (3) for which the sum is bx:

   $(\Box x + \Box)(\Box x + \Box) = ax^2 + bx + c$.

   I
   O
   FOIL

5. Always check by multiplying.

---

360 CHAPTER 5 Polynomials: Factoring

EXAMPLE 1 Factor: $3x^2 - 10x - 8$.

1) First, we check for a common factor. Here there is none (other than 1 or -1).

2) Find two **First** terms whose product is $3x^2$.

   The only possibilities for the **First** terms are $3x$ and x, so any factorization must be of the form

   $(3x + \square)(x + \square)$.

3) Find two **Last** terms whose product is -8.

   Possible factorizations of -8 are

   $(-8) \cdot 1, \quad 8 \cdot (-1), \quad (-2) \cdot 4, \quad \text{and} \quad 2 \cdot (-4)$.

   Since the First terms are not identical, we must also consider

   $1 \cdot (-8), \quad (-1) \cdot 8, \quad 4 \cdot (-2), \quad \text{and} \quad (-4) \cdot 2$.

4) Inspect the **O**utside and **I**nside products resulting from steps (2) and (3). Look for a combination in which the sum of the products is the middle term, $-10x$:

Trial	Product	
$(3x - 8)(x + 1)$	$3x^2 + 3x - 8x - 8$ $= 3x^2 - 5x - 8$	← Wrong middle term
$(3x + 8)(x - 1)$	$3x^2 - 3x + 8x - 8$ $= 3x^2 + 5x - 8$	← Wrong middle term
$(3x - 2)(x + 4)$	$3x^2 + 12x - 2x - 8$ $= 3x^2 + 10x - 8$	← Wrong middle term
$(3x + 2)(x - 4)$	$3x^2 - 12x + 2x - 8$ $= 3x^2 - 10x - 8$	← Correct middle term!
$(3x + 1)(x - 8)$	$3x^2 - 24x + x - 8$ $= 3x^2 - 23x - 8$	← Wrong middle term
$(3x - 1)(x + 8)$	$3x^2 + 24x - x - 8$ $= 3x^2 + 23x - 8$	← Wrong middle term
$(3x + 4)(x - 2)$	$3x^2 - 6x + 4x - 8$ $= 3x^2 - 2x - 8$	← Wrong middle term
$(3x - 4)(x + 2)$	$3x^2 + 6x - 4x - 8$ $= 3x^2 + 2x - 8$	← Wrong middle term

   The correct factorization is $(3x + 2)(x - 4)$.

5) Check: $(3x + 2)(x - 4) = 3x^2 - 10x - 8$.

Two observations can be made from Example 1. First, we listed all possible trials even though we could have stopped after having found the correct factorization. We did this to show that each trial differs only in the middle term of the product. **Second, note that only the sign of the middle term changes when the signs in the binomials are reversed:**

Plus Minus
↓   ↓
$(3x + 4)(x - 2) = 3x^2 - 2x - 8$

Minus Plus
↓   ↓
$(3x - 4)(x + 2) = 3x^2 + 2x - 8$.

↑——— Middle term changes sign

CALCULATOR CORNER

A partial check of a factorization can be performed using a table or a graph. To check the factorization $6x^3 - 9x^2 + 4x - 6 = (2x - 3)(3x^2 + 2)$, for example, we enter $y_1 = 6x^3 - 9x^2 + 4x - 6$ and $y_2 = (2x - 3)(3x^2 + 2)$ on the equation-editor screen. Then we set up a table in AUTO mode. If the factorization is correct, the values of y_1 and y_2 will be the same regardless of the table settings used.

X	Y1	Y2
-3	-261	-261
-2	-98	-98
-1	-25	-25
0	-6	-6
1	-5	-5
2	14	14
3	87	87

X = -3

We can also graph $y_1 = 6x^3 - 9x^2 + 4x - 6$ and $y_2 = (2x - 3)(3x^2 + 2)$. If the graphs appear to coincide, the factorization is probably correct.

$y_1 = 6x^3 - 9x^2 + 4x - 6,$
$y_2 = (2x - 3)(3x^2 + 2)$

Yscl = 2

EXERCISES: Use a table or a graph as a partial check to determine whether each factorization is correct.

1. $24x^2 - 76x + 40 =$
   $4(3x - 2)(2x - 5)$
2. $4x^2 - 5x - 6 =$
   $(4x + 3)(x - 2)$
3. $5x^2 + 17x - 12 =$
   $(5x + 3)(x - 4)$
4. $10x^2 + 37x + 7 =$
   $(5x - 1)(2x + 7)$

Factor.

1. $2x^2 - x - 15$

2. $12x^2 - 17x - 5$

◀ Do Exercises 1 and 2.

EXAMPLE 2 Factor: $24x^2 - 76x + 40$.

1) First, we factor out the largest common factor, 4:

   $4(6x^2 - 19x + 10)$.

   Now we factor the trinomial $6x^2 - 19x + 10$.

2) Because $6x^2$ can be factored as $3x \cdot 2x$ or $6x \cdot x$, we have these possibilities for factorizations:

   $(3x + \square)(2x + \square)$ or $(6x + \square)(x + \square)$.

3) There are four pairs of factors of 10 and each pair can be listed in two ways:

   10, 1   −10, −1   5, 2   −5, −2

   and

   1, 10   −1, −10   2, 5   −2, −5.

4) The two possibilities from step (2) and the eight possibilities from step (3) give $2 \cdot 8$, or 16, possibilities for factorizations. We look for **O**utside and **I**nside products resulting from steps (2) and (3) for which the sum is the middle term, $-19x$. Since the sign of the middle term is negative, but the sign of the last term, 10, is positive, both factors of 10 must be negative. This means only four pairings from step (3) need be considered. We first try these factors with

   $(3x + \square)(2x + \square)$.

   If none gives the correct factorization, we will consider

   $(6x + \square)(x + \square)$.

Trial	Product	
$(3x - 10)(2x - 1)$	$6x^2 - 3x - 20x + 10$ $= 6x^2 - 23x + 10$	← Wrong middle term
$(3x - 1)(2x - 10)$	$6x^2 - 30x - 2x + 10$ $= 6x^2 - 32x + 10$	← Wrong middle term
$(3x - 5)(2x - 2)$	$6x^2 - 6x - 10x + 10$ $= 6x^2 - 16x + 10$	← Wrong middle term
$(3x - 2)(2x - 5)$	$6x^2 - 15x - 4x + 10$ $= 6x^2 - 19x + 10$	← Correct middle term!

   Since we have a correct factorization, we need not consider

   $(6x + \square)(x + \square)$.

   The factorization of $6x^2 - 19x + 10$ is $(3x - 2)(2x - 5)$, but *do not forget the common factor!* We must include it in order to factor the original trinomial:

   $24x^2 - 76x + 40 = 4(6x^2 - 19x + 10)$
   $= 4(3x - 2)(2x - 5)$.

5) Check: $4(3x - 2)(2x - 5) = 4(6x^2 - 19x + 10) = 24x^2 - 76x + 40$.

·················· **Caution!** ··················

When factoring any polynomial, always look for a common factor first. Failure to do so is such a common error that this caution bears repeating.

Answers

1. $(2x + 5)(x - 3)$ **2.** $(4x + 1)(3x - 5)$

In Example 2, look again at the possibility $(3x - 5)(2x - 2)$. Without multiplying, we can reject such a possibility. To see why, consider the following:

$$(3x - 5)(2x - 2) = (3x - 5)(2)(x - 1) = 2(3x - 5)(x - 1).$$

The expression $2x - 2$ has a common factor, 2. But we removed the *largest* common factor in the first step. If $2x - 2$ were one of the factors, then 2 would have to be a common factor in addition to the original 4. Thus, $(2x - 2)$ cannot be part of the factorization of the original trinomial.

> Given that the largest common factor is factored out at the outset, we need not consider factorizations that have a common factor.

Do Exercises 3 and 4.

Factor.

3. $3x^2 - 19x + 20$

4. $20x^2 - 46x + 24$

EXAMPLE 3 Factor: $10x^2 + 37x + 7$.

1) There is no common factor (other than 1 or -1).

2) Because $10x^2$ factors as $10x \cdot x$ or $5x \cdot 2x$, we have these possibilities for factorizations:

   $(10x + \square)(x + \square)$ or $(5x + \square)(2x + \square)$.

3) There are two pairs of factors of 7 and each pair can be listed in two ways:

   1, 7 $-1, -7$ and   7, 1 $-7, -1$.

4) From steps (2) and (3), we see that there are 8 possibilities for factorizations. Look for **O**uter and **I**nner products for which the sum is the middle term. Because all coefficients in $10x^2 + 37x + 7$ are positive, we need consider only positive factors of 7. The possibilities are

   $(10x + 1)(x + 7) = 10x^2 + 71x + 7,$
   $(10x + 7)(x + 1) = 10x^2 + 17x + 7,$
   $(5x + 7)(2x + 1) = 10x^2 + 19x + 7,$
   $(5x + 1)(2x + 7) = 10x^2 + 37x + 7.$ ← Correct middle term

   The factorization is $(5x + 1)(2x + 7)$.

5) Check: $(5x + 1)(2x + 7) = 10x^2 + 37x + 7.$

Do Exercise 5.

5. Factor: $6x^2 + 7x + 2$.

> **TIPS FOR FACTORING** $ax^2 + bx + c, a \neq 1$
>
> - Always factor out the largest common factor first, if one exists.
> - Once the common factor has been factored out of the original trinomial, no binomial factor can contain a common factor (other than 1 or -1).
> - If c is positive, then the signs in both binomial factors must match the sign of b. (This assumes that $a > 0$.)
> - Reversing the signs in the binomials reverses the sign of the middle term of their product.
> - Organize your work so that you can keep track of which possibilities have or have not been checked.
> - Always check by multiplying.

Answers
3. $(3x - 4)(x - 5)$ 4. $2(5x - 4)(2x - 3)$
5. $(2x + 1)(3x + 2)$

SECTION 5.3 Factoring $ax^2 + bx + c, a \neq 1$: The FOIL Method   **363**

Factor.

6. $2 - x - 6x^2$

7. $2x + 8 - 6x^2$

EXAMPLE 4 Factor: $10x + 8 - 3x^2$.

An important problem-solving strategy is to find a way to make new problems look like problems we already know how to solve. The factoring tips on the preceding page apply only to trinomials of the form $ax^2 + bx + c$, with $a > 0$. This leads us to rewrite $10x + 8 - 3x^2$ in descending order:

$$10x + 8 - 3x^2 = -3x^2 + 10x + 8. \quad \text{Writing in descending order}$$

Although $-3x^2 + 10x + 8$ looks similar to the trinomials we have factored, the factoring tips require a positive leading coefficient, so we factor out -1:

$$-3x^2 + 10x + 8 = -1(3x^2 - 10x - 8) \quad \text{Factoring out } -1 \text{ changes the signs of the coefficients.}$$

$$= -1(3x + 2)(x - 4). \quad \text{Using the result from Example 1}$$

The factorization of $10x + 8 - 3x^2$ is $-1(3x + 2)(x - 4)$. Other correct answers are

$$10x + 8 - 3x^2 = (3x + 2)(-x + 4) \quad \text{Multiplying } x - 4 \text{ by } -1$$
$$= (-3x - 2)(x - 4). \quad \text{Multiplying } 3x + 2 \text{ by } -1$$

◀ Do Exercises 6 and 7.

EXAMPLE 5 Factor: $6p^2 - 13pv - 28v^2$.

1) Factor out a common factor, if any.

   There is none (other than 1 or -1).

2) Factor the first term, $6p^2$.

   Possibilities are $2p, 3p$ and $6p, p$. We have these as possibilities for factorizations:

   $$(2p + \square)(3p + \square) \quad \text{or} \quad (6p + \square)(p + \square).$$

3) Factor the last term, $-28v^2$, which has a negative coefficient.

   There are six pairs of factors and each can be listed in two ways:

   $$-28v, v \quad 28v, -v \quad -14v, 2v \quad 14v, -2v \quad -7v, 4v \quad 7v, -4v$$

   and

   $$v, -28v \quad -v, 28v \quad 2v, -14v \quad -2v, 14v \quad 4v, -7v \quad -4v, 7v.$$

4) The coefficient of the middle term is negative, so we look for combinations of factors from steps (2) and (3) such that the sum of their products has a negative coefficient. We try some possibilities:

   $$(2p + v)(3p - 28v) = 6p^2 - 53pv - 28v^2,$$
   $$(2p - 7v)(3p + 4v) = 6p^2 - 13pv - 28v^2. \quad \leftarrow \text{Correct middle term}$$

   The factorization of $6p^2 - 13pv - 28v^2$ is $(2p - 7v)(3p + 4v)$.

5) The check is left to the student.

◀ Do Exercises 8 and 9.

Factor.

8. $6a^2 - 5ab + b^2$

9. $6x^2 + 15xy + 9y^2$

Answers

6. $-1(2x - 1)(3x + 2)$, or $(2x - 1)(-3x - 2)$, or $(-2x + 1)(3x + 2)$ 7. $-2(3x - 4)(x + 1)$, or $2(3x - 4)(-x - 1)$, or $2(-3x + 4)(x + 1)$
8. $(2a - b)(3a - b)$ 9. $3(2x + 3y)(x + y)$

5.3 Exercise Set

✓ Check Your Understanding

Reading Check Determine whether each statement is true or false.

RC1. When factoring a polynomial, we always look for a common factor first.

RC2. We can check any factorization by multiplying.

RC3. When we are factoring $10x^2 + 21x + 2$, the only choices for the First terms in the binomial factors are $2x$ and $5x$.

RC4. The factorization of $10x^2 + 21x + 2$ is $(2x + 1)(5x + 2)$.

Concept Check Use the possible trial factorizations of $6x^2 + 5x - 50$ listed in the columns on the right below to answer the following questions.

CC1. Cross out all trials that contain a common factor. Which trials can be eliminated?

CC2. Find the middle term of each remaining product.

CC3. What is the factorization of $6x^2 + 5x - 50$?

a) $(6x + 1)(x - 50)$
b) $(6x + 50)(x - 1)$
c) $(6x + 2)(x - 25)$
d) $(6x + 25)(x - 2)$
e) $(6x + 5)(x - 10)$
f) $(6x + 10)(x - 5)$
g) $(3x + 1)(2x - 50)$
h) $(3x + 50)(2x - 1)$
i) $(3x + 2)(2x - 25)$
j) $(3x + 25)(2x - 2)$
k) $(3x + 5)(2x - 10)$
l) $(3x + 10)(2x - 5)$

a Factor.

1. $2x^2 - 7x - 4$
2. $3x^2 - x - 4$
3. $5x^2 - x - 18$
4. $4x^2 - 17x + 15$

5. $6x^2 + 23x + 7$
6. $6x^2 - 23x + 7$
7. $3x^2 + 4x + 1$
8. $7x^2 + 15x + 2$

9. $4x^2 + 4x - 15$
10. $9x^2 + 6x - 8$
11. $2x^2 - x - 1$
12. $15x^2 - 19x - 10$

13. $9x^2 + 18x - 16$
14. $2x^2 + 5x + 2$
15. $3x^2 - 5x - 2$
16. $18x^2 - 3x - 10$

17. $12x^2 + 31x + 20$
18. $15x^2 + 19x - 10$
19. $14x^2 + 19x - 3$
20. $35x^2 + 34x + 8$

21. $9x^2 + 18x + 8$
22. $6 - 13x + 6x^2$
23. $49 - 42x + 9x^2$
24. $16 + 36x^2 + 48x$

25. $24x^2 + 47x - 2$
26. $16p^2 - 78p + 27$
27. $35x^2 - 57x - 44$
28. $9a^2 + 12a - 5$

29. $20 + 6x - 2x^2$
30. $15 + x - 2x^2$
31. $12x^2 + 28x - 24$
32. $6x^2 + 33x + 15$

33. $30x^2 - 24x - 54$
34. $18t^2 - 24t + 6$
35. $4y + 6y^2 - 10$
36. $-9 + 18x^2 - 21x$

37. $3x^2 - 4x + 1$
38. $6t^2 + 13t + 6$
39. $12x^2 - 28x - 24$
40. $6x^2 - 33x + 15$

41. $-1 + 2x^2 - x$
42. $-19x + 15x^2 + 6$
43. $9x^2 - 18x - 16$
44. $14y^2 + 35y + 14$

45. $15x^2 - 25x - 10$
46. $18x^2 + 3x - 10$
47. $12p^3 + 31p^2 + 20p$
48. $15x^3 + 19x^2 - 10x$

49. $16 + 18x - 9x^2$
50. $33t - 15 - 6t^2$
51. $-15x^2 + 19x - 6$
52. $1 + p - 2p^2$

53. $14x^4 + 19x^3 - 3x^2$
54. $70x^4 + 68x^3 + 16x^2$
55. $168x^3 - 45x^2 + 3x$
56. $144x^5 + 168x^4 + 48x^3$

57. $15x^4 - 19x^2 + 6$
58. $9x^4 + 18x^2 + 8$
59. $25t^2 + 80t + 64$
60. $9x^2 - 42x + 49$

61. $6x^3 + 4x^2 - 10x$ **62.** $18x^3 - 21x^2 - 9x$ **63.** $25x^2 + 79x + 64$ **64.** $9y^2 + 42y + 47$

65. $6x^2 - 19x - 5$ **66.** $2x^2 + 11x - 9$ **67.** $12m^2 - mn - 20n^2$ **68.** $12a^2 - 17ab + 6b^2$

69. $6a^2 - ab - 15b^2$ **70.** $3p^2 - 16pw - 12w^2$ **71.** $9a^2 + 18ab + 8b^2$ **72.** $10s^2 + 4st - 6t^2$

73. $35p^2 + 34pt + 8t^2$ **74.** $30a^2 + 87ab + 30b^2$ **75.** $18x^2 - 6xy - 24y^2$ **76.** $15a^2 - 5ab - 20b^2$

Skill Maintenance

Graph.

77. $y = \dfrac{2}{5}x - 1$ [3.2a]   **78.** $2x = 6$ [3.3b]   **79.** $x = 4 - 2y$ [3.2a]   **80.** $y = -3$ [3.3b]

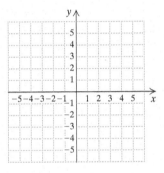 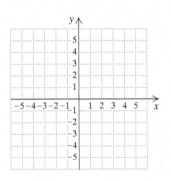 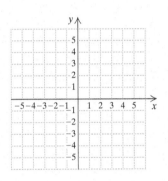 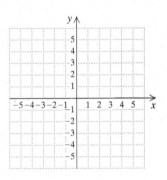

Find the intercepts of each equation. Then graph the equation. [3.3a]

81. $x + y = 4$ **82.** $x - y = 3$ **83.** $5x - 3y = 15$ **84.** $y - 3x = 3$

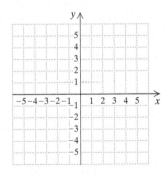 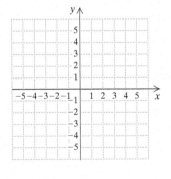 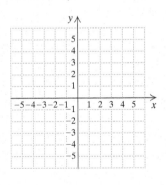 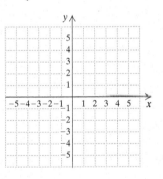

Synthesis

Factor.

85. $20x^{2n} + 16x^n + 3$ **86.** $-15x^{2m} + 26x^m - 8$

87. $3x^{6a} - 2x^{3a} - 1$ **88.** $x^{2n+1} - 2x^{n+1} + x$

89.–94. Use the TABLE feature to check the factoring in Exercises 15–20. (See the Calculator Corner on p. 361.)

5.4 Factoring $ax^2 + bx + c$, $a \neq 1$: The ac-Method

OBJECTIVE

a Factor trinomials of the type $ax^2 + bx + c$, $a \neq 1$, using the ac-method.

SKILL REVIEW

Factor by grouping. [5.1c]

Factor.
1. $2x^3 - 3x^2 - 10x + 15$
2. $n^3 + 3n^2 + n + 3$

Answers:
1. $(2x - 3)(x^2 - 5)$
2. $(n + 3)(n^2 + 1)$

a THE ac-METHOD

Another method for factoring trinomials of the type $ax^2 + bx + c$, $a \neq 1$, involves the product, ac, of the leading coefficient a and the last term c. It is called the **ac-method**. Because it uses factoring by grouping, it is also referred to as the **grouping method**.

We know how to factor the trinomial $x^2 + 5x + 6$. We look for factors of the constant term, 6, whose sum is the coefficient of the middle term, 5. What happens when the leading coefficient is not 1? To factor a trinomial like $3x^2 - 10x - 8$, we can use a method similar to the one that we used for $x^2 + 5x + 6$. That method is outlined as follows.

THE ac-METHOD

To factor a trinomial using the ac-method:

1. Factor out a common factor, if any. We refer to the remaining trinomial as $ax^2 + bx + c$.
2. Multiply the leading coefficient a and the constant c.
3. Try to factor the product ac so that the sum of the factors is b. That is, find integers p and q such that $pq = ac$ and $p + q = b$.
4. Split the middle term, writing it as a sum using the factors found in step (3).
5. Factor by grouping.
6. Check by multiplying.

EXAMPLE 1 Factor: $3x^2 - 10x - 8$.

1) First, we factor out a common factor, if any. There is none (other than 1 or -1).

2) We multiply the leading coefficient, 3, and the constant, -8:

$$3(-8) = -24.$$

3) Then we look for a factorization of -24 in which the sum of the factors is the coefficient of the middle term, -10.

PAIRS OF FACTORS	SUMS OF FACTORS
-1, 24	23
1, -24	-23
-2, 12	10
2, -12	**-10** ← $2 + (-12) = -10$
-3, 8	5
3, -8	-5
-4, 6	2
4, -6	-2

We show these sums for completeness. In practice, we stop when we find the correct sum.

4) Next, we split the middle term as a sum or a difference using the factors found in step (3): $-10x = 2x - 12x$.

5) Finally, we factor by grouping, as follows:

$3x^2 - 10x - 8 = 3x^2 + 2x - 12x - 8$ Substituting $2x - 12x$ for $-10x$

$= (3x^2 + 2x) + (-12x - 8)$
$= x(3x + 2) - 4(3x + 2)$ Factoring by grouping
$= (3x + 2)(x - 4)$.

We can also split the middle term as $-12x + 2x$. We still get the same factorization, although the factors may be in a different order. Note the following:

$3x^2 - 10x - 8 = 3x^2 - 12x + 2x - 8$ Substituting $-12x + 2x$ for $-10x$

$= (3x^2 - 12x) + (2x - 8)$
$= 3x(x - 4) + 2(x - 4)$ Factoring by grouping
$= (x - 4)(3x + 2)$.

6) Check: $(3x + 2)(x - 4) = 3x^2 - 10x - 8$.

Do Exercises 1 and 2. ▶

EXAMPLE 2 Factor: $8x^2 + 8x - 6$.

1) First, we factor out a common factor, if any. The number 2 is common to all three terms, so we factor it out: $2(4x^2 + 4x - 3)$.

2) Next, we factor the trinomial $4x^2 + 4x - 3$. We multiply the leading coefficient and the constant, 4 and -3: $4(-3) = -12$.

3) We try to factor -12 so that the sum of the factors is 4.

PAIRS OF FACTORS	SUMS OF FACTORS	
$-1, 12$	11	
$1, -12$	-11	
$-2, 6$	4 ←	$-2 + 6 = 4$
$2, -6$	•	We have found the correct sum, so there is no need to complete the table.
$-3, 4$	•	
$3, -4$	•	

4) Then we split the middle term, $4x$, as follows: $4x = -2x + 6x$.

5) Finally, we factor by grouping:

$4x^2 + 4x - 3 = 4x^2 - 2x + 6x - 3$ Substituting $-2x + 6x$ for $4x$

$= (4x^2 - 2x) + (6x - 3)$
$= 2x(2x - 1) + 3(2x - 1)$ Factoring by grouping
$= (2x - 1)(2x + 3)$.

The factorization of $4x^2 + 4x - 3$ is $(2x - 1)(2x + 3)$. *But don't forget the common factor!* We must include it to get a factorization of the original trinomial: $8x^2 + 8x - 6 = 2(2x - 1)(2x + 3)$.

6) Check: $2(2x - 1)(2x + 3) = 2(4x^2 + 4x - 3) = 8x^2 + 8x - 6$.

Do Exercises 3 and 4. ▶

Factor.

1. $6x^2 + 7x + 2$

 2. $12x^2 - 17x - 5$

1) There is no common factor.
2) Multiplying the leading coefficient and the constant:
   $12(-5) = \underline{}$.
3) Look for a pair of factors of -60 whose sum is -17. Those factors are 3 and $\underline{}$.
4) Split the middle term:
   $-17x = 3x - \underline{}$.
5) Factor by grouping:
   $12x^2 + 3x - 20x - 5$
   $= 3x(4x + 1) - 5(\underline{})$
   $= (\underline{})(3x - 5)$.
6) Check:
   $(4x + 1)(3x - 5)$
   $= 12x^2 - 17x - 5$.

Factor.

3. $6x^2 + 15x + 9$

4. $20x^2 - 46x + 24$

Answers
1. $(2x + 1)(3x + 2)$ **2.** $(4x + 1)(3x - 5)$
3. $3(2x + 3)(x + 1)$ **4.** $2(5x - 4)(2x - 3)$
Guided Solution:
2. $-60, -20, 20x, 4x + 1, 4x + 1$

SECTION 5.4 Factoring $ax^2 + bx + c$, $a \neq 1$: The ac-Method   369

5.4 Exercise Set

✓ Check Your Understanding

Reading Check Complete each step in the process to factor $10x^2 + 21x + 2$.

RC1. Using the *ac*-method, multiply the _____ 10 and the constant 2. The product is 20.

RC2. Find two integers whose _____ is 20 and whose _____ is 21. The integers are 20 and 1.

RC3. Split the middle term, $21x$, writing it as the _____ of $20x$ and x.

RC4. Factor by _____: $10x^2 + 20x + x + 2 = 10x(x + 2) + 1(x + 2) = (x + 2)(10x + 1)$.

Concept Check Each of the following trinomials is to be factored using the *ac*-method. Choose the option from the column on the right that indicates how the middle term should be split.

CC1. $6x^2 - 7x - 20$

CC2. $2x^2 - 7x - 4$

CC3. $3x^2 - 7x - 10$

CC4. $3x^2 - 7x - 6$

a) $x - 8x$
b) $2x - 9x$
c) $3x - 10x$
d) $8x - 15x$

a Factor. Note that the middle term has already been split.

1. $x^2 + 2x + 7x + 14$

2. $x^2 + 3x + x + 3$

3. $x^2 - 4x - x + 4$

4. $a^2 + 5a - 2a - 10$

5. $6x^2 + 4x + 9x + 6$

6. $3x^2 - 2x + 3x - 2$

7. $3x^2 - 4x - 12x + 16$

8. $24 - 18y - 20y + 15y^2$

9. $35x^2 - 40x + 21x - 24$

10. $8x^2 - 6x - 28x + 21$

11. $4x^2 + 6x - 6x - 9$

12. $2x^4 - 6x^2 - 5x^2 + 15$

13. $2x^4 + 6x^2 + 5x^2 + 15$

14. $9x^4 - 6x^2 - 6x^2 + 4$

Factor using the *ac*-method.

15. $2x^2 + 7x + 6$ **16.** $5x^2 + 17x + 6$ **17.** $3x^2 - 4x - 15$ **18.** $3x^2 + x - 4$

19. $5x^2 + 11x + 2$ **20.** $3x^2 + 16x + 5$ **21.** $3x^2 - 4x + 1$ **22.** $7x^2 - 15x + 2$

23. $6x^2 + 23x + 7$ **24.** $6x^2 + 13x + 6$ **25.** $4x^2 - 4x - 15$ **26.** $9x^2 - 6x - 8$

27. $15x^2 + 19x - 10$ **28.** $6 - 13x + 6x^2$ **29.** $9x^2 - 18x - 16$ **30.** $18x^2 + 3x - 10$

31. $3x^2 + 5x - 2$ **32.** $2x^2 - 5x + 2$ **33.** $12x^2 - 31x + 20$ **34.** $35x^2 - 34x + 8$

35. $14x^2 - 19x - 3$ **36.** $15x^2 - 19x - 10$ **37.** $49 - 42x + 9x^2$ **38.** $25x^2 + 40x + 16$

39. $9x^2 + 18x + 8$ **40.** $24x^2 - 47x - 2$ **41.** $5 - 9a^2 - 12a$ **42.** $17x - 4x^2 + 15$

43. $20 + 6x - 2x^2$ **44.** $15 + x - 2x^2$ **45.** $12x^2 + 28x - 24$ **46.** $6x^2 + 33x + 15$

47. $30x^2 - 24x - 54$ **48.** $18t^2 - 24t + 6$ **49.** $4y + 6y^2 - 10$ **50.** $-9 + 18x^2 - 21x$

51. $3x^2 - 4x + 1$
52. $6t^2 + t - 15$
53. $12x^2 - 28x - 24$
54. $6x^2 - 33x + 15$

55. $-1 + 2x^2 - x$
56. $-19x + 15x^2 + 6$
57. $9x^2 + 18x - 16$
58. $14y^2 + 35y + 14$

59. $15x^2 - 25x - 10$
60. $18x^2 + 3x - 10$
61. $12p^3 + 31p^2 + 20p$
62. $15x^3 + 19x^2 - 10x$

63. $4 - x - 5x^2$
64. $1 - p - 2p^2$
65. $33t - 15 - 6t^2$
66. $-15x^2 - 19x - 6$

67. $14x^4 + 19x^3 - 3x^2$
68. $70x^4 + 68x^3 + 16x^2$
69. $168x^3 - 45x^2 + 3x$
70. $144x^5 + 168x^4 + 48x^3$

71. $15x^4 - 19x^2 + 6$
72. $9x^4 + 18x^2 + 8$
73. $25t^2 + 80t + 64$
74. $9x^2 - 42x + 49$

75. $6x^3 + 4x^2 - 10x$
76. $18x^3 - 21x^2 - 9x$
77. $3x^2 + 9x + 5$
78. $4x^2 + 6x + 3$

79. $6x^2 - 19x - 5$
80. $2x^2 + 11x - 9$
81. $12m^2 - mn - 20n^2$
82. $12a^2 - 17ab + 6b^2$

83. $6a^2 - ab - 15b^2$
84. $3p^2 - 16pq - 12q^2$
85. $9a^2 - 18ab + 8b^2$
86. $10s^2 + 4st - 6t^2$

87. $35p^2 + 34pq + 8q^2$
88. $30a^2 + 87ab + 30b^2$
89. $18x^2 - 6xy - 24y^2$
90. $15a^2 - 5ab - 20b^2$

91. $60x + 18x^2 - 6x^3$
92. $60x + 4x^2 - 8x^3$
93. $35x^5 - 57x^4 - 44x^3$
94. $15x^3 + 33x^4 + 6x^5$

Skill Maintenance

Simplify. Express the result using positive exponents.

95. $(3x^4)^3$ [4.2a, b]
96. $5^{-6} \cdot 5^{-8}$ [4.1d, f]
97. $(x^2y)(x^3y^5)$ [4.1d]
98. $\dfrac{a^{-7}}{a^{-8}}$ [4.1e, f]

99. Convert to scientific notation: 30,080,000,000. [4.2c]

100. Convert to decimal notation: 1.5×10^{-5}. [4.2c]

Solve. [2.6a]

101. Each treehouse at Free Spirit Spheres in Vancouver, British Columbia, is a sphere with an external circumference of about 10 m. Find the radius of a treehouse, in meters and in feet. Use 3.14 for π. (*Hint*: 1 m $\approx$ 3.3 ft.)

    Data: freespiritspheres.com

102. The second angle of a triangle is 10° less than twice the first. The third angle is 15° more than four times the first. Find the measure of the second angle.

Synthesis

Factor.

103. $9x^{10} - 12x^5 + 4$
104. $24x^{2n} + 22x^n + 3$

105. $16x^{10} + 8x^5 + 1$
106. $(a + 4)^2 - 2(a + 4) + 1$

107.–112. Use graphs to check the factoring in Exercises 15–20. (See the Calculator Corner on p. 361.)

Mid-Chapter Review

Concept Reinforcement

Determine whether each statement is true or false.

_____ 1. The greatest common factor (GCF) of a set of natural numbers is at least 1 and always less than or equal to the smallest number in the set. [5.1a]

_____ 2. To factor $x^2 + bx + c$, we use a trial-and-error process that looks for factors of b whose sum is c. [5.2a]

_____ 3. The only numbers that we can factor out of a prime polynomial are 1 and -1. [5.2a]

_____ 4. When factoring $x^2 - 14x + 45$, we need consider only positive pairs of factors of 45. [5.2a]

Guided Solutions

GS Fill in each blank with the number, variable, or expression that creates a correct statement or solution.

5. Factor: $10y^3 - 18y^2 + 12y$. [5.1b]

$$10y^3 - 18y^2 + 12y = \underline{} \cdot 5y^2 - \underline{} \cdot 9y + \underline{} \cdot 6$$
$$= 2y(\underline{})$$

6. Factor $2x^2 - x - 6$ using the ac-method. [5.4a]

$a \cdot c = \underline{} \cdot \underline{} = -12;$ Multiplying the leading coefficient and the constant

$-x = \underline{} + 3x;$ Splitting the middle term

$2x^2 - x - 6 = 2x^2 - 4x + \underline{} - 6$
$= \underline{}(x - 2) + \underline{}(x - 2)$
$= (x - 2)(\underline{})$

Mixed Review

Find the GCF. [5.1a]

7. $x^3, 3x$

8. $5x^4, x^2$

9. $6x^5, -12x^3$

10. $-8x, -12, 16x^2$

11. $15x^3y^2, 5x^2y, 40x^4y^3$

12. $x^2y^4, -x^3y^3, x^3y^2, x^5y^4$

Factor completely. [5.1b, c], [5.2a], [5.3a], [5.4a]

13. $x^3 - 8x$

14. $3x^2 + 12x$

15. $2y^2 + 8y - 4$

16. $3t^6 - 5t^4 - 2t^3$

17. $x^2 + 4x + 3$

18. $z^2 - 4z + 4$

19. $x^3 + 4x^2 + 3x + 12$

20. $8y^5 - 48y^3$

21. $6x^3y + 24x^2y^2 - 42xy^3$

22. $6 - 11t + 4t^2$

23. $z^2 + 4z - 5$

24. $2z^3 + 8z^2 + 5z + 20$

25. $3p^3 - 2p^2 - 9p + 6$

26. $10x^8 - 25x^6 - 15x^5 + 35x^3$

27. $2w^3 + 3w^2 - 6w - 9$

374 CHAPTER 5 Polynomials: Factoring

28. $4x^4 - 5x^3 + 3x^2$
29. $6y^2 + 7y - 10$
30. $3x^2 - 3x - 18$
31. $6x^3 + 4x^2 + 3x + 2$
32. $15 - 8w + w^2$
33. $8x^3 + 20x^2 + 2x + 5$
34. $10z^2 - 21z - 10$
35. $6x^2 + 7x + 2$
36. $x^2 - 10xy + 24y^2$
37. $6z^3 + 3z^2 + 2z + 1$
38. $a^3b^7 + a^4b^5 - a^2b^3 + a^5b^6$
39. $4y^2 - 7yz - 15z^2$
40. $3x^3 + 21x^2 + 30x$
41. $x^3 - 3x^2 - 2x + 6$
42. $9y^2 + 6y + 1$
43. $y^2 + 6y + 8$
44. $6y^2 + 33y + 45$
45. $x^3 - 7x^2 + 4x - 28$
46. $4 + 3y - y^2$
47. $16x^2 - 16x - 60$
48. $10a^2 - 11ab + 3b^2$
49. $6w^3 - 15w^2 - 10w + 25$
50. $y^3 + 9y^2 + 18y$
51. $4x^2 + 11xy + 6y^2$
52. $6 - 5z - 6z^2$
53. $12t^3 + 8t^2 - 9t - 6$
54. $y^2 + yz - 20z^2$
55. $9x^2 - 6xy - 8y^2$
56. $-3 + 8z + 3z^2$
57. $m^2 - 6mn - 16n^2$
58. $2w^2 - 12w + 18$
59. $18t^3 - 18t^2 + 4t$
60. $5z^3 + 15z^2 + z + 3$
61. $-14 + 5t + t^2$
62. $4t^2 - 20t + 25$
63. $t^2 + 4t - 12$
64. $12 + 5z - 2z^2$
65. $12 + 4y - y^2$

Understanding Through Discussion and Writing

66. Explain how one could construct a polynomial with four terms that can be factored by grouping. [5.1c], [5.4a]

67. When searching for a factorization, why do we list pairs of numbers with the correct *product* instead of pairs of numbers with the correct *sum*? [5.2a]

68. Without multiplying $(x - 17)(x - 18)$, explain why it cannot possibly be a factorization of $x^2 + 35x + 306$. [5.2a]

69. A student presents the following work:
$$4x^2 + 28x + 48 = (2x + 6)(2x + 8)$$
$$= 2(x + 3)(x + 4).$$
Is it correct? Explain. [5.3a], [5.4a]

STUDYING FOR SUCCESS   Applications

☐ If you find applied problems challenging, don't give up! Your skill will improve with each problem that you solve.

☐ Make applications real! Look for applications of the math that you are studying in newspapers and magazines.

5.5 Factoring Trinomial Squares and Differences of Squares

OBJECTIVES

a Recognize trinomial squares.

b Factor trinomial squares.

c Recognize differences of squares.

d Factor differences of squares, being careful to factor completely.

In this section, we first learn to factor trinomials that are squares of binomials. Then we factor binomials that are differences of squares.

a RECOGNIZING TRINOMIAL SQUARES

Some trinomials are squares of binomials. For example, the trinomial $x^2 + 10x + 25$ is the square of the binomial $x + 5$. To see this, we can calculate $(x + 5)^2$. It is $x^2 + 2 \cdot x \cdot 5 + 5^2$, or $x^2 + 10x + 25$. A trinomial that is the square of a binomial is called a **trinomial square**, or a **perfect-square trinomial**.

We can use the following special-product rules in reverse to factor trinomial squares:

$$(A + B)^2 = A^2 + 2AB + B^2;$$
$$(A - B)^2 = A^2 - 2AB + B^2.$$

How can we recognize when an expression to be factored is a trinomial square? Look at $A^2 + 2AB + B^2$ and $A^2 - 2AB + B^2$. In order for an expression to be a trinomial square:

a) The two expressions A^2 and B^2 must be squares, such as

   4, x^2, $25x^4$, $16t^2$.

   When the coefficient is a perfect square and the power(s) of the variable(s) is (are) even, then the expression is a perfect square.

b) There must be no minus sign before A^2 or B^2.

c) The remaining term is either twice the product of A and B, $2AB$, or its opposite, $-2AB$.

If a number c can be multiplied by itself to get a number n, then c is a **square root** of n. Thus, 3 is a square root of 9 because $3 \cdot 3$, or 3^2, is 9. Similarly, A is a square root of A^2 and B is a square root of B^2.

EXAMPLE 1 Determine whether $x^2 + 6x + 9$ is a trinomial square.

a) We know that x^2 and 9 are squares.

b) There is no minus sign before x^2 or 9.

c) If we multiply the square roots, x and 3, and double the product, we get the remaining term: $2 \cdot x \cdot 3 = 6x$.

Thus, $x^2 + 6x + 9$ is the square of a binomial. In fact, $x^2 + 6x + 9 = (x + 3)^2$.

It would be helpful to memorize this table of perfect squares.

NUMBER, N	PERFECT SQUARE, N^2
1	1
2	4
3	9
4	16
5	25
6	36
7	49
8	64
9	81
10	100
11	121
12	144
13	169
14	196
15	225
16	256
20	400
25	625

EXAMPLE 2 Determine whether $x^2 + 6x + 11$ is a trinomial square.

The answer is no, because only one term, x^2, is a square.

EXAMPLE 3 Determine whether $16x^2 + 49 - 56x$ is a trinomial square.

It helps to first write the trinomial in descending order:

$$16x^2 - 56x + 49.$$

a) We know that $16x^2$ and 49 are squares.

b) There is no minus sign before $16x^2$ or 49.

c) We multiply the square roots, $4x$ and 7, and double the product to get $2 \cdot 4x \cdot 7 = 56x$. The remaining term, $-56x$, is the opposite of this product.

Thus, $16x^2 + 49 - 56x$ is a trinomial square.

Do Exercises 1–8.

Determine whether each is a trinomial square. Write "yes" or "no."

1. $x^2 + 8x + 16$
2. $25 - x^2 + 10x$
3. $t^2 - 12t + 4$
4. $25 + 20y + 4y^2$
5. $5x^2 + 16 - 14x$
6. $16x^2 + 40x + 25$
7. $p^2 + 6p - 9$
8. $25a^2 + 9 - 30a$

b FACTORING TRINOMIAL SQUARES

We can use the factoring methods from Sections 5.2–5.4 to factor trinomial squares, but there is a faster method using the following equations.

FACTORING TRINOMIAL SQUARES

$A^2 + 2AB + B^2 = (A + B)^2;$
$A^2 - 2AB + B^2 = (A - B)^2$

We use square roots of the squared terms and the sign of the remaining term to factor a trinomial square.

EXAMPLE 4 Factor: $x^2 + 6x + 9$.

$x^2 + 6x + 9 = x^2 + 2 \cdot x \cdot 3 + 3^2 = (x + 3)^2$ The sign of the middle term is positive.

$A^2 + 2 \ A \ B + B^2 = (A + B)^2$

EXAMPLE 5 Factor: $x^2 + 49 - 14x$.

$x^2 + 49 - 14x = x^2 - 14x + 49$ Changing to descending order
$= x^2 - 2 \cdot x \cdot 7 + 7^2$ The sign of the middle term is negative.
$= (x - 7)^2$

EXAMPLE 6 Factor: $16x^2 - 40x + 25$.

$16x^2 - 40x + 25 = (4x)^2 - 2 \cdot 4x \cdot 5 + 5^2 = (4x - 5)^2$

$A^2 - 2 \ A \ B + B^2 = (A - B)^2$

Do Exercises 9–13.

Factor.

9. $x^2 + 2x + 1$
10. $1 - 2x + x^2$
11. $4 + t^2 + 4t$
12. $25x^2 - 70x + 49$

GS 13. $49 - 56y + 16y^2$

Write in descending order: $16y^2 - 56y + 49$.

Factor as a trinomial square:
$(4y)^2 - 2 \cdot 4y \cdot \underline{} + (\underline{})^2$
$= (4y - \underline{})^2$.

Answers
1. Yes   2. No   3. No   4. Yes   5. No
6. Yes   7. No   8. Yes   9. $(x + 1)^2$
10. $(x - 1)^2$, or $(1 - x)^2$ 11. $(t + 2)^2$
12. $(5x - 7)^2$ 13. $(4y - 7)^2$, or $(7 - 4y)^2$

Guided Solution:
13. 7, 7, 7

EXAMPLE 7 Factor: $t^4 + 20t^2 + 100$.
$$t^4 + 20t^2 + 100 = (t^2)^2 + 2(t^2)(10) + 10^2$$
$$= (t^2 + 10)^2$$

EXAMPLE 8 Factor: $75m^3 + 210m^2 + 147m$.

Always look first for a common factor. This time there is one, $3m$:
$$75m^3 + 210m^2 + 147m = 3m(25m^2 + 70m + 49)$$
$$= 3m[(5m)^2 + 2(5m)(7) + 7^2]$$
$$= 3m(5m + 7)^2.$$

Factor.

14. $48m^2 + 75 + 120m$

15. $p^4 + 18p^2 + 81$

16. $4z^5 - 20z^4 + 25z^3$

17. $9a^2 + 30ab + 25b^2$

EXAMPLE 9 Factor: $4p^2 - 12pq + 9q^2$.
$$4p^2 - 12pq + 9q^2 = (2p)^2 - 2(2p)(3q) + (3q)^2$$
$$= (2p - 3q)^2$$

◀ Do Exercises 14–17.

C RECOGNIZING DIFFERENCES OF SQUARES

SKILL REVIEW *Simplify exponential expressions using the power rule.* [4.2a]
Simplify.

1. $(m^5)^2$ 2. $(7a^4)^2$

Answers: 1. m^{10} 2. $49a^8$

A **difference of squares** is an expression in the form $A^2 - B^2$. The polynomials $x^2 - 9$, $4t^2 - 49$, and $a^2 - 25b^2$ are differences of squares. To factor a difference of squares such as $x^2 - 9$, we will use the following special-product rule in reverse:

$$(A + B)(A - B) = A^2 - B^2.$$

How can we recognize such expressions? Look at $A^2 - B^2$. In order for a binomial to be a difference of squares:

a) There must be two expressions, both squares, such as

$$4x^2, \quad 9, \quad 25t^4, \quad 1, \quad x^6, \quad 49y^8.$$

b) The terms must have different signs.

EXAMPLE 10 Is $9x^2 - 64$ a difference of squares?

a) The first expression is a square: $9x^2 = (3x)^2$.
   The second expression is a square: $64 = 8^2$.

b) The terms have different signs, $+9x^2$ and -64.

Thus we have a difference of squares, $(3x)^2 - 8^2$.

EXAMPLE 11 Is $25 - t^3$ a difference of squares?

a) The expression t^3 is not a square.

The expression is not a difference of squares.

Answers
14. $3(4m + 5)^2$ 15. $(p^2 + 9)^2$
16. $z^3(2z - 5)^2$ 17. $(3a + 5b)^2$

EXAMPLE 12 Is $-4x^2 + 16$ a difference of squares?

a) The expressions $4x^2$ and 16 are squares: $4x^2 = (2x)^2$ and $16 = 4^2$.
b) The terms have different signs, $-4x^2$ and $+16$.

Thus we have a difference of squares. We can also see this by rewriting in the equivalent form: $16 - 4x^2$.

Do Exercises 18–24.

d FACTORING DIFFERENCES OF SQUARES

To factor a difference of squares, we use the following equation.

> **FACTORING A DIFFERENCE OF SQUARES**
> $A^2 - B^2 = (A + B)(A - B)$

To factor a difference of squares $A^2 - B^2$, we find A and B, which are square roots of the expressions A^2 and B^2. We then use A and B to form two factors. One is the sum $A + B$, and the other is the difference $A - B$.

EXAMPLE 13 Factor: $x^2 - 4$.

$$x^2 - 4 = x^2 - 2^2 = (x + 2)(x - 2)$$
$$A^2 - B^2 = (A + B)(A - B)$$

EXAMPLE 14 Factor: $9 - 16t^4$.

$$9 - 16t^4 = 3^2 - (4t^2)^2 = (3 + 4t^2)(3 - 4t^2)$$
$$A^2 - B^2 = (A + B)(A - B)$$

EXAMPLE 15 Factor: $m^2 - 4p^2$.

$$m^2 - 4p^2 = m^2 - (2p)^2 = (m + 2p)(m - 2p)$$

EXAMPLE 16 Factor: $x^2 - \frac{1}{9}$.

$$x^2 - \frac{1}{9} = x^2 - \left(\frac{1}{3}\right)^2 = \left(x + \frac{1}{3}\right)\left(x - \frac{1}{3}\right)$$

EXAMPLE 17 Factor: $18x^2 - 50x^6$.

Always look first for a factor common to all terms. This time there is one, $2x^2$.

$$18x^2 - 50x^6 = 2x^2(9 - 25x^4)$$
$$= 2x^2[3^2 - (5x^2)^2]$$
$$= 2x^2(3 + 5x^2)(3 - 5x^2)$$

Determine whether each is a difference of squares. Write "yes" or "no."

18. $x^2 - 25$

19. $t^2 - 24$

20. $y^2 + 36$

21. $4x^2 - 15$

22. $16x^4 - 49$

23. $9w^6 - 1$

24. $-49 + 25t^2$

Answers
18. Yes 19. No 20. No 21. No
22. Yes 23. Yes 24. Yes

SECTION 5.5 Factoring Trinomial Squares and Differences of Squares

Factor.

25. $x^2 - 9$

26. $4t^2 - 64$

27. $a^2 - 25b^2$
$= a^2 - ()^2$
$= (a +)(a -)$ GS

28. $64x^4 - 25x^6$

29. $5 - 20t^6$
[*Hint*: $t^6 = (t^3)^2$.]

EXAMPLE 18 Factor: $36x^{10} - 4x^2$.

Although this expression is a difference of squares, the terms have a common factor. We always begin factoring by factoring out the greatest common factor.

$36x^{10} - 4x^2 = 4x^2(9x^8 - 1)$
$\phantom{36x^{10} - 4x^2} = 4x^2[(3x^4)^2 - 1^2]$ Note that $x^8 = (x^4)^2$ and $1 = 1^2$.
$\phantom{36x^{10} - 4x^2} = 4x^2(3x^4 + 1)(3x^4 - 1)$

◀ Do Exercises 25–29.

······················· **Caution!** ·······················

Note carefully in these examples that a difference of squares is *not* the square of the difference; that is,

$$A^2 - B^2 \neq (A - B)^2.$$

For example,

$$(45 - 5)^2 = 40^2 = 1600,$$

but

$$45^2 - 5^2 = 2025 - 25 = 2000.$$

Factoring Completely

If a factor with more than one term can still be factored, you should do so. When no factor can be factored further, you have **factored completely**. Always factor completely whenever told to factor.

EXAMPLE 19 Factor: $p^4 - 16$.

$p^4 - 16 = (p^2)^2 - 4^2$
$ = (p^2 + 4)(p^2 - 4)$ Factoring a difference of squares
$ = (p^2 + 4)(p + 2)(p - 2)$ Factoring further; $p^2 - 4$ is a difference of squares.

The polynomial $p^2 + 4$ cannot be factored further into polynomials with real coefficients.

······················· **Caution!** ·······················

Apart from possibly removing a common factor, we cannot, in general, factor a sum of squares. In particular,

$$A^2 + B^2 \neq (A + B)^2.$$

Consider $25x^2 + 100$. In this case, a sum of squares has a common factor, 25. Factoring, we get $25(x^2 + 4)$, where $x^2 + 4$ is prime.

Answers

25. $(x + 3)(x - 3)$
26. $4(t + 4)(t - 4)$
27. $(a + 5b)(a - 5b)$
28. $x^4(8 + 5x)(8 - 5x)$
29. $5(1 + 2t^3)(1 - 2t^3)$

Guided Solution:
27. $5b, 5b, 5b$

EXAMPLE 20 Factor: $y^4 - 16x^{12}$.

$$y^4 - 16x^{12} = (y^2 + 4x^6)(y^2 - 4x^6)$$ Factoring a difference of squares

$$= (y^2 + 4x^6)(y + 2x^3)(y - 2x^3)$$ Factoring further. The factor $y^2 - 4x^6$ is a difference of squares.

The polynomial $y^2 + 4x^6$ cannot be factored further into polynomials with real coefficients.

EXAMPLE 21 Factor: $\frac{1}{16}x^8 - 81$.

$$\frac{1}{16}x^8 - 81 = \left(\frac{1}{4}x^4 + 9\right)\left(\frac{1}{4}x^4 - 9\right)$$ Factoring a difference of squares

$$= \left(\frac{1}{4}x^4 + 9\right)\left(\frac{1}{2}x^2 + 3\right)\left(\frac{1}{2}x^2 - 3\right)$$ Factoring further. The factor $\frac{1}{4}x^4 - 9$ is a difference of squares.

TIPS FOR FACTORING

- Always look first for a common factor. If there is one, factor it out.
- Be alert for trinomial squares and differences of squares. Once recognized, they can be factored without trial and error.
- Always factor completely.
- Check by multiplying.

Do Exercises 30–32.

Factor completely.
30. $81x^4 - 1$
31. $16 - \frac{1}{81}y^8$
32. $49p^4 - 25q^6$

Answers
30. $(9x^2 + 1)(3x + 1)(3x - 1)$
31. $\left(4 + \frac{1}{9}y^4\right)\left(2 + \frac{1}{3}y^2\right)\left(2 - \frac{1}{3}y^2\right)$
32. $(7p^2 + 5q^3)(7p^2 - 5q^3)$

5.5 Exercise Set

FOR EXTRA HELP   MyLab Math

✓ Check Your Understanding

Reading Check Determine whether each statement is true or false.

RC1. A trinomial can be considered a trinomial square if only one term is a perfect square.

RC2. A trinomial square is the square of a binomial.

RC3. In order for a binomial to be a difference of squares, the terms in the binomial must have the same sign.

RC4. A binomial cannot have a common factor.

Concept Check Each of the following is in the form $A^2 - B^2$. Determine A and B.

CC1. $x^2 - 64$ **CC2.** $25a^2 - c^2$ **CC3.** $x^{10} - 1$ **CC4.** $9 - \frac{1}{49}y^2$

a Determine whether each of the following is a trinomial square. Answer "yes" or "no."

1. $x^2 - 14x + 49$
2. $x^2 - 16x + 64$
3. $x^2 + 16x - 64$
4. $x^2 - 14x - 49$

5. $x^2 - 2x + 4$
6. $x^2 + 3x + 9$
7. $9x^2 - 24x + 16$
8. $25x^2 + 30x + 9$

b Factor completely. Remember to look first for a common factor and to check by multiplying.

9. $x^2 - 14x + 49$
10. $x^2 - 20x + 100$
11. $x^2 + 16x + 64$
12. $x^2 + 20x + 100$

13. $x^2 - 2x + 1$
14. $x^2 + 2x + 1$
15. $4 + 4x + x^2$
16. $4 + x^2 - 4x$

17. $y^2 + 12y + 36$
18. $y^2 + 18y + 81$
19. $16 + t^2 - 8t$
20. $9 + t^2 - 6t$

21. $q^4 - 6q^2 + 9$
22. $64 + 16a^2 + a^4$
23. $49 + 56y + 16y^2$
24. $75 + 48a^2 - 120a$

25. $2x^2 - 4x + 2$
26. $2x^2 - 40x + 200$
27. $x^3 - 18x^2 + 81x$
28. $x^3 + 24x^2 + 144x$

29. $12q^2 - 36q + 27$
30. $20p^2 + 100p + 125$
31. $49 - 42x + 9x^2$
32. $64 - 112x + 49x^2$

33. $5y^4 + 10y^2 + 5$
34. $a^4 + 14a^2 + 49$
35. $1 + 4x^4 + 4x^2$

36. $1 - 2a^5 + a^{10}$
37. $4p^2 + 12pt + 9t^2$
38. $25m^2 + 20mn + 4n^2$

39. $a^2 - 6ab + 9b^2$
40. $x^2 - 14xy + 49y^2$
41. $81a^2 - 18ab + b^2$

42. $64p^2 + 16pt + t^2$
43. $36a^2 + 96ab + 64b^2$
44. $16m^2 - 40mn + 25n^2$

c Determine whether each of the following is a difference of squares. Answer "yes" or "no."

45. $x^2 - 4$
46. $x^2 - 36$
47. $x^2 + 25$
48. $x^2 + 9$

49. $x^2 - 45$
50. $x^2 - 80y^2$
51. $-25y^2 + 16x^2$
52. $-1 + 36x^2$

d Factor completely. Remember to look first for a common factor.

53. $y^2 - 4$
54. $q^2 - 1$
55. $p^2 - 1$
56. $x^2 - 36$

57. $-49 + t^2$
58. $-64 + m^2$
59. $a^2 - b^2$
60. $p^2 - v^2$

61. $25t^2 - m^2$
62. $w^2 - 49z^2$
63. $100 - k^2$
64. $81 - w^2$

65. $16a^2 - 9$
66. $25x^2 - 4$
67. $4x^2 - 25y^2$
68. $9a^2 - 16b^2$

SECTION 5.5   Factoring Trinomial Squares and Differences of Squares

69. $8x^2 - 98$ **70.** $24x^2 - 54$ **71.** $36x - 49x^3$ **72.** $16x - 81x^3$

73. $\dfrac{1}{16} - 49x^8$ **74.** $\dfrac{1}{625}x^8 - 49$ **75.** $0.09y^2 - 0.0004$ **76.** $0.16p^2 - 0.0025$

77. $49a^4 - 81$ **78.** $25a^4 - 9$ **79.** $a^4 - 16$ **80.** $y^4 - 1$

81. $5x^4 - 405$ **82.** $4x^4 - 64$ **83.** $1 - y^8$ **84.** $x^8 - 1$

85. $x^{12} - 16$ **86.** $x^8 - 81$ **87.** $y^2 - \dfrac{1}{16}$ **88.** $x^2 - \dfrac{1}{25}$

89. $25 - \dfrac{1}{49}x^2$ **90.** $\dfrac{1}{4} - 9q^2$ **91.** $16m^4 - t^4$ **92.** $p^4 t^4 - 1$

Skill Maintenance

Find the intercepts of each equation. [3.3a]

93. $4x + 16y = 64$ **94.** $x - 1.3y = 6.5$ **95.** $y = 2x - 5$

Find the intercepts. Then graph each equation. [3.3a]

96. $y - 5x = 5$ **97.** $2x + 5y = 10$ **98.** $3x - 5y = 15$

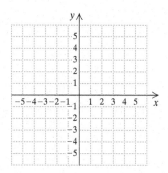
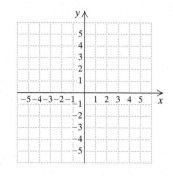
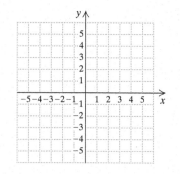

Find a polynomial for the shaded area in each figure. (Leave results in terms of π where appropriate.) [4.4d]

99.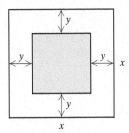

100.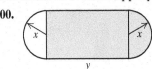

Synthesis

Factor completely, if possible.

101. $49x^2 - 216$
102. $27x^3 - 13x$
103. $x^2 + 22x + 121$
104. $x^2 - 5x + 25$

105. $18x^3 + 12x^2 + 2x$
106. $162x^2 - 82$
107. $x^8 - 2^8$
108. $4x^4 - 4x^2$

109. $3x^5 - 12x^3$
110. $3x^2 - \frac{1}{3}$
111. $18x^3 - \frac{8}{25}x$
112. $x^2 - 2.25$

113. $0.49p - p^3$
114. $3.24x^2 - 0.81$
115. $0.64x^2 - 1.21$
116. $1.28x^2 - 2$

117. $(x + 3)^2 - 9$
118. $(y - 5)^2 - 36q^2$
119. $x^2 - \left(\frac{1}{x}\right)^2$
120. $a^{2n} - 49b^{2n}$

121. $81 - b^{4k}$
122. $9x^{18} + 48x^9 + 64$
123. $9b^{2n} + 12b^n + 4$
124. $(x + 7)^2 - 4x - 24$

125. $(y + 3)^2 + 2(y + 3) + 1$
126. $49(x + 1)^2 - 42(x + 1) + 9$

Find c such that the polynomial is the square of a binomial.
127. $cy^2 + 6y + 1$
128. $cy^2 - 24y + 9$

Use the TABLE feature or graphs to determine whether each factorization is correct. (See the Calculator Corner on p. 361.)
129. $x^2 + 9 = (x + 3)(x + 3)$
130. $x^2 - 49 = (x - 7)(x + 7)$

131. $x^2 + 9 = (x + 3)^2$
132. $x^2 - 49 = (x - 7)^2$

5.6 Factoring Sums or Differences of Cubes

OBJECTIVE

a Factor sums and differences of cubes.

a SUMS OR DIFFERENCES OF CUBES

We can factor the sum or the difference of two expressions that are cubes. Consider the following products:

$$(A + B)(A^2 - AB + B^2) = A(A^2 - AB + B^2) + B(A^2 - AB + B^2)$$
$$= A^3 - A^2B + AB^2 + A^2B - AB^2 + B^3$$
$$= A^3 + B^3$$

and $(A - B)(A^2 + AB + B^2) = A(A^2 + AB + B^2) - B(A^2 + AB + B^2)$
$$= A^3 + A^2B + AB^2 - A^2B - AB^2 - B^3$$
$$= A^3 - B^3.$$

The above equations (reversed) show how we can factor a sum or a difference of two cubes. Each factors as a product of a binomial and a trinomial.

N	N^3
0.2	0.008
0.1	0.001
0	0
1	1
2	8
3	27
4	64
5	125
6	216
7	343
8	512
9	729
10	1000

SUM OR DIFFERENCE OF CUBES

$A^3 + B^3 = (A + B)(A^2 - AB + B^2);$
$A^3 - B^3 = (A - B)(A^2 + AB + B^2)$

Note that what we are considering here is a sum or a difference of cubes. We are not cubing a binomial. For example, $(A + B)^3$ is *not* the same as $A^3 + B^3$. The table of cubes in the margin is helpful.

EXAMPLE 1 Factor: $x^3 - 27$.

We have

$$x^3 - 27 = x^3 - 3^3.$$

In one set of parentheses, we write the cube root of the first term, x. Then we write the cube root of the second term, -3. This gives us the expression $x - 3$:

$(x - 3)(\quad).$

To get the next factor, we think of $x - 3$ and do the following:

- Square the first term: $x \cdot x = x^2$.
- Multiply the terms, $x(-3) = -3x$, and then change the sign: $3x$.
- Square the second term: $(-3)^2 = 9$.

$(x - 3)(x^2 + 3x + 9).$
$(A - B)(A^2 + AB + B^2)$

Note that we cannot factor $x^2 + 3x + 9$. It is not a trinomial square nor can it be factored by trial and error. Check this on your own.

◀ Do Exercises 1 and 2.

Factor.
1. $x^3 - 8$
2. $64 - y^3$

Answers
1. $(x - 2)(x^2 + 2x + 4)$
2. $(4 - y)(16 + 4y + y^2)$

EXAMPLE 2 Factor: $125x^3 + y^3$.

We have
$$125x^3 + y^3 = (5x)^3 + y^3.$$

In one set of parentheses, we write the cube root of the first term, $5x$. Then we write the cube root of the second term, y. This gives us the expression $5x + y$:

$$(5x + y)(\quad).$$

To get the next factor, we think of $5x + y$ and do the following:

- Square the first term: $(5x)(5x) = 25x^2$.
- Multiply the terms, $5x \cdot y = 5xy$, and then change the sign: $-5xy$.
- Square the second term: $y \cdot y = y^2$.

$$(5x + y)(25x^2 - 5xy + y^2).$$
$$(A + B)(A^2 - AB + B^2)$$

Do Exercises 3 and 4. ▶

Factor.
3. $27x^3 + y^3$

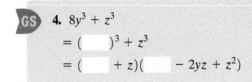
4. $8y^3 + z^3$
$= (\quad)^3 + z^3$
$= (\quad + z)(\quad - 2yz + z^2)$

EXAMPLE 3 Factor: $128y^7 - 250x^6y$.

We first look for the largest common factor:
$$\begin{aligned} 128y^7 - 250x^6y &= 2y(64y^6 - 125x^6) \\ &= 2y[(4y^2)^3 - (5x^2)^3] \\ &= 2y(4y^2 - 5x^2)(16y^4 + 20x^2y^2 + 25x^4). \end{aligned}$$

EXAMPLE 4 Factor: $a^6 - b^6$.

We factor a difference of squares:
$$\begin{aligned} a^6 - b^6 &= (a^3)^2 - (b^3)^2 \\ &= (a^3 + b^3)(a^3 - b^3). \end{aligned}$$

One factor is a sum of two cubes, and the other factor is a difference of two cubes. We factor them:
$$a^6 - b^6 = (a + b)(a^2 - ab + b^2)(a - b)(a^2 + ab + b^2).$$

We have now factored completely.

In Example 4, had we thought of factoring first as a difference of two cubes, we would have had
$$\begin{aligned} (a^2)^3 - (b^2)^3 &= (a^2 - b^2)(a^4 + a^2b^2 + b^4) \\ &= (a + b)(a - b)(a^4 + a^2b^2 + b^4). \end{aligned}$$

In this case, we might have missed some factors; $a^4 + a^2b^2 + b^4$ can be factored as $(a^2 - ab + b^2)(a^2 + ab + b^2)$, but we probably would not have known to do such factoring.

Answers
3. $(3x + y)(9x^2 - 3xy + y^2)$
4. $(2y + z)(4y^2 - 2yz + z^2)$
Guided Solution:
4. $2y, 2y, 4y^2$

> When you can factor as either a difference of squares or a difference of cubes, factor as a difference of squares first.

EXAMPLE 5 Factor: $64a^6 - 729b^6$.

We have

$$64a^6 - 729b^6 = (8a^3)^2 - (27b^3)^2$$
$$= (8a^3 - 27b^3)(8a^3 + 27b^3) \quad \text{Factoring a difference of squares}$$
$$= [(2a)^3 - (3b)^3][(2a)^3 + (3b)^3].$$

Each factor is a sum or a difference of cubes. We factor each:

$$= (2a - 3b)(4a^2 + 6ab + 9b^2)(2a + 3b)(4a^2 - 6ab + 9b^2).$$

Factor.

5. $m^6 - n^6$

6. $16x^7y + 54xy^7$

7. $729x^6 - 64y^6$

8. $x^3 - 0.027$

FACTORING SUMMARY

Sum of cubes: $A^3 + B^3 = (A + B)(A^2 - AB + B^2);$
Difference of cubes: $A^3 - B^3 = (A - B)(A^2 + AB + B^2);$
Difference of squares: $A^2 - B^2 = (A + B)(A - B);$
Sum of squares: $A^2 + B^2$ cannot be factored as the square of a binomial: $A^2 + B^2 \neq (A + B)^2.$

◀ Do Exercises 5–8.

Answers
5. $(m + n)(m^2 - mn + n^2)(m - n)(m^2 + mn + n^2)$
6. $2xy(2x^2 + 3y^2)(4x^4 - 6x^2y^2 + 9y^4)$
7. $(3x + 2y)(9x^2 - 6xy + 4y^2)(3x - 2y)(9x^2 + 6xy + 4y^2)$
8. $(x - 0.3)(x^2 + 0.3x + 0.09)$

5.6 Exercise Set

FOR EXTRA HELP   MyLab Math

✓ Check Your Understanding

Reading and Concept Check Choose from the column on the right the expression that makes each statement correct.

RC1. The cube root of x^3 is _____.

RC2. The cube root of 27 is _____.

RC3. The cube root of -27 is _____.

RC4. The cube root of $27x^3$ is _____.

RC5. The cube root of x^6 is _____.

RC6. The cube root of 1 is _____.

RC7. The cube root of -1 is _____.

RC8. The cube root of $1000x^{12}$ is _____.

a) 1
b) -1
c) 3
d) -3
e) x
f) x^2
g) $3x$
h) $10x^4$

a Factor.

1. $z^3 + 27$
2. $a^3 + 8$
3. $x^3 - 1$
4. $c^3 - 64$

5. $y^3 + 125$
6. $x^3 + 1$
7. $8a^3 + 1$
8. $27x^3 + 1$

9. $y^3 - 8$
10. $p^3 - 27$
11. $8 - 27b^3$
12. $64 - 125x^3$

13. $64y^3 + 1$
14. $125x^3 + 1$
15. $8x^3 + 27$
16. $27y^3 + 64$

17. $a^3 - b^3$
18. $x^3 - y^3$
19. $a^3 + \dfrac{1}{8}$
20. $b^3 + \dfrac{1}{27}$

21. $2y^3 - 128$
22. $3z^3 - 3$
23. $24a^3 + 3$
24. $54x^3 + 2$

25. $rs^3 + 64r$
26. $ab^3 + 125a$
27. $5x^3 - 40z^3$
28. $2y^3 - 54z^3$

29. $x^3 + 0.001$
30. $y^3 + 0.125$
31. $64x^6 - 8t^6$
32. $125c^6 - 8d^6$

33. $2y^4 - 128y$
34. $3z^5 - 3z^2$
35. $z^6 - 1$
36. $t^6 + 1$

37. $t^6 + 64y^6$ **38.** $p^6 - q^6$ **39.** $8w^9 - z^9$ **40.** $a^9 + 64b^9$

41. $\dfrac{1}{8}c^3 + d^3$ **42.** $\dfrac{27}{125}x^3 - y^3$ **43.** $0.001x^3 - 0.008y^3$ **44.** $0.125r^3 - 0.216s^3$

Skill Maintenance

Simplify. [4.1f], [4.2b]

45. $(7y^{-5})^3$ **46.** $(a^{-4}b^{-9})^{-2}$ **47.** $\left(\dfrac{x^3}{4}\right)^{-2}$

Multiply.

48. $(2y^5 + 3)(2y^5 - 3)$ [4.6b] **49.** $\left(w - \dfrac{1}{3}\right)^2$ [4.6c] **50.** $(x - 0.1)(x + 0.5)$ [4.6a]

Synthesis

51. *Volume of Carpeting.* The volume of a carpet that is rolled up can be estimated by the polynomial $\pi R^2 h - \pi r^2 h$.

52. Show how the geometric model below can be used to verify the formula for factoring $a^3 - b^3$.

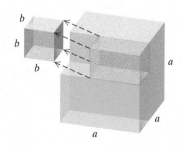

a) Factor the polynomial.
b) Use both the original form and the factored form to find the volume of a roll for which $R = 50$ cm, $r = 10$ cm, and $h = 4$ m. Use 3.14 for π.

Factor. Assume that variables in exponents represent positive integers.

53. $3x^{3a} + 24y^{3b}$ **54.** $\dfrac{8}{27}x^3 + \dfrac{1}{64}y^3$ **55.** $\dfrac{1}{24}x^3y^3 + \dfrac{1}{3}z^3$

56. $7x^3 - \dfrac{7}{8}$ **57.** $(x + y)^3 - x^3$ **58.** $(1 - x)^3 + (x - 1)^6$

59. $(a + 2)^3 - (a - 2)^3$ **60.** $y^4 - 8y^3 - y + 8$

Factoring: A General Strategy

5.7

OBJECTIVE

a Factor polynomials completely using any of the methods considered in this chapter.

a We now combine all of our factoring techniques and consider a general strategy for factoring polynomials. Here we will encounter polynomials of all the types we have considered, in random order, so you will have the opportunity to determine which method to use.

> **FACTORING STRATEGY**
>
> To factor a polynomial:
>
> **a)** Always look first for a common factor. If there is one, factor out the largest common factor.
>
> **b)** Then look at the number of terms.
>
> *Two terms*: Determine whether you have a difference of squares, $A^2 - B^2$, or a sum or difference of cubes, $A^3 + B^3$ or $A^3 - B^3$. Do not try to factor a sum of squares: $A^2 + B^2$.
>
> *Three terms*: Determine whether the trinomial is a square. If it is, you know how to factor. If not, try trial and error, using FOIL or the *ac*-method.
>
> *Four terms*: Try factoring by grouping.
>
> **c)** *Always factor completely.* If a factor with more than one term can still be factored, you should factor it. When no factor can be factored further, you have finished.
>
> **d)** Check by multiplying.

EXAMPLE 1 Factor: $5t^4 - 80$.

a) We look for a common factor. There is one, 5.

$$5t^4 - 80 = 5(t^4 - 16)$$

b) The factor $t^4 - 16$ has only two terms. It is a difference of squares: $(t^2)^2 - 4^2$. We factor $t^4 - 16$ and then include the common factor:

$$5(t^2 + 4)(t^2 - 4).$$

c) We see that one of the factors, $t^2 - 4$, is again a difference of squares. We factor it:

$$5(t^2 + 4)(t + 2)(t - 2).$$
↑
This is a sum of squares. It cannot be factored.

We have factored completely because no factor with more than one term can be factored further.

d) Check: $5(t^2 + 4)(t + 2)(t - 2) = 5(t^2 + 4)(t^2 - 4)$
$= 5(t^4 - 16)$
$= 5t^4 - 80.$

EXAMPLE 2 Factor: $2x^3 + 10x^2 + x + 5$.

a) We look for a common factor. There isn't one.

b) There are four terms. We try factoring by grouping:

$$2x^3 + 10x^2 + x + 5$$
$$= (2x^3 + 10x^2) + (x + 5) \quad \text{Separating into two binomials}$$
$$= 2x^2(x + 5) + 1(x + 5) \quad \text{Factoring each binomial}$$
$$= (x + 5)(2x^2 + 1). \quad \text{Factoring out the common factor } x + 5$$

c) None of these factors can be factored further, so we have factored completely.

d) Check: $(x + 5)(2x^2 + 1) = x \cdot 2x^2 + x \cdot 1 + 5 \cdot 2x^2 + 5 \cdot 1$
$$= 2x^3 + x + 10x^2 + 5, \text{ or}$$
$$2x^3 + 10x^2 + x + 5.$$

EXAMPLE 3 Factor: $x^5 - 2x^4 - 35x^3$.

a) We look first for a common factor. This time there is one, x^3:

$$x^5 - 2x^4 - 35x^3 = x^3(x^2 - 2x - 35).$$

b) The factor $x^2 - 2x - 35$ has three terms, but it is not a trinomial square. We factor it using trial and error:

$$x^5 - 2x^4 - 35x^3 = x^3(x^2 - 2x - 35) = x^3(x - 7)(x + 5).$$

> Don't forget to include the common factor in the final answer!

c) No factor with more than one term can be factored further, so we have factored completely.

d) Check: $x^3(x - 7)(x + 5) = x^3(x^2 - 2x - 35)$
$$= x^5 - 2x^4 - 35x^3.$$

EXAMPLE 4 Factor: $x^4 - 10x^2 + 25$.

a) We look first for a common factor. There isn't one.

b) There are three terms. We see that this polynomial is a trinomial square. We factor it:

$$x^4 - 10x^2 + 25 = (x^2)^2 - 2 \cdot x^2 \cdot 5 + 5^2 = (x^2 - 5)^2.$$

We could use trial and error if we had not recognized that we have a trinomial square.

c) Since $x^2 - 5$ cannot be factored further, we have factored completely.

d) Check: $(x^2 - 5)^2 = (x^2)^2 - 2(x^2)(5) + 5^2 = x^4 - 10x^2 + 25.$

◀ Do Exercises 1–5.

Factor.
1. $3m^4 - 3$
2. $x^6 + 8x^3 + 16$
3. $2x^4 + 8x^3 + 6x^2$
4. $3x^3 + 12x^2 - 2x - 8$

5. $8x^3 - 200x$ **GS**
   a) Factor out the largest common factor:
   $8x^3 - 200x$
   $= \underline{}(x^2 - 25).$
   b) There are two terms inside the parentheses. Factor the difference of squares:
   $8x(x^2 - 25)$
   $= 8x(x + \underline{})(x - \underline{}).$
   c) We have factored completely.
   d) Check:
   $8x(x + 5)(x - 5)$
   $= 8x(x^2 - 25)$
   $= 8x^3 - 200x.$

Answers
1. $3(m^2 + 1)(m + 1)(m - 1)$ 2. $(x^3 + 4)^2$
3. $2x^2(x + 1)(x + 3)$ 4. $(x + 4)(3x^2 - 2)$
5. $8x(x + 5)(x - 5)$

Guided Solution:
5. $8x, 5, 5$

EXAMPLE 5 Factor: $6x^2y^4 - 21x^3y^5 + 3x^2y^6$.

a) We look first for a common factor:
$$6x^2y^4 - 21x^3y^5 + 3x^2y^6 = 3x^2y^4(2 - 7xy + y^2).$$

b) There are three terms in $2 - 7xy + y^2$. Since only y^2 is a square, we do not have a trinomial square. Can the trinomial be factored by trial and error? A key to the answer is that x is in only the term $-7xy$. The polynomial might be in a form like $(1 - y)(2 + y)$, but there would be no x in the middle term. Thus, $2 - 7xy + y^2$ cannot be factored.

c) Have we factored completely? Yes, because no factor with more than one term can be factored further.

d) The check is left to the student.

EXAMPLE 6 Factor: $(p + q)(x + 2) + (p + q)(x + y)$.

a) We look for a common factor:
$$(p + q)(x + 2) + (p + q)(x + y) = (p + q)[(x + 2) + (x + y)]$$
$$= (p + q)(2x + y + 2).$$

b) The trinomial $2x + y + 2$ cannot be factored further.

c) Neither factor can be factored further, so we have factored completely.

d) The check is left to the student.

EXAMPLE 7 Factor: $px + py + qx + qy$.

a) We look first for a common factor. There isn't one.

b) There are four terms. We try factoring by grouping:
$$px + py + qx + qy = p(x + y) + q(x + y)$$
$$= (x + y)(p + q).$$

c) Since neither factor can be factored further, we have factored completely.

d) Check: $(x + y)(p + q) = px + qx + py + qy$, or
$$px + py + qx + qy.$$

EXAMPLE 8 Factor: $25x^2 + 20xy + 4y^2$.

a) We look first for a common factor. There isn't one.

b) There are three terms. We determine whether the trinomial is a square. The first term and the last term are squares:
$$25x^2 = (5x)^2 \quad \text{and} \quad 4y^2 = (2y)^2.$$

Since twice the product of $5x$ and $2y$ is the other term,
$$2 \cdot 5x \cdot 2y = 20xy,$$
the trinomial is a perfect square.

We factor by writing the square roots of the square terms and the sign of the middle term:
$$25x^2 + 20xy + 4y^2 = (5x + 2y)^2.$$

c) Since $5x + 2y$ cannot be factored further, we have factored completely.

d) Check: $(5x + 2y)^2 = (5x)^2 + 2(5x)(2y) + (2y)^2$
$$= 25x^2 + 20xy + 4y^2.$$

EXAMPLE 9 Factor: $p^2q^2 + 7pq + 12$.

a) We look first for a common factor. There isn't one.

b) There are three terms. We determine whether the trinomial is a square. The first term is a square, but neither of the other terms is a square, so we do not have a trinomial square. We factor, thinking of the product pq as a single variable. We consider this possibility for factorization:

$$(pq + \square)(pq + \square).$$

We factor the last term, 12. All the signs are positive, so we consider only positive factors. Possibilities are 1, 12 and 2, 6 and 3, 4. The pair 3, 4 gives a sum of 7 for the coefficient of the middle term. Thus,

$$p^2q^2 + 7pq + 12 = (pq + 3)(pq + 4).$$

c) No factor with more than one term can be factored further, so we have factored completely.

d) Check: $(pq + 3)(pq + 4) = (pq)(pq) + 4 \cdot pq + 3 \cdot pq + 3 \cdot 4$
$= p^2q^2 + 7pq + 12.$

EXAMPLE 10 Factor: $8x^4 - 20x^2y - 12y^2$.

a) We look first for a common factor:

$$8x^4 - 20x^2y - 12y^2 = 4(2x^4 - 5x^2y - 3y^2).$$

b) There are three terms in $2x^4 - 5x^2y - 3y^2$. Since none of the terms is a square, we do not have a trinomial square. The x^2 in the middle term, $-5x^2y$, leads us to factor $2x^4$ as $2x^2 \cdot x^2$. We also factor the last term, $-3y^2$. Possibilities are $3y, -y$ and $-3y, y$ and others. We look for factors such that the sum of their products is the middle term. We try some possibilities:

$$(2x^2 - y)(x^2 + 3y) = 2x^4 + 5x^2y - 3y^2,$$
$$(2x^2 + y)(x^2 - 3y) = 2x^4 - 5x^2y - 3y^2. \quad \text{Correct middle term}$$

c) No factor with more than one term can be factored further, so we have factored completely. The factorization, including the common factor, is

$$4(2x^2 + y)(x^2 - 3y).$$

d) Check: $4(2x^2 + y)(x^2 - 3y) = 4[(2x^2)(x^2) + 2x^2(-3y) + yx^2 + y(-3y)]$
$= 4[2x^4 - 6x^2y + x^2y - 3y^2]$
$= 4(2x^4 - 5x^2y - 3y^2)$
$= 8x^4 - 20x^2y - 12y^2.$

EXAMPLE 11 Factor: $a^4 - 16b^4$.

a) We look first for a common factor. There isn't one.

b) There are two terms. Since $a^4 = (a^2)^2$ and $16b^4 = (4b^2)^2$, we see that we have a difference of squares. Thus,

$$a^4 - 16b^4 = (a^2 + 4b^2)(a^2 - 4b^2).$$

c) The last factor can be factored further. It is also a difference of squares.

$$a^4 - 16b^4 = (a^2 + 4b^2)(a + 2b)(a - 2b)$$

d) Check: $(a^2 + 4b^2)(a + 2b)(a - 2b) = (a^2 + 4b^2)(a^2 - 4b^2)$
$= a^4 - 16b^4.$

◀ Do Exercises 6–10.

Factor.

6. $15x^4 + 5x^2y - 10y^2$

7. $10p^6q^2 + 4p^5q^3 + 2p^4q^4$

8. $(a - b)(x + 5) + (a - b)(x + y^2)$

9. $ax^2 + ay + bx^2 + by$

10. $x^4 + 2x^2y^2 + y^4$ **GS**

  a) There is no common factor.
  b) There are three terms. Factor the trinomial square:
  $x^4 + 2x^2y^2 + y^4 = (x^2 + \underline{})^2.$
  c) We have factored completely.
  d) Check:
  $(x^2 + y^2)^2$
  $= (x^2)^2 + 2(x^2)(y^2) + (y^2)^2$
  $= x^4 + 2x^2y^2 + y^4.$

Answers

6. $5(3x^2 - 2y)(x^2 + y)$
7. $2p^4q^2(5p^2 + 2pq + q^2)$
8. $(a - b)(2x + 5 + y^2)$
9. $(x^2 + y)(a + b)$
10. $(x^2 + y^2)^2$

Guided Solution:
10. y^2

EXAMPLE 12 Factor: $40t^3 - 5s^3$.

a) We look first for a common factor:
$$40t^3 - 5s^3 = 5(8t^3 - s^3).$$

b) The factor $8t^3 - s^3$ has only two terms. It is a difference of cubes. We factor as follows:
$$(2t - s)(4t^2 + 2ts + s^2).$$

c) No factor with more than one term can be factored further, so we have factored completely. The factorization, including the common factor, is
$$5(2t - s)(4t^2 + 2ts + s^2).$$

d) The check is left to the student.

Do Exercises 11–13.

Factor.
11. $x^2y^2 + 5xy + 4$
12. $p^4 - 81q^4$
13. $15a^3 - 120b^3$

Answers
11. $(xy + 1)(xy + 4)$
12. $(p^2 + 9q^2)(p + 3q)(p - 3q)$
13. $15(a - 2b)(a^2 + 2ab + 4b^2)$

5.7 Exercise Set

FOR EXTRA HELP   MyLab Math

✓ Check Your Understanding

Reading Check Choose from the list on the right the appropriate word to complete each step in the following factoring strategy.

RC1. Always look first for a _____ factor.

RC2. If there are two terms, determine whether the binomial is a _____ of squares.

RC3. If there are three terms, determine whether the trinomial is a _____.

RC4. If there are four terms, try factoring by _____.

RC5. Always factor _____.

RC6. Always _____ by multiplying.

check
completely
grouping
sum
common
product
square
difference

Concept Check Choose from the column on the right the most appropriate first step to factor each polynomial. Choices may be used more than once or not at all.

CC1. $x^2 - 81$

CC2. $x^3 - 2x^2 + x$

CC3. $x^3 - 7x^2 + 2x - 14$

CC4. $x^2 + 7x + 12$

CC5. $4x^6 - 16x^4$

a) Factor out the largest common factor.
b) Factor the difference of squares.
c) Factor the trinomial square.
d) Factor using FOIL or the ac-method.
e) Factor by grouping.

a Factor completely.

1. $3x^2 - 192$
2. $2t^2 - 18$
3. $a^2 + 25 - 10a$
4. $y^2 + 49 + 14y$
5. $2x^2 - 11x + 12$
6. $8y^2 - 18y - 5$
7. $x^3 + 24x^2 + 144x$
8. $x^3 - 18x^2 + 81x$
9. $x^3 + 3x^2 - 4x - 12$
10. $x^3 - 5x^2 - 25x + 125$
11. $48x^2 - 3$
12. $50x^2 - 32$
13. $9x^3 + 12x^2 - 45x$
14. $20x^3 - 4x^2 - 72x$
15. $x^2 + 4$
16. $t^2 + 25$
17. $x^4 + 7x^2 - 3x^3 - 21x$
18. $m^4 + 8m^3 + 8m^2 + 64m$
19. $x^5 - 14x^4 + 49x^3$
20. $2x^6 + 8x^5 + 8x^4$
21. $20 - 6x - 2x^2$
22. $45 - 3x - 6x^2$
23. $x^2 - 6x + 1$
24. $x^2 + 8x + 5$
25. $4x^4 - 64$
26. $5x^5 - 80x$
27. $1 - y^8$
28. $t^8 - 1$
29. $x^5 - 4x^4 + 3x^3$
30. $x^6 - 2x^5 + 7x^4$

31. $\frac{1}{81}x^6 - \frac{8}{27}x^3 + \frac{16}{9}$

32. $36a^2 - 15a + \frac{25}{16}$

33. $\frac{1}{1000}m^3 - \frac{1}{27}n^3$

34. $125a^3 - 8b^3$

35. $9x^2y^2 - 36xy$

36. $x^2y - xy^2$

37. $2\pi rh + 2\pi r^2$

38. $10p^4t^4 + 35p^3t^3 + 10p^2t^2$

39. $(a+b)(x-3) + (a+b)(x+4)$

40. $5c(a^3+b) - (a^3+b)$

41. $(x-1)(x+1) - y(x+1)$

42. $3(p-c) - c^2(p-c)$

43. $n^2 + 2n + np + 2p$

44. $a^2 - 3a + ay - 3y$

45. $6w^2 - 3w + 2pw - p$

46. $2x^2 - 4x + xy - 2y$

47. $4b^2 + a^2 - 4ab$

48. $x^2 + y^2 - 2xy$

49. $16x^2 + 24xy + 9y^2$

50. $9c^2 + 6cd + d^2$

51. $49m^4 - 112m^2n + 64n^2$

52. $4x^2y^2 + 12xyz + 9z^2$

53. $y^4 + 10y^2z^2 + 25z^4$

54. $0.01x^4 - 0.1x^2y^2 + 0.25y^4$

55. $\frac{1}{4}a^2 + \frac{1}{3}ab + \frac{1}{9}b^2$

56. $4p^2n + 4pn^2 + n^3$

57. $a^2 - ab - 2b^2$

58. $3b^2 - 17ab - 6a^2$

59. $2mn - 360n^2 + m^2$

60. $15 + x^2y^2 + 8xy$

SECTION 5.7 Factoring: A General Strategy

61. $m^2n^2 - 4mn - 32$ **62.** $x^2z^2 + 7xz + 6$ **63.** $r^5s^2 - 10r^4s + 16r^3$

64. $c^5d^2 + 3c^4d - 10c^3$ **65.** $a^5 + 4a^4b - 5a^3b^2$ **66.** $2s^6t^2 + 10s^3t^3 + 12t^4$

67. $a^2 - \dfrac{1}{25}b^2$ **68.** $p^2 - \dfrac{1}{49}b^2$ **69.** $7x^6 - 7y^6$

70. $16p^3 + 54q^3$ **71.** $16 - c^4d^4$ **72.** $15a^4 - 15b^4$

73. $1 - 16x^{12}y^{12}$ **74.** $81a^4 - b^4$ **75.** $q^3 + 8q^2 - q - 8$

76. $m^3 - 7m^2 - 4m + 28$ **77.** $6a^3b^3 - a^2b^2 - 2ab$ **78.** $4ab^5 - 32b^4 + a^2b^6$

79. $m^4 - 5m^2 + 4$ **80.** $8x^3y^3 - 6x^2y^2 - 5xy$

Skill Maintenance

Compute and simplify. [1.8d]

81. $-50 \div (-5)(-2) - 18 \div (-3)^2$

82. $3(-2) - 2 + |-4 - (-1)|$

83. Evaluate $-x$ when $x = -7$. [1.3b]

84. Use either $<$ or $>$ for $\square$ to write a true sentence:
$-\dfrac{1}{3} \square -\dfrac{1}{2}$. [1.2d]

Synthesis

Factor completely.

85. $t^4 - 2t^2 + 1$ **86.** $x^4 + 9$ **87.** $x^3 + 20 - (5x^2 + 4x)$

88. $\dfrac{1}{5}x^2 - x + \dfrac{4}{5}$ **89.** $12.25x^2 - 7x + 1$ **90.** $x^3 + x^2 - (4x + 4)$

91. $18 + y^3 - 9y - 2y^2$ **92.** $3x^4 - 15x^2 + 12$ **93.** $y^2(y - 1) - 2y(y - 1) + (y - 1)$

Solving Quadratic Equations by Factoring

5.8

OBJECTIVES

a Solve equations (already factored) using the principle of zero products.

b Solve quadratic equations by factoring and then using the principle of zero products.

Second-degree equations like $x^2 + x - 156 = 0$ and $9 - x^2 = 0$ are examples of *quadratic equations*.

QUADRATIC EQUATION

A **quadratic equation** is an equation equivalent to an equation of the type

$$ax^2 + bx + c = 0, \ a \neq 0.$$

In order to solve quadratic equations, we need a new equation-solving principle.

a THE PRINCIPLE OF ZERO PRODUCTS

The product of two numbers is 0 if one or both of the numbers is 0. Furthermore, *if any product is 0, then a factor must be* 0. For example:

If $7x = 0$, then we know that $x = 0$.
If $x(2x - 9) = 0$, then we know that $x = 0$ or $2x - 9 = 0$.
If $(x + 3)(x - 2) = 0$, then we know that $x + 3 = 0$ or $x - 2 = 0$.

.... Caution! ....

In a product such as $ab = 24$, we cannot conclude with certainty that a is 24 or that b is 24, but if $ab = 0$, we can conclude that $a = 0$ or $b = 0$.

SKILL REVIEW Solve equations using both the addition principle and the multiplication principle. [2.3a]
Solve.
1. $3x - 7 = 8$
2. $4y + 5 = 2$

Answers: **1.** 5 **2.** $-\frac{3}{4}$

EXAMPLE 1 Solve: $(x + 3)(x - 2) = 0$.

We have a product of 0. This equation will be true when either factor is 0. Thus it is true when

$x + 3 = 0 \quad or \quad x - 2 = 0.$

Here we have two simple equations that we know how to solve:

$x = -3 \quad or \quad x = 2.$

Each of the numbers -3 and 2 is a solution of the original equation, as we can see in the following checks.

Check: For -3:

$$\frac{(x + 3)(x - 2) = 0}{(-3 + 3)(-3 - 2) \ ? \ 0}$$
$$0(-5) \ \bigg| $$
$$0 \ \bigg| \ \text{TRUE}$$

For 2:

$$\frac{(x + 3)(x - 2) = 0}{(2 + 3)(2 - 2) \ ? \ 0}$$
$$5(0) \ \bigg| $$
$$0 \ \bigg| \ \text{TRUE}$$

CALCULATOR CORNER

Solving Quadratic Equations We can solve quadratic equations graphically. Consider the equation $x^2 + 2x = 8$. First, we write the equation with 0 on one side: $x^2 + 2x - 8 = 0$. Next, we graph $y = x^2 + 2x - 8$ in a window that shows the x-intercepts. The standard window works well in this case.

The solutions of the equation are the values of x for which $x^2 + 2x - 8 = 0$. These are also the first coordinates of the x-intercepts of the graph. We use the ZERO feature from the CALC menu to find these numbers. For each x-intercept, we choose an x-value to the left of the intercept as a Left Bound, an x-value to the right of the intercept as a Right Bound, and an x-value near the intercept as a Guess. Beginning with the intercept on the left, we can read its coordinates, $(-4, 0)$, from the resulting screen.

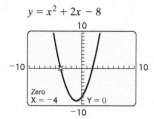

We can repeat this procedure to find the coordinates of the other x-intercept, $(2, 0)$. The solutions of $x^2 + 2x - 8 = 0$ are -4 and 2.

EXERCISE:

1. Solve each of the equations in Examples 4–6 graphically.

Solve using the principle of zero products.

1. $(x - 3)(x + 4) = 0$
2. $(x - 7)(x - 3) = 0$
3. $(4t + 1)(3t - 2) = 0$
4. $y(3y - 17) = 0$

Answers

1. $3, -4$ 2. $7, 3$ 3. $-\dfrac{1}{4}, \dfrac{2}{3}$ 4. $0, \dfrac{17}{3}$

We now have a principle to help in solving quadratic equations.

THE PRINCIPLE OF ZERO PRODUCTS

An equation $ab = 0$ is true if and only if $a = 0$ is true or $b = 0$ is true, or both are true. (A product is 0 if and only if one or both of the factors is 0.)

EXAMPLE 2 Solve: $(5x + 1)(x - 7) = 0$.

We have

$(5x + 1)(x - 7) = 0$

$5x + 1 = 0 \quad \text{or} \quad x - 7 = 0$ Using the principle of zero products

$5x = -1 \quad \text{or} \quad x = 7$ Solving the two equations separately

$x = -\dfrac{1}{5} \quad \text{or} \quad x = 7.$

Check: For $-\dfrac{1}{5}$:

$$\dfrac{(5x + 1)(x - 7) = 0}{(5(-\tfrac{1}{5}) + 1)(-\tfrac{1}{5} - 7) \; ? \; 0}$$
$$(-1 + 1)(-7\tfrac{1}{5})$$
$$0(-7\tfrac{1}{5})$$
$$0 \quad \text{TRUE}$$

For 7:

$$\dfrac{(5x + 1)(x - 7) = 0}{(5(7) + 1)(7 - 7) \; ? \; 0}$$
$$(35 + 1) \cdot 0$$
$$36 \cdot 0$$
$$0 \quad \text{TRUE}$$

The solutions are $-\dfrac{1}{5}$ and 7.

When some factors have only one term, you can still use the principle of zero products.

EXAMPLE 3 Solve: $x(2x - 9) = 0$.

We have

$x(2x - 9) = 0$

$x = 0 \quad \text{or} \quad 2x - 9 = 0$ Using the principle of zero products

$x = 0 \quad \text{or} \quad 2x = 9$

$x = 0 \quad \text{or} \quad x = \dfrac{9}{2}.$

Check: For 0:

$$\dfrac{x(2x - 9) = 0}{0 \cdot (2 \cdot 0 - 9) \; ? \; 0}$$
$$0 \cdot (-9)$$
$$0 \quad \text{TRUE}$$

For $\dfrac{9}{2}$:

$$\dfrac{x(2x - 9) = 0}{\tfrac{9}{2} \cdot (2 \cdot \tfrac{9}{2} - 9) \; ? \; 0}$$
$$\tfrac{9}{2} \cdot (9 - 9)$$
$$\tfrac{9}{2} \cdot 0$$
$$0 \quad \text{TRUE}$$

The solutions are 0 and $\dfrac{9}{2}$.

◀ Do Exercises 1–4.

When you solve an equation using the principle of zero products, a check by substitution will detect errors in solving.

b USING FACTORING TO SOLVE EQUATIONS

Using factoring and the principle of zero products, we can solve some new kinds of equations. Thus we have extended our equation-solving abilities.

EXAMPLE 4 Solve: $x^2 + 5x + 6 = 0$.

There are no like terms to collect, and we have a squared term. We first factor the polynomial. Then we use the principle of zero products.

$x^2 + 5x + 6 = 0$
$(x + 2)(x + 3) = 0$ Factoring
$x + 2 = 0$ or $x + 3 = 0$ Using the principle of zero products
$x = -2$ or $x = -3$

Check: For -2:
$$\begin{array}{c|c} x^2 + 5x + 6 = 0 \\ \hline (-2)^2 + 5(-2) + 6 \; ? \; 0 \\ 4 - 10 + 6 \\ -6 + 6 \\ 0 \end{array} \text{ TRUE}$$

For -3:
$$\begin{array}{c|c} x^2 + 5x + 6 = 0 \\ \hline (-3)^2 + 5(-3) + 6 \; ? \; 0 \\ 9 - 15 + 6 \\ -6 + 6 \\ 0 \end{array} \text{ TRUE}$$

The solutions are -2 and -3.

·········· **Caution!** ··········

Keep in mind that you *must* have 0 on one side of the equation before you can use the principle of zero products. Get all nonzero terms on one side and 0 on the other.

··

Do Exercise 5. ▶

EXAMPLE 5 Solve: $x^2 - 8x = -16$.

We first add 16 to both sides to get 0 on one side:

$x^2 - 8x = -16$
$x^2 - 8x + 16 = 0$ Adding 16 to both sides
$(x - 4)(x - 4) = 0$ Factoring
$x - 4 = 0$ or $x - 4 = 0$ Using the principle of zero products
$x = 4$ or $x = 4$.   Solving each equation

There is only one solution, 4. The check is left to the student.

Do Exercises 6 and 7. ▶

EXAMPLE 6 Solve: $x^2 + 5x = 0$.

$x^2 + 5x = 0$
$x(x + 5) = 0$ Factoring out a common factor
$x = 0$ or $x + 5 = 0$ Using the principle of zero products
$x = 0$ or $x = -5$

The solutions are 0 and -5. The check is left to the student.

MyLab Math
ANIMATION

GS **5.** Solve: $x^2 - x - 6 = 0$.
$x^2 - x - 6 = 0$
$(x + 2)() = 0$
$x + 2 = 0$ or $ = 0$
$x = -2$ or $x = $
Both numbers check.
The solutions are -2 and .

Solve.
6. $x^2 - 3x = 28$

7. $x^2 = 6x - 9$

Answers
5. $-2, 3$ 6. $-4, 7$ 7. 3
Guided Solution:
5. $x - 3, x - 3, 3, 3$

Solve.

8. $x^2 - 4x = 0$
$\square(x - 4) = 0$
$\square = 0 \text{ or } x - 4 = 0$
$x = 0 \text{ or } x = \square$

Both numbers check.
The solutions are 0 and $\square$.

9. $9x^2 = 16$

EXAMPLE 7 Solve: $4x^2 = 25$.

$$4x^2 = 25$$
$$4x^2 - 25 = 0 \quad \text{Subtracting 25 on both sides to get 0 on one side}$$
$$(2x - 5)(2x + 5) = 0 \quad \text{Factoring a difference of squares}$$
$$2x - 5 = 0 \quad or \quad 2x + 5 = 0 \quad \text{Using the principle of zero products}$$
$$2x = 5 \quad or \quad 2x = -5 \quad \text{Solving each equation}$$
$$x = \frac{5}{2} \quad or \quad x = -\frac{5}{2}$$

The solutions are $\frac{5}{2}$ and $-\frac{5}{2}$. The check is left to the student.

◀ Do Exercises 8 and 9.

EXAMPLE 8 Solve: $-5x^2 + 2x + 3 = 0$.

In this case, the leading coefficient of the trinomial is negative. Thus we first multiply by -1 and then proceed as we have in Examples 4–7.

$$-5x^2 + 2x + 3 = 0$$
$$-1(-5x^2 + 2x + 3) = -1 \cdot 0 \quad \text{Multiplying by } -1$$
$$5x^2 - 2x - 3 = 0 \quad \text{Simplifying}$$
$$(5x + 3)(x - 1) = 0 \quad \text{Factoring}$$
$$5x + 3 = 0 \quad or \quad x - 1 = 0 \quad \text{Using the principle of zero products}$$
$$5x = -3 \quad or \quad x = 1$$
$$x = -\frac{3}{5} \quad or \quad x = 1$$

Solve.

10. $-2x^2 + 13x - 21 = 0$

11. $10 - 3x - x^2 = 0$

The solutions are $-\frac{3}{5}$ and 1. The check is left to the student.

◀ Do Exercises 10 and 11.

EXAMPLE 9 Solve: $(x + 2)(x - 2) = 5$.

Be careful with an equation like this one! It might be tempting to set each factor equal to 5. **Remember: We must have 0 on one side.** We first carry out the multiplication on the left. Next, we subtract 5 on both sides to get 0 on one side. Then we proceed using the principle of zero products.

$$(x + 2)(x - 2) = 5$$
$$x^2 - 4 = 5 \quad \text{Multiplying on the left}$$
$$x^2 - 4 - 5 = 5 - 5 \quad \text{Subtracting 5}$$
$$x^2 - 9 = 0 \quad \text{Simplifying}$$
$$(x + 3)(x - 3) = 0 \quad \text{Factoring}$$
$$x + 3 = 0 \quad or \quad x - 3 = 0 \quad \text{Using the principle of zero products}$$
$$x = -3 \quad or \quad x = 3$$

12. Solve: $(x + 1)(x - 1) = 8$.

The solutions are -3 and 3. The check is left to the student.

◀ Do Exercise 12.

Answers

8. 0, 4 **9.** $-\frac{4}{3}, \frac{4}{3}$ **10.** 3, $\frac{7}{2}$ **11.** $-5, 2$
12. $-3, 3$

Guided Solution:
8. $x, x, 4, 4$

CHAPTER 5 Polynomials: Factoring

ALGEBRAIC ► GRAPHICAL CONNECTION

To find the *x*-intercept of a linear equation, we replace *y* with 0 and solve for *x*. This procedure can also be used to find the *x*-intercepts of a quadratic equation.

The graph of $y = ax^2 + bx + c, a \neq 0$, is shaped like one of the following curves. Note that each *x*-intercept represents a solution of $ax^2 + bx + c = 0$.

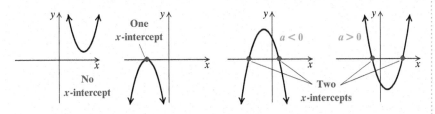

EXAMPLE 10 Find the *x*-intercepts of the graph of $y = x^2 - 4x - 5$, shown at right. (The grid is intentionally not included.)

To find the *x*-intercepts, we let $y = 0$ and solve for *x*:

$y = x^2 - 4x - 5$
$0 = x^2 - 4x - 5$ Substituting 0 for *y*
$0 = (x - 5)(x + 1)$ Factoring
$x - 5 = 0$ *or* $x + 1 = 0$ Using the principle of zero products
$x = 5$ *or* $x = -1$.

The solutions of the equation $0 = x^2 - 4x - 5$ are 5 and -1. Thus the *x*-intercepts of the graph of $y = x^2 - 4x - 5$ are $(5, 0)$ and $(-1, 0)$. We can now label them on the graph.

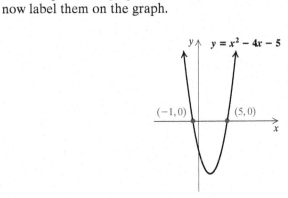

Do Exercises 13 and 14. ▶

13. Find the *x*-intercepts of the following graph.

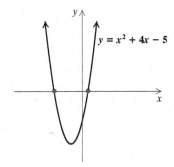

14. Use *only* the following graph to solve $3x - x^2 = 0$.

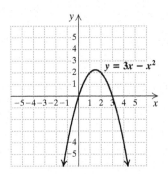

Answers
13. $(-5, 0), (1, 0)$ **14.** $0, 3$

5.8 Exercise Set

✓ Check Your Understanding

Reading Check Determine whether each statement is true or false.

RC1. If $(x + 2)(x + 3) = 10$, then $x + 2 = 10$ or $x + 3 = 10$.

RC2. A quadratic equation always has two different solutions.

RC3. The number 0 is never a solution of a quadratic equation.

RC4. If $ax^2 + bx + c = 0$ has no real-number solution, then the graph of $y = ax^2 + bx + c$ has no x-intercept.

Concept Check For each equation, use the principle of zero products to write two or three linear equations—one for each factor that includes a variable. Do not solve.

CC1. $(x - 5)(x + 4) = 0$

CC2. $(3x + 2)(x - 7) = 0$

CC3. $x(x + 6) = 0$

CC4. $5x(x - 8) = 0$

CC5. $x(x - 1)(x + 3) = 0$

CC6. $9(3x - 7)(x + 1) = 0$

a Solve using the principle of zero products.

1. $(x + 4)(x + 9) = 0$

2. $(x + 2)(x - 7) = 0$

3. $(x + 3)(x - 8) = 0$

4. $(x + 6)(x - 8) = 0$

5. $(x + 12)(x - 11) = 0$

6. $(x - 13)(x + 53) = 0$

7. $x(x + 3) = 0$

8. $y(y + 5) = 0$

9. $0 = y(y + 18)$

10. $0 = x(x - 19)$

11. $(2x + 5)(x + 4) = 0$

12. $(2x + 9)(x + 8) = 0$

13. $(5x + 1)(4x - 12) = 0$

14. $(4x + 9)(14x - 7) = 0$

15. $(7x - 28)(28x - 7) = 0$

16. $(13x + 14)(6x - 5) = 0$

17. $2x(3x - 2) = 0$

18. $55x(8x - 9) = 0$

19. $\left(\frac{1}{5} + 2x\right)\left(\frac{1}{9} - 3x\right) = 0$

20. $\left(\frac{7}{4}x - \frac{1}{16}\right)\left(\frac{2}{3}x - \frac{16}{15}\right) = 0$

21. $(0.3x - 0.1)(0.05x + 1) = 0$

22. $(0.1x + 0.3)(0.4x - 20) = 0$

23. $9x(3x - 2)(2x - 1) = 0$

24. $(x + 5)(x - 75)(5x - 1) = 0$

CHAPTER 5   Polynomials: Factoring

b Solve by factoring and using the principle of zero products. Remember to check.

25. $x^2 + 6x + 5 = 0$
26. $x^2 + 7x + 6 = 0$
27. $x^2 + 7x - 18 = 0$
28. $x^2 + 4x - 21 = 0$

29. $x^2 - 8x + 15 = 0$
30. $x^2 - 9x + 14 = 0$
31. $x^2 - 8x = 0$
32. $x^2 - 3x = 0$

33. $x^2 + 18x = 0$
34. $x^2 + 16x = 0$
35. $x^2 = 16$
36. $100 = x^2$

37. $9x^2 - 4 = 0$
38. $4x^2 - 9 = 0$
39. $0 = 6x + x^2 + 9$
40. $0 = 25 + x^2 + 10x$

41. $x^2 + 16 = 8x$
42. $1 + x^2 = 2x$
43. $5x^2 = 6x$
44. $7x^2 = 8x$

45. $6x^2 - 4x = 10$
46. $3x^2 - 7x = 20$
47. $12y^2 - 5y = 2$
48. $2y^2 + 12y = -10$

49. $t(3t + 1) = 2$
50. $x(x - 5) = 14$
51. $100y^2 = 49$
52. $64a^2 = 81$

53. $x^2 - 5x = 18 + 2x$
54. $3x^2 + 8x = 9 + 2x$
55. $10x^2 - 23x + 12 = 0$
56. $12x^2 + 17x - 5 = 0$

Find the *x*-intercepts of the graph of each equation. (The grids are intentionally not included.)

57.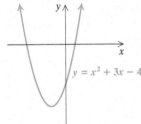
$y = x^2 + 3x - 4$

58.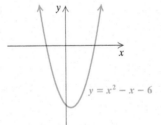
$y = x^2 - x - 6$

59.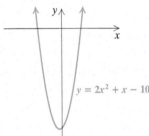
$y = 2x^2 + x - 10$

60.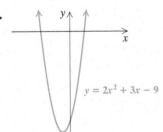
$y = 2x^2 + 3x - 9$

61.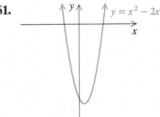
$y = x^2 - 2x - 15$

62.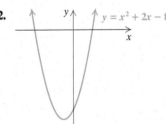
$y = x^2 + 2x - 8$

63. Use the following graph to solve $x^2 - 3x - 4 = 0$.

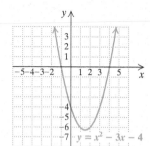

64. Use the following graph to solve $x^2 + x - 6 = 0$.

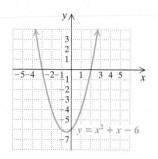

65. Use the following graph to solve $-x^2 + 2x + 3 = 0$.

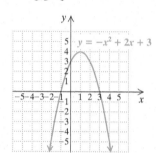

66. Use the following graph to solve $-x^2 - x + 6 = 0$.

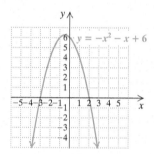

Skill Maintenance

Translate to an algebraic expression. [1.1b]

67. The square of the sum of a and b

68. The sum of the squares of a and b

Solve. [2.7d, e]

69. $-10x > 1000$

70. $6 - 3x \geq -18$

71. $3 - 2x - 4x > -9$

72. $\frac{1}{2}x - 6x + 10 \leq x - 5x$

Synthesis

Solve.

73. $b(b + 9) = 4(5 + 2b)$

74. $y(y + 8) = 16(y - 1)$

75. $(t - 3)^2 = 36$

76. $(t - 5)^2 = 2(5 - t)$

77. $x^2 - \frac{1}{64} = 0$

78. $x^2 - \frac{25}{36} = 0$

79. $\frac{5}{16}x^2 = 5$

80. $\frac{27}{25}x^2 = \frac{1}{3}$

Use a graphing calculator to find the solutions of each equation. Round solutions to the nearest hundredth.

81. $x^2 - 9.10x + 15.77 = 0$

82. $-x^2 + 0.63x + 0.22 = 0$

83. Find an equation that has the given numbers as solutions. For example, 3 and -2 are solutions of $x^2 - x - 6 = 0$.

a) $-3, 4$
b) $-3, -4$
c) $\frac{1}{2}, \frac{1}{2}$
d) $5, -5$
e) $0, 0.1, \frac{1}{4}$

Applications of Quadratic Equations

5.9

a APPLIED PROBLEMS, QUADRATIC EQUATIONS, AND FACTORING

OBJECTIVE

a Solve applied problems involving quadratic equations that can be solved by factoring.

We can solve problems that translate to quadratic equations using the five steps for solving problems.

EXAMPLE 1 *Kitchen Island.* Lisa buys a kitchen island with a butcher-block top as part of a remodeling project. The top of the island is a rectangle that is twice as long as it is wide and that has an area of 800 in². What are the dimensions of the top of the island?

1. **Familiarize.** We first make a drawing. Recall that the area of a rectangle is Length · Width. We let $x =$ the width of the top, in inches. The length is then $2x$.

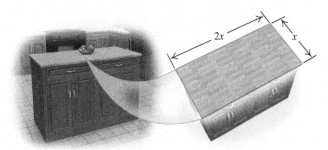

SKILL REVIEW *Solve applied problems.* [2.6a]

Solve.
1. The length of a rectangular garden is twice as long as the width. The perimeter of the garden is 48 ft. Find the length and the width of the garden.
2. The second angle of a triangle is 25° more than the first angle. The third angle is twice as large as the second angle. Find the measures of the angles of the triangle.

Answers:
1. Length: 16 ft; width: 8 ft
2. 26.25°, 51.25°, 102.5°

2. **Translate.** We reword and translate as follows:

   Rewording: The area of the rectangle is 800 in²

   Translating: $2x \cdot x = 800.$

3. **Solve.** We solve the equation as follows:

   $2x \cdot x = 800$

   $2x^2 = 800$

   $2x^2 - 800 = 0$ Subtracting 800 to get 0 on one side

   $2(x^2 - 400) = 0$ Removing a common factor of 2

   $2(x - 20)(x + 20) = 0$ Factoring a difference of squares

   $(x - 20)(x + 20) = 0$ Dividing by 2

   $x - 20 = 0$ or $x + 20 = 0$ Using the principle of zero products

   $x = 20$ or $x = -20.$ Solving each equation

4. **Check.** The solutions of the equation are 20 and -20. Since the width must be positive, -20 cannot be a solution. To check 20 in., we note that if the width is 20 in., then the length is $2 \cdot 20$ in., or 40 in., and the area is 20 in. · 40 in., or 800 in². Thus the solution 20 checks.

5. **State.** The top of the island is 20 in. wide and 40 in. long.

   Do Exercise 1.

1. *Dimensions of a Picture.* A rectangular picture is twice as long as it is wide. If the area of the picture is 288 in², what are its dimensions?

Answer
1. Length: 24 in.; width: 12 in.

SECTION 5.9   Applications of Quadratic Equations   **407**

EXAMPLE 2 *Butterfly Wings.* The *Graphium sarpedon* butterfly has areas of light blue on each wing. When the wings are joined, the blue areas form a triangle, giving rise to the butterfly's common name, Blue Triangle Butterfly. On one butterfly, the base of the blue triangle is 6 cm longer than the height of the triangle. The area of the triangle is 8 cm². Find the base and the height of the triangle.
Data: australianmuseum.net.au

1. **Familiarize.** We first make a drawing, letting $h =$ the height of the triangle, in centimeters. Then $h + 6 =$ the base. We also recall or look up the formula for the area of a triangle: Area $= \frac{1}{2}$(base)(height).

2. **Translate.** We reword the problem and translate:

   Rewording: $\frac{1}{2}$ times Base times Height is 8

   Translating: $\frac{1}{2} \cdot (h + 6) \cdot h = 8$.

3. **Solve.** We solve the quadratic equation using the principle of zero products:

   $\frac{1}{2} \cdot (h + 6) \cdot h = 8$

   $\frac{1}{2}(h^2 + 6h) = 8$ Multiplying $h + 6$ and h

   $2 \cdot \frac{1}{2}(h^2 + 6h) = 2 \cdot 8$ Multiplying by 2

   $h^2 + 6h = 16$ Simplifying

   $h^2 + 6h - 16 = 16 - 16$ Subtracting 16 to get 0 on one side

   $h^2 + 6h - 16 = 0$

   $(h - 2)(h + 8) = 0$ Factoring

   $h - 2 = 0$ or $h + 8 = 0$ Using the principle of zero products

   $h = 2$ or $h = -8$.

4. **Check.** The height of a triangle cannot have a negative length, so -8 cannot be a solution. Suppose that the height is 2 cm. The base is 6 cm more than the height, so the base is 2 cm + 6 cm, or 8 cm, and the area is $\frac{1}{2}(8)(2)$, or 8 cm². The numbers check in the original problem.

5. **State.** The base of the blue triangle is 8 cm and the height is 2 cm.

◀ Do Exercise 2.

2. *Dimensions of a Sail.* The triangular mainsail on Stacey's Lightning sailboat has an area of 125 ft². The height of the sail is 15 ft more than the base. Find the height and the base of the sail.

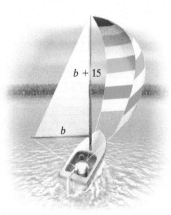

Answer
2. Height: 25 ft; base: 10 ft

EXAMPLE 3 *Apps.* The number of apps a available in the iTunes Store, in thousands, can be estimated by the polynomial

$$a = 40t^2 - 40t + 80,$$

where t is the number of years after 2008. In what year were there approximately 3680 thousand apps available in the iTunes Store?
Data: statista.com

1., 2. Familiarize and **Translate.** We are given that t is the number of years after 2008 and a is the number of apps available in the iTunes Store, in thousands. To translate the problem to an equation, we substitute 3680 for a in the equation:

$$3680 = 40t^2 - 40t + 80. \quad \text{Substituting 3680 for } a$$

3. Solve. We solve the equation for t:

$$3680 = 40t^2 - 40t + 80$$
$$3680 - 3680 = 40t^2 - 40t + 80 - 3680 \quad \text{Subtracting 3680 to get 0 on one side}$$
$$0 = 40t^2 - 40t - 3600$$
$$\left.\begin{array}{l}0 = 40(t^2 - t - 90) \\ 0 = 40(t - 10)(t + 9)\end{array}\right\} \quad \text{Factoring}$$
$$t - 10 = 0 \quad \text{or} \quad t + 9 = 0 \quad \text{Using the principle of zero products}$$
$$t = 10 \quad \text{or} \quad t = -9.$$

4. Check. The solutions of the equation are 10 and -9. Since, in the context of the problem, t is not negative, -9 cannot be a solution. But 10 checks, since

$$40(10)^2 - 40(10) + 80 = 4000 - 400 + 80 = 3680.$$

5. State. There were 3680 thousand apps in the iTunes Store 10 years after 2008, or in 2018.

Do Exercise 3. ▶

3. *Wind Energy.* The cumulative global capacity c of wind power installations, in gigawatts (GW), can be estimated by the polynomial $c = 2t^2 - 4t + 23$, where t is the number of years after 2000. In what year was the cumulative global capacity of wind power installations approximately 263 GW?
Data: World Wind Energy Association

EXAMPLE 4 *Race Numbers.* When Terry and Jody registered their boats in the Lakeport Race, the racing numbers assigned to their boats were consecutive integers, the product of which was 156. Find the integers.

1. Familiarize. Consecutive integers are one unit apart, like 49 and 50. Let $x =$ the first boat number; then $x + 1 =$ the next boat number.

Answer
3. 12 years after 2000, or in 2012

2. **Translate.** We reword the problem before translating:

Rewording: First integer times Second integer is 156

Translating: $x \cdot (x+1) = 156$.

3. **Solve.** We solve the equation as follows:

$$x(x+1) = 156$$
$$x^2 + x = 156 \qquad \text{Multiplying}$$
$$x^2 + x - 156 = 156 - 156 \qquad \text{Subtracting 156 to get 0 on one side}$$
$$x^2 + x - 156 = 0 \qquad \text{Simplifying}$$
$$(x - 12)(x + 13) = 0 \qquad \text{Factoring}$$
$$x - 12 = 0 \quad \text{or} \quad x + 13 = 0 \qquad \text{Using the principle of zero products}$$
$$x = 12 \quad \text{or} \quad x = -13.$$

4. **Page Numbers.** The product of the page numbers on two facing pages of a book is 506. Find the page numbers.

4. **Check.** The solutions of the equation are 12 and −13. Since racing numbers are not negative, −13 must be rejected. On the other hand, if x is 12, then $x + 1$ is 13 and $12 \cdot 13 = 156$. Thus the solution 12 checks.

5. **State.** The boat numbers for Terry and Jody were 12 and 13.

◀ Do Exercise 4.

The Pythagorean Theorem

The Pythagorean theorem states a relationship involving the lengths of the sides of a *right* triangle. A triangle is a **right triangle** if it has a 90°, or *right*, angle. The side opposite the 90° angle is called the **hypotenuse**. The other sides are called **legs**.

THE PYTHAGOREAN THEOREM

In any right triangle, if a and b are the lengths of the legs and c is the length of the hypotenuse, then

$$a^2 + b^2 = c^2.$$

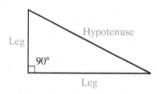

The symbol ⌐ denotes a 90° angle.

Answer

4. 22 and 23

EXAMPLE 5 *Wood Scaffold.* Jonah is building a wood scaffold to use for a home improvement project. The scaffold has diagonal braces that are 5 ft long and that span a distance of 3 ft. How high does each brace reach vertically?

1. **Familiarize.** We make a drawing as shown above and let h = the height, in feet, to which each brace rises vertically.

2. **Translate.** A right triangle is formed, so we can use the Pythagorean theorem:
$$a^2 + b^2 = c^2$$
$$3^2 + h^2 = 5^2. \quad \text{Substituting}$$

3. **Solve.** We solve the equation as follows:

$$3^2 + h^2 = 5^2$$
$$9 + h^2 = 25 \quad \text{Squaring 3 and 5}$$
$$9 + h^2 - 25 = 25 - 25 \quad \text{Subtracting 25 to get 0 on one side}$$
$$h^2 - 16 = 0 \quad \text{Simplifying}$$
$$(h - 4)(h + 4) = 0 \quad \text{Factoring}$$
$$h - 4 = 0 \quad or \quad h + 4 = 0 \quad \text{Using the principle of zero products}$$
$$h = 4 \quad or \quad h = -4.$$

4. **Check.** Since height cannot be negative, -4 cannot be a solution. If the height is 4 ft, we have $3^2 + 4^2 = 9 + 16 = 25$, which is 5^2. Thus, 4 checks and is the solution.

5. **State.** Each brace reaches a height of 4 ft.

Do Exercise 5.

5. *Reach of a Ladder.* Twila has a 26-ft ladder leaning against her house. If the bottom of the ladder is 10 ft from the base of the house, how high does the ladder reach?

Answer
5. 24 ft

EXAMPLE 6 *Ladder Settings.* A ladder of length 13 ft is placed against a building in such a way that the distance from the top of the ladder to the ground is 7 ft more than the distance from the bottom of the ladder to the building. Find both distances.

1. **Familiarize.** We first make a drawing. The ladder and the missing distances form the hypotenuse and the legs of a right triangle. We let $x =$ the length of the side (leg) across the bottom, in feet. Then $x + 7 =$ the length of the other side (leg). The hypotenuse has length 13 ft.

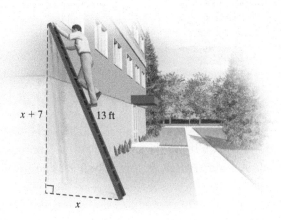

2. **Translate.** Since a right triangle is formed, we can use the Pythagorean theorem:
$$a^2 + b^2 = c^2$$
$$x^2 + (x + 7)^2 = 13^2. \quad \text{Substituting}$$

3. **Solve.** We solve the equation as follows:

$$x^2 + (x^2 + 14x + 49) = 169 \quad \text{Squaring the binomial and 13}$$
$$2x^2 + 14x + 49 = 169 \quad \text{Collecting like terms}$$
$$2x^2 + 14x + 49 - 169 = 169 - 169 \quad \text{Subtracting 169 to get 0 on one side}$$
$$2x^2 + 14x - 120 = 0 \quad \text{Simplifying}$$
$$2(x^2 + 7x - 60) = 0 \quad \text{Factoring out a common factor}$$
$$x^2 + 7x - 60 = 0 \quad \text{Dividing by 2}$$
$$(x + 12)(x - 5) = 0 \quad \text{Factoring}$$
$$x + 12 = 0 \quad \text{or} \quad x - 5 = 0 \quad \text{Using the principle of zero products}$$
$$x = -12 \quad \text{or} \quad x = 5.$$

4. **Check.** The negative integer -12 cannot be the length of a side. When $x = 5$, $x + 7 = 12$, and $5^2 + 12^2 = 13^2$. Thus, 5 and 12 check.

5. **State.** The distance from the top of the ladder to the ground is 12 ft. The distance from the bottom of the ladder to the building is 5 ft.

◀ Do Exercise 6.

6. *Right-Triangle Geometry.* The length of one leg of a right triangle is 1 m longer than the other. The length of the hypotenuse is 5 m. Find the lengths of the legs.

Answers

6. 3 m, 4 m

Translating for Success

1. *Angle Measures.* The measures of the angles of a triangle are three consecutive integers. Find the measures of the angles.

2. *Rectangle Dimensions.* The area of a rectangle is 3599 ft². The length is 2 ft longer than the width. Find the dimensions of the rectangle.

3. *Sales Tax.* Claire paid $40,704 for a new hybrid car. This included 6% for sales tax. How much did the vehicle cost before tax?

4. *Wire Cutting.* A 180-m wire is cut into three pieces. The third piece is 2 m longer than the first. The second is two-thirds as long as the first. How long is each piece?

5. *Perimeter.* The perimeter of a rectangle is 240 ft. The length is 2 ft greater than the width. Find the length and the width.

The goal of these matching questions is to practice step (2), Translate, of the five-step problem-solving process. Translate each word problem to an equation and select a correct translation from equations A–O.

A. $2x \cdot x = 288$

B. $x(x + 60) = 7021$

C. $59 = x \cdot 60$

D. $x^2 + (x + 2)^2 = 3599$

E. $x^2 + (x + 70)^2 = 130^2$

F. $6\% \cdot x = 40{,}704$

G. $2(x + 2) + 2x = 240$

H. $\frac{1}{2}x(x - 1) = 1770$

I. $x + \frac{2}{3}x + (x + 2) = 180$

J. $59\% \cdot x = 60$

K. $x + 6\% \cdot x = 40{,}704$

L. $2x^2 + x = 288$

M. $x(x + 2) = 3599$

N. $x^2 + 60 = 7021$

O. $x + (x + 1) + (x + 2) = 180$

Answers on page A-16

6. *Cell-Phone Tower.* A guy wire on a cell-phone tower is 130 ft long and is attached to the top of the tower. The height of the tower is 70 ft longer than the distance from the point on the ground where the wire is attached to the bottom of the tower. Find the height of the tower.

7. *Sales Meeting Attendance.* PTQ Corporation holds a sales meeting in Tucson. Of the 60 employees, 59 of them attend the meeting. What percent attend the meeting?

8. *Dimensions of a Pool.* A rectangular swimming pool is twice as long as it is wide. The area of the surface is 288 ft². Find the dimensions of the pool.

9. *Dimensions of a Triangle.* The height of a triangle is 1 cm less than the length of the base. The area of the triangle is 1770 cm². Find the height and the length of the base.

10. *Width of a Rectangle.* The length of a rectangle is 60 ft longer than the width. Find the width if the area of the rectangle is 7021 ft².

5.9 Exercise Set

FOR EXTRA HELP — MyLab Math

✓ Check Your Understanding

Reading Check Choose from the column on the right the word to complete each statement.

RC1. The numbers 31 and 32 are _____ integers.

RC2. In a right triangle, the _____ is the side opposite the right angle.

RC3. The area of a triangle is _____ the product of the triangle's base and height.

RC4. The symbol ⌐ indicates a(n) _____ angle.

hypotenuse
leg
obtuse
right
consecutive
even
half
twice

Concept Check Match each statement with an appropriate translation from the column on the right. Choices may be used more than once or not at all.

CC1. The product of two consecutive integers is 20.

CC2. The length of a rectangle is 1 cm longer than the width. The area of the rectangle is 20 cm².

CC3. One leg of a right triangle is 1 cm longer than the other leg. The length of the hypotenuse is 20 cm.

CC4. One leg of a right triangle is 1 cm longer than the other leg. The area of the triangle is 20 cm².

a) $x + (x + 1) = 20$

b) $x(x + 1) = 20$

c) $\frac{1}{2}x(x + 1) = 20$

d) $x^2 + (x + 1)^2 = 20^2$

a Solve.

1. *Dimensions of a Painting.* A rectangular painting is three times as long as it is wide. The area of the picture is 588 in². Find the dimensions of the painting.

2. *Area of a Garden.* The length of a rectangular garden is 4 m greater than the width. The area of the garden is 96 m². Find the length and the width.

414 CHAPTER 5 Polynomials: Factoring

3. *Design.* The screen of the TI-84 Plus graphing calculator is nearly rectangular. The length of the rectangle is 2 cm more than the width. If the area of the rectangle is 24 cm², find the length and the width.

4. *Construction.* The front porch on Trent's new home is five times as long as it is wide. If the area of the porch is 320 ft², find the dimensions.

5. *Dimensions of a Triangle.* A triangle is 10 cm wider than it is tall. The area is 28 cm². Find the height and the base.

6. *Dimensions of a Triangle.* The height of a triangle is 3 cm less than the length of the base. The area of the triangle is 35 cm². Find the height and the length of the base.

7. *Road Design.* A triangular traffic island has a base half as long as its height. The island has an area of 64 m². Find the base and the height.

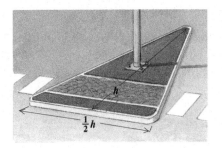

8. *Dimensions of a Sail.* The height of the jib sail on a Lightning sailboat is 5 ft greater than the length of its "foot." The area of the sail is 42 ft². Find the length of the foot and the height of the sail.

Games in a League. In a league of x teams in which each team plays every other team twice, the total number N of games to be played is given by $x^2 - x = N$. Use this equation for Exercises 9–12.

9. A Scrabble league has 14 teams. What is the total number of games to be played if each team plays every other team twice?

10. A chess league has 23 teams. What is the total number of games to be played if each team plays every other team twice?

11. A slow-pitch softball league plays a total of 132 games. How many teams are in the league if each team plays every other team twice?

12. A basketball league plays a total of 90 games. How many teams are in the league if each team plays every other team twice?

Handshakes. Dr. Benton wants to investigate the potential spread of germs by contact. She knows that the number of possible handshakes within a group of x people, assuming each person shakes every other person's hand exactly once, is given by

$$N = \tfrac{1}{2}(x^2 - x).$$

Use this formula for Exercises 13–16.

13. There are 100 people at a party. How many handshakes are possible?

14. There are 40 people at a meeting. How many handshakes are possible?

15. Everyone at a meeting shook hands with each other. There were 300 handshakes in all. How many people were at the meeting?

16. Everyone at a party shook hands with each other. There were 153 handshakes in all. How many people were at the party?

17. *Consecutive Page Numbers.* The product of the page numbers on two facing pages of a book is 210. Find the page numbers.

18. *Consecutive Page Numbers.* The product of the page numbers on two facing pages of a book is 420. Find the page numbers.

19. The product of two consecutive even integers is 168. Find the integers. (Consecutive even integers are two units apart.)

20. The product of two consecutive even integers is 224. Find the integers. (Consecutive even integers are two units apart.)

21. The product of two consecutive odd integers is 255. Find the integers. (Consecutive odd integers are two units apart.)

22. The product of two consecutive odd integers is 143. Find the integers. (Consecutive odd integers are two units apart.)

23. *Roadway Design.* Elliott Street is 24 ft wide when it ends at Main Street in Brattleboro, Vermont. A 40-ft long diagonal crosswalk allows pedestrians to cross Main Street to or from either corner of Elliott Street (see the figure). Determine the width of Main Street.

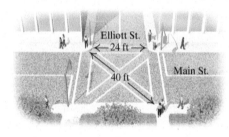

24. *Lookout Tower.* The diagonal braces in a lookout tower are 15 ft long and span a distance of 12 ft. How high does each brace reach vertically?

25. *Right-Triangle Geometry.* The length of one leg of a right triangle is 8 ft. The length of the hypotenuse is 2 ft longer than the other leg. Find the lengths of the hypotenuse and the other leg.

26. *Right-Triangle Geometry.* The length of one leg of a right triangle is 24 ft. The length of the other leg is 16 ft shorter than the hypotenuse. Find the lengths of the hypotenuse and the other leg.

27. *Archaeology.* Archaeologists have discovered that the 18th-century garden of the Charles Carroll House in Annapolis, Maryland, was a right triangle. One leg of the triangle was formed by a 400-ft long sea wall. The hypotenuse of the triangle was 200 ft longer than the other leg. What were the dimensions of the garden?

Data: bsos.umd.edu

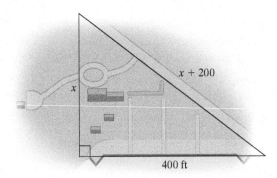

28. *Guy Wire.* The height of a wind power assessment tower is 5 m shorter than the guy wire that supports it. If the guy wire is anchored 15 m from the foot of the tower, how tall is the tower?

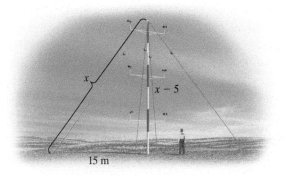

29. *Right Triangle.* The shortest side of a right triangle measures 7 m. The lengths of the other two sides are consecutive integers. Find the lengths of the other two sides.

30. *Right Triangle.* The shortest side of a right triangle measures 8 cm. The lengths of the other two sides are consecutive odd integers. Find the lengths of the other two sides.

31. *Architecture.* An architect has allocated a rectangular space of 264 ft² for a square dining room and a 10-ft wide kitchen, as shown in the figure. Find the dimensions of each room.

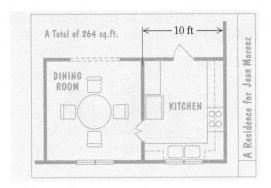

32. *Design.* A window panel for a sun porch consists of a 7-ft tall rectangular window stacked above a square window. The windows have the same width. If the total area of the window panel is 18 ft², find the dimensions of each window.

Height of a Rocket. For Exercises 33 and 34, assume that a water rocket is launched upward with an initial velocity of 48 ft/sec. Its height h, in feet, after t seconds, is given by $h = 48t - 16t^2$.

33. When will the rocket be exactly 32 ft above the ground?

34. When will the rocket crash into the ground?

35. The sum of the squares of two consecutive odd positive integers is 74. Find the integers.

36. The sum of the squares of two consecutive odd positive integers is 130. Find the integers.

SECTION 5.9 Applications of Quadratic Equations

Skill Maintenance

Compute and simplify.

37. $-3.57 + 8.1$ [1.3a]

38. $-\dfrac{2}{3} - \dfrac{1}{6}$ [1.4a]

39. $(-2)(-4)(-5)$ [1.5a]

40. $2 \cdot 6^2 \div (-2) \cdot 3 - 8$ [1.8d]

41. $\dfrac{2 - |3 - 8|}{(-1 - 4)^2}$ [1.8d]

42. $1.2 + (-2)^3 + 3.4$ [1.8d]

Remove parentheses and simplify.

43. $2(y - 7) - (6y - 1)$ [1.8b]

44. $2\{x - 3[4 - (x - 1)] + x\}$ [1.8c]

Synthesis

45. *Pool Sidewalk.* A cement walk of constant width is built around a 20-ft by 40-ft rectangular pool. The total area of the pool and the walk is 1500 ft². Find the width of the walk.

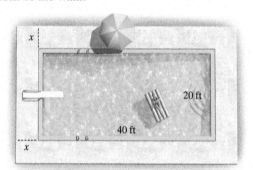

46. *Roofing.* A *square* of shingles covers 100 ft² of surface area. How many squares will be needed to reshingle the roof of the house shown?

47. *Dimensions of an Open Box.* A rectangular piece of cardboard is twice as long as it is wide. A 4-cm square is cut out of each corner, and the sides are turned up to make a box with an open top. The volume of the box is 616 cm³. Find the original dimensions of the cardboard.

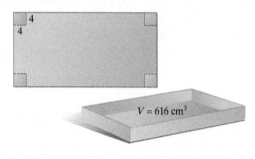

48. *Rain-Gutter Design.* An open rectangular gutter is made by turning up the sides of a piece of metal 20 in. wide. The area of the cross-section of the gutter is 50 in². Find the depth of the gutter.

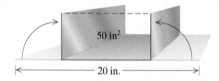

49. *Right Triangle.* The longest side of a right triangle is 5 yd shorter than six times the length of the shortest side. The other side of the triangle is 5 yd longer than five times the length of the shortest side. Find the lengths of the sides of the triangle.

50. Solve for x.

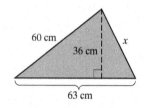

418   CHAPTER 5   Polynomials: Factoring

CHAPTER 5 Summary and Review

Vocabulary Reinforcement

Complete each statement with the correct term from the column on the right. Some of the choices may be used more than once or not at all.

1. To _____ a polynomial is to express it as a product. [5.1a]

2. A(n) _____ of a polynomial P is a polynomial that can be used to express P as a product. [5.1a]

3. A(n) _____ of a polynomial is an expression that names that polynomial as a product. [5.1a]

4. When factoring, always look first for a(n) _____ factor. [5.1b]

5. When factoring a polynomial with four terms, try factoring by _____. [5.7a]

6. A trinomial square is the square of a(n) _____. [5.5a]

7. The principle of _____ products states that if $ab = 0$, then $a = 0$ or $b = 0$. [5.8a]

8. The factorization of a _____ of squares is the product of the sum and the difference of two terms. [5.5d]

common
similar
product
difference
factor
factorization
grouping
monomial
binomial
trinomial
zero

Concept Reinforcement

Determine whether each statement is true or false.

_____ 1. Every polynomial with four terms can be factored by grouping. [5.1c]

_____ 2. When factoring $x^2 + 5x + 6$, we need consider only positive pairs of factors of 6. [5.2a]

_____ 3. A product is 0 if and only if all the factors are 0. [5.8a]

_____ 4. If the principle of zero products is to be used, one side of the equation must be 0. [5.8b]

Study Guide

Objective 5.1a Find the greatest common factor, the GCF, of monomials.

Example Find the GCF of $15x^4y^2$, $-18x$, and $12x^3y$.

$15x^4y^2 = 3 \cdot 5 \cdot x^4 \cdot y^2$;
$-18x = -1 \cdot 2 \cdot 3 \cdot 3 \cdot x$;
$12x^3y = 2 \cdot 2 \cdot 3 \cdot x^3 \cdot y$

The GCF of the coefficients is 3. The GCF of the powers of x is x because 1 is the smallest exponent of x. The GCF of the powers of y is 1 because $-18x$ has no y-factor. Thus the GCF is $3 \cdot x \cdot 1$, or $3x$.

Practice Exercise

1. Find the GCF of $8x^3y^2$, $-20xy^3$, and $32x^2y$.

Objective 5.1b Factor polynomials when the terms have a common factor, factoring out the greatest common factor.

Example Factor: $16y^4 + 8y^3 - 24y^2$.

The *largest* common factor is $8y^2$.

$16y^4 + 8y^3 - 24y^2 = (8y^2)(2y^2) + (8y^2)(y) - (8y^2)(3)$
$= 8y^2(2y^2 + y - 3)$

Practice Exercise

2. Factor $27x^5 - 9x^3 + 18x^2$, factoring out the largest common factor.

Objective 5.1c Factor certain expressions with four terms using factoring by grouping.

Example Factor $6x^3 + 4x^2 - 15x - 10$ by grouping.

$6x^3 + 4x^2 - 15x - 10 = (6x^3 + 4x^2) + (-15x - 10)$
$= 2x^2(3x + 2) - 5(3x + 2)$
$= (3x + 2)(2x^2 - 5)$

Practice Exercise

3. Factor $z^3 - 3z^2 + 4z - 12$ by grouping.

Objective 5.2a Factor trinomials of the type $x^2 + bx + c$ by examining the constant term c.

Example Factor: $x^2 - x - 12$.

Since the constant term, -12, is negative, we look for a factorization of -12 in which one factor is positive and one factor is negative. The sum of the factors must be the coefficient of the middle term, -1, so the negative factor must have the larger absolute value. The possible pairs of factors that meet these criteria are $1, -12$ and $2, -6$ and $3, -4$. The numbers we need are 3 and -4:

$x^2 - x - 12 = (x + 3)(x - 4)$.

Practice Exercise

4. Factor: $x^2 + 6x + 8$.

Objective 5.3a Factor trinomials of the type $ax^2 + bx + c$, $a \neq 1$, using the FOIL method.

Example Factor: $2y^3 + 5y^2 - 3y$.

1) Factor out the largest common factor, y:

$y(2y^2 + 5y - 3)$.

Now we factor $2y^2 + 5y - 3$.

2) Because $2y^2$ factors as $2y \cdot y$, we have this possibility for a factorization:

$(2y + \ )(y + \ )$.

3) There are two pairs of factors of -3 and each can be written in two ways:

$3, -1 \quad \quad -3, 1$
and $-1, 3 \quad \quad 1, -3$.

4) From steps (2) and (3), we see that there are 4 possibilities for factorizations. We look for **O**utside and **I**nside products for which the sum is the middle term, $5y$. We try some possibilities and find that the factorization of $2y^2 + 5y - 3$ is $(2y - 1)(y + 3)$.

We must include the common factor to get a factorization of the original trinomial:

$2y^3 + 5y^2 - 3y = y(2y - 1)(y + 3)$.

Practice Exercise

5. Factor: $6z^2 - 21z - 12$.

Objective 5.4a Factor trinomials of the type $ax^2 + bx + c$, $a \neq 1$, using the ac-method.

Example Factor $5x^2 + 7x - 6$ using the ac-method.
1) There is no common factor (other than 1 or -1).
2) Multiply the leading coefficient 5 and the constant, -6:
$$5(-6) = -30.$$
3) Look for a factorization of -30 in which the sum of the factors is the coefficient of the middle term, 7. One number will be positive and the other will be negative. Since their sum, 7, is positive, the positive number will have the larger absolute value. The numbers we need are 10 and -3.
4) Split the middle term, writing it as a sum or a difference using the factors found in step (3):
$$7x = 10x - 3x.$$
5) Factor by grouping:
$$5x^2 + 7x - 6 = 5x^2 + 10x - 3x - 6$$
$$= 5x(x + 2) - 3(x + 2)$$
$$= (x + 2)(5x - 3).$$
6) Check: $(x + 2)(5x - 3) = 5x^2 + 7x - 6.$

Practice Exercise
6. Factor $6y^2 + 7y - 3$ using the ac-method.

Objective 5.5b Factor trinomial squares.

Example Factor: $9x^2 - 12x + 4$.
$9x^2 - 12x + 4 = (3x)^2 - 2 \cdot 3x \cdot 2 + 2^2 = (3x - 2)^2$

Practice Exercise
7. Factor: $4x^2 + 4x + 1$.

Objective 5.5d Factor differences of squares, being careful to factor completely.

Example Factor: $b^6 - b^2$.
$$b^6 - b^2 = b^2(b^4 - 1) = b^2(b^2 + 1)(b^2 - 1)$$
$$= b^2(b^2 + 1)(b + 1)(b - 1)$$

Practice Exercise
8. Factor $18x^2 - 8$ completely.

Objective 5.6a Factor sums and differences of cubes.

Example Factor: **(a)** $w^3 - 512$; **(b)** $125a^3 + b^3$.
a) $w^3 - 512 = w^3 - 8^3$
$$= (w - 8)(w^2 + 8w + 64)$$
b) $125a^3 + b^3 = (5a)^3 + b^3$
$$= (5a + b)(25a^2 - 5ab + b^2)$$

Practice Exercise
9. Factor: $27 - 125x^3$.

10. Factor: $\dfrac{1}{8}q^3 + 8a^3$.

Objective 5.8b Solve quadratic equations by factoring and then using the principle of zero products.

Example Solve: $x^2 - 3x = 28$.
$$x^2 - 3x = 28$$
$$x^2 - 3x - 28 = 28 - 28$$
$$x^2 - 3x - 28 = 0$$
$$(x + 4)(x - 7) = 0$$
$$x + 4 = 0 \quad \text{or} \quad x - 7 = 0$$
$$x = -4 \quad \text{or} \quad x = 7$$
The solutions are -4 and 7.

Practice Exercise
11. Solve: $x^2 + 4x = 5$.

Review Exercises

Find the GCF. [5.1a]
1. $-15y^2$, $25y^6$

2. $12x^3$, $-60x^2y$, $36xy$

Factor completely. [5.7a]
3. $5 - 20x^6$
4. $x^2 - 3x$

5. $9x^2 - 4$
6. $x^2 + 4x - 12$

7. $x^2 + 14x + 49$
8. $6x^3 + 12x^2 + 3x$

9. $x^3 + x^2 + 3x + 3$
10. $6x^2 - 5x + 1$

11. $x^4 - 81$
12. $9x^3 + 12x^2 - 45x$

13. $2x^2 - 50$
14. $x^4 + 4x^3 - 2x - 8$

15. $16x^4 - 1$
16. $8x^6 - 32x^5 + 4x^4$

17. $75 + 12x^2 + 60x$
18. $x^2 + 9$

19. $x^3 - x^2 - 30x$
20. $4x^2 - 25$

21. $9x^2 + 25 - 30x$
22. $6x^2 - 28x - 48$

23. $x^2 - 6x + 9$
24. $2x^2 - 7x - 4$

25. $18x^2 - 12x + 2$
26. $3x^2 - 27$

27. $15 - 8x + x^2$
28. $25x^2 - 20x + 4$

29. $49b^{10} + 4a^8 - 28a^4b^5$

30. $x^2y^2 + xy - 12$

31. $12a^2 + 84ab + 147b^2$

32. $m^2 + 5m + mt + 5t$

33. $32x^4 - 128y^4z^4$
34. $5y^3 + 40t^3$

Solve. [5.8a, b]
35. $(x - 1)(x + 3) = 0$
36. $x^2 + 2x - 35 = 0$

37. $x^2 + 4x = 0$
38. $3x^2 + 2 = 5x$

39. $x^2 = 64$
40. $16 = x(x - 6)$

Find the *x*-intercepts of the graph of each equation. [5.8b]

41. $y = x^2 + 9x + 20$

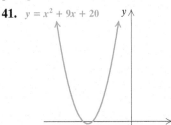

42. $y = 2x^2 - 7x - 15$

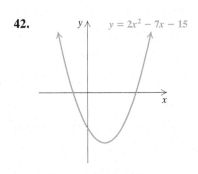

Solve. [5.9a]

43. *Sharks' Teeth.* Sharks' teeth are shaped like triangles. The height of a tooth of a great white shark is 1 cm longer than the base. The area is 15 cm². Find the height and the base.

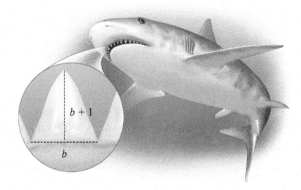

44. The product of two consecutive even integers is 288. Find the integers.

45. *Zipline.* On one zipline in a canopy tour in Costa Rica, riders drop 58 ft while covering a distance of 840 ft along the ground. How long is the zipline?

46. *Tree Supports.* A 5-ft cable is used to support a newly planted maple tree. The distance from the base of the tree to the point on the ground where the cable is anchored is 1 ft more than the distance from the base of the tree to the point where the cable is attached to the tree. Find both distances.

47. If the sides of a square are lengthened by 3 km, the area increases to 81 km². Find the length of a side of the original square.

48. Factor: $x^2 - 9x + 8$. Which of the following is one factor? [5.2a], [5.7a]

A. $(x + 1)$ B. $(x - 1)$
C. $(x + 8)$ D. $(x - 4)$

49. Factor $15x^2 + 5x - 20$ completely. Which of the following is one factor? [5.3a], [5.4a], [5.7a]

A. $(3x + 4)$ B. $(3x - 4)$
C. $(5x - 5)$ D. $(15x + 20)$

Synthesis

Solve. [5.9a]

50. The pages of a book measure 15 cm by 20 cm. Margins of equal width surround the printing on each page and constitute one-half of the area of the page. Find the width of the margins.

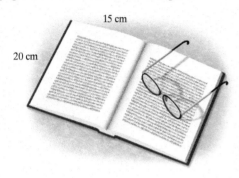

51. The cube of a number is the same as twice the square of the number. Find all such numbers.

52. The length of a rectangle is two times its width. When the length is increased by 20 in. and the width is decreased by 1 in., the area is 160 in². Find the original length and width.

53. Use the information in the following figure to determine the height of the telephone pole.

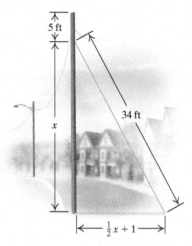

Solve. [5.8b]

54. $x^2 + 25 = 0$

55. $(x - 2)(x + 3)(2x - 5) = 0$

56. $(x - 3)4x^2 + 3x(x - 3) - (x - 3)10 = 0$

Understanding Through Discussion and Writing

1. Gwen factors $x^3 - 8x^2 + 15x$ as $(x^2 - 5x)(x - 3)$. Is she wrong? Why or why not? What advice would you offer? [5.2a]

2. After a test, Josh told a classmate that he was sure he had not written any incorrect factorizations. How could he be certain? [5.7a]

3. Kelly factored $16 - 8x + x^2$ as $(x - 4)^2$, while Tony factored it as $(4 - x)^2$. Evaluate each expression for several values of x. Then explain why both answers are correct. [5.5b]

4. What is wrong with the following? Explain the correct method of solution. [5.8b]

$(x - 3)(x + 4) = 8$
$x - 3 = 8$ or $x + 4 = 8$
$x = 11$ or $x = 4$

5. What is incorrect about solving $x^2 = 3x$ by dividing by x on both sides? [5.8b]

6. An archaeologist has measuring sticks of 3 ft, 4 ft, and 5 ft. Explain how she could draw a 7-ft by 9-ft rectangle on a piece of land being excavated. [5.9a]

CHAPTER 5 Test

1. Find the GCF: $28x^3, 48x^7$.

Factor completely.

2. $x^2 - 7x + 10$

3. $x^2 + 25 - 10x$

4. $6y^2 - 8y^3 + 4y^4$

5. $x^3 + x^2 + 2x + 2$

6. $x^2 - 5x$

7. $x^3 + 2x^2 - 3x$

8. $28x - 48 + 10x^2$

9. $4x^2 - 9$

10. $x^2 - x - 12$

11. $6m^3 + 9m^2 + 3m$

12. $3w^2 - 75$

13. $60x + 45x^2 + 20$

14. $3x^4 - 48$

15. $49x^2 - 84x + 36$

16. $5x^2 - 26x + 5$

17. $x^4 + 2x^3 - 3x - 6$

18. $80 - 5x^4$

19. $6t^3 + 9t^2 - 15t$

20. $4x^2 - 4x - 15$

21. $3m^2 - 9mn - 30n^2$

22. $1000a^3 - 27b^3$

Solve.

23. $x^2 - 3x = 0$

24. $2x^2 = 32$

25. $x^2 - x - 20 = 0$

26. $2x^2 + 7x = 15$

27. $x(x - 3) = 28$

Find the *x*-intercepts of the graph of each equation.

28. 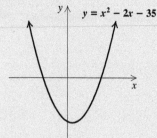 $y = x^2 - 2x - 35$

29. 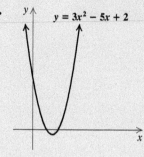 $y = 3x^2 - 5x + 2$

Solve.

30. The length of a rectangle is 2 m more than the width. The area of the rectangle is 48 m². Find the length and the width.

31. The base of a triangle is 6 cm greater than twice the height. The area is 28 cm². Find the height and the base.

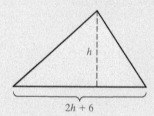

32. *Masonry Corner.* A mason wants to be sure that he has a right-angle corner of a building's foundation. He marks a point 3 ft from the corner along one wall and another point 4 ft from the corner along the other wall. If the corner is a right angle, what should the distance be between the two marked points?

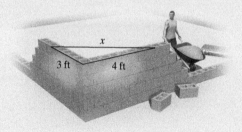

33. Factor $2y^4 - 32$ completely. Which of the following is one factor?
A. $(y + 2)$
B. $(y + 4)$
C. $(y^2 - 4)$
D. $(2y^2 + 8)$

Synthesis

34. The length of a rectangle is five times its width. When the length is decreased by 3 m and the width is increased by 2 m, the area of the new rectangle is 60 m². Find the original length and width.

35. Factor: $(a + 3)^2 - 2(a + 3) - 35$.

36. Solve: $20x(x + 2)(x - 1) = 5x^3 - 24x - 14x^2$.

37. If $x + y = 4$ and $x - y = 6$, then $x^2 - y^2$ equals which of the following?
A. 2
B. 10
C. 34
D. 24

426   CHAPTER 5   Polynomials: Factoring

CHAPTERS 1–5 Cumulative Review

Use either < or > for ☐ to write a true sentence.

1. $\dfrac{2}{3}\ \square\ \dfrac{5}{7}$

2. $-\dfrac{4}{7}\ \square\ -\dfrac{8}{11}$

Compute and simplify.

3. $2.06 + (-4.79) - (-3.08)$

4. $5.652 \div (-3.6)$

5. $\left(\dfrac{2}{9}\right)\left(-\dfrac{3}{8}\right)\left(\dfrac{6}{7}\right)$

6. $\dfrac{21}{5} \div \left(-\dfrac{7}{2}\right)$

Simplify.

7. $[3x + 2(x - 1)] - [2x - (x + 3)]$

8. $1 - [14 + 28 \div 7 - (6 + 9 \div 3)]$

9. $(2x^2 y^{-1})^3$

10. $\dfrac{3x^5}{4x^3} \cdot \dfrac{-2x^{-3}}{9x^2}$

11. Add: $(2x^2 - 3x^3 + x - 4) + (x^4 - x - 5x^2)$.

12. Subtract: $(2x^2 y^2 + xy - 2xy^2) - (2xy - 2xy^2 + x^2 y)$.

13. Divide: $(x^3 + 2x^2 - x + 1) \div (x - 1)$.

Multiply.

14. $(2t - 3)^2$

15. $(x^2 - 3)(x^2 + 3)$

16. $(2x + 4)(3x - 4)$

17. $2x(x^3 + 3x^2 + 4x)$

18. $(2y - 1)(2y^2 + 3y + 4)$

19. $\left(x + \dfrac{2}{3}\right)\left(x - \dfrac{2}{3}\right)$

Factor.

20. $x^2 + 2x - 8$

21. $4x^2 - 25$

22. $3x^3 - 4x^2 + 3x - 4$

23. $x^2 - 26x + 169$

24. $75x^2 - 108y^2$

25. $6x^2 - 13x - 63$

26. $x^4 - 2x^2 - 3$

27. $4y^3 - 6y^2 - 4y + 6$

28. $6p^2 + pq - q^2$

29. $10x^3 + 52x^2 + 10x$

30. $49x^3 - 42x^2 + 9x$

31. $3x^2 + 5x - 4$

32. $75x^3 + 27x$

33. $3x^8 - 48y^8$

34. $14x^2 + 28 + 42x$

35. $2x^5 - 2x^3 + x^2 - 1$

Solve.

36. $3x - 5 = 2x + 10$

37. $3y + 4 > 5y - 8$

38. $(x - 15)\left(x + \frac{1}{4}\right) = 0$

39. $-98x(x + 37) = 0$

40. $x^3 + x^2 = 25x + 25$

41. $2x^2 = 72$

42. $9x^2 + 1 = 6x$

43. $x^2 + 17x + 70 = 0$

44. $14y^2 = 21y$

45. $1.6 - 3.5x = 0.9$

46. $(x + 3)(x - 4) = 8$

47. $1.5x - 3.6 \leq 1.3x + 0.4$

48. $2x - [3x - (2x + 3)] = 3x + [4 - (2x + 1)]$

49. $y = mx + b$, for m

Solve.

50. The sum of two consecutive even integers is 102. Find the integers.

51. The product of two consecutive even integers is 360. Find the integers.

52. The length of a rectangular window is 3 ft longer than the height. The area of the window is 18 ft². Find the length and the height.

53. The length of a rectangular lot is 200 m longer than the width. The perimeter of the lot is 1000 m. Find the dimensions of the lot.

54. Money is borrowed at 6% simple interest. After 1 year, $6890 pays off the loan. How much was originally borrowed?

55. The length of one leg of a right triangle is 15 m. The length of the other leg is 9 m shorter than the length of the hypotenuse. Find the length of the hypotenuse.

56. A 100-m wire is cut into three pieces. The second piece is twice as long as the first piece. The third piece is one-third as long as the first piece. How long is each piece?

57. After a 25% price reduction, a pair of shoes is on sale for $33. What was the price before reduction?

58. The front of a house is a triangle that is as wide as it is tall. Its area is 98 ft². Find the height and the base.

59. Find the intercepts. Then graph the equation.
$3x + 4y = -12$

Synthesis

Solve.

60. $(x + 3)(x - 5) \leq (x + 2)(x - 1)$

61. $\dfrac{x - 3}{2} - \dfrac{2x + 5}{26} = \dfrac{4x + 11}{13}$

62. $(x + 1)^2 = 25$

Factor.

63. $x^2(x - 3) - x(x - 3) - 2(x - 3)$

64. $4a^2 - 4a + 1 - 9b^2 - 24b - 16$

Solve.

65. Find c such that the polynomial will be the square of a binomial: $cx^2 - 40x + 16$.

66. The length of the radius of a circle is increased by 2 cm to form a new circle. The area of the new circle is four times the area of the original circle. Find the length of the radius of the original circle.

CHAPTER 6

Rational Expressions and Equations

- 6.1 Multiplying and Simplifying Rational Expressions
- 6.2 Division and Reciprocals
- 6.3 Least Common Multiples and Denominators
- 6.4 Adding Rational Expressions
- 6.5 Subtracting Rational Expressions

Mid-Chapter Review

- 6.6 Complex Rational Expressions
- 6.7 Solving Rational Equations
- 6.8 Applications Using Rational Equations and Proportions

Translating for Success

- 6.9 Variation and Applications

Summary and Review
Test
Cumulative Review

Laying sod is an instant method of establishing turf. The main uses of sod are home lawns, golf courses, and high-traffic areas. In the United States, there were 320,566 acres of land used for sod production in 2014. That year, 195,497 acres were harvested, and the sales totaled $1,138,465,000. Acreage within only six states provides 58.1% of the total acreage dedicated to sod production.

Data: agcensus.usda.gov

Sod Price List

Sod Type	Amount (in square feet)	Price per square foot
Kentucky bluegrass	250–1200	$0.25
Kentucky bluegrass	1201+	0.20
Turf type tall fescue	250–1200	0.25
Turf type tall fescue	1201+	0.20

DATA: Myers Sod Farm, Seymour IN, 2016 price list

JUST IN TIME Review topics 4, 8, 9, and 12 in the **Just In Time** section at the front of the text. This provides excellent prerequisite skill review for this chapter.

MyLab Math VIDEO

In Example 2 of Section 6.8, we will calculate how long it will take two crews working together to lay 7 skids of sod.

STUDYING FOR SUCCESS   Quiz and Test Follow-Up

- Immediately after completing a chapter quiz or test, write out a step-by-step solution of each question that you missed.
- Visit your instructor or tutor for help with problems that are still giving you trouble.
- Keep your tests and quizzes, along with their corrections, to use as a study guide for the final examination.

6.1 Multiplying and Simplifying Rational Expressions

OBJECTIVES

a Find all numbers for which a rational expression is not defined.

b Multiply a rational expression by 1, using an expression such as A/A.

c Simplify rational expressions by factoring the numerator and the denominator and removing factors of 1.

d Multiply rational expressions and simplify.

a RATIONAL EXPRESSIONS AND REPLACEMENTS

Rational numbers are quotients of integers. Some examples are

$$\frac{2}{3}, \quad \frac{4}{-5}, \quad \frac{-8}{17}, \quad \frac{563}{1}.$$

The following are called **rational expressions**, or **fraction expressions**. They are quotients, or ratios, of polynomials:

$$\frac{3}{4}, \quad \frac{z}{6}, \quad \frac{5}{x+2}, \quad \frac{t^2 + 3t - 10}{7t^2 - 4}.$$

A rational expression is also a division. For example,

$$\frac{3}{4} \text{ means } 3 \div 4 \quad \text{and} \quad \frac{x-8}{x+2} \text{ means } (x-8) \div (x+2).$$

Because rational expressions indicate division, we must be careful to avoid denominators of zero. When a variable is replaced with a number that produces a denominator equal to zero, the rational expression is not defined. For example, in the expression

$$\frac{x-8}{x+2},$$

when x is replaced with -2, the denominator is 0, and the expression is *not* defined:

$$\frac{x-8}{x+2} = \frac{-2-8}{-2+2} = \frac{-10}{0}. \leftarrow \text{Division by 0 is not defined.}$$

When x is replaced with a number other than -2, such as 3, the expression *is* defined because the denominator is nonzero:

$$\frac{x-8}{x+2} = \frac{3-8}{3+2} = \frac{-5}{5} = -1.$$

EXAMPLE 1 Find all numbers for which the rational expression

$$\frac{x+4}{x^2 - 3x - 10}$$

is not defined.

The value of the numerator has no bearing on whether or not a rational expression is defined. To determine which numbers make the rational expression not defined, we set the *denominator* equal to 0 and solve:

$$x^2 - 3x - 10 = 0$$
$$(x - 5)(x + 2) = 0 \quad \text{Factoring}$$
$$x - 5 = 0 \quad \text{or} \quad x + 2 = 0 \quad \text{Using the principle of zero products (See Section 5.8.)}$$
$$x = 5 \quad \text{or} \quad x = -2.$$

The rational expression is not defined for the replacement numbers 5 and −2.

Do Exercises 1–3.

Find all numbers for which the rational expression is not defined.

1. $\dfrac{16}{x - 3}$

GS 2. $\dfrac{2x - 7}{x^2 + 5x - 24}$

$$x^2 + 5x - 24 = 0$$
$$(x + \underline{})(x - 3) = 0$$
$$x + 8 = 0 \quad \text{or} \quad x - \underline{} = 0$$
$$x = \underline{} \quad \text{or} \quad x = 3$$

The rational expression is not defined for replacements −8 and $\underline{}$.

3. $\dfrac{x + 5}{8}$

b MULTIPLYING BY 1

We multiply rational expressions in the same way that we multiply fraction notation in arithmetic. For review, we see that

$$\dfrac{3}{7} \cdot \dfrac{2}{5} = \dfrac{3 \cdot 2}{7 \cdot 5} = \dfrac{6}{35}.$$

MULTIPLYING RATIONAL EXPRESSIONS

To multiply rational expressions, multiply numerators and multiply denominators:

$$\dfrac{A}{B} \cdot \dfrac{C}{D} = \dfrac{AC}{BD}.$$

For example,

$$\dfrac{x - 2}{3} \cdot \dfrac{x + 2}{x + 7} = \dfrac{(x - 2)(x + 2)}{3(x + 7)}. \quad \text{Multiplying the numerators and the denominators}$$

Note that we leave the numerator, $(x - 2)(x + 2)$, and the denominator, $3(x + 7)$, in factored form because it is easier to simplify if we do not multiply. In order to learn to simplify, we first need to consider multiplying the rational expression by 1.

Any rational expression with the same numerator and denominator (except 0/0) is a symbol for 1:

$$\dfrac{19}{19} = 1, \quad \dfrac{x + 8}{x + 8} = 1, \quad \dfrac{3x^2 - 4}{3x^2 - 4} = 1, \quad \dfrac{-1}{-1} = 1.$$

EQUIVALENT EXPRESSIONS

Expressions that have the same value for all allowable (or meaningful) replacements are called **equivalent expressions**.

Answers
1. 3 **2.** −8, 3 **3.** None
Guided Solution:
2. 8, 3, −8, 3

We can multiply by 1 to obtain an *equivalent expression*. At this point, we select expressions for 1 arbitrarily. Later, we will have a system for our choices when we add and subtract.

EXAMPLES Multiply.

2. $\dfrac{3x+2}{x+1} \cdot 1 = \dfrac{3x+2}{x+1} \cdot \dfrac{2x}{2x}$ Using the identity property of 1. We arbitrarily choose $(2x)/(2x)$ as a symbol for 1.

$\phantom{2.\ \dfrac{3x+2}{x+1} \cdot 1} = \dfrac{(3x+2)2x}{(x+1)2x}$

Multiply.

4. $\dfrac{2x+1}{3x-2} \cdot \dfrac{x}{x}$

5. $\dfrac{x+1}{x-2} \cdot \dfrac{x+2}{x+2}$

6. $\dfrac{x-8}{x-y} \cdot \dfrac{-1}{-1}$

3. $\dfrac{x+2}{x-7} \cdot \dfrac{x+3}{x+3} = \dfrac{(x+2)(x+3)}{(x-7)(x+3)}$ We arbitrarily choose $(x+3)/(x+3)$ as a symbol for 1.

4. $\dfrac{2+x}{2-x} \cdot \dfrac{-1}{-1} = \dfrac{(2+x)(-1)}{(2-x)(-1)}$ Using $(-1)/(-1)$ as a symbol for 1

◀ Do Exercises 4–6.

C SIMPLIFYING RATIONAL EXPRESSIONS

SKILL REVIEW Factor trinomials of the type $ax^2 + bx + c$, $a \neq 1$, using the FOIL method. [5.3a]
Factor.
1. $2x^2 - x - 21$ 2. $40x^2 - 43x - 6$

Answers: 1. $(2x-7)(x+3)$ 2. $(5x-6)(8x+1)$

Simplifying rational expressions is similar to simplifying fraction expressions in arithmetic. For example, an expression like can be simplified as follows:

$\dfrac{15}{40} = \dfrac{3 \cdot 5}{8 \cdot 5}$ Factoring the numerator and the denominator. Note the common factor, 5.

$\phantom{\dfrac{15}{40}} = \dfrac{3}{8} \cdot \dfrac{5}{5}$ Factoring the fraction expression

$\phantom{\dfrac{15}{40}} = \dfrac{3}{8} \cdot 1$ $\dfrac{5}{5} = 1$

$\phantom{\dfrac{15}{40}} = \dfrac{3}{8}.$ Using the identity property of 1, or "removing" a factor of 1

Similar steps are followed when simplifying rational expressions: We factor and remove a factor of 1, using the fact that

$$\dfrac{ab}{cb} = \dfrac{a}{c} \cdot \dfrac{b}{b} = \dfrac{a}{c} \cdot 1 = \dfrac{a}{c}.$$

In algebra, instead of simplifying $\tfrac{15}{40}$, we may need to simplify an expression like

$$\dfrac{x^2 - 16}{x + 4}.$$

Just as factoring is important in simplifying in arithmetic, so too is it important in simplifying rational expressions. The factoring that we use most is the factoring of polynomials, which we studied in Chapter 5.

Answers

4. $\dfrac{(2x+1)x}{(3x-2)x}$ 5. $\dfrac{(x+1)(x+2)}{(x-2)(x+2)}$

6. $\dfrac{(x-8)(-1)}{(x-y)(-1)}$

To simplify, we can do the reverse of multiplying. We factor the numerator and the denominator and remove a factor of 1.

EXAMPLE 5 Simplify: $\dfrac{8x^2}{24x}$.

$\dfrac{8x^2}{24x} = \dfrac{8 \cdot x \cdot x}{3 \cdot 8 \cdot x}$ Factoring the numerator and the denominator. Note the common factor, $8x$.

$= \dfrac{8x}{8x} \cdot \dfrac{x}{3}$ Factoring the rational expression

$= 1 \cdot \dfrac{x}{3}$ $\dfrac{8x}{8x} = 1$

$= \dfrac{x}{3}$ We removed a factor of 1.

Do Exercises 7 and 8.

Simplify.

7. $\dfrac{5y}{y}$ 8. $\dfrac{9x^2}{36x}$

EXAMPLES Simplify.

6. $\dfrac{5a + 15}{10} = \dfrac{5(a + 3)}{5 \cdot 2}$ Factoring the numerator and the denominator

$= \dfrac{5}{5} \cdot \dfrac{a + 3}{2}$ Factoring the rational expression

$= 1 \cdot \dfrac{a + 3}{2}$ $\dfrac{5}{5} = 1$

$= \dfrac{a + 3}{2}$ Removing a factor of 1

7. $\dfrac{6a + 12}{7a + 14} = \dfrac{6(a + 2)}{7(a + 2)}$ Factoring the numerator and the denominator

$= \dfrac{6}{7} \cdot \dfrac{a + 2}{a + 2}$ Factoring the rational expression

$= \dfrac{6}{7} \cdot 1$ $\dfrac{a + 2}{a + 2} = 1$

$= \dfrac{6}{7}$ Removing a factor of 1

8. $\dfrac{6x^2 + 4x}{2x^2 + 2x} = \dfrac{2x(3x + 2)}{2x(x + 1)}$ Factoring the numerator and the denominator

$= \dfrac{2x}{2x} \cdot \dfrac{3x + 2}{x + 1}$ Factoring the rational expression

$= 1 \cdot \dfrac{3x + 2}{x + 1}$ $\dfrac{2x}{2x} = 1$

$= \dfrac{3x + 2}{x + 1}$ Removing a factor of 1

↑
············ Caution! ············

Note that you *cannot* simplify further by removing the x's because x is not a *factor* of the entire numerator, $3x + 2$, and the entire denominator, $x + 1$.

Answers

7. 5   8. $\dfrac{x}{4}$

SECTION 6.1   Multiplying and Simplifying Rational Expressions   **433**

9. $\dfrac{x^2 + 3x + 2}{x^2 - 1} = \dfrac{(x + 2)(x + 1)}{(x + 1)(x - 1)}$ Factoring the numerator and the denominator

$= \dfrac{x + 1}{x + 1} \cdot \dfrac{x + 2}{x - 1}$ Factoring the rational expression

$= 1 \cdot \dfrac{x + 2}{x - 1}$ $\dfrac{x + 1}{x + 1} = 1$

$= \dfrac{x + 2}{x - 1}$ Removing a factor of 1

Canceling

You may have encountered canceling when working with rational expressions. With great concern, we mention it as a possible way to speed up your work. Our concern is that canceling be done with care and understanding. Example 9 might have been done faster as follows:

$\dfrac{x^2 + 3x + 2}{x^2 - 1} = \dfrac{(x + 2)(x + 1)}{(x + 1)(x - 1)}$ Factoring the numerator and the denominator

$= \dfrac{(x + 2)\cancel{(x + 1)}}{\cancel{(x + 1)}(x - 1)}$ When a factor of 1 is noted, it is canceled, as shown: $\dfrac{x + 1}{x + 1} = 1$.

$= \dfrac{x + 2}{x - 1}.$ Simplifying

◀ Do Exercises 9–12.

Opposites in Rational Expressions

Expressions of the form $a - b$ and $b - a$ are **opposites** of each other. When either of these binomials is multiplied by -1, the result is the other binomial:

$-1(a - b) = -a + b = b + (-a) = b - a;$
$-1(b - a) = -b + a = a + (-b) = a - b.$ Multiplication by -1 reverses the order in which subtraction occurs.

Consider, for example,

$\dfrac{x - 4}{4 - x}.$

At first glance, it appears as though the numerator and the denominator do not have any common factors other than 1. But $x - 4$ and $4 - x$ are opposites, or additive inverses, of each other. Thus we can rewrite one as the opposite of the other by factoring out a -1.

EXAMPLE 10 Simplify: $\dfrac{x - 4}{4 - x}.$

$\dfrac{x - 4}{4 - x} = \dfrac{x - 4}{-(x - 4)} = \dfrac{1(x - 4)}{-1(x - 4)}$ $4 - x = -(x - 4); 4 - x$ and $x - 4$ are opposites.

$= \dfrac{1}{-1} \cdot \dfrac{x - 4}{x - 4}$

$= -1 \cdot 1$ $1/(-1) = -1$

$= -1$

◀ Do Exercises 13–15.

Simplify.

9. $\dfrac{2x^2 + x}{3x^2 + 2x}$

10. $\dfrac{x^2 - 1}{2x^2 - x - 1}$

11. $\dfrac{7x + 14}{7}$

12. $\dfrac{12y + 24}{48}$

Simplify.

13. $\dfrac{x - 8}{8 - x}$ GS

$= \dfrac{x - 8}{-(x -)}$

$= \dfrac{1(x - 8)}{-1(x - 8)} = \dfrac{1}{} \cdot \dfrac{x - 8}{x - 8}$

$= -1 \cdot =$

14. $\dfrac{c - d}{d - c}$

15. $\dfrac{-x - 7}{x + 7}$

Answers

9. $\dfrac{2x + 1}{3x + 2}$ 10. $\dfrac{x + 1}{2x + 1}$ 11. $x + 2$

12. $\dfrac{y + 2}{4}$ 13. -1 14. -1 15. -1

Guided Solution:
13. $8, -1, 1, -1$

d MULTIPLYING AND SIMPLIFYING

We try to simplify after we multiply. That is why we leave the numerator and the denominator in factored form.

EXAMPLE 11 Multiply and simplify: $\dfrac{5a^3}{4} \cdot \dfrac{2}{5a}$.

$\dfrac{5a^3}{4} \cdot \dfrac{2}{5a} = \dfrac{5a^3(2)}{4(5a)}$ Multiplying the numerators and the denominators

$= \dfrac{5 \cdot a \cdot a \cdot a \cdot 2}{2 \cdot 2 \cdot 5 \cdot a}$ Factoring the numerator and the denominator

$= \dfrac{\cancel{5} \cdot \cancel{a} \cdot a \cdot a \cdot \cancel{2}}{2 \cdot \cancel{2} \cdot \cancel{5} \cdot \cancel{a}}$ Removing a factor of 1: $\dfrac{2 \cdot 5 \cdot a}{2 \cdot 5 \cdot a} = 1$

$= \dfrac{a^2}{2}$ Simplifying

EXAMPLE 12 Multiply and simplify: $\dfrac{x^2 + 6x + 9}{x^2 - 4} \cdot \dfrac{x - 2}{x + 3}$.

$\dfrac{x^2 + 6x + 9}{x^2 - 4} \cdot \dfrac{x - 2}{x + 3} = \dfrac{(x^2 + 6x + 9)(x - 2)}{(x^2 - 4)(x + 3)}$ Multiplying the numerators and the denominators

$= \dfrac{(x + 3)(x + 3)(x - 2)}{(x + 2)(x - 2)(x + 3)}$ Factoring the numerator and the denominator

$= \dfrac{\cancel{(x+3)}(x + 3)\cancel{(x-2)}}{(x + 2)\cancel{(x-2)}\cancel{(x+3)}}$ Removing a factor of 1: $\dfrac{(x + 3)(x - 2)}{(x + 3)(x - 2)} = 1$

$= \dfrac{x + 3}{x + 2}$ Simplifying

Do Exercise 16.

EXAMPLE 13 Multiply and simplify: $\dfrac{x^2 + x - 2}{15} \cdot \dfrac{5}{2x^2 - 3x + 1}$.

$\dfrac{x^2 + x - 2}{15} \cdot \dfrac{5}{2x^2 - 3x + 1} = \dfrac{(x^2 + x - 2)5}{15(2x^2 - 3x + 1)}$ Multiplying the numerators and the denominators

$= \dfrac{(x + 2)(x - 1)5}{5(3)(x - 1)(2x - 1)}$ Factoring the numerator and the denominator

$= \dfrac{(x + 2)\cancel{(x-1)}\cancel{5}}{\cancel{5}(3)\cancel{(x-1)}(2x - 1)}$ Removing a factor of 1: $\dfrac{(x - 1)5}{(x - 1)5} = 1$

$= \underbrace{\dfrac{x + 2}{3(2x - 1)}}_{\uparrow}$ Simplifying

You need not carry out this multiplication.

Do Exercise 17.

GS **16.** Multiply and simplify:
$\dfrac{a^2 - 4a + 4}{a^2 - 9} \cdot \dfrac{a + 3}{a - 2}$.

$\dfrac{a^2 - 4a + 4}{a^2 - 9} \cdot \dfrac{a + 3}{a - 2}$

$= \dfrac{(a^2 - 4a + 4)(a + \boxed{})}{(a^2 - \boxed{})(a - 2)}$

$= \dfrac{(a - \boxed{})(a - 2)(a + 3)}{(a + 3)(a - \boxed{})(a - 2)}$

$= \dfrac{\cancel{(a-2)}(a - 2)\cancel{(a+3)}}{\cancel{(a+3)}(a - 3)\cancel{(a-2)}}$

$= \dfrac{a - \boxed{}}{a - \boxed{}}$

17. Multiply and simplify:
$\dfrac{x^2 - 25}{6} \cdot \dfrac{3}{x + 5}$.

Answers
16. $\dfrac{a - 2}{a - 3}$ **17.** $\dfrac{x - 5}{2}$

Guided Solution:
16. 3, 9, 2, 3, 2, 3

6.1 Exercise Set

FOR EXTRA HELP — MyLab Math

✓ Check Your Understanding

Reading Check Choose the word from below each blank that best completes the statement.

RC1. Expressions that have the same value for all allowable replacements are called _____ expressions.
   rational/equivalent

RC2. A rational expression is undefined when the _____ is zero.
   denominator/numerator

RC3. A rational expression can be written as a _____ of two polynomials.
   product/quotient

RC4. A rational expression is simplified when the numerator and the denominator have no _____ (other than 1) in common.
   factors/terms

Concept Check Choose from the column on the right the expression that illustrates each rational expression described.

CC1. The rational expression is defined for all values of x.

CC2. The rational expression is not defined when $x = -4$.

CC3. The rational expression is not defined when $x = 3$.

CC4. The rational expression is not defined when $x = 0$.

a) $\dfrac{x-3}{x+4}$

b) $\dfrac{x+3}{4}$

c) $\dfrac{x}{x-4}$

d) $\dfrac{x+3}{x-3}$

e) $\dfrac{x-4}{x^2}$

a Find all numbers for which each rational expression is not defined.

1. $\dfrac{-3}{2x}$

2. $\dfrac{24}{-8y}$

3. $\dfrac{5}{x-8}$

4. $\dfrac{y-4}{y+6}$

5. $\dfrac{3}{2y+5}$

6. $\dfrac{x^2-9}{4x-15}$

7. $\dfrac{x^2+11}{x^2-3x-28}$

8. $\dfrac{p^2-9}{p^2-7p+10}$

9. $\dfrac{m^3-2m}{m^2-25}$

10. $\dfrac{7-3x+x^2}{49-x^2}$

11. $\dfrac{x-4}{3}$

12. $\dfrac{x^2-25}{14}$

b Multiply. Do not simplify. Note that in each case you are multiplying by 1.

13. $\dfrac{4x}{4x} \cdot \dfrac{3x^2}{5y}$

14. $\dfrac{5x^2}{5x^2} \cdot \dfrac{6y^3}{3z^4}$

15. $\dfrac{2x}{2x} \cdot \dfrac{x-1}{x+4}$

16. $\dfrac{2a-3}{5a+2} \cdot \dfrac{a}{a}$

17. $\dfrac{3-x}{4-x} \cdot \dfrac{-1}{-1}$

18. $\dfrac{x-5}{5-x} \cdot \dfrac{-1}{-1}$

19. $\dfrac{y+6}{y+6} \cdot \dfrac{y-7}{y+2}$

20. $\dfrac{x^2+1}{x^3-2} \cdot \dfrac{x-4}{x-4}$

C Simplify.

21. $\dfrac{8x^3}{32x}$

22. $\dfrac{4x^2}{20x}$

23. $\dfrac{48p^7q^5}{18p^5q^4}$

24. $\dfrac{-76x^8y^3}{-24x^4y^3}$

25. $\dfrac{4x-12}{4x}$

26. $\dfrac{5a-40}{5}$

27. $\dfrac{3m^2+3m}{6m^2+9m}$

28. $\dfrac{4y^2-2y}{5y^2-5y}$

29. $\dfrac{a^2-9}{a^2+5a+6}$

30. $\dfrac{t^2-25}{t^2+t-20}$

31. $\dfrac{a^2-10a+21}{a^2-11a+28}$

32. $\dfrac{x^2-2x-8}{x^2-x-6}$

33. $\dfrac{x^2-25}{x^2-10x+25}$

34. $\dfrac{x^2+8x+16}{x^2-16}$

35. $\dfrac{a^2-1}{a-1}$

36. $\dfrac{t^2-1}{t+1}$

37. $\dfrac{x^2+1}{x+1}$

38. $\dfrac{m^2+9}{m+3}$

39. $\dfrac{6x^2-54}{4x^2-36}$

40. $\dfrac{8x^2-32}{4x^2-16}$

SECTION 6.1 Multiplying and Simplifying Rational Expressions

41. $\dfrac{6t + 12}{t^2 - t - 6}$

42. $\dfrac{4x + 32}{x^2 + 9x + 8}$

43. $\dfrac{2t^2 + 6t + 4}{4t^2 - 12t - 16}$

44. $\dfrac{3a^2 - 9a - 12}{6a^2 + 30a + 24}$

45. $\dfrac{t^2 - 4}{(t + 2)^2}$

46. $\dfrac{m^2 - 36}{(m - 6)^2}$

47. $\dfrac{6 - x}{x - 6}$

48. $\dfrac{t - 3}{3 - t}$

49. $\dfrac{a - b}{b - a}$

50. $\dfrac{y - x}{-x + y}$

51. $\dfrac{6t - 12}{2 - t}$

52. $\dfrac{5a - 15}{3 - a}$

53. $\dfrac{x^2 - 1}{1 - x}$

54. $\dfrac{a^2 - b^2}{b^2 - a^2}$

55. $\dfrac{6qt - 3t^4}{t^3 - 2q}$

56. $\dfrac{2z - w^5}{5w^{10} - 10zw^5}$

d Multiply and simplify.

57. $\dfrac{4x^3}{3x} \cdot \dfrac{14}{x}$

58. $\dfrac{18}{x^3} \cdot \dfrac{5x^2}{6}$

59. $\dfrac{3c}{d^2} \cdot \dfrac{4d}{6c^3}$

60. $\dfrac{3x^2y}{2} \cdot \dfrac{4}{xy^3}$

61. $\dfrac{x + 4}{x} \cdot \dfrac{x^2 - 3x}{x^2 + x - 12}$

62. $\dfrac{t^2}{t^2 - 4} \cdot \dfrac{t^2 - 5t + 6}{t^2 - 3t}$

63. $\dfrac{a^2 - 9}{a^2} \cdot \dfrac{a^2 - 3a}{a^2 + a - 12}$

64. $\dfrac{x^2 + 10x - 11}{x^2 - 1} \cdot \dfrac{x + 1}{x + 11}$

65. $\dfrac{4a^2}{3a^2 - 12a + 12} \cdot \dfrac{3a - 6}{2a}$

66. $\dfrac{5v + 5}{v - 2} \cdot \dfrac{v^2 - 4v + 4}{v^2 - 1}$

67. $\dfrac{t^4 - 16}{t^4 - 1} \cdot \dfrac{t^2 + 1}{t^2 + 4}$

68. $\dfrac{x^4 - 1}{x^4 - 81} \cdot \dfrac{x^2 + 9}{x^2 + 1}$

69. $\dfrac{(x + 4)^3}{(x + 2)^3} \cdot \dfrac{x^2 + 4x + 4}{x^2 + 8x + 16}$

70. $\dfrac{(t - 2)^3}{(t - 1)^3} \cdot \dfrac{t^2 - 2t + 1}{t^2 - 4t + 4}$

71. $\dfrac{5a^2 - 180}{10a^2 - 10} \cdot \dfrac{20a + 20}{2a - 12}$

72. $\dfrac{2t^2 - 98}{4t^2 - 4} \cdot \dfrac{8t + 8}{16t - 112}$

Skill Maintenance

Graph.

73. $x + y = -1$ [3.3a]

74. $y = -\dfrac{7}{2}$ [3.3b]

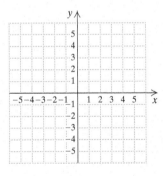
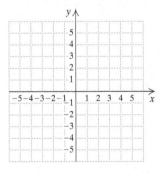

Factor. [5.7a]

75. $x^5 - 2x^4 - 35x^3$

76. $2y^3 - 10y^2 + y - 5$

77. $16 - t^4$

78. $10x^2 + 80x + 70$

Synthesis

Simplify.

79. $\dfrac{x - 1}{x^2 + 1} \cdot \dfrac{x^4 - 1}{(x - 1)^2} \cdot \dfrac{x^2 - 1}{x^4 - 2x^2 + 1}$

80. $\dfrac{(t + 2)^3}{(t + 1)^3} \cdot \dfrac{t^2 + 2t + 1}{t^2 + 4t + 4} \cdot \dfrac{t + 1}{t + 2}$

81. Select any number x, multiply by 2, add 5, multiply by 5, subtract 25, and divide by 10. What do you get? Explain how this procedure can be used for a number trick.

6.2 Division and Reciprocals

OBJECTIVES

a Find the reciprocal of a rational expression.

b Divide rational expressions and simplify.

There is a similarity between what we do with rational expressions and what we do with rational numbers. In fact, after variables have been replaced with rational numbers, a rational expression represents a rational number.

a FINDING RECIPROCALS

Two expressions are **reciprocals** of each other if their product is 1. The reciprocal of a rational expression is found by interchanging the numerator and the denominator.

EXAMPLES

1. The reciprocal of $\frac{2}{5}$ is $\frac{5}{2}$. $\left(\text{This is because } \frac{2}{5} \cdot \frac{5}{2} = \frac{10}{10} = 1.\right)$

2. The reciprocal of $\frac{2x^2 - 3}{x + 4}$ is $\frac{x + 4}{2x^2 - 3}$.

3. The reciprocal of $x + 2$ is $\frac{1}{x + 2}$. $\left(\text{Think of } x + 2 \text{ as } \frac{x + 2}{1}.\right)$

◀ Do Exercises 1–4.

Find the reciprocal.

1. $\frac{7}{2}$

2. $\frac{x^2 + 5}{2x^3 - 1}$

3. $x - 5$

4. $\frac{1}{x^2 - 3}$

SKILL REVIEW *Factor polynomials.* [5.7a]
Factor.
1. $x^2 - 2x$
2. $5y^2 - 11y - 12$

Answers: 1. $x(x - 2)$ 2. $(5y + 4)(y - 3)$

b DIVISION

We divide rational expressions in the same way that we divide fraction notation in arithmetic.

> **DIVIDING RATIONAL EXPRESSIONS**
>
> To divide by a rational expression, multiply by its reciprocal:
>
> $$\frac{A}{B} \div \frac{C}{D} = \frac{A}{B} \cdot \frac{D}{C} = \frac{AD}{BC}.$$
>
> Then factor and, if possible, simplify.

EXAMPLE 4 Divide and simplify: $\frac{3}{4} \div \frac{9}{5}$.

$\frac{3}{4} \div \frac{9}{5} = \frac{3}{4} \cdot \frac{5}{9}$ Multiplying by the reciprocal of the divisor

$= \frac{3 \cdot 5}{4 \cdot 9} = \frac{3 \cdot 5}{2 \cdot 2 \cdot 3 \cdot 3}$ Factoring

$= \frac{3 \cdot 5}{2 \cdot 2 \cdot 3 \cdot 3}$ Removing a factor of 1: $\frac{3}{3} = 1$

$= \frac{5}{12}$ Simplifying

5. Divide and simplify: $\frac{3}{5} \div \frac{7}{10}$.

◀ Do Exercise 5.

Answers

1. $\frac{2}{7}$ 2. $\frac{2x^3 - 1}{x^2 + 5}$ 3. $\frac{1}{x - 5}$
4. $x^2 - 3$ 5. $\frac{6}{7}$

EXAMPLE 5 Divide and simplify: $\dfrac{2}{x} \div \dfrac{3}{x}$.

$$\dfrac{2}{x} \div \dfrac{3}{x} = \dfrac{2}{x} \cdot \dfrac{x}{3} \quad \text{Multiplying by the reciprocal of the divisor}$$

$$= \dfrac{2 \cdot x}{x \cdot 3} = \dfrac{2 \cdot \cancel{x}}{\cancel{x} \cdot 3} \quad \text{Removing a factor of 1: } \dfrac{x}{x} = 1$$

$$= \dfrac{2}{3}$$

Do Exercise 6. ▶

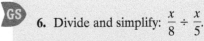

6. Divide and simplify: $\dfrac{x}{8} \div \dfrac{x}{5}$.

$$\dfrac{x}{8} \div \dfrac{x}{5} = \dfrac{x}{8} \cdot \dfrac{5}{\square}$$

$$= \dfrac{x \cdot \square}{8 \cdot x}$$

$$= \dfrac{\cancel{x} \cdot 5}{8 \cdot \cancel{x}}$$

$$= \dfrac{\square}{8}$$

EXAMPLE 6 Divide and simplify: $\dfrac{x+1}{x+2} \div \dfrac{x-1}{x+3}$.

$$\dfrac{x+1}{x+2} \div \dfrac{x-1}{x+3} = \dfrac{x+1}{x+2} \cdot \dfrac{x+3}{x-1} \quad \text{Multiplying by the reciprocal of the divisor}$$

$$= \dfrac{(x+1)(x+3)}{(x+2)(x-1)}$$

⬅ We generally do not carry out the multiplication in the numerator or the denominator. It is not wrong to do so, but the factored form is often more useful.

Do Exercise 7. ▶

7. Divide and simplify:
$\dfrac{x-3}{x+5} \div \dfrac{x+5}{x-2}$.

EXAMPLE 7 Divide and simplify: $\dfrac{4}{x^2 - 7x} \div \dfrac{28x}{x^2 - 49}$.

$$\dfrac{4}{x^2 - 7x} \div \dfrac{28x}{x^2 - 49} = \dfrac{4}{x^2 - 7x} \cdot \dfrac{x^2 - 49}{28x} \quad \text{Multiplying by the reciprocal}$$

$$= \dfrac{4(x^2 - 49)}{(x^2 - 7x)(28x)}$$

$$= \dfrac{2 \cdot 2 \cdot (x-7)(x+7)}{x(x-7) \cdot 2 \cdot 2 \cdot 7 \cdot x} \quad \text{Factoring the numerator and the denominator}$$

$$= \dfrac{2 \cdot 2 \cdot \cancel{(x-7)}(x+7)}{x\cancel{(x-7)} \cdot 2 \cdot 2 \cdot 7 \cdot x} \quad \text{Removing a factor of 1: } \dfrac{2 \cdot 2 \cdot (x-7)}{2 \cdot 2 \cdot (x-7)} = 1$$

$$= \dfrac{x+7}{7x^2}$$

Do Exercise 8. ▶

8. Divide and simplify:
$\dfrac{a^2 + 5a}{6} \div \dfrac{a^2 - 25}{18a}$.

EXAMPLE 8 Divide and simplify: $\dfrac{x+1}{x^2 - 1} \div \dfrac{x+1}{x^2 - 2x + 1}$.

$$\dfrac{x+1}{x^2 - 1} \div \dfrac{x+1}{x^2 - 2x + 1} = \dfrac{x+1}{x^2 - 1} \cdot \dfrac{x^2 - 2x + 1}{x+1} \quad \text{Multiplying by the reciprocal}$$

Answers

6. $\dfrac{5}{8}$ 7. $\dfrac{(x-3)(x-2)}{(x+5)(x+5)}$ 8. $\dfrac{3a^2}{a-5}$

Guided Solution:
6. $x, 5, 5$

SECTION 6.2 Division and Reciprocals   441

Divide and simplify.

9. $\dfrac{x-3}{x+5} \div \dfrac{x+2}{x+5}$

10. $\dfrac{x^2 - 5x + 6}{x+5} \div \dfrac{x+2}{x+5}$

11. $\dfrac{y^2 - 1}{y+1} \div \dfrac{y^2 - 2y + 1}{y+1}$

$= \dfrac{y^2 - 1}{y+1} \cdot \dfrac{\square + 1}{y^2 - \square + 1}$

$= \dfrac{(y^2 - 1)(y+1)}{(y+1)(y^2 - 2y + 1)}$

$= \dfrac{(y+\square)(y-\square)(y+1)}{(y+1)(y-\square)(y-\square)}$

$= \dfrac{\cancel{(y+1)}\cancel{(y-1)}(y+1)}{\cancel{(y+1)}\cancel{(y-1)}(y-1)}$

$= \dfrac{y + \square}{\square - 1}$

Then we multiply numerators and multiply denominators. We have

$= \dfrac{(x+1)(x^2 - 2x + 1)}{(x^2 - 1)(x+1)}$

$= \dfrac{(x+1)(x-1)(x-1)}{(x-1)(x+1)(x+1)}$ Factoring the numerator and the denominator

$= \dfrac{\cancel{(x+1)}\cancel{(x-1)}(x-1)}{\cancel{(x-1)}\cancel{(x+1)}(x+1)}$ Removing a factor of 1: $\dfrac{(x+1)(x-1)}{(x+1)(x-1)} = 1$

$= \dfrac{x-1}{x+1}$.

EXAMPLE 9 Divide and simplify: $\dfrac{x^2 - 2x - 3}{x^2 - 4} \div \dfrac{x+1}{x+5}$.

$\dfrac{x^2 - 2x - 3}{x^2 - 4} \div \dfrac{x+1}{x+5}$

$= \dfrac{x^2 - 2x - 3}{x^2 - 4} \cdot \dfrac{x+5}{x+1}$ Multiplying by the reciprocal

$= \dfrac{(x^2 - 2x - 3)(x+5)}{(x^2 - 4)(x+1)}$

$= \dfrac{(x-3)(x+1)(x+5)}{(x-2)(x+2)(x+1)}$ Factoring the numerator and the denominator

$= \dfrac{(x-3)\cancel{(x+1)}(x+5)}{(x-2)(x+2)\cancel{(x+1)}}$ Removing a factor of 1: $\dfrac{x+1}{x+1} = 1$

$= \dfrac{(x-3)(x+5)}{(x-2)(x+2)}$ ← You need not carry out the multiplications in the numerator and the denominator.

◀ Do Exercises 9–11.

Answers

9. $\dfrac{x-3}{x+2}$ 10. $\dfrac{(x-3)(x-2)}{x+2}$ 11. $\dfrac{y+1}{y-1}$

Guided Solution:
11. $y, 2y, 1, 1, 1, 1, 1, y$

6.2 Exercise Set

FOR EXTRA HELP   MyLab Math

✓ Check Your Understanding

Reading Check Determine whether each statement is true or false.

RC1. To divide an expression by a rational expression, multiply by the reciprocal of the divisor.

RC2. The reciprocal of $2 - x$ is $x - 2$.

RC3. Two expressions are reciprocals if their sum is 1.

RC4. The reciprocal of a rational expression is found by interchanging the numerator and the denominator.

Concept Check Choose from the columns on the right an equivalent expression.

CC1. The reciprocal of $\dfrac{2}{x-2}$

CC2. The reciprocal of $x - 2$

CC3. $\dfrac{2}{x} \cdot \dfrac{x}{2}$

CC4. $\dfrac{x}{2} \cdot \dfrac{2}{x+1}$

CC5. $\dfrac{1}{x} \div \dfrac{1}{2}$

CC6. $\dfrac{1}{x} \div \dfrac{2}{x}$

a) 1

b) $\dfrac{2}{x}$

c) $\dfrac{1}{x} \cdot \dfrac{x}{2}$

d) $\dfrac{x-2}{2}$

e) $\dfrac{1}{x-2}$

f) $\dfrac{x}{2} \div \dfrac{x+1}{2}$

a Find the reciprocal.

1. $\dfrac{4}{x}$

2. $\dfrac{a+3}{a-1}$

3. $x^2 - y^2$

4. $x^2 - 5x + 7$

5. $\dfrac{1}{a+b}$

6. $\dfrac{x^2}{x^2 - 3}$

7. $\dfrac{x^2 + 2x - 5}{x^2 - 4x + 7}$

8. $\dfrac{(a-b)(a+b)}{(a+4)(a-5)}$

b Divide and simplify.

9. $\dfrac{2}{5} \div \dfrac{4}{3}$

10. $\dfrac{3}{10} \div \dfrac{3}{2}$

11. $\dfrac{2}{x} \div \dfrac{8}{x}$

12. $\dfrac{t}{3} \div \dfrac{t}{15}$

13. $\dfrac{a}{b^2} \div \dfrac{a^2}{b^3}$

14. $\dfrac{x^2}{y} \div \dfrac{x^3}{y^3}$

15. $\dfrac{a+2}{a-3} \div \dfrac{a-1}{a+3}$

16. $\dfrac{x-8}{x+9} \div \dfrac{x+2}{x-1}$

17. $\dfrac{x^2 - 1}{x} \div \dfrac{x+1}{x-1}$

18. $\dfrac{4y - 8}{y + 2} \div \dfrac{y - 2}{y^2 - 4}$

19. $\dfrac{x+1}{6} \div \dfrac{x+1}{3}$

20. $\dfrac{a}{a-b} \div \dfrac{b}{a-b}$

21. $\dfrac{5x - 5}{16} \div \dfrac{x - 1}{6}$

22. $\dfrac{4y - 12}{12} \div \dfrac{y - 3}{3}$

23. $\dfrac{-6 + 3x}{5} \div \dfrac{4x - 8}{25}$

24. $\dfrac{-12 + 4x}{4} \div \dfrac{-6 + 2x}{6}$

25. $\dfrac{a+2}{a-1} \div \dfrac{3a+6}{a-5}$

26. $\dfrac{t-3}{t+2} \div \dfrac{4t-12}{t+1}$

27. $\dfrac{x^2 - 4}{x} \div \dfrac{x - 2}{x + 2}$

28. $\dfrac{x+y}{x-y} \div \dfrac{x^2 + y}{x^2 - y^2}$

29. $\dfrac{x^2 - 9}{4x + 12} \div \dfrac{x - 3}{6}$ **30.** $\dfrac{a - b}{2a} \div \dfrac{a^2 - b^2}{8a^3}$ **31.** $\dfrac{c^2 + 3c}{c^2 + 2c - 3} \div \dfrac{c}{c + 1}$ **32.** $\dfrac{y + 5}{2y} \div \dfrac{y^2 - 25}{4y^2}$

33. $\dfrac{2y^2 - 7y + 3}{2y^2 + 3y - 2} \div \dfrac{6y^2 - 5y + 1}{3y^2 + 5y - 2}$ **34.** $\dfrac{x^2 + x - 20}{x^2 - 7x + 12} \div \dfrac{x^2 + 10x + 25}{x^2 - 6x + 9}$

35. $\dfrac{x^2 - 1}{4x + 4} \div \dfrac{2x^2 - 4x + 2}{8x + 8}$ **36.** $\dfrac{5t^2 + 5t - 30}{10t + 30} \div \dfrac{2t^2 - 8}{6t^2 + 36t + 54}$

Skill Maintenance

Solve.

37. Thomas is taking an astronomy course. In order to receive an A, he must average at least 90 after four exams. Thomas scored 96, 98, and 89 on the first three tests. Determine (in terms of an inequality) what scores on the last test will earn him an A. [2.8b]

38. *Triangle Dimensions.* The base of a triangle is 4 in. less than twice the height. The area is 35 in². Find the height and the base. [5.9a]

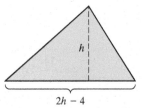

Simplify. [4.2a, b]

39. $(2x^{-3}y^4)^2$ **40.** $(5x^6y^{-4})^3$ **41.** $\left(\dfrac{2x^3}{y^5}\right)^2$ **42.** $\left(\dfrac{a^{-3}}{b^4}\right)^5$

Synthesis

Simplify.

43. $\dfrac{a^2b^2 + 3ab^2 + 2b^2}{a^2b^4 + 4b^4} \div (5a^2 + 10a)$ **44.** $\dfrac{3x + 3y + 3}{9x} \div \dfrac{x^2 + 2xy + y^2 - 1}{x^4 + x^2}$

45. The volume of this rectangular solid is $x - 3$. What is its height?

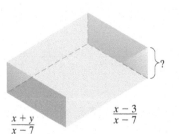

Least Common Multiples and Denominators

6.3

OBJECTIVES

a Find the LCM of several numbers by factoring.

b Add fractions, first finding the LCD.

c Find the LCM of algebraic expressions by factoring.

a LEAST COMMON MULTIPLES

To add when denominators are different, we first find a common denominator. For example, to add $\frac{5}{12}$ and $\frac{7}{30}$, we first look for the **least common multiple, LCM**, of 12 and 30. That number becomes the **least common denominator, LCD**. To find the LCM of 12 and 30, we factor:

$12 = 2 \cdot 2 \cdot 3;$
$30 = 2 \cdot 3 \cdot 5.$

The LCM is the number that has 2 as a factor twice, 3 as a factor once, and 5 as a factor once:

LCM $= 2 \cdot 2 \cdot 3 \cdot 5 = 60.$

— 12 is a factor of the LCM.
— 30 is a factor of the LCM.

> **FINDING LCMs**
>
> To find the LCM, use each factor the greatest number of times that it appears in any one factorization.

EXAMPLE 1 Find the LCM of 24 and 36.

$24 = 2 \cdot 2 \cdot 2 \cdot 3$
$36 = 2 \cdot 2 \cdot 3 \cdot 3$ LCM $= 2 \cdot 2 \cdot 2 \cdot 3 \cdot 3,$ or 72

Do Exercises 1–4. ▶

b ADDING USING THE LCD

Let's finish adding $\frac{5}{12}$ and $\frac{7}{30}$:

$$\frac{5}{12} + \frac{7}{30} = \frac{5}{2 \cdot 2 \cdot 3} + \frac{7}{2 \cdot 3 \cdot 5}.$$

The least common denominator, LCD, is $2 \cdot 2 \cdot 3 \cdot 5$. To get the LCD in the first denominator, we need a 5. To get the LCD in the second denominator, we need another 2. We get these numbers by multiplying by forms of 1:

$$\frac{5}{12} + \frac{7}{30} = \frac{5}{2 \cdot 2 \cdot 3} \cdot \frac{5}{5} + \frac{7}{2 \cdot 3 \cdot 5} \cdot \frac{2}{2} \quad \text{Multiplying by 1}$$

$$= \frac{25}{2 \cdot 2 \cdot 3 \cdot 5} + \frac{14}{2 \cdot 3 \cdot 5 \cdot 2} \quad \text{Each denominator is now the LCD.}$$

$$= \frac{39}{2 \cdot 2 \cdot 3 \cdot 5} \quad \text{Adding the numerators and keeping the LCD}$$

$$= \frac{3 \cdot 13}{2 \cdot 2 \cdot 3 \cdot 5} \quad \text{Factoring the numerator and removing a factor of 1: } \frac{3}{3} = 1$$

$$= \frac{13}{20}. \quad \text{Simplifying}$$

Find the LCM by factoring.

GS 1. 16, 18
$16 = 2 \cdot 2 \cdot 2 \cdot \square$
$18 = 2 \cdot \square \cdot \square$
LCM $= 2 \cdot 2 \cdot 2 \cdot \square \cdot 3 \cdot 3,$
or $\square$

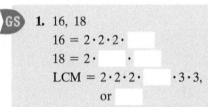

2. 6, 12

3. 2, 5

4. 24, 30, 20

Answers
1. 144 **2.** 12 **3.** 10 **4.** 120
Guided Solution:
1. 2, 3, 3, 2, 144

SECTION 6.3 Least Common Multiples and Denominators 445

Add, first finding the LCD. Simplify, if possible.

5. $\dfrac{3}{16} + \dfrac{1}{18}$ **GS**

$= \dfrac{3}{2 \cdot 2 \cdot 2 \cdot \square} + \dfrac{1}{2 \cdot \square \cdot 3}$

$= \dfrac{3}{2 \cdot 2 \cdot 2 \cdot 2} \cdot \dfrac{3 \cdot 3}{3 \cdot \square}$

$+ \dfrac{1}{2 \cdot 3 \cdot 3} \cdot \dfrac{2 \cdot \square}{2 \cdot 2 \cdot 2}$

$= \dfrac{27 + \square}{2 \cdot 2 \cdot 2 \cdot 2 \cdot 3 \cdot 3}$

$= \dfrac{35}{\square}$

6. $\dfrac{1}{6} + \dfrac{1}{12}$

7. $\dfrac{1}{2} + \dfrac{3}{5}$

8. $\dfrac{1}{24} + \dfrac{1}{30} + \dfrac{3}{20}$

Find the LCM.

9. $12xy^2$, $15x^3y$

10. $y^2 + 5y + 4$, $y^2 + 2y + 1$

11. $t^2 + 16$, $t - 2$, 7

12. $x^2 + 2x + 1$, $3x^2 - 3x$, $x^2 - 1$

Answers

5. $\dfrac{35}{144}$ 6. $\dfrac{1}{4}$ 7. $\dfrac{11}{10}$ 8. $\dfrac{9}{40}$ 9. $60x^3y^2$
10. $(y+1)^2(y+4)$ 11. $7(t^2+16)(t-2)$
12. $3x(x+1)^2(x-1)$

Guided Solution:
5. 2, 3, 3, 2, 8, 144

EXAMPLE 2 Add: $\dfrac{5}{12} + \dfrac{11}{18}$.

$\left.\begin{array}{l} 12 = 2 \cdot 2 \cdot 3 \\ 18 = 2 \cdot 3 \cdot 3 \end{array}\right\}$ LCD $= 2 \cdot 2 \cdot 3 \cdot 3$, or 36

$\dfrac{5}{12} + \dfrac{11}{18} = \dfrac{5}{2 \cdot 2 \cdot 3} \cdot \dfrac{3}{3} + \dfrac{11}{2 \cdot 3 \cdot 3} \cdot \dfrac{2}{2} = \dfrac{15 + 22}{2 \cdot 2 \cdot 3 \cdot 3} = \dfrac{37}{36}$

◀ Do Exercises 5–8.

c LCMs OF ALGEBRAIC EXPRESSIONS

SKILL REVIEW *Factor trinomial squares.* [5.5b]
Factor.
1. $x^2 - 16x + 64$
2. $4x^2 + 12x + 9$

Answers: **1.** $(x-8)^2$ **2.** $(2x+3)^2$

To find the LCM of two or more algebraic expressions, we factor them. Then we use each factor the greatest number of times that it occurs in any one expression. In Section 6.4, each LCM will become an LCD used to add rational expressions.

EXAMPLE 3 Find the LCM of $12x$, $16y$, and $8xyz$.

$\left.\begin{array}{l} 12x = 2 \cdot 2 \cdot 3 \cdot x \\ 16y = 2 \cdot 2 \cdot 2 \cdot 2 \cdot y \\ 8xyz = 2 \cdot 2 \cdot 2 \cdot x \cdot y \cdot z \end{array}\right\}$ LCM $= 2 \cdot 2 \cdot 2 \cdot 2 \cdot 3 \cdot x \cdot y \cdot z$
$= 48xyz$

EXAMPLE 4 Find the LCM of $x^2 + 5x - 6$ and $x^2 - 1$.

$\left.\begin{array}{l} x^2 + 5x - 6 = (x+6)(x-1) \\ x^2 - 1 = (x+1)(x-1) \end{array}\right\}$ LCM $= (x+6)(x-1)(x+1)$

EXAMPLE 5 Find the LCM of $x^2 + 4$, $x + 1$, and 5.

These expressions do not share a common factor other than 1, so the LCM is their product:

$5(x^2+4)(x+1)$.

EXAMPLE 6 Find the LCM of $x^2 - 25$ and $2x - 10$.

$\left.\begin{array}{l} x^2 - 25 = (x+5)(x-5) \\ 2x - 10 = 2(x-5) \end{array}\right\}$ LCM $= 2(x+5)(x-5)$

EXAMPLE 7 Find the LCM of $x^2 - 4y^2$, $x^2 - 4xy + 4y^2$, and $x - 2y$.

$\left.\begin{array}{l} x^2 - 4y^2 = (x-2y)(x+2y) \\ x^2 - 4xy + 4y^2 = (x-2y)(x-2y) \\ x - 2y = x - 2y \end{array}\right\}$ LCM $= (x+2y)(x-2y)(x-2y)$
$= (x+2y)(x-2y)^2$

◀ Do Exercises 9–12.

6.3 Exercise Set

FOR EXTRA HELP: MyLab Math

✓ Check Your Understanding

Reading and Concept Check Choose from the column on the right the best choice to complete each statement. Some choices will not be used.

To add $\frac{5}{16} + \frac{7}{24}$, we begin by finding a
RC1. _____ denominator. We first look for the least common **RC2.** _____ of 16 and 24. That number becomes the least common **RC3.** _____ of the two fractions. We factor 16 and 24: $16 = 2 \cdot 2 \cdot 2 \cdot 2$ and $24 = 2 \cdot 2 \cdot 2 \cdot 3$. Then to find the LCM of 16 and 24, we use each factor the **RC4.** _____ number of times that it appears in any one factorization. The LCM is **RC5.** _____.

- $2 \cdot 2 \cdot 2 \cdot 2 \cdot 3$
- $2 \cdot 2 \cdot 2 \cdot 2 \cdot 2 \cdot 3$
- common
- multiple
- numerator
- denominator
- greatest
- least

a Find the LCM.

1. 12, 27
2. 10, 15
3. 8, 9
4. 12, 18
5. 6, 9, 21

6. 8, 36, 40
7. 24, 36, 40
8. 4, 5, 20
9. 10, 100, 500
10. 28, 42, 60

b Add, first finding the LCD. Simplify, if possible.

11. $\frac{7}{24} + \frac{11}{18}$
12. $\frac{7}{60} + \frac{2}{25}$
13. $\frac{1}{6} + \frac{3}{40}$

14. $\frac{5}{24} + \frac{3}{20}$
15. $\frac{1}{20} + \frac{1}{30} + \frac{2}{45}$
16. $\frac{2}{15} + \frac{5}{9} + \frac{3}{20}$

c Find the LCM.

17. $6x^2$, $12x^3$
18. $2a^2b$, $8ab^3$
19. $2x^2$, $6xy$, $18y^2$

20. p^3q, p^2q, pq^2
21. $2(y - 3)$, $6(y - 3)$
22. $5(m + 2)$, $15(m + 2)$

23. t, $t + 2$, $t - 2$
24. y, $y - 5$, $y + 5$
25. $x^2 - 4$, $x^2 + 5x + 6$

26. $x^2 - 4$, $x^2 - x - 2$
27. $t^3 + 4t^2 + 4t$, $t^2 - 4t$
28. $m^4 - m^2$, $m^3 - m^2$

29. $a + 1$, $(a - 1)^2$, $a^2 - 1$

30. $a^2 - 2ab + b^2$, $a^2 - b^2$, $3a + 3b$

31. $m^2 - 5m + 6$, $m^2 - 4m + 4$

32. $2x^2 + 5x + 2$, $2x^2 - x - 1$

33. $2 + 3x$, $4 - 9x^2$, $2 - 3x$

34. $9 - 4x^2$, $3 + 2x$, $3 - 2x$

35. $10v^2 + 30v$, $5v^2 + 35v + 60$

36. $12a^2 + 24a$, $4a^2 + 20a + 24$

37. $9x^3 - 9x^2 - 18x$, $6x^5 - 24x^4 + 24x^3$

38. $x^5 - 4x^3$, $x^3 + 4x^2 + 4x$

39. $x^5 + 4x^4 + 4x^3$, $3x^2 - 12$, $2x + 4$

40. $x^5 + 2x^4 + x^3$, $2x^3 - 2x$, $5x - 5$

41. $24w^4$, w^2, $10w^3$, w^6

42. t, $6t^4$, t^2, $15t^{15}$, $2t^3$

Skill Maintenance

Complete the following tables, finding the LCM, the GCF, and the product of each pair of expressions. [4.5a], [5.1a], [6.3a]

	Expressions	LCM	GCF	Product
Example	$12x^3$, $8x^2$	$24x^3$	$4x^2$	$96x^5$
43.	$40x^3$, $24x^4$			
45.	$16x^5$, $48x^6$			
47.	$20x^2$, $10x$			

	Expressions	LCM	GCF	Product
44.	$12ab$, $16ab^3$			
46.	$10x^2$, $24x^3$			
48.	a^5, a^{15}			

Synthesis

49. *Running.* Gabriela and Madison leave the starting point of a fitness loop at the same time. Gabriela jogs a lap in 6 min and Madison jogs one in 8 min. Assuming they continue to run at the same pace, after how long will they next meet at the starting point?

Adding Rational Expressions

6.4

OBJECTIVE

a Add rational expressions.

a ADDING RATIONAL EXPRESSIONS

We add rational expressions as we do rational numbers.

> **ADDING RATIONAL EXPRESSIONS WITH LIKE DENOMINATORS**
>
> To add when the denominators are the same, add the numerators and keep the same denominator. Then simplify, if possible.

EXAMPLES Add.

1. $\dfrac{x}{x+1} + \dfrac{2}{x+1} = \dfrac{x+2}{x+1}$

2. $\dfrac{2x^2 + 3x - 7}{2x + 1} + \dfrac{x^2 + x - 8}{2x + 1} = \dfrac{(2x^2 + 3x - 7) + (x^2 + x - 8)}{2x + 1}$

   $= \dfrac{3x^2 + 4x - 15}{2x + 1}$ Factoring the numerator to determine if we can simplify

   $= \dfrac{(x + 3)(3x - 5)}{2x + 1}$

3. $\dfrac{x - 5}{x^2 - 9} + \dfrac{2}{x^2 - 9} = \dfrac{(x - 5) + 2}{x^2 - 9} = \dfrac{x - 3}{x^2 - 9}$

   $= \dfrac{x - 3}{(x - 3)(x + 3)}$ Factoring

   $= \dfrac{1(x - 3)}{(x - 3)(x + 3)}$ Removing a factor of 1: $\dfrac{x-3}{x-3} = 1$

   $= \dfrac{1}{x + 3}$ Simplifying

   Do Exercises 1–3. ▶

When denominators are different, we find the least common denominator, LCD. The procedure we use follows.

> **ADDING RATIONAL EXPRESSIONS WITH DIFFERENT DENOMINATORS**
>
> To add rational expressions with different denominators:
> 1. Find the LCM of the denominators. This is the least common denominator (LCD).
> 2. For each rational expression, find an equivalent expression with the LCD. Multiply by 1 using an expression for 1 made up of factors of the LCD that are missing from the original denominator.
> 3. Add the numerators. Write the sum over the LCD.
> 4. Simplify, if possible.

SKILL REVIEW

Simplify rational expressions by factoring the numerator and the denominator and removing factors of 1. [6.1c]

Simplify.

1. $\dfrac{a^2 - b^2}{a + b}$

2. $\dfrac{x^2 - x - 6}{x^2 + 2x - 15}$

Answers: 1. $a - b$ 2. $\dfrac{x + 2}{x + 5}$

Add.

1. $\dfrac{5}{9} + \dfrac{2}{9}$

2. $\dfrac{3}{x - 2} + \dfrac{x}{x - 2}$

3. $\dfrac{4x + 5}{x - 1} + \dfrac{2x - 1}{x - 1}$

Answers

1. $\dfrac{7}{9}$ 2. $\dfrac{3 + x}{x - 2}$ 3. $\dfrac{2(3x + 2)}{x - 1}$

EXAMPLE 4 Add: $\dfrac{5x^2}{8} + \dfrac{7x}{12}$.

First, we find the LCD:

$$\left.\begin{array}{l} 8 = 2\cdot 2\cdot 2 \\ 12 = 2\cdot 2\cdot 3 \end{array}\right\} \quad \text{LCD} = 2\cdot 2\cdot 2\cdot 3, \text{ or } 24.$$

Compare the factorization $8 = 2\cdot 2\cdot 2$ with the factorization of the LCD, $24 = 2\cdot 2\cdot 2\cdot 3$. The factor of 24 that is missing from 8 is 3. Compare $12 = 2\cdot 2\cdot 3$ and $24 = 2\cdot 2\cdot 2\cdot 3$. The factor of 24 that is missing from 12 is 2.

We multiply each term by a symbol for 1 to get the LCD in each expression, and then add and, if possible, simplify:

$$\dfrac{5x^2}{8} + \dfrac{7x}{12} = \dfrac{5x^2}{2\cdot 2\cdot 2} + \dfrac{7x}{2\cdot 2\cdot 3}$$

$$= \dfrac{5x^2}{2\cdot 2\cdot 2}\cdot\dfrac{3}{3} + \dfrac{7x}{2\cdot 2\cdot 3}\cdot\dfrac{2}{2} \quad \text{Multiplying by 1 to get the same denominators}$$

$$= \dfrac{15x^2}{24} + \dfrac{14x}{24} = \dfrac{15x^2 + 14x}{24} = \dfrac{x(15x+14)}{24}.$$

Add.

4. $\dfrac{3x}{16} + \dfrac{5x^2}{24}$

5. $\dfrac{3}{16x} + \dfrac{5}{24x^2}$

$16x = 2\cdot 2\cdot 2\cdot \boxed{}\cdot x$

$24x^2 = 2\cdot 2\cdot 2\cdot 3\cdot \boxed{}\cdot x$

LCD $= 2\cdot 2\cdot 2\cdot \boxed{}\cdot 3\cdot x\cdot x$, or $48x^2$

$\dfrac{3}{16x}\cdot\dfrac{3x}{\boxed{}} + \dfrac{5}{24x^2}\cdot\dfrac{\boxed{}}{2}$

$= \dfrac{\boxed{}}{48x^2} + \dfrac{10}{\boxed{}}$

$= \dfrac{9x + \boxed{}}{48x^2}$

EXAMPLE 5 Add: $\dfrac{3}{8x} + \dfrac{5}{12x^2}$.

First, we find the LCD:

$$\left.\begin{array}{l} 8x = 2\cdot 2\cdot 2\cdot x \\ 12x^2 = 2\cdot 2\cdot 3\cdot x\cdot x \end{array}\right\} \quad \text{LCD} = 2\cdot 2\cdot 2\cdot 3\cdot x\cdot x, \text{ or } 24x^2.$$

The factors of the LCD missing from $8x$ are 3 and x. The factor of the LCD missing from $12x^2$ is 2. We multiply each term by 1 to get the LCD in each expression, and then add and, if possible, simplify:

$$\dfrac{3}{8x} + \dfrac{5}{12x^2} = \dfrac{3}{8x}\cdot\dfrac{3\cdot x}{3\cdot x} + \dfrac{5}{12x^2}\cdot\dfrac{2}{2}$$

$$= \dfrac{9x}{24x^2} + \dfrac{10}{24x^2} = \dfrac{9x + 10}{24x^2}.$$

◀ Do Exercises 4 and 5.

EXAMPLE 6 Add: $\dfrac{2a}{a^2 - 1} + \dfrac{1}{a^2 + a}$.

First, we find the LCD:

$$\left.\begin{array}{l} a^2 - 1 = (a-1)(a+1) \\ a^2 + a = a(a+1) \end{array}\right\} \quad \text{LCD} = a(a-1)(a+1).$$

We multiply each term by 1 to get the LCD in each expression, and then add and, if possible, simplify:

$$\dfrac{2a}{(a-1)(a+1)}\cdot\dfrac{a}{a} + \dfrac{1}{a(a+1)}\cdot\dfrac{a-1}{a-1}$$

$$= \dfrac{2a^2}{a(a-1)(a+1)} + \dfrac{a-1}{a(a-1)(a+1)}$$

$$= \dfrac{2a^2 + a - 1}{a(a-1)(a+1)}$$

$$= \dfrac{(a+1)(2a-1)}{a(a-1)(a+1)}. \quad \text{Factoring the numerator in order to simplify}$$

Answers

4. $\dfrac{x(9 + 10x)}{48}$ **5.** $\dfrac{9x + 10}{48x^2}$

Guided Solution:
5. 2, x, 2, 3x, 2, 9x, 48x², 10

Then

$$= \frac{(a+1)(2a-1)}{a(a-1)(a+1)} \quad \text{Removing a factor of 1: } \frac{a+1}{a+1} = 1$$

$$= \frac{2a-1}{a(a-1)}.$$

Do Exercise 6. ▶

6. Add:
$$\frac{3}{x^3 - x} + \frac{4}{x^2 + 2x + 1}.$$

EXAMPLE 7 Add: $\dfrac{x+4}{x-2} + \dfrac{x-7}{x+5}$.

First, we find the LCD. It is just the product of the denominators:

$$\text{LCD} = (x-2)(x+5).$$

We multiply by 1 to get the LCD in each expression, and then add and simplify:

$$\frac{x+4}{x-2} \cdot \frac{x+5}{x+5} + \frac{x-7}{x+5} \cdot \frac{x-2}{x-2}$$

$$= \frac{(x+4)(x+5)}{(x-2)(x+5)} + \frac{(x-7)(x-2)}{(x-2)(x+5)}$$

$$= \frac{x^2 + 9x + 20}{(x-2)(x+5)} + \frac{x^2 - 9x + 14}{(x-2)(x+5)}$$

$$= \frac{x^2 + 9x + 20 + x^2 - 9x + 14}{(x-2)(x+5)}$$

$$= \frac{2x^2 + 34}{(x-2)(x+5)} = \frac{2(x^2 + 17)}{(x-2)(x+5)}.$$

Do Exercise 7. ▶

7. Add:
$$\frac{x-2}{x+3} + \frac{x+7}{x+8}.$$

EXAMPLE 8 Add: $\dfrac{x}{x^2 + 11x + 30} + \dfrac{-5}{x^2 + 9x + 20}$.

$$\frac{x}{x^2 + 11x + 30} + \frac{-5}{x^2 + 9x + 20}$$

$$= \frac{x}{(x+5)(x+6)} + \frac{-5}{(x+5)(x+4)} \quad \begin{array}{l}\text{Factoring the denominators} \\ \text{in order to find the LCD. The} \\ \text{LCD is } (x+4)(x+5)(x+6).\end{array}$$

$$= \frac{x}{(x+5)(x+6)} \cdot \frac{x+4}{x+4} + \frac{-5}{(x+5)(x+4)} \cdot \frac{x+6}{x+6} \quad \begin{array}{l}\text{Multiplying} \\ \text{by 1}\end{array}$$

$$= \frac{x(x+4) + (-5)(x+6)}{(x+4)(x+5)(x+6)} = \frac{x^2 + 4x - 5x - 30}{(x+4)(x+5)(x+6)}$$

$$= \frac{x^2 - x - 30}{(x+4)(x+5)(x+6)}$$

$$= \frac{(x-6)(x+5)}{(x+4)(x+5)(x+6)} \quad \left.\begin{array}{l}\text{Always simplify at the end if} \\ \text{possible: } \dfrac{x+5}{x+5} = 1.\end{array}\right.$$

$$= \frac{x-6}{(x+4)(x+6)}$$

Do Exercise 8. ▶

8. Add:
$$\frac{5}{x^2 + 17x + 16} + \frac{3}{x^2 + 9x + 8}.$$

Denominators That Are Opposites

When one denominator is the opposite of the other, we can first multiply either expression by 1 using $-1/-1$.

Answers

6. $\dfrac{4x^2 - x + 3}{x(x-1)(x+1)^2}$ 7. $\dfrac{2x^2 + 16x + 5}{(x+3)(x+8)}$

8. $\dfrac{8(x+11)}{(x+16)(x+1)(x+8)}$

EXAMPLES

9. $\dfrac{x}{2} + \dfrac{3}{-2} = \dfrac{x}{2} + \dfrac{3}{-2} \cdot \dfrac{-1}{-1}$ Multiplying by 1 using $\dfrac{-1}{-1}$

$= \dfrac{x}{2} + \dfrac{-3}{2}$ The denominators are now the same.

$= \dfrac{x + (-3)}{2} = \dfrac{x - 3}{2}$

10. $\dfrac{3x + 4}{x - 2} + \dfrac{x - 7}{2 - x} = \dfrac{3x + 4}{x - 2} + \dfrac{x - 7}{2 - x} \cdot \dfrac{-1}{-1}$

We could have chosen to multiply this expression by $-1/-1$. We multiply only one expression, *not* both.

$= \dfrac{3x + 4}{x - 2} + \dfrac{-x + 7}{x - 2}$ Note: $(2 - x)(-1) = -2 + x = x - 2.$

$= \dfrac{(3x + 4) + (-x + 7)}{x - 2} = \dfrac{2x + 11}{x - 2}$

◀ Do Exercises 9 and 10.

Factors That Are Opposites

Suppose that when we factor to find the LCD, we find factors that are opposites. The easiest way to handle this is to first go back and multiply by $-1/-1$ appropriately to change factors so that they are not opposites.

EXAMPLE 11 Add: $\dfrac{x}{x^2 - 25} + \dfrac{3}{10 - 2x}.$

First, we factor to find the LCD:

$x^2 - 25 = (x - 5)(x + 5);$
$10 - 2x = 2(5 - x).$

We note that $x - 5$ is one factor of $x^2 - 25$ and $5 - x$ is one factor of $10 - 2x$. If the denominator of the second expression were $2x - 10$, then $x - 5$ would be a factor of both denominators. To rewrite the second expression with a denominator of $2x - 10$, we multiply by 1 using $-1/-1$, and then continue as before:

$\dfrac{x}{x^2 - 25} + \dfrac{3}{10 - 2x} = \dfrac{x}{(x - 5)(x + 5)} + \dfrac{3}{10 - 2x} \cdot \dfrac{-1}{-1}$

$= \dfrac{x}{(x - 5)(x + 5)} + \dfrac{-3}{2x - 10}$

$= \dfrac{x}{(x - 5)(x + 5)} + \dfrac{-3}{2(x - 5)}$ LCD $= 2(x - 5)(x + 5)$

$= \dfrac{x}{(x - 5)(x + 5)} \cdot \dfrac{2}{2} + \dfrac{-3}{2(x - 5)} \cdot \dfrac{x + 5}{x + 5}$

$= \dfrac{2x}{2(x - 5)(x + 5)} + \dfrac{-3(x + 5)}{2(x - 5)(x + 5)}$

$= \dfrac{2x - 3(x + 5)}{2(x - 5)(x + 5)} = \dfrac{2x - 3x - 15}{2(x - 5)(x + 5)}$

$= \dfrac{-x - 15}{2(x - 5)(x + 5)}.$ Collecting like terms

◀ Do Exercise 11.

Add.

9. $\dfrac{x}{4} + \dfrac{5}{-4}$

10. $\dfrac{2x + 1}{x - 3} + \dfrac{x + 2}{3 - x}$

$= \dfrac{2x + 1}{x - 3} + \dfrac{x + 2}{3 - x} \cdot \dfrac{-1}{\boxed{}}$

$= \dfrac{2x + 1}{x - 3} + \dfrac{\boxed{} - 2}{x - 3}$

$= \dfrac{(2x + 1) + (-x - 2)}{x - \boxed{}}$

$= \dfrac{\boxed{} - 1}{x - 3}$

11. Add:

$\dfrac{x + 3}{x^2 - 16} + \dfrac{5}{12 - 3x}.$

Answers

9. $\dfrac{x - 5}{4}$ **10.** $\dfrac{x - 1}{x - 3}$ **11.** $\dfrac{-2x - 11}{3(x + 4)(x - 4)}$

Guided Solution:
10. $-1, -x, 3, x$

6.4 Exercise Set

✓ Check Your Understanding

Reading Check Choose from the column on the right the best choice to complete each statement.

RC1. To add rational expressions when the denominators are the same, add the _____ and keep the same _____.

RC2. To add rational expressions with different denominators, first find the _____.

RC3. When factoring to find the LCD and finding factors that are _____, multiply by $-1/-1$.

LCD
LCM
opposites
reciprocals
numerator(s)
denominator(s)

Concept Check From the choices on the right, select the names for 1 that result in the same denominator in each addition. Some choices may be used more than once. Some choices may not be used. Do not complete the addition.

CC1. $\dfrac{2x^2}{15} + \dfrac{3x}{10} = \dfrac{2x^2}{3\cdot 5}\cdot\left(\ \right) + \dfrac{3x}{2\cdot 5}\cdot\left(\ \right)$

CC2. $\dfrac{3}{2} + \dfrac{2x}{x+3} = \dfrac{3}{2}\cdot\left(\ \right) + \dfrac{2x}{x+3}\cdot\left(\ \right)$

CC3. $\dfrac{3x}{x+2} + \dfrac{2}{x-3} = \dfrac{3x}{x+2}\cdot\left(\ \right) + \dfrac{2}{x-3}\cdot\left(\ \right)$

CC4. $\dfrac{2}{15x} + \dfrac{3}{10x^2} = \dfrac{2}{3\cdot 5\cdot x}\cdot\left(\ \right) + \dfrac{3}{2\cdot 5\cdot x\cdot x}\cdot\left(\ \right)$

$\dfrac{x+3}{x+3}$ $\dfrac{2}{2}$

$\dfrac{3x}{3x}$ $\dfrac{x-2}{x-2}$

$\dfrac{x-3}{x-3}$ $\dfrac{3}{3}$

$\dfrac{2x}{2x}$ $\dfrac{x+2}{x+2}$

a   Add. Simplify, if possible.

1. $\dfrac{5}{8} + \dfrac{3}{8}$

2. $\dfrac{3}{16} + \dfrac{5}{16}$

3. $\dfrac{1}{3+x} + \dfrac{5}{3+x}$

4. $\dfrac{x^2+7x}{x^2-5x} + \dfrac{x^2-4x}{x^2-5x}$

5. $\dfrac{4x+6}{2x-1} + \dfrac{5-8x}{-1+2x}$

6. $\dfrac{4}{x+y} + \dfrac{9}{y+x}$

7. $\dfrac{2}{x} + \dfrac{5}{x^2}$

8. $\dfrac{3}{y^2} + \dfrac{6}{y}$

9. $\dfrac{5}{6r} + \dfrac{7}{8r}$

10. $\dfrac{13}{18x} + \dfrac{7}{24x}$

11. $\dfrac{4}{xy^2} + \dfrac{6}{x^2y}$

12. $\dfrac{8}{ab^3} + \dfrac{3}{a^2b}$

13. $\dfrac{2}{9t^3} + \dfrac{1}{6t^2}$

14. $\dfrac{5}{c^2 d^3} + \dfrac{-4}{7cd^2}$

15. $\dfrac{x+y}{xy^2} + \dfrac{3x+y}{x^2 y}$

16. $\dfrac{2c-d}{c^2 d} + \dfrac{c+d}{cd^2}$

17. $\dfrac{3}{x-2} + \dfrac{3}{x+2}$

18. $\dfrac{2}{y+1} + \dfrac{2}{y-1}$

19. $\dfrac{3}{x+1} + \dfrac{2}{3x}$

20. $\dfrac{4}{5y} + \dfrac{7}{y-2}$

21. $\dfrac{2x}{x^2-16} + \dfrac{x}{x-4}$

22. $\dfrac{4x}{x^2-25} + \dfrac{x}{x+5}$

23. $\dfrac{5}{z+4} + \dfrac{3}{3z+12}$

24. $\dfrac{t}{t-3} + \dfrac{5}{4t-12}$

25. $\dfrac{3}{x-1} + \dfrac{2}{(x-1)^2}$

26. $\dfrac{8}{(y+3)^2} + \dfrac{5}{y+3}$

27. $\dfrac{4a}{5a-10} + \dfrac{3a}{10a-20}$

28. $\dfrac{9x}{6x-30} + \dfrac{3x}{4x-20}$

29. $\dfrac{x+4}{x} + \dfrac{x}{x+4}$

30. $\dfrac{a}{a-3} + \dfrac{a-3}{a}$

31. $\dfrac{4}{a^2-a-2} + \dfrac{3}{a^2+4a+3}$

32. $\dfrac{a}{a^2-2a+1} + \dfrac{1}{a^2-5a+4}$

33. $\dfrac{x+3}{x-5} + \dfrac{x-5}{x+3}$

34. $\dfrac{3x}{2y-3} + \dfrac{2x}{3y-2}$

35. $\dfrac{a}{a^2-1} + \dfrac{2a}{a^2-a}$

36. $\dfrac{3x+2}{3x+6} + \dfrac{x-2}{x^2-4}$

37. $\dfrac{7}{8} + \dfrac{5}{-8}$

38. $\dfrac{5}{-3} + \dfrac{11}{3}$

39. $\dfrac{3}{t} + \dfrac{4}{-t}$

40. $\dfrac{5}{-a} + \dfrac{8}{a}$

41. $\dfrac{2x+7}{x-6} + \dfrac{3x}{6-x}$

42. $\dfrac{2x-7}{5x-8} + \dfrac{6+10x}{8-5x}$

43. $\dfrac{y^2}{y-3} + \dfrac{9}{3-y}$

44. $\dfrac{t^2}{t-2} + \dfrac{4}{2-t}$

45. $\dfrac{b-7}{b^2-16} + \dfrac{7-b}{16-b^2}$

46. $\dfrac{a-3}{a^2-25} + \dfrac{a-3}{25-a^2}$

47. $\dfrac{a^2}{a-b} + \dfrac{b^2}{b-a}$

48. $\dfrac{x^2}{x-7} + \dfrac{49}{7-x}$

49. $\dfrac{x+3}{x-5} + \dfrac{2x-1}{5-x} + \dfrac{2(3x-1)}{x-5}$

50. $\dfrac{3(x-2)}{2x-3} + \dfrac{5(2x+1)}{2x-3} + \dfrac{3(x+1)}{3-2x}$

51. $\dfrac{2(4x+1)}{5x-7} + \dfrac{3(x-2)}{7-5x} + \dfrac{-10x-1}{5x-7}$

52. $\dfrac{5(x-2)}{3x-4} + \dfrac{2(x-3)}{4-3x} + \dfrac{3(5x+1)}{4-3x}$

SECTION 6.4 Adding Rational Expressions

53. $\dfrac{x+1}{(x+3)(x-3)} + \dfrac{4(x-3)}{(x-3)(x+3)} + \dfrac{(x-1)(x-3)}{(3-x)(x+3)}$

54. $\dfrac{2(x+5)}{(2x-3)(x-1)} + \dfrac{3x+4}{(2x-3)(1-x)} + \dfrac{x-5}{(3-2x)(x-1)}$

55. $\dfrac{6}{x-y} + \dfrac{4x}{y^2-x^2}$

56. $\dfrac{a-2}{3-a} + \dfrac{4-a^2}{a^2-9}$

57. $\dfrac{4-a}{25-a^2} + \dfrac{a+1}{a-5}$

58. $\dfrac{x+2}{x-7} + \dfrac{3-x}{49-x^2}$

59. $\dfrac{2}{t^2+t-6} + \dfrac{3}{t^2-9}$

60. $\dfrac{10}{a^2-a-6} + \dfrac{3a}{a^2+4a+4}$

Skill Maintenance

Simplify. [4.2a, b]

61. $\left(\dfrac{x^{-4}}{y^7}\right)^3$

62. $(5x^{-2}y^{-3})^2$

Solve.

63. $3x - 7 = 5x + 9$ [2.3b]

64. $x^2 - 7x = 18$ [5.8b]

Graph.

65. $y = \dfrac{1}{2}x - 5$ [3.2a]

66. $2y + x + 10 = 0$ [3.3a]

67. $y = 3$ [3.3b]

68. $x = -5$ [3.3b]

Synthesis

Find the perimeter and the area of each figure.

69. Rectangle with length $\dfrac{y+4}{3}$ and width $\dfrac{y-2}{5}$

70. Rectangle with length $\dfrac{3}{x+4}$ and width $\dfrac{2}{x-5}$

Add. Simplify, if possible.

71. $\dfrac{5}{z+2} + \dfrac{4z}{z^2-4} + 2$

72. $\dfrac{-2}{y^2-9} + \dfrac{4y}{(y-3)^2} + \dfrac{6}{3-y}$

73. $\dfrac{3z^2}{z^4-4} + \dfrac{5z^2-3}{2z^4+z^2-6}$

Subtracting Rational Expressions

6.5

a SUBTRACTING RATIONAL EXPRESSIONS

We subtract rational expressions as we do rational numbers.

OBJECTIVES

a Subtract rational expressions.

b Simplify combined additions and subtractions of rational expressions.

SUBTRACTING RATIONAL EXPRESSIONS WITH LIKE DENOMINATORS

To subtract when the denominators are the same, subtract the numerators and keep the same denominator. Then simplify, if possible.

EXAMPLE 1 Subtract: $\dfrac{8}{x} - \dfrac{3}{x}$.

$$\dfrac{8}{x} - \dfrac{3}{x} = \dfrac{8-3}{x} = \dfrac{5}{x}$$

EXAMPLE 2 Subtract: $\dfrac{3x}{x+2} - \dfrac{x-2}{x+2}$.

$$\dfrac{3x}{x+2} - \dfrac{x-2}{x+2} = \dfrac{3x - (x-2)}{x+2}$$

Caution! The parentheses are important to make sure that you subtract the entire numerator.

$$= \dfrac{3x - x + 2}{x+2} \quad \text{Removing parentheses}$$

$$= \dfrac{2x+2}{x+2} = \dfrac{2(x+1)}{x+2}$$

Do Exercises 1–3. ▶

To subtract rational expressions with different denominators, we use a procedure similar to what we used for addition, except that we subtract numerators and write the difference over the LCD.

SUBTRACTING RATIONAL EXPRESSIONS WITH DIFFERENT DENOMINATORS

To subtract rational expressions with different denominators:

1. Find the LCM of the denominators. This is the least common denominator (LCD).
2. For each rational expression, find an equivalent expression with the LCD. To do so, multiply by 1 using a symbol for 1 made up of factors of the LCD that are missing from the original denominator.
3. Subtract the numerators. Write the difference over the LCD.
4. Simplify, if possible.

SKILL REVIEW

Find an equivalent expression for an opposite without parentheses, where an expression has several terms. [1.8a]

Find an expression without parentheses.

1. $-(3x - 11)$
2. $-(-x + 8)$

Answers:
1. $-3x + 11$ 2. $x - 8$

Subtract.

1. $\dfrac{7}{11} - \dfrac{3}{11}$

2. $\dfrac{7}{y} - \dfrac{2}{y}$

3. $\dfrac{2x^2 + 3x - 7}{2x+1} - \dfrac{x^2 + x - 8}{2x+1}$

Answers

1. $\dfrac{4}{11}$ 2. $\dfrac{5}{y}$ 3. $\dfrac{(x+1)^2}{2x+1}$

4. Subtract:

$$\frac{x-2}{3x} - \frac{2x-1}{5x}.$$

$\dfrac{x-2}{3x} - \dfrac{2x-1}{5x}$

LCD $= 3 \cdot x \cdot 5 = 15x$

$= \dfrac{x-2}{3x} \cdot \dfrac{5}{\boxed{}} - \dfrac{2x-1}{5x} \cdot \dfrac{\boxed{}}{3}$

$= \dfrac{5x - \boxed{}}{15x} - \dfrac{\boxed{} - 3}{15x}$

$= \dfrac{5x - 10 - (6x - \boxed{})}{15x}$

$= \dfrac{5x - 10 - \boxed{} + 3}{15x}$

$= \dfrac{\boxed{} - 7}{15x}$

EXAMPLE 3 Subtract: $\dfrac{x+2}{x-4} - \dfrac{x+1}{x+4}$.

The LCD $= (x-4)(x+4)$.

$\dfrac{x+2}{x-4} \cdot \dfrac{x+4}{x+4} - \dfrac{x+1}{x+4} \cdot \dfrac{x-4}{x-4}$ Multiplying by 1

$= \dfrac{(x+2)(x+4)}{(x-4)(x+4)} - \dfrac{(x+1)(x-4)}{(x-4)(x+4)}$

$= \dfrac{x^2 + 6x + 8}{(x-4)(x+4)} - \dfrac{x^2 - 3x - 4}{(x-4)(x+4)}$

⎯⎯⎯⎯ Subtracting this numerator. Don't forget the parentheses.

$= \dfrac{x^2 + 6x + 8 - (x^2 - 3x - 4)}{(x-4)(x+4)}$

$= \dfrac{x^2 + 6x + 8 - x^2 + 3x + 4}{(x-4)(x+4)}$ Removing parentheses

$= \dfrac{9x + 12}{(x-4)(x+4)} = \dfrac{3(3x+4)}{(x-4)(x+4)}$

◀ Do Exercise 4.

EXAMPLE 4 Subtract: $\dfrac{x}{x^2 + 5x + 6} - \dfrac{2}{x^2 + 3x + 2}$.

$\dfrac{x}{x^2 + 5x + 6} - \dfrac{2}{x^2 + 3x + 2}$

$= \dfrac{x}{(x+2)(x+3)} - \dfrac{2}{(x+2)(x+1)}$ LCD $= (x+1)(x+2)(x+3)$

$= \dfrac{x}{(x+2)(x+3)} \cdot \dfrac{x+1}{x+1} - \dfrac{2}{(x+2)(x+1)} \cdot \dfrac{x+3}{x+3}$

$= \dfrac{x^2 + x}{(x+1)(x+2)(x+3)} - \dfrac{2x + 6}{(x+1)(x+2)(x+3)}$

⎯⎯⎯⎯ Subtracting this numerator. Don't forget the parentheses.

$= \dfrac{x^2 + x - (2x + 6)}{(x+1)(x+2)(x+3)}$

$= \dfrac{x^2 + x - 2x - 6}{(x+1)(x+2)(x+3)} = \dfrac{x^2 - x - 6}{(x+1)(x+2)(x+3)}$

$= \dfrac{(x+2)(x-3)}{(x+1)(x+2)(x+3)}$

$= \dfrac{\cancel{(x+2)}(x-3)}{(x+1)\cancel{(x+2)}(x+3)}$ Simplifying by removing a factor of 1: $\dfrac{x+2}{x+2} = 1$

$= \dfrac{x-3}{(x+1)(x+3)}$

◀ Do Exercise 5.

5. Subtract:

$$\dfrac{x}{x^2 + 15x + 56} - \dfrac{6}{x^2 + 13x + 42}.$$

Denominators That Are Opposites

When one denominator is the opposite of the other, we can first multiply one expression by $-1/-1$ to obtain a common denominator.

Answers

4. $\dfrac{-x - 7}{15x}$ **5.** $\dfrac{x^2 - 48}{(x+7)(x+8)(x+6)}$

Guided Solution:
4. 5, 3, 10, 6x, 3, 6x, −x

EXAMPLE 5 Subtract: $\dfrac{x}{5} - \dfrac{3x-4}{-5}$.

$$\dfrac{x}{5} - \dfrac{3x-4}{-5} = \dfrac{x}{5} - \dfrac{3x-4}{-5} \cdot \dfrac{-1}{-1} \quad \text{Multiplying by 1 using } \dfrac{-1}{-1}$$

This is equal to 1 (not -1).

$$= \dfrac{x}{5} - \dfrac{(3x-4)(-1)}{(-5)(-1)}$$

$$= \dfrac{x}{5} - \dfrac{4-3x}{5}$$

$$= \dfrac{x-(4-3x)}{5} \quad \text{Remember the parentheses!}$$

$$= \dfrac{x-4+3x}{5} = \dfrac{4x-4}{5} = \dfrac{4(x-1)}{5}$$

EXAMPLE 6 Subtract: $\dfrac{5y}{y-5} - \dfrac{2y-3}{5-y}$.

$$\dfrac{5y}{y-5} - \dfrac{2y-3}{5-y} = \dfrac{5y}{y-5} - \dfrac{2y-3}{5-y} \cdot \dfrac{-1}{-1}$$

$$= \dfrac{5y}{y-5} - \dfrac{(2y-3)(-1)}{(5-y)(-1)}$$

$$= \dfrac{5y}{y-5} - \dfrac{3-2y}{y-5}$$

$$= \dfrac{5y-(3-2y)}{y-5} \quad \text{Remember the parentheses!}$$

$$= \dfrac{5y-3+2y}{y-5} = \dfrac{7y-3}{y-5}$$

Do Exercises 6 and 7.

Subtract.

6. $\dfrac{x}{3} - \dfrac{2x-1}{-3}$

7. $\dfrac{3x}{x-2} - \dfrac{x-3}{2-x}$

Factors That Are Opposites

Suppose that when we factor to find the LCD, we find factors that are opposites. Then we multiply by $-1/-1$ appropriately to change factors so that they are not opposites.

EXAMPLE 7 Subtract: $\dfrac{p}{64-p^2} - \dfrac{5}{p-8}$.

Factoring $64 - p^2$, we get $(8-p)(8+p)$. Note that the factors $8-p$ in the first denominator and $p-8$ in the second denominator are opposites. We multiply the first expression by $-1/-1$ to avoid this situation. Then we proceed as before.

$$\dfrac{p}{64-p^2} - \dfrac{5}{p-8} = \dfrac{p}{64-p^2} \cdot \dfrac{-1}{-1} - \dfrac{5}{p-8}$$

$$= \dfrac{-p}{p^2-64} - \dfrac{5}{p-8}$$

$$= \dfrac{-p}{(p-8)(p+8)} - \dfrac{5}{p-8} \quad \text{LCD} = (p-8)(p+8)$$

$$= \dfrac{-p}{(p-8)(p+8)} - \dfrac{5}{p-8} \cdot \dfrac{p+8}{p+8}$$

Answers

6. $\dfrac{3x-1}{3}$ 7. $\dfrac{4x-3}{x-2}$

8. Subtract:

$$\frac{y}{16 - y^2} - \frac{7}{y - 4}.$$

$$\frac{y}{16 - y^2} - \frac{7}{y - 4}$$

$$= \frac{y}{16 - y^2} \cdot \frac{-1}{\boxed{}} - \frac{7}{y - 4}$$

$$= \frac{-y}{\boxed{} - 16} - \frac{7}{y - 4}$$

$$= \frac{-y}{(y + 4)(y - \boxed{})} - \frac{7}{y - 4} \cdot \frac{\boxed{} + 4}{y + 4}$$

$$= \frac{-y}{(y + 4)(y - 4)} - \frac{7y + \boxed{}}{(y + 4)(y - 4)}$$

$$= \frac{-y - (7y + 28)}{(y + 4)(y - 4)} = \frac{-y - 7y - \boxed{}}{(y + 4)(y - 4)}$$

$$= \frac{\boxed{} - 28}{(y + 4)(y - 4)} = \frac{\boxed{}(2y + 7)}{(y + 4)(y - 4)}$$

9. Perform the indicated operations and simplify:

$$\frac{x + 2}{x^2 - 9} - \frac{x - 7}{9 - x^2} + \frac{-8 - x}{x^2 - 9}.$$

10. Perform the indicated operations and simplify:

$$\frac{1}{x} - \frac{5}{3x} + \frac{2x}{x + 1}.$$

Answers

8. $\dfrac{-4(2y + 7)}{(y + 4)(y - 4)}$ 9. $\dfrac{x - 13}{(x + 3)(x - 3)}$

10. $\dfrac{2(3x^2 - x - 1)}{3x(x + 1)}$

Guided Solution:
8. $-1, y^2, 4, y, 28, 28, 8y, -4$

Multiplying, we have

$$\frac{-p}{(p - 8)(p + 8)} - \frac{5p + 40}{(p - 8)(p + 8)}$$

Subtracting this numerator. Don't forget the parentheses.

$$= \frac{-p - (5p + 40)}{(p - 8)(p + 8)}$$

$$= \frac{-p - 5p - 40}{(p - 8)(p + 8)} = \frac{-6p - 40}{(p - 8)(p + 8)} = \frac{-2(3p + 20)}{(p - 8)(p + 8)}.$$

◀ Do Exercise 8.

b COMBINED ADDITIONS AND SUBTRACTIONS

Now let's look at some combined additions and subtractions.

EXAMPLE 8 Perform the indicated operations and simplify:

$$\frac{x + 9}{x^2 - 4} + \frac{5 - x}{4 - x^2} - \frac{2 + x}{x^2 - 4}.$$

$$\frac{x + 9}{x^2 - 4} + \frac{5 - x}{4 - x^2} - \frac{2 + x}{x^2 - 4}$$

$$= \frac{x + 9}{x^2 - 4} + \frac{5 - x}{4 - x^2} \cdot \frac{-1}{-1} - \frac{2 + x}{x^2 - 4}$$

$$= \frac{x + 9}{x^2 - 4} + \frac{x - 5}{x^2 - 4} - \frac{2 + x}{x^2 - 4} = \frac{(x + 9) + (x - 5) - (2 + x)}{x^2 - 4}$$

$$= \frac{x + 9 + x - 5 - 2 - x}{x^2 - 4} = \frac{x + 2}{x^2 - 4} = \frac{(x + 2) \cdot 1}{(x + 2)(x - 2)} = \frac{1}{x - 2}$$

◀ Do Exercise 9.

EXAMPLE 9 Perform the indicated operations and simplify:

$$\frac{1}{x} - \frac{1}{x^2} + \frac{2}{x + 1}.$$

The LCD $= x \cdot x(x + 1)$, or $x^2(x + 1)$.

$$\frac{1}{x} \cdot \frac{x(x + 1)}{x(x + 1)} - \frac{1}{x^2} \cdot \frac{(x + 1)}{(x + 1)} + \frac{2}{x + 1} \cdot \frac{x^2}{x^2}$$

$$= \frac{x(x + 1)}{x^2(x + 1)} - \frac{x + 1}{x^2(x + 1)} + \frac{2x^2}{x^2(x + 1)}$$

Subtracting this numerator. Don't forget the parentheses.

$$= \frac{x(x + 1) - (x + 1) + 2x^2}{x^2(x + 1)}$$

$$= \frac{x^2 + x - x - 1 + 2x^2}{x^2(x + 1)} \quad \text{Removing parentheses}$$

$$= \frac{3x^2 - 1}{x^2(x + 1)}$$

◀ Do Exercise 10.

6.5 Exercise Set

✓ Check Your Understanding

Reading Check When we are subtracting rational expressions, parentheses are important to make sure that we subtract the entire numerator. In the following exercises, complete each numerator by **(a)** filling in the expression in parentheses, **(b)** removing the parentheses, and **(c)** collecting like terms.

RC1. $\dfrac{10x}{x-7} - \dfrac{3x+5}{x-7} = \underbrace{\dfrac{10x-()}{x-7}}_{(a)} = \underbrace{\dfrac{}{x-7}}_{(b)} = \underbrace{\dfrac{}{x-7}}_{(c)}$

RC2. $\dfrac{7}{4+a} - \dfrac{4-9a}{4+a} = \underbrace{\dfrac{7-()}{4+a}}_{(a)} = \underbrace{\dfrac{}{4+a}}_{(b)} = \underbrace{\dfrac{}{4+a}}_{(c)}$

RC3. $\dfrac{9y-2}{y^2-10} - \dfrac{y+1}{y^2-10} = \underbrace{\dfrac{9y-2-()}{y^2-10}}_{(a)} = \underbrace{\dfrac{}{y^2-10}}_{(b)} = \underbrace{\dfrac{}{y^2-10}}_{(c)}$

Concept Check Choose from the column on the right the correct numerator for each difference.

CC1. $\dfrac{4x}{x+5} - \dfrac{3-x}{x+5} = \dfrac{}{x+5}$

CC2. $\dfrac{2(x-3)}{(x+2)(x-3)} - \dfrac{(x-1)(x+2)}{(x-3)(x+2)} = \dfrac{}{(x+2)(x-3)}$

a) $-x^2 + 3x - 1$
b) $x^2 + x - 5$
c) $3x - 3$
d) $-x^2 + x - 4$
e) $5x - 3$
f) $5x + 3$

a Subtract. Simplify, if possible.

1. $\dfrac{7}{x} - \dfrac{3}{x}$

2. $\dfrac{5}{a} - \dfrac{8}{a}$

3. $\dfrac{y}{y-4} - \dfrac{4}{y-4}$

4. $\dfrac{t^2}{t+5} - \dfrac{25}{t+5}$

5. $\dfrac{2x-3}{x^2+3x-4} - \dfrac{x-7}{x^2+3x-4}$

6. $\dfrac{x+1}{x^2-2x+1} - \dfrac{5-3x}{x^2-2x+1}$

7. $\dfrac{a-2}{10} - \dfrac{a+1}{5}$

8. $\dfrac{y+3}{2} - \dfrac{y-4}{4}$

9. $\dfrac{4z-9}{3z} - \dfrac{3z-8}{4z}$

10. $\dfrac{a-1}{4a} - \dfrac{2a+3}{a}$

11. $\dfrac{4x+2t}{3xt^2} - \dfrac{5x-3t}{x^2 t}$

12. $\dfrac{5x+3y}{2x^2 y} - \dfrac{3x+4y}{xy^2}$

13. $\dfrac{5}{x+5} - \dfrac{3}{x-5}$

14. $\dfrac{3t}{t-1} - \dfrac{8t}{t+1}$

15. $\dfrac{3}{2t^2 - 2t} - \dfrac{5}{2t-2}$

16. $\dfrac{11}{x^2 - 4} - \dfrac{8}{x+2}$

17. $\dfrac{2s}{t^2 - s^2} - \dfrac{s}{t-s}$

18. $\dfrac{3}{12 + x - x^2} - \dfrac{2}{x^2 - 9}$

19. $\dfrac{y-5}{y} - \dfrac{3y-1}{4y}$

20. $\dfrac{3x-2}{4x} - \dfrac{3x+1}{6x}$

21. $\dfrac{a}{x+a} - \dfrac{a}{x-a}$

22. $\dfrac{a}{a-b} - \dfrac{a}{a+b}$

23. $\dfrac{11}{6} - \dfrac{5}{-6}$

24. $\dfrac{5}{9} - \dfrac{7}{-9}$

25. $\dfrac{5}{a} - \dfrac{8}{-a}$

26. $\dfrac{8}{x} - \dfrac{3}{-x}$

27. $\dfrac{4}{y-1} - \dfrac{4}{1-y}$

28. $\dfrac{5}{a-2} - \dfrac{3}{2-a}$

29. $\dfrac{3-x}{x-7} - \dfrac{2x-5}{7-x}$

30. $\dfrac{t^2}{t-2} - \dfrac{4}{2-t}$

31. $\dfrac{a-2}{a^2-25} - \dfrac{6-a}{25-a^2}$

32. $\dfrac{x-8}{x^2-16} - \dfrac{x-8}{16-x^2}$

33. $\dfrac{4-x}{x-9} - \dfrac{3x-8}{9-x}$

34. $\dfrac{4x-6}{x-5} - \dfrac{7-2x}{5-x}$

35. $\dfrac{5x}{x^2-9} - \dfrac{4}{3-x}$

36. $\dfrac{8x}{16-x^2} - \dfrac{5}{x-4}$

37. $\dfrac{t^2}{2t^2-2t} - \dfrac{1}{2t-2}$

38. $\dfrac{4}{5a^2-5a} - \dfrac{2}{5a-5}$

39. $\dfrac{x}{x^2+5x+6} - \dfrac{2}{x^2+3x+2}$

40. $\dfrac{a}{a^2+11a+30} - \dfrac{5}{a^2+9a+20}$

b Perform the indicated operations and simplify.

41. $\dfrac{3(2x+5)}{x-1} - \dfrac{3(2x-3)}{1-x} + \dfrac{6x-1}{x-1}$

42. $\dfrac{a-2b}{b-a} - \dfrac{3a-3b}{a-b} + \dfrac{2a-b}{a-b}$

43. $\dfrac{x-y}{x^2-y^2} + \dfrac{x+y}{x^2-y^2} - \dfrac{2x}{x^2-y^2}$

44. $\dfrac{x-3y}{2(y-x)} + \dfrac{x+y}{2(x-y)} - \dfrac{2x-2y}{2(x-y)}$

45. $\dfrac{2(x-1)}{2x-3} - \dfrac{3(x+2)}{2x-3} - \dfrac{x-1}{3-2x}$

46. $\dfrac{5(2y+1)}{2y-3} - \dfrac{3(y-1)}{3-2y} - \dfrac{3(y-2)}{2y-3}$

47. $\dfrac{10}{2y-1} - \dfrac{6}{1-2y} + \dfrac{y}{2y-1} + \dfrac{y-4}{1-2y}$

48. $\dfrac{(x+1)(2x-1)}{(2x-3)(x-3)} - \dfrac{(x-3)(x+1)}{(3-x)(3-2x)} + \dfrac{(2x+1)(x+3)}{(3-2x)(x-3)}$

49. $\dfrac{a+6}{4-a^2} - \dfrac{a+3}{a+2} + \dfrac{a-3}{2-a}$

50. $\dfrac{4t}{t^2-1} - \dfrac{2}{t} - \dfrac{2}{t+1}$

51. $\dfrac{2z}{1-2z} + \dfrac{3z}{2z+1} - \dfrac{3}{4z^2-1}$

52. $\dfrac{1}{x-y} - \dfrac{2x}{x^2-y^2} + \dfrac{1}{x+y}$

53. $\dfrac{1}{x+y} - \dfrac{1}{x-y} + \dfrac{2x}{x^2-y^2}$

54. $\dfrac{2b}{a^2-b^2} - \dfrac{1}{a+b} + \dfrac{1}{a-b}$

Skill Maintenance

Simplify.

55. $(a^2 b^{-5})^{-4}$ [4.2a, b]

56. $\dfrac{54x^{10}}{3x^7}$ [4.1e]

Solve. [2.3b]

57. $\dfrac{4}{7} + 3x = \dfrac{1}{2}x - \dfrac{3}{14}$

58. $6x - 0.5 = 6 - 0.5x$

Find a polynomial for the shaded area of each figure. [4.4d]

59.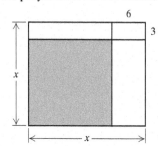

60.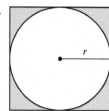

Synthesis

61. The perimeter of the following right triangle is $2a + 5$. Find the missing length of the third side and the area.

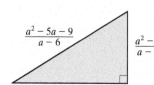

Mid-Chapter Review

Concept Reinforcement

Determine whether each statement is true or false.

_____ 1. The reciprocal of $\dfrac{3-w}{w+2}$ is $\dfrac{w-3}{w+2}$. [6.2a]

_____ 2. The value of the numerator has no bearing on whether a rational expression is defined. [6.1a]

_____ 3. To add or subtract rational expressions when the denominators are the same, add or subtract the numerators and keep the same denominator. [6.4a], [6.5a]

_____ 4. For the rational expression $\dfrac{x(x-2)}{x+3}$, x is a factor of the numerator and a factor of the denominator. [6.1c]

_____ 5. To find the LCM, use each factor the greatest number of times that it appears in any one factorization. [6.3a, c]

Guided Solutions

GS Fill in each blank with the number or expression that creates a correct solution.

6. Subtract: $\dfrac{x-1}{x-2} - \dfrac{x+1}{x+2} - \dfrac{x-6}{4-x^2}$. [6.5b]

$$\dfrac{x-1}{x-2} - \dfrac{x+1}{x+2} - \dfrac{x-6}{4-x^2} = \dfrac{x-1}{x-2} - \dfrac{x+1}{x+2} - \dfrac{x-6}{4-x^2} \cdot \dfrac{\boxed{}}{\boxed{}} = \dfrac{x-1}{x-2} - \dfrac{x+1}{x+2} - \dfrac{6-\boxed{}}{\boxed{}-4}$$

$$= \dfrac{x-1}{x-2} - \dfrac{x+1}{x+2} - \dfrac{6-x}{(x-\boxed{})(\boxed{}+2)}$$

$$= \dfrac{x-1}{x-2} \cdot \dfrac{\boxed{}}{\boxed{}} - \dfrac{x+1}{x+2} \cdot \dfrac{\boxed{}}{\boxed{}} - \dfrac{6-x}{(x-2)(x+2)}$$

$$= \dfrac{x^2 + \boxed{} - 2}{(x-2)(x+2)} - \dfrac{\boxed{} - x - 2}{(x-2)(x+2)} - \dfrac{6-x}{(x-2)(x+2)}$$

$$= \dfrac{x^2 + x - \boxed{} - x^2 + \boxed{} + 2 - \boxed{} + x}{(x-2)(x+2)}$$

$$= \dfrac{\boxed{} - \boxed{}}{(x-2)(x+2)} = \dfrac{3(\boxed{} - \boxed{})}{(x-2)(x+2)} = \dfrac{\boxed{}}{\boxed{}} \cdot \dfrac{3}{x+2} = \dfrac{3}{\boxed{}}$$

Mixed Review

Find all numbers for which the rational expression is not defined. [6.1a]

7. $\dfrac{t^2 - 16}{3}$

8. $\dfrac{x-8}{x^2 - 11x + 24}$

9. $\dfrac{7}{2w-7}$

Simplify. [6.1c]

10. $\dfrac{x^2 + 2x - 3}{x^2 - 9}$

11. $\dfrac{6y^2 + 12y - 48}{3y^2 - 9y + 6}$

12. $\dfrac{r-s}{s-r}$

13. Find the reciprocal of $-x + 3$. [6.2a]

14. Find the LCM of
$x^2 - 100, \quad 10x^3, \quad \text{and} \quad x^2 - 20x + 100.$
[6.3c]

Add, subtract, multiply, or divide and simplify, if possible.

15. $\dfrac{a^2 - a - 2}{a^2 - a - 6} \div \dfrac{a^2 - 2a}{2a + a^2}$ [6.2b]

16. $\dfrac{3y}{y^2 - 7y + 10} - \dfrac{2y}{y^2 - 8y + 15}$ [6.5a]

17. $\dfrac{x^2}{x - 11} + \dfrac{121}{11 - x}$ [6.4a]

18. $\dfrac{x^2 - y^2}{(x - y)^2} \cdot \dfrac{1}{x + y}$ [6.1d]

19. $\dfrac{3a - b}{a^2 b} + \dfrac{a + 2b}{ab^2}$ [6.4a]

20. $\dfrac{5x}{x^2 - 4} - \dfrac{3}{x} + \dfrac{4}{x + 2}$ [6.5b]

Matching. Perform the indicated operation and simplify. Then select the correct answer from selections A–G listed in the second column. [6.1d], [6.2b], [6.4a], [6.5a]

21. $\dfrac{2}{x - 2} \div \dfrac{1}{x + 3}$

22. $\dfrac{1}{x + 3} - \dfrac{2}{x - 2}$

23. $\dfrac{2}{x - 2} - \dfrac{1}{x + 3}$

24. $\dfrac{1}{x + 3} \div \dfrac{2}{x - 2}$

25. $\dfrac{2}{x - 2} + \dfrac{1}{x + 3}$

26. $\dfrac{2}{x - 2} \cdot \dfrac{1}{x + 3}$

A. $\dfrac{-x - 8}{(x - 2)(x + 3)}$

B. $\dfrac{x - 2}{2(x + 3)}$

C. $\dfrac{2}{(x - 2)(x + 3)}$

D. $\dfrac{x + 8}{(x - 2)(x + 3)}$

E. $\dfrac{2(x + 3)}{x - 2}$

F. $\dfrac{3x + 4}{(x - 2)(x + 3)}$

G. $\dfrac{x + 3}{x - 2}$

Understanding Through Discussion and Writing

27. Explain why the product of two numbers is not always their least common multiple. [6.3a]

28. Is the reciprocal of a product the product of the reciprocals? Why or why not? [6.2a]

29. A student insists on finding a common denominator by always multiplying the denominators of the expressions being added. How could this approach be improved? [6.4a]

30. Explain why the expressions
$\dfrac{1}{3 - x}$ and $\dfrac{1}{x - 3}$
are opposites. [6.4a]

31. Explain why 5, −1, and 7 are *not* allowable replacements in the division
$\dfrac{x + 3}{x - 5} \div \dfrac{x - 7}{x + 1}.$ [6.1a], [6.2a, b]

32. If the LCM of a binomial and a trinomial is the trinomial, what relationship exists between the two expressions? [6.3c]

STUDYING FOR SUCCESS *Make the Most of Your Time in Class*

- ☐ Before class, try to glance at the next section in your text to be discussed, so that you can concentrate on the instruction in class.
- ☐ Get a great seat! Sitting near the front will help you both hear the instruction more clearly and avoid distractions.
- ☐ Let your instructor know in advance if you must miss class, and do your best to keep up with any work that you miss.

Complex Rational Expressions

6.6

a SIMPLIFYING COMPLEX RATIONAL EXPRESSIONS

A **complex rational expression**, or **complex fraction expression**, is a rational expression that has one or more rational expressions within its numerator or denominator. Here are some examples:

$$\frac{1+\frac{2}{x}}{3}, \quad \frac{\frac{x+y}{2}}{\frac{2x}{x+1}}, \quad \frac{\frac{1}{3}+\frac{1}{5}}{\frac{2}{x}-\frac{x}{y}}.$$

These are rational expressions within the complex rational expression.

There are two methods used to simplify complex rational expressions.

OBJECTIVE

a Simplify complex rational expressions.

SKILL REVIEW Find the LCM of algebraic expressions by factoring. [6.3c]

1. Find the LCM of 2, 4, 6, and 8.
2. Find the LCM of x, x^2, and $5x$.

Answers: **1.** 24 **2.** $5x^2$

METHOD 1: MULTIPLYING BY THE LCM OF ALL THE DENOMINATORS

To simplify a complex rational expression:

1. First, find the LCM of all the denominators of all the rational expressions occurring *within* both the numerator and the denominator of the complex rational expression.
2. Then multiply by 1 using LCM/LCM.
3. If possible, simplify by removing a factor of 1.

To the instructor and the student: Students can be instructed either to try both methods and then choose the one that works best for them or to use the method chosen by the instructor.

EXAMPLE 1 Simplify: $\dfrac{\frac{1}{2}+\frac{3}{4}}{\frac{5}{6}-\frac{3}{8}}$.

$\dfrac{\frac{1}{2}+\frac{3}{4}}{\frac{5}{6}-\frac{3}{8}}$
 $\begin{cases} \text{The denominators } within \text{ the complex rational expression} \\ \text{are 2, 4, 6, and 8. The LCM of these denominators is 24.} \\ \text{We multiply by 1 using } \frac{24}{24}. \text{ This amounts to multiplying} \\ \text{both the numerator } and \text{ the denominator by 24.} \end{cases}$

$= \dfrac{\frac{1}{2}+\frac{3}{4}}{\frac{5}{6}-\frac{3}{8}} \cdot \dfrac{24}{24}$ Multiplying by 1

SECTION 6.6 Complex Rational Expressions **467**

Using the distributive laws, we carry out the multiplications:

$$= \frac{\left(\frac{1}{2} + \frac{3}{4}\right)24}{\left(\frac{5}{6} - \frac{3}{8}\right)24} = \frac{\frac{1}{2}(24) + \frac{3}{4}(24)}{\frac{5}{6}(24) - \frac{3}{8}(24)}$$

← Multiplying the numerator by 24
← Multiplying the denominator by 24

$$= \frac{12 + 18}{20 - 9} \quad \text{Simplifying}$$

$$= \frac{30}{11}.$$

Multiplying in this manner has the effect of clearing fractions in both the numerator and the denominator of the complex rational expression.

◀ Do Exercise 1.

1. Simplify: $\dfrac{\frac{1}{3} + \frac{4}{5}}{\frac{7}{8} - \frac{5}{6}}$.

EXAMPLE 2 Simplify: $\dfrac{\frac{3}{x} + \frac{1}{2x}}{\frac{1}{3x} - \frac{3}{4x}}$.

The denominators within the complex expression are x, $2x$, $3x$, and $4x$. The LCM of these denominators is $12x$. We multiply by 1 using $12x/12x$.

$$\frac{\frac{3}{x} + \frac{1}{2x}}{\frac{1}{3x} - \frac{3}{4x}} \cdot \frac{12x}{12x} = \frac{\left(\frac{3}{x} + \frac{1}{2x}\right)12x}{\left(\frac{1}{3x} - \frac{3}{4x}\right)12x} = \frac{\frac{3}{x}(12x) + \frac{1}{2x}(12x)}{\frac{1}{3x}(12x) - \frac{3}{4x}(12x)}$$

$$= \frac{36 + 6}{4 - 9} = \frac{42}{-5} = -\frac{42}{5}$$

◀ Do Exercise 2.

2. Simplify: $\dfrac{\frac{x}{2} + \frac{2x}{3}}{\frac{1}{x} - \frac{x}{2}}$.

$\dfrac{\frac{x}{2} + \frac{2x}{3}}{\frac{1}{x} - \frac{x}{2}}$

LCM of denominators = $6x$

$$= \frac{\frac{x}{2} + \frac{2x}{3}}{\frac{1}{x} - \frac{x}{2}} \cdot \frac{6x}{\boxed{}}$$

$$= \frac{\left(\frac{x}{2} + \frac{2x}{3}\right) \cdot \boxed{}}{\left(\frac{1}{x} - \frac{x}{2}\right) \cdot 6x}$$

$$= \frac{3x^2 + \boxed{}}{\boxed{} - 3x^2} = \frac{7x^2}{3(\boxed{} - x^2)}$$

EXAMPLE 3 Simplify: $\dfrac{1 - \frac{1}{x}}{1 - \frac{1}{x^2}}$.

The denominators within the complex expression are x and x^2. The LCM of these denominators is x^2. We multiply by 1 using x^2/x^2. Then, after obtaining a single rational expression, we simplify:

$$\frac{1 - \frac{1}{x}}{1 - \frac{1}{x^2}} \cdot \frac{x^2}{x^2} = \frac{\left(1 - \frac{1}{x}\right)x^2}{\left(1 - \frac{1}{x^2}\right)x^2} = \frac{1(x^2) - \frac{1}{x}(x^2)}{1(x^2) - \frac{1}{x^2}(x^2)} = \frac{x^2 - x}{x^2 - 1}$$

$$= \frac{x(x-1)}{(x+1)(x-1)} = \frac{x}{x+1}.$$

◀ Do Exercise 3.

3. Simplify: $\dfrac{1 + \frac{1}{x}}{1 - \frac{1}{x^2}}$.

Answers
1. $\dfrac{136}{5}$ 2. $\dfrac{7x^2}{3(2 - x^2)}$ 3. $\dfrac{x}{x - 1}$

Guided Solution:
2. $6x, 6x, 4x^2, 6, 2$

METHOD 2: ADDING IN THE NUMERATOR AND THE DENOMINATOR

To simplify a complex rational expression:

1. Add or subtract, as necessary, to get a single rational expression in the numerator.
2. Add or subtract, as necessary, to get a single rational expression in the denominator.
3. Divide the numerator by the denominator.
4. If possible, simplify by removing a factor of 1.

We will redo Examples 1–3 using this method.

EXAMPLE 4 Simplify: $\dfrac{\frac{1}{2} + \frac{3}{4}}{\frac{5}{6} - \frac{3}{8}}$.

The LCM of 2 and 4 in the numerator is 4. The LCM of 6 and 8 in the denominator is 24. We have

$$\dfrac{\frac{1}{2} + \frac{3}{4}}{\frac{5}{6} - \frac{3}{8}} = \dfrac{\frac{1}{2} \cdot \frac{2}{2} + \frac{3}{4}}{\frac{5}{6} \cdot \frac{4}{4} - \frac{3}{8} \cdot \frac{3}{3}} \quad \begin{array}{l} \leftarrow \text{Multiplying } \frac{1}{2} \text{ by 1 to get the common denominator, 4} \\ \leftarrow \text{Multiplying } \frac{5}{6} \text{ and } \frac{3}{8} \text{ by 1 to get the common denominator, 24} \end{array}$$

$$= \dfrac{\frac{2}{4} + \frac{3}{4}}{\frac{20}{24} - \frac{9}{24}}$$

$$= \dfrac{\frac{5}{4}}{\frac{11}{24}} \quad \text{Adding in the numerator; subtracting in the denominator}$$

$$= \frac{5}{4} \div \frac{11}{24}$$

$$= \frac{5}{4} \cdot \frac{24}{11} \quad \text{Multiplying by the reciprocal of the divisor}$$

$$= \frac{5 \cdot 2 \cdot 2 \cdot 2 \cdot 3}{2 \cdot 2 \cdot 11} \quad \text{Factoring}$$

$$= \frac{5 \cdot \cancel{2} \cdot \cancel{2} \cdot 2 \cdot 3}{\cancel{2} \cdot \cancel{2} \cdot 11} \quad \text{Removing a factor of 1: } \frac{2 \cdot 2}{2 \cdot 2} = 1$$

$$= \frac{5 \cdot 2 \cdot 3}{11}$$

$$= \frac{30}{11}.$$

Do Exercise 4.

4. Simplify. Use method 2.

$$\dfrac{\frac{1}{3} + \frac{4}{5}}{\frac{7}{8} - \frac{5}{6}}$$

Answer

4. $\dfrac{136}{5}$

SECTION 6.6 Complex Rational Expressions

5. Simplify. Use method 2.

$$\dfrac{\dfrac{x}{2}+\dfrac{2x}{3}}{\dfrac{1}{x}-\dfrac{x}{2}}$$

$$\dfrac{\dfrac{x}{2}+\dfrac{2x}{3}}{\dfrac{1}{x}-\dfrac{x}{2}} \begin{array}{l}\leftarrow \text{LCD} = 6 \\ \leftarrow \text{LCD} = 2x\end{array}$$

$$= \dfrac{\dfrac{x}{2}\cdot\dfrac{3}{\square}+\dfrac{2x}{3}\cdot\dfrac{\square}{2}}{\dfrac{1}{x}\cdot\dfrac{\square}{2}-\dfrac{x}{2}\cdot\dfrac{x}{\square}}$$

$$= \dfrac{\dfrac{\square}{6}+\dfrac{4x}{6}}{\dfrac{2}{2x}-\dfrac{\square}{2x}} = \dfrac{\dfrac{3x+4x}{6}}{\dfrac{2-x^2}{\square}}$$

$$= \dfrac{7x}{6}\cdot\dfrac{\square}{2-x^2} = \dfrac{7\cdot 2\cdot x\cdot x}{2\cdot 3(2-x^2)}$$

$$= \dfrac{\square}{3(2-x^2)}$$

6. Simplify. Use method 2.

$$\dfrac{1+\dfrac{1}{x}}{1-\dfrac{1}{x^2}}$$

Answers

5. $\dfrac{7x^2}{3(2-x^2)}$ 6. $\dfrac{x}{x-1}$

Guided Solution:
5. $3, 2, 2, x, 3x, x^2, 6, 2x, 2x, 7x^2$

EXAMPLE 5 Simplify: $\dfrac{\dfrac{3}{x}+\dfrac{1}{2x}}{\dfrac{1}{3x}-\dfrac{3}{4x}}$.

$$\dfrac{\dfrac{3}{x}+\dfrac{1}{2x}}{\dfrac{1}{3x}-\dfrac{3}{4x}} = \dfrac{\dfrac{3}{x}\cdot\dfrac{2}{2}+\dfrac{1}{2x}}{\dfrac{1}{3x}\cdot\dfrac{4}{4}-\dfrac{3}{4x}\cdot\dfrac{3}{3}} \begin{array}{l}\leftarrow \text{Finding the LCD, } 2x, \text{and multiplying} \\ \text{by 1 in the numerator} \\ \leftarrow \text{Finding the LCD, } 12x, \text{and multiplying} \\ \text{by 1 in the denominator}\end{array}$$

$$= \dfrac{\dfrac{6}{2x}+\dfrac{1}{2x}}{\dfrac{4}{12x}-\dfrac{9}{12x}} = \dfrac{\dfrac{7}{2x}}{\dfrac{-5}{12x}} = \dfrac{7}{2x}\div\dfrac{-5}{12x} = \dfrac{7}{2x}\cdot\dfrac{12x}{-5}$$

$$= \dfrac{7\cdot 6\cdot \cancel{(2x)}}{\cancel{(2x)}(-5)} = \dfrac{42}{-5} = -\dfrac{42}{5} \quad \text{Removing a factor of 1} \quad \text{}$$

EXAMPLE 6 Simplify: $\dfrac{1-\dfrac{1}{x}}{1-\dfrac{1}{x^2}}$.

$$\dfrac{1-\dfrac{1}{x}}{1-\dfrac{1}{x^2}} = \dfrac{1\cdot\dfrac{x}{x}-\dfrac{1}{x}}{1\cdot\dfrac{x^2}{x^2}-\dfrac{1}{x^2}} \begin{array}{l}\leftarrow \text{Finding the LCD, } x, \text{and} \\ \text{multiplying by 1 in the numerator} \\ \leftarrow \text{Finding the LCD, } x^2, \text{and} \\ \text{multiplying by 1 in the denominator}\end{array}$$

$$= \dfrac{\dfrac{x-1}{x}}{\dfrac{x^2-1}{x^2}} = \dfrac{x-1}{x}\div\dfrac{x^2-1}{x^2} = \dfrac{x-1}{x}\cdot\dfrac{x^2}{x^2-1}$$

$$= \dfrac{\cancel{(x-1)}x\cdot x}{x\cancel{(x-1)}(x+1)} = \dfrac{x}{x+1} \quad \text{Removing a factor of 1}$$

◀ Do Exercises 5 and 6.

6.6 Exercise Set

FOR EXTRA HELP — MyLab Math

✓ Check Your Understanding

Concept Check For each complex rational expression, find the LCM of the denominators in the numerator, the LCM of the denominators in the denominator, and the LCM of all the denominators of all the rational expressions within both the numerator and the denominator of the complex rational expression.

CC1. $\dfrac{\dfrac{3}{5}-\dfrac{7}{10}}{\dfrac{5}{12}+\dfrac{3}{8}}$

CC2. $\dfrac{\dfrac{2}{y}+\dfrac{7}{5y}}{\dfrac{1}{10y}-\dfrac{2}{3y}}$

Reading Check Consider the expression $\dfrac{\dfrac{8}{x} - \dfrac{5}{9}}{\dfrac{2}{x}}$. Choose from the column on the right the correct word(s) to complete each statement.

RC1. The expression given above is a(n) _____ rational expression.

RC2. The expression $\dfrac{8}{x} - \dfrac{5}{9}$ is the _____ of the given expression.

RC3. The _____ of the rational expressions $\dfrac{8}{x}, \dfrac{5}{9},$ and $\dfrac{2}{x}$ is $9x$.

RC4. After subtracting in the numerator to get a single rational expression, $\dfrac{72 - 5x}{9x}$, we can simplify by multiplying the numerator by the _____ of the divisor, $\dfrac{2}{x}$.

numerator
denominator
opposite
reciprocal
complex
least common denominator

a Simplify.

1. $\dfrac{1 + \dfrac{9}{16}}{1 - \dfrac{3}{4}}$

2. $\dfrac{6 - \dfrac{3}{8}}{4 + \dfrac{5}{6}}$

3. $\dfrac{1 - \dfrac{3}{5}}{1 + \dfrac{1}{5}}$

4. $\dfrac{2 + \dfrac{2}{3}}{2 - \dfrac{2}{3}}$

5. $\dfrac{\dfrac{1}{2} + \dfrac{3}{4}}{\dfrac{5}{8} - \dfrac{5}{6}}$

6. $\dfrac{\dfrac{3}{4} + \dfrac{7}{8}}{\dfrac{2}{3} - \dfrac{5}{6}}$

7. $\dfrac{\dfrac{1}{x} + 3}{\dfrac{1}{x} - 5}$

8. $\dfrac{2 - \dfrac{1}{a}}{4 + \dfrac{1}{a}}$

9. $\dfrac{4 - \dfrac{1}{x^2}}{2 - \dfrac{1}{x}}$

10. $\dfrac{\dfrac{2}{y} + \dfrac{1}{2y}}{y + \dfrac{y}{2}}$

11. $\dfrac{8 + \dfrac{8}{d}}{1 + \dfrac{1}{d}}$

12. $\dfrac{3 + \dfrac{2}{t}}{3 - \dfrac{2}{t}}$

13. $\dfrac{\dfrac{x}{8} - \dfrac{8}{x}}{\dfrac{1}{8} + \dfrac{1}{x}}$

14. $\dfrac{\dfrac{2}{m} + \dfrac{m}{2}}{\dfrac{m}{3} - \dfrac{3}{m}}$

15. $\dfrac{1 + \dfrac{1}{y}}{1 - \dfrac{1}{y^2}}$

16. $\dfrac{\dfrac{1}{q^2} - 1}{\dfrac{1}{q} + 1}$

SECTION 6.6 Complex Rational Expressions

17. $\dfrac{\dfrac{1}{5}-\dfrac{1}{a}}{\dfrac{5-a}{5}}$

18. $\dfrac{\dfrac{4}{t}}{4+\dfrac{1}{t}}$

19. $\dfrac{\dfrac{1}{a}+\dfrac{1}{b}}{\dfrac{1}{a^2}-\dfrac{1}{b^2}}$

20. $\dfrac{\dfrac{1}{x^2}-\dfrac{1}{y^2}}{\dfrac{2}{x}-\dfrac{2}{y}}$

21. $\dfrac{\dfrac{p}{q}+\dfrac{q}{p}}{\dfrac{1}{p}+\dfrac{1}{q}}$

22. $\dfrac{x-3+\dfrac{2}{x}}{x-4+\dfrac{3}{x}}$

23. $\dfrac{\dfrac{2}{a}+\dfrac{4}{a^2}}{\dfrac{5}{a^3}-\dfrac{3}{a}}$

24. $\dfrac{\dfrac{5}{x^3}-\dfrac{1}{x^2}}{\dfrac{2}{x}+\dfrac{3}{x^2}}$

25. $\dfrac{\dfrac{2}{7a^4}-\dfrac{1}{14a}}{\dfrac{3}{5a^2}+\dfrac{2}{15a}}$

26. $\dfrac{\dfrac{5}{4x^3}-\dfrac{3}{8x}}{\dfrac{3}{2x}+\dfrac{3}{4x^3}}$

27. $\dfrac{\dfrac{a}{b}+\dfrac{c}{d}}{\dfrac{b}{a}+\dfrac{d}{c}}$

28. $\dfrac{\dfrac{a}{b}-\dfrac{c}{d}}{\dfrac{b}{a}-\dfrac{d}{c}}$

29. $\dfrac{\dfrac{x}{5y^3}+\dfrac{3}{10y}}{\dfrac{3}{10y}+\dfrac{x}{5y^3}}$

30. $\dfrac{\dfrac{a}{6b^3}+\dfrac{4}{9b^2}}{\dfrac{5}{6b}-\dfrac{1}{9b^3}}$

31. $\dfrac{\dfrac{3}{x+1}+\dfrac{1}{x}}{\dfrac{2}{x+1}+\dfrac{3}{x}}$

32. $\dfrac{x-7+\dfrac{5}{x-1}}{x-3+\dfrac{1}{x-1}}$

Skill Maintenance

Solve. [2.7e]

33. $4-\dfrac{1}{6}x \geq -12$

34. $3(b-8) > -2(3b+1)$

35. $1.5x + 19.2 < 4.2 - 3.5x$

Solve. [5.9a]

36. *Ladder Distances.* A ladder of length 13 ft is placed against a building in such a way that the distance from the top of the ladder to the ground is 7 ft more than the distance from the bottom of the ladder to the building. Find these distances.

37. *Perimeter of a Rectangle.* The length of a rectangle is 3 yd greater than the width. The area of the rectangle is 10 yd². Find the perimeter.

Synthesis

Simplify.

38. $\left[\dfrac{\dfrac{x+1}{x-1}+1}{\dfrac{x+1}{x-1}-1}\right]^5$

39. $1 + \dfrac{1}{1+\dfrac{1}{1+\dfrac{1}{1+\dfrac{1}{x}}}}$

40. $\dfrac{\dfrac{z}{1-\dfrac{z}{2+2z}}-2z}{\dfrac{2z}{5z-2}-3}$

Solving Rational Equations

6.7

OBJECTIVE

a. Solve rational equations.

a RATIONAL EQUATIONS

In Sections 6.1–6.6, we studied operations with *rational expressions*. These expressions have no equals signs. We can add, subtract, multiply, or divide and simplify expressions, but we cannot solve if there are no equals signs — as, for example, in

$$\frac{x^2 + 6x + 9}{x^2 - 4} \cdot \frac{x - 2}{x + 3}, \quad \frac{x + y}{x - y} \div \frac{x^2 + y}{x^2 - y^2}, \quad \text{and} \quad \frac{a + 3}{a^2 - 16} + \frac{5}{12 - 3a}.$$

Operation signs occur. There are no equals signs!

Most often, the result of our calculation is another rational expression that has not been cleared of fractions.

Equations *do have* equals signs, and we can clear them of fractions as we did in Section 2.3. A **rational**, or **fraction**, **equation**, is an equation containing one or more rational expressions. Here are some examples:

$$\frac{2}{3} + \frac{5}{6} = \frac{x}{9}, \quad x + \frac{6}{x} = -5, \quad \text{and} \quad \frac{x^2}{x - 1} = \frac{1}{x - 1}.$$

There are equals signs as well as operation signs.

SKILL REVIEW

Solve equations in which like terms may need to be collected. [2.3b]

Solve. Clear fractions first.

1. $4 - \frac{5}{6}y = y + \frac{7}{12}$

2. $\frac{2}{5}x + \frac{1}{3} = \frac{7}{10}x - 2$

Answers: 1. $\frac{41}{22}$ 2. $\frac{70}{9}$

SOLVING RATIONAL EQUATIONS

To solve a rational equation, the first step is to clear the equation of fractions. To do this, multiply all terms on both sides of the equation by the LCM of all the denominators. Then carry out the equation-solving process as we learned it in Chapters 2 and 5.

When clearing an equation of fractions, we use the terminology LCM instead of LCD because we are *not* adding or subtracting rational expressions.

EXAMPLE 1 Solve: $\frac{2}{3} + \frac{5}{6} = \frac{x}{9}$.

The LCM of all denominators is $2 \cdot 3 \cdot 3$, or 18. We multiply all terms on both sides by 18:

$$18\left(\frac{2}{3} + \frac{5}{6}\right) = 18 \cdot \frac{x}{9} \quad \text{Multiplying by the LCM on both sides}$$

$$18 \cdot \frac{2}{3} + 18 \cdot \frac{5}{6} = 18 \cdot \frac{x}{9} \quad \text{Multiplying each term by the LCM to remove parentheses}$$

$$12 + 15 = 2x \quad \text{Simplifying. Note that we have now cleared fractions.}$$

$$27 = 2x$$

$$\frac{27}{2} = x.$$

The check is left to the student. The solution is $\frac{27}{2}$.

Do Exercise 1.

.... Caution! ....

We are introducing a new use of the LCM in this section. We previously used the LCM in adding or subtracting rational expressions. *Now* we have equations with equals signs. We clear fractions by multiplying by the LCM on both sides of the equation. This eliminates the denominators. Do *not* make the mistake of trying to clear fractions when you do not have an equation.

1. Solve: $\frac{3}{4} + \frac{5}{8} = \frac{x}{12}$.

Answer

1. $\frac{33}{2}$

ALGEBRAIC ◆ GRAPHICAL CONNECTION

We can obtain a visual check of the solutions of a rational equation by graphing. For example, consider the equation

$$\frac{x}{4} + \frac{x}{2} = 6.$$

We can examine the solution by graphing the equations

$$y = \frac{x}{4} + \frac{x}{2} \quad \text{and} \quad y = 6$$

using the same set of axes.

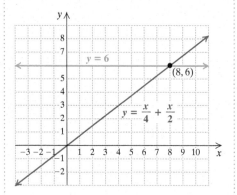

The first coordinate of the point of intersection of the graphs is the value of x for which $\frac{x}{4} + \frac{x}{2} = 6$, so it is the solution of the equation. It appears from the graph that when $x = 8$, the value of $x/4 + x/2$ is 6. We can check by substitution:

$$\frac{x}{4} + \frac{x}{2} = \frac{8}{4} + \frac{8}{2} = 2 + 4 = 6.$$

Thus the solution is 8.

EXAMPLE 2 Solve: $\dfrac{x}{6} - \dfrac{x}{8} = \dfrac{1}{12}$.

The LCM is 24. We multiply all terms on both sides by 24:

$$\frac{x}{6} - \frac{x}{8} = \frac{1}{12}$$

$$24\left(\frac{x}{6} - \frac{x}{8}\right) = 24 \cdot \frac{1}{12} \quad \text{Multiplying by the LCM on both sides}$$

$$24 \cdot \frac{x}{6} - 24 \cdot \frac{x}{8} = 24 \cdot \frac{1}{12} \quad \text{Multiplying to remove parentheses}$$

> Be sure to multiply each term by the LCM.

$$4x - 3x = 2 \quad \text{Simplifying}$$
$$x = 2.$$

Check:
$$\begin{array}{c|c}
\dfrac{x}{6} - \dfrac{x}{8} = \dfrac{1}{12} \\
\hline
\dfrac{2}{6} - \dfrac{2}{8} \;?\; \dfrac{1}{12} \\
\dfrac{1}{3} - \dfrac{1}{4} \\
\dfrac{4}{12} - \dfrac{3}{12} \\
\dfrac{1}{12} \;\bigg|\; \text{TRUE}
\end{array}$$

This checks, so the solution is 2.

EXAMPLE 3 Solve: $\dfrac{1}{x} = \dfrac{1}{4 - x}$.

The LCM is $x(4 - x)$. We multiply all terms on both sides by $x(4 - x)$:

$$\frac{1}{x} = \frac{1}{4 - x}$$

$$x(4 - x) \cdot \frac{1}{x} = x(4 - x) \cdot \frac{1}{4 - x} \quad \begin{array}{l}\text{Multiplying by the LCM}\\ \text{on both sides}\end{array}$$

$$4 - x = x \quad \text{Simplifying}$$
$$4 = 2x$$
$$x = 2.$$

Check:
$$\begin{array}{c|c}
\dfrac{1}{x} = \dfrac{1}{4 - x} \\
\hline
\dfrac{1}{2} \;?\; \dfrac{1}{4 - 2} \\
\dfrac{1}{2} \;\bigg|\; \text{TRUE}
\end{array}$$

This checks, so the solution is 2.

◀ Do Exercises 2 and 3.

Solve.

2. $\dfrac{x}{4} - \dfrac{x}{6} = \dfrac{1}{8}$

3. $\dfrac{1}{x} = \dfrac{1}{6 - x}$

Answers

2. $\dfrac{3}{2}$ 3. 3

EXAMPLE 4 Solve: $\frac{2}{3x} + \frac{1}{x} = 10$.

The LCM is $3x$. We multiply all terms on both sides by $3x$:

$$\frac{2}{3x} + \frac{1}{x} = 10$$

$$3x\left(\frac{2}{3x} + \frac{1}{x}\right) = 3x \cdot 10 \quad \text{Multiplying by the LCM on both sides}$$

$$3x \cdot \frac{2}{3x} + 3x \cdot \frac{1}{x} = 3x \cdot 10 \quad \text{Multiplying to remove parentheses}$$

$$2 + 3 = 30x \quad \text{Simplifying}$$

$$5 = 30x$$

$$\frac{5}{30} = x$$

$$\frac{1}{6} = x.$$

The check is left to the student. The solution is $\frac{1}{6}$.

Do Exercise 4.

4. Solve: $\frac{1}{2x} + \frac{1}{x} = -12$.

$$\frac{1}{2x} + \frac{1}{x} = -12$$

$$\text{LCM} = 2x$$

$$2x\left(\frac{1}{2x} + \frac{1}{x}\right) = \boxed{}(-12)$$

$$2x \cdot \frac{1}{2x} + \boxed{} \cdot \frac{1}{x} = 2x(-12)$$

$$1 + \boxed{} = \boxed{} \cdot x$$

$$\boxed{} = -24x$$

$$\frac{3}{\boxed{}} = x$$

$$-\frac{\boxed{}}{8} = x$$

EXAMPLE 5 Solve: $x + \frac{6}{x} = -5$.

The LCM is x. We multiply all terms on both sides by x:

$$x + \frac{6}{x} = -5$$

$$x\left(x + \frac{6}{x}\right) = x \cdot (-5) \quad \text{Multiplying by } x \text{ on both sides}$$

$$x \cdot x + x \cdot \frac{6}{x} = -5x \quad \text{Note that each rational expression on the left is now multiplied by } x.$$

$$x^2 + 6 = -5x \quad \text{Simplifying}$$

$$x^2 + 5x + 6 = 0 \quad \text{Adding } 5x \text{ to get 0 on one side}$$

$$(x + 3)(x + 2) = 0 \quad \text{Factoring}$$

$$x + 3 = 0 \quad \text{or} \quad x + 2 = 0 \quad \text{Using the principle of zero products}$$

$$x = -3 \quad \text{or} \quad x = -2.$$

Check: For -3:

$$\begin{array}{c|c} x + \frac{6}{x} = -5 \\ \hline -3 + \frac{6}{-3} \:?\: -5 \\ -3 - 2 \\ -5 \end{array} \quad \text{TRUE}$$

For -2:

$$\begin{array}{c|c} x + \frac{6}{x} = -5 \\ \hline -2 + \frac{6}{-2} \:?\: -5 \\ -2 - 3 \\ -5 \end{array} \quad \text{TRUE}$$

Both of these check, so there are two solutions, -3 and -2.

Do Exercise 5.

CHECKING POSSIBLE SOLUTIONS

When we multiply by the LCM on both sides of an equation, the resulting equation might have solutions that are *not* solutions of the original equation. Thus we must *always* check possible solutions in the original equation.

5. Solve: $x + \frac{1}{x} = 2$.

Answers

4. $-\frac{1}{8}$ 5. 1

Guided Solution:
4. $2x, 2x, 2, -24, 3, -24, 1$

CALCULATOR CORNER

Checking Solutions of Rational Equations Consider the equation in Example 6 and the possible solutions that were found, 1 and −1. To check these solutions, we enter $y_1 = x^2/(x - 1)$ and $y_2 = 1/(x - 1)$ on the equation-editor screen. Then, with a table set in ASK mode, we enter $x = 1$. The ERROR messages indicate that 1 is not a solution. Next, we enter $x = -1$. Since y_1 and y_2 have the same value, we know that the equation is true, so −1 is a solution.

$y_1 = x^2/(x - 1)$,
$y_2 = 1/(x - 1)$

X	Y1	Y2
1	ERROR	ERROR
−1	−.5	−.5

EXERCISES: Use a graphing calculator to check the possible solutions.
1. Examples 1, 3, 5, and 7
2. Margin Exercises 1, 3, 6, and 7

Example 6 illustrates the importance of checking all possible solutions.

EXAMPLE 6 Solve: $\dfrac{x^2}{x - 1} = \dfrac{1}{x - 1}$.

The LCM is $x - 1$. We multiply all terms on both sides by $x - 1$:

$$\dfrac{x^2}{x - 1} = \dfrac{1}{x - 1}$$

$(x - 1) \cdot \dfrac{x^2}{x - 1} = (x - 1) \cdot \dfrac{1}{x - 1}$ Multiplying by $x - 1$ on both sides

$x^2 = 1$ Simplifying

$x^2 - 1 = 0$ Subtracting 1 to get 0 on one side

$(x - 1)(x + 1) = 0$ Factoring

$x - 1 = 0$ or $x + 1 = 0$ Using the principle of zero products

$x = 1$ or $x = -1$.

The numbers 1 and −1 are possible solutions.

Check: For 1:

$$\dfrac{x^2}{x - 1} = \dfrac{1}{x - 1}$$

$\dfrac{1^2}{1 - 1} \,?\, \dfrac{1}{1 - 1}$

$\dfrac{1}{0} \,\bigg|\, \dfrac{1}{0}$ NOT DEFINED

For −1:

$$\dfrac{x^2}{x - 1} = \dfrac{1}{x - 1}$$

$\dfrac{(-1)^2}{(-1) - 1} \,?\, \dfrac{1}{(-1) - 1}$

$-\dfrac{1}{2} \,\bigg|\, -\dfrac{1}{2}$ TRUE

We look at the original equation and see that 1 makes a denominator 0 and is thus not a solution. The number −1 checks and is a solution.

Solve.

6. $\dfrac{x^2}{x + 2} = \dfrac{4}{x + 2}$

7. $\dfrac{4}{x - 2} + \dfrac{1}{x + 2} = \dfrac{26}{x^2 - 4}$

$\text{LCM} = (x - 2)(x + 2)$

$(x - 2)(x + \boxed{})\left(\dfrac{4}{x - 2} + \dfrac{1}{x + 2}\right)$

$= (x - 2)(x + 2) \cdot \dfrac{26}{x^2 - 4}$

$\boxed{}(x + 2) + \boxed{}(x - 2) = 26$

$\boxed{} + 8 + x - \boxed{} = 26$

$\boxed{} + 6 = 26$

$5x = \boxed{}$

$x = \boxed{}$

EXAMPLE 7 Solve: $\dfrac{3}{x - 5} + \dfrac{1}{x + 5} = \dfrac{2}{x^2 - 25}$.

The LCM is $(x - 5)(x + 5)$. We multiply all terms on both sides by $(x - 5)(x + 5)$:

$(x - 5)(x + 5)\left(\dfrac{3}{x - 5} + \dfrac{1}{x + 5}\right) = (x - 5)(x + 5)\left(\dfrac{2}{x^2 - 25}\right)$

Multiplying by the LCM on both sides

$(x - 5)(x + 5) \cdot \dfrac{3}{x - 5} + (x - 5)(x + 5) \cdot \dfrac{1}{x + 5}$

$= (x - 5)(x + 5) \cdot \dfrac{2}{x^2 - 25}$

$3(x + 5) + (x - 5) = 2$ Simplifying

$3x + 15 + x - 5 = 2$ Removing parentheses

$4x + 10 = 2$

$4x = -8$

$x = -2$.

The check is left to the student. The number −2 checks and is the solution.

◀ Do Exercises 6 and 7.

Answers
6. 2   7. 4
Guided Solution:
7. 2, 4, 1, 4x, 2, 5x, 20, 4

6.7 Exercise Set

✓ Check Your Understanding

Reading Check One of the common difficulties with this chapter is being sure about the task at hand. Are you combining expressions using operations to get another *rational expression*, or are you solving equations for which the results are numbers that are *solutions* of an equation? To learn to make these decisions, determine for each of the following exercises the type of answer you should get: "Rational expression" or "Solutions." You need not complete the mathematical operations.

RC1. Add: $\dfrac{5a}{a^2 - 1} + \dfrac{a}{a^2 - a}$.

RC2. Solve: $\dfrac{5}{y - 3} - \dfrac{30}{y^2 - 9} = 1$.

RC3. Subtract: $\dfrac{4}{x - 2} - \dfrac{1}{x + 2}$.

RC4. Divide: $\dfrac{x + 4}{x - 2} \div \dfrac{6x}{x^2 - 4}$.

RC5. Solve: $\dfrac{x^2}{x - 1} = \dfrac{1}{x - 1}$.

RC6. Solve: $\dfrac{10}{x} + x = -2$.

RC7. Multiply: $\dfrac{2t^2}{t^2 - 25} \cdot \dfrac{t^2 + 10t + 25}{t^8}$.

RC8. Solve: $\dfrac{7}{x - 4} - \dfrac{2}{x + 4} = \dfrac{1}{x^2 - 16}$.

Concept Check Determine whether each statement is true or false.

CC1. When we multiply by the LCM on both sides of a rational equation, the resulting equation might have solutions that are not solutions of the original equation.

CC2. A rational equation is an equation containing one or more rational expressions.

a Solve. Don't forget to check!

1. $\dfrac{4}{5} - \dfrac{2}{3} = \dfrac{x}{9}$

2. $\dfrac{x}{20} = \dfrac{3}{8} - \dfrac{4}{5}$

3. $\dfrac{3}{5} + \dfrac{1}{8} = \dfrac{1}{x}$

4. $\dfrac{2}{3} + \dfrac{5}{6} = \dfrac{1}{x}$

5. $\dfrac{3}{8} + \dfrac{4}{5} = \dfrac{x}{20}$

6. $\dfrac{3}{5} + \dfrac{2}{3} = \dfrac{x}{9}$

7. $\dfrac{1}{x} = \dfrac{2}{3} - \dfrac{5}{6}$

8. $\dfrac{1}{x} = \dfrac{1}{8} - \dfrac{3}{5}$

9. $\dfrac{1}{6} + \dfrac{1}{8} = \dfrac{1}{t}$

SECTION 6.7 Solving Rational Equations

10. $\dfrac{1}{8} + \dfrac{1}{12} = \dfrac{1}{t}$

11. $x + \dfrac{4}{x} = -5$

12. $\dfrac{10}{x} - x = 3$

13. $\dfrac{x}{4} - \dfrac{4}{x} = 0$

14. $\dfrac{x}{5} - \dfrac{5}{x} = 0$

15. $\dfrac{5}{x} = \dfrac{6}{x} - \dfrac{1}{3}$

16. $\dfrac{4}{x} = \dfrac{5}{x} - \dfrac{1}{2}$

17. $\dfrac{5}{3x} + \dfrac{3}{x} = 1$

18. $\dfrac{5}{2y} + \dfrac{8}{y} = 1$

19. $\dfrac{t-2}{t+3} = \dfrac{3}{8}$

20. $\dfrac{x-7}{x+2} = \dfrac{1}{4}$

21. $\dfrac{2}{x+1} = \dfrac{1}{x-2}$

22. $\dfrac{8}{y-3} = \dfrac{6}{y+4}$

23. $\dfrac{x}{6} - \dfrac{x}{10} = \dfrac{1}{6}$

24. $\dfrac{x}{8} - \dfrac{x}{12} = \dfrac{1}{8}$

25. $\dfrac{t+2}{5} - \dfrac{t-2}{4} = 1$

26. $\dfrac{x+1}{3} - \dfrac{x-1}{2} = 1$

27. $\dfrac{5}{x-1} = \dfrac{3}{x+2}$

28. $\dfrac{x-7}{x-9} = \dfrac{2}{x-9}$

29. $\dfrac{a-3}{3a+2} = \dfrac{1}{5}$

30. $\dfrac{x+7}{8x-5} = \dfrac{2}{3}$

31. $\dfrac{x-1}{x-5} = \dfrac{4}{x-5}$

32. $\dfrac{y+11}{y+8} = \dfrac{3}{y+8}$

33. $\dfrac{2}{x+3} = \dfrac{5}{x}$

34. $\dfrac{6}{y} = \dfrac{5}{y-8}$

35. $\dfrac{x-2}{x-3} = \dfrac{x-1}{x+1}$

36. $\dfrac{t+5}{t-2} = \dfrac{t-2}{t+4}$

37. $\dfrac{1}{x+3} + \dfrac{1}{x-3} = \dfrac{1}{x^2-9}$

38. $\dfrac{4}{x-3} + \dfrac{2x}{x^2-9} = \dfrac{1}{x+3}$

39. $\dfrac{x}{x+4} - \dfrac{4}{x-4} = \dfrac{x^2+16}{x^2-16}$

40. $\dfrac{5}{y-3} - \dfrac{30}{y^2-9} = 1$

41. $\dfrac{4-a}{8-a} = \dfrac{4}{a-8}$

42. $\dfrac{3}{x-7} = \dfrac{x+10}{x-7}$

43. $2 - \dfrac{a-2}{a+3} = \dfrac{a^2-4}{a+3}$

44. $\dfrac{5}{x-1} + x + 1 = \dfrac{5x+4}{x-1}$

45. $\dfrac{x+1}{x+2} = \dfrac{x+3}{x+4}$

46. $\dfrac{x^2}{x^2-4} = \dfrac{x}{x+2} - \dfrac{2x}{2-x}$

47. $4a - 3 = \dfrac{a+13}{a+1}$

48. $\dfrac{3x-9}{x-3} = \dfrac{5x-4}{2}$

49. $\dfrac{4}{y-2} - \dfrac{2y-3}{y^2-4} = \dfrac{5}{y+2}$

50. $\dfrac{y^2-4}{y+3} = 2 - \dfrac{y-2}{y+3}$

Skill Maintenance

Add. [4.4a]

51. $(2x^3 - 4x^2 + x - 7) + (4x^4 + x^3 + 4x^2 + x)$

52. $(2x^3 - 4x^2 + x - 7) + (-2x^3 + 4x^2 - x + 7)$

Factor. [5.7a]

53. $50p^2 - 100$

54. $5p^2 - 40p - 100$

Solve.

55. *Consecutive Even Integers.* The product of two consecutive even integers is 360. Find the integers. [5.9a]

56. *Chemistry.* About 5 L of oxygen can be dissolved in 100 L of water at 0°C. This is 1.6 times the amount that can be dissolved in the same volume of water at 20°C. How much oxygen can be dissolved in 100 L at 20°C? [2.6a]

Synthesis

57. Solve: $\dfrac{x}{x^2+3x-4} + \dfrac{x+1}{x^2+6x+8} = \dfrac{2x}{x^2+x-2}$.

58. Use a graphing calculator to check the solutions to Exercises 13, 15, and 25.

Applications Using Rational Equations and Proportions

6.8

OBJECTIVES

a Solve applied problems using rational equations.

b Solve proportion problems.

In many areas of study, applications involving rates, proportions, or reciprocals translate to rational equations. By using the five steps for problem solving and the skills of Sections 6.1–6.7, we can now solve such problems.

a SOLVING APPLIED PROBLEMS

SKILL REVIEW

Solve a formula for a specified letter. [2.4b]
Solve for the indicated letter.
1. $x = w \cdot y$, for w
2. $A = c - bt$, for t

Answers: 1. $w = \dfrac{x}{y}$ 2. $t = \dfrac{c - A}{b}$, or $\dfrac{A - c}{-b}$

MyLab Math
ANIMATION

Problems Involving Motion

Problems that deal with distance, speed (or rate), and time are called **motion problems**. Translation of these problems involves the distance formula, $d = r \cdot t$, and/or the equivalent formulas $r = d/t$ and $t = d/r$.

MOTION FORMULAS

$d = rt$; Distance = Rate · Time (basic formula)

$r = \dfrac{d}{t}$; Rate = Distance/Time

$t = \dfrac{d}{r}$ Time = Distance/Rate

EXAMPLE 1 *Speed of Sea Animals.* The shortfin Mako shark is known to have the fastest speed of all sharks. The sailfish has the fastest speed of all fish. The top speed recorded for a sailfish is approximately 25 mph faster than the fastest speed of a Mako shark. A sailfish can swim 14 mi in the same time that a Mako shark can swim 9 mi. Find the speed of each sea animal.
Data: International Union for the Conservation of Nature; theshark.dk/en/records.php; thetravelalmanac.com

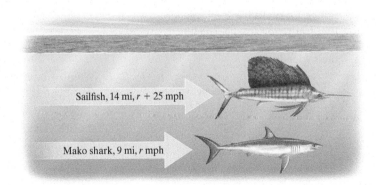

Sailfish, 14 mi, $r + 25$ mph

Mako shark, 9 mi, r mph

1. **Familiarize.** We first make a drawing. We let $r =$ the speed of the shark. Then $r + 25 =$ the speed of the sailfish.

Recall that sometimes we need to use a formula in order to solve an application. As we see above, a formula that relates the notions of distance, speed, and time is $d = rt$, or *Distance = Speed · Time*.

Since each sea animal travels for the same length of time, we can use t for the time of each sea animal. We organize the information in a chart, as follows.

	DISTANCE	SPEED	TIME	
MAKO SHARK	9	r	t	$\rightarrow 9 = rt$
SAILFISH	14	$r + 25$	t	$\rightarrow 14 = (r + 25)t$

2. **Translate.** We can apply the formula $d = rt$ along the rows of the table to obtain two equations:

$$9 = rt \quad \text{and} \quad 14 = (r + 25)t.$$

We know that the sea animals travel for the same length of time. Thus if we solve each equation for t and set the results equal to each other, we get an equation in terms of r.

Solving $9 = rt$ for t: $\quad t = \dfrac{9}{r}$

Solving $14 = (r + 25)t$ for t: $\quad t = \dfrac{14}{r + 25}$

Since the times are the same, we have the following equation:

$$\frac{9}{r} = \frac{14}{r + 25}.$$

3. **Solve.** To solve the equation, we first multiply on both sides by the LCM, which is $r(r + 25)$:

$r(r + 25) \cdot \dfrac{9}{r} = r(r + 25) \cdot \dfrac{14}{r + 25}$ Multiplying on both sides by the LCM, which is $r(r + 25)$

$9(r + 25) = 14r$ Simplifying

$9r + 225 = 14r$ Removing parentheses

$225 = 5r$

$45 = r.$

We now have a possible solution. The speed of the shark is 45 mph, and the speed of the sailfish is $r + 25 = 45 + 25$, or 70 mph.

4. **Check.** We check the speeds of 45 mph for the shark and 70 mph for the sailfish. The sailfish does swim 25 mph faster than the shark. If the sailfish swims 14 mi at 70 mph, the time that it has traveled is $\frac{14}{70}$, or $\frac{1}{5}$ hr. If the shark swims 9 mi at 45 mph, the time that it has traveled is $\frac{9}{45}$, or $\frac{1}{5}$ hr. Since the times are the same, the speeds check.

5. **State.** The speed of the Mako shark is 45 mph, and the speed of the sailfish is 70 mph.

◀ Do Exercise 1.

1. *Driving Speed.* Catherine drives 20 mph faster than her father, Gary. In the same time that Catherine travels 180 mi, her father travels 120 mi. Find their speeds.

Catherine's car
180 mi, $r + 20$ mph

Gary's car
120 mi, r mph

Answer
1. Gary: 40 mph; Catherine: 60 mph

Problems Involving Work

EXAMPLE 2 *Sodding a Yard.* Charlie's Lawn Care has two three-person crews who lay sod. Crew A can lay 7 skids of sod in 4 hr, while crew B requires 6 hr to do the same job. How long would it take the two crews working together to lay 7 skids of sod?

1. **Familiarize.** A common *incorrect* way to translate the problem is to add the two times: $4\,\text{hr} + 6\,\text{hr} = 10\,\text{hr}$. Let's think about this. Crew A can do the job in 4 hr. If crew A and crew B work together, the time that it takes them should be *less* than 4 hr. Thus we reject 10 hr as a solution, but we do have a partial check on any answer we get. The answer should be less than 4 hr.

    We proceed to a translation by considering how much of the job is finished in 1 hr, 2 hr, 3 hr, and so on. It takes crew A 4 hr to do the sodding job alone. Then, in 1 hr, crew A can do $\frac{1}{4}$ of the job. It takes crew B 6 hr to do the job alone. Then, in 1 hr, crew B can do $\frac{1}{6}$ of the job. Working together (see Fig. 1), the crews can do

    $$\frac{1}{4} + \frac{1}{6}, \text{ or } \frac{3}{12} + \frac{2}{12}, \text{ or } \frac{5}{12} \text{ of the job in 1 hr.}$$

    In 2 hr, crew A can do $2(\frac{1}{4})$ of the job and crew B can do $2(\frac{1}{6})$ of the job. Working together (see Fig. 2), they can do

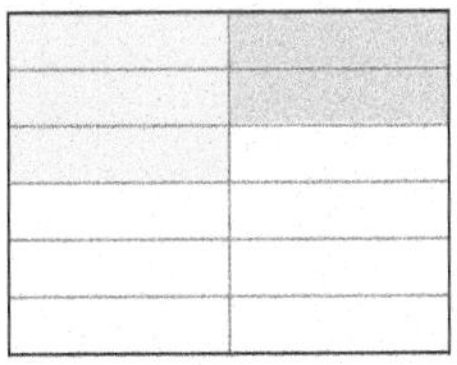

    $$2\left(\frac{1}{4}\right) + 2\left(\frac{1}{6}\right), \text{ or } \frac{6}{12} + \frac{4}{12}, \text{ or } \frac{10}{12}, \text{ or } \frac{5}{6} \text{ of the job in 2 hr.}$$

In 1 hour:
Crew A   Crew B

FIGURE 1

In 2 hours:
Crew A   Crew B

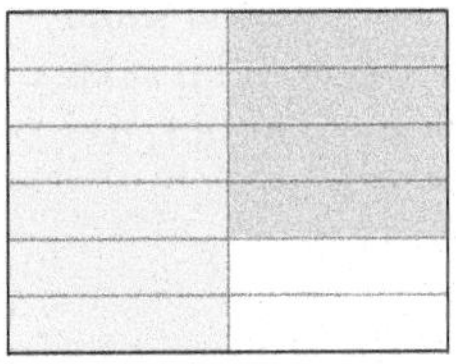

FIGURE 2

TIME	FRACTION OF THE JOB COMPLETED		
	CREW A	CREW B	TOGETHER
1 hr	$\frac{1}{4}$	$\frac{1}{6}$	$\frac{1}{4} + \frac{1}{6}$, or $\frac{5}{12}$
2 hr	$2\left(\frac{1}{4}\right)$	$2\left(\frac{1}{6}\right)$	$2\left(\frac{1}{4}\right) + 2\left(\frac{1}{6}\right)$, or $\frac{5}{6}$
3 hr	$3\left(\frac{1}{4}\right)$	$3\left(\frac{1}{6}\right)$	$3\left(\frac{1}{4}\right) + 3\left(\frac{1}{6}\right)$, or $1\frac{1}{4}$
t hr	$t\left(\frac{1}{4}\right)$	$t\left(\frac{1}{6}\right)$	$t\left(\frac{1}{4}\right) + t\left(\frac{1}{6}\right)$

We see that the answer is somewhere between 2 hr and 3 hr. What we want is a number t such that the fraction of the job that is completed is 1; that is, the job is just completed.

2. **Translate.** From the table, we see that the time we want is some number t for which

    $$t\left(\frac{1}{4}\right) + t\left(\frac{1}{6}\right) = 1, \text{ or } \frac{t}{4} + \frac{t}{6} = 1,$$

    where 1 represents the idea that the entire job is completed in time t.

3. Solve. We solve the equation:

$$12\left(\frac{t}{4} + \frac{t}{6}\right) = 12 \cdot 1 \quad \text{Multiplying by the LCM, which is } 2\cdot 2 \cdot 3, \text{ or } 12$$

$$12 \cdot \frac{t}{4} + 12 \cdot \frac{t}{6} = 12$$

$$3t + 2t = 12$$

$$5t = 12$$

$$t = \frac{12}{5}, \text{ or } 2\frac{2}{5} \text{ hr.}$$

4. Check. In $\frac{12}{5}$ hr, crew A does $\frac{12}{5} \cdot \frac{1}{4}$, or $\frac{3}{5}$, of the job and crew B does $\frac{12}{5} \cdot \frac{1}{6}$, or $\frac{2}{5}$, of the job. Together, they do $\frac{3}{5} + \frac{2}{5}$, or 1 entire job. The answer, $2\frac{2}{5}$ hr, is between 2 hr and 3 hr (see the table), and it is less than 4 hr, the time it takes crew A working alone. The answer checks.

5. State. It takes $2\frac{2}{5}$ hr for crew A and crew B working together to lay 7 skids of sod.

THE WORK PRINCIPLE

Suppose a = the time that it takes A to do a job, b = the time that it takes B to do the same job, and t = the time that it takes them to do the job working together. Then

$$\frac{t}{a} + \frac{t}{b} = 1.$$

◀ Do Exercise 2.

b APPLICATIONS INVOLVING PROPORTIONS

We now consider applications with proportions. A **proportion** involves ratios. A **ratio** of two quantities is their quotient. For example, 73% is the ratio of 73 to 100, $\frac{73}{100}$. The ratio of two different kinds of measure is called a **rate**. Suppose that an animal travels 2720 ft in 2.5 hr. Its **rate**, or **speed**, is then

$$\frac{2720 \text{ ft}}{2.5 \text{ hr}} = 1088 \frac{\text{ft}}{\text{hr}}.$$

◀ Do Exercises 3–6.

PROPORTION

An equality of ratios,

$$\frac{A}{B} = \frac{C}{D},$$

is called a **proportion**. The numbers within a proportion are said to be **proportional** to each other.

---

2. *Work Recycling.* Emma and Evan work as volunteers at a community recycling center. Emma can sort a morning's accumulation of recyclable objects in 3 hr, while Evan requires 5 hr to do the same job. How long would it take them, working together, to sort the recyclable material?

3. Find the ratio of 145 km to 2.5 liters (L).

4. *Batting Average.* Recently, a baseball player got 7 hits in 25 times at bat. What was the rate, or batting average, in number of hits per times at bat?

5. Impulses in nerve fibers travel 310 km in 2.5 hr. What is the rate, or speed, in kilometers per hour?

6. A lake of area 550 yd² contains 1320 fish. What is the population density of the lake, in number of fish per square yard?

Answers

2. $1\frac{7}{8}$ hr   **3.** 58 km/L
4. 0.28 hit per times at bat   **5.** 124 km/h
6. 2.4 fish/yd²

EXAMPLE 3 *Mileage.* A 2017 Jeep Compass Trailhawk can travel 330 mi of highway driving on 11 gal of gas. Find the amount of gas required for 495 mi of highway driving.

Data: *Motor Trend,* June 2017

1. **Familiarize.** We know that the Jeep can travel 330 mi on 11 gal of gas. Thus we can set up a proportion, letting x = the number of gallons of gas required to drive 495 mi.

2. **Translate.** We assume that the car uses gas at the same rate in all highway driving. Thus the ratios are the same and we can write a proportion. Note that the units of *mileage* are in the numerators and the units of *gasoline* are in the denominators.

$$\text{Miles} \rightarrow \frac{330}{11} = \frac{495}{x} \leftarrow \text{Miles}$$
$$\text{Gas} \rightarrow \quad\quad\quad\quad \leftarrow \text{Gas}$$

3. **Solve.** To solve for x, we multiply on both sides by the LCM, which is $11x$:

$$11x \cdot \frac{330}{11} = 11x \cdot \frac{495}{x} \quad \text{Multiplying by } 11x$$
$$330x = 5445 \quad \text{Simplifying}$$
$$\frac{330x}{330} = \frac{5445}{330} \quad \text{Dividing by 330}$$
$$x = 16.5. \quad \text{Simplifying}$$

We can also use cross products to solve the proportion:

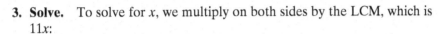

$$\frac{330}{11} = \frac{495}{x} \quad\quad 330 \cdot x \text{ and } 11 \cdot 495 \text{ are cross products.}$$

$$330 \cdot x = 11 \cdot 495 \quad \text{Equating cross products}$$
$$\frac{330x}{330} = \frac{5445}{330} \quad \text{Dividing by 330}$$
$$x = 16.5.$$

4. **Check.** The check is left to the student.
5. **State.** The Jeep will require 16.5 gal of gas for 495 mi of highway driving.

Do Exercise 7. ▶

EXAMPLE 4 *Fruit Quality.* A company that prepares and sells gift boxes and baskets of fruit must order quantities of fruit greater than what they need to allow for selecting fruit that meets their quality standards. The packing-room supervisor keeps records and notes that approximately 87 pears from a shipment of 1000 do not meet the company standards. Over the holidays, a shipment of 3200 pears is ordered. How many pears can the company expect will not meet the quality required?

7. *Mileage.* In city driving, a 2017 Volkswagen Golf Wolfsburg Edition can travel 225 mi on 9 gal of gas. How much gas will be required for 335 mi of city driving?

Data: *Car and Driver,* April 2017

Answer
7. 13.4 gal

1. **Familiarize.** The ratio of the number of pears P that do not meet the standards to the total order of 3200 is $P/3200$. The ratio of the average number of pears that do not meet the standard in an order of 1000 pears is $\frac{87}{1000}$.

2. **Translate.** Assuming that the two ratios are the same, we can translate to a proportion:
$$\frac{P}{3200} = \frac{87}{1000}.$$

3. **Solve.** We solve the proportion. We multiply by the LCM, which is 16,000.
$$16{,}000 \cdot \frac{P}{3200} = 16{,}000 \cdot \frac{87}{1000}$$
$$5 \cdot P = 16 \cdot 87$$
$$P = \frac{16 \cdot 87}{5}$$
$$P = 278.4, \text{ so } P \approx 278.$$

4. **Check.** The check is left to the student.

5. **State.** We estimate that in an order of 3200 pears, there will be about 278 pears that do not meet the quality standards.

◀ Do Exercise 8.

8. *Chlorine for a Pool.* XYZ Pools and Spas, Inc., adds 2 gal of chlorine per 8000 gal of water in a newly constructed pool. How much chlorine is needed for a pool requiring 20,500 gal of water? Round the answer to the nearest tenth of a gallon.

Similar Triangles

Proportions arise in geometry when we are studying *similar triangles*. If two triangles are **similar**, then their corresponding angles have the same measure and their corresponding sides are proportional. To illustrate, if triangle ABC is similar to triangle RST, then angles A and R have the same measure, angles B and S have the same measure, angles C and T have the same measure, and
$$\frac{a}{r} = \frac{b}{s} = \frac{c}{t}.$$

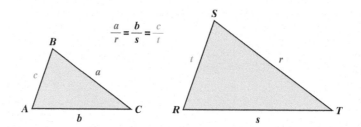

Answer
8. 5.1 gal

SIMILAR TRIANGLES

In **similar triangles**, corresponding angles have the same measure and the lengths of corresponding sides are proportional.

EXAMPLE 5 *Similar Triangles.* Triangles ABC and XYZ below are similar triangles. Solve for z if $a = 8$, $c = 5$, and $x = 10$.

We make a drawing, write a proportion, and then solve. Note that side a is always opposite angle A, side x is always opposite angle X, and so on.

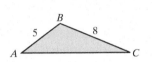 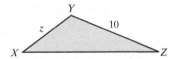

We have

$$\frac{z}{5} = \frac{10}{8}$$ The proportion $\frac{5}{z} = \frac{8}{10}$ could also be used.

$$40 \cdot \frac{z}{5} = 40 \cdot \frac{10}{8}$$ Multiplying by 40

$$8z = 50$$

$$z = \frac{50}{8}$$ Dividing by 8

$$z = \frac{25}{4}, \text{ or } 6.25.$$

Do Exercise 9. ▶

EXAMPLE 6 *Rafters of a House.* Carpenters use similar triangles to determine the lengths of rafters for a house. They first choose the pitch of the roof, or the ratio of the rise over the run. Then using a triangle with that ratio, they calculate the length of the rafter needed for the house. Loren is constructing rafters for a roof with a 6/12 pitch on a house that is 30 ft wide. Using a rafter guide (see the figure at right), Loren knows that the rafter length corresponding to a 6-unit rise and a 12-unit run is 13.4. Find the length x of the rafter of the house.

We have the proportion

Length of rafter in 6/12 triangle → $\dfrac{13.4}{x}$ = $\dfrac{12}{15}$ ← Run in 6/12 triangle
Length of rafter → ← Run in similar
on the house   triangle on the house

Solve: $13.4 \cdot 15 = x \cdot 12$ Equating cross products

$$\frac{13.4 \cdot 15}{12} = \frac{x \cdot 12}{12}$$ Dividing by 12 on both sides

$$\frac{13.4 \cdot 15}{12} = x$$

$$16.75 = x$$

The length of the rafter x of the house is about 16.75 ft, or 16 ft 9 in.

Do Exercise 10. ▶

9. Height of a Flagpole. How high is a flagpole that casts a 45-ft shadow at the same time that a 5.5-ft woman casts a 10-ft shadow?

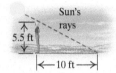

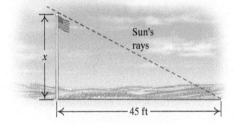

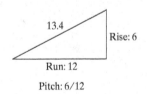

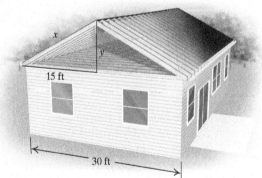

10. Rafters of a House. Refer to Example 6. Find the length y in the house.

Answers
9. 24.75 ft **10.** 7.5 ft

SECTION 6.8 Applications Using Rational Equations and Proportions **487**

Translating for Success

The goal of these matching questions is to practice step (2), Translate, of the five-step problem-solving process. Translate each word problem to an equation and select a correct translation from equations A–O.

1. *Pharmaceutical Marketing.* In 2013, a pharmaceutical firm spent $3.6 million on marketing a new drug. This was a 25% increase over the amount spent for marketing in 2012. How much was spent in 2012?

2. *Cycling Distance.* A bicyclist traveled 197 mi in 7 days. At this rate, how many miles could the cyclist travel in 30 days?

3. *Bicycling.* The speed of one bicyclist is 2 km/h faster than the speed of another bicyclist. The first bicyclist travels 60 km in the same amount of time that it takes the second to travel 50 km. Find the speed of each bicyclist.

4. *Filling Time.* A swimming pool can be filled in 5 hr by hose A alone and in 6 hr by hose B alone. How long would it take to fill the tank if both hoses were working?

5. *Office Budget.* Emma has $36 budgeted for office stationery. Engraved stationery costs $20 for the first 25 sheets and $0.08 for each additional sheet. How many engraved sheets of stationery can Emma order and still stay within her budget?

6. *Sides of a Square.* If each side of a square is increased by 2 ft, the area of the original square plus the area of the enlarged square is 452 ft². Find the length of a side of the original square.

7. *Consecutive Integers.* The sum of two consecutive integers is 613. Find the integers.

8. *Sums of Squares.* The sum of the squares of two consecutive odd integers is 612. Find the integers.

9. *Sums of Squares.* The sum of the squares of two consecutive integers is 613. Find the integers.

10. *Rectangle Dimensions.* The length of a rectangle is 1 ft longer than its width. Find the dimensions of the rectangle such that the perimeter of the rectangle is 613 ft.

A. $2x + 2(x + 1) = 613$

B. $x^2 + (x + 1)^2 = 613$

C. $\dfrac{60}{x + 2} = \dfrac{50}{x}$

D. $20 + 0.08(x - 25) = 36$

E. $\dfrac{197}{7} = \dfrac{x}{30}$

F. $x + (x + 1) = 613$

G. $\dfrac{7}{197} = \dfrac{x}{30}$

H. $x^2 + (x + 2)^2 = 612$

I. $x^2 + (x + 1)^2 = 612$

J. $\dfrac{50}{x + 2} = \dfrac{60}{x}$

K. $x + 25\% \cdot x = 3.6$

L. $t + 5 = 7$

M. $x^2 + (x + 1)^2 = 452$

N. $\dfrac{t}{5} + \dfrac{t}{6} = 1$

O. $x^2 + (x + 2)^2 = 452$

Answers on page A-20

6.8 Exercise Set

FOR EXTRA HELP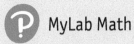

✓ Check Your Understanding

Reading and Concept Check Choose from the column on the right the appropriate word(s) to complete each statement.

RC1. If two triangles are similar, then their _____ angles have the _____ measures and their corresponding sides are _____.

RC2. A ratio of two quantities is their _____.

RC3. An equality of ratios, $\dfrac{A}{B} = \dfrac{C}{D}$, is called a(n) _____.

RC4. Distance equals _____ times time.

RC5. Rate equals _____ divided by time.

RC6. To solve the equation $\dfrac{18}{5} = \dfrac{30}{x}$, we can use the _____, $18x$ and $5 \cdot 30$.

same
product
distance
cross products
proportion
similar
different
quotient
rate
proportional
corresponding

a Solve.

1. *Car Speed.* Rick drives his four-wheel-drive truck 40 km/h faster than Sarah drives her Kia. While Sarah travels 150 km, Rick travels 350 km. Find their speeds.

Complete this table as part of the *Familiarize* step.

	d	=	r	·	t
	DISTANCE		SPEED		TIME
Car	150		r		
Truck	350				t

2. *Train Speed.* The speed of a CSW freight train is 14 mph slower than the speed of an Amtrak passenger train. The freight train travels 330 mi in the same time that it takes the passenger train to travel 400 mi. Find the speed of each train.

Complete this table as part of the *Familiarize* step.

	d	=	r	·	t
	DISTANCE		SPEED		TIME
CSW	330				t
Amtrak	400		r		

3. *Animal Speeds.* An ostrich can run 8 mph faster than a giraffe. An ostrich can run 5 mi in the same time that a giraffe can run 4 mi. Find the speed of each animal.

Data: infoplease.com

4. *Animal Speeds.* A cheetah can run 28 mph faster than a gray fox. A cheetah can run 10 mi in the same time that a gray fox can run 6 mi. Find the speed of each animal.

Data: infoplease.com

SECTION 6.8 Applications Using Rational Equations and Proportions   489

5. *Bicycle Speed.* Hank bicycles 5 km/h slower than Kelly. In the time that it takes Hank to bicycle 42 km, Kelly can bicycle 57 km. How fast does each bicyclist travel?

6. *Driving Speed.* Kaylee's Lexus travels 30 mph faster than Gavin's Harley. In the same time that Gavin travels 75 mi, Kaylee travels 120 mi. Find their speeds.

7. *Trucking Speed.* A long-distance trucker traveled 120 mi in one direction during a snowstorm. The return trip in rainy weather was accomplished at double the speed and took 3 hr less time. Find the speed going.

8. *Car Speed.* After driving 126 mi, Syd found that the drive would have taken 1 hr less time by increasing the speed by 8 mph. What was the actual speed?

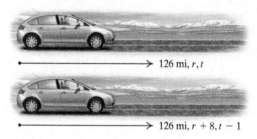

9. *Walking Speed.* Bonnie power walks 3 km/h faster than Ralph. In the time that it takes Ralph to walk 7.5 km, Bonnie walks 12 km. Find their speeds.

10. *Cross-Country Skiing.* Gerard skis cross-country 4 km/h faster than Sally. In the time that it takes Sally to ski 18 km, Gerard skis 24 km. Find their speeds.

11. *Boat Speed.* Tory and Emilio's motorboats travel at the same speed. Tory pilots her boat 40 km before docking. Emilio continues for another 2 hr, traveling a total of 100 km before docking. How long did it take Tory to navigate the 40 km?

12. *Tractor Speed.* Hobart's tractor is just as fast as Evan's. It takes Hobart 1 hr more than it takes Evan to drive to town. If Hobart is 20 mi from town and Evan is 15 mi from town, how long does it take Evan to drive to town?

13. *Gardening.* Nicole can weed her vegetable garden in 50 min. Glen can weed the same garden in 40 min. How long would it take if they worked together?

14. *Harvesting.* Bobbi can pick a quart of raspberries in 20 min. Blanche can pick a quart in 25 min. How long would it take if Bobbi and Blanche worked together?

15. *Shoveling.* Vern can shovel the snow from his driveway in 45 min. Nina can do the same job in 60 min. How long would it take Nina and Vern to shovel the driveway if they worked together?

16. *Raking.* Zoë can rake her yard in 4 hr. Steffi does the same job in 3 hr. How long would it take them, working together, to rake the yard?

17. *Deli Trays.* A grocery needs to prepare a large order of deli trays for Super Bowl weekend. It would take Henry 8.5 hr to prepare the trays. Carly can complete the job in 10.4 hr. How long would it take them, working together, to prepare the trays? Round the time to the nearest tenth of an hour.

18. *School Photos.* Rebecca can take photos for an elementary school with 325 students in 11.5 days. Jack can complete the same job in 9.2 days. How long would it take them working together? Round the time to the nearest tenth of a day.

19. *Wiring.* By checking work records, a contractor finds that Peggyann can wire a home theater in 9 hr. It takes Matthew 7 hr to wire the same room. How long would it take if they worked together?

20. *Plumbing.* By checking work records, a plumber finds that Raul can plumb a house in 48 hr. Mira can do the same job in 36 hr. How long would it take if they worked together?

21. *Office Printers.* The HP Officejet 4215 All-In-One printer, fax, scanner, and copier can print one black-and-white copy of a company's year-end report in 10 min. The HP Officejet 7410 All-In-One can print the same report in 6 min. How long would it take the two printers, working together, to print one copy of the report?

22. *Office Copiers.* The HP Officejet 7410 All-In-One printer, fax, scanner, and copier can make a color copy of a staff training manual in 9 min. The HP Officejet 4215 All-In-One can copy the same manual in 15 min. How long would it take the two copiers, working together, to make one copy of the manual?

b Find the ratio of each of the following. Simplify, if possible.

23. 60 students, 18 teachers

24. 800 mi, 50 gal

25. *Speed of a Black Racer.* A black racer snake travels 4.6 km in 2 hr. What is the speed, in kilometers per hour?

26. *Speed of Light.* Light travels 558,000 mi in 3 sec. What is the speed, in miles per second?

Solve.

27. *Protein Needs.* A 120-lb person should eat a minimum of 44 g of protein each day. How much protein should a 180-lb person eat each day?

28. *Coffee Beans.* The coffee beans from 14 trees are required to produce 7.7 kg of coffee. (This is the amount that the average person in the United States drinks each year.) How many trees are required to produce 320 kg of coffee?

29. *Hemoglobin.* A normal 10-cc specimen of human blood contains 1.2 g of hemoglobin. How much hemoglobin would 16 cc of the same blood contain?

30. *Walking Speed.* Wanda walked 234 km in 14 days. At this rate, how far would she walk in 42 days?

31. *Mileage.* A 2017 Chevrolet Camaro V-61LE can travel 208 mi of city driving on 13 gal of gas. Find the amount of gas required for 112 mi of city driving.

    Data: *Road & Track*, May 2017

32. *Mileage.* A 2017 Honda Civic Sport can travel 624 mi of highway driving on 16 gal of gas. Find the amount of gas required for 897 mi of highway driving.

    Data: *Car and Driver*, April 2017

33. *Estimating a Trout Population.* To determine the number of trout in a lake, a conservationist catches 112 trout, tags them, and throws them back into the lake. Later, 82 trout are caught; 32 of them are tagged. Estimate the number of trout in the lake.

34. *Grass Seed.* It takes 60 oz of grass seed to seed 3000 ft^2 of lawn. At this rate, how much would be needed to seed 5000 ft^2 of lawn?

35. *Quality Control.* A sample of 144 firecrackers contained 9 "duds." How many duds would you expect in a sample of 3200 firecrackers?

36. *Frog Population.* To estimate how many frogs there are in a rain forest, a research team tags 600 frogs and then releases them. Later, the team catches 300 frogs and notes that 25 of them have been tagged. Estimate the total frog population in the rain forest.

37. *College Acceptance.* During the 2016–17 undergraduate academic year, Columbia University received 36,292 applications for admission. Estimate how many students were accepted if the acceptance rate was 5.4 students per 100 applicants.

Data: Columbia University, *Wall Street Journal*, February 23, 2017. "The Short Answer," David Crook and Merrill Sherman

38. *Endangered Bird Species.* Recent data show that approximately 9.4 per 25 North American bird species are considered vulnerable to extinction. Estimate the number of bird species in danger of extinction if there are 1154 bird species in North America.

Data: North American Bird Conservation Initiative's "The State of North American Birds 2016" report

39. *Honey Bees.* Making 1 lb of honey requires 20,000 trips by bees to flowers to gather nectar. How many pounds of honey would 35,000 trips produce?

Data: Tom Turpin, Professor of Entomology, Purdue University

40. *Money.* The ratio of the weight of copper to the weight of zinc in a U.S. penny is $\frac{1}{39}$. If 50 kg of zinc is being turned into pennies, how much copper is needed?

Geometry. For each pair of similar triangles, find the length of the indicated side.

41. b: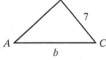

42. a:

43. f:

44. r: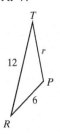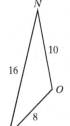

45. h: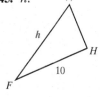

46. n: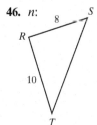

47. *Environmental Science.* The Fish and Wildlife Division of the Indiana Department of Natural Resources recently completed a study that determined the number of largemouth bass in Lake Monroe, near Bloomington, Indiana. For this project, anglers caught 300 largemouth bass, tagged them, and threw them back into the lake. Later, they caught 85 largemouth bass and found that 15 of them were tagged. Estimate how many largemouth bass are in the lake.

Data: Department of Natural Resources, Fish and Wildlife Division, Kevin Hoffman

48. *Environmental Science.* To determine the number of humpback whales in a pod, a marine biologist, using tail markings, identifies 27 members of the pod. Several weeks later, 40 whales from the pod are randomly sighted. Of the 40 sighted, 12 are from the 27 originally identified. Estimate the number of whales in the pod.

Skill Maintenance

Find the slope, if it exists, of the line containing the given pair of points. [3.4a]

49. $(7, -6), (0, -6)$

50. $(3, -11), (-4, 3)$

Simplify. [4.1d, f]

51. $x^5 \cdot x^6$

52. $x^{-5} \cdot x^6$

53. $x^{-5} \cdot x^{-6}$

54. $x^5 \cdot x^{-6}$

Graph.

55. $y = -\dfrac{3}{4}x + 2$ [3.2a]

56. $y = \dfrac{2}{5}x - 4$ [3.2a]

57. $x = -3$ [3.3b]

Synthesis

58. Rachel allows herself 1 hr to reach a sales appointment 50 mi away. After she has driven 30 mi, she realizes that she must increase her speed by 15 mph in order to arrive on time. What was her speed for the first 30 mi?

59. How soon, in minutes, after 5 o'clock will the hands on a clock first be together?

Variation and Applications

6.9

We now extend our study of formulas and functions by considering applications involving variation.

a EQUATIONS OF DIRECT VARIATION

A substitute teacher earns $65 per day. For 1 day, $65 is earned; for 2 days, $130 is earned; for 3 days, $195 is earned; and so on. We plot this information on a graph, using the number of hours as the first coordinate and the amount earned as the second coordinate to form a set of ordered pairs:

(1, 65),   (2, 130),
(3, 195), (4, 260),

and so on.

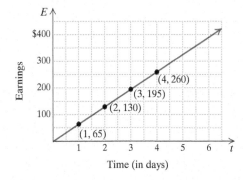

OBJECTIVES

a Find an equation of direct variation given a pair of values of the variables.

b Solve applied problems involving direct variation.

c Find an equation of inverse variation given a pair of values of the variables.

d Solve applied problems involving inverse variation.

e Find equations of other kinds of variation given values of the variables.

f Solve applied problems involving other kinds of variation.

Note that the ratio of the second coordinate to the first coordinate is the same number for each point:

$$\frac{65}{1} = 65, \quad \frac{130}{2} = 65, \quad \frac{195}{3} = 65, \quad \frac{260}{4} = 65, \quad \text{and so on.}$$

Whenever a situation produces pairs of numbers in which the *ratio is constant*, we say that there is **direct variation**. Here the amount earned varies directly as the time:

$$\frac{E}{t} = 65 \text{ (a constant)}, \quad \text{or} \quad E = 65t,$$

or, using function notation, $E(t) = 65t$. The equation is an **equation of direct variation**. The coefficient, 65 in the situation above, is called the **variation constant**. In this case, it is the rate of change of earnings with respect to time.

DIRECT VARIATION

If a situation gives rise to a linear function $f(x) = kx$, or $y = kx$, where k is a positive constant, we say that we have **direct variation**, or that **y varies directly as x**, or that **y is directly proportional to x**. The number k is called the **variation constant**, or the **constant of proportionality**.

1. Find the variation constant and an equation of variation in which y varies directly as x, and $y = 8$ when $x = 20$.

$y = kx$
$8 = k \cdot \boxed{}$
$\dfrac{\boxed{}}{20} = k$
$\dfrac{2}{\boxed{}} = k$

The variation constant is $\boxed{}$. The equation of variation is $y = \boxed{} \cdot x$.

2. Find the variation constant and an equation of variation in which y varies directly as x, and $y = 5.6$ when $x = 8$.

3. *Ohm's Law.* Ohm's Law states that the voltage V in an electric circuit varies directly as the number of amperes I of electric current in the circuit. If the voltage is 10 volts when the current is 3 amperes, what is the voltage when the current is 15 amperes?

EXAMPLE 1 Find the variation constant and an equation of variation in which y varies directly as x, and $y = 32$ when $x = 2$.

We know that $(2, 32)$ is a solution of $y = kx$. Thus,

$y = kx$
$32 = k \cdot 2$ Substituting
$\dfrac{32}{2} = k$, or $k = 16$. Solving for k

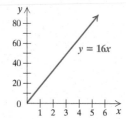

The variation constant, 16, is the rate of change of y with respect to x. The equation of variation is $y = 16x$.

The graph of $y = kx$, $k > 0$, always goes through the origin and rises from left to right. Note that as x increases, y increases. The constant k is also the slope of the line.

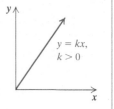

◀ Do Exercises 1 and 2.

b APPLICATIONS OF DIRECT VARIATION

EXAMPLE 2 *Water from Melting Snow.* The number of centimeters W of water produced from melting snow varies directly as S, the number of centimeters of snow. Meteorologists have found that, under certain conditions, 150 cm of snow will melt to 16.8 cm of water. To how many centimeters of water will 200 cm of snow melt?

We first find the variation constant using the data and then find an equation of variation:

$W = kS$ W varies directly as S.
$16.8 = k \cdot 150$ Substituting
$\dfrac{16.8}{150} = k$ Solving for k
$0.112 = k$. This is the variation constant.

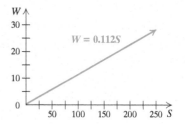

The equation of variation is $W = 0.112S$.

Next, we use the equation to find how many centimeters of water will result from melting 200 cm of snow:

$W = 0.112S$
$ = 0.112(200)$ Substituting
$ = 22.4.$

Thus, 200 cm of snow will melt to 22.4 cm of water.

◀ Do Exercises 3 and 4. (Exercise 4 is on the following page.)

Answers

1. $\dfrac{2}{5}$; $y = \dfrac{2}{5}x$ 2. 0.7; $y = 0.7x$ 3. 50 volts

Guided Solution:
1. $20, 8, 5, \dfrac{2}{5}, \dfrac{2}{5}$

C EQUATIONS OF INVERSE VARIATION

A bus is traveling a distance of 20 mi. At a speed of 5 mph, the trip will take 4 hr; at 20 mph, it will take 1 hr; at 40 mph, it will take $\frac{1}{2}$ hr; and so on. We plot this information on a graph, using speed as the first coordinate and time as the second coordinate to determine a set of ordered pairs:

(5, 4),   (10, 2),
(20, 1), (40, $\frac{1}{2}$),

and so on.

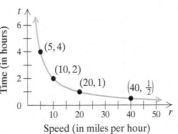

Note that the products of the coordinates are all the same number:

$5 \cdot 4 = 20$, $20 \cdot 1 = 20$, $40 \cdot \frac{1}{2} = 20$,   and so on.

Whenever a situation produces pairs of numbers in which the *product is constant*, we say that there is **inverse variation**. Here the time varies inversely as the speed:

$$rt = 20 \text{ (a constant)}, \quad \text{or} \quad t = \frac{20}{r}.$$

The equation is an **equation of inverse variation**. The constant, 20 in the situation above, is called the **variation constant**. Note that as the first number (speed) increases, the second number (time) decreases.

INVERSE VARIATION

If a situation gives rise to a function $f(x) = k/x$, or $y = k/x$, where k is a positive constant, we say that we have **inverse variation**, or that ***y* varies inversely as *x***, or that ***y* is inversely proportional to *x***. The number k is called the **variation constant**, or the **constant of proportionality**.

EXAMPLE 3 Find the variation constant and an equation of variation in which y varies inversely as x, and $y = 32$ when $x = 0.2$.

We know that (0.2, 32) is a solution of $y = k/x$. We substitute:

$y = \dfrac{k}{x}$

$32 = \dfrac{k}{0.2}$ Substituting

$(0.2)32 = k$ Solving for k

$6.4 = k.$

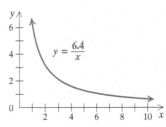

The variation constant is 6.4. The equation of variation is $y = \dfrac{6.4}{x}$.

---

4. *Bees and Honey.* The amount of honey H produced varies directly as the number of bees who produce the honey. It takes 15,000 bees to produce 25 lb of honey. How much honey is produced by 40,000 bees?

Answer

4. $66\dfrac{2}{3}$ lb

5. Find the variation constant and an equation of variation in which y varies inversely as x, and $y = 0.012$ when $x = 50$.

$$y = \frac{k}{x}$$

$$0.012 = \frac{k}{\boxed{}}$$

$$\boxed{} \cdot 50 = k$$

$$\boxed{} = k$$

The variation constant is $\boxed{}$.
The equation of variation is

$$y = \frac{\boxed{}}{x}.$$

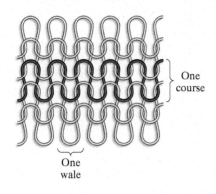

One course

One wale

6. *Cleaning Bleachers.* The time t to do a job varies inversely as the number of people P who work on the job (assuming that all work at the same rate). It takes 4.5 hr for 12 people to clean a section of bleachers after a NASCAR race. How long would it take 15 people to complete the same job?

Answers
5. 0.6; $y = \dfrac{0.6}{x}$ 6. 3.6 hr

Guided Solution:
5. $50, 0.012, 0.6, 0.6, 0.6$

It is helpful to look at the graph of $y = k/x$, $k > 0$. The graph is like the one shown at right for positive values of x. Note that as x increases, y decreases.

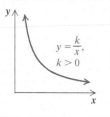

◀ Do Exercise 5.

d APPLICATIONS OF INVERSE VARIATION

EXAMPLE 4 *Fabric Manufacturing.* Knitted cotton fabric is described in terms of wales per inch (WPI) (for the fabric width) and courses per inch (CPI) (for the fabric length). The CPI, C, is inversely proportional to the stitch length l. For a specific fabric with a stitch length of 0.162 in., the CPI is 35.71. What would the CPI be if the stitch length were increased to 0.171 in.?
Data: Cotton Incorporated, "Guidelines for Engineering Cotton Knits"

We first find the variation constant using the data given and then find an equation of variation:

$$C = \frac{k}{l} \qquad \text{C varies inversely as l.}$$

$$35.71 = \frac{k}{0.162} \qquad \text{Substituting}$$

$$5.785 \approx k. \qquad \text{Solving for k, the variation constant}$$

The equation of variation is $C = \dfrac{5.785}{l}$.

Next, we use the equation to find the CPI for a fabric that has a stitch length of 0.171 in.:

$$C = \frac{5.785}{l} \qquad \text{Equation of variation}$$

$$= \frac{5.785}{0.171} \qquad \text{Substituting}$$

$$\approx 33.83.$$

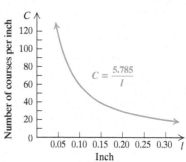

The CPI of a fabric that has a stitch length of 0.171 in. is 33.83.
◀ Do Exercise 6.

e OTHER KINDS OF VARIATION

We now look at other kinds of variation. Consider the equation for the area of a circle, in which A and r are variables and π is a constant:

$$A = \pi r^2, \quad \text{or, as a function,} \quad A(r) = \pi r^2.$$

We say that the area *varies directly* as the square of the radius.

y varies directly as the nth power of x if there is some positive constant k such that
$$y = kx^n.$$

EXAMPLE 5 Find an equation of variation in which y varies directly as the square of x, and $y = 12$ when $x = 2$.

We write an equation of variation and find k:

$y = kx^2$
$12 = k \cdot 2^2$
$12 = k \cdot 4$
$3 = k.$

Thus, $y = 3x^2$.

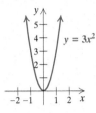

Do Exercise 7.

7. Find an equation of variation in which y varies directly as the square of x, and $y = 175$ when $x = 5$.

$y = kx^2$
$\boxed{} = k \cdot 5^2$
$175 = k \cdot 25$
$\boxed{} = k$

The equation of variation is $y = \boxed{} x^2$.

From the law of gravity, we know that the weight W of an object *varies inversely* as the square of its distance d from the center of the Earth:

$$W = \frac{k}{d^2}.$$

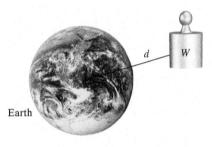

Earth

y varies inversely as the nth power of x if there is some positive constant k such that
$$y = \frac{k}{x^n}.$$

EXAMPLE 6 Find an equation of variation in which W varies inversely as the square of d, and $W = 3$ when $d = 5$.

$W = \dfrac{k}{d^2}$

$3 = \dfrac{k}{5^2}$ Substituting

$3 = \dfrac{k}{25}$

$75 = k$

Thus, $W = \dfrac{75}{d^2}$.

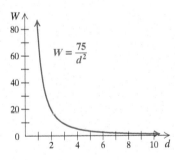

Do Exercise 8.

8. Find an equation of variation in which y varies inversely as the square of x, and $y = \frac{1}{4}$ when $x = 6$.

Consider the equation for the area A of a triangle with height h and base b: $A = \frac{1}{2}bh$. We say that the area **varies jointly** as the height and the base.

Answers

7. $y = 7x^2$ **8.** $y = \dfrac{9}{x^2}$

Guided Solution:
7. 175, 7, 7

9. Find an equation of variation in which y varies jointly as x and z, and $y = 65$ when $x = 10$ and $z = 13$.

$$y = kxz$$
$$65 = k \cdot \boxed{} \cdot 13$$
$$65 = k \cdot 130$$
$$\frac{65}{\boxed{}} = k$$
$$\boxed{} = k$$

The equation of variation is $y = \boxed{} xz.$

> y varies jointly as x and z if there is some positive constant k such that
> $$y = kxz.$$

EXAMPLE 7 Find an equation of variation in which y varies jointly as x and z, and $y = 42$ when $x = 2$ and $z = 3$.

$$y = kxz$$
$$42 = k \cdot 2 \cdot 3 \quad \text{Substituting}$$
$$42 = k \cdot 6$$
$$7 = k$$

Thus, $y = 7xz$.

◀ Do Exercise 9.

Different types of variation can be combined. For example, the equation
$$y = k \cdot \frac{xz^2}{w}$$
asserts that y varies jointly as x and the square of z, and inversely as w.

EXAMPLE 8 Find an equation of variation in which y varies jointly as x and z and inversely as the square of w, and $y = 105$ when $x = 3$, $z = 20$, and $w = 2$.

$$y = k \cdot \frac{xz}{w^2}$$
$$105 = k \cdot \frac{3 \cdot 20}{2^2} \quad \text{Substituting}$$
$$105 = k \cdot 15$$
$$7 = k$$

Thus, $y = 7 \cdot \frac{xz}{w^2}$.

◀ Do Exercise 10.

10. Find an equation of variation in which y varies jointly as x and the square of z and inversely as w, and $y = 80$ when $x = 4$, $z = 10$, and $w = 25$.

f OTHER APPLICATIONS OF VARIATION

EXAMPLE 9 *Volume of a Tree.* The volume of wood V in a tree varies jointly as the height h and the square of the girth g (girth is distance around the trunk). If the volume of a redwood tree is 216 m³ when the height is 30 m and the girth is 1.5 m, what is the height of a tree whose volume is 960 m³ and girth is 2 m?

Answers

9. $y = \frac{1}{2}xz$ 10. $y = \frac{5xz^2}{w}$

Guided Solution:
9. 10, 130, $\frac{1}{2}$, $\frac{1}{2}$

We first find k using the first set of data. Then we solve for h using the second set of data.

$$V = khg^2$$
$$216 = k \cdot 30 \cdot 1.5^2$$
$$3.2 = k$$

Then the equation of variation is $V = 3.2hg^2$. We substitute the second set of data into the equation:

$$960 = 3.2 \cdot h \cdot 2^2$$
$$75 = h.$$

Thus the height of the tree is 75 m.

EXAMPLE 10 *Intensity of a Light.* The intensity of light I from a light bulb varies inversely as the square of the distance d from the bulb. Suppose that I is 90 W/m² (watts per square meter) when the distance is 5 m. How much *further* would it be to a point at which the intensity is 40 W/m²?

We first find k using the first set of data. Then we solve for I using the second set of data.

$$I = \frac{k}{d^2}$$
$$90 = \frac{k}{5^2}$$
$$2250 = k$$

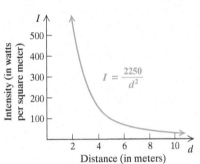

Then the equation of variation is $I = 2250/d^2$. We substitute the second intensity into the equation and solve for d:

$$I = \frac{2250}{d^2}$$
$$40 = \frac{2250}{d^2}$$
$$40d^2 = 2250$$
$$d^2 = 56.25$$
$$d = 7.5.$$

When the intensity is 40 W/m², the distance is 7.5 m. Thus the point at which the intensity is 40 W/m² is 7.5 − 5, or 2.5, m further.

Do Exercises 11 and 12.

11. *Distance of a Dropped Object.* The distance s that an object falls when dropped from some point above the ground varies directly as the square of the time t that it falls. If the object falls 19.6 m in 2 sec, how far will the object fall in 10 sec?

12. *Electrical Resistance.* At a fixed temperature, the resistance R of a wire varies directly as the length l and inversely as the square of its diameter d. If the resistance is 0.1 ohm when the diameter is 1 mm and the length is 50 cm, what is the resistance when the length is 2000 cm and the diameter is 2 mm?

Answers
11. 490 m 12. 1 ohm

6.9 Exercise Set

✓ Check Your Understanding

Reading Check Choose the word from below each blank that best completes the statement.

RC1. If a situation gives rise to a function $f(x) = k/x$, or $y = k/x$, when k is a positive constant, we say that we have _____ variation.
direct/inverse

RC2. If a situation gives rise to a linear function $f(x) = kx$, or $y = kx$, where k is a positive constant, we say that we have _____ variation.
direct/inverse

Concept Check Match each description of variation with the appropriate equation of variation listed on the right.

CC1. a varies directly as z.

CC2. y is inversely proportional to b.

CC3. y varies jointly as z and b and inversely as c.

CC4. x is directly proportional to c.

CC5. b varies directly as z.

CC6. a varies inversely as z.

CC7. c varies inversely as x.

CC8. y varies jointly as c and x.

a) $a = \dfrac{k}{z}$ b) $y = kcx$

c) $b = kz$ d) $y = \dfrac{k}{b}$

e) $x = ky$ f) $a = kz$

g) $c = \dfrac{k}{x}$ h) $y = \dfrac{kzb}{c}$

i) $x = kc$ j) $y = \dfrac{kbc}{z}$

a Find the variation constant and an equation of variation in which y varies directly as x and the following are true.

1. $y = 40$ when $x = 8$

2. $y = 54$ when $x = 12$

3. $y = 4$ when $x = 30$

4. $y = 3$ when $x = 33$

5. $y = 0.9$ when $x = 0.4$

6. $y = 0.8$ when $x = 0.2$

b Solve.

7. *Shipping by Semi Truck.* The number of semi trucks T needed to ship metal varies directly as the weight W of the metal. It takes 75 semi trucks to ship 1500 tons of metal. How many trucks are needed for 3500 tons of metal?

Data: scrappy.com

8. *Shipping by Rail Cars.* The number of rail cars R needed to ship metal varies directly as the weight W of the metal. It takes approximately 21 rail cars to ship 1500 tons of metal. How many rail cars are needed for 3500 tons of metal?

Data: scrappy.com

9. *Fat Intake.* The maximum number of grams of fat that should be in a diet varies directly as a person's weight. A person weighing 120 lb should consume no more than 60 g of fat per day. What is the maximum daily fat intake for a person weighing 180 lb?

10. *Relative Aperture.* The relative aperture, or f-stop, of a 23.5-mm diameter lens is directly proportional to the focal length F of the lens. If a 150-mm focal length has an f-stop of 6.3, find the f-stop of a 23.5-mm diameter lens with a focal length of 80 mm.

11. *Mass of Water in Human Body.* The number of kilograms W of water in a human body varies directly as the mass of the body. A 96-kg person contains 64 kg of water. How many kilograms of water are in a 60-kg person?

12. *Weight on Mars.* The weight M of an object on Mars varies directly as its weight E on Earth. A person who weighs 95 lb on Earth weighs 38 lb on Mars. How much would a 100-lb person weigh on Mars?

13. *Aluminum Usage.* The number of aluminum cans N used each year varies directly as the number of people using them. If 250 people use 60,000 cans in one year, how many cans are used each year in St. Louis, Missouri, which has a population of 318,172?

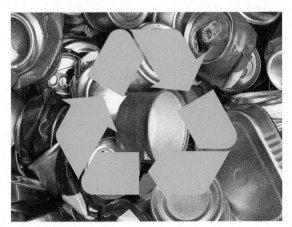

14. *Hooke's Law.* Hooke's law states that the distance d that a spring is stretched by a hanging object varies directly as the weight w of the object. If a spring is stretched 40 cm by a 3-kg barbell, what is the distance stretched by a 5-kg barbell?

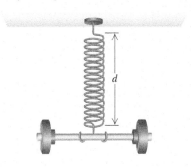

c Find the variation constant and an equation of variation in which y varies inversely as x and the following are true.

15. $y = 14$ when $x = 7$

16. $y = 1$ when $x = 8$

17. $y = 3$ when $x = 12$

18. $y = 12$ when $x = 5$

19. $y = 0.1$ when $x = 0.5$

20. $y = 1.8$ when $x = 0.3$

d Solve.

21. *Work Rate.* The time T required to do a job varies inversely as the number of people P working. It takes 5 hr for 7 bricklayers to build a park wall. How long will it take 10 bricklayers to complete the job?

22. *Pumping Rate.* The time t required to empty a tank varies inversely as the rate r of pumping. If a pump can empty a tank in 45 min at the rate of 600 kL/min, how long will it take the pump to empty the same tank at the rate of 1000 kL/min?

23. *Current and Resistance.* The current I in an electrical conductor varies inversely as the resistance R of the conductor. If the current is $\frac{1}{2}$ ampere when the resistance is 240 ohms, what is the current when the resistance is 540 ohms?

24. *Wavelength and Frequency.* The wavelength W of a radio wave varies inversely as its frequency F. A wave with a frequency of 1200 kilohertz has a length of 300 meters. What is the length of a wave with a frequency of 800 kilohertz?

25. *Beam Weight.* The weight W that a horizontal beam can support varies inversely as the length L of the beam. Suppose that a 12-ft beam can support 1200 lb. How many pounds can a 15-ft beam support?

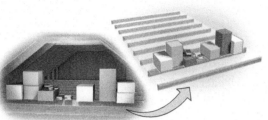

26. *Musical Pitch.* The pitch P of a musical tone varies inversely as its wavelength W. One tone has a pitch of 440 vibrations per second and a wavelength of 2.4 ft. Find the wavelength of another tone that has a pitch of 275 vibrations per second.

27. *Rate of Travel.* The time t required to drive a fixed distance varies inversely as the speed r. It takes 5 hr at a speed of 80 km/h to drive a fixed distance. How long will it take to drive the same distance at a speed of 70 km/h?

28. *Volume and Pressure.* The volume V of a gas varies inversely as the pressure P upon it. The volume of a gas is 200 cm^3 under a pressure of 32 kg/cm^2. What will its volume be under a pressure of 40 kg/cm^2?

e   Find an equation of variation in which the following are true.

29. y varies directly as the square of x, and $y = 0.15$ when $x = 0.1$

30. y varies directly as the square of x, and $y = 6$ when $x = 3$

31. y varies inversely as the square of x, and $y = 0.15$ when $x = 0.1$

32. y varies inversely as the square of x, and $y = 6$ when $x = 3$

33. y varies jointly as x and z, and $y = 56$ when $x = 7$ and $z = 8$

34. y varies directly as x and inversely as z, and $y = 4$ when $x = 12$ and $z = 15$

35. y varies jointly as x and the square of z, and $y = 105$ when $x = 14$ and $z = 5$

36. y varies jointly as x and z and inversely as w, and $y = \frac{3}{2}$ when $x = 2$, $z = 3$, and $w = 4$

37. y varies jointly as x and z and inversely as the product of w and p, and $y = \frac{3}{28}$ when $x = 3$, $z = 10$, $w = 7$, and $p = 8$

38. y varies jointly as x and z and inversely as the square of w, and $y = \frac{12}{5}$ when $x = 16$, $z = 3$, and $w = 5$

f   Solve.

39. *Stopping Distance of a Car.* The stopping distance d of a car after the brakes have been applied varies directly as the square of the speed r. If a car traveling 60 mph can stop in 200 ft, how fast can a car travel and still stop in 72 ft?

40. *Intensity of Light.* The intensity I of light from a light bulb varies inversely as the square of the distance d from the bulb. Suppose that I is 90 W/m^2 (watts per square meter) when the distance is 5 m. How much *further* would it be to a point at which the intensity is 30 W/m^2? Round the answer to the nearest hundredth of a meter.

41. *Weight of a Sphere.* The weight W of a sphere of a given material varies directly as its volume V, and its volume V varies directly as the cube of its diameter.
   a) Find an equation of variation relating the weight W to the diameter d.
   b) An iron ball that is 5 in. in diameter is known to weigh 25 lb. Find the weight of an iron ball that is 8 in. in diameter.

42. *Combined Gas Law.* The volume V of a given mass of a gas varies directly as the temperature T and inversely as the pressure P. If $V = 231 \text{ cm}^3$ when $T = 42°$ and $P = 20 \text{ kg/cm}^2$, what is the volume when $T = 30°$ and $P = 15 \text{ kg/cm}^2$?

43. *Earned-Run Average.* A pitcher's earned-run average E varies directly as the number R of earned runs allowed and inversely as the number I of innings pitched. In 2016, Clayton Kershaw of the Los Angeles Dodgers had an earned-run average of 1.69. He gave up 28 earned runs in 149.0 innings. How many earned runs would he have given up had he pitched 210 innings with the same average?

Data: Major League Baseball

44. *Atmospheric Drag.* Wind resistance, or atmospheric drag, tends to slow down moving objects. Atmospheric drag varies jointly as an object's surface area A and velocity v. If a car traveling at a speed of 40 mph with a surface area of 37.8 ft^2 experiences a drag force of 222 N (newtons), how fast must a car with 51 ft^2 of surface area travel in order to experience a drag force of 430 N?

45. *Water Flow.* The amount Q of water emptied by a pipe varies directly as the square of the diameter d. A pipe 5 in. in diameter will empty 225 gal of water over a fixed time period. If we assume the same kind of flow, how many gallons of water are emptied in the same amount of time by a pipe that is 9 in. in diameter?

46. *Wind Power.* Wind power P from a turbine varies directly as the square of the length r of one of its blades. Two common blade lengths for commercial wind turbines are 35 m and 50 m. When the blade length is 35 m, about 1.5 MW (megawatts) of power is produced under favorable conditions. How much power would be produced, under favorable conditions, by a turbine with 50-m blades?

Data: aweo.org

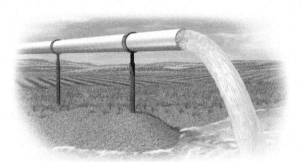

Skill Maintenance

Factor.

47. $x^2 - 14x + 49$ [5.5b]

48. $125y^3 - 27$ [5.6a]

49. $121w^2 - 9$ [5.5d]

Solve.

50. $2(3 - 4t) = -66$ [2.3c]

51. $y^2 - 100 = 0$ [5.8b]

52. $2x - 7 = -17$ [2.3a]

Synthesis

53. In each of the following equations, state whether y varies directly as x, inversely as x, or neither directly nor inversely as x.
   a) $7xy = 14$
   b) $x - 2y = 12$
   c) $-2x + 3y = 0$
   d) $x = \frac{3}{4}y$

54. *Area of a Circle.* The area of a circle varies directly as the square of the length of a diameter. What is the variation constant?

CHAPTER 6 Summary and Review

Vocabulary Reinforcement

Complete each statement with the correct term from the column on the right.

1. A _____ rational expression is a rational expression that has one or more rational expressions within its numerator or its denominator. [6.6a]

2. An equality of ratios, $\dfrac{A}{B} = \dfrac{C}{D}$, is called a(n) _____. [6.8b]

3. Two expressions are _____ of each other if their product is 1. [6.2a]

4. Expressions that have the same value for all allowable replacements are called _____ expressions. [6.1b]

5. Expressions of the form $a - b$ and $b - a$ are _____ of each other. [6.1c]

6. In _____ triangles, corresponding angles have the same measure and the lengths of corresponding sides are proportional. [6.8b]

7. When a situation translates to an equation described by $y = \dfrac{k}{x}$, with k a positive constant, we say that y varies _____ as x. The equation $y = \dfrac{k}{x}$ is called an equation of _____. [6.9c]

8. When a situation translates to an equation described by $y = kx$, with k a positive constant, we say that y varies _____ as x. The equation $y = kx$ is called an equation of _____. [6.9a]

reciprocals
proportion
rational
equivalent
directly
complex
direct variation
similar
inversely
opposites
inverse variation

Concept Reinforcement

Determine whether each statement is true or false.

_____ 1. To determine the numbers for which a rational expression is not defined, we set the denominator equal to 0 and solve. [6.1a]

_____ 2. The expressions $y + 5$ and $y - 5$ are opposites of each other. [6.1c]

_____ 3. The opposite of $2 - x$ is $x - 2$. [6.1c]

Study Guide

Objective 6.1a Find all numbers for which a rational expression is not defined.

Example Find all numbers for which the rational expression $\dfrac{2 - y}{y^2 + 3y - 28}$ is not defined.

$y^2 + 3y - 28 = 0$
$(y + 7)(y - 4) = 0$
$y + 7 = 0 \quad \text{or} \quad y - 4 = 0$
$y = -7 \quad \text{or} \quad y = 4$

The rational expression is not defined for -7 and 4.

Practice Exercise

1. Find all numbers for which the rational expression $\dfrac{c + 8}{c^2 - 11c + 30}$ is not defined.

Objective 6.1c Simplify rational expressions by factoring the numerator and the denominator and removing factors of 1.

Example Simplify: $\dfrac{6y - 12}{2y^2 + y - 10}$.

$$\dfrac{6y - 12}{2y^2 + y - 10} = \dfrac{6(y - 2)}{(2y + 5)(y - 2)}$$
$$= \dfrac{y - 2}{y - 2} \cdot \dfrac{6}{2y + 5}$$
$$= 1 \cdot \dfrac{6}{2y + 5} = \dfrac{6}{2y + 5}$$

Practice Exercise

2. Simplify:
$$\dfrac{2x^2 - 2}{4x^2 + 24x + 20}.$$

Objective 6.1d Multiply rational expressions and simplify.

Example Multiply and simplify:
$$\dfrac{x^2 + 14x + 49}{x^2 - 25} \cdot \dfrac{x + 5}{x + 7}.$$

$$\dfrac{x^2 + 14x + 49}{x^2 - 25} \cdot \dfrac{x + 5}{x + 7} = \dfrac{(x^2 + 14x + 49)(x + 5)}{(x^2 - 25)(x + 7)}$$
$$= \dfrac{(x + 7)\cancel{(x + 7)}\cancel{(x + 5)}}{\cancel{(x + 5)}(x - 5)\cancel{(x + 7)}}$$
$$= \dfrac{x + 7}{x - 5}$$

Practice Exercise

3. Multiply and simplify:
$$\dfrac{2y^2 + 7y - 15}{5y^2 - 45} \cdot \dfrac{y - 3}{2y - 3}.$$

Objective 6.2b Divide rational expressions and simplify.

Example Divide and simplify: $\dfrac{a^2 - 9a}{a^2 - a - 6} \div \dfrac{a}{a + 2}$.

$$\dfrac{a^2 - 9a}{a^2 - a - 6} \div \dfrac{a}{a + 2} = \dfrac{a^2 - 9a}{a^2 - a - 6} \cdot \dfrac{a + 2}{a}$$
$$= \dfrac{(a^2 - 9a)(a + 2)}{(a^2 - a - 6)a}$$
$$= \dfrac{\cancel{a}(a - 9)\cancel{(a + 2)}}{\cancel{(a + 2)}(a - 3)\cancel{a}}$$
$$= \dfrac{a - 9}{a - 3}$$

Practice Exercise

4. Divide and simplify:
$$\dfrac{b^2 + 3b - 28}{b^2 + 5b - 24} \div \dfrac{b - 4}{b - 3}.$$

Objective 6.3b Add fractions, first finding the LCD.

Example Add: $\dfrac{13}{30} + \dfrac{11}{24}$.

$$\left.\begin{array}{l}30 = 2 \cdot 3 \cdot 5 \\ 24 = 2 \cdot 2 \cdot 2 \cdot 3\end{array}\right\} \text{LCD} = 2 \cdot 2 \cdot 2 \cdot 3 \cdot 5, \text{ or } 120$$

$$\dfrac{13}{30} + \dfrac{11}{24} = \dfrac{13}{2 \cdot 3 \cdot 5} \cdot \dfrac{2 \cdot 2}{2 \cdot 2} + \dfrac{11}{2 \cdot 2 \cdot 2 \cdot 3} \cdot \dfrac{5}{5}$$
$$= \dfrac{52 + 55}{2 \cdot 2 \cdot 2 \cdot 3 \cdot 5} = \dfrac{107}{120}$$

Practice Exercise

5. Add: $\dfrac{5}{18} + \dfrac{7}{60}$.

Objective 6.3c Find the LCM of algebraic expressions by factoring.

Example Find the LCM of
$$x^2 - 36 \text{ and } x^2 - 5x - 6.$$
$$x^2 - 36 = (x+6)(x-6)$$
$$x^2 - 5x - 6 = (x-6)(x+1)$$
$$\text{LCM} = (x+6)(x-6)(x+1)$$

Practice Exercise

6. Find the LCM of
$$x^2 - 7x - 18 \text{ and } x^2 - 81.$$

Objective 6.4a Add rational expressions.

Example Add and simplify: $\dfrac{6x-5}{x-1} + \dfrac{x}{1-x}$.

$$\dfrac{6x-5}{x-1} + \dfrac{x}{1-x} = \dfrac{6x-5}{x-1} + \dfrac{x}{1-x} \cdot \dfrac{-1}{-1}$$
$$= \dfrac{6x-5}{x-1} + \dfrac{-x}{x-1}$$
$$= \dfrac{6x-5-x}{x-1}$$
$$= \dfrac{5x-5}{x-1}$$
$$= \dfrac{5(x-1)}{x-1} = 5$$

Practice Exercise

7. Add and simplify:
$$\dfrac{x}{x-4} + \dfrac{2x-4}{4-x}.$$

Objective 6.5a Subtract rational expressions.

Example Subtract: $\dfrac{3}{x^2-1} - \dfrac{2x-1}{x^2+x-2}$.

$$\dfrac{3}{x^2-1} - \dfrac{2x-1}{x^2+x-2}$$
$$= \dfrac{3}{(x+1)(x-1)} - \dfrac{2x-1}{(x+2)(x-1)}$$
$$\qquad \text{The LCM is } (x+1)(x-1)(x+2).$$
$$= \dfrac{3}{(x+1)(x-1)} \cdot \dfrac{x+2}{x+2} - \dfrac{2x-1}{(x+2)(x-1)} \cdot \dfrac{x+1}{x+1}$$
$$= \dfrac{3(x+2)}{(x+1)(x-1)(x+2)} - \dfrac{(2x-1)(x+1)}{(x+2)(x-1)(x+1)}$$
$$= \dfrac{3x+6-(2x^2+x-1)}{(x+1)(x-1)(x+2)}$$
$$= \dfrac{3x+6-2x^2-x+1}{(x+1)(x-1)(x+2)}$$
$$= \dfrac{-2x^2+2x+7}{(x+1)(x-1)(x+2)}$$

Practice Exercise

8. Subtract:
$$\dfrac{x}{x^2+x-2} - \dfrac{5}{x^2-1}.$$

Objective 6.6a Simplify complex rational expressions.

Example Simplify $\dfrac{\frac{1}{3}-\frac{1}{x}}{\frac{1}{x}-\frac{1}{2}}$ using method 1.

The LCM of 3, x, and 2 is $6x$.

$$\dfrac{\frac{1}{3}-\frac{1}{x}}{\frac{1}{x}-\frac{1}{2}} = \dfrac{\frac{1}{3}-\frac{1}{x}}{\frac{1}{x}-\frac{1}{2}} \cdot \dfrac{6x}{6x} = \dfrac{\frac{1}{3}\cdot 6x - \frac{1}{x}\cdot 6x}{\frac{1}{x}\cdot 6x - \frac{1}{2}\cdot 6x}$$

$$= \dfrac{2x-6}{6-3x} = \dfrac{2(x-3)}{3(2-x)}$$

Practice Exercise

9. Simplify: $\dfrac{\frac{2}{5}-\frac{1}{y}}{\frac{3}{y}-\frac{1}{3}}$.

Objective 6.7a Solve rational equations.

Example Solve: $12 = \dfrac{1}{5x} + \dfrac{4}{x}$.

The LCM of the denominators is $5x$. We multiply by $5x$ on both sides.

$$12 = \dfrac{1}{5x} + \dfrac{4}{x}$$

$$5x \cdot 12 = 5x\left(\dfrac{1}{5x} + \dfrac{4}{x}\right)$$

$$5x \cdot 12 = 5x \cdot \dfrac{1}{5x} + 5x \cdot \dfrac{4}{x}$$

$$60x = 1 + 20$$

$$60x = 21$$

$$x = \dfrac{21}{60} = \dfrac{7}{20}$$

This checks, so the solution is $\dfrac{7}{20}$.

Practice Exercise

10. Solve: $\dfrac{1}{x} = \dfrac{2}{3-x}$.

Objective 6.9a Find an equation of direct variation given a pair of values of the variables.

Example Find an equation of variation in which y varies directly as x, and $y = 30$ when $x = 200$. Then find the value of y when $x = \frac{1}{2}$.

$y = kx$ Direct variation

$30 = k \cdot 200$ Substituting 30 for y and 200 for x

$\dfrac{30}{200} = k$, or $k = \dfrac{3}{20}$

The equation of variation is $y = \dfrac{3}{20}x$.
 Next, we substitute $\frac{1}{2}$ for x in $y = \dfrac{3}{20}x$ and solve for y:

$$y = \dfrac{3}{20}x = \dfrac{3}{20}\cdot\dfrac{1}{2} = \dfrac{3}{40}.$$

When $x = \dfrac{1}{2}$, $y = \dfrac{3}{40}$.

Practice Exercise

11. Find an equation of variation in which y varies directly as x, and $y = 60$ when $x = 0.4$. Then find the value of y when $x = 2$.

Objective 6.9c Find an equation of inverse variation given a pair of values of the variables.

Example Find an equation of variation in which y varies inversely as x, and $y = 0.5$ when $x = 20$. Then find the value of y when $x = 6$.

$y = \dfrac{k}{x}$ Inverse variation

$0.5 = \dfrac{k}{20}$ Substituting 0.5 for y and 20 for x

$10 = k$

The equation of variation is $y = \dfrac{10}{x}$.

Next, we substitute 6 for x in $y = 10/x$ and solve for y:

$y = \dfrac{10}{x} = \dfrac{10}{6} = \dfrac{5}{3}$.

When $x = 6$, $y = \dfrac{5}{3}$.

Practice Exercise

12. Find an equation of variation in which y varies inversely as x, and $y = 150$ when $x = 1.5$. Then find the value of y when $x = 10$.

Review Exercises

Find all numbers for which the rational expression is not defined. [6.1a]

1. $\dfrac{3}{x}$

2. $\dfrac{4}{x-6}$

3. $\dfrac{x+5}{x^2-36}$

4. $\dfrac{x^2-3x+2}{x^2+x-30}$

5. $\dfrac{-4}{(x+2)^2}$

6. $\dfrac{x-5}{5}$

Simplify. [6.1c]

7. $\dfrac{4x^2-8x}{4x^2+4x}$

8. $\dfrac{14x^2-x-3}{2x^2-7x+3}$

9. $\dfrac{(y-5)^2}{y^2-25}$

Multiply and simplify. [6.1d]

10. $\dfrac{a^2-36}{10a} \cdot \dfrac{2a}{a+6}$

11. $\dfrac{6t-6}{2t^2+t-1} \cdot \dfrac{t^2-1}{t^2-2t+1}$

Divide and simplify. [6.2b]

12. $\dfrac{10-5t}{3} \div \dfrac{t-2}{12t}$

13. $\dfrac{4x^4}{x^2-1} \div \dfrac{2x^3}{x^2-2x+1}$

Find the LCM. [6.3c]

14. $3x^2$, $10xy$, $15y^2$

15. $a-2$, $4a-8$

16. y^2-y-2, y^2-4

Add and simplify. [6.4a]

17. $\dfrac{x+8}{x+7} + \dfrac{10-4x}{x+7}$

18. $\dfrac{3}{3x-9} + \dfrac{x-2}{3-x}$

19. $\dfrac{2a}{a+1} + \dfrac{4a}{a^2-1}$

20. $\dfrac{d^2}{d-c} + \dfrac{c^2}{c-d}$

510 CHAPTER 6 Rational Expressions and Equations

Subtract and simplify. [6.5a]

21. $\dfrac{6x-3}{x^2-x-12} - \dfrac{2x-15}{x^2-x-12}$

22. $\dfrac{3x-1}{2x} - \dfrac{x-3}{x}$

23. $\dfrac{x+3}{x-2} - \dfrac{x}{2-x}$

24. $\dfrac{1}{x^2-25} - \dfrac{x-5}{x^2-4x-5}$

25. Perform the indicated operations and simplify: [6.5b]

$$\dfrac{3x}{x+2} - \dfrac{x}{x-2} + \dfrac{8}{x^2-4}.$$

Simplify. [6.6a]

26. $\dfrac{\dfrac{1}{z}+1}{\dfrac{1}{z^2}-1}$

27. $\dfrac{\dfrac{c}{d}-\dfrac{d}{c}}{\dfrac{1}{c}+\dfrac{1}{d}}$

Solve. [6.7a]

28. $\dfrac{3}{y} - \dfrac{1}{4} = \dfrac{1}{y}$

29. $\dfrac{15}{x} - \dfrac{15}{x+2} = 2$

Solve. [6.8a]

30. *Highway Work.* In checking records, a contractor finds that crew A can pave a certain length of highway in 9 hr, while crew B can do the same job in 12 hr. How long would it take if they worked together?

31. *Airplane Speed.* One plane travels 80 mph faster than another. While one travels 1750 mi, the other travels 950 mi. Find the speed of each plane.

32. *Train Speed.* A manufacturer is testing two high-speed trains. One train travels 40 km/h faster than the other. While one train travels 70 km, the other travels 60 km. Find the speed of each train.

70 km, $r+40$

60 km, r

Solve. [6.8b]

33. *Quality Control.* A sample of 250 calculators contained 8 defective calculators. How many defective calculators would you expect to find in a sample of 5000?

34. *Pizza Proportions.* At Finnelli's Pizzeria, the following ratios are used: 5 parts sausage to 7 parts cheese, 6 parts onion to 13 parts green pepper, and 9 parts pepperoni to 14 parts cheese.
   a) Finnelli's makes several pizzas with green pepper and onion. They use 2 cups of green pepper. How much onion would they use?
   b) Finnelli's makes several pizzas with sausage and cheese. They use 3 cups of sausage. How much cheese would they use?
   c) Finnelli's makes several pizzas with pepperoni and cheese. They use 6 cups of pepperoni. How much cheese would they use?

35. *Estimating a Whale Population.* To determine the number of blue whales in the world's oceans, marine biologists tag 500 blue whales in various parts of the world. Later, 400 blue whales are checked, and it is found that 20 of them are tagged. Estimate the blue whale population.

36. Triangles *ABC* and *XYZ* below are similar. Find the value of *x*.

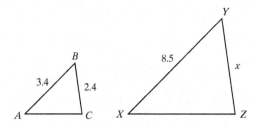

37. Find an equation of variation in which y varies directly as x, and y = 100 when x = 25. [6.9a]

38. Find an equation of variation in which y varies inversely as x, and y = 100 when x = 25. [6.9c]

39. Find an equation of variation in which y varies jointly as x and z, and y = 210 when x = 5 and z = 7. [6.9e]

40. Find an equation of variation in which y varies directly as x and inversely as z, and y = 16 when x = 40 and z = 5. [6.9e]

Solve.

41. *Pumping Time.* The time *t* required to empty a tank varies inversely as the rate *r* of pumping. If a pump can empty a tank in 35 min at the rate of 800 kL per minute, how long will it take the pump to empty the same tank at the rate of 1400 kL per minute? [6.9b]

42. *Test Score.* The score *N* on a test varies directly as the number of correct responses *a*. Ellen answers 28 questions correctly and earns a score of 87. What would Ellen's score have been if she had answered 25 questions correctly? Round the answer to the nearest whole number. [6.9d]

43. *Power of Electric Current.* The power *P* expended by heat in an electric circuit of fixed resistance varies directly as the square of the current C in the circuit. A circuit expends 180 watts when a current of 6 amperes is flowing. What is the amount of power expended when the current is 10 amperes? [6.9f]

44. Find all numbers for which
$$\frac{3x^2 - 2x - 1}{3x^2 + x}$$
is not defined. [6.1a]

A. $1, -\frac{1}{3}$ B. $-\frac{1}{3}$

C. $0, -\frac{1}{3}$ D. $0, \frac{1}{3}$

45. Subtract: $\frac{1}{x-5} - \frac{1}{x+5}$. [6.5a]

A. $\frac{10}{(x-5)(x+5)}$ B. 0

C. $\frac{5}{x-5}$ D. $\frac{10}{x+5}$

Synthesis

46. Simplify: [6.1d], [6.2b]
$$\frac{2a^2 + 5a - 3}{a^2} \cdot \frac{5a^3 + 30a^2}{2a^2 + 7a - 4} \div \frac{a^2 + 6a}{a^2 + 7a + 12}.$$

47. Compare
$$\frac{A+B}{B} = \frac{C+D}{D}$$
with the proportion
$$\frac{A}{B} = \frac{C}{D}. \quad [6.8b]$$

Understanding Through Discussion and Writing

1. Are parentheses as important when adding rational expressions as they are when subtracting? Why or why not? [6.4a], [6.5a]

2. How can a graph be used to determine how many solutions an equation has? [6.7a]

3. How is the process of canceling related to the identity property of 1? [6.1c]

4. Determine whether the following situation represents direct variation, inverse variation, or neither. Give a reason for your answer. [6.9a, c]
The number of plays that it takes to go 80 yd for a touchdown and the average gain per play

5. Explain how a rational expression can be formed for which −3 and 4 are not allowable replacements. [6.1a]

6. Why is it especially important to check the possible solutions to a rational equation? [6.7a]

CHAPTER 6 Test

For Extra Help — For step-by-step test solutions, access the Chapter Test Prep Videos in MyLab Math.

Find all numbers for which the rational expression is not defined.

1. $\dfrac{8}{2x}$

2. $\dfrac{5}{x+8}$

3. $\dfrac{x-7}{x^2-49}$

4. $\dfrac{x^2+x-30}{x^2-3x+2}$

5. $\dfrac{11}{(x-1)^2}$

6. $\dfrac{x+2}{2}$

7. Simplify:
$$\dfrac{6x^2+17x+7}{2x^2+7x+3}.$$

8. Multiply and simplify:
$$\dfrac{a^2-25}{6a}\cdot\dfrac{3a}{a-5}.$$

9. Divide and simplify:
$$\dfrac{25x^2-1}{9x^2-6x}\div\dfrac{5x^2+9x-2}{3x^2+x-2}.$$

10. Find the LCM:
$$y^2-9,\quad y^2+10y+21,\quad y^2+4y-21.$$

Add or subtract. Simplify, if possible.

11. $\dfrac{16+x}{x^3}+\dfrac{7-4x}{x^3}$

12. $\dfrac{5-t}{t^2+1}-\dfrac{t-3}{t^2+1}$

13. $\dfrac{x-4}{x-3}+\dfrac{x-1}{3-x}$

14. $\dfrac{x-4}{x-3}-\dfrac{x-1}{3-x}$

15. $\dfrac{5}{t-1}+\dfrac{3}{t}$

16. $\dfrac{1}{x^2-16}-\dfrac{x+4}{x^2-3x-4}$

17. $\dfrac{1}{x-1}+\dfrac{4}{x^2-1}-\dfrac{2}{x^2-2x+1}$

18. Simplify: $\dfrac{9-\dfrac{1}{y^2}}{3-\dfrac{1}{y}}.$

Solve.

19. $\dfrac{7}{y}-\dfrac{1}{3}=\dfrac{1}{4}$

20. $\dfrac{15}{x}-\dfrac{15}{x-2}=-2$

Find an equation of variation in which y varies directly as x and the following are true. Then find the value of y when $x=25$.

21. $y=6$ when $x=3$

22. $y=1.5$ when $x=3$

Find an equation of variation in which y varies inversely as x and the following are true. Then find the value of y when $x = 100$.

23. $y = 6$ when $x = 3$

24. $y = 11$ when $x = 2$

25. Find an equation of variation in which Q varies jointly as x and y, and $Q = 25$ when $x = 2$ and $y = 5$.

Solve.

26. *Train Travel.* The distance d traveled by a train varies directly as the time t that it travels. The train travels 60 km in $\frac{1}{2}$ hr. How far will it travel in 2 hr?

27. *Concrete Work.* It takes 3 hr for 2 concrete mixers to mix a fixed amount of concrete. The number of hours varies inversely as the number of concrete mixers used. How long would it take 5 concrete mixers to do the same job?

28. *Quality Control.* A sample of 125 spark plugs contained 4 defective spark plugs. How many defective spark plugs would you expect to find in a sample of 500?

29. *Estimating a Zebra Population.* A game warden catches, tags, and then releases 15 zebras. A month later, a sample of 20 zebras is collected and 6 of them have tags. Use this information to estimate the size of the zebra population in that area.

30. *Copying Time.* Kopy Kwik has 2 copiers. One can copy a year-end report in 20 min. The other can copy the same document in 30 min. How long would it take both machines, working together, to copy the report?

31. *Driving Speed.* Craig drives 20 km/h faster than Marilyn. In the same time that Marilyn drives 225 km, Craig drives 325 km. Find the speed of each car.

32. This pair of triangles is similar. Find the missing length x.

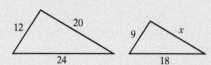

33. Solve: $\dfrac{2}{x-4} + \dfrac{2x}{x^2-16} = \dfrac{1}{x+4}$.

A. -4
B. 4
C. $4, -4$
D. No solution

Synthesis

34. Reggie and Rema work together to mulch the flower beds around an office complex in $2\frac{6}{7}$ hr. Working alone, it would take Reggie 6 hr more than it would take Rema. How long would it take each of them to complete the landscaping working alone?

35. Simplify: $1 + \dfrac{1}{1 + \dfrac{1}{1 + \dfrac{1}{a}}}$.

CHAPTERS 1–6 Cumulative Review

1. Find the absolute value: $|3.5|$.

2. Identify the degree of each term and the degree of the polynomial:
$$x^3 - 2x^2 + x - 1.$$

3. *Millennials Living with Parents.* As of July 2015, the U.S. population was approximately 321,369,000. Millennials comprised 23.4% of this number, and about 32.1% of millennials lived with their parents. For the first time in over a century, the number of millennials living with their parents surpassed the number living with a spouse or a partner. How many millennials lived with their parents?

   Data: marketingcharts.com; *AARP Bulletin*, July–August 2016

4. *Square Footage.* In the third quarter of 2008, the size of new single-family homes averaged 2438 ft², down from 2629 ft² in the second quarter. What was the percent decrease?

   Data: Gopal Ahluwalia, Director of Research, National Association of Home Builders

5. *Principal Borrowed.* Money is borrowed at 6% simple interest. After 1 year, $2650 pays off the loan. How much was originally borrowed?

6. *Car Travel.* One car travels 105 mi in the same time that a car traveling 10 mph slower travels 75 mi. Find the speed of each car.

7. *Areas.* If each side of a square is increased by 2 ft to form another square, the sum of the areas of the two squares is 452 ft². Find the length of a side of the original square.

8. *Muscle Weight.* The number of pounds of muscle M in the human body varies directly as body weight B. A person who weighs 175 lb has a muscle weight of 70 lb.
   a) Write an equation of variation that describes this situation.
   b) Mike weighs 192 lb. What is his muscle weight?

9. Collect like terms: $x^2 - 3x^3 - 4x^2 + 5x^3 - 2$.

Simplify.

10. $\dfrac{1}{2}x - \left[\dfrac{3}{8}x - \left(\dfrac{2}{3} + \dfrac{1}{4}x\right) - \dfrac{1}{3}\right]$

11. $\left(\dfrac{2x^3}{3x^{-1}}\right)^{-2}$

12. $\dfrac{\dfrac{4}{x} - \dfrac{6}{x^2}}{\dfrac{5}{x} + \dfrac{7}{2x}}$

Perform the indicated operations. Simplify, if possible.

13. $(5xy^2 - 6x^2y^2 - 3xy^3) - (-4xy^3 + 7xy^2 - 2x^2y^2)$

14. $(4x^4 + 6x^3 - 6x^2 - 4) + (2x^5 + 2x^4 - 4x^3 - 4x^2 + 3x - 5)$

15. $\dfrac{2y+4}{21} \cdot \dfrac{7}{y^2 + 4y + 4}$

16. $\dfrac{x^2 - 9}{x^2 + 8x + 15} \div \dfrac{x-3}{2x+10}$

17. $\dfrac{x^2}{x-4} + \dfrac{16}{4-x}$

18. $\dfrac{5x}{x^2-4} - \dfrac{-3}{2-x}$

Cumulative Review: Chapters 1–6   **515**

Multiply.
19. $(2.5a + 7.5)(0.4a - 1.2)$

20. $(6x - 5)^2$

21. $(2x^3 + 1)(2x^3 - 1)$

Factor.
22. $9a^2 + 52a - 12$

23. $9x^2 - 30xy + 25y^2$

24. $49x^2 - 1$

Solve.
25. $x - [x - (x - 1)] = 2$

26. $2x^2 + 7x = 4$

27. $x^2 = 10x$

28. $3(x - 2) \leq 4(x + 5)$

29. $\dfrac{5x - 2}{4} - \dfrac{4x - 5}{3} = 1$

30. $t = ax + ay$, for a

Find the slope, if it exists, of the line containing the given pair of points.

31. $(-2, 6)$ and $(-2, -1)$

32. $(-4, 1)$ and $(3, -2)$

33. $\left(-\dfrac{1}{2}, 4\right)$ and $\left(3\dfrac{1}{2}, -5\right)$

34. $\left(-7, \dfrac{3}{4}\right)$ and $\left(-4, \dfrac{3}{4}\right)$

For each equation, find the coordinates of the y-intercept and the x-intercept. Do not graph.

35. $-8x - 24y = 48$

36. $15 - 40x = -120y$

37. $y = 25$

38. $x = -\dfrac{1}{4}$

Graph on a plane.

39. $x = -3$

40. $y = -3$

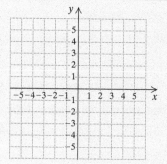 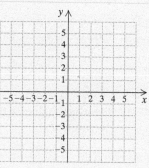

41. $3x - 5y = 15$

42. $2x - 6y = 12$

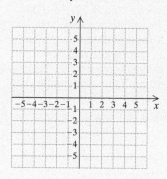 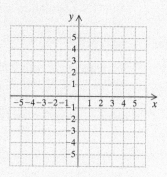

43. $y = -\dfrac{1}{3}x - 2$

44. $x - y = -5$

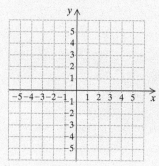 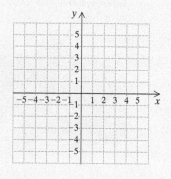

Synthesis

45. Find all numbers for which the following complex rational expression is not defined:

$$\dfrac{\dfrac{1}{x} + x}{2 + \dfrac{1}{x - 3}}.$$

CHAPTER 7

Graphs, Functions, and Applications

7.1 Functions and Graphs
7.2 Finding Domain and Range

Mid-Chapter Review

7.3 Linear Functions: Graphs and Slope
7.4 More on Graphing Linear Equations

Visualizing for Success

7.5 Finding Equations of Lines; Applications

Summary and Review
Test
Cumulative Review

With the increasing numbers of both cars with better gas mileage and electric cars, U.S. gas tax revenue is expected to decrease in the next decade. When planning for needed cash to build and repair roads, states are considering increasing the number of tollways in order to generate the revenue lost due to declining gas tax revenue.

Data: Joint Commission on Taxation

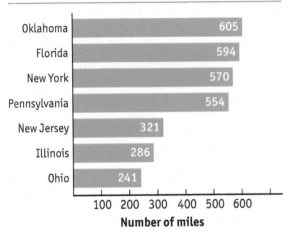

States with the Most Miles of Tollways

- Oklahoma: 605
- Florida: 594
- New York: 570
- Pennsylvania: 554
- New Jersey: 321
- Illinois: 286
- Ohio: 241

Number of miles

DATA: International Bridge, Tunnel, and Tollway Association

In Exercise 52 of Exercise Set 7.5, we will find a linear function that fits the gas tax revenue data and then use the function to determine when the tax revenue will be $19.8 billion.

517

STUDYING FOR SUCCESS   *Doing Your Homework*

- ☐ Prepare for your homework by reading explanations of concepts and by following the step-by-step solutions of examples in the text.
- ☐ Include all the steps. This will keep your work organized, help you to avoid computational errors, and give you a study guide for an exam.
- ☐ Try to do your homework as soon as possible after each class. Avoid waiting until a deadline is near to begin work on an assignment.

7.1 Functions and Graphs

OBJECTIVES

a Determine whether a correspondence is a function.

b Given a function described by an equation, find function values (outputs) for specified values (inputs).

c Draw the graph of a function.

d Determine whether a graph is that of a function using the vertical-line test.

e Solve applied problems involving functions and their graphs.

a IDENTIFYING FUNCTIONS

Consider the equation $y = 2x - 3$. If we substitute a value for x—say, 5—we get a value for y, 7:

$$y = 2x - 3 = 2(5) - 3 = 10 - 3 = 7.$$

The equation $y = 2x - 3$ is an example of a *function*, one of the most important concepts in mathematics.

In much the same way that ordered pairs form correspondences between first and second coordinates, a *function* is a correspondence from one set to another. For example:

To each student in a college, there corresponds his or her student ID.

To each item in a store, there corresponds its price.

To each real number, there corresponds the cube of that number.

In each case, the first set is called the **domain** and the second set is called the **range**. Each of these correspondences is a **function**, because given a member of the domain, there is *just one* member of the range to which it corresponds. Given a student, there is *just one* ID. Given an item, there is *just one* price. Given a real number, there is *just one* cube.

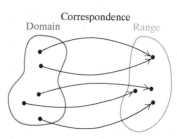
Correspondence
Domain   Range

EXAMPLE 1   Determine whether the correspondence is a function.

f:
Domain → Range
1 → $107.40
2 → $34.10
3 → $29.60
4 → $19.60

g:
Domain → Range
3 → 5
4 → 9
5 → -7
6 →

h:
Domain → Range
Chicago → Cubs
Chicago → White Sox
Baltimore → Orioles
San Diego → Padres

p:
Domain → Range
Cubs → Chicago
White Sox → Chicago
Orioles → Baltimore
Padres → San Diego

The correspondence *f is* a function because each member of the domain is matched to *only one* member of the range.

The correspondence *g is* a function because each member of the domain is matched to *only one* member of the range. Note that a function allows two or more members of the domain to correspond to the same member of the range.

The correspondence *h is not* a function because one member of the domain, Chicago, is matched to *more than one* member of the range.

The correspondence *p is* a function because each member of the domain is matched to *only one* member of the range.

> **FUNCTION; DOMAIN; RANGE**
>
> A **function** is a correspondence between a first set, called the **domain**, and a second set, called the **range**, such that each member of the domain corresponds to **exactly one** member of the range.

Determine whether each correspondence is a function.

1. Domain   Range

   Cheetah ⟶ 70 mph
   Human   ⟶ 28 mph
   Lion   ⟶ 50 mph
   Chicken ⟶ 9 mph

2. Domain   Range

   A ⟶ a
   B ⟶ b
   C ⟶ c
   D ⟶ d
         e

3. Domain   Range

   -2 ⟶ 4
   2
   -3 ⟶ 9
   3
   0 ⟶ 0

4. Domain   Range

   4 ⟶ -2
            2
   9 ⟶ -3
            3
   0 ⟶ 0

Do Exercises 1–4.

EXAMPLE 2 Determine whether each correspondence is a function.

Domain	Correspondence	Range
a) The integers	Each number's square	A set of nonnegative integers
b) A set of presidents (listed below)	Each president's appointees to the Supreme Court	A set of Supreme Court Justices (listed below)

APPOINTING PRESIDENT	SUPREME COURT JUSTICE
George H. W. Bush	Samuel A. Alito, Jr.
William Jefferson Clinton	Stephen G. Breyer
	Ruth Bader Ginsburg
	Neil Gorsuch
George W. Bush	Elena Kagan
	John G. Roberts, Jr.
Barack H. Obama	Sonia M. Sotomayor
Donald J. Trump	Clarence Thomas

a) The correspondence *is* a function because each integer has *only one* square.

b) The correspondence *is not* a function because there is at least one member of the domain who is paired with more than one member of the range (William Jefferson Clinton with Stephen G. Breyer and Ruth Bader Ginsburg; George W. Bush with Samuel A. Alito, Jr., and John G. Roberts, Jr.; Barack H. Obama with Elena Kagan and Sonia M. Sotomayor).

Do Exercises 5–7 on the following page.

Answers
1. Yes   **2.** No   **3.** Yes   **4.** No

Determine whether each correspondence is a function.

5. *Domain*
A set of numbers
Correspondence
Square each number and subtract 10.
Range
A set of numbers

6. *Domain*
A set of polygons
Correspondence
Find the perimeter of each polygon.
Range
A set of numbers

7. Determine whether the correspondence is a function.
Domain
A set of numbers
Correspondence
The area of a rectangle
Range
A set of rectangles

.... **Caution!** ....

The notation $f(x)$ *does not mean "f times x"* and should not be read that way.

SKILL REVIEW *Evaluate algebraic expressions by substitution.* [1.1a]
Evaluate.
1. $-\frac{1}{4}x$, when $x = 40$
2. $y^2 - 2y + 6$, when $y = -1$

Answers: **1.** -10 **2.** 9

MyLab Math VIDEO

When a correspondence between two sets is not a function, it is still an example of a **relation**.

> **RELATION**
>
> A **relation** is a correspondence between a first set, called the **domain**, and a second set, called the **range**, such that each member of the domain corresponds to **at least one** member of the range.

Thus, although the correspondences of Examples 1 and 2 are not all functions, they *are* all relations. A function is a special type of relation—one in which each member of the domain is paired with *exactly one* member of the range.

b FINDING FUNCTION VALUES

Most functions considered in mathematics are described by equations like $y = 2x + 3$ or $y = 4 - x^2$. We graph the function $y = 2x + 3$ by first performing calculations like the following:

for $x = 4, y = 2x + 3 = 2 \cdot 4 + 3 = 8 + 3 = 11$;
for $x = -5, y = 2x + 3 = 2 \cdot (-5) + 3 = -10 + 3 = -7$;
for $x = 0, y = 2x + 3 = 2 \cdot 0 + 3 = 0 + 3 = 3$; and so on.

For $y = 2x + 3$, the **inputs** (members of the domain) are values of x substituted into the equation. The **outputs** (members of the range) are the resulting values of y. If we call the function f, we can use x to represent an arbitrary *input* and $f(x)$—read "f of x," or "f at x," or "the value of f at x"— to represent the corresponding *output*. In this notation, the function given by $y = 2x + 3$ is written as $f(x) = 2x + 3$ and the calculations above can be written more concisely as follows:

$y = f(4) = 2 \cdot 4 + 3 = 8 + 3 = 11$;
$y = f(-5) = 2 \cdot (-5) + 3 = -10 + 3 = -7$;
$y = f(0) = 2 \cdot 0 + 3 = 0 + 3 = 3$; and so on.

Thus instead of writing "when $x = 4$, the value of y is 11," we can simply write "$f(4) = 11$," which can also be read as "f of 4 is 11" or "for the input 4, the output of f is 11."

We can think of a function as a machine. Think of $f(4) = 11$ as putting 4, a member of the domain (an input), into the machine. The machine knows the correspondence $f(x) = 2x + 3$, multiplies 4 by 2 and adds 3, and produces 11, a member of the range (the output).

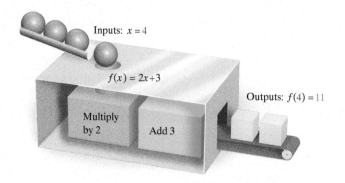

Answers
5. Yes **6.** Yes **7.** No

EXAMPLE 3 A function f is given by $f(x) = 3x^2 - 2x + 8$. Find each of the indicated function values.

a) $f(0)$ **b)** $f(-5)$ **c)** $f(7a)$

One way to find function values when a formula is given is to think of the formula with blanks, or placeholders, replacing the variable as follows:

$$f(\square) = 3\square^2 - 2\square + 8.$$

To find an output for a given input, we think: "Whatever goes in the blank on the left goes in the blank(s) on the right." With this in mind, let's complete the example.

a) $f(0) = 3 \cdot 0^2 - 2 \cdot 0 + 8 = 8$
b) $f(-5) = 3(-5)^2 - 2 \cdot (-5) + 8 = 3 \cdot 25 + 10 + 8 = 75 + 10 + 8 = 93$
c) $f(7a) = 3(7a)^2 - 2(7a) + 8 = 3 \cdot 49a^2 - 2(7a) + 8 = 147a^2 - 14a + 8$

Do Exercise 8.

EXAMPLE 4 Find the indicated function value.

a) $f(5)$, for $f(x) = 3x + 2$ **b)** $g(-2)$, for $g(x) = 7$
c) $F(a + 1)$, for $F(x) = 5x - 8$ **d)** $f(a + h)$, for $f(x) = -2x + 1$

a) $f(5) = 3 \cdot 5 + 2 = 15 + 2 = 17$
b) For the function given by $g(x) = 7$, all inputs share the same output, 7. Thus, $g(-2) = 7$. The function g is an example of a **constant function**.
c) $F(a + 1) = 5(a + 1) - 8 = 5a + 5 - 8 = 5a - 3$
d) $f(a + h) = -2(a + h) + 1 = -2a - 2h + 1$

Do Exercise 9.

8. Find the indicated function values for the function
$f(x) = 2x^2 + 3x - 4.$

a) $f(8) = 2 \cdot \boxed{}^2 + 3 \cdot \boxed{} - 4$
$= 2 \cdot \boxed{} + 3 \cdot 8 - 4$
$= \boxed{} + 24 - 4$
$= 152 - 4$
$= \boxed{}$

b) $f(0)$
c) $f(-5)$
d) $f(2a)$

9. Find the indicated function value.
a) $f(-6)$, for $f(x) = 5x - 3$
b) $g(55)$, for $g(x) = -3$
c) $F(a + 2)$, for $F(x) = -5x + 8$
d) $f(a - h)$, for $f(x) = 6x - 7$

Answers
8. (a) 148; (b) −4; (c) 31; (d) $8a^2 + 6a - 4$ 9. (a) −33; (b) −3; (c) −5a − 2; (d) 6a − 6h − 7
Guided Solution:
8. (a) 8, 8, 64, 128, 148

CALCULATOR CORNER

Finding Function Values We can find function values using a graphing calculator. One method is to substitute inputs directly into the formula. Consider the function $f(x) = x^2 + 3x - 4$. We find that $f(-5) = 6$. See Figure 1.

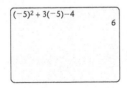
FIGURE 1

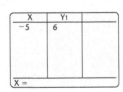
FIGURE 2

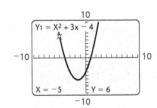
FIGURE 3

After we have entered the function as $y_1 = x^2 + 3x - 4$ on the equation-editor screen, there are other methods that we can use to find function values. We can use a table set in ASK mode and enter $x = -5$. We see that the function value, y_1, is 6. See Figure 2. We can also use the VALUE feature to evaluate the function. To do this, we first graph the function. Then we press 2ND CALC 1 to access the VALUE feature. Next, we supply the desired x-value. Finally, we press ENTER to see X = −5, Y = 6 at the bottom of the screen. See Figure 3. Again we see that the function value is 6. Note that when the VALUE feature is used to find a function value, the x-value must be in the viewing window.

EXERCISES: Find each function value.

1. $f(-5.1)$, for $f(x) = -3x + 2$

2. $f(3)$, for $f(x) = 4x^2 + x - 5$

C GRAPHS OF FUNCTIONS

To graph a function, we find ordered pairs (x, y) or $(x, f(x))$, plot them, and connect the points. Note that y and $f(x)$ are used interchangeably—that is, $y = f(x)$—when we are working with functions and their graphs.

EXAMPLE 5 Graph: $f(x) = x + 2$.

A list of some function values is shown in the following table. We plot the points and connect them. The graph is a straight line. The "y" on the vertical axis could also be labeled "$f(x)$."

x	$f(x)$
-4	-2
-3	-1
-2	0
-1	1
0	2
1	3
2	4
3	5
4	6

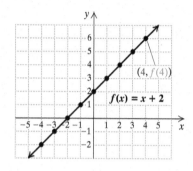

10. Graph: $f(x) = x - 4$.

x	$f(x)$

◀ Do Exercise 10.

EXAMPLE 6 Graph: $g(x) = 4 - x^2$.

We calculate some function values, plot the corresponding points, and draw the curve.

$g(0) = 4 - 0^2 = 4 - 0 = 4,$
$g(-1) = 4 - (-1)^2 = 4 - 1 = 3,$
$g(1) = 4 - (1^2) = 4 - 1 = 3,$
$g(-2) = 4 - (-2)^2 = 4 - 4 = 0,$
$g(2) = 4 - 2^2 = 4 - 4 = 0,$
$g(-3) = 4 - (-3)^2 = 4 - 9 = -5$
$g(3) = 4 - 3^2 = 4 - 9 = -5$

11. Graph: $g(x) = 5 - x^2$.

x	$g(x)$

x	$g(x)$
-3	-5
-2	0
-1	3
0	4
1	3
2	0
3	-5

◀ Do Exercise 11.

Answers

10.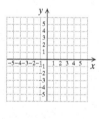
$f(x) = x - 4$

11.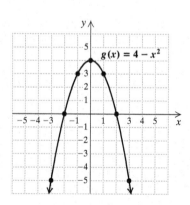
$g(x) = 5 - x^2$

EXAMPLE 7 Graph: $h(x) = |x|$.

A list of some function values is shown in the following table. We plot the points and connect them. The graph is a V-shaped "curve" that rises on either side of the vertical axis.

x	$h(x)$
-3	3
-2	2
-1	1
0	0
1	1
2	2
3	3

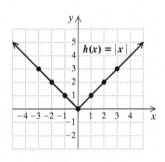

Do Exercise 12.

d THE VERTICAL-LINE TEST

Consider the graph of the function f described by $f(x) = x^2 - 5$ shown at right. It is also the graph of the equation $y = x^2 - 5$.

To find a function value, like $f(3)$, from a graph, we locate the input on the horizontal axis, move directly up or down to the graph of the function, and then move left or right to find the output on the vertical axis. Thus, $f(3) = 4$. Keep in mind that members of the domain are found on the horizontal axis, members of the range are found on the vertical axis, and the y on the vertical axis could also be labeled $f(x)$.

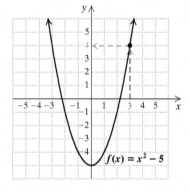

When one member of the domain is paired with two or more different members of the range, the correspondence is not a function. Thus, when a graph contains two or more different points with the same first coordinate, the graph cannot represent a function. Points sharing a common first coordinate are vertically above or below each other. (See the following graph.) This observation leads to the *vertical-line test*.

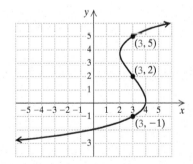

Since 3 is paired with more than one member of the range, the graph does not represent a function.

THE VERTICAL-LINE TEST

If it is possible for a vertical line to cross a graph more than once, then the graph is *not* the graph of a function.

12. Graph: $t(x) = 3 - |x|$.

x	$t(x)$

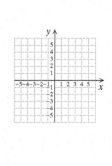

CALCULATOR CORNER

Graphing Functions
To graph a function using a graphing calculator, we replace the function notation with y and proceed as described in the Calculator Corner on p. 211. To graph $f(x) = 2x^2 + x$ in the standard window, for example, we replace $f(x)$ with y and enter $y_1 = 2x^2 + x$ on the Y = screen and then press ZOOM 6.

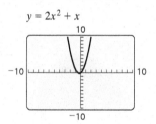

EXERCISES: Graph each function.
1. $f(x) = x - 4$
2. $f(x) = -2x - 3$
3. $h(x) = 1 - x^2$
4. $f(x) = 3x^2 - 4x + 1$
5. $f(x) = x^3$
6. $f(x) = |x + 3|$

Answer
12.
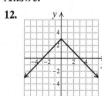
$t(x) = 3 - |x|$

Determine whether each of the following is the graph of a function.

13.

14.
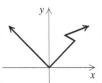

15.
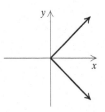

16.
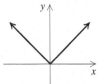

EXAMPLE 8 Determine whether each of the following is the graph of a function.

a)
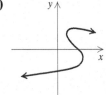

b)
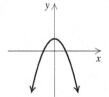

a) The graph *is not* that of a function because a vertical line can cross the graph at more than one point.

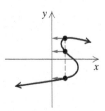

b) The graph *is* that of a function because no vertical line can cross the graph more than once.

◀ Do Exercises 13–16.

e APPLICATIONS OF FUNCTIONS AND THEIR GRAPHS

Functions are often described by graphs, whether or not an equation is given. To use a graph in an application, we note that each point on the graph represents a pair of values.

EXAMPLE 9 *Gift-Card Sales.* The following graph represents worldwide gift-card sales from 2008 to 2018. This amount of sales, in billions of dollars, is a function of the year. Note that no equation is given for the function.

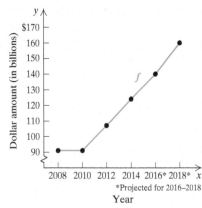

DATA: wallethub.com

a) What was the total amount spent on gift cards in 2012? That is, find $f(2012)$.

b) What is the projected amount spend on gift cards in 2017? That is, find $f(2017)$.

Answers
13. Yes 14. No 15. No 16. Yes

a) To estimate the amount spent on gift cards in 2012, we locate 2012 on the horizontal axis and move directly up until we reach the graph. Then we move across to the vertical axis. We come to a point that is about 107, so we estimate the amount spent on gift cards in 2012 to be $107 billion.

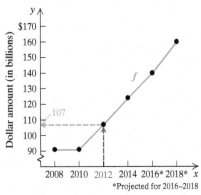

DATA: wallethub.com

DATA: wallethub.com

b) To estimate the amount spent on gift cards in 2017, we locate 2017 on the horizontal axis and move directly up until we reach the graph. Then we move across to the vertical axis. We come to a point that is about 149, so we estimate the amount spent on gift cards in 2017 to be $149 billion.

Do Exercises 17 and 18.

Refer to the graph in Example 9.

17. What was the total spent on gift cards in 2014?

18. What was the total spent on gift cards in 2011?

Answers
17. $124 billion **18.** $99 billion

7.1 Exercise Set

FOR EXTRA HELP — MyLab Math

✓ Check Your Understanding

Reading Check Choose from the list on the right the word(s) that complete(s) each definition. Some words will be used more than once.

RC1. A function is a correspondence between a first set, called the _____, and a second set, called the _____, such that each member of the _____ corresponds to _____ member of the _____.

RC2. A relation is a correspondence between a first set, called the _____, and a second set, called the _____, such that each member of the _____ corresponds to _____ member of the _____.

at least one
exactly one
domain
range

Concept Check Use the graph at right to find the given function value by locating the input on the horizontal axis, moving directly up or down to the graph of the function, and then moving left or right to find the output on the vertical axis. As an example, finding $f(4) = -5$ is illustrated.

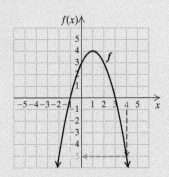

CC1. $f(2)$

CC2. $f(0)$

CC3. $f(-2)$

CC4. $f(3)$

a Determine whether each correspondence is a function.

1. Domain → Range
2 → 9
5 → 8
19 →

2. Domain → Range
5 → 3
−3 → 7
7 →
−7 →

3. Domain → Range
−5 → 1
5 →
8 →

4. Domain → Range
6 → −6
7 → −7
3 → −3

5. Domain → Range
9 → 3, −3
16 → 4, −4
25 → 5, −5

6. Domain → Range
The Color Purple, 1982 (Pulitzer Prize 1983) → Ray Bradbury
East of Eden, 1952 → Pearl Buck
Fahrenheit 451, 1953 → Ernest Hemingway
The Good Earth, 1931 (Pulitzer Prize 1932) → Harper Lee
For Whom the Bell Tolls, 1940 → John Steinbeck
The Grapes of Wrath, 1939 (Pulitzer Prize 1940) → Alice Walker
To Kill a Mockingbird, 1960 (Pulitzer Prize 1961)
The Old Man and the Sea, 1952 (Pulitzer Prize 1953)

7. Domain → Range
Florida ↔ Florida State University, University of Florida, University of Miami
Kansas ↔ Baker University, Kansas State University, University of Kansas

8. Domain → Range
Colorado State University, University of Colorado, University of Denver → Colorado
Gonzaga University, University of Washington, Washington State University → Washington

	Domain	Correspondence	Range
9.	A set of numbers	The area of a triangle	A set of triangles
10.	A family	Each person's height, in inches	A set of positive numbers
11.	The set of U.S. Senators	The state that a Senator represents	The set of all states
12.	The set of all states	Each state's members of the U.S. Senate	The set of U.S. Senators

West Virginia

Shelley Moore Capito

Joe Manchin III

b Find the function values.

13. $f(x) = x + 5$
   a) $f(4)$ b) $f(7)$
   c) $f(-3)$ d) $f(0)$
   e) $f(2.4)$ f) $f(\frac{2}{3})$

14. $g(t) = t - 6$
   a) $g(0)$ b) $g(6)$
   c) $g(13)$ d) $g(-1)$
   e) $g(-1.08)$ f) $g(\frac{7}{8})$

15. $h(p) = 3p$
   a) $h(-7)$ b) $h(5)$
   c) $h(\frac{2}{3})$ d) $h(0)$
   e) $h(6a)$ f) $h(a + 1)$

16. $f(x) = -4x$
   a) $f(6)$ b) $f(-\frac{1}{2})$
   c) $f(0)$ d) $f(-1)$
   e) $f(3a)$ f) $f(a - 1)$

17. $g(s) = 3s + 4$
   a) $g(1)$ b) $g(-7)$
   c) $g(\frac{2}{3})$ d) $g(0)$
   e) $g(a - 2)$ f) $g(a + h)$

18. $h(x) = 19$, a constant function
   a) $h(4)$ b) $h(-6)$
   c) $h(12.5)$ d) $h(0)$
   e) $h(\frac{2}{3})$ f) $h(a + 3)$

19. $f(x) = 2x^2 - 3x$
   a) $f(0)$ b) $f(-1)$
   c) $f(2)$ d) $f(10)$
   e) $f(-5)$ f) $f(4a)$

20. $f(x) = 3x^2 - 2x + 1$
   a) $f(0)$ b) $f(1)$
   c) $f(-1)$ d) $f(10)$
   e) $f(-3)$ f) $f(2a)$

21. $f(x) = |x| + 1$
   a) $f(0)$ b) $f(-2)$
   c) $f(2)$ d) $f(-10)$
   e) $f(a - 1)$ f) $f(a + h)$

22. $g(t) = |t - 1|$
   a) $g(4)$ b) $g(-2)$
   c) $g(-1)$ d) $g(100)$
   e) $g(5a)$ f) $g(a + 1)$

23. $f(x) = x^3$
   a) $f(0)$ b) $f(-1)$
   c) $f(2)$ d) $f(10)$
   e) $f(-5)$ f) $f(-3a)$

24. $f(x) = x^4 - 3$
   a) $f(1)$ b) $f(-1)$
   c) $f(0)$ d) $f(2)$
   e) $f(-2)$ f) $f(-a)$

25. *Gold Production in China.* The function $G(x) = 0.943x + 10.943$ can be used to estimate the production of gold, in thousands of troy ounces, in China from 2010 through 2015. Let $G(x) =$ the amount of gold produced and $x =$ the number of years after 2010. What was the amount of gold produced in China in 2012? in 2015?

Data: Mineral Commodity Summaries 2015, U.S. Geological Survey, U.S. Department of Interior

26. *International Travelers to the United States.* The function $T(x) = 3.72x + 55.54$ can be used to estimate the number of international visitors, in millions, to the United States from 2009 through 2015. Let $T(x) =$ the number of international travelers and $x =$ the number of years after 2009. How many international travelers, to the nearest million, visited the United States in 2011? in 2014?

Data: National Travel and Tourism Office, International Trade Administration, U.S. Department of Commerce; World Tourism Organization

27. *Pressure at Sea Depth.* The function $P(d) = 1 + (d/33)$ gives the pressure, in *atmospheres* (atm), at a depth of d feet in the sea. Note that $P(0) = 1$ atm, $P(33) = 2$ atm, and so on. Find the pressure at 20 ft, 30 ft, and 100 ft.

28. *Temperature as a Function of Depth.* The function $T(d) = 10d + 20$ gives the temperature, in degrees Celsius, inside the earth as a function of the depth d, in kilometers. Find the temperature at 5 km, 20 km, and 1000 km.

29. *Melting Snow.* The function $W(d) = 0.112d$ approximates the amount of water, in centimeters, that results from d centimeters of snow melting. Find the amount of water that results from melting of snow with a depth of 16 cm, 25 cm, and 100 cm.

30. *Temperature Conversions.* The function $C(F) = \frac{5}{9}(F - 32)$ determines the Celsius temperature that corresponds to F degrees Fahrenheit. Find the Celsius temperature that corresponds to 62°F, 77°F, and 23°F.

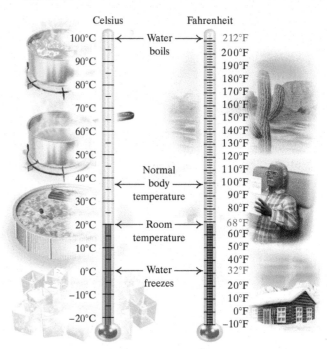

C Graph each function.

31. $f(x) = -2x$

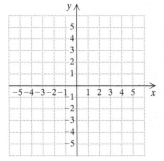

32. $g(x) = 3x$

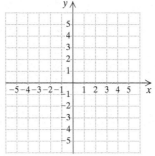

33. $g(x) = 3x - 1$

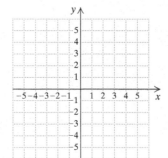

34. $f(x) = 2x + 5$

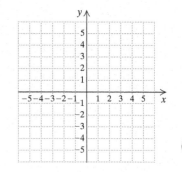

35. $g(x) = -2x + 3$

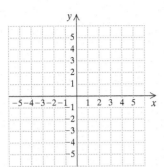

36. $f(x) = -\frac{1}{2}x + 2$

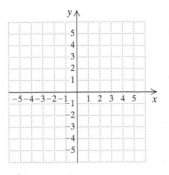

37. $f(x) = \frac{1}{2}x + 1$

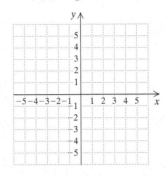

38. $f(x) = -\frac{3}{4}x - 2$

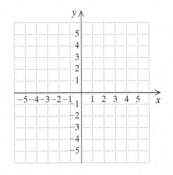

39. $f(x) = 2 - |x|$

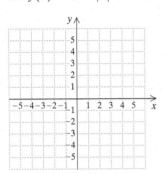

40. $f(x) = |x| - 4$

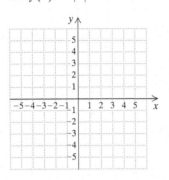

41. $g(x) = |x - 1|$

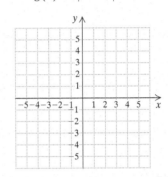

42. $g(x) = |x + 3|$

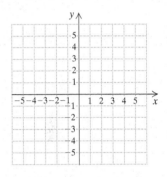

43. $g(x) = x^2 + 2$

x	g(x)
-2	
-1	
0	
1	
2	

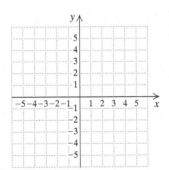

44. $f(x) = x^2 + 1$

x	f(x)
-2	
-1	
0	
1	
2	

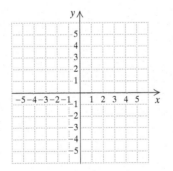

45. $f(x) = x^2 - 2x - 3$

x	f(x)
-2	
-1	
0	
1	
2	
3	
4	

46. $f(x) = x^2 + 6x + 5$

x	f(x)
-6	
-5	
-4	
-3	
-2	
-1	
0	

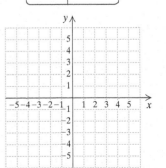

47. $f(x) = -x^2 + 1$ **48.** $f(x) = -x^2 + 2$ **49.** $f(x) = x^3 + 1$ **50.** $f(x) = x^3 - 2$

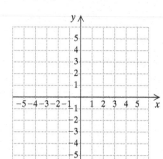 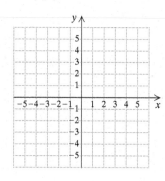 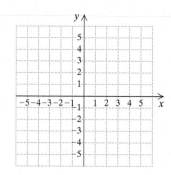 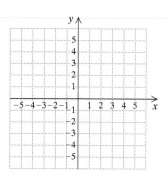

d Determine whether each of the following is the graph of a function.

51. 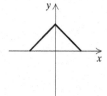   **52.** 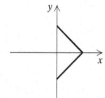   **53.** 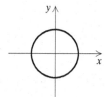   **54.**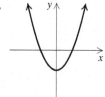

55. 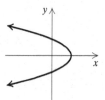   **56.** 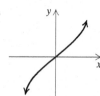   **57.** 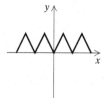   **58.**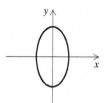

e Solve.

Median Price of a Home. The following graph approximates the median price of a home in the United States from 2011 through 2016. The price is a function f of the year x.

59. Approximate the median price of a home in 2014.

60. Approximate the median price of a home in 2012.

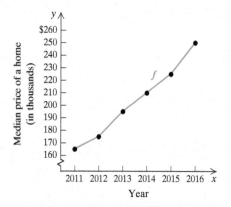

DATA: National Association of REALTORS®

Scheduled Airline Passengers. The following graph approximates the number of scheduled airline passengers in the United States from 2000 through 2016. The number of airline passengers is a function g of the year x.

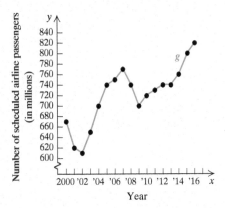

DATA: Airlines for America; airlines.org

61. Approximate the number of passengers in 2010.

62. Approximate the number of passengers in 2016.

63. Approximate the number of passengers in 2003.

64. Approximate the number of passengers in 2011.

Skill Maintenance

Solve.

65. $4 - 7y > 2y - 32$ [2.7e]

66. $\frac{2}{3}(4x - 2) \geq 60$ [2.7e]

67. $7y - 2 = 3 + 7y$ [2.3c]

68. $x^2 + 2x - 80 = 0$ [5.8b]

69. $13x - 5 - x = 2(x + 5)$ [2.3c]

70. $\frac{1}{16}x + 4 = \frac{5}{8}x - 1$ [2.3b]

Synthesis

71. Suppose that for some function g, $g(x - 6) = 10x - 1$. Find $g(-2)$.

72. Suppose that for some function h, $h(x + 5) = x^2 - 4$. Find $h(3)$.

For Exercises 73 and 74, let $f(x) = 3x^2 - 1$ and $g(x) = 2x + 5$.

73. Find $f(g(-4))$ and $g(f(-4))$.

74. Find $f(g(-1))$ and $g(f(-1))$.

75. Suppose that a function g is such that $g(-1) = -7$ and $g(3) = 8$. Find a formula for g if $g(x)$ is of the form $g(x) = mx + b$, where m and b are constants.

7.2 Finding Domain and Range

OBJECTIVE

a Find the domain and the range of a function.

SKILL REVIEW
Solve equations using both the addition principle and the multiplication principle. [2.3a]

Solve.
1. $6x - 3 = 51$
2. $15 - 2x = 0$

Answers: 1. 9 2. $\frac{15}{2}$, or 7.5

1. Find the domain and the range of the function f whose graph is shown below.

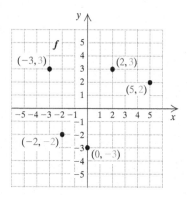

a FINDING DOMAIN AND RANGE

The solutions of an equation in two variables consist of a set of ordered pairs. A set of ordered pairs is called a **relation**. When a set of ordered pairs is such that no two different pairs share a common first coordinate, we have a **function**. The **domain** is the set of all first coordinates, and the **range** is the set of all second coordinates.

EXAMPLE 1 Find the domain and the range of the function f whose graph is shown below.

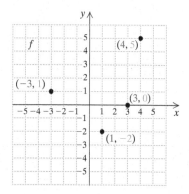

This function contains just four ordered pairs and it can be written as

$$\{(-3, 1), (1, -2), (3, 0), (4, 5)\}.$$

We can determine the domain and the range by reading the x- and y-values directly from the graph.

The domain is the set of all first coordinates, or x-values, $\{-3, 1, 3, 4\}$. The range is the set of all second coordinates, or y-values, $\{1, -2, 0, 5\}$.

◀ Do Exercise 1.

EXAMPLE 2 For the function f whose graph is shown at right, determine each of the following.

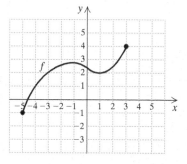

a) The number in the range that is paired with 1 from the domain. That is, find $f(1)$.

b) The domain of f

c) The numbers in the domain that are paired with 1 from the range. That is, find all x such that $f(x) = 1$.

d) The range of f

a) To determine which number in the range is paired with 1 in the domain, we locate 1 on the horizontal axis. Next, we find the point on the graph of f for which 1 is the first coordinate. From that point, we can look to the vertical axis to find the corresponding y-coordinate, 2. The input 1 has the output 2—that is, $f(1) = 2$.

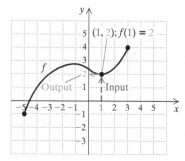

Answer
1. Domain = $\{-3, -2, 0, 2, 5\}$; range = $\{3, -2, -3, 2\}$

b) The domain of the function is the set of all x-values, or inputs, of the points on the graph. These extend from −5 to 3 and can be viewed as the curve's shadow, or projection, onto the x-axis. Thus the domain is $\{x \mid -5 \leq x \leq 3\}$.

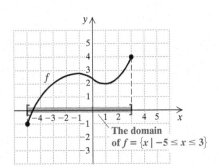

c) To determine which numbers in the domain are paired with 1 in the range, we locate 1 on the vertical axis. From there, we look left and right to the graph of f to find any points for which 1 is the second coordinate (output). One such point exists, (−4, 1). For this function, we note that $x = -4$ is the only member of the domain paired with 1. For other functions, there might be more than one member of the domain paired with a member of the range.

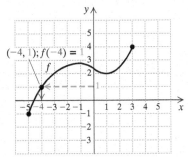

2. For the function f whose graph is shown below, determine each of the following.

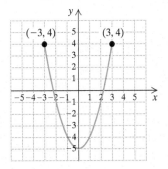

a) The number in the range that is paired with the input 1. That is, find $f(1)$.
b) The domain of f
c) The numbers in the domain that are paired with 4
d) The range of f

d) The range of the function is the set of all y-values, or outputs, of the points on the graph. These extend from −1 to 4 and can be viewed as the curve's shadow, or projection, onto the y-axis. Thus the range is $\{y \mid -1 \leq y \leq 4\}$.

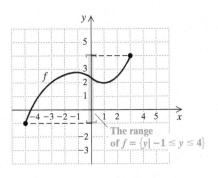

Do Exercise 2.

EXAMPLE 3 Find the domain and the range of the function h whose graph is shown below.

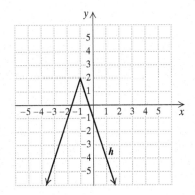

3. Find the domain and the range of the function f whose graph is shown below.

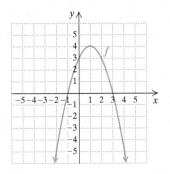

Since no endpoints are indicated, the graph extends indefinitely horizontally. Thus the domain, or the set of inputs, is the set of all real numbers. The range, or the set of outputs, is the set of all y-values of the points on the graph. Thus the range is $\{y \mid y \leq 2\}$.

Do Exercise 3.

Answers

2. (a) −4; (b) $\{x \mid -3 \leq x \leq 3\}$; (c) −3, 3; (d) $\{y \mid -5 \leq y \leq 4\}$
3. Domain: all real numbers; range: $\{y \mid y \leq 4\}$.

SECTION 7.2 Finding Domain and Range

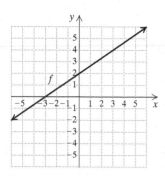

EXAMPLE 4 Find the domain and the range of the function f whose graph is shown at left.

Since no endpoints are indicated, the graph extends indefinitely both horizontally and vertically. Thus the domain is the set of all real numbers. Likewise, the range is the set of all real numbers.

◀ Do Exercise 4.

When a function is given by an equation or a formula, the domain is understood to be the largest set of real numbers (inputs) for which function values (outputs) can be calculated. That is, the domain is the set of all possible allowable inputs into the formula. To find the domain, think, "What can we substitute?"

4. Find the domain and the range of the function f whose graph is shown below.

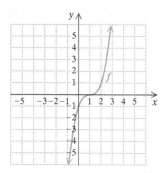

EXAMPLE 5 Find the domain: $f(x) = |x|$.

We ask, "What can we substitute?" Is there any number x for which we cannot calculate $|x|$? The answer is no. Thus the domain of f is the set of all real numbers.

EXAMPLE 6 Find the domain: $f(x) = \dfrac{3}{2x - 5}$.

We ask, "What can we substitute?" Is there any number x for which we cannot calculate $3/(2x - 5)$? Since $3/(2x - 5)$ cannot be calculated when the denominator $2x - 5$ is 0, we solve the following equation to find those real numbers that must be excluded from the domain of f:

$$2x - 5 = 0 \quad \text{Setting the denominator equal to 0}$$
$$2x = 5 \quad \text{Adding 5}$$
$$x = \tfrac{5}{2}. \quad \text{Dividing by 2}$$

Thus, $\tfrac{5}{2}$ is not in the domain, whereas all other real numbers are.
The domain of f is $\{x \mid x \text{ is a real number } and\ x \neq \tfrac{5}{2}\}$.

◀ Do Exercises 5 and 6.

Find the domain.

5. $f(x) = x^3 - |x|$

6. $f(x) = \dfrac{4}{3x + 2}$

Set the denominator equal to 0 and solve for x:

$$3x + 2 = \boxed{}$$
$$3x = -2$$
$$x = \boxed{}.$$

Thus, $\boxed{}$ is not in the domain of $f(x)$; all other real numbers are.

Domain = $\{x \mid x \text{ is a real number } and\ x \neq \boxed{}\}$.

Functions: A Review

The following is a review of the function concepts considered in Sections 7.1 and 7.2. Use the graph below to visualize the concepts.

Function Concepts

- Formula for f: $f(x) = x^2 - 7$
- For every input of f, there is exactly one output.
- When 1 is the input, -6 is the output.
- $f(1) = -6$
- $(1, -6)$ is on the graph.
- Domain = The set of all inputs = The set of all real numbers
- Range = The set of all outputs = $\{y \mid y \geq -7\}$

Graph

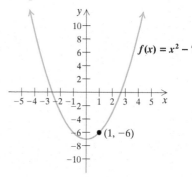

Answers

4. Domain: all real numbers; range: all real numbers   5. All real numbers
6. $\left\{x \mid x \text{ is a real number } and\ x \neq -\dfrac{2}{3}\right\}$.

Guided Solution:

6. $0, -\dfrac{2}{3}, -\dfrac{2}{3}, -\dfrac{2}{3}$

7.2 Exercise Set

✓ Check Your Understanding

Reading and Concept Check Choose from the column on the right the domain of the function. Some choices may be used more than once, others not at all.

RC1. $f(x) = 5 - x$ RC2. $f(x) = \dfrac{-5}{5 - x}$

RC3. $f(x) = |5 - x|$ RC4. $f(x) = \dfrac{5}{|x - 5|}$

RC5. $f(x) = 5 - |x|$ RC6. $f(x) = \dfrac{x - 5}{x + 5}$

a) All real numbers
b) $\{x\,|\,x \text{ is a real number } and\ x \neq 5\}$
c) $\{x\,|\,x \text{ is a real number } and\ x \neq -5\ and\ x \neq 5\}$
d) $\{x\,|\,x \text{ is a real number } and\ x \neq -5\}$

a In Exercises 1–8, the graph is that of a function. Determine for each one **(a)** $f(1)$; **(b)** the domain; **(c)** all x-values such that $f(x) = 2$; and **(d)** the range.

1.
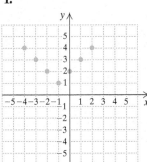

2.
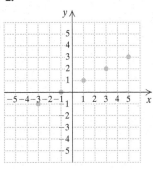

3.
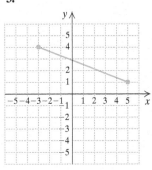

4.
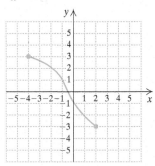

5.
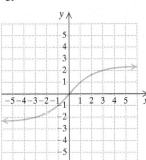

6.
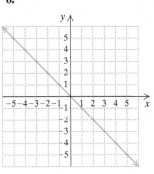

7.
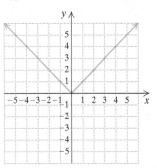

8.
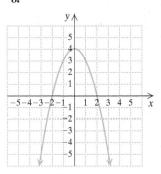

Find the domain.

9. $f(x) = \dfrac{2}{x + 3}$

10. $f(x) = \dfrac{7}{5 - x}$

11. $f(x) = 2x + 1$

12. $f(x) = 4 - 5x$

13. $f(x) = x^2 + 3$

14. $f(x) = x^2 - 2x + 3$

15. $f(x) = \dfrac{8}{5x - 14}$

16. $f(x) = \dfrac{x - 2}{3x + 4}$

SECTION 7.2 Finding Domain and Range   535

17. $f(x) = |x| - 4$
18. $f(x) = |x - 4|$
19. $f(x) = \dfrac{x^2 - 3x}{|4x - 7|}$
20. $f(x) = \dfrac{4}{|2x - 3|}$

21. $g(x) = \dfrac{1}{x - 1}$
22. $g(x) = \dfrac{-11}{4 + x}$
23. $g(x) = x^2 - 2x + 1$
24. $g(x) = 8 - x^2$

25. $g(x) = x^3 - 1$
26. $g(x) = 4x^3 + 5x^2 - 2x$
27. $g(x) = \dfrac{7}{20 - 8x}$
28. $g(x) = \dfrac{2x - 3}{6x - 12}$

29. $g(x) = |x + 7|$
30. $g(x) = |x| + 1$
31. $g(x) = \dfrac{-2}{|4x + 5|}$
32. $g(x) = \dfrac{x^2 + 2x}{|10x - 20|}$

33. For the function f whose graph is shown below, find $f(-1), f(0),$ and $f(1)$.

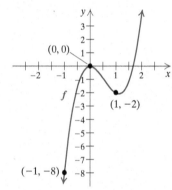

34. For the function g whose graph is shown below, find all the x-values for which $g(x) = 1$.

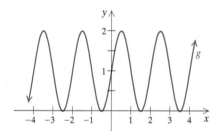

Skill Maintenance

Simplify. [6.1c]

35. $\dfrac{a^2 - 1}{a + 1}$
36. $\dfrac{10y^2 + 10y - 20}{35y^2 + 210y - 245}$
37. $\dfrac{5x - 15}{x^2 - x - 6}$

Divide. [4.8b]

38. $(t^2 + 3t - 28) \div (t + 7)$
39. $(w^2 + 4w + 5) \div (w + 3)$

Synthesis

40. Determine the range of each of the functions in Exercises 9, 14, 17, and 18.

41. Determine the range of each of the functions in Exercises 22, 23, 24, and 30.

Find the domain of each function.

42. $g(x) = \sqrt{2 - x}$
43. $f(x) = \sqrt[3]{x - 1}$

Mid-Chapter Review

Concept Reinforcement

Determine whether each statement is true or false.

_____ 1. Every function is a relation. [7.1a]

_____ 2. It is possible for one input of a function to have two or more outputs. [7.1a]

_____ 3. It is possible for all the inputs of a function to have the same output. [7.1a]

_____ 4. If it is possible for a vertical line to cross a graph more than once, the graph is not the graph of a function. [7.1d]

_____ 5. If the domain of a function is the set of real numbers, then the range is the set of real numbers. [7.2a]

Guided Solutions

 Use the graph to complete the table of ordered pairs that name points on the graph. [7.1c]

6.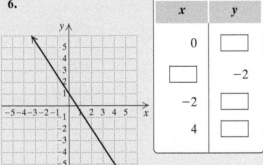

7.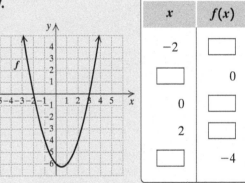

Mixed Review

Determine whether the correspondence is a function. [7.1a]

8. Domain   Range   9. Domain   Range   10. Find the domain and the range. [7.2a]

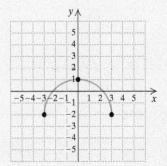

Find the function value. [7.1b]

11. $g(x) = 2 + x$; $g(-5)$

12. $f(x) = x - 7$; $f(0)$

13. $h(x) = 8$; $h(\frac{1}{2})$

14. $f(x) = 3x^2 - x + 5$; $f(-1)$

15. $g(p) = p^4 - p^3$; $g(10)$

16. $f(t) = \frac{1}{2}t + 3$; $f(-6)$

Determine whether each of the following is the graph of a function. [7.1d]

17.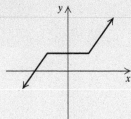

18.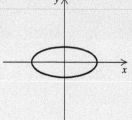

19.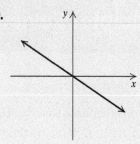

Find the domain. [7.2a]

20. $g(x) = \dfrac{3}{12 - 3x}$

21. $f(x) = x^2 - 10x + 3$

22. $h(x) = \dfrac{x - 2}{x + 2}$

23. $f(x) = |x - 4|$

Graph. [7.1c]

24. $g(x) = -\dfrac{2}{3}x - 2$

25. $f(x) = x - 1$

26. $h(x) = 2x + \dfrac{1}{2}$

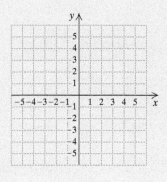
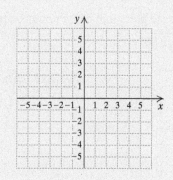
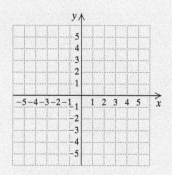

27. $g(x) = |x| - 3$

28. $f(x) = 1 + x^2$

29. $f(x) = -\dfrac{1}{4}x$

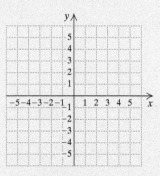
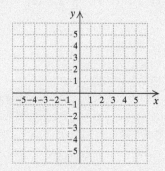
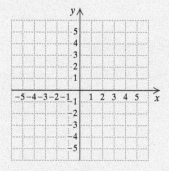

Understanding Through Discussion and Writing

30. Is it possible for a function to have more numbers as outputs than as inputs? Why or why not? [7.1a]

31. Without making a drawing, how can you tell that the graph of $f(x) = x - 30$ passes through three quadrants? [7.1c]

32. For a given function f, it is known that $f(2) = -3$. Give as many interpretations of this fact as you can. [7.1b], [7.2a]

33. Explain the difference between the domain and the range of a function. [7.2a]

STUDYING FOR SUCCESS *Making Positive Choices*
- ☐ Choose to improve your attitude and raise your goals.
- ☐ Choose to make a strong commitment to learning.
- ☐ Choose to place the primary responsibility for learning on yourself.
- ☐ Choose to allocate the proper amount of time to learn.

Linear Functions: Graphs and Slope

7.3

OBJECTIVES

a Find the y-intercept of a line from the equation $y = mx + b$ or $f(x) = mx + b$.

b Given two points on a line, find the slope. Given a linear equation, derive the equivalent slope–intercept equation and determine the slope and the y-intercept.

c Solve applied problems involving slope.

We now turn our attention to functions whose graphs are straight lines. Such functions are called **linear** and can be written in the form $f(x) = mx + b$.

> **LINEAR FUNCTION**
>
> A **linear function** f is any function that can be described by $f(x) = mx + b$.

Compare the two equations $7y + 2x = 11$ and $y = 3x + 5$. Both are linear equations because their graphs are straight lines. Each can be expressed in an equivalent form that is a linear function.

The equation $y = 3x + 5$ can be expressed as $f(x) = mx + b$, where $m = 3$ and $b = 5$.

The equation $7y + 2x = 11$ also has an equivalent form $f(x) = mx + b$. To see this, we solve for y:

$$7y + 2x = 11$$
$$7y + 2x - 2x = -2x + 11 \quad \text{Subtracting } 2x$$
$$7y = -2x + 11$$
$$\frac{7y}{7} = \frac{-2x + 11}{7} \quad \text{Dividing by 7}$$
$$y = -\frac{2}{7}x + \frac{11}{7}. \quad \text{Simplifying}$$

We now have an equivalent function in the form $f(x) = mx + b$:

$$f(x) = -\frac{2}{7}x + \frac{11}{7}, \quad \text{where} \quad m = -\frac{2}{7} \quad \text{and} \quad b = \frac{11}{7}.$$

In this section, we consider the effects of the constants m and b on the graphs of linear functions.

1. Graph $y = 3x$ and $y = 3x - 6$ using the same set of axes. Then compare the graphs.

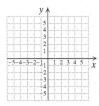

2. Graph $y = -2x$ and $y = -2x + 3$ using the same set of axes. Then compare the graphs.

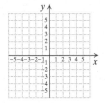

CALCULATOR CORNER

Exploring b We can use a graphing calculator to explore the effect of the constant b on the graph of a function of the form $f(x) = mx + b$. Graph $y_1 = x$ in the standard $[-10, 10, -10, 10]$ viewing window. Then graph $y_2 = x + 4$, followed by $y_3 = x - 3$, in the same viewing window.

EXERCISES:

1. Compare the graph of y_2 with the graph of y_1.
2. Compare the graph of y_3 with the graph of y_1.

a THE CONSTANT b: THE y-INTERCEPT

Let's first explore the effect of the constant b.

EXAMPLE 1 Graph $y = 2x$ and $y = 2x + 3$ using the same set of axes. Then compare the graphs.

We first make a table of solutions of both equations. Next, we plot these points. Drawing a red line for $y = 2x$ and a blue line for $y = 2x + 3$, we note that the graph of $y = 2x + 3$ is simply the graph of $y = 2x$ shifted, or *translated*, up 3 units. The lines are parallel.

x	y $y = 2x$	y $y = 2x + 3$
0	0	3
1	2	5
-1	-2	1
2	4	7
-2	-4	-1

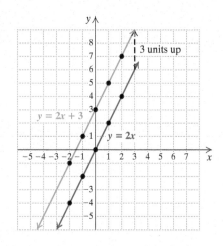

◀ Do Exercises 1 and 2.

EXAMPLE 2 Graph $f(x) = \frac{1}{3}x$ and $g(x) = \frac{1}{3}x - 2$ using the same set of axes. Then compare the graphs.

We first make a table of solutions of both equations. By choosing multiples of 3, we can avoid fractions.

x	$f(x)$ $f(x) = \frac{1}{3}x$	$g(x)$ $g(x) = \frac{1}{3}x - 2$
0	0	-2
3	1	-1
-3	-1	-3
6	2	0

We then plot these points. Drawing a red line for $f(x) = \frac{1}{3}x$ and a blue line for $g(x) = \frac{1}{3}x - 2$, we see that the graph of $g(x) = \frac{1}{3}x - 2$ is simply the graph of $f(x) = \frac{1}{3}x$ shifted, or translated, down 2 units. The lines are parallel.

Answers

1. 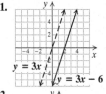 The graph of $y = 3x - 6$ is the graph of $y = 3x$ shifted down 6 units.

2. The graph of $y = -2x + 3$ is the graph of $y = -2x$ shifted up 3 units.

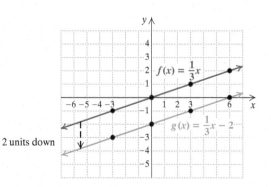

In Example 1, we saw that the graph of $y = 2x + 3$ is parallel to the graph of $y = 2x$ and that it passes through the point $(0, 3)$. Similarly, in Example 2, we saw that the graph of $y = \frac{1}{3}x - 2$ is parallel to the graph of $y = \frac{1}{3}x$ and that it passes through the point $(0, -2)$. In general, the graph of $y = mx + b$ is a line parallel to $y = mx$, passing through the point $(0, b)$. The point $(0, b)$ is called the **y-intercept** because it is the point at which the graph crosses the y-axis. Often it is convenient to refer to the number b as the y-intercept. The constant b has the effect of moving the graph of $y = mx$ up or down $|b|$ units to obtain the graph of $y = mx + b$.

Do Exercise 3. ▶

3. Graph $f(x) = \frac{1}{3}x$ and $g(x) = \frac{1}{3}x + 2$ using the same set of axes. Then compare the graphs.

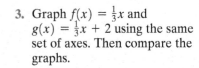

y-INTERCEPT

The y-intercept of the graph of $f(x) = mx + b$ is the point $(0, b)$ or, simply, b.

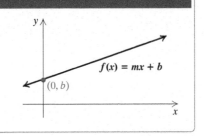

EXAMPLE 3 Find the y-intercept: $y = -5x + 4$.

$y = -5x + 4$
↓
$(0, 4)$ is the y-intercept.

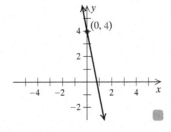

EXAMPLE 4 Find the y-intercept: $f(x) = 6.3x - 7.8$.

$f(x) = 6.3x - 7.8$
↓
$(0, -7.8)$ is the y-intercept.

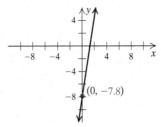

Do Exercises 4 and 5. ▶

Find the y-intercept.

4. $y = 7x + 8$

5. $f(x) = -6x - \frac{2}{3}$

MyLab Math
ANIMATION

b THE CONSTANT m: SLOPE

SKILL REVIEW

Subtract real numbers. [1.4a]
Subtract.

1. $11 - (-8)$
2. $-6 - (-6)$

Answers: 1. 19 **2.** 0

MyLab Math
VIDEO

Answers

3. 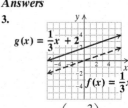 The graph of $g(x)$ is the graph of $f(x)$ shifted up 2 units.

4. $(0, 8)$ 5. $\left(0, -\dfrac{2}{3}\right)$

SECTION 7.3 Linear Functions: Graphs and Slope   **541**

Look again at the graphs in Examples 1 and 2. Note that the slant of each red line seems to match the slant of each blue line. This leads us to believe that the number m in the equation $y = mx + b$ is related to the slant of the line. Let's consider some examples.

Graphs with $m < 0$:

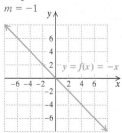
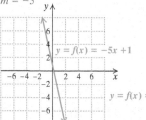
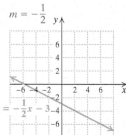

Graphs with $m = 0$:

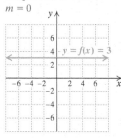
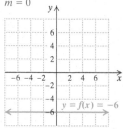

Graphs with $m > 0$:

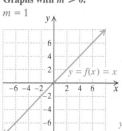
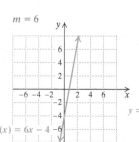
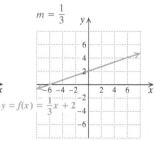

Note that

$m < 0 \longrightarrow$ The graph slants down from left to right;

$m = 0 \longrightarrow$ the graph is horizontal; and

$m > 0 \longrightarrow$ the graph slants up from left to right.

The following definition enables us to visualize the slant and attach a number, a geometric ratio, or *slope*, to the line.

SLOPE

The **slope** m of a line containing points (x_1, y_1) and (x_2, y_2) is given by

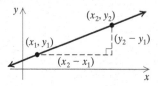

$$m = \frac{\text{rise}}{\text{run}} = \frac{\text{change in } y}{\text{change in } x} = \frac{y_2 - y_1}{x_2 - x_1} = \frac{y_1 - y_2}{x_1 - x_2}.$$

Consider a line with two points marked P_1 and P_2, as follows. As we move from P_1 to P_2, the y-coordinate changes from 1 to 3 and the x-coordinate changes from 2 to 7. The change in y is $3 - 1$, or 2. The change in x is $7 - 2$, or 5.

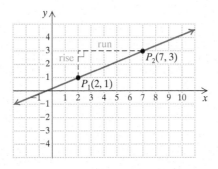

We call the change in y the **rise** and the change in x the **run**. The ratio rise/run is the same for any two points on a line. We call this ratio the **slope**. Slope describes the slant of a line. The slope of the line in the graph above is given by

$$\frac{\text{rise}}{\text{run}}, \quad \text{or} \quad \frac{\text{change in } y}{\text{change in } x}, \quad \text{or} \quad \frac{2}{5}.$$

Whenever x increases by 5 units, y increases by 2 units. Equivalently, whenever x increases by 1 unit, y increases by $\frac{2}{5}$ unit.

EXAMPLE 5 Graph the line containing the points $(-4, 3)$ and $(2, -5)$ and find the slope.

The graph is shown below. Moving from $(-4, 3)$ to $(2, -5)$, we see that the change in y, or the rise, is $-5 - 3$, or -8. The change in x, or the run, is $2 - (-4)$, or 6.

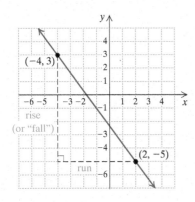

$$\text{Slope} = \frac{\text{rise}}{\text{run}} = \frac{\text{change in } y}{\text{change in } x}$$
$$= \frac{-5 - 3}{2 - (-4)}$$
$$= \frac{-8}{6} = -\frac{8}{6}, \text{ or } -\frac{4}{3}.$$

The formula

$$m = \frac{y_2 - y_1}{x_2 - x_1} = \frac{y_1 - y_2}{x_1 - x_2}$$

tells us that we can subtract in two ways. We must remember, however, to subtract the x-coordinates in the same order that we subtract the y-coordinates.
Let's do Example 5 again:

$$\text{Slope} = \frac{\text{change in } y}{\text{change in } x} = \frac{3 - (-5)}{-4 - 2} = \frac{8}{-6} = -\frac{8}{6} = -\frac{4}{3}.$$

We see that both ways give the same value for the slope.

CALCULATOR CORNER

Visualizing Slope

EXERCISES: Use the window settings $[-6, 6, -4, 4]$, with Xscl = 1 and Yscl = 1.

1. Graph $y = x$, $y = 2x$, and $y = 5x$ in the same window. What do you think the graph of $y = 10x$ will look like?

2. Graph $y = x$, $y = 0.5x$, and $y = 0.1x$ in the same window. What do you think the graph of $y = 0.005x$ will look like?

3. Graph $y = -x$, $y = -2x$, and $y = -5x$ in the same window. What do you think the graph of $y = -10x$ will look like?

4. Graph $y = -x$, $y = -0.5x$, and $y = -0.1x$ in the same window. What do you think the graph of $y = -0.005x$ will look like?

Graph the line through the given points and find its slope.

6. $(-1, -1)$ and $(2, -4)$

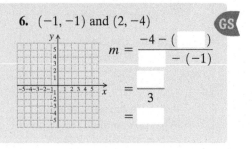

$$m = \frac{-4 - (\quad)}{\quad - (-1)}$$
$$= \frac{\quad}{3}$$
$$= \quad$$

7. $(0, 2)$ and $(3, 1)$

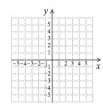

8. Find the slope of the line $f(x) = -\frac{2}{3}x + 1$. Use the points $(9, -5)$ and $(3, -1)$.

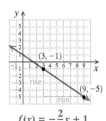

Answers

6. 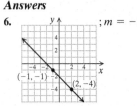 ; $m = -1$

7. 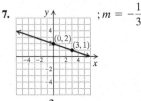 ; $m = -\frac{1}{3}$

8. $m = -\frac{2}{3}$

Guided Solution:
6. $-1, 2, -3, -1$

The slope of a line tells how it slants. A line with positive slope slants up from left to right. The larger the positive number, the steeper the slant. A line with negative slope slants down from left to right. The smaller the negative number, the steeper the line.

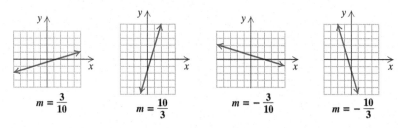

◀ Do Exercises 6 and 7.

How can we find the slope from a given equation? Let's consider the equation $y = 2x + 3$, which is in the form $y = mx + b$. We can find two points by choosing convenient values for x—say, 0 and 1—and substituting to find the corresponding y-values.

If $x = 0$, $y = 2 \cdot 0 + 3 = 3$.
If $x = 1$, $y = 2 \cdot 1 + 3 = 5$.

We find two points on the line to be

$(0, 3)$ and $(1, 5)$.

The slope of the line is found as follows, using the definition of slope:

$$m = \frac{\text{change in } y}{\text{change in } x}$$
$$= \frac{5 - 3}{1 - 0} = \frac{2}{1} = 2.$$

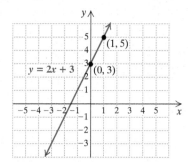

The slope is 2. Note that this is the coefficient of the x-term in the equation $y = 2x + 3$.

If we had chosen different points on the line—say, $(-2, -1)$ and $(4, 11)$—the slope would still be 2, as we see in the following calculation:

$$m = \frac{11 - (-1)}{4 - (-2)} = \frac{11 + 1}{4 + 2} = \frac{12}{6} = 2.$$

◀ Do Exercise 8.

We see that the slope of the line $y = mx + b$ is indeed the constant m, the coefficient of x.

SLOPE

The **slope** of the line $y = mx + b$ is m.

From a linear equation in the form $y = mx + b$, we can read the slope and the y-intercept of the graph directly.

SLOPE–INTERCEPT EQUATION

The equation $y = mx + b$ is called the **slope–intercept equation**. The slope is m and the y-intercept is $(0, b)$.

Note that any graph of an equation $y = mx + b$ passes the vertical-line test and thus represents a function.

EXAMPLE 6 Find the slope and the y-intercept of $y = 5x - 4$.

Since the equation is already in the form $y = mx + b$, we simply read the slope and the y-intercept from the equation:

$$y = 5x - 4.$$

The slope is 5.   The y-intercept is $(0, -4)$.

EXAMPLE 7 Find the slope and the y-intercept of $2x + 3y = 8$.

We first solve for y so we can easily read the slope and the y-intercept:

$$2x + 3y = 8$$
$$3y = -2x + 8 \quad \text{Subtracting } 2x$$
$$\frac{3y}{3} = \frac{-2x + 8}{3} \quad \text{Dividing by 3}$$
$$y = -\frac{2}{3}x + \frac{8}{3}. \quad \text{Finding the form } y = mx + b$$

The slope is $-\frac{2}{3}$.   The y-intercept is $\left(0, \frac{8}{3}\right)$.

Do Exercises 9 and 10.

Find the slope and the y-intercept.

9. $f(x) = -8x + 23$

GS 10. $5x - 10y = 25$

First solve for y:
$$5x - 10y = 25$$
$$-10y = \boxed{} + 25$$
$$y = \frac{-5x + 25}{\boxed{}}$$
$$y = \boxed{} x - \frac{5}{\boxed{}}.$$

Slope is $\boxed{}$; y-intercept is $\left(0, \boxed{}\right)$.

c APPLICATIONS

Slope has many real-world applications. For example, numbers like 2%, 3%, and 6% are often used to represent the *grade* of a road, a measure of how steep a road on a hill or a mountain is. A 3% grade $\left(3\% = \frac{3}{100}\right)$ means that for every horizontal distance of 100 ft that the road runs, the road rises 3 ft, and a -3% grade means that for every horizontal distance of 100 ft, the road drops 3 ft. (Normally, the road-grade signs do not include negative signs, since it is obvious whether you are climbing or descending.)

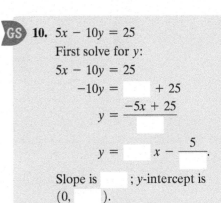

Answers

9. Slope: -8; y-intercept: $(0, 23)$

10. Slope: $\frac{1}{2}$; y-intercept: $\left(0, -\frac{5}{2}\right)$

Guided Solution:

10. $-5x, -10, \frac{1}{2}, 2, \frac{1}{2}, -\frac{5}{2}$

SECTION 7.3   Linear Functions: Graphs and Slope   **545**

An athlete might change the grade of a treadmill during a workout. An escape ramp on an airliner might have a slope of about −0.6.

Architects and carpenters use slope when designing and building stairs, ramps, or roof pitches. Another application occurs in hydrology. The strength or force of a river depends on how far the river falls vertically compared to how far it flows horizontally. Slope can also be considered as a **rate of change**.

EXAMPLE 8 *Cancer Center Advertising.* The total spent by U.S. cancer centers on advertising increased from $54 million in 2005 to $173 million in 2014. Twenty of the 890 cancer centers that spent $173 million in 2014 account for 86% of the total amount spent. Find the rate of change in the amount spent on advertising by cancer centers with respect to time, in years.

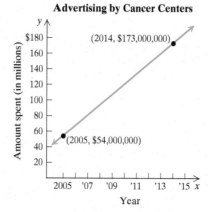

DATA: Indiana University School of Medicine; University of Pittsburgh School of Medicine and Graduate School of Public Health

The rate of change with respect to time, in years, is given by

$$\text{Rate of change} = \frac{\$173{,}000{,}000 - \$54{,}000{,}000}{2014 - 2005}$$

$$= \frac{\$119{,}000{,}000}{9 \text{ years}}$$

$$\approx \$13{,}222{,}222 \text{ per year}.$$

The amount spent on advertising by cancer centers is increasing at a rate of about $13,222,222 per year.

◀ Do Exercise 11.

11. *Holiday Sales.* Retail sales during the holiday season grew from $416.4 billion in 2002 to $655.8 billion in 2016. Find the rate of change in holiday sales with respect to time, in years.

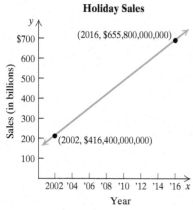

DATA: National Retail Federation

Answer
11. The rate of change in holiday sales is about $17,100,000,000 per year.

EXAMPLE 9 *Consumption of Potatoes.* The annual per-capita consumption of potatoes has been decreasing since 1970. Find the rate of change of the per-capita consumption of potatoes with respect to time, in years.

DATA: U.S. Department of Agriculture

The rate of change with respect to time, in years, is given by

$$\text{Rate of change} = \frac{32.2 - 59.3}{2014 - 1970} = \frac{-27.1}{44 \text{ years}} \approx -0.616 \text{ pound per year.}$$

The per-capita consumption of potatoes is decreasing at a rate of about 0.616 pound per year.

Do Exercise 12.

12. *Newspaper Circulation.* Daily newspaper circulation has decreased dramatically since 2005. The following graph shows the circulation of daily newspapers, in millions, for 2005 and for 2015. Find the rate of change in the circulation of daily newspapers per year.

DATA: Editor & Publisher International Data Book

Answer
12. The rate of change is −1.84 million papers per year.

7.3 Exercise Set

FOR EXTRA HELP — MyLab Math

✓ Check Your Understanding

Reading Check From the words below each blank, choose the one that correctly completes the statement.

RC1. If $m > 0$, the graph slants _____ from left to right.
   up/down

RC2. If $m = 0$, the graph is _____.
   horizontal/vertical

RC3. If $m < 0$, the graph slants _____ from left to right.
   up/down

Concept Check Choose from the columns on the right the slope of each line.

CC1. 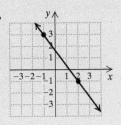   CC2. 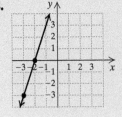   CC3.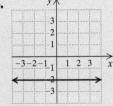

a) $-\dfrac{3}{4}$ d) -4

b) 3 e) $\dfrac{3}{4}$

c) 0 f) $-\dfrac{4}{3}$

a, b Find the slope and the *y*-intercept of each equation.

1. $y = 4x + 5$
2. $y = -5x + 10$
3. $f(x) = -2x - 6$
4. $g(x) = -5x + 7$
5. $y = -\frac{3}{8}x - \frac{1}{5}$
6. $y = \frac{15}{7}x + \frac{16}{5}$
7. $g(x) = 0.5x - 9$
8. $f(x) = -3.1x + 5$
9. $2x - 3y = 8$
10. $-8x - 7y = 24$
11. $9x = 3y + 6$
12. $9y + 36 - 4x = 0$
13. $3 - \frac{1}{4}y = 2x$
14. $5x = \frac{2}{3}y - 10$
15. $17y + 4x + 3 = 7 + 4x$
16. $3y - 2x = 5 + 9y - 2x$

b Find the slope of each line.

17.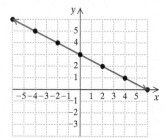
18.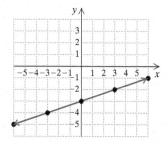
19.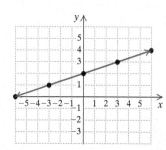
20.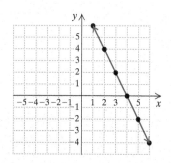

Find the slope of the line containing the given pair of points.

21. $(6, 9)$ and $(4, 5)$
22. $(8, 7)$ and $(2, -1)$
23. $(9, -4)$ and $(3, -8)$
24. $(17, -12)$ and $(-9, -15)$
25. $(-16.3, 12.4)$ and $(-5.2, 8.7)$
26. $(14.4, -7.8)$ and $(-12.5, -17.6)$

c Find the slope (or rate of change).

27. Find the slope (or grade) of the treadmill.

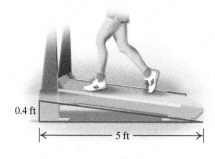

28. Find the slope (or head) of the river.

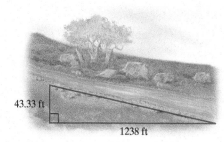

29. Find the slope (or pitch) of the roof.

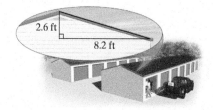

548   CHAPTER 7   Graphs, Functions, and Applications

30. Public buildings regularly include steps with 7-in. risers and 11-in. treads. Find the grade of such a stairway.

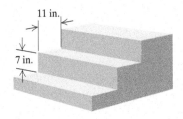

31. *Super Bowl Ad Spending.* Find the rate of change in Super Bowl advertising spending with respect to time, in years.

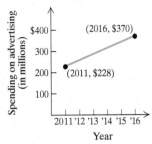

DATA: KANTAR MEDIA

32. *Vehicle Production.* Find the rate of change in U.S. vehicle production with respect to time, in years.

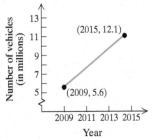

DATA: Automotive News Data Center; R. L. Polk; OICA

Find the rate of change.

33.

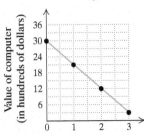

34.

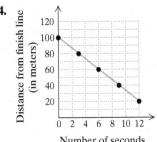

35.

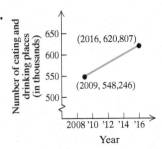

DATA: U.S. Bureau of Labor Statistics

36.

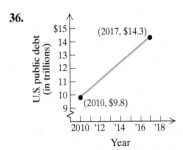

DATA: Budget of the U.S Government; Office of Management and Budget; Executive Office of the President

Skill Maintenance

Simplify. [1.8c, d], [4.1b]

37. $3^2 - 24 \cdot 56 + 144 \div 12$

38. $9\{2x - 3[5x + 2(-3x + y^0 - 2)]\}$

39. $10\{2x + 3[5x - 2(-3x + y^1 - 2)]\}$

40. $5^4 \div 625 \div 5^2 \cdot 5^7 \div 5^3$

Solve. [2.6a]

41. One side of a square is 5 yd less than a side of an equilateral triangle. If the perimeter of the square is the same as the perimeter of the triangle, what is the length of a side of the square? of the triangle?

Factor. [5.6a]

42. $8 - 125x^3$

43. $c^6 - d^6$

44. $56x^3 - 7$

45. Divide: $(a^2 - 11a + 6) \div (a - 1)$. [4.8b]

7.4 More on Graphing Linear Equations

OBJECTIVES

a Graph linear equations using intercepts.

b Given a linear equation in slope–intercept form, use the slope and the y-intercept to graph the line.

c Graph linear equations of the form $x = a$ or $y = b$.

d Given the equations of two lines, determine whether their graphs are parallel or whether they are perpendicular.

a GRAPHING USING INTERCEPTS

The *x*-intercept of the graph of a linear equation or function is the point at which the graph crosses the *x*-axis. The **y-intercept** is the point at which the graph crosses the *y*-axis. We know from geometry that only one line can be drawn through two given points. Thus, if we know the intercepts, we can graph the line. To ensure that a computation error has not been made, it is a good idea to calculate a third point as a check.

Many equations of the type $Ax + By = C$ can be graphed conveniently using intercepts.

> **x-INTERCEPTS AND y-INTERCEPTS**
>
> A **y-intercept** is a point $(0, b)$. To find b, let $x = 0$ and solve for y.
> An **x-intercept** is a point $(a, 0)$. To find a, let $y = 0$ and solve for x.

EXAMPLE 1 Find the intercepts of $3x + 2y = 12$ and then graph the line.

y-intercept: To find the *y*-intercept, we let $x = 0$ and solve for y:

$$3x + 2y = 12$$
$$3 \cdot 0 + 2y = 12 \quad \text{Substituting 0 for } x$$
$$2y = 12$$
$$y = 6.$$

The *y*-intercept is $(0, 6)$.

x-intercept: To find the *x*-intercept, we let $y = 0$ and solve for x:

$$3x + 2y = 12$$
$$3x + 2 \cdot 0 = 12 \quad \text{Substituting 0 for } y$$
$$3x = 12$$
$$x = 4.$$

The *x*-intercept is $(4, 0)$.

We plot these points and draw the line, using a third point as a check. We choose $x = 6$ and solve for y:

$$3(6) + 2y = 12$$
$$18 + 2y = 12$$
$$2y = -6$$
$$y = -3.$$

We plot $(6, -3)$ and note that it is on the line so the graph is probably correct.

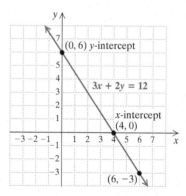

SKILL REVIEW

Plot points associated with ordered pairs of numbers. [3.1a]

1. Plot the following points:
   $A(0, 4), B(0, -1),$
   $C(0, 0), D(3, 0),$ and
   $E\left(-\frac{7}{2}, 0\right).$

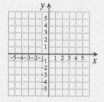

Answer:

1.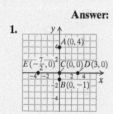

MyLab Math VIDEO

550   CHAPTER 7   Graphs, Functions, and Applications

When both the *x*-intercept and the *y*-intercept are $(0,0)$, as is the case with an equation such as $y = 2x$, whose graph passes through the origin, we would need to calculate another point and use a third point as a check.

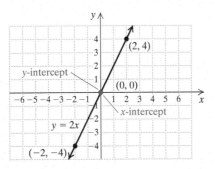

Do Exercise 1.

1. Find the intercepts of $4y - 12 = -6x$ and then graph the line.

To find the *y*-intercept, set $x = 0$ and solve for *y*:

$4y - 12 = -6 \cdot \boxed{}$

$4y - 12 = \boxed{}$

$4y = \boxed{}$

$y = \boxed{}$.

The *y*-intercept is $(0, \boxed{})$.

To find the *x*-intercept, set $y = 0$ and solve for *x*:

$4 \cdot \boxed{} - 12 = -6x$

$\boxed{} = -6x$

$\boxed{} = x$.

The *x*-intercept is $(\boxed{}, 0)$.

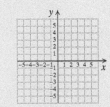

CALCULATOR CORNER

Viewing the Intercepts Knowing the intercepts of a linear equation helps us determine a good viewing window for the graph of the equation. For example, when we graph the equation $y = -x + 15$ in the standard window, we see only a small portion of the graph in the upper right-hand corner of the screen, as shown on the left below.

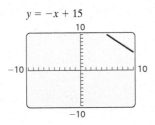 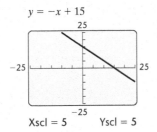

Using algebra, as we did in Example 1, we can find that the intercepts of the graph of this equation are $(0, 15)$ and $(15, 0)$. This tells us that, if we are to see a portion of the graph that includes the intercepts, both Xmax and Ymax should be greater than 15. We can try different window settings until we find one that suits us. One good choice, shown on the right above, is $[-25, 25, -25, 25]$, with Xscl = 5 and Yscl = 5.

EXERCISES: Find the intercepts of the equation algebraically. Then graph the equation on a graphing calculator, choosing window settings that allow the intercepts to be seen clearly. (Settings may vary.)

1. $y = -3.2x - 16$
2. $y - 4.25x = 85$
3. $6x + 5y = 90$
4. $5x - 6y = 30$
5. $8x + 3y = 9$
6. $y = 0.4x - 5$
7. $y = 1.2x - 12$
8. $4x - 5y = 2$

Answer

1.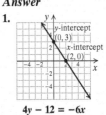

$4y - 12 = -6x$

Guided Solution:
1. 0, 0, 12, 3, 3; 0, −12, 2, 2

Graph using the slope and the y-intercept.

2. $y = \dfrac{3}{2}x + 1$

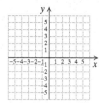

3. $f(x) = \dfrac{3}{4}x - 2$

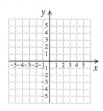

4. $g(x) = -\dfrac{3}{5}x + 5$

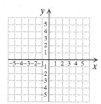

5. $y = -\dfrac{5}{3}x - 4$

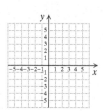

b GRAPHING USING THE SLOPE AND THE y-INTERCEPT

We can also graph a line using its slope and y-intercept.

EXAMPLE 2 Graph: $y = -\dfrac{2}{3}x + 1$.

This equation is in slope–intercept form, $y = mx + b$. The y-intercept is $(0, 1)$. We plot $(0, 1)$. We can think of the slope ($m = -\dfrac{2}{3}$) as $\dfrac{-2}{3}$.

$$m = \dfrac{\text{Rise}}{\text{Run}} = \dfrac{-2}{3} \quad \begin{array}{l}\text{Move 2 units down.}\\ \text{Move 3 units right.}\end{array}$$

Starting at the y-intercept and using the slope, we find another point by moving 2 units down (since the numerator is *negative* and corresponds to the change in y) and 3 units to the right (since the denominator is *positive* and corresponds to the change in x). We get to a new point, $(3, -1)$. In a similar manner, we can move from the point $(3, -1)$ to find another point, $(6, -3)$.

We could also think of the slope ($m = -\dfrac{2}{3}$) as $\dfrac{2}{-3}$.

$$m = \dfrac{\text{Rise}}{\text{Run}} = \dfrac{2}{-3} \quad \begin{array}{l}\text{Move 2 units up.}\\ \text{Move 3 units left.}\end{array}$$

Then we can start again at $(0, 1)$, but this time we move 2 units up (since the numerator is *positive* and corresponds to the change in y) and 3 units to the left (since the denominator is *negative* and corresponds to the change in x). We get another point on the graph, $(-3, 3)$, and from it we can obtain $(-6, 5)$ and others in a similar manner. We plot the points and draw the line.

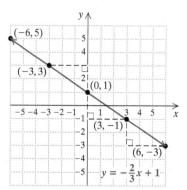

EXAMPLE 3 Graph: $f(x) = \dfrac{2}{5}x + 4$.

First, we plot the y-intercept, $(0, 4)$. We then consider the slope $\dfrac{2}{5}$. A slope of $\dfrac{2}{5}$ tells us that, for every 2 units that the graph rises, it runs 5 units horizontally in the positive direction, or to the right. Thus, starting at the y-intercept and using the slope, we find another point by moving 2 units up (since the numerator is *positive* and corresponds to the change in y) and 5 units to the right (since the denominator is *positive* and corresponds to the change in x). We get to a new point, $(5, 6)$.

Answers

2.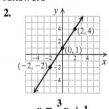
$y = \dfrac{3}{2}x + 1$

3.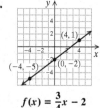
$f(x) = \dfrac{3}{4}x - 2$

4.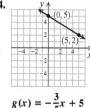
$g(x) = -\dfrac{3}{5}x + 5$

5.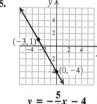
$y = -\dfrac{5}{3}x - 4$

We can also think of the slope $\frac{2}{5}$ as $\frac{-2}{-5}$. A slope of $\frac{-2}{-5}$ tells us that, for every 2 units that the graph drops, it runs 5 units horizontally in the negative direction, or to the left. We again start at the y-intercept, $(0, 4)$. We move 2 units down (since the numerator is *negative* and corresponds to the change in y) and 5 units to the left (since the denominator is *negative* and corresponds to the change in x). We get to another new point, $(-5, 2)$. We plot the points and draw the line.

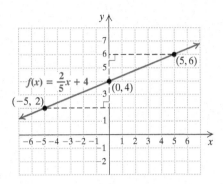

Do Exercises 2–5 on the preceding page.

c HORIZONTAL LINES AND VERTICAL LINES

Some equations have graphs that are parallel to one of the axes. This happens when either A or B is 0 in $Ax + By = C$. These equations have a missing variable; that is, there is only one variable in the equation. In the following example, x is missing.

EXAMPLE 4 Graph: $y = 3$.

Since x is missing, any number for x will do. Thus all ordered pairs $(x, 3)$ are solutions. The graph is a **horizontal line** parallel to the x-axis.

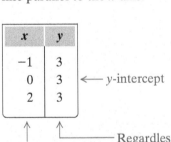

x	y
-1	3
0	3	← y-intercept
2	3

↑ ↑
| └─── Regardless of x, y must be 3.
Choose *any* number for x.

What about the slope of a horizontal line? In Example 4, consider the points $(-1, 3)$ and $(2, 3)$, which are on the line $y = 3$. The change in y is $3 - 3$, or 0. The change in x is $-1 - 2$, or -3. Thus,

$$m = \frac{3 - 3}{-1 - 2} = \frac{0}{-3} = 0.$$

Any two points on a horizontal line have the same y-coordinate. Thus the change in y is always 0, so the slope is 0.

Do Exercises 6 and 7.

Graph and determine the slope.

6. $f(x) = -4$

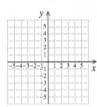

7. $y = 3.6$

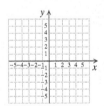

Answers

6. $m = 0$

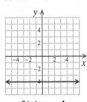

$f(x) = -4$

7. $m = 0$

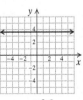

$y = 3.6$

SECTION 7.4 More on Graphing Linear Equations 553

We can also determine the slope by noting that $y = 3$ can be written in slope–intercept form as $y = 0x + 3$, or $f(x) = 0x + 3$. From this equation, we read that the slope is 0. A function of this type is called a **constant function**. We can express it in the form $y = b$, or $f(x) = b$. Its graph is a horizontal line that crosses the y-axis at $(0, b)$.

In the following example, y is missing and the graph is parallel to the y-axis.

EXAMPLE 5 Graph: $x = -2$.

Since y is missing, any number for y will do. Thus all ordered pairs $(-2, y)$ are solutions. The graph is a **vertical line** parallel to the y-axis.

Graph.

8. $x = -5$

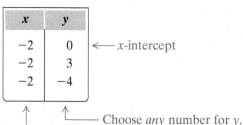

9. $8x - 5 = 19$ (*Hint*: Solve for x.)

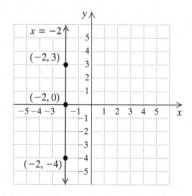

This graph is not the graph of a function because it fails the vertical-line test. The vertical line itself crosses the graph more than once.

◀ Do Exercises 8 and 9.

What about the slope of a vertical line? In Example 5, consider the points $(-2, 3)$ and $(-2, -4)$, which are on the line $x = -2$. The change in y is $3 - (-4)$, or 7. The change in x is $-2 - (-2)$, or 0. Thus,

$$m = \frac{3 - (-4)}{-2 - (-2)} = \frac{7}{0}. \quad \text{Not defined}$$

10. Determine, if possible, the slope of each line.
    a) $x = -12$ b) $y = 6$
    c) $2y + 7 = 11$ d) $x = 0$
    e) $y = -\frac{3}{4}$
    f) $10 - 5x = 15$

Since division by 0 is not defined, the slope of this line is not defined. Any two points on a vertical line have the same x-coordinate. Thus the change in x is always 0, so the slope of any vertical line is not defined.

The following summarizes the characteristics of horizontal lines and vertical lines and their equations.

HORIZONTAL LINE; VERTICAL LINE

The graph of $y = b$, or $f(x) = b$, is a **horizontal line** with y-intercept $(0, b)$. It is the graph of a constant function with slope 0.

The graph of $x = a$ is a **vertical line** with x-intercept $(a, 0)$. The slope is not defined. It is not the graph of a function.

Answers

8.

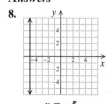

$x = -5$

9. $8x - 5 = 19$

◀ Do Exercise 10.

10. (a) Not defined; (b) 0; (c) 0; (d) not defined; (e) 0; (f) not defined

We have graphed linear equations in several ways in this chapter. Although, in general, you can use any method that works best for you, we list some guidelines in the margin at right.

d PARALLEL LINES AND PERPENDICULAR LINES

Parallel Lines

Parallel lines extend indefinitely without intersecting. If two lines are vertical, they are parallel. How can we tell whether nonvertical lines are parallel? We examine their slopes and y-intercepts.

> **PARALLEL LINES**
>
> Two nonvertical lines are **parallel** if they have the *same* slope and *different* y-intercepts. Vertical lines are parallel.

EXAMPLE 6 Determine if the graphs of $y - 3x = 1$ and $3x + 2y = -2$ are parallel.

To determine if lines are parallel, we first find their slopes. To do this, we find the slope–intercept form of each equation by solving for y:

$$y - 3x = 1 \qquad\qquad 3x + 2y = -2$$
$$y = 3x + 1; \qquad\qquad 2y = -3x - 2$$
$$\qquad\qquad y = \tfrac{1}{2}(-3x - 2)$$
$$\qquad\qquad y = -\tfrac{3}{2}x - 1.$$

The slopes, 3 and $-\tfrac{3}{2}$, are different. Thus the lines are not parallel, as the graphs at right confirm.

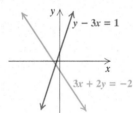

EXAMPLE 7 Determine if the graphs of $3x - y = -5$ and $y - 3x = -2$ are parallel.

We first find the slope–intercept form of each equation by solving for y:

$$3x - y = -5 \qquad\qquad y - 3x = -2$$
$$-y = -3x - 5 \qquad\qquad y = 3x - 2.$$
$$-1(-y) = -1(-3x - 5)$$
$$y = 3x + 5;$$

The slopes, 3, are the same. The y-intercepts, $(0, 5)$ and $(0, -2)$, are different. Thus the lines are parallel, as the graphs appear to confirm.

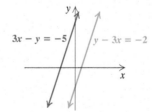

Do Exercises 11–13.

To graph a linear equation:

1. Is the equation of the type $x = a$ or $y = b$? If so, the graph will be a line parallel to an axis; $x = a$ is vertical and $y = b$ is horizontal.
2. If the line is of the type $y = mx$, both intercepts are the origin, $(0, 0)$. Plot $(0, 0)$ and one other point.
3. If the line is of the type $y = mx + b$, plot the y-intercept and one other point.
4. If the equation is of the form $Ax + By = C$, graph using intercepts. If the intercepts are too close together, choose another point farther from the origin.
5. In all cases, use a third point as a check.

Determine whether the graphs of the given pair of lines are parallel.

11. $x + 4 = y,$
$y - x = -3$

Write each equation in the form $y = mx + b$:
$x + 4 = y \rightarrow y = x + 4;$
$y - x = -3 \rightarrow y = \underline{\quad} - 3.$
The slope of each line is $\underline{\quad}$, and the y-intercepts, $(0, 4)$ and $(0, \underline{\quad})$, are different. Thus the lines $\underline{\quad\quad}$ parallel.
$\qquad\qquad$ are/are not

12. $y + 4 = 3x,$
$4x - y = -7$

13. $y = 4x + 5,$
$2y = 8x + 10$

Answers
11. Yes **12.** No **13.** No; they are the same line.
Guided Solution:
11. $x, 1, -3,$ are

Perpendicular Lines

If one line is vertical and another is horizontal, they are perpendicular. For example, the lines $x = 5$ and $y = -3$ are perpendicular. Otherwise, how can we tell whether two lines are perpendicular?

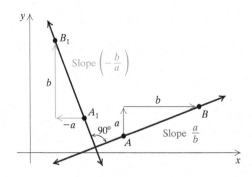

Consider a line $\overleftrightarrow{AB}$, as shown in the figure above, with slope a/b. Then think of rotating the line 90° to get a line $\overleftrightarrow{A_1B_1}$ perpendicular to $\overleftrightarrow{AB}$. For the new line, the rise and the run are interchanged, but the run is now negative. Thus the slope of the new line is $-b/a$, which is the opposite of the reciprocal of the slope of the first line. Also note that when we multiply the slopes, we get

$$\frac{a}{b}\left(-\frac{b}{a}\right) = -1.$$

This is the condition under which lines will be perpendicular.

PERPENDICULAR LINES

Two lines are **perpendicular** if the product of their slopes is -1. (If one line has slope m, then the slope of a line perpendicular to it is $-1/m$. That is, to find the slope of a line perpendicular to a given line, we take the reciprocal of the given slope and change the sign.)

Lines are also perpendicular if one of them is vertical ($x = a$) and one of them is horizontal ($y = b$).

EXAMPLE 8 Determine whether the graphs of $5y = 4x + 10$ and $4y = -5x + 4$ are perpendicular.

In order to determine whether the lines are perpendicular, we determine whether the product of their slopes is -1. We first find the slope–intercept form of each equation by solving for y.

We have

$5y = 4x + 10$ $4y = -5x + 4$
$y = \frac{1}{5}(4x + 10)$ $y = \frac{1}{4}(-5x + 4)$
$y = \frac{4}{5}x + 2;$ $y = -\frac{5}{4}x + 1.$

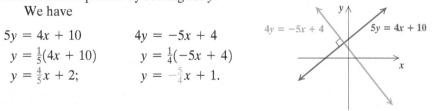

The slope of the first line is $\frac{4}{5}$, and the slope of the second line is $-\frac{5}{4}$. The product of the slopes is $\frac{4}{5} \cdot \left(-\frac{5}{4}\right) = -1$. Thus the lines are perpendicular.

◂ Do Exercises 14 and 15.

Determine whether the graphs of the given pair of lines are perpendicular.

14. $2y - x = 2,$
$y + 2x = 4$

Write each equation in the form $y = mx + b$:
$2y - x = 2 \rightarrow y = \underline{}x + 1;$
$y + 2x = 4 \rightarrow y = \underline{}x + 4.$

The slopes of these lines are $\underline{}$ and -2. The product of the slopes $\frac{1}{2} \cdot (-2) = \underline{}$.
Thus the lines $\underline{}$ are/are not
perpendicular.

15. $3y = 2x + 15,$
$2y = 3x + 10$

Answers
14. Yes   **15.** No
Guided Solution:
14. $\frac{1}{2}, -2, \frac{1}{2}, -1,$ are

Visualizing for Success

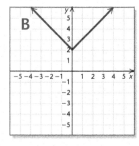

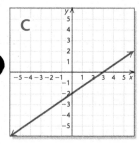

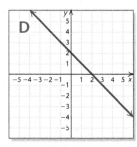

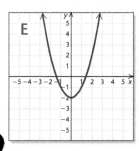

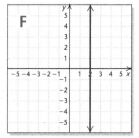

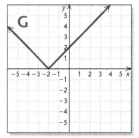

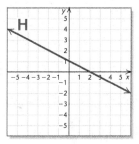

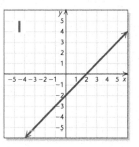

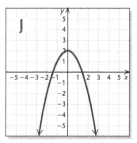

Match each equation with its graph.

1. $y = 2 - x$

2. $x - y = 2$

3. $x + 2y = 2$

4. $2x - 3y = 6$

5. $x = 2$

6. $y = 2$

7. $y = |x + 2|$

8. $y = |x| + 2$

9. $y = x^2 - 2$

10. $y = 2 - x^2$

Answers on page A-23

7.4 Exercise Set

✓ Check Your Understanding

Reading Check Determine whether each statement is true or false.

RC1. Two nonvertical lines are parallel if they have the same slope and the same y-intercepts.

RC2. The x-intercept of $x = a$, a vertical line, is $(a, 0)$.

RC3. The slope of a horizontal line is 0.

RC4. Two lines are perpendicular if the product of their slopes is 1.

Concept Check Choose from the columns on the right the ordered pair, equation, or word that completes the statement.

CC1. The x-intercept of $x = -\frac{2}{7}$ is _____.

CC2. The y-intercept of $y = -2x + 7$ is _____.

CC3. The graphs of the lines $x = -4$ and $y = 5$ are _____.

CC4. The graphs of the lines $y = -13$ and _____ are parallel.

$(0, 7)$ parallel
$(-2, 0)$ perpendicular
$\left(0, -\frac{2}{7}\right)$ $y = \frac{2}{7}$
$\left(-\frac{2}{7}, 0\right)$ $x = -7$

a   Find the intercepts and then graph the line.

1. $x - 2 = y$

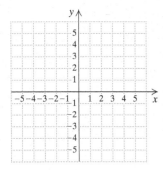

2. $x + 3 = y$

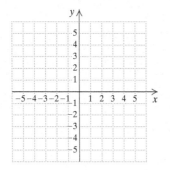

3. $x + 3y = 6$

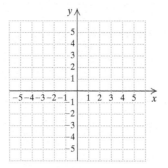

4. $x - 2y = 4$

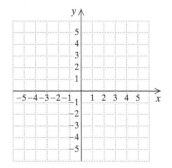

5. $2x + 3y = 6$

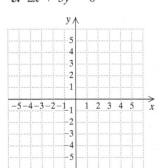

6. $5x - 2y = 10$

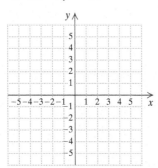

7. $f(x) = -2 - 2x$

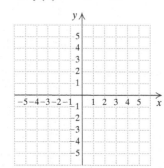

8. $g(x) = 5x - 5$

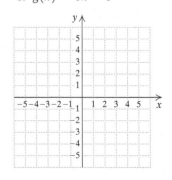

9. $5y = -15 + 3x$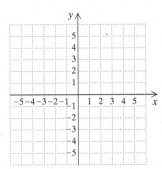

10. $5x - 10 = 5y$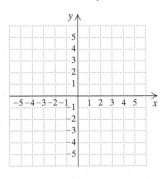

11. $2x - 3y = 6$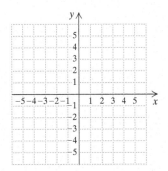

12. $4x + 5y = 20$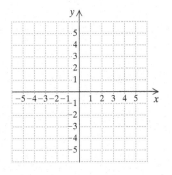

13. $2.8y - 3.5x = -9.8$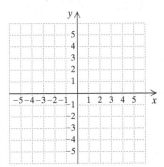

14. $10.8x - 22.68 = 4.2y$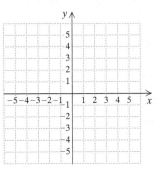

15. $5x + 2y = 7$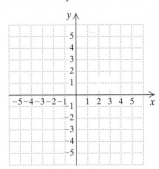

16. $3x - 4y = 10$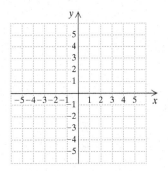

b Graph using the slope and the y-intercept.

17. $y = \dfrac{5}{2}x + 1$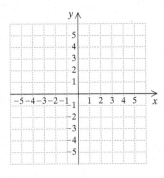

18. $y = \dfrac{2}{5}x - 4$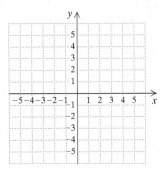

19. $f(x) = -\dfrac{5}{2}x - 4$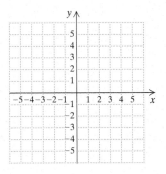

20. $f(x) = \dfrac{2}{5}x + 3$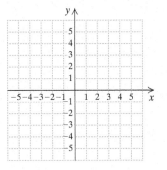

21. $x + 2y = 4$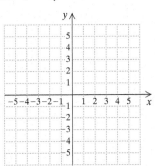

22. $x - 3y = 6$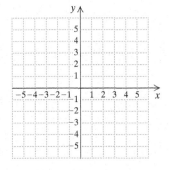

23. $4x - 3y = 12$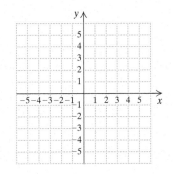

24. $2x + 6y = 12$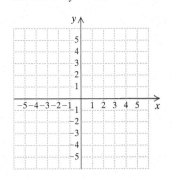

SECTION 7.4 More on Graphing Linear Equations

25. $f(x) = \dfrac{1}{3}x - 4$

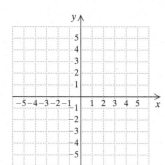

26. $g(x) = -0.25x + 2$

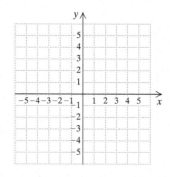

27. $5x + 4 \cdot f(x) = 4$
(*Hint*: Solve for $f(x)$.)

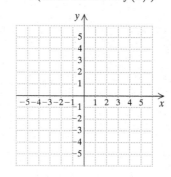

28. $3 \cdot f(x) = 4x + 6$

c Graph and, if possible, determine the slope.

29. $x = 1$

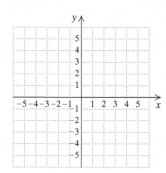

30. $x = -4$

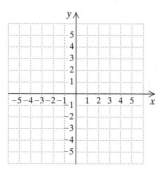

31. $y = -1$

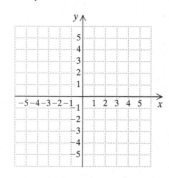

32. $y = \dfrac{3}{2}$

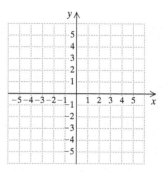

33. $f(x) = -6$

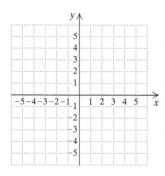

34. $f(x) = 2$

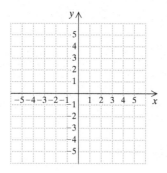

35. $y = 0$

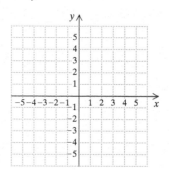

36. $x = 0$

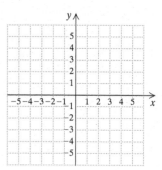

37. $2 \cdot f(x) + 5 = 0$

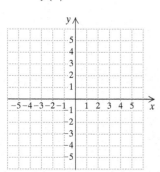

38. $4 \cdot g(x) + 3x = 12 + 3x$

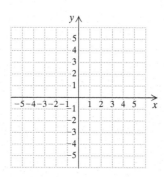

39. $7 - 3x = 4 + 2x$

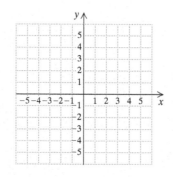

40. $3 - f(x) = 2$

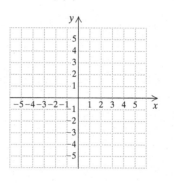

560 CHAPTER 7 Graphs, Functions, and Applications

d Determine whether the graphs of the given pair of lines are parallel.

41. $x + 6 = y$,
$y - x = -2$

42. $2x - 7 = y$,
$y - 2x = 8$

43. $y + 3 = 5x$,
$3x - y = -2$

44. $y + 8 = -6x$,
$-2x + y = 5$

45. $y = 3x + 9$,
$2y = 6x - 2$

46. $y + 7x = -9$,
$-3y = 21x + 7$

47. $12x = 3$,
$-7x = 10$

48. $5y = -2$,
$\frac{3}{4}x = 16$

Determine whether the graphs of the given pair of lines are perpendicular.

49. $y = 4x - 5$,
$4y = 8 - x$

50. $2x - 5y = -3$,
$2x + 5y = 4$

51. $x + 2y = 5$,
$2x + 4y = 8$

52. $y = -x + 7$,
$y = x + 3$

53. $2x - 3y = 7$,
$2y - 3x = 10$

54. $x = y$,
$y = -x$

55. $2x = 3$,
$-3y = 6$

56. $-5y = 10$,
$y = -\frac{4}{9}$

Skill Maintenance

Write in scientific notation. [4.2c]

57. 53,000,000,000

58. 0.000047

59. 0.018

60. 99,902,000

Write in decimal notation. [4.2c]

61. 2.13×10^{-5}

62. 9.01×10^8

63. 2×10^4

64. 8.5677×10^{-2}

Factor. [1.7d]

65. $9x - 15y$

66. $12a + 21ab$

67. $21p - 7pq + 14p$

68. $64x - 128y + 256$

Synthesis

69. Find the value of a such that the graphs of $5y = ax + 5$ and $\frac{1}{4}y = \frac{1}{10}x - 1$ are parallel.

70. Find the value of k such that the graphs of $x + 7y = 70$ and $y + 3 = kx$ are perpendicular.

71. Write an equation of the line that has x-intercept $(-3, 0)$ and y-intercept $(0, \frac{2}{5})$.

72. Find the coordinates of the point of intersection of the graphs of the equations $x = -4$ and $y = 5$.

73. Write an equation for the x-axis. Is this equation a function?

74. Write an equation for the y-axis. Is this equation a function?

75. Find the value of m in $y = mx + 3$ so that the x-intercept of its graph will be $(4, 0)$.

76. Find the value of b in $2y = -7x + 3b$ so that the y-intercept of its graph will be $(0, -13)$.

7.5 Finding Equations of Lines; Applications

OBJECTIVES

a Find an equation of a line when the slope and the y-intercept are given.

b Find an equation of a line when the slope and a point are given.

c Find an equation of a line when two points are given.

d Given a line and a point not on the given line, find an equation of the line parallel to the line and containing the point, and find an equation of the line perpendicular to the line and containing the point.

e Solve applied problems involving linear functions.

---

1. A line has slope 3.4 and y-intercept (0, −8). Find an equation of the line.

MyLab Math
ANIMATION

In this section, we will learn to find an equation of a line for which we have been given two pieces of information.

a FINDING AN EQUATION OF A LINE WHEN THE SLOPE AND THE y-INTERCEPT ARE GIVEN

If we know the slope and the y-intercept of a line, we can find an equation of the line using the slope–intercept equation $y = mx + b$.

EXAMPLE 1 A line has slope -0.7 and y-intercept $(0, 13)$. Find an equation of the line.

We use the slope–intercept equation and substitute -0.7 for m and 13 for b:

$$y = mx + b$$
$$y = -0.7x + 13.$$

◀ Do Exercise 1.

b FINDING AN EQUATION OF A LINE WHEN THE SLOPE AND A POINT ARE GIVEN

Suppose that we know the slope of a line and the coordinates of one point on the line. We can use the slope–intercept equation to find an equation of the line. Or, we can use the **point–slope equation**. We first develop a formula for such a line.

Suppose that a line of slope m passes through the point (x_1, y_1). For any other point (x, y) on this line, we must have

$$\frac{y - y_1}{x - x_1} = m.$$

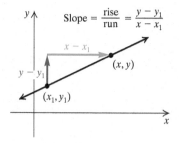

It is tempting to use this last equation as an equation of the line of slope m that passes through (x_1, y_1). The only problem with this form is that when x and y are replaced with x_1 and y_1, we have $\frac{0}{0} = m$, a false equation. To avoid this difficulty, we multiply by $x - x_1$ on both sides and simplify:

$$\frac{y - y_1}{x - x_1}(x - x_1) = m(x - x_1) \quad \text{Multiplying by } x - x_1 \text{ on both sides}$$

$$y - y_1 = m(x - x_1). \quad \text{Removing a factor of 1: } \frac{x - x_1}{x - x_1} = 1$$

This is the *point–slope* form of a linear equation.

Answer
1. $y = 3.4x - 8$

POINT–SLOPE EQUATION

The **point–slope equation** of a line with slope m, passing through (x_1, y_1), is
$$y - y_1 = m(x - x_1).$$

If we know the slope of a line and a point on the line, we can find an equation of the line using either the point–slope equation,
$$y - y_1 = m(x - x_1),$$
or the slope–intercept equation,
$$y = mx + b.$$

EXAMPLE 2 Find an equation of the line with slope 5 and containing the point $\left(\frac{1}{2}, -1\right)$.

Using the Point–Slope Equation: We consider $\left(\frac{1}{2}, -1\right)$ to be (x_1, y_1). We substitute $\frac{1}{2}$ for x_1, -1 for y_1, and 5 for m:

$$y - y_1 = m(x - x_1) \quad \text{Point–slope equation}$$
$$y - (-1) = 5\left(x - \tfrac{1}{2}\right) \quad \text{Substituting}$$
$$y + 1 = 5x - \tfrac{5}{2} \quad \text{Simplifying}$$
$$y = 5x - \tfrac{5}{2} - 1$$
$$y = 5x - \tfrac{5}{2} - \tfrac{2}{2}$$
$$y = 5x - \tfrac{7}{2}.$$

Using the Slope–Intercept Equation: The point $\left(\frac{1}{2}, -1\right)$ is on the line, so it is a solution of the equation. Thus we can substitute $\frac{1}{2}$ for x and -1 for y in $y = mx + b$. We also substitute 5 for m, the slope. Then we solve for b:

$$y = mx + b \quad \text{Slope–intercept equation}$$
$$-1 = 5 \cdot \left(\tfrac{1}{2}\right) + b \quad \text{Substituting}$$
$$-1 = \tfrac{5}{2} + b$$
$$-1 - \tfrac{5}{2} = b$$
$$-\tfrac{2}{2} - \tfrac{5}{2} = b$$
$$-\tfrac{7}{2} = b. \quad \text{Solving for } b$$

We then use the slope–intercept equation $y = mx + b$ again and substitute 5 for m and $-\frac{7}{2}$ for b:
$$y = 5x - \tfrac{7}{2}.$$

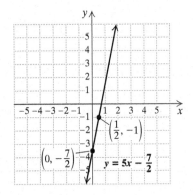

Find an equation of the line with the given slope and containing the given point.

2. $m = -5,\ (-4, 2)$

3. $m = 3,\ (1, -2)$

4. $m = 8,\ (3, 5)$

5. $m = -\dfrac{2}{3},\ (1, 4)$

Do Exercises 2–5.

Answers

2. $y = -5x - 18$ 3. $y = 3x - 5$
4. $y = 8x - 19$ 5. $y = -\dfrac{2}{3}x + \dfrac{14}{3}$

C FINDING AN EQUATION OF A LINE WHEN TWO POINTS ARE GIVEN

We can also use the slope–intercept equation or the point–slope equation to find an equation of a line when two points are given.

EXAMPLE 3 Find an equation of the line containing the points $(2, 3)$ and $(-6, 1)$.

First, we find the slope:

$$m = \frac{3-1}{2-(-6)} = \frac{2}{8}, \text{ or } \frac{1}{4}.$$

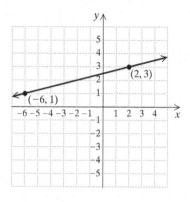

Now we have the slope and two points. We then proceed as we did in Example 2, using either point, and either the point–slope equation or the slope–intercept equation.

Using the Point–Slope Equation: We choose $(2, 3)$ and substitute 2 for x_1, 3 for y_1, and $\frac{1}{4}$ for m:

$$y - y_1 = m(x - x_1) \quad \text{Point–slope equation}$$
$$y - 3 = \tfrac{1}{4}(x - 2) \quad \text{Substituting}$$
$$y - 3 = \tfrac{1}{4}x - \tfrac{1}{2}$$
$$y = \tfrac{1}{4}x - \tfrac{1}{2} + 3$$
$$y = \tfrac{1}{4}x - \tfrac{1}{2} + \tfrac{6}{2}$$
$$y = \tfrac{1}{4}x + \tfrac{5}{2}.$$

Using the Slope–Intercept Equation: We choose $(2, 3)$ and substitute 2 for x, 3 for y, and $\frac{1}{4}$ for m:

$$y = mx + b \quad \text{Slope–intercept equation}$$
$$3 = \tfrac{1}{4} \cdot 2 + b \quad \text{Substituting}$$
$$3 = \tfrac{1}{2} + b$$
$$3 - \tfrac{1}{2} = \tfrac{1}{2} + b - \tfrac{1}{2}$$
$$\tfrac{6}{2} - \tfrac{1}{2} = b$$
$$\tfrac{5}{2} = b. \quad \text{Solving for } b$$

Finally, we use the slope–intercept equation $y = mx + b$ again and substitute $\frac{1}{4}$ for m and $\frac{5}{2}$ for b:

$$y = \tfrac{1}{4}x + \tfrac{5}{2}.$$

◀ Do Exercises 6 and 7.

---

6. Find an equation of the line containing the points $(4, -3)$ and $(1, 2)$.

First, find the slope:

$$m = \frac{\boxed{} - (-3)}{1 - 4} = \frac{\boxed{}}{-3} = -\frac{5}{\boxed{}}.$$

Using the point–slope equation,

$$y - y_1 = m(x - x_1),$$

substitute 4 for x_1, -3 for y_1, and $-\frac{5}{3}$ for m:

$$y - (\boxed{}) = \boxed{}(x - \boxed{})$$
$$y + \boxed{} = -\tfrac{5}{3}x + \boxed{}$$
$$y = -\tfrac{5}{3}x + \tfrac{20}{3} - 3$$
$$y = -\tfrac{5}{3}x + \tfrac{20}{3} - \tfrac{\boxed{}}{3}$$
$$y = -\tfrac{5}{3}x + \tfrac{\boxed{}}{3}.$$

7. Find an equation of the line containing the points $(-3, -5)$ and $(-4, 12)$.

Answers

6. $y = -\dfrac{5}{3}x + \dfrac{11}{3}$ **7.** $y = -17x - 56$

Guided Solution:

6. 2, 5, 3; -3, $-\dfrac{5}{3}$, 4, 3, $\dfrac{20}{3}$, 9, 11

d FINDING AN EQUATION OF A LINE PARALLEL OR PERPENDICULAR TO A GIVEN LINE THROUGH A POINT NOT ON THE LINE

We can also use the methods of Example 2 to find an equation of a line parallel or perpendicular to a given line and containing a point not on the line.

EXAMPLE 4 Find an equation of the line containing the point $(-1, 3)$ and parallel to the line $2x + y = 10$.

A line parallel to the given line $2x + y = 10$ must have the same slope as the given line. To find that slope, we first find the slope–intercept equation by solving for y:

$$2x + y = 10$$
$$y = -2x + 10.$$

Thus the line we want to find through $(-1, 3)$ must also have slope -2.

Using the Point–Slope Equation: We use the point $(-1, 3)$ and the slope -2, substituting -1 for x_1, 3 for y_1, and -2 for m:

$$y - y_1 = m(x - x_1)$$
$$y - 3 = -2(x - (-1)) \quad \text{Substituting}$$
$$y - 3 = -2(x + 1) \quad \text{Simplifying}$$
$$y - 3 = -2x - 2$$
$$y = -2x + 1.$$

Using the Slope–Intercept Equation: We substitute -1 for x, 3 for y, and -2 for m in $y = mx + b$. Then we solve for b:

$$y = mx + b$$
$$3 = -2(-1) + b \quad \text{Substituting}$$
$$3 = 2 + b$$
$$1 = b. \quad \text{Solving for } b$$

We then use the equation $y = mx + b$ again and substitute -2 for m and 1 for b:

$$y = -2x + 1.$$

The given line $2x + y = 10$, or $y = -2x + 10$, and the line $y = -2x + 1$ have the same slope but different y-intercepts. Thus their graphs are parallel.

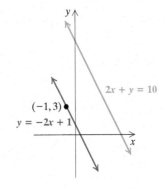

Do Exercise 8.

GS 8. Find an equation of the line containing the point $(2, -1)$ and parallel to the line $9x - 3y = 5$.

Find the slope of the given line:
$$9x - 3y = 5$$
$$-3y = \boxed{} + 5$$
$$y = \boxed{} x - \frac{5}{3}.$$

The slope is $\boxed{}$.

The line parallel to $9x - 3y = 5$ must have slope $\boxed{}$.

Using the slope–intercept equation,
$$y = mx + b,$$
substitute 3 for m, 2 for x, and -1 for y, and solve for b:
$$\boxed{} = 3 \cdot \boxed{} + b$$
$$-1 = 6 + b$$
$$\boxed{} = b.$$

Substitute 3 for m and -7 for b in $y = mx + b$:
$$y = 3x + (\boxed{})$$
$$y = 3x - 7.$$

Answer
8. $y = 3x - 7$
Guided Solution:
8. $-9x, 3, 3, 3; -1, 2, -7, -7$

EXAMPLE 5 Find an equation of the line containing the point $(2, -3)$ and perpendicular to the line $4y - x = 20$.

To find the slope of the given line, we first find its slope–intercept form by solving for y:

$$4y - x = 20$$
$$4y = x + 20$$
$$\frac{4y}{4} = \frac{x + 20}{4} \quad \text{Dividing by 4}$$
$$y = \tfrac{1}{4}x + 5.$$

We know that the slope of the perpendicular line must be the opposite of the reciprocal of $\tfrac{1}{4}$. Thus the new line through $(2, -3)$ must have slope -4.

Using the Point–Slope Equation: We use the point $(2, -3)$ and the slope -4, substituting 2 for x_1, -3 for y_1, and -4 for m:

$$y - y_1 = m(x - x_1)$$
$$y - (-3) = -4(x - 2) \quad \text{Substituting}$$
$$y + 3 = -4x + 8$$
$$y = -4x + 5.$$

Using the Slope–Intercept Equation: We now substitute 2 for x and -3 for y in $y = mx + b$. We also substitute -4 for m, the slope. Then we solve for b:

$$y = mx + b$$
$$-3 = -4(2) + b \quad \text{Substituting}$$
$$-3 = -8 + b$$
$$5 = b. \quad \text{Solving for } b$$

Finally, we use the equation $y = mx + b$ again and substitute -4 for m and 5 for b:

$$y = -4x + 5.$$

The product of the slopes of the lines $4y - x = 20$ and $y = -4x + 5$ is $\tfrac{1}{4} \cdot (-4) = -1$. Thus their graphs are perpendicular.

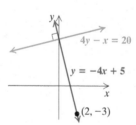

◀ Do Exercise 9.

---

9. Find an equation of the line containing the point $(5, 4)$ and perpendicular to the line $2x - 4y = 9$.
Find the slope of the given line:
$$2x - 4y = 9$$
$$-4y = \boxed{} + 9$$
$$y = \boxed{} x - \frac{9}{4}.$$
The slope is $\boxed{}$.
The slope of a line perpendicular to $2x - 4y = 9$ is the opposite of the reciprocal of $\tfrac{1}{2}$, or $\boxed{}$.
Using the point-slope equation,
$$y - y_1 = m(x - x_1),$$
substitute -2 for m, 5 for x_1, and 4 for y_1:
$$y - \boxed{} = \boxed{}(x - 5)$$
$$y - 4 = -2x + \boxed{}$$
$$y = -2x + 14.$$

Answer
9. $y = -2x + 14$
Guided Solution:
9. $-2x, \tfrac{1}{2}, \tfrac{1}{2}, -2; 4, -2, 10$

e APPLICATIONS OF LINEAR FUNCTIONS

When the essential parts of a problem are described in mathematical language, we say that we have a **mathematical model**. We have already studied many kinds of mathematical models in this text—for example, the formulas in Section 2.4 and the functions in Section 7.1. Here we study linear functions as models.

EXAMPLE 6 *Cost of a Necklace.* Amelia's Beads offers a class in designing necklaces. For a necklace made of 6-mm beads, 4.23 beads per inch are needed. The cost of a necklace of 6-mm gemstone beads that sell for 40¢ each is $7 for the clasp and the crimps and approximately $1.70 per inch.

a) Formulate a linear function that models the total cost of a necklace $C(n)$, where n is the length of the necklace, in inches.
b) Graph the model.
c) Use the model to determine the cost of a 30-in. necklace.

a) The problem describes a situation in which cost per inch is charged in addition to the fixed cost of the clasp and the crimps. The total cost of a 16-in. necklace is

$$\$7 + \$1.70 \cdot 16 = \$34.20.$$

For a 17-in. necklace, the total cost is

$$\$7 + \$1.70 \cdot 17 = \$35.90.$$

These calculations lead us to generalize that for a necklace that is n inches long, the total cost is given by $C(n) = 7 + 1.7n$, where $n \geq 0$ since the length of the necklace cannot be negative. (Actually most necklaces are at least 14 in. long.) The notation $C(n)$ indicates that the cost C is a function of the length n.

b) Before we draw the graph, we rewrite the model in slope–intercept form:

$$C(n) = 1.7n + 7.$$

The y-intercept is $(0, 7)$ and the slope, or rate of change, is $\$1.70$, or $\frac{\$17}{10}$ $\left(\frac{1.7}{1} = \frac{17}{10}\right)$, per inch. We first plot $(0, 7)$; from that point, we move 17 units up and 10 units to the right to the point $(10, 24)$. We then draw a line through these points. We also calculate a third value as a check:

$$C(20) = 1.7 \cdot 20 + 7 = 41.$$

The point $(20, 41)$ lines up with the other two points so the graph is correct.

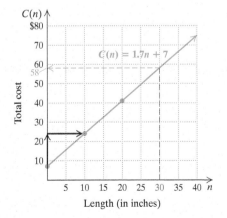

10. *Cost of a Service Call.* For a service call, Belmont Heating and Air Conditioning charges a $65 trip fee and $80 per hour for labor.

a) Formulate a linear function for the total cost of the service call $C(t)$, where t is the length of the call, in hours.

b) Graph the model.

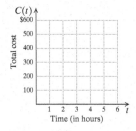

c) Use the model to determine the cost of a $2\frac{1}{2}$-hr service call.

c) To determine the total cost of a 30-in. necklace, we find $C(30)$:

$$C(30) = 1.7 \cdot 30 + 7 = 58.$$

From the graph, we see that the input 30 corresponds to the output 58. Thus we see that a 30-in. necklace costs $58.

◀ Do Exercise 10.

In the following example, we use two points and find an equation for the linear function through these points. Then we use the equation to estimate.

EXAMPLE 7 *Decreasing Number of Banks.* The number of banks in the United States has decreased steadily in recent years. The following table lists data regarding the correspondence between the year and the number of banks.

YEAR, x (in number of years after 2008)	NUMBER OF BANKS, N
2008, 0	8441
2016, 8	6068

Data: Federal Deposit Insurance Company

a) Assuming a constant rate of change, use the two data points to find a linear function that fits the data.

b) Use the function to determine the number of banks in 2013.

c) In what year will the number of banks reach 5200?

a) We let x = the number of years after 2008 and N = the number of banks. The table gives us two ordered pairs, $(0, 8441)$ and $(8, 6068)$. We use them to find a linear function that fits the data. First, we find the slope:

$$m = \frac{6068 - 8441}{8 - 0} = \frac{-2373}{8} = -296.625.$$

Next, we find an equation $N = mx + b$ that fits the data. One of the data points, $(0, 8441)$, is the y-intercept. Thus we know the value of b in the slope–intercept equation, $y = mx + b$. We use the equation $N = mx + b$ and substitute -296.625 for m and 8441 for b:

$$N = -296.625x + 8441.$$

Using function notation, we have

$$N(x) = -296.625x + 8441.$$

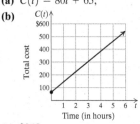

Answer

10. (a) $C(t) = 80t + 65$;

(b)

(c) $265

b) To determine the number of banks in 2013, we substitute 5 for x (2013 is 5 years after 2008) in the function $N(x) = -296.625x + 8441$:

$N(x) = -296.625x + 8441$
$N(5) = -296.625(5) + 8441$ Substituting
$ = -1483.125 + 8441$
$ = 6957.875 \approx 6958.$

There were about 6958 banks in 2013.

c) To find the year in which the number of banks will reach 5200, we substitute 5200 for $N(x)$ and solve for x:

$N(x) = -296.625x + 8441$
$5200 = -296.625x + 8441$ Substituting
$-3241 = -296.625x$ Subtracting 8441
$11 \approx x.$ Dividing by -296.625

The number of banks will reach 5200 about 11 years after 2008, or in 2019.

Do Exercise 11.

11. *Hat Size as a Function of Head Circumference.* The following table lists data relating hat size to head circumference.

HEAD CIRCUMFERENCE, C (in inches)	HAT SIZE, H
21.2	$6\frac{3}{4}$
22	7

Data: Shushan's New Orleans

a) Assuming a constant rate of change, use the two data points to find a linear function that fits the data.
b) Use the function to determine the hat size of a person whose head has a circumference of 24.8 in.
c) Jerome's hat size is 8. What is the circumference of his head?

Answer

11. (a) $H(C) = \frac{5}{16}C + \frac{1}{8}$, or $H(C) = 0.3125C + 0.125$;
(b) $7\frac{7}{8}$, or 7.875; **(c)** 25.2 in.

7.5 Exercise Set

FOR EXTRA HELP — MyLab Math

✓ Check Your Understanding

Reading Check Complete each statement with the correct equation.

RC1. The slope–intercept equation of a line with slope m and passing through (x_1, y_1) is _____.

RC2. The point–slope equation of a line with slope m and passing through (x_1, y_1) is _____.

Concept Check For the given equation, determine the slope of the line **(a)** parallel to the given line and **(b)** perpendicular to the given line.

CC1. $y = \frac{4}{11}x - 2$

CC2. $y = -5$

CC3. $2x - y = -4$

CC4. $y - \frac{4}{3} = -\frac{5}{6}x$

CC5. $x = 3$

CC6. $10x + 5y = 14$

a Find an equation of the line having the given slope and y-intercept.

1. Slope: -8; y-intercept: $(0, 4)$
2. Slope: 5; y-intercept: $(0, -3)$
3. Slope: 2.3; y-intercept: $(0, -1)$
4. Slope: -9.1; y-intercept: $(0, 2)$

Find a linear function $f(x) = mx + b$ whose graph has the given slope and y-intercept.

5. Slope: $-\frac{7}{3}$; y-intercept: $(0, -5)$
6. Slope: $\frac{4}{5}$; y-intercept: $(0, 28)$
7. Slope: $\frac{2}{3}$; y-intercept: $\left(0, \frac{5}{8}\right)$
8. Slope: $-\frac{7}{8}$; y-intercept: $\left(0, -\frac{7}{11}\right)$

b Find an equation of the line having the given slope and containing the given point.

9. $m = 5$, $(4, 3)$
10. $m = 4$, $(5, 2)$
11. $m = -3$, $(9, 6)$
12. $m = -2$, $(2, 8)$

13. $m = 1$, $(-1, -7)$
14. $m = 3$, $(-2, -2)$
15. $m = -2$, $(8, 0)$
16. $m = -3$, $(-2, 0)$

17. $m = 0$, $(0, -7)$
18. $m = 0$, $(0, 4)$
19. $m = \frac{2}{3}$, $(1, -2)$
20. $m = -\frac{4}{5}$, $(2, 3)$

c Find an equation of the line containing the given pair of points.

21. $(1, 4)$ and $(5, 6)$
22. $(2, 5)$ and $(4, 7)$
23. $(-3, -3)$ and $(2, 2)$
24. $(-1, -1)$ and $(9, 9)$

25. $(-4, 0)$ and $(0, 7)$
26. $(0, -5)$ and $(3, 0)$
27. $(-2, -3)$ and $(-4, -6)$
28. $(-4, -7)$ and $(-2, -1)$

29. $(0, 0)$ and $(6, 1)$
30. $(0, 0)$ and $(-4, 7)$
31. $\left(\frac{1}{4}, -\frac{1}{2}\right)$ and $\left(\frac{3}{4}, 6\right)$
32. $\left(\frac{2}{3}, \frac{3}{2}\right)$ and $\left(-3, \frac{5}{6}\right)$

d Write an equation of the line containing the given point and parallel to the given line.

33. $(3, 7)$; $x + 2y = 6$
34. $(0, 3)$; $2x - y = 7$

35. $(2, -1)$; $5x - 7y = 8$
36. $(-4, -5)$; $2x + y = -3$

37. $(-6, 2)$; $3x = 9y + 2$
38. $(-7, 0)$; $2y + 5x = 6$

Write an equation of the line containing the given point and perpendicular to the given line.

39. $(2, 5)$; $2x + y = 3$

40. $(4, 1)$; $x - 3y = 9$

41. $(3, -2)$; $3x + 4y = 5$

42. $(-3, -5)$; $5x - 2y = 4$

43. $(0, 9)$; $2x + 5y = 7$

44. $(-3, -4)$; $6y - 3x = 2$

e Solve.

45. *School Fund-Raiser.* A school club is raising funds by having a "Shred It Day," when residents of the community can bring in their sensitive documents to be shredded. The club is charging $10 for the first three paper bags full of documents and $5 for each additional bag.

a) Formulate a linear function that models the total cost $C(x)$ of shredding x additional bags of documents.
b) Graph the model.
c) Use the model to determine the total cost of shredding 7 bags of documents.

46. *Fitness Club Costs.* A fitness club charges an initiation fee of $165 plus $24.95 per month.

a) Formulate a linear function that models the total cost $C(t)$ of a club membership for t months.
b) Graph the model.
c) Use the model to determine the total cost of a 14-month membership.

47. *Value of a Lawn Mower.* A landscaping business purchased a ZTR commercial lawn mower for $9400. The value $V(t)$ of the mower depreciates (declines) at a rate of $85 per month.

a) Formulate a linear function that models the value $V(t)$ of the mower after t months.
b) Graph the model.
c) Use the model to determine the value of the mower after 18 months.

48. *Value of a Computer.* True Tone Graphics bought a computer for $3800. The value $V(t)$ of the computer depreciates at a rate of $50 per month.

a) Formulate a linear function that models the value $V(t)$ of the computer after t months.
b) Graph the model.
c) Use the model to determine the value of the computer after $10\frac{1}{2}$ months.

In Exercises 49–54, assume that a constant rate of change exists for each model formed.

49. *Urban Population.* The following table lists data regarding the percentage of the world population considered urban in 1960 and in 2014.

YEAR, x (in number of years after 1960)	PERCENTAGE OF WORLD POPULATION THAT WAS URBAN
1960, 0	34%, or 0.34
2014, 54	54%, or 0.54

Data: Global Health Observation Data, World Health Organization

a) Use the two data points to find a linear function that fits the data. Let x = the number of years after 1960 and $U(x)$ = the percentage of the world population considered urban. (Use 0.34 and 0.54 when finding the linear function. Then convert answers to percent notation.)
b) Use the function of part (a) to estimate the percentage of the world population considered urban in 1990. Predict the percentage that will be considered urban in 2020.

50. *Disney Revenue.* The following table lists data regarding revenue from the Disney parks and resorts in 2010 and in 2016.

YEAR, x (in number of years after 2010)	REVENUE FROM PARKS AND RESORTS (in billions of dollars)
2010, 0	$10.667
2016, 6	16.974

Data: The Disney Company

a) Use the two data points to find a linear function that fits the data. Let x = the number of years after 2010 and $R(x)$ = the revenue, in billions of dollars, from the parks and resorts.
b) Use the function of part (a) to estimate the revenue in 2014 and to predict the revenue in 2021.

51. *Bridges Needing Repair.* The number of U.S. bridges in need of repair has been declining recently. In 2000, 89,460 bridges were labeled structurally deficient. By 2016, this number had dropped to 56,000. Let $B(x)$ = the number of bridges in need of repair and x = the number of years after 2000.

Data: U.S. Department of Transportation; *The Wall Street Journal*, "Maligned Bridges Get Needed Repair" by David Harrison, 5/22/17

a) Find a linear function that fits the data.
b) Use the function of part (a) to estimate the number of bridges in need of repair in 2008.
c) At this rate of decrease, when will the number of bridges in need of repair reach 48,000?

52. *Revenue from Gas Taxes.* As a result of cars getting better gas mileage, the increased use of electric vehicles, and fewer miles being driven, revenues from U.S. gas taxes are declining. It is projected that the gas-tax revenue will drop from $24.1 billion in 2017 to $20.3 billion by 2025. Let $G(x)$ = the gas tax revenue, in billions of dollars, and x = the number of years after 2017.

Data: Joint Commission on Taxation

a) Find a linear function that fits the data.
b) Use the function of part (a) to estimate the projected gas tax revenue in 2019 and in 2023.
c) At this rate of decrease, when will the gas tax revenue be $19.8 billion?

53. *Life Expectancy in South Africa.* In 2003, the life expectancy in South Africa was 46.56 years. In 2015, it was 62.34 years. Let $E(t)$ = life expectancy and t = the number of years after 2003.

Data: *CIA World Factbook* 2003–2012

a) Find a linear function that fits the data.
b) Use the function of part (a) to estimate life expectancy in 2020.

54. *Life Expectancy in Monaco.* In 2003, the life expectancy in Monaco was 79.27 years. In 2015, it was 89.52 years. Let $E(t)$ = life expectancy and t = the number of years since 2003.

Data: *CIA World Factbook* 2003–2017

a) Find a linear function that fits the data.
b) Use the function of part (a) to estimate life expectancy in 2020.

Skill Maintenance

Simplify. [6.1c]

55. $\dfrac{w-t}{t-w}$

56. $\dfrac{b^2-1}{b-1}$

57. $\dfrac{3x^2+15x-72}{6x^2+18x-240}$

58. $\dfrac{4y+32}{y^2-y-72}$

Find the slope, if it exists, of the line. [7.3b], [7.4c]

59. $2x-7y=-10$

60. $y=-1$

61. $x=42$

62. $4-6y=12x$

Synthesis

63. Find k such that the line containing the points $(-3, k)$ and $(4, 8)$ is parallel to the line containing the points $(5, 3)$ and $(1, -6)$.

64. Find an equation of the line passing through the point $(4, 5)$ and perpendicular to the line passing through the points $(-1, 3)$ and $(2, 9)$.

CHAPTER 7 Summary and Review

Vocabulary Reinforcement

Complete each statement with the correct term from the column on the right. Some of the choices may be used more than once, and some may not be used at all.

1. The graph of $x = a$ is a(n) _____ line with x-intercept $(a, 0)$. [7.4c]

2. The _____ equation of a line with slope m and passing through (x_1, y_1) is $y - y_1 = m(x - x_1)$. [7.5b]

3. A(n) _____ is a correspondence between a first set, called the _____, and a second set, called the _____, such that each member of the _____ corresponds to _____ member of the _____. [7.1a]

4. The _____ of a line containing points (x_1, y_1) and (x_2, y_2) is given by $m =$ the change in y/the change in x, also described as rise/run. [7.3b]

5. Two lines are _____ if the product of their slopes is -1. [7.4d]

6. The equation $y = mx + b$ is called the _____ equation of a line with slope m and y-intercept $(0, b)$. [7.3b]

7. Lines are _____ if they have the same slope and different y-intercepts. [7.4d]

x-intercept
y-intercept
at least one
exactly one
slope–intercept
point–slope
slope
function
relation
parallel
perpendicular
vertical
horizontal
domain
range

Concept Reinforcement

Determine whether each statement is true or false.

_____ 1. The slope of a vertical line is 0. [7.4c]

_____ 2. A line with slope 1 slants less steeply than a line with slope -5. [7.3b]

_____ 3. Parallel lines have the same slope and y-intercept. [7.4d]

Study Guide

Objective 7.1a Determine whether a correspondence is a function.

Example Determine whether each correspondence is a function.

Domain   Range   Domain   Range
         4   A
f: -4 → 2 g: Q → C
         6 → 0   R → E
         -6 → -2

The correspondence f is a function because each member of the domain is matched to *only one* member of the range. The correspondence g is *not* a function because one member of the domain, Q, is matched to more than one member of the range.

Practice Exercise

1. Determine whether the correspondence is a function.

    Domain   Range
           11 → 20
    h: 15
           19 → 31

Objective 7.1b Given a function described by an equation, find function values (outputs) for specified values (inputs).

Example Find the indicated function value.
a) $f(0)$, for $f(x) = -x + 6$ b) $g(5)$, for $g(x) = -10$
c) $h(-1)$, for $h(x) = 4x^2 + x$

a) $f(x) = -x + 6$: $f(0) = -0 + 6 = 6$
b) $g(x) = -10$: $g(5) = -10$
c) $h(x) = 4x^2 + x$: $h(-1) = 4(-1)^2 + (-1) = 3$

Practice Exercise
2. Find $g(0)$, $g(-2)$, and $g(6)$ for $g(x) = \frac{1}{2}x - 2$.

Objective 7.1c Draw the graph of a function.

Example Graph: $f(x) = -\frac{2}{3}x + 2$.

By choosing multiples of 3 for x, we can avoid fraction values for y. If $x = -3$, then $y = -\frac{2}{3} \cdot (-3) + 2 = 2 + 2 = 4$. We list three ordered pairs in a table, plot the points, draw the line, and label the graph.

x	$f(x)$
3	0
0	2
-3	4

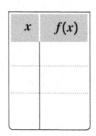

Practice Exercise
3. Graph: $f(x) = \frac{2}{5}x - 3$.

x	$f(x)$

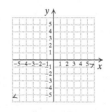

Objective 7.1d Determine whether a graph is that of a function using the vertical-line test.

Example Determine whether each of the following is the graph of a function.

a)   b)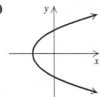

a) The graph is that of a function because no vertical line can cross the graph at more than one point.

b) The graph is not that of a function because a vertical line can cross the graph more than once.

Practice Exercise
4. Determine whether the graph is the graph of a function.

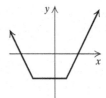

Objective 7.2a Find the domain and the range of a function.

Example For the function f whose graph is shown below, determine the domain and the range.

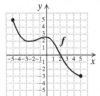 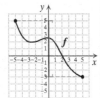

Domain: $[-5, 5]$; range: $[-3, 5]$

Practice Exercises
5. For the function g whose graph is shown below, determine the domain and the range.

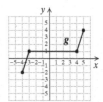

Example Find the domain of $g(x) = \dfrac{x+1}{2x-6}$.

Since $(x+1)/(2x-6)$ cannot be calculated when the denominator $2x - 6$ is 0, we solve $2x - 6 = 0$ to find the real numbers that must be excluded from the domain of g:

$$2x - 6 = 0$$
$$2x = 6$$
$$x = 3.$$

Thus, 3 is not in the domain. The domain of g is $\{x \mid x$ is a real number *and* $x \neq 3\}$.

6. Find the domain of
$$h(x) = \dfrac{x-3}{3x+9}.$$

Objective 7.3b Given two points on a line, find the slope. Given a linear equation, derive the equivalent slope–intercept equation and determine the slope and the y-intercept.

Example Find the slope of the line containing $(-5, 6)$ and $(-1, -4)$.

$$m = \dfrac{\text{change in } y}{\text{change in } x} = \dfrac{6 - (-4)}{-5 - (-1)} = \dfrac{6+4}{-5+1} = \dfrac{10}{-4} = -\dfrac{5}{2}$$

Example Find the slope and the y-intercept of
$$4x - 2y = 20.$$
We first solve for y:
$$4x - 2y = 20$$
$$-2y = -4x + 20 \quad \text{Subtracting } 4x$$
$$y = 2x - 10. \quad \text{Dividing by } -2$$
The slope is 2, and the y-intercept is $(0, -10)$.

Practice Exercises

7. Find the slope of the line containing $(2, -8)$ and $(-3, 2)$.

8. Find the slope and the y-intercept of
$$3x = -6y + 12.$$

Objective 7.4a Graph linear equations using intercepts.

Example Find the intercepts of $x - 2y = 6$ and then graph the line.

To find the y-intercept, we let $x = 0$ and solve for y:
$$0 - 2y = 6 \quad \text{Substituting 0 for } x$$
$$-2y = 6$$
$$y = -3.$$
The y-intercept is $(0, -3)$.

To find the x-intercept, we let $y = 0$ and solve for x:
$$x - 2 \cdot 0 = 6 \quad \text{Substituting 0 for } y$$
$$x = 6.$$
The x-intercept is $(6, 0)$.

We plot these points and draw the line, using a third point as a check. We let $x = -2$ and solve for y:
$$-2 - 2y = 6$$
$$-2y = 8$$
$$y = -4.$$
We plot $(-2, -4)$ and note that it is on the line. Thus the graph is correct.

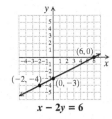

$x - 2y = 6$

Practice Exercise

9. Find the intercepts of $3y - 3 = x$ and then graph the line.

Objective 7.4b Given a linear equation in slope–intercept form, use the slope and the y-intercept to graph the line.

Example Graph using the slope and the y-intercept:
$$y = -\frac{3}{2}x + 5.$$

This equation is in slope–intercept form, $y = mx + b$. The y-intercept is $(0, 5)$. We plot $(0, 5)$. We can think of the slope $(m = -\frac{3}{2})$ as $\frac{-3}{2}$.

Starting at the y-intercept, we use the slope to find another point on the graph. We move 3 units down and 2 units to the right. We get a new point: $(2, 2)$.

To get a third point for a check, we start at $(2, 2)$ and move 3 units down and 2 units to the right to the point $(4, -1)$. We plot the points and draw the line.

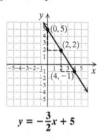

$y = -\frac{3}{2}x + 5$

Practice Exercise

10. Graph using the slope and the y-intercept:
$$y = \frac{1}{4}x - 3.$$

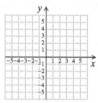

Objective 7.4c Graph linear equations of the form $x = a$ or $y = b$.

Example Graph: $y = -1$.

All ordered pairs $(x, -1)$ are solutions; y is -1 at each point. The graph is a horizontal line that intersects the y-axis at $(0, -1)$.

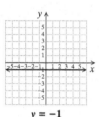

$y = -1$

Example Graph: $x = 2$.

All ordered pairs $(2, y)$ are solutions; x is 2 at each point. The graph is a vertical line that intersects the x-axis at $(2, 0)$.

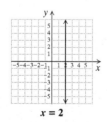

$x = 2$

Practice Exercises

11. Graph: $y = 3$.

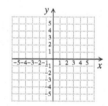

12. Graph: $x = -\frac{5}{2}$.

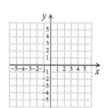

Objective 7.4d Given the equations of two lines, determine whether their graphs are parallel or whether they are perpendicular.

Example Determine whether the graphs of the given pair of lines are parallel, perpendicular, or neither.
a) $2y - x = 16,$
   $x + \frac{1}{2}y = 4$
b) $5x - 3 = 2y,$
   $2y + 12 = 5x$

a) Writing each equation in slope–intercept form, we have $y = \frac{1}{2}x + 8$ and $y = -2x + 8$. The slopes are $\frac{1}{2}$ and -2. The product of the slopes is -1: $\frac{1}{2} \cdot (-2) = -1$. The graphs are perpendicular.

b) Writing each equation in slope–intercept form, we have $y = \frac{5}{2}x - \frac{3}{2}$ and $y = \frac{5}{2}x - 6$. The slopes are the same, $\frac{5}{2}$, and the y-intercepts are different. The graphs are parallel.

Practice Exercises

Determine whether the graphs of the given pair of lines are parallel, perpendicular, or neither.

13. $-3x + 8y = -8,$
    $8y = 3x + 40$

14. $5x - 2y = -8,$
    $2x + 5y = 15$

Objective 7.5a Find an equation of a line when the slope and the y-intercept are given.

Example A line has slope 0.8 and y-intercept $(0, -17)$. Find an equation of the line.

We use the slope–intercept equation and substitute 0.8 for m and -17 for b:

$y = mx + b$ Slope–intercept equation
$y = 0.8x - 17.$

Practice Exercise

15. A line has slope -8 and y-intercept $(0, 0.3)$. Find an equation of the line.

Objective 7.5b Find an equation of a line when the slope and a point are given.

Example Find an equation of the line with slope -2 and containing the point $\left(\frac{1}{3}, -1\right)$.

Using the *point–slope equation*, we substitute -2 for m, $\frac{1}{3}$ for x_1, and -1 for y_1:

$y - (-1) = -2\left(x - \frac{1}{3}\right)$ Using $y - y_1 = m(x - x_1)$

$y + 1 = -2x + \frac{2}{3}$

$y = -2x - \frac{1}{3}.$

Using the *slope–intercept equation*, we substitute -2 for m, $\frac{1}{3}$ for x, and -1 for y, and then solve for b:

$-1 = -2 \cdot \frac{1}{3} + b$ Using $y = mx + b$

$-1 = -\frac{2}{3} + b$

$-\frac{1}{3} = b.$

Then, substituting -2 for m and $-\frac{1}{3}$ for b in the slope–intercept equation $y = mx + b$, we have $y = -2x - \frac{1}{3}$.

Practice Exercise

16. Find an equation of the line with slope -4 and containing the point $\left(\frac{1}{2}, -3\right)$.

Objective 7.5c Find an equation of a line when two points are given.

Example Find an equation of the line containing the points $(-3, 9)$ and $(1, -2)$.

We first find the slope:
$$\frac{9 - (-2)}{-3 - 1} = \frac{11}{-4} = -\frac{11}{4}.$$

Using the slope–intercept equation and the point $(1, -2)$, we substitute $-\frac{11}{4}$ for m, 1 for x, and -2 for y, and then solve for b. We could also have used the point $(-3, 9)$.

$$y = mx + b$$
$$-2 = -\tfrac{11}{4} \cdot 1 + b$$
$$-\tfrac{8}{4} = -\tfrac{11}{4} + b$$
$$\tfrac{3}{4} = b$$

Then substituting $-\frac{11}{4}$ for m and $\frac{3}{4}$ for b in $y = mx + b$, we have $y = -\frac{11}{4}x + \frac{3}{4}$.

Practice Exercise

17. Find an equation of the line containing the points $(-2, 7)$ and $(4, -3)$.

Objective 7.5d Given a line and a point not on the given line, find an equation of the line parallel to the line and containing the point, and find an equation of the line perpendicular to the line and containing the point.

Example Write an equation of the line containing $(-1, 1)$ and parallel to $3y - 6x = 5$.

Solving $3y - 6x = 5$ for y, we get $y = 2x + \frac{5}{3}$. The slope of the given line is 2.

A line parallel to the given line must have the same slope, 2. We substitute 2 for m, -1 for x_1, and 1 for y_1 in the point–slope equation:

$$y - 1 = 2[x - (-1)] \quad \text{Using } y - y_1 = m(x - x_1)$$
$$y - 1 = 2(x + 1)$$
$$y - 1 = 2x + 2$$
$$y = 2x + 3. \quad \text{Line parallel to the given line and passing through } (-1, 1)$$

Practice Exercises

18. Write an equation of the line containing the point $(2, -5)$ and parallel to $4x - 3y = 6$.

Example Write an equation of the line containing the point $(2, -4)$ and perpendicular to $6x + 2y = 13$.

Solving $6x + 2y = 13$ for y, we get $y = -3x + \frac{13}{2}$. The slope of the given line is -3.

The slope of a line perpendicular to the given line is the opposite of the reciprocal of -3, or $\frac{1}{3}$. We substitute $\frac{1}{3}$ for m, 2 for x_1, and -4 for y_1 in the point–slope equation:

$$y - (-4) = \tfrac{1}{3}(x - 2) \quad \text{Using } y - y_1 = m(x - x_1)$$
$$y + 4 = \tfrac{1}{3}x - \tfrac{2}{3}$$
$$y = \tfrac{1}{3}x - \tfrac{14}{3}. \quad \text{Line perpendicular to the given line and passing through } (2, -4)$$

19. Write an equation of the line containing $(2, -5)$ and perpendicular to $4x - 3y = 6$.

Review Exercises

Determine whether each correspondence is a function. [7.1a]

1.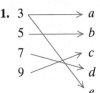
3 → a
5 → b
7 → c
9 → d
    → e

2.
1 → a
2 → b
3 → c
4 → d
5

Find the function values. [7.1b]

3. $g(x) = -2x + 5$; $g(0)$ and $g(-1)$

4. $f(x) = 3x^2 - 2x + 7$; $f(0)$ and $f(-1)$

5. *Gift-Card Sales.* The function $G(t) = 6.22t + 76.22$ can be used to approximate the average amount of gift-card sales, in billions of dollars, where t is the number of years after 2006. Estimate the amount of gift-card sales in 2016. That is, find $G(10)$. [7.1b]

   Data: National Retail Federation

Graph. [7.1c]

6. $f(x) = -3x + 2$

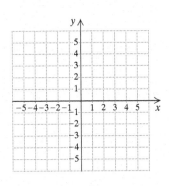

7. $g(x) = \frac{5}{2}x - 3$

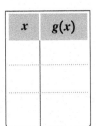
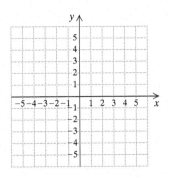

8. $f(x) = |x - 3|$

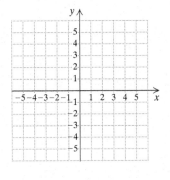

9. $h(x) = 3 - x^2$

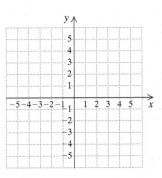

Determine whether each of the following is the graph of a function. [7.1d]

10.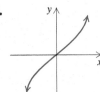

11.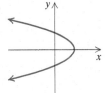

12. For the following graph of a function f, determine (a) $f(2)$; (b) the domain; (c) all x-values such that $f(x) = 2$; and (d) the range. [7.2a]

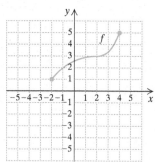

Find the domain. [7.2a]

13. $f(x) = \dfrac{5}{x-4}$

14. $g(x) = x - x^2$

Find the slope and the y-intercept. [7.3a, b]

15. $y = -3x + 2$

16. $4y + 2x = 8$

17. Find the slope, if it exists, of the line containing the points $(13, 7)$ and $(10, -4)$. [7.3b]

Find the intercepts. Then graph the equation. [7.4a]

18. $2y + x = 4$

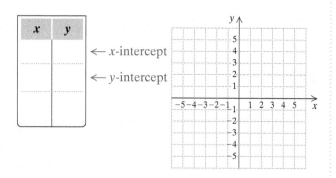

19. $2y = 6 - 3x$

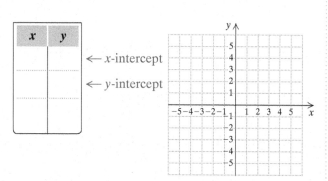

Graph using the slope and the y-intercept. [7.4b]

20. $g(x) = -\tfrac{2}{3}x - 4$

21. $f(x) = \tfrac{5}{2}x + 3$

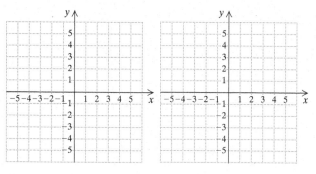

Graph. [7.4c]

22. $x = -3$

23. $f(x) = 4$

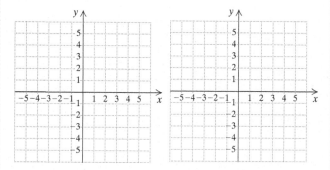

Determine whether the graphs of the given pair of lines are parallel or perpendicular. [7.4d]

24. $y + 5 = -x,$
    $x - y = 2$

25. $3x - 5 = 7y,$
    $7y - 3x = 7$

26. $4y + x = 3,$
    $2x + 8y = 5$

27. $x = 4,$
    $y = -3$

28. Find a linear function $f(x) = mx + b$ whose graph has slope 4.7 and y-intercept $(0, -23)$. [7.5a]

29. Find an equation of the line having slope -3 and containing the point $(3, -5)$. [7.5b]

580   CHAPTER 7   Graphs, Functions, and Applications

30. Find an equation of the line containing the points $(-2, 3)$ and $(-4, 6)$. [7.5c]

31. Find an equation of the line containing the given point and parallel to the given line:
$(14, -1)$; $5x + 7y = 8$. [7.5d]

32. Find an equation of the line containing the given point and perpendicular to the given line:
$(5, 2)$; $3x + y = 5$. [7.5d]

33. *Records in the 400-Meter Run.* The following table lists data regarding the Summer Olympics winning times in the men's 400-m run. [7.5e]

YEAR	SUMMER OLYMPICS WINNING TIME IN MEN'S 400-M RUN (in seconds)
1972	44.66
2016	43.03

a) Use the two data points to find a linear function that fits the data. Let x = the number of years after 1972 and $R(x)$ = the Summer Olympics winning time x years from 1972.

b) Use the function to estimate the winning time in the men's 400-m run in 2000 and in 2010.

34. What is the domain of $f(x) = \dfrac{x + 3}{x - 2}$? [7.2a]

A. $\{x \mid x \geq -3\}$
B. $\{x \mid x \text{ is a real number } and \ x \neq -3 \ and \ x \neq 2\}$
C. $\{x \mid x \text{ is a real number } and \ x \neq 2\}$
D. $\{x \mid x > -3\}$

35. Find an equation of the line containing the point $(-2, 1)$ and perpendicular to $3y - \frac{1}{2}x = 0$. [7.5d]

A. $6x + y = -11$
B. $y = -\dfrac{1}{6}x - 11$
C. $y = -2x - 3$
D. $2x + \dfrac{1}{3} = 0$

Synthesis

36. Homespun Jellies charges $2.49 for each jar of preserves. Shipping charges are $3.75 for handling, plus $0.60 per jar. Find a linear function for determining the cost of buying and shipping x jars of preserves. [7.5e]

Understanding Through Discussion and Writing

1. Under what conditions will the x-intercept and the y-intercept of a line be the same? What would the equation for such a line look like? [7.4a]

2. Explain the usefulness of the concept of slope when describing a line. [7.3b, c], [7.4b], [7.5a, b, c, d]

3. A student makes a mistake when using a graphing calculator to draw $4x + 5y = 12$ and the following screen appears. Use algebra to show that a mistake has been made. What do you think the mistake was? [7.3b]

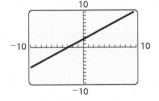

4. *Computer Repair.* The cost $R(t)$, in dollars, of computer repair at PC Pros is given by
$$R(t) = 50t + 35,$$
where t is the number of hours that the repair requires. Determine m and b in this application and explain their meaning. [7.5e]

5. Explain why the slope of a vertical line is not defined but the slope of a horizontal line is 0. [7.4c]

6. A student makes a mistake when using a graphing calculator to draw $5x - 2y = 3$ and the following screen appears. Use algebra to show that a mistake has been made. What do you think the mistake was? [7.3b]

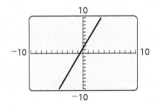

CHAPTER 7 Test

Determine whether each correspondence is a function.

1. cat → dog
   fish → worm
   dog → cat
   tiger → fish
   teacher → student

2. Lake Placid → 1980
   Oslo → 1976
   Squaw Valley → 1960
   Innsbruck → 1952
   → 1932

Find the function values.

3. $f(x) = -3x - 4$; $f(0)$ and $f(-2)$

4. $g(x) = x^2 + 7$; $g(0)$ and $g(-1)$

5. $h(x) = -6$; $h(-4)$ and $h(-6)$

6. $f(x) = |x + 7|$; $f(-10)$ and $f(-7)$

Graph.

7. $h(x) = -2x - 5$

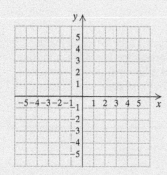

8. $f(x) = -\dfrac{3}{5}x$

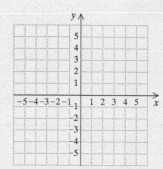

9. $g(x) = 2 - |x|$

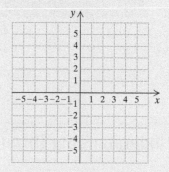

10. $f(x) = x^2 + 2x - 3$

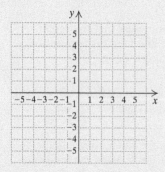

11. $y = f(x) = -3$

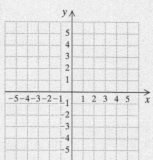

12. $2x = -4$

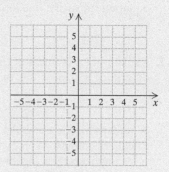

13. *Median Age of Cars.* The function
$$A(t) = 0.22t + 5.87$$
can be used to estimate the median age of cars in the United States t years after 1990. (This means, for example, that if the median age of cars is 4 years, then half the cars are older than 4 years and half are younger.)

Data: The Polk Co; autonews.com

a) Find the median age of cars in 2010.
b) In what year was the median age of cars 11.6 years?

Determine whether each of the following is the graph of a function.

14.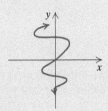

15.

Find the domain.

16. $f(x) = \dfrac{8}{2x + 3}$

17. $g(x) = 5 - x^2$

18. For the following graph of function f, determine (a) $f(1)$; (b) the domain; (c) all x-values such that $f(x) = 2$; and (d) the range.

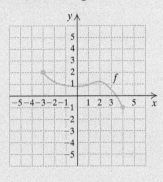

Find the slope and the y-intercept.

19. $f(x) = -\dfrac{3}{5}x + 12$

20. $-5y - 2x = 7$

Find the slope, if it exists, of the line containing the following points.

21. $(-2, -2)$ and $(6, 3)$

22. $(-3.1, 5.2)$ and $(-4.4, 5.2)$

23. Find the slope, or rate of change, of the graph at right.

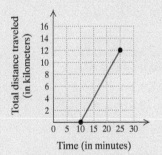

24. Find the intercepts. Then graph the equation.
$2x + 3y = 6$

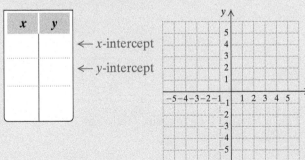

25. Graph using the slope and the y-intercept:
$f(x) = -\dfrac{2}{3}x - 1.$

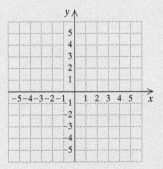

Determine whether the graphs of the given pair of lines are parallel or perpendicular.

26. $4y + 2 = 3x$,
$-3x + 4y = -12$

27. $y = -2x + 5$,
$2y - x = 6$

28. Find an equation of the line that has the given characteristics:

slope: -3; y-intercept: $(0, 4.8)$.

29. Find a linear function $f(x) = mx + b$ whose graph has the given slope and y-intercept:

slope: 5.2; y-intercept: $\left(0, -\frac{5}{8}\right)$.

30. Find an equation of the line having the given slope and containing the given point:

$m = -4$; $(1, -2)$.

31. Find an equation of the line containing the given pair of points:

$(4, -6)$ and $(-10, 15)$.

32. Find an equation of the line containing the given point and parallel to the given line:

$(4, -1)$; $x - 2y = 5$.

33. Find an equation of the line containing the given point and perpendicular to the given line:

$(2, 5)$; $x + 3y = 2$.

34. *Median Age of Men at First Marriage.* The following table lists data regarding the median age of men at first marriage in 1970 and in 2010.

YEAR	MEDIAN AGE OF MEN AT FIRST MARRIAGE
1970	23.2
2015	29.2

Data: U.S. Census Bureau

a) Use the two data points to find a linear function that fits the data. Let $x =$ the number of years after 1970 and $A =$ the median age at first marriage x years after 1970.

b) Use the function to estimate the median age of men at first marriage in 2008 and in 2019.

35. Find an equation of the line having slope -2 and containing the point $(3, 1)$.

A. $y - 1 = 2(x - 3)$ **B.** $y - 1 = -2(x - 3)$
C. $x - 1 = -2(y - 3)$ **D.** $x - 1 = 2(y - 3)$

Synthesis

36. Find k such that the line $3x + ky = 17$ is perpendicular to the line $8x - 5y = 26$.

37. Find a formula for a function f for which $f(-2) = 3$.

584 CHAPTER 7 Graphs, Functions, and Applications

CHAPTERS 1–7 Cumulative Review

Compute and simplify.

1. $-2[1.4 - (-0.8 - 1.2)]$

2. $(1.3 \times 10^8)(2.4 \times 10^{-10})$

3. $\left(-\dfrac{1}{6}\right) \div \left(\dfrac{2}{9}\right)$

4. $\dfrac{2^{12}\, 2^{-7}}{2^8}$

Simplify.

5. $\dfrac{x^2 - 9}{2x^2 - 7x + 3}$

6. $\dfrac{t^2 - 16}{(t + 4)^2}$

7. $\dfrac{x - \dfrac{x}{x+2}}{\dfrac{2}{x} - \dfrac{1}{x+2}}$

Perform the indicated operations and simplify.

8. $(1 - 3x^2)(2 - 4x^2)$

9. $(2a^2b - 5ab^2)^2$

10. $(3x^2 + 4y)(3x^2 - 4y)$

11. $-2x^2(x - 2x^2 + 3x^3)$

12. $(1 + 2x)(4x^2 - 2x + 1)$

13. $\left(8 - \dfrac{1}{3}x\right)\left(8 + \dfrac{1}{3}x\right)$

14. $(-8y^2 - y + 2) - (y^3 - 6y^2 + y - 5)$

15. $(2x^3 - 3x^2 - x - 1) \div (2x - 1)$

16. $\dfrac{7}{5x - 25} + \dfrac{x + 7}{5 - x}$

17. $\dfrac{2x - 1}{x - 2} - \dfrac{2x}{2 - x}$

18. $\dfrac{y^2 + y}{y^2 + y - 2} \cdot \dfrac{y + 2}{y^2 - 1}$

19. $\dfrac{7x + 7}{x^2 - 2x} \div \dfrac{14}{3x - 6}$

Factor completely.

20. $6x^5 - 36x^3 + 9x^2$

21. $16y^4 - 81$

22. $3x^2 + 10x - 8$

23. $4x^4 - 12x^2y + 9y^2$

24. $3m^3 + 6m^2 - 45m$

25. $x^3 + x^2 - x - 1$

Solve.

26. $3x - 4(x + 1) = 5$

27. $x(2x - 5) = 0$

28. $5x + 3 \geq 6(x - 4) + 7$

29. $1.5x - 2.3x = 0.4(x - 0.9)$

30. $2x^2 = 338$

31. $3x^2 + 15 = 14x$

32. $\dfrac{2}{x} - \dfrac{3}{x - 2} = \dfrac{1}{x}$

33. $1 + \dfrac{3}{x} + \dfrac{x}{x + 1} = \dfrac{1}{x^2 + x}$

34. $w - \dfrac{9}{10} = -\dfrac{9}{10} + w$

35. $20 - 3y - 2y \leq 45 + 5y$

36. $9x - 4 - 2x = 5x - 5 + 2x$

37. $N = rx - t$, for x

Solve.

38. *Digital Photo Frame.* Joel paid $37.10, including 6% sales tax, for a digital photo frame. What was the price of the frame itself?

39. *Roofing Time.* It takes David 15 hr to put a roof on a house. It takes Loren 12 hr to put a roof on the same type of house. How long would it take to complete the job if they worked together?

40. *Triangle Dimensions.* The length of one leg of a right triangle is 12 in. The length of the hypotenuse is 8 in. longer than the length of the other leg. Find the lengths of the hypotenuse and the other leg.

41. *Quality Control.* A sample of 120 computer chips contained 5 defective chips. How many defective chips would you expect to find in a batch of 1800 chips?

42. *Triangle Dimensions.* The height of a triangle is 5 ft more than the base. The area is 18 ft². Find the height and the base.

43. *Height of a Parallelogram.* The height h of a parallelogram of fixed area varies inversely as the length of the base b. Suppose that the height is 24 ft when the base is 15 ft. Find the height when the base is 5 ft. What is the variation constant?

44. Find an equation of variation in which y varies directly as x, and $y = 2.4$ when $x = 12$.

Find the function values.

45. $g(x) = -\frac{1}{2}x + 6$; $g(0)$ and $g(-6)$

46. $f(x) = |2x - 3|$; $f(-4)$ and $f(0)$

47. Find the slope of the line containing the points $(2, 3)$ and $(-1, 3)$.

48. Find the slope and the y-intercept of the line $2x + 3y = 6$.

49. Find an equation of the line that contains the points $(-5, 6)$ and $(2, -4)$.

50. Find an equation of the line containing the point $(0, -3)$ and having the slope $m = 6$.

Graph on a plane.

51. $y = -2$ **52.** $2x + 5y = 10$

53. $f(x) = -\frac{3}{4}x + 2$ **54.** $2x - y = 3$

55. $x - 4 = 0$ **56.** $3x - y = -3$

Synthesis

57. Solve: $x^2 + 2 < 0$.

58. Simplify:
$$\frac{x-5}{x+3} - \frac{x^2 - 6x + 5}{x^2 + x - 2} \div \frac{x^2 + 4x + 3}{x^2 + 3x + 2}.$$

59. Find the value of k such that $y - kx = 4$ and $10x - 3y = -12$ are perpendicular.

CHAPTER 8

Systems of Equations

8.1 Systems of Equations in Two Variables
8.2 Solving by Substitution
8.3 Solving by Elimination
8.4 Solving Applied Problems: Two Equations

Translating for Success
Mid-Chapter Review

8.5 Systems of Equations in Three Variables
8.6 Solving Applied Problems: Three Equations

Summary and Review
Test
Cumulative Review

Video gamers are present in 65% of U.S. households and are represented in every age group, with about one-fourth of gamers under 18 and one-fourth over 50. In 2016, U.S. consumers spent $24.5 billion on video game software, an increase of $7 billion since 2010. The delivery format of video games has changed from mostly physical in 2009 to mostly digital in 2016. As the graph indicates, the number of games delivered physically was the same as the number delivered digitally sometime during 2012. The genres of games are changing in popularity as well. In 2016, half of the most popular games were shooting and action games; these genres represented only 32% of the most popular games in 2009.

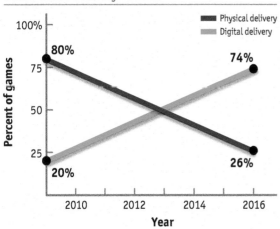

Data: Entertainment Software Association; statista.com

In Exercise 1 of Exercise Set 8.4, we will calculate the number of video-game production locations in the United States, and in Exercise 33 of Exercise Set 8.6, we will calculate the number of games rated for different audiences.

587

STUDYING FOR SUCCESS Working with Others

- Try being a tutor for a fellow student. You may find that you understand concepts better after explaining them to someone else.
- Consider forming a study group.
- Verbalize the math. Often simply talking about a concept with a classmate can help clarify it for you.

8.1 Systems of Equations in Two Variables

OBJECTIVE

a Solve a system of two linear equations or two functions by graphing and determine whether a system is consistent or inconsistent and whether the equations in a system are dependent or independent.

Sets of equations for which we seek a common solution are called *systems of equations*. We can solve systems of two equations in two variables by graphing.

a SOLVING SYSTEMS OF EQUATIONS GRAPHICALLY

SKILL REVIEW *Graph linear equations.* [3.2a]
Graph.
1. $x + y = 3$
2. $y = x - 2$

Answers: 1. 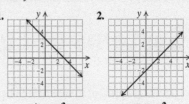 2.

MyLab Math
VIDEO

School Enrollment. In 2016, approximately 50 million children were enrolled in public elementary and secondary schools in the United States. There were 20 million more students enrolled in prekindergarten–grade 8 than there were in grades 9–12. How many were enrolled at each level?
Data: National Center for Education Statistics

To solve, we first let

x = the number enrolled in prekindergarten–grade 8, and
y = the number enrolled in grades 9–12,

where x and y are in millions of students. The problem gives us two statements that can be translated to equations.
First, we consider the total number enrolled:

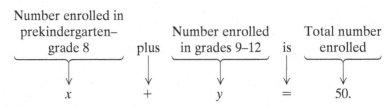

588 CHAPTER 8 Systems of Equations

The second statement of the problem compares the enrollment at the two levels:

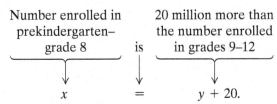

We have now translated the problem to a pair of equations, or a **system of equations**:

$$x + y = 50,$$
$$x = y + 20.$$

A **solution** of a system of two equations in two variables is an ordered pair that makes *both* equations true. If we graph a system of equations, the point at which the graphs intersect will be a solution of *both* equations. To find the solution of the system above, we graph both equations, as shown at the right.

We see that the graphs intersect at the point $(35, 15)$—that is, $x = 35$ and $y = 15$. These numbers check in the statement of the original problem. This tells us that 35 million students were enrolled in prekindergarten–grade 8, and 15 million students were enrolled in grades 9–12.

Systems of two equations may have one solution, no solution, or infinitely many solutions.

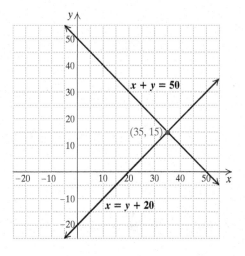

One Solution

EXAMPLE 1 Solve this system graphically:

$$y - x = 1,$$
$$y + x = 3.$$

We draw the graph of each equation and find the coordinates of the point of intersection.

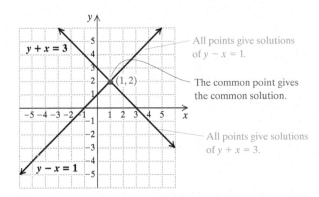

The point of intersection has coordinates that make *both* equations true. The solution seems to be the point $(1, 2)$. However, solving by graphing may give only approximate answers. Thus we check the pair $(1, 2)$ in both equations.

Check:
$$\begin{array}{c|c} y - x = 1 \\ \hline 2 - 1 \;?\; 1 \\ 1 \;|\; \text{TRUE} \end{array} \quad \begin{array}{c|c} y + x = 3 \\ \hline 2 + 1 \;?\; 3 \\ 3 \;|\; \text{TRUE} \end{array}$$

The solution is $(1, 2)$.

Do Exercises 1 and 2.

Solve each system graphically.

1. $-2x + y = 1,$
   $3x + y = 1$

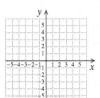

2. $y = \frac{1}{2}x,$
   $y = -\frac{1}{4}x + \frac{3}{2}$

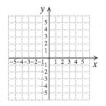

Answers

1. $(0, 1)$ 2. $(2, 1)$

No Solution

Sometimes the equations in a system have graphs that are parallel lines.

EXAMPLE 2 Solve graphically:

$$f(x) = -3x + 5,$$
$$g(x) = -3x - 2.$$

Note that this system is written using function notation. We graph the functions. The graphs have the same slope, -3, and different y-intercepts, so they are parallel. There is no point at which they cross, so the system has no solution. No matter what point we try, it will *not* check in *both* equations. The solution set is thus the empty set, denoted $\emptyset$, or $\{\ \}$.

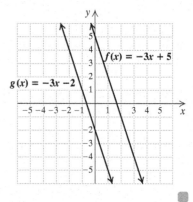

3. Solve graphically:
   $$y + 2x = 3,$$
   $$y + 2x = -4.$$

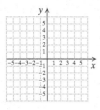

4. Classify each of the systems in Margin Exercises 1–3 as consistent or inconsistent.

   The system in Margin Exercise 1 has a solution, so it is ____.

   The system in Margin Exercise 2 has a solution, so it is ____.

   The system in Margin Exercise 3 does not have a solution, so it is ____.

CONSISTENT SYSTEMS AND INCONSISTENT SYSTEMS

If a system of equations has at least one solution, then it is **consistent**.

If a system of equations has no solution, then it is **inconsistent**.

The system in Example 1 is consistent. The system in Example 2 is inconsistent.

◂ Do Exercises 3 and 4.

Infinitely Many Solutions

Sometimes the equations in a system have the same graph. In such a case, the equations have an *infinite* number of solutions in common.

EXAMPLE 3 Solve graphically:

$$3y - 2x = 6,$$
$$-12y + 8x = -24.$$

We graph the equations and see that the graphs are the same. Thus any solution of one of the equations is a solution of the other. Each equation has an infinite number of solutions, two of which are shown on the graph.

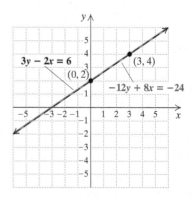

Answers
3. No solution **4.** Consistent: Margin Exercises 1 and 2; inconsistent: Margin Exercise 3

Guided Solution:
4. consistent, consistent, inconsistent

We check one such solution, $(0, 2)$, which is the y-intercept of each equation.

Check:
$$\begin{array}{c|c}
3y - 2x = 6 & -12y + 8x = -24 \\
\hline
3(2) - 2(0) \;?\; 6 & -12(2) + 8(0) \;?\; -24 \\
6 - 0 & -24 + 0 \\
6 \quad \text{TRUE} & -24 \quad \text{TRUE}
\end{array}$$

We leave it to the student to check that $(3, 4)$ is a solution of both equations. If $(0, 2)$ and $(3, 4)$ are solutions, then all points on the line containing them will be solutions. The system has an infinite number of solutions.

> **DEPENDENT EQUATIONS AND INDEPENDENT EQUATIONS**
>
> If for a system of two equations in two variables:
>
> the graphs of the equations are the same line, then the equations are **dependent**.
>
> the graphs of the equations are different lines, then the equations are **independent**.

When we graph a system of two equations, one of the following three situations can occur.

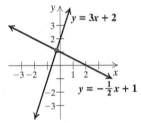

One solution.
Graphs intersect.
The system is *consistent* and *the equations are independent*.

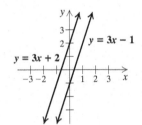

No solution.
Graphs are parallel.
The system is *inconsistent* and *the equations are independent*.

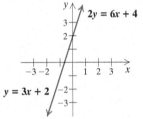

Infinitely many solutions.
Equations have the same graph. The system is *consistent* and *the equations are dependent*.

Let's summarize what we know about the systems of equations shown in Examples 1–3.

	NUMBER OF SOLUTIONS	GRAPHS OF EQUATIONS
EXAMPLE 1	1 System is consistent.	Different Equations are independent.
EXAMPLE 2	0 System is inconsistent.	Different Equations are independent.
EXAMPLE 3	Infinitely many System is consistent.	Same Equations are dependent.

Do Exercises 5 and 6.

5. Solve graphically:

$$2x - 5y = 10,$$
$$-6x + 15y = -30.$$

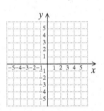

6. Classify the equations in Margin Exercises 1, 2, 3, and 5 as dependent or independent.

In Margin Exercise 1, the graphs are different, so the equations are _____.

In Margin Exercise 2, the graphs are different, so the equations are _____.

In Margin Exercise 3, the graphs are different, so the equations are _____.

In Margin Exercise 5, the graphs are the same, so the equations are _____.

Answers
5. Infinitely many solutions
6. Independent: Margin Exercises 1, 2, and 3; dependent: Margin Exercise 5

Guided Solution:
6. independent, independent, independent, dependent

ALGEBRAIC ▶◀ GRAPHICAL CONNECTION

Consider the equation $-2x + 13 = 4x - 17$. Let's solve it algebraically:

$$-2x + 13 = 4x - 17$$
$$13 = 6x - 17 \quad \text{Adding } 2x$$
$$30 = 6x \quad \text{Adding } 17$$
$$5 = x. \quad \text{Dividing by } 6$$

We can also solve the equation graphically, as we see in the following two methods. Using method 1, we graph two functions. The solution of the original equation is the x-coordinate of the point of intersection. Using method 2, we graph one function. The solution of the original equation is the x-coordinate of the x-intercept of the graph.

Method 1: Solve $-2x + 13 = 4x - 17$ graphically.

We let $f(x) = -2x + 13$ and $g(x) = 4x - 17$. Graphing the system of equations, we get the graph shown below.

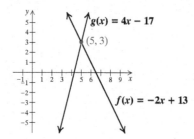

The point of intersection of the two graphs is $(5, 3)$. Note that the x-coordinate of this point is 5. This is the value of x for which $-2x + 13 = 4x - 17$, so it is the solution of the equation.

◀ Do Exercises 7 and 8.

Method 2: Solve $-2x + 13 = 4x - 17$ graphically.

Adding $-4x$ and 17 on both sides, we obtain an equation with 0 on one side: $-6x + 30 = 0$. This time we let $f(x) = -6x + 30$ and $g(x) = 0$. Since the graph of $g(x) = 0$, or $y = 0$, is the x-axis, we need only graph $f(x) = -6x + 30$ and see where it crosses the x-axis.

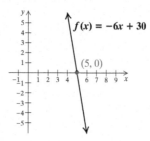

Note that the x-intercept of $f(x) = -6x + 30$ is $(5, 0)$, or just 5. This x-value is the solution of the equation $-2x + 13 = 4x - 17$.

◀ Do Exercises 9 and 10.

---

7. a) Solve $x + 1 = \frac{2}{3}x$ algebraically.

  b) Solve $x + 1 = \frac{2}{3}x$ graphically using method 1.

8. Solve $\frac{1}{2}x + 3 = 2$ graphically using method 1.

9. a) Solve $x + 1 = \frac{2}{3}x$ graphically using method 2.

  b) Compare your answers to Margin Exercises 7(a), 7(b), and 9(a).

10. Solve $\frac{1}{2}x + 3 = 2$ graphically using method 2.

Answers

7. (a) -3; (b) the same: -3 **8.** -2
9. (a) -3; (b) All are -3. **10.** -2

CALCULATOR CORNER

Solving Systems of Equations We can solve a system of two equations in two variables using a graphing calculator. Consider the system of equations in Example 1. First, we solve the equations for y, obtaining $y = x + 1$ and $y = -x + 3$. Next, we enter $y_1 = x + 1$ and $y_2 = -x + 3$ on the equation-editor screen and graph the equations. We can use the standard viewing window, $[-10, 10, -10, 10]$.

We will use the **INTERSECT** feature to find the coordinates of the point of intersection of the lines. To access this feature, we press **2ND** **CALC** **5**. (**CALC** is the second operation associated with the **TRACE** key.) The query "First curve?" appears on the graph screen. The blinking cursor is positioned on the graph of y_1. We press **ENTER** to indicate that this is the first curve involved in the intersection. Next, the query "Second curve?" appears and the blinking cursor is positioned on the graph of y_2. We press **ENTER** to indicate that this is the second curve. Now the query "Guess?" appears. We use the ▷ and ◁ keys to move the cursor close to the point of intersection or we enter an x-value close to the first coordinate of the point of intersection. Then we press **ENTER**. The coordinates of the point of intersection of the graphs, $x = 1$, $y = 2$, appear at the bottom of the screen. Thus the solution of the system of equations is $(1, 2)$.

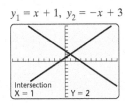

EXERCISES: Use a graphing calculator to solve each system of equations.

1. $x + y = 5,$
   $y = x + 1$

2. $y = x + 3,$
   $2x - y = -7$

3. $x - y = -6,$
   $y = 2x + 7$

4. $x + 4y = -1,$
   $x - y = 4$

8.1 Exercise Set

FOR EXTRA HELP — MyLab Math

✓ Check Your Understanding

Reading Check Determine whether each statement is true or false.

RC1. Every system of equations has one solution.

RC2. A solution of a system of equations in two variables is an ordered pair.

RC3. Graphs of two lines may have one point, no points, or an infinite number of points in common.

RC4. If a system of two equations has only one solution, the system is consistent and the equations in the system are independent.

Concept Check Determine whether the given ordered pair is a solution of the system of equations.

CC1. $(-2, -3);$ $x - y = 1,$
$x + y = -5$

CC2. $(-1, 0);$ $b - a = 1,$
$2a = b - 2$

CC3. $(5, -6);$ $c - d = 11,$
$2c = d + 4$

a Solve each system of equations graphically. Then classify the system as consistent or inconsistent and the equations as dependent or independent. Complete the check for Exercises 1–4.

1. $x + y = 4$,
$x - y = 2$

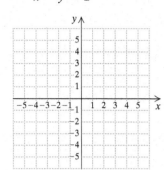

Check: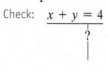

2. $x - y = 3$,
$x + y = 5$

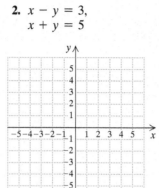

Check: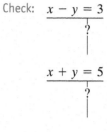

3. $2x - y = 4$,
$2x + 3y = -4$

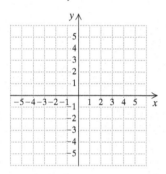

Check: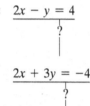

4. $3x + y = 5$,
$x - 2y = 4$

Check:

5. $2x + y = 6$,
$3x + 4y = 4$

6. $2y = 6 - x$,
$3x - 2y = 6$

7. $f(x) = x - 1$,
$g(x) = -2x + 5$

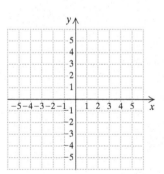

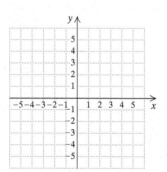

8. $f(x) = x + 1$,
$g(x) = \frac{2}{3}x$

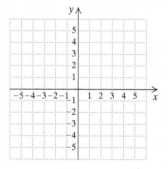

9. $2u + v = 3$,
$2u = v + 7$

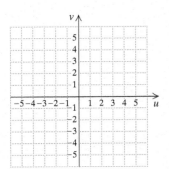

10. $2b + a = 11$,
$a - b = 5$

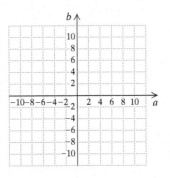

11. $f(x) = -\frac{1}{3}x - 1$,
$g(x) = \frac{4}{3}x - 6$

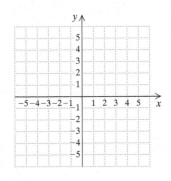

12. $f(x) = -\frac{1}{4}x + 1$,
$g(x) = \frac{1}{2}x - 2$

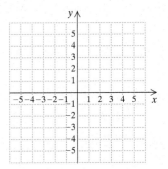

13. $6x - 2y = 2$,
$9x - 3y = 1$

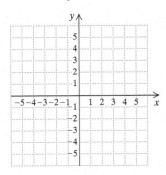

14. $y - x = 5$,
$2x - 2y = 10$

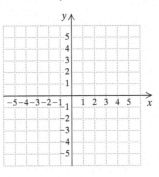

15. $2x - 3y = 6$,
$3y - 2x = -6$

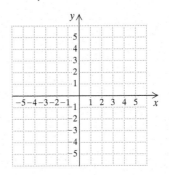

16. $y = 3 - x$,
$2x + 2y = 6$

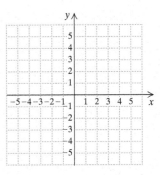

17. $x = 4$,
$y = -5$

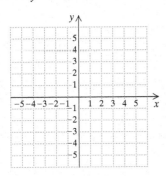

18. $x = -3$,
$y = 2$

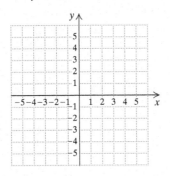

19. $y = -x - 1$,
$x - 2y = 8$

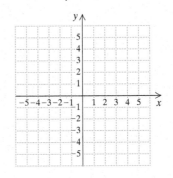

20. $a + 2b = -3$,
$b - a = 6$

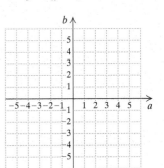

Matching. Each of Exercises 21–26 shows the graph of a system of equations and its solution. First, classify the system as consistent or inconsistent and the equations as dependent or independent. Then match it with one of the appropriate systems of equations (A)–(F), which follow.

21. Solution: $(3, 3)$

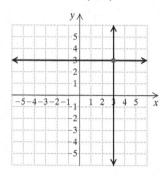

22. Solution: $(1, 1)$

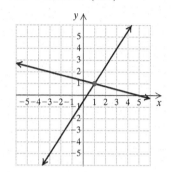

23. Solutions: Infinitely many

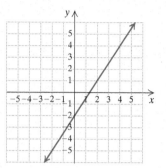

24. Solution: $(4, -3)$

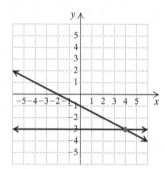

25. Solution: No solution

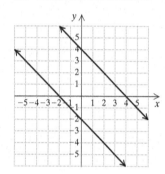

26. Solution: $(-1, 3)$

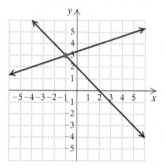

A. $3y - x = 10$,
$x = -y + 2$

B. $9x - 6y = 12$,
$y = \frac{3}{2}x - 2$

C. $2y - 3x = -1$,
$x + 4y = 5$

D. $x + y = 4$,
$y = -x - 2$

E. $\frac{1}{2}x + y = -1$,
$y = -3$

F. $x = 3$,
$y = 3$

Skill Maintenance

Write an equation of the line containing the given point and parallel to the given line. [7.5d]

27. $(-4, 2)$; $3x = 5y - 4$

28. $(-6, 0)$; $8y - 3x = 2$

Write an equation of the line containing the given point and perpendicular to the given line. [7.5d]

29. $(-4, 6)$; $2x = 3y - 12$

30. $(3, -10)$; $8y - 4 = -6x$

Synthesis

Use a graphing calculator to solve each system of equations. Round all answers to the nearest hundredth. You may need to solve for y first.

31. $2.18x + 7.81y = 13.78$,
$5.79x - 3.45y = 8.94$

32. $f(x) = 123.52x + 89.32$,
$g(x) = -89.22x + 33.76$

Solve graphically.

33. $y = |x|$,
$x + 4y = 15$

34. $x - y = 0$,
$y = x^2$

35. $2x - 3 = 4 - 5x$

36. $x + 3 = 2x^2$

8.2 Solving by Substitution

OBJECTIVES

a Solve systems of equations in two variables by the substitution method.

b Solve applied problems by solving systems of two equations using substitution.

Consider this system of equations:

$$5x + 9y = 2,$$
$$4x - 9y = 10.$$

What is the solution? It is rather difficult to tell exactly by graphing. It would appear that fractions are involved. It turns out that the solution is $\left(\frac{4}{3}, -\frac{14}{27}\right)$.

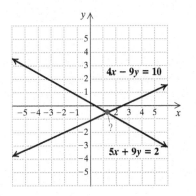

Solving by graphing, though useful in many applied situations, is not always fast or accurate in cases where solutions are not integers. We need techniques involving algebra to determine the solution exactly. Because these techniques use algebra, they are called **algebraic methods**.

a THE SUBSTITUTION METHOD

One nongraphical method for solving systems is known as the **substitution method**.

EXAMPLE 1 Solve this system:

$$x + y = 4, \quad (1)$$
$$x = y + 1. \quad (2)$$

Equation (2) says that x and $y + 1$ name the same number. Thus we can substitute $y + 1$ for x in equation (1):

$$x + y = 4 \quad \text{Equation (1)}$$
$$(y + 1) + y = 4. \quad \text{Substituting } y + 1 \text{ for } x$$

Since this equation has only one variable, we can solve for y using methods learned earlier:

$$(y + 1) + y = 4$$
$$2y + 1 = 4 \quad \text{Removing parentheses and collecting like terms}$$
$$2y = 3 \quad \text{Subtracting 1}$$
$$y = \tfrac{3}{2}. \quad \text{Dividing by 2}$$

We return to the original pair of equations and substitute $\frac{3}{2}$ for y in *either* equation so that we can solve for x. Calculation will be easier if we choose equation (2) since it is already solved for x:

$$x = y + 1 \quad \text{Equation (2)}$$
$$= \tfrac{3}{2} + 1 \quad \text{Substituting } \tfrac{3}{2} \text{ for } y$$
$$= \tfrac{3}{2} + \tfrac{2}{2} = \tfrac{5}{2}.$$

We obtain the ordered pair $\left(\tfrac{5}{2}, \tfrac{3}{2}\right)$. Even though we solved for *y first*, it is still the *second* coordinate since x is before y alphabetically. We check to be sure that the ordered pair is a solution.

SKILL REVIEW

Solve equations. [2.3c]

Solve.

1. $3y - 4 = 2$
2. $2(x + 1) + 5 = 1$

Answers: 1. 2 2. −3

MyLab Math
VIDEO

Check:

$$\frac{x+y=4}{\frac{5}{2}+\frac{3}{2}\stackrel{?}{\,}4}$$
$$\frac{\frac{8}{2}}{4}\quad \text{TRUE}$$

$$\frac{x=y+1}{\frac{5}{2}\stackrel{?}{\,}\frac{3}{2}+1}$$
$$\frac{\frac{3}{2}+\frac{2}{2}}{\frac{5}{2}}\quad \text{TRUE}$$

Since $\left(\frac{5}{2},\frac{3}{2}\right)$ checks, it is the solution. Even though exact fraction solutions are difficult to determine graphically, a graph can help us to visualize whether the solution is reasonable.

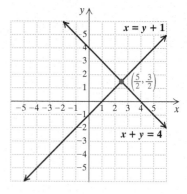

Solve by the substitution method.

1. $x + y = 6,$
   $y = x + 2$

2. $y = 7 - x,$
   $2x - y = 8$

(*Caution*: Use parentheses when you substitute, being careful about removing them. Remember to solve for both variables.)

◀ Do Exercises 1 and 2.

Suppose that neither equation of a pair has a variable alone on one side. We then solve one equation for one of the variables.

EXAMPLE 2 Solve this system:

$2x + y = 6,$ (1)
$3x + 4y = 4.$ (2)

First, we solve one equation for one variable. Since the coefficient of y is 1 in equation (1), it is the easier one to solve for y:

$y = 6 - 2x.$ (3)

Next, we substitute $6 - 2x$ for y in equation (2) and solve for x:

Solve by the substitution method.

3. $2y + x = 1,$
   $y - 2x = 8$

.... **Caution!** ....
Remember to use parentheses when you substitute. Then remove them properly.

$3x + 4(6 - 2x) = 4$ Substituting $6 - 2x$ for y
$3x + 24 - 8x = 4$ Multiplying to remove parentheses
$24 - 5x = 4$ Collecting like terms
$-5x = -20$ Subtracting 24
$x = 4.$ Dividing by -5

In order to find y, we return to either of the original equations, (1) or (2), or equation (3), which we solved for y. It is generally easier to use an equation like (3), where we have solved for the specific variable. We substitute 4 for x in equation (3) and solve for y:

$y = 6 - 2x = 6 - 2(4) = 6 - 8 = -2.$

We obtain the ordered pair $(4, -2)$.

GS 4. $8x - 5y = 12,$ (1)
$x - y = 3$ (2)

Solve for x in equation (2):
$x - y = 3$
$x = \underline{} + 3.$ (3)

Substitute $y + 3$ for $\underline{}$ in equation (1) and solve for $\underline{}$:
$8x - 5y = 12$
$8(y + 3) - 5y = 12$
$8y + \underline{} - 5y = 12$
$3y + 24 = 12$
$3y = -12$
$y = \underline{}.$

Substitute -4 for y in equation (3) and solve for x:
$x = y + 3$
$= -4 + 3$
$= \underline{}.$

The ordered pair checks in both equations. The solution is $(\underline{}, \underline{}).$

Check:

$$\frac{2x+y=6}{2(4)+(-2)\stackrel{?}{\,}6}$$
$$\frac{8-2}{6}\quad \text{TRUE}$$

$$\frac{3x+4y=4}{3(4)+4(-2)\stackrel{?}{\,}4}$$
$$\frac{12-8}{4}\quad \text{TRUE}$$

Since $(4, -2)$ checks, it is the solution.

◀ Do Exercises 3 and 4.

Answers
1. $(2, 4)$ 2. $(5, 2)$ 3. $(-3, 2)$ 4. $(-1, -4)$
Guided Solution:
4. $y, x, y, 24, -4, -1, -1, -4$

EXAMPLE 3 Solve this system of equations:

$$y = -3x + 5, \quad (1)$$
$$y = -3x - 2. \quad (2)$$

The graphs of the equations in the system are shown at right. Since the graphs are parallel, there is no solution. Let's try to solve this system algebraically using substitution. We substitute $-3x - 2$ for y in equation (1):

$$-3x - 2 = -3x + 5 \quad \text{Substituting } -3x - 2 \text{ for } y$$
$$-2 = 5. \quad \text{Adding } 3x$$

We have a false equation. The equation has no solution. This means that the system has **no solution**.

Do Exercise 5.

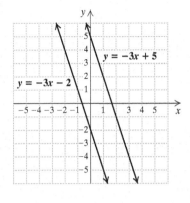

5. Solve using substitution:
$$y + 2x = 3,$$
$$y + 2x = -4.$$

EXAMPLE 4 Solve this system of equations:

$$x = 2y - 1, \quad (1)$$
$$4y - 2x = 2. \quad (2)$$

The graphs of the equations in the system are shown at right. Since the graphs are the same, there is an infinite number of solutions.

Let's try to solve this system algebraically using substitution. We substitute $2y - 1$ for x in equation (2):

$$4y - 2(2y - 1) = 2 \quad \text{Substituting } 2y - 1 \text{ for } x$$
$$4y - 4y + 2 = 2 \quad \text{Removing parentheses}$$
$$2 = 2. \quad \text{Simplifying; } 4y - 4y = 0$$

We have a true equation. Any value of y will make this equation true. This means that the system has **infinitely many solutions**.

Do Exercise 6.

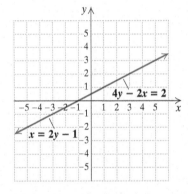

6. Solve using substitution:
$$x - 3y = 1,$$
$$6y - 2x = -2.$$

SPECIAL CASES

When solving a system of two linear equations in two variables:

1. If a false equation is obtained, then the system has no solution.
2. If a true equation is obtained, then the system has an infinite number of solutions.

---

b SOLVING APPLIED PROBLEMS INVOLVING TWO EQUATIONS

Many applied problems are easier to solve if we first translate to a system of two equations rather than to a single equation.

EXAMPLE 5 *Architecture.* The architects who designed the John Hancock Building in Chicago created a visually appealing building that slants on the sides. The ground floor is in the shape of a rectangle that is larger than the rectangle formed by the top floor. The ground floor has a perimeter of 860 ft. The length is 100 ft more than the width. Find the length and the width.

Answers
5. No solution **6.** Infinitely many solutions

1. **Familiarize.** We first make a drawing and label it, using l for length and w for width. We recall or look up the formula for perimeter: $P = 2l + 2w$. This formula can be found at the back of this book.

2. **Translate.** We translate as follows:

   The perimeter is 860 ft.
   $$2l + 2w = 860.$$

   We can also write a second equation:

   The length is 100 ft more than the width.
   $$l = w + 100.$$

   $l = w + 100$, w

   We now have a system of equations:
   $$2l + 2w = 860, \quad (1)$$
   $$l = w + 100. \quad (2)$$

3. **Solve.** We substitute $w + 100$ for l in equation (1) and solve for w:

   $2(w + 100) + 2w = 860$ Substituting in equation (1)
   $2w + 200 + 2w = 860$ Multiplying to remove parentheses
   $4w + 200 = 860$ Collecting like terms
   $4w = 660$
   $w = 165.$ Solving for w

   Next, we substitute 165 for w in equation (2) and solve for l:
   $$l = 165 + 100 = 265.$$

4. **Check.** Consider the dimensions 265 ft and 165 ft. The length is 100 ft more than the width. The perimeter is $2(265 \text{ ft}) + 2(165 \text{ ft})$, or 860 ft. The dimensions 265 ft and 165 ft check in the original problem.

5. **State.** The length is 265 ft, and the width is 165 ft.

◀ Do Exercise 7.

7. *Architecture.* The top floor of the John Hancock Building is also in the shape of a rectangle, but its perimeter is 520 ft. The width is 60 ft less than the length. Find the length and the width.

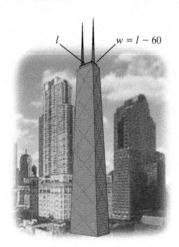
l, $w = l - 60$

Answer
7. Length: 160 ft; width: 100 ft

8.2 Exercise Set

FOR EXTRA HELP MyLab Math

✓ Check Your Understanding

Reading Check Determine whether each statement is true or false.

RC1. The substitution method is an algebraic method for solving systems of equations.

RC2. We can find solutions of systems of equations involving fractions using the substitution method.

RC3. When we are writing the solution of a system, the value that we found first is always the first number in the ordered pair.

RC4. When solving using substitution, if we obtain a false equation, then the system has many solutions.

Concept Check For each system of equations, the value of one variable is given. Find the value of the other variable. Then write the solution as an ordered pair, using alphabetical order for the variables.

CC1. $2a + 5b = 36$,
$2a - b = 12$
$a = 8$

CC2. $3x - 2y = 10$,
$2y - 2x = 9$
$y = \dfrac{47}{2}$

CC3. $5w - p = 7$,
$3w = 8 + p$
$p = -\dfrac{19}{2}$

a Solve each system of equations by the substitution method.

1. $y = 5 - 4x$,
$2x - 3y = 13$

2. $x = 8 - 4y$,
$3x + 5y = 3$

3. $2y + x = 9$,
$x = 3y - 3$

4. $9x - 2y = 3$,
$3x - 6 = y$

5. $3s - 4t = 14$,
$5s + t = 8$

6. $m - 2n = 3$,
$4m + n = 1$

7. $9x - 2y = -6$,
$7x + 8 = y$

8. $t = 4 - 2s$,
$t + 2s = 6$

9. $-5s + t = 11$,
$4s + 12t = 4$

10. $5x + 6y = 14$,
$-3y + x = 7$

11. $2x - 3 = y$,
$y - 2x = 1$

12. $4p - 2w = 16$,
$5p + 7w = 1$

13. $3a - b = 7$,
$2a + 2b = 5$

14. $3x + y = 4$,
$12 - 3y = 9x$

15. $2x - 6y = 4$,
$3y + 2 = x$

16. $5x + 3y = 4$,
$x - 4y = 3$

17. $2x + 2y = 2$,
$3x - y = 1$

18. $4x + 13y = 5$,
$-6x + y = 13$

b Solve.

19. *Archaeology.* The remains of an ancient ball court in Monte Alban, Mexico, include a rectangular playing alley with a perimeter of about 60 m. The length of the alley is five times the width. Find the length and the width of the playing alley.

20. *Soccer Field.* The perimeter of a soccer field is 340 m. The length exceeds the width by 50 m. Find the length and the width.

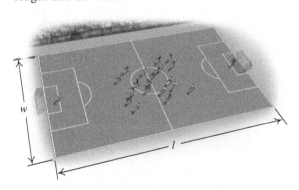

21. *Supplementary Angles.* **Supplementary angles** are angles whose sum is 180°. Two supplementary angles are such that one angle is 12° less than three times the other. Find the measures of the angles.

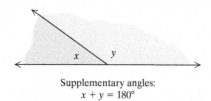

Supplementary angles:
$x + y = 180°$

22. *Complementary Angles.* **Complementary angles** are angles whose sum is 90°. Two complementary angles are such that one angle is 6° more than five times the other. Find the measures of the angles.

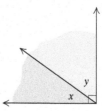

Complementary angles:
$x + y = 90°$

23. *Hockey Points.* At one time, hockey teams received two points when they won a game and one point when they tied. One season, a team won a championship with 60 points. They won 9 more games than they tied. How many wins and how many ties did the team have?

24. *Airplane Seating.* An airplane has a total of 152 seats. The number of coach-class seats is 5 more than six times the number of first-class seats. How many of each type of seat are there on the plane?

Skill Maintenance

25. Find the slope of the line $y = 1.3x - 7$. [7.3b]

26. Simplify: $-9(y + 7) - 6(y - 4)$. [1.8b]

27. Solve $A = \dfrac{pq}{7}$ for p. [2.4b]

28. Find the slope of the line containing the points $(-2, 3)$ and $(-5, -4)$. [7.3b]

Solve. [2.3c]

29. $-4x + 5(x - 7) = 8x - 6(x + 2)$

30. $-12(2x - 3) = 16(4x - 5)$

Synthesis

31. Two solutions of $y = mx + b$ are $(1, 2)$ and $(-3, 4)$. Find m and b.

32. Solve for x and y in terms of a and b:
$$5x + 2y = a,$$
$$x - y = b.$$

33. *Design.* A piece of posterboard has a perimeter of 156 in. If you cut 6 in. off the width, the length becomes four times the width. What are the dimensions of the original piece of posterboard?

34. *Nontoxic Scouring Powder.* A nontoxic scouring powder is made up of 4 parts baking soda and 1 part vinegar. How much of each ingredient is needed for a 16-oz mixture?

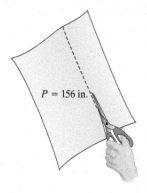

$P = 156$ in.

Solving by Elimination

8.3

a THE ELIMINATION METHOD

OBJECTIVES

a Solve systems of equations in two variables by the elimination method.

b Solve applied problems by solving systems of two equations using elimination.

SKILL REVIEW

Solve equations. [2.3b]
Solve. Clear the fractions or decimals first.

1. $4.2x - 10.4 = 45.4 - 5.1x$
2. $\frac{1}{4}x - \frac{2}{5} + \frac{1}{2}x = \frac{3}{5} + x$

Answers: 1. 6 2. -4

MyLab Math VIDEO

The **elimination method*** for solving systems of equations makes use of the *addition principle* for equations.

EXAMPLE 1 Solve this system:

$2x - 3y = 0,$ **(1)**
$-4x + 3y = -1.$ **(2)**

The elimination method works well to solve this system because the $-3y$ in one equation and the $3y$ in the other are opposites. If we add them, these terms will add to 0 and, in effect, the variable y will have been "eliminated."

We will use the addition principle for equations, adding the same number on both sides of the equation. According to equation (2), $-4x + 3y$ and -1 are the same number. Thus we can use a vertical form and add $-4x + 3y$ on the left side of equation (1) and -1 on the right side:

$$
\begin{aligned}
2x - 3y &= 0 \quad &\textbf{(1)} \\
\underline{-4x + 3y} &= \underline{-1} \quad &\textbf{(2)} \\
-2x + 0y &= -1 \quad &\text{Adding} \\
-2x + 0 &= -1 \\
-2x &= -1.
\end{aligned}
$$

We have eliminated the variable y. We now have an equation with just one variable, which we solve for x:

$$-2x = -1$$
$$x = \tfrac{1}{2}.$$

Next, we substitute $\tfrac{1}{2}$ for x in either equation and solve for y:

$2 \cdot \tfrac{1}{2} - 3y = 0$ Substituting in equation (1)
$1 - 3y = 0$
$-3y = -1$ Subtracting 1
$y = \tfrac{1}{3}.$ Dividing by -3

We obtain the ordered pair $\left(\tfrac{1}{2}, \tfrac{1}{3}\right)$. We can check by solving graphically, as shown at right. To check algebraically, we substitute $\tfrac{1}{2}$ for x and $\tfrac{1}{3}$ for y in both equations.

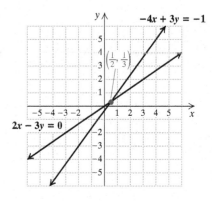

---

*This method is also called the *addition method*.

Solve by the elimination method.
1. $5x + 3y = 17$,
   $-5x + 2y = 3$

2. $-3a + 2b = 0$,
   $3a - 4b = -1$

Check:
$$\begin{array}{c|c} 2x - 3y = 0 & -4x + 3y = -1 \\ \hline 2(\frac{1}{2}) - 3(\frac{1}{3}) \;?\; 0 & -4(\frac{1}{2}) + 3(\frac{1}{3}) \;?\; -1 \\ 1 - 1 & -2 + 1 \\ 0 \quad \text{TRUE} & -1 \quad \text{TRUE} \end{array}$$

Since $(\frac{1}{2}, \frac{1}{3})$ checks, it is the solution.

◀ Do Exercises 1 and 2.

In order to eliminate a variable, we sometimes use the multiplication principle to multiply one or both of the equations by a particular number before adding.

EXAMPLE 2 Solve this system:
$$3x + 3y = 15, \quad (1)$$
$$2x + 6y = 22. \quad (2)$$

If we add directly, we will not eliminate a variable. However, note that if the $3y$ in equation (1) were $-6y$, we could eliminate y. Thus we multiply by -2 on both sides of equation (1) and add:

$$\begin{aligned} -6x - 6y &= -30 & \text{Multiplying by } -2 \text{ on both sides of equation (1)} \\ 2x + 6y &= 22 & \text{Equation (2)} \\ \hline -4x + 0 &= -8 & \text{Adding} \\ -4x &= -8 \\ x &= 2. & \text{Solving for } x \end{aligned}$$

Then
$$\begin{aligned} 2 \cdot 2 + 6y &= 22 & \text{Substituting 2 for } x \text{ in equation (2)} \\ 4 + 6y &= 22 \\ 6y &= 18 \\ y &= 3. \end{aligned} \right\} \text{Solving for } y$$

We obtain $(2, 3)$, or $x = 2$, $y = 3$.

3. Solve by the elimination method:
   $2y + 3x = 12$, **(1)**
   $-4y + 5x = -2$. **(2)**
   Multiply by 2 on both sides of equation (1) and add:
   $4y + 6x = 24$
   $\underline{-4y + 5x = -2}$
   $0 + \boxed{} = \boxed{}$
   $11x = 22$
   $x = \boxed{}$.
   Substitute $\boxed{}$ for x in equation (1) and solve for y:
   $2y + 3x = 12$
   $2y + 3(\boxed{}) = 12$
   $2y + 6 = 12$
   $2y = \boxed{}$
   $y = \boxed{}$.
   The ordered pair checks in both equations, so the solution is ($\boxed{}$, $\boxed{}$).

Check:
$$\begin{array}{c|c} 3x + 3y = 15 & 2x + 6y = 22 \\ \hline 3(2) + 3(3) \;?\; 15 & 2(2) + 6(3) \;?\; 22 \\ 6 + 9 & 4 + 18 \\ 15 \quad \text{TRUE} & 22 \quad \text{TRUE} \end{array}$$

Since $(2, 3)$ checks, it is the solution. We can also see this in the graph at left.

◀ Do Exercise 3.

EXAMPLE 3 Solve this system:
$$2x + 3y = 17, \quad (1)$$
$$5x + 7y = 29. \quad (2)$$

We must multiply twice in order to make one pair of terms with the same variable opposites. We decide to do this with the x-terms in each equation. We multiply equation (1) by 5 and equation (2) by -2. Then we get $10x$ and $-10x$, which are opposites.

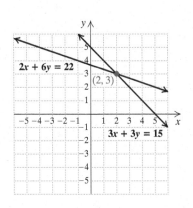

Answers
1. $(1, 4)$ 2. $\left(\frac{1}{3}, \frac{1}{2}\right)$ 3. $(2, 3)$

Guided Solution:
3. $11x$, 22, 2, 2, 2, 6, 3, 2, 3

From equation (1): $10x + 15y = 85$ Multiplying by 5
From equation (2): $\underline{-10x - 14y = -58}$ Multiplying by -2
                        $0 + y = 27$ Adding
                        $y = 27$ Solving for y

Then

$2x + 3 \cdot 27 = 17$ Substituting 27 for y in equation (1)
$2x + 81 = 17$
$2x = -64$ $\}$ Solving for x
$x = -32.$

We check the ordered pair $(-32, 27)$.

Check:
$$\begin{array}{c|c} 2x + 3y = 17 \\ \hline 2(-32) + 3(27) \; ? \; 17 \\ -64 + 81 \\ 17 \; | \; \text{TRUE} \end{array} \qquad \begin{array}{c|c} 5x + 7y = 29 \\ \hline 5(-32) + 7(27) \; ? \; 29 \\ -160 + 189 \\ 29 \; | \; \text{TRUE} \end{array}$$

We obtain $(-32, 27)$, or $x = -32$, $y = 27$, as the solution.

Do Exercises 4 and 5.

When solving a system of equations using the elimination method, it helps to first write the equations in the form $Ax + By = C$. When decimals or fractions occur, it also helps to *clear* before solving.

EXAMPLE 4 Solve this system:

$0.2x + 0.3y = 1.7,$
$\frac{1}{7}x + \frac{1}{5}y = \frac{29}{35}.$

We have

$0.2x + 0.3y = 1.7, \xrightarrow{\text{Multiplying by 10}} 2x + 3y = 17,$
$\frac{1}{7}x + \frac{1}{5}y = \frac{29}{35} \xrightarrow{\text{Multiplying by 35}} 5x + 7y = 29.$
$\text{to clear fractions}$

We multiplied by 10 to clear the decimals. Multiplication by 35, the least common denominator, clears the fractions. The problem is now identical to Example 3. The solution is $(-32, 27)$, or $x = -32$, $y = 27$.

Do Exercises 6 and 7.

To use the elimination method to solve systems of two equations:

1. Write both equations in the form $Ax + By = C$.
2. Clear any decimals or fractions.
3. Choose a variable to eliminate.
4. Make the chosen variable's terms opposites by multiplying one or both equations by appropriate numbers if necessary.
5. Eliminate a variable by adding the corresponding sides of the equations and then solve for the remaining variable.
6. Substitute in either of the original equations to find the value of the other variable.

Solve by the elimination method.

4. $4x + 5y = -8,$
$7x + 9y = 11$

5. $4x - 5y = 38,$
$7x - 8y = -22$

6. Clear the decimals. Then solve.
$0.02x + 0.03y = 0.01,$
$0.3x - 0.1y = 0.7$
(*Hint*: Multiply the first equation by 100 and the second one by 10.)

7. Clear the fractions. Then solve.
$\frac{3}{5}x + \frac{2}{3}y = \frac{1}{3},$
$\frac{3}{4}x - \frac{1}{3}y = \frac{1}{4}$

Answers
4. $(-127, 100)$ **5.** $(-138, -118)$
6. $2x + 3y = 1,$
$3x - y = 7; (2, -1)$
7. $9x + 10y = 5,$
$9x - 4y = 3; \left(\frac{25}{63}, \frac{1}{7}\right)$

Some systems have no solution. How do we recognize such systems if we are solving using elimination?

EXAMPLE 5 Solve this system:

$$y + 3x = 5, \quad (1)$$
$$y + 3x = -2. \quad (2)$$

If we find the slope–intercept equations for this system, we get

$$y = -3x + 5,$$
$$y = -3x - 2.$$

The graphs are parallel lines. The system has no solution.

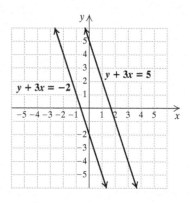

Let's attempt to solve the system by the elimination method:

$$\begin{array}{ll} y + 3x = 5 & \text{Equation (1)} \\ \underline{-y - 3x = 2} & \text{Multiplying equation (2) by } -1 \\ 0 = 7. & \text{Adding, we obtain a false equation.} \end{array}$$

The *x*-terms and the *y*-terms are eliminated and we have a *false* equation. If we obtain a false equation, such as $0 = 7$, when solving algebraically, we know that the system has **no solution**. The system is inconsistent, and the equations are independent.

◀ Do Exercise 8.

Some systems have infinitely many solutions. How can we recognize such a situation when we are solving systems using an algebraic method?

EXAMPLE 6 Solve this system:

$$3y - 2x = 6, \quad (1)$$
$$-12y + 8x = -24. \quad (2)$$

The graphs are the same line. The system has an infinite number of solutions.

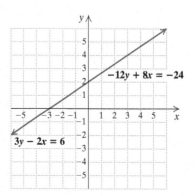

---

8. Solve by the elimination method:

$$y + 2x = 3,$$
$$y + 2x = -1.$$

Multiply the second equation by -1 and add:

$$\begin{array}{l} y + 2x = 3 \\ \underline{-y - 2x = 1} \\ 0 = \underline{}. \end{array}$$

The equation is _____, so
         true/false
the system has no solution.

Answer
8. No solution
Guided Solution:
8. 4, false

Suppose that we try to solve this system by the elimination method:

$12y - 8x = 24$ Multiplying equation (1) by 4
$-12y + 8x = -24$ Equation (2)
$0 = 0.$ Adding, we obtain a true equation.

We have eliminated both variables, and what remains is a true equation, $0 = 0$. It can be expressed as $0 \cdot x + 0 \cdot y = 0$, and is true for all numbers x and y. If an ordered pair is a solution of one of the original equations, then it will be a solution of the other. The system has an **infinite number of solutions**. The system is consistent, and the equations are dependent.

> **SPECIAL CASES**
>
> When solving a system of two linear equations in two variables:
>
> 1. If a false equation is obtained, such as $0 = 7$, then the system has no solution. The system is *inconsistent*, and the equations are *independent*.
> 2. If a true equation is obtained, such as $0 = 0$, then the system has an infinite number of solutions. The system is *consistent*, and the equations are *dependent*.

Do Exercise 9. ▶

9. Solve by the elimination method:
$$2x - 5y = 10,$$
$$-6x + 15y = -30.$$

Comparing Methods

We can solve systems of equations graphically, or we can solve them algebraically using substitution or elimination. When deciding which method to use, consider the information in this table as well as directions from your instructor.

METHOD	STRENGTHS	WEAKNESSES
Graphical	Can "see" solutions.	Inexact when solutions involve numbers that are not integers. Solutions may not appear on the part of the graph drawn.
Substitution	Yields exact solutions. Convenient to use when a variable has a coefficient of 1.	Can introduce extensive computations with fractions. Cannot "see" solutions quickly.
Elimination	Yields exact solutions. Convenient to use when no variable has a coefficient of 1. The preferred method for systems of three or more equations in three or more variables. (See Section 8.5.)	Cannot "see" solutions quickly.

Answer
9. Infinitely many solutions

b SOLVING APPLIED PROBLEMS USING ELIMINATION

Let's now solve an applied problem using the elimination method.

EXAMPLE 7 *Stimulating the Hometown Economy.* To stimulate the economy in his town of Brewton, Alabama, in 2009, Danny Cottrell, co-owner of The Medical Center Pharmacy, gave each of his full-time employees $700 and each part-time employee $300. He asked that each person donate 15% to a charity of his or her choice and spend the rest locally. The money was paid in $2 bills, a rarely used currency, so that the business community could easily see how the money circulated. Cottrell gave away a total of $16,000 to his 24 employees. How many full-time employees and how many part-time employees were there?

Data: *The Press-Register,* March 4, 2009

1. **Familiarize.** We let $f =$ the number of full-time employees and $p =$ the number of part-time employees. Each full-time employee received $700, so a total of $700f$ was paid to them. Similarly, the part-time employees received a total of $300p$. Thus a total of $700f + 300p$ was given away.

2. **Translate.** We translate to two equations.

   Total amount given away is $16,000.
   $$700f + 300p = 16{,}000$$

   Total number of employees is 24.
   $$f + p = 24$$

   We now have a system of equations:
   $$700f + 300p = 16{,}000, \quad (1)$$
   $$f + p = 24. \quad (2)$$

3. **Solve.** First, we multiply by -300 on both sides of equation (2) and add:

$700f + 300p = 16{,}000$	Equation (1)
$-300f - 300p = -7200$	Multiplying by -300 on both sides of equation (2)
$400f = 8800$	Adding
$f = 22.$	Solving for f

   Next, we substitute 22 for f in equation (2) and solve for p:
   $$22 + p = 24$$
   $$p = 2.$$

4. **Check.** If there are 22 full-time employees and 2 part-time employees, there is a total of $22 + 2$, or 24, employees. The 22 full-time employees received a total of $\$700 \cdot 22$, or $\$15{,}400$, and the 2 part-time employees received a total of $\$300 \cdot 2$, or $\$600$. Then a total of $\$15{,}400 + \$600$, or $\$16{,}000$, was given away. The numbers check in the original problem.

5. **State.** There were 22 full-time employees and 2 part-time employees.

◀ Do Exercise 10.

---

10. *Bonuses.* Monica gave each of the full-time employees in her small business a year-end bonus of $500 while each part-time employee received $250. She gave a total of $4000 in bonuses to her 10 employees. How many full-time employees and how many part-time employees did Monica have?

Answer

10. Full-time: 6; part-time: 4

8.3 Exercise Set

✓ Check Your Understanding

Reading Check Choose from the column on the right the word that best completes each sentence. Words may be used more than once.

RC1. If a system of equations has a solution, then it is _____.

RC2. If a system of equations has no solution, then it is _____.

RC3. If a system of equations has infinitely many solutions, then it is _____.

RC4. If the graphs of the equations in a system of two equations in two variables are the same line, then the equations are _____.

RC5. If the graphs of the equations in a system of two equations in two variables are parallel, then the system is _____.

RC6. If the graphs of the equations in a system of two equations in two variables intersect at one point, then the equations are _____.

consistent
inconsistent
dependent
independent

Concept Check For each system, determine by what number the second equation should be multiplied in order to eliminate y by adding.

CC1. $3x + 6y = 7,$
$2x + 2y = 1$

CC2. $5x - 6y = 2,$
$2x + 3y = 5$

CC3. $7x + 3y = 2,$
$4x + 3y = 9$

a Solve each system of equations using the elimination method.

1. $x + 3y = 7,$
$-x + 4y = 7$

2. $x + y = 9,$
$2x - y = -3$

3. $9x + 5y = 6,$
$2x - 5y = -17$

4. $2x - 3y = 18,$
$2x + 3y = -6$

5. $5x + 3y = -11,$
$3x - y = -1$

6. $2x + 3y = -9,$
$5x - 6y = -9$

7. $5r - 3s = 19,$
$2r - 6s = -2$

8. $2a + 3b = 11,$
$4a - 5b = -11$

9. $2x + 3y = 1,$
$4x + 6y = 2$

10. $3x - 2y = 1,$
$-6x + 4y = -2$

11. $5x - 9y = 7,$
$7y - 3x = -5$

12. $5x + 4y = 2,$
$2x - 8y = 4$

13. $3x + 2y = 24,$
    $2x + 3y = 26$

14. $5x + 3y = 25,$
    $3x + 4y = 26$

15. $2x - 4y = 5,$
    $2x - 4y = 6$

16. $3x - 5y = -2,$
    $5y - 3x = 7$

17. $2a + b = 12,$
    $a + 2b = -6$

18. $10x + y = 306,$
    $10y + x = 90$

19. $\frac{1}{3}x + \frac{1}{5}y = 7,$
    $\frac{1}{6}x - \frac{2}{5}y = -4$

20. $\frac{2}{3}x + \frac{1}{7}y = -11,$
    $\frac{1}{7}x - \frac{1}{3}y = -10$

21. $\frac{1}{5}x + \frac{1}{2}y = 6,$
    $\frac{2}{5}x - \frac{3}{2}y = -8$

22. $\frac{2}{3}x + \frac{3}{5}y = -17,$
    $\frac{1}{2}x - \frac{1}{3}y = -1$

23. $\frac{1}{2}x - \frac{1}{3}y = -4,$
    $\frac{1}{4}x + \frac{5}{6}y = 4$

24. $\frac{4}{3}x + \frac{3}{2}y = 4,$
    $\frac{5}{6}x - \frac{1}{8}y = -6$

25. $0.3x - 0.2y = 4,$
    $0.2x + 0.3y = 0.5$

26. $0.7x - 0.3y = 0.5,$
    $-0.4x + 0.7y = 1.3$

27. $0.05x + 0.25y = 22,$
    $0.15x + 0.05y = 24$

28. $1.3x - 0.2y = 12,$
    $0.4x + 17y = 89$

b Solve. Use the elimination method when solving the translated system.

29. *Finding Numbers.* The sum of two numbers is 63. The larger number minus the smaller number is 9. Find the numbers.

30. *Finding Numbers.* The sum of two numbers is 2. The larger number minus the smaller number is 20. Find the numbers.

31. *Finding Numbers.* The sum of two numbers is 3. Three times the larger number plus two times the smaller number is 24. Find the numbers.

32. *Finding Numbers.* The sum of two numbers is 9. Two times the larger number plus three times the smaller number is 2. Find the numbers.

33. *Complementary Angles.* Two angles are complementary. (**Complementary angles** are angles whose sum is 90°.) Their difference is 6°. Find the angles.

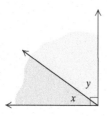

Complementary angles:
$x + y = 90°$

34. *Supplementary Angles.* Two angles are supplementary. (**Supplementary angles** are angles whose sum is 180°.) Their difference is 22°. Find the angles.

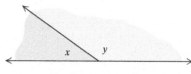

Supplementary angles:
$x + y = 180°$

35. *Basketball Scoring.* In their championship game, the Eastside Golden Eagles scored 60 points on a combination of two-point shots and three-point shots. If they made a total of 27 shots, how many of each kind of shot was made?

36. *Basketball Scoring.* Wilt Chamberlain once scored 100 points, setting a record for points scored in an NBA game. Chamberlain took only two-point shots and (one-point) foul shots and made a total of 64 shots. How many shots of each type did he make?

37. Each course offered during the winter session at New Heights Community College is worth either 3 credits or 4 credits. The members of the Touring Concert Chorale took a total of 33 courses during the winter session, worth a total of 107 credits. How many of each type of class did the chorale members take?

38. Daphne's Lawn and Garden Center offered customers who bought a custom lawn-care package a free ornamental tree, either an Eastern Redbud or a Kousa Dogwood. The center's cost for each Eastern Redbud was $37, and its cost for each Kousa Dogwood was $45. A total of 18 customers took advantage of the offer. The center's total cost for the promotional items was $754. How many patrons chose each type of ornamental tree?

Skill Maintenance

Given the function $f(x) = 3x^2 - x + 1$, find each of the following function values. [7.1b]

39. $f(0)$ **40.** $f(-1)$ **41.** $f(-2)$ **42.** $f(2a)$

43. Find the domain of the function
$$f(x) = \frac{x-5}{x+7}.$$ [7.2a]

44. Find the domain and the range of the function $g(x) = 5 - x^2$. [7.2a]

45. Find an equation of the line with slope $-\frac{3}{5}$ and y-intercept $(0, -7)$. [7.5a]

46. Find an equation of the line containing the points $(-10, 2)$ and $(-2, 10)$. [7.5c]

Synthesis

47. Use the INTERSECT feature to solve the following system of equations. You may need to first solve for y. Round answers to the nearest hundredth.
$$3.5x - 2.1y = 106.2,$$
$$4.1x + 16.7y = -106.28$$

48. Solve:
$$\frac{x+y}{2} - \frac{x-y}{5} = 1,$$
$$\frac{x-y}{2} + \frac{x+y}{6} = -2.$$

49. The solution of this system is $(-5, -1)$. Find A and B.
$$Ax - 7y = -3,$$
$$x - By = -1$$

50. Find an equation to pair with $6x + 7y = -4$ such that $(-3, 2)$ is a solution of the system.

51. The points $(0, -3)$ and $\left(-\frac{3}{2}, 6\right)$ are two of the solutions of the equation $px - qy = -1$. Find p and q.

52. Determine a and b for which $(-4, -3)$ will be a solution of the system
$$ax + by = -26,$$
$$bx - ay = 7.$$

8.4 Solving Applied Problems: Two Equations

OBJECTIVES

a Solve applied problems involving total value and mixture using systems of two equations.

b Solve applied problems involving motion using systems of two equations.

MyLab Math
ANIMATION

a TOTAL-VALUE PROBLEMS AND MIXTURE PROBLEMS

Systems of equations can be a useful tool in solving applied problems. Using systems often makes the *Translate* step easier than using a single equation. The first kind of problem we consider involves quantities of items purchased and the total value, or cost, of the items. We refer to this type of problem as a **total-value problem**.

EXAMPLE 1 *Lunch Orders.* In order to pick up lunch, Cathy collected $181.50 from her co-workers for a total of 21 salads and sandwiches. When she got to the deli, she forgot how many of each were ordered. If salads cost $7.50 and sandwiches cost $9.50, how many of each should she buy?

1. **Familiarize.** Let's begin by guessing that 5 salads were ordered. Since there was a total of 21 orders, this means that 16 sandwiches were ordered. The total cost of the order would then be

$$\underbrace{\$7.50(5)}_{\text{Cost of salads}} + \underbrace{\$9.50(16)}_{\text{Cost of sandwiches}} = \$37.50 + \$152 = \$189.50.$$

The guess is incorrect, but we can use the same process to translate the problem to a system of equations. We also note that our guess resulted in a total that was too high, so there were more salads and fewer sandwiches ordered than we guessed.

We let $d =$ the number of salads and $w =$ the number of sandwiches ordered. The cost of d salads is $\$7.50d$, and the cost of w sandwiches is $\$9.50w$. Organizing the information in a table can help us translate the information to a system of equations.

	SALADS	SANDWICHES	TOTAL	
NUMBER OF ORDERS	d	w	21	$\longrightarrow d + w = 21$
COST PER ORDER	$7.50	$9.50		
TOTAL COST	$7.50d$	$9.50w$	$181.50	$\longrightarrow 7.50d + 9.50w = 181.50$

2. **Translate.** The first row of the table gives us one equation:

   $d + w = 21.$

   The last row of the table gives us a second equation:

   $7.50d + 9.50w = 181.50.$

612   CHAPTER 8   Systems of Equations

Clearing decimals in the second equation gives us the following system of equations:

$$d + w = 21, \quad (1)$$
$$75d + 95w = 1815. \quad (2)$$

3. **Solve.** We use the elimination method to solve the system of equations. We eliminate d by multiplying by -75 on both sides of equation (1) and then adding the result to equation (2):

$$\begin{array}{rl} -75d - 75w = -1575 & \text{Multiplying equation (1) by } -75 \\ \underline{75d + 95w = 1815} & \text{Equation (2)} \\ 20w = 240 & \text{Adding} \\ w = 12. & \text{Dividing by 20} \end{array}$$

Next, we substitute 12 for w in equation (1) and solve for d:

$$\begin{array}{rl} d + w = 21 & \text{Equation (1)} \\ d + 12 = 21 & \text{Substituting 12 for } w \\ d = 9. & \text{Solving for } d \end{array}$$

We obtain $(9, 12)$, or $d = 9$, $w = 12$.

4. **Check.** We check in the original problem.

   Total number of orders: $d + w = 9 + 12 = 21$
   Cost of salads: $\$7.50d = \$7.50(9) = \$67.50$
   Cost of sandwiches: $\$9.50w = \$9.50(12) = \underline{\$114.00}$
   Total $= \$181.50$

   The numbers check.

5. **State.** Cathy should buy 9 salads and 12 sandwiches.

   Do Exercise 1. ▶

The following problem, similar to Example 1, is called a **mixture problem**.

EXAMPLE 2 *Blending Spices.* Spice It Up sells ground turmeric for $1.35 per ounce and ground sumac for $1.85 per ounce. Ethan wants to make a 20-oz seasoning blend of the two spices that sells for $1.65 per ounce. How much of each should he use?

 1. *Retail Sales of Tee Shirts.* A campus bookstore sells college tee shirts. White tee shirts sell for $18.95 each and red ones sell for $19.50 each. If receipts for the sale of 30 tee shirts total $572.90, how many of each color did the shop sell?

Complete the following table, letting $w =$ the number of white tee shirts and $r =$ the number of red tee shirts.

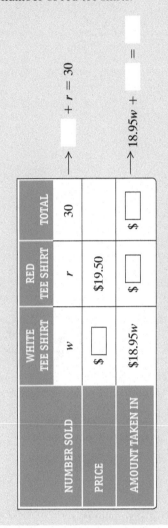

Answer
1. White: 22; red: 8
Guided Solution:
1.

White	Red	Total	
w	r	30	→ $w + r = 30$
$18.95	$19.50		
$18.95w	$19.50r	$572.90	→ $18.95w + 19.50r$ $= 572.90$

1. **Familiarize.** Suppose that Ethan uses 4 oz of sumac. Since he wants a total of 20 oz, he will need 16 oz of turmeric. We compare the value of the spices separately with the desired value of the blend:

   Spices purchased separately: $1.85(4) + $1.35(16), or $29.
   Blend: $1.65(20) = $33

   Since these amounts are not the same, our guess is not correct, but these calculations help us to translate the problem.

   We let s = the number of ounces of sumac and t = the number of ounces of turmeric. Then we organize the information in a table as follows.

	SUMAC	TURMERIC	BLEND	
NUMBER OF OUNCES	s	t	20	$\rightarrow s + t = 20$
PRICE PER OUNCE	$1.85	$1.35	$1.65	
VALUE OF SPICES	$1.85s$	$1.35t$	$1.65(20)$, or $33	$\rightarrow 1.85s + 1.35t = 33$

2. **Translate.** The total number of ounces in the blend is 20, so we have one equation:

   $s + t = 20.$

   The value of the sumac is $1.85s$, and the value of the turmeric is $1.35t$. These amounts are in dollars. Since the total value is to be $1.65(20)$, or 33, we have

   $1.85s + 1.35t = 33.$

   We can multiply by 100 on both sides of the second equation in order to clear the decimals. Thus we have translated to a system of equations:

   $s + t = 20,$ (1)
   $185s + 135t = 3300.$ (2)

3. **Solve.** We will solve this system using substitution, but elimination is also an appropriate method to use. When equation (1) is solved for t, we get $t = 20 - s$. We substitute $20 - s$ for t in equation (2) and solve:

   $185s + 135(20 - s) = 3300$ Substituting
   $185s + 2700 - 135s = 3300$ Using the distributive law
   $50s = 600$ Subtracting 2700 and collecting like terms
   $s = 12.$

   We have $s = 12$. Substituting 12 for s in the equation $t = 20 - s$, we obtain $t = 20 - 12$, or 8.

4. **Check.** We check in a manner similar to our guess in the *Familiarize* step.

   Total number of ounces: $12 + 8 = 20$
   Value of the blend: $1.85(12) + 1.35(8) = 33$

   Thus the number of ounces of each spice checks.

5. **State.** Ethan should use 12 oz of sumac and 8 oz of turmeric.

◀ Do Exercise 2.

2. *Blending Coffees.* The Coffee Counter charges $18.00 per pound for organic Kenyan French Roast coffee and $16.00 per pound for Sumatran coffee. How much of each type should be used in order to make a 20-lb blend that sells for $16.70 per pound?

Answer
2. Kenyan: 7 lb; Sumatran: 13 lb

EXAMPLE 3 *Student Loans.* Jeron's student loans totaled $16,200. Part was a Perkins loan made at 5% interest and the rest was a Stafford loan made at 4% interest. After one year, Jeron's loans accumulated $715 in interest. What was the amount of each loan?

1. **Familiarize.** Listing the given information in a table will help. The columns in the table come from the formula for simple interest: $I = Prt$. We let x = the number of dollars in the Perkins loan and y = the number of dollars in the Stafford loan.

	PERKINS LOAN	STAFFORD LOAN	TOTAL
PRINCIPAL	x	y	$16,200	→ $x + y = 16,200$
RATE OF INTEREST	5%	4%	
TIME	1 year	1 year	
INTEREST	$0.05x$	$0.04y$	$715	→ $0.05x + 0.04y = 715$

2. **Translate.** The total of the amounts of the loans is found in the first row of the table. This gives us one equation:

$$x + y = 16,200.$$

Look at the last row of the table. The interest totals $715. This gives us a second equation:

$$5\%x + 4\%y = 715, \quad \text{or} \quad 0.05x + 0.04y = 715.$$

After we multiply on both sides to clear the decimals, we have

$$5x + 4y = 71,500.$$

3. **Solve.** Using either elimination or substitution, we solve the resulting system:

$$x + y = 16,200,$$
$$5x + 4y = 71,500.$$

We find that $x = 6700$ and $y = 9500$.

4. **Check.** The sum is $6700 + $9500, or $16,200. The interest from $6700 at 5% for one year is 5%($6700), or $335. The interest from $9500 at 4% for one year is 4%($9500), or $380. The total amount of interest is $335 + $380, or $715. The numbers check in the problem.

5. **State.** The Perkins loan was for $6700, and the Stafford loan was for $9500.

Do Exercise 3. ▶

3. *Client Investments.* Infinite Financial Services invested Jasmine's IRA contribution of $3700 for one year at simple interest, yielding $297. Part of the money is invested at 7% and the rest at 9%. How much was invested at each rate?

Do the *Familiarize* and *Translate* steps by completing the following table. Let x = the number of dollars invested at 7% and y = the number of dollars invested at 9%.

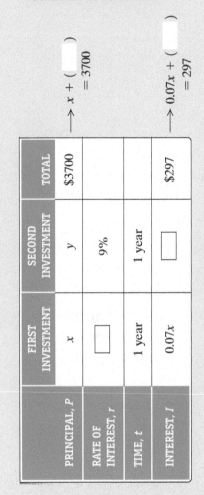

Answer
3. $1800 at 7%; $1900 at 9%
Guided Solution:
3.

First Investment	Second Investment	Total	
x	y	$3700	→ $x + y = 3700$
7%	9%		
1 year	1 year		
$0.07x$	$0.09y$	$297	→ $0.07x + 0.09y = 297$

EXAMPLE 4 *Mixing Fertilizers.* Nature's Landscapes carries two kinds of fertilizer containing nitrogen and water. "Gently Green" is 5% nitrogen and "Sun Saver" is 15% nitrogen. Nature's Landscapes needs to combine the two types of solution in order to make 90 L of a solution that is 12% nitrogen. How much of each brand should be used?

1. **Familiarize.** We first make a drawing and a guess to become familiar with the problem.

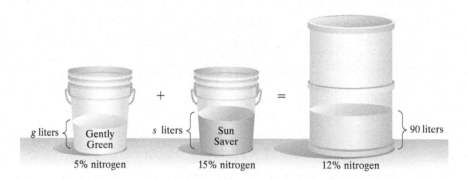

We choose two numbers that total 90 L—say, 40 L of Gently Green and 50 L of Sun Saver—for the amounts of each fertilizer. Will the resulting mixture have the correct percentage of nitrogen? To find out, we multiply as follows:

$5\%(40\,\text{L}) = 2\,\text{L}$ of nitrogen and $15\%(50\,\text{L}) = 7.5\,\text{L}$ of nitrogen.

Thus the total amount of nitrogen in the mixture is 2 L + 7.5 L, or 9.5 L. The final mixture of 90 L is supposed to be 12% nitrogen. Now

$12\%(90\,\text{L}) = 10.8\,\text{L}.$

Since 9.5 L and 10.8 L are not the same, our guess is incorrect. But these calculations help us to make the translation.

We let g = the number of liters of Gently Green and s = the number of liters of Sun Saver in the mixture.

	GENTLY GREEN	SUN SAVER	MIXTURE	
NUMBER OF LITERS	g	s	90	→ $g + s = 90$
PERCENT OF NITROGEN	5%	15%	12%	
AMOUNT OF NITROGEN	$0.05g$	$0.15s$	0.12×90, or 10.8 liters	→ $0.05g + 0.15s = 10.8$

2. **Translate.** If we add g and s in the first row, we get 90, and this gives us one equation:

   $g + s = 90.$

   If we add the amounts of nitrogen listed in the third row, we get 10.8, and this gives us another equation:

   $5\%g + 15\%s = 10.8,$ or $0.05g + 0.15s = 10.8.$

After clearing the decimals, we have the following system:

$g + s = 90,$ (1)
$5g + 15s = 1080.$ (2)

3. **Solve.** We solve the system using elimination. We multiply equation (1) by −5 and add the result to equation (2):

$-5g - 5s = -450$	Multiplying equation (1) by −5
$\underline{5g + 15s = 1080}$	Equation (2)
$10s = 630$	Adding
$s = 63.$	Dividing by 10

Next, we substitute 63 for s in equation (1) and solve for g:

$g + 63 = 90$	Substituting in equation (1)
$g = 27.$	Solving for g

We obtain $(27, 63)$, or $g = 27, s = 63$.

4. **Check.** Remember that g is the number of liters of Gently Green, with 5% nitrogen, and s is the number of liters of Sun Saver, with 15% nitrogen.

Total number of liters of mixture: $g + s = 27 + 63 = 90$ L
Amount of nitrogen: $5\%(27) + 15\%(63) = 1.35 + 9.45 = 10.8$ L
Percentage of nitrogen in mixture: $\dfrac{10.8}{90} = 0.12 = 12\%$

The numbers check in the original problem.

5. **State.** Nature's Landscapes should mix 27 L of Gently Green and 63 L of Sun Saver.

Do Exercise 4. ▶

b MOTION PROBLEMS

When a problem deals with speed, distance, and time, we can expect to use the following *motion formula*.

THE MOTION FORMULA

Distance = Rate (or speed) · Time
$d = rt$

TIPS FOR SOLVING MOTION PROBLEMS

1. Make a drawing using an arrow or arrows to represent distance and the direction of each object in motion.
2. Organize the information in a table or a chart.
3. Look for as many things as you can that are the same, so you can write equations.

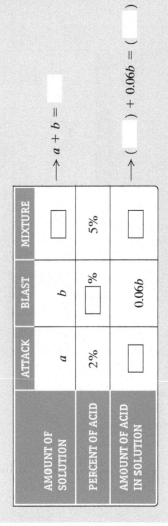

4. **Mixing Cleaning Solutions.** King's Service Station uses two kinds of cleaning solution containing acid and water. "Attack" is 2% acid and "Blast" is 6% acid. They want to mix the two in order to get 60 qt of a solution that is 5% acid. How many quarts of each should they use?

Do the *Familiarize* and *Translate* steps by completing the following table. Let a = the number of quarts of Attack and b = the number of quarts of Blast.

Answer
4. Attack: 15 qt; Blast: 45 qt
Guided Solution:
4.

Attack	Blast	Mixture	
a	b	60	→ $a + b = 60$
2%	6%	5%	
$0.02a$	$0.06b$	0.05×60, or 3	→ $0.02a + 0.06b = 3$

5. Train Travel. A train leaves Barstow traveling east at 35 km/h. One hour later, a faster train leaves Barstow, also traveling east on a parallel track at 40 km/h. How far from Barstow will the faster train catch up with the slower one?

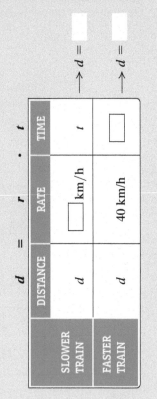

EXAMPLE 5 *Auto Travel.* Keri left Monday morning to drive to a seminar that began Monday evening. An hour after she had left the office, her assistant, Matt, realized that she had forgotten to take a large portfolio needed for a presentation. Knowing Keri would not answer her cell phone when driving, Matt left immediately with the portfolio to try to catch up with her. If Keri drove at a speed of 55 mph and Matt drove at a speed of 65 mph, how long did it take Matt to catch up with her? Assume that neither driver stopped to take a break.

1. **Familiarize.** We first make a drawing. From the drawing, we see that when Matt catches up with Keri, the distances from the office are the same. We let $d=$ the distance, in miles. If we let $t=$ the time, in hours, for Matt to catch Keri, then $t+1=$ the time traveled by Keri at a slower speed.

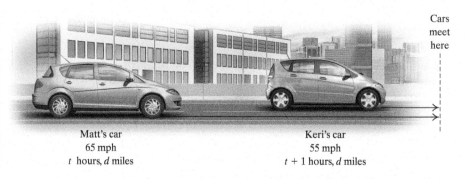

Matt's car
65 mph
t hours, d miles

Keri's car
55 mph
$t+1$ hours, d miles

Cars meet here

We organize the information in a table as follows.

$d = r \cdot t$

	DISTANCE	RATE	TIME	
KERI	d	55	$t+1$	$\rightarrow d = 55(t+1)$
MATT	d	65	t	$\rightarrow d = 65t$

2. **Translate.** Using $d = rt$ in each row of the table, we get an equation. Thus we have a system of equations:

$d = 55(t+1)$,   **(1)**
$d = 65t$.   **(2)**

3. **Solve.** We solve the system using the substitution method:

$65t = 55(t+1)$ Substituting $65t$ for d in equation (1)
$65t = 55t + 55$ Multiplying to remove parentheses on the right
$10t = 55$
$t = 5.5.$ Solving for t

Matt's time is 5.5 hr, which means that Keri's time is $5.5 + 1$, or 6.5 hr.

4. **Check.** At 65 mph, Matt will travel $65 \cdot 5.5$, or 357.5 mi, in 5.5 hr. At 55 mph, Keri will travel $55 \cdot 6.5$, or the same 357.5 mi, in 6.5 hr. The distances are the same, so the numbers check.

5. **State.** Matt caught up with Keri in 5.5 hr.

◀ Do Exercise 5.

Answer
5. 280 km

Guided Solution:
5.

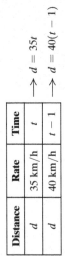

EXAMPLE 6 *Marine Travel.* A Coast Guard patrol boat travels 4 hr on a trip downstream with a 6-mph current. The return trip against the same current takes 5 hr. Find the speed of the boat in still water.

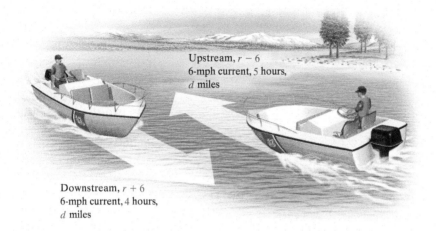

Upstream, $r - 6$
6-mph current, 5 hours,
d miles

Downstream, $r + 6$
6-mph current, 4 hours,
d miles

1. **Familiarize.** We first make a drawing. From the drawing, we see that the distances are the same. We let d = the distance, in miles, and r = the speed of the boat in still water, in miles per hour. Then, when the boat is traveling downstream, its speed is $r + 6$. (The current helps the boat along.) When it is traveling upstream, its speed is $r - 6$. (The current holds the boat back.) We can organize the information in a table. We use the formula $d = rt$.

$$d = r \cdot t$$

	DISTANCE	RATE	TIME	
DOWNSTREAM	d	$r + 6$	4	$\rightarrow d = (r+6)4$
UPSTREAM	d	$r - 6$	5	$\rightarrow d = (r-6)5$

2. **Translate.** From each row of the table, we get an equation, $d = rt$:

   $d = 4r + 24$, **(1)**
   $d = 5r - 30$. **(2)**

3. **Solve.** We solve the system using the substitution method:

   $4r + 24 = 5r - 30$ Substituting $4r + 24$ for d in equation (2)
   $24 = r - 30$
   $54 = r.$ Solving for r

4. **Check.** If $r = 54$, then $r + 6 = 60$; and $60 \cdot 4 = 240$ mi, the distance traveled downstream. If $r = 54$, then $r - 6 = 48$; and $48 \cdot 5 = 240$ mi, the distance traveled upstream. The distances are the same.

   When checking your answer, always ask, "Have I found what the problem asked for?" We could solve for a certain variable but still have not answered the question of the original problem. For example, we might have found speed when the problem wanted distance. In this problem, we want the speed of the boat in still water, and that is r.

5. **State.** The speed in still water is 54 mph.

Do Exercise 6.

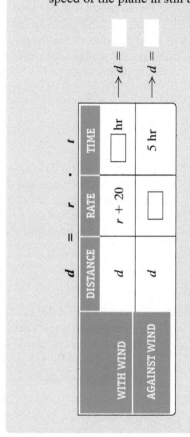

GS 6. *Air Travel.* An airplane flew for 4 hr with a 20-mph tailwind. The return flight against the same wind took 5 hr. Find the speed of the plane in still air.

$d = r \cdot t$

DISTANCE	RATE	TIME	
WITH WIND	d	$r + 20$	☐ hr	$\rightarrow d =$ ☐
AGAINST WIND	d	☐	5 hr	$\rightarrow d =$ ☐

Answer
6. 180 mph
Guided Solution:
6.

Distance	Rate	Time	
d	$r + 20$	4 hr	$\rightarrow d = (r+20)4$
d	$r - 20$	5 hr	$\rightarrow d = (r-20)5$

Translating for Success

The goal of these matching questions is to practice step (2), Translate, of the five-step problem-solving process. Translate each word problem to a system of equations and select a correct translation from systems A–J.

1. **Office Expense.** The monthly telephone expense for an office is $1094 less than the janitorial expense. Three times the janitorial expense minus four times the telephone expense is $248. What is the total of the two expenses?

2. **Dimensions of a Triangle.** The sum of the base and the height of a triangle is 192 in. The height is twice the base. Find the base and the height.

3. **Supplementary Angles.** Two supplementary angles are such that twice one angle is 7° more than the other. Find the measures of the angles.

4. **SAT Scores.** The total of Megan's writing and math scores on the SAT was 1094. Her math score was 248 points higher than her writing score. What were her math and writing SAT scores?

5. **Sightseeing Boat.** A sightseeing boat travels 3 hr on a trip downstream with a 2.5-mph current. The return trip against the same current takes 3.5 hr. Find the speed of the boat in still water.

A. $x = y + 248,$
   $x + y = 1094$

B. $5x = 2y - 3,$
   $y = \frac{2}{3}x + 5$

C. $y = \frac{1}{2}x,$
   $2x + 2y = 192$

D. $2x = 7 + y,$
   $x + y = 180$

E. $x + y = 192,$
   $x = 2y$

F. $x + y = 180,$
   $x = 2y + 7$

G. $x - 1094 = y,$
   $3x - 4y = 248$

H. $3\%x + 2.5\%y = 97.50,$
   $x + y = 2500$

I. $2x = 5 + \frac{2}{3}y,$
   $3y = 15x - 4$

J. $x = (y + 2.5) \cdot 3,$
   $3.5(y - 2.5) = x$

Answers on page A-26

6. **Running Distances.** Each day Tricia runs 5 mi more than two-thirds the distance that Chris runs. Five times the distance that Chris runs is 3 mi less than twice the distance that Tricia runs. How far does Tricia run daily?

7. **Dimensions of a Rectangle.** The perimeter of a rectangle is 192 in. The width is half the length. Find the length and the width.

8. **Mystery Numbers.** Teka asked her students to determine the two numbers that she placed in a sealed envelope. Twice the smaller number is 5 more than two-thirds the larger number. Three times the larger number is 4 less than fifteen times the smaller. Find the numbers.

9. **Supplementary Angles.** Two supplementary angles are such that one angle is 7° more than twice the other. Find the measures of the angles.

10. **Student Loans.** Brandt's student loans totaled $2500. Part was borrowed at 3% interest and the rest at 2.5%. After one year, Brandt had accumulated $97.50 in interest. What was the amount of each loan?

8.4 Exercise Set

FOR EXTRA HELP MyLab Math

✓ Check Your Understanding

Reading and Concept Check Consider the following mixture problem and the table used to translate the problem.

Cherry Breeze is 30% fruit juice and Berry Choice is 15% fruit juice. How much of each should be used in order to make 10 gal of a drink that is 20% fruit juice?

Choose from the options below the expression that best fits each numbered space in the table.

    2   10   15   0.15y

	CHERRY BREEZE	BERRY CHOICE	MIXTURE
GALLONS OF DRINK	x	y	RC1. ____
PERCENT OF FRUIT JUICE	30%	RC2. ____ %	20%
GALLONS OF FRUIT JUICE IN MIXTURE	$0.3x$	RC3. ____	RC4. ____

a Solve.

1. *Video-Game Production.* Video games are developed or published at 2848 locations in the United States. There are 1796 more locations at which video games are developed than locations at which video games are published. At how many locations are video games developed and at how many are they published?

   Data: Entertainment Software Association

2. *Spoken Languages.* Of a total of 422 indigenous and immigrant languages spoken in the United States, 10 more spoken languages are indigenous than are immigrant. How many indigenous languages and how many immigrant languages are spoken in the United States?

   Data: USA Today, 11/8/2015

3. *Entertainment.* For her personal-finance class, Laura was required to estimate her annual entertainment expenditures. She discovered that during the previous year, she spent $225.32 on a total of 68 e-books and game applications. If each book cost $3.99 and each game cost $1.99, how many books and how many games did she purchase?

4. *Flowers.* Kevin's Floral Emporium offers two types of sunflowers for sale by the stem. When in season, the small ones sell for $2.50 per stem, and the large ones sell for $3.95 per stem. One late summer weekend, Kevin sold a total of 118 stems for $376.20. How many of each size did he sell?

5. *Furniture Polish.* A nontoxic furniture polish can be made by combining vinegar and olive oil. The amount of oil should be three times the amount of vinegar. How much of each ingredient is needed in order to make 30 oz of furniture polish?

6. *Nontoxic Floor Wax.* A nontoxic floor wax can be made by combining lemon juice and food-grade linseed oil. The amount of oil should be twice the amount of lemon juice. How much of each ingredient is needed in order to make 32 oz of floor wax? (The mix should be spread with a rag and buffed when dry.)

7. *Catering.* Stella's Catering is planning a wedding reception. The bride and groom would like to serve a nut mixture containing 25% peanuts. Stella has available mixtures that are either 40% or 10% peanuts. How much of each type should be mixed in order to get a 10-lb mixture that is 25% peanuts?

8. *Blending Granola.* Deep Thought Granola is 25% nuts and dried fruit. Oat Dream Granola is 10% nuts and dried fruit. How much of Deep Thought and how much of Oat Dream should be mixed in order to form a 20-lb batch of granola that is 19% nuts and dried fruit?

9. *Ink Remover.* Etch Clean Graphics uses one cleanser that is 25% acid and a second that is 50% acid. How many liters of each should be mixed in order to get 10 L of a solution that is 40% acid?

10. *Livestock Feed.* Soybean meal is 16% protein and corn meal is 9% protein. How many pounds of each should be mixed in order to get a 350-lb mixture that is 12% protein?

11. *Vegetable Seeds.* Tara's website, verdantveggies.com, specializes in the sale of rare or unusual vegetable seeds. Tara sells packets of sweet-pepper seeds for $2.85 each and packets of hot-pepper seeds for $4.29 each. She also offers a 16-packet mixed-pepper assortment combining packets of both types of seeds at $3.30 per packet. How many packets of each type of seed are in the assortment?

12. *Flower Bulbs.* Heritage Bulbs sells heirloom flower bulbs. Acuminata tulip bulbs cost $4.85 each, and Cafe Brun tulip bulbs cost $9.50 each. An assortment of 12 of these bulbs is priced at $7.95 per bulb. How many of each type of bulb are in the assortment?

Sweet peppers   Hot peppers   Assorted

13. *Student Loans.* Sarah's two student loans totaled $12,000. One of her loans was at 6% simple interest and the other at 3%. After one year, Sarah owed $585 in interest. What was the amount of each loan?

14. *Investments.* Ana and Johnny made two investments totaling $45,000. In one year, these investments yielded $2430 in simple interest. Part of the money was invested at 4% and the rest at 6%. How much was invested at each rate?

15. *Food Science.* The following bar graph shows the milk fat percentages in three dairy products. How many pounds each of whole milk and cream should be mixed in order to form 200 lb of milk for cream cheese?

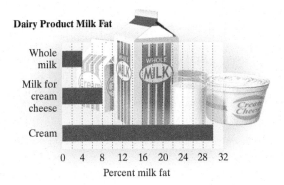

16. *Automotive Maintenance.* Arctic Antifreeze is 18% alcohol and Frost No-More is 10% alcohol. How many liters of Arctic Antifreeze should be mixed with 7.5 L of Frost No-More in order to get a mixture that is 15% alcohol?

17. *Investments.* William opened two investment accounts for his daughter's college fund. The first year, these investments, which totaled $3200, yielded $155 in simple interest. Part of the money was invested at 5.5% and the rest at 4%. How much was invested at each rate?

18. *Student Loans.* Cole's two student loans totaled $31,000. One of his loans was at 2.8% simple interest and the other at 4.5%. After one year, Cole owed $1024.40 in interest. What was the amount of each loan?

19. *Making Change.* Juan goes to a bank and gets change for a $50 bill consisting of all $5 bills and $1 bills. There are 22 bills in all. How many of each kind are there?

20. *Making Change.* Christina makes a $9.25 purchase at a bookstore with a $20 bill. The store has no bills and gives her the change in quarters and dollar coins. There are 19 coins in all. How many of each kind are there?

21. *Super Bowl Advertisements.* For commercials aired during Super Bowl L, advertisers paid an average of $5 million to air a 30-sec commercial. Even at this rate, some commercials were longer than 30 sec. A total of 62 commercials ran for either 30 sec or 1 min. The total running time was $39\frac{1}{2}$ min. How many of each length commercial aired during the Super Bowl?

Data: *Sports Illustrated*

22. *Typing.* It takes Ashleigh 5 min to type a page of her essay and $1\frac{1}{2}$ min to proofread a page. One afternoon she spent 2 hr typing or proofreading a total of 38 pages. How many pages did she type and how many did she proofread?

 Solve.

23. *Train Travel.* A train leaves Danville Junction and travels north at a speed of 75 mph. Two hours later, a second train leaves on a parallel track and travels north at 125 mph. How far from the station will the second train catch up to the first train?

24. *Car Travel.* Max leaves Kansas City and drives east at a speed of 80 km/h. One hour later, Olivia leaves Kansas City traveling in the same direction as Max but at 96 km/h. Assuming that neither driver stops for a break, how far from Kansas City will they be when Olivia catches up with Max?

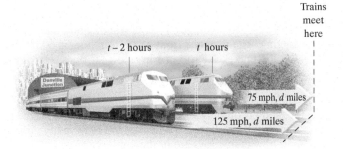

25. *Canoeing.* Darren paddled for 4 hr with a 6-km/h current to reach a campsite. The return trip against the same current took 10 hr. Find the speed of Darren's canoe in still water.

26. *Boating.* Mia's motorboat took 3 hr to make a trip downstream with a 6-mph current. The return trip against the same current took 5 hr. Find the speed of the boat in still water.

27. *Air Travel.* Christie pilots her Cessna 150 plane for 270 mi against a headwind in 3 hr. The flight would take 1 hr and 48 min with a tailwind of the same speed. Find the headwind and the speed of the plane in still air.

28. *Air Travel.* Rod is a pilot for Crossland Airways. He computes his flight time against a headwind for a trip of 2900 mi at 5 hr. The flight would take 4 hr and 50 min if the headwind were half as strong. Find the headwind and the plane's air speed in still air.

29. *Air Travel.* Two airplanes start at the same time and fly toward each other from points 1000 km apart at rates of 420 km/h and 330 km/h. After how many hours will they meet?

30. *Air Travel.* Two planes start at the same time and travel toward each other from cities that are 780 km apart at rates of 190 km/h and 200 km/h. In how many hours will they meet?

31. *Point of No Return.* A plane flying the 3458-mi trip from New York City to London has a 50-mph tailwind. The flight's *point of no return* is the point at which the flight time required to return to New York is the same as the time required to continue to London. If the speed of the plane in still air is 360 mph, how far is New York from the point of no return?

32. *Point of No Return.* A plane is flying the 2553-mi trip from Los Angeles to Honolulu into a 60-mph headwind. If the speed of the plane in still air is 310 mph, how far from Los Angeles is the plane's point of no return? (See Exercise 31.)

Skill Maintenance

Find the domain. [7.2a]

33. $g(x) = x^2 + 5x - 14$

34. $f(x) = \dfrac{2}{x - 14}$

Given the function $f(x) = 4x - 7$, find each of the following function values. [7.1b]

35. $f\left(\dfrac{3}{4}\right)$

36. $f(-2.5)$

37. $f(-3h)$

38. $f(1000)$

Synthesis

39. *Automotive Maintenance.* The radiator in Michelle's car contains 16 L of antifreeze and water. This mixture is 30% antifreeze. How much of this mixture should she drain and replace with pure antifreeze so that there will be a mixture of 50% antifreeze?

40. *Physical Exercise.* Natalie jogs and walks to school each day. She averages 4 km/h walking and 8 km/h jogging. The distance from home to school is 6 km and Natalie makes the trip in 1 hr. How far does she jog in a trip?

41. *Fuel Economy.* Ashlee's SUV gets 18 miles per gallon (mpg) in city driving and 24 mpg in highway driving. The SUV is driven 465 mi on 23 gal of gasoline. How many miles were driven in the city and how many were driven on the highway?

42. *Siblings.* Phil and Maria are siblings. Maria has twice as many brothers as she has sisters. Phil has the same number of brothers as he has sisters. How many girls and how many boys are in the family?

Mid-Chapter Review

Concept Reinforcement

Determine whether each statement is true or false.

_____ 1. If, when solving a system of two linear equations in two variables, a false equation is obtained, the system has infinitely many solutions. [8.2a], [8.3a]

_____ 2. Every system of equations has at least one solution. [8.1a]

_____ 3. If the graphs of two linear equations intersect, then the system is consistent. [8.1a]

_____ 4. The intersection of the graphs of the lines $x = a$ and $y = b$ is (a, b). [8.1a]

Guided Solutions

Fill in each box with the number, the variable, or the expression that creates a correct statement or solution.

GS Solve. [8.2a], [8.3a]

5. $x + 2y = 3$, (1)
   $y = x - 6$ (2)

   $x + 2(\boxed{}) = 3$ Substituting for y in equation (1)
   $x + \boxed{}x - \boxed{} = 3$ Removing parentheses
   $\boxed{}x - 12 = 3$ Collecting like terms
   $3x = \boxed{}$
   $x = \boxed{}$
   $y = \boxed{} - 6$ Substituting in equation (2)
   $y = \boxed{}$ Subtracting
   The solution is ($\boxed{}$, $\boxed{}$).

6. $3x - 2y = 5$, (1)
   $2x + 4y = 14$ (2)

   $\boxed{}x - \boxed{}y = \boxed{}$ Multiplying equation (1) by 2
   $2x + \boxed{}4y = 14$ Equation (2)
   $\boxed{}x = \boxed{}$ Adding
   $x = \boxed{}$
   $2 \cdot \boxed{} + 4y = 14$ Substituting for x in equation (2)
   $\boxed{} + 4y = 14$ Multiplying
   $4y = \boxed{}$
   $y = \boxed{}$
   The solution is ($\boxed{}$, $\boxed{}$).

Mixed Review

Solve each system of equations graphically. Then classify the system as consistent or inconsistent and the equations as dependent or independent. [8.1a]

7. $y = x - 6$,
   $y = 4 - x$

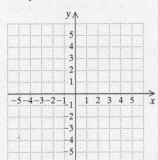

8. $x + y = 3$,
   $3x + y = 3$

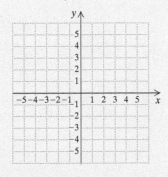

9. $y = 2x - 3$,
   $4x - 2y = 6$

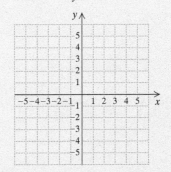

10. $x - y = 3$,
    $2y - 2x = 6$

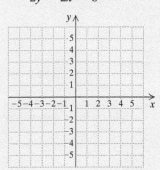

Solve using the substitution method. [8.2a]

11. $x = y + 2,$
$2x - 3y = -2$

12. $y = x - 5,$
$x - 2y = 8$

13. $4x + 3y = 3,$
$y = x + 8$

14. $3x - 2y = 1,$
$x = y + 1$

Solve using the elimination method. [8.3a]

15. $2x + y = 2,$
$x - y = 4$

16. $x - 2y = 13,$
$x + 2y = -3$

17. $3x - 4y = 5,$
$5x - 2y = -1$

18. $3x + 2y = 11,$
$2x + 3y = 9$

19. $x - 2y = 5,$
$3x - 6y = 10$

20. $4x - 6y = 2,$
$-2x + 3y = -1$

21. $\frac{1}{2}x + \frac{1}{3}y = 1,$
$\frac{1}{5}x - \frac{3}{4}y = 11$

22. $0.2x + 0.3y = 0.6,$
$0.1x - 0.2y = -2.5$

Solve.

23. *Garden Dimensions.* A landscape architect designs a garden with a perimeter of 44 ft. The width is 2 ft less than the length. Find the length and the width. [8.2b]

24. *Investments.* Sandy made two investments totaling $5000. Part of the money was invested at 2% and the rest at 3%. In one year, these investments earned $129 in simple interest. How much was invested at each rate? [8.4a]

25. *Mixing Solutions.* A lab technician wants to mix a solution that is 20% acid with a second solution that is 50% acid in order to get 84 L of a solution that is 30% acid. How many liters of each solution should be used? [8.4a]

26. *Boating.* Monica's motorboat took 5 hr to make a trip downstream with a 6-mph current. The return trip against the same current took 8 hr. Find the speed of the boat in still water. [8.4b]

Understanding Through Discussion and Writing

27. Explain how to find the solution of $\frac{3}{4}x + 2 = \frac{2}{5}x - 5$ in two ways graphically and in two ways algebraically. [8.1a], [8.2a], [8.3a]

28. Write a system of equations with the given solution. Answers may vary. [8.1a], [8.2a], [8.3a]
a) $(4, -3)$
b) No solution
c) Infinitely many solutions

29. Describe a method that could be used to create an inconsistent system of equations. [8.1a], [8.2a], [8.3a]

30. Describe a method that could be used to create a system of dependent equations. [8.1a], [8.2a], [8.3a]

STUDYING FOR SUCCESS *Throughout the Semester*

- ☐ Review regularly. A good way to do this is by doing the Skill Maintenance exercises found in each exercise set.
- ☐ Try creating your own glossary. Understanding terminology is essential for success in any math course.
- ☐ Memorizing is a helpful tool in the study of mathematics. Ask your instructor what you are expected to have memorized for tests.

8.5 Systems of Equations in Three Variables

OBJECTIVE

a Solve systems of three equations in three variables.

a SOLVING SYSTEMS IN THREE VARIABLES

SKILL REVIEW Solve systems of equations in two variables by the elimination method. [8.3a]

Solve.

1. $3x + y = 1,$
   $5x - y = 7$

2. $2x + 3y = 9,$
   $3x + 2y = 1$

Answers: 1. $(1, -2)$ **2.** $(-3, 5)$

MyLab Math VIDEO

A **linear equation in three variables** is an equation equivalent to one of the type $Ax + By + Cz = D$. A **solution** of a system of three equations in three variables is an ordered triple (x, y, z) that makes *all three* equations true.

The substitution method can be used to solve systems of three equations, but it is not efficient unless a variable has already been eliminated from one or more of the equations. Therefore, we will use only the elimination method.* The first step is to eliminate a variable and obtain a system of two equations in two variables.

EXAMPLE 1 Solve the following system of equations:

$x + y + z = 4,$ (1)
$x - 2y - z = 1,$ (2)
$2x - y - 2z = -1.$ (3)

a) We first use *any* two of the three equations to get an equation in two variables. In this case, let's use equations (1) and (2) and add to eliminate z:

$x + y + z = 4$ (1)
$\underline{x - 2y - z = 1}$ (2)
$2x - y = 5.$ (4)   Adding to eliminate z

b) We use a *different* pair of equations and eliminate the **same variable** that we did in part (a). Let's use equations (1) and (3) and again eliminate z.

---

*Other methods for solving systems of equations are considered in Appendixes D and E.

> **Caution!**
> A common error is to eliminate a different variable the second time.

$$x + y + z = 4, \quad (1)$$
$$2x - y - 2z = -1; \quad (3)$$

$$\begin{aligned} 2x + 2y + 2z &= 8 &&\text{Multiplying equation (1) by 2}\\ \underline{2x - y - 2z} &= \underline{-1} &&(3)\\ 4x + y &= 7 &&(5) \quad \text{Adding to eliminate } z \end{aligned}$$

c) Now we solve the resulting system of equations, (4) and (5). That solution will give us two of the numbers. Note that we now have two equations in two variables. Had we eliminated two *different* variables in parts (a) and (b), this would not be the case.

$$\begin{aligned} 2x - y &= 5 &&(4)\\ \underline{4x + y} &= \underline{7} &&(5)\\ 6x &= 12 &&\text{Adding}\\ x &= 2 \end{aligned}$$

We can use either equation (4) or (5) to find y. We choose equation (5):

$$\begin{aligned} 4x + y &= 7 &&(5)\\ 4(2) + y &= 7 &&\text{Substituting 2 for } x\\ 8 + y &= 7\\ y &= -1. \end{aligned}$$

d) We now have $x = 2$ and $y = -1$. To find the value for z, we use any of the original three equations, substitute, and solve for z. Let's use equation (1) and substitute our two numbers in it:

$$\begin{aligned} x + y + z &= 4 &&(1)\\ 2 + (-1) + z &= 4 &&\text{Substituting 2 for } x \text{ and } -1 \text{ for } y\\ 1 + z &= 4\\ z &= 3. \end{aligned} \Big\} \text{Solving for } z$$

We have obtained the ordered triple $(2, -1, 3)$. To check, we substitute $(2, -1, 3)$ into each of the three equations using alphabetical order of the variables.

Check:
$$\begin{array}{c|c} x + y + z = 4 \\ \hline 2 + (-1) + 3 \;?\; 4 \\ 4 \;\Big|\; \text{TRUE} \end{array}$$

$$\begin{array}{c|c} x - 2y - z = 1 \\ \hline 2 - 2(-1) - 3 \;?\; 1 \\ 2 + 2 - 3 \\ 1 \;\Big|\; \text{TRUE} \end{array}$$

$$\begin{array}{c|c} 2x - y - 2z = -1 \\ \hline 2(2) - (-1) - 2\cdot 3 \;?\; -1 \\ 4 + 1 - 6 \\ -1 \;\Big|\; \text{TRUE} \end{array}$$

The triple $(2, -1, 3)$ checks and is the solution.

> To use the elimination method to solve systems of three equations:
>
> 1. Write all equations in the standard form, $Ax + By + Cz = D$.
> 2. Clear any decimals or fractions.
> 3. Choose a variable to eliminate. Then use *any* two of the three equations to eliminate that variable, getting an equation in two variables.
> 4. Next, use a different pair of equations and get another equation in *the same two variables*. That is, eliminate the same variable that you did in step (3).
> 5. Solve the resulting system (pair) of equations. That will give two of the numbers.
> 6. Then use any of the original three equations to find the third number.

Do Exercise 1.

1. Solve. Don't forget to check.
$$4x - y + z = 6,$$
$$-3x + 2y - z = -3,$$
$$2x + y + 2z = 3$$

EXAMPLE 2 Solve this system:

$$4x - 2y - 3z = 5, \quad (1)$$
$$-8x - y + z = -5, \quad (2)$$
$$2x + y + 2z = 5. \quad (3)$$

a) The equations are in standard form and do not contain decimals or fractions.

b) We decide to eliminate the variable y since the y-terms are opposites in equations (2) and (3). We add:

$$-8x - y + z = -5 \quad (2)$$
$$\underline{2x + y + 2z = 5} \quad (3)$$
$$-6x + 3z = 0. \quad (4) \quad \text{Adding}$$

c) We use another pair of equations to get an equation in the same two variables, x and z. We use equations (1) and (3) and eliminate y:

$$4x - 2y - 3z = 5, \quad (1)$$
$$2x + y + 2z = 5; \quad (3)$$

$$4x - 2y - 3z = 5 \quad (1)$$
$$\underline{4x + 2y + 4z = 10} \quad \text{Multiplying equation (3) by 2}$$
$$8x + z = 15. \quad (5) \quad \text{Adding}$$

d) Next, we solve the resulting system of equations (4) and (5). That will give us two of the numbers:

$$-6x + 3z = 0, \quad (4)$$
$$8x + z = 15. \quad (5)$$

We multiply equation (5) by -3 and then add:

$$-6x + 3z = 0 \quad (4)$$
$$\underline{-24x - 3z = -45} \quad \text{Multiplying equation (5) by } -3$$
$$-30x = -45 \quad \text{Adding}$$
$$x = \frac{-45}{-30} = \frac{3}{2}.$$

Answer

1. $(2, 1, -1)$

We now use equation (5) to find z:

$$8x + z = 15 \quad (5)$$
$$8(\tfrac{3}{2}) + z = 15 \quad \text{Substituting } \tfrac{3}{2} \text{ for } x$$
$$\left.\begin{array}{r} 12 + z = 15 \\ z = 3. \end{array}\right\} \text{Solving for } z$$

e) Next, we use any of the original equations and substitute to find the third number, y. We choose equation (3) since the coefficient of y there is 1:

$$2x + y + 2z = 5 \quad (3)$$
$$2(\tfrac{3}{2}) + y + 2(3) = 5 \quad \text{Substituting } \tfrac{3}{2} \text{ for } x \text{ and } 3 \text{ for } z$$
$$\left.\begin{array}{r} 3 + y + 6 = 5 \\ y + 9 = 5 \\ y = -4. \end{array}\right\} \text{Solving for } y$$

The solution is $(\tfrac{3}{2}, -4, 3)$. The check is as follows.

Check:
$$\begin{array}{c|c} 4x - 2y - 3z = 5 \\ \hline 4 \cdot \tfrac{3}{2} - 2(-4) - 3(3) \;?\; 5 \\ 6 + 8 - 9 \\ 5 \end{array} \quad \text{TRUE}$$

$$\begin{array}{c|c} -8x - y + z = -5 \\ \hline -8 \cdot \tfrac{3}{2} - (-4) + 3 \;?\; -5 \\ -12 + 4 + 3 \\ -5 \end{array} \quad \text{TRUE}$$

$$\begin{array}{c|c} 2x + y + 2z = 5 \\ \hline 2 \cdot \tfrac{3}{2} + (-4) + 2(3) \;?\; 5 \\ 3 - 4 + 6 \\ 5 \end{array} \quad \text{TRUE}$$

◀ Do Exercise 2.

In Example 3, two of the equations have a missing variable.

EXAMPLE 3 Solve this system:

$$x + y + z = 180, \quad (1)$$
$$x \qquad - z = -70, \quad (2)$$
$$2y - z = 0. \quad (3)$$

We note that there is no y in equation (2). In order to have a system of two equations in the variables x and z, we need to find another equation without a y. We use equations (1) and (3) to eliminate y:

$$x + y + z = 180, \quad (1)$$
$$2y - z = 0; \quad (3)$$

$$\begin{array}{rl} -2x - 2y - 2z = -360 & \text{Multiplying equation (1) by } -2 \\ 2y - z = 0 & (3) \\ \hline -2x - 3z = -360. & (4) \quad \text{Adding} \end{array}$$

2. Solve. Don't forget to check.
$$2x + y - 4z = 0,$$
$$x - y + 2z = 5,$$
$$3x + 2y + 2z = 3$$

Answer
2. $\left(2, -2, \tfrac{1}{2}\right)$

Next, we solve the resulting system of equations (2) and (4):

$$x - z = -70, \quad (2)$$
$$-2x - 3z = -360; \quad (4)$$

$$\begin{array}{ll} 2x - 2z = -140 & \text{Multiplying equation (2) by 2} \\ \underline{-2x - 3z = -360} & (4) \\ -5z = -500 & \text{Adding} \\ z = 100. \end{array}$$

To find x, we substitute 100 for z in equation (2) and solve for x:

$$x - z = -70$$
$$x - 100 = -70$$
$$x = 30.$$

To find y, we substitute 100 for z in equation (3) and solve for y:

$$2y - z = 0$$
$$2y - 100 = 0$$
$$2y = 100$$
$$y = 50.$$

The triple $(30, 50, 100)$ is the solution. The check is left to the student.

Do Exercise 3. ▶

It is possible for a system of three equations to have no solution, that is, to be inconsistent. An example is the system

$$x + y + z = 14,$$
$$x + y + z = 11,$$
$$2x - 3y + 4z = -3.$$

Note the first two equations. It is not possible for a sum of three numbers to be both 14 and 11. Thus the system has no solution. We will not consider such systems here, nor will we consider systems with infinitely many solutions, which also exist.

 3. Solve. Don't forget to check.

$$x + y + z = 100, \quad (1)$$
$$x - y = -10, \quad (2)$$
$$x - z = -30, \quad (3)$$

Add equations (1) and (3):

$$\begin{array}{ll} x + y + z = 100 & (1) \\ \underline{x - z = -30} & (3) \\ 2x + y = \boxed{}. & (4) \end{array}$$

Add equations (2) and (4) and solve for x:

$$\begin{array}{ll} x - y = -10 & (2) \\ \underline{2x + y = 70} & (4) \\ 3x = \boxed{} \\ x = \boxed{}. \end{array}$$

Substitute 20 for x in equation (4) and solve for y:

$$2(20) + y = 70$$
$$y = \boxed{}.$$

Substitute 20 for x and 30 for y in equation (1) and solve for z:

$$20 + 30 + z = 100$$
$$z = \boxed{}.$$

The numbers check. The solution is $(20, 30, \boxed{})$.

Answer
3. $(20, 30, 50)$
Guided Solution:
3. 70, 60, 20, 30, 50, 50

8.5 Exercise Set

FOR EXTRA HELP MyLab Math

✓ Check Your Understanding

Reading Check Choose from the column on the right the option that is an example of each term. Choices may be used more than once.

RC1. A linear equation in three variables

RC2. A system of equations in three variables

RC3. A solution of a linear equation in three variables

RC4. A solution of a system of equations in three variables

a) $(4, -3, 0)$

b) $a + b - c = 1$

c) $a + 3b - c = 1,$
$ 2a + 3b - c = -1,$
$ a - 2b + 3c = 10$

SECTION 8.5 Systems of Equations in Three Variables **631**

Concept Check Determine whether $(2, -1, -3)$ is a solution of each system.

CC1. $x + y - 2z = 7,$
$2x - y - z = 8,$
$-x - 2y - 3z = 9$

CC2. $r - s + t = 0,$
$2r - s + t = 2,$
$r - 2s + t = -3$

CC3. $a - c = 5,$
$3b - c = 0,$
$2a + 5b = -1$

a Solve.

1. $x + y + z = 2,$
$2x - y + 5z = -5,$
$-x + 2y + 2z = 1$

2. $2x - y - 4z = -12,$
$2x + y + z = 1,$
$x + 2y + 4z = 10$

3. $2x - y + z = 5,$
$6x + 3y - 2z = 10,$
$x - 2y + 3z = 5$

4. $x - y + z = 4,$
$3x + 2y + 3z = 7,$
$2x + 9y + 6z = 5$

5. $2x - 3y + z = 5,$
$x + 3y + 8z = 22,$
$3x - y + 2z = 12$

6. $6x - 4y + 5z = 31,$
$5x + 2y + 2z = 13,$
$x + y + z = 2$

7. $3a - 2b + 7c = 13,$
$a + 8b - 6c = -47,$
$7a - 9b - 9c = -3$

8. $x + y + z = 0,$
$2x + 3y + 2z = -3,$
$-x + 2y - 3z = -1$

9. $2x + 3y + z = 17,$
$x - 3y + 2z = -8,$
$5x - 2y + 3z = 5$

10. $2x + y - 3z = -4,$
$4x - 2y + z = 9,$
$3x + 5y - 2z = 5$

11. $2x + y + z = -2,$
$2x - y + 3z = 6,$
$3x - 5y + 4z = 7$

12. $2x + y + 2z = 11,$
$3x + 2y + 2z = 8,$
$x + 4y + 3z = 0$

13. $x - y + z = 4,$
$5x + 2y - 3z = 2,$
$3x - 7y + 4z = 8$

14. $2x + y + 2z = 3,$
$x + 6y + 3z = 4,$
$3x - 2y + z = 0$

15. $4x - y - z = 4,$
$2x + y + z = -1,$
$6x - 3y - 2z = 3$

16. $2r + s + t = 6,$
$3r - 2s - 5t = 7,$
$r + s - 3t = -10$

17. $a - 2b - 5c = -3,$
$3a + b - 2c = -1,$
$2a + 3b + c = 4$

18. $x + 4y - z = 5,$
$2x - y + 3z = -5,$
$4x + 3y + z = 5$

19. $2r + 3s + 12t = 4,$
$4r - 6s + 6t = 1,$
$r + s + t = 1$

20. $10x + 6y + z = 7,$
$5x - 9y - 2z = 3,$
$15x - 12y + 2z = -5$

21. $a + 2b + c = 1,$
$7a + 3b - c = -2,$
$a + 5b + 3c = 2$

22. $3p + 2r = 11,$
$q - 7r = 4,$
$p - 6q = 1$

23. $x + y + z = 57,$
$-2x + y = 3,$
$x - z = 6$

24. $4a + 9b = 8,$
$8a + 6c = -1,$
$6b + 6c = -1$

25. $r + s = 5,$
$3s + 2t = -1,$
$4r + t = 14$

26. $a - 5c = 17,$
$b + 2c = -1,$
$4a - b - 3c = 12$

27. $x + y + z = 105,$
$10y - z = 11,$
$2x - 3y = 7$

Skill Maintenance

Solve for the indicated letter. [2.4b]

28. $F = 3ab$, for a

29. $Q = 4(a + b)$, for a

30. $F = \frac{1}{2}t(c - d)$, for d

31. $F = \frac{1}{2}t(c - d)$, for c

32. $Ax + By = c$, for y

33. $Ax - By = c$, for y

Find the slope and the y-intercept. [7.3b]

34. $y = -\frac{2}{3}x - \frac{5}{4}$

35. $y = 5 - 4x$

36. $2x - 5y = 10$

37. $7x - 6.4y = 20$

Synthesis

Solve.

38. $w + x - y + z = 0,$
$w - 2x - 2y - z = -5,$
$w - 3x - y + z = 4,$
$2w - x - y + 3z = 7$

39. $w + x + y + z = 2,$
$w + 2x + 2y + 4z = 1,$
$w - x + y + z = 6,$
$w - 3x - y + z = 2$

8.6 Solving Applied Problems: Three Equations

OBJECTIVE

a Solve applied problems using systems of three equations.

Solving systems of three or more equations is important in many applications occurring in the natural and social sciences, business, and engineering.

a USING SYSTEMS OF THREE EQUATIONS

SKILL REVIEW *Solve applied problems by translating to equations.* [2.6a]
Solve.

1. The second angle of a triangle is twice as large as the first angle. The third angle is 5° larger than the second angle. Find the measures of the angles.

2. Malika invested her tax refund check in a fund paying 4% interest. After one year, she had earned $50.40 in interest. How much did she invest?

Answers: **1.** 35°, 70°, 75° **2.** $1260

MyLab Math
VIDEO

EXAMPLE 1 *Jewelry Design.* Kim is designing a triangular-shaped pendant for a client of her custom jewelry business. The largest angle of the triangle is 70° greater than the smallest angle. The largest angle is twice as large as the remaining angle. Find the measure of each angle.

1. **Familiarize.** We first make a drawing. We let x = the smallest angle, z = the largest angle, and y = the remaining angle.

2. **Translate.** In order to translate the problem, we use the fact that the sum of the measures of the angles of a triangle is 180°:

$$x + y + z = 180.$$

There are two statements in the problem that we can translate directly.

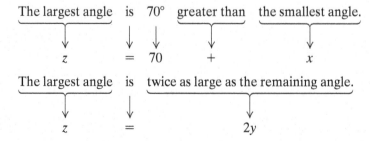

We now have a system of three equations:

$$\begin{aligned} x + y + z &= 180, & x + y + z &= 180, & \text{(1)} \\ x + 70 &= z, & \text{or} \quad x \quad - z &= -70, & \text{(2)} \\ 2y &= z; & 2y - z &= 0. & \text{(3)} \end{aligned}$$

3. **Solve.** We can solve the system using elimination. However, because we can write both x and y in terms of z, we can also use substitution to solve this system.

   We first solve equation (2) for x:

   $$x - z = -70$$
   $$x = z - 70. \quad \text{(4)}$$

   We then solve equation (3) for y:

   $$2y - z = 0$$
   $$y = \frac{z}{2}. \quad \text{(5)}$$

   Finally, we substitute in equation (1) and solve for z:

   $$x + y + z = 180$$
   $$(z - 70) + \frac{z}{2} + z = 180 \quad \text{Substituting}$$
   $$\frac{5}{2}z - 70 = 180$$
   $$\frac{5}{2}z = 250$$
   $$z = 100. \quad \text{Solving for } z$$

   We now substitute 100 for z in equations (4) and (5) and find $x = 30$ and $y = 50$.

4. **Check.** The sum of the numbers is 180. The largest angle measures 100° and the smallest measures 30°, so the largest angle is 70° greater than the smallest. The largest angle is twice as large as 50°, the remaining angle. We have an answer to the problem.

5. **State.** The measures of the angles of the triangle are 30°, 50°, and 100°.

   Do Exercise 1.

EXAMPLE 2 *Airline Tickets.* Some airlines charge fees for baggage that is checked in rather than carried with the passenger. A Spanish class flew to Costa Rica using an airline that charged $25 for one checked bag and $60 for two checked bags. The 30 students paid a total of $1120 in baggage fees for 40 checked bags. How many students checked no bags, how many checked one bag, and how many checked two bags?

1. **Familiarize.** A careful reading of the problem leads us to recognize that there are three relationships given: the total number of students, the total baggage fee, and the total number of bags. These three relationships will give us three equations. We let x = the number of students who checked no bags, y = the number of students who checked one bag, and z = the number of students who checked two bags.

1. *Triangle Measures.* One angle of a triangle is twice as large as a second angle. The remaining angle is 20° greater than the first angle. Find the measure of each angle.

Answer
1. 64°, 32°, 84°

2. Client Investments. Kaufman Financial Corporation makes investments for corporate clients. One year, a client receives $1120 in simple interest from three investments that total $25,000. Part was invested at 3%, part at 4%, and part at 5%. There was $11,000 more invested at 5% than at 4%. How much was invested at each rate?

Let x = the amount invested at 3%, y = the amount invested at 4%, and z = the amount invested at 5%. Complete the following table to help in the translation.

	FIRST INVESTMENT	SECOND INVESTMENT	THIRD INVESTMENT	TOTAL
PRINCIPAL, P	x	y	z	☐ $
RATE OF INTEREST, r	3%	4%	☐ %	
TIME, t	1 year	1 year	1 year	
INTEREST, I	$0.03x$	$0.04y$	$0.05z$	☐ $

2. Translate. We translate the three relationship statements to equations.
- A total of 30 students went on the trip. → $x + y + z = 30$,
- The baggage fee was $1120. → $0 \cdot x + 25 \cdot y + 60 \cdot z = 1120$,
- A total of 40 bags were checked. → $0 \cdot x + 1 \cdot y + 2 \cdot z = 40$

3. Solve. We write the equations in standard form:

$x + y + z = 30$, (1)
$25y + 60z = 1120$, (2)
$y + 2z = 40$. (3)

Equations (2) and (3) form a system of two equations in two unknowns that we can solve for y and z.

$$25y + 60z = 1120,$$
$$\underline{-25y - 50z = -1000} \quad \text{Multiplying equation (3) by } -25$$
$$10z = 120 \quad \text{Adding}$$
$$z = 12 \quad \text{Solving for } z$$

We substitute 12 for z in equation (3) and solve for y:

$y + 2 \cdot 12 = 40$
$y + 24 = 40$
$y = 16$.

Finally, we find x by substituting 16 for y and 12 for z in equation (1):

$x + 16 + 12 = 30$
$x = 2$.

We have $x = 2$, $y = 16$, and $z = 12$.

4. Check. We check our answers in each statement of the problem.
- The total number of students on the trip was $2 + 16 + 12 = 30$.
- The baggage fee was $0 \cdot 2 + 25 \cdot 16 + 60 \cdot 12 = 1120$.
- The total number of bags was $0 \cdot 2 + 1 \cdot 16 + 2 \cdot 12 = 40$.

5. State. There were 2 students who checked no bags, 16 students who checked 1 bag, and 12 students who checked 2 bags.

◀ Do Exercise 2.

Answer
2. $4000 at 3%; $5000 at 4%; $16,000 at 5%

Guided Solution:
2. 25,000, 5, 1120

8.6 Exercise Set

✓ Check Your Understanding

Reading and Concept Check Choose from the column on the right an appropriate translation for each statement.

RC1. The sum of three numbers is 60.

RC2. The first number minus the second number plus the third number is 60.

RC3. The first number is 60 more than the sum of the other two numbers.

RC4. The first number is 60 less than the sum of the other two numbers.

a) $x = y + z + 60$
b) $x = y + z - 60$
c) $x + y + z = 60$
d) $x - y + z = 60$

a Solve.

1. *Advanced Placement Exams.* In 2016, approximately 4.7 million Advanced Placement (AP) exams were taken in the United States. A total of 40 thousand exams was taken in Chinese, French, and German Language and Culture. The number of Chinese exams taken was 8 thousand more than the number of German exams taken. The number of French exams taken was 4 thousand more than the sum of the number of Chinese and German exams. Find the number of exams taken in each language.
   Data: collegeboard.org

2. *Fat Content of Fast Food.* A meal at McDonald's consisting of a Big Mac, a medium order of fries, and a medium vanilla milkshake contains 61 g of fat. The Big Mac has 11 more grams of fat than the milkshake. The total fat content of the fries and the shake exceeds that of the Big Mac by 5 g. Find the fat content of each food item.
   Data: McDonald's

3. *Triangle Measures.* In triangle ABC, the measure of angle B is three times that of angle A. The measure of angle C is 20° more than that of angle A. Find the measure of each angle.

4. *Triangle Measures.* In triangle ABC, the measure of angle B is twice the measure of angle A. The measure of angle C is 80° more than that of angle A. Find the measure of each angle.

5. The sum of three numbers is 55. The difference of the largest and the smallest is 49, and the sum of the two smaller numbers is 13. Find the numbers.

6. *History.* Find the year in which the first U.S. transcontinental railroad was completed. The following are some facts about the number. The sum of the digits in the year is 24. The ones digit is 1 more than the hundreds digit. Both the tens and the ones digits are multiples of 3.

7. *Eggs.* Chicken eggs are sorted by size before they are sold. A dozen small eggs weighs 18 oz, a dozen medium eggs weighs 21 oz, and a dozen large eggs weighs 24 oz. A camp cook needed 480 oz of eggs and purchased 22 dozen eggs at a local convenience store for a total of $43.88. If a dozen small eggs cost $1.79, a dozen medium eggs cost $1.99, and a dozen large eggs cost $2.09, how many dozens of each size egg did the cook purchase?

   Data: thekitchn.com

8. *Coffee.* A coffee shop on campus sells coffee in three sizes: a 12-oz small, a 16-oz medium, and a 20-oz large. A small coffee sells for $1.75, a medium coffee for $1.95, and a large coffee for $2.25. One morning, Brandie served 50 coffees for a total of $98.70. She made the coffee in 80-oz batches, and used exactly 10 of the batches during the morning. How many of each size did she sell?

9. *Cholesterol Levels.* Recent studies indicate that a child's intake of cholesterol should be no more than 300 mg per day. By eating 1 egg, 1 cupcake, and 1 slice of pizza, a child consumes 302 mg of cholesterol. If the child eats 2 cupcakes and 3 slices of pizza, he or she takes in 65 mg of cholesterol. By eating 2 eggs and 1 cupcake, a child consumes 567 mg of cholesterol. How much cholesterol is in each item?

10. *Book Sale.* Katie, Rachel, and Logan went together to a library book sale. Katie bought 22 children's books, 10 paperbacks, and 5 hardbacks for a total of $63.50. Rachel bought 12 paperbacks and 15 hardbacks for a total of $52.50. Logan bought 8 children's books and 6 hardbacks for a total of $29.00. How much did each type of book cost?

11. *Automobile Pricing.* A recent basic model of a particular automobile had a price of $14,685. The basic model with the added features of automatic transmission and power door locks was $16,070. The basic model with air conditioning (AC) and power door locks was $15,580. The basic model with AC and automatic transmission was $15,925. What was the individual cost of each of the three options?

12. *Computer Pricing.* Lindsay plans to buy a new desktop computer for gaming. The base price of the computer is $480. If she upgrades the processor and the memory, the price of the computer is $745. If she upgrades the memory and the graphics card, the price of the computer is $690. If she upgrades the processor and the graphics card, the price of the computer is $805. What is the price of each upgrade?

13. *Pet Ownership.* In 2017, a total of 204 million birds, cats, and dogs lived as pets in U.S. homes. The number of cats was 4 million more than the number of dogs, and the number of birds was 3 million less than one-eighth of the sum of the numbers of cats and dogs. Find the number of each type of pet living in U.S. homes.

    Data: American Pet Productions Association

14. *Glaciers.* The western U.S. states of Washington, California, and Wyoming contain a total of 6366 glaciers. Washington has 164 fewer glaciers than do California and Wyoming combined. California has 262 more glaciers than one-third of the sum of the number of glaciers in Washington and Wyoming. Find the number of glaciers in each state.

    Data: glaciers.research.pdx.edu

15. *Nutrition.* A dietician in a hospital prepares meals under the guidance of a physician. Suppose that for a particular patient a physician prescribes a meal to have 800 calories, 55 g of protein, and 220 mg of vitamin C. The dietician prepares a meal of roast beef, baked potato, and broccoli according to the data in the following table. How many servings of each food are needed in order to satisfy the doctor's orders?

16. *Nutrition.* Repeat Exercise 15 but replace the broccoli with asparagus, for which one 180-g serving contains 50 calories, 5 g of protein, and 44 mg of vitamin C. Which meal would you prefer eating?

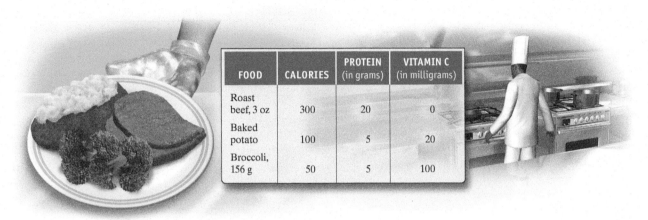

FOOD	CALORIES	PROTEIN (in grams)	VITAMIN C (in milligrams)
Roast beef, 3 oz	300	20	0
Baked potato	100	5	20
Broccoli, 156 g	50	5	100

17. *Investments.* A business class divided an imaginary investment of $80,000 among three mutual funds. The first fund grew by 2%, the second by 6%, and the third by 3%. Total earnings were $2250. The earnings from the first fund were $150 more than the earnings from the third. How much was invested in each fund?

18. *Student Loans.* Terrence owes $32,000 in student loans. The interest rate on his Perkins loan is 5%, the rate on his Stafford loan is 4%, and the rate on his bank loan is 7%. Interest for one year totaled $1500. The interest for one year from the Perkins loan is $220 more than the interest from the bank loan. What is the amount of each loan?

19. *Golf.* On an 18-hole golf course, there are par-3 holes, par-4 holes, and par-5 holes. A golfer who shoots par on every hole has a total of 70. There are twice as many par-4 holes as there are par-5 holes. How many of each type of hole are there on the golf course?

20. *Basketball Scoring.* The New York Knicks once scored a total of 92 points on a combination of 2-point field goals, 3-point field goals, and 1-point foul shots. Altogether, the Knicks made 50 baskets and 19 more 2-pointers than foul shots. How many shots of each kind were made?

21. *Lens Production.* When Sight-Rite's three polishing machines, A, B, and C, are all working, 5700 lenses can be polished in one week. When only A and B are working, 3400 lenses can be polished in one week. When only B and C are working, 4200 lenses can be polished in one week. How many lenses can be polished in a week by each machine alone?

22. *Telemarketing.* Steve, Teri, and Isaiah can process 740 telephone orders per day. Steve and Teri together can process 470 orders, while Teri and Isaiah together can process 520 orders per day. How many orders can each person process alone?

Skill Maintenance

Graph each function. [7.1c]

23. $f(x) = 2x - 3$

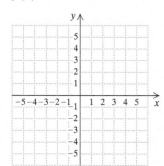

24. $g(x) = |x + 1|$

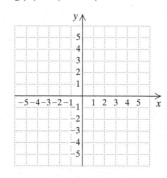

25. $h(x) = x^2 - 2$

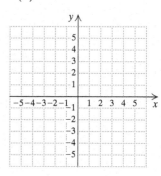

Determine whether each of the following is the graph of a function. [7.1d]

26.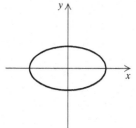

27.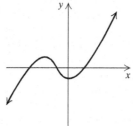

28.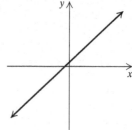

Synthesis

29. Find the sum of the angle measures at the tips of the star in this figure.

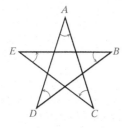

30. *Sharing Raffle Tickets.* Hal gives Tom as many raffle tickets as Tom has and Gary as many as Gary has. In like manner, Tom then gives Hal and Gary as many tickets as each then has. Similarly, Gary gives Hal and Tom as many tickets as each then has. If each finally has 40 tickets, with how many tickets does Tom begin?

31. *Digits.* Find a three-digit positive integer such that the sum of all three digits is 14, the tens digit is 2 more than the ones digit, and if the digits are reversed, the number is unchanged.

32. *Ages.* Tammy's age is the sum of the ages of Carmen and Dennis. Carmen's age is 2 more than the sum of the ages of Dennis and Mark. Dennis's age is four times Mark's age. The sum of all four ages is 42. How old is Tammy?

33. *Video-Game Ratings.* The Entertainment Software Rating Board assigned one of the following ratings to 1476 video games produced in 2016: E (everyone), E10+ (everyone ages 10 and older), T (teens), and M (mature audiences). More games were rated E than any other rating, with only 74 fewer E games than the total number of the other games. The sum of the number of games rated E or E10+ was 45 more than twice the sum of the number of games rated T or M. The number of games rated T was 15 less than twice the number of games rated M. How many games were assigned each rating?

Data: Entertainment Software Association

CHAPTER 8 Summary and Review

Vocabulary Reinforcement

Complete each statement with the correct term from the column on the right. Some of the choices may be used more than once and some may not be used at all.

1. A solution of a system of two equations in two variables is an ordered _____ that makes both equations true. [8.1a]

2. A(n) _____ system of equations has at least one solution. [8.1a]

3. The substitution method is a(n) _____ method for solving systems of equations. [8.2a]

4. A solution of a system of three equations in three variables is an ordered _____ that makes all three equations true. [8.5a]

5. If, for a system of two equations in two variables, the graphs of the equations are different lines, then the equations are _____. [8.1a]

algebraic
graphical
independent
dependent
consistent
inconsistent
pair
triple

Concept Reinforcement

Determine whether each statement is true or false.

_____ 1. A system of equations with infinitely many solutions is inconsistent. [8.1a]

_____ 2. It is not possible for the equations in an inconsistent system of two equations to be dependent. [8.1a]

_____ 3. When $(0, b)$ is a solution of each equation in a system of two equations, the graphs of the two equations have the same y-intercept. [8.1a]

_____ 4. The system of equations $x = 4$ and $y = -4$ is inconsistent. [8.1a]

Study Guide

Objective 8.1a Solve a system of two linear equations or two functions by graphing and determine whether a system is consistent or inconsistent and whether the equations in a system are dependent or independent.

Example Solve this system of equations graphically. Then classify the system as consistent or inconsistent and the equations as dependent or independent.

$$x - y = 3,$$
$$y = 2x - 4$$

We graph the equations. The point of intersection appears to be $(1, -2)$. This checks in both equations, so it is the solution. The system has one solution, so it is consistent and the equations are independent.

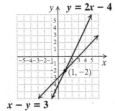

Practice Exercise

1. Solve this system of equations graphically. Then classify the system as consistent or inconsistent and the equations as dependent or independent.

$$x + 3y = 1,$$
$$x + y = 3$$

Objective 8.2a Solve systems of equations in two variables by the substitution method.

Example Solve the system
$$x - 2y = 1, \quad (1)$$
$$2x - 3y = 3. \quad (2)$$

We solve equation (1) for x, since the coefficient of x is 1 in that equation:
$$x - 2y = 1$$
$$x = 2y + 1. \quad (3)$$

Next, we substitute for x in equation (2) and solve for y:
$$2x - 3y = 3$$
$$2(2y + 1) - 3y = 3$$
$$4y + 2 - 3y = 3$$
$$y + 2 = 3$$
$$y = 1.$$

Then we substitute 1 for y in equation (1), (2), or (3) and find x. We choose equation (3) since it is already solved for x:
$$x = 2y + 1 = 2 \cdot 1 + 1 = 2 + 1 = 3$$

Check:

$x - 2y = 1$	$2x - 3y = 3$
$3 - 2 \cdot 1 \; ? \; 1$	$2 \cdot 3 - 3 \cdot 1 \; ? \; 3$
$3 - 2$	$6 - 3$
1 | TRUE	3 | TRUE

The ordered pair $(3, 1)$ checks in both equations, so it is the solution of the system of equations.

Practice Exercise

2. Solve the system
$$2x + y = 2,$$
$$3x + 2y = 5$$
using the substitution method.

Objective 8.3a Solve systems of equations in two variables by the elimination method.

Example Solve the system
$$2a + 3b = -1, \quad (1)$$
$$3a + 2b = 6. \quad (2)$$

We could eliminate either a or b. In this case, we decide to eliminate the a-terms. We multiply equation (1) by 3 and equation (2) by -2 and then add and solve for b:
$$6a + 9b = -3$$
$$\underline{-6a - 4b = -12}$$
$$5b = -15$$
$$b = -3.$$

Next, we substitute -3 for b in either of the original equations:
$$2a + 3b = -1 \quad (1)$$
$$2a + 3(-3) = -1$$
$$2a - 9 = -1$$
$$2a = 8$$
$$a = 4.$$

The ordered pair $(4, -3)$ checks in both equations, so it is a solution of the system of equations.

Practice Exercise

3. Solve the system
$$2x + 3y = 5,$$
$$3x + 4y = 6$$
using the elimination method.

642 CHAPTER 8 Systems of Equations

Objective 8.4a Solve applied problems involving total value and mixture using systems of two equations.

Example To start a small business, Michael took two loans totaling $18,000. One of the loans was at 7% interest and the other at 8%. After one year, Michael owed $1365 in interest. What was the amount of each loan?

1. **Familiarize.** We let x and y represent the amounts of the two loans. Next, we organize the information in a table and use the simple interest formula, $I = Prt$.

	LOAN 1	LOAN 2	TOTAL
PRINCIPAL	x	y	$18,000
RATE OF INTEREST	7%	8%	
TIME	1 year	1 year	
INTEREST	7%x, or 0.07x	8%y, or 0.08y	$1365

2. **Translate.** The total amount of the loans is found in the first row of the table. This gives us one equation:

   $x + y = 18,000$.

   From the last row of the table, we see that the interest totals $1365. This gives us a second equation:

   $0.07x + 0.08y = 1365$, or $7x + 8y = 136,500$.

3. **Solve.** We solve the resulting system of equations:

   $x + y = 18,000$,   (1)
   $7x + 8y = 136,500$.   (2)

   We multiply by -7 on both sides of equation (1) and add:

   $-7x - 7y = -126,000$
   $\underline{7x + 8y = 136,500}$ (2)
   $y = 10,500.$ Adding

   Then

   $x + 10,500 = 18,000$ Substituting 10,500 for y in equation (1)

   $x = 7500.$ Solving for x

   We find that $x = 7500$ and $y = 10,500$.

4. **Check.** The sum is $7500 + $10,500, or $18,000. The interest from $7500 at 7% for one year is 7%($7500), or $525. The interest from $10,500 at 8% for one year is 8%($10,500), or $840. The total amount of interest is $525 + $840, or $1365. The numbers check in the problem.

5. **State.** Michael took loans of $7500 at 7% interest and $10,500 at 8% interest.

Practice Exercise

4. Jaretta made two investments totaling $23,000. In one year, these investments yielded $1237 in simple interest. Part of the money was invested at 6% and the rest at 5%. How much was invested at each rate?

Objective 8.5a Solve systems of three equations in three variables.

Example Solve:
$$x - y - z = -2, \quad (1)$$
$$2x + 3y + z = 3, \quad (2)$$
$$5x - 2y - 2z = -1. \quad (3)$$

The equations are in standard form and do not contain decimals or fractions. We choose to eliminate z since the z-terms in equations (1) and (2) are opposites.

First, we add these two equations:
$$x - y - z = -2$$
$$\underline{2x + 3y + z = 3}$$
$$3x + 2y = 1. \quad (4)$$

Next, we multiply equation (2) by 2 and add it to equation (3) to eliminate z from another pair of equations:
$$4x + 6y + 2z = 6$$
$$\underline{5x - 2y - 2z = -1}$$
$$9x + 4y = 5. \quad (5)$$

Now we solve the system consisting of equations (4) and (5). We multiply equation (4) by -2 and add:
$$-6x - 4y = -2$$
$$\underline{9x + 4y = 5}$$
$$3x = 3$$
$$x = 1.$$

Then we use either equation (4) or (5) to find y:
$$3x + 2y = 1 \quad (4)$$
$$3 \cdot 1 + 2y = 1$$
$$3 + 2y = 1$$
$$2y = -2$$
$$y = -1.$$

Finally, we use one of the original equations to find z:
$$2x + 3y + z = 3 \quad (2)$$
$$2 \cdot 1 + 3(-1) + z = 3$$
$$-1 + z = 3$$
$$z = 4.$$

Check:

$$\begin{array}{c|c}
x - y - z = -2 \\ \hline
1 - (-1) - 4 \; ? \; -2 \\
1 + 1 - 4 \\
-2 \; \bigg| \; \text{TRUE}
\end{array}
\qquad
\begin{array}{c|c}
2x + 3y + z = 3 \\ \hline
2 \cdot 1 + 3(-1) + 4 \; ? \; 3 \\
2 - 3 + 4 \\
3 \; \bigg| \; \text{TRUE}
\end{array}$$

$$\begin{array}{c|c}
5x - 2y - 2z = -1 \\ \hline
5 \cdot 1 - 2(-1) - 2 \cdot 4 \; ? \; -1 \\
5 + 2 - 8 \\
-1 \; \bigg| \; \text{TRUE}
\end{array}$$

The ordered triple $(1, -1, 4)$ checks in all three equations, so it is the solution of the system of equations.

Practice Exercise

5. Solve:
$$x - y + z = 9,$$
$$2x + y + 2z = 3,$$
$$4x + 2y - 3z = -1.$$

Review Exercises

Solve graphically. Then classify the system as consistent or inconsistent and the equations as dependent or independent. [8.1a]

1. $4x - y = -9,$
   $x - y = -3$

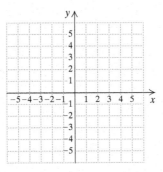

2. $15x + 10y = -20,$
   $3x + 2y = -4$

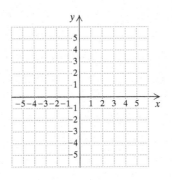

3. $y - 2x = 4,$
   $y - 2x = 5$

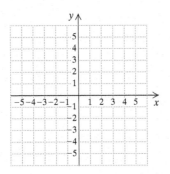

Solve using the substitution method. [8.2a]

4. $2x - 3y = 5,$
   $x = 4y + 5$

5. $y = x + 2,$
   $y - x = 8$

6. $7x - 4y = 6,$
   $y - 3x = -2$

Solve by the elimination method. [8.3a]

7. $x + 3y = -3,$
   $2x - 3y = 21$

8. $3x - 5y = -4,$
   $5x - 3y = 4$

9. $\dfrac{1}{3}x + \dfrac{2}{9}y = 1,$
   $\dfrac{3}{2}x + \dfrac{1}{2}y = 6$

10. $1.5x - 3 = -2y,$
    $3x + 4y = 6$

11. *Air Travel.* An airplane flew for 3 hr with a 30-mph tailwind. The return flight against the same wind took 4.5 hr. Find the speed of the plane in still air. [8.4a]

12. *Retail Sales.* Paint Town sold 45 paintbrushes, one kind at $8.50 each and another at $9.75 each. In all, $398.75 was taken in for the brushes. How many of each kind were sold? [8.4a]

13. *Orange Drink Mixtures.* "Orange Thirst" is 15% orange juice and "Quencho" is 5% orange juice. How many liters of each should be combined in order to get 10 L of a mixture that is 10% orange juice? [8.4a]

14. *Train Travel.* A train leaves Watsonville at noon traveling north at 44 mph. One hour later, another train, going 52 mph, travels north on a parallel track. How many hours will the second train travel before it overtakes the first train? [8.4b]

Solve. [8.5a]

15. $x + 2y + z = 10$,
$2x - y + z = 8$,
$3x + y + 4z = 2$

16. $3x + 2y + z = 1$,
$2x - y - 3z = 1$,
$-x + 3y + 2z = 6$

17. $2x - 5y - 2z = -4$,
$7x + 2y - 5z = -6$,
$-2x + 3y + 2z = 4$

18. $x + y + 2z = 1$,
$x - y + z = 1$,
$x + 2y + z = 2$

19. *Triangle Measure.* In triangle ABC, the measure of angle A is four times the measure of angle C, and the measure of angle B is 45° more than the measure of angle C. What are the measures of the angles of the triangle? [8.6a]

20. *Popcorn.* Paul paid a total of $49 for 1 bag of caramel nut crunch popcorn, 1 bag of plain popcorn, and 1 bag of mocha choco latte popcorn. The price of the caramel nut crunch popcorn was six times the price of the plain popcorn and $16 more than the mocha choco latte popcorn. What was the price of each type of popcorn? [8.6a]

21. Solve using the elimination method:
$x - y = -9$,
$y - 2x = 9$.
The first coordinate of the solution is which of the following? [8.3a]
A. 9
B. −9
C. 0
D. $\frac{9}{2}$

22. The sum of two numbers is −2. The sum of twice one number and the other is 4. One number is which of the following? [8.3b]
A. −6
B. 2
C. 6
D. 8

23. *Motorcycle Travel.* Sally and Elliot travel on motorcycles toward each other from Chicago and Indianapolis, which are about 350 km apart, and they are biking at rates of 110 km/h and 90 km/h. They started at the same time. In how many hours will they meet? [8.4b]
A. 1.75 hr
B. 3.9 hr
C. 3.2 hr
D. 17.5 hr

Synthesis

24. Solve graphically: [7.1c], [8.1a]
$y = x + 2$,
$y = x^2 + 2$.

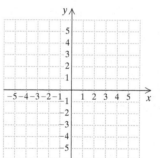

Understanding Through Discussion and Writing

1. Write a problem for a classmate to solve. Design the problem so the answer is "The florist sold 14 hanging baskets and 9 flats of petunias." [8.4a]

2. Exercise 21 in Exercise Set 8.6 can be solved mentally after a careful reading of the problem. Explain how this can be done. [8.6a]

3. *Ticket Revenue.* A pops-concert audience of 100 people consists of adults, senior citizens, and children. The ticket prices are $10 each for adults, $3 each for senior citizens, and $0.50 each for children. The total amount of money taken in is $100. How many adults, senior citizens, and children are in attendance? Does there seem to be some information missing? Do some careful reasoning and explain. [8.6a]

CHAPTER 8 Test

Solve graphically. Then classify the system as consistent or inconsistent and the equations as dependent or independent.

1. $y = 3x + 7,$
   $3x + 2y = -4$

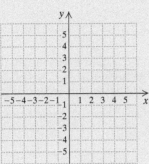

2. $y = 3x + 4,$
   $y = 3x - 2$

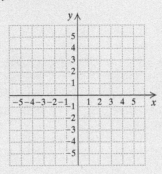

3. $y - 3x = 6,$
   $6x - 2y = -12$

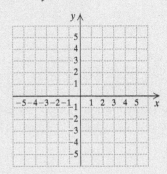

Solve using the substitution method.

4. $4x + 3y = -1,$
   $y = 2x - 7$

5. $x = 3y + 2,$
   $2x - 6y = 4$

6. $x + 2y = 6,$
   $2x + 3y = 7$

7. $t = 2 - r,$
   $3r - 2t = 36$

Solve using the elimination method.

8. $2x + 5y = 3,$
   $-2x + 3y = 5$

9. $x + y = -2,$
   $4x - 6y = -3$

10. $\frac{2}{3}x - \frac{4}{5}y = 1,$
    $\frac{1}{3}x - \frac{2}{5}y = 2$

11. $0.3a - 0.4b = 11,$
    $0.7a + 1.2b = -17$

Solve.

12. *Tennis Court.* The perimeter of a standard tennis court used for playing doubles is 288 ft. The width of the court is 42 ft less than the length. Find the length and the width.

13. *Air Travel.* An airplane flew for 5 hr with a 20-km/h tailwind and returned in 7 hr against the same wind. Find the speed of the plane in still air.

14. *Chicken Dinners.* High Flyin' Wings charges $12 for a bucket of chicken wings and $7 for a chicken dinner. After filling 28 orders for buckets and dinners during a football game, the waiters had collected $281. How many buckets and how many dinners did they sell?

15. *Mixing Solutions.* A chemist has one solution that is 20% salt and a second solution that is 45% salt. How many liters of each should be used in order to get 20 L of a solution that is 30% salt?

16. Solve:
$$6x + 2y - 4z = 15,$$
$$-3x - 4y + 2z = -6,$$
$$4x - 6y + 3z = 8.$$

17. *Repair Rates.* An electrician, a carpenter, and a plumber are hired to work on a house. The electrician earns $21 per hour, the carpenter $19.50 per hour, and the plumber $24 per hour. The first day on the job, they worked a total of 21.5 hr and earned a total of $469.50. If the plumber worked 2 hr more than the carpenter did, how many hours did the electrician work?

18. A business class divided an imaginary $30,000 investment among three funds. The first fund grew 2%, the second grew 3%, and the third grew 5%. Total earnings were $990. The earnings from the third fund were $280 more than the earnings from the first. How much was invested at 5%?
  A. $9000
  B. $10,000
  C. $11,000
  D. $12,000

Synthesis

19. The graph of the function $f(x) = mx + b$ contains the points $(-1, 3)$ and $(-2, -4)$. Find m and b.

CHAPTERS 1–8 Cumulative Review

Perform the indicated operations and simplify.

1. $(3x^4 - 2y^5)(3x^4 + 2y^5)$ 2. $(x^2 + 4)^2$

3. $\left(2x + \dfrac{1}{4}\right)\left(4x - \dfrac{1}{2}\right)$ 4. $\dfrac{x}{2x-1} - \dfrac{3x+2}{1-2x}$

5. $(3x^2 - 2x^3) - (x^3 - 2x^2 + 5) + (3x^2 - 5x + 5)$

6. $\dfrac{2x+2}{3x-9} \cdot \dfrac{x^2 - 8x + 15}{x^2 - 1}$

7. $\dfrac{2x^2 - 2}{2x^2 + 7x + 3} \div \dfrac{4x - 4}{2x^2 - 5x - 3}$

8. $(3x^3 - 2x^2 + x - 5) \div (x - 2)$

Factor completely.

9. $3 - 12x^8$ 10. $12t - 4t^2 - 48t^4$

11. $6x^2 - 28x + 16$ 12. $4x^3 + 4x^2 - x - 1$

13. $16x^4 - 56x^2 + 49$ 14. $x^2 + 3x - 180$

15. Find the slope and the y-intercept of $5y - 4x = 20$.

16. Find an equation of the line with slope -3 and containing the point $(5, 2)$.

17. Find an equation of the line parallel to $3x - 9y = 2$ and containing the point $(-6, 2)$.

18. Determine whether the graphs of the given lines are parallel, perpendicular, or neither.
$$x - 2y = 4,$$
$$4x + 2y = 1$$

Solve.

19. $x^2 = -17x$ 20. $\dfrac{1}{4}x + \dfrac{2}{3}x = \dfrac{2}{3} - \dfrac{3}{4}x$

21. $\dfrac{1}{x} + \dfrac{2}{3} = \dfrac{1}{4}$ 22. $x^2 - 30 = x$

23. $-4(x + 5) \geq 2(x + 5) - 3$

24. $\dfrac{x}{x-1} - \dfrac{x}{x+1} = \dfrac{1}{2x - 2}$

25. Solve $4A = pr + pq$ for p.

Solve.

26. $3x + 4y = 4,$
    $x = 2y + 2$

27. $3x + y = 2,$
    $6x - y = 7$

28. $4x + 3y = 5,$
    $3x + 2y = 3$

29. $x - y + z = 1,$
    $2x + y + z = 3,$
    $x + y - 2z = 4$

Graph on a plane.

30. $3y = 9$

31. $f(x) = -\dfrac{1}{2}x - 3$

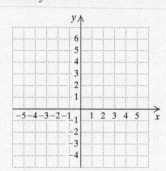

32. $3x - 1 = y$

33. $3x + 5y = 15$

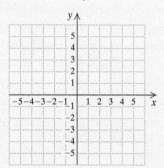 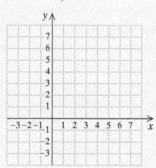

34. For the function f whose graph is shown below, determine **(a)** the domain, **(b)** the range, **(c)** $f(-3)$, and **(d)** any input for which $f(x) = 5$.

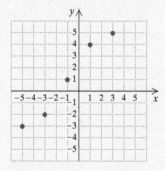

35. Find the domain of the function given by
$$f(x) = \dfrac{7}{2x - 1}.$$

36. Given $g(x) = 1 - 2x^2$, find $g(-1)$, $g(0)$, and $g(3)$.

Solve.

37. *Mixing Solutions.* A technician wants to mix one solution that is 15% alcohol with another solution that is 25% alcohol in order to get 30 L of a solution that is 18% alcohol. How much of each solution should be used?

38. *Utility Cost.* One month Ladi and Bo spent $680 for electricity, rent, and telephone. The electric bill was one-fourth of the rent and the rent was $400 more than the phone bill. How much was the electric bill?

39. *Quality Control.* A sample of 150 resistors contained 12 defective resistors. How many defective resistors would you expect to find in a sample of 250 resistors?

40. *Rectangle Dimensions.* The length of a rectangle is 3 m greater than the width. The area of the rectangle is 180 m². Find the length and the width.

41. *Apparent Size.* The apparent size A of an object varies inversely as the distance d of the object from the eye. You are sitting at a concert 100 ft from the stage. The musicians appear to be 4 ft tall. How tall would they appear to be if you were sitting 1000 ft away in the lawn seats?

42. *Angles of a Triangle.* The second angle of a triangle is twice as large as the first. The third angle is 48° less than the sum of the other two angles. Find the measures of the angles.

Synthesis

43. *Radio Advertising.* An automotive dealer discovers that when $1000 is spent on radio advertising, weekly sales increase by $101,000. When $1250 is spent on radio advertising, weekly sales increase by $126,000. Assuming that sales increase according to a linear function, by what amount would sales increase when $1500 is spent on radio advertising?

44. Given that $f(x) = mx + b$ and that $f(5) = -3$ when $f(-4) = 2$, find m and b.

CHAPTER 9

9.1 Sets, Inequalities, and Interval Notation

Translating for Success

9.2 Intersections, Unions, and Compound Inequalities

Mid-Chapter Review

9.3 Absolute-Value Equations and Inequalities

9.4 Systems of Inequalities in Two Variables

Visualizing for Success

Summary and Review

Test

Cumulative Review

More on Inequalities

The scope of the job of web developer is rapidly expanding, from desktop and laptop computers to mobile devices such as smartphones and tablets. The top six U.S. cities for web developers to live and work in are listed in the table below. Rankings are based on average salary, cost of living, and demand for developers.

Best Cities for Web Developers

Rank	City	Average Salary	Number of Jobs
1	San Francisco, CA	$95,600	3380
2	San Jose, CA	106,580	2590
3	Seattle, WA	87,520	3850
4	Washington, DC	84,990	4580
5	Bethesda, MD	79,280	1220
6	Boston, MA	79,140	2790

DATA: Valuepenguin.com/2015/07/best-cities-web-developers

We will estimate the years after 2014 in which the number of web developers in the United States will be greater than 200,000 in Example 15 of Section 9.1.

651

STUDYING FOR SUCCESS *Make the Most of Your Homework*
- ☐ Double-check that you have copied each exercise from the book correctly.
- ☐ Check that your answer is reasonable.
- ☐ Include correct units with answers.

9.1 Sets, Inequalities, and Interval Notation

OBJECTIVES

a. Determine whether a given number is a solution of an inequality.

b. Write interval notation for the solution set or the graph of an inequality.

c. Solve an inequality using the addition principle and the multiplication principle and then graph the inequality.

d. Solve applied problems by translating to inequalities.

a INEQUALITIES

We begin this chapter with a review of solving inequalities. Here we will write solution sets using both set-builder notation and *interval notation*.

> **INEQUALITY**
> An **inequality** is a sentence containing $<, >, \leq, \geq,$ or $\neq$.

Some examples of inequalities are

$$-2 < a, \quad x > 4, \quad x + 3 \leq 6,$$
$$6 - 7y \geq 10y - 4, \quad \text{and} \quad 5x \neq 10.$$

> **SOLUTION OF AN INEQUALITY**
> Any replacement or value for the variable that makes an inequality true is called a **solution** of the inequality. The set of all solutions is called the **solution set**. When all the solutions of an inequality have been found, we say that we have **solved** the inequality.

EXAMPLES Determine whether the given number is a solution of the inequality.

1. $x + 3 < 6;\ 5$

   We substitute 5 for x and get $5 + 3 < 6$, or $8 < 6$, a *false* sentence. Therefore, 5 is not a solution.

2. $2x - 3 > -3;\ 1$

   We substitute 1 for x and get $2(1) - 3 > -3$, or $-1 > -3$, a *true* sentence. Therefore, 1 is a solution.

3. $4x - 1 \leq 3x + 2;\ -3$

   We substitute -3 for x and get $4(-3) - 1 \leq 3(-3) + 2$, or $-13 \leq -7$, a *true* sentence. Therefore, -3 is a solution.

◀ Do Exercises 1–3.

Determine whether the given number is a solution of the inequality.

1. $3 - x < 2;\ 8$

2. $3x + 2 > -1;\ -2$

3. $3x + 2 \leq 4x - 3;\ 5$

Answers
1. Yes 2. No 3. Yes

b INEQUALITIES AND INTERVAL NOTATION

SKILL REVIEW

Graph an inequality on the number line. [2.7b]

Graph each inequality.

1. $x > -2$
2. $x \leq 1$

Answers:

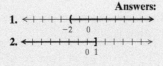

MyLab Math
ANIMATION

The solutions of an inequality can be represented graphically. The solution set can be written using both *set-builder notation* and *interval notation*.

The **graph** of an inequality in one variable is a drawing on the number line that represents its solutions. For example, to graph $x < 4$, we note that the solutions are all real numbers less than 4. We shade all numbers less than 4 on the number line and indicate that 4 is not a solution by using a right parenthesis ")" at 4.

$\{x | x < 4\}$, or $(-\infty, 4)$

We write the solution set for $x < 4$ using **set-builder notation** as $\{x | x < 4\}$. This is read "The set of all x such that x is less than 4." We write the solution set for $x < 4$ using **interval notation** as $(-\infty, 4)$. Interval notation uses parentheses ( ) and brackets [ ] and closely parallels the graph of an inequality on the number line.

If a and b are real numbers such that $a < b$, we define the **open interval** (a, b) as the set of all numbers between but not including a and b—that is, the set of all x for which $a < x < b$. Thus,

$$(a, b) = \{x | a < x < b\}.$$

The points a and b are the **endpoints** of the interval. The parentheses indicate that the endpoints are *not* included in the graph.

The **closed interval** $[a, b]$ is defined as the set of all numbers x for which $a \leq x \leq b$. Thus,

$$[a, b] = \{x | a \leq x \leq b\}.$$

> **Caution!**
>
> The *interval* (a, b) uses the same notation as the *ordered pair* (a, b), which denotes a point in the plane, as we saw in Chapter 3. The context in which the notation appears usually makes the meaning clear.

The brackets indicate that the endpoints *are* included in the graph.*

The following **half-open intervals** include one endpoint and exclude the other:

$(a, b] = \{x | a < x \leq b\}.$ The graph excludes a and includes b.

$[a, b) = \{x | a \leq x < b\}.$ The graph includes a and excludes b.

---

*Some books use the representations ⊸⊸ and ⊢⊣ instead of, respectively, ⟨⟩ and [ ].

SECTION 9.1 Sets, Inequalities, and Interval Notation 653

Some intervals extend without bound in one or both directions. We use the symbols ∞, read "infinity," and $-\infty$, read "negative infinity," to name these intervals. The notation (a, ∞) represents the set of all numbers greater than a—that is,

$$(a, \infty) = \{x \mid x > a\}.$$

Similarly, the notation $(-\infty, a)$ represents the set of all numbers less than a—that is,

$$(-\infty, a) = \{x \mid x < a\}.$$

The notations $[a, \infty)$ and $(-\infty, a]$ are used when we want to include the endpoint a. The interval $(-\infty, \infty)$ names the set of all real numbers.

$$(-\infty, \infty) = \{x \mid x \text{ is a real number}\}$$

Interval notation is summarized in the following table.

Intervals: Notation and Graphs

INTERVAL NOTATION	SET NOTATION	GRAPH
(a, b)	$\{x \mid a < x < b\}$	
$[a, b]$	$\{x \mid a \leq x \leq b\}$	
$[a, b)$	$\{x \mid a \leq x < b\}$	
$(a, b]$	$\{x \mid a < x \leq b\}$	
(a, ∞)	$\{x \mid x > a\}$	
$[a, \infty)$	$\{x \mid x \geq a\}$	
$(-\infty, b)$	$\{x \mid x < b\}$	
$(-\infty, b]$	$\{x \mid x \leq b\}$	
$(-\infty, \infty)$	$\{x \mid x \text{ is a real number}\}$	

························· **Caution!** ·························

Whenever the symbol ∞ is included in interval notation, a right parenthesis ")" is used. Similarly, when $-\infty$ is included, a left parenthesis "(" is used.

EXAMPLES Write interval notation for the given set.

4. $\{x \mid -4 < x < 5\} = (-4, 5)$
5. $\{x \mid x \geq -2\} = [-2, \infty)$
6. $\{x \mid 7 > x \geq 1\} = \{x \mid 1 \leq x < 7\} = [1, 7)$

EXAMPLES Write interval notation for the given graph.

7.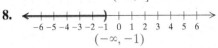
$(-2, 4]$

8.
$(-\infty, -1)$

Do Exercises 4–8.

c SOLVING INEQUALITIES

Two inequalities are **equivalent** if they have the same solution set. For example, the inequalities $x > 4$ and $4 < x$ are equivalent. Just as the addition principle for equations gives us equivalent equations, the addition principle for inequalities gives us equivalent inequalities.

> **THE ADDITION PRINCIPLE FOR INEQUALITIES**
>
> For any real numbers a, b, and c:
>
> $a < b$ is equivalent to $a + c < b + c$;
> $a > b$ is equivalent to $a + c > b + c$.
>
> Similar statements hold for $\leq$ and $\geq$.

Since subtracting c is the same as adding $-c$, there is no need for a separate subtraction principle.

EXAMPLE 9 Solve and graph: $x + 5 > 1$.

We have

$x + 5 > 1$
$x + 5 - 5 > 1 - 5$ Using the addition principle: adding -5 or subtracting 5
$x > -4$.

We used the addition principle to show that the inequalities $x + 5 > 1$ and $x > -4$ are equivalent. The solution set is $\{x | x > -4\}$ and consists of an infinite number of solutions. We cannot possibly check them all. Instead, we can perform a partial check by substituting one member of the solution set (here we use -1) into the original inequality:

$$\frac{x + 5 > 1}{-1 + 5 \ ? \ 1}$$
$$4 \ | \ \text{TRUE}$$

Since $4 > 1$ is true, we have a partial check. The solution set is $\{x | x > -4\}$, or $(-4, \infty)$. The graph is as follows:

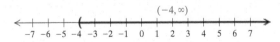

Do Exercises 9 and 10.

Write interval notation for the given set or graph.

4. $\{x | -4 \leq x < 5\}$

5. $\{x | x \leq -2\}$

6. $\{x | 6 \geq x > 2\}$

7.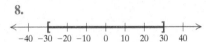

8.

Solve and graph.

9. $x + 6 > 9$

10. $x + 4 \leq 7$

Answers

4. $[-4, 5)$ 5. $(-\infty, -2]$ 6. $(2, 6]$
7. $[10, \infty)$ 8. $[-30, 30]$
9. $\{x | x > 3\}$, or $(3, \infty)$;

10. $\{x | x \leq 3\}$, or $(-\infty, 3]$;

EXAMPLE 10 Solve and graph: $4x - 1 \geq 5x - 2$.

We have

$$4x - 1 \geq 5x - 2$$
$$4x - 1 + 2 \geq 5x - 2 + 2 \quad \text{Adding 2}$$
$$4x + 1 \geq 5x \quad \text{Simplifying}$$
$$4x + 1 - 4x \geq 5x - 4x \quad \text{Subtracting } 4x$$
$$1 \geq x. \quad \text{Simplifying}$$

The inequalities $1 \geq x$ and $x \leq 1$ have the same meaning and the same solutions. The solution set is $\{x | 1 \geq x\}$ or, more commonly, $\{x | x \leq 1\}$. Using interval notation, we write that the solution set is $(-\infty, 1]$. The graph is as follows:

▸ Do Exercise 11.

11. Solve and graph:
$2x - 3 \geq 3x - 1.$

$2x - 3 \geq 3x - 1$
$2x - 3 - \boxed{} \geq 3x - 1 - 2x$
$\boxed{} \geq \boxed{} - 1$
$-3 + \boxed{} \geq x - 1 + 1$
$\boxed{} \geq x$, or
$x \leq -2$

The solution set is
$\{x | x \leq -2\}$, or $(-\infty, \boxed{}]$.

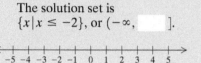

The multiplication principle for inequalities differs from the multiplication principle for equations. Consider the true inequality

$$-4 < 9.$$

If we multiply both numbers by 2, we get another true inequality:

$$-4(2) < 9(2), \quad \text{or} \quad -8 < 18. \quad \text{True}$$

If we multiply both numbers by -3, we get a false inequality:

$$-4(-3) < 9(-3), \quad \text{or} \quad 12 < -27. \quad \text{False}$$

However, if we now *reverse* the inequality symbol above, we get a true inequality:

$$12 > -27. \quad \text{True}$$

> **THE MULTIPLICATION PRINCIPLE FOR INEQUALITIES**
>
> For any real numbers a and b, and any *positive* number c:
>
> $a < b$ is equivalent to $ac < bc$;
> $a > b$ is equivalent to $ac > bc$.
>
> For any real numbers a and b, and any *negative* number c:
>
> $a < b$ is equivalent to $ac > bc$;
> $a > b$ is equivalent to $ac < bc$.
>
> Similar statements hold for $\leq$ and $\geq$.

Since division by c is the same as multiplication by $1/c$, there is no need for a separate division principle.

> The multiplication principle tells us that when we multiply or divide on both sides of an inequality by a negative number, we must reverse the inequality symbol to obtain an equivalent inequality.

Answer
11. $\{x | x \leq -2\}$, or $(-\infty, -2]$;

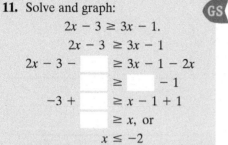

Guided Solution:
11. $2x, -3, x, 1, -2, -2$

EXAMPLE 11 Solve and graph: $3y < \frac{3}{4}$.

We have

$$3y < \frac{3}{4}$$

$$\frac{1}{3} \cdot 3y < \frac{1}{3} \cdot \frac{3}{4} \quad \text{Multiplying by } \tfrac{1}{3}. \text{ Since } \tfrac{1}{3} > 0,\text{ the symbol stays the same.}$$

$$y < \frac{1}{4}. \quad \text{Simplifying}$$

Any number less than $\frac{1}{4}$ is a solution. The solution set is $\{y | y < \frac{1}{4}\}$, or $(-\infty, \frac{1}{4})$. The graph is as follows:

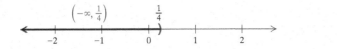

EXAMPLE 12 Solve and graph: $-5x \geq -80$.

We have

$$-5x \geq -80$$

$$\frac{-5x}{-5} \leq \frac{-80}{-5} \quad \text{Dividing by } -5. \text{ Since } -5 < 0,\text{ the inequality symbol must be reversed.}$$

$$x \leq 16.$$

The solution set is $\{x | x \leq 16\}$, or $(-\infty, 16]$. The graph is as follows:

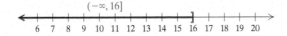

Do Exercises 12–14.

We use the addition and multiplication principles together in solving inequalities in much the same way as in solving equations.

EXAMPLE 13 Solve: $16 - 7y \geq 10y - 4$.

We have

$$16 - 7y \geq 10y - 4$$
$$-16 + 16 - 7y \geq -16 + 10y - 4 \quad \text{Adding } -16$$
$$-7y \geq 10y - 20 \quad \text{Collecting like terms}$$
$$-10y + (-7y) \geq -10y + 10y - 20 \quad \text{Adding } -10y$$
$$-17y \geq -20 \quad \text{Collecting like terms}$$
$$\frac{-17y}{-17} \leq \frac{-20}{-17} \quad \text{Dividing by } -17. \text{ The symbol must be reversed.}$$
$$y \leq \frac{20}{17}. \quad \text{Simplifying}$$

The solution set is $\{y | y \leq \frac{20}{17}\}$ or $(-\infty, \frac{20}{17}]$.

We can avoid multiplying or dividing by a negative number by using the addition principle in a different way. Let's rework Example 13 by adding $7y$ instead of $-10y$.

Solve and graph.

12. $5y \leq \frac{3}{2}$

13. $-2y > 10$

14. $-\frac{1}{3}x \leq -4$

Answers

12. $\{y | y \leq \frac{3}{10}\}$, or $(-\infty, \frac{3}{10}]$

13. $\{y | y < -5\}$, or $(-\infty, -5)$

14. $\{x | x \geq 12\}$, or $[12, \infty)$

$$16 - 7y \geq 10y - 4$$
$$16 - 7y + 7y \geq 10y - 4 + 7y \quad \text{Adding } 7y. \text{ This makes the coefficient of the } y\text{-term positive.}$$
$$16 \geq 17y - 4 \quad \text{Collecting like terms}$$
$$16 + 4 \geq 17y - 4 + 4 \quad \text{Adding 4}$$
$$20 \geq 17y \quad \text{Collecting like terms}$$
$$\frac{20}{17} \geq \frac{17y}{17} \quad \text{Dividing by 17. The symbol stays the same.}$$
$$\frac{20}{17} \geq y, \text{ or } y \leq \frac{20}{17}$$

EXAMPLE 14 Solve: $-3(x + 8) - 5x > 4x - 9$.

$$-3(x + 8) - 5x > 4x - 9$$
$$-3x - 24 - 5x > 4x - 9 \quad \text{Using the distributive law}$$
$$-24 - 8x > 4x - 9 \quad \text{Collecting like terms}$$
$$-24 - 8x + 8x > 4x - 9 + 8x \quad \text{Adding } 8x$$
$$-24 > 12x - 9 \quad \text{Collecting like terms}$$
$$-24 + 9 > 12x - 9 + 9 \quad \text{Adding 9}$$
$$-15 > 12x$$

Dividing by 12. The symbol stays the same.

$$\frac{-15}{12} > \frac{12x}{12}$$
$$-\frac{5}{4} > x.$$

The solution set is $\{x | -\frac{5}{4} > x\}$, or $\{x | x < -\frac{5}{4}\}$, or $\left(-\infty, -\frac{5}{4}\right)$.

◀ Do Exercises 15–17.

Solve.

15. $6 - 5y \geq 7$
$$6 - 5y - 6 \geq 7 - \boxed{}$$
$$\boxed{} \geq 1$$
$$\frac{-5y}{-5} \boxed{} \frac{1}{\boxed{}}$$
$$\boxed{} \leq -\frac{1}{5}$$

The solution set is

$\left\{y | y \leq -\frac{1}{5}\right\}$, or $\left(\boxed{}, -\frac{1}{5}\right]$.

16. $3x + 5x < 4$

17. $17 - 5(y - 2) \leq 45y + 8(2y - 3) - 39y$

d APPLICATIONS AND PROBLEM SOLVING

Many problem-solving and applied situations translate to inequalities.

IMPORTANT WORDS	SAMPLE SENTENCE	TRANSLATION
is at least	Max is at least 5 years old.	$m \geq 5$
is at most	At most 6 people could fit in the elevator.	$n \leq 6$
cannot exceed	Total weight in the elevator cannot exceed 2000 pounds.	$w \leq 2000$
must exceed	The speed must exceed 15 mph.	$s > 15$
is between	Heather's income is between $23,000 and $35,000.	$23{,}000 < h < 35{,}000$
no more than	Bing weighs no more than 90 pounds.	$w \leq 90$
no less than	Saul would accept no less than $4000 for the piano.	$p \geq 4000$

Answers

15. $\left\{y | y \leq -\frac{1}{5}\right\}$, or $\left(-\infty, -\frac{1}{5}\right]$
16. $\left\{x | x < \frac{1}{2}\right\}$, or $\left(-\infty, \frac{1}{2}\right)$
17. $\left\{y | y \geq \frac{17}{9}\right\}$, or $\left[\frac{17}{9}, \infty\right)$

Guided Solution:
15. $6, -5y, \leq, -5, y, -\infty$

The following phrases deserve special attention.

> **TRANSLATING "AT LEAST" AND "AT MOST"**
>
> A quantity x is **at least** some amount q: $x \geq q$.
> (If x is at least q, it cannot be less than q.)
>
> A quantity x is **at most** some amount q: $x \leq q$.
> (If x is at most q, it cannot be more than q.)

Do Exercises 18–24. ▶

EXAMPLE 15 *Web Developers.* As a result of the increase in online activity, the employment demand for web developers is expected to increase 27% from 2014 to 2024. The equation

$$P = 3950t + 148{,}500$$

can be used to estimate the number of web developers in the work force, where t is the number of years after 2014. Determine the years in which the number of web developers will be greater than 200,000.
Data: U.S. Department of Labor

1. **Familiarize.** We already have an equation. To become more familiar with it, we might make a substitution for t. Suppose that we want to know the number of web developers 8 years after 2014, or in 2022. We substitute 8 for t:

    $$P = 3950(8) + 148{,}500 = 180{,}100.$$

    We see that in 2022, the number of web developers will be fewer than 200,000. To find the years in which the number of web developers exceeds 200,000, we could make guesses greater than 8, but it is more efficient to proceed to the next step.

2. **Translate.** The number of web developers is to be more than 200,000. Thus we have

    $$P > 200{,}000.$$

    We replace P with $3950t + 148{,}500$:

    $$3950t + 148{,}500 > 200{,}000.$$

3. **Solve.** We solve the inequality:

    $3950t + 148{,}500 > 200{,}000$
    $\phantom{3950t + 148{,}500} 3950t > 51{,}500$ Subtracting 148,500
    $\phantom{3950t + 148{,}500} t > 13.04.$ Dividing by 3950 and rounding

4. **Check.** As a partial check, we can substitute a value for t that is greater than 13.04, say 14: $P = 3950(14) + 148{,}500 = 203{,}800$.

5. **State.** The number of web developers will be greater than 200,000 in years more than 13.04 years after 2014, so we have $\{t \mid t > 13.04\}$.

Do Exercise 25. ▶

Translate.

18. Russell will pay at most $250 for that plane ticket.

19. Emma scored at least an 88 on her Spanish test.

20. The time of the test was between 50 min and 60 min.

21. The University of Southern Indiana is more than 25 mi away.

22. Sarah's weight is less than 110 lb.

23. That number is greater than −8.

24. The costs of production of that bar-code scanner cannot exceed $135,000.

25. *Web Developers.* Refer to Example 15. Determine, in terms of an inequality, the years in which the number of web developers is greater than 175,000.

Answers
18. $t \leq 250$ **19.** $s \geq 88$ **20.** $50 < t < 60$
21. $d > 25$ **22.** $w < 110$ **23.** $n > -8$
24. $c \leq 135{,}000$ **25.** More than 6.71 years after 2014, or $\{t \mid t > 6.71\}$

EXAMPLE 16 *Salary Plans.* On her new job, Rose can be paid in one of two ways: *Plan A* is a salary of $600 per month, plus a commission of 4% of sales; and *Plan B* is a salary of $800 per month, plus a commission of 6% of sales in excess of $10,000. For what amount of monthly sales is plan A better than plan B, if we assume that sales are always more than $10,000?

1. **Familiarize.** Listing the given information in a table will be helpful.

PLAN A: MONTHLY INCOME	PLAN B: MONTHLY INCOME
$600 salary	$800 salary
4% of sales	6% of sales over $10,000
Total: $600 + 4% of sales	*Total*: $800 + 6% of sales over $10,000

Next, suppose that Rose had sales of $12,000 in one month. Which plan would be better? Under plan A, she would earn $600 plus 4% of $12,000, or

$600 + 0.04(12,000) = \$1080.$

Since with plan B commissions are paid only on sales in excess of $10,000, Rose would earn $800 plus 6% of ($12,000 − $10,000), or

$800 + 0.06(12,000 - 10,000) = 800 + 0.06(2000) = \$920.$

This shows that for monthly sales of $12,000, plan A is better. Similar calculations will show that for sales of $30,000 per month, plan B is better. To determine *all* values for which plan A pays more money, we must solve an inequality that is based on the calculations above.

2. **Translate.** We let S = the amount of monthly sales. If we examine the calculations in the *Familiarize* step, we see that the monthly income from plan A is $600 + 0.04S$ and from plan B is $800 + 0.06(S - 10,000)$. Thus we want to find all values of S for which

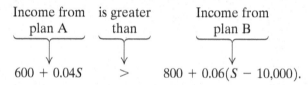

$600 + 0.04S > 800 + 0.06(S - 10,000).$

3. **Solve.** We solve the inequality:

$600 + 0.04S > 800 + 0.06(S - 10,000)$
$600 + 0.04S > 800 + 0.06S - 600$ Using the distributive law
$600 + 0.04S > 200 + 0.06S$ Collecting like terms
$400 > 0.02S$ Subtracting 200 and $0.04S$
$20,000 > S$, or $S < 20,000.$ Dividing by 0.02

4. **Check.** For $S = 20,000$, the income from plan A is

$600 + 4\% \cdot 20,000$, or $\$1400$.

The income from plan B is

$800 + 6\% \cdot (20,000 - 10,000)$, or $\$1400$.   } This confirms that for sales of $20,000, Rose's pay is the same under either plan.

In the *Familiarize* step, we saw that for sales of $12,000, plan A pays more. Since $12,000 < 20,000$, this is a partial check. Since we cannot check all possible values of S, we will stop here.

5. **State.** For monthly sales of less than $20,000, plan A is better.

◀ Do Exercise 26.

26. *Salary Plans.* A painter can be paid in one of two ways:
Plan A: $500 plus $4 per hour;
Plan B: Straight $9 per hour.
Suppose that the job takes n hours. For what values of n is plan A better for the painter?

Answer
26. For $\{n | n < 100\}$, plan A is better.

Translating for Success

The goal of these matching questions is to practice step (2), Translate, of the five-step problem-solving process. Translate each word problem to an equation or an inequality and select a correct translation from A–O.

1. *Consecutive Integers.* The sum of two consecutive even integers is 102. Find the integers.

2. *Salary Increase.* After Susanna earned a 5% raise, her new salary was $25,750. What was her former salary?

3. *Dimensions of a Rectangle.* The length of a rectangle is 6 in. more than the width. The perimeter of the rectangle is 102 in. Find the length and the width.

4. *Population.* The population of Doddville is decreasing at a rate of 5% per year. The current population is 25,750. What was the population the previous year?

5. *Reading Assignment.* Quinn has 6 days to complete a 150-page reading assignment. How many pages must he read the first day so that he has no more than 102 pages left to read on the 5 remaining days?

A. $0.05(25{,}750) = x$

B. $x + 2x = 102$

C. $2x + 2(x + 6) = 102$

D. $150 - x \leq 102$

E. $x - 0.05x = 25{,}750$

F. $x + (x + 2) = 102$

G. $x + (x + 6) > 102$

H. $x + 5x = 150$

I. $x + 0.05x = 25{,}750$

J. $x + (2x + 6) = 102$

K. $x + (x + 1) = 102$

L. $102 + x > 150$

M. $0.05x = 25{,}750$

N. $102 + 5x > 150$

O. $x + (x + 6) = 102$

Answers on page A-28

6. *Numerical Relationship.* One number is 6 more than twice another. The sum of the numbers is 102. Find the numbers.

7. *Clients.* Together Ella and Ken have 102 clients. If Ken has 6 more clients than Ella, how many does each have?

8. *Sales Commissions.* Will earns a commission of 5% on his sales. One year he earned commissions totaling $25,750. What were his total sales for the year?

9. *Fencing.* Jess has 102 ft of fencing that he plans to use to enclose two dog runs. The perimeter of one run is to be twice the perimeter of the other. Into what lengths should the fencing be cut?

10. *Quiz Scores.* Lupe has a total of 102 points on the first 6 quizzes in her sociology class. How many total points must she earn on the 5 remaining quizzes in order to have more than 150 points for the semester?

9.1 Exercise Set

✓ Check Your Understanding

Reading Check Choose from the column on the right the word that best completes each statement. Not every word will be used.

RC1. Because $-5 > -6$ is true, -5 is a(n) _____ of $x > -6$.

RC2. The solution set $\{x | x > 9\}$ is written in _____ notation.

RC3. The interval $[5, 7]$ is a(n) _____ interval.

RC4. The interval $[-3, 10)$ is a(n) _____ interval.

RC5. We reverse the direction of the inequality symbol when we multiply both sides of an inequality by a(n) _____ number.

closed
half-open
interval
negative
open
positive
set-builder
solution

Concept Check For each solution set expressed in set-builder notation, select from the column on the right the equivalent interval notation.

CC1. $\{x | a \le x < b\}$

CC2. $\{x | x < b\}$

CC3. $\{x | x \text{ is a real number}\}$

CC4. $\{x | a < x < b\}$

CC5. $\{x | a \le x \le b\}$

CC6. $\{x | x \ge a\}$

a) (a, b)
b) $[a, b)$
c) $(-\infty, \infty)$
d) $[a, \infty)$
e) $(-\infty, b]$
f) $(a, b]$
g) $[a, b]$
h) $(-\infty, b)$
i) (a, ∞)

a Determine whether the given numbers are solutions of the inequality.

1. $x - 2 \ge 6$; $-4, 0, 4, 8$

2. $3x + 5 \le -10$; $-5, -10, 0, 27$

3. $t - 8 > 2t - 3$; $0, -8, -9, -3, -\frac{7}{8}$

4. $5y - 7 < 8 - y$; $2, -3, 0, 3, \frac{2}{3}$

b Write interval notation for the given set or graph.

5. $\{x | x < 5\}$

6. $\{t | t \ge -5\}$

7. $\{x | -3 \le x \le 3\}$

8. $\{t | -10 < t \le 10\}$

9. $\{x | -4 > x > -8\}$

10. $\{x | 13 > x \ge 5\}$

11. [number line with open interval from -2 to 5, marks at $-6, -5, -4, -3, -2, -1, 0, 1, 2, 3, 4, 5, 6$]

12. [number line with half-open interval from -20 to 30, marks at $-40, -30, -20, -10, 0, 10, 20, 30, 40$]

13. [number line showing open interval starting at $-\sqrt{2}$, marks at $-2, -1, 0, 1, 2$]

14. [number line with half-open interval ending at 8, marks at $-12, -8, -4, 0, 4, 8, 12$]

C Solve and graph.

15. $x + 2 > 1$

16. $x + 8 > 4$

17. $y + 3 < 9$

18. $y + 4 < 10$

19. $a - 9 \leq -31$

20. $a + 6 \leq -14$

21. $t + 13 \geq 9$

22. $x - 8 \leq 17$

23. $y - 8 > -14$

24. $y - 9 > -18$

25. $x - 11 \leq -2$

26. $y - 18 \leq -4$

27. $8x \geq 24$

28. $8t < -56$

29. $0.3x < -18$

30. $0.6x < 30$

31. $\frac{2}{3}x > 2$

32. $\frac{3}{5}x > -3$

Solve.

33. $-9x \geq -8.1$

34. $-5y \leq 3.5$

35. $-\frac{3}{4}x \geq -\frac{5}{8}$

36. $-\frac{1}{8}y \leq -\frac{9}{8}$

37. $2x + 7 < 19$

38. $5y + 13 > 28$

39. $5y + 2y \leq -21$

40. $-9x + 3x \geq -24$

SECTION 9.1 Sets, Inequalities, and Interval Notation **663**

41. $2y - 7 < 5y - 9$

42. $8x - 9 < 3x - 11$

43. $0.4x + 5 \leq 1.2x - 4$

44. $0.2y + 1 > 2.4y - 10$

45. $5x - \frac{1}{12} \leq \frac{5}{12} + 4x$

46. $2x - 3 < \frac{13}{4}x + 10 - 1.25x$

47. $4(4y - 3) \geq 9(2y + 7)$

48. $2m + 5 \geq 16(m - 4)$

49. $3(2 - 5x) + 2x < 2(4 + 2x)$

50. $2(0.5 - 3y) + y > (4y - 0.2)8$

51. $5[3m - (m + 4)] > -2(m - 4)$

52. $[8x - 3(3x + 2)] - 5 \geq 3(x + 4) - 2x$

53. $3(r - 6) + 2 > 4(r + 2) - 21$

54. $5(t + 3) + 9 < 3(t - 2) + 6$

55. $19 - (2x + 3) \leq 2(x + 3) + x$

56. $13 - (2c + 2) \geq 2(c + 2) + 3c$

57. $\frac{1}{4}(8y + 4) - 17 < -\frac{1}{2}(4y - 8)$

58. $\frac{1}{3}(6x + 24) - 20 > -\frac{1}{4}(12x - 72)$

59. $2[4 - 2(3 - x)] - 1 \geq 4[2(4x - 3) + 7] - 25$

60. $5[3(7 - t) - 4(8 + 2t)] - 20 \leq -6[2(6 + 3t) - 4]$

61. $\frac{4}{5}(7x - 6) < 40$

62. $\frac{2}{3}(4x - 3) > 30$

63. $\frac{3}{4}(3 + 2x) + 1 \geq 13$

64. $\frac{7}{8}(5 - 4x) - 17 \geq 38$

65. $\frac{3}{4}(3x - \frac{1}{2}) - \frac{2}{3} < \frac{1}{3}$

66. $\frac{2}{3}(\frac{7}{8} - 4x) - \frac{5}{8} < \frac{3}{8}$

67. $0.7(3x + 6) \geq 1.1 - (x + 2)$

68. $0.9(2x + 8) < 20 - (x + 5)$

69. $a + (a - 3) \leq (a + 2) - (a + 1)$

70. $0.8 - 4(b - 1) > 0.2 + 3(4 - b)$

d Solve.

Body Mass Index. Body mass index I can be used to determine whether an individual has a healthy weight for his or her height. An index in the range 18.5–24.9 indicates a normal weight. Body mass index is given by the formula, or model,

$$I = \frac{703W}{H^2},$$

where W is weight, in pounds, and H is height, in inches. Use this formula for Exercises 71 and 72.

Data: Centers for Disease Control and Prevention

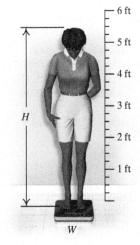

71. *Body Mass Index.* Alexandra's height is 62 in. Determine, in terms of an inequality, those weights W that will keep her body mass index below 25.

72. *Body Mass Index.* Josiah's height is 77 in. Determine, in terms of an inequality, those weights W that will keep his body mass index below 25.

73. *Grades.* David is taking an economics course in which there will be 4 tests, each worth 100 points. He has scores of 89, 92, and 95 on the first three tests. He must make a total of at least 360 in order to get an A. What scores on the last test will give David an A?

74. *Grades.* Elizabeth is taking a mathematics course in which there will be 5 tests, each worth 100 points. She has scores of 94, 90, and 89 on the first three tests. She must make a total of at least 450 in order to get an A. What scores on the fourth test will keep Elizabeth eligible for an A?

75. *Insurance Claims.* After a serious automobile accident, most insurance companies will replace the damaged car with a new one if repair costs exceed 80% of the N.A.D.A., or "blue-book," value of the car. Miguel's car recently sustained $9200 worth of damage but was not replaced. What was the blue-book value of his car?

76. *Delivery Service.* Jay's Express prices cross-town deliveries at $15 for the first 10 miles plus $1.25 for each additional mile. PDQ, Inc., prices its cross-town deliveries at $25 for the first 10 miles plus $0.75 for each additional mile. For what number of miles is PDQ less expensive?

77. *Salary Plans.* Imani can be paid in one of two ways:
   Plan A: A salary of $400 per month plus a commission of 8% of gross sales;
   Plan B: A salary of $610 per month, plus a commission of 5% of gross sales.
   For what amount of gross sales should Imani select plan A?

78. *Salary Plans.* Aiden can be paid for his masonry work in one of two ways:
   Plan A: $300 plus $9.00 per hour;
   Plan B: Straight $12.50 per hour.
   Suppose that the job takes n hours. For what values of n is plan B better for Aiden?

79. *Prescription Coverage.* Low Med offers two prescription-drug insurance plans. With plan 1, James would pay the first $150 of his prescription costs and 30% of all costs after that. With plan 2, James would pay the first $280 of costs, but only 10% of the rest. For what amount of prescription costs will plan 2 save James money? (Assume that his prescription costs exceed $280.)

80. *Insurance Benefits.* Bayside Insurance offers two plans. Under plan A, Giselle would pay the first $50 of her medical bills and 20% of all bills after that. Under plan B, Giselle would pay the first $250 of bills, but only 10% of the rest. For what amount of medical bills will plan B save Giselle money? (Assume that her bills will exceed $250.)

81. *Wedding Costs.* The Arnold Inn offers two plans for wedding parties. Under plan A, the inn charges $30 for each person in attendance. Under plan B, the inn charges $1300 plus $20 for each person in excess of the first 25 who attend. For what size parties will plan B cost less? (Assume that more than 25 guests will attend.)

82. *Investing.* Matthew is about to invest $20,000, part at 3% and the rest at 4%. What is the most that he can invest at 3% and still be guaranteed at least $650 in interest per year?

83. *Renting Office Space.* An investment group is renovating a commercial building and will rent offices to small businesses. The formula
$$R = 2(s + 70)$$
can be used to determine the monthly rent for an office with s square feet. All utilities are included in the monthly payment. For what square footage will the rent be less than $2100?

84. *Temperatures of Solids.* The formula
$$C = \tfrac{5}{9}(F - 32)$$
can be used to convert Fahrenheit temperatures F to Celsius temperatures C.
   a) Gold is a solid at Celsius temperatures less than 1063°C. Find the Fahrenheit temperatures for which gold is a solid.
   b) Silver is a solid at Celsius temperatures less than 960.8°C. Find the Fahrenheit temperatures for which silver is a solid.

85. *Tuition and Fees at Two-Year Colleges.* The equation
$$C = 103t + 1923$$
can be used to estimate the average cost of tuition and fees at two-year public institutions of higher education, where t is the number of years after 2005.

Data: National Center for Education Statistics, U.S. Department of Education

a) What was the average cost of tuition and fees in 2010? in 2016?

b) For what years will the cost of tuition and fees be more than $3500?

86. *Dewpoint Spread.* Pilots use the **dewpoint spread**, or the difference between the current temperature and the dewpoint (the temperature at which dew occurs), to estimate the height of the cloud cover. Each 3° of dewpoint spread corresponds to an increase of 1000 ft in the height of cloud cover. A plane flying with limited instruments must have a cloud cover higher than 3500 ft. What dewpoint spreads will allow the plane to fly?

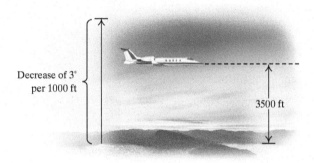

Skill Maintenance

Multiply. [4.6a]

87. $(3x - 4)(x + 8)$

88. $(r - 4s)(6r + s)$

89. $(2a - 5)(3a + 11)$

90. $(t + 2s)(t - 9s)$

Factor. [5.7a]

91. $4x^2 - 36x + 81$

92. $400y^2 - 16$

93. $27w^3 - 8$

94. $80 - 14x - 6x^2$

Find the domain. [7.2a]

95. $f(x) = \dfrac{-3}{x + 8}$

96. $f(x) = 3x - 5$

97. $f(x) = |x| - 4$

98. $f(x) = \dfrac{x + 7}{3x - 2}$

Synthesis

99. *Supply and Demand.* The supply S and demand D for a certain product are given by
$$S = 460 + 94p \quad \text{and} \quad D = 2000 - 60p.$$

a) Find those values of p for which supply exceeds demand.

b) Find those values of p for which supply is less than demand.

Determine whether each statement is true or false. If false, give a counterexample.

100. For any real numbers x and y, if $x < y$, then $x^2 < y^2$.

101. For any real numbers $a, b, c,$ and d, if $a < b$ and $c < d$, then $a + c < b + d$.

102. Determine whether the inequalities
$$x < 3 \quad \text{and} \quad 0 \cdot x < 0 \cdot 3$$
are equivalent. Give reasons to support your answer.

Solve.

103. $x + 5 \leq 5 + x$

104. $x + 8 < 3 + x$

105. $x^2 + 1 > 0$

9.2 Intersections, Unions, and Compound Inequalities

OBJECTIVES

a Find the intersection of two sets. Solve and graph conjunctions of inequalities.

b Find the union of two sets. Solve and graph disjunctions of inequalities.

c Solve applied problems involving conjunctions and disjunctions of inequalities.

Cholesterol is a substance that is found in every cell of the human body. High levels of cholesterol can cause fatty deposits in the blood vessels that increase the risk of heart attack or stroke. A blood test can be used to measure *total cholesterol*. The following table shows the health risk associated with various cholesterol levels.

TOTAL CHOLESTEROL	RISK LEVEL
Less than 200	Normal
From 200 to 239	Borderline high
240 or higher	High

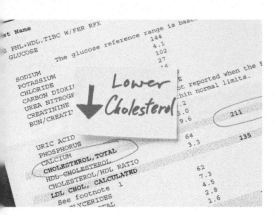

A total-cholesterol level T from 200 to 239 is considered borderline high. We can express this by the sentence

$$200 \leq T \quad \text{and} \quad T \leq 239$$

or more simply by

$$200 \leq T \leq 239.$$

This is an example of a *compound inequality*. **Compound inequalities** consist of two or more inequalities joined by the word *and* or the word *or*. We now "solve" such sentences—that is, we find the set of all solutions.

a INTERSECTIONS OF SETS AND CONJUNCTIONS OF INEQUALITIES

> **INTERSECTION**
>
> The **intersection** of two sets A and B is the set of all members that are common to A and B. We denote the intersection of sets A and B as
>
> $A \cap B.$

The intersection of two sets is often illustrated as shown below.

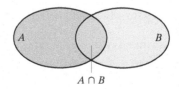

$A \cap B$

EXAMPLE 1 Find the intersection: $\{1, 2, 3, 4, 5\} \cap \{-2, -1, 0, 1, 2, 3\}$.

Only the numbers 1, 2, and 3 are common to the two sets, so the intersection is $\{1, 2, 3\}$.

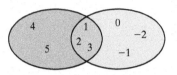

1. Find the intersection:
   $\{0, 3, 5, 7\} \cap \{0, 1, 3, 11\}$.

2. Shade the intersection of sets A and B.

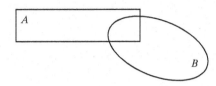

Do Exercises 1 and 2.

CONJUNCTION

When two or more sentences are joined by the word *and* to make a compound sentence, the new sentence is called a **conjunction** of the sentences.

The following is a conjunction of inequalities:

$$-2 < x \quad \text{and} \quad x < 1.$$

A number is a solution of a conjunction if it is a solution of *both* inequalities. For example, 0 is a solution of $-2 < x$ and $x < 1$ because $-2 < 0$ and $0 < 1$. Shown below is the graph of $-2 < x$, followed by the graph of $x < 1$, and then by the graph of the conjunction $-2 < x$ and $x < 1$. As the graphs demonstrate, *the solution set of a conjunction is the intersection of the solution sets of the individual inequalities.*

$\{x \mid -2 < x\}$ $\qquad$ $(-2, \infty)$

$\{x \mid x < 1\}$ $\qquad$ $(-\infty, 1)$

$\{x \mid -2 < x\} \cap \{x \mid x < 1\}$
$= \{x \mid -2 < x \text{ and } x < 1\}$ $\qquad$ $(-2, 1)$

Because there are numbers that are both greater than -2 and less than 1, the conjunction $-2 < x$ and $x < 1$ can be abbreviated as $-2 < x < 1$. Thus the interval $(-2, 1)$ can be represented as $\{x \mid -2 < x < 1\}$, the set of all numbers that are *simultaneously* greater than -2 *and* less than 1. Note that, in general, for $a < b$,

$a < x \quad \text{and} \quad x < b \quad \text{can be abbreviated} \quad a < x < b;$

and $\quad b > x \quad \text{and} \quad x > a \quad \text{can be abbreviated} \quad b > x > a.$

.... **Caution!** ....

"$a > x$ and $x < b$" cannot be abbreviated as "$a > x < b$".

Answers
1. $\{0, 3\}$
2. [shaded intersection diagram]

SECTION 9.2 Intersections, Unions, and Compound Inequalities 669

3. Graph and write interval notation:

   $-1 < x$ and $x < 4$.

   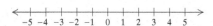

> ### "AND"; "INTERSECTION"
> The word **"and"** corresponds to **"intersection"** and to the symbol "∩". In order for a number to be a solution of a conjunction, it must make each part of the conjunction true.

◀ Do Exercise 3.

EXAMPLE 2 Solve and graph: $-1 \leq 2x + 5 < 13$.

This inequality is an abbreviation for the conjunction

$$-1 \leq 2x + 5 \quad \text{and} \quad 2x + 5 < 13.$$

The word *and* corresponds to set *intersection*, ∩. To solve the conjunction, we solve each of the two inequalities separately and then find the intersection of the solution sets:

$$\begin{array}{lll} -1 \leq 2x + 5 & \text{and} & 2x + 5 < 13 \\ -6 \leq 2x & \text{and} & 2x < 8 \quad \text{Subtracting 5} \\ -3 \leq x & \text{and} & x < 4. \quad \text{Dividing by 2} \end{array}$$

We now abbreviate the result:

$$-3 \leq x < 4.$$

The solution set is $\{x \mid -3 \leq x < 4\}$, or, in interval notation, $[-3, 4)$. The graph is the intersection of the two separate solution sets.

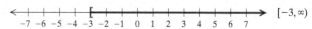

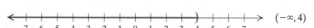

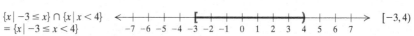

The steps above are generally combined as follows:

$$\begin{array}{ll} -1 \leq 2x + 5 < 13 & 2x + 5 \text{ appears in both inequalities.} \\ -6 \leq 2x < 8 & \text{Subtracting 5} \\ -3 \leq x < 4. & \text{Dividing by 2} \end{array}$$

◀ Do Exercise 4.

4. Solve and graph:

   $-22 < 3x - 7 \leq 23$.

EXAMPLE 3 Solve and graph: $2x - 5 \geq -3$ and $5x + 2 \geq 17$.

We first solve each inequality separately:

$$\begin{array}{lll} 2x - 5 \geq -3 & \text{and} & 5x + 2 \geq 17 \\ 2x \geq 2 & \text{and} & 5x \geq 15 \\ x \geq 1 & \text{and} & x \geq 3. \end{array}$$

Answers

3. ; $(-1, 4)$
4. $\{x \mid -5 < x \leq 10\}$, or $(-5, 10]$;

Next, we find the intersection of the two separate solution sets:

$\{x \mid x \geq 1\}$ [1, ∞)

$\{x \mid x \geq 3\}$ [3, ∞)

$\{x \mid x \geq 1\} \cap \{x \mid x \geq 3\}$
$= \{x \mid x \geq 3\}$ [3, ∞)

The numbers common to both sets are those that are greater than or equal to 3. Thus the solution set is $\{x \mid x \geq 3\}$, or, in interval notation, $[3, \infty)$. You should check that any number in $[3, \infty)$ satisfies the conjunction whereas numbers outside $[3, \infty)$ do not.

Do Exercise 5.

5. Solve and graph:
$3x + 4 \leq 10$ *and* $2x - 7 < -13$.

EMPTY SET; DISJOINT SETS

Sometimes two sets have no elements in common. In such a case, we say that the intersection of the two sets is the **empty set**, denoted $\{\ \}$ or $\emptyset$. Two sets with an empty intersection are said to be **disjoint**.

$A \cap B = \emptyset$ 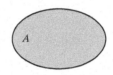 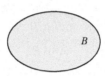

EXAMPLE 4 Solve and graph: $2x - 3 > 1$ *and* $3x - 1 < 2$.

We solve each inequality separately:

$2x - 3 > 1$ and $3x - 1 < 2$
$2x > 4$ and $3x < 3$
$x > 2$ and $x < 1$.

The solution set is the intersection of the solution sets of the individual inequalities.

$\{x \mid x > 2\}$ (2, ∞)

$\{x \mid x < 1\}$ (−∞, 1)

$\{x \mid x > 2\} \cap \{x \mid x < 1\}$
$= \{x \mid x > 2 \text{ and } x < 1\}$
$= \emptyset$ ∅

Since no number is both greater than 2 and less than 1, the solution set is the empty set, $\emptyset$.

Do Exercise 6.

6. Solve and graph:
$3x - 7 \leq -13$ *and* $4x + 3 > 8$.

Answers

5. $\{x \mid x < -3\}$;   **6.** $\emptyset$

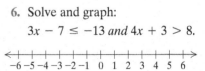

7. Solve: $-4 \leq 8 - 2x \leq 4$.
$$-4 \leq 8 - 2x \leq 4$$
$$-4 - \boxed{} \leq 8 - 2x - 8 \leq 4 - 8$$
$$\boxed{} \leq -2x \leq \boxed{}$$
$$\frac{-12}{-2} \quad \boxed{} \quad \frac{-2x}{-2} \geq \frac{-4}{-2}$$
$$6 \geq \boxed{} \geq 2, \text{ or}$$
$$2 \quad \boxed{} \quad x \leq 6$$

The solution set is $\{x | 2 \leq x \leq 6\}$, or $[\boxed{}, 6]$.

EXAMPLE 5 Solve: $3 \leq 5 - 2x < 7$.

We have
$$3 \leq 5 - 2x < 7$$
$$3 - 5 \leq 5 - 2x - 5 < 7 - 5 \quad \text{Subtracting 5}$$
$$-2 \leq \quad -2x \quad < 2 \quad \text{Simplifying}$$
$$\frac{-2}{-2} \geq \frac{-2x}{-2} > \frac{2}{-2} \quad \text{Dividing by } -2. \text{ The symbols must be reversed.}$$
$$1 \geq x > -1. \quad \text{Simplifying}$$

The solution set is $\{x | 1 \geq x > -1\}$, or $\{x | -1 < x \leq 1\}$, since the inequalities $1 \geq x > -1$ and $-1 < x \leq 1$ are equivalent. The solution, in interval notation, is $(-1, 1]$.

◂ Do Exercise 7.

b UNIONS OF SETS AND DISJUNCTIONS OF INEQUALITIES

UNION

The **union** of two sets A and B is the collection of elements belonging to A and/or B. We denote the union of A and B by

$A \cup B$.

The union of two sets is often illustrated as shown below.

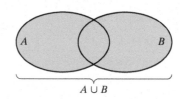

EXAMPLE 6 Find the union: $\{2, 3, 4\} \cup \{3, 5, 7\}$.

The numbers in either or both sets are 2, 3, 4, 5, and 7, so the union is $\{2, 3, 4, 5, 7\}$. We don't list the number 3 twice.

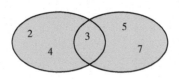

8. Find the union:
$\{0, 1, 3, 4\} \cup \{0, 1, 7, 9\}$.

9. Shade the union of sets A and B.

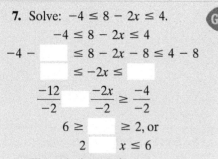

◂ Do Exercises 8 and 9.

DISJUNCTION

When two or more sentences are joined by the word *or* to make a compound sentence, the new sentence is called a **disjunction** of the sentences.

Answers
7. $\{x | 2 \leq x \leq 6\}$, or $[2, 6]$
8. $\{0, 1, 3, 4, 7, 9\}$
9. (shaded figure)

Guided Solution:
7. $8, -12, -4, \geq, x, \leq, 2$

The following is an example of a disjunction:

$x < -3 \quad or \quad x > 3.$

A number is a solution of a disjunction if it is a solution of at least one of the individual inequalities. For example, -7 is a solution of $x < -3$ or $x > 3$ because $-7 < -3$. Similarly, 5 is also a solution because $5 > 3$.

Shown below is the graph of $x < -3$, followed by the graph of $x > 3$, and then by the graph of the disjunction $x < -3$ or $x > 3$. As the graphs demonstrate, *the solution set of a disjunction is the union of the solution sets of the individual sentences.*

$\{x \mid x < -3\}$ $(-\infty, -3)$

$\{x \mid x > 3\}$ $(3, \infty)$

$\{x \mid x < -3\} \cup \{x \mid x > 3\}$
$= \{x \mid x < -3 \, or \, x > 3\}$ $(-\infty, -3) \cup (3, \infty)$

The solution set of

$x < -3 \quad or \quad x > 3$

is written $\{x \mid x < -3 \, or \, x > 3\}$, or, in interval notation, $(-\infty, -3) \cup (3, \infty)$. This cannot be written in a more condensed form.

"OR"; "UNION"

The word "**or**" corresponds to "**union**" and the symbol "∪". In order for a number to be in the solution set of a disjunction, it must be in *at least one* of the solution sets of the individual sentences.

Do Exercise 10. ▶

10. Graph and write interval notation:

$x \leq -2 \, or \, x > 4.$

EXAMPLE 7 Solve and graph: $7 + 2x < -1 \, or \, 13 - 5x \leq 3$.

We solve each inequality separately, retaining the word *or*:

$7 + 2x < -1 \quad or \quad 13 - 5x \leq 3$
$2x < -8 \quad or \quad -5x \leq -10$ ⎯⎯ Dividing by -5. The symbol must be reversed.
$x < -4 \quad or \quad x \geq 2.$

To find the solution set of the disjunction, we consider the individual graphs. We graph $x < -4$ and then $x \geq 2$. Then we take the union of the graphs.

$\{x \mid x < -4\}$ $(-\infty, -4)$

$\{x \mid x \geq 2\}$ $[2, \infty)$

$\{x \mid x < -4 \, or \, x \geq 2\}$ $(-\infty, -4) \cup [2, \infty)$

The solution set is written $\{x \mid x < -4 \, or \, x \geq 2\}$, or, in interval notation, $(-\infty, -4) \cup [2, \infty)$.

Answer
10.
$(-\infty, -2] \cup (4, \infty)$

Solve and graph.

11. $x - 4 < -3 \text{ or } x - 3 \geq 3$

```
<-+-+-+-+-+-+-+-+-+-+-+->
-10 -8 -6 -4 -2 0 2 4 6 8 10
```

12. $-2x + 4 \leq -3 \text{ or } x + 5 < 3$

```
<-+-+-+-+-+-+-+-+-+-+-+->
-6 -5 -4 -3 -2 -1 0 1 2 3 4 5 6
```

13. Solve:
$-3x - 7 < -1 \text{ or } x + 4 < -1$.

$$-3x - 7 < -1 \quad \text{or} \quad x + 4 < -1$$
$$-3x < \boxed{} \quad \text{or} \quad x < \boxed{}$$
$$\frac{-3x}{-3} \; \boxed{} \; \frac{6}{-3} \quad \text{or} \quad x < -5$$
$$x > \boxed{} \quad \text{or} \quad x < -5$$

The solution set is $\{x \mid x < \boxed{} \text{ or } x > -2\}$, or, in interval notation,
$(-\infty, -5) \cup (-2, \boxed{})$.

14. Solve and graph:
$5x - 7 \leq 13 \text{ or } 2x - 1 \geq -7$.

```
<-+-+-+-+-+-+-+-+-+-+-+->
-6 -5 -4 -3 -2 -1 0 1 2 3 4 5 6
```

Answers

11. $\{x \mid x < 1 \text{ or } x \geq 6\}$, or $(-\infty, 1) \cup [6, \infty)$;

```
<-+-+-)-+-+-+[-+-+->
      0 1         6
```

12. $\left\{ x \mid x < -2 \text{ or } x \geq \frac{7}{2} \right\}$, or $(-\infty, -2) \cup \left[\frac{7}{2}, \infty \right)$;

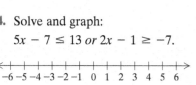

13. $\{x \mid x < -5 \text{ or } x > -2\}$, or $(-\infty, -5) \cup (-2, \infty)$
14. All real numbers;

```
<-+-+-+-+-+-+-+-+->
        0
```

Guided Solution:
13. $6, -5, >, -2, -5, \infty$

> **Caution!**
>
> A compound inequality like
>
> $x < -4 \quad or \quad x \geq 2$,
>
> as in Example 7, *cannot* be expressed as $2 \leq x < -4$ because to do so would be to say that x is *simultaneously* less than -4 and greater than or equal to 2. No number is both less than -4 *and* greater than or equal to 2, but many are less than -4 *or* greater than or equal to 2.

◀ Do Exercises 11 and 12.

EXAMPLE 8 Solve: $-2x - 5 < -2 \text{ or } x - 3 < -10$.

We solve the individual inequalities separately, retaining the word *or*:

$$-2x - 5 < -2 \quad \text{or} \quad x - 3 < -10$$
$$-2x < 3 \quad \text{or} \quad x < -7$$

Reversing the symbol

$$x > -\tfrac{3}{2} \quad \text{or} \quad x < -7.$$

—— Keep the word "or."

The solution set is written $\{x \mid x < -7 \text{ or } x > -\tfrac{3}{2}\}$, or, in interval notation, $(-\infty, -7) \cup \left(-\tfrac{3}{2}, \infty\right)$.

◀ Do Exercise 13.

EXAMPLE 9 Solve: $3x - 11 < 4 \text{ or } 4x + 9 \geq 1$.

We solve the individual inequalities separately, retaining the word *or*:

$$3x - 11 < 4 \quad \text{or} \quad 4x + 9 \geq 1$$
$$3x < 15 \quad \text{or} \quad 4x \geq -8$$
$$x < 5 \quad \text{or} \quad x \geq -2.$$

To find the solution set, we first look at the individual graphs.

$\{x \mid x < 5\}$ ... $(-\infty, 5)$

$\{x \mid x \geq -2\}$ ... $[-2, \infty)$

$\{x \mid x < 5\} \cup \{x \mid x \geq -2\}$
$= \{x \mid x < 5 \text{ or } x \geq -2\}$
$= \{x \mid x \text{ is a real number}\}$ ... $(-\infty, \infty)$
= The set of all real numbers

Since any number is either less than 5 or greater than or equal to -2, the two sets fill the entire number line. Thus the solution set is the set of all real numbers, $(-\infty, \infty)$.

◀ Do Exercise 14.

C APPLICATIONS AND PROBLEM SOLVING

EXAMPLE 10 *Renting Retail Space.* The equation

$$R = 2(s + 70)$$

can be used to determine the monthly rent R for a retail space with s square footage in a renovated commercial building. All utilities are included in the monthly payment. A florist shop has a monthly rental budget between $1720 and $2560. What square footage can be rented and remain within budget?

1. **Familiarize.** We have an equation for calculating the monthly rent. Thus we can substitute a value into the formula. For 720 ft², the rent is found as follows:

   $$R = 2(720 + 70) = 2 \cdot 790 = \$1580.$$

   This familiarizes us with the equation and also tells us that the number of square feet that we are looking for must be larger than 720 since $1580 is less than $1720.

2. **Translate.** We want the monthly rent to be between $1720 and $2560, so we need to find those values of s for which $1720 < R < 2560$. Substituting $2(s + 70)$ for R, we have

   $$1720 < 2(s + 70) < 2560.$$

3. **Solve.** We solve the inequality:

   $$1720 < 2(s + 70) < 2560$$
   $$\frac{1720}{2} < \frac{2(s + 70)}{2} < \frac{2560}{2} \quad \text{Dividing by 2}$$
   $$860 < s + 70 < 1280$$
   $$790 < s < 1210. \quad \text{Subtracting 70}$$

4. **Check.** We substitute some values as we did in the *Familiarize* step.

5. **State.** Square footage between 790 ft² and 1210 ft² can be rented for a budget between $1720 and $2560 per month.

Do Exercise 15.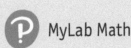

15. *Renting Retail Space.* Refer to Example 10. What square footage can be rented for a budget between $2000 and $3200?

Answer
15. Between 930 ft² and 1530 ft²

9.2 Exercise Set

FOR EXTRA HELP — MyLab Math

✓ Check Your Understanding

Reading Check Determine whether each statement is true or false.

RC1. A compound inequality like $x < 5$ and $x > -2$ can be expressed as $-2 < x < 5$.

RC2. A compound inequality like $x \geq 5$ and $x < -2$ can be expressed as $5 \leq x < -2$.

RC3. The solution set of $x < -4$ and $x > 4$ can be written as $\emptyset$.

RC4. The solution set of $x \leq 3$ and $x \geq 0$ can be written as $[0, 3]$.

Concept Check Determine whether -5 is in the solution set of each compound inequality.

CC1. $-7 < x < 7$
CC2. $-5 \le x < 3$
CC3. $-5 < x \le 3$
CC4. $x > -10$ and $x < 0$
CC5. $x < -10$ or $x > 0$
CC6. $x < -4$ or $x > 1$

a, b Find the intersection or union.

1. $\{9, 10, 11\} \cap \{9, 11, 13\}$
2. $\{1, 5, 10, 15\} \cap \{5, 15, 20\}$
3. $\{a, b, c, d\} \cap \{b, f, g\}$
4. $\{m, n, o, p\} \cap \{m, o, p\}$
5. $\{9, 10, 11\} \cup \{9, 11, 13\}$
6. $\{1, 5, 10, 15\} \cup \{5, 15, 20\}$
7. $\{a, b, c, d\} \cup \{b, f, g\}$
8. $\{m, n, o, p\} \cup \{m, o, p\}$
9. $\{2, 5, 7, 9\} \cap \{1, 3, 4\}$
10. $\{a, e, i, o, u\} \cap \{m, q, w, s, t\}$
11. $\{3, 5, 7\} \cup \emptyset$
12. $\{3, 5, 7\} \cap \emptyset$

a Graph and write interval notation.

13. $-4 < a$ and $a \le 1$
14. $-\frac{5}{2} \le m$ and $m < \frac{3}{2}$
15. $1 < x < 6$
16. $-3 \le y \le 4$

Solve and graph.

17. $-10 \le 3x + 2$ and $3x + 2 < 17$
18. $-11 < 4x - 3$ and $4x - 3 \le 13$
19. $3x + 7 \ge 4$ and $2x - 5 \ge -1$
20. $4x - 7 < 1$ and $7 - 3x > -8$
21. $4 - 3x \ge 10$ and $5x - 2 > 13$
22. $5 - 7x > 19$ and $2 - 3x < -4$

Solve.

23. $-4 < x + 4 < 10$
24. $-6 < x + 6 \leq 8$
25. $6 > -x \geq -2$
26. $3 > -x \geq -5$

27. $2 < x + 3 \leq 9$
28. $-6 \leq x + 1 < 9$
29. $1 < 3y + 4 \leq 19$

30. $5 \leq 8x + 5 \leq 21$
31. $-10 \leq 3x - 5 \leq -1$
32. $-6 \leq 2x - 3 < 6$

33. $-18 \leq -2x - 7 < 0$
34. $4 > -3m - 7 \geq 2$
35. $-\dfrac{1}{2} < \dfrac{1}{4}x - 3 \leq \dfrac{1}{2}$

36. $-\dfrac{2}{3} \leq 4 - \dfrac{1}{4}x < \dfrac{2}{3}$
37. $-4 \leq \dfrac{7 - 3x}{5} \leq 4$
38. $-3 < \dfrac{2x - 5}{4} < 8$

b Graph and write interval notation.

39. $x < -2 \text{ or } x > 1$

40. $x < -4 \text{ or } x > 0$

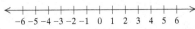

41. $x \leq -3 \text{ or } x > 1$

42. $x \leq -1 \text{ or } x > 3$

Solve and graph.

43. $x + 3 < -2 \text{ or } x + 3 > 2$

44. $x - 2 < -1 \text{ or } x - 2 > 3$

45. $2x - 8 \leq -3 \text{ or } x - 1 \geq 3$

46. $x - 5 \leq -4 \text{ or } 2x - 7 \geq 3$

47. $7x + 4 \geq -17 \text{ or } 6x + 5 \geq -7$

48. $4x - 4 < -8 \text{ or } 4x - 4 < 12$

Solve.

49. $7 > -4x + 5 \text{ or } 10 \leq -4x + 5$

50. $6 > 2x - 1 \text{ or } -4 \leq 2x - 1$

51. $3x - 7 > -10$ *or* $5x + 2 \leq 22$

52. $3x + 2 < 2$ *or* $4 - 2x < 14$

53. $-2x - 2 < -6$ *or* $-2x - 2 > 6$

54. $-3m - 7 < -5$ *or* $-3m - 7 > 5$

55. $\frac{2}{3}x - 14 < -\frac{5}{6}$ *or* $\frac{2}{3}x - 14 > \frac{5}{6}$

56. $\frac{1}{4} - 3x \leq -3.7$ *or* $\frac{1}{4} - 5x \geq 4.8$

57. $\frac{2x - 5}{6} \leq -3$ *or* $\frac{2x - 5}{6} \geq 4$

58. $\frac{7 - 3x}{5} < -4$ *or* $\frac{7 - 3x}{5} > 4$

C Solve.

59. *Pressure at Sea Depth.* The equation
$$P = 1 + \frac{d}{33}$$
gives the pressure P, in atmospheres (atm), at a depth of d feet in the sea. For what depths d is the pressure at least 1 atm and at most 7 atm?

60. *Temperatures of Liquids.* The formula
$$C = \tfrac{5}{9}(F - 32)$$
can be used to convert Fahrenheit temperatures F to Celsius temperatures C.
a) Gold is a liquid for Celsius temperatures C such that $1063° \leq C < 2660°$. Find such an inequality for the corresponding Fahrenheit temperatures.
b) Silver is a liquid for Celsius temperatures C such that $960.8° \leq C < 2180°$. Find such an inequality for the corresponding Fahrenheit temperatures.

61. *Aerobic Exercise.* In order to achieve maximum results from aerobic exercise, one should maintain one's heart rate at a certain level. A 30-year-old woman with a resting heart rate of 60 beats per minute should keep her heart rate between 138 and 162 beats per minute while exercising. She checks her pulse for 10 sec while exercising. What should the number of beats be?

62. *Minimizing Tolls.* A $6.00 toll is charged to cross the bridge from mainland Florida to Sanibel Island. A six-month pass, costing $50.00, reduces the toll to $2.00. A six-month unlimited-trip pass costs $300 and allows for unlimited crossings. How many crossings in six months does it take, on average, for the reduced-toll pass to be the most economical choice?

Data: leewayinfo.com

63. *Body Mass Index.* Refer to Exercises 71 and 72 in Exercise Set 9.1. Alexandra's height is 62 in. What weights W will allow Alexandra to keep her body mass index I in the 18.5–24.9 range?

64. *Body Mass Index.* Refer to Exercises 71 and 72 in Exercise Set 9.1. Josiah's height is 77 in. What weights W will allow Josiah to keep his body mass index in the 18.5–24.9 range?

65. *Young's Rule in Medicine.* Young's rule for determining the amount of a medicine dosage for a child is given by
$$c = \frac{ad}{a + 12},$$
where a is the child's age and d is the usual adult dosage, in milligrams. (*Warning!* Do not apply this formula without checking with a physician!) An 8-year-old child needs medication. What adult dosage can be used if a child's dosage must stay between 100 mg and 200 mg?

Data: Olsen, June L., et al., *Medical Dosage Calculations*, 6th ed. Reading, MA: Addison Wesley Longman, p. A-31.

66. *Young's Rule in Medicine.* Refer to Exercise 65. The dosage of a medication for a 5-year-old child must stay between 50 mg and 100 mg. Find the equivalent adult dosage.

Skill Maintenance

Solve. [8.2a], [8.3a]

67. $3x - 2y = -7,$
$2x + 5y = 8$

68. $4x - 7y = 23,$
$x + 6y = -33$

69. $x + y = 0,$
$x - y = 8$

Find an equation of the line containing the given pair of points. [7.5c]

70. $(2, 7), (3, -4)$

71. $(0, 7), (2, -1)$

72. $(4, -2), (-2, 4)$

Multiply. [4.6a]

73. $(2a - b)(3a + 5b)$

74. $(5y + 6)(5y + 1)$

75. $(7x - 8)(3x - 5)$

76. $(13x - 2y)(x + 3y)$

Synthesis

Solve.

77. $x - 10 < 5x + 6 \leq x + 10$

78. $4m - 8 > 6m + 5 \text{ or } 5m - 8 < -2$

79. $-\frac{2}{15} \leq \frac{2}{3}x - \frac{2}{5} \leq \frac{2}{15}$

80. $2[5(3 - y) - 2(y - 2)] > y + 4$

81. $x + 4 < 2x - 6 \leq x + 12$

82. $2x + 3 \leq x - 6 \text{ or } 3x - 2 \leq 4x + 5$

Determine whether each sentence is true or false for all real numbers a, b, and c.

83. If $-b < -a$, then $a < b$.

84. If $a \leq c$ and $c \leq b$, then $b \geq a$.

85. If $a < c$ and $b < c$, then $a < b$.

86. If $-a < c$ and $-c > b$, then $a > b$.

Mid-Chapter Review

Concept Reinforcement

Determine whether each statement is true or false.

_____ 1. The inequalities $x - 5 > 2$ and $x > 7$ are equivalent. [9.1c]

_____ 2. If a is at most c, then it cannot be less than c. [9.1d]

_____ 3. Sets A and B where $A = \{x | x < 2\}$ and $B = \{x | x \geq 2\}$ are disjoint sets. [9.2a]

_____ 4. The union of two sets A and B is the collection of elements belonging to A and/or B. [9.2b]

Guided Solutions

 Fill in each blank with the number, variable, or symbol that creates a correct solution.

5. Solve: $8 - 5x \leq x + 20$. [9.1c]
$$8 - 5x \leq x + 20$$
$$-5x \leq x + \boxed{}$$
$$\boxed{} x \leq 12$$
$$x \boxed{} -2$$

6. Solve: $-17 < 3 - x < 36$. [9.2a]
$$-17 < 3 - x < 36$$
$$\boxed{} < -x < 33$$
$$20 \boxed{} x \boxed{} -33$$

Mixed Review

Match each graph of a solution with the correct set-builder notation or interval notation from selections A–H. [9.1b], [9.2a, b]

7. [number line from −5 to 5]

8. [number line from −5 to 5]

9. [number line from −5 to 5]

10. [number line from −5 to 5]

11. [number line from −5 to 5]

12. [number line from −5 to 5]

A. $(-\infty, -3]$

B. $\{x | -3 \leq x \leq 3\}$

C. $\{x | x \leq -3 \text{ or } x > 3\}$

D. $[-3, \infty)$

E. $(-\infty, -3) \cup (3, \infty)$

F. $\{x | -3 \leq x < 3\}$

G. $\{x | x > -3\}$

H. $(3, \infty)$

680   CHAPTER 9   More on Inequalities

Find the intersection or union. [9.2a, b]

13. $\{-1, 0, 10, 21, 40\} \cap \{-10, 0, 10\}$

14. $\{e, f, g, h\} \cup \{b, d, e\}$

15. $\left\{\dfrac{1}{4}, \dfrac{3}{8}\right\} \cup \varnothing$

16. $\{3, 6, 9, 12, 15\} \cap \{-12, -6, 7, 8\}$

Solve. Express the answer in both set-builder notation and interval notation.

17. $y - 8 \leq -10$ [9.1c]

18. $-\dfrac{5}{11}x \geq -\dfrac{20}{11}$ [9.1c]

19. $x - 6 < -15 \text{ or } x + 2 > 3$ [9.2b]

20. $-6 \leq x - 9 < 15$ [9.2a]

21. $x + 6 < -4 \text{ or } x + 8 > 9$ [9.2b]

22. $4(3t - 4) > 2(6 - t)$ [9.1c]

23. $3x - 2 \geq -11 \text{ or } 5x + 3 \geq -7$ [9.2b]

24. $0.1y + 3 < 5.6y - 2$ [9.1c]

25. $-6 < \dfrac{2x - 1}{3} < 8$ [9.2a]

26. $20 - (2x - 9) \leq 3(x - 2) + x$ [9.1c]

27. $-\dfrac{1}{2} < 8 - \dfrac{1}{2}x < 6$ [9.2a]

28. $2x - 7 > -18 \text{ or } 3x - 7 \leq 40$ [9.2b]

Solve. [9.1d]

29. Jada is taking a chemistry course in which there will be 5 exams, each worth 100 points. She has scores of 85, 96, 88, and 95 on the first four exams. She must make a total of at least 450 points in order to get an A. What scores on the last exam will give Jada an A?

30. Michael is about to invest $12,500, part at 4.5% and the rest at 5%. What is the most he can invest at 4.5% and still be guaranteed at least $610 in interest per year?

Understanding Through Discussion and Writing

31. Explain in your own words why the inequality symbol must be reversed when both sides of an inequality are multiplied or divided by a negative number. [9.1c]

32. Find the error or errors in each of the following steps: [9.1c]

$7 - 9x + 6x < -9(x + 2) + 10x$
$7 - 9x + 6x < -9x + 2 + 10x$ (1)
$7 + 6x > 2 + 10x$ (2)
$-4x > 8$ (3)
$x > -2.$ (4)

35. Explain why the conjunction $3 < x \text{ and } x < 5$ is equivalent to $3 < x < 5$, but the disjunction $3 < x \text{ or } x < 5$ is not. [9.2a, b]

STUDYING FOR SUCCESS   Your Textbook as a Resource

☐ Study any drawings. Note the details in any sketches or graphs that accompany the explanations.

☐ Note the careful use of color to indicate substitutions and to highlight steps in a multistep problem.

9.3 Absolute-Value Equations and Inequalities

OBJECTIVES

a Simplify expressions containing absolute-value symbols.

b Find the distance between two points on the number line.

c Solve equations with absolute-value expressions.

d Solve equations with two absolute-value expressions.

e Solve inequalities with absolute-value expressions.

a PROPERTIES OF ABSOLUTE VALUE

SKILL REVIEW Find the absolute value of a real number. [1.2e]
Find each absolute value.
1. $|-4|$ 2. $|3.5|$

Answers: 1. 4 2. 3.5

MyLab Math
VIDEO

We can think of the **absolute value** of a number as the number's distance from zero on the number line.

ABSOLUTE VALUE

The **absolute value** of x, denoted $|x|$, is defined as follows:

$$x \geq 0 \longrightarrow |x| = x; \qquad x < 0 \longrightarrow |x| = -x.$$

This definition tells us that, when x is nonnegative, the absolute value of x is x and, when x is negative, the absolute value of x is the opposite of x. For example, $|3| = 3$ and $|-3| = -(-3) = 3$. We see that absolute value is never negative.

Some simple properties of absolute value allow us to manipulate or simplify algebraic expressions.

PROPERTIES OF ABSOLUTE VALUE

a) $|ab| = |a| \cdot |b|$, for any real numbers a and b.
(The absolute value of a product is the product of the absolute values.)

b) $\left|\dfrac{a}{b}\right| = \dfrac{|a|}{|b|}$, for any real numbers a and b and $b \neq 0$.
(The absolute value of a quotient is the quotient of the absolute values.)

c) $|-a| = |a|$, for any real number a.
(The absolute value of the opposite of a number is the same as the absolute value of the number.)

682   CHAPTER 9   More on Inequalities

EXAMPLES Simplify, leaving as little as possible inside the absolute-value signs.

1. $|5x| = |5| \cdot |x| = 5|x|$
2. $|-3y| = |-3| \cdot |y| = 3|y|$
3. $|7x^2| = |7| \cdot |x^2| = 7|x^2| = 7x^2$ Since x^2 is never negative for any number x
4. $\left|\dfrac{6x}{-3x^2}\right| = \left|\dfrac{-2}{x}\right| = \dfrac{|-2|}{|x|} = \dfrac{2}{|x|}$

Do Exercises 1–5.

Simplify, leaving as little as possible inside the absolute-value signs.
1. $|7x|$ 2. $|x^8|$
3. $|5a^2b|$ 4. $\left|\dfrac{7a}{b^2}\right|$
5. $|-9x|$

b DISTANCE ON THE NUMBER LINE

The number line at right shows that the distance between -3 and 2 is 5.

Another way to find the distance between two numbers on the number line is to determine the absolute value of the difference, as follows:

$|-3 - 2| = |-5| = 5$, or $|2 - (-3)| = |5| = 5$.

Note that the order in which we subtract does not matter because we are taking the absolute value after we have subtracted.

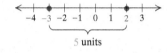

DISTANCE AND ABSOLUTE VALUE

For any real numbers a and b, the **distance** between them is $|a - b|$.

We should note that the distance is also $|b - a|$, because $a - b$ and $b - a$ are opposites and hence have the same absolute value.

EXAMPLE 5 Find the distance between -8 and -92 on the number line.

$|-8 - (-92)| = |84| = 84$, or $|-92 - (-8)| = |-84| = 84$

EXAMPLE 6 Find the distance between x and 0 on the number line.

$|x - 0| = |x|$

Do Exercises 6–8.

Find the distance between the points.

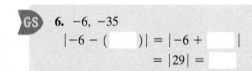

6. $-6, -35$
   $|-6 - ()| = |-6 + |$
   $ = |29| = $

7. 19, 14

8. 0, p

c EQUATIONS WITH ABSOLUTE VALUE

EXAMPLE 7 Solve: $|x| = 4$. Then graph on the number line.

Note that $|x| = |x - 0|$, so that $|x - 0|$ is the distance from x to 0. Thus solutions of the equation $|x| = 4$, or $|x - 0| = 4$ are those numbers x whose distance from 0 is 4. Those numbers are -4 and 4. The solution set is $\{-4, 4\}$. The graph consists of just two points, as shown.

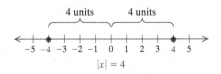

$|x| = 4$

Answers
1. $7|x|$ 2. x^8 3. $5a^2|b|$ 4. $\dfrac{7|a|}{b^2}$
5. $9|x|$ 6. 29 7. 5 8. $|p|$
Guided Solution:
6. $-35, 35, 29$

EXAMPLE 8 Solve: $|x| = 0$.

The only number whose absolute value is 0 is 0 itself. Thus the solution is 0. The solution set is $\{0\}$.

EXAMPLE 9 Solve: $|x| = -7$.

The absolute value of a number is always nonnegative. Thus there is no number whose absolute value is -7; consequently, the equation has no solution. The solution set is $\emptyset$.

Examples 7–9 lead us to the following principle for solving linear equations with absolute value.

THE ABSOLUTE-VALUE PRINCIPLE

For any positive number p and any algebraic expression X:

a) The solution of $|X| = p$ is those numbers that satisfy $X = -p$ or $X = p$.

b) The equation $|X| = 0$ is equivalent to the equation $X = 0$.

c) The equation $|X| = -p$ has no solution.

---

9. Solve: $|x| = 6$. Then graph on the number line.

10. Solve: $|x| = -6$.

11. Solve: $|p| = 0$.

◀ Do Exercises 9–11.

We can use the absolute-value principle with the addition and multiplication principles to solve equations with absolute value.

EXAMPLE 10 Solve: $2|x| + 5 = 9$.

We first use the addition and multiplication principles to get $|x|$ by itself. Then we use the absolute-value principle.

$$2|x| + 5 = 9$$
$$2|x| = 4 \qquad \text{Subtracting 5}$$
$$|x| = 2 \qquad \text{Dividing by 2}$$
$$x = -2 \quad \text{or} \quad x = 2 \qquad \text{Using the absolute-value principle}$$

The solutions are -2 and 2. The solution set is $\{-2, 2\}$.

◀ Do Exercises 12–14.

Solve.
12. $|3x| = 6$

13. $4|x| + 10 = 27$

14. $3|x| - 2 = 10$

EXAMPLE 11 Solve: $|x - 2| = 3$.

We can consider solving this equation in two different ways.

Method 1: This allows us to see the meaning of the solutions graphically. The solution set consists of those numbers that are 3 units from 2 on the number line.

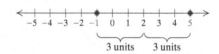

The solutions of $|x - 2| = 3$ are -1 and 5. The solution set is $\{-1, 5\}$.

Answers

9. $\{6, -6\}$;

10. $\emptyset$ 11. $\{0\}$ 12. $\{-2, 2\}$

13. $\left\{-\dfrac{17}{4}, \dfrac{17}{4}\right\}$ 14. $\{-4, 4\}$

Method 2: This method is more efficient. We use the absolute-value principle, replacing X with $x - 2$ and p with 3. Then we solve each equation separately.

$$|X| = p$$
$$|x - 2| = 3$$
$$x - 2 = -3 \quad \text{or} \quad x - 2 = 3 \quad \text{Absolute-value principle}$$
$$x = -1 \quad \text{or} \quad x = 5$$

The solutions are -1 and 5. The solution set is $\{-1, 5\}$.

Do Exercise 15.

15. Solve: $|x - 4| = 1$. Use two methods as in Example 11.

EXAMPLE 12 Solve: $|2x + 5| = 13$.

We use the absolute-value principle, replacing X with $2x + 5$ and p with 13:

$$|X| = p$$
$$|2x + 5| = 13$$
$$2x + 5 = -13 \quad \text{or} \quad 2x + 5 = 13 \quad \text{Absolute-value principle}$$
$$2x = -18 \quad \text{or} \quad 2x = 8$$
$$x = -9 \quad \text{or} \quad x = 4.$$

The solutions are -9 and 4. The solution set is $\{-9, 4\}$.

Do Exercise 16.

GS 16. Solve: $|3x - 4| = 17$.
$$|3x - 4| = 17$$
$$3x - 4 = -17 \quad \text{or} \quad 3x - 4 = \boxed{}$$
$$3x = \boxed{} \quad \text{or} \quad 3x = 21$$
$$x = \boxed{} \quad \text{or} \quad x = \boxed{}$$
The solution set is $\{\boxed{}, 7\}$.

EXAMPLE 13 Solve: $|4 - 7x| = -8$.

Since absolute value is always nonnegative, this equation has no solution. The solution set is $\emptyset$.

Do Exercise 17.

17. Solve: $|6 + 2x| = -3$.

d EQUATIONS WITH TWO ABSOLUTE-VALUE EXPRESSIONS

Sometimes equations have two absolute-value expressions. Consider $|a| = |b|$. This means that a and b are the same distance from 0. If a and b are the same distance from 0, then either they are the same number or they are opposites.

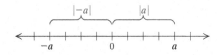

EXAMPLE 14 Solve: $|2x - 3| = |x + 5|$.

Either $2x - 3 = x + 5$ or $2x - 3 = -(x + 5)$. We solve each equation:

$$2x - 3 = x + 5 \quad \text{or} \quad 2x - 3 = -(x + 5)$$
$$x - 3 = 5 \quad \text{or} \quad 2x - 3 = -x - 5$$
$$x = 8 \quad \text{or} \quad 3x - 3 = -5$$
$$x = 8 \quad \text{or} \quad 3x = -2$$
$$x = 8 \quad \text{or} \quad x = -\tfrac{2}{3}.$$

The solutions are 8 and $-\tfrac{2}{3}$. The solution set is $\{8, -\tfrac{2}{3}\}$.

Answers

15. $\{3, 5\}$ **16.** $\left\{-\tfrac{13}{3}, 7\right\}$ **17.** $\emptyset$

Guided Solution:
16. 17, -13, $-\tfrac{13}{3}$, 7, $-\tfrac{13}{3}$

Solve.

18. $|5x - 3| = |x + 4|$

19. $|x - 3| = |x + 10|$

20. Solve: $|x| = 5$. Then graph on the number line.

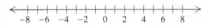

21. Solve: $|x| < 5$. Then graph.

22. Solve: $|x| \geq 5$. Then graph.

Answers

18. $\left\{\frac{7}{4}, -\frac{1}{6}\right\}$ 19. $\left\{-\frac{7}{2}\right\}$

20. $\{-5, 5\}$;

21. $\{x | -5 < x < 5\}$, or $(-5, 5)$;

22. $\{x | x \leq -5 \text{ or } x \geq 5\}$, or $(-\infty, -5] \cup [5, \infty)$;

EXAMPLE 15 Solve: $|x + 8| = |x - 5|$.

$$x + 8 = x - 5 \quad \text{or} \quad x + 8 = -(x - 5)$$
$$8 = -5 \quad \text{or} \quad x + 8 = -x + 5$$
$$8 = -5 \quad \text{or} \quad 2x = -3$$
$$8 = -5 \quad \text{or} \quad x = -\tfrac{3}{2}$$

The first equation has no solution. The solution of the second equation is $-\tfrac{3}{2}$. The solution set is $\left\{-\tfrac{3}{2}\right\}$.

◀ Do Exercises 18 and 19.

e INEQUALITIES WITH ABSOLUTE VALUE

We can extend our methods for solving equations with absolute value to those for solving inequalities with absolute value.

EXAMPLE 16 Solve: $|x| = 4$. Then graph on the number line.

From Example 7, we know that the solutions are -4 and 4. The solution set is $\{-4, 4\}$. The graph consists of just two points, as shown here.

◀ Do Exercise 20.

EXAMPLE 17 Solve: $|x| < 4$. Then graph.

The solutions of $|x| < 4$ are the solutions of $|x - 0| < 4$ and are those numbers x whose distance from 0 is less than 4. We can check by substituting or by looking at the number line to see that numbers like $-3, -2, -1, -\tfrac{1}{2}, -\tfrac{1}{4}, 0, \tfrac{1}{4}, \tfrac{1}{2}, 1, 2$, and 3 are all solutions. In fact, the solutions are all the real numbers x between -4 and 4. The solution set is $\{x | -4 < x < 4\}$ or, in interval notation, $(-4, 4)$. The graph is as follows.

◀ Do Exercise 21.

EXAMPLE 18 Solve: $|x| \geq 4$. Then graph.

The solutions of $|x| \geq 4$ are solutions of $|x - 0| \geq 4$ and are those numbers whose distance from 0 is greater than or equal to 4—in other words, those numbers x such that $x \leq -4$ or $x \geq 4$. The solution set is $\{x | x \leq -4 \text{ or } x \geq 4\}$, or $(-\infty, -4] \cup [4, \infty)$. The graph is as follows.

◀ Do Exercise 22.

Examples 16–18 illustrate three cases of solving equations and inequalities with absolute value. The following is a general principle for solving equations and inequalities with absolute value.

SOLUTIONS OF ABSOLUTE-VALUE EQUATIONS AND INEQUALITIES

For any positive number p and any algebraic expression X:

a) The solutions of $|X| = p$ are those numbers that satisfy
$X = -p$ or $X = p$.

As an example, replacing X with $5x - 1$ and p with 8, we see that the solutions of $|5x - 1| = 8$ are those numbers x for which

$5x - 1 = -8$ or $5x - 1 = 8$
$5x = -7$ or $5x = 9$
$x = -\frac{7}{5}$ or $x = \frac{9}{5}$.

The solution set is $\{-\frac{7}{5}, \frac{9}{5}\}$.

b) The solutions of $|X| < p$ are those numbers that satisfy
$-p < X < p$.

As an example, replacing X with $6x + 7$ and p with 5, we see that the solutions of $|6x + 7| < 5$ are those numbers x for which

$-5 < 6x + 7 < 5$
$-12 < 6x < -2$
$-2 < x < -\frac{1}{3}$.

The solution set is $\{x \mid -2 < x < -\frac{1}{3}\}$, or $\left(-2, -\frac{1}{3}\right)$.

c) The solutions of $|X| > p$ are those numbers that satisfy
$X < -p$ or $X > p$.

As an example, replacing X with $2x - 9$ and p with 4, we see that the solutions of $|2x - 9| > 4$ are those numbers x for which

$2x - 9 < -4$ or $2x - 9 > 4$
$2x < 5$ or $2x > 13$
$x < \frac{5}{2}$ or $x > \frac{13}{2}$.

The solution set is $\{x \mid x < \frac{5}{2} \text{ or } x > \frac{13}{2}\}$, or $\left(-\infty, \frac{5}{2}\right) \cup \left(\frac{13}{2}, \infty\right)$.

Solve. Then graph.

23. $|2x - 3| < 7$

24. $|7 - 3x| \leq 4$
$\boxed{} \leq 7 - 3x \leq 4$
$-11 \leq -3x \leq -3$
$\dfrac{-11}{-3} \boxed{} \dfrac{-3x}{-3} \boxed{} \dfrac{-3}{-3}$
$\dfrac{11}{3} \geq \boxed{} \geq 1$

The solution set is
$\left\{x \mid \boxed{} \leq x \leq \dfrac{11}{3}\right\}$, or $\left[1, \dfrac{11}{3}\right]$.

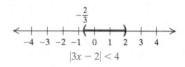

25. $|3x + 2| \geq 5$
$3x + 2 \leq \boxed{}$ or $3x + 2 \geq 5$
$3x \leq -7$ or $3x \geq \boxed{}$
$x \leq \boxed{}$ or $x \geq 1$

The solution set is
$\left\{x \mid x \leq -\dfrac{7}{3} \text{ or } x \geq \boxed{}\right\}$, or
$\left(\boxed{}, -\dfrac{7}{3}\right] \cup [\boxed{}, \infty)$.

EXAMPLE 19 Solve: $|3x - 2| < 4$. Then graph.

We use part (b). In this case, X is $3x - 2$ and p is 4:

$|X| < p$
$|3x - 2| < 4$ Replacing X with $3x - 2$ and p with 4
$-4 < 3x - 2 < 4$
$-2 < 3x < 6$
$-\dfrac{2}{3} < x < 2.$

The solution set is $\left\{x \mid -\dfrac{2}{3} < x < 2\right\}$, or $\left(-\dfrac{2}{3}, 2\right)$. The graph is as follows.

EXAMPLE 20 Solve: $|8 - 4x| \leq 5$. Then graph.

We use part (b). In this case, X is $8 - 4x$ and p is 5:

$|X| \leq p$
$|8 - 4x| \leq 5$ Replacing X with $8 - 4x$ and p with 5
$-5 \leq 8 - 4x \leq 5$
$-13 \leq -4x \leq -3$
$\dfrac{13}{4} \geq x \geq \dfrac{3}{4}.$ Dividing by -4 and reversing the inequality symbols

The solution set is $\left\{x \mid \dfrac{13}{4} \geq x \geq \dfrac{3}{4}\right\}$, or $\left\{x \mid \dfrac{3}{4} \leq x \leq \dfrac{13}{4}\right\}$, or $\left[\dfrac{3}{4}, \dfrac{13}{4}\right]$.

EXAMPLE 21 Solve: $|4x + 2| \geq 6$. Then graph.

We use part (c). In this case, X is $4x + 2$ and p is 6:

$|X| \geq p$
$|4x + 2| \geq 6$ Replacing X with $4x + 2$ and p with 6
$4x + 2 \leq -6$ or $4x + 2 \geq 6$
$4x \leq -8$ or $4x \geq 4$
$x \leq -2$ or $x \geq 1.$

The solution set is $\{x \mid x \leq -2 \text{ or } x \geq 1\}$, or $(-\infty, -2] \cup [1, \infty)$.

◀ Do Exercises 23–25.

Answers

23. $\{x \mid -2 < x < 5\}$, or $(-2, 5)$;

24. $\left\{x \mid 1 \leq x \leq \dfrac{11}{3}\right\}$, or $\left[1, \dfrac{11}{3}\right]$;

25. $\left\{x \mid x \leq -\dfrac{7}{3} \text{ or } x \geq 1\right\}$, or $\left(-\infty, -\dfrac{7}{3}\right] \cup [1, \infty)$;

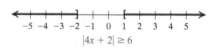

Guided Solutions:
24. $-4, \geq, \geq, x, 1$
25. $-5, 3, -\dfrac{7}{3}, 1, -\infty, 1$

9.3 Exercise Set

✓ Check Your Understanding

Reading Check Determine whether each statement is true or false.

RC1. The number a is $|a|$ units from 0 on the number line.

RC2. $|x|$ is never negative.

RC3. $|x|$ is always positive.

RC4. If two numbers are opposites, their absolute values are equal.

RC5. The distance between -53 and 46 is $|-53 - 46|$.

RC6. There are exactly two solutions of $|2x| = 10$.

Concept Check Choose from the column on the right the graph of the solution set of each equation or inequality.

CC1. $|x| > 3$

CC2. $|x| \geq 3$

CC3. $|x| < 3$

CC4. $|x| = 3$

CC5. $|x| \leq 3$

CC6. $|x| > -3$

a Simplify, leaving as little as possible inside absolute-value signs.

1. $|9x|$
2. $|26x|$
3. $|2x^2|$
4. $|8x^2|$

5. $|-2x^2|$
6. $|-20x^2|$
7. $|-6y|$
8. $|-17y|$

9. $\left|\dfrac{-2}{x}\right|$
10. $\left|\dfrac{y}{3}\right|$
11. $\left|\dfrac{x^2}{-y}\right|$
12. $\left|\dfrac{x^4}{-y}\right|$

13. $\left|\dfrac{-8x^2}{2x}\right|$
14. $\left|\dfrac{-9y^2}{3y}\right|$
15. $\left|\dfrac{4y^3}{-12y}\right|$
16. $\left|\dfrac{5x^3}{-25x}\right|$

b Find the distance between the points on the number line.

17. $-8, -46$
18. $-7, -32$
19. $36, 17$
20. $52, 18$

21. $-3.9, 2.4$
22. $-1.8, -3.7$
23. $-5, 0$
24. $\dfrac{2}{3}, -\dfrac{5}{6}$

c Solve.

25. $|x| = 3$
26. $|x| = 5$
27. $|x| = -3$
28. $|x| = -9$

29. $|q| = 0$
30. $|y| = 7.4$
31. $|x - 3| = 12$
32. $|3x - 2| = 6$

33. $|2x - 3| = 4$
34. $|5x + 2| = 3$
35. $|4x - 9| = 14$
36. $|9y - 2| = 17$

37. $|x| + 7 = 18$
38. $|x| - 2 = 6.3$
39. $574 = 283 + |t|$
40. $-562 = -2000 + |x|$

41. $|5x| = 40$
42. $|2y| = 18$
43. $|3x| - 4 = 17$
44. $|6x| + 8 = 32$

45. $7|w| - 3 = 11$
46. $5|x| + 10 = 26$
47. $\left|\dfrac{2x - 1}{3}\right| = 5$
48. $\left|\dfrac{4 - 5x}{6}\right| = 7$

49. $|m + 5| + 9 = 16$
50. $|t - 7| - 5 = 4$
51. $10 - |2x - 1| = 4$
52. $2|2x - 7| + 11 = 25$

53. $|3x - 4| = -2$
54. $|x - 6| = -8$
55. $\left|\dfrac{5}{9} + 3x\right| = \dfrac{1}{6}$
56. $\left|\dfrac{2}{3} - 4x\right| = \dfrac{4}{5}$

d Solve.

57. $|3x + 4| = |x - 7|$
58. $|2x - 8| = |x + 3|$
59. $|x + 3| = |x - 6|$

60. $|x - 15| = |x + 8|$
61. $|2a + 4| = |3a - 1|$
62. $|5p + 7| = |4p + 3|$

63. $|y - 3| = |3 - y|$

64. $|m - 7| = |7 - m|$

65. $|5 - p| = |p + 8|$

66. $|8 - q| = |q + 19|$

67. $\left|\dfrac{2x - 3}{6}\right| = \left|\dfrac{4 - 5x}{8}\right|$

68. $\left|\dfrac{6 - 8x}{5}\right| = \left|\dfrac{7 + 3x}{2}\right|$

69. $|\tfrac{1}{2}x - 5| = |\tfrac{1}{4}x + 3|$

70. $|2 - \tfrac{2}{3}x| = |4 + \tfrac{7}{8}x|$

e Solve.

71. $|x| < 3$

72. $|x| \leq 5$

73. $|x| \geq 2$

74. $|y| > 12$

75. $|x - 1| < 1$

76. $|x + 4| \leq 9$

77. $5|x + 4| \leq 10$

78. $2|x - 2| > 6$

79. $|2x - 3| \leq 4$

80. $|5x + 2| \leq 3$

81. $|2y - 7| > 10$

82. $|3y - 4| > 8$

83. $|4x - 9| \geq 14$

84. $|9y - 2| \geq 17$

85. $|y - 3| < 12$

86. $|p - 2| < 6$

87. $|2x + 3| \leq 4$

88. $|5x + 2| \leq 13$

89. $|4 - 3y| > 8$

90. $|7 - 2y| > 5$

91. $|9 - 4x| \geq 14$

92. $|2 - 9p| \geq 17$

93. $|3 - 4x| < 21$

94. $|-5 - 7x| \leq 30$

SECTION 9.3 Absolute-Value Equations and Inequalities

95. $\left|\dfrac{1}{2} + 3x\right| \geq 12$ 96. $\left|\dfrac{1}{4}y - 6\right| > 24$ 97. $\left|\dfrac{x-7}{3}\right| < 4$ 98. $\left|\dfrac{x+5}{4}\right| \leq 2$

99. $\left|\dfrac{2-5x}{4}\right| \geq \dfrac{2}{3}$ 100. $\left|\dfrac{1+3x}{5}\right| > \dfrac{7}{8}$ 101. $|m+5| + 9 \leq 16$ 102. $|t-7| + 3 \geq 4$

103. $7 - |3-2x| \geq 5$ 104. $16 \leq |2x-3| + 9$ 105. $\left|\dfrac{2x-1}{3}\right| \leq 1$ 106. $\left|\dfrac{3x-2}{5}\right| \geq 1$

Skill Maintenance

Find the slope, if it exists, of each line. [3.4b]

107. $x - 11y = 22$ 108. $x = 10$ 109. $10x = 5y - 3$ 110. $y = -\dfrac{2}{5}$

Factor. [5.6a]

Divide and simplify. [6.2b]

111. $27w^3 - 1000$ 112. $8 + 125t^3$ 113. $\dfrac{w-z}{3w} \div \dfrac{w^2 - z^2}{9w^3}$ 114. $\dfrac{t}{15} \div \dfrac{t}{25}$

Synthesis

115. *Motion of a Spring.* A weighted spring is bouncing up and down so that its distance d above the ground satisfies the inequality $|d - 6 \text{ ft}| \leq \tfrac{1}{2}$ ft. Find all possible distances d.

116. *Container Sizes.* A container company is manufacturing rectangular boxes of various sizes. The length of any box must exceed the width by at least 3 in., but the perimeter cannot exceed 24 in. What widths are possible?

$l \geq w + 3,$
$2l + 2w \leq 24$

Solve.

117. $|x + 5| > x$ 118. $1 - |\tfrac{1}{4}x + 8| = \tfrac{3}{4}$ 119. $|7x - 2| = x + 4$

120. $|x - 1| = x - 1$ 121. $|x - 6| \leq -8$ 122. $|3x - 4| > -2$

Find an equivalent inequality with absolute value.

123. $-3 < x < 3$ 124. $-5 \leq y \leq 5$ 125. $x \leq -6 \text{ or } x \geq 6$

126. $-5 < x < 1$ 127. $x < -8 \text{ or } x > 2$

Systems of Inequalities in Two Variables

9.4

OBJECTIVES

a Determine whether an ordered pair of numbers is a solution of an inequality in two variables.

b Graph linear inequalities in two variables.

c Graph systems of linear inequalities and find coordinates of any vertices.

A **graph** of an inequality is a drawing that represents its solutions. An inequality in one variable can be graphed on the number line. An inequality in two variables can be graphed on a coordinate plane.

A **linear inequality** is one that we can get from a related linear equation by changing the equals symbol to an inequality symbol. The graph of a linear inequality is the half-plane on one side of the graph of the related equation. The graph sometimes includes the graph of the related line at the boundary of the half-plane.

a SOLUTIONS OF INEQUALITIES IN TWO VARIABLES

The solutions of an inequality in two variables are ordered pairs.

EXAMPLES Determine whether the ordered pair is a solution of the inequality $5x - 4y > 13$.

1. $(-3, 2)$

$$\begin{array}{c|c} 5x - 4y > 13 \\ \hline 5(-3) - 4 \cdot 2 \; ? \; 13 \\ -15 - 8 \\ -23 & \text{FALSE} \end{array}$$

We use alphabetical order to replace x with -3 and y with 2.

Since $-23 > 13$ is false, $(-3, 2)$ is not a solution.

2. $(4, -3)$

$$\begin{array}{c|c} 5x - 4y > 13 \\ \hline 5(4) - 4(-3) \; ? \; 13 \\ 20 + 12 \\ 32 & \text{TRUE} \end{array}$$

Replacing x with 4 and y with -3

Since $32 > 13$ is true, $(4, -3)$ is a solution.

Do Exercises 1 and 2. ▶

1. Determine whether $(1, -4)$ is a solution of $4x - 5y < 12$.

$$\begin{array}{c} 4x - 5y < 12 \\ \hline ? \end{array}$$

2. Determine whether $(4, -3)$ is a solution of $3y - 2x \le 6$.

$$\begin{array}{c} 3y - 2x \le 6 \\ \hline ? \end{array}$$

b GRAPHING INEQUALITIES IN TWO VARIABLES

MyLab Math
ANIMATION

SKILL REVIEW *Graph linear equations using intercepts.* [7.4a]
Find the intercepts. Then graph the equation.
1. $3x - 2y = 6$ **2.** $2x + y = 4$

Answers: **1.** 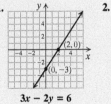   **2.**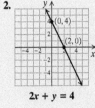

$3x - 2y = 6$ $2x + y = 4$

MyLab Math
VIDEO

Answers
1. No **2.** Yes

SECTION 9.4 Systems of Inequalities in Two Variables **693**

Let's visualize the results of Examples 1 and 2. The equation $5x - 4y = 13$ is represented by the dashed line in the following graphs.

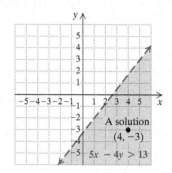

The solutions of the inequality $5x - 4y > 13$ are shaded below that dashed line. As shown in the graph on the left above, the pair $(-3, 2)$ is not a solution of the inequality $5x - 4y > 13$ and is not in the shaded region.

The pair $(4, -3)$ is a solution of the inequality $5x - 4y > 13$ and is in the shaded region. See the graph on the right above.

We now consider how to graph inequalities.

EXAMPLE 3 Graph: $y < x$.

We first graph the line $y = x$. Every solution of $y = x$ is an ordered pair like $(3, 3)$, where the first and second coordinates are the same. The graph of $y = x$ is shown in Figure 1. We draw it dashed because these points are *not* solutions of $y < x$.

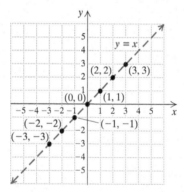

FIGURE 1   **FIGURE 2**

Now look at the graph in Figure 2. Several ordered pairs are plotted on the half-plane below $y = x$. Each is a solution of $y < x$. We can check the pair $(4, 2)$ as follows:

$$\frac{y < x}{2 \; ? \; 4 \quad \text{TRUE}}$$

It turns out that any point on the same side of $y = x$ as $(4, 2)$ is also a solution. Thus, *if you know that one point in a half-plane is a solution of an inequality, then all points in that half-plane are solutions*. In this text, we will usually indicate this by color shading. We shade the half-plane below $y = x$, as shown in Figure 3.

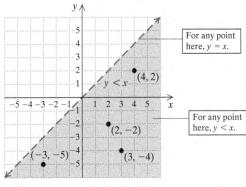

FIGURE 3

EXAMPLE 4 Graph: $8x + 3y \geq 24$.

First, we sketch the line $8x + 3y = 24$. Points on the line $8x + 3y = 24$ are also in the graph of $8x + 3y \geq 24$, so we draw the line solid. This indicates that all points on the line are solutions. The rest of the solutions are in the half-plane either to the left or to the right of the line. To determine which, we select a point that is not on the line and determine whether it is a solution of $8x + 3y \geq 24$. We try $(-3, 4)$ as a test point.

$$8x + 3y \geq 24$$

$$8(-3) + 3(4) \;?\; 24 \quad \text{Using } (-3, 4) \text{ as a test point}$$
$$-24 + 12$$
$$-12 \quad \text{FALSE}$$

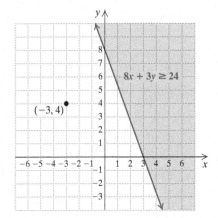

We see that $-12 \geq 24$ is *false*. Since $(-3, 4)$ is not a solution, none of the points in the half-plane containing $(-3, 4)$ is a solution. Thus the points in the opposite half-plane are solutions. We shade that half-plane and obtain the graph shown at right.

To graph an inequality in two variables:

1. Replace the inequality symbol with an equals sign and graph this related equation. This separates points that represent solutions from those that do not.

2. If the inequality symbol is $<$ or $>$, draw the line dashed. If the inequality symbol is $\leq$ or $\geq$, draw the line solid.

3. The graph consists of a half-plane that is either above or below or to the left or to the right of the line and, if the line is solid, the line as well. To determine which half-plane to shade, choose a point not on the line as a test point. Substitute to determine whether that point is a solution. If so, shade the half-plane containing that point. If not, shade the opposite half-plane.

Graph.

3. $6x - 3y < 18$

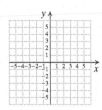

4. $4x + 3y \geq 12$ GS
   1. Graph the related equation
      $4x + 3y$ ____ 12.
   2. Since the inequality symbol is $\geq$, draw a _____ line.
      dashed/solid
   3. Try the test point $(0, 0)$.
      $$\begin{array}{c|c} 4x + 3y \geq 12 \\ \hline 4(0) + 3() \;?\; 12 \\ 0 & \text{FALSE} \end{array}$$
      $(0, 0)$ is not a solution, so we shade the opposite half-plane.

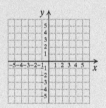

EXAMPLE 5 Graph: $6x - 2y < 12$.

1. We first graph the related equation $6x - 2y = 12$.
2. Since the inequality uses the symbol $<$, points on the line are not solutions of the inequality, so we draw a dashed line.
3. To determine which half-plane to shade, we consider a test point *not* on the line. We try $(0, 0)$ and substitute:
   $$\begin{array}{c|c} 6x - 2y < 12 \\ \hline 6(0) - 2(0) \;?\; 12 \\ 0 - 0 \\ 0 & \text{TRUE} \end{array}$$

Since the inequality $0 < 12$ is *true*, the point $(0, 0)$ is a solution; each point in the half-plane containing $(0, 0)$ is a solution. Thus each point in the opposite half-plane is *not* a solution. The graph is shown below.

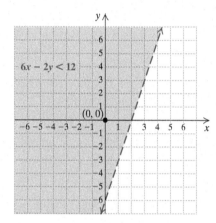

◀ Do Exercises 3 and 4.

EXAMPLE 6 Graph $x > -3$ on a plane.

There is only one variable in this inequality. If we graph the inequality on the number line, its graph is as follows:

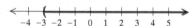

However, we can also write this inequality as $x + 0y > -3$ and consider graphing it in the plane. We first graph the related equation $x = -3$ in the plane. We draw the boundary with a dashed line. The rest of the graph is a half-plane to the right or to the left of the line $x = -3$. To determine which, we consider a test point, $(2, 5)$.
$$\begin{array}{c|c} x + 0y > -3 \\ \hline 2 + 0(5) \;?\; -3 \\ 2 & \text{TRUE} \end{array}$$

Answers

3.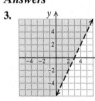
   $6x - 3y < 18$

4.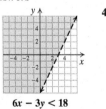
   $4x + 3y \geq 12$

Guided Solution:
4. $=$, solid, 0

Since (2, 5) is a solution, all the points in the half-plane containing (2, 5) are solutions. We shade that half-plane.

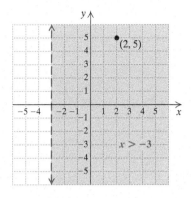

EXAMPLE 7 Graph $y \leq 4$ on a plane.

We first graph $y = 4$ using a solid line. We then use $(-2, 5)$ as a test point and substitute in $0x + y \leq 4$.

$$\begin{array}{c|c} 0x + y \leq 4 \\ \hline 0(-2) + 5 \; ? \; 4 \\ 0 + 5 \\ 5 \end{array} \quad \text{FALSE}$$

We see that $(-2, 5)$ is *not* a solution, so all the points in the half-plane containing $(-2, 5)$ are not solutions. Thus each point in the opposite half-plane is a solution. The graph of the inequality is shown below.

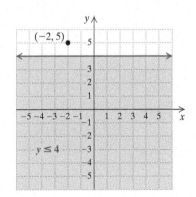

Graph on a plane.

5. $x < 3$

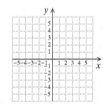

6. $y \geq -4$

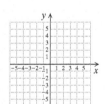

Do Exercises 5 and 6.

c SYSTEMS OF LINEAR INEQUALITIES

The following is an example of a system of two linear inequalities in two variables:

$$x + y \leq 4,$$
$$x - y < 4.$$

A **solution** of a system of linear inequalities is an ordered pair that is a solution of *both* inequalities. To graph solutions of systems of linear inequalities, we graph each inequality and determine where the graphs overlap, or intersect. That will be a region in which the ordered pairs are solutions of both inequalities.

Answers

5.   6.

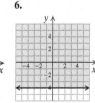

$x < 3$ $y \geq -4$

SECTION 9.4 Systems of Inequalities in Two Variables

EXAMPLE 8 Graph the solutions of the system

$$x + y \leq 4,$$
$$x - y < 4.$$

We graph $x + y \leq 4$ by first graphing the equation $x + y = 4$ using a solid red line. We consider $(0, 0)$ as a test point and find that it is a solution, so we shade all points on that side of the line using red shading. (See the graph on the left below.) The arrows near the ends of the line also indicate the half-plane, or region, that contains the solutions.

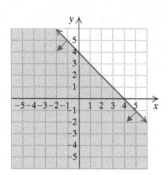 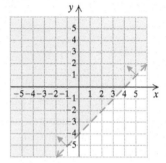

Next, we graph $x - y < 4$. We begin by graphing the equation $x - y = 4$ using a dashed blue line and consider $(0, 0)$ as a test point. Again, $(0, 0)$ is a solution so we shade that side of the line using blue shading. (See the graph on the right above.) The solution set of the system is the region that is shaded both red and blue and part of the line $x + y = 4$.

7. Graph:

$$x + y \geq 1,$$
$$y - x \geq 2.$$

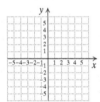

◀ Do Exercise 7.

Answers

7.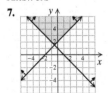

EXAMPLE 9 Graph: $-2 < x \leq 5$.

This is actually a system of inequalities:

$-2 < x,$
$x \leq 5.$

We graph the equation $-2 = x$ and see that the graph of the first inequality is the half-plane to the right of the line $-2 = x$. (See the graph on the left below.)

Next, we graph the second inequality, starting with the line $x = 5$, and find that its graph is the line as well as the half-plane to the left of it. (See the graph on the right below.) Then we shade the intersection of these graphs.

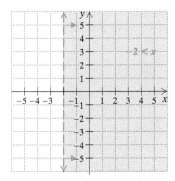
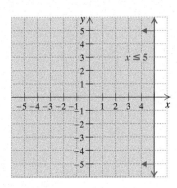

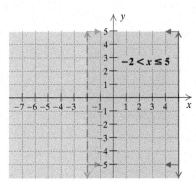

8. Graph: $-3 \leq y < 4$.

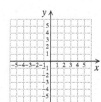

Do Exercise 8.

Answer

8.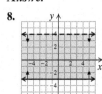

A system of inequalities may have a graph that consists of a polygon and its interior. In *linear programming*, which is a topic rich in application that you may study in a later course, it is important to be able to find the vertices of such a polygon.

EXAMPLE 10 Graph the following system of inequalities. Find the coordinates of any vertices formed.

$$6x - 2y \le 12, \quad (1)$$
$$y - 3 \le 0, \quad (2)$$
$$x + y \ge 0 \quad (3)$$

We graph the lines $6x - 2y = 12$, $y - 3 = 0$, and $x + y = 0$ using solid lines. The regions for each inequality are indicated by the arrows at the ends of the lines. We then note where the regions overlap and shade the region of solutions using one color.

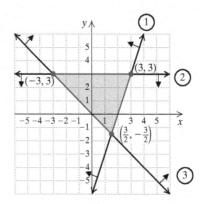

9. Graph the system of inequalities. Find the coordinates of any vertices formed.

$$5x + 6y \le 30,$$
$$0 \le y \le 3,$$
$$0 \le x \le 4$$

To find the vertices, we solve three different systems of equations. The system of equations from inequalities (1) and (2) is

$$6x - 2y = 12, \quad (1)$$
$$y - 3 = 0. \quad (2)$$

Solving, we obtain the vertex $(3, 3)$.
The system of equations from inequalities (1) and (3) is

$$6x - 2y = 12, \quad (1)$$
$$x + y = 0. \quad (3)$$

Solving, we obtain the vertex $\left(\frac{3}{2}, -\frac{3}{2}\right)$.
The system of equations from inequalities (2) and (3) is

$$y - 3 = 0, \quad (2)$$
$$x + y = 0. \quad (3)$$

Solving, we obtain the vertex $(-3, 3)$.

◀ Do Exercise 9.

Answer
9.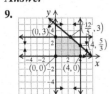

EXAMPLE 11 Graph the following system of inequalities. Find the coordinates of any vertices formed.

$$x + y \leq 16, \quad (1)$$
$$3x + 6y \leq 60, \quad (2)$$
$$x \geq 0, \quad (3)$$
$$y \geq 0 \quad (4)$$

We graph each inequality using solid lines. The regions for each inequality are indicated by the arrows at the ends of the lines. We then note where the regions overlap and shade the region of solutions using one color.

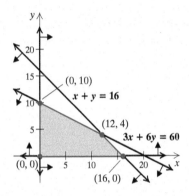

To find the vertices, we solve four different systems of equations. The system of equations from inequalities (1) and (2) is

$$x + y = 16, \quad (1)$$
$$3x + 6y = 60. \quad (2)$$

Solving, we obtain the vertex $(12, 4)$.
The system of equations from inequalities (1) and (4) is

$$x + y = 16, \quad (1)$$
$$y = 0. \quad (4)$$

Solving, we obtain the vertex $(16, 0)$.
The system of equations from inequalities (3) and (4) is

$$x = 0, \quad (3)$$
$$y = 0. \quad (4)$$

The vertex is $(0, 0)$.
The system of equations from inequalities (2) and (3) is

$$3x + 6y = 60, \quad (2)$$
$$x = 0. \quad (3)$$

Solving, we obtain the vertex $(0, 10)$.

Do Exercise 10.

10. Graph the system of inequalities. Find the coordinates of any vertices formed.

$$2x + 4y \leq 8,$$
$$x + y \leq 3,$$
$$x \geq 0,$$
$$y \geq 0$$

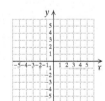

Answer
10.

SECTION 9.4 Systems of Inequalities in Two Variables

Visualizing for Success

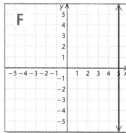
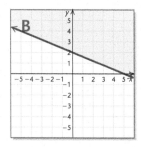
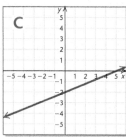
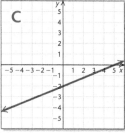
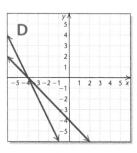
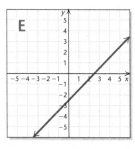
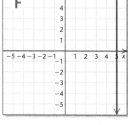
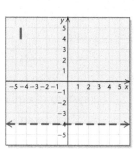

Match the equation, inequality, system of equations, or system of inequalities with its graph.

1. $x + y = -4$,
   $2x + y = -8$

2. $2x + 5y \geq 10$

3. $2x - 2y = 5$

4. $2x - 5y = 10$

5. $-2y < 8$

6. $5x - 2y = 10$

7. $2x = 10$

8. $5x + 2y < 10$,
   $2x - 5y > 10$

9. $5x \geq -10$

10. $y - 2x < 8$

Answers on page A-29

9.4 Exercise Set

FOR EXTRA HELP MyLab Math

✓ Check Your Understanding

Reading Check Choose from the column on the right the word that best completes each statement.

RC1. A(n) _____ of an inequality is a drawing that represents its solutions.

RC2. The sentence $4x - y < 3$ is an example of a linear _____.

RC3. The graph of $4x - y < 3$ is a(n) _____.

RC4. The ordered pair $(1, 6)$ is a(n) _____ of $4x - y < 3$.

RC5. For $4x - y < 3$, the related _____ is $4x - y = 3$.

RC6. To determine which half-plane to shade when graphing an inequality, we can use a(n) _____ point.

equation
graph
half-plane
inequality
solution
test

Concept Check Determine whether the graph of the related equation should be drawn dashed or solid in order to graph the inequality.

CC1. $2x - y < 5$ CC2. $x \geq -3$ CC3. $4y \leq 2 - x$

Determine whether the point $(0, 0)$ is a solution of each inequality.

CC4. $2x + y < 5$ CC5. $y < x - 3$ CC6. $x \leq 0.5$

a Determine whether the given ordered pair is a solution of the given inequality.

1. $(-3, 3)$; $3x + y < -5$
2. $(6, -8)$; $4x + 3y \geq 0$
3. $(5, 9)$; $2x - y > -1$
4. $(5, -2)$; $6y - x > 2$

b Graph each inequality on a plane.

5. $y > 2x$

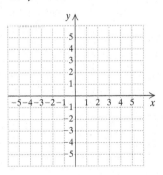

6. $y < 3x$

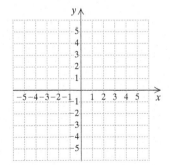

7. $y < x + 1$

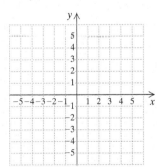

8. $y \leq x - 3$

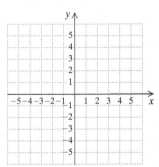

9. $y > x - 2$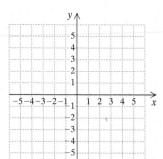

10. $y \geq x + 4$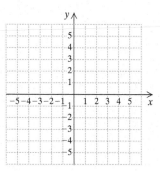

11. $x + y < 4$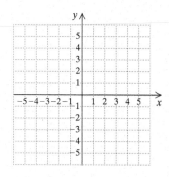

12. $x - y \geq 3$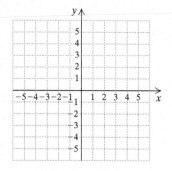

13. $3x + 4y \leq 12$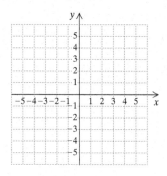

14. $2x + 3y < 6$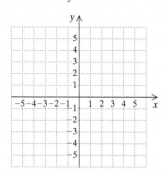

15. $2y - 3x > 6$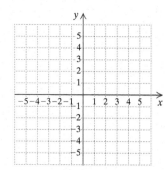

16. $2y - x \leq 4$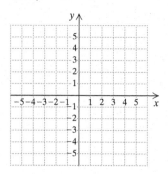

17. $3x - 2 \leq 5x + y$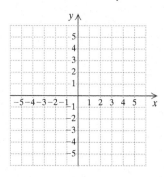

18. $2x - 2y \geq 8 + 2y$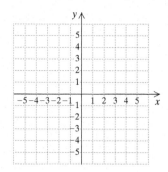

19. $x < 5$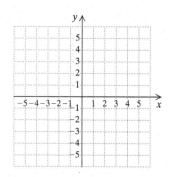

20. $y \geq -2$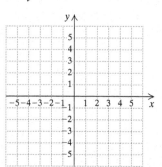

21. $y > 2$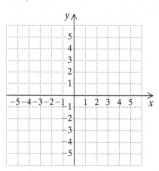

22. $x \leq -4$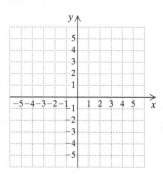

23. $2x + 3y \leq 6$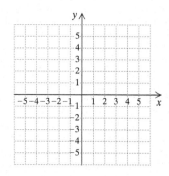

24. $7x + 2y \geq 21$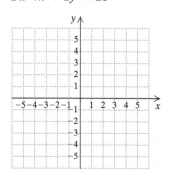

Matching. Each of Exercises 25–30 shows the graph of an inequality. Match the graph with one of the appropriate inequalities (A)–(F), which follow.

25.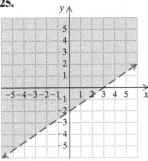

26.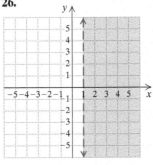

27.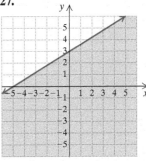

28.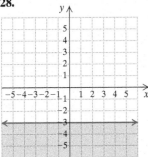

29.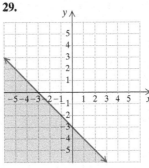

30.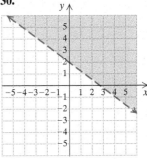

A. $4y > 8 - 3x$ **B.** $3x \geq 5y - 15$ **C.** $y + x \leq -3$ **D.** $x > 1$ **E.** $y \leq -3$ **F.** $2x - 3y < 6$

C Graph each system of inequalities. Find the coordinates of any vertices formed.

31. $y \geq x,$
$y \leq -x + 2$

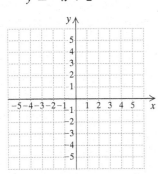

32. $y \geq x,$
$y \leq -x + 4$

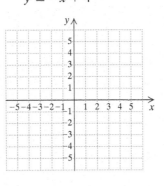

33. $y > x,$
$y < -x + 1$

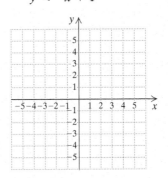

34. $y < x,$
$y > -x + 3$

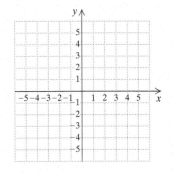

35. $x \leq 3,$
$y \geq -3x + 2$

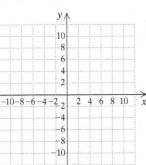

36. $x \geq 2,$
$y \leq -2x + 3$

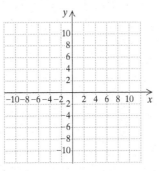

37. $x + y \leq 1,$
$x - y \leq 2$

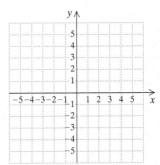

38. $x + y \leq 3,$
$x - y \leq 4$

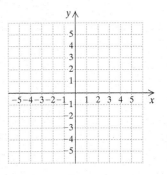

39. $y \le 2x + 1,$
$y \ge -2x + 1,$
$x \le 2$

40. $x - y \le 2,$
$x + 2y \ge 8,$
$y \le 4$

41. $x + 2y \le 12,$
$2x + y \le 12,$
$x \ge 0,$
$y \ge 0$

42. $y - x \ge 1,$
$y - x \le 3,$
$2 \le x \le 5$

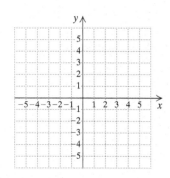
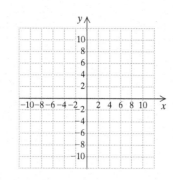
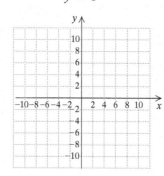
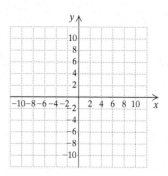

Skill Maintenance

Solve. [2.3b, c]

43. $5(3x - 4) = -2(x + 5)$

44. $4(3x + 4) = 2 - x$

45. $2(x - 1) + 3(x - 2) - 4(x - 5) = 10$

46. $10x - 8(3x - 7) = 2(4x - 1)$

47. $5x + 7x = -144$

48. $0.5x - 2.34 + 2.4x = 7.8x - 9$

Given the function $f(x) = |2 - x|$, find each of the following function values. [7.1b]

49. $f(0)$

50. $f(-1)$

51. $f(10)$

52. $f(2a)$

Synthesis

53. *Waterfalls.* In order for a waterfall to be classified as a classical waterfall, its height must be less than twice its crest width, and its crest width cannot exceed one-and-a-half times its height. The tallest waterfall in the world is about 3200 ft high. Let h represent a waterfall's height, in feet, and w the crest width, in feet. Write and graph a system of inequalities that represents all possible combinations of heights and crest widths of classical waterfalls.

54. *Exercise Danger Zone.* It is dangerous to exercise when the weather is hot and humid. The solutions of the following system of inequalities give a "danger zone" in which it is dangerous to exercise intensely:

$$4H - 3F < 70,$$
$$F + H > 160,$$
$$2F + 3H > 390,$$

where F is the temperature, in degrees Fahrenheit, and H is the humidity.

a) Draw the danger zone by graphing the system of inequalities.

b) Is it dangerous to exercise when $F = 80°$ and $H = 80\%$?

CHAPTER 9 Summary and Review

Vocabulary Reinforcement

Complete each statement with the correct term from the column on the right. Some of the choices may not be used and some may be used more than once.

1. A(n) _____ is a sentence containing $<, \leq, >, \geq,$ or $\neq$. [9.1a]

2. Using _____ notation, we write the solution set for $x < 7$ as $\{x | x < 7\}$. [9.1b]

3. Using _____ notation, we write the solution set of $-5 \leq y < 16$ as $[-5, 16)$. [9.1b]

4. The _____ of two sets A and B is the set of all members that are common to A and B. [9.2a]

5. When two or more sentences are joined by the word *and* to make a compound sentence, the new sentence is called a(n) _____ of the sentences. [9.2a]

6. When two sets have no elements in common, the intersection of the two sets is the _____. [9.2a]

7. Two sets with an empty intersection are said to be _____. [9.2a]

8. The _____ of two sets A and B is the collection of elements belonging to A and/or B. [9.2b]

9. When two or more sentences are joined by the word *or* to make a compound sentence, the new sentence is called a(n) _____ of the sentences. [9.2b]

10. A quantity is _____ some amount q: $x \leq q$. [9.1d]

11. A quantity is _____ some amount q: $x \geq q$. [9.1d]

12. For any real numbers a and b, the _____ between them is $|a - b|$. [9.3b]

union
set-builder
empty set
absolute value
disjunction
at least
at most
inequality
intersection
distance
interval
disjoint sets
compound
conjunction

Concept Reinforcement

Determine whether each statement is true or false.

_____ 1. If one point in a half-plane is a solution of a linear inequality, then all points in that half-plane are solutions. [9.4b]

_____ 2. Every system of linear inequalities has at least one solution. [9.4c]

_____ 3. For any real numbers a, b, and c, $c \neq 0$, $a \leq b$ is equivalent to $ac \leq bc$. [9.1c]

_____ 4. The inequalities $x < 2$ and $x \leq 1$ are equivalent. [9.1c]

_____ 5. If x is negative, $|x| = -x$. [9.3a]

_____ 6. $|x|$ is always positive. [9.3a]

_____ 7. $|a - b| = |b - a|$. [9.3b]

Study Guide

Objective 9.1a Determine whether a given number is a solution of an inequality.

Example Determine whether -3 and 1 are solutions of the inequality $4 - x \geq 2 - 5x$.

We substitute -3 for x and get
$$4 - (-3) \geq 2 - 5(-3), \text{ or } 7 \geq 17,$$
a *false* sentence. Therefore, -3 is not a solution.

We substitute 1 for x and get
$$4 - 1 \geq 2 - 5 \cdot 1, \text{ or } 3 \geq -3,$$
a *true* sentence. Therefore, 1 is a solution.

Practice Exercise

1. Determine whether -2 and 5 are solutions of the inequality $8 - 3x \leq 3x + 6$.

Objective 9.1b Write interval notation for the solution set of an inequality.

Example Write interval notation for the solution set.
a) $\{x \mid x \leq -12\} = (-\infty, -12]$
b) $\{r \mid r > -1\} = (-1, \infty)$
c) $\{y \mid -8 \leq y < 9\} = [-8, 9)$
d) $\{x \mid 0 \geq x \geq -6\} = [-6, 0]$
e) $\{c \mid -25 < c \leq 25\} = (-25, 25]$

Practice Exercise

2. Write interval notation for the solution set.
a) $\{t \mid t < -8\}$
b) $\{x \mid -7 \leq x < 10\}$
c) $\{b \mid b \geq 3\}$

Objective 9.1c Solve an inequality using the addition principle and the multiplication principle and then graph the inequality.

Example Solve and graph: $6x - 7 \leq 3x + 2$.

$6x - 7 \leq 3x + 2$
$3x - 7 \leq 2$ Subtracting $3x$
$3x \leq 9$ Adding 7
$x \leq 3$ Dividing by 3

The solution set is $\{x \mid x \leq 3\}$, or $(-\infty, 3]$. We graph the solution set.

Practice Exercise

3. Solve and graph: $5y + 5 < 2y - 1$.

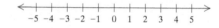

Objective 9.2a Find the intersection of two sets. Solve and graph conjunctions of inequalities.

Example Solve and graph: $-5 < 2x - 3 \leq 3$.

$-5 < 2x - 3 \leq 3$
$-2 < 2x \leq 6$ Adding 3
$-1 < x \leq 3$ Dividing by 2

The solution set is $\{x \mid -1 < x \leq 3\}$, or $(-1, 3]$. We graph the solution set.

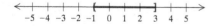

Practice Exercise

4. Solve and graph: $-4 \leq 5z + 6 < 11$.

Objective 9.2b Find the union of two sets. Solve and graph disjunctions of inequalities.

Example Solve and graph:
$$2x + 1 \leq -5 \quad \text{or} \quad 3x + 1 > 7.$$

$$2x + 1 \leq -5 \quad \text{or} \quad 3x + 1 > 7$$
$$2x \leq -6 \quad \text{or} \quad 3x > 6$$
$$x \leq -3 \quad \text{or} \quad x > 2$$

The solution set is $\{x \mid x \leq -3 \text{ or } x > 2\}$, or $(-\infty, -3] \cup (2, \infty)$. We graph the solution set.

Practice Exercise

5. Solve and graph: $z + 4 < 3$ or $4z + 1 \geq 5$.

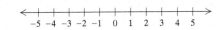

Objective 9.3c Solve equations with absolute-value expressions.

Example Solve: $|y - 2| = 1$.
$$y - 2 = -1 \quad \text{or} \quad y - 2 = 1$$
$$y = 1 \quad \text{or} \quad y = 3$$
The solution set is $\{1, 3\}$.

Practice Exercise

6. Solve: $|5x - 1| = 9$.

Objective 9.3d Solve equations with two absolute-value expressions.

Example Solve: $|4x - 4| = |2x + 8|$.
$$4x - 4 = 2x + 8 \quad \text{or} \quad 4x - 4 = -(2x + 8)$$
$$2x - 4 = 8 \quad \text{or} \quad 4x - 4 = -2x - 8$$
$$2x = 12 \quad \text{or} \quad 6x - 4 = -8$$
$$x = 6 \quad \text{or} \quad 6x = -4$$
$$x = 6 \quad \text{or} \quad x = -\frac{2}{3}$$

The solution set is $\left\{6, -\frac{2}{3}\right\}$.

Practice Exercise

7. Solve: $|z + 4| = |3z - 2|$.

Objective 9.3e Solve inequalities with absolute-value expressions.

Example Solve: **(a)** $|5x + 3| < 2$; **(b)** $|x + 3| \geq 1$.

a) $|5x + 3| < 2$
$$-2 < 5x + 3 < 2$$
$$-5 < 5x < -1$$
$$-1 < x < -\frac{1}{5}$$

The solution set is $\left\{x \mid -1 < x < -\frac{1}{5}\right\}$, or $\left(-1, -\frac{1}{5}\right)$.

b) $|x + 3| \geq 1$
$$x + 3 \leq -1 \quad \text{or} \quad x + 3 \geq 1$$
$$x \leq -4 \quad \text{or} \quad x \geq -2$$

The solution set is $\{x \mid x \leq -4 \text{ or } x \geq -2\}$, or $(-\infty, -4] \cup [-2, \infty)$.

Practice Exercise

8. Solve: **(a)** $|2x + 3| < 5$; **(b)** $|3x + 2| \geq 8$.

Objective 9.4b Graph linear inequalities in two variables.

Example Graph: $2x + y \leq 4$.

First, we graph the line $2x + y = 4$. The intercepts are $(0, 4)$ and $(2, 0)$. We draw the line solid because the inequality symbol is $\leq$. Next, we choose a test point not on the line and determine whether it is a solution of the inequality. We choose $(0, 0)$, since it is usually an easy point to use.

$$\begin{array}{c|c} 2x + y \leq 4 \\ \hline 2 \cdot 0 + 0 \; ? \; 4 \\ 0 \; | \; \text{TRUE} \end{array}$$

Since $(0, 0)$ is a solution, we shade the half-plane that contains $(0, 0)$.

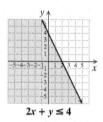
$2x + y \leq 4$

Practice Exercise

9. Graph: $3x - 2y > 6$.

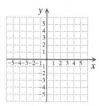

Objective 9.4c Graph systems of linear inequalities and find coordinates of any vertices.

Example Graph this system of inequalities and find the coordinates of any vertices formed:

$$x - 2y \geq -2, \quad (1)$$
$$3x - y \leq 4, \quad (2)$$
$$y \geq -1. \quad (3)$$

We graph the related equations using solid lines. Then we indicate the region for each inequality by arrows at the ends of the line. Next, we shade the region of overlap.

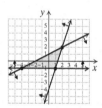

To find the vertices, we solve three different systems of related equations. From (1) and (2), we solve

$$x - 2y = -2,$$
$$3x - y = 4$$

to find the vertex $(2, 2)$. From (1) and (3), we solve

$$x - 2y = -2,$$
$$y = -1$$

to find the vertex $(-4, -1)$. From (2) and (3), we solve

$$3x - y = 4,$$
$$y = -1$$

to find the vertex $(1, -1)$.

Practice Exercise

10. Graph this system of inequalities and find the coordinates of any vertices found:

$$x - 2y \leq 4,$$
$$x + y \leq 4,$$
$$x - 1 \geq 0.$$

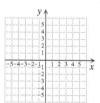

Review Exercises

1. Determine whether -3 and 7 are solutions of the inequality $2(1 - x) \leq 3x + 15$. [9.1a]

Write interval notation for the given set or graph. [9.1b]

2. $\{x \mid -8 \leq x < 9\}$

3.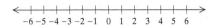

Solve and graph. Write interval notation for the solution set. [9.1c]

4. $x - 2 \leq -4$

5. $x + 5 > 6$

Solve. [9.1c]

6. $a + 7 \leq -14$ 7. $y - 5 \geq -12$

8. $4y > -16$ 9. $-0.3y < 9$

10. $-6x - 5 < 13$

11. $4y + 3 \leq -6y - 9$

12. $-\frac{1}{2}x - \frac{1}{4} > \frac{1}{2} - \frac{1}{4}x$

13. $0.3y - 8 < 2.6y + 15$

14. $2(x - 5) \geq 6(x + 7) - 12$

15. *Moving Costs.* Metro Movers charges $85 plus $40 per hour to move households across town. Champion Moving charges $60 per hour for cross-town moves. For what lengths of time is Champion more expensive? [9.1d]

16. *Investments.* Joe plans to invest $30,000, part at 3% and part at 4%, for one year. What is the most that can be invested at 3% in order to make at least $1100 interest in one year? [9.1d]

Graph and write interval notation. [9.2a, b]

17. $-2 \leq x < 5$

18. $x \leq -2$ or $x > 5$

19. Find the intersection:
$\{1, 2, 5, 6, 9\} \cap \{1, 3, 5, 9\}$. [9.2a]

20. Find the union:
$\{1, 2, 5, 6, 9\} \cup \{1, 3, 5, 9\}$. [9.2b]

Solve. [9.2a, b]

21. $2x - 5 < -7$ and $3x + 8 \geq 14$

22. $-4 < x + 3 \leq 5$

23. $-15 < -4x - 5 < 0$

24. $3x < -9$ or $-5x < -5$

25. $2x + 5 < -17$ or $-4x + 10 \leq 34$

26. $2x + 7 \leq -5$ or $x + 7 \geq 15$

Simplify. [9.3a]

27. $\left| -\dfrac{3}{x} \right|$ 28. $\left| \dfrac{2x}{y^2} \right|$ 29. $\left| \dfrac{12y}{-3y^2} \right|$

30. Find the distance between -23 and 39. [9.3b]

Solve. [9.3c, d]

31. $|x| = 6$ 32. $|x - 2| = 7$

33. $|2x + 5| = |x - 9|$ 34. $|5x + 6| = -8$

Solve. [9.3e]

35. $|2x + 5| < 12$ 36. $|x| \geq 3.5$

37. $|3x - 4| \geq 15$ 38. $|x| < 0$

Graph. [9.4b]

39. $2x + 3y < 12$

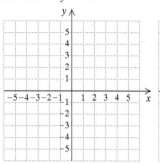

40. $y \leq 0$

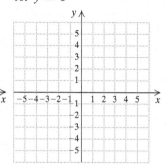

41. $x + y \geq 1$

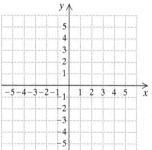

Graph. Find the coordinates of any vertices formed. [9.4c]

42. $y \geq -3$,
$x \geq 2$

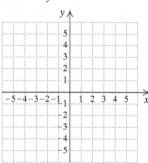

43. $x + 3y \geq -1$,
$x + 3y \leq 4$

44. $x - y \leq 3$,
$x + y \geq -1$,
$y \leq 2$

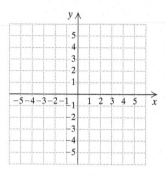

45. Solve: $-2(x + 3) - 1 < -2(2 - x)$. [9.1c]

A. $\left(-\infty, -\dfrac{3}{4}\right)$ **B.** $(-6, \infty)$

C. $(6, \infty)$ **D.** $\left(-\dfrac{3}{4}, \infty\right)$

46. Solve: $\left|-\dfrac{1}{3}x - 10\right| \geq 30$. [9.3e]

A. $(-\infty, -100] \cup [80, \infty)$
B. $(-\infty, -60] \cup [120, \infty)$
C. $(-\infty, -120] \cup [60, \infty)$
D. $(-\infty, -80] \cup [100, \infty)$

Synthesis

47. Solve: $|2x + 5| \leq |x + 3|$. [9.3d, e]

Understanding Through Discussion and Writing

1. Describe the circumstances under which, for intervals, $[a, b] \cup [c, d] = [a, d]$. [9.2b]

2. Explain in your own words why the solutions of the inequality $|x + 5| \leq 2$ can be interpreted as "all those numbers x whose distance from -5 is at most 2 units." [9.3e]

3. When graphing linear inequalities, Ron always shades above the line when he sees a $\geq$ symbol. Is this wise? Why or why not? [9.4a]

4. Explain in your own words why the interval $[6, \infty)$ is only part of the solution set of $|x| \geq 6$. [9.3e]

CHAPTER 9 Test

Write interval notation for the given set or graph.

1. $\{x \mid -3 < x \leq 2\}$

2.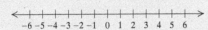

Solve and graph. Write interval notation for the solution set.

3. $x - 2 \leq 4$

4. $-4y - 3 \geq 5$

Solve.

5. $x - 4 \geq 6$

6. $-0.6y < 30$

7. $3a - 5 \leq -2a + 6$

8. $-5y - 1 > -9y + 3$

9. $4(5 - x) < 2x + 5$

10. $-8(2x + 3) + 6(4 - 5x) \geq 2(1 - 7x) - 4(4 + 6x)$

Solve.

11. *Moving Costs.* Mitchell Moving Company charges $105 plus $30 per hour to move households across town. Quick-Pak Moving charges $80 per hour for cross-town moves. For what lengths of time is Quick-Pak more expensive?

12. *Pressure at Sea Depth.* The equation
$$P = 1 + \frac{d}{33}$$
gives the pressure P, in atmospheres (atm), at a depth of d feet in the sea. For what depths d is the pressure at least 2 atm and at most 8 atm?

Graph and write interval notation.

13. $-3 \leq x \leq 4$

14. $x < -3$ or $x > 4$

Solve.

15. $5 - 2x \leq 1$ and $3x + 2 \geq 14$

16. $-3 < x - 2 < 4$

17. $-11 \leq -5x - 2 < 0$

18. $-3x > 12$ or $4x > -10$

19. $x - 7 \leq -5$ or $x - 7 \geq -10$

20. $3x - 2 < 7$ or $x - 2 > 4$

Simplify.

21. $\left|\dfrac{7}{x}\right|$

22. $\left|\dfrac{-6x^2}{3x}\right|$

23. Find the distance between 4.8 and -3.6.

24. Find the intersection:
$\{1, 3, 5, 7, 9\} \cap \{3, 5, 11, 13\}$.

25. Find the union:
$\{1, 3, 5, 7, 9\} \cup \{3, 5, 11, 13\}$.

Solve.

26. $|x| = 9$

27. $|x - 3| = 9$

28. $|x + 10| = |x - 12|$

29. $|2 - 5x| = -10$

30. $|4x - 1| < 4.5$

31. $|x| > 3$

32. $\left|\dfrac{6 - x}{7}\right| \le 15$

33. $|-5x - 3| \ge 10$

Graph. Find the coordinates of any vertices formed.

34. $x - 6y < -6$

35. $x + y \ge 3$,
$x - y \ge 5$

36. $2y - x \ge -4$,
$2y + 3x \le -6$,
$y \le 0$,
$x \le 0$

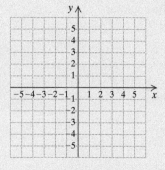
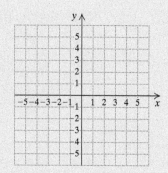
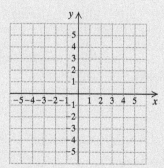

37. Solve: $\left|\dfrac{1}{2}x - 2\right| \ge 2.2$.

 A. $[-0.4, 8.4]$
 B. $(-\infty, -0.4] \cup [8.4, \infty)$
 C. $(-\infty, 8.4]$
 D. $[-0.4, \infty)$

Synthesis

Solve.

38. $|3x - 4| \le -3$

39. $7x < 8 - 3x < 6 + 7x$

Cumulative Review

Chapters 1–9

Graph.

1. $y = -5x + 4$

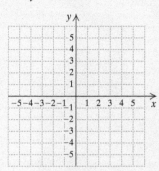

2. $3x - 18 = 0$

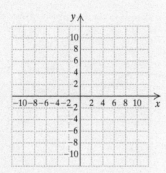

3. $x + 3y < 4$

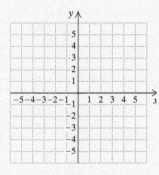

4. $x + y \geq 4,$
   $x - y \geq 1$

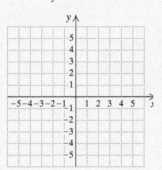

5. Given that $g(x) = |x - 4| + 5$, find $g(-2)$.

6. Given that
$$f(x) = \frac{x-2}{x^2 - 25},$$
find the domain.

7. Find the domain and the range of the function graphed below.

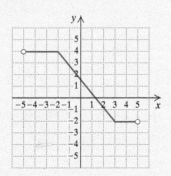

Simplify.

8. $(6m - n)^2$

9. $(3a - 4b)(5a + 2b)$

10. $\dfrac{y^2 - 4}{3y + 33} \cdot \dfrac{y + 11}{y + 2}$

11. $\dfrac{9x^2 - 25}{x^2 - 16} \div \dfrac{3x + 5}{x - 4}$

12. $\dfrac{2x + 1}{4x - 12} - \dfrac{x - 2}{5x - 15}$

13. $\dfrac{1 - \dfrac{2}{y^2}}{1 - \dfrac{1}{y^3}}$

14. $(6p^2 - 2p + 5) - (-10p^2 + 6p + 5)$

15. $\dfrac{2}{x + 2} + \dfrac{3}{x - 2} - \dfrac{x + 1}{x^2 - 4}$

16. $(2x^3 - 7x^2 + x - 3) \div (x + 2)$

Solve.

17. $9y - (5y - 3) = 33$

18. $-3 < -2x - 6 < 0$

19. $\dfrac{3x}{x - 2} - \dfrac{6}{x + 2} = \dfrac{24}{x^2 - 4}$

20. $P = \dfrac{3a}{a + b}$, for a

21. $F = \dfrac{9}{5}C + 32$, for C

22. $|x| \geq 2.1$

23. $\dfrac{6}{x - 5} = \dfrac{2}{2x}$

24. $8x = 1 + 16x^2$

25. $14 + 3x = 2x^2$

Solve.

26. $4x - 2y = 6,$
$6x - 3y = 9$

27. $4x + 5y = -3,$
$x = 1 - 3y$

28. $x + 2y - 2z = 9,$
$2x - 3y + 4z = -4,$
$5x - 4y + 2z = 5$

29. $x + 6y + 4z = -2,$
$4x + 4y + z = 2,$
$3x + 2y - 4z = 5$

Factor.

30. $4x^3 + 18x^2$

31. $8a^3 - 4a^2 - 6a + 3$

32. $x^2 + 8x - 84$

33. $6x^2 + 11x - 10$

34. $16y^2 - 81$

35. $t^2 - 16t + 64$

36. $64x^3 + 8$

37. $0.027b^3 - 0.008c^3$

38. $x^6 - x^2$

39. $20x^2 + 7x - 3$

40. Find an equation of the line with slope $-\frac{1}{2}$ passing through the point $(2, -2)$.

41. Find an equation of the line that is perpendicular to the line $2x + y = 5$ and passes through the point $(3, -1)$.

42. *Hockey Results.* A hockey team played 81 games in a season. They won 1 fewer game than three times the number of ties and lost 8 fewer games than they won. How many games did they win? lose? tie?

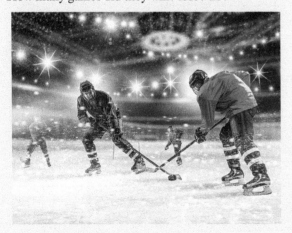

43. *Waste Generation.* The amount of waste generated by a fast-food restaurant varies directly as the number of customers served. A typical restaurant that serves 2000 customers per day generates 238 lb of waste daily. How many pounds of waste would be generated daily by a restaurant that serves 1700 customers per day?

44. Solve: $\dfrac{x}{x-4} - \dfrac{4}{x+3} = \dfrac{28}{x^2 - x - 12}$.

A. No solution
B. 0
C. $-4, 3$
D. $4, -3$

45. Solve: $x^2 - x - 6 = 6$.

A. $4, 9$
B. $3, 8$
C. $4, -3$
D. $0, 1$

46. *Tank Filling.* An oil storage tank can be filled in 10 hr by ship A working alone and in 15 hr by ship B working alone. How many hours will it take to fill the oil storage tank if both ships A and B are working?

A. 8 hr
B. 6 hr
C. $12\frac{1}{2}$ hr
D. 25 hr

Synthesis

47. The graph of $y = ax^2 + bx + c$ contains the three points $(4, 2), (2, 0),$ and $(1, 2)$. Find $a, b,$ and c.

Solve.

48. $16x^3 = x$

49. $\dfrac{18}{x-9} + \dfrac{10}{x+5} = \dfrac{28x}{x^2 - 4x - 45}$

CHAPTER 10

Radical Expressions, Equations, and Functions

10.1 Radical Expressions and Functions
10.2 Rational Numbers as Exponents
10.3 Simplifying Radical Expressions
10.4 Addition, Subtraction, and More Multiplication

Mid-Chapter Review

10.5 More on Division of Radical Expressions
10.6 Solving Radical Equations
10.7 Applications Involving Powers and Roots

Translating for Success

10.8 The Complex Numbers

Summary and Review
Test
Cumulative Review

By 2017, smartphone ownership had increased to approximately 44% of the world population and is expected to grow to 59% by 2022. In the first quarter of 2017, worldwide sales of smartphones totaled 380 million units—an increase of 9.1% over the first quarter of 2016. The graph at right shows the percent of market share for five top vendors. The combined market share of three Chinese manufacturers, Huawei, Oppo, and Vivo, made up nearly 24% of the first-quarter sales.

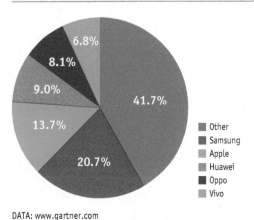

Worldwide Smartphone Sales by Vendor

- Other 41.7%
- Samsung 20.7%
- Apple 13.7%
- Huawei 9.0%
- Oppo 8.1%
- Vivo 6.8%

DATA: www.gartner.com

Data: gartner.com; strategyanalytics.com, "44% of Population Will Own Smartphones in 2017," by Linda Sui, December 21, 2016

In Exercise 15 of Exercise Set 10.7, we will calculate the length of the infinity screen on the Samsung Galaxy S8.

STUDYING FOR SUCCESS Make the Most of Your Time

- Keep your textbook handy. If you find yourself with a few moments of unexpected free time, you can review material or look ahead to the next section.
- Review your schedule and revise it, if necessary, for upcoming final exams.
- Use e-mail to contact your instructor and classmates.

10.1 Radical Expressions and Functions

OBJECTIVES

a Find principal square roots and their opposites, approximate square roots, identify radicands, find outputs of square-root functions, graph square-root functions, and find the domains of square-root functions.

b Simplify radical expressions with perfect-square radicands.

c Find cube roots, simplifying certain expressions, and find outputs of cube-root functions.

d Simplify expressions involving odd roots and even roots.

In this section, we consider roots, such as square roots and cube roots. We define the symbolism associated with roots and consider methods of manipulating symbols.

a SQUARE ROOTS AND SQUARE-ROOT FUNCTIONS

SKILL REVIEW Find the domain of a function. [7.2a]
Find the domain.

1. $f(x) = \dfrac{6}{x - 8}$
2. $f(x) = x^2 + \dfrac{1}{3}$

Answers: 1. $\{x \mid x \text{ is a real number } and \ x \neq 8\}$
2. All real numbers

When we raise a number to the second power, we say that we have **squared** the number. Sometimes we may need to find the number that was squared. We call this process **finding a square root** of a number.

SQUARE ROOT

The number c is a **square root** of a if $c^2 = a$.

For example:

5 is a *square root* of 25 because $5^2 = 5 \cdot 5 = 25$;
-5 is a *square root* of 25 because $(-5)^2 = (-5)(-5) = 25$.

The number -4 does not have a real-number square root because there is no real number c such that $c^2 = -4$.

PROPERTIES OF SQUARE ROOTS

Every positive real number has two real-number square roots.
The number 0 has just one square root, 0 itself.
Negative numbers do not have real-number square roots.*

---

*In Section 10.8, we will consider a number system in which negative numbers do have square roots.

EXAMPLE 1 Find the two square roots of 64.

The square roots of 64 are 8 and -8 because $8^2 = 64$ and $(-8)^2 = 64$.

Do Exercises 1–3.

PRINCIPAL SQUARE ROOT

The **principal square root** of a nonnegative number is its nonnegative square root. The symbol $\sqrt{a}$ represents the principal square root of a. To name the negative square root of a, we can write $-\sqrt{a}$.

EXAMPLES Simplify.

2. $\sqrt{25} = 5$ Remember: $\sqrt{}$ indicates the principal (nonnegative) square root.

3. $-\sqrt{25} = -5$

4. $\sqrt{\dfrac{81}{64}} = \dfrac{9}{8}$ because $\left(\dfrac{9}{8}\right)^2 = \dfrac{9}{8} \cdot \dfrac{9}{8} = \dfrac{81}{64}$.

5. $\sqrt{0.0049} = 0.07$ because $(0.07)^2 = (0.07)(0.07) = 0.0049$.

6. $-\sqrt{0.000001} = -0.001$

7. $\sqrt{0} = 0$

8. $\sqrt{-25}$ Does not exist as a real number. Negative numbers do not have real-number square roots.

Do Exercises 4–13.

We found exact square roots in Examples 1–8. It would be helpful to memorize the table of exact square roots at right. We often need to use rational numbers to *approximate* square roots that are irrational. Such expressions can be found using a calculator with a square-root key.

EXAMPLES Use a calculator to approximate each of the following.

Number	Using a calculator with a 10-digit readout	Rounded to three decimal places
9. $\sqrt{11}$	3.316624790	3.317
10. $\sqrt{487}$	22.06807649	22.068
11. $-\sqrt{7297.8}$	-85.42716196	-85.427
12. $\sqrt{\dfrac{463}{557}}$	.9117229728	0.912

Do Exercises 14–17.

Find the square roots.

1. 9   **2.** 36   **3.** 121

Simplify.

4. a) $\sqrt{16}$ **5. a)** $\sqrt{49}$
 b) $-\sqrt{16}$ **b)** $-\sqrt{49}$
 c) $\sqrt{-16}$ **c)** $\sqrt{-49}$

6. $\sqrt{1}$ **7.** $-\sqrt{36}$

8. $\sqrt{\dfrac{81}{100}}$ **9.** $\sqrt{0.0064}$

10. $-\sqrt{\dfrac{25}{64}}$ **11.** $\sqrt{\dfrac{16}{9}}$

12. $-\sqrt{0.81}$ **13.** $\sqrt{1.44}$

TABLE OF SQUARE ROOTS

$\sqrt{1} = 1$	$\sqrt{196} = 14$
$\sqrt{4} = 2$	$\sqrt{225} = 15$
$\sqrt{9} = 3$	$\sqrt{256} = 16$
$\sqrt{16} = 4$	$\sqrt{289} = 17$
$\sqrt{25} = 5$	$\sqrt{324} = 18$
$\sqrt{36} = 6$	$\sqrt{361} = 19$
$\sqrt{49} = 7$	$\sqrt{400} = 20$
$\sqrt{64} = 8$	$\sqrt{441} = 21$
$\sqrt{81} = 9$	$\sqrt{484} = 22$
$\sqrt{100} = 10$	$\sqrt{529} = 23$
$\sqrt{121} = 11$	$\sqrt{576} = 24$
$\sqrt{144} = 12$	$\sqrt{625} = 25$
$\sqrt{169} = 13$	

Use a calculator to approximate each square root to three decimal places.

14. $\sqrt{17}$ **15.** $\sqrt{1138}$

16. $-\sqrt{867.6}$ **17.** $\sqrt{\dfrac{22}{35}}$

Answers
1. 3, −3   **2.** 6, −6   **3.** 11, −11
4. (a) 4; **(b)** −4; **(c)** does not exist as a real number
5. (a) 7; **(b)** −7; **(c)** does not exist as a real number
6. 1   **7.** −6   **8.** $\dfrac{9}{10}$ **9.** 0.08   **10.** $-\dfrac{5}{8}$
11. $\dfrac{4}{3}$ **12.** −0.9   **13.** 1.2   **14.** 4.123
15. 33.734   **16.** −29.455   **17.** 0.793

RADICAL; RADICAL EXPRESSION; RADICAND

The symbol $\sqrt{}$ is called a **radical**.

An expression written with a radical is called a **radical expression**.

The expression written under the radical is called the **radicand**.

These are radical expressions:

$$\sqrt{5}, \quad \sqrt{a}, \quad -\sqrt{5x}, \quad \sqrt{y^2 + 7}.$$

The radicands in these expressions are 5, a, $5x$, and $y^2 + 7$, respectively.

EXAMPLE 13 Identify the radicand in $x\sqrt{x^2 - 9}$.

The radicand is the expression under the radical, $x^2 - 9$.

◀ Do Exercises 18 and 19.

Identify the radicand.

18. $5\sqrt{28 + x}$

19. $\sqrt{\dfrac{y}{y + 3}}$

Since each nonnegative real number x has exactly one principal square root, the symbol $\sqrt{x}$ represents exactly one real number and thus can be used to define a square-root function:

$$f(x) = \sqrt{x}.$$

The domain of this function is the set of nonnegative real numbers. In interval notation, the domain is $[0, \infty)$.

EXAMPLE 14 For the given function, find the indicated function values:

$$f(x) = \sqrt{3x - 2}; \quad f(1), f(5), \text{ and } f(0).$$

We have

$f(1) = \sqrt{3 \cdot 1 - 2}$ Substituting 1 for x
$ = \sqrt{3 - 2} = \sqrt{1} = 1;$ Simplifying and taking the square root

$f(5) = \sqrt{3 \cdot 5 - 2}$ Substituting 5 for x
$ = \sqrt{13} \approx 3.606;$ Simplifying and approximating

$f(0) = \sqrt{3 \cdot 0 - 2}$ Substituting 0 for x
$ = \sqrt{-2}.$ Negative radicand. No real-number function value exists; 0 is not in the domain of f.

◀ Do Exercises 20 and 21.

For the given function, find the indicated function values.

20. $g(x) = \sqrt{6x + 4}$; $g(0), g(3),$ and $g(-5)$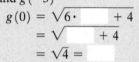

$g(0) = \sqrt{6 \cdot \underline{} + 4}$
$ = \sqrt{\underline{} + 4}$
$ = \sqrt{4} = \underline{}$

$g(3) = \sqrt{6 \cdot \underline{} + 4}$
$ = \sqrt{\underline{} + 4}$
$ = \sqrt{22}$

$g(-5) = \sqrt{6(\underline{}) + 4}$
$ = \sqrt{\underline{} + 4}$
$ = \sqrt{-26}$

-26 is a $\underline{}$ radicand. No real-number function value exists.
  negative/positive

EXAMPLE 15 Find the domain of $g(x) = \sqrt{x + 2}$.

The expression $\sqrt{x + 2}$ is a real number only when $x + 2$ is nonnegative. Thus the domain of $g(x) = \sqrt{x + 2}$ is the set of all x-values for which $x + 2 \geq 0$. We solve as follows:

$x + 2 \geq 0$
$x \geq -2.$ Adding -2

The domain of $g = \{x \mid x \geq -2\}$, or $[-2, \infty)$.

21. $f(x) = -\sqrt{x}$; $f(4), f(7),$ and $f(-3)$

Answers

18. $28 + x$ 19. $\dfrac{y}{y + 3}$ 20. $2; \sqrt{22};$ does not exist as a real number
21. $-2; -\sqrt{7} \approx -2.646;$ does not exist as a real number

Guided Solution:
20. 0, 0, 2; 3, 18; $-5, -30$, negative

EXAMPLE 16 Graph: **(a)** $f(x) = \sqrt{x}$; **(b)** $g(x) = \sqrt{x+2}$.

We first find outputs as we did in Example 14. We can either select inputs that have exact outputs or use a calculator to make approximations. Once ordered pairs have been calculated, a smooth curve can be drawn.

a)

x	$f(x) = \sqrt{x}$	$(x, f(x))$
0	0	(0, 0)
1	1	(1, 1)
3	1.7	(3, 1.7)
4	2	(4, 2)
7	2.6	(7, 2.6)
9	3	(9, 3)

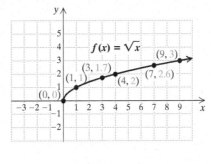

We can see from the table and the graph that the domain of f is $[0, \infty)$. The range is also the set of nonnegative real numbers $[0, \infty)$.

b)

x	$g(x) = \sqrt{x+2}$	$(x, g(x))$
−2	0	(−2, 0)
−1	1	(−1, 1)
0	1.4	(0, 1.4)
3	2.2	(3, 2.2)
5	2.6	(5, 2.6)
10	3.5	(10, 3.5)

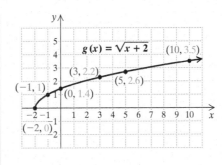

We can see from the table, the graph, and Example 15 that the domain of g is $[−2, \infty)$. The range is the set of nonnegative real numbers $[0, \infty)$.

Do Exercises 22–25. ▶

b FINDING $\sqrt{a^2}$

In the expression $\sqrt{a^2}$, the radicand is a perfect square. It is tempting to think that $\sqrt{a^2} = a$, but we see below that this is not always the case.

Suppose $a = 5$. Then we have $\sqrt{5^2}$, which is $\sqrt{25}$, or 5.

Suppose $a = −5$. Then we have $\sqrt{(−5)^2}$, which is $\sqrt{25}$, or 5.

Suppose $a = 0$. Then we have $\sqrt{0^2}$, which is $\sqrt{0}$, or 0.

The symbol $\sqrt{a^2}$ never represents a negative number. It represents the principal square root of a^2. Note the following.

SIMPLIFYING $\sqrt{a^2}$

$a \geq 0 \longrightarrow \sqrt{a^2} = a$

If a is positive or 0, the principal square root of a^2 is a.

$a < 0 \longrightarrow \sqrt{a^2} = -a$

If a is negative, the principal square root of a^2 is the opposite of a.

Find the domain of each function.

22. $f(x) = \sqrt{x-5}$

23. $g(x) = \sqrt{2x+3}$

Graph.

24. $g(x) = -\sqrt{x}$

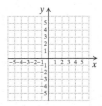

25. $f(x) = 2\sqrt{x+3}$

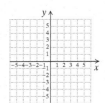

Answers

22. $\{x | x \geq 5\}$, or $[5, \infty)$
23. $\{x | x \geq -\frac{3}{2}\}$, or $[-\frac{3}{2}, \infty)$
24.

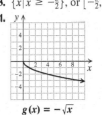

$g(x) = -\sqrt{x}$

25.

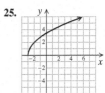

$f(x) = 2\sqrt{x+3}$

In all cases, the radical expression $\sqrt{a^2}$ represents the absolute value of a.

> **PRINCIPAL SQUARE ROOT OF a^2**
>
> For any real number a, $\sqrt{a^2} = |a|$. The principal (nonnegative) square root of a^2 is the absolute value of a.

The absolute value is used to ensure that the principal square root is nonnegative, which is as it is defined.

EXAMPLES Find each of the following. Assume that letters can represent any real number.

17. $\sqrt{(-16)^2} = |-16|$, or 16

18. $\sqrt{(3b)^2} = |3b| = |3| \cdot |b| = 3|b|$

> $|3b|$ can be simplified to $3|b|$ because the absolute value of any product is the product of the absolute values. That is, $|a \cdot b| = |a| \cdot |b|$.

19. $\sqrt{(x-1)^2} = |x-1|$

20. $\sqrt{x^2 + 8x + 16} = \sqrt{(x+4)^2}$
$= |x + 4|$

..... Caution! .....
$|x + 4|$ is *not* the same as $|x| + 4$.

◀ Do Exercises 26–33.

Find each of the following. Assume that letters can represent *any* real number.

26. $\sqrt{y^2}$ **27.** $\sqrt{(-24)^2}$

28. $\sqrt{(5y)^2}$ **29.** $\sqrt{16y^2}$

30. $\sqrt{(x+7)^2}$

31. $\sqrt{4(x-2)^2}$

32. $\sqrt{49(y+5)^2}$

33. $\sqrt{x^2 - 6x + 9}$
$= \sqrt{(x - \underline{})^2}$
$= |\underline{} - 3|$ GS

c CUBE ROOTS

> **CUBE ROOT**
>
> The number c is the **cube root** of a, written $\sqrt[3]{a}$, if the third power of c is a—that is, if $c^3 = a$, then $\sqrt[3]{a} = c$.

For example:

  2 is the *cube root* of 8 because $2^3 = 2 \cdot 2 \cdot 2 = 8$;

  -4 is the *cube root* of -64 because $(-4)^3 = (-4)(-4)(-4) = -64$.

We talk about *the* cube root of a number rather than *a* cube root because of the following.

> Every real number has exactly one cube root in the system of real numbers. The symbol $\sqrt[3]{a}$ represents *the* cube root of a.

Answers
26. $|y|$ **27.** 24 **28.** $5|y|$ **29.** $4|y|$
30. $|x+7|$ **31.** $2|x-2|$ **32.** $7|y+5|$
33. $|x-3|$
Guided Solution:
33. $3, x$

EXAMPLES Find each of the following.

21. $\sqrt[3]{8} = 2$ because $2^3 = 8$.
22. $\sqrt[3]{-27} = -3$
23. $\sqrt[3]{-\dfrac{216}{125}} = -\dfrac{6}{5}$
24. $\sqrt[3]{0.001} = 0.1$
25. $\sqrt[3]{x^3} = x$
26. $\sqrt[3]{-8} = -2$
27. $\sqrt[3]{0} = 0$
28. $\sqrt[3]{-8y^3} = \sqrt[3]{(-2y)^3} = -2y$

When we are determining a cube root, no absolute-value signs are needed because a real number has just one cube root. The real-number cube root of a positive number is positive. The real-number cube root of a negative number is negative. The cube root of 0 is 0. That is, $\sqrt[3]{a^3} = a$ whether $a > 0$, $a < 0$, or $a = 0$.

Do Exercises 34–37. ▶

Since the symbol $\sqrt[3]{x}$ represents exactly one real number, it can be used to define a cube-root function: $f(x) = \sqrt[3]{x}$.

EXAMPLE 29 For the given function, find the indicated function values:

$$f(x) = \sqrt[3]{x}; \quad f(125), f(0), f(-8), \text{ and } f(-10).$$

We have

$f(125) = \sqrt[3]{125} = 5;$
$f(0) = \sqrt[3]{0} = 0;$
$f(-8) = \sqrt[3]{-8} = -2;$
$f(-10) = \sqrt[3]{-10} \approx -2.154.$

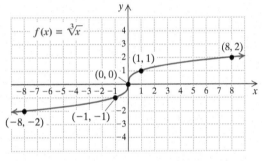

The graph of $f(x) = \sqrt[3]{x}$ is shown above for reference. Note that both the domain and the range consist of the entire set of real numbers, $(-\infty, \infty)$.

Do Exercise 38. ▶

Find each of the following.

34. $\sqrt[3]{-64}$
35. $\sqrt[3]{27y^3}$
36. $\sqrt[3]{8(x+2)^3}$
37. $\sqrt[3]{-\dfrac{343}{64}}$

38. For the given function, find the indicated function values:
$g(x) = \sqrt[3]{x-4};\ g(-23),$
$g(4), g(-1), \text{ and } g(11).$

d ODD AND EVEN kTH ROOTS

In the expression $\sqrt[k]{a}$, we call k the **index** and assume $k \geq 2$.

Odd Roots

The 5th root of a number a is the number c for which $c^5 = a$. There are also 7th roots, 9th roots, and so on. Whenever the number k in $\sqrt[k]{\ }$ is an odd number, we say that we are taking an **odd root**.

Every number has just one real-number odd root. If the number is positive, then the root is positive. If the number is negative, then the root is negative. If the number is 0, then the root is 0. For example, $\sqrt[3]{8} = 2$, $\sqrt[3]{-8} = -2$, and $\sqrt[3]{0} = 0$. Absolute-value signs are *not* needed when we are finding odd roots.

Answers
34. -4 35. $3y$ 36. $2(x+2)$
37. $-\dfrac{7}{4}$ 38. $-3; 0; \sqrt[3]{-5} \approx -1.710;$ $\sqrt[3]{7} \approx 1.913$

If k is an *odd* natural number, then for any real number a,
$$\sqrt[k]{a^k} = a.$$

Find each of the following.
39. $\sqrt[5]{243}$
40. $\sqrt[5]{-243}$
41. $\sqrt[5]{x^5}$
42. $\sqrt[5]{y^7}$
43. $\sqrt[5]{0}$
44. $\sqrt[5]{-32x^5}$
45. $\sqrt[7]{(3x+2)^7}$

EXAMPLES Find each of the following.
30. $\sqrt[5]{32} = 2$
31. $\sqrt[5]{-32} = -2$
32. $-\sqrt[5]{32} = -2$
33. $-\sqrt[5]{-32} = -(-2) = 2$
34. $\sqrt[7]{x^7} = x$
35. $\sqrt[7]{128} = 2$
36. $\sqrt[7]{-128} = -2$
37. $\sqrt[7]{0} = 0$
38. $\sqrt[5]{a^5} = a$
39. $\sqrt[9]{(x-1)^9} = x - 1$

◀ Do Exercises 39–45.

Even Roots

When the index k in $\sqrt[k]{}$ is an even number, we say that we are taking an **even root**. When the index is 2, we do not write it. Every positive real number has two real-number kth roots when k is even. One of those roots is positive and one is negative. Negative real numbers do not have real-number kth roots when k is even. When we are finding even kth roots, absolute-value signs are sometimes necessary, as we have seen with square roots. For example,

$$\sqrt{64} = 8, \quad \sqrt[6]{64} = 2, \quad -\sqrt[6]{64} = -2, \quad \sqrt[6]{64x^6} = \sqrt[6]{(2x)^6} = |2x| = 2|x|.$$

Note that in $\sqrt[6]{64x^6}$, we need absolute-value signs because a variable is involved.

EXAMPLES Find each of the following. Assume that variables can represent any real number.
40. $\sqrt[4]{16} = 2$
41. $-\sqrt[4]{16} = -2$
42. $\sqrt[4]{-16}$ Does not exist as a real number.
43. $\sqrt[4]{81x^4} = \sqrt[4]{(3x)^4} = |3x| = 3|x|$
44. $\sqrt[6]{(y+7)^6} = |y+7|$
45. $\sqrt{81y^2} = \sqrt{(9y)^2} = |9y| = 9|y|$

Find each of the following. Assume that letters can represent any real number.
46. $\sqrt[4]{81}$
47. $-\sqrt[4]{81}$
48. $\sqrt[4]{-81}$
49. $\sqrt[4]{0}$
50. $\sqrt[4]{16(x-2)^4}$
51. $\sqrt[6]{x^6}$
52. $\sqrt[8]{(x+3)^8}$
53. $\sqrt[7]{(x+3)^7}$
54. $\sqrt[5]{243x^5}$

The following is a summary of how absolute value is used when we are taking even roots or odd roots.

SIMPLIFYING

For any real number a:

a) $\sqrt[k]{a^k} = |a|$ when k is an *even* natural number. We use absolute value when k is even unless a is nonnegative.

b) $\sqrt[k]{a^k} = a$ when k is an *odd* natural number greater than 1. We do not use absolute value when k is odd.

◀ Do Exercises 46–54.

Answers
39. 3 40. -3 41. x 42. y 43. 0
44. $-2x$ 45. $3x + 2$ 46. 3 47. -3
48. Does not exist as a real number
49. 0 50. $2|x - 2|$ 51. $|x|$ 52. $|x + 3|$
53. $x + 3$ 54. $3x$

10.1 Exercise Set

✓ Check Your Understanding

Reading Check Determine whether each statement is true or false.

RC1. $-\sqrt{16} = -4$

RC2. $\sqrt{-16} = -4$

RC3. $\sqrt{49}$ represents 7 or -7.

RC4. 36 has two square roots.

RC5. There are no real numbers that when squared yield negative numbers.

RC6. The radicand in the expression $\sqrt{x} - 10$ is $x - 10$.

Concept Check Choose from the columns on the right the domain of the given function. Some of the choices may not be used, and some may be used more than once.

CC1. $f(x) = \sqrt{9 - x}$

CC2. $f(x) = \sqrt{x + 9} + 3$

CC3. $g(x) = \sqrt{x - 3}$

CC4. $h(x) = x + 9$

CC5. $f(x) = 3 - x$

CC6. $g(x) = 3 - \sqrt{3 - x}$

a) $[-3, \infty)$
b) $[-9, \infty)$
c) $(3, \infty)$
d) $(-\infty, -9)$
e) $(-\infty, -3]$
f) $[9, \infty)$
g) $(-\infty, 3]$
h) $[3, \infty)$
i) $(-\infty, \infty)$
j) $(-\infty, 9]$

a Find the square roots.

1. 16

2. 225

3. 144

4. 9

5. 400

6. 81

Simplify.

7. $-\sqrt{\dfrac{49}{36}}$

8. $-\sqrt{\dfrac{361}{9}}$

9. $\sqrt{196}$

10. $\sqrt{441}$

11. $\sqrt{0.0036}$

12. $\sqrt{0.04}$

13. $\sqrt{-225}$

14. $\sqrt{-64}$

Use a calculator to approximate to three decimal places.

15. $\sqrt{347}$

16. $-\sqrt{1839.2}$

17. $\sqrt{\dfrac{285}{74}}$

18. $\sqrt{\dfrac{839.4}{19.7}}$

Identify the radicand.

19. $9\sqrt{y^2 + 16}$

20. $-3\sqrt{p^2 - 10}$

21. $x^4 y^5 \sqrt{\dfrac{x}{y - 1}}$

22. $a^2 b^2 \sqrt{\dfrac{a^2 - b}{b}}$

For the given function, find the indicated function values.

23. $f(x) = \sqrt{5x - 10}$; $f(6), f(2), f(1),$ and $f(-1)$

24. $t(x) = -\sqrt{2x + 1}$; $t(4), t(0), t(-1),$ and $t\left(-\dfrac{1}{2}\right)$

25. $g(x) = \sqrt{x^2 - 25}$; $g(-6), g(3), g(6),$ and $g(13)$

26. $F(x) = \sqrt{x^2 + 1}$; $F(0), F(-1),$ and $F(-10)$

27. Find the domain of the function f in Exercise 23.

28. Find the domain of the function t in Exercise 24.

29. *Parking-Lot Arrival Spaces.* The attendants at a parking lot park cars in temporary spaces before the cars are taken to long-term parking stalls. The number N of such spaces needed is approximated by the function
$$N(a) = 2.5\sqrt{a},$$
where a is the average number of arrivals in peak hours. What is the number of spaces needed when the average number of arrivals is 66? 100?

30. *Body Surface Area.* Body surface area B can be estimated using the Mosteller formula
$$B = \sqrt{\frac{h \times w}{3600}},$$
where B is in square meters, h is height, in centimeters, and w is weight, in kilograms. Estimate the body surface area of **(a)** a woman whose height is 165 cm and whose weight is 63 kg and **(b)** a man whose height is 183 cm and whose weight is 100 kg. Round to the nearest tenth.

Graph.

31. $f(x) = 2\sqrt{x}$

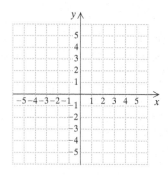

32. $g(x) = 3 - \sqrt{x}$

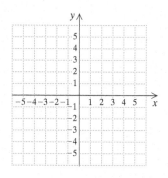

33. $f(x) = -3\sqrt{x}$

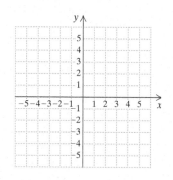

34. $f(x) = 2 + \sqrt{x - 1}$

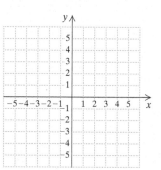

35. $f(x) = \sqrt{x}$

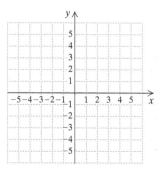

36. $g(x) = -\sqrt{x}$

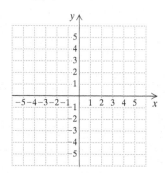

37. $f(x) = \sqrt{x - 2}$

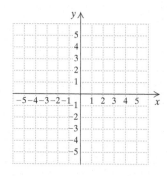

38. $g(x) = \sqrt{x + 3}$

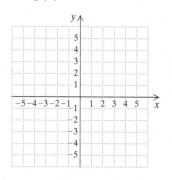

39. $f(x) = \sqrt{12 - 3x}$ **40.** $g(x) = \sqrt{8 - 4x}$ **41.** $g(x) = \sqrt{3x + 9}$ **42.** $f(x) = \sqrt{3x - 6}$

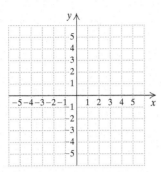
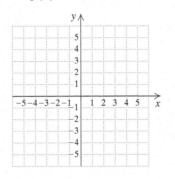
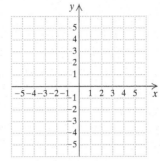
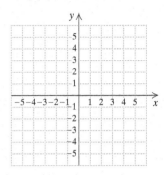

b Find each of the following. Assume that letters can represent *any* real number.

43. $\sqrt{16x^2}$

44. $\sqrt{25t^2}$

45. $\sqrt{(-12c)^2}$

46. $\sqrt{(-9d)^2}$

47. $\sqrt{(p + 3)^2}$

48. $\sqrt{(2 - x)^2}$

49. $\sqrt{x^2 - 4x + 4}$

50. $\sqrt{9t^2 - 30t + 25}$

c Simplify.

51. $\sqrt[3]{27}$

52. $-\sqrt[3]{64}$

53. $\sqrt[3]{-64x^3}$

54. $\sqrt[3]{-125y^3}$

55. $\sqrt[3]{-216}$

56. $-\sqrt[3]{-1000}$

57. $\sqrt[3]{0.343(x + 1)^3}$

58. $\sqrt[3]{0.000008(y - 2)^3}$

For the given function, find the indicated function values.

59. $f(x) = \sqrt[3]{x + 1}$; $f(7), f(26), f(-9),$ and $f(-65)$

60. $g(x) = -\sqrt[3]{2x - 1}$; $g(-62), g(0), g(-13),$ and $g(63)$

61. $f(x) = -\sqrt[3]{3x + 1}$; $f(0), f(-7), f(21),$ and $f(333)$

62. $g(t) = \sqrt[3]{t - 3}$; $g(30), g(-5), g(1),$ and $g(67)$

d Find each of the following. Assume that letters can represent *any* real number.

63. $-\sqrt[4]{625}$
64. $-\sqrt[4]{256}$
65. $\sqrt[5]{-1}$
66. $\sqrt[5]{-32}$

67. $\sqrt[5]{-\dfrac{32}{243}}$
68. $\sqrt[5]{-\dfrac{1}{32}}$
69. $\sqrt[6]{x^6}$
70. $\sqrt[8]{y^8}$

71. $\sqrt[4]{(5a)^4}$
72. $\sqrt[4]{(7b)^4}$
73. $\sqrt[10]{(-6)^{10}}$
74. $\sqrt[12]{(-10)^{12}}$

75. $\sqrt[414]{(a+b)^{414}}$
76. $\sqrt[1999]{(2a+b)^{1999}}$
77. $\sqrt[7]{y^7}$
78. $\sqrt[3]{(-6)^3}$

79. $\sqrt[5]{(x-2)^5}$
80. $\sqrt[9]{(2xy)^9}$

Skill Maintenance

Solve. [5.8b]
81. $4x^2 - 49 = 0$
82. $2x^2 - 26x + 72 = 0$
83. $4x^3 - 20x^2 + 25x = 0$
84. $4x^2 - 20x + 25 = 0$

Simplify.
85. $(a^3b^2c^5)^3$ [4.2a]
86. $(5a^7b^8)(2a^3b)$ [4.1d]

Synthesis

87. Find the domain of
$$f(x) = \dfrac{\sqrt{x+3}}{\sqrt{2-x}}.$$

88. Use only the graph of $f(x) = \sqrt[3]{x}$, shown below, to approximate $\sqrt[3]{4}$, $\sqrt[3]{6}$, and $\sqrt[3]{-5}$. Answers may vary.

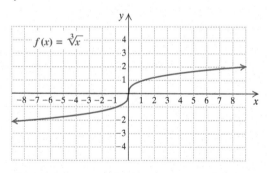

89. Use the TABLE, TRACE, and GRAPH features of a graphing calculator to find the domain and the range of each of the following functions.
   a) $f(x) = \sqrt[3]{x}$
   b) $g(x) = \sqrt[3]{4x - 5}$
   c) $q(x) = 2 - \sqrt{x+3}$
   d) $h(x) = \sqrt[4]{x}$
   e) $t(x) = \sqrt[4]{x-3}$

Rational Numbers as Exponents

10.2

In this section, we give meaning to expressions such as $a^{1/3}, 7^{-1/2}$, and $(3x)^{0.84}$, which have rational numbers as exponents. We will see that using such notation can help simplify certain radical expressions.

OBJECTIVES

a Write expressions with or without rational exponents, and simplify, if possible.

b Write expressions without negative exponents, and simplify, if possible.

c Use the laws of exponents with rational exponents.

d Use rational exponents to simplify radical expressions.

a RATIONAL EXPONENTS

Expressions like $a^{1/2}, 5^{-1/4}$, and $(2y)^{4/5}$ have not yet been defined. We will define such expressions so that the general properties of exponents hold.

Consider $a^{1/2} \cdot a^{1/2}$. If we want to multiply by adding exponents, it must follow that $a^{1/2} \cdot a^{1/2} = a^{1/2+1/2}$, or a^1. Thus we should define $a^{1/2}$ to be a square root of a. Similarly, $a^{1/3} \cdot a^{1/3} \cdot a^{1/3} = a^{1/3+1/3+1/3}$, or a^1, so $a^{1/3}$ should be defined to mean $\sqrt[3]{a}$.

$a^{1/n}$

For any *nonnegative* real number a and any natural number index n ($n \neq 1$),

$$a^{1/n} \quad \text{means} \quad \sqrt[n]{a} \quad \text{(the nonnegative } n\text{th root of } a\text{)}.$$

With rational exponents, we assume that the bases are nonnegative.

EXAMPLES Rewrite without rational exponents, and simplify, if possible.

1. $27^{1/3} = \sqrt[3]{27} = 3$ **2.** $(abc)^{1/5} = \sqrt[5]{abc}$

3. $x^{1/2} = \sqrt{x}$ An index of 2 is not written.

Do Exercises 1–5. ▶

EXAMPLES Rewrite with rational exponents.

4. $\sqrt[5]{7xy} = (7xy)^{1/5}$ We need parentheses around the radicand.

5. $8\sqrt[3]{xy} = 8(xy)^{1/3}$ **6.** $\sqrt[7]{\dfrac{x^3 y}{9}} = \left(\dfrac{x^3 y}{9}\right)^{1/7}$

Do Exercises 6–9. ▶

How should we define $a^{2/3}$? If the general properties of exponents are to hold, we have $a^{2/3} = (a^{1/3})^2$, or $(a^2)^{1/3}$, or $(\sqrt[3]{a})^2$, or $\sqrt[3]{a^2}$. We define this accordingly.

$a^{m/n}$

For any natural numbers m and n ($n \neq 1$) and any nonnegative real number a,

$$a^{m/n} \quad \text{means} \quad \sqrt[n]{a^m}, \quad \text{or} \quad (\sqrt[n]{a})^m.$$

Rewrite without rational exponents, and simplify, if possible.

1. $y^{1/4}$ **2.** $(3a)^{1/2}$

3. $16^{1/4}$ **4.** $(125)^{1/3}$

5. $(a^3 b^2 c)^{1/5}$

Rewrite with rational exponents.

6. $\sqrt[3]{19ab}$ **7.** $19\sqrt[3]{ab}$

8. $\sqrt[5]{\dfrac{x^2 y}{16}}$ **9.** $7\sqrt[4]{2ab}$

Answers
1. $\sqrt[4]{y}$ **2.** $\sqrt{3a}$ **3.** 2 **4.** 5 **5.** $\sqrt[5]{a^3 b^2 c}$
6. $(19ab)^{1/3}$ **7.** $19(ab)^{1/3}$ **8.** $\left(\dfrac{x^2 y}{16}\right)^{1/5}$
9. $7(2ab)^{1/4}$

SECTION 10.2 Rational Numbers as Exponents

Rewrite without rational exponents, and simplify, if possible.
10. $x^{3/5}$ 11. $8^{2/3}$
12. $4^{5/2}$

EXAMPLES Rewrite without rational exponents, and simplify, if possible.

7. $(27)^{2/3} = \sqrt[3]{27^2}$
$= (\sqrt[3]{27})^2$
$= 3^2$
$= 9$

8. $4^{3/2} = \sqrt{4^3}$
$= (\sqrt{4})^3$
$= 2^3$
$= 8$

◀ Do Exercises 10–12.

EXAMPLES Rewrite with rational exponents.

The index becomes the denominator of the rational exponent.

9. $\sqrt[3]{9^4} = 9^{4/3}$

10. $(\sqrt[4]{7xy})^5 = (7xy)^{5/4}$

◀ Do Exercises 13 and 14.

Rewrite with rational exponents.
13. $(\sqrt[3]{7abc})^4$ 14. $\sqrt[5]{6^7}$

b NEGATIVE RATIONAL EXPONENTS

SKILL REVIEW *Rewrite expressions with or without negative integers as exponents.* [4.1f]
Express with positive exponents.
1. $3x^{-2}$
2. cd^{-5}

Answers: 1. $\dfrac{3}{x^2}$ 2. $\dfrac{c}{d^5}$

MyLab Math VIDEO

Negative rational exponents have a meaning similar to that of negative integer exponents.

$a^{-m/n}$

For any rational number m/n and any positive real number a,

$$a^{-m/n} \text{ means } \frac{1}{a^{m/n}};$$

that is, $a^{m/n}$ and $a^{-m/n}$ are reciprocals.

Rewrite with positive exponents, and simplify, if possible.
15. $16^{-1/4}$ 16. $(3xy)^{-7/8}$

17. $81^{-3/4} = \dfrac{1}{81^{3/4}} = \dfrac{1}{(\sqrt[4]{81})}$
$= \dfrac{1}{3} = \dfrac{1}{}$ GS

EXAMPLES Rewrite with positive exponents, and simplify, if possible.

11. $9^{-1/2} = \dfrac{1}{9^{1/2}} = \dfrac{1}{\sqrt{9}} = \dfrac{1}{3}$

12. $(5xy)^{-4/5} = \dfrac{1}{(5xy)^{4/5}}$

13. $64^{-2/3} = \dfrac{1}{64^{2/3}} = \dfrac{1}{(\sqrt[3]{64})^2} = \dfrac{1}{4^2} = \dfrac{1}{16}$

14. $4x^{-2/3}y^{1/5} = 4 \cdot \dfrac{1}{x^{2/3}} \cdot y^{1/5} = \dfrac{4y^{1/5}}{x^{2/3}}$

18. $7p^{3/4}q^{-6/5}$ 19. $\left(\dfrac{11m}{7n}\right)^{-2/3}$

15. $\left(\dfrac{3r}{7s}\right)^{-5/2} = \left(\dfrac{7s}{3r}\right)^{5/2}$ Since $\left(\dfrac{a}{b}\right)^{-n} = \left(\dfrac{b}{a}\right)^n$

◀ Do Exercises 15–19.

Answers
10. $\sqrt[5]{x^3}$ 11. 4   12. 32   13. $(7abc)^{4/3}$
14. $6^{7/5}$ 15. $\dfrac{1}{2}$ 16. $\dfrac{1}{(3xy)^{7/8}}$ 17. $\dfrac{1}{27}$
18. $\dfrac{7p^{3/4}}{q^{6/5}}$ 19. $\left(\dfrac{7n}{11m}\right)^{2/3}$

Guided Solution:
17. 3, 3, 27

CALCULATOR CORNER

Rational Exponents We can use a graphing calculator to approximate rational roots of real numbers, as shown at right. The display indicates that $7^{2/3} \approx 3.659$ and $14^{-1.9} \approx 0.007$.

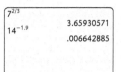

EXERCISES: Approximate each of the following.
1. $5^{3/4}$
2. $8^{4/7}$
3. $29^{-3/8}$
4. $73^{0.56}$
5. $34^{-2.78}$
6. $32^{0.2}$

c LAWS OF EXPONENTS

The same laws hold for rational-number exponents as for integer exponents. We list them for review.

For any real number a and any rational exponents m and n:

1. $a^m \cdot a^n = a^{m+n}$ In multiplying, we add exponents if the bases are the same.

2. $\dfrac{a^m}{a^n} = a^{m-n}$ In dividing, we subtract exponents if the bases are the same.

3. $(a^m)^n = a^{m \cdot n}$ To raise a power to a power, we multiply the exponents.

4. $(ab)^m = a^m b^m$ To raise a product to a power, we raise each factor to the power.

5. $\left(\dfrac{a}{b}\right)^n = \dfrac{a^n}{b^n}$ To raise a quotient to a power, we raise both the numerator and the denominator to the power.

EXAMPLES Use the laws of exponents to simplify.

16. $3^{1/5} \cdot 3^{3/5} = 3^{1/5 + 3/5} = 3^{4/5}$ Adding exponents

17. $\dfrac{7^{1/4}}{7^{1/2}} = 7^{1/4 - 1/2} = 7^{1/4 - 2/4} = 7^{-1/4} = \dfrac{1}{7^{1/4}}$ Subtracting exponents

18. $(7.2^{2/3})^{3/4} = 7.2^{2/3 \cdot 3/4} = 7.2^{6/12} = 7.2^{1/2}$ Multiplying exponents

19. $(a^{-1/3} b^{2/5})^{1/2} = a^{-1/3 \cdot 1/2} \cdot b^{2/5 \cdot 1/2}$ Raising a product to a power and multiplying exponents
$= a^{-1/6} b^{1/5} = \dfrac{b^{1/5}}{a^{1/6}}$

Do Exercises 20–23.

Use the laws of exponents to simplify.
20. $7^{1/3} \cdot 7^{3/5}$
21. $\dfrac{5^{7/6}}{5^{5/6}}$
22. $(9^{3/5})^{2/3}$
23. $(p^{-2/3} q^{1/4})^{1/2}$

d SIMPLIFYING RADICAL EXPRESSIONS

Rational exponents can be used to simplify some radical expressions. The procedure is as follows.

Answers
20. $7^{14/15}$ 21. $5^{1/3}$ 22. $9^{2/5}$ 23. $\dfrac{q^{1/8}}{p^{1/3}}$

> **SIMPLIFYING RADICAL EXPRESSIONS**
>
> 1. Convert radical expressions to exponential expressions.
> 2. Use arithmetic and the laws of exponents to simplify.
> 3. Convert back to radical notation when appropriate.
>
> *Important:* This procedure works only when we assume that a negative number has not been raised to an even power in the radicand. With this assumption, no absolute-value signs will be needed.

EXAMPLES Use rational exponents to simplify.

20. $\sqrt[6]{x^3} = x^{3/6}$ Converting to an exponential expression
    $= x^{1/2}$ Simplifying the exponent
    $= \sqrt{x}$ Converting back to radical notation

21. $\sqrt[6]{4} = 4^{1/6}$ Converting to exponential notation
    $= (2^2)^{1/6}$ Renaming 4 as 2^2
    $= 2^{2/6}$ Using $(a^m)^n = a^{mn}$; multiplying exponents
    $= 2^{1/3}$ Simplifying the exponent
    $= \sqrt[3]{2}$ Converting back to radical notation

22. $\sqrt[8]{a^2 b^4} = (a^2 b^4)^{1/8}$ Converting to exponential notation
    $= a^{2/8} \cdot b^{4/8}$ Using $(ab)^n = a^n b^n$
    $= a^{1/4} \cdot b^{1/2}$ Simplifying the exponents
    $= a^{1/4} \cdot b^{2/4}$ Rewriting $\frac{1}{2}$ with a denominator of 4
    $= (ab^2)^{1/4}$ Using $a^n b^n = (ab)^n$
    $= \sqrt[4]{ab^2}$ Converting back to radical notation

◀ Do Exercises 24–29.

Use rational exponents to simplify.
24. $\sqrt[4]{a^2}$
25. $\sqrt[4]{x^4}$
26. $\sqrt[6]{8}$
27. $\sqrt[12]{x^3 y^6}$
28. $\sqrt[6]{a^{12} b^3}$
29. $\sqrt[5]{a^5 b^{10}}$

We can use properties of rational exponents to write a single radical expression for a product or a quotient.

EXAMPLE 23 Use rational exponents to write a single radical expression for $\sqrt[3]{5} \cdot \sqrt{2}$.

$\sqrt[3]{5} \cdot \sqrt{2} = 5^{1/3} \cdot 2^{1/2}$ Converting to exponential notation
$= 5^{2/6} \cdot 2^{3/6}$ Rewriting so that the exponents have a common denominator
$= (5^2 \cdot 2^3)^{1/6}$ Using $a^n b^n = (ab)^n$
$= \sqrt[6]{5^2 \cdot 2^3}$ Converting back to radical notation
$= \sqrt[6]{25 \cdot 8}$
$= \sqrt[6]{200}$ Multiplying under the radical

◀ Do Exercise 30.

30. Use rational exponents to write a single radical expression for
$\sqrt[4]{7} \cdot \sqrt{3}$.

EXAMPLE 24 Write a single radical expression for $a^{1/2} b^{-1/2} c^{5/6}$.

$a^{1/2} b^{-1/2} c^{5/6} = a^{3/6} b^{-3/6} c^{5/6}$ Rewriting so that the exponents have a common denominator
$= (a^3 b^{-3} c^5)^{1/6}$ Using $a^n b^n = (ab)^n$
$= \sqrt[6]{a^3 b^{-3} c^5}$ Converting to radical notation

Answers
24. $\sqrt{a}$ 25. x 26. $\sqrt{2}$ 27. $\sqrt[4]{xy^2}$
28. $a^2 \sqrt{b}$ 29. ab^2 30. $\sqrt[4]{63}$

EXAMPLE 25 Write a single radical expression for $\dfrac{x^{5/6} \cdot y^{3/8}}{x^{4/9} \cdot y^{1/4}}$.

$$\dfrac{x^{5/6} \cdot y^{3/8}}{x^{4/9} \cdot y^{1/4}} = x^{5/6-4/9} \cdot y^{3/8-1/4}$$ Subtracting exponents
$$= x^{15/18-8/18} \cdot y^{3/8-2/8}$$ Finding common denominators so that exponents can be subtracted
$$= x^{7/18} \cdot y^{1/8}$$ Carrying out the subtraction of exponents
$$= x^{28/72} \cdot y^{9/72}$$ Rewriting so that the exponents have a common denominator
$$= (x^{28}y^9)^{1/72}$$ Using $a^n b^n = (ab)^n$
$$= \sqrt[72]{x^{28}y^9}$$ Converting to radical notation

Do Exercises 31 and 32.

Write a single radical expression.
31. $x^{2/3}y^{1/2}z^{5/6}$

GS 32. $\dfrac{a^{1/2}b^{3/8}}{a^{1/4}b^{1/8}} = a^{1/2-\square} \cdot b^{\square-1/8}$
$= a^{2/4-1/4} \cdot b^{2/8}$
$= a^{\square} \cdot b^{\square}$
$= \sqrt[\square]{ab}$

EXAMPLES Use rational exponents to simplify.

26. $\sqrt[6]{(5x)^3} = (5x)^{3/6}$ Converting to exponential notation
$= (5x)^{1/2}$ Simplifying the exponent
$= \sqrt{5x}$ Converting back to radical notation

27. $\sqrt[5]{t^{20}} = t^{20/5}$ Converting to exponential notation
$= t^4$ Simplifying the exponent

28. $(\sqrt[3]{pq^2c})^{12} = (pq^2c)^{12/3}$ Converting to exponential notation
$= (pq^2c)^4$ Simplifying the exponent
$= p^4 q^8 c^4$ Using $(ab)^n = a^n b^n$

29. $\sqrt{\sqrt[3]{x}} = \sqrt{x^{1/3}}$ Converting the radicand to exponential notation
$= (x^{1/3})^{1/2}$ Try to go directly to this step.
$= x^{1/6}$ Multiplying exponents
$= \sqrt[6]{x}$ Converting back to radical notation

Do Exercises 33–36.

Use rational exponents to simplify.
33. $\sqrt[14]{(5m)^2}$ 34. $\sqrt[18]{m^3}$
35. $(\sqrt[6]{a^5 b^3 c})^{24}$ 36. $\sqrt[5]{\sqrt{x}}$

Answers
31. $\sqrt[6]{x^4 y^3 z^5}$ 32. $\sqrt[4]{ab}$ 33. $\sqrt[7]{5m}$
34. $\sqrt[6]{m}$ 35. $a^{20}b^{12}c^4$ 36. $\sqrt[10]{x}$
Guided Solution:
32. $1/4, 3/8, 1/4, 1/4, 4$

10.2 Exercise Set

FOR EXTRA HELP — MyLab Math

✓ Check Your Understanding

Reading and Concept Check Match the expression with an equivalent expression from the columns on the right.

RC1. $\dfrac{c^2}{c^5}$ RC2. $c^{-2/5}$ a) c^{2+5} b) $\dfrac{1}{c^{2/5}}$

RC3. $\sqrt{c}$ RC4. $c^{2/5}$ c) $c^{1/2}$ d) $c^{2\cdot 5}$

RC5. $\sqrt{c^5}$ RC6. $(c^2)^5$ e) $c^{5/2}$ f) $-c^{5+2}$

RC7. $-c^5 \cdot c^2$ RC8. $c^2 c^5$ g) $(\sqrt[5]{c})^2$ h) c^{2-5}

a Rewrite without rational exponents, and simplify, if possible.

1. $y^{1/7}$
2. $x^{1/6}$
3. $8^{1/3}$
4. $16^{1/2}$
5. $(a^3b^3)^{1/5}$

6. $(x^2y^2)^{1/3}$
7. $16^{3/4}$
8. $4^{7/2}$
9. $49^{3/2}$
10. $27^{4/3}$

Rewrite with rational exponents.

11. $\sqrt{17}$
12. $\sqrt{x^3}$
13. $\sqrt[3]{18}$
14. $\sqrt[3]{23}$
15. $\sqrt[5]{xy^2z}$

16. $\sqrt[7]{x^3y^2z^2}$
17. $(\sqrt{3mn})^3$
18. $(\sqrt[3]{7xy})^4$
19. $(\sqrt[7]{8x^2y})^5$
20. $(\sqrt[6]{2a^5b})^7$

b Rewrite with positive exponents, and simplify, if possible.

21. $27^{-1/3}$
22. $100^{-1/2}$
23. $100^{-3/2}$
24. $16^{-3/4}$
25. $3x^{-1/4}$

26. $8y^{-1/7}$
27. $(2rs)^{-3/4}$
28. $(5xy)^{-5/6}$
29. $2a^{3/4}b^{-1/2}c^{2/3}$
30. $5x^{-2/3}y^{4/5}z$

31. $\left(\dfrac{7x}{8yz}\right)^{-3/5}$
32. $\left(\dfrac{2ab}{3c}\right)^{-5/6}$
33. $\dfrac{1}{x^{-2/3}}$
34. $\dfrac{1}{a^{-7/8}}$
35. $2^{-1/3}x^4y^{-2/7}$

36. $3^{-5/2}a^3b^{-7/3}$
37. $\dfrac{7x}{\sqrt[3]{z}}$
38. $\dfrac{6a}{\sqrt[4]{b}}$
39. $\dfrac{5a}{3c^{-1/2}}$
40. $\dfrac{2z}{5x^{-1/3}}$

c Use the laws of exponents to simplify. Write the answers with positive exponents.

41. $5^{3/4} \cdot 5^{1/8}$
42. $11^{2/3} \cdot 11^{1/2}$
43. $\dfrac{7^{5/8}}{7^{3/8}}$
44. $\dfrac{3^{5/8}}{3^{-1/8}}$
45. $\dfrac{4.9^{-1/6}}{4.9^{-2/3}}$

46. $\dfrac{2.3^{-3/10}}{2.3^{-1/5}}$
47. $(6^{3/8})^{2/7}$
48. $(3^{2/9})^{3/5}$
49. $a^{2/3} \cdot a^{5/4}$
50. $x^{3/4} \cdot x^{2/3}$

51. $(a^{2/3} \cdot b^{5/8})^4$ **52.** $(x^{-1/3} \cdot y^{-2/5})^{-15}$ **53.** $(x^{2/3})^{-3/7}$ **54.** $(a^{-3/2})^{2/9}$

55. $\left(\dfrac{x^{3/4}}{y^{1/2}}\right)^{-2/3}$ **56.** $\left(\dfrac{a^{-3/2}}{b^{-5/3}}\right)^{1/3}$ **57.** $(m^{-1/4} \cdot n^{-5/6})^{-12/5}$ **58.** $(x^{3/8} \cdot y^{5/2})^{4/3}$

d Use rational exponents to simplify. Write the answer in radical notation if appropriate.

59. $\sqrt[6]{a^2}$ **60.** $\sqrt[6]{t^4}$ **61.** $\sqrt[3]{x^{15}}$ **62.** $\sqrt[4]{a^{12}}$ **63.** $\sqrt[6]{x^{-18}}$

64. $\sqrt[5]{a^{-10}}$ **65.** $(\sqrt[3]{ab})^{15}$ **66.** $(\sqrt[7]{cd})^{14}$ **67.** $\sqrt[14]{128}$ **68.** $\sqrt[6]{81}$

69. $\sqrt[6]{4x^2}$ **70.** $\sqrt[3]{8y^6}$ **71.** $\sqrt{x^4y^6}$ **72.** $\sqrt[4]{16x^4y^2}$ **73.** $\sqrt[5]{32c^{10}d^{15}}$

Use rational exponents to write a single radical expression.

74. $\sqrt[3]{3}\sqrt{3}$ **75.** $\sqrt[3]{7} \cdot \sqrt[4]{5}$ **76.** $\sqrt[7]{11} \cdot \sqrt[6]{13}$ **77.** $\sqrt[4]{5} \cdot \sqrt[5]{7}$ **78.** $\sqrt[3]{y}\sqrt[5]{3y}$

79. $\sqrt{x}\sqrt[3]{2x}$ **80.** $(\sqrt[3]{x^2y^5})^{12}$ **81.** $(\sqrt[5]{a^2b^4})^{15}$ **82.** $\sqrt[4]{\sqrt{x}}$ **83.** $\sqrt[3]{\sqrt[6]{m}}$

84. $a^{2/3} \cdot b^{3/4}$ **85.** $x^{1/3} \cdot y^{1/4} \cdot z^{1/6}$ **86.** $\dfrac{x^{8/15} \cdot y^{7/5}}{x^{1/3} \cdot y^{-1/5}}$ **87.** $\left(\dfrac{c^{-4/5}d^{5/9}}{c^{3/10}d^{1/6}}\right)^3$ **88.** $\sqrt[3]{\sqrt[4]{xy}}$

Skill Maintenance

Solve. [9.3c]

89. $|7x - 5| = 9$ **90.** $|3x| = 120$ **91.** $8 - |2x + 5| = -2$ **92.** $\left|\dfrac{1}{2} + x\right| = \dfrac{7}{8}$

Synthesis

93. Use the SIMULTANEOUS mode to graph
$y_1 = x^{1/2}$, $y_2 = 3x^{2/5}$, $y_3 = x^{4/7}$, $y_4 = \tfrac{1}{5}x^{3/4}$.
Then, looking only at coordinates, match each graph with its equation.

94. Simplify:
$\left(\sqrt[10]{\sqrt[5]{x^{15}}}\right)^5 \left(\sqrt[5]{\sqrt[10]{x^{15}}}\right)^5$.

SECTION 10.2 Rational Numbers as Exponents

10.3 Simplifying Radical Expressions

OBJECTIVES

a Multiply and simplify radical expressions.

b Divide and simplify radical expressions.

a MULTIPLYING AND SIMPLIFYING RADICAL EXPRESSIONS

Note that $\sqrt{4}\sqrt{25} = 2 \cdot 5 = 10$. Also $\sqrt{4 \cdot 25} = \sqrt{100} = 10$. Likewise, $\sqrt[3]{27}\sqrt[3]{8} = 3 \cdot 2 = 6$ and $\sqrt[3]{27 \cdot 8} = \sqrt[3]{216} = 6$.

These examples suggest the following.

THE PRODUCT RULE FOR RADICALS

For any nonnegative real numbers a and b and any index k,

$$\sqrt[k]{a} \cdot \sqrt[k]{b} = \sqrt[k]{a \cdot b}, \quad \text{or} \quad a^{1/k} \cdot b^{1/k} = (ab)^{1/k}.$$

The index must be the same throughout.

(To multiply, multiply the radicands.)

SKILL REVIEW Find principal square roots and their opposites. [10.1a]

Find the square roots.
1. 81
2. 100

Answers: 1. 9, −9 2. 10, −10

EXAMPLES Multiply.
1. $\sqrt{3} \cdot \sqrt{5} = \sqrt{3 \cdot 5} = \sqrt{15}$
2. $\sqrt{5a}\sqrt{2b} = \sqrt{5a \cdot 2b} = \sqrt{10ab}$
3. $\sqrt[3]{4}\sqrt[3]{5} = \sqrt[3]{4 \cdot 5} = \sqrt[3]{20}$
4. $\sqrt[4]{\dfrac{y}{5}} \sqrt[4]{\dfrac{7}{x}} = \sqrt[4]{\dfrac{y}{5} \cdot \dfrac{7}{x}} = \sqrt[4]{\dfrac{7y}{5x}}$

Caution! A common error is to omit the index in the answer.

Multiply.
1. $\sqrt{19}\sqrt{7}$
2. $\sqrt{3p}\sqrt{7q}$
3. $\sqrt[4]{403}\sqrt[4]{7}$
4. $\sqrt[3]{\dfrac{5}{p}} \cdot \sqrt[3]{\dfrac{2}{q}}$

◀ Do Exercises 1–4.

Keep in mind that the product rule can be used only when the indexes are the same. When indexes differ, we can use rational exponents as we did in Examples 23 and 24 of Section 10.2.

EXAMPLE 5 Multiply: $\sqrt{5x} \cdot \sqrt[4]{3y}$.

$\sqrt{5x} \cdot \sqrt[4]{3y} = (5x)^{1/2}(3y)^{1/4}$ Converting to exponential notation
$= (5x)^{2/4}(3y)^{1/4}$ Rewriting so that the exponents have a common denominator
$= [(5x)^2(3y)]^{1/4}$ Using $a^n b^n = (ab)^n$
$= [(25x^2)(3y)]^{1/4}$ Squaring $5x$
$= \sqrt[4]{(25x^2)(3y)}$ Converting back to radical notation
$= \sqrt[4]{75x^2y}$ Multiplying under the radical

◀ Do Exercises 5 and 6.

Multiply.
5. $\sqrt{5}\sqrt[3]{2}$
6. $\sqrt[4]{x}\sqrt[3]{2y}$

Answers
1. $\sqrt{133}$ 2. $\sqrt{21pq}$ 3. $\sqrt[4]{2821}$
4. $\sqrt[3]{\dfrac{10}{pq}}$ 5. $\sqrt[6]{500}$ 6. $\sqrt[12]{16x^3y^4}$

We can reverse the product rule to simplify a product. We simplify the root of a product by taking the root of each factor separately.

FACTORING RADICAL EXPRESSIONS

For any nonnegative real numbers a and b and any index k,
$$\sqrt[k]{ab} = \sqrt[k]{a} \cdot \sqrt[k]{b}, \quad \text{or} \quad (ab)^{1/k} = a^{1/k} \cdot b^{1/k}.$$

(Take the kth root of each factor separately.)

Compare the following:
$$\sqrt{50} = \sqrt{10 \cdot 5} = \sqrt{10}\sqrt{5};$$
$$\sqrt{50} = \sqrt{25 \cdot 2} = \sqrt{25}\sqrt{2} = 5\sqrt{2}.$$

In the second case, the radicand is written with the perfect-square factor 25. If you do not recognize any perfect-square factors, try factoring the radicand into its prime factors. For example,
$$\sqrt{50} = \sqrt{2 \cdot \underbrace{5 \cdot 5}} = 5\sqrt{2}.$$

Perfect square (a pair of the same numbers)

Square-root radical expressions in which the radicand has no perfect-square factors, such as $5\sqrt{2}$, are considered to be in simplest form. A procedure for simplifying kth roots follows.

SIMPLIFYING kTH ROOTS

To simplify a radical expression by factoring:

1. Look for the largest factors of the radicand that are perfect kth powers (where k is the index).
2. Then take the kth root of the resulting factors.
3. A radical expression, with index k, is *simplified* when its radicand has no factors that are perfect kth powers.

EXAMPLES Simplify by factoring.

6. $\sqrt{50} = \sqrt{25 \cdot 2} = \sqrt{25} \cdot \sqrt{2} = \sqrt{5 \cdot 5} \cdot \sqrt{2} = 5\sqrt{2}$

This factor is a perfect square.

7. $\sqrt[3]{32} = \sqrt[3]{8 \cdot 4} = \sqrt[3]{8} \cdot \sqrt[3]{4} = \sqrt[3]{2 \cdot 2 \cdot 2} \cdot \sqrt[3]{2 \cdot 2} = 2\sqrt[3]{4}$

This factor is a perfect cube (third power).

8. $\sqrt[4]{48} = \sqrt[4]{16 \cdot 3} = \sqrt[4]{16} \cdot \sqrt[4]{3} = \sqrt[4]{2 \cdot 2 \cdot 2 \cdot 2} \cdot \sqrt[4]{3} = 2\sqrt[4]{3}$

This factor is a perfect fourth power.

Simplify by factoring.

7. $\sqrt{32}$ 8. $\sqrt[3]{80}$

Do Exercises 7 and 8.

Answers
7. $4\sqrt{2}$ 8. $2\sqrt[3]{10}$

Frequently, expressions under radicals do not contain negative numbers raised to even powers. In such cases, absolute-value notation is not necessary. **For this reason, we will no longer use absolute-value notation.**

EXAMPLES Simplify by factoring. Assume that no radicands were formed by raising negative numbers to even powers.

9. $\sqrt{5x^2} = \sqrt{5 \cdot x^2}$ Factoring the radicand
   $= \sqrt{5} \cdot \sqrt{x^2}$ Factoring into two radicals
   $= \sqrt{5} \cdot x$ Taking the square root of x^2

Absolute-value notation is not needed because we assume that x is not negative.

10. $\sqrt{18x^2y} = \sqrt{9 \cdot 2 \cdot x^2 \cdot y}$ Looking for perfect-square factors and factoring the radicand
    $= \sqrt{9 \cdot x^2 \cdot 2 \cdot y}$
    $= \sqrt{9} \cdot \sqrt{x^2} \cdot \sqrt{2 \cdot y}$ Factoring into several radicals
    $= 3x\sqrt{2y}$ Taking square roots

11. $\sqrt{216x^5y^3} = \sqrt{36 \cdot 6 \cdot x^4 \cdot x \cdot y^2 \cdot y}$ Looking for perfect-square factors and factoring the radicand
    $= \sqrt{36 \cdot x^4 \cdot y^2 \cdot 6 \cdot x \cdot y}$
    $= \sqrt{36}\sqrt{x^4}\sqrt{y^2}\sqrt{6xy}$ Factoring into several radicals
    $= 6x^2y\sqrt{6xy}$ Taking square roots

Let's look at this example another way. We do a complete factorization and look for pairs of factors. Each pair of factors makes a square:

$\sqrt{216x^5y^3} = \sqrt{2 \cdot 2 \cdot 2 \cdot 3 \cdot 3 \cdot 3 \cdot x \cdot x \cdot x \cdot x \cdot x \cdot y \cdot y \cdot y}$ Each pair of factors makes a perfect square.

$= 2 \cdot 3 \cdot x \cdot x \cdot y \cdot \sqrt{2 \cdot 3 \cdot x \cdot y}$
$= 6x^2y\sqrt{6xy}.$

12. $\sqrt[3]{16a^7b^{11}} = \sqrt[3]{8 \cdot 2 \cdot a^6 \cdot a \cdot b^9 \cdot b^2}$ Factoring the radicand. The index is 3, so we look for the largest powers that are multiples of 3 because these are perfect cubes.
    $= \sqrt[3]{8} \cdot \sqrt[3]{a^6} \cdot \sqrt[3]{b^9} \cdot \sqrt[3]{2ab^2}$ Factoring into radicals
    $= 2a^2b^3\sqrt[3]{2ab^2}$ Taking cube roots

Let's look at this example another way. We do a complete factorization and look for triples of factors. Each triple of factors makes a cube:

$\sqrt[3]{16a^7b^{11}}$
$= \sqrt[3]{2 \cdot 2 \cdot 2 \cdot 2 \cdot a \cdot a \cdot a \cdot a \cdot a \cdot a \cdot a \cdot b \cdot b \cdot b \cdot b \cdot b \cdot b \cdot b \cdot b \cdot b \cdot b \cdot b}$
    Each triple of factors makes a cube.
$= 2 \cdot a \cdot a \cdot b \cdot b \cdot b \cdot \sqrt[3]{2 \cdot a \cdot b \cdot b}$
$= 2a^2b^3\sqrt[3]{2ab^2}.$

◂ Do Exercises 9–14.

Simplify by factoring. Assume that no radicands were formed by raising negative numbers to even powers.

9. $\sqrt{300}$ 10. $\sqrt{36y^2}$
11. $\sqrt{27zw^2}$ 12. $\sqrt[3]{16}$

13. $\sqrt{12ab^3c^2}$ GS
    $= \sqrt{ \cdot 3 \cdot a \cdot b^2 \cdot \cdot c^2}$
    $= \sqrt{4} \cdot \sqrt{b^2} \cdot \sqrt{c^2} \cdot \sqrt{}$
    $= 2 \cdot \cdot c \sqrt{3ab}$

14. $\sqrt[3]{81x^4y^8}$

Answers
9. $10\sqrt{3}$ 10. $6y$ 11. $3w\sqrt{3z}$
12. $2\sqrt[3]{2}$ 13. $2bc\sqrt{3ab}$ 14. $3xy^2\sqrt[3]{3xy^2}$
Guided Solution:
13. $4, b, 3ab, b$

Sometimes after we have multiplied, we can simplify by factoring.

EXAMPLES Multiply and simplify. Assume that no radicands were formed by raising negative numbers to even powers.

13. $\sqrt{20}\sqrt{8} = \sqrt{20 \cdot 8} = \sqrt{4 \cdot 5 \cdot 4 \cdot 2} = 4\sqrt{10}$

14. $3\sqrt[3]{25} \cdot 2\sqrt[3]{5} = 3 \cdot 2 \cdot \sqrt[3]{25} \cdot \sqrt[3]{5} = 6 \cdot \sqrt[3]{25 \cdot 5}$
$= 6 \cdot \sqrt[3]{5 \cdot 5 \cdot 5}$
$= 6 \cdot 5 = 30$

15. $\sqrt[3]{18y^3}\sqrt[3]{4x^2} = \sqrt[3]{18y^3 \cdot 4x^2}$ Multiplying radicands
$= \sqrt[3]{2 \cdot 3 \cdot 3 \cdot y \cdot y \cdot y \cdot 2 \cdot 2 \cdot x \cdot x}$
$= 2 \cdot y \cdot \sqrt[3]{3 \cdot 3 \cdot x \cdot x}$
$= 2y\sqrt[3]{9x^2}$

Do Exercises 15–18.

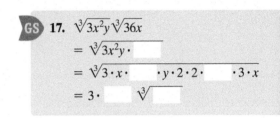

Multiply and simplify. Assume that no radicands were formed by raising negative numbers to even powers.

15. $\sqrt{3}\sqrt{6}$ 16. $\sqrt{18y}\sqrt{14y}$

GS 17. $\sqrt[3]{3x^2y}\sqrt[3]{36x}$
$= \sqrt[3]{3x^2y \cdot }$
$= \sqrt[3]{3 \cdot x \cdot \cdot y \cdot 2 \cdot 2 \cdot \cdot 3 \cdot x}$
$= 3 \cdot \sqrt[3]{}$

18. $\sqrt{7a}\sqrt{21b}$

b DIVIDING AND SIMPLIFYING RADICAL EXPRESSIONS

Note that $\dfrac{\sqrt[3]{27}}{\sqrt[3]{8}} = \dfrac{3}{2}$ and that $\sqrt[3]{\dfrac{27}{8}} = \dfrac{3}{2}$. This example suggests the following.

THE QUOTIENT RULE FOR RADICALS

For any nonnegative number a, any positive number b, and any index k,

$$\dfrac{\sqrt[k]{a}}{\sqrt[k]{b}} = \sqrt[k]{\dfrac{a}{b}}, \quad \text{or} \quad \dfrac{a^{1/k}}{b^{1/k}} = \left(\dfrac{a}{b}\right)^{1/k}.$$

(To divide, divide the radicands. After doing this, you can sometimes simplify by taking roots.)

EXAMPLES Divide and simplify. Assume that no radicands were formed by raising negative numbers to even powers.

16. $\dfrac{\sqrt{80}}{\sqrt{5}} = \sqrt{\dfrac{80}{5}} = \sqrt{16} = 4$ We divide the radicands.

17. $\dfrac{5\sqrt[3]{32}}{\sqrt[3]{2}} = 5\sqrt[3]{\dfrac{32}{2}} = 5\sqrt[3]{16} = 5\sqrt[3]{8 \cdot 2} = 5\sqrt[3]{8}\sqrt[3]{2} = 5 \cdot 2\sqrt[3]{2} = 10\sqrt[3]{2}$

18. $\dfrac{\sqrt{72xy}}{2\sqrt{2}} = \dfrac{1}{2} \cdot \dfrac{\sqrt{72xy}}{\sqrt{2}} = \dfrac{1}{2}\sqrt{\dfrac{72xy}{2}} = \dfrac{1}{2}\sqrt{36xy} = \dfrac{1}{2}\sqrt{36}\sqrt{xy}$
$= \dfrac{1}{2} \cdot 6\sqrt{xy} = 3\sqrt{xy}$

Do Exercises 19–22.

Divide and simplify. Assume that no radicands were formed by raising negative numbers to even powers.

19. $\dfrac{\sqrt{75}}{\sqrt{3}}$ 20. $\dfrac{14\sqrt{128xy}}{2\sqrt{2}}$

21. $\dfrac{\sqrt{50a^3}}{\sqrt{2a}}$ 22. $\dfrac{4\sqrt[3]{250}}{7\sqrt[3]{2}}$

Answers
15. $3\sqrt{2}$ 16. $6y\sqrt{7}$ 17. $3x\sqrt[3]{4y}$
18. $7\sqrt{3ab}$ 19. 5   20. $56\sqrt{xy}$ 21. $5a$
22. $\dfrac{20}{7}$

Guided Solution:
17. $36x, x, 3, x, 4y$

We can simplify the root of a quotient by taking the roots of the numerator and of the denominator separately.

> **kTH ROOTS OF QUOTIENTS**
>
> For any nonnegative number a, any positive number b, and any index k,
>
> $$\sqrt[k]{\frac{a}{b}} = \frac{\sqrt[k]{a}}{\sqrt[k]{b}}, \quad \text{or} \quad \left(\frac{a}{b}\right)^{1/k} = \frac{a^{1/k}}{b^{1/k}}.$$
>
> (Take the kth roots of the numerator and of the denominator separately.)

EXAMPLES Simplify by taking the roots of the numerator and of the denominator. Assume that no radicands were formed by raising negative numbers to even powers.

19. $\sqrt[3]{\dfrac{27}{125}} = \dfrac{\sqrt[3]{27}}{\sqrt[3]{125}} = \dfrac{3}{5}$ We take the cube root of the numerator and of the denominator.

20. $\sqrt{\dfrac{25}{y^2}} = \dfrac{\sqrt{25}}{\sqrt{y^2}} = \dfrac{5}{y}$ We take the square root of the numerator and of the denominator.

21. $\sqrt{\dfrac{16x^3}{y^4}} = \dfrac{\sqrt{16x^3}}{\sqrt{y^4}} = \dfrac{\sqrt{16x^2 \cdot x}}{\sqrt{y^4}} = \dfrac{\sqrt{16x^2} \cdot \sqrt{x}}{\sqrt{y^4}} = \dfrac{4x\sqrt{x}}{y^2}$

22. $\sqrt[3]{\dfrac{27y^5}{343x^3}} = \dfrac{\sqrt[3]{27y^5}}{\sqrt[3]{343x^3}} = \dfrac{\sqrt[3]{27y^3 \cdot y^2}}{\sqrt[3]{343x^3}} = \dfrac{\sqrt[3]{27y^3} \cdot \sqrt[3]{y^2}}{\sqrt[3]{343x^3}} = \dfrac{3y\sqrt[3]{y^2}}{7x}$

We are assuming here that no variable represents 0 or a negative number. Thus we need not be concerned about zero denominators or absolute value.

◀ Do Exercises 23–25.

Simplify by taking the roots of the numerator and of the denominator. Assume that no radicands were formed by raising negative numbers to even powers.

23. $\sqrt{\dfrac{25}{36}}$ **24.** $\sqrt{\dfrac{x^2}{100}}$

25. $\sqrt[3]{\dfrac{54x^5}{125}}$

When indexes differ, we can use rational exponents.

EXAMPLE 23 Divide and simplify: $\dfrac{\sqrt[3]{a^2b^4}}{\sqrt{ab}}$.

$$\dfrac{\sqrt[3]{a^2b^4}}{\sqrt{ab}} = \dfrac{(a^2b^4)^{1/3}}{(ab)^{1/2}}$$ Converting to exponential notation

$$= \dfrac{a^{2/3}b^{4/3}}{a^{1/2}b^{1/2}}$$ Using the product and power rules

$$= a^{2/3 - 1/2}b^{4/3 - 1/2}$$ Subtracting exponents

$$= a^{4/6 - 3/6}b^{8/6 - 3/6}$$ Finding common denominators so exponents can be subtracted

$$= a^{1/6}b^{5/6}$$

$$= (ab^5)^{1/6}$$ Using $a^n b^n = (ab)^n$

$$= \sqrt[6]{ab^5}$$ Converting back to radical notation

26. Divide and simplify:

$$\dfrac{\sqrt[4]{x^3y^2}}{\sqrt[3]{x^2y}}.$$

◀ Do Exercise 26.

Answers

23. $\dfrac{5}{6}$ **24.** $\dfrac{x}{10}$ **25.** $\dfrac{3x\sqrt[3]{2x^2}}{5}$ **26.** $\sqrt[12]{xy^2}$

10.3 Exercise Set

✓ Check Your Understanding

Reading Check Determine whether each statement is true or false.

RC1. For any nonnegative real numbers a and b and any index k, $\sqrt[k]{a} \cdot \sqrt[k]{b} = \sqrt[k]{ab}$.

RC2. For $q > 0$, $\sqrt{q^2 - 100} = q + 10$.

RC3. The expression $\sqrt{Y}$ is not simplified if Y contains a factor that contains a perfect square.

RC4. For any nonnegative number a, any positive number b, and any index k, $\dfrac{\sqrt[k]{a}}{\sqrt[k]{b}} = \sqrt[k]{\dfrac{a}{b}}$.

Concept Check Determine whether each radical expression is in simplest form.

CC1. $\sqrt[4]{162x^2}$ **CC2.** $3\sqrt{121}$ **CC3.** $\sqrt[3]{15y^2}$

CC4. $\sqrt{320}$ **CC5.** $\sqrt{210xy}$ **CC6.** $\sqrt[3]{6y^6}$

a Simplify by factoring. Assume that no radicands were formed by raising negative numbers to even powers.

1. $\sqrt{24}$
2. $\sqrt{20}$
3. $\sqrt{90}$

4. $\sqrt{18}$
5. $\sqrt[3]{250}$
6. $\sqrt[3]{108}$

7. $\sqrt{180x^4}$
8. $\sqrt{175y^6}$
9. $\sqrt[3]{54x^8}$

10. $\sqrt[3]{40y^3}$
11. $\sqrt[3]{80t^8}$
12. $\sqrt[3]{108x^5}$

13. $\sqrt[4]{80}$
14. $\sqrt[4]{32}$
15. $\sqrt{32a^2b}$

16. $\sqrt{75p^3q^4}$
17. $\sqrt[4]{243x^8y^{10}}$
18. $\sqrt[4]{162c^4d^6}$

19. $\sqrt[5]{96x^7y^{15}}$
20. $\sqrt[5]{p^{14}q^9r^{23}}$

Multiply and simplify. Assume that no radicands were formed by raising negative numbers to even powers.

21. $\sqrt{10}\sqrt{5}$
22. $\sqrt{6}\sqrt{3}$
23. $\sqrt{15}\sqrt{6}$
24. $\sqrt{2}\sqrt{32}$

25. $\sqrt[3]{2}\sqrt[3]{4}$
26. $\sqrt[3]{9}\sqrt[3]{3}$
27. $\sqrt{45}\sqrt{60}$
28. $\sqrt{24}\sqrt{75}$

29. $\sqrt{3x^3}\sqrt{6x^5}$
30. $\sqrt{5a^7}\sqrt{15a^3}$
31. $\sqrt{5b^3}\sqrt{10c^4}$
32. $\sqrt{2x^3y}\sqrt{12xy}$

33. $\sqrt[3]{5a^2}\sqrt[3]{2a}$
34. $\sqrt[3]{7x}\sqrt[3]{3x^2}$
35. $\sqrt[3]{y^4}\sqrt[3]{16y^5}$
36. $\sqrt[3]{s^2t^4}\sqrt[3]{s^4t^6}$

37. $\sqrt[4]{16}\sqrt[4]{64}$
38. $\sqrt[5]{64}\sqrt[5]{16}$
39. $\sqrt{12a^3b}\sqrt{8a^4b^2}$
40. $\sqrt{30x^3y^4}\sqrt{18x^2y^5}$

41. $\sqrt{2}\sqrt[3]{5}$
42. $\sqrt{6}\sqrt[3]{5}$
43. $\sqrt[4]{3}\sqrt{2}$
44. $\sqrt[3]{5}\sqrt[4]{2}$

45. $\sqrt{a}\sqrt[4]{a^3}$

46. $\sqrt[3]{x^2}\sqrt[6]{x^5}$

47. $\sqrt[5]{b^2}\sqrt{b^3}$

48. $\sqrt[4]{a^3}\sqrt[3]{a^2}$

49. $\sqrt{xy^3}\sqrt[3]{x^2y}$

50. $\sqrt{y^5z}\sqrt[3]{yz^4}$

51. $\sqrt{2a^3b}\sqrt[4]{8ab^2}$

52. $\sqrt[4]{9ab^3}\sqrt{3a^4b}$

b Divide and simplify. Assume that all expressions under radicals represent positive numbers.

53. $\dfrac{\sqrt{90}}{\sqrt{5}}$

54. $\dfrac{\sqrt{98}}{\sqrt{2}}$

55. $\dfrac{\sqrt{35q}}{\sqrt{7q}}$

56. $\dfrac{\sqrt{30x}}{\sqrt{10x}}$

57. $\dfrac{\sqrt[3]{54}}{\sqrt[3]{2}}$

58. $\dfrac{\sqrt[3]{40}}{\sqrt[3]{5}}$

59. $\dfrac{\sqrt{56xy^3}}{\sqrt{8x}}$

60. $\dfrac{\sqrt{52ab^3}}{\sqrt{13a}}$

61. $\dfrac{\sqrt[3]{96a^4b^2}}{\sqrt[3]{12a^2b}}$

62. $\dfrac{\sqrt[3]{189x^5y^7}}{\sqrt[3]{7x^2y^2}}$

63. $\dfrac{\sqrt{128xy}}{2\sqrt{2}}$

64. $\dfrac{\sqrt{48ab}}{2\sqrt{3}}$

65. $\dfrac{\sqrt[4]{48x^9y^{13}}}{\sqrt[4]{3xy^5}}$

66. $\dfrac{\sqrt[5]{64a^{11}b^{28}}}{\sqrt[5]{2ab^2}}$

67. $\dfrac{\sqrt[3]{a}}{\sqrt{a}}$

68. $\dfrac{\sqrt{x}}{\sqrt[4]{x}}$

69. $\dfrac{\sqrt[3]{a^2}}{\sqrt[4]{a}}$

70. $\dfrac{\sqrt[3]{x^2}}{\sqrt[5]{x}}$

71. $\dfrac{\sqrt[4]{x^2y^3}}{\sqrt[3]{xy}}$

72. $\dfrac{\sqrt[5]{a^4b^2}}{\sqrt[3]{ab^2}}$

Simplify.

73. $\sqrt{\dfrac{25}{36}}$

74. $\sqrt{\dfrac{49}{64}}$

75. $\sqrt{\dfrac{16}{49}}$

76. $\sqrt{\dfrac{100}{81}}$

77. $\sqrt[3]{\dfrac{125}{27}}$ 78. $\sqrt[3]{\dfrac{343}{1000}}$ 79. $\sqrt{\dfrac{49}{y^2}}$ 80. $\sqrt{\dfrac{121}{x^2}}$

81. $\sqrt{\dfrac{25y^3}{x^4}}$ 82. $\sqrt{\dfrac{36a^5}{b^6}}$ 83. $\sqrt[3]{\dfrac{81y^5}{64}}$ 84. $\sqrt[3]{\dfrac{8z^7}{125}}$

85. $\sqrt[3]{\dfrac{27a^4}{8b^3}}$ 86. $\sqrt[3]{\dfrac{64x^7}{216y^6}}$ 87. $\sqrt[4]{\dfrac{81x^4}{16}}$ 88. $\sqrt[4]{\dfrac{256}{81x^8}}$

89. $\sqrt[4]{\dfrac{16a^{12}}{b^4c^{16}}}$ 90. $\sqrt[4]{\dfrac{81x^4}{y^8z^4}}$ 91. $\sqrt[5]{\dfrac{32x^8}{y^{10}}}$ 92. $\sqrt[5]{\dfrac{32b^{10}}{243a^{20}}}$

93. $\sqrt[5]{\dfrac{w^7}{z^{10}}}$ 94. $\sqrt[5]{\dfrac{z^{11}}{w^{20}}}$ 95. $\sqrt[6]{\dfrac{x^{13}}{y^6z^{12}}}$ 96. $\sqrt[6]{\dfrac{p^9q^{24}}{r^{18}}}$

Skill Maintenance

Solve. [5.9a]

97. The sum of a number and its square is 90. Find the number.

98. *Triangle Dimensions.* The base of a triangle is 2 in. longer than the height. The area is 12 in². Find the height and the base.

Solve. [6.7a]

99. $\dfrac{12x}{x-4} - \dfrac{3x^2}{x+4} = \dfrac{384}{x^2-16}$

100. $\dfrac{4x}{x+5} + \dfrac{20}{x} = \dfrac{100}{x^2+5x}$

Synthesis

101. *Pendulums.* The **period** of a pendulum is the time it takes to complete one cycle, swinging to and fro. For a pendulum that is L centimeters long, the period T is given by the function

$$T(L) = 2\pi\sqrt{\dfrac{L}{980}},$$

where T is in seconds. Find, to the nearest hundredth of a second, the period of a pendulum of length **(a)** 65 cm; **(b)** 98 cm; **(c)** 120 cm. Use a calculator's π key if possible.

Simplify.

102. $\dfrac{\sqrt[3]{x^3 - y^3}}{\sqrt[3]{x - y}}$

103. $\dfrac{\sqrt{44x^2y^9z}\,\sqrt{22y^9z^6}}{(\sqrt{11xy^8z^2})^2}$

Addition, Subtraction, and More Multiplication

10.4

OBJECTIVES

a Add or subtract with radical notation and simplify.

b Multiply expressions involving radicals in which some factors contain more than one term.

a ADDITION AND SUBTRACTION

Any two real numbers can be added. For example, the sum of 7 and $\sqrt{3}$ can be expressed as $7 + \sqrt{3}$. We cannot simplify this sum. However, when we have **like radicals** (radicals having the same index and radicand), we can use the distributive laws to simplify by collecting like radical terms. For example,

$$7\sqrt{3} + \sqrt{3} = 7\sqrt{3} + 1\sqrt{3} = (7 + 1)\sqrt{3} = 8\sqrt{3}.$$

EXAMPLES Add or subtract.

1. $6\sqrt{7} + 4\sqrt{7} = (6 + 4)\sqrt{7}$ Using a distributive law
   $= 10\sqrt{7}$

2. $8\sqrt[3]{2} - 7x\sqrt[3]{2} + 5\sqrt[3]{2} = (8 - 7x + 5)\sqrt[3]{2}$ Factoring out $\sqrt[3]{2}$
   $= (13 - 7x)\sqrt[3]{2}$

   These parentheses are necessary!

3. $3\sqrt[5]{4x} + 7\sqrt[5]{4x} - \sqrt[3]{4x} = (3 + 7)\sqrt[5]{4x} - \sqrt[3]{4x}$
   $= 10\sqrt[5]{4x} - \sqrt[3]{4x}$

   Note that these expressions have the same *radicand*, but they are not like radicals because they do not have the same *index*.

Do Exercises 1 and 2.

Sometimes, in order to determine whether there are like radical terms, we need to simplify radicals by factoring.

EXAMPLES Add or subtract.

4. $3\sqrt{8} - 5\sqrt{2} = 3\sqrt{4 \cdot 2} - 5\sqrt{2}$ Factoring 8
   $= 3\sqrt{4} \cdot \sqrt{2} - 5\sqrt{2}$ Factoring $\sqrt{4 \cdot 2}$ into two radicals
   $= 3 \cdot 2\sqrt{2} - 5\sqrt{2}$ Taking the square root of 4
   $= 6\sqrt{2} - 5\sqrt{2}$
   $= (6 - 5)\sqrt{2}$ Collecting like radical terms
   $= \sqrt{2}$

5. $5\sqrt{2} - 4\sqrt{3}$ No simplification is possible.

6. $5\sqrt[3]{16y^4} + 7\sqrt[3]{2y} = 5\sqrt[3]{8y^3 \cdot 2y} + 7\sqrt[3]{2y}$ ⎫ Factoring the first radical
   $= 5\sqrt[3]{8y^3} \cdot \sqrt[3]{2y} + 7\sqrt[3]{2y}$ ⎭
   $= 5 \cdot 2y \cdot \sqrt[3]{2y} + 7\sqrt[3]{2y}$ Taking the cube root of $8y^3$
   $= 10y\sqrt[3]{2y} + 7\sqrt[3]{2y}$
   $= (10y + 7)\sqrt[3]{2y}$ Collecting like radical terms

Do Exercises 3–5.

SKILL REVIEW

Simplify an expression by collecting like terms. [1.7e]

Collect like terms.
1. $2x + 5x$
2. $y + 3 - 4y + 1$

Answers: **1.** $7x$ **2.** $-3y + 4$

MyLab Math VIDEO

Add or subtract.
1. $5\sqrt{2} + 8\sqrt{2}$
2. $7\sqrt[4]{5x} + 3\sqrt[4]{5x} - \sqrt{7}$

Add or subtract.
3. $7\sqrt{45} - 2\sqrt{5}$
4. $3\sqrt[3]{y^5} + 4\sqrt[3]{y^2} + \sqrt[3]{8y^6}$

GS 5. $\sqrt{25x - 25} - \sqrt{9x - 9}$
   $= \sqrt{(x - 1)} - \sqrt{(x - 1)}$
   $= 5\sqrt{x - 1} - \sqrt{x - 1}$
   $= \sqrt{x - 1}$

Answers
1. $13\sqrt{2}$ 2. $10\sqrt[4]{5x} - \sqrt{7}$ 3. $19\sqrt{5}$
4. $(3y + 4)\sqrt[3]{y^2} + 2y^2$ 5. $2\sqrt{x - 1}$
Guided Solution:
5. 25, 9, 3, 2

SECTION 10.4   Addition, Subtraction, and More Multiplication   745

b MORE MULTIPLICATION

To multiply expressions in which some factors contain more than one term, we use the procedures for multiplying polynomials.

EXAMPLES Multiply.

7. $\sqrt{3}(x - \sqrt{5}) = \sqrt{3} \cdot x - \sqrt{3} \cdot \sqrt{5}$ Using a distributive law
$= x\sqrt{3} - \sqrt{15}$ Multiplying radicals

8. $\sqrt[3]{y}(\sqrt[3]{y^2} + \sqrt[3]{2}) = \sqrt[3]{y} \cdot \sqrt[3]{y^2} + \sqrt[3]{y} \cdot \sqrt[3]{2}$ Using a distributive law
$= \sqrt[3]{y^3} + \sqrt[3]{2y}$ Multiplying radicals
$= y + \sqrt[3]{2y}$ Simplifying $\sqrt[3]{y^3}$

◀ Do Exercises 6 and 7.

Multiply. Assume that no radicands were formed by raising negative numbers to even powers.

6. $\sqrt{2}(5\sqrt{3} + 3\sqrt{7})$

7. $\sqrt[3]{a^2}(\sqrt[3]{3a} - \sqrt[3]{2})$

EXAMPLE 9 Multiply: $(4\sqrt{3} + \sqrt{2})(\sqrt{3} - 5\sqrt{2})$.

$$(4\sqrt{3} + \sqrt{2})(\sqrt{3} - 5\sqrt{2}) = 4(\sqrt{3})^2 - 20\sqrt{3} \cdot \sqrt{2} + \sqrt{2} \cdot \sqrt{3} - 5(\sqrt{2})^2$$

F O I L

$= 4 \cdot 3 - 20\sqrt{6} + \sqrt{6} - 5 \cdot 2$
$= 12 - 20\sqrt{6} + \sqrt{6} - 10$
$= 2 - 19\sqrt{6}$ Collecting like terms

EXAMPLE 10 Multiply: $(\sqrt{a} + \sqrt{3})(\sqrt{b} + \sqrt{3})$. Assume that all expressions under radicals represent nonnegative numbers.

$(\sqrt{a} + \sqrt{3})(\sqrt{b} + \sqrt{3}) = \sqrt{a}\sqrt{b} + \sqrt{a}\sqrt{3} + \sqrt{3}\sqrt{b} + \sqrt{3}\sqrt{3}$
$= \sqrt{ab} + \sqrt{3a} + \sqrt{3b} + 3$

EXAMPLE 11 Multiply: $(\sqrt{5} + \sqrt{7})(\sqrt{5} - \sqrt{7})$.

$(\sqrt{5} + \sqrt{7})(\sqrt{5} - \sqrt{7}) = (\sqrt{5})^2 - (\sqrt{7})^2$ This is now a difference of two squares:
$(A - B)(A + B) = A^2 - B^2$.
$= 5 - 7 = -2$

Multiply. Assume that no radicands were formed by raising negative numbers to even powers.

8. $(\sqrt{3} - 5\sqrt{2})(2\sqrt{3} + \sqrt{2})$

9. $(\sqrt{a} + 2\sqrt{3})(3\sqrt{b} - 4\sqrt{3})$

10. $(\sqrt{2} + \sqrt{5})(\sqrt{2} - \sqrt{5})$
$= ()^2 - (\sqrt{5})^2$
$= 2 - $
$= $

EXAMPLE 12 Multiply: $(\sqrt{a} + \sqrt{b})(\sqrt{a} - \sqrt{b})$. Assume that no radicands were formed by raising negative numbers to even powers.

$(\sqrt{a} + \sqrt{b})(\sqrt{a} - \sqrt{b}) = (\sqrt{a})^2 - (\sqrt{b})^2$
$= a - b$ No radicals

Expressions of the form $\sqrt{a} + \sqrt{b}$ and $\sqrt{a} - \sqrt{b}$ are called **conjugates**. Their product is always an expression that has no radicals.

◀ Do Exercises 8–11.

11. $(\sqrt{p} - \sqrt{q})(\sqrt{p} + \sqrt{q})$

Multiply.

12. $(2\sqrt{5} - y)^2$

13. $(3\sqrt{6} + 2)^2$

EXAMPLE 13 Multiply: $(\sqrt{3} + x)^2$.

$(\sqrt{3} + x)^2 = (\sqrt{3})^2 + 2 \cdot \sqrt{3} \cdot x + x^2$ Squaring a binomial
$= 3 + 2x\sqrt{3} + x^2$

◀ Do Exercises 12 and 13.

Answers
6. $5\sqrt{6} + 3\sqrt{14}$
7. $a\sqrt[3]{3} - \sqrt[3]{2a^2}$ 8. $-4 - 9\sqrt{6}$
9. $3\sqrt{ab} - 4\sqrt{3a} + 6\sqrt{3b} - 24$
10. -3 11. $p - q$ 12. $20 - 4y\sqrt{5} + y^2$
13. $58 + 12\sqrt{6}$

Guided Solution:
10. $\sqrt{2}, 5, -3$

10.4 Exercise Set

✓ Check Your Understanding

Reading Check Determine whether each statement is true or false.

RC1. The expressions $3\sqrt{7y}$ and $\sqrt[3]{7y}$ are like radicals.

RC2. The expression $3 - \sqrt{3}$ cannot be simplified.

RC3. The expressions $\sqrt{x} - \sqrt{y}$ and $\sqrt{x} + \sqrt{y}$ are conjugates.

RC4. The product of expressions that are conjugates is always an expression that has no radicals.

Concept Check Like radical terms have the *same* index and the *same* radicand. Determine whether the given pair of terms are like radicals. Answer "yes" or "no."

CC1. $4\sqrt[3]{5y}, \ 2\sqrt[3]{5y}$

CC2. $5, \ 5\sqrt{2}$

CC3. $\sqrt[7]{x^2 y^3}, \ \sqrt[7]{x^2 y^2}$

CC4. $q\sqrt[4]{q^3}, \ 2\sqrt[4]{q^3}$

CC5. $-4\sqrt{3}, \ \sqrt{3}$

CC6. $x\sqrt[3]{y}, \ y\sqrt[3]{x}$

CC7. $3\sqrt[5]{a-b}, \ 3\sqrt[4]{a-b}$

CC8. $\dfrac{1}{4}\sqrt[3]{\dfrac{x^2}{y}}, \ 4\sqrt[3]{\dfrac{x^2}{y}}$

a Add or subtract. Assume that no radicands were formed by raising negative numbers to even powers.

1. $7\sqrt{5} + 4\sqrt{5}$

2. $2\sqrt{3} + 9\sqrt{3}$

3. $6\sqrt[3]{7} - 5\sqrt[3]{7}$

4. $13\sqrt[5]{3} - 8\sqrt[5]{3}$

5. $4\sqrt[3]{y} + 9\sqrt[3]{y}$

6. $6\sqrt[4]{t} - 3\sqrt[4]{t}$

7. $5\sqrt{6} - 9\sqrt{6} - 4\sqrt{6}$

8. $3\sqrt{10} - 8\sqrt{10} + 7\sqrt{10}$

9. $4\sqrt[3]{3} - \sqrt{5} + 2\sqrt[3]{3} + \sqrt{5}$

10. $5\sqrt{7} - 8\sqrt[4]{11} + \sqrt{7} + 9\sqrt[4]{11}$

11. $8\sqrt{27} - 3\sqrt{3}$

12. $9\sqrt{50} - 4\sqrt{2}$

13. $8\sqrt{45} + 7\sqrt{20}$

14. $9\sqrt{12} + 16\sqrt{27}$

15. $18\sqrt{72} + 2\sqrt{98}$

16. $12\sqrt{45} - 8\sqrt{80}$

17. $3\sqrt[3]{16} + \sqrt[3]{54}$

18. $\sqrt[3]{27} - 5\sqrt[3]{8}$

19. $2\sqrt{128} - \sqrt{18} + 4\sqrt{32}$

20. $5\sqrt{50} - 2\sqrt{18} + 9\sqrt{32}$

21. $\sqrt{5a} + 2\sqrt{45a^3}$

22. $4\sqrt{3x^3} - \sqrt{12x}$

23. $\sqrt[3]{24x} - \sqrt[3]{3x^4}$

24. $\sqrt[3]{54x} - \sqrt[3]{2x^4}$

25. $7\sqrt{27x^3} + \sqrt{3x}$

26. $2\sqrt{45x^3} - \sqrt{5x}$

27. $\sqrt{4} + \sqrt{18}$

28. $\sqrt[3]{8} - \sqrt[3]{24}$

29. $5\sqrt[3]{32} - \sqrt[3]{108} + 2\sqrt[3]{256}$

30. $3\sqrt[3]{8x} - 4\sqrt[3]{27x} + 2\sqrt[3]{64x}$

31. $\sqrt[3]{6x^4} + \sqrt[3]{48x} - \sqrt[3]{6x}$

32. $\sqrt[4]{80x^5} - \sqrt[4]{405x^9} + \sqrt[4]{5x}$

33. $\sqrt{4a - 4} + \sqrt{a - 1}$

34. $\sqrt{9y + 27} + \sqrt{y + 3}$

35. $\sqrt{x^3 - x^2} + \sqrt{9x - 9}$

36. $\sqrt{4x - 4} + \sqrt{x^3 - x^2}$

b Multiply. Assume that no radicands were formed by raising negative numbers to even powers.

37. $\sqrt{5}(4 - 2\sqrt{5})$

38. $\sqrt{6}(2 + \sqrt{6})$

39. $\sqrt{3}(\sqrt{2} - \sqrt{7})$

40. $\sqrt{2}(\sqrt{5} - \sqrt{2})$

41. $\sqrt{3}(-4\sqrt{3} + 6)$

42. $\sqrt{2}(-5\sqrt{2} - 7)$

43. $\sqrt{3}(2\sqrt{5} - 3\sqrt{4})$

44. $\sqrt{2}(3\sqrt{10} - 2\sqrt{2})$

45. $\sqrt[3]{2}(\sqrt[3]{4} - 2\sqrt[3]{32})$

46. $\sqrt[3]{3}(\sqrt[3]{9} - 4\sqrt[3]{21})$

47. $3\sqrt[3]{y}(2\sqrt[3]{y^2} - 4\sqrt[3]{y})$

48. $2\sqrt[3]{y^2}(5\sqrt[3]{y} + 4\sqrt[3]{y^2})$

49. $\sqrt[3]{a}(\sqrt[3]{2a^2} + \sqrt[3]{16a^2})$

50. $\sqrt[3]{x}(\sqrt[3]{3x^2} - \sqrt[3]{81x^2})$

51. $(\sqrt{3} - \sqrt{2})(\sqrt{3} + \sqrt{2})$

52. $(\sqrt{5} + \sqrt{6})(\sqrt{5} - \sqrt{6})$

53. $(\sqrt{8} + 2\sqrt{5})(\sqrt{8} - 2\sqrt{5})$

54. $(\sqrt{18} + 3\sqrt{7})(\sqrt{18} - 3\sqrt{7})$

55. $(7 + \sqrt{5})(7 - \sqrt{5})$

56. $(4 - \sqrt{3})(4 + \sqrt{3})$

57. $(2 - \sqrt{3})(2 + \sqrt{3})$

58. $(11 - \sqrt{2})(11 + \sqrt{2})$

59. $(\sqrt{8} + \sqrt{5})(\sqrt{8} - \sqrt{5})$

60. $(\sqrt{6} - \sqrt{7})(\sqrt{6} + \sqrt{7})$

61. $(3 + 2\sqrt{7})(3 - 2\sqrt{7})$

62. $(6 - 3\sqrt{2})(6 + 3\sqrt{2})$

63. $(\sqrt{c} + \sqrt{d})(\sqrt{c} - \sqrt{d})$

64. $(\sqrt{x} - \sqrt{y})(\sqrt{x} + \sqrt{y})$

65. $(3 - \sqrt{5})(2 + \sqrt{5})$

66. $(2 + \sqrt{6})(4 - \sqrt{6})$

67. $(\sqrt{3} + 1)(2\sqrt{3} + 1)$

68. $(4\sqrt{3} + 5)(\sqrt{3} - 2)$

69. $(2\sqrt{7} - 4\sqrt{2})(3\sqrt{7} + 6\sqrt{2})$

70. $(4\sqrt{5} + 3\sqrt{3})(3\sqrt{5} - 4\sqrt{3})$

71. $(\sqrt{a} + \sqrt{2})(\sqrt{a} + \sqrt{3})$

72. $(2 - \sqrt{x})(1 - \sqrt{x})$

73. $(2\sqrt[3]{3} + \sqrt[3]{2})(\sqrt[3]{3} - 2\sqrt[3]{2})$

74. $(3\sqrt[3]{7} + \sqrt[3]{6})(2\sqrt[3]{7} - 3\sqrt[3]{6})$

75. $(2 + \sqrt{3})^2$

76. $(\sqrt{5} + 1)^2$

77. $(\sqrt[5]{9} - \sqrt[5]{3})(\sqrt[5]{8} + \sqrt[5]{27})$

78. $(\sqrt[3]{8x} - \sqrt[3]{5y})^2$

Skill Maintenance

Multiply or divide and simplify.

79. $\dfrac{a^3 + 8}{a^2 - 4} \cdot \dfrac{a^2 - 4a + 4}{a^2 - 2a + 4}$ [6.1d]

80. $\dfrac{a^2 - 4}{a} \div \dfrac{a - 2}{a + 4}$ [6.2b]

Simplify. [6.6a]

81. $\dfrac{x - \dfrac{1}{3}}{x + \dfrac{1}{4}}$

82. $\dfrac{1 - \dfrac{1}{x}}{1 - \dfrac{1}{x^2}}$

83. $\dfrac{\dfrac{1}{p} - \dfrac{1}{q}}{\dfrac{1}{p^2} - \dfrac{1}{q^2}}$

84. $\dfrac{\dfrac{1}{a} + \dfrac{1}{b}}{\dfrac{1}{a^3} + \dfrac{1}{b^3}}$

Solve. [9.3c, d, e]

85. $|3x + 7| = 22$

86. $|3x + 7| < 22$

87. $|3x + 7| \geq 22$

88. $|3x + 7| = |2x - 5|$

Synthesis

89. Graph the function $f(x) = \sqrt{(x - 2)^2}$. What is the domain?

90. Use a graphing calculator to check your answers to Exercises 5, 22, and 72.

Multiply and simplify.

91. $\sqrt{9 + 3\sqrt{5}}\sqrt{9 - 3\sqrt{5}}$

92. $(\sqrt{x + 2} - \sqrt{x - 2})^2$

93. $(\sqrt{3} + \sqrt{5} - \sqrt{6})^2$

94. $\sqrt[3]{y}(1 - \sqrt[3]{y})(1 + \sqrt[3]{y})$

95. $(\sqrt[3]{9} - 2)(\sqrt[3]{9} + 4)$

96. $[\sqrt{3 + \sqrt{2} + \sqrt{1}}]^4$

Mid-Chapter Review

Concept Reinforcement

Determine whether each statement is true or false.

_____ 1. Every real number has two real-number square roots. [10.1a]

_____ 2. If $\sqrt[3]{q}$ is negative, then q is negative. [10.1c]

_____ 3. $a^{m/n}$ and $a^{n/m}$ are reciprocals. [10.2b]

_____ 4. To multiply radicals with the same index, we multiply the radicands. [10.3a]

Guided Solutions

GS Fill in each blank with the number that creates a correct statement or solution.
Perform the indicated operations and simplify. [10.3a], [10.4a]

5. $\sqrt{6}\sqrt{10} = \sqrt{6 \cdot \underline{}} = \sqrt{2 \cdot \underline{} \cdot 2 \cdot \underline{}} = \underline{}\sqrt{\underline{}}$

6. $5\sqrt{32} - 3\sqrt{18} = 5\sqrt{\underline{} \cdot 2} - 3\sqrt{\underline{} \cdot 2}$

   $= 5 \cdot 4\sqrt{2} - 3 \cdot \underline{}\sqrt{2}$

   $= \underline{}\sqrt{2} - \underline{}\sqrt{2}$

   $= \underline{}\sqrt{2}$

Mixed Review

Simplify. [10.1a]

7. $\sqrt{81}$

8. $-\sqrt{144}$

9. $\sqrt{\dfrac{16}{25}}$

10. $\sqrt{-9}$

11. For $f(x) = \sqrt{2x+3}$, find $f(3)$ and $f(-2)$. [10.1a]

12. Find the domain of $f(x) = \sqrt{4-x}$. [10.1a]

Graph. [10.1a]

13. $f(x) = -2\sqrt{x}$

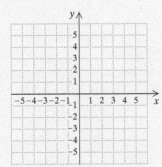

14. $g(x) = \sqrt{x+1}$

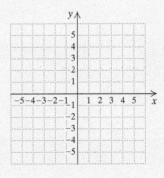

Find each of the following. Assume that letters can represent *any* real number. [10.1b, c, d]

15. $\sqrt{36z^2}$

16. $\sqrt{x^2 - 8x + 16}$

17. $\sqrt[3]{-64}$

18. $-\sqrt[3]{27a^3}$

19. $\sqrt[5]{32}$

20. $\sqrt[10]{y^{10}}$

Rewrite without rational exponents and simplify, if possible. [10.2a]

21. $125^{1/3}$

22. $(a^3b)^{1/4}$

Rewrite with rational exponents. [10.2a]

23. $\sqrt[5]{16}$

24. $\sqrt[3]{6m^2n}$

Simplify. Write the answer with positive exponents. [10.2b, c]

25. $3^{1/4} \cdot 3^{-5/8}$

26. $\dfrac{7^{6/5}}{7^{2/5}}$

27. $(x^{3/4}y^{-2/3})^2$

28. $(n^{-3/5})^{5/4}$

Use rational exponents to simplify. Write the answer in radical notation. [10.2d]

29. $\sqrt[6]{16}$

30. $(\sqrt[10]{ab})^5$

Use rational exponents to write a single radical expression. [10.2d]

31. $\sqrt{y}\,\sqrt[3]{y}$

32. $a^{2/3}b^{3/5}$

Perform the indicated operation and simplify. Assume that no radicands were formed by raising negative numbers to even powers. [10.3a, b], [10.4a, b]

33. $\sqrt{5}\sqrt{15}$

34. $\sqrt[3]{4x^2y}\,\sqrt[3]{6xy^4}$

35. $\dfrac{\sqrt[3]{80}}{\sqrt[3]{2}}$

36. $\sqrt{\dfrac{49a^5}{b^8}}$

37. $5\sqrt{7} + 6\sqrt{7}$

38. $3\sqrt{18x^3} - 6\sqrt{32x}$

39. $\sqrt{3}(2 - 5\sqrt{3})$

40. $(1 - \sqrt{x})(3 - \sqrt{x})$

41. $(\sqrt{m} - \sqrt{n})(\sqrt{m} + \sqrt{n})$

42. $(\sqrt{7} + 2)^2$

43. $(2\sqrt{3} + 3\sqrt{5})(3\sqrt{3} - 4\sqrt{5})$

Understanding Through Discussion and Writing

44. Does the nth root of x^2 always exist? Why or why not? [10.1a]

45. Explain how to formulate a radical expression that can be used to define a function f with a domain of $\{x | x \leq 5\}$. [10.1a]

46. Explain why $\sqrt[3]{x^6} = x^2$ for any value of x, but $\sqrt{x^6} = x^3$ only when $x \geq 0$. [10.2d]

47. Is the quotient of two irrational numbers always an irrational number? Why or why not? [10.3b]

STUDYING FOR SUCCESS *Make Your Text Your Own*
- ☐ Make notes on the corresponding page of your text as your instructor presents concepts.
- ☐ Mark any difficult exercises for review as you study for a quiz or a test. Aim for mastery.
- ☐ Highlight trouble spots so that you can ask questions during class or in a tutoring session.

10.5 More on Division of Radical Expressions

OBJECTIVES

a Rationalize the denominator of a radical expression having one term in the denominator.

b Rationalize the denominator of a radical expression having two terms in the denominator.

a RATIONALIZING DENOMINATORS

Sometimes in mathematics it is useful to find an equivalent expression without a radical in the denominator. This provides a standard notation for expressing results. The procedure for finding such an expression is called **rationalizing the denominator**. We carry this out by multiplying by 1.

EXAMPLE 1 Rationalize the denominator: $\sqrt{\dfrac{7}{3}}$.

We multiply by 1, using $\sqrt{3}/\sqrt{3}$. We do this so that the denominator of the radicand will be a perfect square.

$$\sqrt{\frac{7}{3}} = \frac{\sqrt{7}}{\sqrt{3}} \cdot \frac{\sqrt{3}}{\sqrt{3}} = \frac{\sqrt{7}\cdot\sqrt{3}}{\sqrt{3}\cdot\sqrt{3}}$$

$$= \frac{\sqrt{21}}{\sqrt{3^2}} = \frac{\sqrt{21}}{3}$$

↑ The radicand is a perfect square.

Do Exercise 1. ▶

EXAMPLE 2 Rationalize the denominator: $\sqrt[3]{\dfrac{7}{25}}$.

We first factor the denominator:

$$\sqrt[3]{\frac{7}{25}} = \sqrt[3]{\frac{7}{5\cdot 5}}.$$

To eliminate the radical in the denominator, we consider the index 3 and the factors of the denominator. We have 2 factors of 5, and we need 3 factors of 5 in order to have a perfect cube. We achieve this by multiplying by 1, using $\sqrt[3]{5}/\sqrt[3]{5}$.

$$\sqrt[3]{\frac{7}{25}} = \frac{\sqrt[3]{7}}{\sqrt[3]{25}} = \frac{\sqrt[3]{7}}{\sqrt[3]{5\cdot 5}} \cdot \frac{\sqrt[3]{5}}{\sqrt[3]{5}}$$

$$= \frac{\sqrt[3]{7}\cdot\sqrt[3]{5}}{\sqrt[3]{5\cdot 5}\cdot\sqrt[3]{5}}$$

$$= \frac{\sqrt[3]{35}}{\sqrt[3]{5^3}} = \frac{\sqrt[3]{35}}{5}$$

↑ The radicand is a perfect cube.

Do Exercise 2. ▶

SKILL REVIEW *Use a rule to multiply a sum and a difference of the same two terms.* [4.6b]

Multiply.
1. $(x + 3)(x - 3)$
2. $(2y + 5)(2y - 5)$

Answers: **1.** $x^2 - 9$ **2.** $4y^2 - 25$

MyLab Math VIDEO

1. Rationalize the denominator:

$$\sqrt{\frac{2}{5}}.$$

2. Rationalize the denominator:

$$\sqrt[3]{\frac{5}{4}}.$$

Answers
1. $\dfrac{\sqrt{10}}{5}$ **2.** $\dfrac{\sqrt[3]{10}}{2}$

3. Rationalize the denominator: $\sqrt{\dfrac{4a}{3b}}$.

$\sqrt{\dfrac{4a}{3b}} = \dfrac{\sqrt{4a}}{\sqrt{3b}} \cdot \dfrac{\boxed{\phantom{\sqrt{3b}}}}{\sqrt{3b}}$

$= \dfrac{\sqrt{ab}}{\sqrt{3^2 \cdot \boxed{}}}$

$= \dfrac{\boxed{}\sqrt{3ab}}{\boxed{}}$

EXAMPLE 3 Rationalize the denominator: $\sqrt{\dfrac{2a}{5b}}$. Assume that no radicands were formed by raising negative numbers to even powers.

$\sqrt{\dfrac{2a}{5b}} = \dfrac{\sqrt{2a}}{\sqrt{5b}}$ Converting to a quotient of radicals

$\phantom{\sqrt{\dfrac{2a}{5b}}} = \dfrac{\sqrt{2a}}{\sqrt{5b}} \cdot \dfrac{\sqrt{5b}}{\sqrt{5b}}$ Multiplying by 1

$\phantom{\sqrt{\dfrac{2a}{5b}}} = \dfrac{\sqrt{10ab}}{\sqrt{5^2 b^2}}$ The radicand in the denominator is a perfect square.

$\phantom{\sqrt{\dfrac{2a}{5b}}} = \dfrac{\sqrt{10ab}}{5b}$

◀ Do Exercise 3.

EXAMPLE 4 Rationalize the denominator: $\dfrac{\sqrt[3]{a}}{\sqrt[3]{9x}}$.

We factor the denominator:

$\dfrac{\sqrt[3]{a}}{\sqrt[3]{9x}} = \dfrac{\sqrt[3]{a}}{\sqrt[3]{3 \cdot 3 \cdot x}}$.

To choose the symbol for 1, we look at $3 \cdot 3 \cdot x$. To make it a cube, we need another 3 and two more x's. Thus we multiply by 1, using $\sqrt[3]{3x^2}/\sqrt[3]{3x^2}$:

$\dfrac{\sqrt[3]{a}}{\sqrt[3]{9x}} = \dfrac{\sqrt[3]{a}}{\sqrt[3]{3 \cdot 3 \cdot x}} \cdot \dfrac{\sqrt[3]{3x^2}}{\sqrt[3]{3x^2}}$ Multiplying by 1

$\phantom{\dfrac{\sqrt[3]{a}}{\sqrt[3]{9x}}} = \dfrac{\sqrt[3]{3ax^2}}{\sqrt[3]{3^3 x^3}}$ The radicand in the denominator is a perfect cube.

$\phantom{\dfrac{\sqrt[3]{a}}{\sqrt[3]{9x}}} = \dfrac{\sqrt[3]{3ax^2}}{3x}$.

Rationalize the denominator.

4. $\dfrac{\sqrt[4]{7}}{\sqrt[4]{2}}$ 5. $\sqrt[3]{\dfrac{3x^5}{2y}}$

◀ Do Exercises 4 and 5.

EXAMPLE 5 Rationalize the denominator: $\dfrac{3x}{\sqrt[5]{2x^2 y^3}}$.

$\dfrac{3x}{\sqrt[5]{2x^2 y^3}} = \dfrac{3x}{\sqrt[5]{2 \cdot x \cdot x \cdot y \cdot y \cdot y}}$

$\phantom{\dfrac{3x}{\sqrt[5]{2x^2 y^3}}} = \dfrac{3x}{\sqrt[5]{2x^2 y^3}} \cdot \dfrac{\sqrt[5]{2^4 x^3 y^2}}{\sqrt[5]{2^4 x^3 y^2}}$

$\phantom{\dfrac{3x}{\sqrt[5]{2x^2 y^3}}} = \dfrac{3x\sqrt[5]{16x^3 y^2}}{\sqrt[5]{2^5 x^5 y^5}}$ The radicand in the denominator is a perfect fifth power.

$\phantom{\dfrac{3x}{\sqrt[5]{2x^2 y^3}}} = \dfrac{3x\sqrt[5]{16x^3 y^2}}{2xy}$

$\phantom{\dfrac{3x}{\sqrt[5]{2x^2 y^3}}} = \dfrac{x}{x} \cdot \dfrac{3\sqrt[5]{16x^3 y^2}}{2y}$

$\phantom{\dfrac{3x}{\sqrt[5]{2x^2 y^3}}} = \dfrac{3\sqrt[5]{16x^3 y^2}}{2y}$

6. Rationalize the denominator:

$\dfrac{7x}{\sqrt[3]{4xy^5}}$.

◀ Do Exercise 6.

Answers

3. $\dfrac{2\sqrt{3ab}}{3b}$ 4. $\dfrac{\sqrt[4]{56}}{2}$ 5. $\dfrac{x\sqrt[3]{12x^2 y^2}}{2y}$

6. $\dfrac{7\sqrt[3]{2x^2 y}}{2y^2}$

Guided Solution:
3. $\sqrt{3b}, 12, b^2, 2, 3b$

b RATIONALIZING WHEN THERE ARE TWO TERMS

Certain pairs of expressions containing square roots, such as $c - \sqrt{b}$, $c + \sqrt{b}$ and $\sqrt{a} - \sqrt{b}$, $\sqrt{a} + \sqrt{b}$, are called **conjugates**. The product of such a pair of conjugates has no radicals in it. (See Example 12 of Section 10.4.) Thus when we wish to rationalize a denominator that has two terms and one or more of them involves a square-root radical, we multiply by 1 using the conjugate of the denominator to write a symbol for 1.

▶ Do Exercises 7 and 8.

Multiply.

7. $(c - \sqrt{b})(c + \sqrt{b})$

8. $(\sqrt{a} + \sqrt{b})(\sqrt{a} - \sqrt{b})$

EXAMPLES In each of the following, what symbol for 1 would you use to rationalize the denominator?

	Expression	Symbol for 1
6.	$\dfrac{3}{x + \sqrt{7}}$	$\dfrac{x - \sqrt{7}}{x - \sqrt{7}}$
7.	$\dfrac{\sqrt{7} + 4}{3 - 2\sqrt{5}}$	$\dfrac{3 + 2\sqrt{5}}{3 + 2\sqrt{5}}$

Change the operation sign in the denominator to obtain the conjugate. Use the conjugate for the numerator and denominator of the symbol for 1.

▶ Do Exercises 9 and 10.

What symbol for 1 would you use to rationalize the denominator?

9. $\dfrac{\sqrt{5} + 1}{\sqrt{3} - y}$

10. $\dfrac{1}{\sqrt{2} + \sqrt{3}}$

EXAMPLE 8 Rationalize the denominator: $\dfrac{4}{\sqrt{3} + x}$.

$$\dfrac{4}{\sqrt{3} + x} = \dfrac{4}{\sqrt{3} + x} \cdot \dfrac{\sqrt{3} - x}{\sqrt{3} - x}$$

$$= \dfrac{4(\sqrt{3} - x)}{(\sqrt{3} + x)(\sqrt{3} - x)}$$

$$= \dfrac{4\sqrt{3} - 4x}{3 - x^2}$$

EXAMPLE 9 Rationalize the denominator: $\dfrac{4 + \sqrt{2}}{\sqrt{5} - \sqrt{2}}$.

$$\dfrac{4 + \sqrt{2}}{\sqrt{5} - \sqrt{2}} = \dfrac{4 + \sqrt{2}}{\sqrt{5} - \sqrt{2}} \cdot \dfrac{\sqrt{5} + \sqrt{2}}{\sqrt{5} + \sqrt{2}}$$ Multiplying by 1, using the conjugate of $\sqrt{5} - \sqrt{2}$, which is $\sqrt{5} + \sqrt{2}$

$$= \dfrac{(4 + \sqrt{2})(\sqrt{5} + \sqrt{2})}{(\sqrt{5} - \sqrt{2})(\sqrt{5} + \sqrt{2})}$$ Multiplying numerators and denominators

$$= \dfrac{4\sqrt{5} + 4\sqrt{2} + \sqrt{2}\sqrt{5} + (\sqrt{2})^2}{(\sqrt{5})^2 - (\sqrt{2})^2}$$ Using $(A - B)(A + B) = A^2 - B^2$ in the denominator

$$= \dfrac{4\sqrt{5} + 4\sqrt{2} + \sqrt{10} + 2}{5 - 2}$$

$$= \dfrac{4\sqrt{5} + 4\sqrt{2} + \sqrt{10} + 2}{3}$$

Rationalize the denominator.

11. $\dfrac{14}{3 + \sqrt{2}}$

$$= \dfrac{14}{3 + \sqrt{2}} \cdot \dfrac{3 - \boxed{}}{\boxed{} - \sqrt{2}}$$

$$= \dfrac{14(3 - \sqrt{2})}{9 - \boxed{}}$$

$$= \dfrac{14(3 - \sqrt{2})}{\boxed{}}$$

$$= \boxed{}(3 - \sqrt{2})$$

$$= 6 - 2\sqrt{2}$$

12. $\dfrac{5 + \sqrt{2}}{1 - \sqrt{2}}$

Answers

7. $c^2 - b$ 8. $a - b$ 9. $\dfrac{\sqrt{3} + y}{\sqrt{3} + y}$

10. $\dfrac{\sqrt{2} - \sqrt{3}}{\sqrt{2} - \sqrt{3}}$ 11. $6 - 2\sqrt{2}$

12. $-7 - 6\sqrt{2}$

Guided Solution:
11. $\sqrt{2}, 3, 2, 7, 2$

▶ Do Exercises 11 and 12.

10.5 Exercise Set

✓ Check Your Understanding

Reading Check Determine whether the two given expressions are conjugates. Answer "yes" or "no."

RC1. $8 - \sqrt{10}$, $\sqrt{10} - 8$

RC2. $\sqrt{4} - \sqrt{15}$, $\sqrt{4} + \sqrt{15}$

RC3. $-\sqrt{2} + 9$, $\sqrt{2} + 9$

RC4. $\dfrac{\sqrt{11}}{\sqrt{3}}$, $\dfrac{\sqrt{3}}{\sqrt{11}}$

Concept Check Choose from the columns on the right the symbol for 1 that you would use to rationalize the denominator. Some choices may be used more than once and others may not be used.

CC1. $\dfrac{2}{x - \sqrt{3}}$

CC2. $\dfrac{\sqrt{5}}{\sqrt{3}}$

a) $\dfrac{\sqrt{3x}}{\sqrt{3x}}$

b) $\dfrac{x - \sqrt{3}}{x - \sqrt{3}}$

CC3. $\dfrac{\sqrt[4]{9x^2}}{\sqrt[4]{x^3}}$

CC4. $\dfrac{\sqrt[3]{3y}}{\sqrt[3]{9y^2}}$

c) $\dfrac{\sqrt{3}}{\sqrt{3}}$

d) $\dfrac{\sqrt{x}}{\sqrt{x}}$

e) $\dfrac{\sqrt[4]{x}}{\sqrt[4]{x}}$

f) $\dfrac{\sqrt{9}}{\sqrt{9}}$

CC5. $\dfrac{x + \sqrt{3}}{\sqrt{3}}$

CC6. $\dfrac{1}{3\sqrt{x}}$

g) $\dfrac{x + \sqrt{3}}{x + \sqrt{3}}$

h) $\dfrac{\sqrt[3]{3y}}{\sqrt[3]{3y}}$

a Rationalize the denominator. Assume that no radicands were formed by raising negative numbers to even powers.

1. $\sqrt{\dfrac{5}{3}}$

2. $\sqrt{\dfrac{8}{7}}$

3. $\sqrt{\dfrac{11}{2}}$

4. $\sqrt{\dfrac{17}{6}}$

5. $\dfrac{2\sqrt{3}}{7\sqrt{5}}$

6. $\dfrac{3\sqrt{5}}{8\sqrt{2}}$

7. $\sqrt[3]{\dfrac{16}{9}}$

8. $\sqrt[3]{\dfrac{1}{3}}$

9. $\dfrac{\sqrt[3]{3a}}{\sqrt[3]{5c}}$

10. $\dfrac{\sqrt[3]{7x}}{\sqrt[3]{3y}}$

11. $\dfrac{\sqrt[3]{2y^4}}{\sqrt[3]{6x^4}}$

12. $\dfrac{\sqrt[3]{3a^4}}{\sqrt[3]{7b^2}}$

13. $\dfrac{1}{\sqrt[4]{st}}$

14. $\dfrac{1}{\sqrt[3]{yz}}$

15. $\sqrt{\dfrac{3x}{20}}$

16. $\sqrt{\dfrac{7a}{32}}$

17. $\sqrt[3]{\dfrac{4}{5x^5y^2}}$
18. $\sqrt[3]{\dfrac{7c}{100ab^5}}$
19. $\sqrt[4]{\dfrac{1}{8x^7y^3}}$
20. $\dfrac{2x}{\sqrt[5]{18x^8y^6}}$

b Rationalize the denominator. Assume that no radicands were formed by raising negative numbers to even powers.

21. $\dfrac{9}{6 - \sqrt{10}}$
22. $\dfrac{3}{8 + \sqrt{5}}$
23. $\dfrac{-4\sqrt{7}}{\sqrt{5} + \sqrt{3}}$
24. $\dfrac{-5\sqrt{2}}{\sqrt{7} - \sqrt{5}}$

25. $\dfrac{6\sqrt{3}}{3\sqrt{2} - \sqrt{5}}$
26. $\dfrac{34\sqrt{5}}{2\sqrt{5} - \sqrt{3}}$
27. $\dfrac{3 + \sqrt{5}}{\sqrt{2} + \sqrt{5}}$
28. $\dfrac{2 + \sqrt{3}}{\sqrt{3} + \sqrt{5}}$

29. $\dfrac{\sqrt{3} - \sqrt{2}}{\sqrt{3} - \sqrt{7}}$
30. $\dfrac{\sqrt{5} - \sqrt{3}}{\sqrt{5} - \sqrt{2}}$
31. $\dfrac{\sqrt{5} - 2\sqrt{6}}{\sqrt{3} - 4\sqrt{5}}$
32. $\dfrac{\sqrt{6} - 3\sqrt{5}}{\sqrt{3} - 2\sqrt{7}}$

33. $\dfrac{2 - \sqrt{a}}{3 + \sqrt{a}}$
34. $\dfrac{5 + \sqrt{x}}{8 - \sqrt{x}}$
35. $\dfrac{2 + 3\sqrt{x}}{3 + 2\sqrt{x}}$
36. $\dfrac{5 + 2\sqrt{y}}{4 + 3\sqrt{y}}$

37. $\dfrac{5\sqrt{3} - 3\sqrt{2}}{3\sqrt{2} - 2\sqrt{3}}$
38. $\dfrac{7\sqrt{2} + 4\sqrt{3}}{4\sqrt{3} - 3\sqrt{2}}$
39. $\dfrac{\sqrt{x} - \sqrt{y}}{\sqrt{x} + \sqrt{y}}$
40. $\dfrac{\sqrt{a} + \sqrt{b}}{\sqrt{a} - \sqrt{b}}$

Skill Maintenance

Solve. [6.7a]

41. $\dfrac{1}{2} - \dfrac{1}{3} = \dfrac{5}{t}$

42. $\dfrac{5}{x - 1} + \dfrac{9}{x^2 + x + 1} = \dfrac{15}{x^3 - 1}$

Divide and simplify. [6.2b]

43. $\dfrac{1}{x^3 - y^3} \div \dfrac{1}{(x - y)(x^2 + xy + y^2)}$

44. $\dfrac{2x^2 - x - 6}{x^2 + 4x + 3} \div \dfrac{2x^2 + x - 3}{x^2 - 1}$

Synthesis

45. Simplify. (*Hint*: Rationalize the denominator.)

$$\sqrt{a^2 - 3} - \dfrac{a^2}{\sqrt{a^2 - 3}}$$

46. Express each of the following as the product of two radical expressions.

   a) $x - 5$
   b) $x - a$

10.6 Solving Radical Equations

OBJECTIVES

a Solve radical equations with one radical term.

b Solve radical equations with two radical terms.

c Solve applied problems involving radical equations.

a THE PRINCIPLE OF POWERS

SKILL REVIEW

Solve quadratic equations by first factoring and then using the principle of zero products. [5.8b]

Solve.

1. $x^2 - x = 6$
2. $x^2 - x = 2x + 4$

Answers: **1.** $-2, 3$ **2.** $-1, 4$

A **radical equation** has variables in one or more radicands. For example,

$$\sqrt[3]{2x+1} = 5 \quad \text{and} \quad \sqrt{x} + \sqrt{4x-2} = 7$$

are radical equations. To solve such an equation, we need a new equation-solving principle. Suppose that an equation $a = b$ is true. If we square both sides, we get another true equation: $a^2 = b^2$. This can be generalized.

THE PRINCIPLE OF POWERS

For any natural number n, if an equation $a = b$ is true, then $a^n = b^n$ is true.

However, if an equation $a^n = b^n$ is true, it *may not* be true that $a = b$, if n is even. For example, $3^2 = (-3)^2$ is true, but $3 = -3$ is not true. Thus we *must check* the possible solutions when we solve an equation using the principle of powers.

To solve an equation with a radical term, we first isolate the radical term on one side of the equation. Then we use the principle of powers.

EXAMPLE 1 Solve: $\sqrt{x} - 3 = 4$.

We have

$\sqrt{x} - 3 = 4$

$\sqrt{x} = 7$ Adding 3 to isolate the radical

$(\sqrt{x})^2 = 7^2$ Using the principle of powers (squaring)

$x = 49$. $\sqrt{x} \cdot \sqrt{x} = x$

The number 49 is a possible solution. But we *must* check in order to be sure!

Check: $\sqrt{x} - 3 = 4$

$\sqrt{49} - 3 \;?\; 4$

$7 - 3$

4 | TRUE

Caution!

The principle of powers does not always give equivalent equations. For this reason, a check is a must!

The solution is 49.

758 CHAPTER 10 Radical Expressions, Equations, and Functions

EXAMPLE 2 Solve: $\sqrt{x} = -3$.

We might note at the outset that this equation has no solution because the principal square root of a number is never negative. Let's continue as above for comparison.

$$\sqrt{x} = -3$$
$$(\sqrt{x})^2 = (-3)^2$$
$$x = 9$$

Check:
$$\sqrt{x} = -3$$
$$\sqrt{9} \;?\; -3$$
$$3 \;|\; \text{FALSE}$$

The number 9 does *not* check. Thus the equation $\sqrt{x} = -3$ has no real-number solution. Note that the solution of the equation $x = 9$ is 9, but the equation $\sqrt{x} = -3$ has *no* solution. Thus the equations $x = 9$ and $\sqrt{x} = -3$ are *not* equivalent equations.

Do Exercises 1 and 2. ▷

Solve.
1. $\sqrt{x} - 7 = 3$
2. $\sqrt{x} = -2$

EXAMPLE 3 Solve: $x - 7 = 2\sqrt{x+1}$.

The radical term is already isolated. We proceed with the principle of powers:

$$x - 7 = 2\sqrt{x+1}$$
$$(x-7)^2 = (2\sqrt{x+1})^2 \quad \text{Using the principle of powers (squaring)}$$
$$(x-7)(x-7) = (2\sqrt{x+1})(2\sqrt{x+1})$$
$$x^2 - 14x + 49 = 2^2(\sqrt{x+1})^2$$
$$x^2 - 14x + 49 = 4(x+1)$$
$$x^2 - 14x + 49 = 4x + 4$$
$$x^2 - 18x + 45 = 0$$
$$(x-3)(x-15) = 0 \quad \text{Factoring}$$
$$x - 3 = 0 \;\; or \;\; x - 15 = 0 \quad \text{Using the principle of zero products}$$
$$x = 3 \;\; or \;\; x = 15.$$

The possible solutions are 3 and 15.

Check:

For 3:
$$x - 7 = 2\sqrt{x+1}$$
$$3 - 7 \;?\; 2\sqrt{3+1}$$
$$-4 \;|\; 2\sqrt{4}$$
$$\;|\; 2(2)$$
$$\;|\; 4 \quad \text{FALSE}$$

For 15:
$$x - 7 = 2\sqrt{x+1}$$
$$15 - 7 \;?\; 2\sqrt{15+1}$$
$$8 \;|\; 2\sqrt{16}$$
$$\;|\; 2(4)$$
$$\;|\; 8 \quad \text{TRUE}$$

The number 3 does *not* check, but the number 15 does check. The solution is 15.

The number 3 in Example 3 is what is sometimes called an *extraneous solution*, but such terminology is risky to use at best because the number 3 is in *no way* a solution of the original equation.

Do Exercises 3 and 4. ▷

Solve.

GS 3. $x + 2 = \sqrt{2x+7}$
$$(x+2) = (\sqrt{2x+7})^2$$
$$x^2 + x + 4 = + 7$$
$$x^2 + 2x - = 0$$
$$(x +)(x-1) = 0$$
$$x + 3 = 0 \;\; or \;\; x - 1 = 0$$
$$x = -3 \;\; or \;\; x = $$
The number ___ does not check, but the number 1 does check. The solution is ___.

4. $x + 1 = 3\sqrt{x-1}$

Answers
1. 100 2. No solution 3. 1 4. 2, 5
Guided Solution:
3. 2, 4, 2x, 3, 3, 1, −3, 1

SECTION 10.6 Solving Radical Equations

ALGEBRAIC ▶◀ GRAPHICAL CONNECTION

We can visualize or check the solutions of a radical equation graphically. Consider the equation of Example 3: $x - 7 = 2\sqrt{x + 1}$. We can examine the solutions by graphing the equations

$$y = x - 7 \quad \text{and} \quad y = 2\sqrt{x + 1}$$

using the same set of axes. A hand-drawn graph of $y = 2\sqrt{x + 1}$ would involve approximating square roots on a calculator.

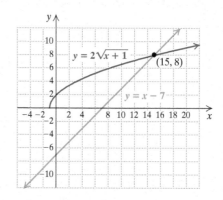

It appears from the graph that when $x = 15$, the values of $y = x - 7$ and $y = 2\sqrt{x + 1}$ are the same, 8. We can check this as we did in Example 3. Note too that the graphs *do not* intersect at $x = 3$, the possible solution that did not check.

CALCULATOR CORNER

Solving Radical Equations We can solve radical equations graphically. Consider the equation in Example 3, $x - 7 = 2\sqrt{x + 1}$.

We first graph each side of the equation. We enter $y_1 = x - 7$ and $y_2 = 2\sqrt{x + 1}$ on the equation-editor screen and graph the equations using the window $[-5, 20, -10, 10]$. Note that there is one point of intersection. We use the **INTERSECT** feature to find its coordinates. (See the Calculator Corner on p. 593 for the procedure.) The first coordinate, 15, is the value of x for which $y_1 = y_2$, or $x - 7 = 2\sqrt{x + 1}$. It is the solution of the equation. Note that the graph shows a single solution whereas the algebraic solution in Example 3 yields two possible solutions, 3 and 15, that must be checked. The algebraic check shows that 15 is the only solution.

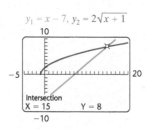

EXERCISES:
1. Solve the equations in Examples 1 and 4 graphically.
2. Solve the equations in Margin Exercises 1, 3, and 4 graphically.

EXAMPLE 4 Solve: $x = \sqrt{x+7} + 5$.

We have

$$x = \sqrt{x+7} + 5$$
$$x - 5 = \sqrt{x+7} \quad \text{Subtracting 5 to isolate the radical term}$$
$$(x-5)^2 = (\sqrt{x+7})^2 \quad \text{Using the principle of powers (squaring both sides)}$$
$$x^2 - 10x + 25 = x + 7$$
$$x^2 - 11x + 18 = 0$$
$$(x-9)(x-2) = 0 \quad \text{Factoring}$$
$$x = 9 \quad \text{or} \quad x = 2. \quad \text{Using the principle of zero products}$$

The possible solutions are 9 and 2.

Check: For 9:
$$x = \sqrt{x+7} + 5$$
$$9 \;?\; \sqrt{9+7} + 5$$
$$\sqrt{16} + 5$$
$$4 + 5$$
$$9 \qquad \text{TRUE}$$

For 2:
$$x = \sqrt{x+7} + 5$$
$$2 \;?\; \sqrt{2+7} + 5$$
$$\sqrt{9} + 5$$
$$3 + 5$$
$$8 \qquad \text{FALSE}$$

Since 9 checks but 2 does not, the solution is 9.

EXAMPLE 5 Solve: $\sqrt[3]{2x+1} + 5 = 0$.

We have

$$\sqrt[3]{2x+1} + 5 = 0$$
$$\sqrt[3]{2x+1} = -5 \quad \text{Subtracting 5. This isolates the radical term.}$$
$$(\sqrt[3]{2x+1})^3 = (-5)^3 \quad \text{Using the principle of powers (raising to the third power)}$$
$$2x + 1 = -125$$
$$2x = -126 \quad \text{Subtracting 1}$$
$$x = -63.$$

Check:
$$\sqrt[3]{2x+1} + 5 = 0$$
$$\sqrt[3]{2 \cdot (-63) + 1} + 5 \;?\; 0$$
$$\sqrt[3]{-126 + 1} + 5$$
$$\sqrt[3]{-125} + 5$$
$$-5 + 5$$
$$0 \qquad \text{TRUE}$$

The solution is -63.

Do Exercises 5 and 6.

Solve.
5. $x = \sqrt{x+5} + 1$

6. $\sqrt[4]{x-1} - 2 = 0$
$\sqrt[4]{x-1} = \boxed{}$
$(\sqrt[4]{x-1})^4 = 2$
$x - 1 = \boxed{}$
$x = \boxed{}$

Answers
5. 4   6. 17
Guided Solution:
6. 2, 4, 16, 17

b EQUATIONS WITH TWO RADICAL TERMS

A general strategy for solving radical equations, including those with two radical terms, is as follows.

> **SOLVING RADICAL EQUATIONS**
>
> To solve radical equations:
> 1. Isolate one of the radical terms.
> 2. Use the principle of powers.
> 3. If a radical remains, perform steps (1) and (2) again.
> 4. Check possible solutions.

EXAMPLE 6 Solve: $\sqrt{x-3} + \sqrt{x+5} = 4$.

$$\sqrt{x-3} + \sqrt{x+5} = 4$$

$\sqrt{x-3} = 4 - \sqrt{x+5}$ Subtracting $\sqrt{x+5}$. This isolates one of the radical terms.

$(\sqrt{x-3})^2 = (4 - \sqrt{x+5})^2$ Using the principle of powers (squaring both sides)

$x - 3 = 16 - 8\sqrt{x+5} + (x+5)$ Using $(A - B)^2 = A^2 - 2AB + B^2$. See this rule in Section 4.6.

$x - 3 = -8\sqrt{x+5} + x + 21$ Collecting like terms

$-3 = 21 - 8\sqrt{x+5}$ Subtracting x

$-24 = -8\sqrt{x+5}$ Subtracting 21 to isolate the radical term

$3 = \sqrt{x+5}$ Dividing by -8

$3^2 = (\sqrt{x+5})^2$ Squaring

$9 = x + 5$

$4 = x$

The number 4 checks and is the solution.

EXAMPLE 7 Solve: $\sqrt{2x-5} = 1 + \sqrt{x-3}$.

$\sqrt{2x-5} = 1 + \sqrt{x-3}$

$(\sqrt{2x-5})^2 = (1 + \sqrt{x-3})^2$ One radical is already isolated. We square both sides.

$2x - 5 = 1 + 2\sqrt{x-3} + (x-3)$

$2x - 5 = 2\sqrt{x-3} + x - 2$

$x - 3 = 2\sqrt{x-3}$ Isolating the remaining radical term

$(x-3)^2 = (2\sqrt{x-3})^2$ Squaring both sides

$x^2 - 6x + 9 = 4(x-3)$

$x^2 - 6x + 9 = 4x - 12$

$x^2 - 10x + 21 = 0$

$(x-7)(x-3) = 0$ Factoring

$x = 7$ or $x = 3$ Using the principle of zero products

The possible solutions are 7 and 3.

Check: For 7:
$$\sqrt{2x-5} = 1 + \sqrt{x-3}$$
$$\sqrt{2(7)-5} \;?\; 1 + \sqrt{7-3}$$
$$\sqrt{14-5} \;\big|\; 1 + \sqrt{4}$$
$$\sqrt{9} \;\big|\; 1 + 2$$
$$3 \;\big|\; 3 \quad \text{TRUE}$$

For 3:
$$\sqrt{2x-5} = 1 + \sqrt{x-3}$$
$$\sqrt{2(3)-5} \;?\; 1 + \sqrt{3-3}$$
$$\sqrt{6-5} \;\big|\; 1 + \sqrt{0}$$
$$\sqrt{1} \;\big|\; 1 + 0$$
$$1 \;\big|\; 1 \quad \text{TRUE}$$

The numbers 7 and 3 check and are the solutions.

Do Exercises 7 and 8. ▶

Solve.

7. $\sqrt{x} - \sqrt{x-5} = 1$

8. $\sqrt{2x-5} - 2 = \sqrt{x-2}$

EXAMPLE 8 Solve: $\sqrt{x+2} - \sqrt{2x+2} + 1 = 0.$

We first isolate one radical.

$$\sqrt{x+2} - \sqrt{2x+2} + 1 = 0$$
$$\sqrt{x+2} + 1 = \sqrt{2x+2} \quad \text{Adding } \sqrt{2x+2} \text{ to isolate a radical term}$$
$$(\sqrt{x+2} + 1)^2 = (\sqrt{2x+2})^2 \quad \text{Squaring both sides}$$
$$x + 2 + 2\sqrt{x+2} + 1 = 2x + 2$$
$$2\sqrt{x+2} = x - 1$$
$$(2\sqrt{x+2})^2 = (x-1)^2$$
$$4(x+2) = x^2 - 2x + 1$$
$$4x + 8 = x^2 - 2x + 1$$
$$0 = x^2 - 6x - 7$$
$$0 = (x-7)(x+1) \quad \text{Factoring}$$
$$x - 7 = 0 \quad \text{or} \quad x + 1 = 0 \quad \text{Using the principle of zero products}$$
$$x = 7 \quad \text{or} \quad x = -1$$

The possible solutions are 7 and −1.

Check: For 7:
$$\sqrt{x+2} - \sqrt{2x+2} + 1 = 0$$
$$\sqrt{7+2} - \sqrt{2 \cdot 7 + 2} + 1 \;?\; 0$$
$$\sqrt{9} - \sqrt{16} + 1$$
$$3 - 4 + 1$$
$$0 \quad \text{TRUE}$$

For −1:
$$\sqrt{x+2} - \sqrt{2x+2} + 1 = 0$$
$$\sqrt{-1+2} - \sqrt{2 \cdot (-1) + 2} + 1 \;?\; 0$$
$$\sqrt{1} - \sqrt{0} + 1$$
$$1 - 0 + 1$$
$$2 \quad \text{FALSE}$$

The number 7 checks, but −1 does not. The solution is 7.

Do Exercise 9. ▶

9. Solve:
$\sqrt{3x+1} - 1 - \sqrt{x+4} = 0.$

Answers

7. 9   **8.** 27   **9.** 5

SECTION 10.6 Solving Radical Equations   **763**

c APPLICATIONS

Speed of Sound. Many applications translate to radical equations. For example, at a temperature of t degrees Fahrenheit, sound travels at a rate of S feet per second, where

$$S = 21.9\sqrt{5t + 2457}.$$

EXAMPLE 9 *Capitol Concert.* Each year, a free concert, know as "A Capitol Fourth," on the west lawn of the Capitol in Washington, DC, precedes the July 4th fireworks display on the National Mall. A few of the performers at the 2017 concert included the Beach Boys, the Blues Brothers, Trace Adkins, Yolanda Adams, and *The Voice* winner Chris Blue. A scientific instrument determined that the sound of the music was traveling at a rate of 1170 ft/sec. What was the air temperature at the concert?

We substitute 1170 for S in the formula $S = 21.9\sqrt{5t + 2457}$:

$$1170 = 21.9\sqrt{5t + 2457}.$$

Then we solve the equation for t:

$$1170 = 21.9\sqrt{5t + 2457}$$

$$\frac{1170}{21.9} = \sqrt{5t + 2457} \qquad \text{Dividing by 21.9}$$

$$\left(\frac{1170}{21.9}\right)^2 = \left(\sqrt{5t + 2457}\right)^2 \qquad \text{Squaring both sides}$$

$$2854.2 \approx 5t + 2457 \qquad \text{Simplifying}$$

$$397.2 \approx 5t \qquad \text{Subtracting 2457}$$

$$79 \approx t. \qquad \text{Dividing by 5}$$

The temperature at the concert was about 79°F.

◀ Do Exercise 10.

10. *Marching Band Performance.* When the Fulton High School marching band performed at half-time of a football game, the speed of sound of the music was measured by a scientific instrument to be 1162 ft/sec. What was the air temperature?

Answer
10. About 72°F

10.6 Exercise Set

FOR EXTRA HELP   MyLab Math

✓ Check Your Understanding

Reading Check Choose from the list on the right the term that best completes each statement. Not every word will be used.

RC1. The equation $\sqrt{4 - 11x} = 3$ is a(n) _____ equation.

RC2. When we square both sides of an equation, we are using the principle of _____.

RC3. To solve an equation with a radical term, we first _____ the radical term on one side of the equation.

RC4. A radical equation has variables in one or more _____.

RC5. A check is essential when we raise both sides of an equation to a(n) _____ power.

even
radical
isolate
odd
radicands
square roots
powers
raise
rational
principle

Concept Check Determine the number of times that the principle of squaring must be used in order to solve the equation. Do not solve the equation.

CC1. $\sqrt{x-2} = x+4$

CC2. $\sqrt{12y-5} = \sqrt{y+24}$

CC3. $3 - \sqrt{x+3} = \sqrt{x-3}$

CC4. $\sqrt{c} + 2 = \sqrt{c+7}$

a Solve.

1. $\sqrt{2x-3} = 4$
2. $\sqrt{5x+2} = 7$
3. $\sqrt{6x+1} = 8$
4. $\sqrt{3x-4} = 6$

5. $\sqrt{y+7} - 4 = 4$
6. $\sqrt{x-1} - 3 = 9$
7. $\sqrt{5y+8} = 10$
8. $\sqrt{2y+9} = 5$

9. $\sqrt[3]{x} = -1$
10. $\sqrt[3]{y} = -2$
11. $\sqrt{x+2} = -4$
12. $\sqrt{y-3} = -2$

13. $\sqrt[3]{x+5} = 2$
14. $\sqrt[3]{x-2} = 3$
15. $\sqrt[4]{y-3} = 2$
16. $\sqrt[4]{x+3} = 3$

17. $\sqrt[3]{6x+9} + 8 = 5$
18. $\sqrt[3]{3y+6} + 2 = 3$
19. $8 = \dfrac{1}{\sqrt{x}}$
20. $\dfrac{1}{\sqrt{y}} = 3$

21. $x - 7 = \sqrt{x-5}$
22. $x - 5 = \sqrt{x+7}$
23. $2\sqrt{x+1} + 7 = x$
24. $\sqrt{2x+7} - 2 = x$

25. $3\sqrt{x-1} - 1 = x$
26. $x - 1 = \sqrt{x+5}$
27. $x - 3 = \sqrt{27-3x}$
28. $x - 1 = \sqrt{1-x}$

b Solve.

29. $\sqrt{3y+1} = \sqrt{2y+6}$

30. $\sqrt{5x-3} = \sqrt{2x+3}$

31. $\sqrt{y-5} + \sqrt{y} = 5$

32. $\sqrt{x-9} + \sqrt{x} = 1$

33. $3 + \sqrt{z-6} = \sqrt{z+9}$

34. $\sqrt{4x-3} = 2 + \sqrt{2x-5}$

35. $\sqrt{20-x} + 8 = \sqrt{9-x} + 11$

36. $4 + \sqrt{10-x} = 6 + \sqrt{4-x}$

37. $\sqrt{4y+1} - \sqrt{y-2} = 3$

38. $\sqrt{y+15} - \sqrt{2y+7} = 1$

39. $\sqrt{x+2} + \sqrt{3x+4} = 2$

40. $\sqrt{6x+7} - \sqrt{3x+3} = 1$

41. $\sqrt{3x-5} + \sqrt{2x+3} + 1 = 0$

42. $\sqrt{2m-3} + 2 - \sqrt{m+7} = 0$

43. $2\sqrt{t-1} - \sqrt{3t-1} = 0$

44. $3\sqrt{2y+3} - \sqrt{y+10} = 0$

C Solve.

Sightings to the Horizon. How far can you see to the horizon from a given height? The function $D = 1.2\sqrt{h}$ can be used to approximate the distance D, in miles, that a person can see to the horizon from a height h, in feet.

45. *Observation Deck at Jeddah Tower.* Jeddah Tower, which will be the tallest building in the world, in Jeddah, Saudi Arabia, is projected to be completed in 2020. An observation deck with a 98-ft diameter outdoor balcony near the top of the tower is 2080 ft high. How far can a tourist see to the horizon from the observation deck?

Data: Council on Tall Buildings and Urban Habitat (CTBUH); Global Construction News

46. *Observation Deck at Washington Monument.* An observation deck near the top of the Washington Monument is 499 ft high. How far can a tourist see to the horizon from the top of the Washington Monument?

Data: Council on Tall Buildings and Urban Habitat (CTBUH); Global Construction News

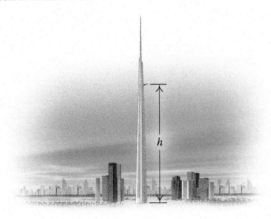

47. Sarah can see 31.3 mi to the horizon from the top of a cliff. What is the height of Sarah's eyes?

48. A steeplejack can see 13 mi to the horizon from the top of a building. What is the height of the steeplejack's eyes?

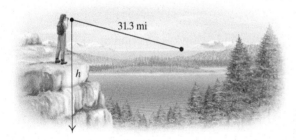

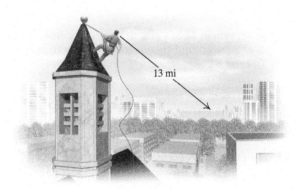

49. A technician can see 30.4 mi to the horizon from the top of a radio tower. How high is the tower?

50. A person can see 230 mi to the horizon from an airplane window. How high is the airplane?

Speed of a Skidding Car. After an automobile accident, how do police determine the speed at which a car had been traveling? The formula
$$r = 2\sqrt{5L}$$
can be used to approximate the speed r, in miles per hour, of a car that has left a skid mark of length L, in feet. Use this formula for Exercises 51 and 52.

51. How far will a car skid at 55 mph? at 75 mph?

52. How far will a car skid at 65 mph? at 100 mph?

53. *Water Flow and Nozzle Pressure.* For accuracy and safety, firefighters rely on nozzle pressure and water flow. They can calculate water flow using the formula
$$F = 29.72D^2\sqrt{P},$$
where F is the water flow, in gallons per minute, D is the nozzle diameter, in inches, and P is the nozzle pressure, in pounds per square inch. What is the nozzle pressure when the nozzle diameter is $1\frac{3}{4}$ in. and the water flow is 815 gallons per minute?

Data: Manchaca Fire Rescue (mvfd.org); Truckee Meadow Fire Protection District (tmfire.us)

54. *Temperature and the Speed of Sound.* Avalanche blasting creates controlled slides. Squaw Valley, a Tahoe ski resort, is using the Gazex Inertia Explorer, a new blasting technology that is safer and creates a larger blast. At a temperature of t degrees Fahrenheit, sound travels at a rate of S feet per second, where
$$S = 21.9\sqrt{5t + 2457}.$$
If at a recent blast the sound traveled at a rate of 1115 ft/sec, what was the temperature at that time?

Data: kolotv.com; Squaw-Unveils-New-Avalanche-Control-Technology-351977181.html, by Colin Lygren, November 19, 2015

Period of a Swinging Pendulum. The formula $T = 2\pi\sqrt{L/32}$ can be used to find the period T, in seconds, of a pendulum of length L, in feet.

55. What is the length of a pendulum that has a period of 1.0 sec? Use 3.14 for π.

56. What is the length of a pendulum that has a period of 2.0 sec? Use 3.14 for π.

57. The pendulum in Jean's grandfather clock has a period of 2.2 sec. Find the length of the pendulum. Use 3.14 for π.

58. A playground swing has a period of 3.1 sec. Find the length of the swing's chain. Use 3.14 for π.

Skill Maintenance

Solve.

59. *Bicycle Travel.* A cyclist traveled 702 mi in 14 days. At this same rate, how far would the cyclist have traveled in 56 days? [6.8b]

60. *Delivering Leaflets.* Jeff can distribute leaflets to homes three times as fast as Grace can. If they work together, it takes them 1 hr to complete the job. How long would it take each to deliver the leaflets alone? [6.8a]

Solve. [5.8b]

61. $x^2 + 2.8x = 0$

62. $3x^2 - 5x = 0$

63. $x^2 - 64 = 0$

64. $2x^2 = x + 21$

For each of the following functions, find and simplify $f(a - 1)$. [7.1b]

65. $f(x) = x^2$

66. $f(x) = x^2 - x$

67. $f(x) = 2x^2 - 3x$

68. $f(x) = 2x^2 + 3x - 7$

Synthesis

Solve.

69. $\sqrt{\sqrt{y + 49} - \sqrt{y}} = \sqrt{7}$

70. $\sqrt[3]{x^2 + x + 15} - 3 = 0$

71. $\sqrt{\sqrt{x^2 + 9x + 34}} = 2$

72. $6\sqrt{y} + 6y^{-1/2} = 37$

73. $\sqrt{a^2 + 30a} = a + \sqrt{5a}$

74. $\sqrt{\sqrt{x} + 4} = \sqrt{x} - 2$

75. $\sqrt{y + 1} - \sqrt{2y - 5} = \sqrt{y - 2}$

76. $\sqrt{x + 1} - \dfrac{2}{\sqrt{x + 1}} = 1$

Applications Involving Powers and Roots

10.7

OBJECTIVE

a Solve applied problems involving the Pythagorean theorem and powers and roots.

a APPLICATIONS

SKILL REVIEW

Approximate square roots of numbers using a calculator. [1.2d]
Use a calculator to approximate each square root. Round to three decimal places.

1. $\sqrt{43}$
2. $\sqrt{310}$

Answers: **1.** 6.557 **2.** 17.607

There are many kinds of applied problems that involve powers and roots. Many also make use of right triangles and the Pythagorean theorem: $a^2 + b^2 = c^2$.

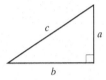

EXAMPLE 1 *Computer Screen Size.* The viewable image size of a widescreen computer measures 21 in. diagonally and has a width of 18.3 in. What is its height?

Using the Pythagorean theorem, $a^2 + b^2 = c^2$, we substitute 18.3 for b and 21 for c and then solve for a:

$$a^2 + b^2 = c^2$$
$$a^2 + 18.3^2 = 21^2 \quad \text{Substituting}$$
$$a^2 + 334.89 = 441$$
$$a^2 = 106.11$$

Exact answer: $a = \sqrt{106.11}$ We consider only the positive root
Approximation: $a \approx 10.3$.   since length cannot be negative.

The height of the viewable image is about 10.3 in.

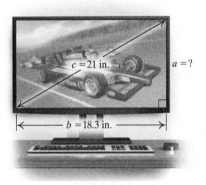

EXAMPLE 2 Find the length of the hypotenuse of this right triangle. Give an exact answer and an approximation to three decimal places.

$$7^2 + 4^2 = c^2 \quad \text{Substituting}$$
$$49 + 16 = c^2$$
$$65 = c^2$$

Exact answer: $c = \sqrt{65}$
Approximation: $c \approx 8.062$ Using a calculator

EXAMPLE 3 Find the missing length b in this right triangle. Give an exact answer and an approximation to three decimal places.

$$1^2 + b^2 = (\sqrt{11})^2 \quad \text{Substituting}$$
$$1 + b^2 = 11$$
$$b^2 = 10$$

Exact answer: $b = \sqrt{10}$
Approximation: $b \approx 3.162$ Using a calculator

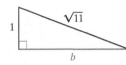

1. Find the length of the hypotenuse of this right triangle. Give an exact answer and an approximation to three decimal places.

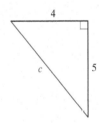

2. Find the length of the leg of this right triangle. Give an exact answer and an approximation to three decimal places.

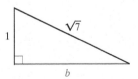

Do Exercises 1 and 2.

Answers
1. $\sqrt{41}$; 6.403 **2.** $\sqrt{6}$; 2.449

EXAMPLE 4 *Construction.* Darla is laying out the footing of a house. To see if the corner is square, she measures 16 ft from the corner along one wall and 12 ft from the corner along the other wall. How long should the diagonal be between those two points if the corner is a right angle?

3. *Baseball Diamond.* A baseball diamond is actually in the shape of a square 90 ft on a side. Suppose that a catcher fields a bunt along the third-base line 10 ft from home plate. How far is the catcher's throw to first base? Give an exact answer and an approximation to three decimal places.

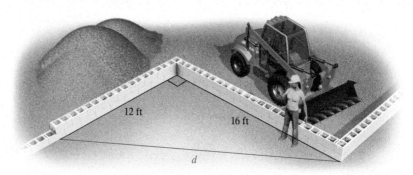

We make a drawing and let d = the length of the diagonal. It is the length of the hypotenuse of a right triangle whose legs are 12 ft and 16 ft. We substitute these values in the Pythagorean theorem to find d:

$$d^2 = 12^2 + 16^2$$
$$d^2 = 144 + 256$$
$$d^2 = 400$$
$$d = \sqrt{400} = 20.$$

The length of the diagonal should be 20 ft.

◀ Do Exercise 3.

EXAMPLE 5 *Road-Pavement Messages.* In a psychological study, it was determined that the ideal length L of the letters of a word painted on pavement is given by

$$L = \frac{0.000169 d^{2.27}}{h},$$

where d is the distance of a car from the lettering and h is the height of the eye above the road. All units are in feet. For a person h feet above the road, a message d feet away will be the most readable if the length of the letters is L. Find L, given that $h = 4$ ft and $d = 180$ ft.

4. Refer to Example 5. Find L given that $h = 3$ ft and $d = 180$ ft. You will need a calculator with an exponentiation key $\boxed{y^x}$, or ⌃.

We substitute 4 for h and 180 for d and calculate L using a calculator with an exponentiation key $\boxed{y^x}$, or ⌃:

$$L = \frac{0.000169(180)^{2.27}}{4} \approx 5.6 \text{ ft.}$$

◀ Do Exercise 4.

Answers

3. $\sqrt{8200}$ ft; 90.554 ft 4. About 7.4 ft

Translating for Success

1. Angles of a Triangle. The second angle of a triangle is four times as large as the first. The third is 27° less than the sum of the other angles. Find the measures of the angles.

2. Lengths of a Rectangle. The area of a rectangle is 180 ft². The length is 26 ft greater than the width. Find the length and the width.

3. Boat Travel. The speed of a river is 3 mph. A boat can go 72 mi upstream and 24 mi downstream in a total time of 16 hr. Find the speed of the boat in still water.

4. Coin Mixture. A collection of nickels and quarters is worth $13.85. There are 85 coins in all. How many of each coin are there?

5. Perimeter. The perimeter of a rectangle is 180 ft. The length is 26 ft greater than the width. Find the length and the width.

Translate each word problem to an equation or a system of equations and select a correct translation from equations A–O.

A. $12^2 + 12^2 = x^2$

B. $x(x + 26) = 180$

C. $10{,}311 + 5\%x = x$

D. $x + y = 85,$
   $5x + 25y = 13.85$

E. $x^2 + 4^2 = 12^2$

F. $\dfrac{240}{x - 18} = \dfrac{384}{x}$

G. $x + 5\%x = 10{,}311$

H. $\dfrac{x}{65} + 1 = \dfrac{x}{85}$

I. $\dfrac{x}{65} + \dfrac{x}{85} = 1$

J. $x + y + z = 180,$
   $y = 4x,$
   $z = x + y - 27$

K. $2x + 2(x + 26) = 180$

L. $\dfrac{384}{x - 18} = \dfrac{240}{x}$

M. $x + y = 85,$
   $0.05x + 0.25y = 13.85$

N. $2x + 2(x + 24) = 240$

O. $\dfrac{72}{x - 3} + \dfrac{24}{x + 3} = 16$

Answers on page A-33

6. Shoveling Time. It takes Marv 65 min to shovel 4 in. of snow from his driveway. It takes Elaine 85 min to do the same job. How long would it take if they worked together?

7. Money Borrowed. Claire borrows some money at 5% simple interest. After 1 year, $10,311 pays off her loan. How much did she originally borrow?

8. Plank Height. A 12-ft plank is leaning against a shed. The bottom of the plank is 4 ft from the building. How high up the side of the shed is the top of the plank?

9. Train Speeds. The speed of train A is 18 mph slower than the speed of train B. Train A travels 240 mi in the same time that it takes train B to travel 384 mi. Find the speed of train A.

10. Diagonal of a Square. Find the length of a diagonal of a square swimming pool whose sides are 12 ft long.

10.7 Exercise Set

✓ Check Your Understanding

Reading Check Complete each of the following statements.

RC1. The _____ theorem states that if a and b are the lengths of the legs of a right triangle and c is the length of the third side of a triangle, then $a^2 + b^2 = c^2$.

RC2. In any right triangle, the two shortest sides are called the _____.

RC3. In any right triangle, the longest side is opposite the right angle and is called the _____.

Concept Check For each right triangle, choose from the column on the right the equation that can be used to find the missing side.

CC1.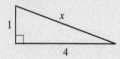

CC2.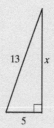

CC3.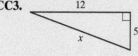

CC4.

a) $12^2 + x^2 = 5^2$
b) $12^2 + 13^2 = x^2$
c) $5^2 + x^2 = 13^2$
d) $x^2 + 4^2 = (\sqrt{17})^2$
e) $1^2 + 4^2 = x^2$
f) $5^2 + 13^2 = x^2$
g) $5^2 + 12^2 = x^2$
h) $(\sqrt{17})^2 + x^2 = 4^2$

a In a right triangle, find the length of the side not given. Give an exact answer and, where appropriate, an approximation to three decimal places.

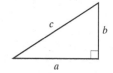

1. $a = 3, b = 5$
2. $a = 8, b = 10$
3. $a = 15, b = 15$
4. $a = 8, b = 8$

5. $b = 12, c = 13$
6. $a = 5, c = 12$
7. $c = 7, a = \sqrt{6}$
8. $c = 10, a = 4\sqrt{5}$

9. $b = 1, c = \sqrt{13}$
10. $a = 1, c = \sqrt{12}$
11. $a = 1, c = \sqrt{n}$
12. $c = 2, a = \sqrt{n}$

In the following problems, give an exact answer and, where appropriate, an approximation to three decimal places.

13. *Road-Pavement Messages.* Using the formula of Example 5, find the length L of a road-pavement message when $h = 4$ ft and $d = 200$ ft.

14. *Road-Pavement Messages.* Using the formula of Example 5, find the length L of a road-pavement message when $h = 8$ ft and $d = 300$ ft.

15. *Length of a Smartphone Screen.* The Samsung Galaxy S8 phone has less bezel at the top and the bottom and a longer screen than previous models. The new infinity screen measures 5.8 in. diagonally and is 2.7 in. wide. Find the length of the display screen.

Data: Samsung

16. *Guy Wire.* How long is a guy wire reaching from the top of a 10-ft pole to a point on the ground 4 ft from the pole?

17. *Pyramide du Louvre.* A large glass and metal pyramid designed by I. M. Pei attracts visitors to the entrance of the Louvre Museum in Paris, France. If a plumb line from the highest point of the pyramid to the center of the base is 71 ft long and each side of the four equilateral triangles measures 100 ft, how far is it from the center of the base to a corner of the base?

Data: glassonweb.com

18. *Central Park.* New York City's rectangular Central Park in Manhattan runs 13,725 ft from 59th Street to 110th Street. A diagonal of the park is 13,977 ft. Find the width of the park.

19. Find all ordered pairs on the x-axis of a Cartesian coordinate system that are 5 units from the point $(0, 4)$.

20. Find all ordered pairs on the y-axis of a Cartesian coordinate system that are 5 units from the point $(3, 0)$.

21. *Bridge Expansion.* During the summer heat, a 2-mi bridge expands 2 ft in length. If we assume that the bulge occurs straight up the middle, how high is the bulge? (The answer may surprise you. In reality, bridges are built with expansion spaces to avoid such buckling.)

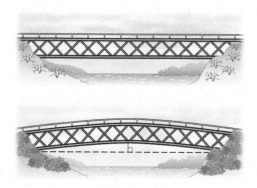

22. *Triangle Areas.* Triangle ABC has sides of lengths 25 ft, 25 ft, and 30 ft. Triangle PQR has sides of lengths 25 ft, 25 ft, and 40 ft. Which triangle has the greater area and by how much?

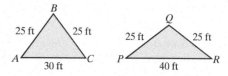

23. *A Baseball Throw.* A baseball diamond is actually in the shape of a square 90 ft on a side. Suppose that a third baseman fields a ball while standing on the third-base line 8 ft from third base, as shown. How far is the third baseman's throw to first base?

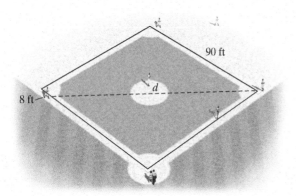

24. *Distance Over Water.* To determine the distance between two points on opposite sides of a pond, a surveyor locates two stakes at those two points and uses instrumentation to place a third stake so that the distance across the pond is the length of a hypotenuse. If the third stake is 90 m from one stake and 70 m from the other, how wide is the pond?

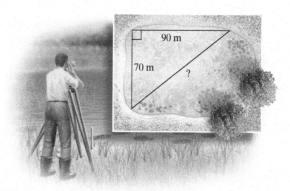

25. *Plumbing.* Plumbers use the Pythagorean theorem to calculate pipe length. If a pipe is to be offset, as shown in the figure, the *travel*, or length, of the pipe, is calculated using the lengths of the *advance* and the *offset*. Find the travel if the offset is 17.75 in. and the advance is 10.25 in.

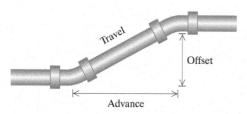

26. *Ramps for the Disabled.* Laws regarding access ramps for the disabled state that a ramp must be in the form of a right triangle, where every vertical length (leg) of 1 ft has a horizontal length (leg) of 12 ft. What is the length of a ramp with a 12-ft horizontal leg and a 1-ft vertical leg?

774 CHAPTER 10 Radical Expressions, Equations, and Functions

27. The length and the width of a rectangle are given by consecutive integers. The area of the rectangle is 90 cm². Find the length of a diagonal of the rectangle.

28. The diagonal of a square has length $8\sqrt{2}$ ft. Find the length of a side of the square.

29. Each side of a regular octagon has length s. Find a formula for the distance d between the parallel sides of the octagon.

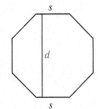

30. The two equal sides of an isosceles right triangle are of length s. Find a formula for the length of the hypotenuse.

Skill Maintenance

Solve. [6.8a]

31. *Commuter Travel.* The speed of the Zionsville Flash commuter train is 14 mph faster than that of the Carmel Crawler. The Flash travels 290 mi in the same time that it takes the Crawler to travel 230 mi. Find the speed of each train.

32. *Marine Travel.* A motor boat travels three times as fast as the current in the Saskatee River. A trip up the river and back takes 10 hr, and the total distance of the trip is 100 mi. Find the speed of the current.

Solve.

33. $2x^2 + 11x - 21 = 0$ [5.8b]

34. $x^2 + 24 = 11x$ [5.8b]

35. $\dfrac{x+2}{x+3} = \dfrac{x-4}{x-5}$ [6.7a]

36. $3x^2 - 12 = 0$ [5.8b]

37. $\dfrac{x-5}{x-7} = \dfrac{4}{3}$ [6.7a]

38. $\dfrac{x-1}{x-3} = \dfrac{6}{x-3}$ [6.7a]

Synthesis

39. *Roofing.* Kit's cottage, which is 24 ft wide and 32 ft long, needs a new roof. By counting clapboards that are 4 in. apart, Kit determines that the peak of the roof is 6 ft higher than the sides. If one packet of shingles covers $33\tfrac{1}{3}$ sq ft, how many packets will the job require?

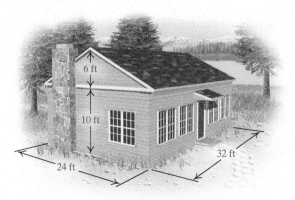

40. *Wind Chill Temperature.* Because wind enhances the loss of heat from the skin, we feel colder when there is wind than when there is not. The *wind chill temperature* is what the temperature would have to be with no wind in order to give the same chilling effect as with the wind. A formula for finding the wind chill temperature, T_W, is

$$T_W = 35.74 + 0.6215T - 35.75V^{0.16} + 0.4275TV^{0.16},$$

where T is the actual temperature given by a thermometer, in degrees Fahrenheit, and V is the wind speed, in miles per hour. This formula can be used only when the wind speed is *above* 3 mph. Use a calculator to find the wind chill temperature in each case. Round to the nearest degree.

Data: National Weather Service

a) $T = 40°F$, $V = 25$ mph
b) $T = 20°F$, $V = 25$ mph
c) $T = 10°F$, $V = 20$ mph
d) $T = 10°F$, $V = 40$ mph
e) $T = -5°F$, $V = 35$ mph
f) $T = -15°F$, $V = 35$ mph

10.8 The Complex Numbers

OBJECTIVES

a Express imaginary numbers as bi, where b is a nonzero real number, and complex numbers as $a + bi$, where a and b are real numbers.

b Add and subtract complex numbers.

c Multiply complex numbers.

d Write expressions involving powers of i in the form $a + bi$.

e Find conjugates of complex numbers and divide complex numbers.

f Determine whether a given complex number is a solution of an equation.

a IMAGINARY NUMBERS AND COMPLEX NUMBERS

Negative numbers do not have square roots in the real-number system. However, mathematicians have described a larger number system that contains the real-number system and in which negative numbers have square roots. That system is called the **complex-number system**. We begin by defining a number that is a square root of -1. We call this new number i.

THE COMPLEX NUMBER i

We define the number i to be $\sqrt{-1}$. That is,
$$i = \sqrt{-1} \quad \text{and} \quad i^2 = -1.$$

To express roots of negative numbers in terms of i, we can use the fact that in the complex-number system, $\sqrt{-p} = \sqrt{-1 \cdot p} = \sqrt{-1}\sqrt{p}$ when p is a positive real number.

EXAMPLES Express in terms of i.

1. $\sqrt{-7} = \sqrt{-1 \cdot 7} = \sqrt{-1} \cdot \sqrt{7} = i\sqrt{7}$, or $\sqrt{7}i$ ← i is *not* under the radical.
2. $\sqrt{-16} = \sqrt{-1 \cdot 16} = \sqrt{-1} \cdot \sqrt{16} = i \cdot 4 = 4i$
3. $-\sqrt{-13} = -\sqrt{-1 \cdot 13} = -\sqrt{-1} \cdot \sqrt{13} = -i\sqrt{13}$, or $-\sqrt{13}i$
4. $-\sqrt{-64} = -\sqrt{-1 \cdot 64} = -\sqrt{-1} \cdot \sqrt{64} = -i \cdot 8 = -8i$
5. $\sqrt{-48} = \sqrt{-1 \cdot 48} = \sqrt{-1} \cdot \sqrt{48} = i\sqrt{48} = i\sqrt{16 \cdot 3}$
$= i \cdot 4\sqrt{3} = 4i\sqrt{3}$, or $4\sqrt{3}i$

◄ Do Exercises 1–5.

Express in terms of i.
1. $\sqrt{-5}$ 2. $\sqrt{-25}$
3. $-\sqrt{-11}$ 4. $-\sqrt{-36}$

5. $\sqrt{-54}$
$= \sqrt{} \cdot 54$
$= \sqrt{-1} \cdot \sqrt{54} = i\sqrt{9 \cdot 6}$
$= i \cdot \sqrt{6}$, or $3\sqrt{6}$

IMAGINARY NUMBER

An **imaginary*** number is a number that can be named
$$bi,$$
where b is some real number and $b \neq 0$.

To form the system of **complex numbers**, we take the imaginary numbers and the real numbers and all possible sums of real and imaginary numbers. These are complex numbers:

$$7 - 4i, \quad -\pi + 19i, \quad 37, \quad i\sqrt{6}.$$

Answers
1. $i\sqrt{5}$, or $\sqrt{5}i$ 2. $5i$ 3. $-i\sqrt{11}$, or $-\sqrt{11}i$
4. $-6i$ 5. $3i\sqrt{6}$, or $3\sqrt{6}i$

Guided Solution:
5. $-1, 3, i$

*Don't let the name "imaginary" fool you. The imaginary numbers are very important in such fields as engineering and the physical sciences.

> **COMPLEX NUMBER**
>
> A **complex number** is any number that can be named
>
> $a + bi$,
>
> where a and b are any real numbers. (Note that either a or b or both can be 0.)

Since $0 + bi = bi$, every imaginary number is a complex number. Similarly, $a + 0i = a$, so every real number is a complex number. The relationships among various real and complex numbers are shown in the following diagram.

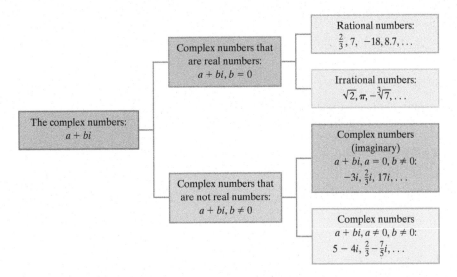

It is important to keep in mind some comparisons between numbers that have real-number roots and those that have complex-number roots that are not real. For example, $\sqrt{-48}$ is a complex number that is not a real number because we are taking the square root of a negative number. *But*, $\sqrt[3]{-125}$ is a real number because we are taking the cube root of a negative number and *any* real number has a cube root that is a real number.

b ADDITION AND SUBTRACTION

The complex numbers follow the commutative and associative laws of addition. Thus we can add and subtract them as we do binomials with real-number coefficients; that is, we collect like terms.

EXAMPLES Add or subtract.

6. $(8 + 6i) + (3 + 2i) = (8 + 3) + (6 + 2)i = 11 + 8i$
7. $(3 + 2i) - (5 - 2i) = (3 - 5) + [2 - (-2)]i = -2 + 4i$

Do Exercises 6–9.

Add or subtract.
6. $(7 + 4i) + (8 - 7i)$
7. $(-5 - 6i) + (-7 + 12i)$
8. $(8 + 3i) - (5 + 8i)$
9. $(5 - 4i) - (-7 + 3i)$

Answers
6. $15 - 3i$ **7.** $-12 + 6i$
8. $3 - 5i$ **9.** $12 - 7i$

C MULTIPLICATION

> **SKILL REVIEW** Use the FOIL method to multiply two binomials. [4.6a]
> Multiply.
> 1. $(w + 4)(w - 6)$
> 2. $(2x + 3y)(3x - 5y)$
>
> Answers: 1. $w^2 - 2w - 24$ 2. $6x^2 - xy - 15y^2$
>
>

The complex numbers obey the commutative, associative, and distributive laws. But although the property $\sqrt{a}\sqrt{b} = \sqrt{ab}$ does *not* hold for complex numbers in general, it does hold when $a = -1$ and b is a positive real number.

To multiply square roots of negative real numbers, we first express them in terms of i. For example,

$$\sqrt{-2} \cdot \sqrt{-5} = \sqrt{-1} \cdot \sqrt{2} \cdot \sqrt{-1} \cdot \sqrt{5} = i\sqrt{2} \cdot i\sqrt{5}$$
$$= i^2\sqrt{10} = -\sqrt{10} \quad \text{is correct!}$$

 But $\sqrt{-2} \cdot \sqrt{-5} = \sqrt{(-2)(-5)} = \sqrt{10}$ is wrong!

.... **Caution!** ....

The rule $\sqrt{a}\sqrt{b} = \sqrt{ab}$ holds only for nonnegative real numbers.

Keeping this and the fact that $i^2 = -1$ in mind, we multiply in much the same way that we do with real numbers.

EXAMPLES Multiply.

8. $\sqrt{-49} \cdot \sqrt{-16} = \sqrt{-1} \cdot \sqrt{49} \cdot \sqrt{-1} \cdot \sqrt{16}$
$= i \cdot 7 \cdot i \cdot 4 = i^2(28)$
$= (-1)(28) \quad i^2 = -1$
$= -28$

9. $\sqrt{-3} \cdot \sqrt{-7} = \sqrt{-1} \cdot \sqrt{3} \cdot \sqrt{-1} \cdot \sqrt{7}$
$= i \cdot \sqrt{3} \cdot i \cdot \sqrt{7} = i^2(\sqrt{21})$
$= (-1)\sqrt{21} \quad i^2 = -1$
$= -\sqrt{21}$

Multiply.

10. $\sqrt{-25} \cdot \sqrt{-4}$

11. $\sqrt{-2} \cdot \sqrt{-17}$

12. $-6i \cdot 7i$

10. $-2i \cdot 5i = -10 \cdot i^2$
$= (-10)(-1) \quad i^2 = -1$
$= 10$

13. $-3i(4 - 3i)$

14. $5i(-5 + 7i)$

11. $(-4i)(3 - 5i) = (-4i) \cdot 3 - (-4i)(5i)$ Using a distributive law
$= -12i + 20i^2 = -12i + 20(-1)$
$= -12i - 20 = -20 - 12i$

15. $(1 + 3i)(1 + 5i)$
$= 1 + \underline{\quad} + 3i + \underline{\quad} i^2$
$= 1 + \underline{\quad} i + 15(-1)$
$= 1 + 8i - 15$
$= \underline{\quad} + 8i$

12. $(1 + 2i)(1 + 3i) = 1 + 3i + 2i + 6i^2$ Using FOIL
$= 1 + 3i + 2i + 6(-1) \quad i^2 = -1$
$= 1 + 3i + 2i - 6$
$= -5 + 5i$ Collecting like terms

16. $(3 - 2i)(1 + 4i)$

17. $(3 + 2i)^2$

13. $(3 - 2i)^2 = 3^2 - 2(3)(2i) + (2i)^2$ Squaring the binomial
$= 9 - 12i + 4i^2 = 9 - 12i + 4(-1)$
$= 9 - 12i - 4 = 5 - 12i$

◀ Do Exercises 10–17.

Answers
10. -10 11. $-\sqrt{34}$ 12. 42 13. $-9 - 12i$
14. $-35 - 25i$ 15. $-14 + 8i$ 16. $11 + 10i$
17. $5 + 12i$

Guided Solution:
15. $5i, 15, 8, -14$

d POWERS OF i

We now want to simplify certain expressions involving powers of i. To do so, we first see how to simplify powers of i. Simplifying powers of i can be done by using the fact that $i^2 = -1$ and expressing the given power of i in terms of even powers, and then in terms of powers of i^2. Consider the following:

i,
$i^2 = -1$,
$i^3 = i^2 \cdot i = (-1)i = -i$,
$i^4 = (i^2)^2 = (-1)^2 = 1$,
$i^5 = i^4 \cdot i = (i^2)^2 \cdot i = (-1)^2 \cdot i = i$,
$i^6 = (i^2)^3 = (-1)^3 = -1$.

Note that the powers of i cycle through the values i, -1, $-i$, and 1.

EXAMPLES Simplify.

14. $i^{37} = i^{36} \cdot i = (i^2)^{18} \cdot i = (-1)^{18} \cdot i = 1 \cdot i = i$
15. $i^{58} = (i^2)^{29} = (-1)^{29} = -1$
16. $i^{75} = i^{74} \cdot i = (i^2)^{37} \cdot i = (-1)^{37} \cdot i = -1 \cdot i = -i$
17. $i^{80} = (i^2)^{40} = (-1)^{40} = 1$

Do Exercises 18–21. ▶

EXAMPLES Simplify to the form $a + bi$.

18. $8 - i^2 = 8 - (-1) = 8 + 1 = 9$
19. $17 + 6i^3 = 17 + 6 \cdot i^2 \cdot i = 17 + 6(-1)i = 17 - 6i$
20. $i^{22} - 67i^2 = (i^2)^{11} - 67(-1) = (-1)^{11} + 67 = -1 + 67 = 66$
21. $i^{23} + i^{48} = (i^{22}) \cdot i + (i^2)^{24} = (i^2)^{11} \cdot i + (-1)^{24} = (-1)^{11} \cdot i + (-1)^{24}$
$= -i + 1 = 1 - i$

Do Exercises 22–25. ▶

e CONJUGATES AND DIVISION

Conjugates of complex numbers are defined as follows.

> **CONJUGATE**
>
> The **conjugate** of a complex number $a + bi$ is $a - bi$, and the **conjugate** of $a - bi$ is $a + bi$.

EXAMPLES Find the conjugate.

22. $5 + 7i$ The conjugate is $5 - 7i$.
23. $-3 - 9i$ The conjugate is $-3 + 9i$.
24. $4i$ The conjugate is $-4i$.

Do Exercises 26–28. ▶

Simplify.
18. i^{47} **19.** i^{68}
20. i^{85} **21.** i^{90}

Simplify.
22. $8 - i^5$
23. $7 + 4i^2$
24. $i^{34} - i^{55}$

GS 25. $6i^{11} + 7i^{14}$
$= 6 \cdot i^{} \cdot i + 7 \cdot i^{14}$
$= 6(i^2)^{} \cdot i + 7(i^2)^7$
$= 6()^5 \cdot i + 7(-1)^7$
$= 6()i + 7()$
$= -6i - $
$= -7 - 6i$

Find the conjugate.
26. $6 + 3i$
27. $-9 - 5i$
28. $-\frac{1}{4}i$

Answers
18. $-i$ **19.** 1 **20.** i **21.** -1 **22.** $8 - i$
23. 3 **24.** $-1 + i$ **25.** $-7 - 6i$
26. $6 - 3i$ **27.** $-9 + 5i$ **28.** $\frac{1}{4}i$

Guided Solution:
25. $10, 5, -1, -1, -1, 7$

When we multiply a complex number by its conjugate, we get a real number.

EXAMPLES Multiply.

25. $(5 + 7i)(5 - 7i) = 5^2 - (7i)^2$ Using $(A + B)(A - B) = A^2 - B^2$
$= 25 - 49i^2$
$= 25 - 49(-1)$ $i^2 = -1$
$= 25 + 49$
$= 74$

26. $(2 - 3i)(2 + 3i) = 2^2 - (3i)^2$
$= 4 - 9i^2$
$= 4 - 9(-1)$ $i^2 = -1$
$= 4 + 9$
$= 13$

Multiply.
29. $(7 - 2i)(7 + 2i)$

30. $(-3 - i)(-3 + i)$

◀ Do Exercises 29 and 30.

We use conjugates when dividing complex numbers.

EXAMPLE 27 Divide and simplify to the form $a + bi$: $\dfrac{-5 + 9i}{1 - 2i}$.

$\dfrac{-5 + 9i}{1 - 2i} \cdot \dfrac{1 + 2i}{1 + 2i} = \dfrac{(-5 + 9i)(1 + 2i)}{(1 - 2i)(1 + 2i)}$ Multiplying by 1 using the conjugate of the denominator in the symbol for 1

$= \dfrac{-5 - 10i + 9i + 18i^2}{1^2 - 4i^2}$

$= \dfrac{-5 - i + 18(-1)}{1 - 4(-1)}$ $i^2 = -1$

$= \dfrac{-5 - i - 18}{1 + 4}$

$= \dfrac{-23 - i}{5}$

$= -\dfrac{23}{5} - \dfrac{1}{5}i$

Note the similarity between the preceding example and rationalizing denominators. In both cases, we used the conjugate of the denominator to write another name for 1. In Example 27, the symbol for the number 1 was chosen using the conjugate of the divisor, $1 - 2i$.

EXAMPLE 28 What symbol for 1 would you use to divide?

Division to be done   *Symbol for 1*

$\dfrac{3 + 5i}{4 + 3i}$ $\dfrac{4 - 3i}{4 - 3i}$

Answers
29. 53   **30.** 10

EXAMPLE 29 Divide and simplify to the form $a + bi$: $\dfrac{3 + 5i}{4 + 3i}$.

$$\dfrac{3 + 5i}{4 + 3i} \cdot \dfrac{4 - 3i}{4 - 3i} = \dfrac{(3 + 5i)(4 - 3i)}{(4 + 3i)(4 - 3i)} \quad \text{Multiplying by 1}$$

$$= \dfrac{12 - 9i + 20i - 15i^2}{4^2 - 9i^2}$$

$$= \dfrac{12 + 11i - 15(-1)}{16 - 9(-1)} \quad i^2 = -1$$

$$= \dfrac{12 + 11i + 15}{16 + 9}$$

$$= \dfrac{27 + 11i}{25} = \dfrac{27}{25} + \dfrac{11}{25}i$$

Do Exercises 31 and 32. ▶

Divide and simplify to the form $a + bi$.

31. $\dfrac{6 + 2i}{1 - 3i}$

32. $\dfrac{2 + 3i}{-1 + 4i}$

CALCULATOR CORNER

Complex Numbers We can perform operations on complex numbers on a graphing calculator. To do so, we first set the calculator in complex, or $a + bi$, mode by pressing **MODE**, using the ⌄ and ⌃ keys to position the blinking cursor over $a + bi$, and then pressing **ENTER**. We press **2ND** **QUIT** to go to the home screen. Now we can add, subtract, multiply, and divide complex numbers; i is the second operation associated with the · key.

To find $(3 + 4i) - (7 - i)$, we note that the parentheses around $3 + 4i$ are optional, but those around $7 - i$ are necessary. To find $\dfrac{5 - 2i}{-1 + 3i}$, we see that parentheses must be used to group the numerator and the denominator. To find $\sqrt{-4} \cdot \sqrt{-9}$, note that the calculator supplies the left parenthesis in each radicand and we supply the right parenthesis. The results of these operations are shown at right.

```
(3+4i)-(7-i)
            -4+5i
(5-2i)/(-1+3i)▶Frac
       -11/10-13/10i
√(-4)*√(-9)
              -6
```

EXERCISES: Carry out each operation.

1. $(9 + 4i) + (-11 - 13i)$
2. $(9 + 4i) - (-11 - 13i)$
3. $(9 + 4i) \cdot (-11 - 13i)$
4. $(9 + 4i) \div (-11 - 13i)$
5. $\sqrt{-16} \cdot \sqrt{-25}$
6. $\sqrt{-23} \cdot \sqrt{-35}$
7. $\dfrac{4 - 5i}{-6 + 8i}$
8. $(-3i)^4$
9. $(1 - i)^3 - (2 + 3i)^4$

---

f SOLUTIONS OF EQUATIONS

The equation $x^2 + 1 = 0$ has no real-number solution, but it has *two* nonreal complex solutions.

EXAMPLE 30 Determine whether i is a solution of the equation $x^2 + 1 = 0$.

We substitute i for x in the equation.

$$\begin{array}{c|c} x^2 + 1 = 0 \\ \hline i^2 + 1 \ ?\ 0 \\ -1 + 1 \\ 0 & \text{TRUE} \end{array}$$

The number i is a solution.

Do Exercise 33. ▶

33. Determine whether $-i$ is a solution of $x^2 + 1 = 0$.

$$\begin{array}{c|c} x^2 + 1 = 0 \\ \hline ? \end{array}$$

Answers

31. $2i$ 32. $\dfrac{10}{17} - \dfrac{11}{17}i$ 33. Yes

Any equation consisting of a polynomial in one variable on one side and 0 on the other has complex-number solutions. (Some may be real.) It is not always easy to find the solutions, but they always exist.

EXAMPLE 31 Determine whether $1 + i$ is a solution of the equation $x^2 - 2x + 2 = 0$.

We substitute $1 + i$ for x in the equation.

$$\begin{array}{r|l} x^2 - 2x + 2 = 0 \\ \hline (1+i)^2 - 2(1+i) + 2 \:?\: 0 \\ 1 + 2i + i^2 - 2 - 2i + 2 \\ 1 + 2i - 1 - 2 - 2i + 2 \\ (1 - 1 - 2 + 2) + (2 - 2)i \\ 0 + 0i \\ 0 & \text{TRUE} \end{array}$$

The number $1 + i$ is a solution.

EXAMPLE 32 Determine whether $2i$ is a solution of $x^2 + 3x - 4 = 0$.

$$\begin{array}{r|l} x^2 + 3x - 4 = 0 \\ \hline (2i)^2 + 3(2i) - 4 \:?\: 0 \\ 4i^2 + 6i - 4 \\ -4 + 6i - 4 \\ -8 + 6i & \text{FALSE} \end{array}$$

The number $2i$ is not a solution.

◀ Do Exercise 34.

34. Determine whether $1 - i$ is a solution of $x^2 - 2x + 2 = 0$.

$$\begin{array}{c} x^2 - 2x + 2 = 0 \\ \hline \\ ? \\ \end{array}$$

Answer
34. Yes

10.8 Exercise Set

FOR EXTRA HELP — MyLab Math

✓ Check Your Understanding

Reading and Concept Check Determine whether each statement is true or false.

RC1. Every real number is a complex number, but not every complex number is a real number.

RC2. The conjugate of the complex number $3 - 7i$ is $3 + 7i$.

RC3. All complex numbers are imaginary.

RC4. The square of a complex number is always a real number.

RC5. The product of a complex number and its conjugate is always a real number.

RC6. The imaginary number i raised to an even power is always 1.

RC7. The quotient of two complex numbers is always a complex number.

RC8. We add complex numbers by combining the real parts and combining the imaginary parts.

a Express in terms of i.

1. $\sqrt{-35}$
2. $\sqrt{-21}$
3. $\sqrt{-16}$
4. $\sqrt{-36}$

5. $-\sqrt{-12}$
6. $-\sqrt{-20}$
7. $\sqrt{-3}$
8. $\sqrt{-4}$

9. $\sqrt{-81}$
10. $\sqrt{-27}$
11. $\sqrt{-98}$
12. $-\sqrt{-18}$

13. $-\sqrt{-49}$
14. $-\sqrt{-125}$
15. $4 - \sqrt{-60}$
16. $6 - \sqrt{-84}$

17. $\sqrt{-4} + \sqrt{-12}$
18. $-\sqrt{-76} + \sqrt{-125}$

b Add or subtract and simplify.

19. $(7 + 2i) + (5 - 6i)$
20. $(-4 + 5i) + (7 + 3i)$
21. $(4 - 3i) + (5 - 2i)$

22. $(-2 - 5i) + (1 - 3i)$
23. $(9 - i) + (-2 + 5i)$
24. $(6 + 4i) + (2 - 3i)$

25. $(6 - i) - (10 + 3i)$
26. $(-4 + 3i) - (7 + 4i)$
27. $(4 - 2i) - (5 - 3i)$

28. $(-2 - 3i) - (1 - 5i)$
29. $(9 + 5i) - (-2 - i)$
30. $(6 - 3i) - (2 + 4i)$

c Multiply.

31. $\sqrt{-36} \cdot \sqrt{-9}$
32. $\sqrt{-16} \cdot \sqrt{-64}$
33. $\sqrt{-7} \cdot \sqrt{-2}$
34. $\sqrt{-11} \cdot \sqrt{-3}$

35. $-3i \cdot 7i$
36. $8i \cdot 5i$
37. $-3i(-8 - 2i)$
38. $4i(5 - 7i)$

39. $(3 + 2i)(1 + i)$
40. $(4 + 3i)(2 + 5i)$
41. $(2 + 3i)(6 - 2i)$
42. $(5 + 6i)(2 - i)$

43. $(6 - 5i)(3 + 4i)$
44. $(5 - 6i)(2 + 5i)$
45. $(7 - 2i)(2 - 6i)$
46. $(-4 + 5i)(3 - 4i)$

47. $(3 - 2i)^2$
48. $(5 - 2i)^2$
49. $(1 + 5i)^2$
50. $(6 + 2i)^2$

51. $(-2 + 3i)^2$
52. $(-5 - 2i)^2$

d Simplify.

53. i^7
54. i^{11}
55. i^{24}
56. i^{35}

57. i^{42}
58. i^{64}
59. i^9
60. $(-i)^{71}$

61. i^6
62. $(-i)^4$
63. $(5i)^3$
64. $(-3i)^5$

Simplify to the form $a + bi$.

65. $7 + i^4$ **66.** $-18 + i^3$ **67.** $i^{28} - 23i$ **68.** $i^{29} + 33i$

69. $i^2 + i^4$ **70.** $5i^5 + 4i^3$ **71.** $i^5 + i^7$ **72.** $i^{84} - i^{100}$

73. $1 + i + i^2 + i^3 + i^4$ **74.** $i - i^2 + i^3 - i^4 + i^5$ **75.** $5 - \sqrt{-64}$

76. $\sqrt{-12} + 36i$ **77.** $\dfrac{8 - \sqrt{-24}}{4}$ **78.** $\dfrac{9 + \sqrt{-9}}{3}$

e Divide and simplify to the form $a + bi$.

79. $\dfrac{4 + 3i}{3 - i}$ **80.** $\dfrac{5 + 2i}{2 + i}$ **81.** $\dfrac{3 - 2i}{2 + 3i}$ **82.** $\dfrac{6 - 2i}{7 + 3i}$

83. $\dfrac{8 - 3i}{7i}$ **84.** $\dfrac{3 + 8i}{5i}$ **85.** $\dfrac{4}{3 + i}$ **86.** $\dfrac{6}{2 - i}$

87. $\dfrac{2i}{5 - 4i}$ **88.** $\dfrac{8i}{6 + 3i}$ **89.** $\dfrac{4}{3i}$

90. $\dfrac{5}{6i}$ **91.** $\dfrac{2 - 4i}{8i}$ **92.** $\dfrac{5 + 3i}{i}$

93. $\dfrac{6 + 3i}{6 - 3i}$ **94.** $\dfrac{4 - 5i}{4 + 5i}$

f Determine whether the complex number is a solution of the equation.

95. $1 - 2i$;
    $x^2 - 2x + 5 = 0$

96. $1 + 2i$;
    $x^2 - 2x + 5 = 0$

97. $2 + i$;
    $x^2 - 4x - 5 = 0$

98. $1 - i$;
    $x^2 + 2x + 2 = 0$

Skill Maintenance

For each equation, find the intercepts. [3.3a]

99. $2x - y = -30$

100. $4y - 3x = 72$

101. $5x = 10 - 2y$

Find the slope of each line. [3.4b]

102. $y = -\frac{7}{3}x$

103. $x - 3y = 15$

104. $6x = -12$

Multiply or divide and write scientific notation for the result. [4.2d]

105. $\dfrac{3.6 \times 10^{-5}}{1.2 \times 10^{-8}}$

106. $(4.1 \times 10^{-2})(6.5 \times 10^6)$

Synthesis

107. A complex function g is given by
$$g(z) = \frac{z^4 - z^2}{z - 1}.$$
Find $g(2i)$, $g(1 + i)$, and $g(-1 + 2i)$.

108. Evaluate $\dfrac{1}{w - w^2}$ when $w = \dfrac{1 - i}{10}$.

Express in terms of i.

109. $\frac{1}{8}(-24 - \sqrt{-1024})$

110. $12\sqrt{-\dfrac{1}{32}}$

111. $7\sqrt{-64} - 9\sqrt{-256}$

Simplify.

112. $\dfrac{i^5 + i^6 + i^7 + i^8}{(1 - i)^4}$

113. $(1 - i)^3(1 + i)^3$

114. $\dfrac{5 - \sqrt{5}i}{\sqrt{5}i}$

115. $\dfrac{6}{1 + \frac{3}{i}}$

116. $\left(\dfrac{1}{2} - \dfrac{1}{3}i\right)^2 - \left(\dfrac{1}{2} + \dfrac{1}{3}i\right)^2$

117. $\dfrac{i - i^{38}}{1 + i}$

118. Find all numbers a for which the opposite of a is the same as the reciprocal of a.

CHAPTER 10 Summary and Review

Vocabulary Reinforcement

Complete each statement with the correct term from the column on the right. Some of the choices may not be used and some may be used more than once.

1. The number c is the _____ root of a, written $\sqrt[3]{a}$, if the third power of c is a—that is, if $c^3 = a$, then $\sqrt[3]{a} = c$. [10.1c]

2. A(n) _____ number is any number that can be named $a + bi$, where a and b are any real numbers. [10.8a]

3. For any real number a, $\sqrt{a^2} = |a|$. The _____ (nonnegative) square root of a^2 is the absolute value of a. [10.1b]

4. Finding an equivalent expression without a radical in the denominator is called _____ the denominator. [10.5a]

5. The symbol $\sqrt{}$ is called a(n) _____. [10.1a]

6. A(n) _____ number is a number that can be named bi, where b is some real number and $b \neq 0$. [10.8a]

7. The number c is a(n) _____ root of a if $c^2 = a$. [10.1a]

8. The expression written under the radical is called the _____. [10.1a]

9. The _____ of a complex number $a + bi$ is $a - bi$. [10.8e]

10. In the expression $\sqrt[k]{a}$, we call the k the _____. [10.1d]

i
radicand
index
square
complex
cube
conjugate
radical
principal
imaginary
rationalizing
odd

Concept Reinforcement

Determine whether each statement is true or false.

_____ 1. For any negative number a, we have $\sqrt{a^2} = -a$. [10.1a]

_____ 2. For any real numbers $\sqrt[m]{a}$ and $\sqrt[n]{b}$, $\sqrt[m]{a} \cdot \sqrt[n]{b} = \sqrt[mn]{ab}$. [10.3a]

_____ 3. For any real numbers $\sqrt[n]{a}$ and $\sqrt[n]{b}$, $\sqrt[n]{a} + \sqrt[n]{b} = \sqrt[n]{a+b}$. [10.4a]

_____ 4. If $x^2 = 4$, then $x = 2$. [10.6a]

_____ 5. All real numbers are complex numbers, but not every complex number is a real number. [10.8a]

_____ 6. The product of a complex number and its conjugate is always a real number. [10.8e]

Study Guide

Objective 10.1b Simplify radical expressions with perfect-square radicands.

Example Simplify: $\sqrt{16x^2}$.
$\sqrt{16x^2} = \sqrt{(4x)^2} = |4x| = |4| \cdot |x| = 4|x|$

Example Simplify: $\sqrt{x^2 - 6x + 9}$.
$\sqrt{x^2 - 6x + 9} = \sqrt{(x-3)^2} = |x - 3|$

Practice Exercises

1. Simplify: $\sqrt{36y^2}$.

2. Simplify: $\sqrt{a^2 + 4a + 4}$.

Objective 10.2a Write expressions with or without rational exponents, and simplify, if possible.

Example Rewrite $x^{1/4}$ without a rational exponent.
Recall that $a^{1/n}$ means $\sqrt[n]{a}$. Then
$$x^{1/4} = \sqrt[4]{x}.$$

Practice Exercises

3. Rewrite $z^{3/5}$ without a rational exponent.

Example Rewrite $(\sqrt[3]{4xy^2})^4$ with a rational exponent.
Recall that $(\sqrt[n]{a})^m$ means $a^{m/n}$. Then
$$(\sqrt[3]{4xy^2})^4 = (4xy^2)^{4/3}.$$

4. Rewrite $(\sqrt{6ab})^5$ with a rational exponent.

Objective 10.2b Write expressions without negative exponents, and simplify, if possible.

Example Rewrite $8^{-2/3}$ with a positive exponent, and simplify, if possible.

Recall that $a^{-m/n}$ means $\dfrac{1}{a^{m/n}}$. Then

$$8^{-2/3} = \frac{1}{8^{2/3}} = \frac{1}{(\sqrt[3]{8})^2} = \frac{1}{2^2} = \frac{1}{4}.$$

Practice Exercise

5. Rewrite $9^{-3/2}$ with a positive exponent, and simplify, if possible.

Objective 10.2d Use rational exponents to simplify radical expressions.

Example Use rational exponents to simplify: $\sqrt[6]{x^2y^4}$.
$$\begin{aligned}\sqrt[6]{x^2y^4} &= (x^2y^4)^{1/6}\\ &= x^{2/6}y^{4/6}\\ &= x^{1/3}y^{2/3}\\ &= (xy^2)^{1/3}\\ &= \sqrt[3]{xy^2}\end{aligned}$$

Practice Exercise

6. Use rational exponents to simplify: $\sqrt[8]{a^6b^2}$.

Objective 10.3a Multiply and simplify radical expressions.

Example Multiply and simplify: $\sqrt[3]{6xy^2}\sqrt[3]{9y}$.
$$\begin{aligned}\sqrt[3]{6xy^2}\sqrt[3]{9y} &= \sqrt[3]{6xy^2 \cdot 9y}\\ &= \sqrt[3]{54xy^3}\\ &= \sqrt[3]{27y^3 \cdot 2x}\\ &= \sqrt[3]{27y^3}\sqrt[3]{2x}\\ &= 3y\sqrt[3]{2x}\end{aligned}$$

Practice Exercise

7. Multiply and simplify. Assume that all expressions under radicals represent nonnegative numbers.
$$\sqrt{5y}\sqrt{30y}$$

Objective 10.3b Divide and simplify radical expressions.

Example Divide and simplify: $\dfrac{\sqrt{24x^5}}{\sqrt{6x}}$.
$$\frac{\sqrt{24x^5}}{\sqrt{6x}} = \sqrt{\frac{24x^5}{6x}} = \sqrt{4x^4} = 2x^2$$

Practice Exercise

8. Divide and simplify: $\dfrac{\sqrt{20a}}{\sqrt{5}}$.

Objective 10.4a Add or subtract with radical notation and simplify.

Example Subtract: $5\sqrt{2} - 4\sqrt{8}$.
$$5\sqrt{2} - 4\sqrt{8} = 5\sqrt{2} - 4\sqrt{4 \cdot 2}$$
$$= 5\sqrt{2} - 4\sqrt{4}\sqrt{2}$$
$$= 5\sqrt{2} - 4 \cdot 2\sqrt{2} = 5\sqrt{2} - 8\sqrt{2}$$
$$= (5 - 8)\sqrt{2} = -3\sqrt{2}$$

Practice Exercise
9. Subtract: $\sqrt{48} - 2\sqrt{3}$.

Objective 10.4b Multiply expressions involving radicals in which some factors contain more than one term.

Example Multiply: $(3 - \sqrt{6})(2 + 4\sqrt{6})$.
We use FOIL:
$$(3 - \sqrt{6})(2 + 4\sqrt{6})$$
$$= 3 \cdot 2 + 3 \cdot 4\sqrt{6} - \sqrt{6} \cdot 2 - \sqrt{6} \cdot 4\sqrt{6}$$
$$= 6 + 12\sqrt{6} - 2\sqrt{6} - 4 \cdot 6$$
$$= 6 + 12\sqrt{6} - 2\sqrt{6} - 24$$
$$= -18 + 10\sqrt{6}.$$

Practice Exercise
10. Multiply: $(5 - \sqrt{x})^2$.

Objective 10.6a Solve radical equations with one radical term.

Example Solve: $x = \sqrt{x - 2} + 4$.
First, we subtract 4 on both sides to isolate the radical. Then we square both sides of the equation.
$$x = \sqrt{x - 2} + 4$$
$$x - 4 = \sqrt{x - 2}$$
$$(x - 4)^2 = (\sqrt{x - 2})^2$$
$$x^2 - 8x + 16 = x - 2$$
$$x^2 - 9x + 18 = 0$$
$$(x - 3)(x - 6) = 0$$
$$x - 3 = 0 \quad \text{or} \quad x - 6 = 0$$
$$x = 3 \quad \text{or} \quad x = 6$$

We must check both possible solutions. When we do, we find that 6 checks, but 3 does not. Thus the solution is 6.

Practice Exercise
11. Solve: $3 + \sqrt{x - 1} = x$.

Objective 10.6b Solve radical equations with two radical terms.

Example Solve: $1 = \sqrt{x + 9} - \sqrt{x}$.
$$1 = \sqrt{x + 9} - \sqrt{x}$$
$$\sqrt{x} + 1 = \sqrt{x + 9} \quad \text{Isolating one radical}$$
$$(\sqrt{x} + 1)^2 = (\sqrt{x + 9})^2 \quad \text{Squaring both sides}$$
$$x + 2\sqrt{x} + 1 = x + 9$$
$$2\sqrt{x} = 8 \quad \text{Isolating the remaining radical}$$
$$\sqrt{x} = 4$$
$$(\sqrt{x})^2 = 4^2$$
$$x = 16$$

The number 16 checks. It is the solution.

Practice Exercise
12. Solve: $\sqrt{x + 3} - \sqrt{x - 2} = 1$.

Objective 10.8c Multiply complex numbers.

Example Multiply: $(3 - 2i)(4 + i)$.
$(3 - 2i)(4 + i) = 12 + 3i - 8i - 2i^2$ Using FOIL
$= 12 + 3i - 8i - 2(-1)$
$= 12 + 3i - 8i + 2$
$= 14 - 5i$

Practice Exercise

13. Multiply: $(2 - 5i)^2$.

Objective 10.8e Find conjugates of complex numbers and divide complex numbers.

Example Divide and simplify to the form $a + bi$:
$$\frac{5 - i}{4 + 3i}.$$
The conjugate of the denominator is $4 - 3i$, so we multiply by 1 using $\frac{4 - 3i}{4 - 3i}$:

$$\frac{5 - i}{4 + 3i} = \frac{5 - i}{4 + 3i} \cdot \frac{4 - 3i}{4 - 3i}$$
$$= \frac{20 - 15i - 4i + 3i^2}{16 - 9i^2}$$
$$= \frac{20 - 19i + 3(-1)}{16 - 9(-1)}$$
$$= \frac{20 - 19i - 3}{16 + 9}$$
$$= \frac{17 - 19i}{25} = \frac{17}{25} - \frac{19}{25}i.$$

Practice Exercise

14. Divide and simplify to the form $a + bi$: $\dfrac{3 - 2i}{2 + i}$.

Review Exercises

Use a calculator to approximate to three decimal places. [10.1a]

1. $\sqrt{778}$

2. $\sqrt{\dfrac{963.2}{23.68}}$

3. For the given function, find the indicated function values. [10.1a]
$f(x) = \sqrt{3x - 16}$; $f(0), f(-1), f(1)$, and $f\left(\frac{41}{3}\right)$

4. Find the domain of the function f in Exercise 3. [10.1a]

Simplify. Assume that letters represent any real number. [10.1b]

5. $\sqrt{81a^2}$

6. $\sqrt{(-7z)^2}$

7. $\sqrt{(6 - b)^2}$

8. $\sqrt{x^2 + 6x + 9}$

Simplify. [10.1c]

9. $\sqrt[3]{-1000}$

10. $\sqrt[3]{-\dfrac{1}{27}}$

11. For the given function, find the indicated function values. [10.1c]
$f(x) = \sqrt[3]{x + 2}$; $f(6), f(-10)$, and $f(25)$

Simplify. Assume that letters represent any real number. [10.1d]

12. $\sqrt[10]{x^{10}}$

13. $-\sqrt[13]{(-3)^{13}}$

Rewrite without rational exponents, and simplify, if possible. [10.2a]

14. $a^{1/5}$

15. $64^{3/2}$

790   CHAPTER 10   Radical Expressions, Equations, and Functions

Rewrite with rational exponents. [10.2a]

16. $\sqrt{31}$ **17.** $\sqrt[5]{a^2b^3}$

Rewrite with positive exponents, and simplify, if possible. [10.2b]

18. $49^{-1/2}$ **19.** $(8xy)^{-2/3}$

20. $5a^{-3/4}b^{1/2}c^{-2/3}$ **21.** $\dfrac{3a}{\sqrt[4]{t}}$

Use the laws of exponents to simplify. Write answers with positive exponents. [10.2c]

22. $(x^{-2/3})^{3/5}$ **23.** $\dfrac{7^{-1/3}}{7^{-1/2}}$

Use rational exponents to simplify. Write the answer in radical notation if appropriate. [10.2d]

24. $\sqrt[3]{x^{21}}$ **25.** $\sqrt[3]{27x^6}$

Use rational exponents to write a single radical expression. [10.2d]

26. $x^{1/3}y^{1/4}$ **27.** $\sqrt[4]{x}\sqrt[3]{x}$

Simplify by factoring. Assume that all expressions under radicals represent nonnegative numbers. [10.3a]

28. $\sqrt{245}$ **29.** $\sqrt[3]{-108}$

30. $\sqrt[3]{250a^2b^6}$

Simplify. Assume that no radicands were formed by raising negative numbers to even powers. [10.3b]

31. $\sqrt{\dfrac{49}{36}}$ **32.** $\sqrt[3]{\dfrac{64x^6}{27}}$

33. $\sqrt[4]{\dfrac{16x^8}{81y^{12}}}$

Perform the indicated operations and simplify. Assume that no radicands were formed by raising negative numbers to even powers. [10.3a, b], [10.4a]

34. $\sqrt{5x}\sqrt{3y}$ **35.** $\sqrt[3]{a^5b}\sqrt[3]{27b}$

36. $\sqrt[3]{a}\sqrt[5]{b^3}$ **37.** $\dfrac{\sqrt[3]{60xy^3}}{\sqrt[3]{10x}}$

38. $\dfrac{\sqrt{75x}}{2\sqrt{3}}$ **39.** $\dfrac{\sqrt[3]{x^2}}{\sqrt[4]{x}}$

40. $5\sqrt[3]{x} + 2\sqrt[3]{x}$ **41.** $2\sqrt{75} - 7\sqrt{3}$

42. $\sqrt{50} + 2\sqrt{18} + \sqrt{32}$ **43.** $\sqrt[3]{8x^4} + \sqrt[3]{xy^6}$

Multiply. [10.4b]
44. $(\sqrt{5} - 3\sqrt{8})(\sqrt{5} + 2\sqrt{8})$

45. $(1 - \sqrt{7})^2$

46. $(\sqrt[3]{27} - \sqrt[3]{2})(\sqrt[3]{27} + \sqrt[3]{2})$

Rationalize the denominator. [10.5a, b]

47. $\sqrt{\dfrac{8}{3}}$ **48.** $\dfrac{2}{\sqrt{a} + \sqrt{b}}$

Solve. [10.6a, b]
49. $x - 3 = \sqrt{5-x}$ **50.** $\sqrt[4]{x+3} = 2$

51. $\sqrt{x+8} - \sqrt{3x+1} = 1$

Automotive Repair. For an engine with a displacement of 2.8 L, the function given by
$$d(n) = 0.75\sqrt{2.8n}$$
can be used to determine the diameter of the carburetor's opening, $d(n)$, in millimeters, where n is the number of rpm at which the engine achieves peak performance. [10.6c]
Data: macdizzy.com

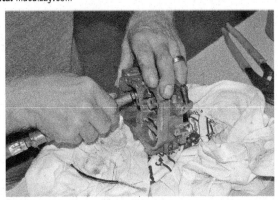

52. ▦ If a carburetor's opening is 81 mm, for what number of rpm will the engine produce peak power?

53. ▦ If a carburetor's opening is 84 mm, for what number of rpm will the engine produce peak power?

54. *Length of a Side of a Square.* The diagonal of a square has length $9\sqrt{2}$ cm. Find the length of a side of the square. [10.7a]

55. *Bookcase Width.* A bookcase is 5 ft tall and has a 7-ft diagonal brace. How wide is the bookcase? [10.7a]

In a right triangle, find the length of the side not given. Give an exact answer and an answer to three decimal places. [10.7a]

56. $a = 7$, $b = 24$ 57. $a = 2$, $c = 5\sqrt{2}$

58. Express in terms of i: $\sqrt{-25} + \sqrt{-8}$. [10.8a]

Add or subtract. [10.8b]
59. $(-4 + 3i) + (2 - 12i)$

60. $(4 - 7i) - (3 - 8i)$

Multiply. [10.8c, d]
61. $(2 + 5i)(2 - 5i)$ 62. i^{13}

63. $(6 - 3i)(2 - i)$

Divide. [10.8e]
64. $\dfrac{-3 + 2i}{5i}$ 65. $\dfrac{1 - 2i}{3 + i}$

66. Graph: $f(x) = \sqrt{x}$. [10.1a]

67. Which of the following is a solution of $x^2 + 4x + 5 = 0$? [10.8f]
    A. $1 - i$ B. $1 + i$
    C. $2 + i$ D. $-2 + i$

Synthesis

68. Simplify: $i \cdot i^2 \cdot i^3 \cdots i^{99} \cdot i^{100}$. [10.8c, d]

69. Solve: $\sqrt{11x + \sqrt{6 + x}} = 6$. [10.6a]

Understanding Through Discussion and Writing

1. Find the domain of
$$f(x) = (x + 5)^{1/2}(x + 7)^{-1/2}$$
and explain how you found your answer. [10.1a], [10.2b]

2. ⌇⌇ Ron is puzzled. When he uses a graphing calculator to graph $y = \sqrt{x} \cdot \sqrt{x}$, he gets the following screen. Explain why Ron did not get the complete line $y = x$. [10.1a], [10.3a]

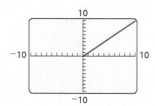

3. In what way(s) is collecting like radical terms the same as collecting like monomial terms? [10.4a]

4. Is checking solutions of equations necessary when the principle of powers is used with an odd power n? Why or why not? [10.1d], [10.6a, b]

5. A student *incorrectly* claims that
$$\frac{5 + \sqrt{2}}{\sqrt{18}} = \frac{5 + \sqrt{1}}{\sqrt{9}} = \frac{5 + 1}{3} = 2.$$
How could you convince the student that a mistake has been made? How would you explain the correct way of rationalizing the denominator? [10.5a]

6. How are conjugates of complex numbers similar to the conjugates used in Section 10.5? [10.8e]

CHAPTER 10 Test

1. Use a calculator to approximate $\sqrt{148}$ to three decimal places.

2. For the given function, find the indicated function values.
$$f(x) = \sqrt{8 - 4x}; \quad f(1) \text{ and } f(3)$$

3. Find the domain of the function f in Exercise 2.

Simplify. Assume that letters represent *any* real number.

4. $\sqrt{(-3q)^2}$

5. $\sqrt{x^2 + 10x + 25}$

6. $\sqrt[3]{-\dfrac{1}{1000}}$

7. $\sqrt[5]{x^5}$

8. $\sqrt[10]{(-4)^{10}}$

Rewrite without rational exponents, and simplify, if possible.

9. $a^{2/3}$

10. $32^{3/5}$

Rewrite with rational exponents.

11. $\sqrt{37}$

12. $(\sqrt{5xy^2})^5$

Rewrite with positive exponents, and simplify, if possible.

13. $1000^{-1/3}$

14. $8a^{3/4}b^{-3/2}c^{-2/5}$

Use the laws of exponents to simplify. Write answers with positive exponents.

15. $(x^{2/3}y^{-3/4})^{12/5}$

16. $\dfrac{2.9^{-5/8}}{2.9^{2/3}}$

Use rational exponents to simplify. Write the answer in radical notation if appropriate. Assume that no radicands were formed by raising negative numbers to even powers.

17. $\sqrt[8]{x^2}$

18. $\sqrt[4]{16x^6}$

Use rational exponents to write a single radical expression.

19. $a^{2/5}b^{1/3}$

20. $\sqrt[4]{2y}\,\sqrt[3]{y}$

Simplify by factoring. Assume that no radicands were formed by raising negative numbers to even powers.

21. $\sqrt{148}$

22. $\sqrt[4]{80}$

23. $\sqrt[3]{24a^{11}b^{13}}$

Simplify. Assume that no radicands were formed by raising negative numbers to even powers.

24. $\sqrt[3]{\dfrac{16x^5}{y^6}}$

25. $\sqrt{\dfrac{25x^2}{36y^4}}$

Perform the indicated operations and simplify. Assume that no radicands were formed by raising negative numbers to even powers.

26. $\sqrt[3]{2x}\sqrt[3]{5y^2}$

27. $\sqrt[4]{x^3y^2}\sqrt[4]{xy}$

28. $\dfrac{\sqrt[5]{x^3y^4}}{\sqrt[5]{xy^2}}$

29. $\dfrac{\sqrt{300a}}{5\sqrt{3}}$

30. Add: $3\sqrt{128} + 2\sqrt{18} + 2\sqrt{32}$.

Multiply.

31. $(\sqrt{20} + 2\sqrt{5})(\sqrt{20} - 3\sqrt{5})$

32. $(3 + \sqrt{x})^2$

33. Rationalize the denominator: $\dfrac{1 + \sqrt{2}}{3 - 5\sqrt{2}}$.

Solve.

34. $\sqrt[5]{x - 3} = 2$

35. $\sqrt{x - 6} = \sqrt{x + 9} - 3$

36. $\sqrt{x - 1} + 3 = x$

37. *Length of a Side of a Square.* The diagonal of a square has length $7\sqrt{2}$ ft. Find the length of a side of the square.

38. *Sighting to the Horizon.* A person can see 72 mi to the horizon from an airplane window. How high is the airplane? Use the formula $D = 1.2\sqrt{h}$, where D is in miles and h is in feet.

In a right triangle, find the length of the side not given. Give an exact answer and an approximation to three decimal places.

39. $a = 7,\ b = 7$

40. $a = 1,\ c = \sqrt{5}$

41. Express in terms of i: $\sqrt{-9} + \sqrt{-64}$.

42. Subtract: $(5 + 8i) - (-2 + 3i)$.

Multiply.

43. $(3 - 4i)(3 + 7i)$

44. i^{95}

45. Divide: $\dfrac{-7 + 14i}{6 - 8i}$.

46. Determine whether $1 + 2i$ is a solution of $x^2 + 2x + 5 = 0$.

47. Which of the following describes the solution(s) of the equation $x - 4 = \sqrt{x - 2}$?
  A. There is exactly one solution, and it is positive.
  B. There are one positive solution and one negative solution.
  C. There are two positive solutions.
  D. There is no solution.

Synthesis

48. Simplify: $\dfrac{1 - 4i}{4i(1 + 4i)^{-1}}$.

49. Solve: $\sqrt{2x - 2} + \sqrt{7x + 4} = \sqrt{13x + 10}$.

CHAPTERS 1–10 Cumulative Review

Simplify. Assume that no radicands were formed by raising negative numbers to even powers.

1. $(2x^2 - 3x + 1) + (6x - 3x^3 + 7x^2 - 4)$

2. $(2x^2 - y)^2$

3. $(5x^2 - 2x + 1)(3x^2 + x - 2)$

4. $\dfrac{x^3 + 64}{x^2 - 49} \cdot \dfrac{x^2 - 14x + 49}{x^2 - 4x + 16}$

5. $\dfrac{\dfrac{y^2 - 5y - 6}{y^2 - 7y - 18}}{\dfrac{y^2 + 3y + 2}{y^2 + 4y + 4}}$

6. $\dfrac{x}{x + 2} + \dfrac{1}{x - 3} - \dfrac{x^2 - 2}{x^2 - x - 6}$

7. $(y^3 + 3y^2 - 5) \div (y + 2)$

8. $\sqrt[3]{-8x^3}$

9. $\sqrt{16x^2 - 32x + 16}$

10. $9\sqrt{75} + 6\sqrt{12}$

11. $\sqrt{2xy^2} \cdot \sqrt{8xy^3}$

12. $\dfrac{3\sqrt{5}}{\sqrt{6} - \sqrt{3}}$

13. $\sqrt[6]{\dfrac{m^{12}n^{24}}{64}}$

14. $6^{2/9} \cdot 6^{2/3}$

15. $(6 + i) - (3 - 4i)$

16. $\dfrac{2 - i}{6 + 5i}$

Solve.

17. $\dfrac{1}{5} + \dfrac{3}{10}x = \dfrac{4}{5}$

18. $M = \dfrac{1}{8}(c - 3)$, for c

19. $3a - 4 < 10 + 5a$

20. $-8 < x + 2 < 15$

21. $|3x - 6| = 2$

22. $625 = 49y^2$

23. $3x + 5y = 30,$
    $5x + 3y = 34$

24. $3x + 2y - z = -7,$
    $-x + y + 2z = 9,$
    $5x + 5y + z = -1$

25. $\dfrac{6x}{x - 5} - \dfrac{300}{x^2 + 5x + 25} = \dfrac{2250}{x^3 - 125}$

26. $\dfrac{3x^2}{x + 2} + \dfrac{5x - 22}{x - 2} = \dfrac{-48}{x^2 - 4}$

27. $I = \dfrac{nE}{R + nr}$, for R

28. $\sqrt{4x + 1} - 2 = 3$

29. $2\sqrt{1 - x} = \sqrt{5}$

30. $13 - x = 5 + \sqrt{x + 4}$

Graph.

31. $f(x) = -\dfrac{2}{3}x + 2$

32. $4x - 2y = 8$

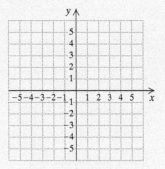 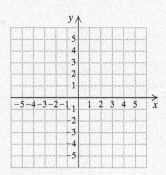

33. $4x \geq 5y + 20$

34. $y \geq -3,$
    $y \leq 2x + 3$

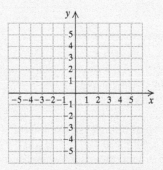 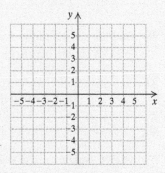

35. $g(x) = x^2 - x - 2$ **36.** $f(x) = |x + 4|$

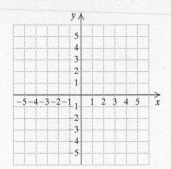
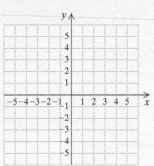

37. $g(x) = \dfrac{4}{x - 3}$ **38.** $f(x) = 2 - \sqrt{x}$

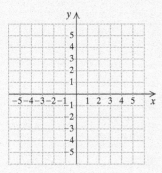
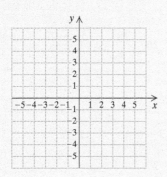

Factor.

39. $12x^2y^2 - 30xy^3$ **40.** $3x^2 - 17x - 28$

41. $y^2 - y - 132$ **42.** $27y^3 + 8$

43. $4x^2 - 625$

Find the domain and the range of each function.

44.   **45.**

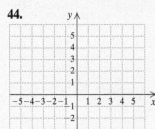
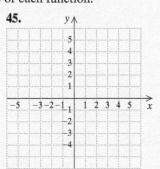

46. Find the slope and the y-intercept of the line $3x - 2y = 8$.

47. Find an equation for the line perpendicular to the line $3x - y = 5$ and passing through $(1, 4)$.

48. *Triangle Area.* The height h of triangles of fixed area varies inversely as the base b. Suppose that the height is 100 ft when the base is 20 ft. Find the height when the base is 16 ft. What is the fixed area?

Solve.

49. *Harvesting Time.* One combine can harvest a field in 3 hr. Another combine can harvest the same field in 1.5 hr. How long should it take them to harvest the field together?

50. *Warning Dye.* A warning dye is used by people in lifeboats to aid search planes. The volume V of the dye used varies directly as the square of the diameter d of the circular area formed by the dye in the water. If 4 L of dye is required for a 10-m wide circle, how much dye is needed for a 40-m wide circle?

51. Rewrite with rational exponents: $\sqrt[5]{xy^4}$.

  A. $\dfrac{1}{(xy^4)^5}$ **B.** $(xy^4)^5$
  C. $(xy)^{4/5}$ **D.** $(xy^4)^{1/5}$

52. A grain bin can be filled in 3 hr if the grain enters through spout A alone or in 15 hr if the grain enters through spout B alone. If grain is entering through both spouts at the same time, how many hours will it take to fill the bin?

  A. $\tfrac{5}{2}$ hr   **B.** 9 hr
  C. $22\tfrac{1}{2}$ hr   **D.** $10\tfrac{1}{2}$ hr

53. Divide: $(x^3 - x^2 + 2x + 4) \div (x - 3)$.

  A. $x^2 + 2x + 8$, R 28   **B.** $x^2 + 2x - 4$, R -8
  C. $x^2 - 4x - 10$, R -26 **D.** $x^2 - 4x + 14$, R 46

54. Solve: $2x + 6 = 8 + \sqrt{5x + 1}$.

  A. $\tfrac{1}{4}$ **B.** 3
  C. $3, \tfrac{1}{4}$ **D.** 4, 3

Synthesis

55. Solve: $\dfrac{x + \sqrt{x + 1}}{x - \sqrt{x + 1}} = \dfrac{5}{11}$.

CHAPTER 11

Quadratic Equations and Functions

11.1 The Basics of Solving Quadratic Equations
11.2 The Quadratic Formula
11.3 Applications Involving Quadratic Equations

Translating for Success

11.4 More on Quadratic Equations

Mid-Chapter Review

11.5 Graphing $f(x) = a(x - h)^2 + k$
11.6 Graphing $f(x) = ax^2 + bx + c$

Visualizing for Success

11.7 Mathematical Modeling with Quadratic Functions
11.8 Polynomial Inequalities and Rational Inequalities

Summary and Review
Test
Cumulative Review

On average, Americans spend about one-fourth of their income on housing. Since 1995, average annual rent has risen from $6000 to over $10,000, and the median price for a new home has risen from $135,000 to over $300,000. Income has not kept pace with rising home prices; while both median income and average annual rent have risen about 65% since 1995, new-home prices have risen over 120%. Rising home prices relative to income may be, in part, responsible for the decrease in homeownership rate from 2005 to 2016, as illustrated in the accompanying graph.

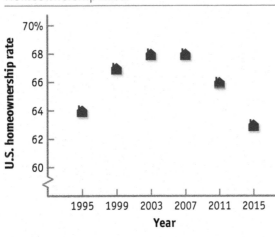

DATA: U.S Census Bureau

Data: pewtrusts.org; areavibes.com; Federal Reserve Bank of St. Louis; U.S. Census Bureau

In Exercise 35 of Exercise Set 11.7, we will model the U.S. homeownership rate using a *quadratic function*.

STUDYING FOR SUCCESS   *Beginning to Study for the Final Exam*

- ☐ Take a few minutes each week to review highlighted information.
- ☐ Prepare a few pages of notes for the course and then try to condense the notes to just one page.
- ☐ Use the Mid-Chapter Reviews, Summary and Reviews, Chapter Tests, and Cumulative Reviews.

11.1 The Basics of Solving Quadratic Equations

OBJECTIVES

a Solve quadratic equations using the principle of square roots, and find the *x*-intercepts of the graph of a related function.

b Solve quadratic equations by completing the square.

c Solve applied problems using quadratic equations.

ALGEBRAIC ▶◀ GRAPHICAL CONNECTION

The graph of the function $f(x) = x^2 + 6x + 8$ and its *x*-intercepts are shown below.

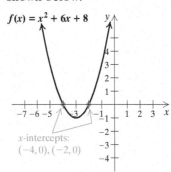

x-intercepts: $(-4, 0), (-2, 0)$

The *x*-intercepts are $(-4, 0)$ and $(-2, 0)$. These pairs are also the points of intersection of the graphs of $f(x) = x^2 + 6x + 8$ and $g(x) = 0$ (the *x*-axis). We will analyze the graphs of quadratic functions in greater detail in Sections 11.5–11.7.

We can solve $x^2 + 6x + 8 = 0$ using factoring:

$$x^2 + 6x + 8 = 0$$
$$(x + 4)(x + 2) = 0 \qquad \text{Factoring}$$
$$x + 4 = 0 \quad or \quad x + 2 = 0 \qquad \text{Using the principle of zero products}$$
$$x = -4 \quad or \quad x = -2.$$

We see that the solutions of $x^2 + 6x + 8 = 0$, -4 and -2, are the first coordinates of the *x*-intercepts, $(-4, 0)$ and $(-2, 0)$, of the graph of $f(x) = x^2 + 6x + 8$.

a   THE PRINCIPLE OF SQUARE ROOTS

SKILL REVIEW   *Solve quadratic equations using the principle of zero products.* [5.8b]
Solve.
1. $x^2 + 6x - 16 = 0$
2. $6x^2 - 13x - 5 = 0$

Answers: 1. $-8, 2$ 2. $-\dfrac{1}{3}, \dfrac{5}{2}$

The quadratic equation $5x^2 + 8x - 2 = 0$ is said to be written in **standard form**.

> **QUADRATIC EQUATION**
>
> An equation of the type $ax^2 + bx + c = 0$, where a, b, and c are real-number constants and $a > 0$, is called the **standard form of a quadratic equation**.

To write standard form for the quadratic equation $-5x^2 + 4x - 7 = 0$, we find an equivalent equation by multiplying by -1 on both sides:

$$-1(-5x^2 + 4x - 7) = -1(0)$$
$$5x^2 - 4x + 7 = 0. \quad \text{Writing in standard form}$$

To solve a quadratic equation using the principle of zero products, we first write the equation in standard form and then factor.

EXAMPLE 1
a) Solve: $x^2 = 25$.
b) Find the x-intercepts of $f(x) = x^2 - 25$.

a) We first find standard form and then factor:

$$x^2 - 25 = 0 \quad \text{Subtracting 25}$$
$$(x - 5)(x + 5) = 0 \quad \text{Factoring}$$
$$x - 5 = 0 \text{ or } x + 5 = 0 \quad \text{Using the principle of zero products}$$
$$x = 5 \text{ or } x = -5.$$

The solutions are 5 and -5.

b) The x-intercepts of $f(x) = x^2 - 25$ are $(-5, 0)$ and $(5, 0)$. The solutions of the equation $x^2 = 25$ are the first coordinates of the x-intercepts of the graph of $f(x) = x^2 - 25$.

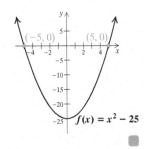

EXAMPLE 2 Solve: $6x^2 - 15x = 0$.

We factor and use the principle of zero products:

$$6x^2 - 15x = 0$$
$$3x(2x - 5) = 0$$
$$3x = 0 \text{ or } 2x - 5 = 0$$
$$x = 0 \text{ or } 2x = 5$$
$$x = 0 \text{ or } x = \tfrac{5}{2}.$$

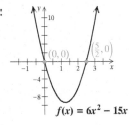

The solutions are 0 and $\tfrac{5}{2}$. The check is left to the student.

Do Exercises 1–3.

---

1. Below is the graph of $f(x) = x^2 - 6x + 8$.

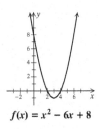

a) What are the x-intercepts of the graph?
b) What are the solutions of $x^2 - 6x + 8 = 0$?
c) What relationship exists between the answers to parts (a) and (b)?

2. a) Solve: $x^2 = 16$.
b) Find the x-intercepts of $f(x) = x^2 - 16$.

3. a) Solve: $4x^2 + 14x = 0$.
b) Find the x-intercepts of $f(x) = 4x^2 + 14x$.

Answers
1. (a) $(2, 0), (4, 0)$; **(b)** 2, 4; **(c)** The solutions of $x^2 - 6x + 8 = 0$, 2 and 4, are the first coordinates of the x-intercepts, $(2, 0)$ and $(4, 0)$, of the graph of $f(x) = x^2 - 6x + 8$.
2. (a) $-4, 4$; **(b)** $(-4, 0), (4, 0)$
3. (a) $0, -\dfrac{7}{2}$; **(b)** $\left(-\dfrac{7}{2}, 0\right), (0, 0)$

EXAMPLE 3

a) Solve: $3x^2 = 2 - x$.

b) Find the x-intercepts of $f(x) = 3x^2 + x - 2$.

a) We first find standard form. Then we factor and use the principle of zero products.

$$3x^2 = 2 - x$$
$$3x^2 + x - 2 = 0 \qquad \text{Adding } x \text{ and subtracting 2}$$
$$(x + 1)(3x - 2) = 0 \qquad \text{Factoring}$$
$$x + 1 = 0 \quad \text{or} \quad 3x - 2 = 0 \qquad \text{Using the principle of zero products}$$
$$x = -1 \quad \text{or} \quad 3x = 2$$
$$x = -1 \quad \text{or} \quad x = \tfrac{2}{3}$$

Check: For -1:

$$\begin{array}{c|c} 3x^2 = 2 - x \\ \hline 3(-1)^2 \;?\; 2 - (-1) \\ 3 \cdot 1 \;\bigm|\; 2 + 1 \\ 3 \;\bigm|\; 3 \qquad \text{TRUE} \end{array}$$

For $\tfrac{2}{3}$:

$$\begin{array}{c|c} 3x^2 = 2 - x \\ \hline 3(\tfrac{2}{3})^2 \;?\; 2 - (\tfrac{2}{3}) \\ 3 \cdot \tfrac{4}{9} \;\bigm|\; \tfrac{6}{3} - \tfrac{2}{3} \\ \tfrac{4}{3} \;\bigm|\; \tfrac{4}{3} \qquad \text{TRUE} \end{array}$$

The solutions are -1 and $\tfrac{2}{3}$.

b) The x-intercepts of $f(x) = 3x^2 + x - 2$ are $(-1, 0)$ and $(\tfrac{2}{3}, 0)$. The solutions of the equation $3x^2 = 2 - x$ are the first coordinates of the x-intercepts of the graph of $f(x) = 3x^2 + x - 2$.

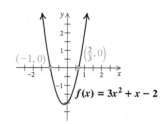

◀ Do Exercise 4.

4. a) Solve: $5x^2 = 8x - 3$.

b) Find the x-intercepts of $f(x) = 5x^2 - 8x + 3$.

$f(x) = 5x^2 - 8x + 3$

Solving Equations of the Type $x^2 = d$

Consider the equation $x^2 = 25$ again. The number 25 has two real-number square roots, namely, 5 and -5. Note that these are the solutions of the equation in Example 1. This illustrates the principle of square roots, which provides a quick method for solving equations of the type $x^2 = d$.

THE PRINCIPLE OF SQUARE ROOTS

The solutions of the equation $x^2 = d$ are $\sqrt{d}$ and $-\sqrt{d}$.

When $d > 0$, the solutions are two real numbers.

When $d = 0$, the only solution is 0.

When $d < 0$, the solutions are two imaginary numbers.

We often use the notation $\pm\sqrt{d}$ to represent both $\sqrt{d}$ and $-\sqrt{d}$.

Answer

4. (a) $\tfrac{3}{5}, 1$; **(b)** $\left(\tfrac{3}{5}, 0\right), (1, 0)$

EXAMPLE 4 Solve: $3x^2 = 6$. Give the exact solutions and approximate the solutions to three decimal places.

We have
$$3x^2 = 6$$
$$x^2 = 2$$
$$x = \sqrt{2} \quad \text{or} \quad x = -\sqrt{2}.$$

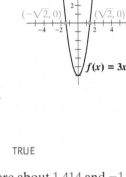

Check: For $\sqrt{2}$:
$$\frac{3x^2 = 6}{3(\sqrt{2})^2 \; ? \; 6}$$
$$3 \cdot 2$$
$$6 \quad \text{TRUE}$$

For $-\sqrt{2}$:
$$\frac{3x^2 = 6}{3(-\sqrt{2})^2 \; ? \; 6}$$
$$3 \cdot 2$$
$$6 \quad \text{TRUE}$$

The solutions are $\sqrt{2}$ and $-\sqrt{2}$, or $\pm\sqrt{2}$, which are about 1.414 and -1.414, or ± 1.414, when rounded to three decimal places.

Do Exercise 5.

Sometimes we rationalize denominators when writing solutions.

EXAMPLE 5 Solve: $-5x^2 + 2 = 0$. Give the exact solutions and approximate the solutions to three decimal places.

$$-5x^2 + 2 = 0$$
$$x^2 = \frac{2}{5} \quad \text{Subtracting 2 and dividing by } -5$$
$$x = \sqrt{\frac{2}{5}} \quad \text{or} \quad x = -\sqrt{\frac{2}{5}} \quad \text{Using the principle of square roots}$$
$$x = \sqrt{\frac{2}{5} \cdot \frac{5}{5}} \quad \text{or} \quad x = -\sqrt{\frac{2}{5} \cdot \frac{5}{5}} \quad \text{Rationalizing the denominators}$$
$$x = \frac{\sqrt{10}}{5} \quad \text{or} \quad x = -\frac{\sqrt{10}}{5}$$

Check: Since there is no x-term in the equation, we can check both numbers at once.

$$\frac{-5x^2 + 2 = 0}{-5\left(\pm\frac{\sqrt{10}}{5}\right)^2 + 2 \; ? \; 0}$$
$$-5\left(\frac{10}{25}\right) + 2$$
$$-2 + 2$$
$$0 \quad \text{TRUE}$$

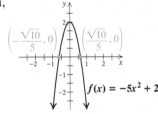

The solutions are $\frac{\sqrt{10}}{5}$ and $-\frac{\sqrt{10}}{5}$, or $\pm\frac{\sqrt{10}}{5}$. The approximate solutions, rounded to three decimal places, are 0.632 and -0.632, or ± 0.632.

Do Exercise 6.

GS 5. Solve: $5x^2 = 15$. Give the exact solutions and approximate the solutions to three decimal places.

$$5x^2 = 15$$
$$x^2 = \boxed{}$$
$$x = \sqrt{\boxed{}} \quad \text{or} \quad x = -\sqrt{\boxed{}}$$

The solutions can also be written $\pm\sqrt{3}$. If we round to three decimal places, the solutions are $\pm\boxed{}$.

6. Solve: $-3x^2 + 8 = 0$. Give the exact solutions and approximate the solutions to three decimal places.

Answers
5. $\sqrt{3}$ and $-\sqrt{3}$, or $\pm\sqrt{3}$; 1.732 and -1.732, or ± 1.732
6. $\frac{2\sqrt{6}}{3}$ and $-\frac{2\sqrt{6}}{3}$, or $\pm\frac{2\sqrt{6}}{3}$; 1.633 and -1.633, or ± 1.633
Guided Solution:
5. 3, 3, 3, 1.732

Sometimes we get solutions that are imaginary numbers.

EXAMPLE 6 Solve: $4x^2 + 9 = 0$.

$$4x^2 + 9 = 0$$
$$x^2 = -\frac{9}{4} \quad \text{Subtracting 9 and dividing by 4}$$
$$x = \sqrt{-\frac{9}{4}} \quad \text{or} \quad x = -\sqrt{-\frac{9}{4}} \quad \text{Using the principle of square roots}$$
$$x = \frac{3}{2}i \quad \text{or} \quad x = -\frac{3}{2}i \quad \text{Simplifying; recall that } \sqrt{-1} = i.$$

Check:
$$\begin{array}{c|c} 4x^2 + 9 = 0 \\ \hline 4\left(\pm\frac{3}{2}i\right)^2 + 9 \;?\; 0 \\ 4\left(-\frac{9}{4}\right) + 9 \\ -9 + 9 \\ 0 \mid \text{TRUE} \end{array}$$

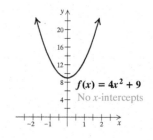

$f(x) = 4x^2 + 9$
No x-intercepts

The solutions are $\frac{3}{2}i$ and $-\frac{3}{2}i$, or $\pm\frac{3}{2}i$.

We see that the graph of $f(x) = 4x^2 + 9$ does not cross the x-axis. This is true because the equation $4x^2 + 9 = 0$ has *imaginary* complex-number solutions. Only real-number solutions correspond to x-intercepts.

◀ Do Exercise 7.

CALCULATOR CORNER

Imaginary Solutions of Quadratic Equations What happens when you use the ZERO feature to solve the equation in Example 6? Explain why this happens.

7. Solve: $2x^2 + 1 = 0$.

Solving Equations of the Type $(x + c)^2 = d$

Equations like $(x - 2)^2 = 7$ can also be solved using the principle of square roots.

EXAMPLE 7

a) Solve: $(x - 2)^2 = 7$.

b) Find the x-intercepts of $f(x) = (x - 2)^2 - 7$.

a) We have
$$(x - 2)^2 = 7$$
$$x - 2 = \sqrt{7} \quad \text{or} \quad x - 2 = -\sqrt{7} \quad \text{Using the principle of square roots}$$
$$x = 2 + \sqrt{7} \quad \text{or} \quad x = 2 - \sqrt{7}.$$
The solutions are $2 + \sqrt{7}$ and $2 - \sqrt{7}$, or $2 \pm \sqrt{7}$.

b) The x-intercepts of $f(x) = (x - 2)^2 - 7$ are $(2 - \sqrt{7}, 0)$ and $(2 + \sqrt{7}, 0)$.

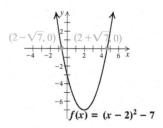

$f(x) = (x - 2)^2 - 7$

◀ Do Exercise 8.

8. a) Solve: $(x - 1)^2 = 5$.
b) Find the x-intercepts of $f(x) = (x - 1)^2 - 5$.

$f(x) = (x - 1)^2 - 5$

Answers

7. $\frac{\sqrt{2}}{2}i$ and $-\frac{\sqrt{2}}{2}i$, or $\pm\frac{\sqrt{2}}{2}i$

8. (a) $1 \pm \sqrt{5}$; **(b)** $(1 - \sqrt{5}, 0), (1 + \sqrt{5}, 0)$

If we can express the left side of an equation as the square of a binomial, we can proceed as we did in Example 7.

EXAMPLE 8 Solve: $x^2 + 6x + 9 = 2$.

We have

$x^2 + 6x + 9 = 2$ The left side is the square of a binomial.

$(x + 3)^2 = 2$

$x + 3 = \sqrt{2}$ or $x + 3 = -\sqrt{2}$ Using the principle of square roots

$x = -3 + \sqrt{2}$ or $x = -3 - \sqrt{2}$.

The solutions are $-3 + \sqrt{2}$ and $-3 - \sqrt{2}$, or $-3 \pm \sqrt{2}$.

Do Exercise 9.

> **GS 9.** Solve: $x^2 + 16x + 64 = 11$.
>
> $x^2 + 16x + 64 = 11$
>
> $(x +)^2 = 11$
>
> $x + 8 = \sqrt{}$ or $x + 8 = -\sqrt{}$
>
> $x = + \sqrt{11}$ or $x = - \sqrt{11}$
>
> The solutions can also be written $-8 \pm \sqrt{11}$.

b COMPLETING THE SQUARE

Consider the equation $x^2 + 14x - 4 = 0$. Since $x^2 + 14x - 4$ cannot be factored and since it is not in the form $(x + c)^2 = d$, we need a new procedure to solve this equation. This procedure is called **completing the square**. *It can be used to solve any quadratic equation.*

Suppose that we have the quadratic equation

$$x^2 + 14x = 4.$$

If we could add on both sides of the equation a constant that would make the expression on the left the square of a binomial, we could then solve the equation using the principle of square roots.

How can we determine what to add to $x^2 + 14x$ in order to construct the square of a binomial? We want to find a number a such that the following equation is satisfied:

$$x^2 + 14x + a^2 = (x + a)(x + a) = x^2 + 2ax + a^2.$$

Thus, $2a = 14$. Solving, we get $a = 7$. Since $a^2 = 7^2 = 49$, we add 49 to our original expression:

$$x^2 + 14x + 49 = (x + 7)^2. \quad x^2 + 14x + 49 \text{ is the square of } x + 7.$$

Note that $7 = \frac{14}{2}$. Thus, a is half of the coefficient of x in $x^2 + 14x$.

Returning to solving our original equation, we first add 49 on *both* sides to *complete the square* on the left. Then we solve:

$x^2 + 14x = 4$ Original equation

$x^2 + 14x + 49 = 4 + 49$ Adding 49: $\left(\frac{14}{2}\right)^2 = 7^2 = 49$

$(x + 7)^2 = 53$

$x + 7 = \sqrt{53}$ or $x + 7 = -\sqrt{53}$ Using the principle of square roots

$x = -7 + \sqrt{53}$ or $x = -7 - \sqrt{53}$.

The solutions are $-7 \pm \sqrt{53}$.

Answer
9. $-8 \pm \sqrt{11}$
Guided Solution:
9. 8, 11, 11, −8, −8

COMPLETING THE SQUARE

When solving an equation, to **complete the square** of an expression like $x^2 + bx$, we take half the x-coefficient, which is $b/2$, and square it. Then we add that number, $(b/2)^2$, on both sides of the equation.

We have seen that a quadratic equation $(x + c)^2 = d$ can be solved using the principle of square roots. Any equation, such as $x^2 - 6x + 8 = 0$, can be put in this form by completing the square. Then we can solve as before.

EXAMPLE 9 Solve: $x^2 - 6x + 8 = 0$.

We have

$$x^2 - 6x + 8 = 0$$
$$x^2 - 6x = -8. \quad \text{Subtracting 8}$$

We take half of -6 and square it, to get 9. Then we add 9 on *both* sides of the equation. This makes the left side the square of a binomial, $x - 3$. We have now *completed the square*.

$$x^2 - 6x + 9 = -8 + 9 \quad \text{Adding 9: } \left(\frac{-6}{2}\right)^2 = (-3)^2 = 9$$
$$(x - 3)^2 = 1$$
$$x - 3 = 1 \quad \text{or} \quad x - 3 = -1 \quad \text{Using the principle of square roots}$$
$$x = 4 \quad \text{or} \quad x = 2$$

The solutions are 2 and 4.

◀ Do Exercises 10 and 11.

EXAMPLE 10 Solve $x^2 + 4x - 7 = 0$ by completing the square.

We have

$$x^2 + 4x - 7 = 0$$
$$x^2 + 4x = 7 \quad \text{Adding 7}$$
$$x^2 + 4x + 4 = 7 + 4 \quad \text{Adding 4: } \left(\frac{4}{2}\right)^2 = (2)^2 = 4$$
$$(x + 2)^2 = 11$$
$$x + 2 = \sqrt{11} \quad \text{or} \quad x + 2 = -\sqrt{11} \quad \text{Using the principle of square roots}$$
$$x = -2 + \sqrt{11} \quad \text{or} \quad x = -2 - \sqrt{11}.$$

The solutions are $-2 \pm \sqrt{11}$.

◀ Do Exercise 12.

Solve.

10. $x^2 + 6x + 8 = 0$

11. $x^2 - 8x - 20 = 0$ (GS)
$$x^2 - 8x = \boxed{}$$
$$x^2 - 8x + \boxed{} = 20 + \boxed{}$$
$$(x - \boxed{})^2 = 36$$
$$x - 4 = 6 \quad \text{or} \quad x - 4 = \boxed{}$$
$$x = 10 \quad \text{or} \quad x = \boxed{}$$

12. Solve by completing the square:
$$x^2 + 6x - 1 = 0.$$

Answers
10. $-2, -4$ **11.** $10, -2$ **12.** $-3 \pm \sqrt{10}$
Guided Solution:
11. 20, 16, 16, 4, -6, -2

When the coefficient of x^2 is not 1, we can multiply to make it 1.

EXAMPLE 11 Solve $3x^2 + 7x = 2$ by completing the square.

We have

$$3x^2 + 7x = 2$$

$$\frac{1}{3}(3x^2 + 7x) = \frac{1}{3} \cdot 2 \qquad \text{Multiplying by } \tfrac{1}{3} \text{ to make the } x^2\text{-coefficient 1}$$

$$x^2 + \frac{7}{3}x = \frac{2}{3} \qquad \text{Multiplying and simplifying}$$

$$x^2 + \frac{7}{3}x + \frac{49}{36} = \frac{2}{3} + \frac{49}{36} \qquad \text{Adding } \tfrac{49}{36}: \left[\tfrac{1}{2} \cdot \tfrac{7}{3}\right]^2 = \tfrac{49}{36}$$

$$\left(x + \frac{7}{6}\right)^2 = \frac{24}{36} + \frac{49}{36} \qquad \text{Finding a common denominator}$$

$$\left(x + \frac{7}{6}\right)^2 = \frac{73}{36}$$

$$x + \frac{7}{6} = \sqrt{\frac{73}{36}} \quad \text{or} \quad x + \frac{7}{6} = -\sqrt{\frac{73}{36}} \qquad \text{Using the principle of square roots}$$

$$x + \frac{7}{6} = \frac{\sqrt{73}}{6} \quad \text{or} \quad x + \frac{7}{6} = -\frac{\sqrt{73}}{6}$$

$$x = -\frac{7}{6} + \frac{\sqrt{73}}{6} \quad \text{or} \quad x = -\frac{7}{6} - \frac{\sqrt{73}}{6}.$$

The solutions are $-\frac{7}{6} \pm \frac{\sqrt{73}}{6}$.

The graph at right shows the x-intercepts of the graph of the related function $f(x) = 3x^2 + 7x - 2$.

Do Exercises 13 and 14. ▶

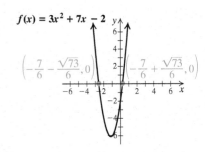

Solve by completing the square.

13. $2x^2 + 6x = 5$

14. $3x^2 - 2x = 7$

SOLVING BY COMPLETING THE SQUARE

To solve an equation $ax^2 + bx + c = 0$ by completing the square:

1. If $a \neq 1$, multiply by $1/a$ so that the x^2-coefficient is 1.
2. Once the x^2-coefficient is 1, add or subtract so that the equation is in the form

   $$x^2 + bx = -c, \quad \text{or} \quad x^2 + \frac{b}{a}x = -\frac{c}{a} \text{ if step (1) has been applied.}$$

3. Take half of the x-coefficient and square it. Add the result on both sides of the equation.
4. Express the side with the variables as the square of a binomial.
5. Use the principle of square roots and complete the solution.

Answers

13. $-\dfrac{3}{2} \pm \dfrac{\sqrt{19}}{2}$ 14. $\dfrac{1}{3} \pm \dfrac{\sqrt{22}}{3}$

EXAMPLE 12 Solve $2x^2 = 3x - 7$ by completing the square.

We have
$$2x^2 = 3x - 7$$
$$2x^2 - 3x = -7 \quad \text{Subtracting } 3x$$
$$\frac{1}{2}(2x^2 - 3x) = \frac{1}{2} \cdot (-7) \quad \text{Multiplying by } \tfrac{1}{2} \text{ to make the } x^2\text{-coefficient 1}$$
$$x^2 - \frac{3}{2}x = -\frac{7}{2} \quad \text{Multiplying and simplifying}$$
$$x^2 - \frac{3}{2}x + \frac{9}{16} = -\frac{7}{2} + \frac{9}{16} \quad \text{Adding } \tfrac{9}{16}: \left[\tfrac{1}{2}\left(-\tfrac{3}{2}\right)\right]^2 = \left[-\tfrac{3}{4}\right]^2 = \tfrac{9}{16}$$
$$\left(x - \frac{3}{4}\right)^2 = -\frac{56}{16} + \frac{9}{16} \quad \text{Finding a common denominator}$$
$$\left(x - \frac{3}{4}\right)^2 = -\frac{47}{16}$$
$$x - \frac{3}{4} = \sqrt{-\frac{47}{16}} \quad \text{or} \quad x - \frac{3}{4} = -\sqrt{-\frac{47}{16}} \quad \text{Using the principle of square roots}$$
$$x - \frac{3}{4} = \frac{\sqrt{47}}{4}i \quad \text{or} \quad x - \frac{3}{4} = -\frac{\sqrt{47}}{4}i \quad \sqrt{-1} = i$$
$$x = \frac{3}{4} + \frac{\sqrt{47}}{4}i \quad \text{or} \quad x = \frac{3}{4} - \frac{\sqrt{47}}{4}i.$$

The solutions are $\frac{3}{4} \pm \frac{\sqrt{47}}{4}i$.

We see at left that the graph of $f(x) = 2x^2 - 3x + 7$ does not cross the x-axis. This is true because the equation $2x^2 = 3x - 7$ has nonreal complex-number solutions.

◀ **Do Exercise 15.**

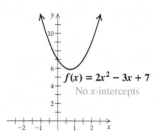
$f(x) = 2x^2 - 3x + 7$
No x-intercepts

15. Solve by completing the square: $3x^2 = 2x - 1$.

C APPLICATIONS AND PROBLEM SOLVING

EXAMPLE 13 *Hang Time.* One of the most exciting plays in basketball is the dunk shot. The amount of time T that passes from the moment a player leaves the ground, goes up, makes the shot, and arrives back on the ground is called *hang time*. A function relating an athlete's vertical leap V, in inches, to hang time T, in seconds, is given by
$$V(T) = 48T^2.$$

Answer

15. $\frac{1}{3} \pm \frac{\sqrt{2}}{3}i$

a) Hall-of-Famer Michael Jordan had a hang time of about 0.889 sec. What was his vertical leap?

b) Although his height is only 5 ft 7 in., Spud Webb, formerly of the Sacramento Kings, had a vertical leap of about 44 in. What was his hang time?

a) To find Jordan's vertical leap, we substitute 0.889 for T in the function and compute V:

$$V(0.889) = 48(0.889)^2 \approx 37.9 \text{ in.}$$

Jordan's vertical leap was about 37.9 in.

b) To find Webb's hang time, we substitute 44 for V and solve for T:

$$44 = 48T^2 \quad \text{Substituting 44 for } V$$
$$\frac{44}{48} = T^2 \quad \text{Solving for } T^2$$
$$0.91\overline{6} = T^2$$
$$\sqrt{0.91\overline{6}} = T \quad \text{Hang time is positive.}$$
$$0.957 \approx T. \quad \text{Using a calculator}$$

Webb's hang time was about 0.957 sec. Note that his hang time was greater than Jordan's.

Do Exercises 16 and 17.

16. *Vertical Leap.* Blake Griffin of the Los Angeles Clippers has a hang time of about 0.878 sec. What is his vertical leap?

17. *Hang Time.* Russell Westbrook of the Oklahoma Thunder has a vertical leap of 40 in. What is his hang time?

Answers
16. About 37.0 in. **17.** About 0.913 sec

11.1 Exercise Set

FOR EXTRA HELP — MyLab Math

✔ Check Your Understanding

Reading Check Determine whether each statement is true or false.

RC1. The quadratic equation $8x^2 - 11x + 50 = 0$ is in standard form.

RC2. Any quadratic equation can be solved by completing the square.

RC3. A quadratic equation may have solutions that are imaginary numbers.

RC4. The notation $\pm\sqrt{7}$ represents two real numbers.

RC5. To solve $5x^2 = 2x$, we can divide by x on both sides.

RC6. If $(x - 6)^2 = \sqrt{7}$, then $x = \sqrt{7}$ or $x = -\sqrt{7}$.

Concept Check Replace the blanks in each equation with constants in order to complete the square and form a true equation.

CC1. $x^2 + 10x + \underline{} = (x + \underline{})^2$

CC2. $t^2 - 12t + \underline{} = (t - \underline{})^2$

CC3. $m^2 - 5m + \underline{} = (m - \underline{})^2$

CC4. $x^2 + x + \underline{} = (x + \underline{})^2$

CC5. $r^2 + \frac{2}{5}r + \underline{} = (r + \underline{})^2$

CC6. $y^2 - \frac{3}{2}y + \underline{} = (y - \underline{})^2$

a

1. a) Solve: $6x^2 = 30$.
   b) Find the x-intercepts of $f(x) = 6x^2 - 30$.

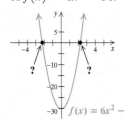

2. a) Solve: $5x^2 = 35$.
   b) Find the x-intercepts of $f(x) = 5x^2 - 35$.

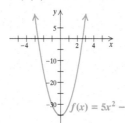

3. a) Solve: $9x^2 + 25 = 0$.
   b) Find the x-intercepts of $f(x) = 9x^2 + 25$.

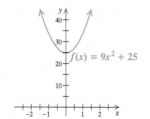

4. a) Solve: $36x^2 + 49 = 0$.
   b) Find the x-intercepts of $f(x) = 36x^2 + 49$.

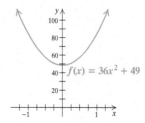

Solve. Give the exact solutions and approximate solutions to three decimal places, when appropriate.

5. $2x^2 - 3 = 0$
6. $3x^2 - 7 = 0$
7. $(x + 2)^2 = 49$
8. $(x - 1)^2 = 6$

9. $(x - 4)^2 = 16$
10. $(x + 3)^2 = 9$
11. $(x - 11)^2 = 7$
12. $(x - 9)^2 = 34$

13. $(x - 7)^2 = -4$
14. $(x + 1)^2 = -9$
15. $(x - 9)^2 = 81$
16. $(t - 2)^2 = 25$

17. $\left(x - \frac{3}{2}\right)^2 = \frac{7}{2}$
18. $\left(y + \frac{3}{4}\right)^2 = \frac{17}{16}$
19. $x^2 + 6x + 9 = 64$

20. $x^2 + 10x + 25 = 100$
21. $y^2 - 14y + 49 = 4$
22. $p^2 - 8p + 16 = 1$

b Solve by completing the square. Show your work.

23. $x^2 + 4x = 2$
24. $x^2 + 2x = 5$
25. $x^2 - 22x = 11$
26. $x^2 - 18x = 10$

27. $x^2 + x = 1$
28. $x^2 - x = 3$
29. $t^2 - 5t = 7$
30. $y^2 + 9y = 8$

31. $x^2 + \frac{3}{2}x = 3$
32. $x^2 - \frac{4}{3}x = \frac{2}{3}$
33. $m^2 - \frac{9}{2}m = \frac{3}{2}$
34. $r^2 + \frac{2}{5}r = \frac{4}{5}$

35. $x^2 + 6x - 16 = 0$
36. $x^2 - 8x + 15 = 0$
37. $x^2 + 22x + 102 = 0$
38. $x^2 + 18x + 74 = 0$

39. $x^2 - 10x - 4 = 0$
40. $x^2 + 10x - 4 = 0$

41. a) Solve:
   $x^2 + 7x - 2 = 0$.
   b) Find the x-intercepts of $f(x) = x^2 + 7x - 2$.

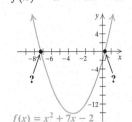

42. a) Solve:
   $x^2 - 7x - 2 = 0$.
   b) Find the x-intercepts of $f(x) = x^2 - 7x - 2$.

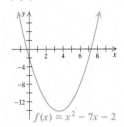

43. a) Solve:
   $2x^2 - 5x + 8 = 0$.
   b) Find the x-intercepts of $f(x) = 2x^2 - 5x + 8$.

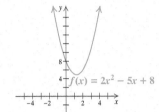

44. a) Solve:
   $2x^2 - 3x + 9 = 0$.
   b) Find the x-intercepts of $f(x) = 2x^2 - 3x + 9$.

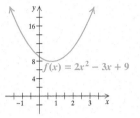

Solve by completing the square. Show your work.

45. $x^2 - \frac{3}{2}x - \frac{1}{2} = 0$

46. $x^2 + \frac{3}{2}x - 2 = 0$

47. $2x^2 - 3x - 17 = 0$

48. $2x^2 + 3x - 1 = 0$

49. $3x^2 - 4x - 1 = 0$

50. $3x^2 + 4x - 3 = 0$

51. $x^2 + x + 2 = 0$

52. $x^2 - x + 1 = 0$

53. $x^2 - 4x + 13 = 0$

54. $x^2 - 6x + 13 = 0$

C *Hang Time.* For Exercises 55 and 56, use the hang-time function $V(T) = 48T^2$, relating vertical leap to hang time.

55. The NBA's Kobe Bryant of the Los Angeles Lakers has a vertical leap of about 38 in. What is his hang time?

56. The NBA's Darrell Griffith had a record vertical leap of 48 in. What was his hang time?

Free-Falling Objects. The function $s(t) = 16t^2$ is used to approximate the distance s, in feet, that an object falls freely from rest in t seconds. Use the formula for Exercises 57–62.

57. The tallest roller coaster in the world is the Kingda Ka, located at Six Flags Great Adventure amusement park, in Jackson, New Jersey. It is 456 ft high. How long would it take an object to fall freely from the top?

58. The Gateway Arch in St. Louis, Missouri, is 630 ft high. How long would it take an object to fall freely from the top?

59. The Bunda Cliffs in Australia extend 62 mi along the coastline and vary in height from 200 ft to 390 ft. How long would it take an object to fall freely from the highest point on these cliffs?

60. Suspended 1854 ft above the water, the Beipanjiang River Bridge in China is the world's highest bridge. How long would it take an object to fall freely from the bridge?

61. The Washington Monument, near the west end of the National Mall in Washington, D.C., is the world's tallest stone structure and the world's tallest obelisk. It is 555.427 ft tall. How long would it take an object to fall freely from that height?

62. Completed in 2010, the Burj Khalifa, in downtown Dubai, is the tallest building in the world. It is 2720 ft tall. How long would it take an object to fall freely from that height?

Skill Maintenance

63. *Marathon Times.* The following table lists the record marathon times in 1981 and in 2011. [7.5e]

NUMBER OF YEARS AFTER 1981	RECORD MARATHON TIME (in minutes)
0	128
30	124

DATA: marathonguide.com

a) Use the two data points in the table to find a linear function $R(t) = mt + b$ that fits the data.
b) Use the function to estimate the record marathon time in 2020.
c) In what year will the marathon record be 122 min?

810 CHAPTER 11 Quadratic Equations and Functions

Graph. [7.1c], [7.4a]

64. $f(x) = 5 - 2x^2$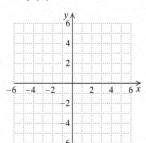

65. $f(x) = 5 - 2x$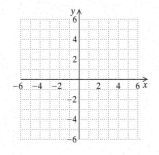

66. $2x - 5y = 10$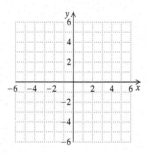

67. $f(x) = |5 - 2x|$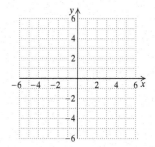

Simplify.

68. $\sqrt{88}$ [10.3a]

69. $\sqrt[5]{32x^5}$ [10.1d]

70. $\dfrac{t^3 - 8}{t^2 - 5t + 6}$ [6.1c]

71. $\dfrac{4x^3 - 6x^2 - 10x}{3x^3 - 3x}$ [6.1c]

72. $\dfrac{\dfrac{1}{x}}{\dfrac{1}{2x} - \dfrac{1}{3x}}$ [6.6a]

73. $\dfrac{\dfrac{t}{t+1}}{t - \dfrac{1}{t}}$ [6.6a]

Synthesis

74. Use a graphing calculator to solve each of the following equations.
  a) $25.55x^2 - 1635.2 = 0$
  b) $-0.0644x^2 + 0.0936x + 4.56 = 0$
  c) $2.101x + 3.121 = 0.97x^2$

75. Problems such as those in Exercises 17, 21, and 25 can be solved without first finding standard form by using the INTERSECT feature on a graphing calculator. We let y_1 = the left side of the equation and y_2 = the right side. Use a graphing calculator to solve Exercises 17, 21, and 25 in this manner.

Find b such that the trinomial is a square.

76. $x^2 + bx + 75$

77. $x^2 + bx + 64$

Solve.

78. $\left(x - \tfrac{1}{3}\right)\left(x - \tfrac{1}{3}\right) + \left(x - \tfrac{1}{3}\right)\left(x + \tfrac{2}{9}\right) = 0$

79. $x(2x^2 + 9x - 56)(3x + 10) = 0$

80. *Boating.* A barge and a fishing boat leave a dock at the same time, traveling at right angles to each other. The barge travels 7 km/h slower than the fishing boat. After 4 hr, the boats are 68 km apart. Find the speed of each vessel.

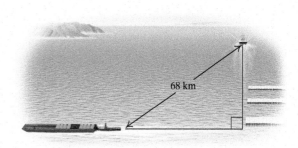

11.2

The Quadratic Formula

OBJECTIVE

a. Solve quadratic equations using the quadratic formula, and approximate solutions using a calculator.

a SOLVING USING THE QUADRATIC FORMULA

SKILL REVIEW

Express imaginary numbers as bi and complex numbers as $a + bi$. [10.8a]

Express in terms of i.

1. $\sqrt{-100}$
2. $10 - \sqrt{-68}$

Answers: **1.** $10i$ **2.** $10 - 2\sqrt{17}i$

MyLab Math VIDEO

Each time you solve by completing the square, the procedure is the same. When we do the same kind of procedure many times, we look for a formula to speed up our work. Consider

$$ax^2 + bx + c = 0, \quad a > 0.$$

Note that if $a < 0$, we can get an equivalent form with $a > 0$ by first multiplying by -1.

Let's solve by *completing the square*. As we carry out the steps, compare them with those in Example 12 in the preceding section.

$$x^2 + \frac{b}{a}x + \frac{c}{a} = 0 \qquad \text{Multiplying by } \frac{1}{a}$$

$$x^2 + \frac{b}{a}x = -\frac{c}{a} \qquad \text{Subtracting } \frac{c}{a}$$

Half of $\frac{b}{a}$ is $\frac{b}{2a}$. The square is $\frac{b^2}{4a^2}$. We add $\frac{b^2}{4a^2}$ on both sides:

$$x^2 + \frac{b}{a}x + \frac{b^2}{4a^2} = -\frac{c}{a} + \frac{b^2}{4a^2} \qquad \text{Adding } \frac{b^2}{4a^2}$$

$$\left(x + \frac{b}{2a}\right)^2 = -\frac{4ac}{4a^2} + \frac{b^2}{4a^2} \qquad \text{Factoring the left side and finding a common denominator on the right}$$

$$\left(x + \frac{b}{2a}\right)^2 = \frac{b^2 - 4ac}{4a^2}$$

$$x + \frac{b}{2a} = \sqrt{\frac{b^2 - 4ac}{4a^2}} \quad \text{or} \quad x + \frac{b}{2a} = -\sqrt{\frac{b^2 - 4ac}{4a^2}}. \qquad \text{Using the principle of square roots}$$

Since $a > 0$, $\sqrt{4a^2} = 2a$, so we can simplify as follows:

$$x + \frac{b}{2a} = \frac{\sqrt{b^2 - 4ac}}{2a} \quad \text{or} \quad x + \frac{b}{2a} = -\frac{\sqrt{b^2 - 4ac}}{2a}.$$

Thus,

$$x = -\frac{b}{2a} \pm \frac{\sqrt{b^2 - 4ac}}{2a}, \quad \text{or} \quad x = \frac{-b \pm \sqrt{b^2 - 4ac}}{2a}.$$

This result is called the **quadratic formula**.

THE QUADRATIC FORMULA

The solutions of $ax^2 + bx + c = 0$ are given by
$$x = \frac{-b \pm \sqrt{b^2 - 4ac}}{2a}.$$

A similar proof would show that the formula also holds when $a < 0$.

ALGEBRAIC ▶◀ GRAPHICAL CONNECTION

The Quadratic Formula (Algebraic). The solutions of $ax^2 + bx + c = 0, a \neq 0$, are given by
$$x = \frac{-b \pm \sqrt{b^2 - 4ac}}{2a}.$$

The Quadratic Formula (Graphical).
The x-intercepts of the graph of the function $f(x) = ax^2 + bx + c$, $a \neq 0$, if they exist, are given by
$\left(\dfrac{-b \pm \sqrt{b^2 - 4ac}}{2a}, 0 \right)$.

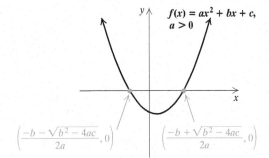

$\left(\dfrac{-b - \sqrt{b^2 - 4ac}}{2a}, 0 \right)$ $\left(\dfrac{-b + \sqrt{b^2 - 4ac}}{2a}, 0 \right)$

EXAMPLE 1 Solve $5x^2 + 8x = -3$ using the quadratic formula.

We first find standard form and determine a, b, and c:
$$5x^2 + 8x + 3 = 0;$$
$$a = 5, \quad b = 8, \quad c = 3.$$

We then use the quadratic formula, $x = \dfrac{-b \pm \sqrt{b^2 - 4ac}}{2a}$:

$x = \dfrac{-8 \pm \sqrt{8^2 - 4 \cdot 5 \cdot 3}}{2 \cdot 5}$ Substituting

$x = \dfrac{-8 \pm \sqrt{64 - 60}}{10}$ Be sure to write the fraction bar all the way across.

$x = \dfrac{-8 \pm \sqrt{4}}{10}$

$x = \dfrac{-8 \pm 2}{10}$

$x = \dfrac{-8 + 2}{10}$ or $x = \dfrac{-8 - 2}{10}$

$x = \dfrac{-6}{10}$ or $x = \dfrac{-10}{10}$

$x = -\dfrac{3}{5}$ or $x = -1$.

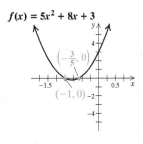

The solutions are $-\dfrac{3}{5}$ and -1.

CALCULATOR CORNER

Solving Quadratic Equations We can use the INTERSECT feature to solve a quadratic equation. Consider the equation
$4x(x - 2) - 5x(x - 1) = 2.$
First, we enter the equations
$y_1 = 4x(x - 2) - 5x(x - 1)$
and $y_2 = 2$. We then graph the equations in a window that shows the point(s) of intersection of the graphs. Next, we use the INTERSECT feature to find the coordinates of the left-hand point of intersection.

The first coordinate of this point, -2, is one solution of the equation. We use the INTERSECT feature again to find the other solution, -1.

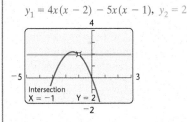

EXERCISES: Solve.
1. $5x^2 = -11x + 12$
2. $2x^2 - 15 = 7x$
3. $6(x - 3) = (x - 3)(x - 2)$
4. $(x + 1)(x - 4) = 3(x - 4)$

It turns out that we could have solved the equation in Example 1 more easily by factoring, as follows:

$$5x^2 + 8x + 3 = 0$$
$$(5x + 3)(x + 1) = 0$$
$$5x + 3 = 0 \quad \text{or} \quad x + 1 = 0$$
$$5x = -3 \quad \text{or} \quad x = -1$$
$$x = -\tfrac{3}{5} \quad \text{or} \quad x = -1.$$

1. Consider the equation $2x^2 = 4 + 7x$.
   a) Solve using the quadratic formula.
   b) Solve by factoring.

> To solve a quadratic equation:
> 1. If the equation is in the form $x^2 = d$ or $(x + c)^2 = d$, use the principle of square roots as in Section 11.1.
> 2. If it is not in the form of step (1), write it in standard form $ax^2 + bx + c = 0$ with a and b nonzero.
> 3. Then try factoring.
> 4. If it is not possible to factor or if factoring seems difficult, use the quadratic formula.
>
> Although the solutions of a quadratic equation cannot always be found by factoring, they can *always* be found using the quadratic formula.

CALCULATOR CORNER

Approximating Solutions of Quadratic Equations In Example 2, we find that the solutions of the equation $5x^2 - 8x = 3$ are $\dfrac{4 + \sqrt{31}}{5}$ and $\dfrac{4 - \sqrt{31}}{5}$. We can use a calculator to approximate these solutions. Parentheses must be used carefully. For example, to approximate $\dfrac{4 + \sqrt{31}}{5}$, we press

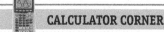
.

The solutions are approximately 1.914 and −0.314.

```
(4+√31)/5
           1.913552873
(4−√31)/5
           −.3135528726
```

EXERCISES: Use a calculator to approximate the solutions in each of the following. Round to three decimal places.
1. Example 4
2. Margin Exercise 2
3. Margin Exercise 4

The solutions to all the exercises in this section could also be found by completing the square. However, the quadratic formula is the preferred method because it is faster.

◀ Do Exercise 1.

We will see in Example 2 that we cannot always rely on factoring.

EXAMPLE 2 Solve: $5x^2 - 8x = 3$. Give the exact solutions and approximate the solutions to three decimal places.

We first find standard form and determine a, b and c:
$$5x^2 - 8x - 3 = 0;$$
$$a = 5, \quad b = -8, \quad c = -3.$$

We then use the quadratic formula, $x = \dfrac{-b \pm \sqrt{b^2 - 4ac}}{2a}$:

$$x = \dfrac{-(-8) \pm \sqrt{(-8)^2 - 4 \cdot 5 \cdot (-3)}}{2 \cdot 5} \quad \text{Substituting}$$

$$= \dfrac{8 \pm \sqrt{64 + 60}}{10} = \dfrac{8 \pm \sqrt{124}}{10} = \dfrac{8 \pm \sqrt{4 \cdot 31}}{10}$$

$$= \dfrac{8 \pm 2\sqrt{31}}{10} = \dfrac{2(4 \pm \sqrt{31})}{2 \cdot 5} = \dfrac{2}{2} \cdot \dfrac{4 \pm \sqrt{31}}{5} = \dfrac{4 \pm \sqrt{31}}{5}.$$

↑ Caution!

To avoid a common error in simplifying, remember to *factor the numerator and the denominator* and then remove a factor of 1.

Answer
1. (a) $-\tfrac{1}{2}, 4$; (b) $-\tfrac{1}{2}, 4$

We can use a calculator to approximate the solutions:

$$\frac{4 + \sqrt{31}}{5} \approx 1.914; \quad \frac{4 - \sqrt{31}}{5} \approx -0.314.$$

Check: Checking the exact solutions $(4 \pm \sqrt{31})/5$ can be quite cumbersome. It could be done on a calculator or by using the approximations. Here we check 1.914; the check for -0.314 is left to the student.

For 1.914:

$$\begin{array}{c|c} 5x^2 - 8x = 3 \\ \hline 5(1.914)^2 - 8(1.914) \;?\; 3 \\ 5(3.663396) - 15.312 \\ 3.00498 \end{array}$$

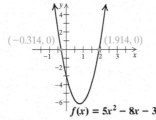

We do not have a perfect check due to the rounding error, but our check seems to confirm the solutions.

Do Exercise 2.

Some quadratic equations have solutions that are nonreal complex numbers.

EXAMPLE 3 Solve: $x^2 + x + 1 = 0$.

We have $a = 1, b = 1, c = 1$. We use the quadratic formula:

$$x = \frac{-1 \pm \sqrt{1^2 - 4 \cdot 1 \cdot 1}}{2 \cdot 1}$$

$$= \frac{-1 \pm \sqrt{1 - 4}}{2}$$

$$= \frac{-1 \pm \sqrt{-3}}{2}$$

$$= \frac{-1 \pm \sqrt{3}i}{2}.$$

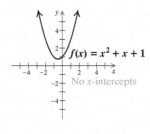

The solutions are

$$\frac{-1 + \sqrt{3}i}{2} \text{ and } \frac{-1 - \sqrt{3}i}{2}, \text{ or } -\frac{1}{2} + \frac{\sqrt{3}}{2}i \text{ and } -\frac{1}{2} - \frac{\sqrt{3}}{2}i.$$

Do Exercise 3.

EXAMPLE 4 Solve: $2 + \frac{7}{x} = \frac{5}{x^2}$. Give the exact solutions and approximate solutions to three decimal places.

We first find an equivalent quadratic equation in standard form:

$$x^2\left(2 + \frac{7}{x}\right) = x^2 \cdot \frac{5}{x^2} \quad \text{Multiplying by } x^2 \text{ to clear fractions, noting that } x \neq 0$$

$$2x^2 + 7x = 5$$

$$2x^2 + 7x - 5 = 0. \quad \text{Subtracting 5}$$

---

 2. Solve using the quadratic formula:

$$3x^2 + 2x = 7.$$

Give the exact solutions and approximate solutions to three decimal places. Write the equation in standard form.

$$3x^2 + 2x - 7 = \boxed{}$$

$$a = \boxed{}, b = \boxed{}, c = \boxed{};$$

$$x = \frac{-\boxed{} \pm \sqrt{2^2 - 4 \cdot 3 \cdot (\boxed{})}}{2 \cdot 3}$$

$$= \frac{-2 \pm \sqrt{\boxed{}}}{6}$$

$$= \frac{-2 \pm 2\sqrt{\boxed{}}}{6}$$

$$= \frac{2(-1 \pm \sqrt{\boxed{}})}{2 \cdot 3}$$

$$= \frac{-1 \pm \sqrt{22}}{\boxed{}}$$

Approximate the solutions and round to three decimal places.

$$\frac{-1 + \sqrt{22}}{3} \approx \boxed{}$$

$$\frac{-1 - \sqrt{22}}{3} \approx \boxed{}$$

3. Solve: $x^2 - x + 2 = 0$.

Answers

2. $\dfrac{-1 \pm \sqrt{22}}{3}$; 1.230, -1.897

3. $\dfrac{1 \pm \sqrt{7}i}{2}$, or $\dfrac{1}{2} \pm \dfrac{\sqrt{7}}{2}i$

Guided Solution:
2. 0, 3, 2, -7, 2, -7, 88, 22, 22, 3, 1.230, -1.897

Then

$$a = 2, \quad b = 7, \quad c = -5;$$

$$x = \frac{-7 \pm \sqrt{7^2 - 4 \cdot 2 \cdot (-5)}}{2 \cdot 2} \quad \text{Substituting}$$

$$x = \frac{-7 \pm \sqrt{49 + 40}}{4}$$

$$x = \frac{-7 \pm \sqrt{89}}{4}$$

$$x = \frac{-7 + \sqrt{89}}{4} \quad \text{or} \quad x = \frac{-7 - \sqrt{89}}{4}.$$

Since we began with a rational equation, we need to check. We cleared the fractions before obtaining a quadratic equation in standard form, and this step could introduce numbers that do not check in the original rational equation. We need to show that neither of the numbers makes a denominator 0. Since neither of them does, the solutions are

$$\frac{-7 + \sqrt{89}}{4} \quad \text{and} \quad \frac{-7 - \sqrt{89}}{4}.$$

We can use a calculator to approximate the solutions:

$$\frac{-7 + \sqrt{89}}{4} \approx 0.608;$$

$$\frac{-7 - \sqrt{89}}{4} \approx -4.108.$$

◀ Do Exercise 4.

4. Solve:

$$3 = \frac{5}{x} + \frac{4}{x^2}.$$

Give the exact solutions and approximate solutions to three decimal places.

Answer

4. $\frac{5 \pm \sqrt{73}}{6}$; 2.257, −0.591

11.2 Exercise Set

FOR EXTRA HELP — MyLab Math

✓ Check Your Understanding

Reading Check Determine whether each statement is true or false.

RC1. The quadratic formula can be used to solve any quadratic equation.

RC2. The quadratic formula does not work if the solutions are imaginary numbers.

RC3. Using the quadratic formula is always the fastest way to solve a quadratic equation.

RC4. In the quadratic formula, if $b^2 - 4ac$ is positive but not a perfect square, the solutions will be irrational.

Concept Check Complete each statement with the correct number or expression.

CC1. When we are using the quadratic formula to solve $3x^2 - x - 8 = 0$, the value of a is _____.

CC2. When we are using the quadratic formula to solve $3x^2 - x - 8 = 0$, the value of b is _____.

CC3. Standard form for the quadratic equation $5x^2 = 9 - x$ is _____ = 0.

CC4. When we are using the quadratic formula to solve $3x^2 = 10x$, the value of c is _____.

a Solve.

1. $x^2 + 8x + 2 = 0$
2. $x^2 - 6x - 4 = 0$
3. $3p^2 = -8p - 1$
4. $3u^2 = 18u - 6$

5. $x^2 - x + 1 = 0$
6. $x^2 + x + 2 = 0$
7. $x^2 + 13 = 4x$
8. $x^2 + 13 = 6x$

9. $r^2 + 3r = 8$
10. $h^2 + 4 = 6h$
11. $1 + \dfrac{2}{x} + \dfrac{5}{x^2} = 0$
12. $1 + \dfrac{5}{x^2} = \dfrac{2}{x}$

13. **a)** Solve: $3x + x(x - 2) = 0$.
    b) Find the x-intercepts of $f(x) = 3x + x(x - 2)$.

14. **a)** Solve: $4x + x(x - 3) = 0$.
    b) Find the x-intercepts of $f(x) = 4x + x(x - 3)$.

15. **a)** Solve: $11x^2 - 3x - 5 = 0$.
    b) Find the x-intercepts of $f(x) = 11x^2 - 3x - 5$.

16. **a)** Solve: $7x^2 + 8x = -2$.
    b) Find the x-intercepts of $f(x) = 7x^2 + 8x + 2$.

17. **a)** Solve: $25x^2 = 20x - 4$.
    b) Find the x-intercepts of $f(x) = 25x^2 - 20x + 4$.

18. **a)** Solve: $49x^2 - 14x + 1 = 0$.
    b) Find the x-intercepts of $f(x) = 49x^2 - 14x + 1$.

Solve.

19. $4x(x - 2) - 5x(x - 1) = 2$
20. $3x(x + 1) - 7x(x + 2) = 6$

21. $14(x - 4) - (x + 2) = (x + 2)(x - 4)$
22. $11(x - 2) + (x - 5) = (x + 2)(x - 6)$

23. $5x^2 = 17x - 2$
24. $15x = 2x^2 + 16$
25. $x^2 + 5 = 4x$
26. $x^2 + 5 = 2x$

27. $x + \dfrac{1}{x} = \dfrac{13}{6}$
28. $\dfrac{3}{x} + \dfrac{x}{3} = \dfrac{5}{2}$
29. $\dfrac{1}{y} + \dfrac{1}{y + 2} = \dfrac{1}{3}$
30. $\dfrac{1}{x} + \dfrac{1}{x + 4} = \dfrac{1}{7}$

31. $(2t - 3)^2 + 17t = 15$

32. $2y^2 - (y + 2)(y - 3) = 12$

33. $(x - 2)^2 + (x + 1)^2 = 0$

34. $(x + 3)^2 + (x - 1)^2 = 0$

35. $x^3 - 1 = 0$
(*Hint:* Factor the difference of cubes. Then use the quadratic formula.)

36. $x^3 + 27 = 0$

Solve. Give the exact solutions and approximate solutions to three decimal places.

37. $x^2 + 6x + 4 = 0$

38. $x^2 + 4x - 7 = 0$

39. $x^2 - 6x + 4 = 0$

40. $x^2 - 4x + 1 = 0$

41. $2x^2 - 3x - 7 = 0$

42. $3x^2 - 3x - 2 = 0$

43. $5x^2 = 3 + 8x$

44. $2y^2 + 2y - 3 = 0$

Skill Maintenance

Solve.

45. $x = \sqrt{x + 2}$ [10.6a]

46. $\sqrt{x + 1} + 2 = \sqrt{3x + 1}$ [10.6b]

47. $\sqrt{2x - 6} + 11 = 2$ [10.6a]

48. $\sqrt[3]{4x - 7} = 2$ [10.6a]

49. $2x^2 = x + 3$ [5.8b]

50. $100x^2 + 1 = 20x$ [5.8b]

51. $\dfrac{3}{x} - \dfrac{1}{4} = \dfrac{7}{2x}$ [6.7a]

52. $\dfrac{3}{x - 2} = \dfrac{5}{6x}$ [6.7a]

Synthesis

53. Use a graphing calculator to solve the equations in Exercises 3, 16, 17, and 43 using the INTERSECT feature, letting $y_1 = $ the left side and $y_2 = $ the right side. Then solve $2.2x^2 + 0.5x - 1 = 0$.

54. Use a graphing calculator to solve the equations in Exercises 9, 27, and 30. Then solve $5.33x^2 = 8.23x + 3.24$.

Solve.

55. $2x^2 - x - \sqrt{5} = 0$

56. $\dfrac{5}{x} + \dfrac{x}{4} = \dfrac{11}{7}$

57. $ix^2 - x - 1 = 0$

58. $\sqrt{3}x^2 + 6x + \sqrt{3} = 0$

59. $\dfrac{x}{x + 1} = 4 + \dfrac{1}{3x^2 - 3}$

60. $(1 + \sqrt{3})x^2 - (3 + 2\sqrt{3})x + 3 = 0$

61. Let $f(x) = (x - 3)^2$. Find all inputs x such that $f(x) = 13$.

62. Let $f(x) = x^2 + 14x + 49$. Find all inputs x such that $f(x) = 36$.

Applications Involving Quadratic Equations

11.3

OBJECTIVES

a Solve applied problems involving quadratic equations.

b Solve a formula for a given letter.

a APPLICATIONS AND PROBLEM SOLVING

Sometimes when we translate a problem to mathematical language, the result is a quadratic equation.

EXAMPLE 1 *Beach Volleyball.* The beach volleyball court at Lake Jean State Park measures 24 m by 16 m. The playing area is surrounded by a free zone of uniform width. The area of the playing area is one-third of the area of the entire court. How wide is the free zone?

1. **Familiarize.** We let $x =$ the width of the free zone and make a drawing.

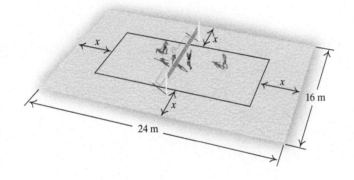

2. **Translate.** The area of a rectangle is lw (length times width). Then:

   Area of entire court $= 24 \cdot 16$;
   Area of playing area $= (24 - 2x)(16 - 2x)$.

   Since the playing area is one-third of the area of the court, we have

   $$(24 - 2x)(16 - 2x) = \frac{1}{3} \cdot 24 \cdot 16.$$

3. **Solve.** We solve the equation:

$384 - 80x + 4x^2 = 128$	Using FOIL on the left
$4x^2 - 80x + 256 = 0$	Finding standard form
$x^2 - 20x + 64 = 0$	Dividing by 4
$(x - 4)(x - 16) = 0$	Factoring
$x = 4$ or $x = 16$.	Using the principle of zero products

4. **Check.** We check in the original problem. We see that 16 is not a solution because a 24-m by 16-m court cannot have a 16-m free zone.
   If the free zone is 4 m wide, then the playing area will have length $24 - 2 \cdot 4$, or 16 m. The width will be $16 - 2 \cdot 4$, or 8 m. The area of the playing area is thus $16 \cdot 8$, or 128 m². The area of the entire court is $24 \cdot 16$, or 384 m². The area of the playing area is one-third of 384 m², so the number 4 checks.

5. **State.** The free zone is 4 m wide.

Do Exercise 1.

1. *Landscaping.* A rectangular garden is 60 ft by 80 ft. Part of the garden is torn up to install a sidewalk of uniform width around it. The area of the new garden is one-half of the old area. How wide is the sidewalk?

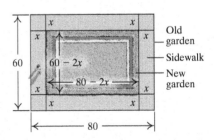

Answer
1. 10 ft

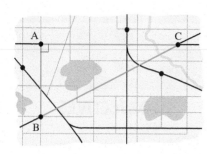

EXAMPLE 2 *Town Planning.* Three towns A, B, and C are situated as shown in the figure at left. The roads at A form a right angle. The distance from A to B is 2 mi less than the distance from A to C. The distance from B to C is 10 mi. Find the distance from A to B and the distance from A to C.

1. **Familiarize.** We first make a drawing and label it. We let $d=$ the distance from A to C. Then the distance from A to B is $d - 2$.

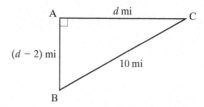

2. **Ladder Location.** A ladder leans against a building, as shown below. The ladder is 20 ft long. The distance to the top of the ladder is 4 ft greater than the distance d from the building. Find the distance d and the distance to the top of the ladder.

2. **Translate.** We see that a right triangle is formed. We can use the Pythagorean equation, $c^2 = a^2 + b^2$:

$$10^2 = d^2 + (d-2)^2.$$

3. **Solve.** We solve the equation:

$$10^2 = d^2 + (d-2)^2$$
$$100 = d^2 + d^2 - 4d + 4 \quad \text{Squaring}$$
$$2d^2 - 4d - 96 = 0 \quad \text{Finding standard form}$$
$$d^2 - 2d - 48 = 0 \quad \text{Dividing by 2}$$
$$(d-8)(d+6) = 0 \quad \text{Factoring}$$
$$d = 8 \quad \text{or} \quad d = -6. \quad \text{Using the principle of zero products}$$

4. **Check.** We know that -6 cannot be a solution because distances are not negative. If $d = 8$, then $d - 2 = 6$, and $8^2 + 6^2 = 64 + 36 = 100$. Since $10^2 = 100$, the distance 8 mi checks.

5. **State.** The distance from A to C is 8 mi, and the distance from A to B is 6 mi.

◀ Do Exercise 2.

EXAMPLE 3 *Landscape Design.* Melanie plans to build a fire pit in her backyard at a safe distance from both her gardening shed and her daughter's playset. The three structures will form a right triangle, with the distance of the fire pit from the gardening shed 2 m less than the distance from the fire pit to the playset. The distance from the playset to the gardening shed is 20 m and forms the longest side of the triangle. How far is the fire pit from the gardening shed and from the playset?

1. **Familiarize.** We make a drawing and label it. We let $d =$ the distance from the fire pit to the playset. Then the distance from the fire pit to the gardening shed is $d - 2$.

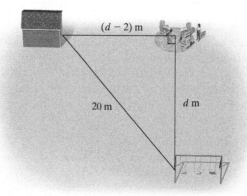

Answer
2. The distance d is 12 ft; the distance to the top of the ladder is 16 ft.

2. **Translate.** We use the Pythagorean equation as in Example 2:
$$20^2 = d^2 + (d-2)^2.$$

3. **Solve.** We solve the equation using the quadratic formula:

$400 = d^2 + d^2 - 4d + 4$ Squaring
$0 = 2d^2 - 4d - 396$ Finding standard form
$0 = d^2 - 2d - 198.$ Dividing by 2 (or multiplying by $\frac{1}{2}$)

Then

$$d = \frac{-b \pm \sqrt{b^2 - 4ac}}{2a}$$

$$= \frac{-(-2) \pm \sqrt{(-2)^2 - 4(1)(-198)}}{2(1)} \qquad a = 1, b = -2, c = -198$$

$$= \frac{2 \pm \sqrt{796}}{2} = \frac{2 \pm \sqrt{4 \cdot 199}}{2} = \frac{2 \pm 2\sqrt{199}}{2} = 1 \pm \sqrt{199}.$$

4. **Check.** Since $1 - \sqrt{199}$ is negative, it cannot be the distance between the structures. Using a calculator, we find that $1 + \sqrt{199} \approx 15.1$. Thus we have $d \approx 15.1$ and $d - 2 \approx 13.1$. The square of the hypotenuse in the triangle is 20^2, or 400. Since $(15.1)^2 + (13.1)^2 = 399.62 \approx 400$, the numbers check.

5. **State.** The fire pit is 13.1 m from the gardening shed and 15.1 m from the playset.

Do Exercise 3.

3. *Ladder Location.* Refer to Margin Exercise 2. Suppose that the ladder has length 10 ft. Find the distance d and the distance $d + 4$.

EXAMPLE 4 *Motorcycle Travel.* Karin's motorcycle traveled 300 mi at a certain speed. Had she gone 10 mph faster, she could have made the trip in 1 hr less time. Find her speed.

1. **Familiarize.** We make a drawing, labeling it with known and unknown information, and organize the information in a table. We let $r =$ the speed, in miles per hour, and $t =$ the time, in hours.

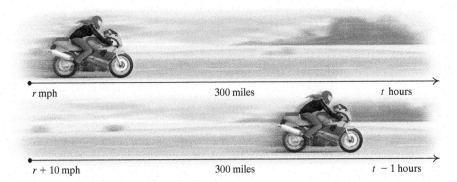

r mph 300 miles t hours

$r + 10$ mph 300 miles $t - 1$ hours

DISTANCE	SPEED	TIME
300	r	t
300	$r + 10$	$t - 1$

$\longrightarrow r = \dfrac{300}{t}$

$\longrightarrow r + 10 = \dfrac{300}{t-1}$

Recalling the motion formula $d = rt$ and solving for r, we get $r = d/t$. From the rows of the table, we obtain

$$r = \frac{300}{t} \quad \text{and} \quad r + 10 = \frac{300}{t-1}.$$

Answer

3. The distance d is about 4.8 ft; the distance to the top of the ladder is about 8.8 ft.

4. *Marine Travel.* Two ships make the same voyage of 3000 nautical miles. The faster ship travels 10 knots faster than the slower one. (A *knot* is 1 nautical mile per hour.) The faster ship makes the voyage in 50 hr less time than the slower one. Find the speeds of the two ships.

Complete this table to help with the familiarization.

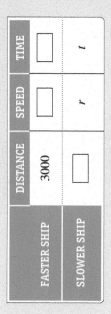

2. **Translate.** We substitute for r from the first equation into the second and get a translation:
$$\frac{300}{t} + 10 = \frac{300}{t-1}.$$

3. **Solve.** We solve as follows:
$$\frac{300}{t} + 10 = \frac{300}{t-1}$$
$$t(t-1)\left[\frac{300}{t} + 10\right] = t(t-1) \cdot \frac{300}{t-1} \quad \text{Multiplying by the LCM}$$
$$t(t-1) \cdot \frac{300}{t} + t(t-1) \cdot 10 = t(t-1) \cdot \frac{300}{t-1}$$
$$300(t-1) + 10(t^2 - t) = 300t$$
$$300t - 300 + 10t^2 - 10t = 300t$$
$$10t^2 - 10t - 300 = 0 \quad \text{Standard form}$$
$$t^2 - t - 30 = 0 \quad \text{Dividing by 10}$$
$$(t-6)(t+5) = 0 \quad \text{Factoring}$$
$$t = 6 \quad \text{or} \quad t = -5. \quad \text{Using the principle of zero products}$$

4. **Check.** Since negative time has no meaning in this problem, we try 6 hr. Remembering that $r = d/t$, we get $r = 300/6 = 50$ mph.
   To check, we take the speed 10 mph faster, which is 60 mph, and see how long the trip would have taken at that speed:
$$t = \frac{d}{r} = \frac{300}{60} = 5 \text{ hr.}$$
This is 1 hr less than the trip actually took, so we have an answer.

5. **State.** Karin's speed was 50 mph.

◀ Do Exercise 4.

b SOLVING FORMULAS

Recall that to solve a formula for a certain letter, we use the principles for solving equations to get that letter alone on one side.

EXAMPLE 5 *Period of a Pendulum.* The time T required for a pendulum of length L to swing back and forth (complete one period) is given by the formula $T = 2\pi\sqrt{L/g}$, where g is the gravitational constant. Solve for L.

$$T = 2\pi\sqrt{\frac{L}{g}} \quad \text{This is a radical equation.}$$
$$T^2 = \left(2\pi\sqrt{\frac{L}{g}}\right)^2 \quad \text{Principle of powers (squaring)}$$
$$T^2 = 2^2\pi^2\frac{L}{g}$$
$$gT^2 = 4\pi^2 L \quad \text{Clearing fractions and simplifying}$$
$$\frac{gT^2}{4\pi^2} = L \quad \text{Multiplying by } \frac{1}{4\pi^2}$$

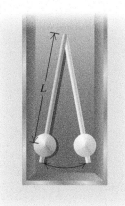

Answer
4. 20 knots, 30 knots
Guided Solution:
4.

	Distance	Speed	Time
Faster ship	3000	$r + 10$	$t - 50$
Slower ship	3000	r	t

We now have L alone on one side and L does not appear on the other side, so the formula is solved for L.

Do Exercise 5. ▶

In most formulas, variables represent nonnegative numbers, so we need only the positive root when taking square roots.

EXAMPLE 6 *Hang Time.* An athlete's *hang time* is the amount of time that the athlete can remain airborne when jumping. A formula relating an athlete's vertical leap V, in inches, to hang time T, in seconds, is $V = 48T^2$. Solve for T.

We have

$$48T^2 = V$$

$$T^2 = \frac{V}{48} \quad \text{Multiplying by } \tfrac{1}{48} \text{ to get } T^2 \text{ alone}$$

$$T = \sqrt{\frac{V}{48}} \quad \text{Using the principle of square roots; note that } T \geq 0.$$

$$= \sqrt{\frac{V}{2 \cdot 2 \cdot 2 \cdot 2 \cdot 3} \cdot \frac{3}{3}}$$

$$= \frac{\sqrt{3V}}{2 \cdot 2 \cdot 3}$$

$$= \frac{\sqrt{3V}}{12}.$$

Do Exercise 6. ▶

EXAMPLE 7 *Falling Distance.* An object that is tossed downward with an initial speed (velocity) of v_0 will travel a distance of s meters, where $s = 4.9t^2 + v_0 t$ and t is measured in seconds. Solve for t.

5. Solve $A = \sqrt{\dfrac{w_1}{w_2}}$ for w_2.

GS 6. Solve $V = \pi r^2 h$ for r. (Volume of a right circular cylinder)

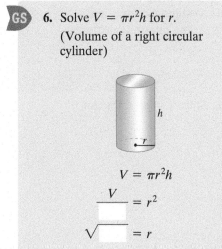

$$V = \pi r^2 h$$

$$\frac{V}{\boxed{}} = r^2$$

$$\sqrt{\boxed{}} = r$$

Answers

5. $w_2 = \dfrac{w_1}{A^2}$ 6. $r = \sqrt{\dfrac{V}{\pi h}}$

Guided Solution:

6. $\pi h, \dfrac{V}{\pi h}$

To solve a formula for a letter, say, t:

1. Clear the fractions and use the principle of powers, as needed, until t does not appear in any radicand or denominator. (In some cases, you may clear the fractions first, and in some cases, you may use the principle of powers first.)
2. Collect all terms with t^2 in them. Also collect all terms with t in them.
3. If t^2 does not appear, you can finish by using just the addition and multiplication principles.
4. If t^2 appears but t does not, solve the equation for t^2. Then take square roots on both sides.
5. If there are terms containing both t and t^2, write the equation in standard form and use the quadratic formula.

Since t is squared in one term and raised to the first power in the other term, the equation is quadratic in t. The variable is t; v_0 and s are treated as constants.

We have

$$4.9t^2 + v_0 t = s$$
$$4.9t^2 + v_0 t - s = 0 \quad \text{Writing standard form}$$
$$a = 4.9, \quad b = v_0, \quad c = -s$$

$$t = \frac{-v_0 \pm \sqrt{(v_0)^2 - 4(4.9)(-s)}}{2(4.9)} \quad \text{Using the quadratic formula: } t = \frac{-b \pm \sqrt{b^2 - 4ac}}{2a}$$

$$= \frac{-v_0 \pm \sqrt{(v_0)^2 + 19.6s}}{9.8}.$$

Since the negative square root would yield a negative value for t, we use only the positive root:

$$t = \frac{-v_0 + \sqrt{(v_0)^2 + 19.6s}}{9.8}.$$

◀ Do Exercise 7.

The steps listed in the margin should help you when solving formulas for a given letter. Try to remember that, when solving a formula, you do the same things you would do to solve an equation.

EXAMPLE 8 Solve $t = \dfrac{a}{\sqrt{a^2 + b^2}}$ for a.

In this case, we could either clear the fractions first or use the principle of powers first. Let's clear the fractions. Multiplying by $\sqrt{a^2 + b^2}$, we have

$$t\sqrt{a^2 + b^2} = a.$$

Now we square both sides and then continue:

$$(t\sqrt{a^2+b^2})^2 = a^2 \quad \text{Squaring}$$

Caution!
Don't forget to square both t and $\sqrt{a^2+b^2}$.

$$t^2(\sqrt{a^2+b^2})^2 = a^2$$
$$t^2(a^2+b^2) = a^2$$
$$t^2 a^2 + t^2 b^2 = a^2$$
$$t^2 b^2 = a^2 - t^2 a^2 \quad \text{Getting all } a^2\text{-terms together}$$
$$t^2 b^2 = a^2(1 - t^2) \quad \text{Factoring out } a^2$$
$$\frac{t^2 b^2}{1 - t^2} = a^2 \quad \text{Dividing by } 1 - t^2$$
$$\sqrt{\frac{t^2 b^2}{1 - t^2}} = a \quad \text{Taking the square root}$$
$$\frac{tb}{\sqrt{1 - t^2}} = a. \quad \text{Simplifying}$$

You need not rationalize denominators in situations such as this.

◀ Do Exercise 8.

7. Solve $s = gt + 16t^2$ for t.

8. Solve $\dfrac{b}{\sqrt{a^2 - b^2}} = t$ for b.

Answers

7. $t = \dfrac{-g + \sqrt{g^2 + 64s}}{32}$

8. $b = \dfrac{ta}{\sqrt{1 + t^2}}$

Translating for Success

Translate each word problem to an equation or a system of equations and select a correct translation from equations A–O.

A. $(80 - 2x)(100 - 2x) = \frac{1}{3} \cdot 80 \cdot 100$

B. $\dfrac{800}{x} + 10 = \dfrac{800}{x-2}$

C. $x + 18\% \cdot x = 3.24$

D. $x + 25y = 26.95,$
$x + y = 117$

E. $2x + 2(x - 7) = 537$

F. $x + (x - 7) + \frac{1}{2}x = 537$

G. $0.10x + 0.25y = 26.95,$
$x + y = 117$

H. $3.24 - 18\% \cdot 3.24 = x$

I. $\dfrac{4}{x+2} + \dfrac{4}{x-2} = 3$

J. $x^2 + (x+1)^2 = 7^2$

K. $75^2 + x^2 = 78^2$

L. $\dfrac{3}{x+2} + \dfrac{3}{x-2} = 4$

M. $75^2 + 78^2 = x^2$

N. $x + (x+1) + (x+2) = 537$

O. $\dfrac{800}{x} + \dfrac{800}{x-2} = 10$

Answers on page A-35

1. *Car Travel.* Sarah drove her car 800 mi to see her friend. The return trip was 2 hr faster at a speed that was 10 mph more. Find her return speed.

2. *Coin Mixture.* A collection of dimes and quarters is worth $26.95. There are 117 coins in all. How many of each coin are there?

3. *Wire Cutting.* A 537-in. wire is cut into three pieces. The second piece is 7 in. shorter than the first. The third is half as long as the first. How long is each piece?

4. *Marine Travel.* The Columbia River flows at a rate of 2 mph for the length of a popular boating route. In order for a motorized dinghy to travel 3 mi upriver and return in a total of 4 hr, how fast must the boat be able to travel in still water?

5. *Locker Numbers.* The numbers on three adjoining lockers are consecutive integers whose sum is 537. Find the integers.

6. *Gasoline Prices.* One day the price of gasoline was increased 18% to a new price of $3.24 per gallon. What was the original price?

7. *Triangle Dimensions.* The hypotenuse of a right triangle is 7 ft. The length of one leg is 1 ft longer than the other. Find the lengths of the legs.

8. *Rectangle Dimensions.* The perimeter of a rectangle is 537 ft. The width of the rectangle is 7 ft shorter than the length. Find the length and the width.

9. *Guy Wire.* A guy wire is 78 ft long. It is attached to the top of a 75-ft cell-phone tower. How far is it from the base of the pole to the point where the wire is attached to the ground?

10. *Landscaping.* A rectangular garden is 80 ft by 100 ft. Part of the garden is torn up to install a sidewalk of uniform width around it. The area of the new garden is $\frac{1}{3}$ of the old area. How wide is the sidewalk?

11.3 Exercise Set

✓ Check Your Understanding

Reading and Concept Check Match each formula with the appropriate description from the column on the right.

RC1. _____ $s = 4.9t^2 + v_0 t$

RC2. _____ $V = 48T^2$

RC3. _____ $a^2 + b^2 = c^2$

RC4. _____ $A = lw$

RC5. _____ $T = 2\pi \sqrt{\dfrac{l}{g}}$

RC6. _____ $t = \dfrac{d}{r}$

a) Area of a rectangle
b) Pythagorean theorem
c) Motion
d) Period of a pendulum
e) Vertical leap
f) Distance

a Solve.

1. *Flower Bed.* The width of a rectangular flower bed is 7 ft less than the length. The area is 18 ft². Find the length and the width.

2. *Feed Lot.* The width of a rectangular feed lot is 8 m less than the length. The area is 20 m². Find the length and the width.

3. *Parking Lot.* The length of a rectangular parking lot is twice the width. The area is 162 yd². Find the length and the width.

4. *Flag Dimensions.* The length of an American flag that is displayed at a government office is 3 in. less than twice its width. The area is 1710 in². Find the length and the width of the flag.

5. *Easter Island.* Easter Island is roughly triangular in shape. The height of the triangle is 7 mi less than the base. The area is 60 mi². Find the base and the height of the triangular-shaped island.

6. *Sailing.* The base of a triangular sail is 9 m less than the height. The area is 56 m². Find the base and the height of the sail.

Area = 56 m²

826 CHAPTER 11 Quadratic Equations and Functions

7. *Parking Lot.* The width of a rectangular parking lot is 51 ft less than its length. Determine the dimensions of the parking lot if it measures 250 ft diagonally.

8. *Sailing.* The base of a triangular sail is 8 ft less than the height. The area is 56 ft^2. Find the base and the height of the sail.

9. *Mirror Framing.* The outside of a mosaic mirror frame measures 14 in. by 20 in., and 160 in^2 of mirror shows. Find the width of the frame.

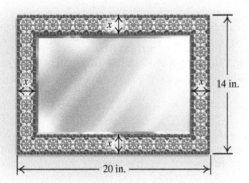

10. *Quilt Dimensions.* Michelle is making a quilt for a wall hanging at the entrance to a state museum. The finished quilt will measure 8 ft by 6 ft. The quilt has a border of uniform width around it. The area of the interior rectangular section is one-half of the area of the entire quilt. How wide is the border?

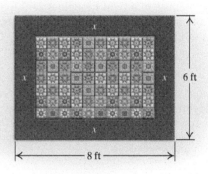

11. *Landscaping.* A landscaper is designing a flower garden in the shape of a right triangle. She wants 10 ft of a perennial border to form the hypotenuse of the triangle, and one leg is to be 2 ft longer than the other. Find the lengths of the legs.

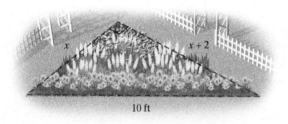

12. *Flag Dimensions.* The diagonal of a Papua New Guinea flag displayed in a school is 60 in. The length of the flag is 12 in. longer than the width. Find the dimensions of the flag.

13. *Raffle Tickets.* Margaret and Zane purchased consecutively numbered raffle tickets at a charity auction. The product of the ticket numbers was 552. Find the ticket numbers.

14. *Box Construction.* An open box is to be made from a 10-ft by 20-ft rectangular piece of cardboard by cutting a square from each corner. The area of the bottom of the box is to be 96 ft^2. What is the length of the sides of the squares that are cut from the corners?

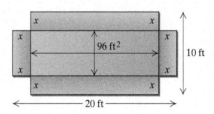

Solve. Find exact answers and approximate answers rounded to three decimal places.

15. The width of a rectangle is 4 ft less than the length. The area is 10 ft². Find the length and the width.

16. The length of a rectangle is twice the width. The area is 328 cm². Find the length and the width.

17. *Page Dimensions.* The outside of an oversized book page measures 14 in. by 20 in., and 100 in² of printed text shows. Find the width of the margin.

18. *Picture Framing.* The outside of a picture frame measures 13 cm by 20 cm, and 80 cm² of picture shows. Find the width of the frame.

19. The hypotenuse of a right triangle is 24 ft long. The length of one leg is 14 ft more than the other. Find the lengths of the legs.

20. The hypotenuse of a right triangle is 22 m long. The length of one leg is 10 m less than the other. Find the lengths of the legs.

21. *Car Trips.* During the first part of a trip, Sam's Toyota Prius Hybrid traveled 120 mi. Sam then drove another 100 mi at a speed that was 10 mph slower. If the total time for Sam's trip was 4 hr, what was his speed on each part of the trip?

DISTANCE	SPEED	TIME

22. *Canoeing.* During the first part of a canoe trip, Doug covered 60 km. He then traveled 24 km at a speed that was 4 km/h slower. If the total time for Doug's trip was 8 hr, what was his speed on each part of the trip?

DISTANCE	SPEED	TIME

23. *Skiing.* Kingdom Trails is a nonprofit conservation organization in Vermont, working with private landowners to manage outdoor recreation opportunities and to preserve and protect trails. In January, Colleen skied 24 km along part of the trails. If she had gone 2 km/h faster, the trip would have taken 1 hr less. Find Colleen's speed.

24. *Bicycling.* In July, Art bicycled 120 mi in Vermont. If he had gone 5 mph faster, the trip would have taken 4 hr less. Find Art's speed.

25. *Air Travel.* A Cessna flies 600 mi. A Beechcraft flies 1000 mi at a speed that is 50 mph faster, but takes 1 hr longer. Find the speed of each plane.

26. *Air Travel.* A turbo-jet flies 50 mph faster than a super-prop plane. If a turbo-jet goes 2000 mi in 3 hr less time than it takes the super-prop to go 2800 mi, find the speed of each plane.

27. *Bicycling.* Naoki bikes 40 mi to Hillsboro. The return trip is made at a speed that is 6 mph slower. Total travel time for the round trip is 14 hr. Find Naoki's speed on each part of the trip.

28. *Car Speed.* On a sales trip, Gail drives 600 mi to Richmond. The return trip is made at a speed that is 10 mph slower. Total travel time for the round trip is 22 hr. How fast did Gail travel on each part of the trip?

29. *Navigation.* The current in a typical Mississippi River shipping route flows at a rate of 4 mph. In order for a barge to travel 24 mi upriver and then return in a total of 5 hr, approximately how fast must the barge be able to travel in still water?

30. *Navigation.* The Hudson River flows at a rate of 3 mph. A patrol boat travels 60 mi upriver and returns in a total time of 9 hr. What is the speed of the boat in still water?

b Solve each formula for the given letter. Assume that all variables represent nonnegative numbers.

31. $A = 6s^2$, for s
(Surface area of a cube)

32. $A = 4\pi r^2$, for r
(Surface area of a sphere)

33. $F = \dfrac{Gm_1 m_2}{r^2}$, for r

34. $N = \dfrac{kQ_1 Q_2}{s^2}$, for s
(Number of phone calls between two cities)

35. $E = mc^2$, for c
(Einstein's energy–mass relationship)

36. $V = \frac{1}{3} s^2 h$, for s
(Volume of a pyramid)

37. $a^2 + b^2 = c^2$, for b
(Pythagorean formula in two dimensions)

38. $a^2 + b^2 + c^2 = d^2$, for c
(Pythagorean formula in three dimensions)

39. $N = \dfrac{k^2 - 3k}{2}$, for k
(Number of diagonals of a polygon of k sides)

40. $s = v_0 t + \dfrac{gt^2}{2}$, for t
(A motion formula)

41. $A = 2\pi r^2 + 2\pi rh$, for r
(Surface area of a cylinder)

42. $A = \pi r^2 + \pi rs$, for r
(Surface area of a cone)

43. $T = 2\pi\sqrt{\dfrac{L}{g}}$, for g
(A pendulum formula)

44. $W = \sqrt{\dfrac{1}{LC}}$, for L
(An electricity formula)

45. $I = \dfrac{703W}{H^2}$, for H
(Body mass index)

46. $N + p = \dfrac{6.2A^2}{pR^2}$, for R

47. $m = \dfrac{m_0}{\sqrt{1 - \dfrac{v^2}{c^2}}}$, for v
(A relativity formula)

48. Solve the formula given in Exercise 47 for c.

Skill Maintenance

Add or subtract.

49. $\dfrac{1}{x-1} + \dfrac{1}{x^2 - 3x + 2}$ [6.4a]

50. $\dfrac{x+1}{x-1} - \dfrac{x+1}{x^2 + x + 1}$ [6.5a]

51. $\dfrac{2}{x+3} - \dfrac{x}{x-1} + \dfrac{x^2+2}{x^2+2x-3}$ [6.5b]

52. Multiply and simplify: $\sqrt{3x^2}\sqrt{3x^3}$. [10.3a]

53. Express in terms of i: $\sqrt{-20}$. [10.8a]

Synthesis

54. Solve: $\dfrac{4}{2x+i} - \dfrac{1}{x-i} = \dfrac{2}{x+i}$.

55. Find a when the reciprocal of $a - 1$ is $a + 1$.

56. *Pizza Crusts.* At Pizza Perfect, Ron can make 100 large pizza crusts in 1.2 hr less than Chad. Together they can do the job in 1.8 hr. How long does it take each to do the job alone?

57. *Surface Area.* A sphere is inscribed in a cube as shown in the following figure. Express the surface area of the sphere as a function of the surface area S of the cube.

58. *Bungee Jumping.* Jesse is tied to one end of a 40-m elasticized (bungee) cord. The other end of the cord is tied to the middle of a train trestle. If Jesse steps off the bridge, for how long will he fall before the cord begins to stretch? (See Example 7 and let $v_0 = 0$.)

59. *The Golden Rectangle.* For over 2000 years, the proportions of a "golden" rectangle have been considered visually appealing. A rectangle of width w and length l is considered "golden" if

$$\dfrac{w}{l} = \dfrac{l}{w + l}.$$

Solve for l.

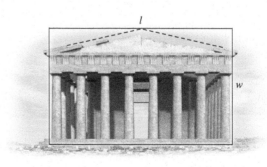

More on Quadratic Equations

11.4

a THE DISCRIMINANT

From the quadratic formula, we know that the solutions x_1 and x_2 of a quadratic equation are given by

$$x_1 = \frac{-b + \sqrt{b^2 - 4ac}}{2a} \quad \text{and} \quad x_2 = \frac{-b - \sqrt{b^2 - 4ac}}{2a}.$$

The expression $b^2 - 4ac$ is called the **discriminant**. When we are using the quadratic formula, it is helpful to compute the discriminant first. If it is 0, there will be just one real solution. If it is positive, there will be two real solutions. If it is negative, we will be taking the square root of a negative number; hence there will be two nonreal complex-number solutions, and they will be complex conjugates.

OBJECTIVES

a Determine the nature of the solutions of a quadratic equation.

b Write a quadratic equation having two given numbers as solutions.

c Solve equations that are quadratic in form.

DISCRIMINANT $b^2 - 4ac$	NATURE OF SOLUTIONS	x-INTERCEPTS
0	Only one solution; it is a real number	Only one
Positive	Two different real-number solutions	Two different
Negative	Two different nonreal complex-number solutions (complex conjugates)	None

If the discriminant is a perfect square, we can solve the equation by factoring and do not need the quadratic formula.

EXAMPLE 1 Determine the nature of the solutions of $9x^2 - 12x + 4 = 0$.

We have

$$a = 9, \quad b = -12, \quad c = 4.$$

We compute the discriminant:

$$b^2 - 4ac = (-12)^2 - 4 \cdot 9 \cdot 4$$
$$= 144 - 144$$
$$= 0.$$

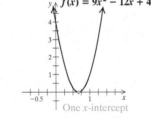
One x-intercept

There is just one solution, and it is a real number. Since 0 is a perfect square, the equation can be solved by factoring.

EXAMPLE 2 Determine the nature of the solutions of $x^2 + 5x + 8 = 0$.

We have

$$a = 1, \quad b = 5, \quad c = 8.$$

We compute the discriminant:

$$b^2 - 4ac = 5^2 - 4 \cdot 1 \cdot 8$$
$$= 25 - 32$$
$$= -7.$$

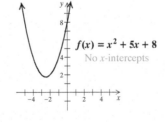
No x-intercepts

Since the discriminant is negative, there are two nonreal complex-number solutions.

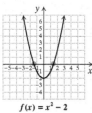

$f(x) = x^2 - 2$
$b^2 - 4ac = 8 > 0$
Two real solutions
Two x-intercepts

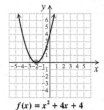

$f(x) = x^2 + 4x + 4$
$b^2 - 4ac = 0$
One real solution
One x-intercept

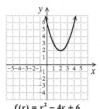

$f(x) = x^2 - 4x + 6$
$b^2 - 4ac = -8 < 0$
No real solutions
No x-intercept

Determine the nature of the solutions without solving.

1. $x^2 + 5x - 3 = 0$

2. $9x^2 - 6x + 1 = 0$

3. $3x^2 - 2x + 1 = 0$
   $a = 3, b = \underline{}, c = 1;$
   $b^2 - 4ac = (\underline{})^2 - 4 \cdot 3 \cdot 1$
   $= \underline{}$

   Since the discriminant is negative, there are $\underline{}$ nonreal solutions.

EXAMPLE 3 Determine the nature of the solutions of $x^2 + 5x + 6 = 0$.

We have
$$a = 1, \quad b = 5, \quad c = 6;$$
$$b^2 - 4ac = 5^2 - 4 \cdot 1 \cdot 6 = 1.$$

Since the discriminant is positive, there are two solutions, and they are real numbers. The equation can be solved by factoring since the discriminant is a perfect square.

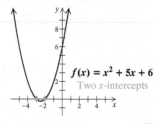

$f(x) = x^2 + 5x + 6$
Two x-intercepts

EXAMPLE 4 Determine the nature of the solutions of $5x^2 + x - 3 = 0$.

We have
$$a = 5, \quad b = 1, \quad c = -3;$$
$$b^2 - 4ac = 1^2 - 4 \cdot 5 \cdot (-3) = 1 + 60 = 61.$$

Since the discriminant is positive, there are two solutions, and they are real numbers. The equation cannot be solved by factoring because 61 is not a perfect square.

The discriminant, $b^2 - 4ac$, tells us how many real-number solutions the equation $ax^2 + bx + c = 0$ has, so it also indicates how many x-intercepts the graph of $f(x) = ax^2 + bx + c$ has. Compare the graphs at left.

◀ Do Exercises 1–3.

b WRITING EQUATIONS FROM SOLUTIONS

We know by the principle of zero products that $(x - 2)(x + 3) = 0$ has solutions 2 and -3. If we know the solutions of an equation, we can write the equation, using this principle in reverse.

EXAMPLE 5 Find a quadratic equation whose solutions are 3 and $-\frac{2}{5}$.

We have

$x = 3$ or $x = -\frac{2}{5}$
$x - 3 = 0$ or $x + \frac{2}{5} = 0$ Getting the 0's on one side
$x - 3 = 0$ or $5x + 2 = 0$ Clearing the fraction
$(x - 3)(5x + 2) = 0$ Using the principle of zero products in reverse
$5x^2 - 13x - 6 = 0.$ Using FOIL

EXAMPLE 6 Write a quadratic equation whose solutions are $2i$ and $-2i$.

We have

$x = 2i$ or $x = -2i$
$x - 2i = 0$ or $x + 2i = 0$ Getting the 0's on one side
$(x - 2i)(x + 2i) = 0$ Using the principle of zero products in reverse
$x^2 - (2i)^2 = 0$ Using $(A - B)(A + B) = A^2 - B^2$
$x^2 - 4i^2 = 0$
$x^2 - 4(-1) = 0$
$x^2 + 4 = 0.$

Answers
1. Two real 2. One real 3. Two nonreal
Guided Solution:
3. $-2, -2, -8$; two

EXAMPLE 7 Write a quadratic equation whose solutions are $\sqrt{3}$ and $-2\sqrt{3}$.

We have

$$x = \sqrt{3} \quad \text{or} \quad x = -2\sqrt{3}$$
$$x - \sqrt{3} = 0 \quad \text{or} \quad x + 2\sqrt{3} = 0 \quad \text{Getting the 0's on one side}$$
$$(x - \sqrt{3})(x + 2\sqrt{3}) = 0 \quad \text{Using the principle of zero products}$$
$$x^2 + 2\sqrt{3}x - \sqrt{3}x - 2(\sqrt{3})^2 = 0 \quad \text{Using FOIL}$$
$$x^2 + \sqrt{3}x - 6 = 0. \quad \text{Collecting like terms}$$

EXAMPLE 8 Write a quadratic equation whose solutions are $-12i$ and $12i$.

We have

$$x = -12i \quad \text{or} \quad x = 12i$$
$$x + 12i = 0 \quad \text{or} \quad x - 12i = 0 \quad \text{Getting the 0's on one side}$$
$$(x + 12i)(x - 12i) = 0 \quad \text{Using the principle of zero products}$$
$$x^2 - 12ix + 12ix - 144i^2 = 0 \quad \text{Using FOIL}$$
$$x^2 - 144(-1) = 0 \quad \text{Collecting like terms; substituting } -1 \text{ for } i^2$$
$$x^2 + 144 = 0.$$

Do Exercises 4–8.

Find a quadratic equation having the following solutions.

4. 7 and -2

5. -4 and $\dfrac{5}{3}$

6. $5i$ and $-5i$

7. $-2\sqrt{2}$ and $\sqrt{2}$

8. $-7i$ and $7i$

C EQUATIONS QUADRATIC IN FORM

SKILL REVIEW *Solve radical equations.* [10.6a]
Solve.
1. $\sqrt{x} = 10$ 2. $\sqrt{x} = -10$

Answers: **1.** 100 **2.** No solution

Certain equations that are not really quadratic can still be solved as quadratic. Consider this fourth-degree equation.

$$x^4 - 9x^2 + 8 = 0$$
$$\downarrow \quad \downarrow \quad \downarrow \quad \downarrow$$
$$(x^2)^2 - 9(x^2) + 8 = 0 \quad \text{Thinking of } x^4 \text{ as } (x^2)^2$$
$$\downarrow \quad \downarrow \quad \downarrow \quad \downarrow$$
$$u^2 - 9u + 8 = 0 \quad \text{To make this clearer, write } u \text{ instead of } x^2.$$

The equation $u^2 - 9u + 8 = 0$ can be solved by factoring or by the quadratic formula. After that, we can find x by remembering that $x^2 = u$. Equations that can be solved like this are said to be **quadratic in form**, or **reducible to quadratic**.

Answers
4. $x^2 - 5x - 14 = 0$
5. $3x^2 + 7x - 20 = 0$
6. $x^2 + 25 = 0$ **7.** $x^2 + \sqrt{2}x - 4 = 0$
8. $x^2 + 49 = 0$

EXAMPLE 9 Solve: $x^4 - 9x^2 + 8 = 0$.

Let $u = x^2$. Then we solve the equation found by substituting u for x^2:

$$u^2 - 9u + 8 = 0$$
$$(u - 8)(u - 1) = 0 \qquad \text{Factoring}$$
$$u - 8 = 0 \quad \text{or} \quad u - 1 = 0 \qquad \text{Using the principle of zero products}$$
$$u = 8 \quad \text{or} \quad u = 1.$$

Next, we substitute x^2 for u and solve these equations:

$$x^2 = 8 \qquad \text{or} \quad x^2 = 1$$
$$x = \pm\sqrt{8} \quad \text{or} \quad x = \pm 1$$
$$x = \pm 2\sqrt{2} \quad \text{or} \quad x = \pm 1.$$

Note that when a number and its opposite are raised to an even power, the results are the same. Thus we can make one check for $\pm 2\sqrt{2}$ and one for ± 1.

Check:

For $\pm 2\sqrt{2}$:

$$\begin{array}{c|c} x^4 - 9x^2 + 8 = 0 \\ \hline (\pm 2\sqrt{2})^4 - 9(\pm 2\sqrt{2})^2 + 8 \;?\; 0 \\ 64 - 9 \cdot 8 + 8 \\ 0 & \text{TRUE} \end{array}$$

For ± 1:

$$\begin{array}{c|c} x^4 - 9x^2 + 8 = 0 \\ \hline (\pm 1)^4 - 9(\pm 1)^2 + 8 \;?\; 0 \\ 1 - 9 + 8 \\ 0 & \text{TRUE} \end{array}$$

The solutions are $1, -1, 2\sqrt{2}$, and $-2\sqrt{2}$.

.... **Caution!** ....

A common error is to solve for u and then forget to solve for x. Remember that you *must* find values for the *original* variable!

◀ Do Exercise 9.

Solving equations quadratic in form can sometimes introduce numbers that are not solutions of the original equation. Thus a check by substitution in the original equation is necessary.

EXAMPLE 10 Solve: $x - 3\sqrt{x} - 4 = 0$.

Let $u = \sqrt{x}$. Then we solve the equation found by substituting u for $\sqrt{x}$ and u^2 for x:

$$u^2 - 3u - 4 = 0$$
$$(u - 4)(u + 1) = 0$$
$$u = 4 \quad \text{or} \quad u = -1.$$

Next, we substitute $\sqrt{x}$ for u and solve these equations:

$$\sqrt{x} = 4 \quad \text{or} \quad \sqrt{x} = -1.$$

Squaring, we get $x = 16$ or $x = 1$. We check both possible solutions.

---

9. Solve: $x^4 - 10x^2 + 9 = 0$. [GS]

Let $u = $ ____ . Then $u^2 = $ ____ .

$$x^4 - 10x^2 + 9 = 0$$
$$u^2 - 10u + 9 = 0$$
$$(u - 1)() = 0$$
$$u - 1 = 0 \quad \text{or} \quad = 0$$
$$u = 1 \quad \text{or} \quad u = $$
$$x^2 = 1 \quad \text{or} \quad x^2 = $$
$$x = \pm 1 \quad \text{or} \quad x = $$

All four numbers check. The solutions are $-1, 1, -3,$ and ____ .

Answer

9. $\pm 1, \pm 3$

Guided Solution:

9. $x^2, x^4, u - 9, u - 9, 9, 9, \pm 3, 3$

Check:

For 16:
$$x - 3\sqrt{x} - 4 = 0$$
$$\overline{16 - 3\sqrt{16} - 4 \;?\; 0}$$
$$16 - 3 \cdot 4 - 4$$
$$16 - 12 - 4$$
$$0 \quad \text{TRUE}$$

For 1:
$$x - 3\sqrt{x} - 4 = 0$$
$$\overline{1 - 3\sqrt{1} - 4 \;?\; 0}$$
$$1 - 3 \cdot 1 - 4$$
$$1 - 3 - 4$$
$$-6 \quad \text{FALSE}$$

Since 16 checks but 1 does not, the solution is 16. Had we noted that 1 is not a solution of $\sqrt{x} = -1$, we could have eliminated 1 at that point as a possible solution.

Do Exercise 10.

10. Solve: $x + 3\sqrt{x} - 10 = 0$. Be sure to check.

EXAMPLE 11 Solve: $y^{-2} - y^{-1} - 2 = 0$.

Let $u = y^{-1}$. Then we solve the equation found by substituting u for y^{-1} and u^2 for y^{-2}:

$$u^2 - u - 2 = 0$$
$$(u - 2)(u + 1) = 0$$
$$u = 2 \quad \text{or} \quad u = -1.$$

Next, we substitute y^{-1} or $1/y$ for u and solve these equations:

$$\frac{1}{y} = 2 \quad \text{or} \quad \frac{1}{y} = -1$$

$$y = \frac{1}{2} \quad \text{or} \quad y = \frac{1}{(-1)} = -1.$$

The numbers $\frac{1}{2}$ and -1 both check. They are the solutions.

Do Exercise 11.

11. Solve: $x^{-2} + x^{-1} - 6 = 0$.

EXAMPLE 12 Find the x-intercepts of the graph of

$$f(x) = (x^2 - 1)^2 - (x^2 - 1) - 2.$$

The x-intercepts occur where $f(x) = 0$, so we must have

$$(x^2 - 1)^2 - (x^2 - 1) - 2 = 0.$$

Let $u = x^2 - 1$. Then we solve the equation found by substituting u for $x^2 - 1$:

$$u^2 - u - 2 = 0$$
$$(u - 2)(u + 1) = 0$$
$$u = 2 \quad \text{or} \quad u = -1.$$

Next, we substitute $x^2 - 1$ for u and solve these equations:

$$x^2 - 1 = 2 \quad \text{or} \quad x^2 - 1 = -1$$
$$x^2 = 3 \quad \text{or} \quad x^2 = 0$$
$$x = \pm\sqrt{3} \quad \text{or} \quad x = 0.$$

Since the numbers $\sqrt{3}, -\sqrt{3}$, and 0 check, they are the solutions of $(x^2 - 1)^2 - (x^2 - 1) - 2 = 0$. Thus the x-intercepts of the graph of $f(x)$ are $(-\sqrt{3}, 0), (0, 0),$ and $(\sqrt{3}, 0)$.

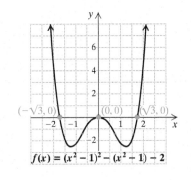

Do Exercise 12.

12. Find the x-intercepts of $f(x) = (x^2 - x)^2 - 14(x^2 - x) + 24$.

Answers

10. 4   11. $-\frac{1}{3}, \frac{1}{2}$

12. $(-3, 0), (-1, 0), (2, 0), (4, 0)$

11.4 Exercise Set

✔ Check Your Understanding

Reading Check Match each discriminant with the appropriate description of the solution(s) from the column on the right. Answers may be used more than once.

RC1. ____ $b^2 - 4ac = 9$

RC2. ____ $b^2 - 4ac = 0$

RC3. ____ $b^2 - 4ac = -1$

RC4. ____ $b^2 - 4ac = 1$

RC5. ____ $b^2 - 4ac = 8$

RC6. ____ $b^2 - 4ac = -12$

a) One real-number solution
b) Two different real-number solutions
c) Two different nonreal complex-number solutions

Concept Check Determine the substitution that could be used to make each equation quadratic in u.

CC1. For $3x - 4\sqrt{x} + 1 = 0$, use $u = $ _____.

CC2. For $p^4 - p^2 - 6 = 0$, use $u = $ _____.

CC3. For $x^{-2} + 5x^{-1} - 14 = 0$, use $u = $ _____.

CC4. For $(2y + 7)^2 - 3(2y + 7) - 40 = 0$, use $u = $ _____.

a   Determine the nature of the solutions of each equation.

1. $x^2 - 8x + 16 = 0$
2. $x^2 + 12x + 36 = 0$
3. $x^2 + 1 = 0$
4. $x^2 + 6 = 0$

5. $x^2 - 6 = 0$
6. $x^2 - 3 = 0$
7. $4x^2 - 12x + 9 = 0$
8. $4x^2 + 8x - 5 = 0$

9. $x^2 - 2x + 4 = 0$
10. $x^2 + 3x + 4 = 0$
11. $9t^2 - 3t = 0$
12. $4m^2 + 7m = 0$

13. $y^2 = \frac{1}{2}y + \frac{3}{5}$
14. $y^2 + \frac{9}{4} = 4y$

15. $4x^2 - 4\sqrt{3}x + 3 = 0$
16. $6y^2 - 2\sqrt{3}y - 1 = 0$

b   Write a quadratic equation having the given numbers as solutions.

17. -4 and 4
18. -11 and 9
19. $-4i$ and $4i$

20. $-i$ and i

21. 8, only solution
[*Hint*: It must be a double solution, that is, $(x-8)(x-8) = 0$.]

22. -3, only solution

23. $-\dfrac{2}{5}$ and $\dfrac{6}{5}$

24. $-\dfrac{1}{4}$ and $-\dfrac{1}{2}$

25. $\dfrac{k}{3}$ and $\dfrac{m}{4}$

26. $\dfrac{c}{2}$ and $\dfrac{d}{2}$

27. $-\sqrt{3}$ and $2\sqrt{3}$

28. $\sqrt{2}$ and $3\sqrt{2}$

29. $6i$ and $-6i$

30. $8i$ and $-8i$

C Solve.

31. $x^4 - 6x^2 + 9 = 0$

32. $x^4 - 7x^2 + 12 = 0$

33. $x - 10\sqrt{x} + 9 = 0$

34. $2x - 9\sqrt{x} + 4 = 0$

35. $(x^2 - 6x)^2 - 2(x^2 - 6x) - 35 = 0$

36. $(x^2 + 5x)^2 + 2(x^2 + 5x) - 24 = 0$

37. $x^{-2} - 5x^{-1} - 36 = 0$

38. $3x^{-2} - x^{-1} - 14 = 0$

39. $(1 + \sqrt{x})^2 + (1 + \sqrt{x}) - 6 = 0$

40. $(2 + \sqrt{x})^2 - 3(2 + \sqrt{x}) - 10 = 0$

41. $(y^2 - 5y)^2 - 2(y^2 - 5y) - 24 = 0$

42. $(2t^2 + t)^2 - 4(2t^2 + t) + 3 = 0$

43. $w^4 - 29w^2 + 100 = 0$

44. $t^4 - 10t^2 + 9 = 0$

45. $2x^{-2} + x^{-1} - 1 = 0$

46. $m^{-2} + 9m^{-1} - 10 = 0$

47. $6x^4 - 19x^2 + 15 = 0$

48. $6x^4 - 17x^2 + 5 = 0$

49. $x^{2/3} - 4x^{1/3} - 5 = 0$

50. $x^{2/3} + 2x^{1/3} - 8 = 0$

51. $\left(\dfrac{x-4}{x+1}\right)^2 - 2\left(\dfrac{x-4}{x+1}\right) - 35 = 0$

52. $\left(\dfrac{x+3}{x-3}\right)^2 - \left(\dfrac{x+3}{x-3}\right) - 6 = 0$

53. $9\left(\dfrac{x+2}{x+3}\right)^2 - 6\left(\dfrac{x+2}{x+3}\right) + 1 = 0$

54. $16\left(\dfrac{x-1}{x-8}\right)^2 + 8\left(\dfrac{x-1}{x-8}\right) + 1 = 0$

55. $\left(\dfrac{x^2-2}{x}\right)^2 - 7\left(\dfrac{x^2-2}{x}\right) - 18 = 0$

56. $\left(\dfrac{y^2-1}{y}\right)^2 - 4\left(\dfrac{y^2-1}{y}\right) - 12 = 0$

Find the *x*-intercepts of the graph of each function.

57. $f(x) = 5x + 13\sqrt{x} - 6$

58. $f(x) = 3x + 10\sqrt{x} - 8$

59. $f(x) = (x^2 - 3x)^2 - 10(x^2 - 3x) + 24$

60. $f(x) = (x^2 - x)^2 - 8(x^2 - x) + 12$

61. $f(x) = x^{2/3} + x^{1/3} - 2$

62. $f(x) = x^{2/5} + x^{1/5} - 6$

Skill Maintenance

Solve. [8.4a]

63. *Coffee Beans.* Twin Cities Roasters sells Kenyan coffee worth $9.75 per pound and Peruvian coffee worth $13.25 per pound. How many pounds of each kind should be mixed in order to obtain a 50-lb mixture that is worth $11.15 per pound?

64. *Solution Mixtures.* Solution A is 18% alcohol and solution B is 45% alcohol. How many liters of each should be mixed in order to get 12 L of a solution that is 36% alcohol?

Multiply and simplify. Assume that no radicands were formed by raising negative numbers to even powers. [10.3a]

65. $\sqrt{8x}\,\sqrt{2x}$

66. $\sqrt[3]{x^2}\,\sqrt[3]{27x^4}$

67. $\sqrt[4]{9a^2}\,\sqrt[4]{18a^3}$

68. $\sqrt[5]{16}\,\sqrt[5]{64}$

Graph. [7.1c], [7.4a, c]

69. $f(x) = -\frac{3}{5}x + 4$

70. $5x - 2y = 8$

71. $y = 4$

72. $f(x) = -x - 3$

Synthesis

73. Use a graphing calculator to check your answers to Exercises 32, 34, 36, and 39.

74. Use a graphing calculator to solve each of the following equations.
a) $6.75x - 35\sqrt{x} - 5.26 = 0$
b) $\pi x^4 - \pi^2 x^2 = \sqrt{99.3}$
c) $x^4 - x^3 - 13x^2 + x + 12 = 0$

For each equation under the given condition, (a) find k and (b) find the other solution.

75. $kx^2 - 2x + k = 0$; one solution is -3.

76. $kx^2 - 17x + 33 = 0$; one solution is 3.

77. Find a quadratic equation for which the sum of the solutions is $\sqrt{3}$ and the product is 8.

78. Find k given that $kx^2 - 4x + (2k - 1) = 0$ and the product of the solutions is 3.

79. The graph of a function of the form
$$f(x) = ax^2 + bx + c$$
is a curve similar to the one shown below. Determine a, b, and c from the information given.

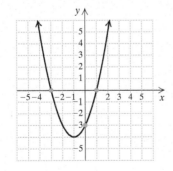

80. While solving a quadratic equation of the form $ax^2 + bx + c = 0$ with a graphing calculator, Shawn-Marie gets the following screen. How could the discriminant help her check the graph?

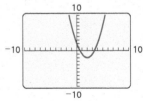

Solve.

81. $\dfrac{x}{x-1} - 6\sqrt{\dfrac{x}{x-1}} - 40 = 0$

82. $\dfrac{x}{x-3} - 24 = 10\sqrt{\dfrac{x}{x-3}}$

83. $\sqrt{x-3} - \sqrt[4]{x-3} = 12$

84. $a^3 - 26a^{3/2} - 27 = 0$

85. $x^6 - 28x^3 + 27 = 0$

86. $x^6 + 7x^3 - 8 = 0$

Mid-Chapter Review

Concept Reinforcement

Determine whether each statement is true or false.

_____ 1. Every quadratic equation has exactly two real-number solutions. [11.4a]

_____ 2. The quadratic formula can be used to find all the solutions of any quadratic equation. [11.2a]

_____ 3. If the graph of a quadratic equation crosses the x-axis, then it has exactly two real-number solutions. [11.4a]

_____ 4. The x-intercepts of $f(x) = x^2 - t$ are $(0, \sqrt{t})$ and $(0, -\sqrt{t})$. [11.1a]

Guided Solutions

GS Fill in each blank with the number that creates a correct solution.

5. Solve $5x^2 + 3x = 4$ by completing the square. [11.1b]

$$5x^2 + 3x = 4$$
$$\boxed{}(5x^2 + 3x) = \boxed{} \cdot 4$$
$$x^2 + \frac{3}{\boxed{}}x = \frac{4}{\boxed{}}$$
$$x^2 + \frac{3}{5}x + \boxed{} = \frac{4}{5} + \boxed{}$$
$$(x + \boxed{})^2 = \frac{\boxed{}}{100}$$

$$x + \frac{3}{10} = \sqrt{\boxed{}} \quad \text{or} \quad x + \frac{3}{10} = -\sqrt{\boxed{}}$$

$$x + \frac{3}{10} = \frac{\sqrt{\boxed{}}}{10} \quad \text{or} \quad x + \frac{3}{10} = -\frac{\sqrt{\boxed{}}}{10}$$

$$x = -\frac{\boxed{}}{10} + \frac{\sqrt{\boxed{}}}{10} \quad \text{or} \quad x = -\frac{\boxed{}}{10} - \frac{\sqrt{\boxed{}}}{10}.$$

The solutions are $-\dfrac{\boxed{}}{10} \pm \dfrac{\sqrt{\boxed{}}}{10}$.

6. Use the quadratic formula to solve $5x^2 + 3x = 4$. [11.2a]

$$5x^2 + 3x = 4$$
$$5x^2 + 3x - \boxed{} = 0$$
$$a = \boxed{}, \quad b = \boxed{}, \quad c = \boxed{}$$

$$x = \frac{-b \pm \sqrt{b^2 - 4ac}}{2a}$$

$$x = \frac{-\boxed{} \pm \sqrt{\boxed{}^2 - 4 \cdot \boxed{} \cdot \boxed{}}}{2 \cdot \boxed{}}$$

$$x = \frac{-3 \pm \sqrt{9 + \boxed{}}}{\boxed{}}$$

$$x = \frac{-3 \pm \sqrt{\boxed{}}}{10}$$

$$x = -\frac{3}{10} \pm \frac{\sqrt{\boxed{}}}{10}$$

Mixed Review

Solve by completing the square. [11.1b]

7. $x^2 + 1 = -4x$

8. $2x^2 + 5x - 3 = 0$

9. $x^2 + 10x - 6 = 0$

10. $x^2 - x = 5$

Determine the nature of the solutions of each equation $ax^2 + bx + c = 0$ and the number of x-intercepts of the graph of the function $f(x) = ax^2 + bx + c$. [11.4a]

11. $x^2 - 10x + 25 = 0$

12. $x^2 - 11 = 0$

13. $y^2 = \dfrac{1}{3}y - \dfrac{4}{7}$

14. $x^2 + 5x + 9 = 0$

15. $x^2 - 4 = 2x$

16. $x^2 - 8x = 0$

840   CHAPTER 11   Quadratic Equations and Functions

Write a quadratic equation having the given numbers as solutions. [11.4b]

17. -1 and 10

18. -13 and 13

19. $-\sqrt{5}$ and $3\sqrt{5}$

20. $-4i$ and $4i$

21. -6, only solution

22. $-\dfrac{4}{3}$ and $\dfrac{2}{7}$

Solve.

23. Jacob traveled 780 mi by car. Had he gone 5 mph faster, he could have made the trip in 1 hr less time. Find his speed. [11.3a]

24. $R = as^2$, for s [11.3b]

Solve. [11.1a], [11.2a], [11.4c]

25. $3x^2 + x = 4$

26. $x^4 - 8x^2 + 15 = 0$

27. $4x^2 = 15x - 5$

28. $7x^2 + 2 = -9x$

29. $2x + x(x - 1) = 0$

30. $(x + 3)^2 = 64$

31. $49x^2 + 16 = 0$

32. $(x^2 - 2)^2 + 2(x^2 - 2) - 24 = 0$

33. $r^2 + 5r = 12$

34. $s^2 + 12s + 37 = 0$

35. $\left(x - \dfrac{5}{2}\right)^2 = \dfrac{11}{4}$

36. $x + \dfrac{1}{x} = \dfrac{7}{3}$

37. $4x + 1 = 4x^2$

38. $(x - 3)^2 + (x + 5)^2 = 0$

39. $b^2 - 16b + 64 = 3$

40. $(x - 3)^2 = -10$

41. $\dfrac{1}{x} + \dfrac{1}{x + 2} = \dfrac{1}{5}$

42. $x - \sqrt{x} - 6 = 0$

Understanding Through Discussion and Writing

43. Given the solutions of a quadratic equation, is it possible to reconstruct the original equation? Why or why not? [11.4b]

44. Explain how the quadratic formula can be used to factor a quadratic polynomial into two binomials. Use it to factor $5x^2 + 8x - 3$. [11.2a]

45. Describe a procedure that could be used to write an equation having the first seven natural numbers as solutions. [11.4b]

46. Describe a procedure that could be used to write an equation that is quadratic in $3x^2 + 1$ and has real-number solutions. [11.4c]

STUDYING FOR SUCCESS   Take Good Care of Yourself

- ☐ Get plenty of rest, especially the night before a test.
- ☐ Try an exercise break when studying. Often a walk or a bike ride will improve your concentration.
- ☐ Plan leisure time in your schedule. Rest and a change of pace will help you avoid burn-out.

11.5 Graphing $f(x) = a(x - h)^2 + k$

OBJECTIVES

a Graph quadratic functions of the type $f(x) = ax^2$ and then label the vertex and the line of symmetry.

b Graph quadratic functions of the type $f(x) = a(x - h)^2$ and then label the vertex and the line of symmetry.

c Graph quadratic functions of the type $f(x) = a(x - h)^2 + k$, finding the vertex, the line of symmetry, and the maximum or minimum function value, or y-value.

a   GRAPHS OF $f(x) = ax^2$

The most basic quadratic function is $f(x) = x^2$.

EXAMPLE 1 Graph: $f(x) = x^2$.

We choose some values for x and compute $f(x)$ for each. Then we plot the ordered pairs and connect them with a smooth curve.

x	$f(x) = x^2$	$(x, f(x))$
−3	9	(−3, 9)
−2	4	(−2, 4)
−1	1	(−1, 1)
0	0	(0, 0)
1	1	(1, 1)
2	4	(2, 4)
3	9	(3, 9)

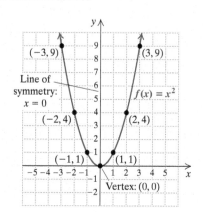

Let's compare the graphs of $g(x) = \frac{1}{2}x^2$ and $h(x) = 2x^2$ with the graph of $f(x) = x^2$. We choose x-values and plot points for both functions.

x	$g(x) = \frac{1}{2}x^2$
−3	$\frac{9}{2}$
−2	2
−1	$\frac{1}{2}$
0	0
1	$\frac{1}{2}$
2	2
3	$\frac{9}{2}$

x	$h(x) = 2x^2$
−3	18
−2	8
−1	2
0	0
1	2
2	8
3	18

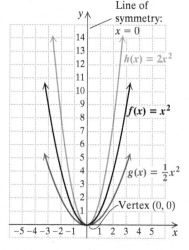

Note the symmetry: For equal increments to the left and right of the vertex, the y-values are the same.

842   CHAPTER 11   Quadratic Equations and Functions

Graphs of quadratic functions are called **parabolas**. They are cup-shaped curves that are symmetric with respect to a vertical line known as the parabola's **line of symmetry**, or **axis of symmetry**. In the graphs shown on the preceding page, the y-axis (or the line $x = 0$) is the line of symmetry. If the paper were to be folded on this line, the two halves of the curve would coincide. The point $(0, 0)$ is the **vertex** of each of the parabolas shown on the preceding page.

Note that the graph of $g(x) = \frac{1}{2}x^2$ is a wider parabola than the graph of $f(x) = x^2$, and the graph of $h(x) = 2x^2$ is narrower.

When we consider the graph of $k(x) = -\frac{1}{2}x^2$, we see that the parabola opens down and has the same shape as the graph of $g(x) = \frac{1}{2}x^2$.

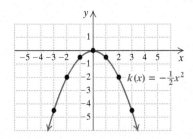

GRAPHS OF $f(x) = ax^2$

The graph of $f(x) = ax^2$, or $y = ax^2$, is a parabola with $x = 0$ as its line of symmetry; its vertex is the origin.

For $a > 0$, the parabola opens up; for $a < 0$, the parabola opens down.

If $|a|$ is greater than 1, the parabola is narrower than $y = x^2$.

If $|a|$ is between 0 and 1, the parabola is wider than $y = x^2$.

Graph.

1. $f(x) = -\frac{1}{3}x^2$

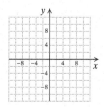

2. $f(x) = 3x^2$

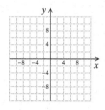

3. $f(x) = -2x^2$

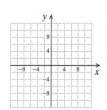

Do Exercises 1–3.

b GRAPHS OF $f(x) = a(x - h)^2$

EXAMPLE 2 Graph: $g(x) = (x - 3)^2$.

We choose some values for x and compute $g(x)$. Then we plot the points and draw the curve.

x	$g(x) = (x - 3)^2$	
3	0	← Vertex
4	1	
5	4	
6	9	
2	1	
1	4	
0	9	

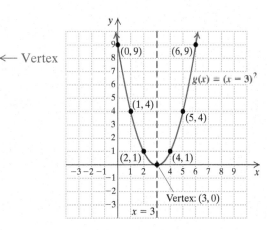

For an x-value of 3, $g(3) = (3 - 3)^2 = 0$. As we increase x-values from 3, the corresponding y-values increase. Then as we decrease x-values from 3, the corresponding y-values increase again. The line $x = 3$ is the line of symmetry.

Answers

1.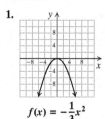

$f(x) = -\frac{1}{3}x^2$

2.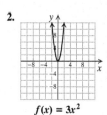

$f(x) = 3x^2$

3.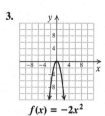

$f(x) = -2x^2$

4. Graph: $g(x) = (x + 2)^2$.

Compute $g(x)$ for each value of x shown.

x	$g(x) = (x + 2)^2$
-2	
-1	1
0	4
1	
-3	1
-4	4
-5	

Plot the points and draw the curve.

Answer

4.

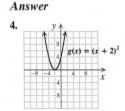

Guided Solution:
4. 0, 9, 9

EXAMPLE 3 Graph: $t(x) = (x + 3)^2$.

We choose some values for x and compute $t(x)$. Then we plot the points and draw the curve.

x	$t(x) = (x + 3)^2$	
-3	0	← Vertex
-2	1	
-1	4	
0	9	
-4	1	
-5	4	
-6	9	

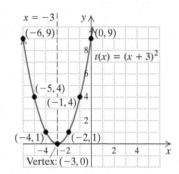

For an x-value of -3, $t(-3) = (-3 + 3)^2 = 0$. As we increase x-values from -3, the corresponding y-values increase. Then as we decrease x-values from -3, the y-values increase again. The line $x = -3$ is the line of symmetry.

◀ Do Exercise 4.

The graph of $g(x) = (x - 3)^2$ in Example 2 looks just like the graph of $f(x) = x^2$ in Example 1, except that it is moved, or translated, 3 units to the right. Comparing the pairs for $g(x)$ with those for $f(x)$, we see that when an input for $g(x)$ is 3 more than an input for $f(x)$, the outputs are the same.

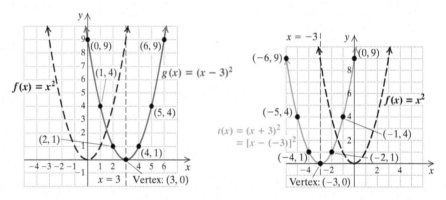

The graph of $t(x) = (x + 3)^2 = [x - (-3)]^2$ in Example 3 looks just like the graph of $f(x) = x^2$ in Example 1, except that it is moved, or translated, 3 units to the left. Comparing the pairs for $t(x)$ with those for $f(x)$, we see that when an input for $t(x)$ is 3 less than an input for $f(x)$, the outputs are the same.

GRAPHS OF $f(x) = a(x - h)^2$

The graph of $f(x) = a(x - h)^2$ has the same shape as the graph of $y = ax^2$.

If h is positive, the graph of $y = ax^2$ is shifted h units to the right.

If h is negative, the graph of $y = ax^2$ is shifted $|h|$ units to the left.

The vertex is $(h, 0)$, and the line of symmetry is $x = h$.

EXAMPLE 4 Graph: $f(x) = -2(x + 3)^2$.

We first rewrite the equation as $f(x) = -2[x - (-3)]^2$. In this case, $a = -2$ and $h = -3$, so the graph looks like that of $g(x) = 2x^2$ translated 3 units to the left and, since $-2 < 0$, the graph opens down. The vertex is $(-3, 0)$, and the line of symmetry is $x = -3$.

x	$f(x) = -2(x + 3)^2$
-3	0 ← Vertex
-2	-2
-1	-8
-4	-2
-5	-8

Graph. Find and label the vertex and the line of symmetry.

5. $f(x) = \frac{1}{2}(x - 4)^2$

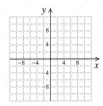

6. $f(x) = -\frac{1}{2}(x - 4)^2$

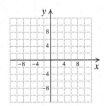

Do Exercises 5 and 6.

C GRAPHS OF $f(x) = a(x - h)^2 + k$

Given a graph of $f(x) = a(x - h)^2$, if we add a positive constant k, each function value $f(x)$ is increased by k, so the curve is moved up. If k is negative, the curve is moved down. The line of symmetry for the parabola remains $x = h$, but the vertex will be at (h, k).

Note that if a parabola opens up ($a > 0$), the function value, or y-value, at the vertex is a least, or **minimum**, value. That is, it is less than the y-value at any other point on the graph. If the parabola opens down ($a < 0$), the function value at the vertex is a greatest, or **maximum**, value.

MyLab Math
ANIMATION

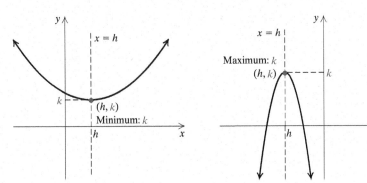

GRAPHS OF $f(x) = a(x - h)^2 + k$

The graph of $f(x) = a(x - h)^2 + k$ has the same shape as the graph of $y = a(x - h)^2$.

If k is positive, the graph of $y = a(x - h)^2$ is shifted k units up.

If k is negative, the graph of $y = a(x - h)^2$ is shifted $|k|$ units down.

The vertex is (h, k), and the line of symmetry is $x = h$.

For $a > 0$, k is the minimum function value. For $a < 0$, k is the maximum function value.

Answers

5.

$f(x) = \frac{1}{2}(x - 4)^2$

6.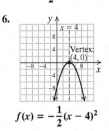

$f(x) = -\frac{1}{2}(x - 4)^2$

7. Graph $g(x) = (x - 1)^2 - 3$, and find the minimum function value.

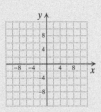

Compute $g(x)$ for each value of x shown.

x	$g(x) = (x - 1)^2 - 3$
1	-3
2	☐
3	1
4	6
0	☐
-1	1
-2	6

The vertex is $(1, -3)$. The line of symmetry is $x = $ ☐, and the minimum function value is ☐.

Plot the points and draw the curve.

EXAMPLE 5 Graph $f(x) = (x - 3)^2 - 5$, and find the minimum function value.

The graph will look like that of $g(x) = (x - 3)^2$ (see Example 2) but translated 5 units down. You can confirm this by plotting some points. For instance,

$$f(4) = (4 - 3)^2 - 5 = -4,$$

whereas in Example 2,

$$g(4) = (4 - 3)^2 = 1.$$

Note that the vertex is $(h, k) = (3, -5)$, so we begin calculating points on both sides of $x = 3$. The line of symmetry is $x = 3$, and the minimum function value is -5.

x	$f(x) = (x - 3)^2 - 5$	
3	-5	← Vertex
4	-4	
5	-1	
6	4	
2	-4	
1	-1	
0	4	

◀ Do Exercise 7.

EXAMPLE 6 Graph $t(x) = \frac{1}{2}(x - 3)^2 + 5$, and find the minimum function value.

The graph looks just like that of $f(x) = \frac{1}{2}x^2$ but moved 3 units to the right and 5 units up. The vertex is $(3, 5)$, and the line of symmetry is $x = 3$. We draw $f(x) = \frac{1}{2}x^2$ and then shift the curve over and up. The minimum function value is 5. By plotting some points, we have a check.

x	$t(x) = \frac{1}{2}(x - 3)^2 + 5$	
3	5	← Vertex
4	$5\frac{1}{2}$	
5	7	
6	$9\frac{1}{2}$	
2	$5\frac{1}{2}$	
1	7	
0	$9\frac{1}{2}$	

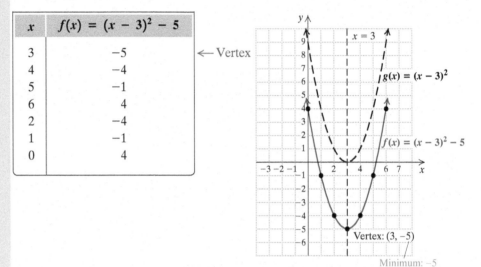

Answer
7.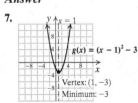

Guided Solution:
7. $-2, -2, 1, -3$

EXAMPLE 7 Graph $f(x) = -2(x + 3)^2 + 5$. Find the vertex, the line of symmetry, and the maximum or minimum value.

We first express the equation in the equivalent form
$$f(x) = -2[x - (-3)]^2 + 5.$$

The graph looks like that of $g(x) = -2x^2$ translated 3 units to the left and 5 units up. The vertex is $(-3, 5)$, and the line of symmetry is $x = -3$. Since $-2 < 0$, we know that the graph opens down so 5, the second coordinate of the vertex, is the maximum y-value. We compute points and draw the graph.

x	$f(x) = -2(x + 3)^2 + 5$	
-3	5	← Vertex
-2	3	
-1	-3	
-4	3	
-5	-3	

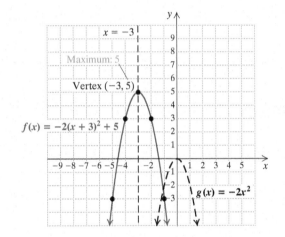

Do Exercises 8 and 9.

Graph. Find the vertex, the line of symmetry, and the maximum or minimum y-value.

8. $f(x) = \frac{1}{2}(x + 2)^2 - 4$

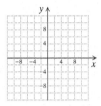

9. $f(x) = -2(x - 5)^2 + 3$

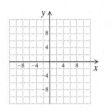

Answers

8.

$f(x) = \frac{1}{2}(x + 2)^2 - 4$

9.

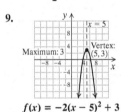

$f(x) = -2(x - 5)^2 + 3$

11.5 Exercise Set

FOR EXTRA HELP — MyLab Math

✔ Check Your Understanding

Reading Check Determine whether each statement is true or false.

RC1. The graph of a quadratic function may be a straight line or a parabola.

RC2. The graph of every quadratic function is symmetric with respect to a vertical line.

RC3. Every quadratic function has either a maximum value or a minimum value.

RC4. The graph of $f(x) = 5x^2$ is wider than the graph of $f(x) = 3x^2$.

Concept Check Determine whether the graph of each function opens up or down.

CC1. $f(x) = -6x^2$ **CC2.** $g(x) = 8(x - 5)^2$ **CC3.** $f(x) = \frac{1}{2}(x - 2)^2 - 9$ **CC4.** $g(x) = -\frac{1}{2}(x + 11)^2 + 24$

a, b Graph. Find and label the vertex and the line of symmetry.

1. $f(x) = 4x^2$

x	$f(x)$
0	
1	
2	
−1	
−2	

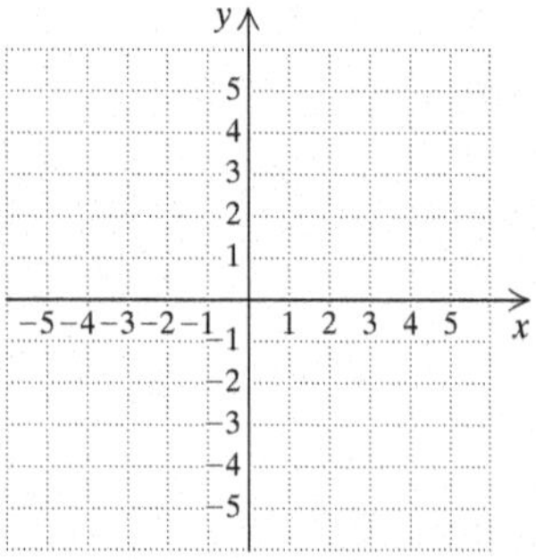

Vertex: (___, ___)
Line of symmetry: $x = $ ___

2. $f(x) = 5x^2$

x	$f(x)$
0	
1	
2	
−1	
−2	

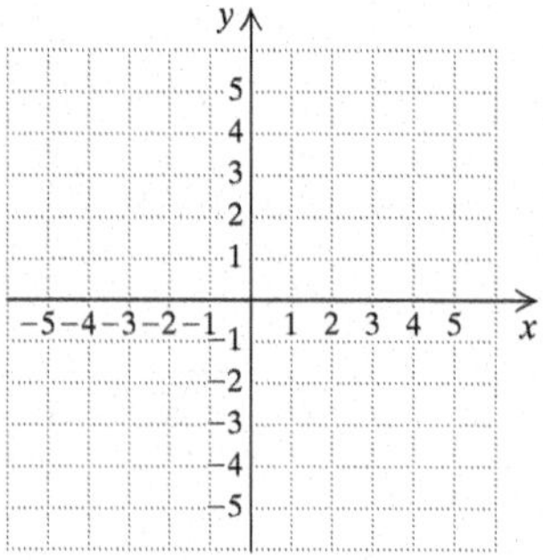

Vertex: (___, ___)
Line of symmetry: $x = $ ___

3. $f(x) = \frac{1}{3}x^2$

x	$f(x)$
0	
1	
2	
−1	
−2	

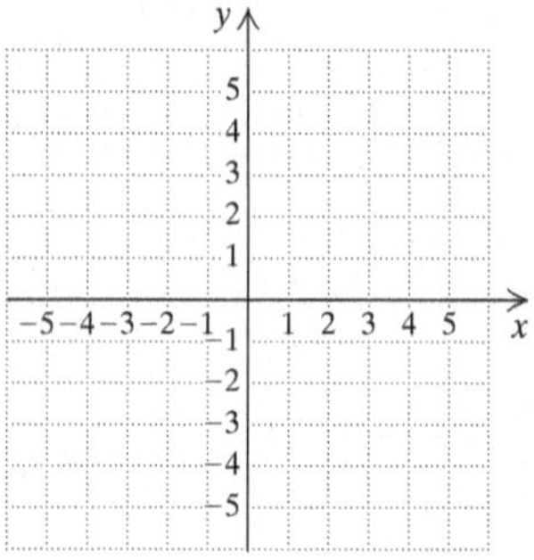

Vertex: (___, ___)
Line of symmetry: $x = $ ___

4. $f(x) = \frac{1}{4}x^2$

x	$f(x)$
0	
1	
2	
−1	
−2	

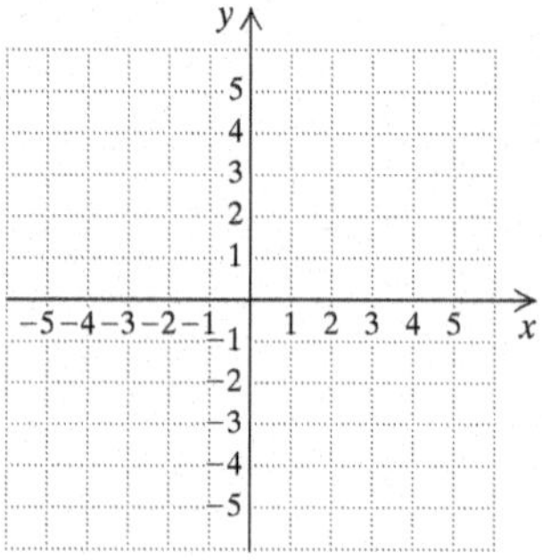

Vertex: (___, ___)
Line of symmetry: $x = $ ___

5. $f(x) = (x + 3)^2$

x	$f(x)$
−3	
−2	
−1	
−4	
−5	

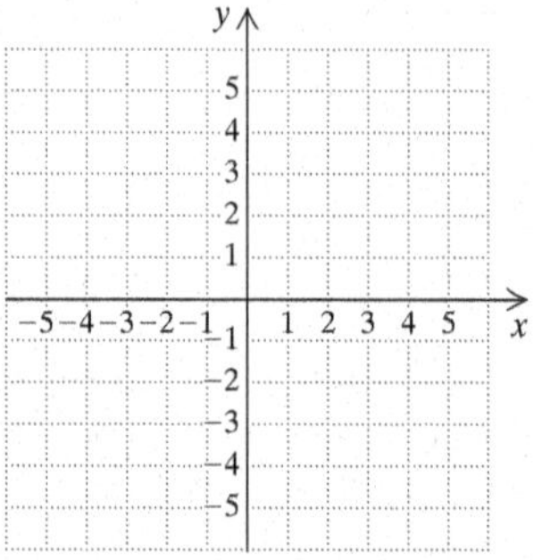

Vertex: (___, ___)
Line of symmetry: $x = $ ___

6. $f(x) = (x + 1)^2$

x	$f(x)$
−1	
0	
1	
−2	
−3	

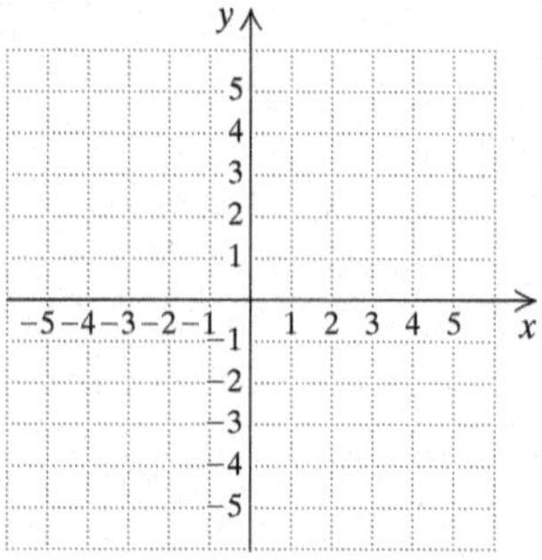

Vertex: (___, ___)
Line of symmetry: $x = $ ___

7. $f(x) = -4x^2$

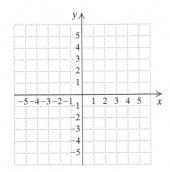

Vertex: (___, ___)
Line of symmetry: $x = $ ___

8. $f(x) = -3x^2$

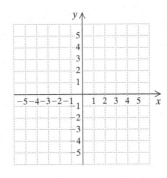

Vertex: (___, ___)
Line of symmetry: $x = $ ___

9. $f(x) = -\frac{1}{2}x^2$

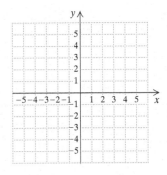

Vertex: (___, ___)
Line of symmetry: $x = $ ___

10. $f(x) = -\frac{1}{4}x^2$

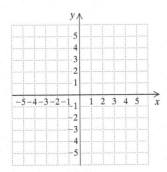

Vertex: (___, ___)
Line of symmetry: $x = $ ___

11. $f(x) = 2(x - 4)^2$

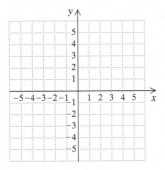

Vertex: (___, ___)
Line of symmetry: $x = $ ___

12. $f(x) = 4(x - 1)^2$

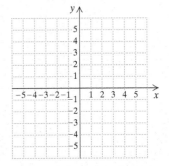

Vertex: (___, ___)
Line of symmetry: $x = $ ___

13. $f(x) = -2(x + 2)^2$

x	$f(x)$
−2	
−3	
−1	
−4	
0	

Vertex: (___, ___)
Line of symmetry: $x = $ ___

14. $f(x) = -2(x + 4)^2$

x	$f(x)$
−4	
−5	
−3	
−6	
−2	

Vertex: (___, ___)
Line of symmetry: $x = $ ___

15. $f(x) = 3(x - 1)^2$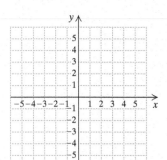

16. $f(x) = 4(x - 2)^2$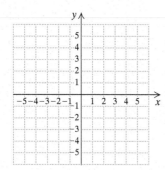

17. $f(x) = -\frac{3}{2}(x + 2)^2$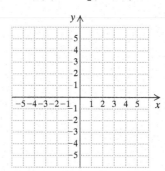

18. $f(x) = -\frac{5}{2}(x + 3)^2$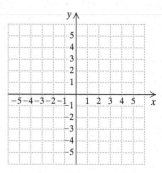

c Graph. Find and label the vertex and the line of symmetry. Find the maximum or minimum value.

19. $f(x) = (x - 3)^2 + 1$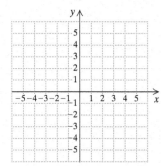

20. $f(x) = (x + 2)^2 - 3$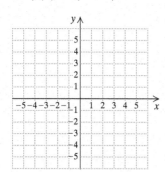

21. $f(x) = -3(x + 4)^2 + 1$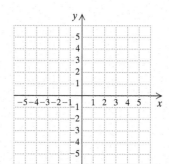

22. $f(x) = -\frac{1}{2}(x - 1)^2 - 3$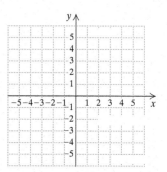

Vertex: (____, ____)
Line of symmetry: $x =$ ____
Minimum value: ____

Vertex: (____, ____)
Line of symmetry: $x =$ ____
Minimum value: ____

Vertex: (____, ____)
Line of symmetry: $x =$ ____
Maximum value: ____

Vertex: (____, ____)
Line of symmetry: $x =$ ____
Maximum value: ____

23. $f(x) = \frac{1}{2}(x + 1)^2 + 4$

24. $f(x) = -2(x - 5)^2 - 3$

25. $f(x) = -(x + 1)^2 - 2$

26. $f(x) = 3(x - 4)^2 + 2$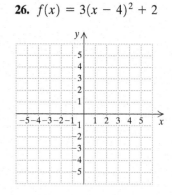

Vertex: (____, ____)
Line of symmetry: $x =$ ____
_____ value: ____

Vertex: (____, ____)
Line of symmetry: $x =$ ____
_____ value: ____

Vertex: (____, ____)
Line of symmetry: $x =$ ____
_____ value: ____

Vertex: (____, ____)
Line of symmetry: $x =$ ____
_____ value: ____

Skill Maintenance

Find the slope, if it exists, of the line containing the following points. [3.4a]

27. $(-10, 0)$ and $(5, -6)$

28. $(7, -3)$ and $(15, -3)$

11.6 Graphing $f(x) = ax^2 + bx + c$

a ANALYZING AND GRAPHING $f(x) = ax^2 + bx + c$

By *completing the square*, we can begin with any quadratic polynomial $ax^2 + bx + c$ and find an equivalent expression $a(x - h)^2 + k$. This allows us to analyze and graph any quadratic function $f(x) = ax^2 + bx + c$.

EXAMPLE 1 For $f(x) = x^2 - 6x + 4$, find the vertex, the line of symmetry, and the maximum or the minimum value. Then graph.

We first find the vertex and the line of symmetry. To do so, we find the equivalent form $a(x - h)^2 + k$ by completing the square, beginning as follows:

$$f(x) = x^2 - 6x + 4 = (x^2 - 6x \quad) + 4.$$

We complete the square inside the parentheses, but in a different manner than we did before. We take half the x-coefficient, $-6/2 = -3$, and square it: $(-3)^2 = 9$. Then we add 0, or $9 - 9$, inside the parentheses. (Instead of adding $(b/2)^2$ on both sides of an equation, we add and subtract it on the same side, effectively adding 0 and not changing the value of the expression.)

$$\begin{aligned}
f(x) &= (x^2 - 6x + 0) + 4 & \text{Adding 0} \\
&= (x^2 - 6x + 9 - 9) + 4 & \text{Substituting } 9 - 9 \text{ for } 0 \\
&= (x^2 - 6x + 9) + (-9 + 4) & \text{Using the associative law of addition to regroup} \\
&= (x - 3)^2 - 5 & \text{Factoring and simplifying}
\end{aligned}$$

The vertex is $(3, -5)$, and the line of symmetry is $x = 3$. The coefficient of x^2 is 1, which is positive, so the graph opens up. This tells us that -5 is the minimum value. We plot the vertex and draw the line of symmetry. We choose some x-values on both sides of the vertex and graph the parabola. Suppose we compute the pair $(5, -1)$:

$$f(5) = 5^2 - 6(5) + 4 = 25 - 30 + 4 = -1.$$

We note that it is 2 units to the right of the line of symmetry. There will also be a pair with the same y-coordinate on the graph 2 units *to the left* of the line of symmetry. Thus we get a second point, $(1, -1)$, without making another calculation.

x	$f(x)$	
3	-5	← Vertex
4	-4	
5	-1	
6	4	
2	-4	
1	-1	
0	4	

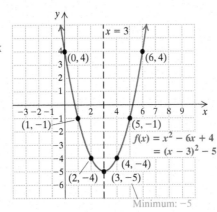

OBJECTIVES

a For a quadratic function, find the vertex, the line of symmetry, and the maximum or the minimum value, and then graph the function.

b Find the intercepts of a quadratic function.

1. For $f(x) = x^2 - 4x + 7$, find the vertex, the line of symmetry, and the maximum or the minimum value. Then graph.

x	f(x)

Vertex: (____, ____)
Line of symmetry:
$x =$ ____
Minimum value: ____

◀ Do Exercise 1.

EXAMPLE 2 For $f(x) = 3x^2 + 12x + 13$, find the vertex, the line of symmetry, and the maximum or the minimum value. Then graph.

Since the coefficient of x^2 is not 1, we factor out 3 from only the *first two* terms of the expression. Remember that we want to write the function in the form $f(x) = a(x - h)^2 + k$:

$$f(x) = 3x^2 + 12x + 13$$
$$= 3(x^2 + 4x) + 13. \quad \text{Factoring 3 out of the first two terms}$$

Next, we complete the square inside the parentheses:

$$f(x) = 3(x^2 + 4x \quad) + 13.$$

We take half the *x*-coefficient, $\frac{1}{2} \cdot 4 = 2$, and square it: $2^2 = 4$. Then we add 0, or $4 - 4$, inside the parentheses:

$$f(x) = 3(x^2 + 4x + 0) + 13 \quad \text{Adding 0}$$
$$= 3(x^2 + 4x + 4 - 4) + 13 \quad \text{Substituting } 4 - 4 \text{ for } 0$$
$$= 3(x^2 + 4x + 4 - 4) + 13$$
$$= 3(x^2 + 4x + 4) + 3(-4) + 13 \quad \left.\begin{matrix}\text{Using the distributive law to}\\\text{separate } -4 \text{ from the trinomial}\end{matrix}\right.$$
$$= 3(x^2 + 4x + 4) - 12 + 13$$
$$= 3(x + 2)^2 + 1 \quad \text{Factoring and simplifying}$$
$$= 3[x - (-2)]^2 + 1.$$

The vertex is $(-2, 1)$, and the line of symmetry is $x = -2$. The coefficient of x^2 is 3, so the graph is narrow and opens up. This tells us that 1 is the minimum value of the function. We choose a few *x*-values on one side of the line of symmetry, compute *y*-values, and use the resulting coordinates to find more points on the other side of the line of symmetry. We plot points and graph the parabola.

x	f(x)	
-2	1	← Vertex
-1	4	
-3	4	
0	13	
-4	13	

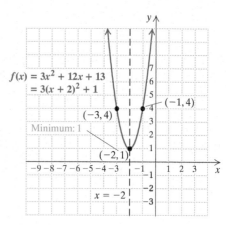

2. For $f(x) = 3x^2 - 24x + 43$, find the vertex, the line of symmetry, and the maximum or the minimum value. Then graph.

x	f(x)

Vertex: (____, ____)
Line of symmetry:
$x =$ ____
Minimum value: ____

◀ Do Exercise 2.

Answers

1.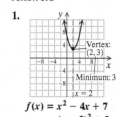
$f(x) = x^2 - 4x + 7$
$= (x - 2)^2 + 3$

2.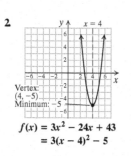
$f(x) = 3x^2 - 24x + 43$
$= 3(x - 4)^2 - 5$

EXAMPLE 3 For $f(x) = -2x^2 + 10x - 7$, find the vertex, the line of symmetry, and the maximum or the minimum value. Then graph.

Again, the coefficient of x^2 is not 1. We factor out -2 from only the *first two* terms of the expression. This makes the coefficient of x^2 inside the parentheses 1:

$$f(x) = -2x^2 + 10x - 7$$
$$= -2(x^2 - 5x) - 7.$$

Next, we complete the square as before:

$$f(x) = -2(x^2 - 5x \quad) - 7.$$

We take half the x-coefficient, $\frac{1}{2}(-5) = -\frac{5}{2}$, and square it: $\left(-\frac{5}{2}\right)^2 = \frac{25}{4}$. Then we add 0, or $\frac{25}{4} - \frac{25}{4}$, inside the parentheses:

$$f(x) = -2\left(x^2 - 5x + \frac{25}{4} - \frac{25}{4}\right) - 7$$
$$= -2\left(x^2 - 5x + \frac{25}{4} - \frac{25}{4}\right) - 7 \qquad \text{Adding 0, or } \frac{25}{4} - \frac{25}{4}$$

Using the distributive law to separate $-\frac{25}{4}$ from the trinomial

$$= -2\left(x^2 - 5x + \frac{25}{4}\right) + (-2)\left(-\frac{25}{4}\right) - 7$$
$$= -2\left(x^2 - 5x + \frac{25}{4}\right) + \frac{25}{2} - 7$$
$$= -2\left(x - \frac{5}{2}\right)^2 + \frac{11}{2}. \qquad \text{Factoring and simplifying}$$

The vertex is $\left(\frac{5}{2}, \frac{11}{2}\right)$, and the line of symmetry is $x = \frac{5}{2}$. The coefficient of x^2 is -2, so the graph is narrow and opens down. This tells us that $\frac{11}{2}$ is the maximum value of the function. We choose a few x-values on one side of the line of symmetry, compute y-values, and use the resulting coordinates to find more points on the other side of the line of symmetry. We plot points and graph the parabola.

x	$f(x)$	
$\frac{5}{2}$	$\frac{11}{2}$, or $5\frac{1}{2}$	← Vertex
3	5	
4	1	
5	-7	
2	5	
1	1	
0	-7	

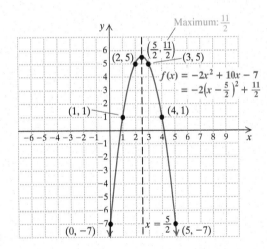

3. For $f(x) = -4x^2 + 12x - 5$, find the vertex, the line of symmetry, and the maximum or the minimum value. Then graph.

x	$f(x)$

Vertex: (___, ___)
Line of symmetry:
$x = $ ___
Maximum value: ___

Do Exercise 3. ▶

Answer

3.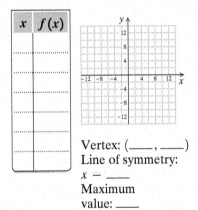

$f(x) = -4x^2 + 12x - 5$
$= -4\left(x - \frac{3}{2}\right)^2 + 4$

The method used in Examples 1–3 can be generalized to find a formula for locating the vertex. We complete the square as follows:

$$f(x) = ax^2 + bx + c$$
$$= a\left(x^2 + \frac{b}{a}x\right) + c. \quad \text{Factoring } a \text{ out of the first two terms. Check by multiplying.}$$

Half of the x-coefficient, $\frac{b}{a}$, is $\frac{b}{2a}$. We square it to get $\frac{b^2}{4a^2}$ and add $\frac{b^2}{4a^2} - \frac{b^2}{4a^2}$ inside the parentheses. Then we distribute a:

$$f(x) = a\left(x^2 + \frac{b}{a}x + \frac{b^2}{4a^2} - \frac{b^2}{4a^2}\right) + c$$

$$= a\left(x^2 + \frac{b}{a}x + \frac{b^2}{4a^2}\right) + a\left(-\frac{b^2}{4a^2}\right) + c \quad \text{Using the distributive law}$$

$$= a\left(x + \frac{b}{2a}\right)^2 + \frac{-b^2}{4a} + \frac{4ac}{4a} \quad \text{Factoring and finding a common denominator}$$

$$= a\left[x - \left(-\frac{b}{2a}\right)\right]^2 + \frac{4ac - b^2}{4a}.$$

Thus we have the following.

VERTEX; LINE OF SYMMETRY

The **vertex** of a parabola given by $f(x) = ax^2 + bx + c$ is

$$\left(-\frac{b}{2a}, \frac{4ac - b^2}{4a}\right), \quad \text{or} \quad \left(-\frac{b}{2a}, f\left(-\frac{b}{2a}\right)\right).$$

The x-coordinate of the vertex is $-b/(2a)$. The **line of symmetry** is $x = -b/(2a)$. The second coordinate of the vertex is easiest to find by computing $f\left(-\frac{b}{2a}\right)$.

Let's reexamine Example 3 to see how we could have found the vertex directly. From the formula above,

the x-coordinate of the vertex is $-\frac{b}{2a} = -\frac{10}{2(-2)} = \frac{5}{2}$.

Substituting $\frac{5}{2}$ into $f(x) = -2x^2 + 10x - 7$, we find the second coordinate of the vertex:

$$f\left(\tfrac{5}{2}\right) = -2\left(\tfrac{5}{2}\right)^2 + 10\left(\tfrac{5}{2}\right) - 7$$
$$= -2\left(\tfrac{25}{4}\right) + 25 - 7$$
$$= -\tfrac{25}{2} + 25 - 7 = -\tfrac{25}{2} + \tfrac{50}{2} - \tfrac{14}{2} = \tfrac{11}{2}.$$

The vertex is $\left(\tfrac{5}{2}, \tfrac{11}{2}\right)$. The line of symmetry is $x = \tfrac{5}{2}$.

We have developed two methods for finding the vertex. One is by completing the square and the other is by using a formula. You should check with your instructor about which method to use.

◀ Do Exercises 4–6.

Find the vertex of each parabola using the formula.

4. $f(x) = x^2 - 6x + 4$

5. $f(x) = 3x^2 - 24x + 43$

6. $f(x) = -4x^2 + 12x - 5$ **GS**

The x-coordinate of the vertex is

$$-\frac{b}{2a} = -\frac{\square}{2(-4)} = \frac{\square}{2}.$$

The second coordinate of the vertex is $f\left(\tfrac{3}{2}\right)$:

$$f\left(\tfrac{3}{2}\right) = -4(\quad)^2 + 12(\quad) - 5$$
$$= -4(\quad) + \square - 5$$
$$= \square + 18 - 5$$
$$= \square.$$

The vertex is ($\square$, $\square$).

Answers

4. $(3, -5)$ **5.** $(4, -5)$ **6.** $\left(\tfrac{3}{2}, 4\right)$

Guided Solution:

6. $12, 3, \tfrac{3}{2}, \tfrac{3}{2}, \tfrac{9}{4}, 18, -9, 4, \tfrac{3}{2}, 4$

b FINDING THE INTERCEPTS OF A QUADRATIC FUNCTION

SKILL REVIEW

Graph linear equations using intercepts. [7.4a]

Find the intercepts and then graph the line.

1. $3x - y = 3$
2. $2x + 4y = -8$

Answers: 1. y-intercept: $(0, -3)$; x-intercept: $(1, 0)$ 2. y-intercept: $(0, -2)$; x-intercept: $(-4, 0)$

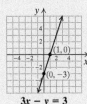
$3x - y = 3$

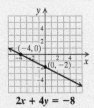
$2x + 4y = -8$

The points at which a graph crosses an axis are called **intercepts**. We determine the y-intercept by finding $f(0)$. For $f(x) = ax^2 + bx + c$, $f(0) = a \cdot 0^2 + b \cdot 0 + c = c$, so the y-intercept is $(0, c)$.

To find the x-intercepts, we look for values of x for which $f(x) = 0$. For $f(x) = ax^2 + bx + c$, we solve

$$0 = ax^2 + bx + c.$$

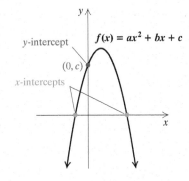

EXAMPLE 4 Find the intercepts of $f(x) = x^2 - 2x - 2$.

The y-intercept is $(0, f(0))$. Since $f(0) = 0^2 - 2 \cdot 0 - 2 = -2$, the y-intercept is $(0, -2)$. To find the x-intercepts, we solve

$$0 = x^2 - 2x - 2.$$

Using the quadratic formula, we have $x = 1 \pm \sqrt{3}$. Thus the x-intercepts are $(1 - \sqrt{3}, 0)$ and $(1 + \sqrt{3}, 0)$, or, approximately, $(-0.732, 0)$ and $(2.732, 0)$.

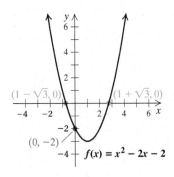

Do Exercises 7–9.

Find the intercepts.

GS 7. $f(x) = x^2 + 2x - 3$

The y-intercept is $(0, f(0))$.

$f(0) = \boxed{}^2 + 2 \cdot \boxed{} - 3 = -3$

The y-intercept is $(0, \boxed{})$. To find the x-intercepts, we solve $0 = x^2 + 2x - 3$:

$0 = x^2 + 2x - 3$
$0 = (x - 1)(\boxed{})$
$x - 1 = 0$ or $\boxed{} = 0$
$x = 1$ or $x = \boxed{}$.

The x-intercepts are $(1, 0)$ and $(\boxed{}, 0)$.

8. $f(x) = x^2 + 8x + 16$

9. $f(x) = x^2 - 4x + 1$

Answers

7. y-intercept: $(0, -3)$; x-intercepts: $(-3, 0), (1, 0)$
8. y-intercept: $(0, 16)$; x-intercept: $(-4, 0)$
9. y-intercept: $(0, 1)$; x-intercepts: $(2 - \sqrt{3}, 0), (2 + \sqrt{3}, 0)$, or $(0.268, 0), (3.732, 0)$

Guided Solution:
7. $0, 0, -3, x + 3, x + 3, -3, -3$

Visualizing for Success

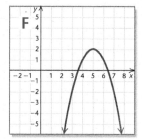

Match each equation or inequality with its graph.

1. $y = -(x - 5)^2 + 2$

2. $2x + 5y = 10$

3. $5x - 2y = 10$

4. $2x - 5y = 10$

5. $y = (x - 5)^2 - 2$

6. $y = x^2 - 5$

7. $5x + 2y \geq 10$

8. $5x + 2y = -10$

9. $y < 5x$

10. $y = -(x + 5)^2 + 2$

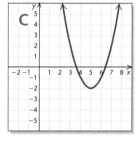
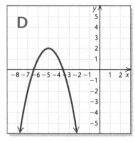
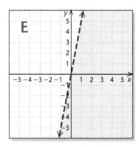
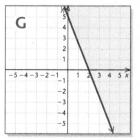
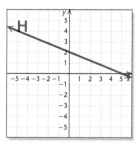
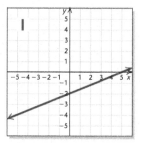
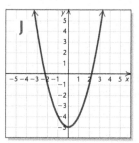

Answers on page A-37

11.6 Exercise Set

✓ Check Your Understanding

Reading and Concept Check Choose from beneath the blank the word that best completes each statement.

RC1. The graph of $f(x) = 5x^2 - 10x - 3$ opens _____.
   downward/upward

RC2. The function given by $f(x) = 2x^2 - x - 7$ has a _____ value.
   maximum/minimum

RC3. The graph of $g(x) = (x + 3)^2 - 2$ has its _____ at $(-3, -2)$.
   vertex/x-intercept

RC4. The _____ of the graph of $g(x) = -x^2 + 6x + 9$ is $(0, 9)$.
   x-intercept/y-intercept

---

a For each quadratic function, find **(a)** the vertex, **(b)** the line of symmetry, and **(c)** the maximum or the minimum value. Then **(d)** graph the function.

1. $f(x) = x^2 - 2x - 3$

   a) Vertex: (____, ____)
   b) Line of symmetry: $x =$ ____
   c) _____ value: ____
   d)

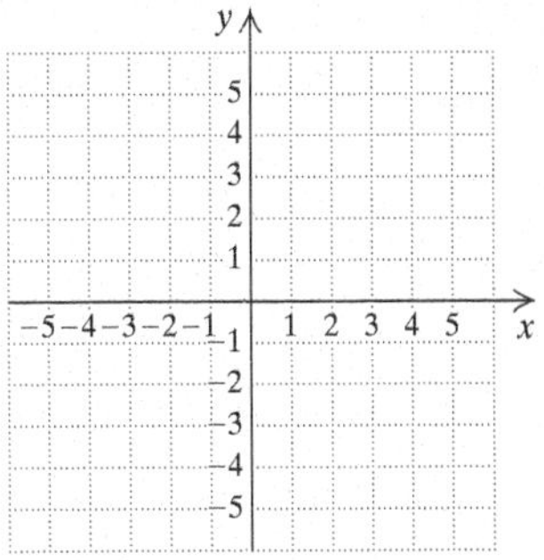

2. $f(x) = x^2 + 2x - 5$

   a) Vertex: (____, ____)
   b) Line of symmetry: $x =$ ____
   c) _____ value: ____
   d)

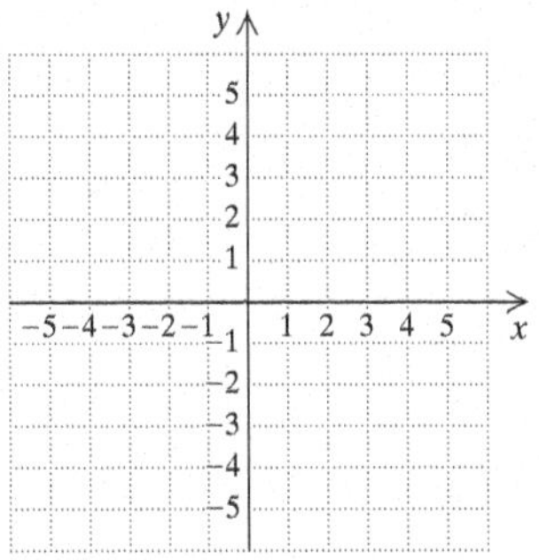

3. $f(x) = -x^2 - 4x - 2$

   a) Vertex: (____, ____)
   b) Line of symmetry: $x =$ ____
   c) _____ value: ____
   d)

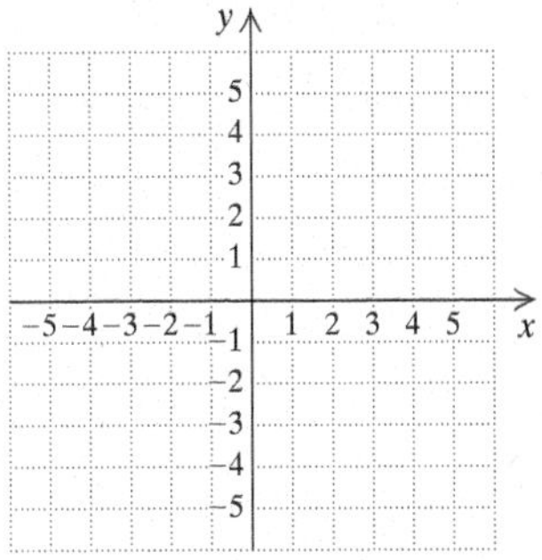

4. $f(x) = -x^2 + 4x + 1$

   a) Vertex: (____, ____)
   b) Line of symmetry: $x =$ ____
   c) _____ value: ____
   d)

SECTION 11.6 Graphing $f(x) = ax^2 + bx + c$

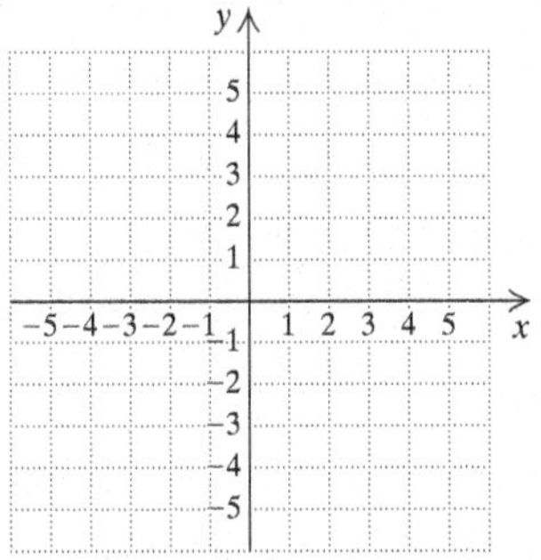

5. $f(x) = 3x^2 - 24x + 50$
   a) Vertex: (____, ____)
   b) Line of symmetry: $x =$ ____
   c) _____ value: ____
   d)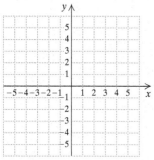

6. $f(x) = 4x^2 + 8x + 1$
   a) Vertex: (____, ____)
   b) Line of symmetry: $x =$ ____
   c) _____ value: ____
   d)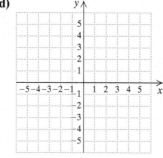

7. $f(x) = -2x^2 - 2x + 3$
   a) Vertex: (____, ____)
   b) Line of symmetry: $x =$ ____
   c) _____ value: ____
   d)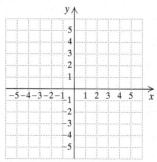

8. $f(x) = -2x^2 + 2x + 1$
   a) Vertex: (____, ____)
   b) Line of symmetry: $x =$ ____
   c) _____ value: ____
   d)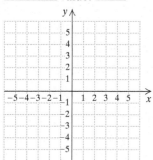

9. $f(x) = 5 - x^2$
   a) Vertex: (____, ____)
   b) Line of symmetry: $x =$ ____
   c) _____ value: ____
   d)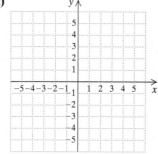

10. $f(x) = x^2 - 3x$
    a) Vertex: (____, ____)
    b) Line of symmetry: $x =$ ____
    c) _____ value: ____
    d)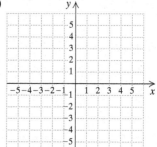

11. $f(x) = 2x^2 + 5x - 2$
    a) Vertex: (____, ____)
    b) Line of symmetry: $x =$ ____
    c) _____ value: ____
    d)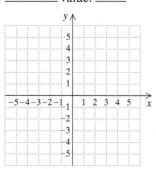

12. $f(x) = -4x^2 - 7x + 2$
    a) Vertex: (____, ____)
    b) Line of symmetry: $x =$ ____
    c) _____ value: ____
    d)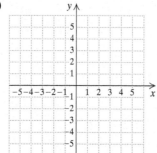

b Find the x- and y-intercepts.

13. $f(x) = x^2 - 6x + 1$ **14.** $f(x) = x^2 + 2x + 12$ **15.** $f(x) = -x^2 + x + 20$ **16.** $f(x) = -x^2 + 5x + 24$

17. $f(x) = 4x^2 + 12x + 9$ **18.** $f(x) = 3x^2 - 6x + 1$ **19.** $f(x) = 4x^2 - x + 8$ **20.** $f(x) = 2x^2 + 4x - 1$

Skill Maintenance

Solve. [6.9b]

21. *Determining Medication Dosage.* A child's dosage D, in milligrams, of a medication varies directly as the child's weight w, in kilograms. To control a fever, a doctor suggests that a child who weighs 28 kg be given 420 mg of analgesic medication. Find an equation of variation.

22. *Calories Burned.* The number C of calories burned while exercising varies directly as the time t, in minutes, spent exercising. Harold exercises for 24 min on a StairMaster and burns 356 calories. Find an equation of variation.

Find the variation constant and an equation of variation in which y varies inversely as x and the following are true. [6.9c]

23. $y = 125$ when $x = 2$

24. $y = 2$ when $x = 125$

Find the variation constant and an equation of variation in which y varies directly as x and the following are true. [6.9a]

25. $y = 125$ when $x = 2$

26. $y = 2$ when $x = 125$

Synthesis

27. Use the TRACE and/or TABLE features of a graphing calculator to estimate the maximum or minimum values of the following functions.
  a) $f(x) = 2.31x^2 - 3.135x - 5.89$
  b) $f(x) = -18.8x^2 + 7.92x + 6.18$

28. Use the INTERSECT feature of a graphing calculator to find the points of intersection of the graphs of the functions.
$f(x) = x^2 + 2x + 1, \quad g(x) = -2x^2 - 4x + 1$

Graph.

29. $f(x) = |x^2 - 1|$

30. $f(x) = |x^2 + 6x + 4|$

31. $f(x) = |x^2 - 3x - 4|$

32. $f(x) = |2(x - 3)^2 - 5|$

33. A quadratic function has $(-1, 0)$ as one of its intercepts and $(3, -5)$ as its vertex. Find an equation for the function.

34. A quadratic function has $(4, 0)$ as one of its intercepts and $(-1, 7)$ as its vertex. Find an equation for the function.

35. Consider
$$f(x) = \frac{x^2}{8} + \frac{x}{4} - \frac{3}{8}.$$
Find the vertex, the line of symmetry, and the maximum or the minimum value. Then draw the graph.

36. Use only the graph in Exercise 35 to approximate the solutions of each of the following equations.
  a) $\frac{x^2}{8} + \frac{x}{4} - \frac{3}{8} = 0$
  b) $\frac{x^2}{8} + \frac{x}{4} - \frac{3}{8} = 1$
  c) $\frac{x^2}{8} + \frac{x}{4} - \frac{3}{8} = 2$

11.7 Mathematical Modeling with Quadratic Functions

OBJECTIVES

a Solve maximum–minimum problems involving quadratic functions.

b Fit a quadratic function to a set of data to form a mathematical model, and solve related applied problems.

MyLab Math
ANIMATION

We now consider some of the many situations in which quadratic functions can serve as mathematical models.

a MAXIMUM–MINIMUM PROBLEMS

We have seen that for any quadratic function $f(x) = ax^2 + bx + c$, the value of $f(x)$ at the vertex is either a maximum or a minimum, meaning that either all outputs are smaller than that value for a maximum or larger than that value for a minimum.

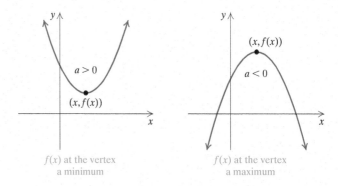

$f(x)$ at the vertex a minimum

$f(x)$ at the vertex a maximum

There are many applied problems in which we want to find a maximum or a minimum value. If a quadratic function can be used as a model, we can find such maximums or minimums by finding coordinates of the vertex.

EXAMPLE 1 *Landscape Design.* As part of his backyard design, Christopher plans to create an outdoor dining area by planting boxwood shrubs around the perimeter of a rectangle. If he has enough shrubs to form a perimeter of 64 ft (after allowing for an opening for entry into the area), what are the dimensions of the largest rectangle that Christopher can enclose?

1. **Familiarize.** We first make a drawing and label it. We let $l =$ the length of the dining area and $w =$ the width. Recall the following formulas:

   Perimeter: $2l + 2w$;
   Area: $l \cdot w$.

To become familiar with the problem, let's choose some dimensions (shown at right) for which $2l + 2w = 64$ and then calculate the corresponding areas. What choice of l and w will maximize A?

l	w	A
22	10	220
20	12	240
18	14	252
18.5	13.5	249.75
12.4	19.6	243.04
15	17	255

2. **Translate.** We have two equations, one for perimeter and one for area:

$$2l + 2w = 64,$$
$$A = l \cdot w.$$

Let's use them to express A as a function of l or w, but not both. To express A in terms of w, for example, we solve for l in the first equation:

$$2l + 2w = 64$$
$$2l = 64 - 2w$$
$$l = \frac{64 - 2w}{2}$$
$$= 32 - w.$$

Substituting $32 - w$ for l, we get a quadratic function $A(w)$, or just A:

$$A = lw = (32 - w)w = 32w - w^2 = -w^2 + 32w.$$

3. **Solve.** To find the vertex, we complete the square:

$$A = -w^2 + 32w \qquad \text{This is a parabola opening down, so a maximum exists.}$$
$$= -1(w^2 - 32w) \qquad \text{Factoring out } -1$$
$$= -1(w^2 - 32w + 256 - 256) \qquad \tfrac{1}{2}(-32) = -16; (-16)^2 = 256.$$
$$\qquad\qquad\qquad\qquad\qquad\qquad \text{We add 0, or } 256 - 256.$$
$$= -1(w^2 - 32w + 256) + (-1)(-256) \qquad \text{Using the distributive law}$$
$$= -(w - 16)^2 + 256.$$

The vertex is $(16, 256)$. Thus the maximum value is 256. It occurs when $w = 16$ and $l = 32 - w = 32 - 16 = 16$.

4. **Check.** We note that 256 is larger than any of the values found in the *Familiarize* step. To be more certain, we could make more calculations. We can also use the graph of the function to check the maximum value.

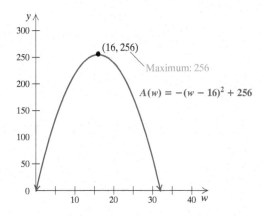

5. **State.** The largest rectangular dining area that can be enclosed is 16 ft by 16 ft; that is, it is a square with sides of 16 ft.

Do Exercise 1.

1. **Stained-Glass Window Design.** An artist is designing a rectangular stained-glass window with a perimeter of 84 in. What dimensions will yield the maximum area?

To familiarize yourself with the problem, complete the following table.

l	w	A
10	32	320
12	30	
15	27	
20	22	
20.5	21.5	

Answer
1. 21 in. by 21 in.

CALCULATOR CORNER

Maximum and Minimum Values We can use a graphing calculator to find the maximum or the minimum value of a quadratic function. Consider the quadratic function in Example 1, $A = -w^2 + 32w$. First, we replace w with x and A with y and graph the function in a window that displays the vertex of the graph. We choose [0, 40, 0, 300], with Xscl = 5 and Yscl = 20. Now, we select the MAXIMUM feature from the CALC menu. We are prompted to select a left bound for the maximum point. This means that we must choose an x-value that is to the left of the x-value of the point where the maximum occurs. This can be done by using the left- and right-arrow keys to move the cursor to a point to the left of the maximum point or by keying in an appropriate value. Once this is done, we press ENTER. Now, we are prompted to select a right bound. We move the cursor to a point to the right of the maximum point or key in an appropriate value.

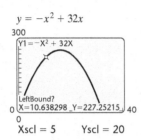
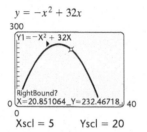

We press ENTER again. Finally, we are prompted to guess the x-value at which the maximum occurs. We move the cursor close to the maximum or key in an x-value. We press ENTER a third time and see that the maximum function value of 256 occurs when $x = 16$. (One or both coordinates of the maximum point might be approximations of the actual values, as shown with the x-value below.)

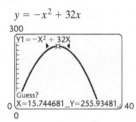
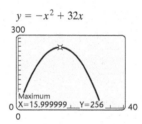

EXERCISES: Use the maximum or minimum feature on a graphing calculator to find the maximum or minimum value of each function.

1. $y = 3x^2 - 6x + 4$
2. $y = 2x^2 + x + 5$
3. $y = -x^2 + 4x + 2$
4. $y = -4x^2 + 5x - 1$

b FITTING QUADRATIC FUNCTIONS TO DATA

SKILL REVIEW
Find an equation of a line when two points are given. [7.5c]
Find an equation of the line containing the given points.
1. $(2, 9)$ and $(0, 8)$
2. $(-3, -1)$ and $(-5, 7)$

Answers: **1.** $y = \frac{1}{2}x + 8$ **2.** $y = -4x - 13$

As we move through our study of mathematics, we develop a library of functions. These functions can serve as models for many applications. Some of them are graphed below. We have not considered the cubic or quartic functions in detail (we leave that discussion to a later course), but we show them here for reference.

Linear function:
$f(x) = mx + b$

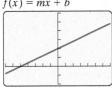

Quadratic function:
$f(x) = ax^2 + bx + c, a > 0$

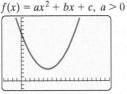

Quadratic function:
$f(x) = ax^2 + bx + c, a < 0$

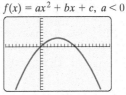

Absolute-value function:
$f(x) = |x|$

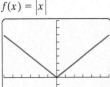

Cubic function:
$f(x) = ax^3 + bx^2 + cx + d, a > 0$

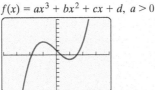

Quartic function:
$f(x) = ax^4 + bx^3 + cx^2 + dx + e, a > 0$

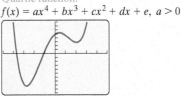

Now let's consider some real-world data. How can we decide which type of function might fit the data of a particular application? One simple way is to graph the data and look for a pattern resembling one of the graphs above. For example, data might be modeled by a linear function if the graph resembles a straight line. The data might be modeled by a quadratic function if the graph rises and then falls, or falls and then rises, in a curved manner resembling a parabola. For a quadratic, it might also just rise or fall in a curved manner as if following only one part of the parabola.

Let's now use our library of functions to see which, if any, might fit certain data situations.

EXAMPLES *Choosing Models.* For the scatterplots and graphs below, determine which, if any, of the following functions might be used as a model for the data.

Linear, $f(x) = mx + b$;
Quadratic, $f(x) = ax^2 + bx + c, a > 0$;
Quadratic, $f(x) = ax^2 + bx + c, a < 0$;
Polynomial, neither quadratic nor linear

2.

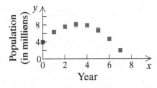

The data rise and then fall in a curved manner fitting a quadratic function $f(x) = ax^2 + bx + c, a < 0$.

Choosing Models. For the scatterplots in Margin Exercises 2–5, determine which, if any, of the following functions might be used as a model for the data:

Linear, $f(x) = mx + b$;

Quadratic, $f(x) = ax^2 + bx + c$, $a > 0$;

Quadratic, $f(x) = ax^2 + bx + c$, $a < 0$;

Polynomial, neither quadratic nor linear.

2.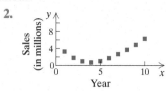

3.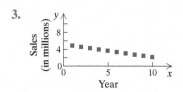

4.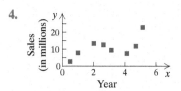

5.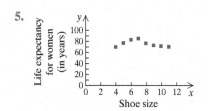

DATA: *Orthopedic Quarterly*

3.

The data seem to fit a linear function $f(x) = mx + b$.

4.

The data rise in a manner fitting the right side of a quadratic function $f(x) = ax^2 + bx + c$, $a > 0$.

5. Precipitation in Sonoma, California

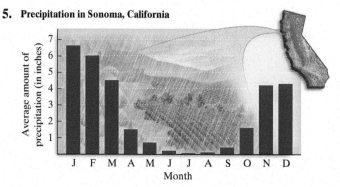

The data fall and then rise in a curved manner fitting a quadratic function $f(x) = ax^2 + bx + c$, $a > 0$.

6. Average Number of Motorcyclists Killed per Hour on the Weekend

DATA: FARS and GES

The data fall, then rise, then fall again. They do not appear to fit a linear function or a quadratic function but might fit a polynomial function that is neither quadratic nor linear.

◀ Do Exercises 2–5.

Answers
2. $f(x) = ax^2 + bx + c$, $a > 0$
3. $f(x) = mx + b$ **4.** Polynomial, neither quadratic nor linear **5.** Polynomial, neither quadratic nor linear

Whenever a quadratic function seems to fit a data situation, that function can be determined if at least three inputs and their outputs are known.

EXAMPLE 7 *Suspension Bridge.* The Clifton Suspension Bridge in Bristol, UK, completed in 1864, is one of the longest bridges suspended by chains. The data in the table at right show the height of the chains above the road surface at different points along the bridge. Here, 0 represents the center of the bridge, negative values of x represent distances in one direction from the center, and positive values of x represent distances in the other direction.

DATA: cliftonbridge.org.uk

LOCATION ON BRIDGE (in meters from center)	HEIGHT OF CHAIN (in meters)
0	1
20	1.7
40	4.0
−40	4.0
−60	7.6

a) Make a scatterplot of the data.
b) Decide whether the data seem to fit a quadratic function.
c) Use the data points $(40, 4)$, $(0, 1)$, and $(-40, 4)$ to find a quadratic function that fits the data.
d) Use the function to estimate the height of the chains at the towers, which are each 107 m from the center of the bridge.

a) The scatterplot is shown at right.

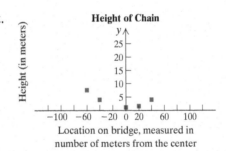

Height of Chain

Location on bridge, measured in number of meters from the center

b) The data seem to fall and rise in a manner similar to a quadratic function. Note that there may not be a function that exactly fits all the data.

c) We are looking for a quadratic function

$$H(x) = ax^2 + bx + c.$$

We need to determine the constants a, b, and c. We use the three data points $(40, 4)$, $(0, 1)$, and $(-40, 4)$ and substitute as follows:

$$4 = a \cdot 40^2 + b \cdot 40 + c,$$
$$1 = a \cdot 0^2 + b \cdot 0 + c,$$
$$4 = a \cdot (-40)^2 + b \cdot (-40) + c.$$

After simplifying, we see that we need to solve the system

$$4 = 1600a + 40b + c,$$
$$1 = c,$$
$$4 = 1600a - 40b + c.$$

6. **Ticket Profits.** Valley Community College is presenting a series of plays. The profit P, in dollars, after x days is listed in the following table.

DAYS x	PROFIT P
0	$-100
90	560
180	872
270	870
360	548
450	-100

a) Make a scatterplot of the data.
b) Decide whether the data can be modeled by a quadratic function.
c) Use the data points $(0, -100)$, $(180, 872)$, and $(360, 548)$ to find a quadratic function that fits the data.
d) Use the function to estimate the profit after 225 days.

Since $c = 1$, we need to solve a system of two equations in two variables:

$$4 = 1600a + 40b + 1 \quad \text{or} \quad 3 = 1600a + 40b \quad (1)$$
$$4 = 1600a - 40b + 1, \quad\quad 3 = 1600a - 40b. \quad (2)$$

We add equations (1) and (2) and solve for a:

$$3 = 1600a + 40b$$
$$3 = 1600a - 40b$$
$$6 = 3200a \quad \text{Adding}$$
$$\frac{3}{1600} = a. \quad \text{Solving for } a$$

Next, we substitute $\frac{3}{1600}$ for a in equation (1) and solve for b:

$$3 = 1600\left(\frac{3}{1600}\right) + 40b$$
$$3 = 3 + 40b$$
$$0 = 40b$$
$$0 = b.$$

This gives us the quadratic function:

$$H(x) = \frac{3}{1600}x^2 + 1, \quad \text{or} \quad H(x) = 0.001875x^2 + 1.$$

d) The towers are 107 m from the center of the bridge. Since the heights of the towers are the same, we need find only $H(107)$:

$$H(107) = 0.001875(107)^2 + 1 \approx 22.5.$$

At a distance of 107 m from the center of the bridge, the chains are about 22.5 m long.

◀ **Do Exercise 6.**

Answer

6. (a)

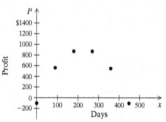

(b) yes; (c) $P(x) = -0.02x^2 + 9x - 100$;
(d) $912.50

11.7 Exercise Set

FOR EXTRA HELP MyLab Math

✓ Check Your Understanding

Reading Check Determine whether each statement is true or false.

RC1. Sometimes we can solve a problem without solving an equation or an inequality.

RC2. Every quadratic function has a maximum value.

RC3. A scatterplot can help us decide what type of function might fit a set of data.

RC4. When fitting a quadratic function to a set of data, the function must go through all the points on the scatterplot.

Concept Check Write an equation indicating that the ordered pair (x, y) is a solution of $y = ax^2 + bx + c$. (See Example 7.)

CC1. $(2, 4)$ **CC2.** $(5, 0)$ **CC3.** $(-6, 10)$

a Solve.

1. *Architecture.* An architect is designing a hotel with a central atrium. Each floor is to be rectangular and is allotted 720 ft of security piping around walls outside the rooms. What dimensions will allow the atrium to have the maximum area?

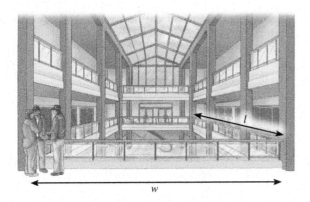

2. *Fenced-In Land.* A farmer has 100 yd of fencing. What are the dimensions of the largest rectangular pen that the farmer can enclose?

3. *Molding Plastics.* Economite Plastics plans to produce a one-compartment vertical file by bending the long side of an 8-in. by 14-in. sheet of plastic along two lines to form a U shape. How tall should the file be in order to maximize the volume that the file can hold?

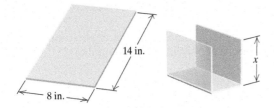

4. *Patio Design.* A stone mason has enough stones to enclose a rectangular patio with a perimeter of 60 ft, assuming that the attached house forms one side of the rectangle. What is the maximum area that the mason can enclose? What should the dimensions of the patio be in order to yield this area?

5. *Minimizing Cost.* Aki's Bicycle Designs has determined that when x hundred bicycles are built, the average cost per bicycle is given by
$$C(x) = 0.1x^2 - 0.7x + 2.425,$$
where $C(x)$ is in hundreds of dollars. How many bicycles should the shop build in order to minimize the average cost per bicycle?

6. *Corral Design.* A rancher needs to enclose two adjacent rectangular corrals, one for sheep and one for cattle. If a river forms one side of the corrals and 180 yd of fencing is available, what is the largest total area that can be enclosed?

7. *Garden Design.* A farmer decides to enclose a rectangular garden, using the side of a barn as one side of the rectangle. What is the maximum area that the farmer can enclose with 40 ft of fence? What should the dimensions of the garden be in order to yield this area?

8. *Composting.* A rectangular compost container is to be formed in a corner of a fenced yard, with 8 ft of chicken wire completing the other two sides of the rectangle. If the chicken wire is 3 ft high, what dimensions of the base will maximize the volume of the container?

9. *Ticket Sales.* The number of tickets sold each day for an upcoming performance of Handel's *Messiah* is given by
$$N(x) = -0.4x^2 + 9x + 11,$$
where x is the number of days after the concert was first announced. When will daily ticket sales peak and how many tickets will be sold that day?

10. *Stock Prices.* The value of a share of a particular stock, in dollars, can be represented by $V(x) = x^2 - 6x + 13$, where x is the number of months after January 2018. What is the lowest value that $V(x)$ will reach, and when will that occur?

Maximizing Profit. Total profit P is the difference between total revenue R and total cost C. Given the following total-revenue and total-cost functions, find the total profit, the maximum value of the total profit, and the value of x at which it occurs.

11. $R(x) = 1000x - x^2$,
    $C(x) = 3000 + 20x$

12. $R(x) = 200x - x^2$,
    $C(x) = 5000 + 8x$

13. What is the maximum product of two numbers whose sum is 22? What numbers yield this product?

14. What is the maximum product of two numbers whose sum is 45? What numbers yield this product?

15. What is the minimum product of two numbers whose difference is 4? What are the numbers?

16. What is the minimum product of two numbers whose difference is 6? What are the numbers?

17. What is the maximum product of two numbers that add to -12? What numbers yield this product?

18. What is the minimum product of two numbers that differ by 9? What are the numbers?

b *Choosing Models.* For the scatterplots and graphs in Exercises 19–26, determine which, if any, of the following functions might be used as a model for the data: Linear, $f(x) = mx + b$; quadratic, $f(x) = ax^2 + bx + c, a > 0$; quadratic, $f(x) = ax^2 + bx + c, a < 0$; polynomial, neither quadratic nor linear.

19.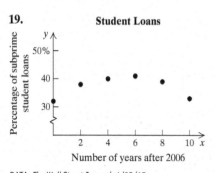
Student Loans
DATA: *The Wall Street Journal*, 4/25/17

20.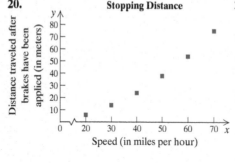
Stopping Distance

21.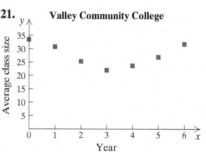
Valley Community College

22.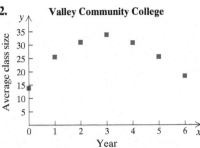

23.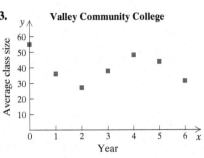

24.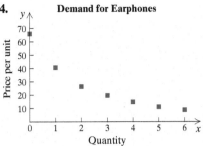

25. **High School Graduation Rate**

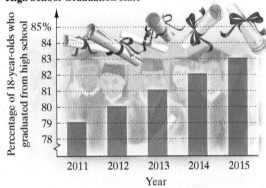

DATA: National Center for Education Statistics

26. **Hurricanes in the Atlantic Basin**

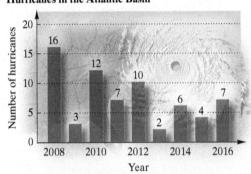

DATA: National Oceanic and Atmospheric Administration/ Hurricane Research Division

Find a quadratic function that fits the set of data points.

27. $(1, 4), (-1, -2), (2, 13)$

28. $(1, 4), (-1, 6), (-2, 16)$

29. $(2, 0), (4, 3), (12, -5)$

30. $(-3, -30), (3, 0), (6, 6)$

31. *Nighttime Accidents.*
    a) Find a quadratic function that fits the following data.

TRAVEL SPEED (in kilometers per hour)	NUMBER OF NIGHTTIME ACCIDENTS (for every 200 million kilometers driven)
60	400
80	250
100	250

    b) Use the function to estimate the number of nighttime accidents that occur at 50 km/h.

32. *Daytime Accidents.*
    a) Find a quadratic function that fits the following data.

TRAVEL SPEED (in kilometers per hour)	NUMBER OF DAYTIME ACCIDENTS (for every 200 million kilometers driven)
60	100
80	130
100	200

    b) Use the function to estimate the number of daytime accidents that occur at 50 km/h.

33. *River Depth.* Typically, rivers are deepest in the middle, with the depth decreasing to zero at the edges. A hydrologist measures the depths D, in feet, of a river at distances x, in feet, from one bank. The results are listed in the following table. Use the data points $(0, 0)$, $(50, 20)$, and $(100, 0)$ to find a quadratic function that fits the data. Then use the function to estimate the depth of the river at 75 ft from the bank.

DISTANCE x FROM THE RIVERBANK (in feet)	DEPTH D OF THE RIVER (in feet)
0	0
15	10.2
25	17
50	20
90	7.2
100	0

34. *Canoe Depth.* The following figure shows the cross section of a canoe. Canoes are deepest at the middle of the center line, with the depth decreasing to zero at the edges. The following table lists measures of the depths D, in inches, along the center line of one canoe at distances x, in inches, from the edge. Use the data points $(0, 0)$, $(18, 14)$, and $(36, 0)$ to find a quadratic function that fits the data. Then use the function to estimate the depth of the canoe 10 in. from the edge along the center line.

DISTANCE x FROM THE EDGE OF THE CANOE ALONG THE CENTER LINE (in inches)	DEPTH D OF THE CANOE (in inches)
0	0
9	10.5
18	14
30	7.75
36	0

35. *U.S. Homeownership.* Between 1995 and 2015, homeownership rates in the United States rose and then fell in a pattern that can be modeled by a quadratic function, as shown in the following graph. Use the data points $(4, 67)$, $(8, 68)$, and $(12, 68)$ to find a quadratic function that fits the data. Then use the function to estimate the homeownership rate in 2005.

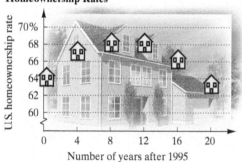

Homeownership Rates
DATA: U.S. Census Bureau

36. *Cake Servings.* Tyler provides customers with the following chart showing the number of servings for various sizes of round cakes.

DIAMETER	NUMBER OF SERVINGS
6 in.	12
8 in.	24
10 in.	35
12 in.	56
14 in.	78

The number of servings should probably be a function of diameter because it should be proportional to the area, and the area is a quadratic function of the diameter. (The area of a circular region is given by $A = \pi r^2$ or $(\pi/4)d^2$.)

a) Express the number of servings as a quadratic function of diameter using the data points $(6, 12)$, $(8, 24)$, and $(12, 56)$.
b) Use the function to find the number of servings in a 9-in. cake.

Skill Maintenance

Add, subtract, or multiply.

37. $(-3x^2 - x - 2) + (x^2 + 3x - 7)$ [4.4a]

38. $(2mn^2 - n^2 - 3m^2n) - (m^2n + 2mn^2 - n^2)$ [4.4c]

39. $(c^2d + 2y)(c^2d - 2y)$ [4.6b]

Factor.

40. $100t^2 - 81$ [5.5d]

41. $12x^3 - 60x^2 + 75x$ [5.5b]

42. $6y^2 + y - 12$ [5.3a], [5.4a]

Polynomial Inequalities and Rational Inequalities

11.8

OBJECTIVES

a Solve quadratic inequalities and other polynomial inequalities.

b Solve rational inequalities.

a QUADRATIC AND OTHER POLYNOMIAL INEQUALITIES

Inequalities like the following are called **quadratic inequalities**:

$$x^2 + 3x - 10 < 0, \quad 5x^2 - 3x + 2 \geq 0.$$

In each case, we have a polynomial of degree 2 on the left. We will solve such inequalities in two ways. The first method provides understanding and the second yields the more efficient method.

We can solve a quadratic inequality, such as $ax^2 + bx + c > 0$, by considering the graph of a related function, $f(x) = ax^2 + bx + c$.

EXAMPLE 1 Solve: $x^2 + 3x - 10 > 0$.

Consider the function $f(x) = x^2 + 3x - 10$ and its graph. The graph opens up since the leading coefficient ($a = 1$) is positive. We find the x-intercepts by setting the polynomial equal to 0 and solving:

$$x^2 + 3x - 10 = 0$$
$$(x + 5)(x - 2) = 0$$
$$x + 5 = 0 \quad \text{or} \quad x - 2 = 0$$
$$x = -5 \quad \text{or} \quad x = 2.$$

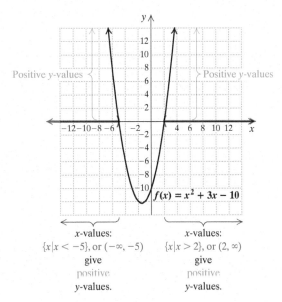

x-values: $\{x | x < -5\}$, or $(-\infty, -5)$ give positive y-values.

x-values: $\{x | x > 2\}$, or $(2, \infty)$ give positive y-values.

Values of y will be positive to the left and right of the intercepts, as shown. Thus the solution set of the inequality is

$$\{x | x < -5 \text{ or } x > 2\}, \quad \text{or} \quad (-\infty, -5) \cup (2, \infty).$$

Do Exercise 1. ▶

We can solve any inequality by considering the graph of a related function and finding x-intercepts, as in Example 1. In some cases, we may need to use the quadratic formula to find the intercepts.

SKILL REVIEW

Write interval notation for the solution set or graph of an inequality. [9.1b]

Write interval notation for the given set.

1. $\{x | -3 < x \leq 10\}$
2. $\{y | y > -\frac{1}{2}\}$

Answers: 1. $(-3, 10]$
2. $\left(-\frac{1}{2}, \infty\right)$

MyLab Math
VIDEO

1. Solve by graphing:
   $x^2 + 2x - 3 > 0$.

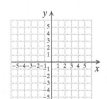

Answer
1. $\{x | x < -3 \text{ or } x > 1\}$, or $(-\infty, -3) \cup (1, \infty)$

EXAMPLE 2 Solve: $x^2 + 3x - 10 < 0$.

Looking again at the graph of $f(x) = x^2 + 3x - 10$ or at least visualizing it tells us that y-values are negative for those x-values between -5 and 2.

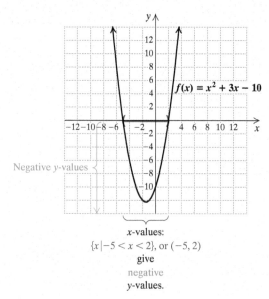

The solution set is $\{x | -5 < x < 2\}$, or $(-5, 2)$.

When an inequality contains $\leq$ or $\geq$, the x-values of the x-intercepts must be included. Thus the solution set of the inequality $x^2 + 3x - 10 \leq 0$ is $\{x | -5 \leq x \leq 2\}$, or $[-5, 2]$.

◀ Do Exercises 2 and 3.

Solve by graphing.

2. $x^2 + 2x - 3 < 0$

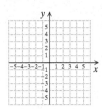

3. $x^2 + 2x - 3 \leq 0$

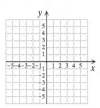

We now consider a more efficient method for solving polynomial inequalities. The preceding discussion provides the understanding for this method. In Examples 1 and 2, we see that the x-intercepts divide the number line into intervals.

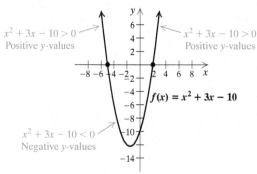

If a function has a positive output for one number in an interval, it will be positive for all the numbers in the interval. The same is true for negative outputs. Thus we can merely make a test substitution in each interval to solve the inequality. This is very similar to our method of using test points to graph a linear inequality in a plane.

Answers

2. $\{x | -3 < x < 1\}$, or $(-3, 1)$
3. $\{x | -3 \leq x \leq 1\}$, or $[-3, 1]$

EXAMPLE 3 Solve: $x^2 + 3x - 10 < 0$.

We set the polynomial equal to 0 and solve. The solutions of $x^2 + 3x - 10 = 0$, or $(x + 5)(x - 2) = 0$, are -5 and 2. We locate the solutions on the number line as follows. Note that the numbers divide the number line into three intervals, which we will call A, B, and C. Within each interval, the values of the function $f(x) = x^2 + 3x - 10$ will be all positive or will be all negative.

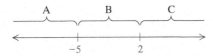

We choose a test number in interval A, say -7, and substitute -7 for x in the function $f(x) = x^2 + 3x - 10$:

$$f(-7) = (-7)^2 + 3(-7) - 10 = 49 - 21 - 10 = 18.$$

Since $f(-7) = 18$ and $18 > 0$, the function values will be positive for any number in interval A.

Next, we try a test number in interval B, say 1, and find the corresponding function value:

$$f(1) = 1^2 + 3(1) - 10 = 1 + 3 - 10 = -6.$$

Since $f(1) = -6$ and $-6 < 0$, the function values will be negative for any number in interval B.

Next, we try a test number in interval C, say 4, and find the corresponding function value:

$$f(4) = 4^2 + 3(4) - 10 = 16 + 12 - 10 = 18.$$

Since $f(4) = 18$ and $18 > 0$, the function values will be positive for any number in interval C.

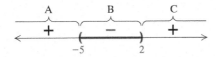

We are looking for numbers x for which $f(x) = x^2 + 3x - 10 < 0$. Thus any number x in interval B is a solution. The solution set is $\{x | -5 < x < 2\}$, or the interval $(-5, 2)$. If the inequality had been $\leq$, it would have been necessary to include the endpoints -5 and 2 in the solution set as well.

Do Exercises 4 and 5.

Solve using the method of Example 3.

4. $x^2 + 3x > 4$

5. $x^2 + 3x \leq 4$

To solve a polynomial inequality:

1. Get 0 on one side, set the expression on the other side equal to 0, and solve to find the x-intercepts.
2. Use the numbers found in step (1) to divide the number line into intervals.
3. Substitute a number from each interval into the related function. If the function value is positive, then the expression will be positive for all numbers in the interval. If the function value is negative, then the expression will be negative for all numbers in the interval.
4. Select the intervals for which the inequality is satisfied and write set-builder notation or interval notation for the solution set.

Answers

4. $\{x | x < -4 \text{ or } x > 1\}$, or $(-\infty, -4) \cup (1, \infty)$
5. $\{x | -4 \leq x \leq 1\}$, or $[-4, 1]$

EXAMPLE 4 Solve: $5x(x + 3)(x - 2) \geq 0$.

The solutions of $f(x) = 0$, or $5x(x + 3)(x - 2) = 0$, are 0, -3, and 2. They divide the number line into four intervals, as shown below.

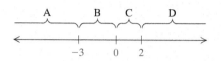

We try test numbers in each interval:

A: Test -5, $f(-5) = 5(-5)(-5 + 3)(-5 - 2) = -350 < 0$.
B: Test -2, $f(-2) = 5(-2)(-2 + 3)(-2 - 2) = 40 > 0$.
C: Test 1, $f(1) = 5(1)(1 + 3)(1 - 2) = -20 < 0$.
D: Test 3, $f(3) = 5(3)(3 + 3)(3 - 2) = 90 > 0$.

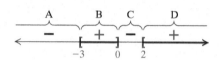

The expression is positive for values of x in intervals B and D. Since the inequality symbol is $\geq$, we must include the x-intercepts. The solution set of the inequality is

$$\{x \mid -3 \leq x \leq 0 \text{ or } x \geq 2\}, \quad \text{or} \quad [-3, 0] \cup [2, \infty).$$

We visualize this with the following graph.

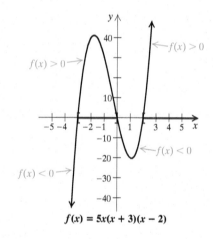

◀ Do Exercise 6.

---

6. Solve: $6x(x + 1)(x - 1) < 0$.

The solutions of $6x(x + 1)(x - 1) = 0$ are 0, -1, and ____. Divide the number line into four intervals and test values of $f(x) = 6x(x + 1)(x - 1)$.

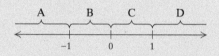

A: Test -2.
$f(-2) = 6()(-2 + 1)(-2 - 1)$
$= $

B: Test $-\dfrac{1}{2}$.
$f\left(-\dfrac{1}{2}\right) = 6()\left(-\dfrac{1}{2} + 1\right)\left(-\dfrac{1}{2} - 1\right)$
$= $

C: Test $\dfrac{1}{2}$.
$f\left(\dfrac{1}{2}\right) = 6()\left(\dfrac{1}{2} + 1\right)\left(\dfrac{1}{2} - 1\right)$
$= $

D: Test 2.
$f(2) = 6()(2 + 1)(2 - 1)$
$= $

The expression is negative for values of x in intervals A and ____. The solution set is $\{x \mid x < -1 \text{ or } 0 < x < 1\}$, or $(-\infty,) \cup (0,)$.

---

b RATIONAL INEQUALITIES

We adapt the preceding method for inequalities that involve rational expressions. We call these **rational inequalities**.

Answer
6. $\{x \mid x < -1 \text{ or } 0 < x < 1\}$, or $(-\infty, -1) \cup (0, 1)$

Guided Solution:
6. $1, -2, -36, -\dfrac{1}{2}, \dfrac{9}{4}, \dfrac{1}{2}, -\dfrac{9}{4}, 2, 36, \text{C}, -1, 1$

EXAMPLE 5 Solve: $\dfrac{x-3}{x+4} \geq 2$.

We write a related equation by changing the $\geq$ symbol to $=$:

$$\dfrac{x-3}{x+4} = 2.$$

Then we solve this related equation. First, we multiply on both sides of the equation by the LCM, which is $x + 4$:

$$(x+4) \cdot \dfrac{x-3}{x+4} = (x+4) \cdot 2$$
$$x - 3 = 2x + 8$$
$$-11 = x.$$

With rational inequalities, we must also determine those numbers for which the rational expression is not defined—that is, those numbers that make the denominator 0. We set the denominator equal to 0 and solve: $x + 4 = 0$, or $x = -4$. Next, we use the numbers -11 and -4 to divide the number line into intervals, as shown below.

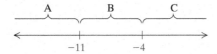

We try test numbers in each interval to see if each satisfies the original inequality.

A: Test -15, $\quad \dfrac{x-3}{x+4} \geq 2$

$$\dfrac{-15-3}{-15+4} \; ? \; 2$$
$$\dfrac{18}{11} \quad \text{FALSE}$$

Since the inequality is false for $x = -15$, the number -15 is not a solution of the inequality. Interval A *is not* part of the solution set.

B: Test -8, $\quad \dfrac{x-3}{x+4} \geq 2$

$$\dfrac{-8-3}{-8+4} \; ? \; 2$$
$$\dfrac{11}{4} \quad \text{TRUE}$$

Since the inequality is true for $x = -8$, the number -8 is a solution of the inequality. Interval B *is* part of the solution set.

C: Test 1, $\quad \dfrac{x-3}{x+4} \geq 2$

$$\dfrac{1-3}{1+4} \; ? \; 2$$
$$-\dfrac{2}{5} \quad \text{FALSE}$$

Since the inequality is false for $x = 1$, the number 1 is not a solution of the inequality. Interval C *is not* part of the solution set.

Solve.

7. $\dfrac{x+1}{x-2} \geq 3$

8. $\dfrac{x}{x-5} < 2$

The solution set includes interval B. The number -11 is also included since the inequality symbol is $\geq$ and -11 is a solution of the related equation. The number -4 is not included; it is not an allowable replacement because it results in division by 0. Thus the solution set of the original inequality is

$$\{x \mid -11 \leq x < -4\}, \quad \text{or} \quad [-11, -4).$$

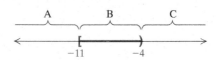

◀ Do Exercises 7 and 8.

To solve a rational inequality:

1. Change the inequality symbol to an equals sign and solve the related equation.
2. Find the numbers for which any denominator in the inequality is not defined.
3. Use the numbers found in steps (1) and (2) to divide the number line into intervals.
4. Substitute a number from each interval into the inequality. If the number is a solution, then the interval to which it belongs is part of the solution set.
5. Select the intervals for which the inequality is satisfied and write set-builder notation or interval notation for the solution set.

Answers

7. $\left\{x \mid 2 < x \leq \dfrac{7}{2}\right\}$, or $\left(2, \dfrac{7}{2}\right]$

8. $\{x \mid x < 5 \text{ or } x > 10\}$, or $(-\infty, 5) \cup (10, \infty)$

11.8 Exercise Set

FOR EXTRA HELP   MyLab Math

✓ Check Your Understanding

Reading Check Complete each statement using either "positive" or "negative."

RC1. To solve $x^2 - 2 < 0$, we look for intervals for which $f(x) = x^2 - 2$ is _____.

RC2. To solve $\dfrac{x}{x+1} > 0$, we look for intervals for which $f(x) = \dfrac{x}{x+1}$ is _____.

RC3. To solve $3x < 5 + x^2$, we look for intervals for which $f(x) = x^2 - 3x + 5$ is _____.

RC4. To solve $(x-1)(x+2) > 3x^2$, we look for intervals for which $f(x) = 2x^2 - x + 2$ is _____.

Concept Check Determine whether -1 is a solution of each inequality.

CC1. $x^2 + 3x < 0$

CC2. $x^2 + x < 0$

CC3. $\dfrac{2x}{x-2} \geq 1$

CC4. $\dfrac{x+1}{x-1} \leq 0$

a Solve algebraically and verify results from the graph.

1. $(x - 6)(x + 2) > 0$
2. $(x - 5)(x + 1) > 0$
3. $4 - x^2 \geq 0$
4. $9 - x^2 \leq 0$

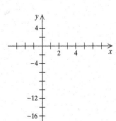 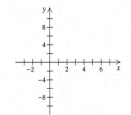 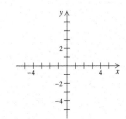 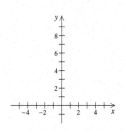

Solve.

5. $3(x + 1)(x - 4) \leq 0$
6. $(x - 7)(x + 3) \leq 0$
7. $x^2 - x - 2 < 0$
8. $x^2 + x - 2 < 0$

9. $x^2 - 2x + 1 \geq 0$
10. $x^2 + 6x + 9 < 0$
11. $x^2 + 8 < 6x$
12. $x^2 - 12 > 4x$

13. $3x(x + 2)(x - 2) < 0$
14. $5x(x + 1)(x - 1) > 0$
15. $(x + 9)(x - 4)(x + 1) > 0$

16. $(x - 1)(x + 8)(x - 2) < 0$
17. $(x + 3)(x + 2)(x - 1) < 0$
18. $(x - 2)(x - 3)(x + 1) < 0$

b Solve.

19. $\dfrac{1}{x - 6} < 0$
20. $\dfrac{1}{x + 4} > 0$
21. $\dfrac{x + 1}{x - 3} > 0$
22. $\dfrac{x - 2}{x + 5} < 0$

23. $\dfrac{3x + 2}{x - 3} \leq 0$
24. $\dfrac{5 - 2x}{4x + 3} \leq 0$
25. $\dfrac{x - 1}{x - 2} > 3$
26. $\dfrac{x + 1}{2x - 3} < 1$

SECTION 11.8 Polynomial Inequalities and Rational Inequalities

27. $\dfrac{(x-2)(x+1)}{x-5} < 0$
28. $\dfrac{(x+4)(x-1)}{x+3} > 0$
29. $\dfrac{x+3}{x} \le 0$
30. $\dfrac{x}{x-2} \ge 0$

31. $\dfrac{x}{x-1} > 2$
32. $\dfrac{x-5}{x} < 1$
33. $\dfrac{x-1}{(x-3)(x+4)} < 0$
34. $\dfrac{x+2}{(x-2)(x+7)} > 0$

35. $3 < \dfrac{1}{x}$
36. $\dfrac{1}{x} \le 2$
37. $\dfrac{x^2+x-2}{x^2-x-12} > 0$
38. $\dfrac{x^2-11x+30}{x^2-8x-9} \ge 0$

Skill Maintenance

Simplify. [10.3b]

39. $\sqrt[3]{\dfrac{125}{27}}$
40. $\sqrt{\dfrac{25}{4a^2}}$
41. $\sqrt{\dfrac{16a^3}{b^4}}$
42. $\sqrt[3]{\dfrac{27c^5}{343d^3}}$

Add or subtract. [10.4a]

43. $3\sqrt{8} - 5\sqrt{2}$

44. $7\sqrt{45} - 2\sqrt{20}$

45. $5\sqrt[3]{16a^4} + 7\sqrt[3]{2a}$

46. $3\sqrt{10} + 8\sqrt{20} - 5\sqrt{80}$

Synthesis

47. Use a graphing calculator to solve Exercises 11, 22, and 25 by graphing two curves, one for each side of the inequality.

48. Use a graphing calculator to solve each of the following.
    a) $x + \dfrac{1}{x} < 0$
    b) $x - \sqrt{x} \ge 0$
    c) $\tfrac{1}{3}x^3 - x + \tfrac{2}{3} \le 0$

Solve.

49. $x^2 - 2x \le 2$
50. $x^2 + 2x > 4$
51. $x^4 + 2x^2 > 0$

52. $x^4 + 3x^2 \le 0$
53. $\left|\dfrac{x+2}{x-1}\right| < 3$

54. *Total Profit.* A company determines that its total profit from the production and sale of x units of a product is given by
$$P(x) = -x^2 + 812x - 9600.$$
  a) The company makes a profit for those nonnegative values of x for which $P(x) > 0$. Find the values of x for which the company makes a profit.
  b) The company loses money for those nonnegative values of x for which $P(x) < 0$. Find the values of x for which the company loses money.

55. *Height of a Thrown Object.* The function
$$H(t) = -16t^2 + 32t + 1920$$
gives the height H of an object thrown from a cliff 1920 ft high, after time t seconds.
  a) For what times is the height greater than 1920 ft?
  b) For what times is the height less than 640 ft?

CHAPTER 11 Summary and Review

Formulas and Principles

Principle of Square Roots: $x^2 = d$ has solutions $\sqrt{d}$ and $-\sqrt{d}$.

Quadratic Formula: $x = \dfrac{-b \pm \sqrt{b^2 - 4ac}}{2a}$

Discriminant: $b^2 - 4ac$

The *vertex* of the graph of $f(x) = ax^2 + bx + c$ is $\left(-\dfrac{b}{2a}, \dfrac{4ac - b^2}{4a}\right)$, or $\left(-\dfrac{b}{2a}, f\left(-\dfrac{b}{2a}\right)\right)$.

The *line of symmetry* of the graph of $f(x) = ax^2 + bx + c$ is $x = -\dfrac{b}{2a}$.

Vocabulary Reinforcement

Complete each statement with the correct term from the column on the right. Some of the choices may be used more than once and some may not be used at all.

1. The equation $x^2 = 2x - 8$ is an example of a(n) _____ equation. [11.1a]

2. The inequality $\dfrac{1}{x} < 7$ is an example of a(n) _____ inequality. [11.8b]

3. We can _____ the square for $y^2 - 8y$ by adding 16. [11.1b]

4. The expression $b^2 - 4ac$ in the quadratic formula is called the _____. [11.4a]

5. The equation $m^6 - m^3 - 12 = 0$ is _____ in form. [11.4c]

6. The graph of a quadratic function is a(n) _____. [11.5a]

7. The vertical line $x = 0$ is the line of _____ for the graph of $y = x^2$. [11.5a]

8. The maximum or minimum value of a quadratic function is the y-coordinate of the _____. [11.5c]

complete
parabola
line
discriminant
symmetry
linear
polynomial
quadratic
rational
vertex
x-intercept
y-intercept

Concept Reinforcement

Determine whether each statement is true or false.

_____ 1. The graph of $f(x) = -(-x^2 - 8x - 3)$ opens downward. [11.5a]

_____ 2. If $(-5, 7)$ is the vertex of a parabola, then $x = -5$ is the line of symmetry. [11.6a]

_____ 3. The graph of $f(x) = -3(x + 2)^2 - 5$ is a translation to the right of the graph of $f(x) = -3x^2 - 5$. [11.5b]

Study Guide

Objective 11.1a Solve quadratic equations using the principle of square roots.

Example Solve: $(x - 3)^2 = -36$.

$x - 3 = \sqrt{-36}$ or $x - 3 = -\sqrt{-36}$
$x - 3 = 6i$ or $x - 3 = -6i$
$x = 3 + 6i$ or $x = 3 - 6i$

The solutions are $3 \pm 6i$.

Practice Exercise

1. Solve: $(x - 2)^2 = -9$.

Objective 11.1b Solve quadratic equations by completing the square.

Example Solve by completing the square:
$x^2 - 8x + 13 = 0$.

$x^2 - 8x = -13$
$x^2 - 8x + 16 = -13 + 16$
$(x - 4)^2 = 3$
$x - 4 = \sqrt{3}$ or $x - 4 = -\sqrt{3}$
$x = 4 + \sqrt{3}$ or $x = 4 - \sqrt{3}$

The solutions are $4 \pm \sqrt{3}$.

Practice Exercise

2. Solve by completing the square:
$x^2 - 12x + 31 = 0$.

Objective 11.2a Solve quadratic equations using the quadratic formula, and approximate solutions using a calculator.

Example Solve: $x^2 - 2x = 2$. Give the exact solutions and approximate solutions to three decimal places.

$x^2 - 2x - 2 = 0$ Standard form
$a = 1, \ b = -2, \ c = -2$

$x = \dfrac{-(-2) \pm \sqrt{(-2)^2 - 4 \cdot 1 \cdot (-2)}}{2 \cdot 1}$ Using the quadratic formula

$= \dfrac{2 \pm \sqrt{4 + 8}}{2}$

$= \dfrac{2 \pm \sqrt{12}}{2}$

$= \dfrac{2 \pm 2\sqrt{3}}{2}$

$= 1 \pm \sqrt{3}$, or 2.732 and -0.732

Practice Exercise

3. Solve: $x^2 - 10x = -23$. Give the exact solutions and approximate solutions to three decimal places.

Objective 11.4a Determine the nature of the solutions of a quadratic equation.

Example Determine the nature of the solutions of the quadratic equation $x^2 - 7x = 1$.

In standard form, we have $x^2 - 7x - 1 = 0$. Thus, $a = 1$, $b = -7$, and $c = -1$. The discriminant, $b^2 - 4ac$, is $(-7)^2 - 4 \cdot 1 \cdot (-1)$, or 53. Since the discriminant is positive, there are two real solutions.

Practice Exercise

4. Determine the nature of the solutions of each quadratic equation.
   a) $x^2 - 3x = 7$
   b) $2x^2 - 5x + 5 = 0$

Objective 11.4b Write a quadratic equation having two given numbers as solutions.

Example Write a quadratic equation whose solutions are 7 and $-\frac{1}{4}$.

$x = 7$ or $x = -\frac{1}{4}$
$x - 7 = 0$ or $x + \frac{1}{4} = 0$
$x - 7 = 0$ or $4x + 1 = 0$ Clearing the fraction
$(x - 7)(4x + 1) = 0$ Using the principle of zero products in reverse
$4x^2 - 27x - 7 = 0$ Using FOIL

Practice Exercise

5. Write a quadratic equation whose solutions are $-\frac{2}{5}$ and 3.

Objective 11.4c Solve equations that are quadratic in form.

Example Solve: $x - 8\sqrt{x} - 9 = 0$.

Let $u = \sqrt{x}$. Then we substitute u for $\sqrt{x}$ and u^2 for x and solve for u:
$$u^2 - 8u - 9 = 0$$
$$(u - 9)(u + 1) = 0$$
$$u = 9 \quad \text{or} \quad u = -1.$$

Next, we substitute $\sqrt{x}$ for u and solve for x:
$$\sqrt{x} = 9 \quad \text{or} \quad \sqrt{x} = -1.$$

Squaring each equation, we get
$$x = 81 \quad \text{or} \quad x = 1.$$

Checking both 81 and 1 in $x - 8\sqrt{x} - 9 = 0$, we find that 81 checks but 1 does not. The solution is 81.

Practice Exercise

6. Solve:
$$(x^2 - 3)^2 - 5(x^2 - 3) - 6 = 0.$$

Objective 11.6a For a quadratic function, find the vertex, the line of symmetry, and the maximum or minimum value, and then graph the function.

Example For $f(x) = -2x^2 + 4x + 1$, find the vertex, the line of symmetry, and the maximum or minimum value. Then graph.

We factor out -2 from only the first two terms:
$$f(x) = -2(x^2 - 2x) + 1.$$

Next, we complete the square, factor, and simplify:
$$f(x) = -2(x^2 - 2x \quad) + 1.$$
$$= -2(x^2 - 2x + 1 - 1) + 1$$
$$= -2(x^2 - 2x + 1) + (-2)(-1) + 1$$
$$= -2(x - 1)^2 + 3.$$

The vertex is $(1, 3)$. The line of symmetry is $x = 1$. The coefficient of x^2 is negative, so the graph opens down. Thus, 3 is the maximum value of the function.

We plot points and graph the parabola.

x	y
1	3
2	1
0	1
3	−5
−1	−5

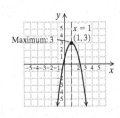

Practice Exercise

7. For $f(x) = -x^2 - 2x - 3$, find the vertex, the line of symmetry, and the maximum or minimum value. Then graph.

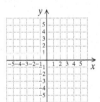

Objective 11.6b Find the intercepts of a quadratic function.

Example Find the intercepts of $f(x) = x^2 - 8x + 14$.

Since $f(0) = 0^2 - 8 \cdot 0 + 14 = 14$, the y-intercept is $(0, 14)$. To find the x-intercepts, we solve $0 = x^2 - 8x + 14$. Using the quadratic formula, we have $x = 4 \pm \sqrt{2}$. Thus the x-intercepts are $(4 - \sqrt{2}, 0)$ and $(4 + \sqrt{2}, 0)$.

Practice Exercise

8. Find the intercepts of $f(x) = x^2 - 6x + 4$.

Objective 11.8a Solve quadratic inequalities and other polynomial inequalities.

Example Solve: $x^2 - 15 > 2x$.

$x^2 - 2x - 15 > 0$ Subtracting $2x$

We set the polynomial equal to 0 and solve. The solutions of $x^2 - 2x - 15 = 0$, or $(x + 3)(x - 5) = 0$, are -3 and 5. They divide the number line into three intervals.

$\leftarrow\!+\!+\!+\!+\!+\!+\!+\!+\!+\!+\!+\!+\!+\!\rightarrow$
$-6\;-5\;-4\;-3\;-2\;-1\;0\;1\;2\;3\;4\;5\;6$

We try a test point in each interval:

Test -5: $(-5)^2 - 2(-5) - 15 = 20 > 0$;
Test 0: $(0)^2 - 2 \cdot 0 - 15 = -15 < 0$;
Test 6: $(6)^2 - 2 \cdot 6 - 15 = 9 > 0$.

The expression $x^2 - 2x - 15$ is positive for values of x in the intervals $(-\infty, -3)$ and $(5, \infty)$. The inequality symbol is $>$, so -3 and 5 are not solutions. The solution set is $\{x | x < -3 \text{ or } x > 5\}$, or $(-\infty, -3) \cup (5, \infty)$.

Practice Exercise

9. Solve: $x^2 + 40 > 14x$.

Objective 11.8b Solve rational inequalities.

Example Solve: $\dfrac{x + 3}{x - 6} \geq 2$.

We first solve the related equation $\dfrac{x + 3}{x - 6} = 2$. The solution is 15. We must also determine those numbers for which the rational expression is not defined. We set the denominator equal to 0 and solve: $x - 6 = 0$, or $x = 6$. The numbers 6 and 15 divide the number line into three intervals. We test a point in each interval.

$\leftarrow\!+\!+\!+\!+\!+\!+\!+\!+\!+\!+\!+\!+\!+\!\rightarrow$
$4\;5\;6\;7\;8\;9\;10\;11\;12\;13\;14\;15\;16$

Test 5: $\dfrac{5 + 3}{5 - 6} \geq 2$, or $-8 \geq 2$, which is false.

Test 9: $\dfrac{9 + 3}{9 - 6} \geq 2$, or $4 \geq 2$, which is true.

Test 17: $\dfrac{17 + 3}{17 - 6} \geq 2$, or $\dfrac{20}{11} \geq 2$, which is false.

The solution set includes the interval $(6, 15)$ and the number 15, the solution of the related equation. The number 6 is not included. It is not an allowable replacement because it results in division by 0. The solution set is $\{x | 6 < x \leq 15\}$, or $(6, 15]$.

Practice Exercise

10. Solve: $\dfrac{x + 7}{x - 5} \geq 3$.

Review Exercises

1. **a)** Solve: $2x^2 - 7 = 0$. [11.1a]
   b) Find the x-intercepts of $f(x) = 2x^2 - 7$. [11.1a]

Solve. [11.2a]

2. $14x^2 + 5x = 0$
3. $x^2 - 12x + 27 = 0$
4. $x^2 - 7x + 13 = 0$
5. $4x^2 + 6x = 1$
6. $4x(x - 1) + 15 = x(3x + 4)$
7. $x^2 + 4x + 1 = 0$. Give exact solutions and approximate solutions to three decimal places.
8. $\dfrac{x}{x-2} + \dfrac{4}{x-6} = 0$
9. $\dfrac{x}{4} - \dfrac{4}{x} = 2$
10. $15 = \dfrac{8}{x+2} - \dfrac{6}{x-2}$
11. Solve $x^2 + 6x + 2 = 0$ by completing the square. Show your work. [11.1b]
12. *Hang Time.* A basketball player has a vertical leap of 39 in. What is his hang time? Use the function $V(T) = 48T^2$. [11.1c]
13. *Notebook.* The width of a pocket-sized rectangular notebook is 5 cm less than the length. The area is 126 cm². Find the length and the width. [11.3a]
14. *Picture Matting.* A picture mat measures 12 in. by 16 in., and 140 in² of picture shows. Find the width of the mat. [11.3a]
15. *Motorcycle Travel.* During the first part of a trip, a motorcyclist travels 50 mi. She travels 80 mi on the second part of the trip at a speed that is 10 mph slower. The total time for the trip is 3 hr. What is the speed on each part of the trip? [11.3a]

Determine the nature of the solutions of each equation. [11.4a]

16. $x^2 + 3x - 6 = 0$
17. $x^2 + 2x + 5 = 0$

Write a quadratic equation having the given solutions. [11.4b]

18. $\frac{1}{5}, -\frac{3}{5}$
19. -4, only solution

Solve for the indicated letter. [11.3b]

20. $N = 3\pi\sqrt{\dfrac{1}{p}}$, for p
21. $2A = \dfrac{3B}{T^2}$, for T

Solve. [11.4c]

22. $x^4 - 13x^2 + 36 = 0$
23. $15x^{-2} - 2x^{-1} - 1 = 0$
24. $(x^2 - 4)^2 - (x^2 - 4) - 6 = 0$
25. $x - 13\sqrt{x} + 36 = 0$

For each quadratic function in Exercises 26–28, find and label **(a)** the vertex, **(b)** the line of symmetry, and **(c)** the maximum or minimum value. Then **(d)** graph the function. [11.5c], [11.6a]

26. $f(x) = -\frac{1}{2}(x - 1)^2 + 3$

x	$f(x)$

a) Vertex: (____, ____)
b) Line of symmetry: $x = $ ____
c) _____ value: ____
d)

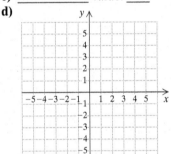

27. $f(x) = x^2 - x + 6$

x	$f(x)$

a) Vertex: (____, ____)
b) Line of symmetry: $x = $ ____
c) _____ value: ____
d)

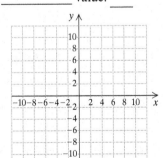

28. $f(x) = -3x^2 - 12x - 8$

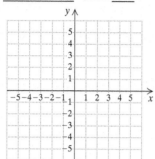

a) Vertex: (___, ___)
b) Line of symmetry: $x = $ ___
c) _____ value: ___
d)

Find the x- and y-intercepts. [11.6b]

29. $f(x) = x^2 - 9x + 14$

30. $g(x) = x^2 - 4x - 3$

31. What is the minimum product of two numbers whose difference is 22? What numbers yield this product? [11.7a]

32. Find a quadratic function that fits the data points $(0, -2)$, $(1, 3)$, and $(3, 7)$. [11.7b]

Solve. [11.8a, b]

33. $(x + 2)(x - 1)(x - 2) > 0$

34. $\dfrac{(x + 4)(x - 1)}{(x + 2)} < 0$

35. *Online Sales Growth.* Sales of products and services through the Internet have increased quadratically since 2010, as illustrated in the following graph. [11.7b]

eCommerce Sales, Real and Projected

DATA: *USA Today*

a) Use the data points $(0, 170)$, $(4, 300)$, and $(10, 680)$ to fit a quadratic function to the data.
b) Use the quadratic function to estimate the amount of online sales in 2017.

36. Determine the nature of the solutions:
$$x^2 - 10x + 25 = 0. \quad [11.4a]$$
A. Infinite number of solutions
B. One real solution
C. Two real solutions
D. No real solutions

37. Solve: $2x^2 - 6x + 5 = 0$. [11.2a]

A. $\dfrac{3}{2} \pm \dfrac{\sqrt{19}}{2}$

B. $3 \pm i$

C. $3 \pm \sqrt{19}$

D. $\dfrac{3}{2} \pm \dfrac{1}{2}i$

Synthesis

38. The sum of the base and the height of a triangle is 38 cm. Find the dimensions for which the area is a maximum, and find the maximum area. [11.7a]

39. The average of two numbers is 171. One of the numbers is the square root of the other. Find the numbers. [11.3a]

Understanding Through Discussion and Writing

1. Does the graph of every quadratic function have a y-intercept? Why or why not? [11.6b]

2. Explain how the leading coefficient of a quadratic function can be used to determine whether a maximum or minimum function value exists. [11.7a]

3. Explain, without plotting points, why the graph of $f(x) = (x + 3)^2 - 4$ looks like the graph of $f(x) = x^2$ translated 3 units to the left and 4 units down. [11.5c]

4. Describe a method that could be used to create quadratic inequalities that have no solution. [11.8a]

5. Is it possible for the graph of a quadratic function to have only one x-intercept if the vertex is off the x-axis? Why or why not? [11.6b]

6. Explain how the x-intercepts of a quadratic function can be used to help find the vertex of the function. What piece of information would still be missing? [11.6a, b]

CHAPTER 11 Test

1. **a)** Solve: $3x^2 - 4 = 0$.
   b) Find the x-intercepts of $f(x) = 3x^2 - 4$.

Solve.

2. $x^2 + x + 1 = 0$

3. $x - 8\sqrt{x} + 7 = 0$

4. $4x(x - 2) - 3x(x + 1) = -18$

5. $4x^4 - 17x^2 + 15 = 0$

6. $x^2 + 4x = 2$. Give exact solutions and approximate solutions to three decimal places.

7. $\dfrac{1}{4 - x} + \dfrac{1}{2 + x} = \dfrac{3}{4}$

8. Solve $x^2 - 4x + 1 = 0$ by completing the square. Show your work.

9. *Free-Falling Objects.* The Peachtree Plaza in Atlanta, Georgia, is 723 ft tall. Use the function $s(t) = 16t^2$ to approximate how long it would take an object to fall from the top.

10. *Marine Travel.* The Columbia River flows at a rate of 2 mph for the length of a popular boating route. In order for a motorized dinghy to travel 3 mi upriver and then return in a total of 4 hr, how fast must the boat be able to travel in still water?

11. *Memory Board.* A computer-parts company wants to make a rectangular memory board that has a perimeter of 28 cm. What dimensions will allow the board to have the maximum area?

12. *Hang Time.* Professional basketball player Nate Robinson has a vertical leap of 43 in. What is his hang time? Use the function $V(T) = 48T^2$.

13. Determine the nature of the solutions of the equation $x^2 + 5x + 17 = 0$.

14. Write a quadratic equation having the solutions $\sqrt{3}$ and $3\sqrt{3}$.

15. Solve $V = 48T^2$ for T.

For the quadratic functions in Exercises 16 and 17, find and label **(a)** the vertex, **(b)** the line of symmetry, and **(c)** the maximum or minimum value. Then **(d)** graph the function.

16. $f(x) = -x^2 - 2x$

 a) Vertex: (___, ___)
 b) Line of symmetry: $x =$ ___
 c) _____ value: ___
 d)

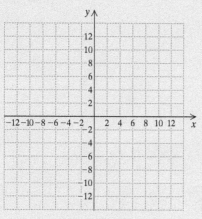

17. $f(x) = 4x^2 - 24x + 41$

 a) Vertex: (___, ___)
 b) Line of symmetry: $x =$ ___
 c) _____ value: ___
 d)

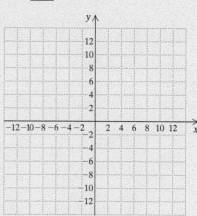

18. Find the x- and y-intercepts:
$$f(x) = -x^2 + 4x - 1.$$

19. What is the minimum product of two numbers whose difference is 8? What numbers yield this product?

20. Find the quadratic function that fits the data points $(0, 0)$, $(3, 0)$, and $(5, 2)$.

21. *Pizza Prices.* Pizza Unlimited has the following prices for pizzas.

DIAMETER	PRICE
8 in.	$10.00
12 in.	$12.50
16 in.	$15.50

Price should probably be a quadratic function of diameter because it should be proportional to the area, and the area is a quadratic function of the diameter. (The area of a circular region is given by $A = \pi r^2$ or $(\pi/4)d^2$.)

 a) Express price as a quadratic function of diameter using the data points $(8, 10)$, $(12, 12.50)$, and $(16, 15.50)$.
 b) Use the function to find the price of a 14-in. pizza.

Solve.

22. $x^2 < 6x + 7$

23. $\dfrac{x-5}{x+3} < 0$

24. $\dfrac{x-2}{(x+3)(x-1)} \geq 0$

25. Write a quadratic equation whose solutions are $\dfrac{1}{2}i$ and $-\dfrac{1}{2}i$.

 A. $4x^2 - 4ix - 1 = 0$ **B.** $x^2 - \dfrac{1}{4} = 0$ **C.** $4x^2 + 1 = 0$ **D.** $x^2 - ix + 1 = 0$

Synthesis

26. A quadratic function has x-intercepts $(-2, 0)$ and $(7, 0)$ and y-intercept $(0, 8)$. Find an equation for the function. What is its maximum or minimum value?

27. One solution of $kx^2 + 3x - k = 0$ is -2. Find the other solution.

CHAPTERS 1–11 Cumulative Review

1. *Golf Courses.* Most golf courses have a hole such as the one shown here, where the safe way to the hole is to hit straight out on a first shot (the distance a) and then make subsequent shots at a right angle to cover the distance b. Golfers are often lured, however, into taking a shortcut over trees, houses, or lakes. If a golfer makes a hole in one on this hole, how long is the shot?

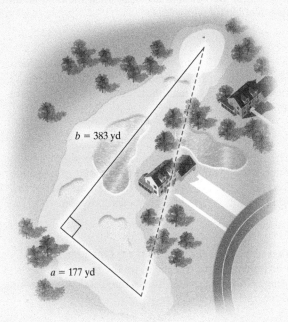

$b = 383$ yd
$a = 177$ yd

Simplify.

2. $(4 + 8x^2 - 5x) - (-2x^2 + 3x - 2)$

3. $(2x^2 - x + 3)(x - 4)$

4. $\dfrac{a^2 - 16}{5a - 15} \cdot \dfrac{2a - 6}{a + 4}$

5. $\dfrac{y}{y^2 - y - 42} \div \dfrac{y^2}{y - 7}$

6. $\dfrac{2}{m + 1} + \dfrac{3}{m - 5} - \dfrac{m^2 - 1}{m^2 - 4m - 5}$

7. $(9x^3 + 5x^2 + 2) \div (x + 2)$

8. $\dfrac{\dfrac{1}{x} - \dfrac{1}{y}}{x + y}$

9. $\sqrt{0.36}$

10. $\sqrt{9x^2 - 36x + 36}$

11. $6\sqrt{45} - 3\sqrt{20}$

12. $\dfrac{2\sqrt{3} - 4\sqrt{2}}{\sqrt{2} - 3\sqrt{6}}$

13. $(8^{2/3})^4$

14. $(3 + 2i)(5 - i)$

15. $\dfrac{6 - 2i}{3i}$

Factor.

16. $2t^2 - 7t - 30$

17. $a^2 + 3a - 54$

18. $-3a^3 + 12a^2$

19. $64a^2 - 9b^2$

20. $3a^2 - 36a + 108$

21. $\dfrac{1}{27}a^3 - 1$

22. $24a^3 + 18a^2 - 20a - 15$

23. $(x + 1)(x - 1) + (x + 1)(x + 2)$

Solve.

24. $3(4x - 5) + 6 = 3 - (x + 1)$

25. $F = \dfrac{mv^2}{r}$, for r

26. $5 - 3(2x + 1) \le 8x - 3$

27. $3x - 2 < -6 \text{ or } x + 3 > 9$

28. $|4x - 1| \leq 14$

29. $5x + 10y = -10,$
    $-2x - 3y = 5$

30. $2x + y - z = 9,$
    $4x - 2y + z = -9,$
    $2x - y + 2z = -12$

31. $10x^2 + 28x - 6 = 0$

32. $\dfrac{2}{n} - \dfrac{7}{n} = 3$

33. $\dfrac{1}{2x - 1} = \dfrac{3}{5x}$

34. $A = \dfrac{mh}{m + a}$, for m

35. $\sqrt{2x - 1} = 6$

36. $\sqrt{x - 2} + 1 = \sqrt{2x - 6}$

37. $16(t - 1) = t(t + 8)$

38. $x^2 - 3x + 16 = 0$

39. $\dfrac{18}{x + 1} - \dfrac{12}{x} = \dfrac{1}{3}$

40. $P = \sqrt{a^2 - b^2}$, for a

41. $\dfrac{(x + 3)(x + 2)}{(x - 1)(x + 1)} < 0$

42. Solve: $4x^2 - 25 > 0$.

Graph.

43. $x + y = 2$

44. $y \geq 6x - 5$

45. $x < -3$

46. $3x - y > 6,$
    $4x + y \leq 3$

47. $f(x) = x^2 - 1$

48. $f(x) = -2x^2 + 3$

49. Find an equation of the line with slope $\tfrac{1}{2}$ and through the point $(-4, 2)$.

50. Find an equation of the line parallel to the line $3x + y = 4$ and through the point $(0, 1)$.

51. *Marine Travel.* The Connecticut River flows at a rate of 4 km/h for the length of a popular scenic route. In order for a cruiser to travel 60 km upriver and then return in a total of 8 hr, how fast must the boat be able to travel in still water?

52. *Architecture.* An architect is designing a rectangular family room with a perimeter of 56 ft. What dimensions will yield the maximum area? What is the maximum area?

53. The perimeter of a hexagon with all six sides the same length is the same as the perimeter of a square. One side of the hexagon is 3 less than the side of the square. Find the perimeter of each polygon.

54. Two pipes can fill a tank in $1\tfrac{1}{2}$ hr. One pipe requires 4 hr longer running alone to fill the tank than the other. How long would it take the faster pipe, working alone, to fill the tank?

55. Complete the square: $f(x) = 5x^2 - 20x + 15$.
    A. $f(x) = 5(x - 2)^2 - 5$
    B. $f(x) = 5(x + 2)^2 + 15$
    C. $f(x) = 5(x + 2)^2 + 6$
    D. $f(x) = 5(x + 2)^2 + 11$

56. How many times does the graph of $f(x) = x^4 - 6x^2 - 16$ cross the x-axis?
    A. 1
    B. 2
    C. 3
    D. 4

Synthesis

57. Solve: $\dfrac{2x + 1}{x} = 3 + 7\sqrt{\dfrac{2x + 1}{x}}$.

58. Factor: $\dfrac{a^3}{8} + \dfrac{8b^3}{729}$.

CHAPTER 12

Exponential Functions and Logarithmic Functions

12.1 Exponential Functions
12.2 Composite Functions and Inverse Functions
12.3 Logarithmic Functions
12.4 Properties of Logarithmic Functions

Mid-Chapter Review

12.5 Natural Logarithmic Functions

Visualizing for Success

12.6 Solving Exponential Equations and Logarithmic Equations
12.7 Mathematical Modeling with Exponential Functions and Logarithmic Functions

Translating for Success
Summary and Review
Test
Cumulative Review

In 2016, over 600 million megawatt hours of electricity was generated in the United States from sources considered renewable, such as the sun, the earth, and waste products. This amount accounted for about 17% of the electricity used in 2016, up from 9% in 2007. As the accompanying graph indicates, the amounts of solar power, geothermal power, and power generated from waste products are all increasing. Geothermal power is increasing linearly at a slow rate, power from waste is increasing linearly at a more rapid rate, and solar power is increasing *exponentially*.

Data: *U.S. Energy Information Administration*

In Example 6 of Section 12.7, we will develop an *exponential function* to model solar power generation.

STUDYING FOR SUCCESS   *Preparing for the Final Exam*

- ☐ Browse through each chapter, reviewing highlighted or boxed information and noting important formulas.
- ☐ Attend any exam tutoring sessions offered by your college or university.
- ☐ Retake the chapter tests that you took in class, or take the chapter tests in the text.
- ☐ Work through the Cumulative Review for Chapters 1–12 as a sample final exam.

12.1 Exponential Functions

OBJECTIVES

a Graph exponential equations and functions.

b Graph exponential equations in which x and y have been interchanged.

c Solve applied problems involving applications of exponential functions and their graphs.

The following graph approximates the graph of an *exponential function*. We will consider such functions and some of their applications.

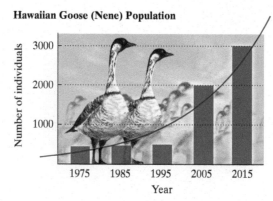

Hawaiian Goose (Nene) Population

DATA: esasuccess.org

a GRAPHING EXPONENTIAL FUNCTIONS

SKILL REVIEW *Evaluate expressions containing negative exponents.* [4.1f]
Evaluate.

1. 2^{-3}
2. $\left(\dfrac{1}{2}\right)^{-3}$

Answers: 1. $\dfrac{1}{8}$ 2. 8

We have defined exponential expressions with rational-number exponents, such as

$$8^{1/4}, \quad 3^{-3/4}, \quad 7^{2.34}, \quad 5^{1.73}.$$

For example, $5^{1.73}$, or $5^{173/100}$, or $\sqrt[100]{5^{173}}$, means to raise 5 to the 173rd power and then take the 100th root. We now develop the meaning of exponential expressions with irrational exponents, such as

$$5^{\sqrt{3}}, \quad 7^{\pi}, \quad 9^{-\sqrt{2}}.$$

Since we can approximate irrational numbers with decimal approximations, we can also approximate expressions with irrational exponents. For example, consider $5^{\sqrt{3}}$. As rational values of r get close to $\sqrt{3}$, 5^r gets close to some real number.

CHAPTER 12   Exponential Functions and Logarithmic Functions

r closes in on $\sqrt{3}$.	5^r closes in on some real number p.
r	5^r
$1 < \sqrt{3} < 2$	$5 = 5^1 < p < 5^2 = 25$
$1.7 < \sqrt{3} < 1.8$	$15.426 \approx 5^{1.7} < p < 5^{1.8} \approx 18.119$
$1.73 < \sqrt{3} < 1.74$	$16.189 \approx 5^{1.73} < p < 5^{1.74} \approx 16.452$
$1.732 < \sqrt{3} < 1.733$	$16.241 \approx 5^{1.732} < p < 5^{1.733} \approx 16.267$

As r closes in on $\sqrt{3}$, 5^r closes in on some real number p. We define $5^{\sqrt{3}}$ to be that number p. To seven decimal places, we have $5^{\sqrt{3}} \approx 16.2424508$.

Any positive irrational exponent can be defined in a similar way. Negative irrational exponents are then defined in the same way as negative integer exponents. Thus the expression a^x has meaning for any real number x. The general laws of exponents still hold, but we will not prove that here.

We now define exponential functions.

EXPONENTIAL FUNCTION

The function $f(x) = a^x$, where a is a positive constant different from 1, is called an **exponential function**, base a.

We restrict the base a to being positive to avoid the possibility of taking even roots of negative numbers such as the square root of -1, $(-1)^{1/2}$, which is not a real number. We restrict the base from being 1 because for $a = 1$, $t(x) = 1^x = 1$, which is a constant. The following are examples of exponential functions:

$$f(x) = 2^x, \quad f(x) = \left(\tfrac{1}{2}\right)^x, \quad f(x) = (0.4)^x.$$

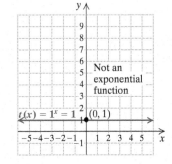

Note that in contrast to polynomial functions like $f(x) = x^2$ and $f(x) = x^3$, the variable is *in the exponent*. Let's consider graphs of exponential functions.

EXAMPLE 1 Graph the exponential function $f(x) = 2^x$.

We compute some function values and list the results in a table. It is a good idea to begin by letting $x = 0$.

$f(0) = 2^0 = 1;$
$f(1) = 2^1 = 2;$
$f(2) = 2^2 = 4;$
$f(3) = 2^3 = 8;$
$f(-1) = 2^{-1} = \dfrac{1}{2^1} = \dfrac{1}{2};$
$f(-2) = 2^{-2} = \dfrac{1}{2^2} = \dfrac{1}{4};$
$f(-3) = 2^{-3} = \dfrac{1}{2^3} = \dfrac{1}{8}.$

x	$f(x)$
0	1
1	2
2	4
3	8
-1	$\tfrac{1}{2}$
-2	$\tfrac{1}{4}$
-3	$\tfrac{1}{8}$

1. Graph: $f(x) = 3^x$. Complete this table. Then plot the points from the table and connect them with a smooth curve.

x	$f(x)$
0	
1	
2	
3	
−1	
−2	
−3	

2. Graph: $f(x) = \left(\dfrac{1}{3}\right)^x$.

Complete this table. Then plot the points from the table and connect them with a smooth curve.

x	$f(x)$
0	
1	
2	
3	
−1	
−2	
−3	

Next, we plot these points and connect them with a smooth curve.

In graphing, be sure to plot enough points to determine how steeply the curve rises.

The curve comes very close to the x-axis, but does not touch or cross it.

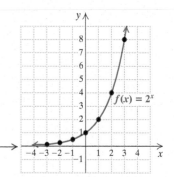

Note that as x increases, the function values increase indefinitely. As x decreases, the function values decrease, getting very close to 0. The x-axis, or the line $y = 0$, is an *asymptote*, meaning here that as x gets very small, the curve comes very close to but never touches the axis.

◀ Do Exercise 1.

EXAMPLE 2 Graph the exponential function $f(x) = \left(\dfrac{1}{2}\right)^x$.

We compute some function values and list the results in a table. Before we do so, note that
$$f(x) = \left(\tfrac{1}{2}\right)^x = (2^{-1})^x = 2^{-x}.$$
Then we have

$f(0) = 2^{-0} = 1;$

$f(1) = 2^{-1} = \dfrac{1}{2^1} = \dfrac{1}{2};$

$f(2) = 2^{-2} = \dfrac{1}{2^2} = \dfrac{1}{4};$

$f(3) = 2^{-3} = \dfrac{1}{2^3} = \dfrac{1}{8};$

$f(-1) = 2^{-(-1)} = 2^1 = 2;$

$f(-2) = 2^{-(-2)} = 2^2 = 4;$

$f(-3) = 2^{-(-3)} = 2^3 = 8.$

x	$f(x)$
0	1
1	$\frac{1}{2}$
2	$\frac{1}{4}$
3	$\frac{1}{8}$
−1	2
−2	4
−3	8

Next, we plot these points and draw the curve. Note that this graph is a reflection across the y-axis of the graph in Example 1. The line $y = 0$ is again an asymptote.

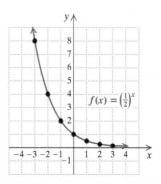

◀ Do Exercise 2.

Answers

1.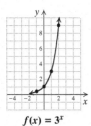

x	$f(x)$
0	1
1	3
2	9
3	27
−1	$\frac{1}{3}$
−2	$\frac{1}{9}$
−3	$\frac{1}{27}$

$f(x) = 3^x$

2.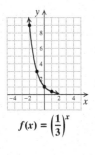

x	$f(x)$
0	1
1	$\frac{1}{3}$
2	$\frac{1}{9}$
3	$\frac{1}{27}$
−1	3
−2	9
−3	27

$f(x) = \left(\dfrac{1}{3}\right)^x$

The preceding examples illustrate exponential functions with various bases. Let's list some of their characteristics. Keep in mind that the definition of an exponential function, $f(x) = a^x$, requires that the base be positive and different from 1.

When $a > 1$, the function $f(x) = a^x$ increases from left to right. The greater the value of a, the steeper the curve. As x gets smaller and smaller, the curve gets closer to the line $y = 0$: It is an asymptote.

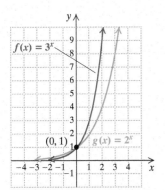
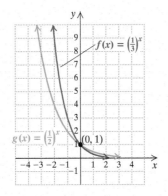

Graph.

3. $f(x) = 4^x$

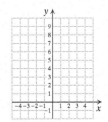

When $0 < a < 1$, the function $f(x) = a^x$ decreases from left to right. As a approaches 1, the curve becomes less steep. As x gets larger and larger, the curve gets closer to the line $y = 0$: It is an asymptote.

4. $f(x) = \left(\dfrac{1}{4}\right)^x$

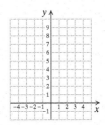

y-INTERCEPT OF AN EXPONENTIAL FUNCTION

All functions $f(x) = a^x$ go through the point $(0, 1)$. That is, the y-intercept is $(0, 1)$.

Do Exercises 3 and 4.

EXAMPLE 3 Graph: $f(x) = 2^{x-2}$.

We construct a table of values. Then we plot the points and connect them with a smooth curve. Be sure to note that $x - 2$ is the *exponent*.

$f(0) = 2^{0-2} = 2^{-2} = \dfrac{1}{2^2} = \dfrac{1}{4}$;

$f(1) = 2^{1-2} = 2^{-1} = \dfrac{1}{2^1} = \dfrac{1}{2}$;

$f(2) = 2^{2-2} = 2^0 = 1$;

$f(3) = 2^{3-2} = 2^1 = 2$;

$f(4) = 2^{4-2} = 2^2 = 4$;

$f(-1) = 2^{-1-2} = 2^{-3} = \dfrac{1}{2^3} = \dfrac{1}{8}$;

$f(-2) = 2^{-2-2} = 2^{-4} = \dfrac{1}{2^4} = \dfrac{1}{16}$

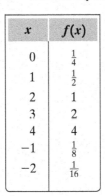

x	$f(x)$
0	$\frac{1}{4}$
1	$\frac{1}{2}$
2	1
3	2
4	4
-1	$\frac{1}{8}$
-2	$\frac{1}{16}$

Answers

3.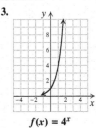
   $f(x) = 4^x$

4.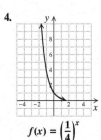
   $f(x) = \left(\dfrac{1}{4}\right)^x$

5. Graph: $f(x) = 2^{x+2}$.

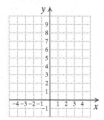

The graph has the same shape as the graph of $g(x) = 2^x$, but it is translated 2 units to the right.

The y-intercept of $g(x) = 2^x$ is $(0, 1)$.
The y-intercept of $f(x) = 2^{x-2}$ is $(0, \frac{1}{4})$.
The line $y = 0$ is still an asymptote.

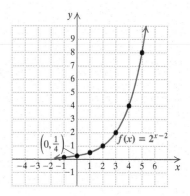

◀ Do Exercise 5.

EXAMPLE 4 Graph: $f(x) = 2^x - 3$.

We construct a table of values. Then we plot the points and connect them with a smooth curve. Note that the only expression in the exponent is x.

$f(0) = 2^0 - 3 = 1 - 3 = -2;$
$f(1) = 2^1 - 3 = 2 - 3 = -1;$
$f(2) = 2^2 - 3 = 4 - 3 = 1;$
$f(3) = 2^3 - 3 = 8 - 3 = 5;$
$f(4) = 2^4 - 3 = 16 - 3 = 13;$
$f(-1) = 2^{-1} - 3 = \frac{1}{2} - 3 = -\frac{5}{2};$
$f(-2) = 2^{-2} - 3 = \frac{1}{4} - 3 = -\frac{11}{4}$

x	$f(x)$
0	-2
1	-1
2	1
3	5
4	13
-1	$-\frac{5}{2}$
-2	$-\frac{11}{4}$

6. Graph: $f(x) = 2^x - 4$.

x	$f(x)$
0	
1	
2	
3	
4	
-1	
-2	

The graph has the same shape as the graph of $g(x) = 2^x$, but it is translated 3 units down. The y-intercept is $(0, -2)$. The line $y = -3$ is an asymptote. The curve gets closer to this line as x gets smaller and smaller.

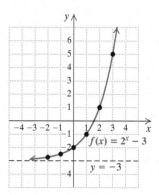

◀ Do Exercise 6.

Answers
Answers to Margin Exercises 5 and 6 are on p. 895.

CALCULATOR CORNER

Graphing Exponential Functions We can use a graphing calculator to graph exponential functions. It might be necessary to try several sets of window dimensions in order to find the ones that give a good view of the curve.

To graph $f(x) = 3^x - 1$, we enter the equation as y_1. We can begin graphing with the standard window $[-10, 10, -10, 10]$. Although this window gives a good view of the curve, we might want to adjust it to show more of the curve in the first quadrant. Changing the dimensions to $[-10, 10, -5, 15]$ accomplishes this.

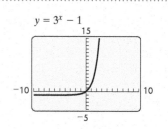

EXERCISE:

1. Use a graphing calculator to graph the functions in Examples 1–4.

b EQUATIONS WITH x AND y INTERCHANGED

It will be helpful in later work to be able to graph an equation in which the x and the y in $y = a^x$ are interchanged.

EXAMPLE 5 Graph: $x = 2^y$.

Note that x is alone on one side of the equation. We can find ordered pairs that are solutions more easily by choosing values for y and then computing the x-values.

For $y = 0$, $x = 2^0 = 1$.
For $y = 1$, $x = 2^1 = 2$.
For $y = 2$, $x = 2^2 = 4$.
For $y = 3$, $x = 2^3 = 8$.
For $y = -1$, $x = 2^{-1} = \dfrac{1}{2^1} = \dfrac{1}{2}$.
For $y = -2$, $x = 2^{-2} = \dfrac{1}{2^2} = \dfrac{1}{4}$.
For $y = -3$, $x = 2^{-3} = \dfrac{1}{2^3} = \dfrac{1}{8}$.

x	y
1	0
2	1
4	2
8	3
$\frac{1}{2}$	-1
$\frac{1}{4}$	-2
$\frac{1}{8}$	-3

(1) Choose values for y.
(2) Compute values for x.

We plot the points and connect them with a smooth curve. What happens as y-values become smaller?

This curve does not touch or cross the y-axis.

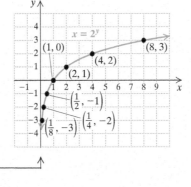

7. Graph: $x = 3^y$.

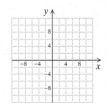

Note that the curve $x = 2^y$ has the same shape as the graph of $y = 2^x$, except that it is reflected, or flipped, across the line $y = x$, as shown below.

$y = 2^x$		$x = 2^y$	
x	y	x	y
0	1	1	0
1	2	2	1
2	4	4	2
3	8	8	3
-1	$\frac{1}{2}$	$\frac{1}{2}$	-1
-2	$\frac{1}{4}$	$\frac{1}{4}$	-2
-3	$\frac{1}{8}$	$\frac{1}{8}$	-3

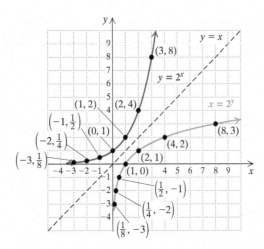

Do Exercise 7.

Answers

5.

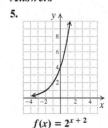

$f(x) = 2^{x+2}$

6.

x	f(x)
0	-3
1	-2
2	0
3	4
4	12
-1	$-\frac{7}{2}$
-2	$-\frac{15}{4}$

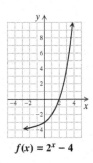

$f(x) = 2^x - 4$

7.

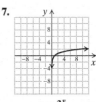

$x = 3^y$

c APPLICATIONS OF EXPONENTIAL FUNCTIONS

When interest is paid on interest, we call it **compound interest**. This is the type of interest paid on investments and loans. Suppose that you have $100,000 in a savings account at an annual interest rate of 4%. This means that in 1 year, the account will contain the original $100,000 plus 4% of $100,000. Thus the total in the account after 1 year will be

$100,000 plus $100,000 × 0.04.

This can also be expressed as

$100,000 + $100,000 × 0.04 = $100,000 × 1 + $100,000 × 0.04
= $100,000(1 + 0.04) Factoring out $100,000 using the distributive law
= $100,000(1.04)
= $104,000.

Now suppose that the total of $104,000 remains in the account for another year. At the end of the second year, the account will contain the $104,000 plus 4% of $104,000. The total in the account will be

$104,000 plus $104,000 × 0.04,

or

$104,000(1.04) = [$100,000(1.04)](1.04) = $100,000(1.04)^2
= $108,160.

Note that in the second year, interest is earned on the first year's interest as well as the original amount. When this happens, we say that the interest is **compounded annually**. If the original amount of $100,000 earned only simple interest for 2 years, the interest would be

$100,000 × 0.04 × 2, or $8000,

and the amount in the account would be

$100,000 + $8000 = $108,000,

less than the $108,160 when interest is compounded annually.

◀ Do Exercise 8.

8. **Interest Compounded Annually.** Find the amount in an account after 1 year and after 2 years if $40,000 is invested at 2%, compounded annually.

Amount after 1 year:
$40,000 + $40,000 × ☐
= $40,000(1.02)
= ☐

Amount after 2 years:
$40,800 + $40,800 × ☐
= $40,800(1.02)
= ☐

The following table shows how the computation continues over 4 years.

$100,000 in an Account

YEAR	WITH INTEREST COMPOUNDED ANNUALLY	WITH SIMPLE INTEREST
Beginning of 1st year	$100,000	
End of 1st year	$100,000(1.04)^1 = $104,000	$104,000
Beginning of 2nd year	$104,000	
End of 2nd year	$100,000(1.04)^2 = $108,160	$108,000
Beginning of 3rd year	$108,160	
End of 3rd year	$100,000(1.04)^3 = $112,486.40	$112,000
Beginning of 4th year	$112,486.40	
End of 4th year	$100,000(1.04)^4 ≈ $116,985.86	$116,000

Answers
8. $40,800; $41,616
Guided Solution:
8. 0.02, $40,800, 0.02, $41,616

We can express interest compounded annually using an exponential function.

EXAMPLE 6 *Interest Compounded Annually.* The amount of money A that a principal P will grow to after t years at interest rate r, compounded annually, is given by the formula

$$A = P(1 + r)^t.$$

Suppose that $100,000 is invested at 4% interest, compounded annually.

a) Find a function for the amount in the account after t years.

b) Find the amount of money in the account at $t = 0$, $t = 4$, $t = 8$, and $t = 10$.

c) Graph the function.

a) If $P = \$100,000$ and $r = 4\% = 0.04$, we can substitute these values and form the following function:

$$A(t) = \$100,000(1 + 0.04)^t = \$100,000(1.04)^t.$$

b) To find the function values, you might find a calculator with a power key helpful.

$A(0) = \$100,000(1.04)^0 = \$100,000;$
$A(4) = \$100,000(1.04)^4 \approx \$116,985.86;$
$A(8) = \$100,000(1.04)^8 \approx \$136,856.91;$
$A(10) = \$100,000(1.04)^{10} \approx \$148,024.43$

c) We use the function values computed in (b) with others, if we wish, to draw the graph as follows. Note that the axes are scaled differently because of the large values of A.

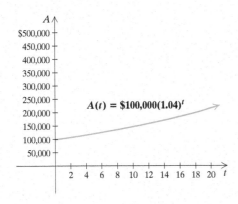

Do Exercise 9.

Suppose that the principal of $100,000 that we just considered were **compounded semiannually**—that is, every half year. Interest would then be calculated twice a year at a rate of 4% ÷ 2, or 2%, each time. The computations are as follows:

After the first $\frac{1}{2}$ year, the account will contain 102% of $100,000:

$$\$100,000 \times 1.02 = \$102,000.$$

After a second $\frac{1}{2}$ year (1 full year), the account will contain 102% of $102,000:

$$\$102,000 \times 1.02 = \$100,000 \times (1.02)^2 = \$104,040.$$

After a third $\frac{1}{2}$ year ($1\frac{1}{2}$ full years), the account will contain 102% of $104,040:

$$\$104,040 \times 1.02 = \$100,000 \times (1.02)^3 = \$106,120.80.$$

9. *Interest Compounded Annually.* Suppose that $40,000 is invested at 5% interest, compounded annually.

a) Find a function for the amount in the account after t years.

b) Find the amount of money in the account at $t = 0$, $t = 4$, $t = 8$, and $t = 10$.

c) Graph the function.

Answers

9. (a) $A(t) = \$40,000(1.05)^t;$
(b) $40,000; $48,620.25; $59,098.22; $65,155.79
(c)

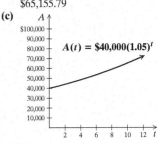

SECTION 12.1 Exponential Functions

After a fourth $\frac{1}{2}$ year (2 full years), the account will contain 102% of $106,120.80:

$$\$106{,}120.80 \times 1.02 = \$100{,}000 \times (1.02)^4$$
$$\approx \$108{,}243.22. \quad \text{Rounded to the nearest cent}$$

Comparing these results with those in the table on p. 896, we can see that by having more compounding periods, we increase the amount in the account at the end of the second year. We have illustrated the following result.

> **COMPOUND-INTEREST FORMULA**
>
> If a principal P has been invested at interest rate r, compounded n times per year, in t years it will grow to an amount A given by
> $$A = P \cdot \left(1 + \frac{r}{n}\right)^{n \cdot t}.$$

EXAMPLE 7 The Ibsens invest $4000 in an account paying $2\frac{5}{8}\%$, compounded quarterly. Find the amount in the account after $2\frac{1}{2}$ years.

The compounding is quarterly—that is, four times per year—so in $2\frac{1}{2}$ years, there are ten $\frac{1}{4}$-year periods. We substitute $4000 for P, $2\frac{5}{8}\%$, or 0.02625, for r, 4 for n, and $2\frac{1}{2}$, or $\frac{5}{2}$, for t and compute A:

$$A = P \cdot \left(1 + \frac{r}{n}\right)^{n \cdot t}$$
$$= 4000 \cdot \left(1 + \frac{2\frac{5}{8}\%}{4}\right)^{4 \cdot \frac{5}{2}}$$
$$= 4000 \cdot \left(1 + \frac{0.02625}{4}\right)^{10}$$
$$= 4000(1.0065625)^{10} \quad \text{Using a calculator}$$
$$\approx \$4270.39.$$

The amount in the account after $2\frac{1}{2}$ years is $4270.39.

◀ **Do Exercise 10.**

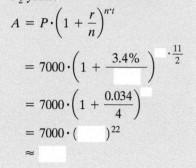

10. A couple invests $7000 in an account paying 3.4%, compounded quarterly. Find the amount in the account after $5\frac{1}{2}$ years.

$$A = P \cdot \left(1 + \frac{r}{n}\right)^{n \cdot t}$$
$$= 7000 \cdot \left(1 + \frac{3.4\%}{\square}\right)^{\square \cdot \frac{11}{2}}$$
$$= 7000 \cdot \left(1 + \frac{0.034}{4}\right)^{\square}$$
$$= 7000 \cdot (\square)^{22}$$
$$\approx \square$$

Answer
10. $8432.72
Guided Solution:
10. 4, 4, 22, 1.0085, $8432.72

12.1 Exercise Set

FOR EXTRA HELP MyLab Math

✓ Check Your Understanding

Reading Check Determine whether each statement is true or false.

RC1. In an exponential function, the variable is in the exponent.

RC2. The graph of $f(x) = 3^x$ goes through the point $(1, 0)$.

RC3. If x and y are interchanged in an equation, then the graph of the new equation will be the reflection of the graph of the original equation across the y-axis.

RC4. For a given interest rate, the amount of interest earned in a given period of time will be greater with semiannual compounding than with annual compounding.

Concept Check Each of the following exercises shows the graph of a function $f(x) = a^x$. Determine from the graph whether $a > 1$ or $0 < a < 1$.

CC1. 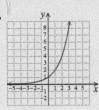   CC2. 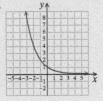   CC3.   CC4.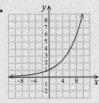

a Graph.

1. $f(x) = 2^x$

x	$f(x)$
0	
1	
2	
3	
−1	
−2	
−3	

2. $f(x) = 3^x$

x	$f(x)$
0	
1	
2	
3	
−1	
−2	
−3	

3. $f(x) = 5^x$

4. $f(x) = 6^x$

5. $f(x) = 2^{x+1}$

6. $f(x) = 2^{x-1}$

7. $f(x) = 3^{x-2}$

8. $f(x) = 3^{x+2}$

9. $f(x) = 2^x - 3$

10. $f(x) = 2^x + 1$

11. $f(x) = 5^{x+3}$

12. $f(x) = 6^{x-4}$

13. $f(x) = \left(\dfrac{1}{2}\right)^x$

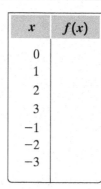

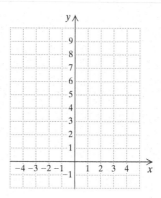

14. $f(x) = \left(\dfrac{1}{3}\right)^x$

15. $f(x) = \left(\dfrac{1}{5}\right)^x$

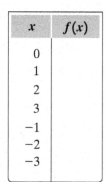

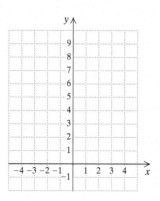

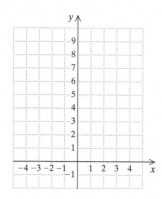

16. $f(x) = \left(\dfrac{1}{4}\right)^x$

17. $f(x) = 2^{2x-1}$

18. $f(x) = 3^{3-x}$

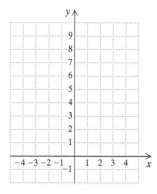

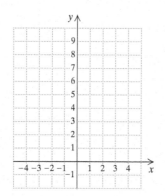

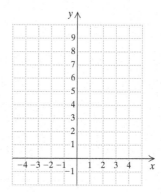

b Graph.

19. $x = 2^y$

20. $x = 6^y$

21. $x = \left(\dfrac{1}{2}\right)^y$

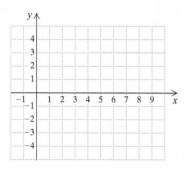

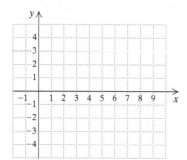

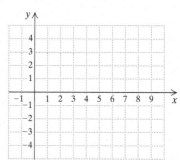

900 CHAPTER 12 Exponential Functions and Logarithmic Functions

22. $x = \left(\dfrac{1}{3}\right)^y$

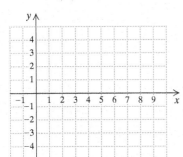

23. $x = 5^y$

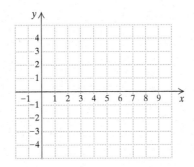

24. $x = \left(\dfrac{2}{3}\right)^y$

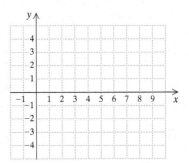

Graph both equations using the same set of axes.

25. $y = 2^x,\ x = 2^y$

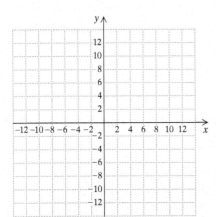

26. $y = \left(\dfrac{1}{2}\right)^x,\ x = \left(\dfrac{1}{2}\right)^y$

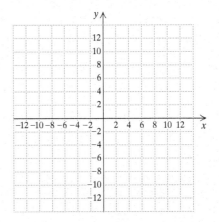

C Solve.

27. *Interest Compounded Annually.* Suppose that $50,000 is invested at 2% interest, compounded annually.
  a) Find a function A for the amount in the account after t years.
  b) Complete the following table of function values.

t	$A(t)$
0	
1	
2	
4	
8	
10	
20	

  c) Graph the function.

28. *Interest Compounded Annually.* Suppose that $50,000 is invested at 3% interest, compounded annually.
  a) Find a function A for the amount in the account after t years.
  b) Complete the following table of function values.

t	$A(t)$
0	
1	
2	
4	
8	
10	
20	

  c) Graph the function.

29. *Interest Compounded Semiannually.* Jesse deposits $2000 in an account paying 2.6%, compounded semiannually. Find the amount in the account after 3 years.

30. *Interest Compounded Semiannually.* Rory deposits $3500 in an account paying 3.2%, compounded semiannually. Find the amount in the account after 2 years.

31. *Interest Compounded Quarterly.* The Jansens invest $4500 in an account paying 3.6%, compounded quarterly. Find the amount in the account after $4\frac{1}{2}$ years.

32. *Interest Compounded Quarterly.* The Gemmers invest $4000 in an account paying 2.8%, compounded quarterly. Find the amount in the account after $3\frac{1}{2}$ years.

33. *Endangered Species.* After the Hawaiian goose, or nene, was placed on the endangered species list, its population began to increase from a critically low number. The number of nene $N(t)$ in Hawaii t years after 1975 can be approximated by
$$N(t) = 374(1.054)^t,$$
where $t = 0$ corresponds to 1975.

Data: esasuccess.org

a) How many nene were in Hawaii in 1975? in 2005? in 2015?
b) Graph the function.

34. *Snapchat.* The popularity of the social media platform Snapchat has increased exponentially since 2014. The average number of daily active Snapchat users $S(t)$, in millions, t years after January 2014 can be approximated by
$$S(t) = 48(1.58)^t,$$
where $t = 0$ corresponds to January 2014.

Data: The Wall Street Journal

a) How many daily active users were there, on average, in January 2014? in January 2015? in January 2016?
b) Graph the function.

35. *TV prices.* The average price of an OLED panel for a TV can be approximated by
$$P(t) = 6553(0.612)^t,$$
where t is the number of years after 2013.

Data: smarttvradar.com

a) What was the average price of an OLED panel in 2013? in 2015? in 2017?
b) Graph the function.

36. *Salvage Value.* An office machine is purchased for $5200. Its value each year is about 80% of the value the preceding year. Its value after t years is given by the exponential function
$$V(t) = \$5200(0.8)^t.$$

a) Find the value of the machine after 0 year, 1 year, 2 years, 5 years, and 10 years.
b) Graph the function.

37. *Recycling Aluminum Cans.* Although Americans discard $1 billion worth of aluminum cans every year, 67% of the aluminum is recycled. If a beverage company distributes 500,000 cans, the amount of aluminum still in use after t years can be made into $N(t)$ cans, where
$$N(t) = 500,000(0.67)^t.$$
Data: The Aluminum Association

a) How many cans can be made from the original 500,000 cans after 1 year? after 3 years? after 7 years?

b) Graph the function.

38. *Growth of Bacteria.* Bladder infections are often caused when the bacteria *Escherichia coli* reach the human bladder. Suppose that 3000 of the bacteria are present at time $t = 0$. Then t minutes later, the number of bacteria present will be
$$N(t) = 3000(2)^{t/20}.$$
Data: Hayes, Chris, "Detecting a Human Health Risk: E. coli," *Laboratory Medicine* **29**, no. 6, pp. 347–355, June 1998

a) How many bacteria will be present after 10 min? 20 min? 30 min? 40 min? 60 min?

b) Graph the function.

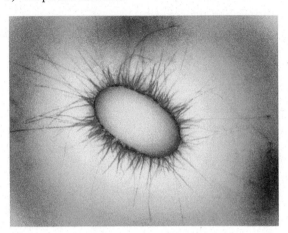

Skill Maintenance

39. Multiply and simplify: $x^{-5} \cdot x^3$. [4.1d, f]

40. Simplify: $(x^{-3})^4$. [4.1f], [4.2a]

Simplify. [4.1b]

41. 9^0

42. $\left(\frac{2}{3}\right)^0$

43. $\left(\frac{2}{3}\right)^1$

44. 2.7^1

Divide and simplify. [4.1e, f]

45. $\dfrac{x^{-3}}{x^4}$

46. $\dfrac{x}{x^{11}}$

47. $\dfrac{x}{x^0}$

48. $\dfrac{x^{-3}}{x^{-4}}$

Synthesis

49. Simplify: $(5^{\sqrt{2}})^{2\sqrt{2}}$.

50. Which is larger: $\pi^{\sqrt{2}}$ or $(\sqrt{2})^\pi$?

Graph.

51. $y = 2^x + 2^{-x}$

52. $y = |2^x - 2|$

53. $y = \left|\left(\frac{1}{2}\right)^x - 1\right|$

54. $y = 2^{-x^2}$

Graph both equations using the same set of axes.

55. $y = 3^{-(x-1)}$, $x = 3^{-(y-1)}$

56. $y = 1^x$, $x = 1^y$

57. Use a graphing calculator to graph each of the equations in Exercises 51–54.

12.2 Composite Functions and Inverse Functions

OBJECTIVES

a Find the composition of functions and express certain functions as a composition of functions.

b Find the inverse of a relation if it is described as a set of ordered pairs or as an equation.

c Given a function, determine whether it is one-to-one and has an inverse that is a function.

d Find a formula for the inverse of a function, if it exists, and graph inverse relations and functions.

e Determine whether a function is an inverse by checking its composition with the original function.

Later in this chapter, we discuss two closely related types of functions: exponential functions and logarithmic functions. In order to properly understand the link between these functions, we must first understand composite functions and inverse functions.

a COMPOSITE FUNCTIONS

Functions frequently occur in which some quantity depends on a variable that, in turn, depends on another variable. For instance, a firm's profits may be a function of the number of items the firm produces, which may in turn be a function of the number of employees hired. In this case, the firm's profits may be considered a **composite function**.

Let's consider an example of a profit function. Tea Mug Collective sells hand-painted tee shirts. Suppose that the monthly profit p, in dollars, from the sale of m shirts is given by $p = 15m - 1200$, and the number of shirts m produced in a month by x employees is given by $m = 40x$.

If Tea Mug Collective employs 10 people, then in one month they can produce $m = 40(10) = 400$ shirts. The profit from selling these 400 shirts would be $p = 15(400) - 1200 = 4800$ dollars. Can we find an equation that would allow us to calculate the monthly profit on the basis of the number of employees? We begin with the profit equation and substitute:

$$p = 15m - 1200$$
$$= 15(40x) - 1200 \quad \text{Substituting } 40x \text{ for } m$$
$$= 600x - 1200.$$

The equation $p = 600x - 1200$ gives the monthly profit when Tea Mug Collective has x employees.

To find a composition of functions, we follow the same reasoning described above using function notation:

$p(m) = 15m - 1200,$ Profit as a function of the number of shirts produced

$m(x) = 40x;$ Number of shirts as a function of the number of employees

$$p(m(x)) = p(40x)$$
$$= 15(40x) - 1200$$
$$= 600x - 1200.$$

If we call this new function P, then $P(x) = 600x - 1200$. This gives profit as a function of the number of employees.

We call P the *composition* of p and m. In general, the composition of f and g is written $f \circ g$ and is read "the composition of f and g," "f composed with g," or "f circle g."

It is not uncommon to use the same variable to represent the input in more than one function.

Throughout this chapter, keep in mind that equations such as $m(x) = 40x$ and $m(t) = 40t$ describe the same function. Both equations tell us to find a function value by multiplying the input by 40.

Tea Mug Collective's Shane Kimberlin, Alaskan artist

COMPOSITE FUNCTION

The **composite function** $f \circ g$, the **composition** of f and g, is defined as
$$(f \circ g)(x) = f(g(x)).$$

We can visualize the composition of functions as follows.

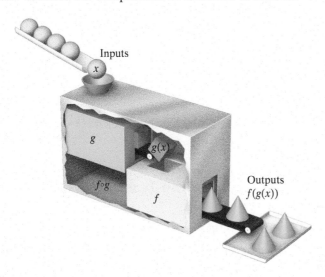

EXAMPLE 1 Given $f(x) = 3x$ and $g(x) = 1 + x^2$:

a) Find $(f \circ g)(5)$ and $(g \circ f)(5)$.
b) Find $(f \circ g)(x)$ and $(g \circ f)(x)$.

We consider each function separately:

$f(x) = 3x$ This function multiplies each input by 3.

and $g(x) = 1 + x^2$.   This function adds 1 to the square of each input.

a) $(f \circ g)(5) = f(g(5)) = f(1 + 5^2) = f(26) = 3(26) = 78;$
$(g \circ f)(5) = g(f(5)) = g(3 \cdot 5) = g(15) = 1 + 15^2 = 1 + 225 = 226$

b) $(f \circ g)(x) = f(g(x))$
$= f(1 + x^2)$ Substituting $1 + x^2$ for $g(x)$
$= 3(1 + x^2)$
$= 3 + 3x^2;$

$(g \circ f)(x) = g(f(x))$
$= g(3x)$ Substituting $3x$ for $f(x)$
$= 1 + (3x)^2$
$= 1 + 9x^2$

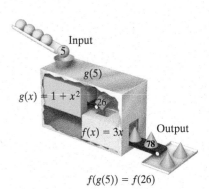

A composition machine for Example 1

We can check the values in part (a) with the formulas found in part (b):

$(f \circ g)(x) = 3 + 3x^2$ $(g \circ f)(x) = 1 + 9x^2$
$(f \circ g)(5) = 3 + 3 \cdot 5^2$ $(g \circ f)(5) = 1 + 9 \cdot 5^2$
$\qquad = 3 + 3 \cdot 25$ $\qquad = 1 + 9 \cdot 25$
$\qquad = 3 + 75$ $\qquad = 1 + 225$
$\qquad = 78;$ $\qquad = 226.$

Do Exercise 1.

1. Given $f(x) = x + 5$ and $g(x) = x^2 - 1$, find $(f \circ g)(x)$ and $(g \circ f)(x)$.

Answer
1. $x^2 + 4; x^2 + 10x + 24$

2. Given $f(x) = 4x + 5$ and $g(x) = \sqrt[3]{x}$, find $(f \circ g)(x)$ and $(g \circ f)(x)$.

$(f \circ g)(x) = f(g(x))$
$= f(\sqrt[3]{})$
$= 4() + 5;$

$(g \circ f)(x) = g(f(x))$
$= g(4x +)$
$= \sqrt[3]{}$

Example 1 shows that $(f \circ g)(5) \neq (g \circ f)(5)$ and, in general,
$$(f \circ g)(x) \neq (g \circ f)(x).$$

EXAMPLE 2 Given $f(x) = \sqrt{x}$ and $g(x) = x - 1$, find $(f \circ g)(x)$ and $(g \circ f)(x)$.

$(f \circ g)(x) = f(g(x)) = f(x - 1) = \sqrt{x - 1};$
$(g \circ f)(x) = g(f(x)) = g(\sqrt{x}) = \sqrt{x} - 1$

◀ Do Exercise 2.

It is important to be able to recognize how a function can be expressed, or "broken down," as a composition. Such a situation can occur in a study of calculus.

EXAMPLE 3 Find $f(x)$ and $g(x)$ such that $h(x) = (f \circ g)(x)$:
$$h(x) = (7x + 3)^2.$$

This is $7x + 3$ to the 2nd power. Two functions that can be used for the composition are $f(x) = x^2$ and $g(x) = 7x + 3$. We can check by forming the composition:
$$h(x) = (f \circ g)(x) = f(g(x)) = f(7x + 3) = (7x + 3)^2.$$

This is the most "obvious" answer to the question. There can be other less obvious answers. For example, if
$$f(x) = (x - 1)^2 \quad \text{and} \quad g(x) = 7x + 4,$$
then
$$h(x) = (f \circ g)(x) = f(g(x)) = f(7x + 4) = (7x + 4 - 1)^2 = (7x + 3)^2.$$

3. Find $f(x)$ and $g(x)$ such that $h(x) = (f \circ g)(x)$. Answers may vary.

a) $h(x) = \sqrt[3]{x^2 + 1}$

b) $h(x) = \dfrac{1}{(x + 5)^4}$

◀ Do Exercise 3.

b INVERSES

A set of ordered pairs is called a **relation**. A function is a special kind of relation in which to each first coordinate there corresponds one and only one second coordinate.

Consider the relation h given as follows:
$$h = \{(-7, 4), (3, -1), (-6, 5), (0, 2)\}.$$

Suppose that we *interchange* the first and second coordinates. The relation we obtain is called the **inverse** of the relation h and is given as follows:

Inverse of $h = \{(4, -7), (-1, 3), (5, -6), (2, 0)\}.$

> **INVERSE RELATION (ORDERED PAIRS)**
>
> Interchanging the coordinates of the ordered pairs in a relation produces the **inverse relation**.

Answers
2. $4\sqrt[3]{x} + 5$; $\sqrt[3]{4x + 5}$
3. (a) $f(x) = \sqrt[3]{x}$; $g(x) = x^2 + 1$;
(b) $f(x) = \dfrac{1}{x^4}$; $g(x) = x + 5$

Guided Solution:
2. x, $\sqrt[3]{x}$, 5, $4x + 5$

EXAMPLE 4 Consider the relation g given by

$$g = \{(2, 4), (-1, 3), (-2, 0)\}.$$

In the following figure, the relation g is shown in red. The inverse of the relation is

$$\{(4, 2), (3, -1), (0, -2)\}$$

and is shown in blue.

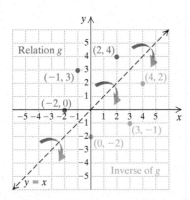

4. Consider the relation g given by

$$g = \{(2, 5), (-1, 4), (-2, 1)\}.$$

The graph of the relation is shown below in red. Find the inverse and draw its graph.

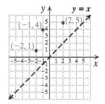

5. Find an equation of the inverse relation. Then complete the table and graph both the original relation and its inverse.

Relation:
$y = 6 - 2x$

x	y
0	6
2	2
3	0
5	-4

Inverse:
_____

x	y
	0
	2
	3
	5

Do Exercise 4.

INVERSE RELATION (EQUATION)

If a relation is defined by an equation, interchanging the variables produces an equation of the **inverse relation**.

EXAMPLE 5 Find an equation of the inverse of $y = 3x - 4$. Then graph both the relation and its inverse.

We interchange x and y and obtain an equation of the inverse:

$$x = 3y - 4.$$

Relation: $y = 3x - 4$ ⟶   Inverse: $x = 3y - 4$

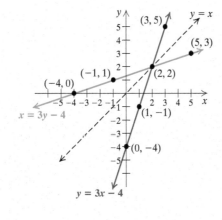

Do Exercise 5.

Note in Example 5 that the relation $y = 3x - 4$ is a function and its inverse relation $x = 3y - 4$ is also a function. Each graph passes the vertical-line test. (See Section 7.1.)

Answer
4. Inverse of $g = \{(5, 2), (4, -1), (1, -2)\}$

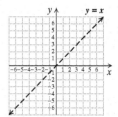

Answer to Margin Exercise 5 is on p. 908.

6. Find an equation of the inverse relation. Then complete the table and graph both the original relation and its inverse.

Relation:
$y = x^2 - 4x + 7$

x	y
0	7
1	4
2	3
3	4
4	7

Inverse:
_____

x	y
0	
1	
2	
3	
4	

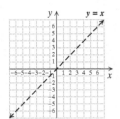

EXAMPLE 6 Find an equation of the inverse of $y = 6x - x^2$. Then graph both the original relation and its inverse.

We interchange x and y and obtain an equation of the inverse:

$$x = 6y - y^2.$$

Relation: $y = 6x - x^2$ ⟶⟶⟶ Inverse: $x = 6y - y^2$

x	y
−1	−7
0	0
1	5
3	9
5	5

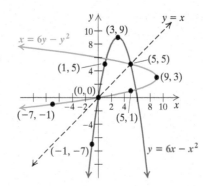

x	y
−7	−1
0	0
5	1
9	3
5	5

Note in Example 6 that the relation $y = 6x - x^2$ is a function because it passes the vertical-line test. However, its inverse relation $x = 6y - y^2$ is not a function because its graph fails the vertical-line test. Therefore, the inverse of a function is *not* always a function.

◀ Do Exercise 6.

C INVERSES AND ONE-TO-ONE FUNCTIONS

Let's consider the following two functions.

NUMBER (Domain)	CUBE (Range)
−3 ⟶	−27
−2 ⟶	−8
−1 ⟶	−1
0 ⟶	0
1 ⟶	1
2 ⟶	8
3 ⟶	27

VIDEO GAME CONSOLE (Domain)	MANUFACTURER (Range)
Xbox ⟶	Microsoft
DS	
Wii ⟶	Nintendo
Switch	
PlayStation ⟶	Sony

Suppose that we reverse the arrows. Are these inverse relations functions?

CUBE ROOT (Range)	NUMBER (Domain)
−3 ⟵	−27
−2 ⟵	−8
−1 ⟵	−1
0 ⟵	0
1 ⟵	1
2 ⟵	8
3 ⟵	27

VIDEO GAME CONSOLE (Range)	MANUFACTURER (Domain)
Xbox ⟵	Microsoft
DS	
Wii ⟵	Nintendo
Switch	
PlayStation ⟵	Sony

Answers

5. Inverse:
$x = 6 - 2y$

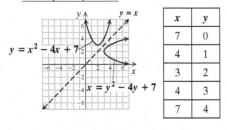

6. Inverse:
$x = y^2 - 4y + 7$

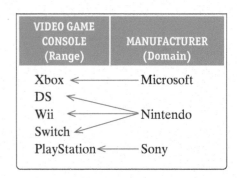

We see that the inverse of the cubing function is a function. The inverse of the game platform function is not a function, however, because the input Nintendo has *three* outputs: DS, Wii, and Switch. Recall that for a function, each input has exactly one output. However, it can happen that the same output comes from two or more different inputs. If this is the case, the inverse cannot be a function. When this possibility is excluded, the inverse is also a function.

In the cubing function, different inputs have different outputs. Thus its inverse is also a function. The cubing function is what is called a **one-to-one function**.

ONE-TO-ONE FUNCTION AND INVERSES

A function f is **one-to-one** if different inputs have different outputs— that is,

$$\text{if } a \neq b, \text{ then } f(a) \neq f(b). \text{ Or,}$$

A function f is **one-to-one** if when the outputs are the same, the inputs are the same—that is,

$$\text{if } f(a) = f(b), \text{ then } a = b.$$

If a function is one-to-one, then its inverse is a function.

How can we tell graphically whether a function is one-to-one and thus has an inverse that is a function?

EXAMPLE 7 The graph of the exponential function $f(x) = 2^x$, or $y = 2^x$, is shown on the left below. The graph of the inverse $x = 2^y$ is shown on the right. How can we tell by examining only the graph on the left whether it has an inverse that is a function?

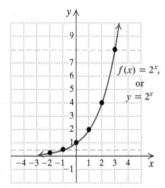
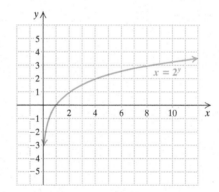

We see that the graph on the right passes the vertical-line test, so we know that it is the graph of a function. However, if we look only at the graph on the left, we think as follows:

A function is one-to-one if different inputs have different outputs. That is, no two x-values will have the same y-value. For this function, no horizontal line can be drawn that will cross the graph more than once. The function is thus one-to-one and its inverse is a function.

THE HORIZONTAL-LINE TEST

If it is possible for a horizontal line to intersect the graph of a function more than once, then the function is not one-to-one and therefore its inverse is not a function.

Determine whether the function is one-to-one and thus has an inverse that is also a function.

7. $f(x) = 4 - x$

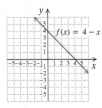

8. $f(x) = x^2 - 1$

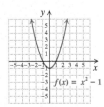

9. $f(x) = 4^x$
(Sketch this graph yourself.)

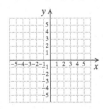

10. $f(x) = |x| - 3$
(Sketch this graph yourself.)

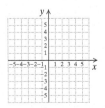

MyLab Math
ANIMATION

A graph is that of a function if no vertical line crosses the graph more than once. A function has an inverse that is also a function if no horizontal line crosses the graph more than once.

EXAMPLE 8 Determine whether the function $f(x) = x^2$ is one-to-one and has an inverse that is also a function.

The graph of $f(x) = x^2$, or $y = x^2$, is shown on the left below. There are many horizontal lines that cross the graph more than once, so this function is not one-to-one and does not have an inverse that is a function.

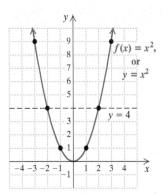 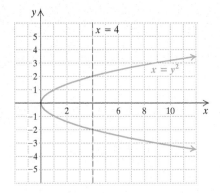

The inverse of the function $y = x^2$ is the relation $x = y^2$. The graph of $x = y^2$ is shown on the right above. It fails the vertical-line test and is not a function.

◀ Do Exercises 7–10.

d INVERSE FORMULAS AND GRAPHS

If the inverse of a function f is also a function, then it is named f^{-1} (read "f-inverse").

.... Caution! ....

The -1 in f^{-1} is *not* an exponent and f^{-1} does *not* represent a reciprocal!

Suppose that a function is described by a formula. If it has an inverse that is a function, how do we find a formula for the inverse function? If for any equation with two variables such as x and y we interchange the variables, we obtain an equation of the inverse relation. We proceed as follows to find a formula for f^{-1}.

> If a function f is one-to-one, a formula for its inverse f^{-1} can be found as follows:
>
> 1. Replace $f(x)$ with y.
> 2. Interchange x and y. (This gives the inverse relation.)
> 3. Solve for y.
> 4. Replace y with $f^{-1}(x)$.

Answers
7. Yes   8. No   9. Yes   10. No

EXAMPLE 9 Given $f(x) = x + 1$:

a) Determine whether the function is one-to-one.
b) If it is one-to-one, find a formula for $f^{-1}(x)$.
c) Graph the inverse function, if it exists.

a) The graph of $f(x) = x + 1$ is shown below. It passes the horizontal-line test, so it is one-to-one. Thus its inverse is a function.

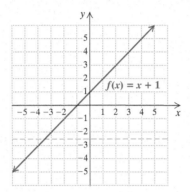

b) 1. Replace $f(x)$ with y: $y = x + 1$.
    2. Interchange x and y: $x = y + 1$.   This gives the inverse relation.
    3. Solve for y: $x - 1 = y$.
    4. Replace y with $f^{-1}(x)$: $f^{-1}(x) = x - 1$.

c) We graph $f^{-1}(x) = x - 1$, or $y = x - 1$. The graph is shown below.

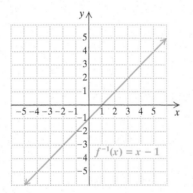

EXAMPLE 10 Given $f(x) = 2x - 3$:

a) Determine whether the function is one-to-one.
b) If it is one-to-one, find a formula for $f^{-1}(x)$.
c) Graph the inverse function, if it exists.

CALCULATOR CORNER

Graphing an Inverse Function The DRAWINV operation can be used to graph a function and its inverse on the same screen. A formula for the inverse function need not be found in order to do this. The graphing calculator must be set in FUNC mode when this operation is used.

To graph $f(x) = 2x - 3$ and $f^{-1}(x)$ using the same set of axes, we first clear any existing equations on the equation-editor screen and then enter $y_1 = 2x - 3$. We graph the function and then use the DRAWINV operation to graph its inverse. The graphs are shown here in a squared window.

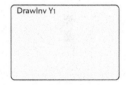

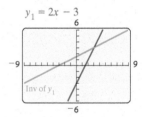

EXERCISES: Use the DRAWINV operation on a graphing calculator to graph each function with its inverse on the same screen.

1. $f(x) = x - 5$
2. $f(x) = \frac{2}{3}x$
3. $f(x) = x^2 + 2$
4. $f(x) = x^3 - 3$

a) The graph of $f(x) = 2x - 3$ is shown below. It passes the horizontal-line test and is one-to-one.

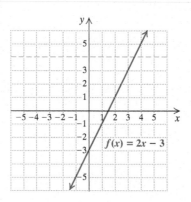

Given each function:
a) Determine whether it is one-to-one.
b) If it is one-to-one, find a formula for the inverse.
c) Graph the inverse function, if it exists.

11. $f(x) = 3 - x$

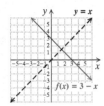

12. $g(x) = 3x - 2$

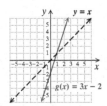

b) 1. Replace $f(x)$ with y: $\quad y = 2x - 3$.
2. Interchange x and y: $\quad x = 2y - 3$.
3. Solve for y: $\quad x + 3 = 2y$

$$\frac{x + 3}{2} = y.$$

4. Replace y with $f^{-1}(x)$: $\quad f^{-1}(x) = \dfrac{x + 3}{2}.$

c) We graph

$$f^{-1}(x) = \frac{x + 3}{2}, \quad \text{or}$$

$$y = \frac{1}{2}x + \frac{3}{2}.$$

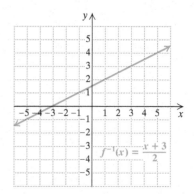

◀ **Do Exercises 11 and 12.**

Let's consider inverses of functions in terms of a function machine. Suppose that a one-to-one function f is programmed into a machine. If the machine is run in reverse, it will perform the inverse function f^{-1}. Inputs then enter at the opposite end, and the entire process is reversed.

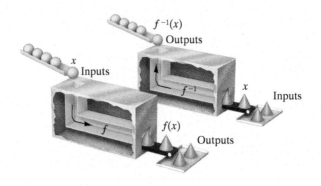

Answers
11. (a) Yes; (b) $f^{-1}(x) = 3 - x$;
(c)

$f^{-1}(x) = 3 - x$

12. (a) Yes; (b) $g^{-1}(x) = \dfrac{x + 2}{3}$;
(c)

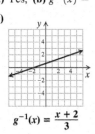

$g^{-1}(x) = \dfrac{x + 2}{3}$

Consider $f(x) = 2x - 3$ and $f^{-1}(x) = (x + 3)/2$ from Example 10. For the input 5,

$$f(5) = 2 \cdot 5 - 3 = 10 - 3 = 7.$$

The output is 7. Now we use 7 for the input in the inverse:

$$f^{-1}(7) = \frac{7 + 3}{2} = \frac{10}{2} = 5.$$

The function f takes 5 to 7. The inverse function f^{-1} takes the number 7 back to 5.

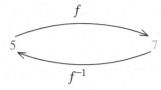

How do the graphs of a function and its inverse compare?

EXAMPLE 11 Graph $f(x) = 2x - 3$ and $f^{-1}(x) = (x + 3)/2$ using the same set of axes. Then compare.

The graph of each function follows. Note that the graph of f^{-1} can be drawn by reflecting the graph of f across the line $y = x$. That is, if we graph $f(x) = 2x - 3$ in wet ink and fold the paper along the line $y = x$, the graph of $f^{-1}(x) = (x + 3)/2$ will appear as the impression made by f.

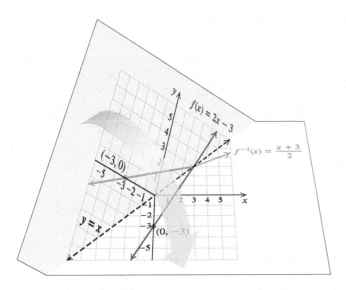

When x and y are interchanged to find a formula for the inverse, we are, in effect, flipping the graph of $f(x) = 2x - 3$ over the line $y = x$. For example, when the coordinates of the y-intercept of the graph of f, $(0, -3)$, are reversed, we get the x-intercept of the graph of f^{-1}, $(-3, 0)$.

The graph of f^{-1} is a reflection of the graph of f across the line $y = x$.

Do Exercise 13.

13. Graph $g(x) = 3x - 2$ and $g^{-1}(x) = (x + 2)/3$ using the same set of axes.

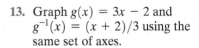

Answer

13.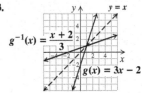

EXAMPLE 12 Consider $g(x) = x^3 + 2$.

a) Determine whether the function is one-to-one.
b) If it is one-to-one, find a formula for its inverse.
c) Graph the inverse, if it exists.

a) The graph of $g(x) = x^3 + 2$ is shown at right in red. It passes the horizontal-line test and thus is one-to-one.

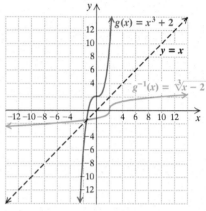

14. Given $f(x) = x^3 + 1$:

a) Determine whether the function is one-to-one.

b) If it is one-to-one, find a formula for its inverse.

c) Graph the function and its inverse using the same set of axes.

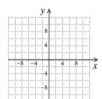

b) **1.** Replace $g(x)$ with y: $y = x^3 + 2$.
   2. Interchange x and y: $x = y^3 + 2$.
   3. Solve for y: $x - 2 = y^3$
                     $\sqrt[3]{x - 2} = y$.   Since a number has only one cube root, we can solve for y.

   4. Replace y with $g^{-1}(x)$: $g^{-1}(x) = \sqrt[3]{x - 2}$.

c) To find the graph, we reflect the graph of $g(x) = x^3 + 2$ across the line $y = x$, as we did in Example 11. It can also be found by substituting into $g^{-1}(x) = \sqrt[3]{x - 2}$ and plotting points. The graph of $g^{-1}(x) = \sqrt[3]{x - 2}$ is shown above in blue.

◀ Do Exercise 14.

We can now see why we exclude 1 as a base for an exponential function. Consider

$$f(x) = a^x = 1^x = 1.$$

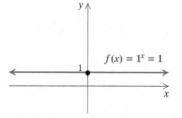

The graph of f is the horizontal line $y = 1$. The graph is not one-to-one. The function does not have an inverse that is a function. All other positive bases yield exponential functions that are one-to-one.

If a function f is one-to-one, then the domain of f is the range of f^{-1}, and the range of f is the domain of f^{-1}.

DOMAIN AND RANGE OF INVERSE FUNCTIONS

The domain of a one-to-one function f is the range of the inverse f^{-1}.
The range of a one-to-one function f is the domain of the inverse f^{-1}.

Answer

14. (a) Yes; (b) $f^{-1}(x) = \sqrt[3]{x - 1}$;
(c)

e INVERSE FUNCTIONS AND COMPOSITION

Suppose that we used some input x for the function f and found its output, $f(x)$. The function f^{-1} would then take that output back to x. Similarly, if we began with an input x for the function f^{-1} and found its output, $f^{-1}(x)$, the original function f would then take that output back to x.

> If a function f is one-to-one, then f^{-1} is the unique function for which
> $$(f^{-1} \circ f)(x) = x \quad \text{and} \quad (f \circ f^{-1})(x) = x.$$

EXAMPLE 13 Let $f(x) = 2x - 3$. Use composition to show that
$$f^{-1}(x) = \frac{x + 3}{2}. \quad \text{(See Example 10.)}$$

We find $(f^{-1} \circ f)(x)$ and $(f \circ f^{-1})(x)$ and check to see that each is x.

$$(f^{-1} \circ f)(x) = f^{-1}(f(x))$$
$$= f^{-1}(2x - 3)$$
$$= \frac{(2x - 3) + 3}{2}$$
$$= \frac{2x}{2}$$
$$= x;$$

$$(f \circ f^{-1})(x) = f(f^{-1}(x))$$
$$= f\left(\frac{x + 3}{2}\right)$$
$$= 2 \cdot \frac{x + 3}{2} - 3$$
$$= x + 3 - 3$$
$$= x$$

Do Exercise 15.

15. Let $f(x) = \frac{2}{3}x - 4$.
Use composition to show that
$$f^{-1}(x) = \frac{3x + 12}{2}.$$

Answer

15. $(f^{-1} \circ f)(x) = f^{-1}(f(x)) = f^{-1}(\frac{2}{3}x - 4)$
$$= \frac{3(\frac{2}{3}x - 4) + 12}{2}$$
$$= \frac{2x - 12 + 12}{2}$$
$$= \frac{2x}{2} = x;$$
$(f \circ f^{-1})(x) = f(f^{-1}(x)) = f\left(\frac{3x + 12}{2}\right)$
$$= \frac{2}{3}\left(\frac{3x + 12}{2}\right) - 4$$
$$= \frac{6x + 24}{6} - 4$$
$$= x + 4 - 4 = x$$

12.2 Exercise Set

FOR EXTRA HELP — MyLab Math

✓ Check Your Understanding

Reading Check Choose from the list on the right the word that best completes each statement. Words may be used more than once or not at all.

RC1. Any set of ordered pairs is a(n) _____.

RC2. The relation $\{(3, 6), (0, -1), (2, 5)\}$ is the _____ of $\{(6, 3), (-1, 0), (5, 2)\}$.

RC3. If the graph of a function passes the _____-line test, the inverse of the function is also a function.

RC4. A function whose inverse is also a function is called a(n) _____ function.

RC5. The function $g(x) = x - 10$ is the _____ of $f(x) = x + 10$.

RC6. The function $g(x) = (x - 1)^2$ is the _____ $(f \circ h)(x)$ of the functions given by $f(x) = x^2$ and $h(x) = x - 1$.

composition
inverse
function
horizontal
one-to-one
relation
vertical

Concept Check Find each of the following, if possible, given $f(x) = 2x + 3$, $g(x) = x^2 - 1$, and $h(x) = \dfrac{2}{x}$.

CC1. $(f \circ g)(1)$ **CC2.** $(g \circ f)(1)$ **CC3.** $(f \circ h)(-2)$

CC4. $(h \circ f)(-2)$ **CC5.** $(g \circ h)(-1)$ **CC6.** $(h \circ g)(-1)$

a Find $(f \circ g)(x)$ and $(g \circ f)(x)$.

1. $f(x) = 2x - 3$,
   $g(x) = 6 - 4x$

2. $f(x) = 9 - 6x$,
   $g(x) = 0.37x + 4$

3. $f(x) = 3x^2 + 2$,
   $g(x) = 2x - 1$

4. $f(x) = 4x + 3$,
   $g(x) = 2x^2 - 5$

5. $f(x) = 4x^2 - 1$,
   $g(x) = \dfrac{2}{x}$

6. $f(x) = \dfrac{3}{x}$,
   $g(x) = 2x^2 + 3$

7. $f(x) = x^2 + 5$,
   $g(x) = x^2 - 5$

8. $f(x) = \dfrac{1}{x^2}$,
   $g(x) = x - 1$

Find $f(x)$ and $g(x)$ such that $h(x) = (f \circ g)(x)$. Answers may vary.

9. $h(x) = (5 - 3x)^2$
10. $h(x) = 4(3x - 1)^2 + 9$
11. $h(x) = \sqrt{5x + 2}$
12. $h(x) = (3x^2 - 7)^5$

13. $h(x) = \dfrac{1}{x - 1}$
14. $h(x) = \dfrac{3}{x} + 4$
15. $h(x) = \dfrac{1}{\sqrt{7x + 2}}$
16. $h(x) = \sqrt{x - 7} - 3$

17. $h(x) = (\sqrt{x} + 5)^4$
18. $h(x) = \dfrac{x^3 + 1}{x^3 - 1}$

b Find the inverse of each relation. Graph the original relation in red and then graph the inverse relation in blue.

19. $\{(1, 2), (6, -3), (-3, -5)\}$

20. $\{(3, -1), (5, 2), (5, -3), (2, 0)\}$

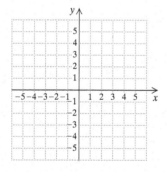
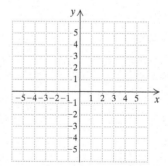

916 CHAPTER 12 Exponential Functions and Logarithmic Functions

Find an equation of the inverse of the relation. Then complete the second table and graph both the original relation and its inverse.

21. $y = 2x + 6$

x	y
−1	4
0	6
1	8
2	10
3	12

x	y
4	
6	
8	
10	
12	

22. $y = \frac{1}{2}x^2 - 8$

x	y
−4	0
−2	−6
0	−8
2	−6
4	0

x	y
0	
−6	
−8	
−6	
0	

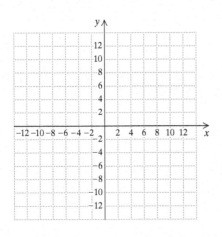
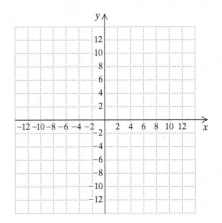

c Determine whether each function is one-to-one.

23. $f(x) = x - 5$ **24.** $f(x) = 3 - 6x$ **25.** $f(x) = x^2 - 2$ **26.** $f(x) = 4 - x^2$

27. $f(x) = |x| - 3$ **28.** $f(x) = |x - 2|$ **29.** $f(x) = 3^x$ **30.** $f(x) = \left(\frac{1}{2}\right)^x$

d Determine whether each function is one-to-one. If it is, find a formula for its inverse.

31. $f(x) = 5x - 2$ **32.** $f(x) = 4 + 7x$ **33.** $f(x) = \dfrac{-2}{x}$ **34.** $f(x) = \dfrac{1}{x}$

35. $f(x) = \frac{4}{3}x + 7$ **36.** $f(x) = -\frac{7}{8}x + 2$ **37.** $f(x) = \dfrac{2}{x+5}$ **38.** $f(x) = \dfrac{1}{x-8}$

39. $f(x) = 5$ **40.** $f(x) = -2$ **41.** $f(x) = \dfrac{2x+1}{5x+3}$ **42.** $f(x) = \dfrac{2x-1}{5x+3}$

43. $f(x) = x^3 - 1$ **44.** $f(x) = x^3 + 5$ **45.** $f(x) = \sqrt[3]{x}$ **46.** $f(x) = \sqrt[3]{x-4}$

Graph each function and its inverse using the same set of axes.

47. $f(x) = \frac{1}{2}x - 3$,
$f^{-1}(x) =$ _____

x	f(x)
-4	
0	
2	
4	

x	$f^{-1}(x)$
-5	
-3	
-2	
-1	

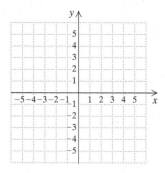

48. $g(x) = x + 4$,
$g^{-1}(x) =$ _____

x	g(x)
-4	
-2	
0	
1	

x	$g^{-1}(x)$
0	
2	
4	
5	

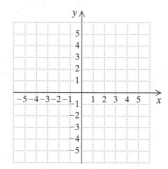

49. $f(x) = x^3$,
$f^{-1}(x) =$ _____

x	f(x)
0	
1	
2	
3	
-1	
-2	
-3	

x	$f^{-1}(x)$

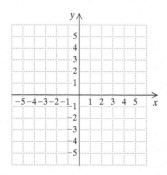

50. $f(x) = x^3 - 1$,
$f^{-1}(x) =$ _____

x	f(x)
0	
1	
2	
3	
-1	
-2	
-3	

x	$f^{-1}(x)$

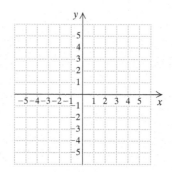

e For each function, use composition to show that the inverse is correct.

51. $f(x) = \frac{4}{5}x$,
$f^{-1}(x) = \frac{5}{4}x$

52. $f(x) = x - 3$,
$f^{-1}(x) = x + 3$

53. $f(x) = \frac{x + 7}{2}$,
$f^{-1}(x) = 2x - 7$

54. $f(x) = \frac{3}{4}x - 1$,
$f^{-1}(x) = \frac{4x + 4}{3}$

55. $f(x) = \frac{1 - x}{x}$,
$f^{-1}(x) = \frac{1}{x + 1}$

56. $f(x) = x^3 - 5$,
$f^{-1}(x) = \sqrt[3]{x + 5}$

Find the inverse of the given function by thinking about the operations of the function and then reversing, or undoing, them. You can check your answer using composition.

Function	Inverse	Function	Inverse
57. $f(x) = 3x$	$f^{-1}(x) = $ _____	**58.** $f(x) = \frac{1}{4}x + 7$	$f^{-1}(x) = $ _____
59. $f(x) = -x$	$f^{-1}(x) = $ _____	**60.** $f(x) = \sqrt[3]{x} - 5$	$f^{-1}(x) = $ _____
61. $f(x) = \sqrt[3]{x - 5}$	$f^{-1}(x) = $ _____	**62.** $f(x) = x^{-1}$	$f^{-1}(x) = $ _____

63. *Dress Sizes in the United States and France.* A size-6 dress in the United States is size 38 in France. A function that converts dress sizes in the United States to those in France is
$$f(x) = x + 32.$$
a) Find the dress sizes in France that correspond to sizes of 8, 10, 14, and 18 in the United States.
b) Determine whether this function has an inverse that is a function. If so, find a formula for the inverse.
c) Use the inverse function to find dress sizes in the United States that correspond to sizes of 40, 42, 46, and 50 in France.

64. *Dress Sizes in the United States and Italy.* A size-6 dress in the United States is size 36 in Italy. A function that converts dress sizes in the United States to those in Italy is
$$f(x) = 2(x + 12).$$
a) Find the dress sizes in Italy that correspond to sizes of 8, 10, 14, and 18 in the United States.
b) Determine whether this function has an inverse that is a function. If so, find a formula for the inverse.
c) Use the inverse function to find dress sizes in the United States that correspond to sizes of 40, 44, 52, and 60 in Italy.

Skill Maintenance

Use rational exponents to simplify. [10.2d]

65. $\sqrt[6]{a^2}$
66. $\sqrt[8]{81}$
67. $\sqrt[12]{64x^6y^6}$
68. $\sqrt[4]{81a^8b^8}$

Simplify.

69. i^{79} [10.8d]
70. $(125x^3y^{-2}z^6)^{-2/3}$ [10.2c]
71. $\sqrt{2400}$ [10.3a]

Multiply.

72. $(x + 1)(x^2 - 7)$ [4.6a]
73. $(2y - 5)^2$ [4.6c]
74. $(7a^2 + c)(7a^2 - c)$ [4.6b]

Synthesis

In Exercises 75–78, use a graphing calculator to help determine whether or not the given functions are inverses of each other.

75. $f(x) = 0.75x^2 + 2$; $g(x) = \sqrt{\dfrac{4(x-2)}{3}}$

76. $f(x) = 1.4x^3 + 3.2$; $g(x) = \sqrt[3]{\dfrac{x - 3.2}{1.4}}$

77. $f(x) = \sqrt{2.5x + 9.25}$; $g(x) = 0.4x^2 - 3.7$, $x \geq 0$

78. $f(x) = 0.8x^{1/2} + 5.23$; $g(x) = 1.25(x^2 - 5.23)$, $x \geq 0$

79. Use a graphing calculator to help match each function in Column A with its inverse from Column B.

Column A

(1) $y = 5x^3 + 10$

(2) $y = (5x + 10)^3$

(3) $y = 5(x + 10)^3$

(4) $y = (5x)^3 + 10$

Column B

A. $y = \dfrac{\sqrt[3]{x} - 10}{5}$

B. $y = \sqrt[3]{\dfrac{x}{5}} - 10$

C. $y = \sqrt[3]{\dfrac{x - 10}{5}}$

D. $y = \dfrac{\sqrt[3]{x - 10}}{5}$

In Exercises 80 and 81, graph the inverse of f.

80.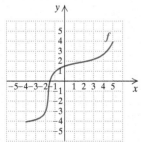

81.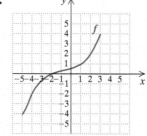

82. Examine the following table. Does it appear that f and g could be inverses of each other? Why or why not?

x	$f(x)$	$g(x)$
6	6	6
7	6.5	8
8	7	10
9	7.5	12
10	8	14
11	8.5	16
12	9	18

83. Refer to Exercise 82. Assume that f and g are both linear functions. Find equations for $f(x)$ and $g(x)$. Are f and g inverses of each other?

Logarithmic Functions

12.3

OBJECTIVES

a Graph logarithmic functions.

b Convert from exponential equations to logarithmic equations and from logarithmic equations to exponential equations.

c Solve logarithmic equations.

d Find common logarithms on a calculator.

We are now ready to study inverses of exponential functions. These functions have many applications and are referred to as *logarithm*, or *logarithmic*, functions.

a GRAPHING LOGARITHMIC FUNCTIONS

Consider the exponential function $f(x) = 2^x$. Like all exponential functions, f is one-to-one. Can a formula for f^{-1} be found? To answer this, we use the method of Section 12.2:

1. Replace $f(x)$ with y: $y = 2^x$.
2. Interchange x and y: $x = 2^y$.
3. Solve for y: $y =$ the power to which we raise 2 to get x.
4. Replace y with $f^{-1}(x)$: $f^{-1}(x) =$ the power to which we raise 2 to get x.

We now define a new symbol to replace the words "the power to which we raise 2 to get x."

MyLab Math ANIMATION

MEANING OF LOGARITHMS

$\log_2 x$, read "the logarithm, base 2, of x," or "log, base 2, of x," means "the power to which we raise 2 to get x."

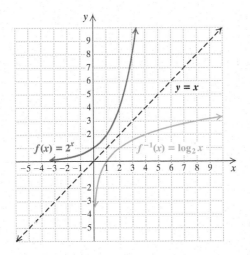

Thus if $f(x) = 2^x$, then $f^{-1}(x) = \log_2 x$. Note that $f^{-1}(8) = \log_2 8 = 3$, because 3 is the *power to which we raise 2 to get* 8; that is, $2^3 = 8$.

Although expressions like $\log_2 13$ can be only approximated, remember that $\log_2 13$ represents the **power** to which we raise 2 to get 13. That is, $2^{\log_2 13} = 13$.

Do Exercise 1.

---

1. Write the meaning of $\log_2 64$. Then find $\log_2 64$.

Answer

1. $\log_2 64$ is the power to which we raise 2 to get 64; 6

For any exponential function $f(x) = a^x$, the inverse is called a **logarithmic function, base a**. The graph of the inverse can, of course, be drawn by reflecting the graph of $f(x) = a^x$ across the line $y = x$. It will be helpful to remember that the inverse of $f(x) = a^x$ is given by $f^{-1}(x) = \log_a x$. Normally, we use a number a that is greater than 1 for the logarithm base.

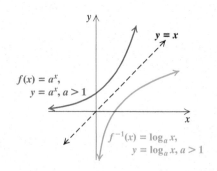

LOGARITHMS

The inverse of $f(x) = a^x$ is given by
$$f^{-1}(x) = \log_a x.$$

We read "$\log_a x$" as "the logarithm, base a, of x." We define $y = \log_a x$ as that number y such that $a^y = x$, where $x > 0$ and a is a positive constant other than 1.

It is helpful in dealing with logarithmic functions to remember that the logarithm of a number is an **exponent**. For instance, $\log_a x$ is the exponent y in $x = a^y$. Keep thinking, "The logarithm, base a, of a number x is the power to which a must be raised in order to get x."

EXPONENTIAL FUNCTION	LOGARITHMIC FUNCTION
$y = a^x$	$x = a^y$
$f(x) = a^x$	$f^{-1}(x) = \log_a x$
$a > 0, a \neq 1$	$a > 0, a \neq 1$
Domain = The set of real numbers	Range = The set of real numbers
Range = The set of positive numbers	Domain = The set of positive numbers

Why do we exclude 1 from being a logarithm base? See the following graph. If we allow 1 as a logarithm base, the graph of the relation $y = \log_1 x$, or $x = 1^y = 1$, is a vertical line, which is not a function and therefore not a logarithmic function.

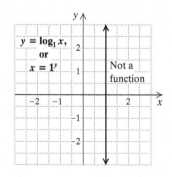

EXAMPLE 1 Graph: $y = f(x) = \log_5 x$.

The equation $y = \log_5 x$ is equivalent to $5^y = x$. We can find ordered pairs that are solutions by choosing values for y and computing the corresponding x-values.

For $y = 0$, $x = 5^0 = 1$.
For $y = 1$, $x = 5^1 = 5$.
For $y = 2$, $x = 5^2 = 25$.
For $y = 3$, $x = 5^3 = 125$.
For $y = -1$, $x = 5^{-1} = \dfrac{1}{5}$.
For $y = -2$, $x = 5^{-2} = \dfrac{1}{25}$.

x, or 5^y	y
1	0
5	1
25	2
125	3
$\frac{1}{5}$	-1
$\frac{1}{25}$	-2

(1) Select y.
(2) Compute x.

The table shows the following:

$\left.\begin{array}{l}\log_5 1 = 0; \\ \log_5 5 = 1; \\ \log_5 25 = 2; \\ \log_5 125 = 3; \\ \log_5 \frac{1}{5} = -1; \\ \log_5 \frac{1}{25} = -2.\end{array}\right\}$ These can all be checked using the equations above.

We plot the ordered pairs and connect them with a smooth curve. The graph of $y = 5^x$ has been shown only for reference.

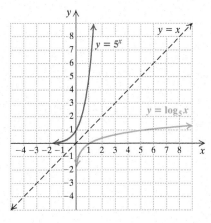

2. Graph: $y = f(x) = \log_3 x$.

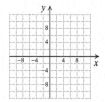

Do Exercise 2.

b CONVERTING BETWEEN EXPONENTIAL EQUATIONS AND LOGARITHMIC EQUATIONS

We use the definition of logarithms to convert from exponential equations to logarithmic equations.

> **CONVERTING BETWEEN EXPONENTIAL EQUATIONS AND LOGARITHMIC EQUATIONS**
>
> $y = \log_a x \longrightarrow a^y = x$; $a^y = x \longrightarrow y = \log_a x$
>
> Be sure to memorize this relationship! It is probably the most important definition in the chapter. Often this definition will be a justification for a proof or a procedure that we are considering.

Answer

2.

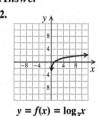

$y = f(x) = \log_3 x$

Convert to a logarithmic equation.
3. $6^0 = 1$
4. $10^{-3} = 0.001$
5. $16^{0.25} = 2$
6. $m^T = P$

EXAMPLES Convert to a logarithmic equation.

2. $8 = 2^x \longrightarrow x = \log_2 8$ The exponent is the logarithm.
                                             The base remains the same.
3. $y^{-1} = 4 \longrightarrow -1 = \log_y 4$
4. $a^b = c \longrightarrow b = \log_a c$

◀ Do Exercises 3–6.

We also use the definition of logarithms to convert from logarithmic equations to exponential equations.

EXAMPLES Convert to an exponential equation.

Convert to an exponential equation.
7. $\log_2 32 = 5$
8. $\log_{10} 1000 = 3$
9. $\log_a Q = 7$
10. $\log_t M = x$

5. $y = \log_3 5 \longrightarrow 3^y = 5$ The logarithm is the exponent.
                                         The base does not change.
6. $-2 = \log_a 7 \longrightarrow a^{-2} = 7$
7. $a = \log_b d \longrightarrow b^a = d$

◀ Do Exercises 7–10.

C SOLVING CERTAIN LOGARITHMIC EQUATIONS

Certain equations involving logarithms can be solved by first converting to exponential equations. We will solve more complicated equations later.

EXAMPLE 8 Solve: $\log_2 x = -3$.

$$\log_2 x = -3$$
$$2^{-3} = x \qquad \text{Converting to an exponential equation}$$
$$\frac{1}{2^3} = x$$
$$\frac{1}{8} = x$$

Check: $\log_2 \frac{1}{8}$ is the exponent to which we raise 2 to get $\frac{1}{8}$. Since $2^{-3} = \frac{1}{8}$, we know that $\frac{1}{8}$ checks and is the solution.

Solve.
11. $\log_{10} x = 4$
    $10^{\boxed{}} = x$
    $\boxed{} = x$

12. $\log_x 81 = 4$
13. $\log_2 x = -2$

EXAMPLE 9 Solve: $\log_x 16 = 2$.

$$\log_x 16 = 2$$
$$x^2 = 16 \qquad \text{Converting to an exponential equation}$$
$$x = 4 \quad or \quad x = -4 \qquad \text{Using the principle of square roots}$$

Check: $\log_4 16 = 2$ because $4^2 = 16$. Thus, 4 is a solution. Since all logarithm bases must be positive, $\log_{-4} 16$ is not defined. Therefore, -4 is not a solution.

◀ Do Exercises 11–13.

Answers
3. $0 = \log_6 1$ 4. $-3 = \log_{10} 0.001$
5. $0.25 = \log_{16} 2$ 6. $T = \log_m P$
7. $2^5 = 32$ 8. $10^3 = 1000$ 9. $a^7 = Q$
10. $t^x = M$ 11. $10,000$ 12. 3 13. $\dfrac{1}{4}$

Guided Solution:
11. $4, 10{,}000$

To think of finding logarithms as solving equations may help in some cases.

EXAMPLE 10 Find $\log_{10} 1000$.

Method 1: Let $\log_{10} 1000 = x$. Then

$10^x = 1000$ Converting to an exponential equation
$10^x = 10^3$
$x = 3$.   The exponents are the same.

Therefore, $\log_{10} 1000 = 3$.

Method 2: Think of the meaning of $\log_{10} 1000$. It is the exponent to which we raise 10 to get 1000. That exponent is 3. Therefore, $\log_{10} 1000 = 3$.

EXAMPLE 11 Find $\log_{10} 0.01$.

Method 1: Let $\log_{10} 0.01 = x$. Then

$10^x = 0.01$ Converting to an exponential equation
$10^x = \dfrac{1}{100}$
$10^x = 10^{-2}$
$x = -2$.   The exponents are the same.

Therefore, $\log_{10} 0.01 = -2$.

Method 2: $\log_{10} 0.01$ is the exponent to which we raise 10 to get 0.01. Noting that

$$0.01 = \dfrac{1}{100} = \dfrac{1}{10^2} = 10^{-2},$$

we see that the exponent is -2. Therefore, $\log_{10} 0.01 = -2$.

EXAMPLE 12 Find $\log_5 1$.

Method 1: Let $\log_5 1 = x$. Then

$5^x = 1$ Converting to an exponential equation
$5^x = 5^0$
$x = 0$.   The exponents are the same.

Therefore, $\log_5 1 = 0$.

Method 2: $\log_5 1$ is the exponent to which we raise 5 to get 1. That exponent is 0. Therefore, $\log_5 1 = 0$.

Do Exercises 14–16.

THE LOGARITHM OF 1

For any base a,
$$\log_a 1 = 0.$$
The logarithm, base a, of 1 is always 0.

Find each of the following.

14. $\log_{10} 10{,}000$

Let $\log_{10} 10{,}000 = x$.
$\boxed{}^x = 10{,}000$
$10^x = \boxed{}^4$
$x = \boxed{}$

15. $\log_{10} 0.0001$

16. $\log_7 1$

Answers
14. 4   **15.** -4 **16.** 0
Guided Solution:
14. 10, 10, 4

The proof follows from the fact that $a^0 = 1$. This is equivalent to the logarithmic equation $\log_a 1 = 0$.

Another property follows similarly. We know that $a^1 = a$ for any real number a. In particular, it holds for any positive number a. This is equivalent to the logarithmic equation $\log_a a = 1$.

> **THE LOGARITHM, BASE a, OF a**
>
> For any base a,
>
> $$\log_a a = 1.$$

EXAMPLE 13 Simplify: $\log_m 1$ and $\log_t t$.

$$\log_m 1 = 0; \qquad \log_t t = 1$$

◀ Do Exercises 17–20.

d FINDING COMMON LOGARITHMS ON A CALCULATOR

Base-10 logarithms are called **common logarithms**. Before calculators became so widely available, common logarithms were used extensively to do complicated calculations. The abbreviation **log**, with no base written, is used for the common logarithm, base-10. Thus,

$$\log 29 \quad \text{means} \quad \log_{10} 29.$$

> Be sure to memorize $\log a = \log_{10} a$.

We can approximate $\log 29$. Note the following:

$$\left.\begin{array}{l}\log 100 = \log_{10} 100 = \mathbf{2}; \\ \log 29 = \mathbf{?}; \\ \log 10 = \log_{10} 10 = \mathbf{1}.\end{array}\right\}$$ It seems reasonable to conclude that $\log 29$ is between 1 and 2.

The calculator key for common logarithms is generally marked **LOG**. We find that

$$\log 29 \approx 1.462397998 \approx 1.4624,$$

rounded to four decimal places. This also tells us that $10^{1.4624} \approx 29$.

On some scientific calculators, the keystrokes for doing such a calculation might be

(2)(9)(LOG)(=).   The display would then read   1.462398.

If we are using a graphing calculator, the keystrokes might be

(LOG)(2)(9)(ENTER).   The display would then read   1.462397998.

EXAMPLES Find the common logarithm, to four decimal places, on a scientific calculator or a graphing calculator.

	Function Value	Readout	Rounded
14.	$\log 287{,}523$	5.458672591	5.4587
15.	$\log 0.000486$	-3.313363731	-3.3134
16.	$\log (-5)$	NONREAL ANS	Does not exist as a real number

Simplify.
17. $\log_3 1$
18. $\log_3 3$
19. $\log_c c$
20. $\log_c 1$

Find the common logarithm, to four decimal places, on a scientific calculator or a graphing calculator.

21. $\log 78{,}235.4$

22. $\log 0.0000309$

23. $\log (-3)$

24. Find

   $\log 1000$ and $\log 10{,}000$

   without using a calculator. Between what two whole numbers is $\log 9874$? Then on a calculator, approximate $\log 9874$, rounded to four decimal places.

Answers

17. 0   **18.** 1   **19.** 1   **20.** 0   **21.** 4.8934
22. -4.5100 **23.** Does not exist as a real number   **24.** $\log 1000 = 3$, $\log 10{,}000 = 4$; between 3 and 4; 3.9945

In Example 16, $\log(-5)$ does not exist as a real number because there is no real-number power to which we can raise 10 to get -5. The number 10 raised to any power is nonnegative. The logarithm of a negative number does not exist as a real number (though it can be defined as a complex number).

Do Exercises 21–24 on the preceding page.

We can use common logarithms to express any positive number as a power of 10. Considering very large or very small numbers as powers of 10 might be a helpful way to compare those numbers.

EXAMPLE 17 Complete the following table to express each number in the first column as a power of 10. Round each exponent to the nearest ten-thousandth.

We simply find the common logarithm of the number using a calculator.

NUMBER	EXPRESSED AS A POWER OF 10
4	$4 \approx 10^{0.6021}$
625	$625 \approx 10^{2.7959}$
134,567	$134{,}567 \approx 10^{5.1289}$
0.00567	$0.00567 \approx 10^{-2.2464}$
0.000374859	$0.000374859 \approx 10^{-3.4261}$
186,000	$186{,}000 \approx 10^{5.2695}$
186,000,000	$186{,}000{,}000 \approx 10^{8.2695}$

25. Complete the following table to express each number in the first column as a power of 10. Round each exponent to the nearest ten-thousandth.

NUMBER	EXPRESSED AS A POWER OF 10
8	
947	
634,567	
0.00708	
0.000778899	
18,600,000	
1860	

Do Exercise 25.

The inverse of a logarithmic function is an exponential function. Thus, if $f(x) = \log x$, then $f^{-1}(x) = 10^x$. Because of this, on many calculators, the **LOG** key doubles as the **10^x** key after a **2ND** or **SHIFT** key has been pressed. To find $10^{5.4587}$ on a scientific calculator, we might enter 5.4587 and press **10^x**. On many graphing calculators, we press **2ND** **10^x**, followed by 5.4587. In either case, we get the approximation $10^{5.4587} \approx 287{,}541.1465$. Compare this computation to that in Example 14. Note that, apart from the rounding error, $10^{5.4587}$ takes us back to about 287,523.

Do Exercise 26.

26. Find $10^{4.8934}$ using a calculator. (Compare your computation to that of Margin Exercise 21.)

Using the scientific keys on a calculator would allow us to construct a graph of $f(x) = \log_{10} x = \log x$ by finding function values directly, rather than converting to exponential form as we did in Example 1.

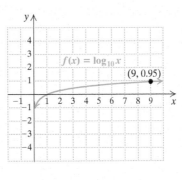

x	$f(x)$
0.5	-0.3010
1	0
2	0.3010
3	0.4771
5	0.6990
9	0.9542
10	1

Answers
25. $10^{0.9031}$; $10^{2.9763}$; $10^{5.8025}$; $10^{-2.1500}$; $10^{-3.1085}$; $10^{7.2695}$; $10^{3.2695}$ **26.** 78,234.8042

12.3 Exercise Set

✓ Check Your Understanding

Reading Check Determine whether each statement is true or false.

RC1. The expression $\log_5 12$ represents the power to which we raise 12 to get 5.

RC2. A logarithm base cannot be negative.

RC3. The logarithm of a negative number does not exist.

RC4. The number 1 is not a valid logarithm base.

RC5. If the base of a logarithm is not written, it is assumed to be 2.

RC6. A LOG key on a calculator represents a base-10 logarithm.

RC7. Any positive number can be written as a power of 10.

RC8. Any positive number can be written as a power of 5.

Concept Check Use the powers in the list on the right to find each logarithm.

CC1. $\log_{10} 100 = $ _____

CC2. $\log_3 9 = $ _____

CC3. $\log_2 8 = $ _____

CC4. $\log_2 1024 = $ _____

$2^3 = 8$
$3^2 = 9$
$10^2 = 100$
$2^{10} = 1024$

a Graph.

1. $f(x) = \log_2 x$, or $y = \log_2 x$
$y = \log_2 x \longrightarrow x = $ _____

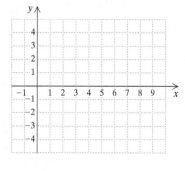

2. $f(x) = \log_{10} x$, or $y = \log_{10} x$
$y = \log_{10} x \longrightarrow x = $ _____

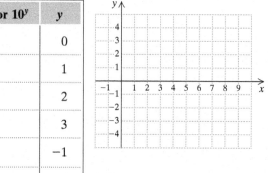

928 CHAPTER 12 Exponential Functions and Logarithmic Functions

3. $f(x) = \log_{1/3} x$

4. $f(x) = \log_{1/2} x$

Graph both functions using the same set of axes.

5. $f(x) = 3^x$, $f^{-1}(x) = \log_3 x$

6. $f(x) = 4^x$, $f^{-1}(x) = \log_4 x$

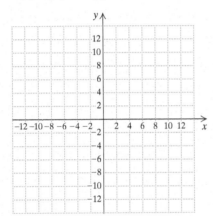
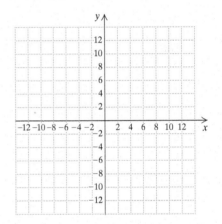

b Convert to a logarithmic equation.

7. $10^3 = 1000$

8. $10^2 = 100$

9. $5^{-3} = \dfrac{1}{125}$

10. $4^{-5} = \dfrac{1}{1024}$

11. $8^{1/3} = 2$

12. $16^{1/4} = 2$

13. $10^{0.3010} = 2$

14. $10^{0.4771} = 3$

15. $e^2 = t$

16. $p^k = 3$

17. $Q^t = x$

18. $P^m = V$

19. $e^2 = 7.3891$

20. $e^3 = 20.0855$

21. $e^{-2} = 0.1353$

22. $e^{-4} = 0.0183$

SECTION 12.3 Logarithmic Functions

Convert to an exponential equation.

23. $w = \log_4 10$
24. $t = \log_5 9$
25. $\log_6 36 = 2$
26. $\log_7 7 = 1$

27. $\log_{10} 0.01 = -2$
28. $\log_{10} 0.001 = -3$
29. $\log_{10} 8 = 0.9031$
30. $\log_{10} 2 = 0.3010$

31. $\log_e 100 = 4.6052$
32. $\log_e 10 = 2.3026$
33. $\log_t Q = k$
34. $\log_m P = a$

C Solve.

35. $\log_3 x = 2$
36. $\log_4 x = 3$
37. $\log_x 16 = 2$
38. $\log_x 64 = 3$

39. $\log_2 16 = x$
40. $\log_5 25 = x$
41. $\log_3 27 = x$
42. $\log_4 16 = x$

43. $\log_x 25 = 1$
44. $\log_x 9 = 1$
45. $\log_3 x = 0$
46. $\log_2 x = 0$

47. $\log_2 x = -1$
48. $\log_3 x = -2$
49. $\log_8 x = \dfrac{1}{3}$
50. $\log_{32} x = \dfrac{1}{5}$

Find each of the following.

51. $\log_{10} 100$
52. $\log_{10} 100{,}000$
53. $\log_{10} 0.1$
54. $\log_{10} 0.001$

55. $\log_{10} 1$
56. $\log_{10} 10$
57. $\log_5 625$
58. $\log_2 64$

59. $\log_7 49$
60. $\log_5 125$
61. $\log_2 8$
62. $\log_8 64$

63. $\log_9 \dfrac{1}{81}$
64. $\log_5 \dfrac{1}{125}$
65. $\log_8 1$
66. $\log_6 6$

67. $\log_e e$
68. $\log_e 1$
69. $\log_{27} 9$
70. $\log_8 2$

d Find the common logarithm, to four decimal places, on a calculator.

71. log 78,889.2 **72.** log 9,043,788 **73.** log 0.67 **74.** log 0.0067

75. log(−97) **76.** log 0 **77.** $\log\left(\dfrac{289}{32.7}\right)$ **78.** $\log\left(\dfrac{23}{86.2}\right)$

79. Complete the following table to express each number in the first column as a power of 10. Round each exponent to the nearest ten-thousandth.

NUMBER	EXPRESSED AS A POWER OF 10
6	
84	
987,606	
0.00987606	
98,760.6	
70,000,000	
7000	

80. Complete the following table to express each number in the first column as a power of 10. Round each exponent to the nearest ten-thousandth.

NUMBER	EXPRESSED AS A POWER OF 10
7	
314	
31.4	
31,400,000	
0.000314	
3.14	
0.0314	

Skill Maintenance

Perform the indicated operations and simplify.

81. $\dfrac{t^2 - 9}{t^2 - 4t + 4} \cdot \dfrac{3t - 6}{3t + 9}$ [6.1d]

82. $\dfrac{c^2}{c - p} + \dfrac{p^2}{p - c}$ [6.4a]

83. $\dfrac{3}{x^2 - x - 2} - \dfrac{1}{2x^2 + 3x + 1}$ [6.5a]

Factor completely.

84. $100x^2 - 9$ [5.5d]

85. $3x^3 - 24x^2 + 48x$ [5.5b]

86. $64a^3 - y^3$ [5.6a]

87. $2x^3 - 3x^2 + 2x - 3$ [5.1c]

88. $10t^6 + 19t^3 - 15$ [5.3a], [5.4a]

Synthesis

Graph.

89. $f(x) = \log_3 |x + 1|$

90. $f(x) = \log_2(x - 1)$

Solve.

91. $\log_{125} x = \tfrac{2}{3}$

92. $|\log_3 x| = 3$

93. $\log_{128} x = \tfrac{5}{7}$

94. $\log_4(3x - 2) = 2$

95. $\log_8(2x + 1) = -1$

96. $\log_{10}(x^2 + 21x) = 2$

Simplify.

97. $\log_{1/4} \tfrac{1}{64}$

98. $\log_{81} 3 \cdot \log_3 81$

99. $\log_{10}(\log_4(\log_3 81))$

100. $\log_2(\log_2(\log_4 256))$

101. $\log_{1/5} 25$

12.4 Properties of Logarithmic Functions

OBJECTIVES

a Express the logarithm of a product as a sum of logarithms, and conversely.

b Express the logarithm of a power as a product.

c Express the logarithm of a quotient as a difference of logarithms, and conversely.

d Convert from logarithms of products, quotients, and powers to expressions in terms of individual logarithms, and conversely.

e Simplify expressions of the type $\log_a a^k$.

We now establish some basic properties that are useful in manipulating logarithmic expressions.

a LOGARITHMS OF PRODUCTS

PROPERTY 1: THE PRODUCT RULE

For any positive numbers M and N and any logarithm base a,
$$\log_a (M \cdot N) = \log_a M + \log_a N.$$
(The logarithm of a product is the sum of the logarithms of the factors.)

EXAMPLE 1 Express as a sum of logarithms: $\log_2 (4 \cdot 16)$.

$\log_2 (4 \cdot 16) = \log_2 4 + \log_2 16$ By Property 1

EXAMPLE 2 Express as a single logarithm: $\log_{10} 0.01 + \log_{10} 1000$.

$\log_{10} 0.01 + \log_{10} 1000 = \log_{10} (0.01 \times 1000)$ By Property 1
$= \log_{10} 10$

◀ Do Exercises 1–4.

Express as a sum of logarithms.
1. $\log_5 (25 \cdot 5)$ 2. $\log_b (PQ)$

Express as a single logarithm.
3. $\log_3 7 + \log_3 5$
4. $\log_a J + \log_a A + \log_a M$

A Proof of Property 1 (Optional): We let $\log_a M = x$ and $\log_a N = y$. Converting to exponential equations, we have $a^x = M$ and $a^y = N$. Then we multiply to obtain
$$M \cdot N = a^x \cdot a^y = a^{x+y}.$$
Converting $M \cdot N = a^{x+y}$ back to a logarithmic equation, we get
$$\log_a (M \cdot N) = x + y.$$
Remembering what x and y represent, we get
$$\log_a (M \cdot N) = \log_a M + \log_a N.$$

b LOGARITHMS OF POWERS

PROPERTY 2: THE POWER RULE

For any positive number M, any real number k, and any logarithm base a,
$$\log_a M^k = k \cdot \log_a M.$$
(The logarithm of a power of M is the exponent times the logarithm of M.)

Answers
1. $\log_5 25 + \log_5 5$ 2. $\log_b P + \log_b Q$
3. $\log_3 35$ 4. $\log_a (JAM)$

EXAMPLES Express as a product.

3. $\log_a 9^{-5} = -5 \log_a 9$ By Property 2
4. $\log_a \sqrt[4]{5} = \log_a 5^{1/4}$ Writing exponential notation
   $= \frac{1}{4} \log_a 5$ By Property 2

Do Exercises 5 and 6. ▶

A Proof of Property 2 (Optional): We let $x = \log_a M$. Then we convert to an exponential equation to get $a^x = M$. Raising both sides to the kth power, we obtain

$$(a^x)^k = M^k, \text{ or } a^{xk} = M^k.$$

Converting back to a logarithmic equation with base a, we get $\log_a M^k = xk$. But $x = \log_a M$, so

$$\log_a M^k = (\log_a M)k = k \cdot \log_a M.$$

c   LOGARITHMS OF QUOTIENTS

PROPERTY 3: THE QUOTIENT RULE

For any positive numbers M and N and any logarithm base a,

$$\log_a \frac{M}{N} = \log_a M - \log_a N.$$

(The logarithm of a quotient is the logarithm of the numerator minus the logarithm of the denominator.)

EXAMPLE 5 Express as a difference of logarithms: $\log_t \frac{6}{U}$.

$$\log_t \frac{6}{U} = \log_t 6 - \log_t U \quad \text{By Property 3}$$

EXAMPLE 6 Express as a single logarithm: $\log_b 17 - \log_b 27$.

$$\log_b 17 - \log_b 27 = \log_b \frac{17}{27} \quad \text{By Property 3}$$

EXAMPLE 7 Express as a single logarithm: $\log_{10} 10{,}000 - \log_{10} 100$.

$$\log_{10} 10{,}000 - \log_{10} 100 = \log_{10} \frac{10{,}000}{100} = \log_{10} 100$$

Do Exercises 7 and 8. ▶

A Proof of Property 3 (Optional): The proof makes use of Property 1 and Property 2.

$$\log_a \frac{M}{N} = \log_a M \cdot \frac{1}{N} = \log_a MN^{-1} \quad \frac{1}{N} = N^{-1}$$
$$= \log_a M + \log_a N^{-1} \quad \text{By Property 1}$$
$$= \log_a M + (-1) \log_a N \quad \text{By Property 2}$$
$$= \log_a M - \log_a N$$

Express as a product.

5. $\log_7 4^5$
6. $\log_a \sqrt{5}$

CALCULATOR CORNER

Properties of Logarithms   Use a table or a graph to determine whether each of the following is correct.

1. $\log(5x) = \log 5 \cdot \log x$
2. $\log(5x) = \log 5 + \log x$
3. $\log x^2 = \log x \cdot \log x$
4. $\log x^2 = 2 \log x$
5. $\log\left(\frac{x}{3}\right) = \frac{\log x}{\log 3}$
6. $\log\left(\frac{x}{3}\right) = \log x - \log 3$
7. $\log(x + 2) = \log x + \log 2$
8. $\log(x + 2) = \log x \cdot \log 2$

7. Express as a difference of logarithms:
$$\log_b \frac{P}{Q}.$$

8. Express as a single logarithm:
$$\log_2 125 - \log_2 25.$$

Answers

5. $5 \log_7 4$ **6.** $\frac{1}{2} \log_a 5$ **7.** $\log_b P - \log_b Q$
8. $\log_2 5$

d USING THE PROPERTIES TOGETHER

EXAMPLES Express in terms of logarithms of w, x, y, and z.

8. $\log_a \dfrac{x^2 y^3}{z^4} = \log_a (x^2 y^3) - \log_a z^4$ Using Property 3

$\phantom{\log_a \dfrac{x^2 y^3}{z^4}} = \log_a x^2 + \log_a y^3 - \log_a z^4$ Using Property 1

$\phantom{\log_a \dfrac{x^2 y^3}{z^4}} = 2\log_a x + 3\log_a y - 4\log_a z$ Using Property 2

9. $\log_a \sqrt[4]{\dfrac{xy}{z^3}} = \log_a \left(\dfrac{xy}{z^3}\right)^{1/4}$ Writing exponential notation

$\phantom{\log_a \sqrt[4]{\dfrac{xy}{z^3}}} = \tfrac{1}{4} \log_a \dfrac{xy}{z^3}$ Using Property 2

$\phantom{\log_a \sqrt[4]{\dfrac{xy}{z^3}}} = \tfrac{1}{4}(\log_a (xy) - \log_a z^3)$ Using Property 3 (note the parentheses)

$\phantom{\log_a \sqrt[4]{\dfrac{xy}{z^3}}} = \tfrac{1}{4}(\log_a x + \log_a y - 3\log_a z)$ Using Properties 1 and 2

$\phantom{\log_a \sqrt[4]{\dfrac{xy}{z^3}}} = \tfrac{1}{4}\log_a x + \tfrac{1}{4}\log_a y - \tfrac{3}{4}\log_a z$ Distributive law

10. $\log_b \dfrac{xy}{w^3 z^4} = \log_b (xy) - \log_b (w^3 z^4)$ Using Property 3

$\phantom{\log_b \dfrac{xy}{w^3 z^4}} = (\log_b x + \log_b y) - (\log_b w^3 + \log_b z^4)$ Using Property 1

$\phantom{\log_b \dfrac{xy}{w^3 z^4}} = \log_b x + \log_b y - \log_b w^3 - \log_b z^4$ Removing parentheses

$\phantom{\log_b \dfrac{xy}{w^3 z^4}} = \log_b x + \log_b y - 3\log_b w - 4\log_b z$ Using Property 2

◀ Do Exercises 9–11.

EXAMPLES Express as a single logarithm.

11. $\dfrac{1}{2}\log_a x - 7\log_a y + \log_a z$

$= \log_a x^{1/2} - \log_a y^7 + \log_a z$ Using Property 2

$= \log_a \dfrac{\sqrt{x}}{y^7} + \log_a z$ Using Property 3

$= \log_a \dfrac{z\sqrt{x}}{y^7}$ Using Property 1

12. $\log_a \dfrac{b}{\sqrt{x}} + \log_a \sqrt{bx}$

$= \log_a b - \log_a \sqrt{x} + \log_a \sqrt{bx}$ Using Property 3

$= \log_a b - \tfrac{1}{2}\log_a x + \tfrac{1}{2}\log_a (bx)$ Using Property 2

$= \log_a b - \tfrac{1}{2}\log_a x + \tfrac{1}{2}(\log_a b + \log_a x)$ Using Property 1

$= \log_a b - \tfrac{1}{2}\log_a x + \tfrac{1}{2}\log_a b + \tfrac{1}{2}\log_a x$

$= \tfrac{3}{2}\log_a b$ Collecting like terms

$= \log_a b^{3/2}$ Using Property 2

Example 12 could also be done as follows:

$\log_a \dfrac{b}{\sqrt{x}} + \log_a \sqrt{bx} = \log_a \left(\dfrac{b}{\sqrt{x}} \sqrt{bx}\right)$ Using Property 1

$= \log_a \left(\dfrac{b}{\sqrt{x}} \cdot \sqrt{b} \cdot \sqrt{x}\right)$

$= \log_a (b\sqrt{b}), \text{ or } \log_a b^{3/2}.$

Express in terms of logarithms of w, x, y, and z.

9. $\log_a \sqrt{\dfrac{z^3}{xy}}$

10. $\log_a \dfrac{x^2}{y^3 z}$

11. $\log_a \dfrac{x^3 y^4}{z^5 w^9}$ **GS**

$= \log_a x^3 y^4 - \log_a \boxed{}$

$= (\log_a x^3 + \log_a \boxed{}) -$

$ (\log_a z^5 + \log_a \boxed{})$

$= \log_a x^3 + \log_a y^4 -$

$ \log_a z^5 - \boxed{}$

$= 3\log_a x + 4\log_a y -$

$ \boxed{} \log_a z - \boxed{} \log_a w$

Express as a single logarithm.

12. $5\log_a x - \log_a y + \dfrac{1}{4}\log_a z$

13. $\log_a \dfrac{\sqrt{x}}{b} - \log_a \sqrt{bx}$

Answers

9. $\dfrac{3}{2}\log_a z - \dfrac{1}{2}\log_a x - \dfrac{1}{2}\log_a y$
10. $2\log_a x - 3\log_a y - \log_a z$
11. $3\log_a x + 4\log_a y - 5\log_a z - 9\log_a w$
12. $\log_a \dfrac{x^5 z^{1/4}}{y}$, or $\log_a \dfrac{x^5 \sqrt[4]{z}}{y}$
13. $\log_a \dfrac{1}{b\sqrt{b}}$, or $\log_a b^{-3/2}$

Guided Solution:
11. $z^5 w^9$, y^4, w^9, $\log_a w^9$, 5, 9

Do Exercises 12 and 13 on the preceding page.

Caution!

Keep in mind that, in general,
$$\log_a (M + N) \neq \log_a M + \log_a N,$$
$$\log_a (M - N) \neq \log_a M - \log_a N,$$
$$\log_a (MN) \neq (\log_a M)(\log_a N),$$
and
$$\log_a (M/N) \neq (\log_a M) \div (\log_a N).$$

EXAMPLES Given $\log_a 2 = 0.301$ and $\log_a 3 = 0.477$, find each of the following, if possible.

13. $\log_a 6 = \log_a (2 \cdot 3) = \log_a 2 + \log_a 3$ Property 1
$= 0.301 + 0.477 = 0.778$

14. $\log_a \frac{2}{3} = \log_a 2 - \log_a 3$ Property 3
$= 0.301 - 0.477 = -0.176$

15. $\log_a 81 = \log_a 3^4 = 4 \log_a 3$ Property 2
$= 4(0.477) = 1.908$

16. $\log_a \frac{1}{3} = \log_a 1 - \log_a 3$ Property 3
$= 0 - 0.477 = -0.477$

17. $\log_a \sqrt{a} = \log_a a^{1/2} = \frac{1}{2} \log_a a = \frac{1}{2} \cdot 1 = \frac{1}{2}$ Property 2

18. $\log_a 2a = \log_a 2 + \log_a a$ Property 1
$= 0.301 + 1 = 1.301$

19. $\log_a 5$ There is no way to find this using these properties
$(\log_a 5 \neq \log_a 2 + \log_a 3).$

20. $\frac{\log_a 3}{\log_a 2} = \frac{0.477}{0.301} \approx 1.58$ We simply divide the logarithms, not using any property of logarithms.

Do Exercises 14–22.

e THE LOGARITHM OF THE BASE TO A POWER

PROPERTY 4

For any base a,
$$\log_a a^k = k.$$
(The logarithm, base a, of a to a power is the power.)

A Proof of Property 4 (Optional): The proof involves Property 2 and the fact that $\log_a a = 1$:

$\log_a a^k = k(\log_a a)$ Using Property 2
$= k \cdot 1$ Using $\log_a a = 1$
$= k.$

EXAMPLES Simplify.

21. $\log_3 3^7 = 7$
22. $\log_{10} 10^{5.6} = 5.6$
23. $\log_e e^{-t} = -t$

Do Exercises 23–25.

Given
$$\log_a 2 = 0.301,$$
$$\log_a 5 = 0.699,$$
find each of the following, if possible.

14. $\log_a 4$ **15.** $\log_a 10$

16. $\log_a \frac{2}{5}$ **17.** $\log_a \frac{5}{2}$

18. $\log_a \frac{1}{5}$ **19.** $\log_a \sqrt{a^3}$

20. $\log_a 5a$ **21.** $\log_a 7$

GS **22.** $\log_a 16$

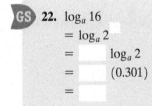

Simplify.

23. $\log_2 2^6$ **24.** $\log_{10} 10^{3.2}$

25. $\log_e e^{12}$

Answers
14. 0.602 **15.** 1 **16.** −0.398 **17.** 0.398
18. −0.699 **19.** $\frac{3}{2}$ **20.** 1.699 **21.** Cannot be found using the properties of logarithms
22. 1.204 **23.** 6 **24.** 3.2 **25.** 12
Guided Solution:
22. 4, 4, 4, 1.204

12.4 Exercise Set

✓ Check Your Understanding

Reading Check Choose from the column on the right the option that best completes each statement.

RC1. $\log_2 (8 \cdot 4) =$ _____

RC2. $\log_2 \left(\dfrac{8}{4}\right) =$ _____

RC3. $\log_2 4^8 =$ _____

RC4. $\log_2 (3 \cdot 4) =$ _____

RC5. $\log_2 4^3 =$ _____

RC6. $\log_2 3^4 =$ _____

a) $\log_2 3 + \log_2 4$
b) $\log_2 8 + \log_2 4$
c) $\log_2 8 - \log_2 4$
d) $8 \log_2 4$
e) $3 \log_2 4$
f) $4 \log_2 3$

Concept Check Choose from the column on the right the option that makes each statement true. Not all choices will be used.

CC1. $\log_5 12 + \log_5 2 = \log_5$ _____

CC2. $\log_5 12 - \log_5 2 = \log_5$ _____

CC3. $\log_5 5 =$ _____

CC4. $\log_5 5^{10} =$ _____

a) 0
b) 1
c) 5
d) 6
e) 10
f) 24

a Express as a sum of logarithms.

1. $\log_2 (32 \cdot 8)$
2. $\log_3 (27 \cdot 81)$
3. $\log_4 (64 \cdot 16)$
4. $\log_5 (25 \cdot 125)$
5. $\log_a Qx$
6. $\log_r 8Z$

Express as a single logarithm.

7. $\log_b 3 + \log_b 84$
8. $\log_a 75 + \log_a 5$
9. $\log_c K + \log_c y$
10. $\log_t H + \log_t M$

b Express as a product.

11. $\log_c y^4$
12. $\log_a x^3$
13. $\log_b t^6$
14. $\log_{10} y^7$
15. $\log_b C^{-3}$
16. $\log_c M^{-5}$

c Express as a difference of logarithms.

17. $\log_a \dfrac{67}{5}$
18. $\log_t \dfrac{T}{7}$
19. $\log_b \dfrac{2}{5}$
20. $\log_a \dfrac{z}{y}$

Express as a single logarithm.

21. $\log_c 22 - \log_c 3$
22. $\log_d 54 - \log_d 9$

d Express in terms of logarithms of a single variable or a number.

23. $\log_a x^2 y^3 z$

24. $\log_a 5xy^4 z^3$

25. $\log_b \dfrac{xy^2}{z^3}$

26. $\log_b \dfrac{p^2 q^5}{m^4 n^7}$

27. $\log_c \sqrt[3]{\dfrac{x^4}{y^3 z^2}}$

28. $\log_a \sqrt{\dfrac{x^6}{p^5 q^8}}$

29. $\log_a \sqrt[4]{\dfrac{m^8 n^{12}}{a^3 b^5}}$

30. $\log_a \sqrt{\dfrac{a^6 b^8}{a^2 b^5}}$

Express as a single logarithm and, if possible, simplify.

31. $\dfrac{2}{3}\log_a x - \dfrac{1}{2}\log_a y$

32. $\dfrac{1}{2}\log_a x + 3\log_a y - 2\log_a x$

33. $\log_a 2x + 3(\log_a x - \log_a y)$

34. $\log_a x^2 - 2\log_a \sqrt{x}$

35. $\log_a \dfrac{a}{\sqrt{x}} - \log_a \sqrt{ax}$

36. $\log_a (x^2 - 4) - \log_a (x - 2)$

Given $\log_b 3 = 1.099$ and $\log_b 5 = 1.609$, find each of the following.

37. $\log_b 15$

38. $\log_b 8$

39. $\log_b \dfrac{5}{3}$

40. $\log_b \dfrac{3}{5}$

41. $\log_b \dfrac{1}{5}$

42. $\log_b \dfrac{1}{3}$

43. $\log_b \sqrt{b}$

44. $\log_b \sqrt{b^3}$

45. $\log_b 5b$

46. $\log_b 3b$

47. $\log_b 2$

48. $\log_b 75$

e Simplify.

49. $\log_e e^t$

50. $\log_w w^8$

51. $\log_p p^5$

52. $\log_Y Y^{-4}$

Solve for x.

53. $\log_2 2^7 = x$

54. $\log_9 9^4 = x$

55. $\log_e e^x = -7$

56. $\log_a a^x = 2.7$

Skill Maintenance

Compute and simplify. Express answers in the form $a + bi$, where $i^2 = -1$. [10.8b, c, d, e]

57. i^{29}

58. i^{34}

59. $(2 + i)(2 - i)$

60. $\dfrac{2 + i}{2 - i}$

61. $(7 - 8i) - (-16 + 10i)$

62. $2i^2 \cdot 5i^3$

63. $(8 + 3i)(-5 - 2i)$

64. $(2 - i)^2$

Synthesis

Determine whether each is true or false.

65. $\dfrac{\log_a P}{\log_a Q} = \log_a \dfrac{P}{Q}$

66. $\dfrac{\log_a P}{\log_a Q} = \log_a P - \log_a Q$

67. $\log_a 3x = \log_a 3 + \log_a x$

68. $\log_a 3x = 3 \log_a x$

69. $\log_a (P + Q) = \log_a P + \log_a Q$

70. $\log_a x^2 = 2 \log_a x$

Mid-Chapter Review

Concept Reinforcement

Determine whether each statement is true or false.

_____ 1. The graph of an exponential function never crosses the x-axis. [12.1a]

_____ 2. A function f is one-to-one if different inputs have different outputs. [12.2c]

_____ 3. $\log_a 0 = 1$ [12.3c]

_____ 4. $\log_a \dfrac{m}{n} = \log_a m - \log_a n$ [12.4c]

Guided Solutions

Fill in each box with the number and/or symbol that creates a correct statement or solution.

5. Solve: $\log_5 x = 3$. [12.3c]

   $\log_5 x = 3$

   $\boxed{} = x$ Converting to an exponential equation

   $\boxed{} = x$ Simplifying

6. Given $\log_a 2 = 0.648$ and $\log_a 9 = 2.046$, find $\log_a 18$. [12.4d]

   $\log_a 18 = \log_a (\boxed{} \cdot \boxed{})$

   $= \log_a 2 \boxed{} \log_a \boxed{}$

   $= 0.648 + \boxed{} = \boxed{}$

Mixed Review

Graph. [12.1a], [12.3a]

7. $f(x) = 3^{x-1}$

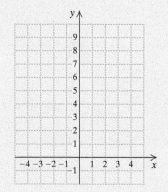

8. $f(x) = \left(\dfrac{3}{4}\right)^x$

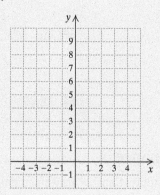

9. $f(x) = \log_4 x$

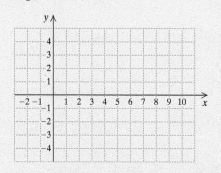

10. $f(x) = \log_{1/4} x$

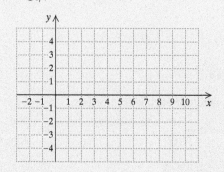

938　CHAPTER 12　Exponential Functions and Logarithmic Functions

11. *Interest Compounded Annually.* Lucas invests $500 at 4% interest, compounded annually. [12.1c]
  a) Find a function A for the amount in the account after t years.
  b) Find the amount in the account at $t = 0$, at $t = 4$, and at $t = 10$.

12. *Interest Compounded Quarterly.* The Currys invest $1500 in an account paying 3.5% interest, compounded quarterly. Find the amount in the account after $1\frac{1}{2}$ years. [12.1c]

Determine whether each function is one-to-one. If it is, find a formula for its inverse. [12.2c]

13. $f(x) = 3x + 1$ **14.** $f(x) = x^3 + 2$

Find $(f \circ g)(x)$ and $(g \circ f)(x)$. [12.2a]

15. $f(x) = 2x - 5$, $g(x) = 3 - x$

16. $f(x) = x^2 + 1$, $g(x) = 3x - 1$

Find $f(x)$ and $g(x)$ such that $h(x) = (f \circ g)(x)$. Answers may vary. [12.2a]

17. $h(x) = \dfrac{3}{x+4}$ **18.** $h(x) = \sqrt{6x - 7}$

For each function, use composition to show that the inverse is correct. [12.2e]

19. $f(x) = \dfrac{x}{3}$, $f^{-1}(x) = 3x$

20. $f(x) = \sqrt[3]{x + 4}$, $f^{-1}(x) = x^3 - 4$

Convert to a logarithmic equation. [12.3b]

21. $7^3 = 343$ **22.** $3^{-4} = \dfrac{1}{81}$

Convert to an exponential equation. [12.3b]

23. $\log_6 12 = t$ **24.** $\log_n T = m$

Solve. [12.3c]

25. $\log_4 64 = x$ **26.** $\log_x \dfrac{1}{4} = -2$

Find each of the following. [12.3c]

27. $\log_7 49$ **28.** $\log_2 32$

Use a calculator to find the logarithm, to four decimal places. [12.3d]

29. $\log 243.7$ **30.** $\log 0.23$

Express in terms of logarithms of a single variable or number. [12.4d]

31. $\log_b \dfrac{2xy^2}{z^3}$ **32.** $\log_a \sqrt[3]{\dfrac{x^2 y^5}{z^4}}$

Express as a single logarithm and, if possible, simplify. [12.4d]

33. $\log_a x - 2\log_a y + \dfrac{1}{2}\log_a z$

34. $\log_m (b^2 - 16) - \log_m (b + 4)$

Simplify. [12.3c], [12.4e]

35. $\log_8 1$ **36.** $\log_3 3$ **37.** $\log_a a^{-3}$ **38.** $\log_c c^5$

Understanding Through Discussion and Writing

39. The function $V(t) = 750(1.2)^t$ is used to predict the value V of a certain rare stamp t years after 2010. Do not calculate $V^{-1}(t)$ but explain how V^{-1} could be used. [12.2c]

40. Explain in your own words what is meant by $\log_a b = c$. [12.3b]

41. Find a way to express $\log_a (x/5)$ as a difference of logarithms without using the quotient rule. Explain your work. [12.4a, b]

42. A student incorrectly reasons that
$$\log_b \dfrac{1}{x} = \log_b \dfrac{x}{x \cdot x}$$
$$= \log_b x - \log_b x + \log_b x$$
$$= \log_b x.$$
What mistake has the student made? Explain what the answer should be. [12.4a, c]

STUDYING FOR SUCCESS *Looking Ahead*

- [ ] As you register for next semester's courses, evaluate your work and family commitments.
- [ ] If you are registering for another math course, consider keeping your notes, tests, and text from this course as a resource.

12.5 Natural Logarithmic Functions

OBJECTIVES

a Find logarithms or powers, base *e*, using a calculator.

b Use the change-of-base formula to find logarithms with bases other than *e* or 10.

c Graph exponential functions and logarithmic functions, base *e*.

Any positive number other than 1 can serve as the base of a logarithmic function. Common, or base-10, logarithms, which were introduced in Section 12.3, are useful because they have the same base as our "commonly" used decimal system of naming numbers.

Today, another base is widely used. It is an irrational number named *e*. We now consider *e* and **natural logarithms**, or logarithms base *e*.

a THE BASE *e* AND NATURAL LOGARITHMS

When interest is computed *n* times per year, the compound-interest formula is

$$A = P\left(1 + \frac{r}{n}\right)^{nt},$$

where *A* is the amount that an initial investment *P* will grow to after *t* years at interest rate *r*. Suppose that $1 could be invested at 100% interest for 1 year. (In reality, no financial institution would pay such an interest rate.) The preceding formula becomes a function *A* defined in terms of the number of compounding periods *n*:

$$A(n) = \left(1 + \frac{1}{n}\right)^n.$$

Let's find some function values, using a calculator and rounding to six decimal places. The numbers in the table shown here approach a very important number called *e*. It is an irrational number, so its decimal representation neither terminates nor repeats.

n	$A(n) = \left(1 + \frac{1}{n}\right)^n$
1 (compounded annually)	$2.00
2 (compounded semiannually)	$2.25
3	$2.370370
4 (compounded quarterly)	$2.441406
5	$2.488320
100	$2.704814
365 (compounded daily)	$2.714567
8760 (compounded hourly)	$2.718127

THE NUMBER *e*

$e \approx 2.7182818284\ldots$

Logarithms, base *e*, are called **natural logarithms**, or **Naperian logarithms**, in honor of John Napier (1550–1617), a Scotsman who invented logarithms.

The abbreviation **ln** is commonly used with natural logarithms. Thus,

ln 29   means $\log_e 29$.

We generally read "ln 29" as "the natural log of 29," or simply "el en of 29."

On a calculator, the key for natural logarithms is generally marked **LN**. Using that key, we find that

$$\ln 29 \approx 3.36729583 \approx 3.3673, \quad \text{Be sure to memorize } \ln a = \log_e a.$$

rounded to four decimal places. This also tells us that $e^{3.3673} \approx 29$.

On some scientific calculators, the keystrokes for doing such a calculation might be **2 9 LN =**. If we were to use a graphing calculator, the keystrokes might be **LN 2 9 ENTER**.

EXAMPLES Find the natural logarithm, to four decimal places, on a calculator.

	Function Value	Readout	Rounded
1.	ln 287,523	12.56905814	12.5691
2.	ln 0.000486	−7.629301934	−7.6293
3.	ln −5	NONREAL ANS	Does not exist as a real number
4.	ln e	1	1
5.	ln 1	0	0

Do Exercises 1–5. ▶

Find the natural logarithm, to four decimal places, on a calculator.
1. ln 78,235.4
2. ln 0.0000309
3. ln (−3)
4. ln 0
5. ln 10

$\ln 1 = 0$ and $\ln e = 1$, for the logarithmic base e.

The inverse of a logarithmic function is an exponential function. Thus, if $f(x) = \ln x$, then $f^{-1}(x) = e^x$. Because of this relationship, on many calculators, the **LN** key doubles as the **e^x** key after a **2ND** or $\boxed{\text{SHIFT}}$ key has been pressed.

EXAMPLE 6 Find $e^{12.5691}$ using a calculator.

On a scientific calculator, we might enter 12.5691 and press **e^x**. On a graphing calculator, we might press **2ND e^x**, followed by 12.5691 **ENTER**. In either case, we get the approximation

$$e^{12.5691} \approx 287,535.0371.$$

Do Exercises 6 and 7. ▶

6. Find $e^{11.2675}$ using a calculator. (Compare this computation to that of Margin Exercise 1.)

7. Find e^{-2} using a calculator.

b CHANGING LOGARITHM BASES

Find common logarithms on a calculator. [12.3d]
Find each of the following using a calculator. Round to four decimal places.

1. $\log \dfrac{8}{3}$

2. $\dfrac{\log 8}{\log 3}$

Answers: **1.** 0.4260 **2.** 1.8928

Answers
1. 11.2675 **2.** −10.3848 **3.** Does not exist as a real number **4.** Does not exist **5.** 2.3026 **6.** 78,237.1596 **7.** 0.1353

Most calculators give the values of both common logarithms and natural logarithms. To find a logarithm with some other base, we can use the following conversion formula.

> **THE CHANGE-OF-BASE FORMULA**
>
> For any logarithm bases a and b and any positive number M,
> $$\log_b M = \frac{\log_a M}{\log_a b}.$$

A Proof of the Change-of-Base Formula (*Optional*): We let $x = \log_b M$. Then, writing an equivalent exponential equation, we have $b^x = M$. Next, we take the logarithm base a on both sides. This gives us

$$\log_a b^x = \log_a M.$$

By Property 2, the Power Rule,

$$x \log_a b = \log_a M,$$

and solving for x, we obtain

$$x = \frac{\log_a M}{\log_a b}.$$

But $x = \log_b M$, so we have

$$\log_b M = \frac{\log_a M}{\log_a b},$$

which is the change-of-base formula.

EXAMPLE 7 Find $\log_4 7$ using common logarithms.

We let $a = 10$, $b = 4$, and $M = 7$. Then we substitute into the change-of-base formula:

$$\log_b M = \frac{\log_a M}{\log_a b}$$

$$\log_4 7 = \frac{\log_{10} 7}{\log_{10} 4} \quad \text{Substituting 10 for } a, \text{ 4 for } b, \text{ and 7 for } M$$

$$= \frac{\log 7}{\log 4}$$

$$\approx 1.4037.$$

To check, we use a calculator with a power key $\boxed{y^x}$ or ⌃ to verify that

$$4^{1.4037} \approx 7.$$

We can also use base e for a conversion.

EXAMPLE 8 Find $\log_4 7$ using natural logarithms.

$$\log_b M = \frac{\log_a M}{\log_a b}$$

$$\log_4 7 = \frac{\log_e 7}{\log_e 4} \quad \text{Substituting } e \text{ for } a, \text{ 4 for } b, \text{ and 7 for } M$$

$$= \frac{\ln 7}{\ln 4}$$

$$\approx 1.4037 \quad \text{Note that this is the same as the answer found in Example 7.}$$

CALCULATOR CORNER

The Change-of-Base Formula To find a logarithm with a base other than 10 or e, we can use the change-of-base formula. For example, we can find $\log_5 8$ using common logarithms.

We let $a = 10$, $b = 5$, and $M = 8$ and substitute in the change-of-base formula. We press `LOG` `8` `)` `÷` `LOG` `5` `)` `ENTER`. Note that the parentheses must be closed in the numerator in order to enter the expression correctly. We also close the parentheses in the denominator for completeness. The result is about 1.2920. We could have let $a = e$ and used natural logarithms to find $\log_5 8$ as well.

```
log(8)/log(5)
           1.292029674
```

Some calculators allow us to find a logarithm with any base directly.

```
log₅(8)
           1.292029674
```

EXAMPLE 9 Find $\log_5 29$ using natural logarithms.

Substituting e for a, 5 for b, and 29 for M, we have

$$\log_5 29 = \frac{\log_e 29}{\log_e 5} \quad \text{Using the change-of-base formula}$$

$$= \frac{\ln 29}{\ln 5} \approx 2.0922.$$

Do Exercises 8 and 9.

C GRAPHS OF EXPONENTIAL FUNCTIONS AND LOGARITHMIC FUNCTIONS, BASE e

EXAMPLE 10 Graph $f(x) = e^x$ and $g(x) = e^{-x}$.

We use a calculator with an $\boxed{e^x}$ key to find approximate values of e^x and e^{-x}. Using these values, we can graph the functions.

x	e^x	e^{-x}
0	1	1
1	2.7	0.4
2	7.4	0.1
−1	0.4	2.7
−2	0.1	7.4

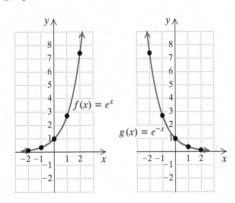

Note that each graph is the image of the other reflected across the y-axis.

EXAMPLE 11 Graph: $f(x) = e^{-0.5x}$.

We find some solutions with a calculator, plot them, and then draw the graph. For example, $f(2) = e^{-0.5(2)} = e^{-1} \approx 0.4$.

x	$e^{-0.5x}$
0	1
1	0.6
2	0.4
3	0.2
−1	1.6
−2	2.7
−3	4.5

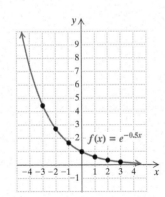

Do Exercises 10 and 11.

8. a) Find $\log_6 7$ using common logarithms.

$$\log_6 7 = \frac{\log \underline{}}{\log \underline{}}$$

$$\approx \underline{}$$

b) Find $\log_6 7$ using natural logarithms.

$$\log_6 7 = \frac{\ln \underline{}}{\ln \underline{}}$$

$$\approx \underline{}$$

9. Find $\log_2 46$ using natural logarithms.

Graph.

10. $f(x) = e^{2x}$

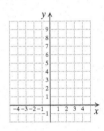

11. $g(x) = \frac{1}{2}e^{-x}$

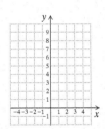

Answers
8. (a) 1.0860; (b) 1.0860   9. 5.5236
10.   11.

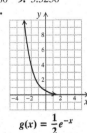

$f(x) = e^{2x}$ $g(x) = \frac{1}{2}e^{-x}$

Guided Solution:
8. (a) 7, 6, 1.0860;   (b) 7, 6, 1.0860

Graph.

12. $f(x) = 2 \ln x$

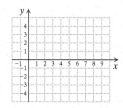

13. $g(x) = \ln(x - 2)$

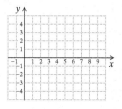

EXAMPLE 12 Graph: $g(x) = \ln x$.

We find some solutions with a calculator and then draw the graph. As expected, the graph is a reflection across the line $y = x$ of the graph of $y = e^x$.

x	$\ln x$
1	0
4	1.4
7	1.9
0.5	−0.7

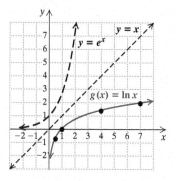

EXAMPLE 13 Graph: $f(x) = \ln(x + 3)$.

We find some solutions with a calculator, and then draw the graph. The graph of $y = \ln(x + 3)$ is the graph of $y = \ln x$ translated 3 units to the left.

x	$\ln(x + 3)$
0	1.1
1	1.4
2	1.6
3	1.8
4	1.9
−1	0.7
−2	0
−2.5	−0.7

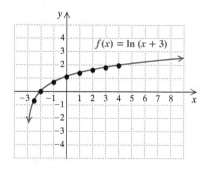

◀ Do Exercises 12 and 13.

Answers

12.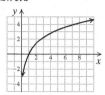
$f(x) = 2 \ln x$

13.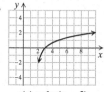
$g(x) = \ln(x - 2)$

CALCULATOR CORNER

Graphing Logarithmic Functions We can graph logarithmic functions with base 10 or base e by entering the function on the equation-editor screen using the **LOG** or **LN** key. To graph a logarithmic function with a base other than 10 or e, we must first use the change-of-base formula to change the base to 10 or e, unless our calculator can find other logarithms directly.

We can graph the function $y = \log_5 x$ on a graphing calculator by first changing the base to e. We let $a = e$, $b = 5$, and $M = x$ and substitute in the change-of-base formula. We enter $y_1 = \ln(x)/\ln(5)$ on the equation-editor screen, select a window, and press **GRAPH**. We could have let $a = 10$ and used base-10 logarithms to graph this function.

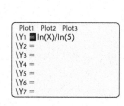 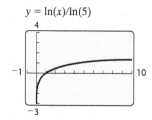

EXERCISES: Graph each of the following on a graphing calculator.

1. $y = \log_2 x$ **2.** $y = \log_3 x$ **3.** $y = \log_{1/2} x$ **4.** $y = \log_{2/3} x$

Visualizing for Success

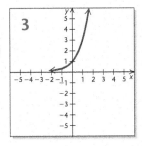

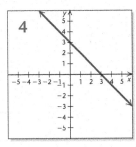

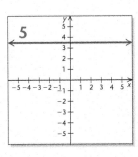

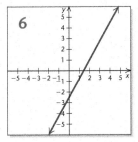

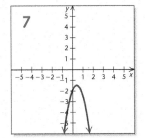

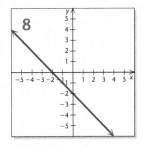

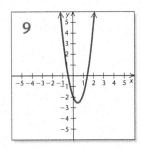

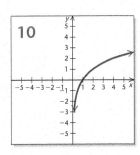

Match each graph with its function.

A. $f(x) = ax^2 + bx + c$, $a < 0, c < 0$

B. $f(x) = a^x, 0 < a < 1$

C. $f(x) = a^x, a < 0$

D. $f(x) = \log_a x, 0 < a < 1$

E. $f(x) = \log_a x, a < 0$

F. $f(x) = mx + b, m > 0, b < 0$

G. $f(x) = mx + b, m < 0, b > 0$

H. $f(x) = mx + b, m < 0, b < 0$

I. $f(x) = ax^2 + bx + c$, $a > 0, c < 0$

J. $f(x) = ax^2 + bx + c$, $a < 0, c > 0$

K. $f(x) = \log_a x, a > 1$

L. $f(x) = ax^2 + bx + c$, $a > 0, c > 0$

M. $f(x) = mx + b, m < 0, b = 0$

N. $f(x) = mx + b, m = 0, b > 0$

O. $f(x) = a^x, a > 1$

Answers on page A-43

12.5 Exercise Set

✓ Check Your Understanding

Reading Check Determine whether each phrase describes either a base-10 logarithm or a base-e logarithm.

RC1. A common logarithm

RC2. A natural logarithm

RC3. $\log x$

RC4. $\ln x$

Concept Check Choose from the list below the option that best completes each statement. Choices may be used more than once.

a) $\log_5 7$
b) $\log_7 5$

CC1. $\dfrac{\log 5}{\log 7} = $ _____

CC2. $\dfrac{\log 7}{\log 5} = $ _____

CC3. $\dfrac{\ln 5}{\ln 7} = $ _____

CC4. $\dfrac{\ln 7}{\ln 5} = $ _____

a Find each of the following logarithms or powers, base e, using a calculator. Round answers to four decimal places.

1. $\ln 2$
2. $\ln 5$
3. $\ln 62$
4. $\ln 30$
5. $\ln 4365$
6. $\ln 901.2$

7. $\ln 0.0062$
8. $\ln 0.00073$
9. $\ln 0.2$
10. $\ln 0.04$
11. $\ln 0$
12. $\ln(-4)$

13. $\ln\left(\dfrac{97.4}{558}\right)$
14. $\ln\left(\dfrac{786.2}{77.2}\right)$
15. $\ln e$
16. $\ln e^2$
17. $e^{2.71}$
18. $e^{3.06}$

19. $e^{-3.49}$
20. $e^{-2.64}$
21. $e^{4.7}$
22. $e^{1.23}$
23. $\ln e^5$
24. $e^{\ln 7}$

b Find each of the following logarithms using the change-of-base formula.

25. $\log_6 100$
26. $\log_3 100$
27. $\log_2 100$
28. $\log_7 100$
29. $\log_7 65$
30. $\log_5 42$

31. $\log_{0.5} 5$
32. $\log_{0.1} 3$
33. $\log_2 0.2$
34. $\log_2 0.08$
35. $\log_\pi 200$
36. $\log_\pi \pi$

c Graph.

37. $f(x) = e^x$

x	$f(x)$
0	
1	
2	
-1	
-2	

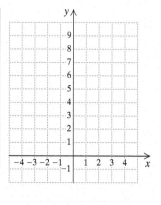

38. $f(x) = e^{0.5x}$

x	$f(x)$
0	
1	
2	
-1	
-2	

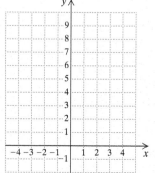

39. $f(x) = e^{-0.5x}$

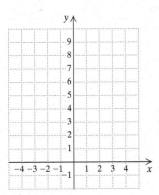

40. $f(x) = e^{-x}$

41. $f(x) = e^{x-1}$

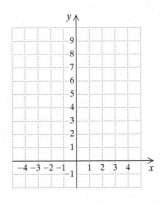

42. $f(x) = e^{-x} + 3$

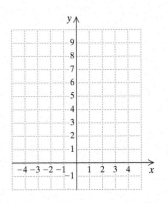

43. $f(x) = e^{x+2}$

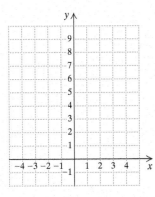

44. $f(x) = e^{x-2}$

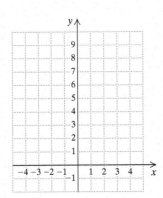

45. $f(x) = e^x - 1$

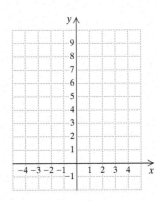

46. $f(x) = 2e^{0.5x}$

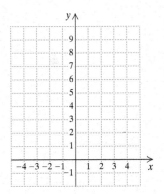

47. $f(x) = \ln(x + 2)$

x	f(x)
0	
1	
2	
3	
−0.5	
−1	
−1.5	

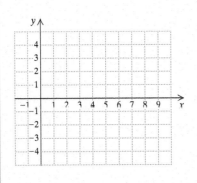

48. $f(x) = \ln(x + 1)$

x	f(x)
0	
1	
2	
3	
4	
−0.5	
−0.75	

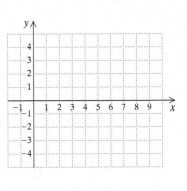

SECTION 12.5 Natural Logarithmic Functions

49. $f(x) = \ln(x - 3)$

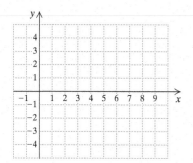

50. $f(x) = 2\ln(x - 2)$

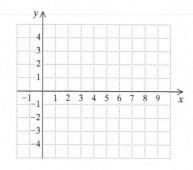

51. $f(x) = 2\ln x$

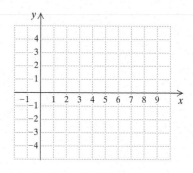

52. $f(x) = \ln x - 3$

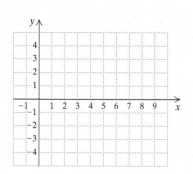

53. $f(x) = \frac{1}{2}\ln x + 1$

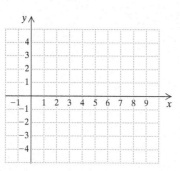

54. $f(x) = \ln x^2$

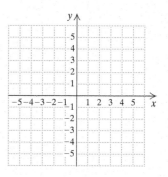

55. $f(x) = |\ln x|$

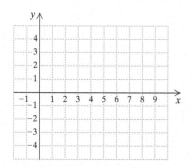

56. $f(x) = \ln |x|$

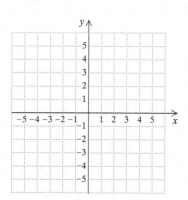

Skill Maintenance

Solve. [11.4c]

57. $x^{1/2} - 6x^{1/4} + 8 = 0$ **58.** $2y - 7\sqrt{y} + 3 = 0$ **59.** $x - 18\sqrt{x} + 77 = 0$ **60.** $x^4 - 25x^2 + 144 = 0$

Synthesis

Use the graph of the function to find the domain and the range.

61. $f(x) = 10x^2 e^{-x}$ **62.** $f(x) = 7.4 e^x \ln x$ **63.** $f(x) = 100(1 - e^{-0.3x})$

Find the domain.

64. $f(x) = \log_3 x^2$ **65.** $f(x) = \log(2x - 5)$

Solving Exponential Equations and Logarithmic Equations

12.6

OBJECTIVES

a Solve exponential equations.
b Solve logarithmic equations.

a SOLVING EXPONENTIAL EQUATIONS

Equations with variables in exponents, such as $5^x = 12$ and $2^{7x} = 64$, are called **exponential equations**. Sometimes we can solve an exponential equation by writing each side as a power of the *same* number. We use the following property, which is true because exponential functions are one-to-one.

> **THE PRINCIPLE OF EXPONENTIAL EQUALITY**
>
> For any $a > 0$, $a \neq 1$,
>
> $a^x = a^y \longrightarrow x = y$.
>
> (When powers are equal, the exponents are equal.)

EXAMPLE 1 Solve: $2^{3x-5} = 16$.

Because $16 = 2^4$, we can write each side as a power of the same number:

$$2^{3x-5} = 2^4.$$

Since the base is the same, 2, the exponents must be the same. Thus,

$$3x - 5 = 4$$
$$3x = 9$$
$$x = 3.$$

Check:
$$\begin{array}{c|c} 2^{3x-5} = 16 \\ \hline 2^{3 \cdot 3 - 5} \;?\; 16 \\ 2^{9-5} \\ 2^4 \\ 16 & \text{TRUE} \end{array}$$

The solution is 3.

Do Exercises 1 and 2.

Solve.
1. $3^{2x} = 9$
2. $4^{2x-3} = 64$

> **ALGEBRAIC** ▶◀ **GRAPHICAL CONNECTION**
>
> The solution, 3, of the equation $2^{3x-5} = 16$ in Example 1 is the x-coordinate of the point of intersection of the graphs of $y = 2^{3x-5}$ and $y = 16$.
>
>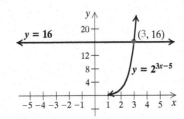

Answers
1. 1   2. 3

When it does not seem possible to write both sides of an equation as powers of the same base, we can use the following principle along with the properties developed in Section 12.4.

> **THE PRINCIPLE OF LOGARITHMIC EQUALITY**
>
> For any logarithm base a, and for $x, y > 0$,
> $$\log_a x = \log_a y \longrightarrow x = y.$$
> (If the logarithms, base a, of two expressions are the same, then the expressions are the same.)

Because calculators can generally find only common or natural logarithms (without resorting to the change-of-base formula), we usually take the common or natural logarithm on both sides of the equation.

The principle of logarithmic equality is useful any time a variable appears as an exponent.

EXAMPLE 2 Solve: $5^x = 12$.

$$5^x = 12$$
$$\log 5^x = \log 12 \quad \text{Taking the common logarithm on both sides}$$
$$x \log 5 = \log 12 \quad \text{Property 2}$$
$$x = \frac{\log 12}{\log 5} \quad \longleftarrow \text{Caution!} \quad \text{This is not } \log \tfrac{12}{5}!$$

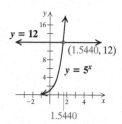

This is an exact answer. We cannot simplify further, but we can approximate using a calculator:

$$x = \frac{\log 12}{\log 5} \approx 1.5440.$$

We can also partially check this answer by finding $5^{1.5440}$ using a calculator:

$$5^{1.5440} \approx 12.00078587.$$

We get an answer close to 12, due to the rounding. This checks.

◀ Do Exercise 3.

If the base is e, we can make our work easier by taking the logarithm, base e, on both sides.

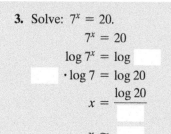

3. Solve: $7^x = 20$.
$$7^x = 20$$
$$\log 7^x = \log \boxed{}$$
$$\boxed{} \cdot \log 7 = \log 20$$
$$x = \frac{\log 20}{\boxed{}}$$
$$x \approx \boxed{}$$

Answer
3. 1.5395
Guided Solution:
3. 20, x, log 7, 1.5395

EXAMPLE 3 Solve: $e^{0.06t} = 1500$.

We take the natural logarithm on both sides:

$$e^{0.06t} = 1500$$
$$\ln e^{0.06t} = \ln 1500 \quad \text{Taking ln on both sides}$$
$$\log_e e^{0.06t} = \ln 1500 \quad \text{Definition of natural logarithms}$$
$$0.06t = \ln 1500 \quad \text{Here we use Property 4: } \log_a a^k = k.$$
$$t = \frac{\ln 1500}{0.06}.$$

We can approximate using a calculator:

$$t = \frac{\ln 1500}{0.06} \approx \frac{7.3132}{0.06} \approx 121.89.$$

We can also partially check this answer using a calculator.

Check:
$$e^{0.06t} = 1500$$
$$e^{0.06(121.89)} \;?\; 1500$$
$$e^{7.3134} \;\Big|$$
$$1500.269444 \;\Big| \quad \text{TRUE}$$

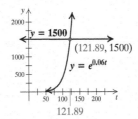

The solution is $\dfrac{\ln 1500}{0.06}$, or about 121.89.

Do Exercise 4. ▶ **4.** Solve: $e^{0.3t} = 80$.

b SOLVING LOGARITHMIC EQUATIONS

SKILL REVIEW
Solve quadratic equations using the principle of zero products. [5.8b]
Solve.
1. $y^2 - y - 6 = 0$
2. $x^2 - 3x = 4$

Answers: **1.** $-2, 3$ **2.** $-1, 4$

MyLab Math VIDEO

Equations containing logarithmic expressions are called **logarithmic equations**. We can solve logarithmic equations by converting to equivalent exponential equations.

EXAMPLE 4 Solve: $\log_2 x = 3$.

We obtain an equivalent exponential equation:
$$x = 2^3$$
$$x = 8.$$

The solution is 8.

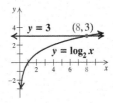

Do Exercise 5. ▶ **5.** Solve: $\log_5 x = 2$.

Answers
4. 14.6068 **5.** 25

> To solve a logarithmic equation, first try to obtain a single logarithmic expression on one side and then write an equivalent exponential equation.

EXAMPLE 5 Solve: $\log_4(8x - 6) = 3$.

We already have a single logarithmic expression, so we write an equivalent exponential equation:

$$8x - 6 = 4^3 \quad \text{Writing an equivalent exponential equation}$$
$$8x - 6 = 64$$
$$8x = 70$$
$$x = \tfrac{70}{8}, \text{ or } \tfrac{35}{4}.$$

Check:
$$\begin{array}{c|c}
\log_4(8x - 6) = 3 & \\
\hline
\log_4\left(8 \cdot \tfrac{35}{4} - 6\right) \stackrel{?}{=} 3 & \\
\log_4(70 - 6) & \\
\log_4 64 & \\
3 & \text{TRUE}
\end{array}$$

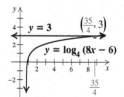

The solution is $\tfrac{35}{4}$.

◀ **Do Exercise 6.**

6. Solve: $\log_3(5x + 7) = 2$.

EXAMPLE 6 Solve: $\log x + \log(x - 3) = 1$.

Here we have common logarithms. It helps to first write in the 10's before we obtain a single logarithmic expression on the left.

$$\log_{10} x + \log_{10}(x - 3) = 1$$
$$\log_{10}[x(x - 3)] = 1 \quad \text{Using Property 1 to obtain a single logarithm}$$
$$x(x - 3) = 10^1 \quad \text{Writing an equivalent exponential expression}$$
$$x^2 - 3x = 10$$
$$x^2 - 3x - 10 = 0$$
$$(x + 2)(x - 5) = 0 \quad \text{Factoring}$$
$$x + 2 = 0 \quad \text{or} \quad x - 5 = 0 \quad \text{Using the principle of zero products}$$
$$x = -2 \quad \text{or} \quad x = 5$$

Check: For -2:
$$\begin{array}{c|c}
\log x + \log(x - 3) = 1 & \text{The number } -2 \text{ does } \textit{not} \text{ check} \\
\hline
\log(-2) + \log(-2 - 3) \stackrel{?}{=} 1 & \text{because negative numbers do not have logarithms.}
\end{array}$$

For 5:
$$\begin{array}{c|c}
\log x + \log(x - 3) = 1 & \\
\hline
\log 5 + \log(5 - 3) \stackrel{?}{=} 1 & \\
\log 5 + \log 2 & \\
\log(5 \cdot 2) & \\
\log 10 & \\
1 & \text{TRUE}
\end{array}$$

The solution is 5.

◀ **Do Exercise 7.**

7. Solve: $\log x + \log(x + 3) = 1$.

Answers

6. $\tfrac{2}{5}$ 7. 2

EXAMPLE 7 Solve: $\log_2(x+7) - \log_2(x-7) = 3$.

$\log_2(x+7) - \log_2(x-7) = 3$

$\log_2 \dfrac{x+7}{x-7} = 3$ Using Property 3 to obtain a single logarithm

$\dfrac{x+7}{x-7} = 2^3$ Writing an equivalent exponential expression

$\dfrac{x+7}{x-7} = 8$

$x + 7 = 8(x-7)$ Multiplying by the LCM, $x-7$

$x + 7 = 8x - 56$ Using a distributive law

$63 = 7x$

$\dfrac{63}{7} = x$

$9 = x$

Check:
$$\begin{array}{c|c}\log_2(x+7) - \log_2(x-7) = 3 \\ \hline \log_2(9+7) - \log_2(9-7) \;?\; 3 \\ \log_2 16 - \log_2 2 \\ \log_2 \tfrac{16}{2} \\ \log_2 8 \\ 3 \;\big|\; \text{TRUE} \end{array}$$

The solution is 9.

Do Exercise 8.

GS 8. Solve:
$$\log_3(2x-1) - \log_3(x-4) = 2.$$

$\log_3(2x-1) - \log_3(x-4) = 2$

$\log_3 \dfrac{2x-1}{\boxed{}} = 2$

$\dfrac{2x-1}{x-4} = \boxed{}^2$

$\dfrac{2x-1}{x-4} = \boxed{}$

$2x - 1 = \boxed{}(x-4)$

$2x - 1 = 9x - \boxed{}$

$\boxed{} = 7x$

$\boxed{} = x$

Answer
8. 5
Guided Solution:
8. $x-4, 3, 9, 9, 36, 35, 5$

12.6 Exercise Set

FOR EXTRA HELP   MyLab Math

✓ Check Your Understanding

Reading Check Determine whether each statement is true or false.

RC1. To solve an exponential equation, we can take the common logarithm on both sides of the equation.

RC2. To solve an exponential equation, we can take the natural logarithm on both sides of the equation.

RC3. We cannot calculate the logarithm of a negative number.

RC4. The solution of a logarithmic equation is never a negative number.

Concept Check Determine whether each statement is true or false.

CC1. The solution of $2^x = 6$ is 3.

CC2. The solution of $2^x = 16$ is 4.

CC3. The solution of $\log_2 4 = x$ is 16.

CC4. The solution of $\log_x 8 = 3$ is 2.

CC5. The solution of $\log_8 x = 1$ is 8.

CC6. The solution of $\log_2 1 = x$ is 0.

a Solve.

1. $2^x = 8$
2. $3^x = 81$
3. $4^x = 256$
4. $5^x = 125$

5. $2^{2x} = 32$
6. $4^{3x} = 64$
7. $3^{5x} = 27$
8. $5^{7x} = 625$

9. $2^x = 11$
10. $2^x = 20$
11. $2^x = 43$
12. $2^x = 55$

13. $5^{4x-7} = 125$
14. $4^{3x+5} = 16$
15. $3^{x^2} \cdot 3^{4x} = \dfrac{1}{27}$
16. $3^{5x} \cdot 9^{x^2} = 27$

17. $4^x = 8$
18. $6^x = 10$
19. $e^t = 100$
20. $e^t = 1000$

21. $e^{-t} = 0.1$
22. $e^{-t} = 0.01$
23. $e^{-0.02t} = 0.06$
24. $e^{0.07t} = 2$

25. $2^x = 3^{x-1}$
26. $3^{x+2} = 5^{x-1}$
27. $(3.6)^x = 62$
28. $(5.2)^x = 70$

b Solve.

29. $\log_4 x = 4$
30. $\log_7 x = 3$
31. $\log_2 x = -5$
32. $\log_9 x = \dfrac{1}{2}$

33. $\log x = 1$
34. $\log x = 3$
35. $\log x = -2$
36. $\log x = -3$

37. $\ln x = 2$

38. $\ln x = 1$

39. $\ln x = -1$

40. $\ln x = -3$

41. $\log_3 (2x + 1) = 5$

42. $\log_2 (8 - 2x) = 6$

43. $\log x + \log (x - 9) = 1$

44. $\log x + \log (x + 9) = 1$

45. $\log x - \log (x + 3) = -1$

46. $\log (x + 9) - \log x = 1$

47. $\log_2 (x + 1) + \log_2 (x - 1) = 3$

48. $\log_2 x + \log_2 (x - 2) = 3$

49. $\log_4 (x + 6) - \log_4 x = 2$

50. $\log_4 (x + 3) - \log_4 (x - 5) = 2$

51. $\log_4 (x + 3) + \log_4 (x - 3) = 2$

52. $\log_5 (x + 4) + \log_5 (x - 4) = 2$

53. $\log_3 (2x - 6) - \log_3 (x + 4) = 2$

54. $\log_4 (2 + x) - \log_4 (3 - 5x) = 3$

Skill Maintenance

Solve.

55. $-3 \leq x - 12 < 4$ [9.2a]

56. $|2x - 5| > 3$ [9.3e]

57. $x^2 - x = 12$ [5.8a]

58. $x^2 - x \leq 12$ [11.8a]

59. $\sqrt{n - 1} = 8$ [10.6a]

60. $\sqrt{y - 3} = y - 5$ [10.6a]

Synthesis

61. Find the value of x for which the natural logarithm is the same as the common logarithm.

62. Use a graphing calculator to check your answers to Exercises 4, 20, 36, and 54.

63. Use a graphing calculator to solve each of the following equations.
  a) $e^{7x} = 14$
  b) $8e^{0.5x} = 3$
  c) $xe^{3x-1} = 5$
  d) $4 \ln (x + 3.4) = 2.5$

64. Use the INTERSECT feature of a graphing calculator to find the points of intersection of the graphs of each pair of functions.
  a) $f(x) = e^{0.5x-7}, g(x) = 2x + 6$
  b) $f(x) = \ln 3x, g(x) = 3x - 8$
  c) $f(x) = \ln x^2, g(x) = -x^2$

Solve.

65. $2^{2x} + 128 = 24 \cdot 2^x$

66. $27^x = 81^{2x-3}$

67. $8^x = 16^{3x+9}$

68. $\log_x (\log_3 27) = 3$

69. $\log_6 (\log_2 x) = 0$

70. $x \log \frac{1}{8} = \log 8$

71. $\log_5 \sqrt{x^2 - 9} = 1$

72. $2^{x^2+4x} = \frac{1}{8}$

73. $\log (\log x) = 5$

74. $\log_5 |x| = 4$

75. $\log x^2 = (\log x)^2$

76. $\log_3 |5x - 7| = 2$

77. $\log_a a^{x^2+4x} = 21$

78. $\sqrt{x} \cdot \sqrt[3]{x} \cdot \sqrt[4]{x} \cdot \sqrt[5]{x} = 146$

79. $3^{2x} - 8 \cdot 3^x + 15 = 0$

80. If $x = (\log_{125} 5)^{\log_5 125}$, what is the value of $\log_3 x$?

12.7 Mathematical Modeling with Exponential Functions and Logarithmic Functions

OBJECTIVES

a Solve applied problems involving logarithmic functions.

b Solve applied problems involving exponential functions.

Exponential functions and logarithmic functions can now be added to our library of functions that can serve as models for many kinds of applications. Let's review some of their graphs.

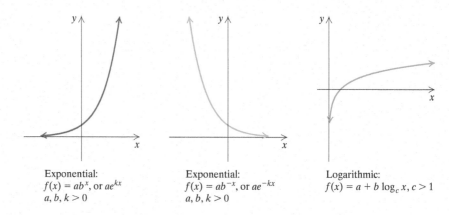

Exponential:
$f(x) = ab^x$, or ae^{kx}
$a, b, k > 0$

Exponential:
$f(x) = ab^{-x}$, or ae^{-kx}
$a, b, k > 0$

Logarithmic:
$f(x) = a + b \log_c x, c > 1$

a APPLICATIONS OF LOGARITHMIC FUNCTIONS

EXAMPLE 1 *Sound Levels.* To measure the "loudness" of any particular sound, the decibel scale is used. The loudness L, in decibels (dB), of a sound is given by

$$L = 10 \cdot \log \frac{I}{I_0},$$

where I is the intensity of the sound, in watts per square meter (W/m²), and $I_0 = 10^{-12}$ W/m². (I_0 is approximately the intensity of the softest sound that can be heard.)

a) An iPod can produce sounds of more than $10^{-0.5}$ W/m², a volume that can damage the hearing of a person exposed to the sound for more than 28 sec. How loud, in decibels, is this sound level?

b) Audiologists and physicians recommend that earplugs be worn when one is exposed to sounds in excess of 90 dB. What is the intensity of such sounds?

Data: American Speech–Language-Hearing Association

a) To find the loudness, in decibels, we use the above formula:

$$L = 10 \cdot \log \frac{I}{I_0}$$

$= 10 \cdot \log \dfrac{10^{-0.5}}{10^{-12}}$ Substituting

$= 10 \cdot \log 10^{11.5}$ Subtracting exponents

$= 10 \cdot 11.5$ $\log 10^a = a$

$= 115.$

The sound level is 115 decibels.

b) We substitute and solve for I:

$$L = 10 \cdot \log \frac{I}{I_0}$$

$$90 = 10 \cdot \log \frac{I}{10^{-12}} \quad \text{Substituting}$$

$$9 = \log \frac{I}{10^{-12}} \quad \text{Dividing by 10}$$

$$9 = \log I - \log 10^{-12} \quad \text{Using Property 3}$$

$$9 = \log I - (-12) \quad \log 10^a = a$$

$$-3 = \log I \quad \text{Adding } -12$$

$$10^{-3} = I. \quad \text{Converting to an exponential equation}$$

Earplugs are recommended for sounds with intensities that exceed $10^{-3}\,\text{W/m}^2$.

Do Exercises 1 and 2. ▶

EXAMPLE 2 *pH.* In chemistry, the pH of a liquid is defined as

$$\text{pH} = -\log\,[\text{H}^+],$$

where $[\text{H}^+]$ is the hydrogen ion concentration in moles per liter.

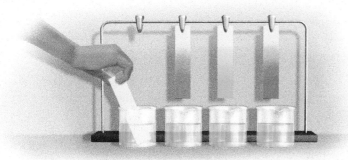

a) The hydrogen ion concentration of human blood is normally about 3.98×10^{-8} moles per liter. Find the pH.

b) The pH of seawater is about 8.3. Find the hydrogen ion concentration.

a) To find the pH of human blood, we use the above formula:

$$\text{pH} = -\log\,[\text{H}^+] = -\log\,[3.98 \times 10^{-8}]$$

$$\approx -(-7.400117) \approx 7.4. \quad \text{Using a calculator and rounding}$$

The pH of human blood is normally about 7.4.

b) We substitute and solve for $[\text{H}^+]$:

$$8.3 = -\log\,[\text{H}^+] \quad \text{Using pH} = -\log\,[\text{H}^+]$$

$$-8.3 = \log\,[\text{H}^+] \quad \text{Dividing by } -1$$

$$10^{-8.3} = [\text{H}^+] \quad \text{Converting to an exponential equation}$$

$$5.01 \times 10^{-9} \approx [\text{H}^+]. \quad \text{Using a calculator; writing scientific notation}$$

The hydrogen ion concentration of seawater is about 5.01×10^{-9} moles per liter.

Do Exercises 3 and 4. ▶

---

 1. *Acoustics.* The intensity of sound in normal conversation is about $3.2 \times 10^{-6}\,\text{W/m}^2$. How high is this sound level in decibels?

$$L = 10 \cdot \log \frac{I}{I_0}$$

$$= 10 \cdot \log \frac{\boxed{}}{10^{-12}}$$

$$= 10 \cdot \log\,(3.2 \times 10^{\boxed{}})$$

$$= 10 \cdot (\log 3.2 + \log 10^6)$$

$$\approx 10 \cdot (0.5051 + \boxed{})$$

$$\approx 65$$

The sound level is about $\boxed{}$ decibels.

2. *Audiology.* Overexposure to excessive sound levels can diminish one's hearing to the point where the softest sound that is audible is 28 dB. What is the intensity of such a sound?

3. *Coffee.* The hydrogen ion concentration of freshly brewed coffee is about 1.3×10^{-5} moles per liter. Find the pH.

 4. *Acidosis.* When the pH of a patient's blood drops below 7.4, a condition called *acidosis* sets in. Acidosis can be fatal at a pH level of 7. What would the hydrogen ion concentration of the patient's blood be at that point?

$$\text{pH} = -\log\,[\text{H}^+]$$

$$\boxed{} = -\log\,[\text{H}^+]$$

$$\boxed{} = \log\,[\text{H}^+]$$

$$10^{\boxed{}} = [\text{H}^+]$$

Answers
1. About 65 decibels 2. $10^{-9.2}\,\text{W/m}^2$
3. About 4.9 4. 10^{-7} moles per liter
Guided Solutions:
1. $3.2 \times 10^{-6}, 6, 6, 65$ 4. $7, -7, -7$

b APPLICATIONS OF EXPONENTIAL FUNCTIONS

> **SKILL REVIEW**
>
> Solve logarithmic equations. [12.6b]
> Solve.
> 1. $\log x = 2$
> 2. $\ln x = -2$
>
> Answers: **1.** 100 **2.** $e^{-2} \approx 0.1353$

EXAMPLE 3 *Interest Compounded Annually.* Suppose that $30,000 is invested at 4% interest, compounded annually. In t years, it will grow to the amount A given by the function

$$A(t) = 30{,}000(1.04)^t.$$

(See Example 6 in Section 12.1.)

a) How long will it take to accumulate $150,000 in the account?

b) Let $T =$ the amount of time it takes for the $30,000 to double itself; T is called the **doubling time**. Find the doubling time.

a) We set $A(t) = 150{,}000$ and solve for t:

$$150{,}000 = 30{,}000(1.04)^t$$

$$\frac{150{,}000}{30{,}000} = (1.04)^t \qquad \text{Dividing by 30,000}$$

$$5 = (1.04)^t$$

$$\log 5 = \log(1.04)^t \qquad \text{Taking the common logarithm on both sides}$$

$$\log 5 = t \log 1.04 \qquad \text{Using Property 2}$$

$$\frac{\log 5}{\log 1.04} = t \qquad \text{Dividing by log 1.04}$$

$$41.04 \approx t. \qquad \text{Using a calculator}$$

It will take about 41 years for the $30,000 to grow to $150,000.

b) To find the *doubling time* T, we replace $A(t)$ with 60,000 and t with T and solve for T:

$$60{,}000 = 30{,}000(1.04)^T$$

$$2 = (1.04)^T \qquad \text{Dividing by 30,000}$$

$$\log 2 = \log(1.04)^T \qquad \text{Taking the common logarithm on both sides}$$

$$\log 2 = T \log 1.04 \qquad \text{Using Property 2}$$

$$T = \frac{\log 2}{\log 1.04} \approx 17.7. \qquad \text{Using a calculator}$$

The doubling time is about 17.7 years.

◀ Do Exercise 5.

5. *Interest Compounded Annually.* Suppose that $40,000 is invested at 4.3% interest, compounded annually.
  a) After what amount of time will there be $250,000 in the account?
  b) Find the doubling time.

Answers
5. (a) 43.5 years; (b) 16.5 years

The function in Example 3 illustrates exponential growth. Populations often grow exponentially according to the following model.

EXPONENTIAL GROWTH MODEL

An **exponential growth model** is a function of the form

$$P(t) = P_0 e^{kt}, \quad k > 0,$$

where P_0 is the population at time 0, $P(t)$ is the population at time t, and k is the **exponential growth rate** for the situation. The **doubling time** is the amount of time necessary for the population to double in size.

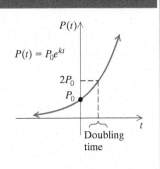

The exponential growth rate is the rate of growth of a population at any *instant* in time. Since the population is continually growing, the percent of total growth after one year will exceed the exponential growth rate.

EXAMPLE 4 *Population Growth.* In 2016, Luxembourg's population was 0.6 million, and the exponential growth rate was 2.05% per year.
Data: Central Intelligence Agency

a) Find the exponential growth function.
b) What will the population be in 2030?

a) The given information allows us to create a model. At $t = 0$ (2016), the population was 0.6 million. We substitute 0.6 for P_0 and 2.05%, or 0.0205, for k to obtain the exponential growth function:

$$P(t) = P_0 e^{kt}$$
$$= 0.6 e^{0.0205t}.$$

Here $P(t)$ is in millions and t is the number of years after 2016.

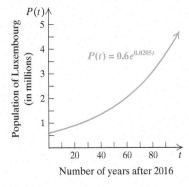

b) In 2030, we have $t = 14$. That is, 14 years have passed since 2016. To find the population in 2030, we substitute 14 for t:

$P(14) = 0.6 e^{0.0205(14)}$ Substituting 14 for t
$\approx 0.8.$ Using a calculator

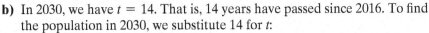

The population of Luxembourg will be about 0.8 million in 2030.

Do Exercise 6.

6. *Population Growth.* Refer to Example 4. What will the population of Luxembourg be in 2040?

EXAMPLE 5 *Interest Compounded Continuously.* Suppose that an amount of money P_0 is invested in a savings account at interest rate k, compounded continuously. That is, suppose that interest is computed every "instant" and added to the amount in the account. The balance $P(t)$, after t years, is given by the exponential growth model

$$P(t) = P_0 e^{kt}.$$

a) Suppose that $30,000 is invested and grows to $34,855.03 in 5 years. Find the interest rate and then the exponential growth function.

Answer
6. About 1 million

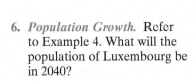

b) What is the balance after 10 years?

c) What is the doubling time?

a) We have $P_0 = 30{,}000$. Thus the exponential growth function is

$$P(t) = 30{,}000e^{kt},$$

where k must still be determined. We know that $P(5) = 34{,}855.03$. We substitute and solve for k:

$$34{,}855.03 = 30{,}000e^{k(5)} = 30{,}000e^{5k}$$

$$\frac{34{,}855.03}{30{,}000} = e^{5k} \qquad \text{Dividing by 30,000}$$

$$1.161834 \approx e^{5k}$$

$$\ln 1.161834 = \ln e^{5k} \qquad \text{Taking the natural logarithm on both sides}$$

$$0.15 \approx 5k \qquad \text{Finding } \ln 1.161834 \text{ on a calculator and simplifying } \ln e^{5k}$$

$$\frac{0.15}{5} = 0.03 \approx k.$$

The interest rate is about 0.03, or 3%, compounded continuously. Note that since interest is being compounded continuously, the interest earned each year is more than 3%. The exponential growth function is

$$P(t) = 30{,}000e^{0.03t}.$$

b) We substitute 10 for t:

$$P(10) = 30{,}000e^{0.03(10)} \approx \$40{,}495.76.$$

The balance in the account after 10 years will be $40,495.76.

c) To find the doubling time T, we replace $P(t)$ with 60,000 and solve for T:

$$60{,}000 = 30{,}000e^{0.03T}$$

$$2 = e^{0.03T} \qquad \text{Dividing by 30,000}$$

$$\ln 2 = \ln e^{0.03T} \qquad \text{Taking the natural logarithm on both sides}$$

$$\ln 2 = 0.03T$$

$$\frac{\ln 2}{0.03} = T \qquad \text{Dividing}$$

$$23.1 \approx T.$$

Thus the original investment of $30,000 will double in about 23.1 years, as shown in the following graph of the growth function.

7. *Interest Compounded Continuously.*

a) Suppose that $5000 is invested and grows to $6356.25 in 4 years. Find the interest rate and then the exponential growth function.

b) What is the balance after 1 year? 2 years? 10 years?

c) What is the doubling time?

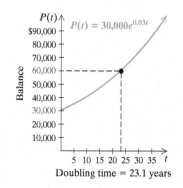

Doubling time = 23.1 years

◀ Do Exercise 7.

Answers

7. (a) $k = 6\%$, $P(t) = 5000e^{0.06t}$;
(b) $5309.18; $5637.48; $9110.59;
(c) about 11.6 years

EXAMPLE 6 *Solar Power.* The amount of electricity generated from the sun by utilities in the United States has increased exponentially from 864 thousand MWh (megawatt hours) in 2008 to 36,755 thousand MWh in 2016, as shown in the following graph.

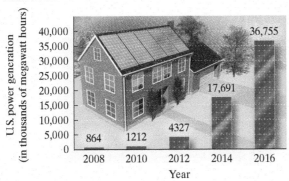

Solar Power

DATA: U.S. Energy Information Administration

a) Let t = the number of years after 2008. Then $t = 0$ corresponds to 2008 and $t = 8$ corresponds to 2016. Use the data points (0, 864) and (8, 36,755) to find the exponential growth rate and then the exponential growth function.

b) Use the function found in part (a) to predict the amount of solar power generated in the United States in 2020.

c) Use the function to determine the year in which there will be about 1 billion MWh of solar power generated.

a) We use the equation $P(t) = P_0 e^{kt}$, where $P(t)$ is the amount of solar power generated t years after 2008, in thousands of MWh. In 2008, at $t = 0$, there was 864 thousand MWh of solar power generated. Thus we substitute 864 for P_0:

$$P(t) = 864e^{kt}.$$

To find the exponential growth rate k, we note that 8 years later, in 2016, there was 36,755 thousand MWh of solar power generated. We substitute and solve for k:

$P(8) = 864e^{k(8)}$ Substituting
$36{,}755 = 864e^{k(8)}$
$\dfrac{36{,}755}{864} = e^{8k}$ Dividing by 864
$\ln \dfrac{36{,}755}{864} = \ln e^{8k}$ Taking the natural logarithm on both sides
$3.750456781 = 8k$ $\ln e^a = a$
$0.469 \approx k.$

The exponential growth rate is 0.469, or 46.9%, and the exponential growth function is $P(t) = 864e^{0.469t}$, where t is the number of years after 2008.

b) Since 2020 is 12 years after 2008, we substitute 12 for t:

$$P(12) = 864e^{0.469(12)} \approx 240{,}283.$$

The amount of solar power generated in 2020 will be about 240,283 thousand MWh.

8. **Global Mobile Data Traffic.** The amount of data transferred using mobile devices is expected to increase exponentially from 7.2 exabytes per month in 2016 to 49 exabytes per month in 2021.

Data: Cisco

a) Let $P(t) = P_0 e^{kt}$, where $P(t)$ is the global mobile data traffic, in exabytes per month, t years after 2016. Then $t = 0$ corresponds to 2016 and $t = 5$ corresponds to 2021. Use the data points $(0, 7.2)$ and $(5, 49)$ to find the exponential growth rate and then the exponential growth function.

b) Use the function found in part (a) to estimate the global mobile data traffic per month in 2019.

c) Assuming exponential growth continues at the same rate, predict the year in which the global mobile data traffic will be 100 exabytes per month.

c) To determine the year in which 1 billion MWh of solar power will be generated, we first note that 1 billion is 1 million · 1 thousand. Thus we can substitute 1,000,000 for $P(t)$ and solve for t:

$$1{,}000{,}000 = 864 e^{0.469t}$$

$$\frac{1{,}000{,}000}{864} = e^{0.469t} \quad \text{Dividing by 864}$$

$$\ln \frac{1{,}000{,}000}{864} = \ln e^{0.469t} \quad \text{Taking the natural logarithm on both sides}$$

$$7.053937789 \approx 0.469t \quad \ln e^a = a$$

$$15 \approx t.$$

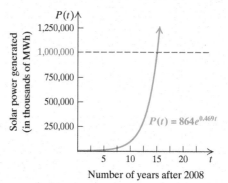

We see that, according to this model, 15 years after 2008, or in 2023, there will be about 1,000,000 thousand, or 1 billion, MWh of solar power generated in the United States.

◀ **Do Exercise 8.**

In some real-life situations, a quantity or a population is *decreasing* or *decaying* exponentially.

EXPONENTIAL DECAY MODEL

An **exponential decay model** is a function of the form

$$P(t) = P_0 e^{-kt}, \quad k > 0,$$

where P_0 is the quantity present at time 0, $P(t)$ is the amount present at time t, and k is the **decay rate**. The **half-life** is the amount of time necessary for half of the quantity to decay.

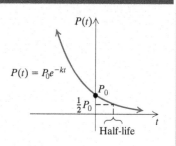

Answers
8. (a) $k \approx 0.384$; $P(t) = 7.2 e^{0.384t}$;
(b) about 22.8 exabytes per month;
(c) 2023

EXAMPLE 7 *Carbon Dating.* The radioactive element carbon-14 has a half-life of 5750 years. The percentage of carbon-14 in the remains of organic matter can be used to determine the age of that material. Recently, near Patuxent River, Maryland, archaeologists discovered charcoal that had lost 8.1% of its carbon-14. The age of this charcoal was evidence that this is the oldest dwelling ever discovered in Maryland. What was the age of the charcoal?

Data: Roylance, Frank D., "Digging Where Indians Camped Before Columbus," *The Baltimore Sun*, July 2, 2009

Reprinted with permission of The Baltimore Sun Media Group. All rights reserved.

We first find k. To do so, we use the concept of half-life. When $t = 5750$ (the half-life), $P(t)$ will be half of P_0. Then

$$0.5P_0 = P_0 e^{-k(5750)}$$
$$0.5 = e^{-5750k} \quad \text{Dividing by } P_0 \text{ on both sides}$$
$$\ln 0.5 = \ln e^{-5750k} \quad \text{Taking the natural logarithm on both sides}$$
$$\ln 0.5 = -5750k$$
$$\frac{\ln 0.5}{-5750} = k$$
$$0.00012 \approx k.$$

Now we have a function for the decay of carbon-14:

$$P(t) = P_0 e^{-0.00012t}. \quad \text{This completes the first part of our solution.}$$

(*Note:* This equation can be used for any subsequent carbon-dating problem.)

If the charcoal has lost 8.1% of its carbon-14 from an initial amount P_0, then 100% − 8.1%, or 91.9%, of P_0 is still present. To find the age t of the charcoal, we solve this equation for t:

$$0.919 P_0 = P_0 e^{-0.00012t} \quad \text{We want to find } t \text{ for which } P(t) = 0.919 P_0.$$
$$0.919 = e^{-0.00012t} \quad \text{Dividing by } P_0 \text{ on both sides}$$
$$\ln 0.919 = \ln e^{-0.00012t} \quad \text{Taking the natural logarithm on both sides}$$
$$\ln 0.919 = -0.00012t \quad \ln e^a = a$$
$$\frac{\ln 0.919}{-0.00012} = t \quad \text{Dividing by } -0.00012 \text{ on both sides}$$
$$700 \approx t. \quad \text{Using a calculator and rounding}$$

The charcoal is about 700 years old.

Do Exercise 9.

9. *Carbon Dating.* In Chaco Canyon, New Mexico, archaeologists found corn pollen that had lost 38.1% of its carbon-14. What was the age of the pollen?

Answer
9. About 4000 years

Translating for Success

The goal of these matching questions is to practice step (2), Translate, of the five-step problem-solving process. Translate each word problem to an equation or a system of equations and select a correct translation from equations A–O.

A. $M(t) = 30{,}000e^{0.207t}$

B. $40x = 50(x - 3)$

C. $x^2 + (x - 10)^2 = 50^2$

D. $\dfrac{8}{x} + \dfrac{8}{4x} = 1$

E. $x + 8\%x = 500$

F. $\dfrac{500}{x} + \dfrac{500}{x - 2} = 8$

G. $x + y = 90,$
$0.1x + 0.25y = 16.50$

H. $x + (x + 1) + (x + 2) = 39$

I. $x + (x + 8) + \dfrac{4}{5}x = 1086$

J. $x + (x + 2) + (x + 4) = 39$

K. $M(t) = 20.7e^{30{,}000t}$

L. $x^2 + (x + 8)^2 = 1086$

M. $2x + 2(x - 10) = 50$

N. $\dfrac{500}{x} = \dfrac{500}{x + 2} + 8$

O. $x + y = 90,$
$0.1x + 0.25y = 1650$

Answers on page A-43

1. *Grain Flow.* Grain flows through spout A four times as fast as through spout B. When grain flows through both spouts, a grain bin is filled in 8 hr. How many hours would it take to fill the bin if grain flows through spout B alone?

2. *Rectangle Dimensions.* The perimeter of a rectangle is 50 ft. The width of the rectangle is 10 ft shorter than the length. Find the length and the width.

3. *Wire Cutting.* A 1086-in. wire is cut into three pieces. The second piece is 8 in. longer than the first. The third is four-fifths as long as the first. How long is each piece?

4. *Movies.* In 2015, there were 30,000 movie screens in China, and this number was increasing exponentially at a rate of 20.7% per year. Write an exponential growth function M for which $M(t)$ approximates the number of movie screens in China t years after 2015.

5. *Charitable Contributions.* In 2017, Kylie donated $500 to charities. This was an 8% increase over her donations in 2015. How much did Kylie donate to charities in 2015?

6. *Uniform Numbers.* The numbers on three baseball uniforms are consecutive integers whose sum is 39. Find the integers.

7. *Triangle Dimensions.* The hypotenuse of a right triangle is 50 ft. The length of one leg is 10 ft shorter than the other. Find the lengths of the legs.

8. *Coin Mixture.* A collection of dimes and quarters is worth $16.50. There are 90 coins in all. How many of each coin are there?

9. *Car Travel.* Emma drove her car 500 mi to see her friend. The return trip was 2 hr faster at a speed that was 8 mph more. Find her return speed.

10. *Train Travel.* An Amtrak train leaves a station and travels east at 40 mph. Three hours later, a second train leaves on a parallel track traveling east at 50 mph. After what amount of time will the second train overtake the first?

12.7 Exercise Set

✓ Check Your Understanding

Reading Check For the exponential growth model $P(t) = P_0 e^{kt}$, $k > 0$, match each variable with its description.

RC1. ____ k

RC2. ____ $P(t)$

RC3. ____ P_0

RC4. ____ T, where $2P_0 = P_0 e^{kT}$

a) Doubling time
b) Exponential growth rate
c) Population at time 0
d) Population at time t

Concept Check A population numbering 550 in 2018 grew at an annual exponential growth rate of 13%. Use the model $P(t) = P_0 e^{kt}$ to find the population in 2020.

CC1. What value should you substitute for P_0?

CC2. What value should you substitute for k?

CC3. What value should you substitute for t?

CC4. What is the population in 2020?

a Solve.

Sound Levels. Use the decibel formula from Example 1 for Exercises 1–4.

1. *Blue Whale.* The blue whale is not only the largest animal on earth, it is also the loudest. The call of a blue whale can reach an intensity of $10^{6.8}$ W/m². What is this sound level, in decibels?

2. *Jackhammer Noise.* A jackhammer can generate sound measurements of 130 dB. What is the intensity of such sounds?

3. *Dishwasher Noise.* A top-of-the-line dishwasher, built to muffle noise, has a sound measurement of 45 dB. A less-expensive dishwasher can have a sound measurement of 60 dB. What is the intensity of each sound?

4. *Sound of an Alarm Clock.* The intensity of sound of an alarm clock is 10^{-4} W/m². What is this sound level, in decibels?

pH. Use the pH formula from Example 2 for Exercises 5–8.

5. *Milk.* The hydrogen ion concentration of milk is about 1.6×10^{-7} moles per liter. Find the pH.

6. *Mouthwash.* The hydrogen ion concentration of mouthwash is about 6.3×10^{-7} moles per liter. Find the pH.

7. *Alkalosis.* When the pH of a person's blood rises above 7.4, a condition called *alkalosis* sets in. Alkalosis can be fatal at a pH level above 7.8. What would the hydrogen ion concentration of the person's blood be at that point?

8. *Orange Juice.* The pH of orange juice is 3.2. What is its hydrogen ion concentration?

Walking Speed. In a study by psychologists Bornstein and Bornstein, it was found that the average walking speed w, in feet per second, of a person living in a city of population P, in thousands, is given by the function

$$w(P) = 0.37 \ln P + 0.05.$$

Data: *International Journal of Psychology*

In Exercises 9–12, various cities and their populations are given. Find the walking speed of people in each city.

9. Minneapolis, Minnesota: 382,600

10. Tallahassee, Florida: 189,900

11. New York City, New York: 8,537,700

12. Phoenix, Arizona: 1,615,100

 Solve.

13. *Otter Population.* Due primarily to the irresponsible use of chemicals, England's otter population declined dramatically during the twentieth century until otters could be found on only a small percentage of riverbanks. After chemical bans were put in place, the number of English riverside sites $R(t)$ occupied by otters rose significantly and can be approximated by the exponential function

$$R(t) = 294(1.074)^t,$$

where t is the number of years since 1985.
Data: Nicolson, Adam, "The Sultans of Streams," *National Geographic* 223(2), February 2013, pp. 124–134.

a) How many riverside sites were occupied by otters in 2010?
b) In what year were there otters on 1000 sites?
c) What is the doubling time for the number of English riverside sites occupied by otters?

14. *Code-School Graduates.* A need for software developers has led to the rise of code schools, which typically offer 12-week training courses in coding skills. The annual number of code-school graduates $G(t)$ in the United States is increasing exponentially and can be approximated by the exponential function

$$G(t) = 9000(2)^t,$$

where t is the number of years after 2015.
Data: *The Wall Street Journal*, 2/26/2017

a) How many code-school graduates were there in 2017?
b) In what year will there be 144,000 code-school graduates?
c) What is the doubling time for the annual number of code-school graduates?

15. *Drug Overdose.* Although primarily used for relief of pain, opioids have addictive properties that often make withdrawal difficult. Deaths through unintentional opioid overdose have risen sharply across the United States since the early 2000s. Ohio, in particular, has an exceptionally high rate of opioid overdose death. The number of deaths in Ohio in year t, where t is the number of years after 2003, can be approximated by

$$D(t) = 318(1.19)^t.$$

Data: *The Wall Street Journal*, 6/7/2017

a) How many unintentional opioid overdose deaths were there in Ohio in 2013?
b) In what year were there about 3000 unintentional opioid overdose deaths in Ohio?
c) What is the doubling time of unintentional opioid overdose deaths in Ohio?

16. *Salvage Value.* A color photocopier is purchased for $4800. Its value each year is about 70% of its value in the preceding year. Its salvage value, in dollars, after t years is given by the exponential function

$$V(t) = 4800(0.7)^t.$$

a) Find the salvage value of the copier after 3 years.
b) After what amount of time will the salvage value be $1200?
c) After what amount of time will the salvage value be half the original value?

Growth. Use the exponential growth model $P(t) = P_0 e^{kt}$ for Exercises 17–22.

17. *Interest Compounded Continuously.* Suppose that P_0 is invested in a savings account in which interest is compounded continuously at 3% per year.
  a) Express $P(t)$ in terms of P_0 and 0.03.
  b) Suppose that $5000 is invested. What is the balance after 1 year? 2 years? 10 years?
  c) When will the investment of $5000 double itself?

18. *Interest Compounded Continuously.* Suppose that P_0 is invested in a savings account in which interest is compounded continuously at 5.4% per year.
  a) Express $P(t)$ in terms of P_0 and 0.054.
  b) Suppose that $10,000 is invested. What is the balance after 1 year? 2 years? 10 years?
  c) When will the investment of $10,000 double itself?

19. *World Population Growth.* In 2017, the population of the world reached 7.5 billion, and the exponential growth rate was 1.06% per year.

  Data: *CIA World Factbook*

  a) Find the exponential growth function.
  b) What will the world population be in 2025?
  c) In what year will the world population reach 15 billion?
  d) What is the doubling time of the world population?

20. *Population Growth.* In 2017, the population of the United States was 324 million, and the exponential growth rate was 0.81% per year.

  Data: *CIA World Factbook*

  a) Find the exponential growth function.
  b) What will the U.S. population be in 2025?
  c) In what year will the U.S. population reach 350 million?
  d) What is the doubling time of the U.S. population?

21. *Employment.* The number of full-time and part-time employees of the online retailer Amazon increased exponentially from 56 thousand in 2011 to 306.8 thousand in 2016.

  Data: *USA Today,* 1/17/2017

  a) Let $t = 0$ correspond to 2011 and $t = 5$ correspond to 2016. Then t is the number of years after 2011. Use the data points $(0, 56)$ and $(5, 306.8)$ to find the exponential growth rate and fit an exponential growth function $P(t) = P_0 e^{kt}$ to the data, where $P(t)$ is the number of Amazon employees, in thousands, t years after 2011.
  b) Use the function found in part (a) to estimate the number of Amazon employees in 2018.
  c) According to this model, when will there be one million Amazon employees?

22. *U.S. Tax Code.* Over the years, the U.S. tax code has become increasingly complex. For example, the length of the instruction book for the 1040 tax form increased exponentially from 2 pages in 1935 to 211 pages in 2015.

  Data: National Taxpayers Union

  a) Let $t = 0$ correspond to 1935 and $t = 80$ correspond to 2015. Then t is the number of years after 1935. Use the data points $(0, 2)$ and $(80, 211)$ to find the exponential growth rate and fit an exponential growth function $C(t) = C_0 e^{kt}$ to the data, where $C(t)$ is the number of pages in the instruction book.
  b) Use the function found in part (a) to estimate the total number of pages in the instruction book in 2020.
  c) When will there be 300 pages in the instruction book?

Carbon Dating. Use the carbon-14 decay function $P(t) = P_0 e^{-0.00012t}$ for Exercises 23 and 24.

23. *Carbon Dating.* When archaeologists found the Dead Sea scrolls, they determined that the linen wrapping had lost 22.3% of its carbon-14. How old was the linen wrapping?

24. *Carbon Dating.* In 2005, researchers were able to start a date tree from a seed found at Masada, in Israel. The seed had lost 21.3% of its carbon-14. How old was the seed?

Decay. Use the exponential decay function $P(t) = P_0 e^{-kt}$ for Exercises 25 and 26.

25. *Chemistry.* The decay rate of iodine-131 is 9.6% per day. What is the half-life?

26. *Chemistry.* The decay rate of krypton-85 is 6.3% per day. What is the half-life?

27. *Home Construction.* The chemical urea formaldehyde was found in some insulation used in houses built during the mid to late 1960s. Unknown at the time was the fact that urea formaldehyde emitted toxic fumes as it decayed. The half-life of urea formaldehyde is 1 year. What is its decay rate?

28. *Plumbing.* Lead pipes and solder are often found in older buildings. Unfortunately, as lead decays, toxic chemicals can get in the water resting in the pipes. The half-life of lead is 22 years. What is its decay rate?

29. *Physical Music Sales.* The number of albums sold in a physical format (such as a compact disc) has decreased exponentially from 451 million albums in 2007 to 118 million albums in 2016.

Data: Nielsen Music Industry Reports

a) Find the exponential decay rate, and write an exponential function that represents the number of albums $A(t)$, in millions, sold in a physical format t years after 2007.
b) Estimate the number of albums sold in a physical format in 2020.
c) In what year were 300 million albums sold in a physical format?

30. *Covered Bridges.* There were as many as 15,000 covered bridges in the United States in the 1800s. Now their number is decreasing exponentially, partly as a result of vandalism. In 1965, there were 1156 covered bridges, but by 2017, only 804 covered bridges remained.

Data: National Society for the Preservation of Covered Bridges; woodcenter.org

a) Find the exponential decay rate, and write an exponential function B that represents the number of covered bridges t years after 1965.
b) Estimate the number of covered bridges in 2002.
c) In what year were there 1000 covered bridges?

31. *Population Decline.* The population of Cleveland, Ohio, declined from 396,000 in 2010 to 390,000 in 2017. Assume that the population decreases according to the exponential decay model.

Data: worldpopulationreview.com

a) Find the exponential decay rate, and write an exponential function that represents the population of Cleveland t years after 2010.
b) Estimate the population of Cleveland in 2022.
c) In what year will the population of Cleveland reach 380,000?

32. *Online Shopping.* Revenue from online grocery shopping in the United States is expected to increase exponentially from $35 billion in 2014 to $175 billion in 2021.

Data: The Wall Street Journal, 6/19/2017

a) Find the exponential growth rate, and write a function that represents revenue from online grocery shopping t years after 2014.
b) Estimate the revenue from online grocery shopping in 2018.
c) In what year will revenue from online grocery shopping reach $300 billion?

33. *Contemporary Art.* In 2017, Jean-Michel Basquiat's "Untitled" was sold at auction for $110.5 million. The painting had previously been sold for $19,000 in 1984. Assume that the painting's value increases exponentially.

Data: news.artnet.com, 4/19/2017; *The New York Times*, 5/18/2017

a) Find the exponential growth rate k, and determine the exponential growth function that can be used to estimate the painting's value $V(t)$, in dollars, t years after 1984.
b) Estimate the value of the painting in 2020.
c) What is the doubling time for the value of the painting?
d) How long after 1984 will the value of the painting be $1 billion?

34. *Classic Cars.* In 2014, a 1962 Ferrari 250 GTO (chassis 3851GT) was sold at auction for a record $38 million. The car had previously been sold for $4000 in 1965. Assume that the car's value increases exponentially.

Data: *Los Angeles Times*, 8/14/2014

a) Find the exponential growth rate k, and determine the exponential growth function that can be used to estimate the car's value $V(t)$, in dollars, t years after 1965.
b) Estimate the value of the car in 2020.
c) What is the doubling time for the value of the car?
d) How many years after 1965 will the value of the car be $150 million?

Skill Maintenance

Solve.

35. $5x + 6y = -2,$
$3x + 10y = 2$ [8.2a], [8.3a]

36. $x + y - z = 0,$
$3x + y + z = 6,$
$x - y + 2z = 5$ [8.5a]

37. $x^2 + 2x + 3 = 0$ [11.2a]

38. $\dfrac{6}{x} + \dfrac{6}{x+2} = \dfrac{5}{2}$ [6.7a]

39. $\dfrac{7}{x^2 - 5x} - \dfrac{2}{x-5} = \dfrac{4}{x}$ [6.7a]

40. $15x^2 + 45 = 0$ [11.1a]

Synthesis

Use a graphing calculator to solve each of the following equations.

41. $2^x = x^{10}$

42. $(\ln 2)x = 10 \ln x$

43. $x^2 = 2^x$

44. $x^3 = e^x$

45. *Nuclear Energy.* Plutonium-239 (Pu-239) is used in nuclear energy plants. The half-life of Pu-239 is 24,360 years. How long will it take for a fuel rod of Pu-239 to lose 90% of its radioactivity?

Data: *Microsoft Encarta 97 Encyclopedia*

46. *Population Growth.* In 2016, the population of China was 1.4 billion, and the exponential growth rate was 0.43%. In the same year, the population of India was 1.3 billion, and the exponential growth rate was 1.19%.

Data: *CIA World Factbook*

a) Find an exponential growth function $C(t)$ that can be used to model the population of China t years after 2016.
b) Find an exponential growth function $I(t)$ that can be used to model the population of India t years after 2016.
c) Use a graphing calculator to estimate the year in which the population of China and the population of India will be the same.

CHAPTER 12 Summary and Review

Key Formulas

Exponential Growth:	$P(t) = P_0 e^{kt}$
Exponential Decay:	$P(t) = P_0 e^{-kt}$
Carbon Dating:	$P(t) = P_0 e^{-0.00012t}$
Interest Compounded Annually:	$A = P(1 + r)^t$
Interest Compounded n Times per Year:	$A = P\left(1 + \dfrac{r}{n}\right)^{nt}$
Interest Compounded Continuously:	$P(t) = P_0 e^{kt}$, where P_0 dollars are invested for t years at interest rate k

Vocabulary Reinforcement

Complete each statement with the correct term from the column on the right. Some of the choices may not be used.

1. The function given by $f(x) = 6^x$ is an example of a(n) _____ function. [12.1a]

2. The _____ of a function given by a set of ordered pairs is found by interchanging the first and second coordinates in each ordered pair. [12.2a]

3. When interest is paid on interest previously earned, it is called _____ interest. [12.1c]

4. Base-10 logarithms are called _____ logarithms. [12.3d]

5. The logarithm of a number is a(n) _____. [12.3a]

6. A quantity's _____ is the amount of time necessary for half of the quantity to decay. [12.7b]

exponential
logarithmic
common
natural
composition
compound
inverse
doubling time
half-life
base
exponent

Concept Reinforcement

Determine whether each statement is true or false.

_____ 1. The y-intercept of a function $f(x) = a^x$ is $(0, 1)$. [12.1a]

_____ 2. If it is possible for a horizontal line to intersect the graph of a function more than once, its inverse is a function. [12.2c]

_____ 3. $\log_a 1 = 0$, $a > 0$ [12.3c]

_____ 4. If we find that $\log(78) \approx 1.8921$ on a calculator, we also know that $10^{1.8921} \approx 78$. [12.3d]

_____ 5. $\ln(35) = \ln 7 \cdot \ln 5$ [12.4a]

_____ 6. The functions $f(x) = e^x$ and $g(x) = \ln x$ are inverses of each other. [12.5a]

Study Guide

Objective 12.1a Graph exponential equations and functions.

Example Graph: $f(x) = 4^x$.

We compute some function values and list the results in a table:

$f(-2) = 4^{-2} = \dfrac{1}{4^2} = \dfrac{1}{16};$

$f(-1) = 4^{-1} = \dfrac{1}{4};$

$f(0) = 4^0 = 1;$

$f(1) = 4^1 = 4;$

$f(2) = 4^2 = 16.$

x	$f(x)$
-2	$\dfrac{1}{16}$
-1	$\dfrac{1}{4}$
0	1
1	4
2	16

Now we plot the points $(x, f(x))$ and connect them with a smooth curve.

Practice Exercise

1. Graph: $f(x) = 2^x$.

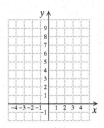

Objective 12.2a Find the composition of functions and express certain functions as a composition of functions.

Example Given $f(x) = x - 2$ and $g(x) = x^2$, find $(f \circ g)(x)$ and $(g \circ f)(x)$.

$(f \circ g)(x) = f(g(x))$
$= f(x^2) = x^2 - 2;$

$(g \circ f)(x) = g(f(x))$
$= g(x - 2) = (x - 2)^2$
$= x^2 - 4x + 4$

Example Find $f(x)$ and $g(x)$ such that $h(x) = (f \circ g)(x)$:

$h(x) = \sqrt[3]{x - 5}.$

Two functions that can be used are $f(x) = \sqrt[3]{x}$ and $g(x) = x - 5$. There are other correct answers.

Practice Exercises

2. Given $f(x) = 2x$ and $g(x) = 4x + 1$, find $(f \circ g)(x)$ and $(g \circ f)(x)$.

3. Find $f(x)$ and $g(x)$ such that $h(x) = (f \circ g)(x)$:

$h(x) = \dfrac{1}{3x + 2}.$

Objective 12.2c Given a function, determine whether it is one-to-one and has an inverse that is a function.

Example Determine whether the function $f(x) = x + 5$ is one-to-one and thus has an inverse that is also a function.

The graph of $f(x) = x + 5$ is shown below. No horizontal line crosses the graph more than once, so the function is one-to-one and has an inverse that is a function.

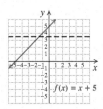

If there is a horizontal line that crosses the graph of a function more than once, the function is not one-to-one and does not have an inverse that is a function.

Practice Exercise

4. Determine whether the function $f(x) = 3^x$ is one-to-one.

Objective 12.2d Find a formula for the inverse of a function, if it exists, and graph inverse relations and functions.

Example Determine whether the function $f(x) = 3x - 1$ is one-to-one. If it is, find a formula for its inverse.

The graph of $f(x) = 3x - 1$ passes the horizontal-line test, so it is one-to-one. Now we find a formula for $f^{-1}(x)$.

1. Replace $f(x)$ with y: $y = 3x - 1$.
2. Interchange x and y: $x = 3y - 1$.
3. Solve for y: $x + 1 = 3y$

$$\frac{x + 1}{3} = y.$$

4. Replace y with $f^{-1}(x)$: $f^{-1}(x) = \frac{x + 1}{3}$.

Example Graph the one-to-one function $g(x) = x - 3$ and its inverse using the same set of axes.

We graph $g(x) = x - 3$ and then draw its reflection across the line $y = x$.

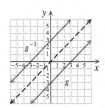

Practice Exercises

5. Determine whether the function $g(x) = 4 - x$ is one-to-one. If it is, find a formula for its inverse.

6. Graph the one-to-one function $f(x) = 2x + 1$ and its inverse using the same set of axes.

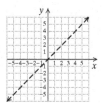

Objective 12.3a Graph logarithmic functions.

Example Graph: $y = f(x) = \log_4 x$.

The equation $y = \log_4 x$ is equivalent to $4^y = x$.

For $y = -2$, $x = 4^{-2} = \dfrac{1}{4^2} = \dfrac{1}{16}$.

For $y = -1$, $x = 4^{-1} = \dfrac{1}{4}$.

For $y = 0$, $x = 4^0 = 1$.

For $y = 1$, $x = 4^1 = 4$.

For $y = 2$, $x = 4^2 = 16$.

x	y
$\frac{1}{16}$	-2
$\frac{1}{4}$	-1
1	0
4	1
16	2

Now we plot these points and connect them with a smooth curve.

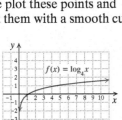

Practice Exercise

7. Graph: $y = \log_5 x$.

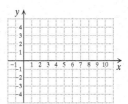

Objective 12.4d Convert from logarithms of products, quotients, and powers to expressions in terms of individual logarithms, and conversely.

Example Express

$$\log_a \frac{x^2 y}{z^3}$$

in terms of logarithms of x, y, and z.

$$\log_a \frac{x^2 y}{z^3} = \log_a (x^2 y) - \log_a z^3$$
$$= \log_a x^2 + \log_a y - \log_a z^3$$
$$= 2\log_a x + \log_a y - 3\log_a z$$

Example Express

$$4\log_a x - \frac{1}{2}\log_a y$$

as a single logarithm.

$$4\log_a x - \frac{1}{2}\log_a y = \log_a x^4 - \log_a y^{1/2}$$
$$= \log_a \frac{x^4}{y^{1/2}}, \text{ or } \log_a \frac{x^4}{\sqrt{y}}$$

Practice Exercises

8. Express $\log_a \sqrt[5]{\dfrac{x^3}{y^2}}$ in terms of logarithms of x and y.

9. Express $\dfrac{1}{2}\log_a x - 3\log_a y$ as a single logarithm.

Objective 12.5c Graph exponential functions and logarithmic functions, base e.

Example Graph: $f(x) = e^{x-1}$.

x	$f(x)$
-1	0.1
0	0.4
1	1
2	2.7

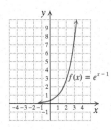

Practice Exercises

10. Graph: $f(x) = e^x - 1$.

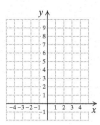

Example Graph: $g(x) = \ln x + 3$.

x	$g(x)$
0.5	2.3
1	3
3	4.1
5	4.6
8	5.1
10	5.3

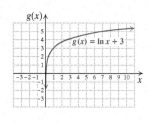

11. Graph: $f(x) = \ln(x + 3)$.

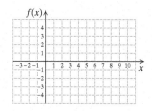

Objective 12.6a Solve exponential equations.

Example Solve: $3^{x-1} = 81$.

$$3^{x-1} = 81$$
$$3^{x-1} = 3^4$$

Since the bases are the same, the exponents must be the same:

$$x - 1 = 4$$
$$x = 5.$$

The solution is 5.

Practice Exercise

12. Solve: $2^{3x} = 16$.

Objective 12.6b Solve logarithmic equations.

Example Solve: $\log x + \log(x + 3) = 1$.

$$\log x + \log(x + 3) = 1$$
$$\log_{10}[x(x + 3)] = 1$$
$$x(x + 3) = 10^1$$
$$x^2 + 3x = 10$$
$$x^2 + 3x - 10 = 0$$
$$(x + 5)(x - 2) = 0$$
$$x + 5 = 0 \quad \text{or} \quad x - 2 = 0$$
$$x = -5 \quad \text{or} \quad x = 2$$

The number -5 does not check, but 2 does. The solution is 2.

Practice Exercise

13. Solve: $\log_3(2x + 3) = 2$.

Review Exercises

1. Find the inverse of the relation
$\{(-4, 2), (5, -7), (-1, -2), (10, 11)\}$. [12.2b]

Determine whether each function is one-to-one. If it is, find a formula for its inverse. [12.2c, d]

2. $f(x) = 4 - x^2$

3. $g(x) = \dfrac{2x - 3}{7}$

4. $f(x) = 8x^3$

5. $f(x) = \dfrac{4}{3 - 2x}$

6. Graph the function $f(x) = x^3 + 1$ and its inverse using the same set of axes. [12.2d]

Graph.

7. $f(x) = 3^{x-1}$ [12.1a]

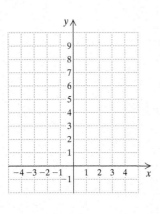

8. $f(x) = \log_3 x$, or $y = \log_3 x$ [12.3a]
$y = \log_3 x \rightarrow x = $ _____

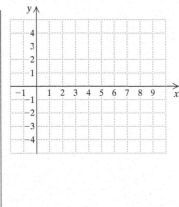

9. $f(x) = e^{x+1}$ [12.5c]

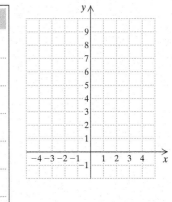

10. $f(x) = \ln(x - 1)$ [12.5c]

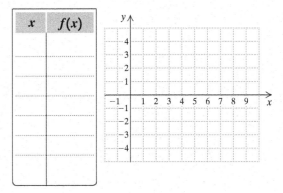

11. Find $(f \circ g)(x)$ and $(g \circ f)(x)$ if $f(x) = x^2$ and $g(x) = 3x - 5$. [12.2a]

12. If $h(x) = \sqrt{4 - 7x}$, find $f(x)$ and $g(x)$ such that $h(x) = (f \circ g)(x)$. [12.2a]

Convert to a logarithmic equation. [12.3b]

13. $10^4 = 10{,}000$

14. $25^{1/2} = 5$

Convert to an exponential equation. [12.3b]

15. $\log_4 16 = x$

16. $\log_{1/2} 8 = -3$

Find each of the following. [12.3c]

17. $\log_3 9$

18. $\log_{10} \dfrac{1}{10}$

19. $\log_m m$

20. $\log_m 1$

Find the common logarithm, to four decimal places, using a calculator. [12.3d]

21. $\log\left(\dfrac{78}{43,112}\right)$ **22.** $\log(-4)$

Express in terms of logarithms of x, y, and z. [12.4d]

23. $\log_a x^4 y^2 z^3$ **24.** $\log \sqrt[4]{\dfrac{z^2}{x^3 y}}$

Express as a single logarithm. [12.4d]

25. $\log_a 8 + \log_a 15$

26. $\dfrac{1}{2} \log a - \log b - 2 \log c$

Simplify. [12.4e]

27. $\log_m m^{17}$ **28.** $\log_m m^{-7}$

Given $\log_a 2 = 1.8301$ and $\log_a 7 = 5.0999$, find each of the following. [12.4d]

29. $\log_a 28$ **30.** $\log_a 3.5$

31. $\log_a \sqrt{7}$ **32.** $\log_a \dfrac{1}{4}$

Find each of the following, to four decimal places, using a calculator. [12.5a]

33. $\ln 0.06774$ **34.** $e^{-0.98}$

35. $e^{2.91}$ **36.** $\ln 1$

37. $\ln 0$ **38.** $\ln e$

Find each logarithm using the change-of-base formula. [12.5b]

39. $\log_5 2$ **40.** $\log_{12} 70$

Solve. Where appropriate, give approximations to four decimal places. [12.6a, b]

41. $\log_3 x = -2$ **42.** $\log_x 32 = 5$

43. $\log x = -4$ **44.** $3 \ln x = -6$

45. $4^{2x-5} = 16$ **46.** $2^{x^2} \cdot 2^{4x} = 32$

47. $4^x = 8.3$ **48.** $e^{-0.1t} = 0.03$

49. $\log_4 16 = x$

50. $\log_4 x + \log_4 (x - 6) = 2$

51. $\log_2 (x + 3) - \log_2 (x - 3) = 4$

52. $\log_3 (x - 4) = 3 - \log_3 (x + 4)$

Solve. [12.7a, b]

53. *Sound Level.* The intensity of sound of a symphony orchestra playing at its peak can reach $10^{1.7}\ \text{W/m}^2$. How high is this sound level, in decibels? (Use $L = 10 \cdot \log(I/I_0)$ and $I_0 = 10^{-12}\ \text{W/m}^2$.)

54. Internet Traffic. Internet traffic is increasing exponentially. The number of petabytes (PB) of global consumer Internet usage per month $U(t)$ can be approximated by the exponential function
$$U(t) = 5204(1.47)^t,$$
where t is the number of years after 2010.

Data: Cisco

a) Estimate the global consumer Internet usage per month in 2010, in 2015, and in 2017.
b) In what year will the global consumer Internet usage per month be 250,000 PB?
c) What is the doubling time for global consumer Interner usage per month?
d) Graph the function.

55. Investment. In 2015, Lucy invested $40,000 in a mutual fund. By 2018, the value of her investment was $53,000. Assume that the value of her investment increased exponentially.

a) Find the value of k, and write an exponential function that describes the value of Lucy's investment t years after 2015.
b) Predict the value of her investment in 2025.
c) In what year will the value of her investment first reach $85,000?

56. The population of a colony of bacteria doubled in 3 days. What was the exponential growth rate?

57. How long will it take $7600 to double itself if it is invested at 3.4%, compounded continuously?

58. How old is a skeleton that has lost 34% of its carbon-14? (Use $P(t) = P_0 e^{-0.00012t}$.)

59. What is the inverse of the function $f(x) = 5^x$, if it exists? [12.3a]

A. $f^{-1}(x) = x^5$ **B.** $f^{-1}(x) = \log_x 5$
C. $f^{-1}(x) = \log_5 x$ **D.** Does not exist

60. Solve: $\log(x^2 - 9) - \log(x + 3) = 1$. [12.6b]

A. 4 **B.** 5
C. 7 **D.** 13

Synthesis

Solve. [12.6a, b]

61. $\ln(\ln x) = 3$

62. $5^{x+y} = 25$, $2^{2x-y} = 64$

Understanding Through Discussion and Writing

1. Explain how the graph of $f(x) = e^x$ could be used to obtain the graph of $g(x) = 1 + \ln x$. [12.2d], [12.5a]

2. Christina first determines that the solution of $\log_3(x + 4) = 1$ is -1, but then rejects it. What mistake do you think she might be making? [12.6b]

3. An organization determines that the cost per person of chartering a bus is given by the function
$$C(x) = \frac{100 + 5x}{x},$$
where x is the number of people in the group and $C(x)$ is in dollars. Determine $C^{-1}(x)$ and explain how this inverse function could be used. [12.2d]

4. Explain how the equation $\ln x = 3$ could be solved using the graph of $f(x) = \ln x$. [12.6b]

5. Explain why you cannot take the logarithm of a negative number. [12.3a]

6. Write a problem for a classmate to solve in which data that seem to fit an exponential growth function are provided. Try to find data in a newspaper to make the problem as realistic as possible. [12.7b]

CHAPTER 12 Test

Graph.

1. $f(x) = 2^{x+1}$

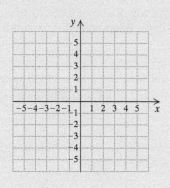

2. $y = \log_2 x$

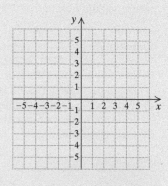

3. $f(x) = e^{x-2}$

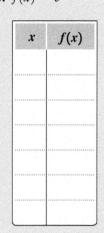 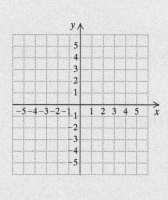

4. $f(x) = \ln(x - 4)$

 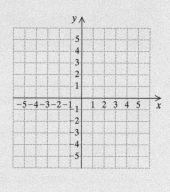

5. Find $(f \circ g)(x)$ and $(g \circ f)(x)$ if $f(x) = x + x^2$ and $g(x) = 5x - 2$.

6. Find the inverse of the relation $\{(-4, 3), (5, -8), (-1, -3), (10, 12)\}$.

Determine whether each function is one-to-one. If it is, find a formula for its inverse.

7. $f(x) = 4x - 3$

8. $f(x) = (x + 1)^3$

9. $f(x) = 2 - |x|$

10. Convert to a logarithmic equation:
$256^{1/2} = 16$.

11. Convert to an exponential equation:
$m = \log_7 49$.

Find each of the following.

12. $\log_5 125$

13. $\log_t t^{23}$

14. $\log_p 1$

Find the common logarithm, to four decimal places, using a calculator.

15. $\log 0.0123$

16. $\log (-5)$

17. Express in terms of logarithms of a, b, and c:
$$\log \frac{a^3 b^{1/2}}{c^2}.$$

18. Express as a single logarithm:
$\frac{1}{3} \log_a x - 3 \log_a y + 2 \log_a z$.

Given $\log_a 2 = 0.301$, $\log_a 6 = 0.778$, and $\log_a 7 = 0.845$, find each of the following.

19. $\log_a \frac{2}{7}$

20. $\log_a 12$

Find each of the following, to four decimal places, using a calculator.

21. $\ln 807.39$

22. $e^{4.68}$

23. $\ln 1$

24. Find $\log_{18} 31$ using the change-of-base formula.

Solve. Where appropriate, give approximations to four decimal places.

25. $\log_x 25 = 2$

26. $\log_4 x = \frac{1}{2}$

27. $\log x = 4$

28. $\ln x = \frac{1}{4}$

29. $7^x = 1.2$

30. $\log(x^2 - 1) - \log(x - 1) = 1$

31. $\log_5 x + \log_5 (x + 4) = 1$

32. *Tomatoes.* What is the pH of tomatoes if the hydrogen ion concentration is 6.3×10^{-5} moles per liter? (Use pH $= -\log[\text{H}^+]$.)

33. *Spread of a Rumor.* The number of people who hear a rumor increases exponentially. If 20 people start a rumor and if each person who hears the rumor repeats it to two people per day, the number of people N who have heard the rumor after t days is given by the function

$$N(t) = 20(3)^t.$$

a) How many people have heard the rumor after 5 days?
b) After what amount of time will 1000 people have heard the rumor?
c) What is the doubling time for the number of people who have heard the rumor?

34. *Interest Compounded Continuously.* Suppose that a $1000 investment, compounded continuously, grows to $1150.27 in 5 years.

a) Find the interest rate and the exponential growth function.
b) What is the balance after 8 years?
c) When will the balance be $1439?
d) What is the doubling time?

35. The population of Masonville grew exponentially and doubled in 23 years. What was the exponential growth rate?

36. How old is an animal bone that has lost 43% of its carbon-14? (Use $P(t) = P_0 e^{-0.00012t}$.)

37. Solve: $\log(3x - 1) + \log x = 1$.

   A. There are one positive solution and one negative solution.
   B. There is exactly one solution, and it is positive.
   C. There is exactly one solution, and it is negative.
   D. There is no solution.

Synthesis

38. Solve: $\log_3 |2x - 7| = 4$.

39. If $\log_a x = 2$, $\log_a y = 3$, and $\log_a z = 4$, find

$$\log_a \frac{\sqrt[3]{x^2 z}}{\sqrt[3]{y^2 z^{-1}}}.$$

CHAPTERS 1–12 Cumulative Review

Solve.

1. $\dfrac{1}{3}x - \dfrac{1}{5} \geq \dfrac{1}{5}x - \dfrac{1}{3}$

2. $|x| > 6.4$

3. $17 - |4 - x| = 2$

4. $3x + y = 4,$
   $-6x - y = -3$

5. $x^4 - 13x^2 + 36 = 0$

6. $2x^2 = x + 3$

7. $3x - \dfrac{6}{x} = 7$

8. $|x + 6| \leq 13$

9. $x(x + 10) = -21$

10. $2x^2 + x + 1 = 0$

11. $4(y - 1) + 6 = -8 - y$

12. $\dfrac{x + 1}{x - 2} > 0$

13. $\log_3 x = 2$

14. $x^2 - 1 \geq 0$

15. $\log_2 x + \log_2(x + 7) = 3$

16. $\sqrt{x + 5} = x - 1$

17. $|x - 2| = |3x + 1|$

18. $7^x = 30$

19. $x - y + 2z = 3,$
    $-x + z = 4,$
    $2x + y - z = -3$

20. $3 \leq 4x + 7 < 31$

21. $\dfrac{3}{x - 3} - \dfrac{x + 2}{x^2 + 2x - 15} = \dfrac{1}{x + 5}$

22. $P = \dfrac{3}{4}(M + 2N)$, for N

Solve.

23. *Oil.* Worldwide demand for oil is expected to grow exponentially. The amount of oil $N(t)$, in millions of barrels per day, demanded t years after 2000, can be approximated by
$$N(t) = 77(1.019)^t,$$
where $t = 0$ corresponds to 2000.

Data: euractiv.com

a) How much oil is projected to be demanded in 2016? in 2025?
b) What is the doubling time?
c) Graph the function.

24. *Interest Compounded Annually.* Suppose that $50,000 is invested at 4% interest, compounded annually.
  a) Find a function A for the amount in the account after t years.
  b) Find the amount of money in the account at $t = 0$, $t = 4$, $t = 8$, and $t = 10$.
  c) Graph the function.

Simplify.

25. $(2x + 3)(x^2 - 2x - 1)$

26. $(3x^2 + x^3 - 1) - (2x^3 + x + 5)$

27. $\dfrac{2m^2 + 11m - 6}{m^3 + 1} \cdot \dfrac{m^2 - m + 1}{m + 6}$

28. $\dfrac{x}{x - 1} + \dfrac{2}{x + 1} - \dfrac{2x}{x^2 - 1}$

29. $\dfrac{1 - \dfrac{5}{x}}{x - 4 - \dfrac{5}{x}}$

30. $(x^4 + 3x^3 - x + 4) \div (x + 1)$

31. $\dfrac{\sqrt{75x^5y^2}}{\sqrt{3xy}}$

32. $4\sqrt{50} - 3\sqrt{18}$

33. $(16^{3/2})^{1/2}$

34. $(2 - i\sqrt{2})(5 + 3i\sqrt{2})$

35. $\dfrac{5 + i}{2 - 4i}$

36. $\left|\dfrac{2}{3} - \dfrac{4}{5}\right|$

37. $\dfrac{63x^2y^3}{-7x^{-4}y}$

38. $1000 \div 10^2 \cdot 25 \div 4$

39. $5x - 3[4(x - 2) - 2(x + 1)]$

40. Find the x- and y-intercepts of the graph of $7x - 14 = 28y$.

41. Find the slope, if it exists, of the line containing the given points.
  a) $(-3, 13), (8, -3)$
  b) $(-1, 2), \left(-1, \dfrac{2}{5}\right)$

42. Find an equation of the line containing the points $(1, 4)$ and $(-1, 0)$.

48. $h(x) = 2 + x - x^2$

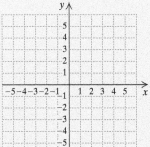

49. $x + y \leq 0,$
    $x \geq -4,$
    $y \geq -1$

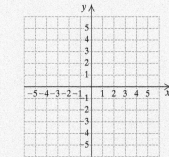

43. Find an equation of the line containing the point $(1, 2)$ and perpendicular to the line whose equation is $2x - y = 3$.

Graph.

44. $y - x < -2$

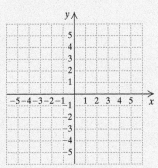

45. $y = -\dfrac{1}{2}x - 3$

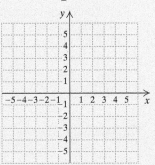

50. $x = 3.5$

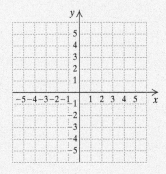

51. $f(x) = 2x^2 - 8x + 9$

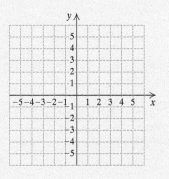

46. $4y - 3x = 12$

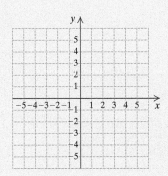

47. $y < -2$

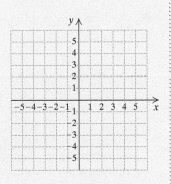

52. $f(x) = e^{-x}$

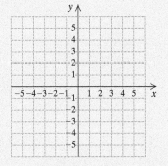

53. $f(x) = \log_2 x$

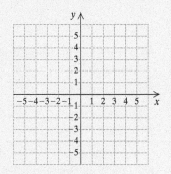

Factor.

54. $2x^4 - 12x^3 + x - 6$ **55.** $3a^2 - 12ab - 135b^2$

56. $x^2 - 17x + 72$ **57.** $81m^4 - n^4$

58. $16x^2 - 16x + 4$ **59.** $81a^3 - 24$

60. $10x^2 + 66x - 28$ **61.** $6x^3 + 27x^2 - 15x$

62. Find $f^{-1}(x)$ when $f(x) = 2x - 3$.

63. z varies directly as x and inversely as the cube of y, and $z = 5$ when $x = 4$ and $y = 2$. What is z when $x = 10$ and $y = 5$?

64. Given the function f described by $f(x) = x^3 - 2$, find $f(-2)$.

65. Given the function g described by $g(x) = -30$, find $g(0)$.

66. *Medicaid Enrollment.* The projected Medicaid enrollment, in millions, can be modeled by the linear function
$$M(t) = 1.05t + 72.65,$$
where t is the number of years after 2016.

Data: Department of Health and Human Services, 2016 Actuarial Report of the Financial Outlook of Medicaid

a) Find the projected Medicaid enrollment in 2018, in 2021, and in 2025.
b) Graph the function.

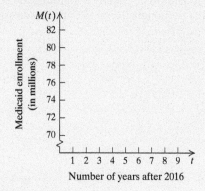

c) Find the y-intercept.
d) Find the slope.
e) Find the rate of change.

67. Rationalize the denominator: $\dfrac{5 + \sqrt{a}}{3 - \sqrt{a}}$.

68. Find the domain: $f(x) = x^2 - 2x + 3$.

69. Find the domain: $h(x) = \dfrac{2 - x}{x + 2}$.

Solve.

70. *Book Club.* A book club offers two types of membership. Limited members pay a fee of $20 per year and can buy books for $20 each. Preferred members pay $40 per year and can buy books for $15 each. For what numbers of annual book purchases would it be less expensive to be a preferred member?

71. *Train Travel.* A passenger train travels at twice the speed of a freight train. The freight train leaves a station at 2 A.M. and travels north at 34 mph. The passenger train leaves the station at 11 A.M, traveling north on a parallel track. How far from the station will the passenger train overtake the freight train?

72. *Perimeters of Polygons.* A pentagon with all five sides the same length has a perimeter equal to that of an octagon in which all eight sides are the same length. One side of the pentagon is 2 less than three times one side of the octagon. What is the perimeter of each figure?

73. *Ammonia Solutions.* A chemist has two solutions of ammonia and water. Solution A is 6% ammonia and solution B is 2% ammonia. How many liters of each solution are needed in order to obtain 80 L of a solution that is 3.2% ammonia?

74. *Air Travel.* An airplane can fly 190 mi with the wind in the same time that it takes to fly 160 mi against the wind. The speed of the wind is 30 mph. How fast can the plane fly in still air?

75. *Work.* Christy can do a certain job in 21 min. Madeline can do the same job in 14 min. How long would it take to do the job if the two worked together?

76. *Centripetal Force.* The centripetal force F of an object moving in a circle varies directly as the square of the velocity v and inversely as the radius r of the circle. If $F = 8$ when $v = 1$ and $r = 10$, what is F when $v = 2$ and $r = 16$?

77. *Maximizing Area.* A farmer wants to fence in a rectangular area next to a river. (Note that no fence will be needed along the river.) What is the area of the largest region that can be fenced in with 100 ft of fencing?

78. *Carbon Dating.* Use the function $P(t) = P_0 e^{-0.00012t}$ to find the age of a bone that has lost 25% of its carbon-14.

79. *Beam Load.* The weight W that a horizontal beam can support varies inversely as the length L of the beam. If a 14-m beam can support 1440 kg, what weight can a 6-m beam support?

80. Fit a linear function to the data points $(2, -3)$ and $(5, -4)$.

81. Fit a quadratic function to the data points $(-2, 4)$, $(-5, -6)$, and $(1, -3)$.

82. Convert to a logarithmic equation: $10^6 = r$.

83. Convert to an exponential equation: $\log_3 Q = x$.

84. Express as a single logarithm:
$\frac{1}{5}(7 \log_b x - \log_b y - 8 \log_b z)$.

85. Express in terms of logarithms of x, y, and z:
$\log_b \left(\frac{xy^5}{z}\right)^{-6}$.

86. What is the maximum product of two numbers whose sum is 26?

87. Determine whether the function $f(x) = 4 - x^2$ is one-to-one.

88. For the graph of function f shown here, determine (a) $f(2)$; (b) the domain; (c) all x-values such that $f(x) = -5$; and (d) the range.

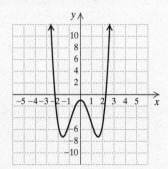

89. *Population Growth of Nevada.* Of the 50 states, Nevada had the greatest percent change in population from 2000 to 2015. In 2015, the population of Nevada was 2,890,845. It had grown from a population of 1,998,257 in 2000. Assume that the population growth increases according to an exponential growth function.

Data: U.S. Census Bureau

a) Let $t = 0$ correspond to 2000 and $t = 15$ correspond to 2015. Then t is the number of years after 2000. Use the data points $(0, 1{,}998{,}257)$ and $(15, 2{,}890{,}845)$ to find the exponential growth rate and fit an exponential growth function $P(t) = P_0 e^{kt}$ to the data, where $P(t)$ is the population of Nevada t years after 2020.
b) Use the function found in part (a) to predict the population of Nevada in 2020.
c) If growth continues at this rate, when will the population reach 4.0 million?

Synthesis

90. Solve: $\dfrac{9}{x} - \dfrac{9}{x+12} = \dfrac{108}{x^2 + 12x}$.

91. Solve: $\log_2(\log_3 x) = 2$.

92. Solve $ax^2 - bx = 0$ for x.

93. Factor: $\dfrac{a^3}{8} + \dfrac{8b^3}{729}$.

94. Simplify: $\left[\dfrac{1}{(-3)^{-2}} - (-3)^1\right] \cdot [(-3)^2 + (-3)^{-2}]$.

Appendixes

A Introductory Algebra Review
B Mean, Median, and Mode
C Synthetic Division
D Determinants and Cramer's Rule
E Elimination Using Matrices
F The Algebra of Functions
G Distance, Midpoints, and Circles

A | Introductory Algebra Review

This text is appropriate for a two-semester course that combines the study of introductory and intermediate algebra. Students who take only the second-semester course (which generally begins with Chapter 7) often need a review of the topics covered in the first-semester course. This appendix is a guide for a review of the first six chapters of this text. Below is a syllabus of selected exercises that can be used as a condensed review of the main objectives in the first half of the text. For extra help, consult the *Student's Solutions Manual,* which contains fully worked-out solutions with step-by-step annotations for all the odd-numbered exercises in the exercise sets.

SECTION/OBJECTIVE	EXAMPLES	EXERCISES IN EXERCISE SET
1.3a	5–13	3, 11, 19, 25, 33, 39, 43
1.4a	6–12	5, 9, 15, 19, 21, 51, 55, 69, 81
1.5a	4–10, 15	15, 27, 31, 35, 45, 65
1.6c	19–23	49, 53, 55, 57
1.7c	14–17	47, 49, 59
1.7d	28–31	73, 89, 91
1.7e	32–38	99, 107, 111, 117
1.8b	7, 11, 12	15, 21
1.8c	16	29, 35
1.8d	20–23	41, 51, 61, 81
2.1b	7, 8	19, 39, 47
2.2a	2, 3, 6	3, 7, 31, 37
2.3a	2	5, 15
2.3b	7, 9	23, 43, 51
2.3c	11, 12, 13	71, 77, 83, 87
2.5a	4, 5, 6, 8	7, 15, 25, 35
2.6a	1, 2, 5, 8	1, 7, 15, 27
2.7e	13, 16	53, 61, 75
3.2a	2, 3, 5	5, 13, 21
3.3a	1	21, 25
3.3b	3, 4	43, 49
3.4a	1	9, 11
3.4b	5, 6, 7, 8	27, 39, 43
4.1b	2	15, 17
4.1c	5, 6	27, 35
4.1d, e, f	8, 10, 12, 15, 22–27	63, 71, 81, 87, 93, 97
4.2a, b	1–4, 8, 12, 14	3, 35, 41, 45
4.4a	3, 4	7, 17
4.4c	8, 9	33, 41, 49
4.5d	9	63
4.6a	2–6	9, 29
4.6b	11–13	41, 45
4.6c	16–18	61, 65, 69
4.7c	4–6	21, 27

(continued)

SECTION/OBJECTIVE	EXAMPLES	EXERCISES IN EXERCISE SET
4.7f	11, 12	41, 57
4.8b	7, 8, 10	25, 29, 33, 43
5.1b	8–11	25, 33
5.1c	13, 14, 17	49, 57
5.2a	1, 2, 3	1, 21, 27
5.3a	1, 2	3, 11, 25, 39, 71
5.4a	1	19, 33
5.5b	4, 6	13, 23, 33
5.5d	13, 17, 19	55, 59, 67, 81
5.6a	1, 2	1, 9, 23, 27
5.8b	4	25, 29, 37
5.9a	6	25
6.1d	11, 12	59, 65
6.2b	6, 8	13, 31
6.3c	4, 6	25, 31
6.4a	5, 6, 8	11, 17, 27, 59
6.5a	3, 4	9, 17, 39
6.6a	6	15
6.7a	1, 3, 5	5, 11, 21, 35
6.8a	1	1
6.8b	3	27

Two other features of the text that can be used for review of the first six chapters are as follows.

- At the end of each chapter is a *Summary and Review* that provides an extensive set of review exercises. Reference codes beside each exercise or the direction line preceding it allow the student to easily return to the objective being reviewed. Answers to all of these exercises appear at the back of the book.
- The *Cumulative Review* that follows every chapter beginning with Chapter 2 can also be used for review. Each reviews material from all preceding chapters. At the back of the text are answers to all Cumulative Review exercises, together with section and objective references, so that students know exactly what material to study if they miss an exercise.

The extensive supplements package that accompanies this text also includes material appropriate for a structured review of the first six chapters. Consult the preface in the text for detailed descriptions of each of the following.

- Chapter Test Prep videos
- *MyMathGuide: Notes, Practice, and Video Path* workbook
- MyLab Math

B OBJECTIVE

a Find the mean (average), the median, and the mode of a set of data and solve related applied problems.

Mean, Median, and Mode

a MEAN, MEDIAN, AND MODE

One way to analyze data is to look for a single representative number, called a **center point** or **measure of central tendency**. Those most often used are the **mean** (or **average**), the **median**, and the **mode**.

Mean

> **MEAN, OR AVERAGE**
>
> The **mean**, or **average**, of a set of numbers is the sum of the numbers divided by the number of addends.

EXAMPLE 1 *Movies Released.* Consider the number of movies released annually in the United States for the years 2011–2016:

602, 668, 689, 706, 705, 735.

What is the mean, or average, of the numbers?
Data: boxofficemojo.com

First, we add the numbers:

$$602 + 668 + 689 + 706 + 705 + 735 = 4105.$$

Then we divide by the number of addends, 6:

$$\frac{4105}{6} \approx 684. \quad \text{Rounding to the nearest one}$$

The mean, or average, number of movies released annually in the United States in those six years is about 684.

Note that if the number of movies had been the average (same) for each of the six years, we would have

$$684 + 684 + 684 + 684 + 684 + 684 = 4104 \approx 4105.$$

The number 684 is called the mean, or average, of the set of numbers.

◀ Do Exercises 1–3.

Find the mean. Round to the nearest tenth.

1. 28, 103, 39

2. 85, 46, 105.7, 22.1

3. A student scored the following on five tests:

    78, 95, 84, 100, 82.

    What was the average score?

Median

The *median* is useful when we wish to de-emphasize extreme values. For example, suppose that five workers in a technology company manufactured the following number of computers during one month's work:

Sarah: 88   Jen: 94   Matt: 92
Mark: 91   Pat: 66

Let's first list the values in order from smallest to largest:

66 88 **91** 92 94.
             ↑
        Middle number

The middle number—in this case, 91—is the **median**.

Answers
1. 56.7   2. 64.7   3. 87.8

MEDIAN

Once a set of data has been arranged from smallest to largest, the **median** of the set of data is the middle number if there is an odd number of data numbers. If there is an even number of data numbers, then there are two middle numbers and the median is the *average* of the two middle numbers.

EXAMPLE 2 What is the median of the following set of yearly salaries?

$76,000, $58,000, $87,000, $32,500, $64,800, $62,500

We first rearrange the numbers in order from smallest to largest.

$32,500, $58,000, $62,500, $64,800, $76,000, $87,000
↑
Median

There is an even number of numbers. We look for the middle two, which are $62,500 and $64,800. In this case, the median is the average of $62,500 and $64,800:

$$\frac{\$62{,}500 + \$64{,}800}{2} = \$63{,}650.$$

Do Exercises 4–6.

Find the median.
4. 17, 13, 18, 14, 19

5. 17, 18, 16, 19, 13, 14

6. 122, 102, 103, 91, 83, 81, 78, 119, 88

Mode

The last center point we consider is called the *mode*. A number that occurs most often in a set of data can be considered a representative number, or center point.

MODE

The **mode** of a set of data is the number or numbers that occur most often. If each number occurs the same number of times, there is *no* mode.

EXAMPLE 3 Find the mode of the following data:

23, 24, 27, 18, 19, 27

The number that occurs most often is 27. Thus the mode is 27.

EXAMPLE 4 Find the mode of the following data:

83, 84, 84, 84, 85, 86, 87, 87, 87, 88, 89, 90.

There are two numbers that occur most often, 84 and 87. Thus the modes are 84 and 87.

EXAMPLE 5 Find the mode of the following data:

115, 117, 211, 213, 219.

Each number occurs the same number of times. The set of data has *no* mode.

Do Exercises 7–10.

Find any modes that exist.
7. 33, 55, 55, 88, 55

8. 90, 54, 88, 87, 87, 54

9. 23.7, 27.5, 54.9, 17.2, 20.1

10. In conducting laboratory tests, Carole discovers bacteria in different lab dishes grew to the following areas, in square millimeters:

    25, 19, 29, 24, 28.

a) What is the mean?
b) What is the median?
c) What is the mode?

Answers
4. 17 **5.** 16.5 **6.** 91 **7.** 55 **8.** 54, 87
9. No mode exists. **10. (a)** 25 mm^2;
(b) 25 mm^2; **(c)** no mode exists.

APPENDIX B   Mean, Median, and Mode

B Exercise Set

a For each set of numbers, find the mean (average), the median, and any modes that exist.

1. 17, 19, 29, 18, 14, 29

2. 13, 32, 25, 27, 13

3. 4.3, 7.4, 1.2, 5.7, 8.3

4. 13.4, 13.4, 12.6, 42.9

5. 234, 228, 234, 229, 234, 278

6. $29.95, $28.79, $30.95, $28.79

7. *Tornadoes.* The following bar graph shows the number of tornado deaths by month for 2010–2016. What is the average number of tornado deaths for the 12 months? the median? the mode?

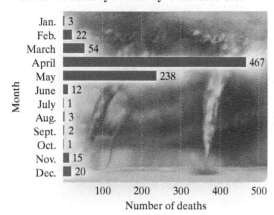

Number of Deaths by Tornado by Month 2010–2016

Jan. 3
Feb. 22
March 54
April 467
May 238
June 12
July 1
Aug. 3
Sept. 2
Oct. 1
Nov. 15
Dec. 20

DATA: National Weather Service's Storm Prediction Center

8. *Hurricanes.* The following bar graph shows the number of hurricanes that made landfall in the United States in various months from 1851 to 2015. What is the average number for the six months given? the median? the mode?

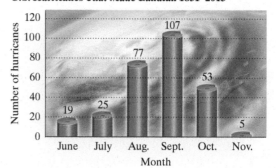

U.S. Hurricanes That Made Landfall 1851–2015

June 19
July 25
Aug. 77
Sept. 107
Oct. 53
Nov. 5

DATA: National Oceanic & Atmospheric Administration

9. *Brussels Sprouts.* The following prices per stalk of Brussels sprouts were found at five supermarkets:

$3.99, $4.49, $4.99, $3.99, $3.49.

What was the average price per stalk? the median price? the mode?

10. *Cheddar Cheese Prices.* The following prices per pound of sharp cheddar cheese were found at five supermarkets:

$5.99, $6.79, $5.99, $6.99, $6.79.

What was the average price per pound? the median price? the mode?

Synthetic Division

a SYNTHETIC DIVISION

To divide a polynomial by a binomial of the type $x - a$, we can streamline the general procedure by a process called **synthetic division**.

Compare the following. In **A**, we perform a division. In **B**, we also divide but we do not write the variables. In **B**, we eliminate some duplication of writing.

OBJECTIVE

a Use synthetic division to divide a polynomial by a binomial of the type $x - a$.

A.
$$
\begin{array}{r}
4x^2 + 5x + 11 \\
x - 2 \overline{) 4x^3 - 3x^2 + x + 7} \\
\underline{4x^3 - 8x^2} \\
5x^2 + x \\
\underline{5x^2 - 10x} \\
11x + 7 \\
\underline{11x - 22} \\
29
\end{array}
$$

B.
$$
\begin{array}{r}
4 + 5 + 11 \\
1 - 2 \overline{) 4 - 3 + 1 + 7} \\
\underline{4 - 8} \\
5 + 1 \\
\underline{5 - 10} \\
11 + 7 \\
\underline{11 - 22} \\
29
\end{array}
$$

In **B**, there is still some duplication of writing. Also, since we can subtract by adding the opposite, we can use 2 instead of -2 and then add instead of subtracting.

C. *Synthetic Division*

a) $2 \underline{|} 4 -3 1 7$

$ 4$

Write 2, the number that is subtracted in the divisor $x - 2$, and the coefficients of the dividend.

Bring down the first coefficient.

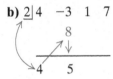

Multiply 4 by 2 to get 8. Add 8 and -3.

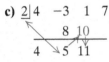

Multiply 5 by 2 to get 10. Add 10 and 1.

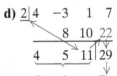

Multiply 11 by 2 to get 22. Add 22 and 7.

Quotient Remainder

The last number, 29, is the remainder. The other numbers are the coefficients of the quotient with that of the term of highest degree first. Note that the degree of the term of highest degree is 1 less than the degree of the dividend.

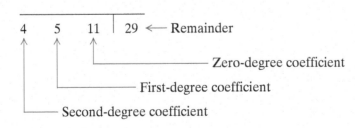

It is important to remember that in order for synthetic division to work, the divisor must be of the form $x - a$, that is, a variable minus a constant. The coefficient of the variable must be 1.

The answer is $4x^2 + 5x + 11$, R 29; or $4x^2 + 5x + 11 + \dfrac{29}{x - 2}$.

EXAMPLE 1 Use synthetic division to divide:
$$(x^3 + 6x^2 - x - 30) \div (x - 2).$$
We have

$$\underline{2 |}\begin{array}{cccc} 1 & 6 & -1 & -30 \\ & 2 & 16 & 30 \\ \hline 1 & 8 & 15 & \big| \ 0 \end{array}$$

The answer is $x^2 + 8x + 15$, R 0; or just $x^2 + 8x + 15$.

◀ Do Exercise 1.

1. Use synthetic division to divide:
$(2x^3 - 4x^2 + 8x - 8) \div (x - 3)$.

When there are missing terms, be sure to write 0's for their coefficients.

EXAMPLES Use synthetic division to divide.

2. $(2x^3 + 7x^2 - 5) \div (x + 3)$

There is no x-term, so we must write a 0 for its coefficient. Note that $x + 3 = x - (-3)$, so we write -3 at the left.

$$\underline{-3 |}\begin{array}{cccc} 2 & 7 & 0 & -5 \\ & -6 & -3 & 9 \\ \hline 2 & 1 & -3 & \big| \ 4 \end{array}$$

The answer is $2x^2 + x - 3$, R 4; or $2x^2 + x - 3 + \dfrac{4}{x + 3}$.

3. $(x^3 + 4x^2 - x - 4) \div (x + 4)$

Note that $x + 4 = x - (-4)$, so we write -4 at the left.

$$\underline{-4 |}\begin{array}{cccc} 1 & 4 & -1 & -4 \\ & -4 & 0 & 4 \\ \hline 1 & 0 & -1 & \big| \ 0 \end{array}$$

The answer is $x^2 - 1$.

4. $(x^4 - 1) \div (x - 1)$

The divisor is $x - 1$, so we write 1 at the left.

$$\underline{1 |}\begin{array}{ccccc} 1 & 0 & 0 & 0 & -1 \\ & 1 & 1 & 1 & 1 \\ \hline 1 & 1 & 1 & 1 & \big| \ 0 \end{array}$$ There are 3 missing terms.

The answer is $x^3 + x^2 + x + 1$.

5. $(8x^5 - 6x^3 + x - 8) \div (x + 2)$

Note that $x + 2 = x - (-2)$, so we write -2 at the left.

$$\underline{-2 |}\begin{array}{cccccc} 8 & 0 & -6 & 0 & 1 & -8 \\ & -16 & 32 & -52 & 104 & -210 \\ \hline 8 & -16 & 26 & -52 & 105 & \big| \ -218 \end{array}$$

The answer is $8x^4 - 16x^3 + 26x^2 - 52x + 105$, R -218; or

$$8x^4 - 16x^3 + 26x^2 - 52x + 105 + \dfrac{-218}{x + 2}.$$

◀ Do Exercises 2 and 3.

Use synthetic division to divide.

2. $(x^3 - 2x^2 + 5x - 4) \div (x + 2)$

3. $(y^3 + 1) \div (y + 1)$

Answers

1. $2x^2 + 2x + 14$, R 34; or
$2x^2 + 2x + 14 + \dfrac{34}{x - 3}$
2. $x^2 - 4x + 13$, R -30; or
$x^2 - 4x + 13 + \dfrac{-30}{x + 2}$
3. $y^2 - y + 1$

C Exercise Set   FOR EXTRA HELP   MyLab Math

a Use synthetic division to divide.

1. $(x^3 - 2x^2 + 2x - 5) \div (x - 1)$
2. $(x^3 - 2x^2 + 2x - 5) \div (x + 1)$
3. $(a^2 + 11a - 19) \div (a + 4)$

4. $(a^2 + 11a - 19) \div (a - 4)$
5. $(x^3 - 7x^2 - 13x + 3) \div (x - 2)$
6. $(x^3 - 7x^2 - 13x + 3) \div (x + 2)$

7. $(3x^3 + 7x^2 - 4x + 3) \div (x + 3)$
8. $(3x^3 + 7x^2 - 4x + 3) \div (x - 3)$
9. $(y^3 - 3y + 10) \div (y - 2)$

10. $(x^3 - 2x^2 + 8) \div (x + 2)$
11. $(3x^4 - 25x^2 - 18) \div (x - 3)$
12. $(6y^4 + 15y^3 + 28y + 6) \div (y + 3)$

13. $(x^3 - 8) \div (x - 2)$
14. $(y^3 + 125) \div (y + 5)$
15. $(y^4 - 16) \div (y - 2)$

16. $(x^5 - 32) \div (x - 2)$
17. $(y^8 - 1) \div (y + 1)$
18. $(y^6 - 2) \div (y - 1)$

APPENDIX C Synthetic Division

D Determinants and Cramer's Rule

OBJECTIVES

a Evaluate second-order determinants.

b Evaluate third-order determinants.

c Solve systems of equations using Cramer's rule.

The elimination method concerns itself primarily with the coefficients and constants of the equations. We now introduce a method for solving a system of equations using just the coefficients and constants. This method involves *determinants*.

a EVALUATING DETERMINANTS

The following symbolism represents a **second-order determinant**:

$$\begin{vmatrix} a_1 & b_1 \\ a_2 & b_2 \end{vmatrix}.$$

To evaluate a determinant, we do two multiplications and subtract.

EXAMPLE 1 Evaluate:

$$\begin{vmatrix} 2 & -5 \\ 6 & 7 \end{vmatrix}.$$

We multiply and subtract as follows:

$$\begin{vmatrix} 2 & -5 \\ 6 & 7 \end{vmatrix} = 2 \cdot 7 - 6 \cdot (-5) = 14 + 30 = 44.$$

Determinants are defined according to the pattern shown in Example 1.

> **SECOND-ORDER DETERMINANT**
>
> The determinant $\begin{vmatrix} a_1 & b_1 \\ a_2 & b_2 \end{vmatrix}$ is defined to mean $a_1b_2 - a_2b_1$.

The value of a determinant is a *number*. In Example 1, the value is 44.

◀ Do Exercises 1 and 2.

Evaluate.

1. $\begin{vmatrix} 3 & 2 \\ 4 & 1 \end{vmatrix}$

2. $\begin{vmatrix} 5 & -2 \\ -1 & -1 \end{vmatrix}$

b THIRD-ORDER DETERMINANTS

A **third-order determinant** is defined as follows.

$$\begin{vmatrix} a_1 & b_1 & c_1 \\ a_2 & b_2 & c_2 \\ a_3 & b_3 & c_3 \end{vmatrix} = a_1 \begin{vmatrix} b_2 & c_2 \\ b_3 & c_3 \end{vmatrix} - a_2 \begin{vmatrix} b_1 & c_1 \\ b_3 & c_3 \end{vmatrix} + a_3 \begin{vmatrix} b_1 & c_1 \\ b_2 & c_2 \end{vmatrix}$$

Note the minus sign here.

Note that the *a*'s come from the first column.

Answers
1. -5 2. -7

Note that the second-order determinants on the right in the preceding definition can be obtained by crossing out the row and the column in which each a occurs.

For a_1:
$$\begin{vmatrix} a_1 & b_1 & c_1 \\ a_2 & b_2 & c_2 \\ a_3 & b_3 & c_3 \end{vmatrix}$$

For a_2:
$$\begin{vmatrix} a_1 & b_1 & c_1 \\ a_2 & b_2 & c_2 \\ a_3 & b_3 & c_3 \end{vmatrix}$$

For a_3:
$$\begin{vmatrix} a_1 & b_1 & c_1 \\ a_2 & b_2 & c_2 \\ a_3 & b_3 & c_3 \end{vmatrix}$$

EXAMPLE 2 Evaluate this third-order determinant:

$$\begin{vmatrix} -1 & 0 & 1 \\ -5 & 1 & -1 \\ 4 & 8 & 1 \end{vmatrix} = -1 \begin{vmatrix} 1 & -1 \\ 8 & 1 \end{vmatrix} - (-5) \begin{vmatrix} 0 & 1 \\ 8 & 1 \end{vmatrix} + 4 \begin{vmatrix} 0 & 1 \\ 1 & -1 \end{vmatrix}.$$

We calculate as follows:

$$-1 \begin{vmatrix} 1 & -1 \\ 8 & 1 \end{vmatrix} - (-5) \begin{vmatrix} 0 & 1 \\ 8 & 1 \end{vmatrix} + 4 \begin{vmatrix} 0 & 1 \\ 1 & -1 \end{vmatrix}$$
$$= -1[1 \cdot 1 - 8(-1)] + 5(0 \cdot 1 - 8 \cdot 1) + 4[0 \cdot (-1) - 1 \cdot 1]$$
$$= -1(9) + 5(-8) + 4(-1)$$
$$= -9 - 40 - 4$$
$$= -53.$$

Do Exercises 3 and 4. ▶

Evaluate.

3. $\begin{vmatrix} 2 & -1 & 1 \\ 1 & 2 & -1 \\ 3 & 4 & -3 \end{vmatrix}$

4. $\begin{vmatrix} 3 & 2 & 2 \\ -2 & 1 & 4 \\ 4 & -3 & 3 \end{vmatrix}$

c SOLVING SYSTEMS USING DETERMINANTS

Here is a system of two equations in two variables:

$$a_1 x + b_1 y = c_1,$$
$$a_2 x + b_2 y = c_2.$$

We form three determinants, which we call D, D_x, and D_y.

$D = \begin{vmatrix} a_1 & b_1 \\ a_2 & b_2 \end{vmatrix}$ In D, we have the coefficients of x and y.

$D_x = \begin{vmatrix} c_1 & b_1 \\ c_2 & b_2 \end{vmatrix}$ To form D_x, we replace the x-coefficients in D with the constants on the right side of the equations.

$D_y = \begin{vmatrix} a_1 & c_1 \\ a_2 & c_2 \end{vmatrix}$ To form D_y, we replace the y-coefficients in D with the constants on the right.

It is important that the replacement be done *without changing the order of the columns*. Then the solution of the system can be found as follows. This is known as **Cramer's rule**.

Answers

3. -6 4. 93

CRAMER'S RULE

$$x = \frac{D_x}{D}, \quad y = \frac{D_y}{D}$$

EXAMPLE 3 Solve using Cramer's rule:

$$3x - 2y = 7,$$
$$3x + 2y = 9.$$

We compute D, D_x, and D_y:

$$D = \begin{vmatrix} 3 & -2 \\ 3 & 2 \end{vmatrix} = 3 \cdot 2 - 3 \cdot (-2) = 6 + 6 = 12;$$

$$D_x = \begin{vmatrix} 7 & -2 \\ 9 & 2 \end{vmatrix} = 7 \cdot 2 - 9(-2) = 14 + 18 = 32;$$

$$D_y = \begin{vmatrix} 3 & 7 \\ 3 & 9 \end{vmatrix} = 3 \cdot 9 - 3 \cdot 7 = 27 - 21 = 6.$$

Then

$$x = \frac{D_x}{D} = \frac{32}{12}, \text{ or } \frac{8}{3} \quad \text{and} \quad y = \frac{D_y}{D} = \frac{6}{12} = \frac{1}{2}.$$

The solution is $\left(\frac{8}{3}, \frac{1}{2}\right)$.

◀ Do Exercise 5.

5. Solve using Cramer's rule:
$$4x - 3y = 15,$$
$$x + 3y = 0.$$

Cramer's rule for three equations is very similar to that for two.

$$a_1 x + b_1 y + c_1 z = d_1,$$
$$a_2 x + b_2 y + c_2 z = d_2,$$
$$a_3 x + b_3 y + c_3 z = d_3$$

$$D = \begin{vmatrix} a_1 & b_1 & c_1 \\ a_2 & b_2 & c_2 \\ a_3 & b_3 & c_3 \end{vmatrix} \quad D_x = \begin{vmatrix} d_1 & b_1 & c_1 \\ d_2 & b_2 & c_2 \\ d_3 & b_3 & c_3 \end{vmatrix}$$

$$D_y = \begin{vmatrix} a_1 & d_1 & c_1 \\ a_2 & d_2 & c_2 \\ a_3 & d_3 & c_3 \end{vmatrix}$$

D is again the determinant of the coefficients of x, y, and z. This time we have one more determinant, D_z. We get it by replacing the z-coefficients in D with the constants on the right:

$$D_z = \begin{vmatrix} a_1 & b_1 & d_1 \\ a_2 & b_2 & d_2 \\ a_3 & b_3 & d_3 \end{vmatrix}.$$

Answer

5. $(3, -1)$

The solution of the system is given by the following.

CRAMER'S RULE

$$x = \frac{D_x}{D}, \quad y = \frac{D_y}{D}, \quad z = \frac{D_z}{D}$$

EXAMPLE 4 Solve using Cramer's rule:

$$x - 3y + 7z = 13,$$
$$x + y + z = 1,$$
$$x - 2y + 3z = 4.$$

We compute D, D_x, D_y, and D_z:

$$D = \begin{vmatrix} 1 & -3 & 7 \\ 1 & 1 & 1 \\ 1 & -2 & 3 \end{vmatrix} = -10; \quad D_x = \begin{vmatrix} 13 & -3 & 7 \\ 1 & 1 & 1 \\ 4 & -2 & 3 \end{vmatrix} = 20;$$

$$D_y = \begin{vmatrix} 1 & 13 & 7 \\ 1 & 1 & 1 \\ 1 & 4 & 3 \end{vmatrix} = -6; \quad D_z = \begin{vmatrix} 1 & -3 & 13 \\ 1 & 1 & 1 \\ 1 & -2 & 4 \end{vmatrix} = -24.$$

Then

$$x = \frac{D_x}{D} = \frac{20}{-10} = -2;$$
$$y = \frac{D_y}{D} = \frac{-6}{-10} = \frac{3}{5};$$
$$z = \frac{D_z}{D} = \frac{-24}{-10} = \frac{12}{5}.$$

The solution is $\left(-2, \frac{3}{5}, \frac{12}{5}\right)$.

In Example 4, we would not have needed to evaluate D_z. Once we found x and y, we could have substituted them into one of the equations to find z. In practice, it is faster to use determinants to find only two of the numbers; then we find the third by substitution into an equation.

Do Exercise 6.

6. Solve using Cramer's rule:
$$x - 3y - 7z = 6,$$
$$2x + 3y + z = 9,$$
$$4x + y = 7.$$

In using Cramer's rule, we divide by D. If D were 0, we could not do so.

INCONSISTENT SYSTEMS; DEPENDENT EQUATIONS

If $D = 0$ and at least one of the other determinants is not 0, then the system does not have a solution, and we say that it is *inconsistent*.

If $D = 0$ and all the other determinants are also 0, then there is an infinite set of solutions. In that case, we say that the equations in the system are *dependent*.

Answer
6. $(1, 3, -2)$

D Exercise Set

a Evaluate.

1. $\begin{vmatrix} 3 & 7 \\ 2 & 8 \end{vmatrix}$

2. $\begin{vmatrix} 5 & 4 \\ 4 & -5 \end{vmatrix}$

3. $\begin{vmatrix} -3 & -6 \\ -5 & -10 \end{vmatrix}$

4. $\begin{vmatrix} 4 & 5 \\ -7 & 9 \end{vmatrix}$

5. $\begin{vmatrix} 8 & 2 \\ 12 & -3 \end{vmatrix}$

6. $\begin{vmatrix} 1 & 1 \\ 9 & 8 \end{vmatrix}$

7. $\begin{vmatrix} 2 & -7 \\ 0 & 0 \end{vmatrix}$

8. $\begin{vmatrix} 0 & -4 \\ 0 & -6 \end{vmatrix}$

b Evaluate.

9. $\begin{vmatrix} 0 & 2 & 0 \\ 3 & -1 & 1 \\ 1 & -2 & 2 \end{vmatrix}$

10. $\begin{vmatrix} 3 & 0 & -2 \\ 5 & 1 & 2 \\ 2 & 0 & -1 \end{vmatrix}$

11. $\begin{vmatrix} -1 & -2 & -3 \\ 3 & 4 & 2 \\ 0 & 1 & 2 \end{vmatrix}$

12. $\begin{vmatrix} 1 & 2 & 2 \\ 2 & 1 & 0 \\ 3 & 3 & 1 \end{vmatrix}$

13. $\begin{vmatrix} 3 & 2 & -2 \\ -2 & 1 & 4 \\ -4 & -3 & 3 \end{vmatrix}$

14. $\begin{vmatrix} 2 & -1 & 1 \\ 1 & 2 & -1 \\ 3 & 4 & -3 \end{vmatrix}$

15. $\begin{vmatrix} 3 & 2 & 4 \\ 1 & 1 & 1 \\ 1 & 1 & 1 \end{vmatrix}$

16. $\begin{vmatrix} -1 & 6 & -5 \\ 2 & 4 & 4 \\ 5 & 3 & 10 \end{vmatrix}$

c Solve using Cramer's rule.

17. $3x - 4y = 6,$
    $5x + 9y = 10$

18. $5x + 8y = 1,$
    $3x + 7y = 5$

19. $-2x + 4y = 3,$
    $3x - 7y = 1$

20. $5x - 4y = -3,$
    $7x + 2y = 6$

21. $4x + 2y = 11,$
    $3x - y = 2$

22. $3x - 3y = 11,$
    $9x - 2y = 5$

23. $x + 4y = 8,$
    $3x + 5y = 3$

24. $x + 4y = 5,$
    $-3x + 2y = 13$

25. $2x - 3y + 5z = 27,$
    $x + 2y - z = -4,$
    $5x - y + 4z = 27$

26. $x - y + 2z = -3,$
    $x + 2y + 3z = 4,$
    $2x + y + z = -3$

27. $r - 2s + 3t = 6,$
    $2r - s - t = -3,$
    $r + s + t = 6$

28. $a - 3c = 6,$
    $b + 2c = 2,$
    $7a - 3b - 5c = 14$

29. $4x - y - 3z = 1,$
    $8x + y - z = 5,$
    $2x + y + 2z = 5$

30. $3x + 2y + 2z = 3,$
    $x + 2y - z = 5,$
    $2x - 4y + z = 0$

31. $p + q + r = 1,$
    $p - 2q - 3r = 3,$
    $4p + 5q + 6r = 4$

32. $x + 2y - 3z = 9,$
    $2x - y + 2z = -8,$
    $3x - y - 4z = 3$

Elimination Using Matrices

E

OBJECTIVE

a Solve systems of two or three equations using matrices.

The elimination method concerns itself primarily with the coefficients and constants of the equations. In what follows, we learn a method for solving systems using just the coefficients and the constants. This procedure involves what are called *matrices*.

a In solving systems of equations, we perform computations with the constants. The variables play no important role until the end. Thus we can simplify writing a system by omitting the variables. For example, the system

$$\begin{array}{l} 3x + 4y = 5, \\ x - 2y = 1 \end{array} \quad \text{simplifies to} \quad \begin{array}{ccc} 3 & 4 & 5 \\ 1 & -2 & 1 \end{array}$$

if we omit the variables, the operation of addition, and the equals signs. The result is a rectangular array of numbers. Such an array is called a **matrix** (plural, **matrices**). We ordinarily write brackets around matrices. The following are matrices.

$$\begin{bmatrix} 4 & 1 & 3 & 5 \\ 1 & 0 & 1 & 2 \\ 6 & 3 & -2 & 0 \end{bmatrix}, \quad \begin{bmatrix} 6 & 2 & 1 & 4 & 7 \\ 1 & 2 & 1 & 3 & 1 \\ 4 & 0 & -2 & 0 & -3 \end{bmatrix}, \quad \begin{bmatrix} 1 & 2 \\ 145 & 0 \\ -7 & 9 \\ 8 & 1 \\ 0 & 0 \end{bmatrix}.$$

The **rows** of a matrix are horizontal, and the **columns** are vertical.

$$\begin{bmatrix} 5 & -2 & 2 \\ 1 & 0 & 1 \\ 0 & 1 & 2 \end{bmatrix} \begin{array}{l} \leftarrow \text{row 1} \\ \leftarrow \text{row 2} \\ \leftarrow \text{row 3} \end{array}$$

$\uparrow \qquad \uparrow \qquad \uparrow$
column 1 column 2 column 3

Let's now use matrices to solve systems of linear equations.

EXAMPLE 1 Solve the system

$$5x - 4y = -1,$$
$$-2x + 3y = 2.$$

We write a matrix using only the coefficients and the constants, keeping in mind that x corresponds to the first column and y to the second. A dashed line separates the coefficients from the constants at the end of each equation:

$$\begin{bmatrix} 5 & -4 & \vdots & -1 \\ -2 & 3 & \vdots & 2 \end{bmatrix}. \quad \text{The individual numbers are called } \textit{elements}, \text{ or } \textit{entries}.$$

Our goal is to transform this matrix into one of the form

$$\begin{bmatrix} a & b & \vdots & c \\ 0 & d & \vdots & e \end{bmatrix}.$$

The variables can then be reinserted to form equations from which we can complete the solution.

We do calculations that are similar to those that we would do if we wrote the entire equations. The first step, if possible, is to multiply and/or interchange the rows so that each number in the first column below the first number is a multiple of that number. In this case, we do so by multiplying Row 2 by 5. This corresponds to multiplying the second equation by 5.

$$\begin{bmatrix} 5 & -4 & \vdots & -1 \\ -10 & 15 & \vdots & 10 \end{bmatrix} \quad \text{New Row } 2 = 5(\text{Row } 2)$$

Next, we multiply the first row by 2 and add the result to the second row. This corresponds to multiplying the first equation by 2 and adding the result to the second equation. Although we write the calculations out here, we generally try to do them mentally:

$$2 \cdot 5 + (-10) = 0; \quad 2(-4) + 15 = 7; \quad 2(-1) + 10 = 8.$$

$$\begin{bmatrix} 5 & -4 & \vdots & -1 \\ 0 & 7 & \vdots & 8 \end{bmatrix} \quad \text{New Row } 2 = 2(\text{Row } 1) + (\text{Row } 2)$$

If we now reinsert the variables, we have

$$5x - 4y = -1, \quad (1)$$
$$7y = 8. \quad (2)$$

We can now proceed as before, solving equation (2) for y:

$$7y = 8 \quad (2)$$
$$y = \tfrac{8}{7}.$$

Next, we substitute $\tfrac{8}{7}$ for y back in equation (1). This procedure is called *back-substitution*.

$$5x - 4y = -1 \quad (1)$$
$$5x - 4 \cdot \tfrac{8}{7} = -1 \quad \text{Substituting } \tfrac{8}{7} \text{ for } y \text{ in equation (1)}$$
$$x = \tfrac{5}{7} \quad \text{Solving for } x$$

The solution is $\left(\tfrac{5}{7}, \tfrac{8}{7}\right)$.

◀ Do Exercise 1.

1. Solve using matrices:
$$5x - 2y = -44,$$
$$2x + 5y = -6.$$

EXAMPLE 2 Solve the system

$$2x - y + 4z = -3,$$
$$x - 4z = 5,$$
$$6x - y + 2z = 10.$$

We first write a matrix, using only the coefficients and the constants. Where there are missing terms, we must write 0's:

$$\begin{bmatrix} 2 & -1 & 4 & \vdots & -3 \\ 1 & 0 & -4 & \vdots & 5 \\ 6 & -1 & 2 & \vdots & 10 \end{bmatrix} \quad \begin{array}{l} \textbf{(P1)} \\ \textbf{(P2)} \\ \textbf{(P3)} \end{array} \quad \begin{array}{l} \text{(P1), (P2), and (P3) designate the} \\ \text{equations that are in the first,} \\ \text{second, and third position,} \\ \text{respectively.} \end{array}$$

Our goal is to find an equivalent matrix of the form

$$\begin{bmatrix} a & b & c & \vdots & d \\ 0 & e & f & \vdots & g \\ 0 & 0 & h & \vdots & i \end{bmatrix}.$$

A matrix of this form can be rewritten as a system of equations from which a solution can be found easily.

Answer
1. $(-8, 2)$

The first step, if possible, is to interchange the rows so that each number in the first column below the first number is a multiple of that number. In this case, we do so by interchanging Rows 1 and 2:

$$\begin{bmatrix} 1 & 0 & -4 & | & 5 \\ 2 & -1 & 4 & | & -3 \\ 6 & -1 & 2 & | & 10 \end{bmatrix}.$$ This corresponds to interchanging the first two equations.

Next, we multiply the first row by -2 and add it to the second row:

$$\begin{bmatrix} 1 & 0 & -4 & | & 5 \\ 0 & -1 & 12 & | & -13 \\ 6 & -1 & 2 & | & 10 \end{bmatrix}.$$ This corresponds to multiplying new equation (P1) by -2 and adding it to new equation (P2). The result replaces the former (P2). We perform the calculations mentally.

Now we multiply the first row by -6 and add it to the third row:

$$\begin{bmatrix} 1 & 0 & -4 & | & 5 \\ 0 & -1 & 12 & | & -13 \\ 0 & -1 & 26 & | & -20 \end{bmatrix}.$$ This corresponds to multiplying equation (P1) by -6 and adding it to equation (P3).

Next, we multiply Row 2 by -1 and add it to the third row:

$$\begin{bmatrix} 1 & 0 & -4 & | & 5 \\ 0 & -1 & 12 & | & -13 \\ 0 & 0 & 14 & | & -7 \end{bmatrix}.$$ This corresponds to multiplying equation (P2) by -1 and adding it to equation (P3).

Reinserting the variables gives us

$$x \quad - 4z = 5, \quad \textbf{(P1)}$$
$$-y + 12z = -13, \quad \textbf{(P2)}$$
$$14z = -7. \quad \textbf{(P3)}$$

We now solve (P3) for z:

$14z = -7$ **(P3)**
$z = -\frac{7}{14}$ Solving for z
$z = -\frac{1}{2}$.

Next, we back-substitute $-\frac{1}{2}$ for z in (P2) and solve for y:

$-y + 12z = -13$ **(P2)**
$-y + 12\left(-\frac{1}{2}\right) = -13$ Substituting $-\frac{1}{2}$ for z in equation (P2)
$-y - 6 = -13$
$-y = -7$
$y = 7$. Solving for y

Since there is no y-term in (P1), we need only substitute $-\frac{1}{2}$ for z in (P1) and solve for x:

$x - 4z = 5$ **(P1)**
$x - 4\left(-\frac{1}{2}\right) = 5$ Substituting $-\frac{1}{2}$ for z in equation (P1)
$x + 2 = 5$
$x = 3$. Solving for x

The solution is $\left(3, 7, -\frac{1}{2}\right)$.

Do Exercise 2.

2. Solve using matrices:
$$x - 2y + 3z = 4,$$
$$2x - y + z = -1,$$
$$4x + y + z = 1.$$

Answer
2. $(-1, 2, 3)$

APPENDIX E Elimination Using Matrices

The best overall method of solving systems of equations is by row-equivalent matrices; graphing calculators and computers are programmed to use them. Matrices are part of a branch of mathematics known as linear algebra. They are also studied in more detail in many courses in finite mathematics.

All the operations used in the preceding example correspond to operations with the equations and produce equivalent systems of equations. We call the matrices **row-equivalent** and the operations that produce them **row-equivalent operations**.

ROW-EQUIVALENT OPERATIONS

Each of the following row-equivalent operations produces an equivalent matrix:

a) Interchanging any two rows.
b) Multiplying each element of a row by the same nonzero number.
c) Multiplying each element of a row by a nonzero number and adding the result to another row.

E Exercise Set

FOR EXTRA HELP — MyLab Math

a Solve using matrices.

1. $4x + 2y = 11,$
   $3x - y = 2$

2. $3x - 3y = 11,$
   $9x - 2y = 5$

3. $x + 4y = 8,$
   $3x + 5y = 3$

4. $x + 4y = 5,$
   $-3x + 2y = 13$

5. $5x - 3y = -2,$
   $4x + 2y = 5$

6. $3x + 4y = 7,$
   $-5x + 2y = 10$

7. $2x - 3y = 50,$
   $5x + y = 40$

8. $4x + 5y = -8,$
   $7x + 9y = 11$

9. $4x - y - 3z = 1,$
   $8x + y - z = 5,$
   $2x + y + 2z = 5$

10. $3x + 2y + 2z = 3,$
    $x + 2y - z = 5,$
    $2x - 4y + z = 0$

11. $p + q + r = 1,$
    $p - 2q - 3r = 3,$
    $4p + 5q + 6r = 4$

12. $x + 2y - 3z = 9,$
    $2x - y + 2z = -8,$
    $3x - y - 4z = 3$

13. $x - y + 2z = 0,$
    $x - 2y + 3z = -1,$
    $2x - 2y + z = -3$

14. $4a + 9b = 8,$
    $8a + 6c = -1,$
    $6b + 6c = -1$

15. $3p + 2r = 11,$
    $q - 7r = 4,$
    $p - 6q = 1$

16. $m + n + t = 6,$
    $m - n - t = 0,$
    $m + 2n + t = 5$

1004 APPENDIXES

The Algebra of Functions

a THE SUM, DIFFERENCE, PRODUCT, AND QUOTIENT OF FUNCTIONS

Suppose that a is in the domain of two functions, f and g. The input a is paired with $f(a)$ by f and with $g(a)$ by g. The outputs can then be added to get $f(a) + g(a)$.

EXAMPLE 1 Let $f(x) = x + 4$ and $g(x) = x^2 + 1$. Find $f(2) + g(2)$.

We visualize two function machines. Because 2 is in the domain of each function, we can compute $f(2)$ and $g(2)$.

OBJECTIVE

a Given two functions f and g, find their sum, difference, product, and quotient.

MyLab Math
ANIMATION

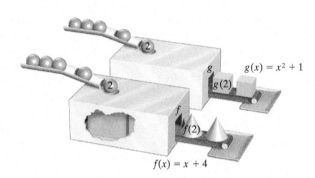

Since
$$f(2) = 2 + 4 = 6 \quad \text{and} \quad g(2) = 2^2 + 1 = 5,$$
we have
$$f(2) + g(2) = 6 + 5 = 11.$$

In Example 1, suppose that we were to write $f(x) + g(x)$ as $(x + 4) + (x^2 + 1)$, or $f(x) + g(x) = x^2 + x + 5$. This could then be regarded as a "new" function: $(f + g)(x) = x^2 + x + 5$. We can alternatively find $f(2) + g(2)$ with $(f + g)(x)$:

$$(f + g)(x) = x^2 + x + 5$$
$$(f + g)(2) = 2^2 + 2 + 5 \quad \text{Substituting 2 for } x$$
$$= 4 + 2 + 5$$
$$= 11.$$

Similar notations exist for subtraction, multiplication, and division of functions.

THE SUM, DIFFERENCE, PRODUCT, AND QUOTIENT OF FUNCTIONS

For any functions f and g, we can form new functions defined as:

1. The **sum** $f + g$: $(f + g)(x) = f(x) + g(x)$;
2. The **difference** $f - g$: $(f - g)(x) = f(x) - g(x)$;
3. The **product** fg: $(f \cdot g)(x) = f(x) \cdot g(x)$;
4. The **quotient** f/g: $(f/g)(x) = f(x)/g(x)$, where $g(x) \neq 0$.

1. Given $f(x) = x^2 + 3$ and $g(x) = x^2 - 3$, find each of the following.
   a) $(f + g)(x)$
   b) $(f - g)(x)$
   c) $(f \cdot g)(x)$
   d) $(f/g)(x)$
   e) $(f \cdot f)(x)$

EXAMPLE 2 Given f and g described by $f(x) = x^2 - 5$ and $g(x) = x + 7$, find $(f + g)(x), (f - g)(x), (f \cdot g)(x), (f/g)(x),$ and $(g \cdot g)(x)$.

$(f + g)(x) = f(x) + g(x) = (x^2 - 5) + (x + 7) = x^2 + x + 2;$
$(f - g)(x) = f(x) - g(x) = (x^2 - 5) - (x + 7) = x^2 - x - 12;$
$(f \cdot g)(x) = f(x) \cdot g(x) = (x^2 - 5)(x + 7) = x^3 + 7x^2 - 5x - 35;$
$(f/g)(x) = f(x)/g(x) = \dfrac{x^2 - 5}{x + 7};$
$(g \cdot g)(x) = g(x) \cdot g(x) = (x + 7)(x + 7) = x^2 + 14x + 49$

Note that the sum, the difference, and the product of polynomials are also polynomial functions, but the quotient may not be.

◀ Do Exercise 1.

EXAMPLE 3 For $f(x) = x^2 - x$ and $g(x) = x + 2$, find $(f + g)(3), (f - g)(-1), (f \cdot g)(5),$ and $(f/g)(-4)$.

We first find $(f + g)(x), (f - g)(x), (f \cdot g)(x),$ and $(f/g)(x)$.

$(f + g)(x) = f(x) + g(x) = x^2 - x + x + 2$
$\qquad = x^2 + 2;$

$(f - g)(x) = f(x) - g(x) = x^2 - x - (x + 2)$
$\qquad = x^2 - x - x - 2$
$\qquad = x^2 - 2x - 2;$

$(f \cdot g)(x) = f(x) \cdot g(x) = (x^2 - x)(x + 2)$
$\qquad = x^3 + 2x^2 - x^2 - 2x$
$\qquad = x^3 + x^2 - 2x;$

$(f/g)(x) = \dfrac{f(x)}{g(x)} = \dfrac{x^2 - x}{x + 2}.$

Then we substitute.

$(f + g)(3) = 3^2 + 2$ Using $(f + g)(x) = x^2 + 2$
$\qquad = 9 + 2 = 11;$

$(f - g)(-1) = (-1)^2 - 2(-1) - 2$ Using $(f - g)(x) = x^2 - 2x - 2$
$\qquad = 1 + 2 - 2 = 1;$

$(f \cdot g)(5) = 5^3 + 5^2 - 2 \cdot 5$ Using $(f \cdot g)(x) = x^3 + x^2 - 2x$
$\qquad = 125 + 25 - 10 = 140;$

$(f/g)(-4) = \dfrac{(-4)^2 - (-4)}{-4 + 2}$ Using $(f/g)(x) = (x^2 - x)/(x + 2)$
$\qquad = \dfrac{16 + 4}{-2} = \dfrac{20}{-2} = -10$

◀ Do Exercise 2.

2. Given $f(x) = x^2 + x$ and $g(x) = 2x - 3$, find each of the following.
   a) $(f + g)(-2)$
   b) $(f - g)(4)$
   c) $(f \cdot g)(-3)$
   d) $(f/g)(2)$

Answers

1. (a) $2x^2$; (b) 6; (c) $x^4 - 9$; (d) $\dfrac{x^2 + 3}{x^2 - 3}$; (e) $x^4 + 6x^2 + 9$ **2.** (a) -5; (b) 15; (c) -54; (d) 6

F Exercise Set

a Let $f(x) = -3x + 1$ and $g(x) = x^2 + 2$. Find each of the following.

1. $f(2) + g(2)$
2. $f(-1) + g(-1)$
3. $f(5) - g(5)$
4. $f(4) - g(4)$
5. $f(-1) \cdot g(-1)$
6. $f(-2) \cdot g(-2)$
7. $f(-4)/g(-4)$
8. $f(3)/g(3)$
9. $g(1) - f(1)$
10. $g(2)/f(2)$
11. $g(0)/f(0)$
12. $g(6) - f(6)$

Let $f(x) = x^2 - 3$ and $g(x) = 4 - x$. Find each of the following.

13. $(f + g)(x)$
14. $(f - g)(x)$
15. $(f + g)(-4)$
16. $(f + g)(-5)$
17. $(f - g)(3)$
18. $(f - g)(2)$
19. $(f \cdot g)(x)$
20. $(f/g)(x)$
21. $(f \cdot g)(-3)$
22. $(f \cdot g)(-4)$
23. $(f/g)(0)$
24. $(f/g)(1)$
25. $(f/g)(-2)$
26. $(f/g)(-1)$

For each pair of functions f and g, find $(f + g)(x)$, $(f - g)(x)$, $(f \cdot g)(x)$, and $(f/g)(x)$.

27. $f(x) = x^2$,
    $g(x) = 3x - 4$

28. $f(x) = 5x - 1$,
    $g(x) = 2x^2$

29. $f(x) = \dfrac{1}{x - 2}$,
    $g(x) = 4x^3$

30. $f(x) = 3x^2$,
    $g(x) = \dfrac{1}{x - 4}$

31. $f(x) = \dfrac{3}{x - 2}$,
    $g(x) = \dfrac{5}{4 - x}$

32. $f(x) = \dfrac{5}{x - 3}$,
    $g(x) = \dfrac{1}{x - 2}$

G

OBJECTIVES

a Use the distance formula to find the distance between two points whose coordinates are known.

b Use the midpoint formula to find the midpoint of a segment when the coordinates of its endpoints are known.

c Given an equation of a circle, find its center and radius and graph it. Given the center and the radius of a circle, write an equation of the circle and graph the circle.

Distance, Midpoints, and Circles

In carpentry, surveying, engineering, and other fields, it is often necessary to determine distances and midpoints and to produce accurately drawn circles.

a THE DISTANCE FORMULA

Suppose that two points are on a horizontal line, and thus have the same second coordinate. We can find the distance between them by subtracting their first coordinates. This difference may be negative, depending on the order in which we subtract. So, to make sure that we get a positive number, we take the absolute value of this difference. The distance between two points on a horizontal line (x_1, y_1) and (x_2, y_1) is thus $|x_2 - x_1|$. Similarly, the distance between two points on a vertical line (x_2, y_1) and (x_2, y_2) is $|y_2 - y_1|$.

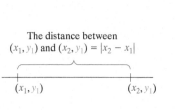

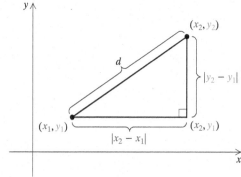

Now consider *any* two points (x_1, y_1) and (x_2, y_2). If $x_1 \neq x_2$ and $y_1 \neq y_2$, these points are vertices of a right triangle, as shown. The other vertex is then (x_2, y_1). The lengths of the legs are $|x_2 - x_1|$ and $|y_2 - y_1|$. We find d, the length of the hypotenuse, by using the Pythagorean equation:

$$d^2 = |x_2 - x_1|^2 + |y_2 - y_1|^2.$$

Since the square of a number is the same as the square of its opposite, we don't need these absolute-value signs. Thus,

$$d^2 = (x_2 - x_1)^2 + (y_2 - y_1)^2.$$

Taking the principal square root, we obtain the formula for the distance between two points.

THE DISTANCE FORMULA

The distance between any two points (x_1, y_1) and (x_2, y_2) is given by
$$d = \sqrt{(x_2 - x_1)^2 + (y_2 - y_1)^2}.$$

This formula holds even when the two points *are* on a vertical line or a horizontal line.

EXAMPLE 1 Find the distance between $(4, -3)$ and $(-5, 4)$. Give an exact answer and an approximation to three decimal places.

We substitute into the distance formula:

$$d = \sqrt{(-5-4)^2 + [4-(-3)]^2} \quad \text{Substituting}$$
$$= \sqrt{(-9)^2 + 7^2}$$
$$= \sqrt{130} \approx 11.402. \quad \text{Using a calculator}$$

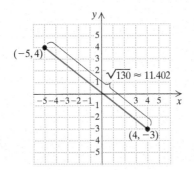

Do Exercises 1 and 2. ▶

Find the distance between each pair of points. Where appropriate, give an approximation to three decimal places.

1. $(2, 6)$ and $(-4, -2)$

2. $(-2, 1)$ and $(4, 2)$

b MIDPOINTS OF SEGMENTS

The distance formula can be used to derive a formula for finding the midpoint of a segment when the coordinates of the endpoints are known.

THE MIDPOINT FORMULA

If the endpoints of a segment are (x_1, y_1) and (x_2, y_2), then the coordinates of the midpoint are

$$\left(\frac{x_1 + x_2}{2}, \frac{y_1 + y_2}{2}\right).$$

(To locate the midpoint, determine the average of the x-coordinates and the average of the y-coordinates.)

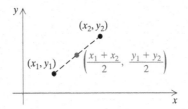

EXAMPLE 2 Find the midpoint of the segment with endpoints $(-2, 3)$ and $(4, -6)$.

Using the midpoint formula, we obtain

$$\left(\frac{-2 + 4}{2}, \frac{3 + (-6)}{2}\right), \quad \text{or} \quad \left(\frac{2}{2}, \frac{-3}{2}\right), \quad \text{or} \quad \left(1, -\frac{3}{2}\right).$$

Do Exercises 3 and 4. ▶

Find the midpoint of the segment with the given endpoints.

3. $(-3, 1)$ and $(6, -7)$

4. $(10, -7)$ and $(8, -3)$

Answers

1. 10 **2.** $\sqrt{37} \approx 6.083$ **3.** $\left(\frac{3}{2}, -3\right)$
4. $(9, -5)$

C CIRCLES

A **circle** is defined as the set of all points in a plane that are a fixed distance from a point in that plane.

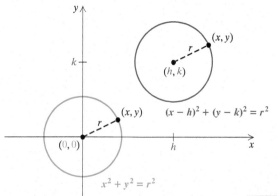

Let's find an equation for a circle. We call the center (h, k) and let the radius have length r. Suppose that (x, y) is any point on the circle. By the distance formula, we have

$$\sqrt{(x - h)^2 + (y - k)^2} = r.$$

Squaring both sides gives an equation of the circle in standard form:

$$(x - h)^2 + (y - k)^2 = r^2.$$

When $h = 0$ and $k = 0$, the circle is centered at the origin. Otherwise, we can think of that circle being translated $|h|$ units horizontally and $|k|$ units vertically from the origin.

EQUATIONS OF CIRCLES

A circle centered at the origin with radius r has equation

$$x^2 + y^2 = r^2.$$

A circle with center (h, k) and radius r has equation

$$(x - h)^2 + (y - k)^2 = r^2. \qquad \text{(Standard form)}$$

EXAMPLE 3 Find the center and the radius and graph this circle:

$$(x + 2)^2 + (y - 3)^2 = 16.$$

First, we find an equivalent equation in standard form:

$$[x - (-2)]^2 + (y - 3)^2 = 4^2.$$

Thus the center is $(-2, 3)$ and the radius is 4. We draw the graph, shown below, by locating the center and then using a compass, setting the radius at 4, to draw the circle.

5. Find the center and the radius of the circle

$$(x - 5)^2 + \left(y + \tfrac{1}{2}\right)^2 = 9.$$

Then graph the circle.

6. Find the center and the radius of the circle $x^2 + y^2 = 64$.

◀ Do Exercises 5 and 6.

Answers

5.   Center: $\left(5, -\tfrac{1}{2}\right)$; Radius: 3

6. $(0, 0); r = 8$

EXAMPLE 4 Write an equation of a circle with center $(9, -5)$ and radius $\sqrt{2}$.

We use standard form $(x - h)^2 + (y - k)^2 = r^2$ and substitute:

$(x - 9)^2 + [y - (-5)]^2 = (\sqrt{2})^2$ Substituting
$(x - 9)^2 + (y + 5)^2 = 2.$ Simplifying

Do Exercise 7. ▶

7. Find an equation of a circle with center $(-3, 1)$ and radius 6.

With certain equations not in standard form, we can complete the square to show that the equations are equations of circles.

EXAMPLE 5 Find the center and the radius and graph this circle:

$x^2 + y^2 + 8x - 2y + 15 = 0.$

First, we regroup the terms and then complete the square twice, once with $x^2 + 8x$ and once with $y^2 - 2y$:

$x^2 + y^2 + 8x - 2y + 15 = 0$

$(x^2 + 8x) + (y^2 - 2y) = -15$ Regrouping and subtracting 15

$(x^2 + 8x + 0) + (y^2 - 2y + 0) = -15$ Adding 0

$(x^2 + 8x + 16 - 16) + (y^2 - 2y + 1 - 1) = -15$ $(\frac{8}{2})^2 = 4^2 = 16;$ $(\frac{-2}{2})^2 = (-1)^2 = 1;$ substituting $16 - 16$ and $1 - 1$ for 0

$(x^2 + 8x + 16) + (y^2 - 2y + 1) - 16 - 1 = -15$ Regrouping

$(x^2 + 8x + 16) + (y^2 - 2y + 1) = -15 + 16 + 1$ Adding 16 and 1 on both sides

$(x + 4)^2 + (y - 1)^2 = 2$ Factoring and simplifying

$[x - (-4)]^2 + (y - 1)^2 = (\sqrt{2})^2.$ Writing standard form

The center is $(-4, 1)$ and the radius is $\sqrt{2}$.

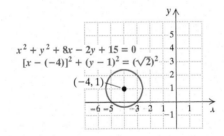

Do Exercise 8. ▶

8. Find the center and the radius of the circle
$x^2 + 2x + y^2 - 4y + 2 = 0.$
Then graph the circle.

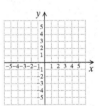

Answers
7. $(x + 3)^2 + (y - 1)^2 = 36$
8. Center: $(-1, 2)$; radius: $\sqrt{3}$;

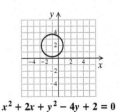

$x^2 + 2x + y^2 - 4y + 2 = 0$

APPENDIX G   Distance, Midpoints, and Circles   **1011**

G Exercise Set

a Find the distance between each pair of points. Where appropriate, give an approximation to three decimal places.

1. $(6, -4)$ and $(2, -7)$
2. $(1, 2)$ and $(-4, 14)$
3. $(0, -4)$ and $(5, -6)$
4. $(8, 3)$ and $(8, -3)$
5. $(9, 9)$ and $(-9, -9)$
6. $(2, 22)$ and $(-8, 1)$
7. $(2.8, -3.5)$ and $(-4.3, -3.5)$
8. $(6.1, 2)$ and $(5.6, -4.4)$
9. $\left(\dfrac{5}{7}, \dfrac{1}{14}\right)$ and $\left(\dfrac{1}{7}, \dfrac{11}{14}\right)$
10. $(0, \sqrt{7})$ and $(\sqrt{6}, 0)$
11. $(-23, 10)$ and $(56, -17)$
12. $(34, -18)$ and $(-46, -38)$
13. (a, b) and $(0, 0)$
14. $(0, 0)$ and (p, q)
15. $(\sqrt{2}, -\sqrt{3})$ and $(-\sqrt{7}, \sqrt{5})$
16. $(\sqrt{8}, \sqrt{3})$ and $(-\sqrt{5}, -\sqrt{6})$
17. $(1000, -240)$ and $(-2000, 580)$
18. $(-3000, 560)$ and $(-430, -640)$

b Find the midpoint of the segment with the given endpoints.

19. $(-1, 9)$ and $(4, -2)$
20. $(5, 10)$ and $(2, -4)$
21. $(3, 5)$ and $(-3, 6)$
22. $(7, -3)$ and $(4, 11)$
23. $(-10, -13)$ and $(8, -4)$
24. $(6, -2)$ and $(-5, 12)$
25. $(-3.4, 8.1)$ and $(2.9, -8.7)$
26. $(4.1, 6.9)$ and $(5.2, -6.9)$
27. $\left(\dfrac{1}{6}, -\dfrac{3}{4}\right)$ and $\left(-\dfrac{1}{3}, \dfrac{5}{6}\right)$

28. $\left(-\dfrac{4}{5}, -\dfrac{2}{3}\right)$ and $\left(\dfrac{1}{8}, \dfrac{3}{4}\right)$

29. $(\sqrt{2}, -1)$ and $(\sqrt{3}, 4)$

30. $(9, 2\sqrt{3})$ and $(-4, 5\sqrt{3})$

C Find the center and the radius of each circle. Then graph the circle.

31. $(x + 1)^2 + (y + 3)^2 = 4$

32. $(x - 2)^2 + (y + 3)^2 = 1$

33. $(x - 3)^2 + y^2 = 2$

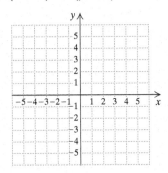 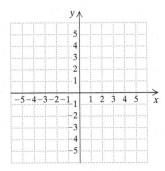 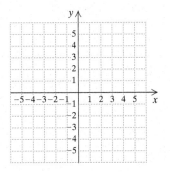

34. $x^2 + (y - 1)^2 = 3$

35. $x^2 + y^2 = 25$

36. $x^2 + y^2 = 9$

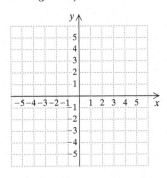 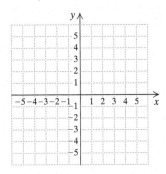 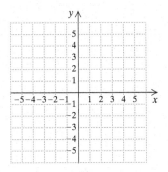

Find an equation of the circle having the given center and radius.

37. Center $(0, 0)$, radius 7

38. Center $(0, 0)$, radius 4

39. Center $(-5, 3)$, radius $\sqrt{7}$

40. Center $(4, 1)$, radius $3\sqrt{2}$

Find the center and the radius of each circle.

41. $x^2 + y^2 + 8x - 6y - 15 = 0$

42. $x^2 + y^2 + 6x - 4y - 15 = 0$

43. $x^2 + y^2 - 8x + 2y + 13 = 0$

44. $x^2 + y^2 + 6x + 4y + 12 = 0$

45. $x^2 + y^2 - 4x = 0$

46. $x^2 + y^2 + 10y - 75 = 0$

Answers

JUST-IN-TIME REVIEW

1 All Factors of a Number
1. 1, 2, 4, 5, 10, 20   **2.** 1, 3, 13, 39   **3.** 1, 3, 9, 27, 81
4. 1, 47   **5.** 1, 2, 4, 5, 8, 10, 16, 20, 32, 40, 80, 160   **6.** 1, 3, 5, 9, 15, 45   **7.** 1, 2, 4, 7, 14, 28   **8.** 1, 2, 3, 4, 6, 9, 12, 18, 36

2 Prime Factorizations
1. $3 \cdot 11$ **2.** $11 \cdot 11$ **3.** $2 \cdot 3 \cdot 3$ **4.** $2 \cdot 2 \cdot 2 \cdot 7$
5. $2 \cdot 2 \cdot 2 \cdot 3 \cdot 5$ **6.** $2 \cdot 3 \cdot 3 \cdot 5$ **7.** $2 \cdot 3 \cdot 5 \cdot 7$ **8.** $7 \cdot 13$

3 Greatest Common Factor
1. 12   **2.** 13   **3.** 1   **4.** 18   **5.** 6   **6.** 15   **7.** 20
8. 14   **9.** 5   **10.** 35

4 Least Common Multiple
1. 216   **2.** 15   **3.** 300   **4.** 299   **5.** 360   **6.** 180
7. 84   **8.** 176   **9.** 60   **10.** 2520

5 Equivalent Expressions and Fraction Notation
1. $\frac{21}{24}$ **2.** $\frac{40}{48}$ **3.** $\frac{20}{16}$ **4.** $\frac{12}{54}$ **5.** $\frac{21}{77}$ **6.** $\frac{65}{80}$

6 Mixed Numerals
1. $\frac{17}{3}$ **2.** $\frac{91}{10}$ **3.** $\frac{154}{5}$ **4.** $\frac{13}{8}$ **5.** $\frac{200}{3}$ **6.** $3\frac{3}{5}$ **7.** $4\frac{5}{6}$ **8.** $5\frac{7}{10}$
9. $13\frac{1}{3}$ **10.** $7\frac{57}{100}$

7 Simplify Fraction Notation
1. $\frac{2}{3}$ **2.** $\frac{7}{8}$ **3.** $\frac{5}{12}$ **4.** $\frac{2}{5}$ **5.** 4   **6.** $\frac{3}{4}$ **7.** 12   **8.** $\frac{1}{7}$ **9.** $\frac{11}{27}$
10. $\frac{13}{7}$ **11.** $\frac{2}{3}$ **12.** $\frac{13}{25}$

8 Multiply and Divide Fraction Notation
1. $\frac{27}{70}$ **2.** $\frac{10}{3}$ **3.** $\frac{7}{4}$ **4.** $\frac{3}{2}$ **5.** $\frac{6}{35}$ **6.** $\frac{10}{3}$ **7.** $\frac{1}{4}$ **8.** $\frac{2}{7}$ **9.** 1
10. $\frac{3}{32}$

Calculator Corner, p. 12
1. $\frac{41}{24}$ **2.** $\frac{27}{112}$ **3.** $\frac{35}{16}$ **4.** $\frac{2}{3}$

9 Add and Subtract Fraction Notation
1. $\frac{8}{11}$ **2.** 2   **3.** $\frac{13}{24}$ **4.** $\frac{7}{6}$ **5.** $\frac{5}{6}$ **6.** $\frac{19}{144}$ **7.** $\frac{13}{20}$ **8.** $\frac{25}{48}$
9. $\frac{31}{60}$ **10.** $\frac{7}{12}$ **11.** $\frac{41}{24}$ **12.** $\frac{47}{50}$

10 Convert from Decimal Notation to Fraction Notation
1. $\frac{53}{10}$ **2.** $\frac{67}{100}$ **3.** $\frac{40,008}{10,000}$ **4.** $\frac{11,223}{10}$ **5.** $\frac{14,703}{1000}$ **6.** $\frac{9}{10}$ **7.** $\frac{18,342}{100}$
8. $\frac{6}{1000}$

Calculator Corner, p. 15
1. 40.42   **2.** 3.33   **3.** 0.69324   **4.** 2.38

11 Add and Subtract Decimal Notation
1. 444.94   **2.** 170.844   **3.** 63.79   **4.** 2008.243   **5.** 32.234
6. 26.835   **7.** 1.52   **8.** 1.9193

12 Multiply and Divide Decimal Notation
1. 13.212   **2.** 0.7998   **3.** 0.000036   **4.** 1.40756   **5.** 9.3
6. 2.3   **7.** 660   **8.** 0.26

13 Convert from Fraction Notation to Decimal Notation
1. 0.34375   **2.** 0.875   **3.** $1.\overline{18}$ **4.** $1.41\overline{6}$ **5.** $0.\overline{5}$ **6.** $0.8\overline{3}$
7. $2.\overline{1}$ **8.** $0.\overline{81}$

14 Rounding with Decimal Notation
1. 745.07; 745.1; 745; 750; 700   **2.** 6780.51; 6780.5; 6781; 6780; 6800   **3.** $17.99; $18   **4.** $20.49; $20   **5.** 0.4167; 0.417; 0.42; 0.4; 0   **6.** 12.3457; 12.346; 12.35; 12.3; 12

15 Convert between Percent Notation and Decimal Notation
1. 0.63   **2.** 0.941   **3.** 2.4   **4.** 0.0081   **5.** 0.023   **6.** 1
7. 76%   **8.** 500%   **9.** 9.3%   **10.** 0.47%   **11.** 67.5%
12. 134%

16 Convert between Percent Notation and Fraction Notation
1. $\frac{60}{100}$ **2.** $\frac{289}{1000}$ **3.** $\frac{110}{100}$ **4.** $\frac{42}{100,000}$ **5.** $\frac{320}{100}$ **6.** $\frac{347}{10,000}$
7. 70%   **8.** 56%   **9.** 3.17%   **10.** 34%   **11.** 37.5%, or $37\frac{1}{2}$%
12. $16.\overline{6}$%, or $16\frac{2}{3}$%

Calculator Corner, p. 23
1. 40,353,607   **2.** 10.4976   **3.** 12,812.904   **4.** $\frac{64}{729}$

17 Exponential Notation
1. 5^4 **2.** 3^5 **3.** 4.2^3 **4.** 9^2 **5.** $\left(\frac{2}{11}\right)^3$ **6.** 64   **7.** 1
8. 6.25   **9.** 1,000,000   **10.** $\frac{27}{8}$

Calculator Corner, p. 24
1. 81   **2.** 2   **3.** 5932   **4.** 743.027   **5.** 783   **6.** 228,112.96

18 Order of Operations
1. 25   **2.** 33   **3.** 28   **4.** 24   **5.** 4   **6.** 24   **7.** 80   **8.** 9
9. 50,000   **10.** 1   **11.** 24   **12.** 15

CHAPTER 1

Exercise Set 1.1, p. 32
RC1. Algebraic expression **RC2.** Algebraic equation
RC3. Algebraic expression **RC4.** Algebraic equation
CC1. Division **CC2.** Multiplication **CC3.** Multiplication
CC4. Division
1. 32 min; 69 min; 81 min **3.** 260 mi **5.** 576 in^2 **7.** 1935 m^2
9. 56 **11.** 8 **13.** 1 **15.** 6 **17.** 2 **19.** $b + 7$, or $7 + b$
21. $c - 12$ **23.** $a + b$, or $b + a$ **25.** $x \div y$, or $\frac{x}{y}$, or x/y
27. $x + w$, or $w + x$ **29.** $n - m$ **31.** $2z$ **33.** $3m$, or $m \cdot 3$
35. $4a + 6$, or $6 + 4a$ **37.** $xy - 8$ **39.** $2t - 5$ **41.** $3n + 11$, or $11 + 3n$ **43.** $4x + 3y$, or $3y + 4x$ **45.** $s + 0.05s$
47. $65t$ miles **49.** $50 - x$ **51.** $12.50n$ **53.** $2 \cdot 2 \cdot 3 \cdot 3 \cdot 3$
54. $2 \cdot 2 \cdot 2 \cdot 2 \cdot 2 \cdot 2 \cdot 3$ **55.** $\frac{41}{56}$ **56.** $\frac{31}{54}$ **57.** 0.0515 **58.** 43,500
59. 96 **60.** 396 **61.** $\frac{1}{4}$ **63.** 0

Calculator Corner, p. 39
1. 8.717797887 **2.** 17.80449381 **3.** 67.08203932
4. 35.4807407 **5.** 3.141592654 **6.** 91.10618695
7. 530.9291585 **8.** 138.8663978

Calculator Corner, p. 40
1. -0.75 **2.** -0.125 **3.** -0.6875 **4.** -3.5

Calculator Corner, p. 42
1. 5 **2.** 17 **3.** 0 **4.** 6.48 **5.** 12.7 **6.** $\frac{5}{7}$

Exercise Set 1.2, p. 43
RC1. True **RC2.** False **RC3.** True **CC1.** H **CC2.** E
CC3. J **CC4.** D **CC5.** B **CC6.** G
1. 24; -2 **3.** 7,200,000,000,000; -460 **5.** 2073; -282

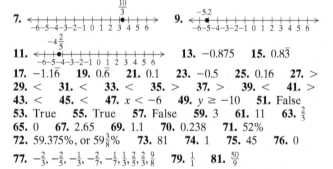

17. $-1.1\overline{6}$ **19.** $0.\overline{6}$ **21.** 0.1 **23.** -0.5 **25.** 0.16 **27.** >
29. < **31.** < **33.** < **35.** > **37.** > **39.** < **41.** >
43. < **45.** < **47.** $x < -6$ **49.** $y \geq -10$ **51.** False
53. True **55.** True **57.** False **59.** 3 **61.** 11 **63.** $\frac{2}{3}$
65. 0 **67.** 2.65 **69.** 1.1 **70.** 0.238 **71.** 52%
72. 59.375%, or $59\frac{3}{8}$% **73.** 81 **74.** 1 **75.** 45 **76.** 0
77. $-\frac{2}{3}, -\frac{2}{5}, -\frac{1}{3}, -\frac{2}{7}, -\frac{1}{7}, \frac{1}{5}, \frac{2}{3}, \frac{9}{8}$ **79.** $\frac{1}{1}$ **81.** $\frac{50}{9}$

Exercise Set 1.3, p. 51
RC1. add; negative **RC2.** subtract; negative
RC3. opposites **RC4.** identity **CC1.** right; right
CC2. left; left **CC3.** right; left **CC4.** left; right
1. -7 **3.** -6 **5.** 0 **7.** -8 **9.** -7 **11.** -27 **13.** 0
15. -42 **17.** 0 **19.** 0 **21.** 3 **23.** -9 **25.** 7 **27.** 0
29. 35 **31.** -3.8 **33.** -8.1 **35.** $-\frac{1}{5}$ **37.** $-\frac{7}{9}$ **39.** $-\frac{3}{8}$
41. $-\frac{19}{24}$ **43.** $\frac{1}{24}$ **45.** $\frac{8}{15}$ **47.** $\frac{16}{45}$ **49.** 37 **51.** 50 **53.** -24
55. 26.9 **57.** -8 **59.** $\frac{13}{8}$ **61.** -43 **63.** $\frac{4}{3}$ **65.** 24 **67.** $\frac{3}{8}$
69. 13,796 ft **71.** -3°F **73.** Profit of $4300 **75.** He owes $85.
77. 39 **78.** $y > -3$ **79.** $-0.08\overline{3}$ **80.** 0.625 **81.** 0
82. 21.4 **83.** All positive numbers **85.** B

Exercise Set 1.4, p. 57
RC1. opposite **RC2.** opposite **RC3.** difference
CC1. (c) **CC2.** (b) **CC3.** (d) **CC4.** (a)
1. -7 **3.** -6 **5.** 0 **7.** -4 **9.** -7 **11.** -6 **13.** 0
15. 14 **17.** 11 **19.** -14 **21.** 5 **23.** -1 **25.** 18 **27.** -3
29. -21 **31.** 5 **33.** -8 **35.** 12 **37.** -23 **39.** -68

41. -73 **43.** 116 **45.** 0 **47.** -1 **49.** $\frac{1}{12}$ **51.** $-\frac{17}{12}$
53. $\frac{1}{8}$ **55.** 19.9 **57.** -8.6 **59.** -0.01 **61.** -193 **63.** 500
65. -2.8 **67.** -3.53 **69.** $-\frac{1}{2}$ **71.** $\frac{6}{7}$ **73.** $-\frac{41}{30}$ **75.** $-\frac{2}{15}$
77. $-\frac{1}{48}$ **79.** $-\frac{43}{60}$ **81.** 37 **83.** -62 **85.** -139 **87.** 6
89. 108.5 **91.** $\frac{1}{4}$ **93.** 30,383 ft **95.** $347.94 **97.** 3780 m
99. 381 ft **101.** 1130°F **103.** $y + 7$, or $7 + y$ **104.** $t - 41$
105. $a - h$ **106.** $6c$, or $c \cdot 6$ **107.** $r + s$, or $s + r$ **108.** $y - x$
109. False; $3 - 0 \neq 0 - 3$ **111.** True **113.** True

Mid-Chapter Review: Chapter 1, p. 61
1. True **2.** False **3.** True **4.** False **5.** $-x = -(-4) = 4$; $-(-x) = -(-(-4)) = -(4) = -4$
6. $5 - 13 = 5 + (-13) = -8$ **7.** $-6 - 7 = -6 + (-7) = -13$
8. 4 **9.** 11 **10.** $3y$ **11.** $n - 5$ **12.** 450; -79
13. [number line with point at -3.5, from -6 to 6] **14.** -0.8 **15.** $2.\overline{3}$ **16.** <
17. > **18.** False **19.** True **20.** $5 > y$ **21.** $t \leq -3$
22. 15.6 **23.** 18 **24.** 0 **25.** $\frac{12}{5}$ **26.** 5.6 **27.** $-\frac{7}{4}$
28. 0 **29.** 49 **30.** 19 **31.** 2.3 **32.** -2 **33.** $-\frac{1}{8}$
34. 0 **35.** -17 **36.** $-\frac{11}{24}$ **37.** -8.1 **38.** -9 **39.** -2
40. -10.4 **41.** 16 **42.** $\frac{7}{20}$ **43.** -12 **44.** -4 **45.** $-\frac{4}{3}$
46. -1.8 **47.** 13 **48.** 9 **49.** -23 **50.** 75 **51.** 14
52. 33°C **53.** $54.80 **54.** Answers may vary. Three examples are $\frac{6}{13}$, -23.8, and $\frac{43}{5}$. These are rational numbers because they can be named in the form $\frac{a}{b}$, where a and b are integers and b is not 0. They are not integers, however, because they are neither whole numbers nor the opposites of whole numbers.
55. Answers may vary. Three examples are π, $-\sqrt{7}$, and $0.31311311131111\ldots$. Irrational numbers cannot be written as the quotient of two integers. Real numbers that are not rational are irrational. Decimal notation for rational numbers either terminates or repeats. Decimal notation for irrational numbers neither terminates nor repeats. **56.** Answers may vary. If we think of the addition on the number line, we start at 0, move to the left to a negative number, and then move to the left again. This always brings us to a point on the negative portion of the number line. **57.** Yes; consider $m - (-n)$, where both m and n are positive. Then $m - (-n) = m + n$. Now $m + n$, the sum of two positive numbers, is positive.

Exercise Set 1.5, p. 66
RC1. negative **RC2.** positive **RC3.** positive
RC4. negative **CC1.** -9 **CC2.** 9 **CC3.** $-\frac{1}{4}$ **CC4.** $-\frac{1}{4}$
1. -8 **3.** -24 **5.** -72 **7.** 16 **9.** 42 **11.** -120
13. -238 **15.** 1200 **17.** 98 **19.** -72 **21.** -12.4
23. 30 **25.** 21.7 **27.** $-\frac{2}{5}$ **29.** $\frac{1}{12}$ **31.** -17.01 **33.** 420
35. $\frac{2}{7}$ **37.** -60 **39.** 150 **41.** 50.4 **43.** $\frac{10}{189}$ **45.** -960
47. 17.64 **49.** $-\frac{5}{784}$ **51.** 0 **53.** -720 **55.** $-30,240$
57. 1 **59.** 16, -16; 16, -16 **61.** $\frac{4}{25}, -\frac{4}{25}; \frac{4}{25}, -\frac{4}{25}$
63. $-9, -9; -9, -9$ **65.** 441, -147; 441, -147 **67.** 20; 20
69. $-2; 2$ **71.** -24°C **73.** -20 lb **75.** $12.71 **77.** -32 m
79. 38°F **81.** 2 **82.** $\frac{4}{15}$ **83.** $-\frac{1}{3}$ **84.** -4.3 **85.** 44
86. $-\frac{1}{12}$ **87.** True **88.** False **89.** False **90.** False
91. A **93.** Largest quotient: $10 \div \frac{1}{5} = 50$; smallest quotient: $-5 \div \frac{1}{5} = -25$

Calculator Corner, p. 74
1. 2.7 **2.** -9.5 **3.** -0.8 **4.** 14.44

Exercise Set 1.6, p. 75
RC1. opposites **RC2.** 1 **RC3.** 0 **RC4.** reciprocals
CC1. 0 **CC2.** 1 **CC3.** 0 **CC4.** 0 **CC5.** 1
1. -8 **3.** -14 **5.** -3 **7.** 3 **9.** -8 **11.** 2 **13.** -12
15. -8 **17.** Not defined **19.** 0 **21.** $\frac{7}{15}$ **23.** $-\frac{13}{47}$ **25.** $\frac{1}{13}$
27. $-\frac{1}{32}$ **29.** -7.1 **31.** 9 **33.** $4y$ **35.** $\frac{3b}{2a}$ **37.** $4 \cdot \left(\frac{1}{17}\right)$

39. $8 \cdot \left(-\frac{1}{13}\right)$ **41.** $13.9 \cdot \left(-\frac{1}{1.5}\right)$ **43.** $\frac{2}{3} \cdot \left(-\frac{5}{4}\right)$ **45.** $x \cdot y$
47. $(3x+4)\left(\frac{1}{5}\right)$ **49.** $-\frac{9}{8}$ **51.** $\frac{5}{3}$ **53.** $\frac{9}{14}$ **55.** $\frac{9}{64}$ **57.** $-\frac{5}{4}$
59. $-\frac{27}{5}$ **61.** $\frac{11}{13}$ **63.** -2 **65.** -16.2 **67.** -2.5
69. -1.25 **71.** Not defined **73.** The percent increase is about 43%. **75.** The percent decrease is about 3%.
77. $-2.125°$F per minute **79.** $-\frac{1}{4}$ **80.** 5 **81.** -42
82. -48 **83.** 8.5 **84.** $-\frac{1}{8}$ **85.** $-0.0\overline{9}$ **86.** $0.91\overline{6}$
87. 3.75 **88.** $-3.\overline{3}$ **89.** $\frac{1}{-10.5}$; -10.5, the reciprocal of the reciprocal, is the original number. **91.** Negative **93.** Positive
95. Negative

Exercise Set 1.7, p. 87

RC1. (g) **RC2.** (h) **RC3.** (f) **RC4.** (e) **RC5.** (d)
RC6. (a) **RC7.** (b) **CC1.** 5 **CC2.** 7 **CC3.** 8 **CC4.** d
1. $\frac{3y}{5y}$ **3.** $\frac{10x}{15x}$ **5.** $\frac{2x}{x^2}$ **7.** $-\frac{3}{2}$ **9.** $-\frac{7}{6}$ **11.** $\frac{4s}{3}$ **13.** $8+y$
15. nm **17.** $xy+9$, or $9+yx$, or $yx+9$ **19.** $c+ab$, or $ba+c$, or $c+ba$ **21.** $(a+b)+2$ **23.** $8(xy)$
25. $a+(b+3)$ **27.** $(3a)b$ **29.** $2+(b+a), (2+a)+b, (b+2)+a$; answers may vary **31.** $(5+w)+v, (v+5)+w, (w+v)+5$; answers may vary **33.** $(3x)y, y(x \cdot 3), 3(yx)$; answers may vary **35.** $a(7b), b(7a), (7b)a$; answers may vary
37. $2b+10$ **39.** $7+7t$ **41.** $30x+12$ **43.** $7x+28+42y$
45. $7x-21$ **47.** $-3x+21$ **49.** $\frac{2}{3}b-4$ **51.** $7.3x-14.6$
53. $-\frac{3}{5}x+\frac{3}{5}y-6$ **55.** $45x+54y-72$ **57.** $-4x+12y+8z$
59. $-3.72x+9.92y-3.41$ **61.** $4x, 3z$ **63.** $7x, 8y, -9z$
65. $2(x+2)$ **67.** $5(6+y)$ **69.** $7(2x+3y)$ **71.** $7(2t-1)$
73. $8(x-3)$ **75.** $6(3a-4b)$ **77.** $-4(y-8)$, or $4(-y+8)$
79. $5(x+2+3y)$ **81.** $8(2m-4n+1)$ **83.** $4(3a+b-6)$
85. $2(4x+5y-11)$ **87.** $a(x-1)$ **89.** $a(x-y-z)$
91. $-6(3x-2y-1)$, or $6(-3x+2y+1)$ **93.** $\frac{1}{3}(2x-5y+1)$
95. $6(6x-y+3z)$ **97.** $19a$ **99.** $9a$ **101.** $8x+9z$
103. $7x+15y^2$ **105.** $-19a+88$ **107.** $4t+6y-4$
109. b **111.** $\frac{13}{4}y$ **113.** $8x$ **115.** $5n$ **117.** $-16y$
119. $17a-12b-1$ **121.** $4x+2y$ **123.** $7x+y$
125. $0.8x+0.5y$ **127.** $\frac{35}{6}a+\frac{3}{2}b-42$ **129.** 38
130. -4.9 **131.** 4 **132.** 10 **133.** $-\frac{4}{49}$ **134.** -106
135. 1 **136.** -34 **137.** 180 **138.** $\frac{4}{13}$ **139.** True
140. False **141.** True **142.** True **143.** Not equivalent; $3 \cdot 2 + 5 \neq 3 \cdot 5 + 2$ **145.** Equivalent; commutative law of addition **147.** $q(1+r+rs+rst)$

Calculator Corner, p. 96

1. -16 **2.** 9 **3.** 117,649 **4.** $-1,419,857$ **5.** $-117,649$
6. $-1,419,857$ **7.** -4 **8.** -2

Exercise Set 1.8, p. 97

RC1. (d) **RC2.** (b) **RC3.** (a) **RC4.** (c)
CC1. Multiplication **CC2.** Addition **CC3.** Subtraction
CC4. Division **CC5.** Division **CC6.** Multiplication
1. $2x-7$ **3.** $-8+x$ **5.** $-4a+3b-7c$
7. $-6x+8y-5$ **9.** $-3x+5y+6$ **11.** $8x+6y+43$
13. $5x-3$ **15.** $-3a+9$ **17.** $5x-6$ **19.** $-19x+2y$
21. $9y-25z$ **23.** $-7x+10y$ **25.** $37a-23b+35c$
27. 7 **29.** -40 **31.** 19 **33.** $12x+30$ **35.** $3x+30$
37. $9x-18$ **39.** $-4x-64$ **41.** -7 **43.** -1 **45.** -16
47. -334 **49.** 14 **51.** 1880 **53.** 12 **55.** 8 **57.** -86
59. 37 **61.** -1 **63.** -10 **65.** -67 **67.** -7988 **69.** -3000
71. 60 **73.** 1 **75.** 10 **77.** $-\frac{13}{45}$ **79.** $-\frac{23}{18}$ **81.** -122
83. 18 **84.** 35 **85.** 0.4 **86.** $\frac{15}{2}$ **87.** $-\frac{1}{9}$ **88.** $\frac{3}{7}$ **89.** -25
90. -35 **91.** 25 **92.** 35 **93.** $-2x-f$ **95.** (a) 52; 52; 28.130169; (b) $-24; -24; -108.307025$ **97.** -6

Summary and Review: Chapter 1, p. 101

Vocabulary Reinforcement

1. integers **2.** additive inverses **3.** commutative law
4. identity property of 1 **5.** associative law
6. multiplicative inverses **7.** identity property of 0

Concept Reinforcement

1. True **2.** True **3.** False **4.** False

Study Guide

1. 14 **2.** $<$ **3.** $\frac{5}{4}$ **4.** -8.5 **5.** -2 **6.** 56 **7.** -8
8. $\frac{9}{20}$ **9.** $\frac{5}{3}$ **10.** $5x+15y-20z$ **11.** $9(3x+y-4z)$
12. $5a-2b$ **13.** $4a-4b$ **14.** -2

Review Exercises

1. 4 **2.** $19\%x$, or $0.19x$ **3.** $620, -125$ **4.** 38 **5.** 126
6. [number line with point at -2.5, from -6 to 6]   **7.** [number line with point at $\frac{8}{9}$, from -6 to 6]
8. $<$ **9.** $>$ **10.** $>$ **11.** $<$ **12.** $x > -3$ **13.** True
14. False **15.** -3.8 **16.** $\frac{3}{4}$ **17.** $\frac{8}{3}$ **18.** $-\frac{1}{7}$ **19.** 34
20. 5 **21.** -3 **22.** -4 **23.** -5 **24.** 1 **25.** $-\frac{7}{5}$
26. -7.9 **27.** 54 **28.** -9.18 **29.** $-\frac{2}{7}$ **30.** -210 **31.** -7
32. -3 **33.** $\frac{3}{4}$ **34.** 24.8 **35.** -2 **36.** 2 **37.** -2
38. 8-yd gain **39.** $-\$360$ **40.** $\$4.64$ **41.** The percent increase is about 64%. **42.** $15x-35$ **43.** $-8x+10$
44. $4x+15$ **45.** $-24+48x$ **46.** $2(x-7)$ **47.** $-6(x-1)$, or $6(-x+1)$ **48.** $5(x+2)$ **49.** $-3(x-4y+4)$, or $3(-x+4y-4)$ **50.** $7a-3b$ **51.** $-2x+5y$ **52.** $5x-y$
53. $-a+8b$ **54.** $-3a+9$ **55.** $-2b+21$ **56.** 6
57. $12y-34$ **58.** $5x+24$ **59.** $-15x+25$ **60.** D **61.** B
62. $-\frac{5}{8}$ **63.** -2.1 **64.** 1000 **65.** $4a+2b$

Understanding Through Discussion and Writing

1. The sum of each pair of opposites such as -50 and 50, -49 and 49, and so on, is 0. The sum of these sums and the remaining integer, 0, is 0. **2.** The product of an even number of negative numbers is positive, and the product of an odd number of negative numbers is negative. Now $(-7)^8$ is the product of 8 factors of -7 so it is positive, and $(-7)^{11}$ is the product of 11 factors of -7 so it is negative. **3.** Consider $\frac{a}{b}=q$, where a and b are both negative numbers. Then $q \cdot b = a$, so q must be a positive number in order for the product to be negative.
4. Consider $\frac{a}{b}=q$, where a is a negative number and b is a positive number. Then $q \cdot b = a$, so q must be a negative number in order for the product to be negative. **5.** We use the distributive law when we collect like terms even though we might not always write this step. **6.** Jake expects the calculator to multiply 2 and 3 first and then divide 18 by that product. This procedure does not follow the rules for order of operations.

Test: Chapter 1, p. 107

1. [1.1a] 6 **2.** [1.1b] $x-9$ **3.** [1.2d] $>$ **4.** [1.2d] $<$
5. [1.2d] $>$ **6.** [1.2d] $-2 > x$ **7.** [1.2d] True **8.** [1.2e] 7
9. [1.2e] $\frac{9}{4}$ **10.** [1.2e] 2.7 **11.** [1.3b] $-\frac{2}{3}$ **12.** [1.3b] 1.4
13. [1.6b] $-\frac{1}{2}$ **14.** [1.6b] $\frac{7}{4}$ **15.** [1.3b] 8 **16.** [1.4a] 7.8
17. [1.3a] -8 **18.** [1.3a] $\frac{7}{40}$ **19.** [1.4a] 10 **20.** [1.4a] -2.5
21. [1.4a] $\frac{7}{8}$ **22.** [1.5a] -48 **23.** [1.5a] $\frac{3}{16}$ **24.** [1.6a] -9
25. [1.6c] $\frac{3}{4}$ **26.** [1.6c] -9.728 **27.** [1.2e], [1.8d] -173
28. [1.8d] -5 **29.** [1.3c], [1.4b] Up 15 points **30.** [1.4b] 2244 m
31. [1.5b] 16,080 **32.** [1.6d] $-0.75°$C each minute
33. [1.7c] $18-3x$ **34.** [1.7c] $-5y+5$ **35.** [1.7d] $2(6-11x)$
36. [1.7d] $7(x+3+2y)$ **37.** [1.4a] 12 **38.** [1.8b] $2x+7$
39. [1.8b] $9a-12b-7$ **40.** [1.8c] $68y-8$ **41.** [1.8d] -4
42. [1.8d] 448 **43.** [1.2d] B **44.** [1.2e], [1.8d] 15
45. [1.8c] $4a$ **46.** [1.7e] $4x+4y$

CHAPTER 2

Exercise Set 2.1, p. 114
RC1. (c) **RC2.** (a) **RC3.** (b) **RC4.** (d) **CC1.** (f) **CC2.** (c) **CC3.** (e) **CC4.** (a)
1. Yes **3.** No **5.** No **7.** Yes **9.** Yes **11.** No **13.** 4
15. -20 **17.** -14 **19.** -18 **21.** 15 **23.** -14 **25.** 2
27. 20 **29.** -6 **31.** $6\frac{1}{2}$ **33.** 19.9 **35.** $\frac{7}{3}$ **37.** $-\frac{7}{4}$ **39.** $\frac{41}{24}$
41. $-\frac{1}{20}$ **43.** 5.1 **45.** 12.4 **47.** -5 **49.** $1\frac{5}{6}$ **51.** $-\frac{10}{21}$
53. $-\frac{3}{2}$ **54.** -5.2 **55.** $-\frac{1}{24}$ **56.** 172.72 **57.** $\$83 - x$
58. $65t$ miles **59.** $-\frac{26}{15}$ **61.** -10 **63.** All real numbers

Exercise Set 2.2, p. 119
RC1. (d) **RC2.** (c) **RC3.** (a) **RC4.** (b) **CC1.** (f) **CC2.** (d) **CC3.** (a) **CC4.** (b)
1. 6 **3.** 9 **5.** 12 **7.** -40 **9.** 1 **11.** -7 **13.** -6
15. 6 **17.** -63 **19.** -48 **21.** 36 **23.** -9 **25.** -21
27. $-\frac{3}{5}$ **29.** $-\frac{3}{2}$ **31.** $\frac{9}{2}$ **33.** 7 **35.** -7 **37.** 8 **39.** 15.9
41. -50 **43.** -14 **45.** $7x$ **46.** $-x + 5$ **47.** $8x + 11$
48. $-32y$ **49.** $x - 4$ **50.** $-5x - 23$ **51.** $-10y - 42$
52. $-22a + 4$ **53.** $8r$ miles **54.** $\frac{1}{2}b \cdot 10$ m², or $5b$ m²
55. -8655 **57.** No solution **59.** No solution **61.** $\frac{b}{3a}$
63. $\frac{4b}{a}$

Calculator Corner, p. 125
1. Left to the student

Exercise Set 2.3, p. 129
RC1. collect **RC2.** clear **RC3.** distributive
RC4. multiplication **CC1.** (d) **CC2.** (a) **CC3.** (c)
CC4. (e) **CC5.** (b)
1. 5 **3.** 8 **5.** 10 **7.** 14 **9.** -8 **11.** -8 **13.** -7
15. 12 **17.** 6 **19.** 4 **21.** 6 **23.** -3 **25.** 1 **27.** 6
29. -20 **31.** 7 **33.** 2 **35.** 5 **37.** 2 **39.** 10 **41.** 4
43. 0 **45.** -1 **47.** $-\frac{4}{3}$ **49.** $\frac{2}{5}$ **51.** -2 **53.** -4 **55.** $\frac{4}{5}$
57. $-\frac{28}{27}$ **59.** 6 **61.** 2 **63.** No solution **65.** All real numbers **67.** 6 **69.** 8 **71.** 1 **73.** 17 **75.** $-\frac{5}{3}$
77. All real numbers **79.** No solution **81.** -3 **83.** 2
85. $\frac{4}{7}$ **87.** No solution **89.** All real numbers **91.** $-\frac{51}{31}$
93. -6.5 **94.** -75.14 **95.** $7(x - 3 - 2y)$ **96.** $8(y - 11x + 1)$
97. -160 **98.** $-17x + 18$ **99.** $91x - 242$ **100.** 0.25
101. $-\frac{5}{32}$ **103.** $\frac{52}{45}$

Exercise Set 2.4, p. 136
RC1. True **RC2.** False **RC3.** True **CC1.** Yes **CC2.** No **CC3.** Yes
1. $14\frac{1}{3}$ meters per cycle **3.** 10.5 calories per ounce
5. (a) 337.5 mi; **(b)** $t = \frac{d}{r}$ **7. (a)** 1423 students; **(b)** $n = 15f$
9. $x = \frac{y}{5}$ **11.** $c = \frac{a}{b}$ **13.** $m = n - 11$ **15.** $x = y + \frac{3}{5}$
17. $x = y - 13$ **19.** $x = y - b$ **21.** $x = 5 - y$ **23.** $x = a - y$
25. $y = \frac{5x}{8}$, or $\frac{5}{8}x$ **27.** $x = \frac{By}{A}$ **29.** $t = \frac{W - b}{m}$
31. $x = \frac{y - c}{b}$ **33.** $h = \frac{A}{b}$ **35.** $w = \frac{P - 2l}{2}$, or $\frac{1}{2}P - l$
37. $a = 2A - b$ **39.** $b = 3A - a - c$ **41.** $t = \frac{A - b}{a}$
43. $x = \frac{c - By}{A}$ **45.** $a = \frac{F}{m}$ **47.** $c^2 = \frac{E}{m}$ **49.** $t = \frac{3k}{v}$
51. 7 **52.** $-21a + 12b$ **53.** -13.2 **54.** $-\frac{3}{2}$ **55.** $-35\frac{1}{2}$

56. $-\frac{1}{6}$ **57.** -9.325 **58.** $3\frac{3}{4}$ **59.** $\frac{11}{8}$ **60.** -1 **61.** -3
62. $\frac{9}{7}$ **63.** $b = \frac{Ha - 2}{H}$, or $a - \frac{2}{H}$; $a = \frac{2 + Hb}{H}$, or $\frac{2}{H} + b$
65. A quadruples. **67.** A increases by $2h$ units.

Mid-Chapter Review: Chapter 2, p. 141
1. False **2.** True **3.** True **4.** False
5. $\quad x + 5 = -3$ **6.** $\quad -6x = 42$
$\quad x + 5 - 5 = -3 - 5 \qquad \quad \frac{-6x}{-6} = \frac{42}{-6}$
$\quad x + 0 = -8 \qquad\qquad\qquad 1 \cdot x = -7$
$\quad x = -8 \qquad\qquad\qquad\quad\; x = -7$

7. $\quad 5y + z = t$
$\quad 5y + z - z = t - z$
$\quad 5y = t - z$
$\quad \frac{5y}{5} = \frac{t - z}{5}$
$\quad y = \frac{t - z}{5}$

8. 6 **9.** -12 **10.** 7 **11.** -10 **12.** 20 **13.** 5 **14.** $\frac{3}{4}$
15. -1.4 **16.** 6 **17.** -17 **18.** -9 **19.** 17 **20.** 21
21. 18 **22.** -15 **23.** $-\frac{3}{2}$ **24.** 1 **25.** -3 **26.** $\frac{3}{2}$ **27.** -1
28. 3 **29.** -7 **30.** 4 **31.** 2 **32.** $\frac{9}{8}$ **33.** $-\frac{21}{5}$ **34.** 9
35. -2 **36.** 0 **37.** All real numbers **38.** No solution
39. $-\frac{13}{2}$ **40.** All real numbers **41.** $b = \frac{A}{4}$ **42.** $x = y + 1.5$
43. $m = s - n$ **44.** $t = \frac{9w}{4}$ **45.** $t = \frac{B + c}{a}$
46. $y = 2M - x - z$ **47.** Equivalent expressions have the same value for all possible replacements for the variable(s). Equivalent equations have the same solution(s). **48.** The equations are not equivalent because they do not have the same solutions. Although 5 is a solution of both equations, -5 is a solution of $x^2 = 25$ but not of $x = 5$. **49.** For an equation $x + a = b$, add the opposite of a (or subtract a) on both sides of the equation. **50.** The student probably added $\frac{1}{3}$ on both sides of the equation rather than adding $-\frac{1}{3}$ (or subtracting $\frac{1}{3}$) on both sides. The correct solution is -2. **51.** For an equation $ax = b$, multiply by $1/a$ (or divide by a) on both sides of the equation. **52.** Answers may vary. A walker who knows how far and how long she walks each day wants to know her average speed each day.

Exercise Set 2.5, p. 147
RC1. percent **RC2.** of **RC3.** base **RC4.** percent
CC1. (d) **CC2.** (b) **CC3.** (e) **CC4.** (a) **CC5.** (f)
CC6. (c)
1. 20% **3.** 150 **5.** 546 **7.** 24% **9.** 2.5 **11.** 5%
13. 25% **15.** 84 **17.** 24% **19.** 16% **21.** $46.\overline{6}$, or $46\frac{2}{3}$
23. 0.8 **25.** 5 **27.** 40 **29.** 811 million **31.** 5274 million
33. 1764 million **35.** 46.4 million bags **37.** $\$244.40$ **39.** 21%
41. (a) 12.5%; **(b)** $\$13.50$ **43. (a)** $\$31$; **(b)** $\$35.65$ **45.** About 85,821 acres **47.** The percent increase is about 8.6%. **49.** The percent decrease is about 33.2%. **51.** The percent increase is about 9.7%. **53.** The percent decrease is about 17.0%.
55. $12 + 3q$ **56.** $5x - 21$ **57.** $\frac{15w}{8}$ **58.** $-\frac{3}{2}$ **59.** 44
60. $x + 8$ **61.** 6 ft 7 in.

Translating for Success, p. 162
1. B **2.** H **3.** G **4.** N **5.** J **6.** C **7.** L **8.** E
9. F **10.** D

Exercise Set 2.6, p. 163
RC1. Familiarize **RC2.** Translate **RC3.** Solve
RC4. Check **RC5.** State **CC1.** $x + 2$; $x + 2$; 32
CC2. $0.06x$, or 6% x; $0.06x$; 36.57

A-4   Answers

1. 91,179 students **3.** 180 in.; 60 in. **5.** 1522 Medals of Honor **7.** 4.37 mi **9.** 1204 and 1205 **11.** 41, 42, 43 **13.** 61, 63, 65 **15.** 36 in. × 110 in. **17.** $63 **19.** $24.95 **21.** 11 visits **23.** 28°, 84°, 68° **25.** 33°, 38°, 109° **27.** $350 **29.** $852.94 **31.** 18 mi **33.** $38.60 **35.** 89 and 96 **37.** −12 **39.** $-\frac{47}{40}$ **40.** $-\frac{17}{40}$ **41.** $-\frac{3}{10}$ **42.** $-\frac{32}{15}$ **43.** −10 **44.** 1.6 **45.** 409.6 **46.** −9.6 **47.** −41.6 **48.** 0.1 **49.** $yz + 12$, $zy + 12$, or $12 + zy$ **50.** $c + (4 + d)$ **51.** 120 apples **53.** About 0.65 in.

Exercise Set 2.7, p. 176

RC1. Not equivalent **RC2.** Not equivalent **RC3.** Equivalent **RC4.** Equivalent **CC1.** ≥ **CC2.** < **CC3.** ≥ **CC4.** <
1. (a) Yes; (b) yes; (c) no; (d) yes; (e) yes **3.** (a) No; (b) no; (c) no; (d) yes; (e) no
5. $x > 4$ **7.** $t < -3$
9. $m \geq -1$ **11.** $-3 < x \leq 4$
13. $0 < x < 3$ **15.** $\{x | x > -5\}$
17. $\{x | x \leq -18\}$; **19.** $\{y | y > -5\}$
21. $\{x | x > 2\}$ **23.** $\{x | x \leq -3\}$ **25.** $\{x | x < 4\}$
27. $\{t | t > 14\}$ **29.** $\{y | y \leq \frac{1}{4}\}$ **31.** $\{x | x > \frac{7}{12}\}$
33. $\{x | x < 7\}$; **35.** $\{x | x < 3\}$;
37. $\{y | y \geq -\frac{2}{5}\}$ **39.** $\{x | x \geq -6\}$ **41.** $\{y | y \leq 4\}$
43. $\{x | x > \frac{17}{3}\}$ **45.** $\{y | y < -\frac{1}{14}\}$ **47.** $\{x | x \leq \frac{3}{10}\}$
49. $\{x | x < 8\}$ **51.** $\{x | x \leq 6\}$ **53.** $\{x | x < -3\}$
55. $\{x | x > -3\}$ **57.** $\{x | x \leq 7\}$ **59.** $\{x | x > -10\}$
61. $\{y | y > 2\}$ **63.** $\{y | y \geq 3\}$ **65.** $\{y | y > -2\}$
67. $\{x | x > -4\}$ **69.** $\{x | x \leq 9\}$ **71.** $\{y | y \leq -3\}$
73. $\{y | y < 6\}$ **75.** $\{m | m \geq 6\}$ **77.** $\{t | t < -\frac{5}{3}\}$
79. $\{r | r > -3\}$ **81.** $\{x | x \geq -\frac{57}{34}\}$ **83.** $\{x | x > -2\}$
85. $-\frac{5}{8}$ **86.** −1.11 **87.** −9.4 **88.** $-\frac{7}{8}$ **89.** 140 **90.** 41
91. $-2x - 23$ **92.** $37x - 1$ **93.** (a) Yes; (b) yes; (c) no; (d) no; (e) no; (f) yes; (g) yes **95.** No solution

Exercise Set 2.8, p. 182

RC1. $r \leq q$ **RC2.** $q \leq r$ **RC3.** $r < q$ **RC4.** $q \leq r$
RC5. $r < q$ **RC6.** $r \leq q$ **CC1.** No **CC2.** Yes
CC3. Yes **CC4.** No
1. $n \geq 7$ **3.** $w > 2$ kg **5.** 90 mph $< s <$ 110 mph **7.** $w \leq 20$ hr
9. $c \geq \$3.20$ **11.** $x > 8$ **13.** $y \leq -4$ **15.** $n \geq 1300$
17. $W \leq 500$ L **19.** $3x + 2 < 13$ **21.** $\{x | x \leq 84\}$
23. $\{C | C < 1063°\}$ **25.** $\{Y | Y \geq 1935\}$
27. 15 or fewer copies **29.** $\{L | L \geq 5 \text{ in.}\}$ **31.** 5 min or more
33. 2 courses **35.** 4 servings or more **37.** Lengths greater than or equal to 92 ft; lengths less than or equal to 92 ft
39. Lengths less than 21.5 cm **41.** The blue-book value is greater than or equal to $10,625. **43.** It has at least 16 g of fat. **45.** Heights greater than or equal to 4 ft **47.** Dates at least 6 weeks after July 1 **49.** 21 calls or more **51.** 40
52. −22 **53.** 12 **54.** 6 **55.** All real numbers **56.** No solution **57.** 7.5% **58.** 31 **59.** 1250 **60.** $83.\overline{3}\%$, or $83\frac{1}{3}\%$
61. Temperatures between −15°C and $-9\frac{4}{9}$°C **63.** They contain at least 7.5 g of fat per serving.

Summary and Review: Chapter 2, p. 188

Vocabulary Reinforcement

1. solution **2.** addition principle **3.** multiplication principle **4.** inequality **5.** equivalent

Concept Reinforcement

1. True **2.** True **3.** False **4.** True

Study Guide

1. −12 **2.** All real numbers **3.** No solution **4.** $b = \frac{2A}{h}$
5. $x > 1$ **6.** $x \leq -1$
7. $\{y | y > -4\}$

Review Exercises

1. −22 **2.** 1 **3.** 25 **4.** 9.99 **5.** $\frac{1}{4}$ **6.** 7 **7.** −192
8. $-\frac{7}{3}$ **9.** $-\frac{15}{64}$ **10.** −8 **11.** 4 **12.** −5 **13.** $-\frac{1}{3}$
14. 3 **15.** 4 **16.** 16 **17.** All real numbers **18.** 6
19. −3 **20.** 28 **21.** 4 **22.** No solution **23.** Yes **24.** No
25. Yes **26.** $\{y | y \geq -\frac{1}{2}\}$ **27.** $\{x | x \geq 7\}$ **28.** $\{y | y > 2\}$
29. $\{y | y \leq -4\}$ **30.** $\{x | x < -11\}$ **31.** $\{y | y > -7\}$
32. $\{x | x > -\frac{9}{11}\}$ **33.** $\{x | x \geq -\frac{1}{12}\}$
34. $x < 3$ **35.** $-2 < x \leq 5$
36. $y > 0$ **37.** $d = \frac{C}{\pi}$ **38.** $B = \frac{3V}{h}$
39. $a = 2A - b$ **40.** $x = \frac{y - b}{m}$ **41.** Length: 365 mi; width: 275 mi **42.** 345, 346 **43.** $2117 **44.** 27 subscriptions
45. 35°, 85°, 60° **46.** 15 **47.** 18.75% **48.** 600
49. The percent increase is about 16.5%. **50.** The percent decrease is about 39.1%. **51.** $220 **52.** $72,500
53. $138.95 **54.** 86 **55.** $\{w | w > 17 \text{ cm}\}$ **56.** C **57.** A
58. 23, −23 **59.** 20, −20 **60.** $a = \frac{y - 3}{2 - b}$

Understanding Through Discussion and Writing

1. The end result is the same either way. If s is the original salary, the new salary after a 5% raise followed by an 8% raise is $1.08(1.05s)$. If the raises occur the other way around, the new salary is $1.05(1.08s)$. By the commutative and associative laws of multiplication, we see that these are equal. However, it would be better to receive the 8% raise first, because this increase yields a higher salary initially than a 5% raise. **2.** No; Erin paid 75% of the original price and was offered credit for 125% of this amount, not to be used on sale items. Now, 125% of 75% is 93.75%, so Erin would have a credit of 93.75% of the original price. Since this credit can be applied only to nonsale items, she has less purchasing power than if the amount she paid were refunded and she could spend it on sale items. **3.** The inequalities are equivalent by the multiplication principle for inequalities. If we multiply on both sides of one inequality by −1, the other inequality results. **4.** For any pair of numbers, their relative position on the number line is reversed when both are multiplied by the same negative number. For example, −3 is to the left of 5 on the number line $(-3 < 5)$, but 12 is to the right of -20 $(-3(-4) > 5(-4))$. **5.** Answers may vary. Fran is more than 3 years older than Todd. **6.** Let n represent "a number." Then "five more than a number" translates to the *expression* $n + 5$, or $5 + n$, and "five is more than a number" translates to the *inequality* $5 > n$.

Test: Chapter 2, p. 193

1. [2.1b] 8 **2.** [2.1b] 26 **3.** [2.2a] −6 **4.** [2.2a] 49
5. [2.3b] −12 **6.** [2.3a] 2 **7.** [2.3a] −8 **8.** [2.1b] $-\frac{7}{20}$
9. [2.3c] 7 **10.** [2.3c] $\frac{5}{3}$ **11.** [2.3b] $\frac{5}{2}$ **12.** [2.3c] No solution
13. [2.3c] All real numbers **14.** [2.7c] $\{x | x \leq -4\}$
15. [2.7c] $\{x | x > -13\}$ **16.** [2.7d] $\{x | x \leq 5\}$
17. [2.7d] $\{y | y \leq -13\}$ **18.** [2.7d] $\{y | y \geq 8\}$

19. [2.7d] $\{x|x \le -\frac{1}{20}\}$ **20.** [2.7e] $\{x|x < -6\}$
21. [2.7e] $\{x|x \le -1\}$
22. [2.7b]   **23.** [2.7b, e]

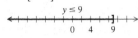

24. [2.7b]   **25.** [2.5a] 18

26. [2.5a] 16.5%   **27.** [2.5a] 40,000   **28.** [2.5a] The percent increase is about 29.3%.   **29.** [2.6a] Width: 7 cm; length: 11 cm   **30.** [2.5a] About $230,556   **31.** [2.6a] 2509, 2510, 2511   **32.** [2.6a] $880   **33.** [2.6a] 3 m, 5 m
34. [2.8b] $\{l|l \ge 174 \text{ yd}\}$ **35.** [2.8b] $\{b|b \le \$105\}$
36. [2.8b] $\{c|c \le 119{,}531\}$ **37.** [2.4b] $r = \dfrac{A}{2\pi h}$
38. [2.4b] $x = \dfrac{y-b}{8}$ **39.** [2.5a] D   **40.** [2.4b] $d = \dfrac{1-ca}{-c}$, or $\dfrac{ca-1}{c}$ **41.** [1.2e], [2.3a] 15, −15   **42.** [2.6a] 60 tickets

Cumulative Review: Chapters 1–2, p. 195

1. [1.1a] $\frac{3}{2}$ **2.** [1.1a] $\frac{15}{4}$ **3.** [1.1a] 0   **4.** [1.1b] $2w-4$
5. [1.2d] >   **6.** [1.2d] >   **7.** [1.2d] <   **8.** [1.3b], [1.6b] $-\frac{2}{5}, \frac{5}{2}$
9. [1.2e] 3   **10.** [1.2e] $\frac{3}{4}$ **11.** [1.2e] 0   **12.** [1.3a] −4.4
13. [1.4a] $-\frac{5}{2}$ **14.** [1.5a] $\frac{5}{6}$ **15.** [1.5a] −105   **16.** [1.6a] −9
17. [1.6c] −3   **18.** [1.6c] $\frac{32}{125}$ **19.** [1.7c] $15x+25y+10z$
20. [1.7c] $-12x-8$ **21.** [1.7c] $-12y+24x$
22. [1.7d] $2(32+9x+12y)$ **23.** [1.7d] $8(2y-7)$
24. [1.7d] $5(a-3b+5)$ **25.** [1.7e] $15b+22y$
26. [1.7e] $4+9y+6z$ **27.** [1.7e] $1-3a-9d$
28. [1.7e] $-2.6x-5.2y$ **29.** [1.8b] $3x-1$ **30.** [1.8b] $-2x-y$
31. [1.8b] $-7x+6$ **32.** [1.8b] $8x$ **33.** [1.8c] $5x-13$
34. [2.1b] 4.5   **35.** [2.2a] $\frac{4}{25}$ **36.** [2.1b] 10.9   **37.** [2.1b] $3\frac{5}{6}$
38. [2.2a] −48   **39.** [2.2a] $-\frac{3}{8}$ **40.** [2.2a] −6.2   **41.** [2.3a] −3
42. [2.3b] $-\frac{12}{5}$ **43.** [2.3b] 8   **44.** [2.3c] 7   **45.** [2.3b] $-\frac{4}{5}$
46. [2.3b] $-\frac{10}{3}$ **47.** [2.3c] All real numbers
48. [2.3c] No solution   **49.** [2.7c] $\{x|x<2\}$
50. [2.7e] $\{y|y<-3\}$ **51.** [2.7e] $\{y|y \ge 4\}$
52. [2.4b] $m = 65 - H$ **53.** [2.4b] $t = \dfrac{I}{Pr}$ **54.** [2.5a] 25.2
55. [2.5a] 45%   **56.** [2.5a] $363   **57.** [2.6a] $24.60
58. [2.6a] $45   **59.** [2.6a] $1050   **60.** [2.6a] 50 m, 53 m, 40 m
61. [2.8b] $\{s|s \ge 84\}$ **62.** [1.8d] C   **63.** [2.6a] $45,200
64. [2.6a] 30%   **65.** [1.2e], [2.3a] 4, −4   **66.** [2.3b] 3
67. [2.4b] $Q = \dfrac{2-pm}{p}$

CHAPTER 3

Exercise Set 3.1, p. 202

RC1. True   **RC2.** False   **RC3.** False   **RC4.** True
CC1. (c)   **CC2.** (d)   **CC3.** (b)   **CC4.** (a)
1.   **3.** II   **5.** IV   **7.** III
9. On an axis, not in a quadrant
11. II   **13.** IV   **15.** II
17. I, IV   **19.** I, III

21. A: (3, 3); B: (0, −4); C: (−5, 0); D: (−1, −1); E: (2, 0)
23. No   **25.** No   **27.** Yes

29.
$$\begin{array}{c} y = x - 5 \\ \hline -1 \;?\; 4 - 5 \\ \;|\; -1 \quad \text{TRUE} \end{array}$$
$$\begin{array}{c} y = x - 5 \\ \hline -4 \;?\; 1 - 5 \\ \;|\; -4 \quad \text{TRUE} \end{array}$$

31. $y = \frac{1}{2}x + 3$
$$\begin{array}{c} 5 \;?\; \frac{1}{2} \cdot 4 + 3 \\ \;|\; 2 + 3 \\ \;|\; 5 \quad \text{TRUE} \end{array}$$
$$\begin{array}{c} y = \frac{1}{2}x + 3 \\ \hline 2 \;?\; \frac{1}{2}(-2) + 3 \\ \;|\; -1 + 3 \\ \;|\; 2 \quad \text{TRUE} \end{array}$$

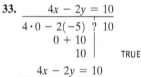

33.
$$\begin{array}{c} 4x - 2y = 10 \\ \hline 4 \cdot 0 - 2(-5) \;?\; 10 \\ 0 + 10 \\ 10 \quad \text{TRUE} \end{array}$$
$$\begin{array}{c} 4x - 2y = 10 \\ \hline 4 \cdot 4 - 2 \cdot 3 \;?\; 10 \\ 16 - 6 \\ 10 \quad \text{TRUE} \end{array}$$

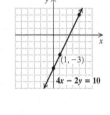

35. 8   **36.** $\frac{7}{4}$ **37.** All real numbers
38. No solution   **39.** $\frac{17}{10}$ **40.** $\frac{1}{3}$
41.   **42.**

43. $3.57   **44.** $48.60   **45.** 15%
46. $-\frac{3}{20}$ **47.** 0   **48.** −455   **49.** (−1, −5)
51. Second axis

53. 26 linear units

Calculator Corner, p. 205
1. Left to the student

Calculator Corner, p. 211

1. $y = -5x + 3$ 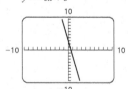   **2.** $y = 4x - 5$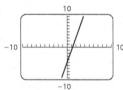

3. $y = \frac{4}{5}x + 2$ 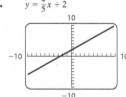   **4.** $y = -\frac{3}{5}x - 1$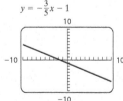

Exercise Set 3.2, p. 211

RC1. (c)   **RC2.** (b)   **RC3.** (d)   **RC4.** (a)
CC1. (e)   **CC2.** (b)   **CC3.** (f)   **CC4.** (c)

1.

x	y
-2	-1
-1	0
0	1
1	2
2	3
3	4

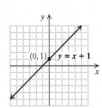

3.

x	y
-2	-2
-1	-1
0	0
1	1
2	2
3	3

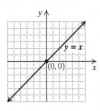

5.

x	y
-2	-1
0	0
4	2

7. 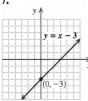 **9.** **11.**

13. **15.** **17.**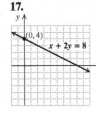

19. **21.** **23.**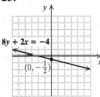

25. (a) 2000: 6.081 billion; 2015: 7.251 billion; 2030: 8.421 billion;
(b)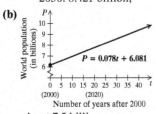
about 7.5 billion;
(c) 35 years after 2000, or in 2035

27. (a) 2000: 10.9%; 2010: 7.5%; 2018: 4.78%;
(b)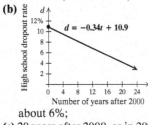
about 6%;
(c) 20 years after 2000, or in 2020

29. 12 **30.** 4.89 **31.** 0 **32.** $\frac{4}{5}$ **33.** $\frac{43}{2}$ **34.** −54 **35.** −10
36. 4 **37.** 16.6 million books **38.** $780 billion **39.** $-\frac{3}{25}$
40. 3 **41.** $-\frac{15}{16}$ **42.** −420 **43.** $\frac{32}{7}$ **44.** −9

Calculator Corner, p. 218

1. y-intercept: $(0, -15)$; x-intercept: $(-2, 0)$; $y = -7.5x - 15$

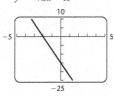
Xscl = 1   Yscl = 5

2. y-intercept: $(0, 43)$; x-intercept: $(-20, 0)$; $y = 2.15x + 43$
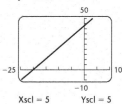
Xscl = 5   Yscl = 5

3. y-intercept: $(0, -30)$; x-intercept: $(25, 0)$; $y = (6x - 150)/5$
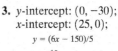
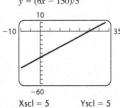
Xscl = 5   Yscl = 5

4. y-intercept: $(0, -4)$; x-intercept: $(20, 0)$; $y = 0.2x - 4$
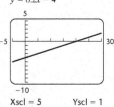
Xscl = 5   Yscl = 1

5. y-intercept: $(0, -15)$; x-intercept: $(10, 0)$; $y = 1.5x - 15$
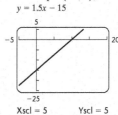
Xscl = 5   Yscl = 5

6. y-intercept: $\left(0, -\frac{1}{2}\right)$; x-intercept: $\left(\frac{2}{5}, 0\right)$; $y = (5x - 2)/4$
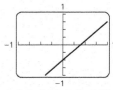
Xscl = 0.25   Yscl = 0.25

Visualizing for Success, p. 221

1. E **2.** C **3.** G **4.** A **5.** I **6.** D **7.** F **8.** J
9. B **10.** H

Exercise Set 3.3, p. 222

RC1. horizontal, y-intercept **RC2.** x-axis **RC3.** $(0, 0)$
RC4. vertical, x-intercept **RC5.** y-axis **RC6.** origin
CC1. False **CC2.** False **CC3.** True **CC4.** True
1. (a) $(0, 5)$; **(b)** $(2, 0)$ **3. (a)** $(0, -4)$; **(b)** $(3, 0)$
5. (a) $(0, 3)$; **(b)** $(5, 0)$ **7. (a)** $(0, -14)$; **(b)** $(4, 0)$
9. (a) $\left(0, \frac{10}{3}\right)$; **(b)** $\left(-\frac{5}{2}, 0\right)$ **11. (a)** $\left(0, -\frac{1}{3}\right)$; **(b)** $\left(\frac{1}{2}, 0\right)$
13. **15.** 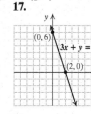 **17.**

19. **21.** 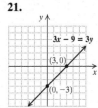 **23.**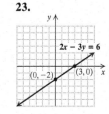

25. 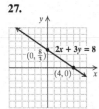 **27.** 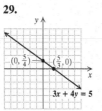 **29.**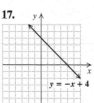

17. **18.**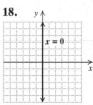

31. 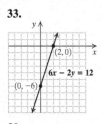 **33.** 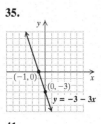 **35.**

19. D **20.** C **21.** B **22.** E **23.** A **24.** No; an equation of the form $x = a$, $a \neq 0$, does not have a y-intercept.
25. Most would probably say that the second equation would be easier to graph because it has been solved for y. This makes it more efficient to find the y-value that corresponds to a given x-value. **26.** $A = 0$. If the line is horizontal, then the equation is of the form $y =$ a constant. Thus, Ax must be 0 and, hence, $A = 0$. **27.** Any ordered pair $(7, y)$ is a solution of $x = 7$. Thus all points on the graph are 7 units to the right of the y-axis, so they lie on a vertical line.

37. 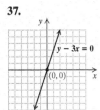 **39.** 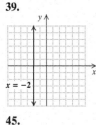 **41.**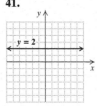

Calculator Corner, p. 232

1. This line will pass through the origin and slant up from left to right. This line will be steeper than $y = 10x$. **2.** This line will pass through the origin and slant up from left to right. This line will be less steep than $y = \frac{5}{32}x$. **3.** This line will pass through the origin and slant down from left to right. This line will be steeper than $y = -10x$. **4.** This line will pass through the origin and slant down from left to right. This line will be less steep than $y = -\frac{5}{32}x$.

43. 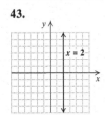 **45.** **47.**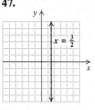

Exercise Set 3.4, p. 235

RC1. (d) **RC2.** (f) **RC3.** (b) **RC4.** (e) **RC5.** (c)
RC6. (a) **CC1.** (c) **CC2.** (b) **CC3.** (a) **CC4.** (d)
1. $-\frac{3}{7}$ **3.** $\frac{2}{3}$ **5.** $\frac{3}{4}$ **7.** 0
9. $-\frac{4}{5}$; 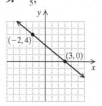 **11.** 3;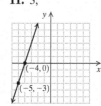

49. 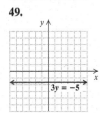 **51.** 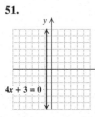 **53.**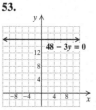

13. $-\frac{2}{3}$; 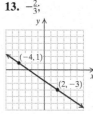 **15.** $\frac{7}{8}$;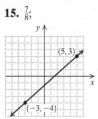

55. $y = -1$ **57.** $x = 4$ **59.** $\{x \mid x < 1\}$ **60.** $\{x \mid x \geq 2\}$
61. $\{x \mid x \leq 7\}$ **62.** $\{x \mid x > 1\}$ **63.** $y = -4$ **65.** $k = 12$

Mid-Chapter Review: Chapter 3, p. 227

1. True **2.** False **3.** True **4.** False
5. (a) The y-intercept is $(0, -3)$. (b) The x-intercept is $(-3, 0)$.
6. (a) The x-intercept is $(c, 0)$. (b) The y-intercept is $(0, d)$.
7. $A: (-1, 0); B: (2, 5); C: (-5, -4); D: (6, -2); E: (-4, 2)$
8. $F: (0, -3); G: (5, 0); H: (1, 5); I: (-4, 4); J: (3, -6)$
9. No **10.** Yes **11.** x-intercept: $(-6, 0)$; y-intercept: $(0, 9)$
12. x-intercept: $\left(\frac{1}{2}, 0\right)$; y-intercept: $\left(0, -\frac{1}{20}\right)$
13. x-intercept: $(40, 0)$; y-intercept: $(0, -2)$
14. x-intercept: $(-42, 0)$; y-intercept: $(0, 105)$

17. $\frac{2}{3}$ **19.** Not defined **21.** $-\frac{5}{13}$ **23.** 0 **25.** -10
27. 3.78 **29.** 3 **31.** $-\frac{1}{5}$ **33.** $-\frac{3}{2}$ **35.** Not defined
37. -1 **39.** 3 **41.** $\frac{5}{4}$ **43.** 0 **45.** $\frac{4}{3}$ **47.** $-\frac{21}{8}$ **49.** $\frac{12}{41}$
51. $\frac{28}{129}$ **53.** 3.0%; yes **55.** 21,375 MWh per year
57. About -500 people per year **59.** 0.12 million, or 120,000, students per year **61.** $\frac{15}{2}$ **62.** -12 **63.** $-\frac{2}{3}p$
64. $5t - 1$ **65.** $y = -x + 5$ **67.** $y = x + 2$

15. 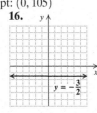 **16.**

Summary and Review: Chapter 3, p. 240

Vocabulary Reinforcement
1. not defined **2.** horizontal; $(0, b)$ **3.** coordinates **4.** y-intercept **5.** vertical; $(a, 0)$ **6.** 0

Concept Reinforcement
1. True **2.** False **3.** True **4.** True **5.** False **6.** False

Study Guide
1. F: $(2, 4)$; G: $(-2, 0)$; H: $(-3, -5)$
2.   **3.**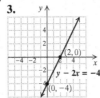
4. 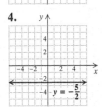 **5.**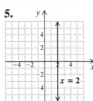
6. m is not defined. **7.** $\frac{3}{2}$ **8.** 0 **9.** m is not defined.
10. -2 **11.** 0 **12.** About 27,400 people per year

Review Exercises
1.–3.   **4.** $(-5, -1)$ **5.** $(-2, 5)$ **6.** $(3, 0)$
7. IV **8.** III **9.** I **10.** No **11.** Yes

12.
$$\begin{array}{c|c} 2x - y = 3 \\ \hline 2 \cdot 0 - (-3) \; ? \; 3 \\ 0 + 3 \\ 3 \end{array} \text{TRUE}$$
$$\begin{array}{c|c} 2x - y = 3 \\ \hline 2 \cdot 2 - 1 \; ? \; 3 \\ 4 - 1 \\ 3 \end{array} \text{TRUE}$$
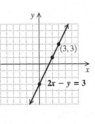

13.   **14.**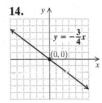
15.   **16.**

17. (a) $14\frac{1}{2}$ ft³, 16 ft³, $20\frac{1}{2}$ ft³, 28 ft³;
(b) 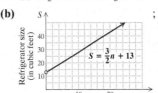 ; 19 ft³; (c) 6 residents

18.   **19.**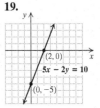
20. 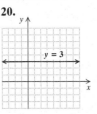 **21.**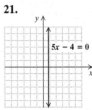
22. $\frac{1}{3}$ **23.** $-\frac{1}{3}$
24. $\frac{3}{5}$; 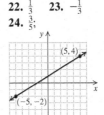 **25.** -1;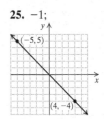

26. $-\frac{5}{8}$ **27.** $\frac{1}{2}$ **28.** Not defined **29.** 0 **30.** $\frac{1}{10}$ **31.** $\frac{3}{2}$
32. (a) 2.4 driveways per hour; (b) 25 minutes per driveway
33. 7% **34.** $19.5 billion per year **35.** D **36.** C
37. 45 square units; 28 linear units **38.** (a) $239.58\overline{3}$ ft per minute;
(b) about 0.004 min per foot

Understanding Through Discussion and Writing
1. With slope $\frac{5}{3}$, for each horizontal change of 3 units, there is a vertical change of 5 units. With slope $\frac{4}{3}$, for each horizontal change of 3 units, there is a vertical change of 4 units. Since $5 > 4$, the line with slope $\frac{5}{3}$ has a steeper slant. **2.** No; the equation $y = b, b \neq 0$, does not have an x-intercept. **3.** The y-intercept is the point at which the graph crosses the y-axis. Since a point on the y-axis is neither left nor right of the origin, the first or x-coordinate of the point is 0. **4.** Any ordered pair $(x, -2)$ is a solution of $y = -2$. All points on the graph are 2 units below the x-axis, so they lie on a horizontal line.

Test: Chapter 3, p. 246
1. [3.1a] II **2.** [3.1a] III **3.** [3.1b] $(-5, 1)$
4. [3.1b] $(0, -4)$
5. [3.1c]
$$\begin{array}{c|c} y - 2x = 5 \\ \hline -3 - 2(-4) \; ? \; 5 \\ -3 + 8 \\ 5 \end{array} \text{TRUE}$$
$$\begin{array}{c|c} y - 2x = 5 \\ \hline 3 - 2(-1) \; ? \; 5 \\ 3 + 2 \\ 5 \end{array} \text{TRUE}$$
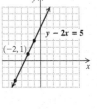

6. [3.2a] 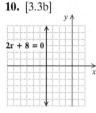   **7.** [3.2a] 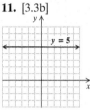   **8.** [3.3a]

9. [3.3a] 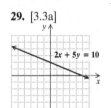   **10.** [3.3b] 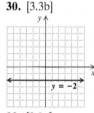   **11.** [3.3b]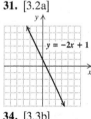

12. [3.4a] -2
13. [3.4a] $\frac{3}{8}$;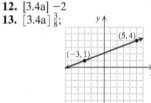

14. [3.4b] $\frac{2}{5}$ **15.** [3.4b] not defined   **16.** [3.4b] 0
17. [3.4b] -11 **18.** [3.4c] $-\frac{1}{20}$, or -0.05
19. [3.2b] **(a)** 2007: $8593; 2009: $9805; 2012: $11,623;
(b) 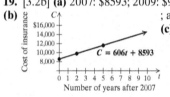 ; approximately $14,000; **(c)** 11 years after 2007, or in 2018

20. [3.4c] **(a)** 14.5 floors per minute; **(b)** $4\frac{4}{29}$ seconds per floor
21. [3.4c] 87.5 mph   **22.** [3.3a], [3.4b] B   **23.** [3.3b] $y = 3$
24. [3.1a] 25 square units; 20 linear units

Cumulative Review: Chapters 1–3, p. 249

1. [1.1a] $-\frac{4}{5}$ **2.** [1.7c] $-\frac{2}{3}x + 4y - 2$ **3.** [1.7d] $3(6w - 8 + 3y)$
4. [1.2c] $-0.\overline{7}$ **5.** [1.2e] $2\frac{1}{5}$ **6.** [1.3b] -8.17 **7.** [1.6b] $-\frac{7}{8}$
8. [1.7e] $-x - y$ **9.** [1.3a] -3 **10.** [1.8d] -6 **11.** [1.6c] $-\frac{2}{3}$
12. [1.8c] $11x + 9$ **13.** [1.5a] 2.64   **14.** [1.8d] -2
15. [2.2a] -81 **16.** [2.3c] No solution   **17.** [2.3b] 3
18. [2.3c] All real numbers   **19.** [2.7e] $\{x | x \leq -\frac{11}{8}\}$
20. [2.1b] $\frac{4}{3}$ **21.** [3.4b] $\frac{3}{4}$ **22.** [3.4b] Not defined
23. [3.4b] 0   **24.** [2.4b] $s = \frac{7t}{z}$ **25.** [2.4b] $h = \frac{2A}{b+c}$
26. [3.1a] IV   **27.** [3.3a] y-intercept: $(0, -3)$; x-intercept: $\left(\frac{21}{2}, 0\right)$
28. [2.7b]
29. [3.3a] 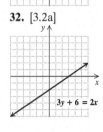   **30.** [3.3b] 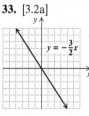   **31.** [3.2a]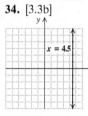

32. [3.2a]   **33.** [3.2a]   **34.** [3.3b]

35. [3.4a] $-\frac{1}{4}$ **36.** [2.6a] 20.2 million Americans
37. [2.8b] $\{x | x \leq 8\}$ **38.** [2.6a] First: 50 m; second: 53 m; third: 40 m   **39.** [1.8d] A   **40.** [3.4a] D
41. [1.2e], [2.3a] $-4, 4$ **42.** [2.3b] 3
43. [2.4b] $Q = \frac{2 - pm}{p}$, or $\frac{2}{p} - m$

CHAPTER 4

Exercise Set 4.1, p. 258

RC1. base   RC2. cubed   RC3. add
CC1. (c)   CC2. (a)   CC3. (e)   CC4. (c)   CC5. (d)
CC6. (c)
1. $3 \cdot 3 \cdot 3 \cdot 3$ **3.** $(-1.1)(-1.1)(-1.1)(-1.1)(-1.1)$
5. $\left(\frac{2}{3}\right)\left(\frac{2}{3}\right)\left(\frac{2}{3}\right)\left(\frac{2}{3}\right)$ **7.** $(7p)(7p)$ **9.** $8 \cdot k \cdot k \cdot k$ **11.** $-6 \cdot y \cdot y \cdot y \cdot y$
13. 1   **15.** b **17.** 1   **19.** -7.03 **21.** 1   **23.** ab **25.** a
27. 27   **29.** 19   **31.** -81 **33.** 256   **35.** 93   **37.** 136
39. 10; 4   **41.** 3629.84 ft^2 **43.** $\frac{1}{3^2} = \frac{1}{9}$ **45.** $\frac{1}{10^3} = \frac{1}{1000}$
47. $\frac{1}{a^3}$ **49.** $8^2 = 64$ **51.** y^4 **53.** $\frac{5}{z^4}$ **55.** $\frac{x}{y^2}$ **57.** 4^{-3}
59. x^{-3} **61.** a^{-5} **63.** 2^7 **65.** 9^{38} **67.** x^5 **69.** x^{17}
71. $(3y)^{12}$ **73.** $(7y)^{17}$ **75.** 3^3 **77.** 1   **79.** $\frac{1}{x^{13}}$ **81.** $\frac{1}{a^{10}}$
83. $x^6 y^{15}$ **85.** $s^3 t^7$ **87.** 7^3 **89.** y^8 **91.** $\frac{1}{16^6}$ **93.** $\frac{1}{m^6}$
95. $\frac{1}{(8x)^4}$ **97.** x^2 **99.** $\frac{1}{z^4}$ **101.** x^3 **103.** 1   **105.** $a^3 b^2$
107. $5^2 = 25; 5^{-2} = \frac{1}{25}; \left(\frac{1}{5}\right)^2 = \frac{1}{25}; \left(\frac{1}{5}\right)^{-2} = 25; -5^2 = -25;$
$(-5)^2 = 25; -\left(-\frac{1}{5}\right)^2 = -\frac{1}{25}; \left(-\frac{1}{5}\right)^{-2} = 25$ **109.** 8 in.; 4 in.
110. 51°, 27°, 102°   **111.** 45%; 37.5%; 17.5%   **112.** Lengths less than 2.5 ft   **113.** $\frac{7}{4}$ **114.** 2   **115.** $\frac{23}{14}$ **116.** $\frac{11}{10}$ **117.** No
119. No   **121.** y^{5x} **123.** a^{4t} **125.** 1   **127.** >   **129.** <
131. $-\frac{1}{10,000}$ **133.** No; for example, $(3 + 4)^2 = 49$, but $3^2 + 4^2 = 25$.

Calculator Corner, p. 266

1. 1.3545×10^{-4} **2.** 3.2×10^5 **3.** 3×10^{-6} **4.** 8×10^{-26}

Exercise Set 4.2, p. 268

RC1. multiply   RC2. nth   RC3. right   RC4. positive
CC1. Positive power of 10   CC2. Negative power of 10
CC3. Negative power of 10   CC4. Positive power of 10
1. 2^6 **3.** $\frac{1}{5^6}$ **5.** x^{12} **7.** $\frac{1}{a^{18}}$ **9.** t^{18} **11.** $\frac{1}{t^{12}}$ **13.** x^8
15. $a^3 b^3$ **17.** $\frac{1}{a^3 b^3}$ **19.** $\frac{1}{m^3 n^6}$ **21.** $16x^6$ **23.** $\frac{9}{x^8}$ **25.** $\frac{1}{x^{12} y^{15}}$
27. $x^{24} y^8$ **29.** $\frac{a^{10}}{b^{35}}$ **31.** $\frac{25t^6}{r^8}$ **33.** $\frac{b^{21}}{a^{15} c^6}$ **35.** $\frac{9x^6}{y^{16} z^6}$
37. $\frac{16x^6}{y^4}$ **39.** $a^{12} b^8$ **41.** $\frac{y^6}{4}$ **43.** $\frac{a^8}{b^{12}}$ **45.** $\frac{8}{y^6}$ **47.** $49x^6$
49. $\frac{x^6 y^3}{z^3}$ **51.** $\frac{c^2 d^6}{a^4 b^2}$ **53.** 2.8×10^{10} **55.** 9.07×10^{17}
57. 3.04×10^{-6} **59.** 1.8×10^{-8} **61.** 1×10^{11}, or 10^{11}
63. 4.19854×10^8 **65.** 7×10^{-9} **67.** 87,400,000
69. 0.00000005704   **71.** 10,000,000   **73.** 0.00001
75. 6×10^9 **77.** 3.38×10^4 **79.** 8.1477×10^{-13} **81.** 2.5×10^{13}
83. 5.0×10^{-4} **85.** 3.0×10^{-21} **87.** Approximately 1.325×10^{14} ft^3 **89.** 7×10^{22} stars   **91.** The mass of Jupiter is 3.18×10^2 times the mass of Earth.   **93.** 1.12×10^2 videos
95. 3.1×10^2 sheets   **97.** 4.375×10^2 days

99. 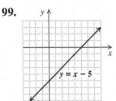 **100.**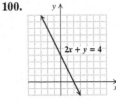

101. 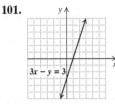 **102.**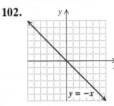

103. 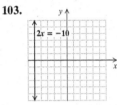 **104.**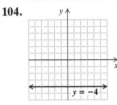

105. 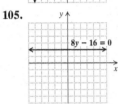 **106.**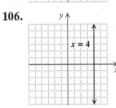

107. 2.478125×10^{-1} **109.** $\frac{1}{5}$ **111.** 3^{11} **113.** 7
115. $\frac{1}{0.4}$, or 2.5 **117.** False **119.** False **121.** True

Calculator Corner, p. 276
1. $3; 2.25; -27$ **2.** $44; 0; 9.28$

Exercise Set 4.3, p. 281
RC1. (b) **RC2.** (f) **RC3.** (c) **RC4.** (d) **RC5.** (a)
RC6. (e) **CC1.** 2 **CC2.** 0 **CC3.** 4 **CC4.** 5 **CC5.** 2
CC6. Ascending order **CC7.** x
1. $-18; 7$ **3.** $19; 14$ **5.** $-12; -7$ **7.** $\frac{13}{3}; 5$ **9.** $9; 1$
11. $56; -2$ **13.** 1112 ft **15.** 55 oranges **17.** $-4, 4, 5, 2.75, 1$
19. (a) 3400 MW; (b) left to the student **21.** 9 words **23.** 6
25. 15 **27.** $2, -3x, x^2$ **29.** $-2x^4, \frac{1}{3}x^3, -x, 3$ **31.** Trinomial
33. None of these **35.** Binomial **37.** Monomial **39.** $-3, 6$
41. $5, \frac{3}{4}, 3$ **43.** $-5, 6, -2.7, 1, -2$ **45.** $1, 0; 1$ **47.** $2, 1, 0; 2$
49. $3, 2, 1, 0; 3$ **51.** $2, 1, 6, 4; 6$
53.

Term	Coefficient	Degree of the Term	Degree of the Polynomial
$7x^4$	7	4	
$6x^3$	6	3	
$-x^2$	-1	2	4
$8x$	8	1	
-2	-2	0	

55. $6x^2$ and $-3x^2$ **57.** $2x^4$ and $-3x^4$; $5x$ and $-7x$ **59.** $3x^5$ and $14x^5$; $-7x$ and $-2x$; 8 and -9 **61.** $-3x$ **63.** $-8x$
65. $11x^3 + 4$ **67.** $x^3 - x$ **69.** $4b^5$ **71.** $\frac{3}{4}x^5 - 2x - 42$
73. x^4 **75.** $\frac{15}{16}x^3 - \frac{7}{6}x^2$ **77.** $x^5 + 6x^3 + 2x^2 + x + 1$
79. $15y^9 + 7y^8 + 5y^3 - y^2 + y$ **81.** $x^6 + x^4$ **83.** $13x^3 - 9x + 8$
85. $-5x^2 + 9x$ **87.** $12x^4 - 2x + \frac{1}{4}$ **89.** x^2, x **91.** x^3, x^2, x^0

93. None missing **95.** $x^3 + 0x^2 + 0x - 27; x^3 - 27$
97. $x^4 + 0x^3 + 0x^2 - x + 0x^0; x^4 - x$
99. $5x^2 + 0x + 0; 5x^2$ **101.** -19 **102.** -1 **103.** -2.25
104. -2.6 **105.** $-\frac{17}{24}$ **106.** $\frac{5}{8}$ **107.** $\frac{1}{3}$ **108.** -0.6
109. -24 **110.** $-\frac{8}{5}$ **111.** 0 **112.** Not defined
113. $10x^6 + 52x^5$ **115.** $4x^5 - 3x^3 + x^2 - 7x$; answers may vary
117. $-4, 4, 5, 2.75, 1$ **119.** 9

Exercise Set 4.4, p. 290
RC1. False **RC2.** True **RC3.** False **RC4.** False
CC1. 0 **CC2.** $2x^2$ **CC3.** x **CC4.** 0 **CC5.** $2x$
CC6. $2x^2$
1. $-x + 5$ **3.** $x^2 - \frac{11}{2}x - 1$ **5.** $2x^2$ **7.** $5x^2 + 3x - 30$
9. $-2.2x^3 - 0.2x^2 - 3.8x + 23$ **11.** $6 + 12x^2$
13. $-\frac{1}{2}x^4 + \frac{2}{3}x^3 + x^2$ **15.** $0.01x^5 + x^4 - 0.2x^3 + 0.2x + 0.06$
17. $9x^8 + 8x^7 - 6x^4 + 8x^2 + 4$
19. $1.05x^4 + 0.36x^3 + 14.22x^2 + x + 0.97$ **21.** $5x$
23. $x^2 - \frac{3}{2}x + 2$ **25.** $-12x^4 + 3x^3 - 3$ **27.** $-3x + 7$
29. $-4x^2 + 3x - 2$ **31.** $4x^4 - 6x^2 - \frac{3}{4}x + 8$ **33.** $7x - 1$
35. $-x^2 - 7x + 5$ **37.** -18 **39.** $6x^4 + 3x^3 - 4x^2 + 3x - 4$
41. $4.6x^3 + 9.2x^2 - 3.8x - 23$ **43.** $\frac{3}{4}x^3 - \frac{1}{2}x$
45. $0.06x^3 - 0.05x^2 + 0.01x + 1$ **47.** $3x + 6$
49. $11x^4 + 12x^3 - 9x^2 - 8x - 9$ **51.** $x^4 - x^3 + x^2 - x$
53. $\frac{23}{2}a + 12$ **55.** $5x^2 + 4x$
57. $(r + 11)(r + 9); 9r + 99 + 11r + r^2$, or $r^2 + 20r + 99$
59. $(x + 3)(x + 3)$, or $(x + 3)^2; x^2 + 3x + 9 + 3x$, or $x^2 + 6x + 9$
61. $\pi r^2 - 25\pi$ **63.** $18z - 64$ **65.** 6 **66.** -19 **67.** $-\frac{7}{22}$
68. 5 **69.** 5 **70.** 1 **71.** $\frac{39}{2}$ **72.** $\frac{37}{2}$ **73.** $\{x | x \geq -10\}$
74. $\{x | x < 0\}$ **75.** $20w + 42$ **77.** $2x^2 + 20x$
79. $y^2 - 4y + 4$ **81.** $12y^2 - 23y + 21$ **83.** $-3y^4 - y^3 + 5y - 2$

Mid-Chapter Review: Chapter 4, p. 295
1. True **2.** False **3.** False **4.** True
5. $4w^3 + 6w - 8w^3 - 3w = (4 - 8)w^3 + (6 - 3)w = -4w^3 + 3w$ **6.** $(3y^4 - y^2 + 11) - (y^4 - 4y^2 + 5) = 3y^4 - y^2 + 11 - y^4 + 4y^2 - 5 = 2y^4 + 3y^2 + 6$ **7.** z **8.** 1
9. -32 **10.** 1 **11.** 5^7 **12.** $(3a)^9$ **13.** $\frac{1}{x^3}$ **14.** 1
15. 7^4 **16.** $\frac{1}{x^2}$ **17.** w^8 **18.** $\frac{1}{y^4}$ **19.** 3^{15} **20.** $\frac{x^{18}}{y^{12}}$
21. $\frac{a^{24}}{5^6}$ **22.** $\frac{x^2 z^4}{4y^6}$ **23.** 2.543×10^7 **24.** 1.2×10^{-4}
25. 0.000036 **26.** 144,000,000 **27.** 6×10^3 **28.** 5×10^{-7}
29. $16; 1$ **30.** $-16; 9$ **31.** $-2x^5 - 5x^2 + 4x + 2$
32. $8x^6 + 2x^3 - 8x^2$ **33.** $3, 1, 0; 3$ **34.** $1, 4, 6; 6$
35. Binomial **36.** Trinomial **37.** $8x^2 + 5$
38. $5x^3 - 2x^2 + 2x - 11$ **39.** $-4x - 10$
40. $-0.4x^2 - 3.4x + 9$ **41.** $3y + 3y^2$ **42.** The area of the smaller square is x^2, and the area of the larger square is $(3x)^2$, or $9x^2$, so the area of the larger square is nine times the area of the smaller square. **43.** The volume of the smaller cube is r^3, and the volume of the larger cube is $(2x)^3$, or $8x^3$, so the volume of the larger cube is eight times the volume of the smaller cube.
44. Exponents are added when powers with like bases are multiplied. Exponents are multiplied when a power is raised to a power.
45. $3^{-29} = \frac{1}{3^{29}}$ and $2^{-29} = \frac{1}{2^{29}}$. Since $3^{29} > 2^{29}$, we have $\frac{1}{3^{29}} < \frac{1}{2^{29}}$.
46. It is better to evaluate a polynomial after like terms have been collected, because there are fewer terms to evaluate. **47.** Yes; consider the following: $(x^2 + 4) + (4x - 7) = x^2 + 4x - 3$.

Calculator Corner, p. 300
1. Correct **2.** Correct **3.** Not correct **4.** Not correct

Exercise Set 4.5, p. 301

RC1. True   **RC2.** False   **RC3.** True   **RC4.** True
CC1. (a)   **CC2.** (a)   **CC3.** (d)   **CC4.** (e)
1. $40x^2$ **3.** x^3 **5.** $32x^8$ **7.** $0.03x^{11}$ **9.** $\frac{1}{15}x^4$ **11.** 0
13. $-24x^{11}$ **15.** $-2x^2 + 10x$ **17.** $-5x^2 + 5x$ **19.** $x^5 + x^2$
21. $6x^3 - 18x^2 + 3x$ **23.** $-6x^4 - 6x^3$ **25.** $18y^6 + 24y^5$
27. $x^2 + 9x + 18$ **29.** $x^2 + 3x - 10$ **31.** $x^2 + 3x - 4$
33. $x^2 - 7x + 12$ **35.** $x^2 - 9$ **37.** $x^2 - 16$
39. $3x^2 + 11x + 10$ **41.** $25 - 15x + 2x^2$ **43.** $4x^2 + 20x + 25$
45. $x^2 - 6x + 9$ **47.** $x^2 - \frac{21}{10}x - 1$ **49.** $x^2 + 2.4x - 10.81$
51. $(x + 2)(x + 6)$, or $x^2 + 8x + 12$ **53.** $(x + 1)(x + 6)$, or $x^2 + 7x + 6$

55.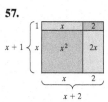
$x + 5$

57.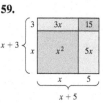
$x + 2$

59.
$x + 5$

61. $x^3 - 1$ **63.** $4x^3 + 14x^2 + 8x + 1$
65. $3y^4 - 6y^3 - 7y^2 + 18y - 6$ **67.** $x^6 + 2x^5 - x^3$
69. $-10x^5 - 9x^4 + 7x^3 + 2x^2 - x$ **71.** $-1 - 2x - x^2 + x^4$
73. $6t^4 + t^3 - 16t^2 - 7t + 4$ **75.** $x^9 - x^5 + 2x^3 - x$
77. $x^4 + 8x^3 + 12x^2 + 9x + 4$ **79.** $2x^4 - 5x^3 + 5x^2 - \frac{19}{10}x + \frac{1}{5}$
81. 47 **82.** 96 **83.** $4(4x - 6y + 9)$ **84.** $-3(3x + 15y - 5)$
85. $75y^2 - 45y$ **87.** $(x^3 + 2x^2 - 210)$ m^3 **89.** 0 **91.** 0

Visualizing for Success, p. 310

1. E, F   **2.** B, O   **3.** K, S   **4.** G, R   **5.** D, M   **6.** J, P
7. C, L   **8.** N, Q   **9.** A, H   **10.** I, T

Exercise Set 4.6, p. 311

RC1. outside; last   **RC2.** descending   **RC3.** difference
RC4. square; binomial   **RC5.** binomials   **RC6.** difference
CC1. (c)   **CC2.** (b)   **CC3.** (a)   **CC4.** (d)
1. $x^3 + x^2 + 3x + 3$ **3.** $x^4 + x^3 + 2x + 2$ **5.** $y^2 - y - 6$
7. $9x^2 + 12x + 4$ **9.** $5x^2 + 4x - 12$ **11.** $9t^2 - 1$
13. $4x^2 - 6x + 2$ **15.** $p^2 - \frac{1}{16}$ **17.** $x^2 - 0.01$
19. $2x^3 + 2x^2 + 6x + 6$ **21.** $-2x^2 - 11x + 6$
23. $a^2 + 14a + 49$ **25.** $1 - x - 6x^2$ **27.** $\frac{9}{64}y^2 - \frac{5}{8}y + \frac{25}{36}$
29. $x^5 + 3x^3 - x^2 - 3$ **31.** $3x^6 - 2x^4 - 6x^2 + 4$
33. $13.16x^2 + 18.99x - 13.95$ **35.** $6x^7 + 18x^5 + 4x^2 + 12$
37. $4x^3 - 12x^2 + 3x - 9$ **39.** $4y^6 + 4y^5 + y^4 + y^3$
41. $x^2 - 16$ **43.** $4x^2 - 1$ **45.** $25m^2 - 4$ **47.** $4x^4 - 9$
49. $9x^8 - 16$ **51.** $x^{12} - x^4$ **53.** $x^8 - 9x^2$ **55.** $x^{24} - 9$
57. $4y^{16} - 9$ **59.** $\frac{25}{64}x^2 - 18.49$ **61.** $x^2 + 4x + 4$
63. $9x^4 + 6x^2 + 1$ **65.** $a^2 - a + \frac{1}{4}$ **67.** $9 + 6x + x^2$
69. $x^4 + 2x^2 + 1$ **71.** $4 - 12x^4 + 9x^8$ **73.** $25 + 60t^2 + 36t^4$
75. $x^2 - \frac{5}{4}x + \frac{25}{64}$ **77.** $9 - 12x^3 + 4x^6$ **79.** $4x^3 + 24x^2 - 12x$
81. $4x^4 - 2x^2 + \frac{1}{4}$ **83.** $9p^2 - 1$ **85.** $15t^5 - 3t^4 + 3t^3$
87. $36x^8 + 48x^4 + 16$ **89.** $12x^3 + 8x^2 + 15x + 10$
91. $64 - 96x^4 + 36x^8$ **93.** $t^3 - 1$ **95.** $25; 49$ **97.** $56; 16$
99. $a^2 + 2a + 1$ **101.** $t^2 + 10t + 24$ **103.** $\frac{28}{27}$ **104.** $-\frac{41}{7}$
105. $\frac{27}{4}$ **106.** $y = \frac{3x - 12}{2}$, or $y = \frac{3}{2}x - 6$ **107.** $b = \frac{C + r}{a}$
108. $a = \frac{5d + 4}{3}$, or $a = \frac{5}{3}d + \frac{4}{3}$ **109.** $30x^3 + 35x^2 - 15x$
111. $a^4 - 50a^2 + 625$ **113.** $81t^{16} - 72t^8 + 16$ **115.** -7
117. First row: 90, -432, -63; second row: 7, -18, -36, -14, 12, -6, -21, -11; third row: 9, -2, -2, 10, -8, -8, -8, -10, 21; fourth row: -19, -6   **119.** Yes   **121.** No

Exercise Set 4.7, p. 319

RC1. True   **RC2.** False   **RC3.** False   **RC4.** False
CC1. a, x, y **CC2.** 5   **CC3.** $-axy$ **CC4.** $3ax^2, -axy, 7ax^2$
CC5. $3ax^2$ and $7ax^2$

1. -1 **3.** -15 **5.** 240 **7.** -145 **9.** 3.715 L
11. 2322 calories   **13.** 44.4624 in^2 **15.** 73.005 in^2
17. Coefficients: 1, -2, 3, -5; degrees: 4, 2, 2, 0; 4
19. Coefficients: 17, -3, -7; degrees: 5, 5, 0; 5   **21.** $-a - 2b$
23. $3x^2y - 2xy^2 + x^2$ **25.** $20au + 10av$ **27.** $8u^2v - 5uv^2$
29. $x^2 - 4xy + 3y^2$ **31.** $3r + 7$ **33.** $-b^2a^3 - 3b^3a^2 + 5ba + 3$
35. $ab^2 - a^2b$ **37.** $2ab - 2$ **39.** $-2a + 10b - 5c + 8d$
41. $6z^2 + 7zu - 3u^2$ **43.** $a^4b^2 - 7a^2b + 10$ **45.** $a^6 - b^2c^2$
47. $y^6x + y^4x + y^4 + 2y^2 + 1$ **49.** $12x^2y^2 + 2xy - 2$
51. $12 - c^2d^2 - c^4d^4$ **53.** $m^3 + m^2n - mn^2 - n^3$
55. $x^9y^9 - x^6y^6 + x^5y^5 - x^2y^2$ **57.** $x^2 + 2xh + h^2$
59. $9a^2 + 12ab + 4b^2$ **61.** $r^6t^4 - 8r^3t^2 + 16$
63. $p^8 + 2m^2n^2p^4 + m^4n^4$ **65.** $3a^3 - 12a^2b + 12ab^2$
67. $m^2 + 2mn + n^2 - 6m - 6n + 9$ **69.** $a^2 - b^2$ **71.** $4a^2 - b^2$
73. $c^4 - d^2$ **75.** $a^2b^2 - c^2d^4$ **77.** $x^2 + 2xy + y^2 - 9$
79. $x^2 - y^2 - 2yz - z^2$ **81.** $a^2 + 2ab + b^2 - c^2$
83. $3x^4 - 7x^2y + 3x^2 - 20y^2 + 22y - 6$ **85.** IV   **86.** III   **87.** I
88. II   **89.** 39 **90.** 1.125 **91.** $<$ **92.** -3 **93.** $4xy - 4y^2$
95. $2xy + \pi x^2$ **97.** $2\pi nh + 2\pi mh + 2\pi n^2 - 2\pi m^2$ **99.** 16 gal
101. $\$12{,}351.94$

Exercise Set 4.8, p. 328

RC1. subtract; divide   **RC2.** divide   **RC3.** multiply; subtract
RC4. multiply; add   **CC1.** $x - 1\overline{)x^2 + 5x - 6}$
CC2. $x - 3\overline{)x^2 + x + 1}$ **CC3.** $x - 2\overline{)x^3 + 0x^2 + 0x - 4}$
1. $3x^4$ **3.** $5x$ **5.** $18x^3$ **7.** $4a^3b$ **9.** $3x^4 - \frac{1}{2}x^3 + \frac{1}{8}x^2 - 2$
11. $1 - 2u - u^4$ **13.** $5t^2 + 8t - 2$ **15.** $-4x^4 + 4x^2 + 1$
17. $6x^2 - 10x + \frac{3}{2}$ **19.** $9x^2 - \frac{5}{2}x + 1$ **21.** $6x^2 + 13x + 4$
23. $3rs + r - 2s$ **25.** $x + 2$ **27.** $x - 5 + \dfrac{-50}{x - 5}$
29. $x - 2 + \dfrac{-2}{x + 6}$ **31.** $x - 3$ **33.** $x^4 - x^3 + x^2 - x + 1$
35. $2x^2 - 7x + 4$ **37.** $x^3 - 6$ **39.** $t^2 + 1$
41. $y^2 - 3y + 1 + \dfrac{-5}{y + 2}$ **43.** $3x^2 + x + 2 + \dfrac{10}{5x + 1}$
45. $6y^2 - 5 + \dfrac{-6}{2y + 7}$ **47.** -1 **48.** $\frac{10}{3}$ **49.** $\frac{23}{19}$
50. $\{r | r \leq -15\}$ **51.** 140% **52.** $\{x | x \geq 95\}$
53. $25{,}543.75$ ft^2 **54.** $228, 229$ **55.** $x^2 + 5$
57. $a + 3 + \dfrac{5}{5a^2 - 7a - 2}$ **59.** $2x^2 + x - 3$
61. $a^5 + a^4b + a^3b^2 + a^2b^3 + ab^4 + b^5$ **63.** -5 **65.** 1

Summary and Review: Chapter 4, p. 331

Vocabulary Reinforcement

1. exponent   **2.** product   **3.** monomial   **4.** trinomial
5. quotient   **6.** descending   **7.** degree   **8.** scientific

Concept Reinforcement

1. True   **2.** False   **3.** False   **4.** True

Study Guide

1. z^8 **2.** a^2b^6 **3.** $\dfrac{y^6}{27x^{12}z^9}$ **4.** 7.63×10^5 **5.** 0.0003
6. 6×10^4 **7.** $2x^4 - 4x^2 - 3$ **8.** $3x^4 + x^3 - 2x^2 + 2$
9. $x^6 - 6x^4 + 11x^2 - 6$ **10.** $2y^2 + 11y + 12$ **11.** $x^2 - 25$
12. $9w^2 + 24w + 16$ **13.** $-2a^3b^2 - 5a^2b + ab^2 - 2ab$
14. $y^2 - 4y + \frac{8}{5}$ **15.** $x - 9 + \dfrac{48}{x + 5}$

Review Exercises

1. $\dfrac{1}{7^2}$ **2.** y^{11} **3.** $(3x)^{14}$ **4.** t^8 **5.** 4^3 **6.** $\dfrac{1}{a^3}$ **7.** 1

8. $9t^8$ **9.** $36x^8$ **10.** $\dfrac{y^3}{8x^3}$ **11.** t^{-5} **12.** $\dfrac{1}{y^4}$ **13.** 3.28×10^{-5}
14. 8,300,000 **15.** 2.09×10^4 **16.** 5.12×10^{-5}
17. 1.564×10^{10} slices **18.** 10 **19.** $-4y^5, 7y^2, -3y, -2$
20. x^2, x^0 **21.** 3, 2, 1, 0; 3 **22.** Binomial **23.** None of these
24. Monomial **25.** $-2x^2 - 3x + 2$ **26.** $10x^4 - 7x^2 - x - \tfrac{1}{2}$
27. $x^5 - 2x^4 + 6x^3 + 3x^2 - 9$ **28.** $-2x^5 - 6x^4 - 2x^3 - 2x^2 + 2$
29. $2x^2 - 4x$ **30.** $x^5 - 3x^3 - x^2 + 8$ **31.** Perimeter: $4w + 6$; area: $w^2 + 3w$ **32.** $(t+3)(t+4); t^2 + 7t + 12$
33. $x^2 + \tfrac{7}{6}x + \tfrac{1}{3}$ **34.** $49x^2 + 14x + 1$
35. $12x^3 - 23x^2 + 13x - 2$ **36.** $9x^4 - 16$
37. $15x^7 - 40x^6 + 50x^5 + 10x^4$ **38.** $x^2 - 3x - 28$
39. $9y^4 - 12y^3 + 4y^2$ **40.** $2t^4 - 11t^2 - 2$ **41.** 49
42. Coefficients: 1, -7, 9, -8; degrees: 6, 2, 2, 0; 6
43. $-y + 9w - 5$ **44.** $m^6 - 2m^2n + 2m^2n^2 + 8n^2m - 6m^3$
45. $-9xy - 2y^2$ **46.** $11x^3y^2 - 8x^2y - 6x^2 - 6x + 6$
47. $p^3 - q^3$ **48.** $9a^8 - 2a^4b^3 + \tfrac{1}{9}b^6$ **49.** $5x^2 - \tfrac{1}{2}x + 3$
50. $3x^2 - 7x + 4 + \dfrac{1}{2x+3}$ **51.** 0, 3.75, -3.75, 0 **52.** B
53. D **54.** $\tfrac{1}{2}x^2 - \tfrac{1}{2}y^2$ **55.** $400 - 4a^2$ **56.** $-28x^8$
57. $\tfrac{94}{13}$ **58.** $x^4 + x^3 + x^2 + x + 1$ **59.** 80 ft by 40 ft

Understanding Through Discussion and Writing

1. 578.6×10^{-7} is not in scientific notation because 578.6 is not a number greater than or equal to 1 and less than 10.
2. When evaluating polynomials, it is essential to know the order in which the operations are to be performed.
3. We label the figure as shown.

Then we see that the area of the figure is $(x+3)^2$, or $x^2 + 3x + 3x + 9 \neq x^2 + 9$. **4.** Emma did not divide *each* term of the polynomial by the divisor. The first term was divided by $3x$, but the second was not. Multiplying Emma's "quotient" by the divisor $3x$, we get $12x^3 - 18x^2 \neq 12x^3 - 6x$. This should convince her that a mistake has been made. **5.** Yes; for example, $(x^2 + xy + 1) + (3x - xy + 2) = x^2 + 3x + 3$.
6. Yes; consider $a + b + c + d$. This is a polynomial in 4 variables but it has degree 1.

Test: Chapter 4, p. 337

1. [4.1d, f] $\dfrac{1}{6^5}$ **2.** [4.1d] x^9 **3.** [4.1d] $(4a)^{11}$ **4.** [4.1e] 3^3
5. [4.1e, f] $\dfrac{1}{x^5}$ **6.** [4.1b, e] 1 **7.** [4.2a] x^6 **8.** [4.2a, b] $-27y^6$
9. [4.2a, b] $16a^{12}b^4$ **10.** [4.2b] $\dfrac{a^3b^3}{c^3}$ **11.** [4.1d], [4.2a, b] $-216x^{21}$
12. [4.1d], [4.2a, b] $-24x^{21}$ **13.** [4.1d], [4.2a, b] $162x^{10}$
14. [4.1d], [4.2a, b] $324x^{10}$ **15.** [4.1f] $\dfrac{1}{5^3}$ **16.** [4.1f] y^{-8}
17. [4.2c] 3.9×10^9 **18.** [4.2c] 0.00000005
19. [4.2d] 1.75×10^{17} **20.** [4.2d] 1.296×10^{22}
21. [4.2e] The mass of Saturn is 9.5×10 times the mass of Earth.
22. [4.3a] -43 **23.** [4.3c] $\tfrac{1}{3}, -1, 7$ **24.** [4.3c] 3, 0, 1, 6; 6
25. [4.3b] Binomial **26.** [4.3d] $5a^2 - 6$ **27.** [4.3d] $\tfrac{7}{4}y^2 - 4y$
28. [4.3e] $x^5 + 2x^3 + 4x^2 - 8x + 3$
29. [4.4a] $4x^5 + x^4 + 2x^3 - 8x^2 + 2x - 7$
30. [4.4a] $5x^4 + 5x^2 + x + 5$ **31.** [4.4c] $-4x^4 + x^3 - 8x - 3$
32. [4.4c] $-x^5 + 0.7x^3 - 0.8x^2 - 21$
33. [4.5b] $-12x^4 + 9x^3 + 15x^2$ **34.** [4.6c] $x^2 - \tfrac{2}{3}x + \tfrac{1}{9}$
35. [4.6b] $9x^2 - 100$ **36.** [4.6a] $3b^2 - 4b - 15$
37. [4.6a] $x^{14} - 4x^8 + 4x^6 - 16$ **38.** [4.6a] $48 + 34y - 5y^2$
39. [4.5d] $6x^3 - 7x^2 - 11x - 3$ **40.** [4.6c] $25t^2 + 20t + 4$
41. [4.7c] $-5x^3y - y^3 + xy^3 - x^2y^2 + 19$
42. [4.7e] $8a^2b^2 + 6ab - 4b^3 + 6ab^2 + ab^3$
43. [4.7f] $9x^{10} - 16y^{10}$ **44.** [4.8a] $4x^2 + 3x - 5$
45. [4.8b] $2x^2 - 4x - 2 + \dfrac{17}{3x+2}$
46. [4.3a] 3, 1.5, -3.5, -5, -5.25
47. [4.4d] $(t+2)(t+2); t^2 + 4t + 4$ **48.** [4.4d] B
49. [4.5b], [4.6a] $V = l^3 - 3l^2 + 2l$ **50.** [2.3b], [4.6b, c] $-\tfrac{61}{12}$

Cumulative Review: Chapters 1–4, p. 339

1. [1.1a] $\tfrac{5}{2}$ **2.** [4.3a] -4 **3.** [4.7a] -14 **4.** [1.2e] 4 **5.** [1.6b] $\tfrac{1}{5}$
6. [1.3a] $-\tfrac{11}{60}$ **7.** [1.4a] 4.2 **8.** [1.5a] 7.28 **9.** [1.6c] $-\tfrac{5}{12}$
10. [4.2d] 2.2×10^{22} **11.** [4.2d] 4×10^{-5} **12.** [1.7a] -3
13. [1.8b] $-2y - 7$ **14.** [1.8c] $5x + 11$ **15.** [1.8d] -2
16. [4.4a] $2x^5 - 2x^4 + 3x^3 + 2$ **17.** [4.7d] $3x^2 + xy - 2y^2$
18. [4.4c] $x^3 + 5x^2 - x - 7$ **19.** [4.4c] $-\tfrac{1}{3}x^2 - \tfrac{3}{4}x$
20. [1.7c] $12x - 15y + 21$ **21.** [4.5a] $6x^8$
22. [4.5b] $2x^5 - 4x^4 + 8x^3 - 10x^2$
23. [4.5d] $3y^4 + 5y^3 - 10y - 12$
24. [4.7f] $2p^4 + 3p^3q + 2p^2q^2 - 2p^4q - p^3q^2 - p^2q^3 + pq^3$
25. [4.6a] $6x^2 + 13x + 6$ **26.** [4.6c] $9x^4 + 6x^2 + 1$
27. [4.6b] $t^2 - \tfrac{1}{4}$ **28.** [4.6b] $4y^4 - 25$
29. [4.6a] $4x^6 + 6x^4 - 6x^2 - 9$ **30.** [4.6c] $t^2 - 4t^3 + 4t^4$
31. [4.7f] $15p^2 - pq - 2q^2$ **32.** [4.8a] $6x^2 + 2x - 3$
33. [4.8b] $3x^2 - 2x - 7$ **34.** [2.1b] -1.2 **35.** [2.2a] -21
36. [2.3a] 9 **37.** [2.2a] $-\tfrac{20}{3}$ **38.** [2.3b] 2 **39.** [2.1b] $\tfrac{13}{8}$
40. [2.3c] $-\tfrac{17}{21}$ **41.** [2.3b] -17 **42.** [2.3b] 2
43. [2.7e] $\{x | x < 16\}$ **44.** [2.7e] $\{x | x \leq -\tfrac{11}{8}\}$
45. [2.4b] $x = \dfrac{A-P}{Q}$ **46.** [2.5a] $3.50
47. [4.4d] $(\pi r^2 - 18)$ ft^2 **48.** [2.6a] 18 and 19
49. [2.6a] 20 ft, 24 ft **50.** [2.6a] $10°$ **51.** [4.1d, f] y^4
52. [4.1e, f] $\dfrac{1}{x}$ **53.** [4.2a, b] $-\dfrac{27x^9}{y^6}$ **54.** [4.1d, e, f] x^3
55. [3.3a]

56. [4.1a, f] $3^2 = 9, 3^{-2} = \tfrac{1}{9}, \left(\tfrac{1}{3}\right)^2 = \tfrac{1}{9}, \left(\tfrac{1}{3}\right)^{-2} = 9, -3^2 = -9,$ $(-3)^2 = 9, \left(-\tfrac{1}{3}\right)^2 = \tfrac{1}{9}, \left(-\tfrac{1}{3}\right)^{-2} = 9$ **57.** [4.4d] $(4x-4)$ in^2
58. [4.1d], [4.2a, b], [4.4a] $12x^5 - 15x^4 - 27x^3 + 4x^2$
59. [4.4a], [4.6c] $5x^2 - 2x + 10$ **60.** [2.3b], [4.6a, c] $\tfrac{11}{7}$
61. [2.3b], [4.8b] 1 **62.** [1.2e], [2.3a] $-5, 5$
63. [2.3b], [4.6a], [4.8b] All real numbers except 5

CHAPTER 5

Exercise Set 5.1, p. 347

RC1. (b) **RC2.** (c) **RC3.** (d) **RC4.** (a)
CC1. Yes **CC2.** No **CC3.** Yes **CC4.** No
1. 6 **3.** 24 **5.** 1 **7.** x **9.** x^2 **11.** 2 **13.** $17xy$
15. x **17.** x^2y^2 **19.** $x(x-6)$ **21.** $2x(x+3)$
23. $x^2(x+6)$ **25.** $8x^2(x^2-3)$ **27.** $2(x^2+x-4)$
29. $17xy(x^4y^2 + 2x^2y + 3)$ **31.** $x^2(6x^2 - 10x + 3)$
33. $x^2y^2(x^3y^3 + x^2y + xy - 1)$ **35.** $2x^3(x^4 - x^3 - 32x^2 + 2)$
37. $0.8x(2x^3 - 3x^2 + 4x + 8)$ **39.** $\tfrac{1}{3}x^3(5x^3 + 4x^2 + x + 1)$
41. $(x+3)(x^2+2)$ **43.** $(3z-1)(4z^2+7)$
45. $(3x+2)(2x^2+1)$ **47.** $(2a-7)(5a^3-1)$
49. $(x+3)(x^2+2)$ **51.** $(x+3)(2x^2+1)$
53. $(2x-3)(4x^2+3)$ **55.** $(3p-4)(4p^2+1)$
57. $(x-1)(5x^2-1)$ **59.** $(x+8)(x^2-3)$

61. $(x-4)(2x^2-9)$ **63.** $y^2+12y+35$
64. $y^2+14y+49$ **65.** y^2-49 **66.** $y^2-14y+49$
67. $16x^3-48x^2+8x$ **68.** $28w^2-53w-66$
69. $49w^2+84w+36$ **70.** $16w^2-88w+121$
71. $16w^2-121$ **72.** y^3-3y^2+5y **73.** $6x^2+11xy-35y^2$
74. $25x^2-10xt+t^2$ **75.** $(2x^2+3)(2x^3+3)$
77. $(x^5+1)(x^7+1)$ **79.** Not factorable by grouping

Exercise Set 5.2, p. 355

RC1. True   **RC2.** True   **RC3.** True   **RC4.** False
CC1. 1, 18; 2, 9; 3, 6   **CC2.** 1, 42; 2, 21; 3, 14; 6, 7
CC3. 1, 96; 2, 48; 3, 32; 4, 24; 6, 16; 8, 12
CC4. 1, 150; 2, 75; 3, 50; 5, 30; 6, 25; 10, 15

1.

Pairs of Factors	Sums of Factors
1, 15	16
−1, −15	−16
3, 5	8
−3, −5	−8

$(x+3)(x+5)$

3.

Pairs of Factors	Sums of Factors
1, 12	13
−1, −12	−13
2, 6	8
−2, −6	−8
3, 4	7
−3, −4	−7

$(x+3)(x+4)$

5.

Pairs of Factors	Sums of Factors
1, 9	10
−1, −9	−10
3, 3	6
−3, −3	−6

$(x-3)^2$

7.

Pairs of Factors	Sums of Factors
−1, 14	13
1, −14	−13
−2, 7	5
2, −7	−5

$(x+2)(x-7)$

9.

Pairs of Factors	Sums of Factors
1, 4	5
−1, −4	−5
2, 2	4
−2, −2	−4

$(b+1)(b+4)$

11.

Pairs of Factors	Sums of Factors
−1, 18	17
1, −18	−17
−2, 9	7
2, −9	−7
−3, 6	3
3, −6	−3

$(t-3)(t+6)$

13. $(d-2)(d-5)$ **15.** $(y-1)(y-10)$ **17.** Prime
19. $(x-9)(x+2)$ **21.** $x(x-8)(x+2)$
23. $y(y-9)(y+5)$ **25.** $(x-11)(x+9)$
27. $(c^2+8)(c^2-7)$ **29.** $(a^2+7)(a^2-5)$
31. $(x-6)(x+7)$ **33.** Prime   **35.** $(x+10)^2$
37. $2z(z-4)(z+3)$ **39.** $3t^2(t^2+t+1)$
41. $x^2(x-25)(x+4)$ **43.** $(x-24)(x+3)$
45. $(x-9)(x-16)$ **47.** $(a+12)(a-11)$
49. $3(t+1)^2$ **51.** $w^2(w-4)^2$ **53.** $-1(x-10)(x+3)$,
or $(-x+10)(x+3)$, or $(x-10)(-x-3)$
55. $-1(a-2)(a+12)$, or $(-a+2)(a+12)$, or
$(a-2)(-a-12)$ **57.** $(x-15)(x-8)$
59. $-1(x+12)(x-9)$, or $(-x-12)(x-9)$, or
$(x+12)(-x+9)$ **61.** $(y-0.4)(y+0.2)$
63. $(p+5q)(p-2q)$ **65.** $-1(t+14)(t-6)$, or
$(-t-14)(t-6)$, or $(t+14)(-t+6)$
67. $(m+4n)(m+n)$ **69.** $(s+3t)(s-5t)$
71. $6a^8(a+2)(a-7)$ **73.** 12   **74.** −1   **75.** $\frac{5}{4}$
76. No solution   **77.** $\{x\,|\,x>-24\}$
78. $\{x\,|\,x\le\frac{14}{5}\}$ **79.** $p=2A-w$ **80.** $x=\dfrac{y-b}{m}$
81. 0.756 billion min, or 756 million min   **82.** 73.6 million
83. 15, −15, 27, −27, 51, −51   **85.** $(x-\frac{1}{2})(x+\frac{1}{4})$
87. $(x+5)(x-\frac{5}{7})$ **89.** $(b^n+5)(b^n+2)$ **91.** $2x^2(4-\pi)$

Calculator Corner, p. 361

1. Correct   **2.** Correct   **3.** Not correct   **4.** Not correct

Exercise Set 5.3, p. 365

RC1. True   **RC2.** True   **RC3.** False   **RC4.** False
CC1. (b), (c), (f), (g), (j), (k)   **CC2.** (a) $-299x$; (d) $13x$; (e) $-55x$;
(h) $97x$; (i) $-71x$; (l) $5x$ **CC3.** (l) $(3x+10)(2x-5)$
1. $(2x+1)(x-4)$ **3.** $(5x+9)(x-2)$ **5.** $(3x+1)(2x+7)$
7. $(3x+1)(x+1)$ **9.** $(2x-3)(2x+5)$ **11.** $(2x+1)(x-1)$
13. $(3x-2)(3x+8)$ **15.** $(3x+1)(x-2)$
17. $(3x+4)(4x+5)$ **19.** $(7x-1)(2x+3)$
21. $(3x+2)(3x+4)$ **23.** $(3x-7)^2$, or $(7-3x)^2$
25. $(24x-1)(x+2)$ **27.** $(5x-11)(7x+4)$
29. $-2(x-5)(x+2)$, or $2(-x+5)(x+2)$, or $2(x-5)(-x-2)$
31. $4(3x-2)(x+3)$ **33.** $6(5x-9)(x+1)$
35. $2(3y+5)(y-1)$ **37.** $(3x-1)(x-1)$
39. $4(3x+2)(x-3)$ **41.** $(2x+1)(x-1)$
43. $(3x+2)(3x-8)$ **45.** $5(3x+1)(x-2)$
47. $p(3p+4)(4p+5)$ **49.** $-1(3x+2)(3x-8)$, or
$(-3x-2)(3x-8)$, or $(3x+2)(-3x+8)$
51. $-1(5x-3)(3x-2)$, or $(-5x+3)(3x-2)$, or
$(5x-3)(-3x+2)$ **53.** $x^2(7x-1)(2x+3)$
55. $3x(8x-1)(7x-1)$ **57.** $(5x^2-3)(3x^2-2)$ **59.** $(5t+8)^2$

61. $2x(3x + 5)(x - 1)$ **63.** Prime **65.** Prime
67. $(4m + 5n)(3m - 4n)$ **69.** $(2a + 3b)(3a - 5b)$
71. $(3a + 2b)(3a + 4b)$ **73.** $(5p + 2t)(7p + 4t)$
75. $6(3x - 4y)(x + y)$

77. 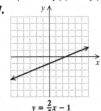   **78.**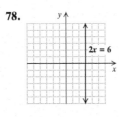

79. 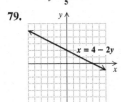   **80.**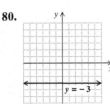

81. 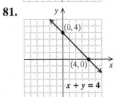   **82.**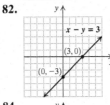

83. 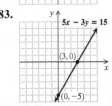   **84.**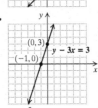

85. $(2x^n + 1)(10x^n + 3)$ **87.** $(x^{3a} - 1)(3x^{3a} + 1)$
89.–93. Left to the student

Exercise Set 5.4, p. 370

RC1. leading coefficient **RC2.** product; sum **RC3.** sum
RC4. grouping **CC1.** (d) **CC2.** (a) **CC3.** (c) **CC4.** (b)
1. $(x + 2)(x + 7)$ **3.** $(x - 4)(x - 1)$ **5.** $(3x + 2)(2x + 3)$
7. $(3x - 4)(x - 4)$ **9.** $(7x - 8)(5x + 3)$
11. $(2x + 3)(2x - 3)$ **13.** $(x^2 + 3)(2x^2 + 5)$
15. $(2x + 3)(x + 2)$ **17.** $(3x + 5)(x - 3)$
19. $(5x + 1)(x + 2)$ **21.** $(3x - 1)(x - 1)$
23. $(2x + 7)(3x + 1)$ **25.** $(2x + 3)(2x - 5)$
27. $(5x - 2)(3x + 5)$ **29.** $(3x + 2)(3x - 8)$
31. $(3x - 1)(x + 2)$ **33.** $(3x - 4)(4x - 5)$
35. $(7x + 1)(2x - 3)$ **37.** $(3x - 7)^2$, or $(7 - 3x)^2$
39. $(3x + 2)(3x + 4)$ **41.** $-1(3a - 1)(3a + 5)$, or $(-3a + 1)(3a + 5)$, or $(3a - 1)(-3a - 5)$
43. $-2(x - 5)(x + 2)$, or $2(-x + 5)(x + 2)$, or $2(x - 5)(-x - 2)$ **45.** $4(3x - 2)(x + 3)$
47. $6(5x - 9)(x + 1)$ **49.** $2(3y + 5)(y - 1)$
51. $(3x - 1)(x - 1)$ **53.** $4(3x + 2)(x - 3)$
55. $(2x + 1)(x - 1)$ **57.** $(3x - 2)(3x + 8)$
59. $5(3x + 1)(x - 2)$ **61.** $p(3p + 4)(4p + 5)$
63. $-1(5x - 4)(x + 1)$, or $(-5x + 4)(x + 1)$, or $(5x - 4)(-x - 1)$ **65.** $-3(2t - 1)(t - 5)$, or $3(-2t + 1)(t - 5)$, or $3(2t - 1)(-t + 5)$
67. $x^2(7x - 1)(2x + 3)$ **69.** $3x(8x - 1)(7x - 1)$
71. $(5x^2 - 3)(3x^2 - 2)$ **73.** $(5t + 8)^2$
75. $2x(3x + 5)(x - 1)$ **77.** Prime **79.** Prime
81. $(4m + 5n)(3m - 4n)$ **83.** $(2a + 3b)(3a - 5b)$
85. $(3a - 2b)(3a - 4b)$ **87.** $(5p + 2q)(7p + 4q)$
89. $6(3x - 4y)(x + y)$ **91.** $-6x(x - 5)(x + 2)$, or $6x(-x + 5)(x + 2)$, or $6x(x - 5)(-x - 2)$

93. $x^3(5x - 11)(7x + 4)$ **95.** $27x^{12}$ **96.** $\dfrac{1}{5^{14}}$ **97.** x^5y^6
98. a **99.** 3.008×10^{10} **100.** 0.000015 **101.** About 1.6 m, or 5.3 ft **102.** $40°$ **103.** $(3x^5 - 2)^2$ **105.** $(4x^5 + 1)^2$
107.–111. Left to the student

Mid-Chapter Review: Chapter 5, p. 374

1. True **2.** False **3.** True **4.** False
5. $10y^3 - 18y^2 + 12y = 2y \cdot 5y^2 - 2y \cdot 9y + 2y \cdot 6$
$\qquad = 2y(5y^2 - 9y + 6)$
6. $a \cdot c = 2 \cdot (-6) = -12;$
$\quad -x = -4x + 3x;$
$\quad 2x^2 - x - 6 = 2x^2 - 4x + 3x - 6$
$\qquad = 2x(x - 2) + 3(x - 2)$
$\qquad = (x - 2)(2x + 3)$
7. x **8.** x^2 **9.** $6x^3$ **10.** 4 **11.** $5x^2y$ **12.** x^2y^2
13. $x(x^2 - 8)$ **14.** $3x(x + 4)$ **15.** $2(y^2 + 4y - 2)$
16. $t^3(3t^3 - 5t - 2)$ **17.** $(x + 1)(x + 3)$ **18.** $(z - 2)^2$
19. $(x + 4)(x^2 + 3)$ **20.** $8y^3(y^2 - 6)$
21. $6xy(x^2 + 4xy - 7y^2)$ **22.** $(4t - 3)(t - 2)$
23. $(z - 1)(z + 5)$ **24.** $(z + 4)(2z^2 + 5)$
25. $(3p - 2)(p^2 - 3)$ **26.** $5x^3(2x^5 - 5x^3 - 3x^2 + 7)$
27. $(2w + 3)(w^2 - 3)$ **28.** $x^2(4x^2 - 5x + 3)$
29. $(6y - 5)(y + 2)$ **30.** $3(x - 3)(x + 2)$
31. $(3x + 2)(2x^2 + 1)$ **32.** $(w - 5)(w - 3)$
33. $(2x + 5)(4x^2 + 1)$ **34.** $(5z + 2)(2z - 5)$
35. $(2x + 1)(3x + 2)$ **36.** $(x - 6y)(x - 4y)$
37. $(2z + 1)(3z^2 + 1)$ **38.** $a^2b^3(ab^4 + a^2b^2 - 1 + a^3b^3)$
39. $(4y + 5z)(y - 3z)$ **40.** $3x(x + 2)(x + 5)$
41. $(x - 3)(x^2 - 2)$ **42.** $(3y + 1)^2$ **43.** $(y + 2)(y + 4)$
44. $3(2y + 5)(y + 3)$ **45.** $(x - 7)(x^2 + 4)$
46. $-1(y - 4)(y + 1)$, or $(-y + 4)(y + 1)$, or $(y - 4)(-y - 1)$
47. $4(2x + 3)(2x - 5)$ **48.** $(5a - 3b)(2a - b)$
49. $(2w - 5)(3w^2 - 5)$ **50.** $y(y + 6)(y + 3)$
51. $(4x + 3y)(x + 2y)$ **52.** $-1(3z - 2)(2z + 3)$, or $(-3z + 2)(2z + 3)$, or $(3z - 2)(-2z - 3)$
53. $(3t + 2)(4t^2 - 3)$ **54.** $(y - 4z)(y + 5z)$
55. $(3x - 4y)(3x + 2y)$ **56.** $(3z - 1)(z + 3)$
57. $(m - 8n)(m + 2n)$ **58.** $2(w - 3)^2$
59. $2t(3t - 2)(3t - 1)$ **60.** $(z + 3)(5z^2 + 1)$
61. $(t - 2)(t + 7)$ **62.** $(2t - 5)^2$ **63.** $(t - 2)(t + 6)$
64. $-1(2z + 3)(z - 4)$, or $(-2z - 3)(z - 4)$, or $(2z + 3)(-z + 4)$ **65.** $-1(y - 6)(y + 2)$, or $(-y + 6)(y + 2)$, or $(y - 6)(-y - 2)$ **66.** Find the product of two binomials. For example, $(2x^2 + 3)(x - 4) = 2x^3 - 8x^2 + 3x - 12$.
67. There is a finite number of pairs of numbers with the correct product, but there are infinitely many pairs with the correct sum. **68.** Since both constants are negative, the middle term will be negative, so $(x - 17)(x - 18)$ cannot be a factorization of $x^2 + 35x + 306$. **69.** No; both $2x + 6$ and $2x + 8$ contain a factor of 2, so $2 \cdot 2$, or 4, must be factored out to obtain the complete factorization. In other words, the largest common factor is 4, not 2.

Exercise Set 5.5, p. 381

RC1. False **RC2.** True **RC3.** False **RC4.** False
CC1. $A = x, B = 8$ **CC2.** $A = 5a, B = c$
CC3. $A = x^5, B = 1$ **CC4.** $A = 3, B = \frac{1}{7}y$
1. Yes **3.** No **5.** No **7.** Yes **9.** $(x - 7)^2$
11. $(x + 8)^2$ **13.** $(x - 1)^2$ **15.** $(x + 2)^2$ **17.** $(y + 6)^2$
19. $(t - 4)^2$ **21.** $(q^2 - 3)^2$ **23.** $(4y + 7)^2$
25. $2(x - 1)^2$ **27.** $x(x - 9)^2$ **29.** $3(2q - 3)^2$
31. $(7 - 3x)^2$, or $(3x - 7)^2$ **33.** $5(y^2 + 1)^2$ **35.** $(1 + 2x^2)^2$
37. $(2p + 3t)^2$ **39.** $(a - 3b)^2$ **41.** $(9a - b)^2$
43. $4(3a + 4b)^2$ **45.** Yes **47.** No **49.** No **51.** Yes
53. $(y + 2)(y - 2)$ **55.** $(p + 1)(p - 1)$ **57.** $(t + 7)(t - 7)$
59. $(a + b)(a - b)$ **61.** $(5t + m)(5t - m)$
63. $(10 + k)(10 - k)$ **65.** $(4a + 3)(4a - 3)$

67. $(2x + 5y)(2x - 5y)$ **69.** $2(2x + 7)(2x - 7)$
71. $x(6 + 7x)(6 - 7x)$ **73.** $(\frac{1}{4} + 7x^4)(\frac{1}{4} - 7x^4)$
75. $(0.3y + 0.02)(0.3y - 0.02)$ **77.** $(7a^2 + 9)(7a^2 - 9)$
79. $(a^2 + 4)(a + 2)(a - 2)$ **81.** $5(x^2 + 9)(x + 3)(x - 3)$
83. $(1 + y^4)(1 + y^2)(1 + y)(1 - y)$
85. $(x^6 + 4)(x^3 + 2)(x^3 - 2)$ **87.** $(y + \frac{1}{4})(y - \frac{1}{4})$
89. $(5 + \frac{1}{7}x)(5 - \frac{1}{7}x)$ **91.** $(4m^2 + t^2)(2m + t)(2m - t)$
93. y-intercept: $(0, 4)$; x-intercept: $(16, 0)$
94. y-intercept: $(0, -5)$; x-intercept: $(6.5, 0)$
95. y-intercept: $(0, -5)$; x-intercept: $(\frac{5}{2}, 0)$
96. 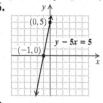 **97.**
98.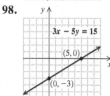
99. $x^2 - 4xy + 4y^2$ **100.** $\frac{1}{2}\pi x^2 + 2xy$ **101.** Prime
103. $(x + 11)^2$ **105.** $2x(3x + 1)^2$
107. $(x^4 + 2^4)(x^2 + 2^2)(x + 2)(x - 2)$ **109.** $3x^3(x + 2)(x - 2)$
111. $2x(3x + \frac{2}{5})(3x - \frac{2}{5})$ **113.** $p(0.7 + p)(0.7 - p)$
115. $(0.8x + 1.1)(0.8x - 1.1)$ **117.** $x(x + 6)$
119. $(x + \frac{1}{x})(x - \frac{1}{x})$ **121.** $(9 + b^{2k})(3 - b^k)(3 + b^k)$
123. $(3b^n + 2)^2$ **125.** $(y + 4)^2$ **127.** 9 **129.** Not correct
131. Not correct

Exercise Set 5.6, p. 388
RC1. (e) **RC2.** (c) **RC3.** (d) **RC4.** (g) **RC5.** (f)
RC6. (a) **RC7.** (b) **RC8.** (h) **1.** $(z + 3)(z^2 - 3z + 9)$
3. $(x - 1)(x^2 + x + 1)$ **5.** $(y + 5)(y^2 - 5y + 25)$
7. $(2a + 1)(4a^2 - 2a + 1)$ **9.** $(y - 2)(y^2 + 2y + 4)$
11. $(2 - 3b)(4 + 6b + 9b^2)$ **13.** $(4y + 1)(16y^2 - 4y + 1)$
15. $(2x + 3)(4x^2 - 6x + 9)$ **17.** $(a - b)(a^2 + ab + b^2)$
19. $(a + \frac{1}{2})(a^2 - \frac{1}{2}a + \frac{1}{4})$ **21.** $2(y - 4)(y^2 + 4y + 16)$
23. $3(2a + 1)(4a^2 - 2a + 1)$ **25.** $r(s + 4)(s^2 - 4s + 16)$
27. $5(x - 2z)(x^2 + 2xz + 4z^2)$ **29.** $(x + 0.1)(x^2 - 0.1x + 0.01)$
31. $8(2x^2 - t^2)(4x^4 + 2x^2t^2 + t^4)$ **33.** $2y(y - 4)(y^2 + 4y + 16)$
35. $(z - 1)(z^2 + z + 1)(z + 1)(z^2 - z + 1)$
37. $(t^2 + 4y^2)(t^4 - 4t^2y^2 + 16y^4)$
39. $(2w^3 - z^3)(4w^6 + 2w^3z^3 + z^6)$
41. $(\frac{1}{2}c + d)(\frac{1}{4}c^2 - \frac{1}{2}cd + d^2)$
43. $(0.1x - 0.2y)(0.01x^2 + 0.02xy + 0.04y^2)$ **45.** $\frac{343}{y^{15}}$
46. a^8b^{18} **47.** $\frac{16}{x^6}$ **48.** $4y^{10} - 9$ **49.** $w^2 - \frac{2}{3}w + \frac{1}{9}$
50. $x^2 + 0.4x - 0.05$ **51.** (a) $\pi h(R + r)(R - r)$;
(b) 3,014,400 cm³, or 3.0144 m³
53. $3(x^a + 2y^b)(x^{2a} - 2x^ay^b + 4y^{2b})$
55. $\frac{1}{3}(\frac{1}{2}xy + z)(\frac{1}{4}x^2y^2 - \frac{1}{2}xyz + z^2)$
57. $y(3x^2 + 3xy + y^2)$ **59.** $4(3a^2 + 4)$

Exercise Set 5.7, p. 395
RC1. common **RC2.** difference **RC3.** square
RC4. grouping **RC5.** completely **RC6.** check
CC1. (b) **CC2.** (a) **CC3.** (e) **CC4.** (d) **CC5.** (a)

1. $3(x + 8)(x - 8)$ **3.** $(a - 5)^2$ **5.** $(2x - 3)(x - 4)$
7. $x(x + 12)^2$ **9.** $(x + 3)(x + 2)(x - 2)$
11. $3(4x + 1)(4x - 1)$ **13.** $3x(3x - 5)(x + 3)$ **15.** Prime
17. $x(x^2 + 7)(x - 3)$ **19.** $x^3(x - 7)^2$ **21.** $-2(x - 2)(x + 5)$,
or $2(-x + 2)(x + 5)$, or $2(x - 2)(-x - 5)$
23. Prime **25.** $4(x^2 + 4)(x + 2)(x - 2)$
27. $(1 + y^4)(1 + y^2)(1 + y)(1 - y)$
29. $x^3(x - 3)(x - 1)$ **31.** $\frac{1}{9}(\frac{1}{3}x^3 - 4)^2$
33. $(\frac{1}{10}m - \frac{1}{3}n)(\frac{1}{100}m^2 + \frac{1}{30}mn + \frac{1}{9}n^2)$
35. $9xy(xy - 4)$ **37.** $2\pi r(h + r)$ **39.** $(a + b)(2x + 1)$
41. $(x + 1)(x - 1 - y)$ **43.** $(n + 2)(n + p)$
45. $(2w - 1)(3w + p)$ **47.** $(2b - a)^2$, or $(a - 2b)^2$
49. $(4x + 3y)^2$ **51.** $(7m^2 - 8n)^2$ **53.** $(y^2 + 5z^2)^2$
55. $(\frac{1}{2}a + \frac{1}{3}b)^2$ **57.** $(a + b)(a - 2b)$
59. $(m + 20n)(m - 18n)$ **61.** $(mn - 8)(mn + 4)$
63. $r^3(rs - 2)(rs - 8)$ **65.** $a^3(a - b)(a + 5b)$
67. $(a + \frac{1}{5}b)(a - \frac{1}{5}b)$
69. $7(x + y)(x^2 - xy + y^2)(x - y)(x^2 + xy + y^2)$
71. $(4 + c^2d^2)(2 + cd)(2 - cd)$
73. $(1 + 4x^6y^6)(1 + 2x^3y^3)(1 - 2x^3y^3)$
75. $(q + 8)(q + 1)(q - 1)$ **77.** $ab(2ab + 1)(3ab - 2)$
79. $(m + 1)(m - 1)(m + 2)(m - 2)$ **81.** -22 **82.** -5
83. 7 **84.** > **85.** $(t + 1)^2(t - 1)^2$
87. $(x - 5)(x + 2)(x - 2)$ **89.** $(3.5x - 1)^2$
91. $(y - 2)(y + 3)(y - 3)$ **93.** $(y - 1)^3$

Calculator Corner, p. 400
1. Left to the student

Exercise Set 5.8, p. 404
RC1. False **RC2.** False **RC3.** False **RC4.** True
CC1. $x - 5 = 0; x + 4 = 0$ **CC2.** $3x + 2 = 0; x - 7 = 0$
CC3. $x = 0; x + 6 = 0$ **CC4.** $5x = 0$ (or $x = 0$); $x - 8 = 0$
CC5. $x = 0; x - 1 = 0; x + 3 = 0$
CC6. $3x - 7 = 0; x + 1 = 0$
1. $-4, -9$ **3.** $-3, 8$ **5.** $-12, 11$ **7.** $0, -3$ **9.** $0, -18$
11. $-\frac{5}{2}, -4$ **13.** $-\frac{1}{5}, 3$ **15.** $4, \frac{1}{4}$ **17.** $0, \frac{2}{3}$ **19.** $-\frac{1}{10}, \frac{1}{27}$
21. $\frac{1}{3}, -20$ **23.** $0, \frac{2}{3}, \frac{1}{2}$ **25.** $-5, -1$ **27.** $-9, 2$ **29.** $3, 5$
31. $0, 8$ **33.** $0, -18$ **35.** $-4, 4$ **37.** $-\frac{2}{3}, \frac{2}{3}$ **39.** -3 **41.** 4
43. $0, \frac{6}{5}$ **45.** $-1, \frac{5}{3}$ **47.** $-\frac{1}{4}, \frac{2}{3}$ **49.** $-1, \frac{2}{3}$ **51.** $-\frac{7}{10}, \frac{7}{10}$
53. $-2, 9$ **55.** $\frac{4}{5}, \frac{3}{2}$ **57.** $(-4, 0), (1, 0)$ **59.** $(-\frac{5}{2}, 0), (2, 0)$
61. $(-3, 0), (5, 0)$ **63.** $-1, 4$ **65.** $-1, 3$ **67.** $(a + b)^2$
68. $a^2 + b^2$ **69.** $\{x | x < -100\}$ **70.** $\{x | x \leq 8\}$
71. $\{x | x < 2\}$ **72.** $\{x | x \geq \frac{20}{3}\}$ **73.** $-5, 4$ **75.** $-3, 9$
77. $-\frac{1}{8}, \frac{1}{8}$ **79.** $-4, 4$ **81.** $2.33, 6.77$ **83.** Answers may vary.
(a) $x^2 - x - 12 = 0$; (b) $x^2 + 7x + 12 = 0$; (c) $4x^2 - 4x + 1 = 0$;
(d) $x^2 - 25 = 0$; (e) $40x^3 - 14x^2 + x = 0$

Translating for Success, p. 413
1. O **2.** M **3.** K **4.** I **5.** G **6.** E **7.** C **8.** A
9. H **10.** B

Exercise Set 5.9, p. 414
RC1. consecutive **RC2.** hypotenuse **RC3.** half
RC4. right **CC1.** (b) **CC2.** (b) **CC3.** (d) **CC4.** (c)
1. Length: 42 in.; width: 14 in. **3.** Length: 6 cm; width: 4 cm
5. Height: 4 cm; base: 14 cm **7.** Base: 8 m; height: 16 m
9. 182 games **11.** 12 teams **13.** 4950 handshakes
15. 25 people **17.** 14 and 15 **19.** 12 and 14; -12 and -14
21. 15 and 17; -15 and -17 **23.** 32 ft **25.** Hypotenuse: 17 ft;
leg: 15 ft **27.** 300 ft by 400 ft by 500 ft **29.** 24 m, 25 m

31. Dining room: 12 ft by 12 ft; kitchen: 12 ft by 10 ft **33.** 1 sec, 2 sec **35.** 5 and 7 **37.** 4.53 **38.** $-\frac{5}{6}$ **39.** -40 **40.** -116 **41.** $-\frac{3}{25}$ **42.** -3.4 **43.** $-4y - 13$ **44.** $10x - 30$ **45.** 5 ft **47.** 30 cm by 15 cm **49.** 11 yd, 60 yd, 61 yd

Summary and Review: Chapter 5, p. 419

Vocabulary Reinforcement
1. factor **2.** factor **3.** factorization **4.** common **5.** grouping **6.** binomial **7.** zero **8.** difference

Concept Reinforcement
1. False **2.** True **3.** False **4.** True

Study Guide
1. $4xy$ **2.** $9x^2(3x^3 - x + 2)$ **3.** $(z - 3)(z^2 + 4)$ **4.** $(x + 2)(x + 4)$ **5.** $3(z - 4)(2z + 1)$ **6.** $(3y - 1)(2y + 3)$ **7.** $(2x + 1)^2$ **8.** $2(3x + 2)(3x - 2)$ **9.** $(3 - 5x)(9 + 15x + 25x^2)$ **10.** $\left(\frac{1}{2}q + 2a\right)\left(\frac{1}{4}q^2 - qa + 4a^2\right)$ **11.** $-5, 1$

Review Exercises
1. $5y^2$ **2.** $12x$ **3.** $5(1 + 2x^3)(1 - 2x^3)$ **4.** $x(x - 3)$ **5.** $(3x + 2)(3x - 2)$ **6.** $(x + 6)(x - 2)$ **7.** $(x + 7)^2$ **8.** $3x(2x^2 + 4x + 1)$ **9.** $(x + 1)(x^2 + 3)$ **10.** $(3x - 1)(2x - 1)$ **11.** $(x^2 + 9)(x + 3)(x - 3)$ **12.** $3x(3x - 5)(x + 3)$ **13.** $2(x + 5)(x - 5)$ **14.** $(x + 4)(x^3 - 2)$ **15.** $(4x^2 + 1)(2x + 1)(2x - 1)$ **16.** $4x^4(2x^2 - 8x + 1)$ **17.** $3(2x + 5)^2$ **18.** Prime **19.** $x(x - 6)(x + 5)$ **20.** $(2x + 5)(2x - 5)$ **21.** $(3x - 5)^2$ **22.** $2(3x + 4)(x - 6)$ **23.** $(x - 3)^2$ **24.** $(2x + 1)(x - 4)$ **25.** $2(3x - 1)^2$ **26.** $3(x + 3)(x - 3)$ **27.** $(x - 5)(x - 3)$ **28.** $(5x - 2)^2$ **29.** $(7b^5 - 2a^4)^2$ **30.** $(xy + 4)(xy - 3)$ **31.** $3(2a + 7b)^2$ **32.** $(m + 5)(m + t)$ **33.** $32(x^2 - 2y^2z^2)(x^2 + 2y^2z^2)$ **34.** $5(y + 2t)(y^2 - 2yt + 4t^2)$ **35.** $1, -3$ **36.** $-7, 5$ **37.** $-4, 0$ **38.** $\frac{2}{3}, 1$ **39.** $-8, 8$ **40.** $-2, 8$ **41.** $(-5, 0), (-4, 0)$ **42.** $\left(-\frac{3}{2}, 0\right), (5, 0)$ **43.** Height: 6 cm; base: 5 cm **44.** -18 and -16; 16 and 18 **45.** 842 ft **46.** On the ground: 4 ft; on the tree: 3 ft **47.** 6 km **48.** B **49.** A **50.** 2.5 cm **51.** $0, 2$ **52.** Length: 12 in.; width: 6 in. **53.** 35 ft **54.** No solution **55.** $2, -3, \frac{5}{2}$ **56.** $-2, \frac{5}{4}, 3$

Understanding Through Discussion and Writing
1. Although $x^3 - 8x^2 + 15x$ can be factored as $(x^2 - 5x)(x - 3)$, this is not a complete factorization of the polynomial since $x^2 - 5x = x(x - 5)$. Gwen should always look for a common factor first. **2.** Josh is correct, because answers can easily be checked by multiplying. **3.** For $x = -3$:
$$(x - 4)^2 = (-3 - 4)^2 = (-7)^2 = 49;$$
$$(4 - x)^2 = [4 - (-3)]^2 = 7^2 = 49.$$
For $x = 1$:
$$(x - 4)^2 = (1 - 4)^2 = (-3)^2 = 9;$$
$$(4 - x)^2 = (4 - 1)^2 = 3^2 = 9.$$
In general, $(x - 4)^2 = [-(-x + 4)]^2 = [-(4 - x)]^2 = (-1)^2(4 - x)^2 = (4 - x)^2$.

4. The equation is not in the form $ab = 0$. The correct procedure is
$$(x - 3)(x + 4) = 8$$
$$x^2 + x - 12 = 8$$
$$x^2 + x - 20 = 0$$
$$(x + 5)(x - 4) = 0$$
$$x + 5 = 0 \quad \text{or} \quad x - 4 = 0$$
$$x = -5 \quad \text{or} \quad x = 4.$$
The solutions are -5 and 4.
5. One solution of the equation is 0. Dividing both sides of the equation by x, leaving the solution $x = 3$, is equivalent to dividing by 0. **6.** She could use the measuring sticks to draw a right angle as shown below. Then she could use the 4-ft stick to extend one leg to 7 ft and the 5-ft stick to extend the other leg to 9 ft.

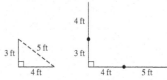

Next, she could draw another right angle with either the 7-ft side or the 9-ft side as a side.

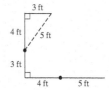

Then she could use the sticks to extend the other side to the appropriate length. Finally, she would draw the remaining side of the rectangle.

Test: Chapter 5, p. 425
1. [5.1a] $4x^3$ **2.** [5.2a] $(x - 5)(x - 2)$ **3.** [5.5b] $(x - 5)^2$ **4.** [5.1b] $2y^2(2y^2 - 4y + 3)$ **5.** [5.1c] $(x + 1)(x^2 + 2)$ **6.** [5.1b] $x(x - 5)$ **7.** [5.2a] $x(x + 3)(x - 1)$ **8.** [5.3a], [5.4a] $2(5x - 6)(x + 4)$ **9.** [5.5d] $(2x + 3)(2x - 3)$ **10.** [5.2a] $(x - 4)(x + 3)$ **11.** [5.3a], [5.4a] $3m(2m + 1)(m + 1)$ **12.** [5.5d] $3(w + 5)(w - 5)$ **13.** [5.5b] $5(3x + 2)^2$ **14.** [5.5d] $3(x^2 + 4)(x + 2)(x - 2)$ **15.** [5.5b] $(7x - 6)^2$ **16.** [5.3a], [5.4a] $(5x - 1)(x - 5)$ **17.** [5.1c] $(x + 2)(x^3 - 3)$ **18.** [5.5d] $5(4 + x^2)(2 + x)(2 - x)$ **19.** [5.3a], [5.4a] $3t(2t + 5)(t - 1)$ **20.** [5.3a], [5.4a] $(2x + 3)(2x - 5)$ **21.** [5.2a] $3(m + 2n)(m - 5n)$ **22.** [5.6a] $(10a - 3b)(100a^2 + 30ab + 9b^2)$ **23.** [5.8b] $0, 3$ **24.** [5.8b] $-4, 4$ **25.** [5.8b] $-4, 5$ **26.** [5.8b] $-5, \frac{3}{2}$ **27.** [5.8b] $-4, 7$ **28.** [5.8b] $(-5, 0), (7, 0)$ **29.** [5.8b] $\left(\frac{2}{3}, 0\right), (1, 0)$ **30.** [5.9a] Length: 8 m; width: 6 m **31.** [5.9a] Height: 4 cm; base: 14 cm **32.** [5.9a] 5 ft **33.** [5.5d] A **34.** [5.9a] Length: 15 m; width: 3 m **35.** [5.2a] $(a - 4)(a + 8)$ **36.** [5.8b] $-\frac{8}{3}, 0, \frac{2}{5}$ **37.** [4.6b], [5.5d] D

Cumulative Review: Chapters 1-5, p. 427

1. [1.2d] <   2. [1.2d] >   3. [1.4a] 0.35   4. [1.6c] -1.57
5. [1.5a] $-\frac{1}{14}$ 6. [1.6c] $-\frac{6}{5}$ 7. [1.8c] $4x+1$ 8. [1.8d] -8
9. [4.2a, b] $\frac{8x^6}{y^3}$ 10. [4.1d, e] $-\frac{1}{6x^3}$
11. [4.4a] $x^4-3x^3-3x^2-4$ 12. [4.7e] $2x^2y^2-x^2y-xy$
13. [4.8b] $x^2+3x+2+\frac{3}{x-1}$ 14. [4.6c] $4t^2-12t+9$
15. [4.6b] x^4-9 16. [4.6a] $6x^2+4x-16$
17. [4.5b] $2x^4+6x^3+8x^2$ 18. [4.5d] $4y^3+4y^2+5y-4$
19. [4.6b] $x^2-\frac{4}{9}$ 20. [5.2a] $(x+4)(x-2)$
21. [5.5d] $(2x+5)(2x-5)$ 22. [5.1c] $(3x-4)(x^2+1)$
23. [5.5b] $(x-13)^2$ 24. [5.5d] $3(5x+6y)(5x-6y)$
25. [5.3a], [5.4a] $(3x+7)(2x-9)$ 26. [5.2a] $(x^2-3)(x^2+1)$
27. [5.7a] $2(2y-3)(y-1)(y+1)$
28. [5.3a], [5.4a] $(3p-q)(2p+q)$
29. [5.3a], [5.4a] $2x(5x+1)(x+5)$ 30. [5.5b] $x(7x-3)^2$
31. [5.3a], [5.4a] Prime   32. [5.1b] $3x(25x^2+9)$
33. [5.5d] $3(x^4+4y^4)(x^2+2y^2)(x^2-2y^2)$
34. [5.2a] $14(x+2)(x+1)$ 35. [5.7a] $(x+1)(x-1)(2x^3+1)$
36. [2.3b] 15   37. [2.7e] $\{y | y < 6\}$ 38. [5.8a] 15, $-\frac{1}{4}$
39. [5.8a] 0, -37 40. [5.8a] 5, -5, -1 41. [5.8b] 6, -6
42. [5.8b] $\frac{1}{3}$ 43. [5.8b] -10, -7 44. [5.8b] 0, $\frac{3}{2}$
45. [2.3a] 0.2   46. [5.8b] -4, 5   47. [2.7e] $\{x | x \le 20\}$
48. [2.3c] All real numbers   49. [2.4b] $m = \frac{y-b}{x}$
50. [2.6a] 50, 52   51. [5.9a] -20 and -18; 18 and 20
52. [5.9a] Length: 6 ft; height: 3 ft   53. [2.6a] 150 m by 350 m
54. [2.5a] $6500   55. [5.9a] 17 m   56. [2.6a] 30 m, 60 m, 10 m
57. [2.5a] $44   58. [5.9a] Height: 14 ft; base: 14 ft
59. [3.3a]

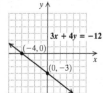

60. [2.7e], [4.6a] $\{x | x \ge -\frac{13}{3}\}$ 61. [2.3b] 22
62. [5.8b] -6, 4   63. [5.7a] $(x-3)(x-2)(x+1)$
64. [5.7a] $(2a+3b+3)(2a-3b-5)$ 65. [5.5a] 25
66. [5.9a] 2 cm

CHAPTER 6

Exercise Set 6.1, p. 436

RC1. equivalent   RC2. denominator   RC3. quotient
RC4. factors   CC1. (b)   CC2. (a)   CC3. (d)   CC4. (e)
1. 0   3. 8   5. $-\frac{5}{2}$ 7. -4, 7   9. -5, 5   11. None
13. $\frac{(4x)(3x^2)}{(4x)(5y)}$ 15. $\frac{2x(x-1)}{2x(x+4)}$ 17. $\frac{(3-x)(-1)}{(4-x)(-1)}$
19. $\frac{(y+6)(y-7)}{(y+6)(y+2)}$ 21. $\frac{x^2}{4}$ 23. $\frac{8p^2q}{3}$ 25. $\frac{x-3}{x}$
27. $\frac{m+1}{2m+3}$ 29. $\frac{a-3}{a+2}$ 31. $\frac{a-3}{a-4}$ 33. $\frac{x+5}{x-5}$ 35. $a+1$
37. $\frac{x^2+1}{x+1}$ 39. $\frac{3}{2}$ 41. $\frac{6}{t-3}$ 43. $\frac{t+2}{2(t-4)}$ 45. $\frac{t-2}{t+2}$
47. -1 49. -1 51. -6 53. $-x-1$ 55. $-3t$ 57. $\frac{56x}{3}$
59. $\frac{2}{dc^2}$ 61. 1   63. $\frac{(a+3)(a-3)}{a(a+4)}$ 65. $\frac{2a}{a-2}$
67. $\frac{(t+2)(t-2)}{(t+1)(t-1)}$ 69. $\frac{x+4}{x+2}$ 71. $\frac{5(a+6)}{a-1}$

73.

74.

75. $x^3(x-7)(x+5)$ 76. $(2y^2+1)(y-5)$
77. $(2+t)(2-t)(4+t^2)$ 78. $10(x+7)(x+1)$
79. $\frac{1}{x-1}$ 81. $\frac{5(2x+5)-25}{10} = \frac{10x+25-25}{10} = \frac{10x}{10} = x$

You get the same number you selected. To do a number trick, ask someone to select a number and then perform these operations. The person will probably be surprised that the result is the original number.

Exercise Set 6.2, p. 442

RC1. True   RC2. False   RC3. False   RC4. True
CC1. (d)   CC2. (e)   CC3. (a)   CC4. (f)   CC5. (b)
CC6. (c)
1. $\frac{x}{4}$ 3. $\frac{1}{x^2-y^2}$ 5. $a+b$ 7. $\frac{x^2-4x+7}{x^2+2x-5}$ 9. $\frac{3}{10}$
11. $\frac{1}{4}$ 13. $\frac{b}{a}$ 15. $\frac{(a+2)(a+3)}{(a-3)(a-1)}$ 17. $\frac{(x-1)^2}{x}$
19. $\frac{1}{2}$ 21. $\frac{15}{8}$ 23. $\frac{15}{4}$ 25. $\frac{a-5}{3(a-1)}$ 27. $\frac{(x+2)^2}{x}$
29. $\frac{3}{2}$ 31. $\frac{c+1}{c-1}$ 33. $\frac{y-3}{2y-1}$ 35. $\frac{x+1}{x-1}$ 37. $\{x | x \ge 77\}$
38. Height: 7 in.; base: 10 in.   39. $\frac{4y^8}{x^6}$ 40. $\frac{125x^{18}}{y^{12}}$ 41. $\frac{4x^6}{y^{10}}$
42. $\frac{1}{a^{15}b^{20}}$ 43. $\frac{a+1}{5ab^2(a^2+4)}$ 45. $\frac{(x-7)^2}{x+y}$

Exercise Set 6.3, p. 447

RC1. common   RC2. multiple   RC3. denominator
RC4. greatest   RC5. $2 \cdot 2 \cdot 2 \cdot 2 \cdot 3$
1. 108   3. 72   5. 126   7. 360   9. 500   11. $\frac{65}{72}$
13. $\frac{29}{120}$ 15. $\frac{23}{180}$ 17. $12x^3$ 19. $18x^2y^2$ 21. $6(y-3)$
23. $t(t+2)(t-2)$ 25. $(x+2)(x-2)(x+3)$
27. $t(t+2)^2(t-4)$ 29. $(a+1)(a-1)^2$ 31. $(m-3)(m-2)^2$
33. $(2+3x)(2-3x)$ 35. $10v(v+4)(v+3)$
37. $18x^3(x-2)^2(x+1)$ 39. $6x^3(x+2)^2(x-2)$
41. $120w^6$ 43. $120x^4$; $8x^3$; $960x^7$ 44. $48ab^3$; $4ab$; $192a^2b^4$
45. $48x^6$; $16x^5$; $768x^{11}$ 46. $120x^3$; $2x^2$; $240x^5$
47. $20x^2$; $10x$; $200x^3$ 48. a^{15}; a^5; a^{20} 49. 24 min

Exercise Set 6.4, p. 453

RC1. numerators; denominator   RC2. LCD   RC3. opposites
CC1. $\frac{2}{2}$; $\frac{3}{3}$ CC2. $\frac{x+3}{x+3}$; $\frac{2}{2}$ CC3. $\frac{x-3}{x-3}$; $\frac{x+2}{x+2}$ CC4. $\frac{2x}{2x}$; $\frac{3}{3}$
1. 1   3. $\frac{6}{3+x}$ 5. $\frac{-4x+11}{2x-1}$ 7. $\frac{2x+5}{x^2}$ 9. $\frac{41}{24r}$
11. $\frac{2(2x+3y)}{x^2y^2}$ 13. $\frac{4+3t}{18t^3}$ 15. $\frac{x^2+4xy+y^2}{x^2y^2}$
17. $\frac{6x}{(x-2)(x+2)}$ 19. $\frac{11x+2}{3x(x+1)}$ 21. $\frac{x(x+6)}{(x+4)(x-4)}$
23. $\frac{6}{z+4}$ 25. $\frac{3x-1}{(x-1)^2}$ 27. $\frac{11a}{10(a-2)}$ 29. $\frac{2(x^2+4x+8)}{x(x+4)}$
31. $\frac{7a+6}{(a-2)(a+1)(a+3)}$ 33. $\frac{2(x^2-2x+17)}{(x-5)(x+3)}$

A-18   Answers

35. $\dfrac{3a+2}{(a+1)(a-1)}$ **37.** $\tfrac{1}{4}$ **39.** $-\dfrac{1}{t}$ **41.** $\dfrac{-x+7}{x-6}$, or $\dfrac{7-x}{x-6}$, or $\dfrac{x-7}{6-x}$ **43.** $y+3$ **45.** $\dfrac{2(b-7)}{(b+4)(b-4)}$ **47.** $a+b$ **49.** $\dfrac{5x+2}{x-5}$ **51.** -1 **53.** $\dfrac{-x^2+9x-14}{(x-3)(x+3)}$ **55.** $\dfrac{2(x+3y)}{(x+y)(x-y)}$ **57.** $\dfrac{a^2+7a+1}{(a+5)(a-5)}$ **59.** $\dfrac{5t-12}{(t+3)(t-3)(t-2)}$ **61.** $\dfrac{1}{x^{12}y^{21}}$ **62.** $\dfrac{25}{x^4 y^6}$ **63.** -8 **64.** $-2, 9$
65. 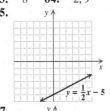 **66.**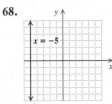
67. 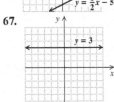 **68.**

69. Perimeter: $\dfrac{16y+28}{15}$; area: $\dfrac{y^2+2y-8}{15}$ **71.** $\dfrac{(z+6)(2z-3)}{(z+2)(z-2)}$ **73.** $\dfrac{11z^4-22z^2+6}{(z^2+2)(z^2-2)(2z^2-3)}$

Exercise Set 6.5, p. 461
RC1. (a) $3x+5$; (b) $10x-3x-5$; (c) $7x-5$
RC2. (a) $4-9a$; (b) $7-4+9a$; (c) $3+9a$
RC3. (a) $y+1$; (b) $9y-2-y-1$; (c) $8y-3$
CC1. $5x-3$ **CC2.** $-x^2+x-4$
1. $\dfrac{4}{x}$ **3.** 1 **5.** $\dfrac{1}{x-1}$ **7.** $\dfrac{-a-4}{10}$ **9.** $\dfrac{7z-12}{12z}$ **11.** $\dfrac{4x^2-13xt+9t^2}{3x^2t^2}$ **13.** $\dfrac{2(x-20)}{(x+5)(x-5)}$ **15.** $\dfrac{3-5t}{2t(t-1)}$ **17.** $\dfrac{2s-st-s^2}{(t+s)(t-s)}$ **19.** $\dfrac{y-19}{4y}$ **21.** $\dfrac{-2a^2}{(x+a)(x-a)}$ **23.** $\tfrac{8}{3}$ **25.** $\dfrac{13}{a}$ **27.** $\dfrac{8}{y-1}$ **29.** $\dfrac{x-2}{x-7}$ **31.** $\dfrac{4}{(a+5)(a-5)}$ **33.** $\dfrac{2(x-2)}{x-9}$ **35.** $\dfrac{3(3x+4)}{(x+3)(x-3)}$ **37.** $\tfrac{1}{2}$ **39.** $\dfrac{x-3}{(x+3)(x+1)}$ **41.** $\dfrac{18x+5}{x-1}$ **43.** 0 **45.** $\dfrac{-9}{2x-3}$ **47.** $\dfrac{20}{2y-1}$ **49.** $\dfrac{2a-3}{2-a}$ **51.** $\dfrac{z-3}{2z-1}$ **53.** $\dfrac{2}{x+y}$ **55.** $\dfrac{b^{20}}{a^8}$ **56.** $18x^3$ **57.** $-\tfrac{11}{35}$ **58.** 1 **59.** $x^2-9x+18$ **60.** $(4-\pi)r^2$ **61.** Missing length: $\dfrac{-2a-15}{a-6}$; area: $\dfrac{-2a^3-15a^2+12a+90}{2(a-6)^2}$

Mid-Chapter Review: Chapter 6, p. 465
1. False **2.** True **3.** True **4.** False **5.** True
6. $\dfrac{x-1}{x-2} - \dfrac{x+1}{x+2} - \dfrac{x-6}{4-x^2}$
$= \dfrac{x-1}{x-2} - \dfrac{x+1}{x+2} - \dfrac{x-6}{4-x^2} \cdot \dfrac{-1}{-1}$
$= \dfrac{x-1}{x-2} - \dfrac{x+1}{x+2} - \dfrac{6-x}{x^2-4}$
$= \dfrac{x-1}{x-2} - \dfrac{x+1}{x+2} - \dfrac{6-x}{(x-2)(x+2)}$
$= \dfrac{x-1}{x-2}\cdot\dfrac{x+2}{x+2} - \dfrac{x+1}{x+2}\cdot\dfrac{x-2}{x-2} - \dfrac{6-x}{(x-2)(x+2)}$
$= \dfrac{x^2+x-2}{(x-2)(x+2)} - \dfrac{x^2-x-2}{(x-2)(x+2)} - \dfrac{6-x}{(x-2)(x+2)}$
$= \dfrac{x^2+x-2-x^2+x+2-6+x}{(x-2)(x+2)}$
$= \dfrac{3x-6}{(x-2)(x+2)}$
$= \dfrac{3(x-2)}{(x-2)(x+2)} = \dfrac{x-2}{x-2}\cdot\dfrac{3}{x+2}$
$= \dfrac{3}{x+2}$

7. None **8.** $3, 8$ **9.** $\tfrac{7}{2}$ **10.** $\dfrac{x-1}{x-3}$ **11.** $\dfrac{2(y+4)}{y-1}$ **12.** -1 **13.** $\dfrac{1}{-x+3}$, or $\dfrac{1}{3-x}$ **14.** $10x^3(x-10)^2(x+10)$ **15.** $\dfrac{a+1}{a-3}$ **16.** $\dfrac{y}{(y-2)(y-3)}$ **17.** $x+11$ **18.** $\dfrac{1}{x-y}$ **19.** $\dfrac{a^2+5ab-b^2}{a^2 b^2}$ **20.** $\dfrac{2(3x^2-4x+6)}{x(x+2)(x-2)}$ **21.** E **22.** A **23.** D **24.** B **25.** F **26.** C **27.** If the numbers have a common factor, then their product contains that factor more than the greatest number of times it occurs in any one factorization. In this case, their product is not their least common multiple. **28.** Yes; consider the product $\dfrac{a}{b}\cdot\dfrac{c}{d} = \dfrac{ac}{bd}$. The reciprocal of the product is $\dfrac{bd}{ac}$. This is equal to the product of the reciprocals of the two original factors: $\dfrac{bd}{ac} = \dfrac{b}{a}\cdot\dfrac{d}{c}$. **29.** Although multiplying the denominators of the expressions being added results in a common denominator, it is often not the *least* common denominator. Using a common denominator other than the LCD makes the expressions more complicated, requires additional simplification after the addition has been performed, and leaves more room for error. **30.** Their sum is 0. Another explanation is that $-\left(\dfrac{1}{3-x}\right) = \dfrac{1}{-(3-x)} = \dfrac{1}{x-3}$. **31.** $\dfrac{x+3}{x-5}$ is not defined for $x=5$, $\dfrac{x-7}{x+1}$ is not defined for $x=-1$, and $\dfrac{x+1}{x-7}$ (the reciprocal of $\dfrac{x-7}{x+1}$) is not defined for $x=7$. **32.** The binomial is a factor of the trinomial.

Exercise Set 6.6, p. 470
CC1. 10; 24; 120 **CC2.** $5y$; $30y$; $30y$ **RC1.** complex **RC2.** numerator **RC3.** least common denominator **RC4.** reciprocal
1. $\tfrac{25}{4}$ **3.** $\tfrac{1}{3}$ **5.** -6 **7.** $\dfrac{1+3x}{1-5x}$ **9.** $\dfrac{2x+1}{x}$ **11.** 8 **13.** $x-8$ **15.** $\dfrac{y}{y-1}$ **17.** $-\dfrac{1}{a}$ **19.** $\dfrac{ab}{b-a}$ **21.** $\dfrac{p^2+q^2}{q+p}$ **23.** $\dfrac{2a(a+2)}{5-3a^2}$ **25.** $\dfrac{15(4-a^3)}{14a^2(9+2a)}$ **27.** $\dfrac{ac}{bd}$ **29.** 1 **31.** $\dfrac{4x+1}{5x+3}$ **33.** $\{x\,|\,x\le 96\}$ **34.** $\{b\,|\,b>\tfrac{22}{9}\}$ **35.** $\{x\,|\,x<-3\}$ **36.** 12 ft, 5 ft **37.** 14 yd **39.** $\dfrac{5x+3}{3x+2}$

Calculator Corner, p. 476
1.–2. Left to the student

Exercise Set 6.7, p. 477

RC1. Rational expression **RC2.** Solutions
RC3. Rational expression **RC4.** Rational expression
RC5. Solutions **RC6.** Solutions **RC7.** Rational expression **RC8.** Solutions **CC1.** True **CC2.** True
1. $\frac{6}{5}$ **3.** $\frac{40}{29}$ **5.** $\frac{47}{2}$ **7.** -6 **9.** $\frac{24}{7}$ **11.** $-4, -1$ **13.** $-4, 4$
15. 3 **17.** $\frac{14}{3}$ **19.** 5 **21.** 5 **23.** $\frac{5}{2}$ **25.** -2 **27.** $-\frac{13}{2}$
29. $\frac{17}{2}$ **31.** No solution **33.** -5 **35.** $\frac{5}{3}$ **37.** $\frac{1}{2}$
39. No solution **41.** No solution **43.** 4 **45.** No solution
47. $-2, 2$ **49.** 7 **51.** $4x^4 + 3x^3 + 2x - 7$ **52.** 0
53. $50(p^2 - 2)$ **54.** $5(p + 2)(p - 10)$ **55.** 18 and 20; -20 and -18 **56.** 3.125 L **57.** $-\frac{1}{6}$

Translating for Success, p. 488

1. K **2.** E **3.** C **4.** N **5.** D **6.** O **7.** F **8.** H **9.** B **10.** A

Exercise Set 6.8, p. 489

RC1. corresponding; same; proportional **RC2.** quotient
RC3. proportion **RC4.** rate **RC5.** distance
RC6. cross products
1. Sarah: 30 km/h; Rick: 70 km/h
3. Ostrich: 40 mph; giraffe: 32 mph
5. Hank: 14 km/h; Kelly: 19 km/h **7.** 20 mph
9. Ralph: 5 km/h; Bonnie: 8 km/h **11.** $1\frac{1}{3}$ hr **13.** $22\frac{2}{9}$ min
15. $25\frac{5}{7}$ min **17.** About 4.7 hr **19.** $3\frac{15}{16}$ hr **21.** $3\frac{3}{4}$ min
23. $\frac{10}{3}$ students/teacher **25.** 2.3 km/h **27.** 66 g **29.** 1.92 g
31. 7 gal **33.** 287 trout **35.** 200 duds **37.** 1960 students
39. 1.75 lb **41.** $\frac{21}{2}$ **43.** $\frac{8}{3}$ **45.** $\frac{35}{2}$ **47.** About 1700 largemouth bass **49.** 0 **50.** -2 **51.** x^{11} **52.** x
53. $\frac{1}{x^{11}}$ **54.** $\frac{1}{x}$
55.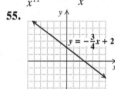
56.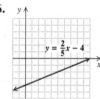
57.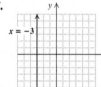
59. $27\frac{3}{11}$ min

Exercise Set 6.9, p. 502

RC1. inverse **RC2.** direct
CC1. (f) **CC2.** (d) **CC3.** (h) **CC4.** (i)
CC5. (c) **CC6.** (a) **CC7.** (g) **CC8.** (b)
1. $5; y = 5x$ **3.** $\frac{2}{15}; y = \frac{2}{15}x$ **5.** $\frac{9}{4}; y = \frac{9}{4}x$
7. 175 semi trucks **9.** 90g **11.** 40 kg
13. 76,361,280 cans **15.** $98; y = \frac{98}{x}$ **17.** $36; y = \frac{36}{x}$
19. $0.05; y = \frac{0.05}{x}$ **21.** 3.5 hr **23.** $\frac{2}{9}$ ampere **25.** 960 lb
27. $5\frac{5}{7}$ hr **29.** $y = 15x^2$ **31.** $y = \frac{0.0015}{x^2}$ **33.** $y = xz$
35. $y = \frac{3}{10}xz^2$ **37.** $y = \frac{xz}{5wp}$ **39.** 36 mph **41.** (a) $W = kd^3$;
(b) $102\frac{2}{5}$ lb **43.** 39 earned runs **45.** 729 gal **47.** $(x - 7)^2$

48. $(5y - 3)(25y^2 + 15y + 9)$ **49.** $(11w + 3)(11w - 3)$
50. 9 **51.** $-10, 10$ **52.** -5 **53.** (a) Inversely; (b) neither; (c) directly; (d) directly

Summary and Review: Chapter 6, p. 506

Vocabulary Reinforcement

1. complex **2.** proportion **3.** reciprocals **4.** equivalent
5. opposites **6.** similar **7.** inversely; inverse variation
8. directly; direct variation

Concept Reinforcement

1. True **2.** False **3.** True

Study Guide

1. 5, 6 **2.** $\frac{x - 1}{2(x + 5)}$ **3.** $\frac{y + 5}{5(y + 3)}$ **4.** $\frac{b + 7}{b + 8}$ **5.** $\frac{71}{180}$
6. $(x + 2)(x - 9)(x + 9)$ **7.** -1 **8.** $\frac{x^2 - 4x - 10}{(x + 2)(x + 1)(x - 1)}$
9. $\frac{3(2y - 5)}{5(9 - y)}$ **10.** 1 **11.** $y = 150x; 300$ **12.** $y = \frac{225}{x}; 22.5$

Review Exercises

1. 0 **2.** 6 **3.** $-6, 6$ **4.** $-6, 5$ **5.** -2 **6.** None
7. $\frac{x - 2}{x + 1}$ **8.** $\frac{7x + 3}{x - 3}$ **9.** $\frac{y - 5}{y + 5}$ **10.** $\frac{a - 6}{5}$ **11.** $\frac{6}{2t - 1}$
12. $-20t$ **13.** $\frac{2x(x - 1)}{x + 1}$ **14.** $30x^2y^2$ **15.** $4(a - 2)$
16. $(y - 2)(y + 2)(y + 1)$ **17.** $\frac{-3(x - 6)}{x + 7}$ **18.** -1
19. $\frac{2a}{a - 1}$ **20.** $d + c$ **21.** $\frac{4}{x - 4}$ **22.** $\frac{x + 5}{2x}$ **23.** $\frac{2x + 3}{x - 2}$
24. $\frac{-x^2 + x + 26}{(x - 5)(x + 5)(x + 1)}$ **25.** $\frac{2(x - 2)}{x + 2}$ **26.** $\frac{z}{1 - z}$
27. $c - d$ **28.** 8 **29.** $-5, 3$ **30.** $5\frac{1}{7}$ hr **31.** 95 mph, 175 mph **32.** 240 km/h, 280 km/h **33.** 160 defective calculators **34.** (a) $\frac{12}{13}$ c; (b) $4\frac{1}{5}$ c; (c) $9\frac{1}{3}$ c
35. 10,000 blue whales **36.** 6 **37.** $y = 4x$
38. $y = \frac{2500}{x}$ **39.** $y = 6xz$ **40.** $y = \frac{2x}{z}$
41. 20 min **42.** 78 **43.** 500 watts **44.** C
45. A **46.** $\frac{5(a + 3)^2}{a}$ **47.** They are equivalent proportions.

Understanding Through Discussion and Writing

1. No; when we are adding, no sign changes are required so the result is the same regardless of whether parentheses are used. When we are subtracting, however, the sign of each term of the expression being subtracted must be changed and parentheses are needed to make sure this is done. **2.** Graph each side of the equation and determine the number of points of intersection of the graphs. **3.** Canceling removes a factor of 1, allowing us to rewrite $a \cdot 1$ as a. **4.** Inverse variation; the greater the average gain per play, the smaller the number of plays required. **5.** Form a rational expression that has factors of $x + 3$ and $x - 4$ in the denominator. **6.** If we multiply both sides of a rational equation by a variable expression in order to clear fractions, it is possible that the variable expression is equal to 0. Thus an equivalent equation might not be produced.

Test: Chapter 6, p. 513

1. [6.1a] 0   2. [6.1a] -8 3. [6.1a] $-7, 7$ 4. [6.1a] 1, 2
5. [6.1a] 1   6. [6.1a] None   7. [6.1c] $\dfrac{3x+7}{x+3}$
8. [6.1d] $\dfrac{a+5}{2}$ 9. [6.2b] $\dfrac{(5x+1)(x+1)}{3x(x+2)}$
10. [6.3a] $(y-3)(y+3)(y+7)$ 11. [6.4a] $\dfrac{23-3x}{x^3}$
12. [6.5a] $\dfrac{2(4-t)}{t^2+1}$ 13. [6.4a] $\dfrac{-3}{x-3}$ 14. [6.5a] $\dfrac{2x-5}{x-3}$
15. [6.4a] $\dfrac{8t-3}{t(t-1)}$ 16. [6.5a] $\dfrac{-x^2-7x-15}{(x+4)(x-4)(x+1)}$
17. [6.5b] $\dfrac{x^2+2x-7}{(x-1)^2(x+1)}$ 18. [6.6a] $\dfrac{3y+1}{y}$
19. [6.7a] 12   20. [6.7a] $-3, 5$
21. [6.9a] $y = 2x$; 50   22. [6.9a] $y = 0.5x$; 12.5
23. [6.9c] $y = \dfrac{18}{x}$; $\dfrac{9}{50}$ 24. [6.9c] $y = \dfrac{22}{x}$; $\dfrac{11}{50}$
25. [6.9e] $Q = 2.5xy$ 26. [6.9b] 240 km   27. [6.9d] $1\dfrac{1}{5}$ hr
28. [6.8b] 16 defective spark plugs   29. [6.8b] 50 zebras
30. [6.8a] 12 min   31. [6.8a] Craig: 65 km/h; Marilyn: 45 km/h
32. [6.8b] 15   33. [6.7a] D   34. [6.8a] Rema: 4 hr; Reggie: 10 hr
35. [6.6a] $\dfrac{3a+2}{2a+1}$

Cumulative Review: Chapters 1–6, p. 515

1. [1.2e] 3.5   2. [4.3c] 3, 2, 1, 0; 3
3. [2.5a] 24,139,311 millennials   4. [2.5a] About 7.3%
5. [2.5a] $2500   6. [6.8a] 35 mph, 25 mph   7. [5.9a] 14 ft
8. (a) [6.9b] $M = 0.4B$; (b) [6.9b] 76.8 lb
9. [4.3d] $2x^3 - 3x^2 - 2$ 10. [1.8c] $\dfrac{3}{8}x + 1$
11. [4.1e], [4.2a, b] $\dfrac{9}{4x^8}$ 12. [6.6a] $\dfrac{4(2x-3)}{17x}$
13. [4.7e] $-2xy^2 - 4x^2y^2 + xy^3$
14. [4.4a] $2x^5 + 6x^4 + 2x^3 - 10x^2 + 3x - 9$
15. [6.1d] $\dfrac{2}{3(y+2)}$ 16. [6.2b] 2
17. [6.4a] $x + 4$ 18. [6.5a] $\dfrac{2(x-3)}{(x+2)(x-2)}$
19. [4.6a] $a^2 - 9$ 20. [4.6c] $36x^2 - 60x + 25$
21. [4.6b] $4x^6 - 1$ 22. [5.3a], [5.4a] $(9a-2)(a+6)$
23. [5.5b] $(3x - 5y)^2$ 24. [5.5d] $(7x-1)(7x+1)$
25. [2.3c] 3   26. [5.8b] $-4, \dfrac{1}{2}$ 27. [5.8b] 0, 10
28. [2.7e] $\{x | x \geq -26\}$ 29. [6.7a] 2
30. [2.4b] $a = \dfrac{t}{x+y}$ 31. [3.4a] Not defined   32. [3.4a] $-\dfrac{3}{7}$
33. [3.4a] $-\dfrac{9}{4}$ 34. [3.4a] 0
35. [3.3a] y-intercept: $(0, -2)$; x-intercept: $(-6, 0)$
36. [3.3a] y-intercept: $\left(0, -\dfrac{1}{8}\right)$; x-intercept: $\left(\dfrac{3}{8}, 0\right)$
37. [3.3b] y-intercept: $(0, 25)$; x-intercept: none
38. [3.3b] y-intercept: none; x-intercept: $\left(-\dfrac{1}{4}, 0\right)$
39. [3.3b] 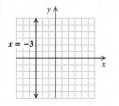   40. [3.3b]
41. [3.3a]   42. [3.3a]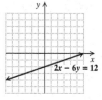
43. [3.2a]
44. [3.2a], [3.3a] 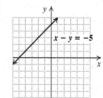   45. [6.1a], [6.7a] $0, 3, \dfrac{5}{2}$

CHAPTER 7

Calculator Corner, p. 521
1. 17.3   2. 34

Calculator Corner, p. 523
1. $y = x - 4$ 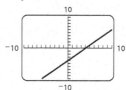   2. $y = -2x - 3$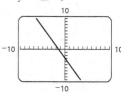
3. $y = 1 - x^2$ 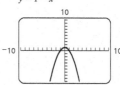   4. $y = 3x^2 - 4x + 1$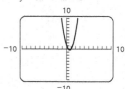
5. $y = x^3$ 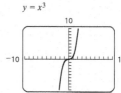   6. $y = |x + 3|$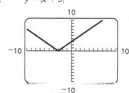

Exercise Set 7.1, p. 525

RC1. domain, range, domain, exactly one, range
RC2. domain, range, domain, at least one, range
CC1. $f(2) = 3$ **CC2.** $f(0) = 3$ **CC3.** $f(-2) = -5$
CC4. $f(3) = 0$
1. Yes   3. Yes   5. No   7. No   9. No   11. Yes
13. (a) 9; (b) 12; (c) 2; (d) 5; (e) 7.4; (f) $5\dfrac{2}{3}$ 15. (a) -21;
(b) 15; (c) 2; (d) 0; (e) $18a$; (f) $3a + 3$ 17. (a) 7; (b) -17;
(c) 6; (d) 4; (e) $3a - 2$; (f) $3a + 3h + 4$

19. (a) 0; (b) 5; (c) 2; (d) 170; (e) 65; (f) $32a^2 - 12a$
21. (a) 1; (b) 3; (c) 3; (d) 11; (e) $|a - 1| + 1$; (f) $|a + h| + 1$
23. (a) 0; (b) -1; (c) 8; (d) 1000; (e) -125; (f) $-27a^3$
25. 2012: about 12.8 million troy ounces; 2015: about 15.7 million troy ounces   **27.** $1\frac{20}{33}$ atm; $1\frac{10}{11}$ atm; $4\frac{1}{33}$ atm
29. 1.792 cm; 2.8 cm; 11.2 cm

31. 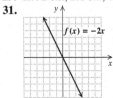   **33.**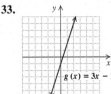

35. 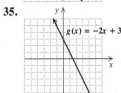   **37.**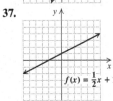

39. 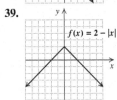   **41.**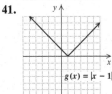

43. 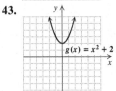   **45.**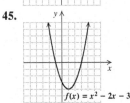

47. 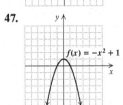   **49.**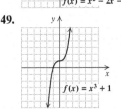

51. Yes **53.** No **55.** No **57.** Yes **59.** $210,000
61. 720 million passengers **63.** 650 million passengers
65. $\{y | y < 4\}$ **66.** $\{x | x \geq 23\}$ **67.** No solution
68. $-10, 8$ **69.** $\frac{3}{2}$ **70.** $\frac{80}{9}$ **71.** $g(-2) = 39$ **73.** 26; 99
75. $g(x) = \frac{15}{4}x - \frac{13}{4}$

Exercise Set 7.2, p. 535
RC1. (a) **RC2.** (b) **RC3.** (a) **RC4.** (b) **RC5.** (a)
RC6. (d)
1. (a) 3; (b) $\{-4, -3, -2, -1, 0, 1, 2\}$; (c) $-2, 0$; (d) $\{1, 2, 3, 4\}$
3. (a) $2\frac{1}{2}$; (b) $\{x | -3 \leq x \leq 5\}$ (c) $2\frac{1}{4}$; (d) $\{y | 1 \leq y \leq 4\}$
5. (a) 1; (b) all real numbers; (c) 3; (d) all real numbers
7. (a) 1; (b) all real numbers; (c) $-2, 2$; (d) $\{y | y \geq 0\}$
9. $\{x | x$ is a real number and $x \neq -3\}$ **11.** All real numbers
13. All real numbers   **15.** $\{x | x$ is a real number and $x \neq \frac{14}{5}\}$
17. All real numbers   **19.** $\{x | x$ is a real number and $x \neq \frac{7}{4}\}$
21. $\{x | x$ is a real number and $x \neq 1\}$ **23.** All real numbers
25. All real numbers   **27.** $\{x | x$ is a real number and $x \neq \frac{5}{2}\}$
29. All real numbers   **31.** $\{x | x$ is a real number and $x \neq -\frac{5}{4}\}$
33. $-8; 0; -2$ **35.** $a - 1$ **36.** $\dfrac{2(y + 2)}{7(y + 7)}$ **37.** $\dfrac{5}{x + 2}$

38. $t - 4$ **39.** $w + 1, R\ 2;$ or $w + 1 + \dfrac{2}{w + 3}$

41. $\{y | y$ is a real number and $y \neq 0\}$;
$\{y | y \geq 0\}$; $\{y | y \leq 8\}$; $\{y | y \geq 1\}$
43. All real numbers

Mid-Chapter Review: Chapter 7, p. 537
1. True **2.** False **3.** True **4.** True **5.** False

6.

x	y
0	1
2	-2
-2	4
4	-5

7.

x	f(x)
-2	0
3	0
0	-6
2	-4
-1	-4

8. Yes **9.** No **10.** Domain: $\{x | -3 \leq x \leq 3\}$, range: $\{y | -2 \leq y \leq 1\}$ **11.** -3 **12.** -7 **13.** 8   **14.** 9
15. 9000   **16.** 0   **17.** Yes   **18.** No   **19.** Yes   **20.** $\{x | x$ is a real number and $x \neq 4\}$ **21.** All real numbers
22. $\{x | x$ is a real number and $x \neq -2\}$ **23.** All real numbers

24. 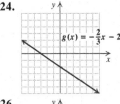   **25.**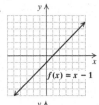

26. 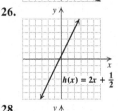   **27.**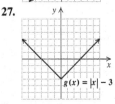

28. 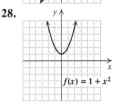   **29.**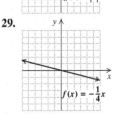

30. No; since each input has exactly one output, the number of outputs cannot exceed the number of inputs. **31.** When $x < 0$, then $y < 0$, and the graph contains points in quadrant III. When $0 < x < 30$, then $y < 0$, and the graph contains points in quadrant IV. When $x > 30$, then $y > 0$, and the graph contains points in quadrant I. Thus the graph passes through three quadrants.   **32.** The output -3 corresponds to the input 2. The number -3 in the range is paired with the number 2 in the domain. The point $(2, -3)$ is on the graph of the function.
33. The domain of a function is the set of all inputs, and the range is the set of all outputs.

Calculator Corner, p. 540
1. The graph of $y_2 = x + 4$ is the graph of $y_1 = x$ shifted up 4 units.   **2.** The graph of $y_3 = x - 3$ is the graph of $y_1 = x$ shifted down 3 units.

Calculator Corner, p. 543
1. The graph of $y = 10x$ will slant up from left to right. It will be steeper than the other graphs.   **2.** The graph of $y = 0.005x$ will slant up from left to right. It will be less steep than the other graphs.   **3.** The graph of $y = -10x$ will slant down from left to right. It will be steeper than the other graphs.   **4.** The graph of $y = -0.005x$ will slant down from left to right. It will be less steep than the other graphs.

Exercise Set 7.3, p. 547

RC1. up **RC2.** horizontal **RC3.** down **CC1.** (f)
CC2. (b) **CC3.** (c)
1. $m = 4$; y-intercept: $(0, 5)$ **3.** $m = -2$; y-intercept: $(0, -6)$
5. $m = -\frac{3}{8}$; y-intercept: $(0, -\frac{1}{5})$ **7.** $m = 0.5$; y-intercept: $(0, -9)$
9. $m = \frac{2}{3}$; y-intercept: $(0, -\frac{8}{3})$ **11.** $m = 3$; y-intercept: $(0, -2)$
13. $m = -8$; y-intercept: $(0, 12)$ **15.** $m = 0$; y-intercept: $(0, \frac{4}{17})$
17. $m = -\frac{1}{2}$ **19.** $m = \frac{1}{3}$ **21.** $m = 2$ **23.** $m = \frac{2}{3}$
25. $m = -\frac{1}{3}$ **27.** $\frac{2}{25}$, or 8% **29.** $\frac{13}{41}$, or about 31.7%
31. The rate of change is about $28.4 million per year.
33. The rate of change is -$900 per year.
35. The rate of change is about 10,366 eating and drinking places per year. **37.** -1323 **38.** $45x + 54$
39. $350x - 60y + 120$ **40.** 25 **41.** Square: 15 yd; triangle: 20 yd
42. $(2 - 5x)(4 + 10x + 25x^2)$
43. $(c - d)(c^2 + cd + d^2)(c + d)(c^2 - cd + d^2)$
44. $7(2x - 1)(4x^2 + 2x + 1)$
45. $a - 10, R\ -4$, or $a - 10 + \dfrac{-4}{a - 1}$

Calculator Corner, p. 551

1. $y = -3.2x - 16$
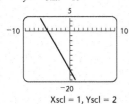
Xscl = 1, Yscl = 2

2. $y = 4.25x + 85$
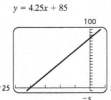
Xscl = 5, Yscl = 5

3. $y = (-6x + 90)/5$
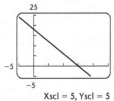
Xscl = 5, Yscl = 5

4. $y = (5x - 30)/6$
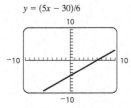

5. $y = (-8x + 9)/3$
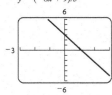

6. $y = 0.4x - 5$
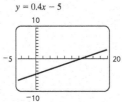
Xscl = 2, Yscl = 1

7. $y = 1.2x - 12$
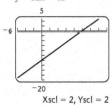
Xscl = 2, Yscl = 2

8. $y = (4x - 2)/5$
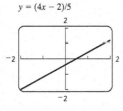

Visualizing for Success, p. 557

1. D **2.** I **3.** H **4.** C **5.** F **6.** A **7.** G **8.** B
9. E **10.** J

Exercise Set 7.4, p. 558

RC1. False **RC2.** True **RC3.** True **RC4.** False
CC1. $\left(-\frac{2}{7}, 0\right)$ **CC2.** $(0, 7)$ **CC3.** perpendicular
CC4. $y = \frac{2}{7}$

1.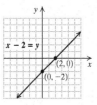
3.
5.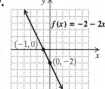
7.
9.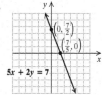
11.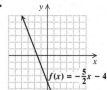
13.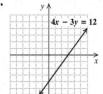
15.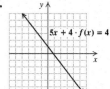
17.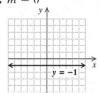
19.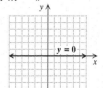
21.
23.
25.
27.
29. Not defined
31. $m = 0$
33. $m = 0$
35. $m = 0$

37. $m = 0$ **39.** Not defined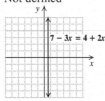

41. Yes **43.** No **45.** Yes **47.** Yes **49.** Yes **51.** No
53. No **55.** Yes **57.** 5.3×10^{10} **58.** 4.7×10^{-5}
59. 1.8×10^{-2} **60.** 9.9902×10^{7} **61.** 0.0000213
62. $901{,}000{,}000$ **63.** $20{,}000$ **64.** 0.085677
65. $3(3x - 5y)$ **66.** $3a(4 + 7b)$ **67.** $7p(3 - q + 2)$
68. $64(x - 2y + 4)$ **69.** $a = 2$ **71.** $y = \frac{2}{15}x + \frac{2}{5}$
73. $y = 0$; yes **75.** $m = -\frac{3}{4}$

Exercise Set 7.5, p. 569

RC1. $y = mx + b$ **RC2.** $y - y_1 = m(x - x_1)$ **CC1. (a)** $\frac{4}{11}$;
(b) $-\frac{11}{4}$ **CC2. (a)** 0; **(b)** not defined **CC3. (a)** 2;
(b) $-\frac{1}{2}$ **CC4. (a)** $-\frac{5}{6}$; **(b)** $\frac{6}{5}$ **CC5. (a)** Not defined;
(b) 0 **CC6. (a)** -2; **(b)** $\frac{1}{2}$
1. $y = -8x + 4$ **3.** $y = 2.3x - 1$ **5.** $f(x) = -\frac{7}{3}x - 5$
7. $f(x) = \frac{2}{3}x + \frac{5}{8}$ **9.** $y = 5x - 17$ **11.** $y = -3x + 33$
13. $y = x - 6$ **15.** $y = -2x + 16$ **17.** $y = -7$
19. $y = \frac{2}{3}x - \frac{8}{3}$ **21.** $y = \frac{1}{2}x + \frac{7}{2}$ **23.** $y = x$
25. $y = \frac{7}{4}x + 7$ **27.** $y = \frac{3}{2}x$ **29.** $y = \frac{1}{6}x$
31. $y = 13x - \frac{15}{4}$ **33.** $y = -\frac{1}{2}x + \frac{17}{2}$ **35.** $y = \frac{5}{7}x - \frac{17}{7}$
37. $y = \frac{1}{3}x + 4$ **39.** $y = \frac{1}{2}x + 4$ **41.** $y = \frac{4}{3}x - 6$
43. $y = \frac{5}{2}x + 9$
45. (a) $C(x) = 5x + 10$; **(b)** 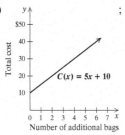 ; **(c)** $30

47. (a) $V(t) = 9400 - 85t$;
(b) 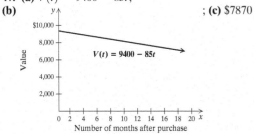 ; **(c)** $7870

49. (a) $U(x) = 0.0037x + 0.34$; **(b)** 1990: 45.1%; 2020: 56.2% **51. (a)** $B(x) = -2091.25x + 89{,}460$; **(b)** 72,730 bridges; **(c)** about 20 years after 2000, or in 2020
53. (a) $E(t) = 1.315t + 46.56$; **(b)** about 68.92 years
55. -1 **56.** $b + 1$ **57.** $\dfrac{x - 3}{2(x - 5)}$ **58.** $\dfrac{4}{y - 9}$ **59.** $\dfrac{2}{7}$
60. 0 **61.** Not defined **62.** -2 **63.** -7.75

Summary and Review: Chapter 7, p. 573

Vocabulary Reinforcement
1. vertical **2.** point–slope **3.** function, domain, range, domain, exactly one, range **4.** slope **5.** perpendicular
6. slope–intercept **7.** parallel

Concept Reinforcement
1. False **2.** True **3.** False

Study Guide
1. No **2.** $g(0) = -2; g(-2) = -3; g(6) = 1$
3. 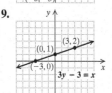 **4.** Yes **5.** Domain: $[-4, 5]$; range: $[-2, 4]$ **6.** $\{x \mid x \text{ is a real number and } x \neq -3\}$ **7.** -2 **8.** Slope: $-\frac{1}{2}$; y-intercept: $(0, 2)$
9. 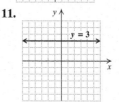 **10.**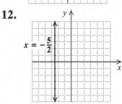
11. 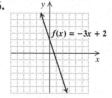 **12.**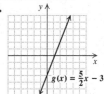
13. Parallel **14.** Perpendicular **15.** $y = -8x + 0.3$
16. $y = -4x - 1$ **17.** $y = -\frac{5}{3}x + \frac{11}{3}$ **18.** $y = \frac{4}{3}x - \frac{23}{3}$
19. $y = -\frac{3}{4}x - \frac{7}{2}$

Review Exercises
1. No **2.** Yes **3.** $g(0) = 5; g(-1) = 7$
4. $f(0) = 7; f(-1) = 12$ **5.** About $138 billion
6. 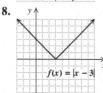 **7.**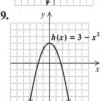
8. 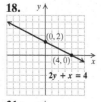 **9.**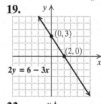
10. Yes **11.** No **12. (a)** $f(2) = 3$; **(b)** $[-2, 4]$; **(c)** -1; **(d)** $[1, 5]$
13. $\{x \mid x \text{ is a real number and } x \neq 4\}$
14. All real numbers **15.** Slope: -3; y-intercept: $(0, 2)$
16. Slope: $-\frac{1}{2}$; y-intercept: $(0, 2)$ **17.** $\frac{11}{3}$
18. **19.** 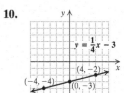 **20.**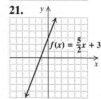
21. **22.** 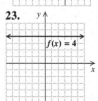 **23.** (see image)
24. Perpendicular **25.** Parallel **26.** Parallel
27. Perpendicular **28.** $f(x) = 4.7x - 23$ **29.** $y = -3x + 4$
30. $y = -\frac{3}{2}x$ **31.** $y = -\frac{5}{7}x + 9$ **32.** $y = \frac{1}{3}x + \frac{1}{3}$

33. (a) $R(x) = -0.037x + 44.66$; (b) 2000: about 43.62 sec; 2010: about 43.25 sec **34.** C **35.** A **36.** $f(x) = 3.09x + 3.75$

Understanding Through Discussion and Writing

1. A line's x- and y-intercepts are the same only when the line passes through the origin. The equation for such a line is of the form $y = mx$. **2.** The concept of slope is useful in describing how a line slants. A line with positive slope slants up from left to right. A line with negative slope slants down from left to right. The larger the absolute value of the slope, the steeper the slant. **3.** Find the slope–intercept form of the equation:

$$4x + 5y = 12$$
$$5y = -4x + 12$$
$$y = -\tfrac{4}{5}x + \tfrac{12}{5}.$$

This form of the equation indicates that the line has a negative slope and thus should slant down from left to right. The student may have graphed $y = \tfrac{4}{5}x + \tfrac{12}{5}$. **4.** For $R(t) = 50t + 35$, $m = 50$ and $b = 35$; 50 signifies that the cost per hour of a repair is $50; 35 signifies that the minimum cost of a repair job is $35.

5. $m = \dfrac{\text{change in } y}{\text{change in } x}$

As we move from one point to another on a vertical line, the y-coordinate changes but the x-coordinate does not. Thus the change in x is 0. Since division by 0 is not defined, the slope of a vertical line is not defined. As we move from one point to another on a horizontal line, the y-coordinate does not change but the x-coordinate does. Thus the change in y is 0 whereas the change in x is a nonzero number, so the slope is 0. **6.** Using algebra, we find that the slope–intercept form of the equation is $y = \tfrac{5}{2}x - \tfrac{3}{2}$. This indicates that the y-intercept is $\left(0, -\tfrac{3}{2}\right)$, so a mistake has been made. It appears that the student graphed $y = \tfrac{5}{2}x + \tfrac{3}{2}$.

Test: Chapter 7, p. 582

1. [7.1a] Yes **2.** [7.1a] No **3.** [7.1b] −4; 2 **4.** [7.1b] 7; 8
5. [7.1b] −6; −6 **6.** [7.1b] 3; 0
7. [7.1c] **8.** [7.1c] **9.** [7.1c]

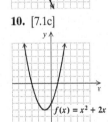 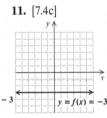 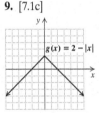

10. [7.1c] **11.** [7.4c] **12.** [7.4c]

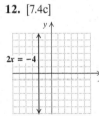

13. [7.1e] (a) About 10.3 years; (b) 2016 **14.** [7.1d] Yes
15. [7.1d] No **16.** [7.2a] $\{x \mid x \text{ is a real number } and\ x \ne -\tfrac{3}{2}\}$
17. [7.2a] All real numbers **18.** [7.2a] (a) 1;
(b) $\{x \mid -3 \le x \le 4\}$; (c) −3; (d) $\{y \mid -1 \le y \le 2\}$
19. [7.3b] Slope: $-\tfrac{3}{5}$; y-intercept: $(0, 12)$ **20.** [7.3b] Slope: $-\tfrac{2}{5}$; y-intercept: $\left(0, -\tfrac{7}{5}\right)$ **21.** [7.3b] $\tfrac{5}{8}$ **22.** [7.3b] 0
23. [7.3c] $\tfrac{4}{5}$ km/min

24. [7.4a] **25.** [7.4b]

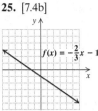

26. [7.4d] Parallel **27.** [7.4d] Perpendicular
28. [7.5a] $y = -3x + 4.8$ **29.** [7.5a] $f(x) = 5.2x - \tfrac{5}{8}$
30. [7.5b] $y = -4x + 2$ **31.** [7.5c] $y = -\tfrac{3}{2}x$
32. [7.5d] $y = \tfrac{1}{2}x - 3$ **33.** [7.5d] $y = 3x - 1$
34. [7.5e] (a) $A(x) = 0.133x + 23.2$; (b) 28.25 years; 29.72 years
35. [7.5b] B **36.** [7.4d] $\tfrac{24}{5}$ **37.** [7.1b] $f(x) = 3$; answers may vary

Cumulative Review: Chapters 1–7, p. 585

1. [1.8d] −6.8 **2.** [4.2d] 3.12×10^{-2} **3.** [1.6c] $-\tfrac{3}{4}$
4. [4.1d, e] $\tfrac{1}{8}$ **5.** [6.1c] $\dfrac{x+3}{2x-1}$ **6.** [6.1c] $\dfrac{t-4}{t+4}$
7. [6.6a] $\dfrac{x^2(x+1)}{x+4}$ **8.** [4.6a] $2 - 10x^2 + 12x^4$
9. [4.7f] $4a^4b^2 - 20a^3b^3 + 25a^2b^4$ **10.** [4.6b], [4.7f] $9x^4 - 16y^2$
11. [4.5b] $-2x^3 + 4x^4 - 6x^5$ **12.** [4.5d] $8x^3 + 1$
13. [4.6b] $64 - \tfrac{1}{9}x^2$ **14.** [4.4c] $-y^3 - 2y^2 - 2y + 7$
15. [4.8b] $x^2 - x - 1 + \dfrac{-2}{2x-1}$ **16.** [6.4a] $\dfrac{-5x-28}{5(x-5)}$
17. [6.5a] $\dfrac{4x-1}{x-2}$ **18.** [6.1d] $\dfrac{y}{(y-1)^2}$
19. [6.2b] $\dfrac{3(x+1)}{2x}$ **20.** [5.1b] $3x^2(2x^3 - 12x + 3)$
21. [5.5d] $(4y^2 + 9)(2y + 3)(2y - 3)$
22. [5.3a], [5.4a] $(3x - 2)(x + 4)$ **23.** [5.5b] $(2x^2 - 3y)^2$
24. [5.3a] $3m(m+5)(m-3)$
25. [5.1c] $(x+1)^2(x-1)$ **26.** [2.3c] −9 **27.** [5.8a] $0, \tfrac{5}{2}$
28. [2.7e] $\{x \mid x \le 20\}$ **29.** [2.3b, c] 0.3 **30.** [5.8b] 13, −13
31. [5.8b] $\tfrac{5}{3}$, 3 **32.** [6.7a] −1 **33.** [6.7a] No solution
34. [2.3c] All real numbers **35.** [2.7e] $\{y \mid y \ge -\tfrac{5}{2}\}$
36. [2.3c] No solution **37.** [2.4b] $x = \dfrac{N+t}{r}$ **38.** [2.6a] $35
39. [6.8a] $6\tfrac{2}{3}$ hr **40.** [5.9a] Hypotenuse: 13 in.; leg: 5 in.
41. [6.8b] 75 chips **42.** [5.9a] Height: 9 ft; base: 4 ft
43. [6.9d] 72 ft; 360 **44.** [6.9a] $y = 0.2x$ **45.** [7.1b] 6; 9
46. [7.1b] 11; 3 **47.** [3.4a] 0 **48.** [7.3b] $-\tfrac{2}{3}, (0, 2)$
49. [7.5c] $y = -\tfrac{10}{7}x - \tfrac{8}{7}$ **50.** [7.5a] $y = 6x - 3$
51. [3.3b]

52. [3.3a] **53.** [7.1c]

54. [3.2a] 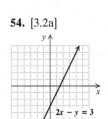   **55.** [3.3b] 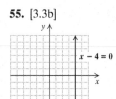   **56.** [3.3a]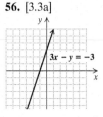

57. [2.7e] No solution   **58.** [1.8d], [6.2b], [6.5a] 0   **59.** [7.4d] $-\frac{3}{10}$

CHAPTER 8

Calculator Corner, p. 593
1. $(2,3)$ **2.** $(-4,-1)$ **3.** $(-1,5)$ **4.** $(3,-1)$

Exercise Set 8.1, p. 593
RC1. False   **RC2.** True   **RC3.** True   **RC4.** True
CC1. Yes   **CC2.** Yes   **CC3.** No
1. $(3,1)$; consistent; independent   **3.** $(1,-2)$; consistent; independent   **5.** $(4,-2)$; consistent; independent
7. $(2,1)$; consistent; independent   **9.** $\left(\frac{5}{2},-2\right)$; consistent; independent   **11.** $(3,-2)$; consistent; independent
13. No solution; inconsistent; independent   **15.** Infinitely many solutions; consistent; dependent   **17.** $(4,-5)$; consistent; independent   **19.** $(2,-3)$; consistent; independent
21. Consistent; independent; F   **23.** Consistent; dependent; B
25. Inconsistent; independent; D   **27.** $y = \frac{3}{5}x + \frac{22}{5}$
28. $y = \frac{3}{8}x + \frac{9}{4}$ **29.** $y = -\frac{3}{2}x$ **30.** $y = \frac{4}{3}x - 14$
31. $(2.23, 1.14)$ **33.** $(3,3), (-5,5)$ **35.** 1

Exercise Set 8.2, p. 600
RC1. True   **RC2.** True   **RC3.** False   **RC4.** False
CC1. $b = 4; (8,4)$ **CC2.** $x = 19; \left(19, \frac{47}{2}\right)$
CC3. $w = -\frac{1}{2}; \left(-\frac{19}{2}, -\frac{1}{2}\right)$
1. $(2,-3)$ **3.** $\left(\frac{21}{5}, \frac{12}{5}\right)$ **5.** $(2,-2)$ **7.** $(-2,-6)$
9. $(-2,1)$ **11.** No solution   **13.** $\left(\frac{19}{8}, \frac{1}{8}\right)$ **15.** Infinitely many solutions   **17.** $\left(\frac{1}{2}, \frac{1}{2}\right)$ **19.** Length: 25 m; width: 5 m
21. 48° and 132°   **23.** Wins: 23; ties: 14   **25.** 1.3
26. $-15y - 39$ **27.** $p = \dfrac{7A}{q}$ **28.** $\frac{7}{3}$ **29.** -23 **30.** $\frac{29}{22}$
31. $m = -\frac{1}{2}; b = \frac{5}{2}$ **33.** Length: 57.6 in.; width: 20.4 in.

Exercise Set 8.3, p. 609
RC1. consistent   **RC2.** inconsistent   **RC3.** consistent
RC4. dependent   **RC5.** inconsistent   **RC6.** independent
CC1. -3 **CC2.** 2   **CC3.** -1
1. $(1,2)$ **3.** $(-1,3)$ **5.** $(-1,-2)$ **7.** $(5,2)$ **9.** Infinitely many solutions   **11.** $\left(\frac{1}{2},-\frac{1}{2}\right)$ **13.** $(4,6)$ **15.** No solution
17. $(10,-8)$ **19.** $(12,15)$ **21.** $(10,8)$ **23.** $(-4,6)$
25. $(10,-5)$ **27.** $(140,60)$ **29.** 36 and 27   **31.** 18 and -15 **33.** 48° and 42°   **35.** Two-point shots: 21; three-point shots: 6   **37.** 3-credit courses: 25; 4-credit courses: 8
39. 1   **40.** 5   **41.** 15   **42.** $12a^2 - 2a + 1$ **43.** $\{x \mid x$ is a real number $and\ x \neq -7\}$ **44.** Domain: all real numbers; range: $\{y \mid y \leq 5\}$ **45.** $y = -\frac{3}{5}x - 7$ **46.** $y = x + 12$
47. $(23.12, -12.04)$ **49.** $A = 2, B = 4$ **51.** $p = 2, q = -\frac{1}{3}$

Translating for Success, p. 620
1. G   **2.** E   **3.** D   **4.** A   **5.** J   **6.** B   **7.** C   **8.** I
9. F   **10.** H

Exercise Set 8.4, p. 621
RC1. 10   **RC2.** 15   **RC3.** $0.15y$ **RC4.** 2
1. Developed: 2322 locations; published: 526 locations
3. Books: 45; games: 23   **5.** Olive oil: $22\frac{1}{2}$ oz; vinegar: $7\frac{1}{2}$ oz
7. 5 lb of each   **9.** 25% acid: 4 L; 50% acid: 6 L
11. Sweet-pepper packets: 11; hot-pepper packets: 5
13. $7500 at 6%; $4500 at 3%   **15.** Whole milk: $169\frac{3}{13}$ lb; cream: $30\frac{10}{13}$ lb   **17.** $1800 at 5.5%; $1400 at 4%   **19.** $5 bills: 7; $1 bills: 15   **21.** 30-sec commercials: 45; 1-min commercials: 17
23. 375 mi   **25.** 14 km/h   **27.** Headwind: 30 mph; plane: 120 mph   **29.** $1\frac{1}{3}$ hr   **31.** About 1489 mi
33. All real numbers   **34.** $\{x \mid x$ is a real number and $x \neq 14\}$
35. -4 **36.** -17 **37.** $-12h - 7$ **38.** 3993   **39.** $4\frac{4}{7}$ L
41. City: 261 mi; highway: 204 mi

Mid-Chapter Review: Chapter 8, p. 625
1. False   **2.** False   **3.** True   **4.** True
5. $x + 2(x - 6) = 3$
$x + 2x - 12 = 3$
$3x - 12 = 3$
$3x = 15$
$x = 5$
$y = 5 - 6$
$y = -1$
The solution is $(5, -1)$.
6. $6x - 4y = 10$
$\underline{2x + 4y = 14}$
$8x = 24$
$x = 3$
$2 \cdot 3 + 4y = 14$
$6 + 4y = 14$
$4y = 8$
$y = 2$
The solution is $(3, 2)$.
7. $(5, -1)$; consistent; independent   **8.** $(0, 3)$; consistent; independent   **9.** Infinitely many solutions; consistent; dependent
10. No solution; inconsistent; independent   **11.** $(8, 6)$
12. $(2, -3)$ **13.** $(-3, 5)$ **14.** $(-1, -2)$ **15.** $(2, -2)$
16. $(5, -4)$ **17.** $(-1, -2)$ **18.** $(3, 1)$ **19.** No solution
20. Infinitely many solutions   **21.** $(10, -12)$ **22.** $(-9, 8)$
23. Length: 12 ft; width: 10 ft   **24.** $2100 at 2%; $2900 at 3%
25. 20% acid: 56 L; 50% acid: 28 L   **26.** 26 mph
27. *Graphically*: **1.** Graph $y = \frac{3}{4}x + 2$ and $y = \frac{2}{5}x - 5$ and find the point of intersection. The first coordinate of this point is the solution of the original equation. **2.** Rewrite the equation as $\frac{7}{20}x + 7 = 0$. Then graph $y = \frac{7}{20}x + 7$ and find the x-intercept. The first coordinate of this point is the solution of the original equation. *Algebraically*: **1.** Use the addition and multiplication principles for equations. **2.** Multiply by 20 to clear the fractions and then use the addition and multiplication principles for equations.
28. (a) Answers may vary.
$x + y = 1,$
$x - y = 7$
(b) Answers may vary.
$x + 2y = 5,$
$3x + 6y = 10$
(c) Answers may vary.
$x - 2y = 3,$
$3x - 6y = 9$
29. Answers may vary. Form a linear expression in two variables and set it equal to two different constants. See Exercises 10 and 19 in this review for examples.   **30.** Answers may vary. Let any linear equation be one equation in the system. Multiply by a constant on both sides of that equation to get the second equation in the system. See Exercises 9 and 20 in this review for examples.

Exercise Set 8.5, p. 631

RC1. (b) **RC2.** (c) **RC3.** (a) **RC4.** (a) **CC1.** Yes
CC2. No **CC3.** Yes
1. $(1, 2, -1)$ **3.** $(2, 0, 1)$ **5.** $(3, 1, 2)$ **7.** $(-3, -4, 2)$
9. $(2, 4, 1)$ **11.** $(-3, 0, 4)$ **13.** $(2, 2, 4)$ **15.** $(\frac{1}{2}, 4, -6)$
17. $(-2, 3, -1)$ **19.** $(\frac{1}{2}, \frac{1}{3}, \frac{1}{6})$ **21.** $(3, -5, 8)$ **23.** $(15, 33, 9)$
25. $(4, 1, -2)$ **27.** $(17, 9, 79)$ **28.** $a = \dfrac{F}{3b}$
29. $a = \dfrac{Q - 4b}{4}$, or $\dfrac{Q}{4} - b$ **30.** $d = \dfrac{tc - 2F}{t}$, or $c - \dfrac{2F}{t}$
31. $c = \dfrac{2F + td}{t}$, or $\dfrac{2F}{t} + d$ **32.** $y = \dfrac{c - Ax}{B}$
33. $y = \dfrac{Ax - c}{B}$ **34.** Slope: $-\frac{2}{3}$; y-intercept: $(0, -\frac{5}{4})$
35. Slope: -4; y-intercept: $(0, 5)$ **36.** Slope: $\frac{2}{5}$; y-intercept: $(0, -2)$ **37.** Slope: 1.09375; y-intercept: $(0, -3.125)$
39. $(1, -2, 4, -1)$

Exercise Set 8.6, p. 637

RC1. (c) **RC2.** (d) **RC3.** (a) **RC4.** (b)
1. Chinese: 13,000 exams; French: 22,000 exams; German: 5000 exams **3.** $32°, 96°, 52°$ **5.** $-7, 20, 42$ **7.** Small: 5 dozen; medium: 6 dozen; large: 11 dozen **9.** Egg: 274 mg; cupcake: 19 mg; pizza: 9 mg **11.** Automatic transmission: $865; power door locks: $520; air conditioning: $375 **13.** Birds: 20 million; cats: 94 million; dogs: 90 million **15.** Roast beef: 2; baked potato: 1; broccoli: 2 **17.** First fund: $45,000; second fund: $10,000; third fund: $25,000 **19.** Par-3: 6 holes; par-4: 8 holes; par-5: 4 holes **21.** A: 1500 lenses; B: 1900 lenses; C: 2300 lenses

23. 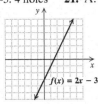 **24.**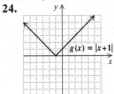

25.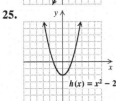

26. No **27.** Yes **28.** Yes **29.** $180°$ **31.** 464
33. E: 701 games; E10+: 298 games; T: 313 games; M: 164 games

Summary and Review: Chapter 8, p. 641

Vocabulary Reinforcement
1. pair **2.** consistent **3.** algebraic **4.** triple **5.** independent

Concept Reinforcement
1. False **2.** True **3.** True **4.** False

Study Guide
1. $(4, -1)$; consistent; independent **2.** $(-1, 4)$ **3.** $(-2, 3)$
4. $8700 at 6%; $14,300 at 5% **5.** $(3, -5, 1)$

Review Exercises
1. $(-2, 1)$; consistent; independent **2.** Infinitely many solutions; consistent; dependent **3.** No solution; inconsistent; independent **4.** $(1, -1)$ **5.** No solution **6.** $(\frac{2}{5}, -\frac{4}{5})$
7. $(6, -3)$ **8.** $(2, 2)$ **9.** $(5, -3)$ **10.** Infinitely many solutions **11.** 150 mph **12.** 32 brushes at $8.50; 13 brushes at $9.75 **13.** 5 L of each **14.** $5\frac{1}{2}$ hr **15.** $(10, 4, -8)$
16. $(-1, 3, -2)$ **17.** $(2, 0, 4)$ **18.** $(2, \frac{1}{3}, -\frac{2}{3})$
19. $90°, 67\frac{1}{2}°, 22\frac{1}{2}°$ **20.** Caramel nut crunch: $30; plain: $5; mocha choco latte: $14 **21.** C **22.** C **23.** A
24. $(0, 2)$ and $(1, 3)$

Understanding Through Discussion and Writing

1. Answers may vary. One day, a florist sold a total of 23 hanging baskets and flats of petunias. Hanging baskets cost $10.95 each and flats of petunias cost $12.95 each. The sales totaled $269.85. How many of each were sold? **2.** We know that machines A, B, and C can polish 5700 lenses in one week when working together. We also know that A and B together can polish 3400 lenses in one week, so C can polish $5700 - 3400$, or 2300, lenses in one week alone. We also know that B and C together can polish 4200 lenses in one week, so A can polish $5700 - 4200$, or 1500, lenses in one week alone. Also, B can polish $4200 - 2300$, or 1900, lenses in one week alone.
3. Let $x =$ the number of adults in the audience, $y =$ the number of senior citizens, and $z =$ the number of children. The total attendance is 100, so we have equation (1), $x + y + z = 100$. The amount taken in was $100, so equation (2) is $10x + 3y + 0.5z = 100$. There is no other information that can be translated to an equation. Clearing decimals in equation (2) and then eliminating z gives us equation (3), $95x + 25y = 500$. Dividing by 5 on both sides, we have equation (4), $19x + 5y = 100$. Since we have only two equations, it is not possible to eliminate z from another pair of equations. However, in $19x + 5y = 100$, note that 5 is a factor of both $5y$ and 100. Therefore, 5 must also be a factor of $19x$, and hence of x, since 5 is not a factor of 19. Then for some positive integer n, $x = 5n$. (We require n to be positive, since the number of adults clearly cannot be negative and must also be nonzero since the exercise states that the audience consists of adults, senior citizens, and children.) We have

$$19 \cdot 5n + 5y = 100$$
$$19n + y = 20. \quad \text{Dividing by 5}$$

Since n and y must both be positive, $n = 1$. (If $n > 1$, then $19n + y > 20$.) Then $x = 5 \cdot 1$, or 5.

$$19 \cdot 5 + 5y = 100 \quad \text{Substituting in (4)}$$
$$y = 1$$
$$5 + 1 + z = 100 \quad \text{Substituting in (1)}$$
$$z = 94$$

There were 5 adults, 1 senior citizen, and 94 children in the audience.

Test: Chapter 8, p. 647

1. [8.1a] $(-2, 1)$; consistent; independent **2.** [8.1a] No solution; inconsistent; independent **3.** [8.1a] Infinitely many solutions; consistent; dependent **4.** [8.2a] $(2, -3)$ **5.** [8.2a] Infinitely many solutions **6.** [8.2a] $(-4, 5)$ **7.** [8.2a] $(8, -6)$
8. [8.3a] $(-1, 1)$ **9.** [8.3a] $(-\frac{3}{2}, -\frac{1}{2})$ **10.** [8.3a] No solution
11. [8.3a] $(10, -20)$ **12.** [8.2b] Length: 93 ft; width: 51 ft
13. [8.4b] 120 km/h **14.** [8.3b], [8.4a] Buckets: 17; dinners: 11
15. [8.4a] 20% solution: 12 L; 45% solution: 8 L
16. [8.5a] $(2, -\frac{1}{2}, -1)$ **17.** [8.6a] 3.5 hr **18.** [8.6a] B
19. [8.3a] $m = 7; b = 10$

Cumulative Review: Chapters 1–8, p. 649
1. [4.7f] $9x^8 - 4y^{10}$ **2.** [4.6c] $x^4 + 8x^2 + 16$ **3.** [4.6a] $8x^2 - \frac{1}{8}$
4. [6.5a] $\dfrac{2(2x+1)}{2x-1}$ **5.** [4.4a, c] $-3x^3 + 8x^2 - 5x$
6. [6.1d] $\dfrac{2(x-5)}{3(x-1)}$ **7.** [6.2b] $\dfrac{(x+1)(x-3)}{2(x+3)}$
8. [4.8b] $3x^2 + 4x + 9 + \dfrac{13}{x-2}$ **9.** [5.5d] $3(1+2x^4)(1-2x^4)$
10. [5.1b] $4t(3 - t - 12t^3)$ **11.** [5.3a], [5.4a] $2(3x - 2)(x - 4)$
12. [5.1c], [5.7a] $(2x+1)(2x-1)(x+1)$
13. [5.5b] $(4x^2 - 7)^2$ **14.** [5.2a] $(x+15)(x-12)$
15. [7.3b] Slope: $\frac{4}{5}$; y-intercept: $(0, 4)$ **16.** [7.5b] $y = -3x + 17$
17. [7.5d] $y = \frac{1}{3}x + 4$ **18.** [7.4d] Perpendicular
19. [5.8b] $-17, 0$ **20.** [2.3b] $\frac{2}{5}$ **21.** [6.7a] $-\frac{12}{5}$
22. [5.8b] $-5, 6$ **23.** [2.7e] $\{x \mid x \leq -\frac{9}{2}\}$ **24.** [6.7a] $\frac{1}{3}$
25. [2.4b] $p = \dfrac{4A}{r+q}$ **26.** [8.2a] $\left(\frac{8}{5}, -\frac{1}{5}\right)$ **27.** [8.3a] $(1, -1)$
28. [8.3a] $(-1, 3)$ **29.** [8.5a] $(2, 0, -1)$
30. [7.4c] **31.** [7.1c]

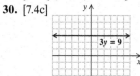

32. [7.4b] **33.** [7.4a]

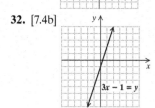
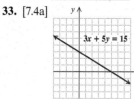

34. [7.2a] **(a)** $\{-5, -3, -1, 1, 3\}$ **(b)** $\{-3, -2, 1, 4, 5\}$ **(c)** -2
(d) 3 **35.** [7.2a] $\{x \mid x \text{ is a real number } and\ x \neq \frac{1}{2}\}$
36. [7.1b] $-1; 1; -17$ **37.** [8.4a] 15%: 21 L; 25%: 9 L
38. [8.6a] $120 **39.** [6.8b] 20 defective resistors
40. [5.9a] Length: 15 m; width: 12 m **41.** [6.9d] 0.4 ft
42. [2.6a] 38°, 76°, 66° **43.** [7.5e] $151,000
44. [8.3a] $m = -\frac{5}{9}; b = -\frac{2}{9}$

CHAPTER 9

Translating for Success, p. 661
1. F **2.** I **3.** C **4.** E **5.** D **6.** J **7.** O **8.** M
9. B **10.** L

Exercise Set 9.1, p. 662
RC1. solution **RC2.** set-builder **RC3.** closed
RC4. half-open **RC5.** negative **CC1.** (b) **CC2.** (h)
CC3. (c) **CC4.** (a) **CC5.** (g) **CC6.** (d)
1. No, no, no, yes **3.** No, yes, yes, no, no **5.** $(-\infty, 5)$
7. $[-3, 3]$ **9.** $(-8, -4)$ **11.** $(-2, 5)$ **13.** $(-\sqrt{2}, \infty)$
15. $\{x \mid x > -1\}$, or $(-1, \infty)$ **17.** $\{y \mid y < 6\}$, or $(-\infty, 6)$
19. $\{a \mid a \leq -22\}$, or $(-\infty, -22]$
21. $\{t \mid t \geq -4\}$, or $[-4, \infty)$ **23.** $\{y \mid y > -6\}$, or $(-6, \infty)$
25. $\{x \mid x \leq 9\}$, or $(-\infty, 9]$ **27.** $\{x \mid x \geq 3\}$, or $[3, \infty)$
29. $\{x \mid x < -60\}$, or $(-\infty, -60)$ **31.** $\{x \mid x > 3\}$, or $(3, \infty)$
33. $\{x \mid x \leq 0.9\}$, or $(-\infty, 0.9]$ **35.** $\{x \mid x \leq \frac{5}{6}\}$, or $(-\infty, \frac{5}{6}]$
37. $\{x \mid x < 6\}$, or $(-\infty, 6)$ **39.** $\{y \mid y \geq -3\}$, or $(-\infty, -3]$
41. $\{y \mid y > \frac{2}{3}\}$, or $(\frac{2}{3}, \infty)$ **43.** $\{x \mid x \geq 11.25\}$, or $[11.25, \infty)$
45. $\{x \mid x \leq \frac{1}{2}\}$, or $(-\infty, \frac{1}{2}]$ **47.** $\{y \mid y \leq -\frac{75}{2}\}$, or $(-\infty, -\frac{75}{2}]$
49. $\{x \mid x > -\frac{2}{17}\}$, or $(-\frac{2}{17}, \infty)$ **51.** $\{m \mid m > \frac{7}{3}\}$, or $(\frac{7}{3}, \infty)$
53. $\{r \mid r < -3\}$, or $(-\infty, -3)$ **55.** $\{x \mid x \geq 2\}$, or $[2, \infty)$
57. $\{y \mid y < 5\}$, or $(-\infty, 5)$ **59.** $\{x \mid x \leq \frac{4}{7}\}$, or $(-\infty, \frac{4}{7}]$
61. $\{x \mid x < 8\}$, or $(-\infty, 8)$ **63.** $\{x \mid x \geq \frac{13}{2}\}$, or $[\frac{13}{2}, \infty)$
65. $\{x \mid x < \frac{11}{18}\}$, or $(-\infty, \frac{11}{18})$ **67.** $\{x \mid x \geq -\frac{51}{31}\}$, or $[-\frac{51}{31}, \infty)$
69. $\{a \mid a \leq 2\}$, or $(-\infty, 2]$
71. $\{W \mid W < (\text{approximately}) \ 136.7 \ \text{lb}\}$ **73.** $\{S \mid S \geq 84\}$
75. $\{B \mid B \geq \$11,500\}$ **77.** $\{S \mid S > \$7000\}$ **79.** $\{c \mid c > \$735\}$
81. $\{p \mid p > 80\}$ **83.** $\{s \mid s < 980 \ \text{ft}^2\}$
85. **(a)** 2010: $2438; 2016: $3056; **(b)** More than 15.31 years
after 2005, or $\{t \mid t \geq 15.31\}$ **87.** $3x^2 + 20x - 32$
88. $6r^2 - 23rs - 4s^2$ **89.** $6a^2 + 7a - 55$ **90.** $t^2 - 7st - 18s^2$
91. $(2x-9)^2$ **92.** $16(5y+1)(5y-1)$
93. $(3w-2)(9w^2 + 6w + 4)$ **94.** $2(8 - 3x)(5 + x)$
95. $\{x \mid x \text{ is a real number } and\ x \neq -8\}$ **96.** All real numbers
97. All real numbers **98.** $\{x \mid x \text{ is a real number } and\ x \neq \frac{2}{3}\}$
99. **(a)** $\{p \mid p > 10\}$; **(b)** $\{p \mid p < 10\}$ **101.** True
103. All real numbers **105.** All real numbers

Exercise Set 9.2, p. 675
RC1. True **RC2.** False **RC3.** True **RC4.** True
CC1. Yes **CC2.** Yes **CC3.** No **CC4.** Yes **CC5.** No
CC6. Yes
1. $\{9, 11\}$ **3.** $\{b\}$ **5.** $\{9, 10, 11, 13\}$ **7.** $\{a, b, c, d, f, g\}$
9. $\varnothing$ **11.** $\{3, 5, 7\}$ **13.** $(-4, 1]$
15. $(1, 6)$
17. $\{x \mid -4 \leq x < 5\}$, or $[-4, 5)$
19. $\{x \mid x \geq 2\}$, or $[2, \infty)$
21. $\varnothing$ **23.** $\{x \mid -8 < x < 6\}$, or $(-8, 6)$
25. $\{x \mid -6 < x \leq 2\}$, or $(-6, 2]$
27. $\{x \mid -1 < x \leq 6\}$, or $(-1, 6]$
29. $\{y \mid -1 < y \leq 5\}$, or $(-1, 5]$
31. $\{x \mid -\frac{5}{3} \leq x \leq \frac{4}{3}\}$, or $[-\frac{5}{3}, \frac{4}{3}]$
33. $\{x \mid -\frac{7}{2} < x \leq \frac{11}{2}\}$, or $(-\frac{7}{2}, \frac{11}{2}]$
35. $\{x \mid 10 < x \leq 14\}$, or $(10, 14]$
37. $\{x \mid -\frac{13}{3} \leq x \leq 9\}$, or $[-\frac{13}{3}, 9]$
39. $(-\infty, -2) \cup (1, \infty)$
41. $(-\infty, -3] \cup (1, \infty)$
43. $\{x \mid x < -5 \text{ or } x > -1\}$, or $(-\infty, -5) \cup (-1, \infty)$;
45. $\{x \mid x \leq \frac{5}{2} \text{ or } x \geq 4\}$, or $(-\infty, \frac{5}{2}] \cup [4, \infty)$;
47. $\{x \mid x \geq -3\}$, or $[-3, \infty)$;
49. $\{x \mid x \leq -\frac{5}{4} \text{ or } x > -\frac{1}{2}\}$, or $(-\infty, -\frac{5}{4}] \cup (-\frac{1}{2}, \infty)$
51. All real numbers, or $(-\infty, \infty)$ **53.** $\{x \mid x < -4 \text{ or } x > 2\}$,
or $(-\infty, -4) \cup (2, \infty)$ **55.** $\{x \mid x < \frac{79}{4} \text{ or } x > \frac{89}{4}\}$, or
$(-\infty, \frac{79}{4}) \cup (\frac{89}{4}, \infty)$ **57.** $\{x \mid x \leq -\frac{13}{2} \text{ or } x \geq \frac{29}{2}\}$, or
$(-\infty, -\frac{13}{2}] \cup [\frac{29}{2}, \infty)$ **59.** $\{d \mid 0 \ \text{ft} \leq d \leq 198 \ \text{ft}\}$
61. Between 23 beats and 27 beats
63. $\{W \mid 101.2 \ \text{lb} \leq W \leq 136.2 \ \text{lb}\}$

65. $\{d \mid 250 \text{ mg} < d < 500 \text{ mg}\}$ **67.** $(-1, 2)$ **68.** $(-3, -5)$
69. $(4, -4)$ **70.** $y = -11x + 29$ **71.** $y = -4x + 7$
72. $y = -x + 2$ **73.** $6a^2 + 7ab - 5b^2$ **74.** $25y^2 + 35y + 6$
75. $21x^2 - 59x + 40$ **76.** $13x^2 + 37xy - 6y^2$
77. $\{x \mid -4 < x \leq 1\}$, or $(-4, 1]$ **79.** $\{x \mid \frac{2}{5} \leq x \leq \frac{4}{5}\}$, or $[\frac{2}{5}, \frac{4}{5}]$
81. $\{x \mid 10 < x \leq 18\}$, or $(10, 18]$ **83.** True **85.** False

Mid-Chapter Review: Chapter 9, p. 680

1. True **2.** False **3.** True **4.** True
5. $8 - 5x \leq x + 20$ **6.** $-17 < 3 - x < 36$
$ -5x \leq x + 12$ $\phantom{-17 <} -20 < -x < 33$
$ -6x \leq 12$ $\phantom{-17 <} 20 > x > -33$
$ x \geq -2$

7. G **8.** B **9.** H **10.** E **11.** F **12.** A **13.** $\{0, 10\}$
14. $\{b, d, e, f, g, h\}$ **15.** $\{\frac{1}{4}, \frac{3}{8}\}$ **16.** $\emptyset$
17. $\{y \mid y \leq -2\}$, or $(-\infty, -2]$ **18.** $\{x \mid x \leq 4\}$, or $(-\infty, 4]$
19. $\{x \mid x < -9 \text{ or } x > 1\}$, or $(-\infty, -9) \cup (1, \infty)$
20. $\{x \mid 3 \leq x \leq 24\}$, or $[3, 24)$ **21.** $\{x \mid x < -10 \text{ or } x > 1\}$, or $(-\infty, -10) \cup (1, \infty)$ **22.** $\{t \mid t > 2\}$, or $(2, \infty)$
23. $\{x \mid x \geq -3\}$, or $[-3, \infty)$ **24.** $\{y \mid y \geq \frac{10}{11}\}$, or $(\frac{10}{11}, \infty)$
25. $\{x \mid -\frac{17}{2} < x < \frac{25}{2}\}$, or $(-\frac{17}{2}, \frac{25}{2})$ **26.** $\{x \mid x \geq \frac{35}{6}\}$, or $[\frac{35}{6}, \infty)$
27. $\{x \mid 4 < x < 17\}$, or $(4, 17)$ **28.** $\{x \mid x \text{ is a real number}\}$, or $(-\infty, \infty)$ **29.** $\{S \mid S \geq 86\}$ **30.** $\$3000$
31. When the signs of the quantities on either side of the inequality symbol are changed, their relative positions on the number line are reversed. **32.** (1) $-9(x + 2) = -9x - 18$, not $-9x + 2$. (2) The left side should be $7 - 3x$. The inequality symbol should not have been reversed. Using the incorrect right side in step (1), this should now be $2 + x$. (3) If (2) were correct, the right-hand side would be -5, not 8. (4) The inequality symbol should be reversed. The correct solution is

$$7 - 9x + 6x < -9(x + 2) + 10x$$
$$7 - 9x + 6x < -9x - 18 + 10x$$
$$7 - 3x < x - 18$$
$$-4x < -25$$
$$x > \tfrac{25}{4}.$$

33. By definition, the notation $3 < x < 5$ indicates that $3 < x$ and $x < 5$. A solution of the disjunction $3 < x$ or $x < 5$ must be in at least one of these sets but not necessarily in both, so the disjunction cannot be written as $3 < x < 5$.

Exercise Set 9.3, p. 689

RC1. True **RC2.** True **RC3.** False **RC4.** True
RC5. True **RC6.** True **CC1.** (f) **CC2.** (b) **CC3.** (e)
CC4. (c) **CC5.** (a) **CC6.** (d)

1. $9|x|$ **3.** $2x^2$ **5.** $2x^2$ **7.** $6|y|$ **9.** $\dfrac{2}{|x|}$ **11.** $\dfrac{x^2}{|y|}$
13. $4|x|$ **15.** $\dfrac{y^2}{3}$ **17.** 38 **19.** 19 **21.** 6.3
23. 5 **25.** $\{-3, 3\}$ **27.** $\emptyset$ **29.** $\{0\}$ **31.** $\{-9, 15\}$
33. $\{-\tfrac{1}{2}, \tfrac{7}{2}\}$ **35.** $\{-\tfrac{5}{4}, \tfrac{23}{4}\}$ **37.** $\{-11, 11\}$
39. $\{-291, 291\}$ **41.** $\{-8, 8\}$ **43.** $\{-7, 7\}$ **45.** $\{-2, 2\}$
47. $\{-7, 8\}$ **49.** $\{-12, 2\}$ **51.** $\{-\tfrac{5}{2}, \tfrac{7}{2}\}$ **53.** $\emptyset$
55. $\{-\tfrac{13}{54}, -\tfrac{7}{54}\}$ **57.** $\{-\tfrac{11}{2}, \tfrac{3}{4}\}$ **59.** $\{\tfrac{3}{2}\}$ **61.** $\{5, -\tfrac{3}{5}\}$
63. All real numbers **65.** $\{-\tfrac{3}{2}\}$ **67.** $\{\tfrac{24}{23}, 0\}$
69. $\{32, \tfrac{8}{3}\}$ **71.** $\{x \mid -3 < x < 3\}$, or $(-3, 3)$
73. $\{x \mid x \leq -2 \text{ or } x \geq 2\}$, or $(-\infty, -2] \cup [2, \infty)$
75. $\{x \mid 0 < x < 2\}$, or $(0, 2)$
77. $\{x \mid -6 \leq x \leq -2\}$, or $[-6, -2]$
79. $\{x \mid -\tfrac{1}{2} \leq x \leq \tfrac{7}{2}\}$, or $[-\tfrac{1}{2}, \tfrac{7}{2}]$
81. $\{y \mid y < -\tfrac{3}{2} \text{ or } y > \tfrac{17}{2}\}$, or $(-\infty, -\tfrac{3}{2}) \cup (\tfrac{17}{2}, \infty)$
83. $\{x \mid x \leq -\tfrac{5}{4} \text{ or } x \geq \tfrac{23}{4}\}$, or $(-\infty, -\tfrac{5}{4}] \cup [\tfrac{23}{4}, \infty)$
85. $\{y \mid -9 < y < 15\}$, or $(-9, 15)$ **87.** $\{x \mid -\tfrac{7}{2} \leq x \leq \tfrac{1}{2}\}$, or $[-\tfrac{7}{2}, \tfrac{1}{2}]$ **89.** $\{y \mid y < -\tfrac{4}{3} \text{ or } y > 4\}$, or $(-\infty, -\tfrac{4}{3}) \cup (4, \infty)$
91. $\{x \mid x \leq -\tfrac{5}{4} \text{ or } x \geq \tfrac{23}{4}\}$, or $(-\infty, -\tfrac{5}{4}] \cup [\tfrac{23}{4}, \infty)$
93. $\{x \mid -\tfrac{9}{2} < x < 6\}$, or $(-\tfrac{9}{2}, 6)$ **95.** $\{x \mid x \leq -\tfrac{25}{6} \text{ or } x \geq \tfrac{23}{6}\}$, or $(-\infty, -\tfrac{25}{6}] \cup [\tfrac{23}{6}, \infty)$ **97.** $\{x \mid -5 < x < 19\}$, or $(-5, 19)$
99. $\{x \mid x \leq -\tfrac{2}{15} \text{ or } x \geq \tfrac{14}{15}\}$, or $(-\infty, -\tfrac{2}{15}] \cup [\tfrac{14}{15}, \infty)$
101. $\{m \mid -12 \leq m \leq 2\}$, or $[-12, 2]$ **103.** $\{x \mid \tfrac{1}{2} \leq x \leq \tfrac{5}{2}\}$, or $[\tfrac{1}{2}, \tfrac{5}{2}]$ **105.** $\{x \mid -1 \leq x \leq 2\}$, or $[-1, 2]$ **107.** $\tfrac{1}{11}$
108. Not defined **109.** 2 **110.** 0
111. $(3w - 10)(9w^2 + 30w + 100)$
112. $(2 + 5t)(4 - 10t + 25t^2)$ **113.** $\dfrac{3w^2}{w + z}$
114. $\tfrac{5}{3}$ **115.** $\{d \mid 5\tfrac{1}{2} \text{ ft} \leq d \leq 6\tfrac{1}{2} \text{ ft}\}$
117. All real numbers **119.** $\{1, -\tfrac{1}{4}\}$ **121.** $\emptyset$
123. $|x| < 3$ **125.** $|x| \geq 6$ **127.** $|x + 3| > 5$

Visualizing for Success, p. 702

1. D **2.** B **3.** E **4.** C **5.** I **6.** G **7.** F **8.** H
9. A **10.** J

Exercise Set 9.4, p. 703

RC1. graph **RC2.** inequality **RC3.** half-plane
RC4. solution **RC5.** equation **RC6.** test **CC1.** Dashed
CC2. Solid **CC3.** Solid **CC4.** Yes **CC5.** No **CC6.** Yes
1. Yes **3.** Yes

5. 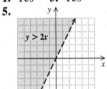 **7.**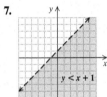

9. 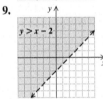 **11.**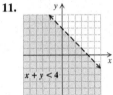

13. 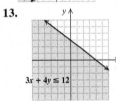 **15.**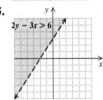

17. 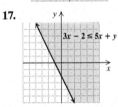 **19.**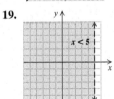

21. 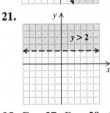 **23.**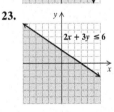

25. F **27.** B **29.** C

31. **33.**

35. 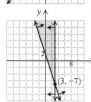 **37.**

39. **41.**

43. $\frac{10}{17}$ **44.** $-\frac{14}{13}$ **45.** -2 **46.** $\frac{29}{11}$ **47.** -12 **48.** $\frac{333}{245}$
49. 2 **50.** 3 **51.** 8 **52.** $|2 - 2a|$, or $2|1 - a|$
53. $h < 2w$,
$w \leq 1.5h$,
$h \leq 3200$,
$h \geq 0$,
$w \geq 0$

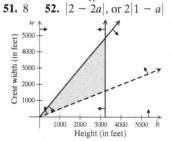

Summary and Review: Chapter 9, p. 707

Vocabulary Reinforcement
1. inequality **2.** set-builder **3.** interval **4.** intersection
5. conjunction **6.** empty set **7.** disjoint sets **8.** union
9. disjunction **10.** at most **11.** at least **12.** distance

Concept Reinforcement
1. True **2.** False **3.** False **4.** False **5.** True **6.** False
7. True

Study Guide
1. -2 is not a solution; 5 is a solution. **2. (a)** $(-\infty, -8)$;
(b) $[-7, 10)$; **(c)** $[3, \infty)$ **3.** $\{y | y < -2\}$, or $(-\infty, -2)$;

4. $\{z | -2 \leq z < 1\}$, or $[-2, 1)$;

5. $\{z | z < -1 \text{ or } z \geq 1\}$, or $(-\infty, -1) \cup [1, \infty)$;

6. $\{-\frac{8}{5}, 2\}$ **7.** $\{3, -\frac{1}{2}\}$ **8. (a)** $\{x | -4 < x < 1\}$, or $(-4, 1)$;
(b) $\{x | x \leq -\frac{10}{3} \text{ or } x \geq 2\}$, or $(-\infty, -\frac{10}{3}] \cup [2, \infty)$
9.   **10.**

Review Exercises
1. -3 is not a solution; 7 is a solution. **2.** $[-8, 9)$
3. $(-\infty, 40]$ **4.** ⟵⫯⫯⫯⫯⟶ ; $(-\infty, -2]$
5. ⟵⫯⫯⫯⫯⟶ ; $(1, \infty)$
6. $\{a | a \leq -21\}$, or $(-\infty, -21]$ **7.** $\{y | y \geq -7\}$, or $[-7, \infty)$
8. $\{y | y > -4\}$, or $(-4, \infty)$ **9.** $\{y | y > -30\}$, or $(-30, \infty)$
10. $\{x | x > -3\}$, or $(-3, \infty)$ **11.** $\{y | y \leq -\frac{6}{5}\}$, or $(-\infty, -\frac{6}{5}]$
12. $\{x | x < -3\}$, or $(-\infty, -3)$ **13.** $\{y | y > -10\}$, or $(-10, \infty)$
14. $\{x | x \leq -\frac{5}{2}\}$, or $(-\infty, -\frac{5}{2}]$ **15.** $\{t | t > 4\frac{1}{4} \text{ hr}\}$
16. \$10,000 **17.** ⟵⫯⫯⫯⫯⟶ ; $[-2, 5)$
18. ⟵⫯⫯⫯⫯⟶ ; $(-\infty, -2] \cup (5, \infty)$
19. $\{1, 5, 9\}$ **20.** $\{1, 2, 3, 5, 6, 9\}$ **21.** $\varnothing$
22. $\{x | -7 < x \leq 2\}$, or $(-7, 2]$
23. $\{x | -\frac{5}{4} < x < \frac{5}{2}\}$, or $(-\frac{5}{4}, \frac{5}{2})$
24. $\{x | x < -3 \text{ or } x > 1\}$, or $(-\infty, -3) \cup (1, \infty)$
25. $\{x | x < -11 \text{ or } x \geq -6\}$, or $(-\infty, -11) \cup [-6, \infty)$
26. $\{x | x \leq -6 \text{ or } x \geq 8\}$, or $(-\infty, -6] \cup [8, \infty)$
27. $\frac{3}{|x|}$ **28.** $\frac{2|x|}{y^2}$ **29.** $\frac{4}{|y|}$ **30.** 62 **31.** $\{-6, 6\}$
32. $\{-5, 9\}$ **33.** $\{-14, \frac{4}{3}\}$ **34.** $\varnothing$
35. $\{x | -\frac{17}{2} < x < \frac{7}{2}\}$, or $(-\frac{17}{2}, \frac{7}{2})$
36. $\{x | x \leq -3.5 \text{ or } x \geq 3.5\}$, or $(-\infty, -3.5] \cup [3.5, \infty)$
37. $\{x | x \leq -\frac{11}{3} \text{ or } x \geq \frac{19}{3}\}$, or $(-\infty, -\frac{11}{3}] \cup [\frac{19}{3}, \infty)$ **38.** $\varnothing$
39. 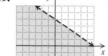 **40.**

41.   **42.**

43.   **44.**

45. D **46.** C **47.** $\{x | -\frac{8}{3} \leq x \leq -2\}$, or $[-\frac{8}{3}, -2]$

Understanding Through Discussion and Writing
1. When $b \geq c$, then the intervals overlap and $[a, b] \cup [c, d] = [a, d]$. **2.** The distance between x and -5 is $|x - (-5)|$, or $|x + 5|$. Then the solutions of the inequality $|x + 5| \leq 2$ can be interpreted as "all those numbers x whose distance from -5 is at most 2 units." **3.** No; the symbol $\geq$ does not always yield a graph in which the half-plane above the line is shaded. For the inequality $-y \geq 3$, for example, the half-plane below the line $y = -3$ is shaded. **4.** The solutions of $|x| \geq 6$ are those numbers whose distance from 0 is greater than or equal to 6. In addition to the numbers in $[6, \infty)$, the distance from 0 of the numbers in $(-\infty, -6]$ is also greater than or equal to 6. Thus, $[6, \infty)$ is only part of the solution of the inequality.

Test: Chapter 9, p. 713

1. [9.1b] $(-3, 2)$ 2. [9.1b] $(-4, \infty)$
3. [9.1c] ⟵———————⟶; $(-\infty, 6]$
4. [9.1c] ⟵———————⟶; $(-\infty, -2]$
5. [9.1c] $\{x | x \geq 10\}$, or $[10, \infty)$ 6. [9.1c] $\{y | y > -50\}$, or $(-50, \infty)$ 7. [9.1c] $\{a | a \leq \frac{11}{5}\}$, or $(-\infty, \frac{11}{5}]$
8. [9.1c] $\{y | y > 1\}$, or $(1, \infty)$ 9. [9.1c] $\{x | x > \frac{5}{2}\}$, or $(\frac{5}{2}, \infty)$
10. [9.1c] $\{x | x \leq \frac{7}{4}\}$, or $(-\infty, \frac{7}{4}]$ 11. [9.1d] $\{h | h > 2\frac{1}{10} \text{ hr}\}$
12. [9.2c] $\{d | 33 \text{ ft} \leq d \leq 231 \text{ ft}\}$
13. [9.2a] ⟵———————⟶; $[-3, 4]$
14. [9.2b] ⟵———————⟶; $(-\infty, -3) \cup (4, \infty)$
15. [9.2a] $\{x | x \geq 4\}$, or $[4, \infty)$ 16. [9.2a] $\{x | -1 < x < 6\}$, or $(-1, 6)$ 17. [9.2a] $\{x | -\frac{2}{5} < x \leq \frac{9}{5}\}$, or $(-\frac{2}{5}, \frac{9}{5}]$
18. [9.2b] $\{x | x < -4 \text{ or } x > -\frac{5}{2}\}$, or $(-\infty, -4) \cup (-\frac{5}{2}, \infty)$
19. [9.2b] All real numbers, or $(-\infty, \infty)$
20. [9.2b] $\{x | x < 3 \text{ or } x > 6\}$, or $(-\infty, 3) \cup (6, \infty)$
21. [9.3a] $\frac{7}{|x|}$ 22. [9.3a] $2|x|$ 23. [9.3b] 8.4
24. [9.2a] $\{3, 5\}$ 25. [9.2b] $\{1, 3, 5, 7, 9, 11, 13\}$
26. [9.3c] $\{-9, 9\}$ 27. [9.3c] $\{-6, 12\}$ 28. [9.3d] $\{1\}$
29. [9.3c] $\emptyset$ 30. [9.3e] $\{x | -0.875 < x < 1.375\}$, or $(-0.875, 1.375)$ 31. [9.3e] $\{x | x < -3 \text{ or } x > 3\}$, or $(-\infty, -3) \cup (3, \infty)$ 32. [9.3e] $\{x | -99 \leq x \leq 111\}$, or $[-99, 111]$ 33. [9.3e] $\{x | x \leq -\frac{13}{5} \text{ or } x \geq \frac{7}{5}\}$, or $(-\infty, -\frac{13}{5}] \cup [\frac{7}{5}, \infty)$

34. [9.4b]

35. [9.4c] 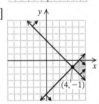 36. [9.4c]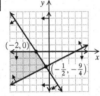

37. [9.3e] B 38. [9.3e] $\emptyset$ 39. [9.2a] $\{x | \frac{1}{5} < x < \frac{4}{5}\}$, or $(\frac{1}{5}, \frac{4}{5})$

Cumulative Review: Chapters 1–9, p. 715

1. [3.2a]   2. [3.3b]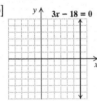

3. [9.4b]   4. [9.4c]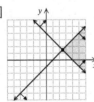

5. [7.1b] 11
6. [7.2a] $\{x | x \text{ is a real number } and \ x \neq -5 \ and \ x \neq 5\}$
7. [7.2a] Domain: $\{x | -5 < x < 5\}$; range: $\{y | -2 \leq y \leq 4\}$

8. [4.6c] $36m^2 - 12mn + n^2$ 9. [4.6a] $15a^2 - 14ab - 8b^2$
10. [6.1d] $\dfrac{y - 2}{3}$ 11. [6.2b] $\dfrac{3x - 5}{x + 4}$ 12. [6.5a] $\dfrac{6x + 13}{20(x - 3)}$
13. [6.6a] $\dfrac{y(y^2 - 2)}{(y - 1)(y^2 + y + 1)}$ 14. [4.4c] $16p^2 - 8p$
15. [6.5b] $\dfrac{4x + 1}{(x + 2)(x - 2)}$ 16. [4.8b] $2x^2 - 11x + 23 + \dfrac{-49}{x + 2}$
17. [2.3c] $\frac{15}{2}$ 18. [9.2a] $\{x | -3 < x < -\frac{3}{2}\}$, or $(-3, -\frac{3}{2})$
19. [6.7a] No solution 20. [2.4b] $a = \dfrac{bP}{3 - P}$
21. [2.4b] $C = \frac{5}{9}(F - 32)$ 22. [9.3e] $\{x | x \leq -2.1 \text{ or } x \geq 2.1\}$, or $(-\infty, -2.1] \cup [2.1, \infty)$ 23. [6.7a] -1 24. [5.8b] $\frac{1}{4}$
25. [5.8b] $-2, \frac{7}{2}$ 26. [8.2a], [8.3a] Infinite number of solutions
27. [8.2a], [8.3a] $(-2, 1)$ 28. [8.5a] $(3, 2, -1)$
29. [8.5a] $(\frac{5}{8}, \frac{1}{16}, -\frac{3}{4})$ 30. [5.1b] $2x^2(2x + 9)$
31. [5.1c] $(2a - 1)(4a^2 - 3)$ 32. [5.2a] $(x - 6)(x + 14)$
33. [5.3a], [5.4a] $(2x + 5)(3x - 2)$ 34. [5.5d] $(4y + 9)(4y - 9)$
35. [5.5b] $(t - 8)^2$ 36. [5.6a] $8(2x + 1)(4x^2 - 2x + 1)$
37. [5.6a] $(0.3b - 0.2c)(0.09b^2 + 0.06bc + 0.04c^2)$
38. [5.7a] $x^2(x^2 + 1)(x + 1)(x - 1)$
39. [5.3a], [5.4a] $(4x - 1)(5x + 3)$ 40. [7.5b] $y = -\frac{1}{2}x - 1$
41. [7.5d] $y = \frac{1}{2}x - \frac{5}{2}$ 42. [8.6a] Win: 38 games; lose: 30 games; tie: 13 games 43. [6.9b] 202.3 lb 44. [6.7a] A
45. [5.8b] C 46. [6.8a] B 47. [8.5a] $a = 1, b = -5, c = 6$
48. [5.8b] $0, \frac{1}{4}, -\frac{1}{4}$ 49. [6.7a] All real numbers except 9 and -5

CHAPTER 10

Exercise Set 10.1, p. 725

RC1. True **RC2.** False **RC3.** False **RC4.** True
RC5. True **RC6.** False **CC1.** (j) **CC2.** (b)
CC3. (h) **CC4.** (i) **CC5.** (i) **CC6.** (g)
1. $4, -4$ **3.** $12, -12$ **5.** $20, -20$ **7.** $-\frac{7}{6}$ **9.** 14 **11.** 0.06
13. Does not exist as a real number **15.** 18.628 **17.** 1.962
19. $y^2 + 16$ **21.** $\dfrac{x}{y - 1}$ **23.** $\sqrt{20} \approx 4.472$; 0; does not exist as a real number; does not exist as a real number
25. $\sqrt{11} \approx 3.317$; does not exist as a real number; $\sqrt{11} \approx 3.317$; 12
27. $\{x | x \geq 2\}$, or $[2, \infty)$ **29.** 21 spaces; 25 spaces

31.   **33.** 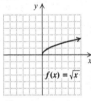 **35.**

37.   **39.** **41.**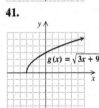

43. $4|x|$ **45.** $12|c|$ **47.** $|p + 3|$ **49.** $|x - 2|$ **51.** 3
53. $-4x$ **55.** -6 **57.** $0.7(x + 1)$ **59.** $2; 3; -2; -4$
61. $-1; -\sqrt[3]{-20} \approx 2.714; -4; -10$ **63.** -5 **65.** -1
67. $-\frac{2}{3}$ **69.** $|x|$ **71.** $5|a|$ **73.** 6 **75.** $|a + b|$ **77.** y
79. $x - 2$ **81.** $-\frac{7}{2}, \frac{7}{2}$ **82.** $4, 9$ **83.** $0, \frac{5}{2}$ **84.** $\frac{5}{2}$ **85.** $a^9 b^6 c^{15}$
86. $10a^{10} b^9$ **87.** $\{x | -3 \leq x < 2\}$, or $[-3, 2)$

89. (a) Domain: $(-\infty, \infty)$; range: $(-\infty, \infty)$;
(b) domain: $(-\infty, \infty)$; range: $(-\infty, \infty)$;
(c) domain: $[-3, \infty)$; range: $(-\infty, 2]$; (d) domain: $[0, \infty)$; range: $[0, \infty)$; (e) domain: $[3, \infty)$; range: $[0, \infty)$

Calculator Corner, p. 731
1. 3.344 **2.** 3.281 **3.** 0.283 **4.** 11.053 **5.** 5.527×10^{-5}
6. 2

Exercise Set 10.2, p. 733
RC1. (h) **RC2.** (b) **RC3.** (c) **RC4.** (g) **RC5.** (e)
RC6. (d) **RC7.** (f) **RC8.** (a)
1. $\sqrt[7]{y}$ **3.** 2 **5.** $\sqrt[5]{a^3b^3}$ **7.** 8 **9.** 343 **11.** $17^{1/2}$
13. $18^{1/3}$ **15.** $(xy^2z)^{1/5}$ **17.** $(3mn)^{3/2}$ **19.** $(8x^2y)^{5/7}$
21. $\frac{1}{3}$ **23.** $\frac{1}{1000}$ **25.** $\frac{3}{x^{1/4}}$ **27.** $\frac{1}{(2rs)^{3/4}}$ **29.** $\frac{2a^{3/4}c^{2/3}}{b^{1/2}}$
31. $\left(\frac{8yz}{7x}\right)^{3/5}$ **33.** $x^{2/3}$ **35.** $\frac{x^4}{2^{1/3}y^{2/7}}$ **37.** $\frac{7x}{z^{1/3}}$ **39.** $\frac{5ac^{1/2}}{3}$
41. $5^{7/8}$ **43.** $7^{1/4}$ **45.** $4.9^{1/2}$ **47.** $6^{3/28}$ **49.** $a^{23/12}$
51. $a^{8/3}b^{5/2}$ **53.** $\frac{1}{x^{2/7}}$ **55.** $\frac{y^{1/3}}{x^{1/2}}$ **57.** $m^{3/5}n^2$ **59.** $\sqrt[3]{a}$
61. x^5 **63.** $\frac{1}{x^3}$ **65.** a^5b^5 **67.** $\sqrt{2}$ **69.** $\sqrt[3]{2x}$ **71.** x^2y^3
73. $2c^2d^3$ **75.** $\sqrt[12]{7^4 \cdot 5^3}$ **77.** $\sqrt[20]{5^5 \cdot 7^4}$ **79.** $\sqrt[6]{4x^5}$
81. a^6b^{12} **83.** $\sqrt[18]{m}$ **85.** $\sqrt[12]{x^4y^3z^2}$ **87.** $\sqrt[30]{\frac{d^{35}}{c^{99}}}$
89. $\{-\frac{4}{7}, 2\}$ **90.** $\{-40, 40\}$ **91.** $\{-\frac{15}{2}, \frac{5}{2}\}$ **92.** $\{-\frac{11}{8}, \frac{3}{8}\}$
93. Left to the student

Exercise Set 10.3, p. 741
RC1. True **RC2.** False **RC3.** True **RC4.** True
CC1. No **CC2.** No **CC3.** Yes **CC4.** No **CC5.** Yes
CC6. No
1. $2\sqrt{6}$ **3.** $3\sqrt{10}$ **5.** $5\sqrt[3]{2}$ **7.** $6x^2\sqrt{5}$ **9.** $3x^2\sqrt[3]{2x^2}$
11. $2t^2\sqrt[3]{10t^2}$ **13.** $2\sqrt[4]{5}$ **15.** $4a\sqrt{2b}$ **17.** $3x^2y^2\sqrt[4]{3y^2}$
19. $2xy^3\sqrt[5]{3x^2}$ **21.** $5\sqrt{2}$ **23.** $3\sqrt{10}$ **25.** 2 **27.** $30\sqrt{3}$
29. $3x^4\sqrt{2}$ **31.** $5bc^2\sqrt{2b}$ **33.** $a\sqrt[3]{10}$ **35.** $2y^3\sqrt[3]{2}$
37. $4\sqrt[4]{4}$ **39.** $4a^3b\sqrt{6ab}$ **41.** $\sqrt[6]{200}$ **43.** $\sqrt[6]{12}$ **45.** $a\sqrt[4]{a}$
47. $b^{10}\sqrt[6]{b^9}$ **49.** $xy\sqrt[6]{xy^5}$ **51.** $2ab\sqrt[6]{2a^3}$ **53.** $3\sqrt{2}$ **55.** $\sqrt{5}$
57. 3 **59.** $y\sqrt{7y}$ **61.** $2\sqrt[3]{a^2b}$ **63.** $4\sqrt{xy}$ **65.** $2x^2y^2$
67. $\frac{1}{\sqrt[6]{a}}$ **69.** $\sqrt[12]{a^5}$ **71.** $\sqrt[12]{x^2y^5}$ **73.** $\frac{5}{6}$ **75.** $\frac{4}{7}$ **77.** $\frac{5}{3}$
79. $\frac{7}{y}$ **81.** $\frac{5y\sqrt{y}}{x^2}$ **83.** $\frac{3y\sqrt[3]{3y^2}}{4}$ **85.** $\frac{3a\sqrt[3]{a}}{2b}$ **87.** $\frac{3x}{2}$
89. $\frac{2a^3}{bc^4}$ **91.** $\frac{2x\sqrt[5]{x^3}}{y^2}$ **93.** $\frac{w\sqrt[5]{w^2}}{z^2}$ **95.** $\frac{x^2\sqrt[6]{x}}{yz}$ **97.** $-10, 9$
98. Height: 4 in.; base: 6 in. **99.** 8 **100.** No solution
101. (a) 1.62 sec; (b) 1.99 sec; (c) 2.20 sec **103.** $2yz\sqrt{2z}$

Exercise Set 10.4, p. 747
RC1. False **RC2.** True **RC3.** True **RC4.** True
CC1. Yes **CC2.** No **CC3.** No **CC4.** Yes
CC5. Yes **CC6.** No **CC7.** No **CC8.** Yes
1. $11\sqrt{5}$ **3.** $\sqrt[3]{7}$ **5.** $13\sqrt[3]{y}$ **7.** $-8\sqrt{6}$ **9.** $6\sqrt[3]{3}$
11. $21\sqrt{3}$ **13.** $38\sqrt{5}$ **15.** $122\sqrt{2}$ **17.** $9\sqrt[3]{2}$ **19.** $29\sqrt{2}$
21. $(1 + 6a)\sqrt{5a}$ **23.** $(2 - x)\sqrt[3]{3x}$ **25.** $(21x + 1)\sqrt{3x}$
27. $2 + 3\sqrt{2}$ **29.** $15\sqrt[3]{4}$ **31.** $(x + 1)\sqrt[3]{6x}$ **33.** $3\sqrt{a - 1}$
35. $(x + 3)\sqrt{x - 1}$ **37.** $4\sqrt{5} - 10$ **39.** $\sqrt{6} - \sqrt{21}$
41. $-12 + 6\sqrt{3}$ **43.** $2\sqrt{15} - 6\sqrt{3}$ **45.** -6
47. $6y - 12\sqrt[3]{y^2}$ **49.** $3a\sqrt[3]{2}$ **51.** 1 **53.** -12 **55.** 44
57. 1 **59.** 3 **61.** -19 **63.** $c - d$ **65.** $1 + \sqrt{5}$

67. $7 + 3\sqrt{3}$ **69.** -6 **71.** $a + \sqrt{3a} + \sqrt{2a} + \sqrt{6}$
73. $2\sqrt[3]{9} - 3\sqrt[3]{6} - 2\sqrt[3]{4}$ **75.** $7 + 4\sqrt{3}$
77. $\sqrt[5]{72} + 3 - \sqrt[5]{24} - \sqrt[5]{81}$ **79.** $a - 2$ **80.** $\frac{(a+2)(a+4)}{a}$
81. $\frac{4(3x-1)}{3(4x+1)}$ **82.** $\frac{x}{x+1}$ **83.** $\frac{pq}{q+p}$ **84.** $\frac{a^2b^2}{b^2 - ab + a^2}$
85. $-\frac{29}{3}, 5$ **86.** $\{x | -\frac{29}{3} < x < 5\}$, or $\left(-\frac{29}{3}, 5\right)$
87. $\{x | x \le -\frac{29}{3} \text{ or } x \ge 5\}$, or $\left(-\infty, -\frac{29}{3}\right] \cup [5, \infty)$
88. $-12, -\frac{2}{5}$ **89.** $(-\infty, \infty)$ **91.** 6
93. $14 + 2\sqrt{15} - 6\sqrt{2} - 2\sqrt{30}$ **95.** $3\sqrt[3]{3} + 2\sqrt[3]{9} - 8$

Mid-Chapter Review: Chapter 10, p. 751
1. False **2.** True **3.** False **4.** True
5. $\sqrt{6}\sqrt{10} = \sqrt{6 \cdot 10} = \sqrt{2 \cdot 3 \cdot 2 \cdot 5} = 2\sqrt{15}$
6. $5\sqrt{32} - 3\sqrt{18} = 5\sqrt{16 \cdot 2} - 3\sqrt{9 \cdot 2} = 5 \cdot 4\sqrt{2} - 3 \cdot 3\sqrt{2} = 20\sqrt{2} - 9\sqrt{2} = 11\sqrt{2}$ **7.** 9 **8.** -12 **9.** $\frac{4}{5}$
10. Does not exist as a real number **11.** 3; does not exist as a real number **12.** $\{x | x \le 4\}$, or $(-\infty, 4]$
13.   **14.**

15. $6|z|$ **16.** $|x - 4|$ **17.** -4 **18.** $-3a$ **19.** 2
20. $|y|$ **21.** 5 **22.** $\sqrt[4]{a^3b}$ **23.** $16^{1/5}$ **24.** $(6m^2n)^{1/3}$
25. $\frac{1}{3^{3/8}}$ **26.** $7^{4/5}$ **27.** $\frac{x^{3/2}}{y^{4/3}}$ **28.** $\frac{1}{n^{3/4}}$ **29.** $\sqrt[3]{4}$ **30.** $\sqrt{ab}$
31. $\sqrt[6]{y^5}$ **32.** $\sqrt[15]{a^{10}b^9}$ **33.** $5\sqrt{3}$ **34.** $2xy\sqrt[3]{3y^2}$
35. $2\sqrt[3]{5}$ **36.** $\frac{7a^2\sqrt{a}}{b^4}$ **37.** $11\sqrt{7}$ **38.** $(9x - 24)\sqrt{2x}$
39. $2\sqrt{3} - 15$ **40.** $3 - 4\sqrt{x} + x$ **41.** $m - n$
42. $11 + 4\sqrt{7}$ **43.** $-42 + \sqrt{15}$ **44.** Yes; since x^2 is nonnegative for any value of x, the nth root of x^2 exists regardless of whether n is even or odd. Thus the nth root of x^2 always exists. **45.** Formulate an expression containing a radical term with an even index and a radicand R such that the solution of the inequality $R \ge 0$ is $\{x | x \le 5\}$. One expression is $\sqrt{5 - x}$. Other expressions could be formulated as $a\sqrt[k]{b(5 - x)} + c$, where $a \ne 0$, $b > 0$, and k is an even integer.
46. Since $x^6 \ge 0$ and $x^2 \ge 0$ for any value of x, then $\sqrt[3]{x^6} = x^2$. However, $x^3 \ge 0$ only for $x \ge 0$, so $\sqrt{x^6} = x^3$ only when $x \ge 0$.
47. No; for example, $\frac{\sqrt{8}}{\sqrt{2}} = \sqrt{\frac{8}{2}} = \sqrt{4} = 2$.

Exercise Set 10.5, p. 756
RC1. No **RC2.** Yes **RC3.** Yes **RC4.** No **CC1.** (g)
CC2. (c) **CC3.** (e) **CC4.** (h) **CC5.** (c) **CC6.** (d)
1. $\frac{\sqrt{15}}{3}$ **3.** $\frac{\sqrt{22}}{2}$ **5.** $\frac{2\sqrt{15}}{35}$ **7.** $\frac{2\sqrt{6}}{3}$ **9.** $\frac{\sqrt[3]{75ac^2}}{5c}$
11. $\frac{y\sqrt[3]{9yx^2}}{3x^2}$ **13.** $\frac{\sqrt[4]{s^3t^3}}{st}$ **15.** $\frac{\sqrt{15x}}{10}$ **17.** $\frac{\sqrt[3]{100xy}}{5x^2y}$
19. $\frac{\sqrt[4]{2xy}}{2x^2y}$ **21.** $\frac{54 + 9\sqrt{10}}{26}$ **23.** $-2\sqrt{35} + 2\sqrt{21}$
25. $\frac{18\sqrt{6} + 6\sqrt{15}}{13}$ **27.** $\frac{3\sqrt{2} - 3\sqrt{5} + \sqrt{10} - 5}{-3}$
29. $\frac{3 + \sqrt{21} - \sqrt{6} - \sqrt{14}}{-4}$ **31.** $\frac{\sqrt{15} + 20 - 6\sqrt{2} - 8\sqrt{30}}{-77}$
33. $\frac{6 - 5\sqrt{a} + a}{9 - a}$ **35.** $\frac{6 + 5\sqrt{x} - 6x}{9 - 4x}$ **37.** $\frac{3\sqrt{6} + 4}{2}$

39. $\dfrac{x - 2\sqrt{xy} + y}{x - y}$ **41.** 30   **42.** $-\dfrac{19}{5}$ **43.** 1   **44.** $\dfrac{x - 2}{x + 3}$
45. $-\dfrac{3\sqrt{a^2 - 3}}{a^2 - 3}$

Calculator Corner, p. 760
1. Left to the student   **2.** Left to the student

Exercise Set 10.6, p. 764
RC1. radical   **RC2.** powers   **RC3.** isolate
RC4. radicands   **RC5.** even   **CC1.** 1   **CC2.** 1
CC3. 2   **CC4.** 2
1. $\dfrac{19}{2}$ **3.** $\dfrac{49}{6}$ **5.** 57   **7.** $\dfrac{92}{5}$ **9.** -1
11. No solution   **13.** 3   **15.** 19   **17.** -6 **19.** $\dfrac{1}{64}$ **21.** 9
23. 15   **25.** 2, 5   **27.** 6   **29.** 5   **31.** 9   **33.** 7   **35.** $\dfrac{80}{9}$
37. 2, 6   **39.** -1 **41.** No solution   **43.** 3
45. About 54.7 mi   **47.** About 680 ft   **49.** About 642 ft
51. 151.25 ft; 281.25 ft   **53.** About 80.2 pounds per square inch
55. About 0.81 ft   **57.** About 3.9 ft   **59.** 2808 mi
60. Jeff: $1\tfrac{1}{3}$ hr; Grace: 4 hr   **61.** $0, -2.8$ **62.** $0, \dfrac{5}{3}$ **63.** $-8, 8$
64. $-3, \dfrac{7}{2}$ **65.** $2ah + h^2$ **66.** $2ah + h^2 - h$
67. $4ah + 2h^2 - 3h$ **68.** $4ah + 2h^2 + 3h$ **69.** 0
71. $-6, -3$ **73.** $0, \dfrac{125}{4}$ **75.** 3

Translating for Success, p. 771
1. J   **2.** B   **3.** O   **4.** M   **5.** K   **6.** I   **7.** G   **8.** E
9. F   **10.** A

Exercise Set 10.7, p. 772
RC1. Pythagorean   **RC2.** legs   **RC3.** hypotenuse
CC1. (e)   **CC2.** (c)   **CC3.** (g)   **CC4.** (d)
1. $\sqrt{34}$; 5.831   **3.** $\sqrt{450}$; 21.213   **5.** 5   **7.** $\sqrt{43}$; 6.557
9. $\sqrt{12}$; 3.464   **11.** $\sqrt{n - 1}$ **13.** About 7.066 ft
15. $\sqrt{26.35}$ in.; about 5.133 in.   **17.** $\sqrt{4959}$ ft; about 70.420 ft
19. $(3, 0), (-3, 0)$ **21.** $\sqrt{10{,}561}$ ft; 102.767 ft
23. $\sqrt{14{,}824}$ ft; 121.754 ft   **25.** $\sqrt{420.125}$ in.; 20.497 in.
27. $\sqrt{181}$ cm; 13.454 cm   **29.** $s + s\sqrt{2}$ **31.** Flash:
$67\tfrac{2}{3}$ mph; Crawler: $53\tfrac{2}{3}$ mph   **32.** $3\tfrac{3}{4}$ mph   **33.** $-7, \dfrac{3}{2}$
34. 3, 8   **35.** 1   **36.** $-2, 2$ **37.** 13   **38.** 7   **39.** 26 packets

Calculator Corner, p. 781
1. $-2 - 9i$ **2.** $20 + 17i$ **3.** $-47 - 161i$ **4.** $-\dfrac{151}{290} + \dfrac{73}{290}i$
5. -20 **6.** -28.373 **7.** $-\dfrac{16}{25} - \dfrac{1}{50}i$ **8.** 81   **9.** $117 + 118i$

Exercise Set 10.8, p. 782
RC1. True   **RC2.** True   **RC3.** False   **RC4.** False
RC5. True   **RC6.** False   **RC7.** True   **RC8.** True
1. $i\sqrt{35}$, or $\sqrt{35}i$ **3.** $4i$ **5.** $2i\sqrt{3}$, or $2\sqrt{3}i$
7. $i\sqrt{3}$, or $\sqrt{3}i$ **9.** $9i$ **11.** $7i\sqrt{2}$, or $7\sqrt{2}i$ **13.** $-7i$
15. $4 - 2i\sqrt{15}$, or $4 - 2\sqrt{15}i$ **17.** $(2 + 2\sqrt{3})i$
19. $12 - 4i$ **21.** $9 - 5i$ **23.** $7 + 4i$ **25.** $-4 - 4i$
27. $-1 + i$ **29.** $11 + 6i$ **31.** -18 **33.** $-\sqrt{14}$ **35.** 21
37. $-6 + 24i$ **39.** $1 + 5i$ **41.** $18 + 14i$ **43.** $38 + 9i$
45. $2 - 46i$ **47.** $5 - 12i$ **49.** $-24 + 10i$ **51.** $-5 - 12i$
53. $-i$ **55.** 1   **57.** -1 **59.** i **61.** -1 **63.** $-125i$
65. 8   **67.** $1 - 23i$ **69.** 0   **71.** 0   **73.** 1   **75.** $5 - 8i$
77. $2 - \dfrac{\sqrt{6}}{2}i$ **79.** $\dfrac{9}{10} + \dfrac{13}{10}i$ **81.** $-i$ **83.** $-\dfrac{3}{7} - \dfrac{8}{7}i$ **85.** $\dfrac{6}{5} - \dfrac{2}{5}i$
87. $-\dfrac{8}{41} + \dfrac{10}{41}i$ **89.** $-\dfrac{4}{3}i$ **91.** $-\dfrac{1}{2} - \dfrac{1}{4}i$ **93.** $\dfrac{3}{5} + \dfrac{4}{5}i$
95.
$$\begin{array}{c|c} x^2 - 2x + 5 = 0 & \\ \hline (1 - 2i)^2 - 2(1 - 2i) + 5 \;?\; 0 & \\ 1 - 4i + 4i^2 - 2 + 4i + 5 & \\ 1 - 4i - 4 - 2 + 4i + 5 & \\ 0 & \text{TRUE} \end{array}$$
Yes

97.
$$\begin{array}{c|c} x^2 - 4x - 5 = 0 & \\ \hline (2 + i)^2 - 4(2 + i) - 5 \;?\; 0 & \\ 4 + 4i + i^2 - 8 - 4i - 5 & \\ 4 + 4i - 1 - 8 - 4i - 5 & \\ -10 & \text{FALSE} \end{array}$$
No
99. x-intercept: $(-15, 0)$, y-intercept: $(0, 30)$
100. x-intercept: $(-24, 0)$, y-intercept: $(0, 18)$
101. x-intercept: $(2, 0)$, y-intercept: $(0, 5)$ **102.** $-\dfrac{7}{3}$ **103.** $\dfrac{1}{3}$
104. Not defined   **105.** 3.0×10^3 **106.** 2.665×10^5
107. $-4 - 8i$; $-2 + 4i$; $8 - 6i$ **109.** $-3 - 4i$ **111.** $-88i$
113. 8   **115.** $\dfrac{3}{5} + \dfrac{9}{5}i$ **117.** 1

Summary and Review: Chapter 10, p. 787

Vocabulary Reinforcement
1. cube   **2.** complex   **3.** principal   **4.** rationalizing
5. radical   **6.** imaginary   **7.** square   **8.** radicand
9. conjugate   **10.** index

Concept Reinforcement
1. True   **2.** False   **3.** False   **4.** False   **5.** True   **6.** True

Study Guide
1. $6|y|$ **2.** $|a + 2|$ **3.** $\sqrt[5]{z^3}$ **4.** $(6ab)^{5/2}$ **5.** $\dfrac{1}{9^{3/2}} = \dfrac{1}{27}$
6. $\sqrt[4]{a^3b}$ **7.** $5y\sqrt{6}$ **8.** $2\sqrt{a}$ **9.** $2\sqrt{3}$ **10.** $25 - 10\sqrt{x} + x$
11. 5   **12.** 6   **13.** $-21 - 20i$ **14.** $\dfrac{4}{5} - \dfrac{7}{5}i$

Review Exercises
1. 27.893   **2.** 6.378   **3.** $f(0), f(-1)$, and $f(1)$ do not exist as
real numbers; $f\!\left(\tfrac{41}{3}\right) = 5$ **4.** $\{x \,|\, x \geq \tfrac{16}{3}\}$, or $\left[\tfrac{16}{3}, \infty\right)$
5. $9|a|$ **6.** $7|z|$ **7.** $|6 - b|$ **8.** $|x + 3|$ **9.** -10 **10.** $-\dfrac{1}{3}$
11. $2; -2; 3$ **12.** $|x|$ **13.** 3   **14.** $\sqrt[5]{a}$ **15.** 512   **16.** $31^{1/2}$
17. $(a^2 b^3)^{1/5}$ **18.** $\dfrac{1}{7}$ **19.** $\dfrac{1}{4x^{2/3} y^{2/3}}$ **20.** $\dfrac{5b^{1/2}}{a^{3/4} c^{2/3}}$ **21.** $\dfrac{3a}{t^{1/4}}$
22. $\dfrac{1}{x^{2/5}}$ **23.** $7^{1/6}$ **24.** x^7 **25.** $3x^2$ **26.** $\sqrt[12]{x^4 y^3}$ **27.** $\sqrt[12]{x^7}$
28. $7\sqrt{5}$ **29.** $-3\sqrt[3]{4}$ **30.** $5b^2 \sqrt[3]{2a^2}$ **31.** $\dfrac{7}{6}$ **32.** $\dfrac{4x^2}{3}$
33. $\dfrac{2x^2}{3y^3}$ **34.** $\sqrt{15xy}$ **35.** $3a\sqrt[3]{a^2 b^2}$ **36.** $\sqrt[15]{a^5 b^9}$
37. $y\sqrt[3]{6}$ **38.** $\dfrac{5}{2}\sqrt{x}$ **39.** $\sqrt[12]{x^5}$ **40.** $7\sqrt[3]{x}$ **41.** $3\sqrt{3}$
42. $15\sqrt{2}$ **43.** $(2x + y^2)\sqrt[3]{x}$ **44.** $-43 - 2\sqrt{10}$
45. $8 - 2\sqrt{7}$ **46.** $9 - \sqrt[3]{4}$ **47.** $\dfrac{2\sqrt{6}}{3}$ **48.** $\dfrac{2\sqrt{a} - 2\sqrt{b}}{a - b}$
49. 4   **50.** 13   **51.** 1   **52.** About 4166 rpm   **53.** 4480 rpm
54. 9 cm   **55.** $\sqrt{24}$ ft; 4.899 ft   **56.** 25   **57.** $\sqrt{46}$; 6.782
58. $(5 + 2\sqrt{2})i$ **59.** 2   **9i**   **60.** $1 + i$ **61.** 29
62. i **63.** $9 - 12i$ **64.** $\dfrac{2}{5} + \dfrac{3}{5}i$ **65.** $\dfrac{1}{10} - \dfrac{7}{10}i$
66.   **67.** D   **68.** -1 **69.** 3

Understanding Through Discussion and Writing
1. $f(x) = (x + 5)^{1/2}(x + 7)^{-1/2}$. Consider $(x + 5)^{1/2}$.
Since the exponent is $\tfrac{1}{2}$, $x + 5$ must be nonnegative. Then
$x + 5 \geq 0$, or $x \geq -5$. Consider $(x + 7)^{-1/2}$. Since the exponent is
$-\tfrac{1}{2}$, $x + 7$ must be positive. Then $x + 7 > 0$, or $x > -7$. Then the
domain of $f = \{x \,|\, x \geq -5 \text{ and } x > -7\}$, or $\{x \,|\, x \geq -5\}$.

2. Since $\sqrt{x}$ exists only for $\{x | x \geq 0\}$, this is the domain of $y = \sqrt{x} \cdot \sqrt{x}$.   **3.** The distributive law is used to collect radical expressions with the same indexes and radicands just as it is used to collect monomials with the same variables and exponents.
4. No; when n is odd, it is true that if $a^n = b^n$, then $a = b$.
5. Use a calculator to show that $\dfrac{5 + \sqrt{2}}{\sqrt{18}} \neq 2$. Explain that we multiply by 1 to rationalize a denominator. In this case, we would write 1 as $\sqrt{2}/\sqrt{2}$.   **6.** When two radical expressions are conjugates, their product contains no radicals. Similarly, the product of a complex number and its conjugate does not contain i.

Test: Chapter 10, p. 793

1. [10.1a] 12.166   **2.** [10.1a] 2; does not exist as a real number
3. [10.1a] $\{x | x \leq 2\}$, or $(-\infty, 2]$ **4.** [10.1b] $3|q|$
5. [10.1b] $|x + 5|$ **6.** [10.1c] $-\frac{1}{10}$ **7.** [10.1d] x
8. [10.1d] 4   **9.** [10.2a] $\sqrt[3]{a^2}$ **10.** [10.2a] 8
11. [10.2a] $37^{1/2}$ **12.** [10.2a] $(5xy^2)^{5/2}$ **13.** [10.2b] $\frac{1}{10}$
14. [10.2b] $\dfrac{8a^{3/4}}{b^{3/2}c^{2/5}}$ **15.** [10.2c] $\dfrac{x^{8/5}}{y^{9/5}}$ **16.** [10.2c] $\dfrac{1}{2 \cdot 9^{31/24}}$
17. [10.2d] $\sqrt[4]{x}$ **18.** [10.2d] $2x\sqrt{x}$ **19.** [10.2d] $\sqrt[15]{a^6b^5}$
20. [10.2d] $\sqrt[12]{8y^7}$ **21.** [10.3a] $2\sqrt{37}$ **22.** [10.3a] $2\sqrt[4]{5}$
23. [10.3a] $2a^3b^4\sqrt[3]{3a^2b}$ **24.** [10.3b] $\dfrac{2x\sqrt[3]{2x^2}}{y^2}$ **25.** [10.3b] $\dfrac{5x}{6y^2}$
26. [10.3a] $\sqrt[3]{10xy^2}$ **27.** [10.3a] $xy\sqrt[4]{x}$ **28.** [10.3b] $\sqrt[5]{x^2y^2}$
29. [10.3b] $2\sqrt{a}$ **30.** [10.4a] $38\sqrt{2}$ **31.** [10.4b] -20
32. [10.4b] $9 + 6\sqrt{x} + x$ **33.** [10.5b] $\dfrac{13 + 8\sqrt{2}}{-41}$
34. [10.6a] 35   **35.** [10.6b] 7   **36.** [10.6a] 5   **37.** [10.7a] 7 ft
38. [10.6c] 3600 ft   **39.** [10.7a] $\sqrt{98}$; 9.899   **40.** [10.7a] 2
41. [10.8a] $11i$ **42.** [10.8b] $7 + 5i$ **43.** [10.8c] $37 + 9i$
44. [10.8d] $-i$ **45.** [10.8e] $-\frac{77}{50} + \frac{7}{25}i$ **46.** [10.8f] No
47. [10.6a] A   **48.** [10.8c, e] $-\frac{17}{4}i$ **49.** [10.6b] 3

Cumulative Review: Chapters 1–10, p. 795

1. [4.4a] $-3x^3 + 9x^2 + 3x - 3$ **2.** [4.6c] $4x^4 - 4x^2y + y^2$
3. [4.5d] $15x^4 - x^3 - 9x^2 + 5x - 2$ **4.** [6.1d] $\dfrac{(x + 4)(x - 7)}{x + 7}$
5. [6.2b] $\dfrac{y - 6}{y - 9}$ **6.** [6.5b] $\dfrac{-2x + 4}{(x + 2)(x - 3)}$, or $\dfrac{-2(x - 2)}{(x + 2)(x - 3)}$
7. [4.8b] $y^2 + y - 2 + \dfrac{-1}{y + 2}$ **8.** [10.1c] $-2x$
9. [10.1b], [10.3a] $4(x - 1)$ **10.** [10.4a] $57\sqrt{3}$
11. [10.3a] $4xy^2\sqrt{y}$ **12.** [10.5b] $\sqrt{30} + \sqrt{15}$
13. [10.1d], [10.3b] $\dfrac{m^2n^4}{2}$ **14.** [10.2c] $6^{8/9}$
15. [10.8b] $3 + 5i$ **16.** [10.8e] $\frac{7}{61} - \frac{16}{61}i$
17. [2.3a] 2   **18.** [2.4b] $c = 8M + 3$ **19.** [9.1c] $\{a | a > -7\}$, or $(-7, \infty)$ **20.** [9.2a] $\{x | -10 < x < 13\}$, or $(-10, 13)$
21. [9.3c] $\frac{4}{3}, \frac{8}{3}$ **22.** [5.8b] $\frac{25}{7}, -\frac{25}{7}$ **23.** [8.3a] $(5, 3)$
24. [8.5a] $(-1, 0, 4)$ **25.** [6.7a] -5 **26.** [6.7a] $\frac{1}{3}$
27. [2.4b], [6.7a] $R = \dfrac{nE - nrI}{I}$ **28.** [10.6a] 6
29. [10.6b] $-\frac{1}{4}$ **30.** [10.6a] 5
31. [7.1c]   **32.** [7.4a]   **33.** [9.4b]

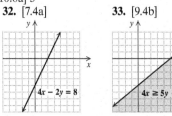

34. [9.4c]   **35.** [7.1c]   **36.** [7.1c]

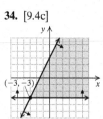 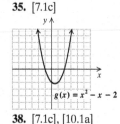 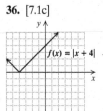

37. [7.1c]   **38.** [7.1c], [10.1a]

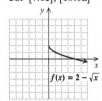

39. [5.1b] $6xy^2(2x - 5y)$ **40.** [5.3a], [5.4a], $(3x + 4)(x - 7)$
41. [5.2a] $(y + 11)(y - 12)$ **42.** [5.6a] $(3y + 2)(9y^2 - 6y + 4)$
43. [5.5d] $(2x + 25)(2x - 25)$ **44.** [7.2a] Domain: $[-5, 5]$; range: $[-3, 4]$ **45.** [7.2a] Domain: $(-\infty, \infty)$; range: $[-5, \infty)$
46. [7.3b] Slope: $\frac{3}{2}$; y-intercept: $(0, -4)$ **47.** [7.5d] $y = -\frac{1}{3}x + \frac{13}{3}$
48. [6.9d] 125 ft; 1000 ft^2 **49.** [6.8a] 1 hr   **50.** [6.9f] 64 L
51. [10.2a] D   **52.** [6.8a] A   **53.** [4.8b] A   **54.** [10.6a] B
55. [10.6b] $-\frac{8}{9}$

CHAPTER 11

Calculator Corner, p. 802

The calculator returns an ERROR message because the graph of $y = 4x^2 + 9$ has no x-intercepts. This indicates that the equation $4x^2 + 9 = 0$ has no real-number solutions.

Exercise Set 11.1, p. 807

RC1. True   **RC2.** True   **RC3.** True   **RC4.** True
RC5. False   **RC6.** False   **CC1.** 25; 5   **CC2.** 36; 6
CC3. $\frac{25}{4}; \frac{5}{2}$ **CC4.** $\frac{1}{4}; \frac{1}{2}$ **CC5.** $\frac{1}{25}; \frac{1}{5}$ **CC6.** $\frac{9}{16}; \frac{3}{4}$
1. (a) $\sqrt{5}, -\sqrt{5}$, or $\pm\sqrt{5}$; **(b)** $(-\sqrt{5}, 0), (\sqrt{5}, 0)$
3. (a) $\frac{5}{3}i, -\frac{5}{3}i$, or $\pm\frac{5}{3}i$; **(b)** no x-intercepts
5. $\pm\dfrac{\sqrt{6}}{2}$; ± 1.225 **7.** $5, -9$ **9.** $8, 0$
11. $11 \pm \sqrt{7}$; 13.646, 8.354   **13.** $7 \pm 2i$ **15.** $18, 0$
17. $\dfrac{3}{2} \pm \dfrac{\sqrt{14}}{2}$; 3.371, -0.371 **19.** $5, -11$ **21.** $9, 5$
23. $-2 \pm \sqrt{6}$ **25.** $11 \pm 2\sqrt{33}$ **27.** $-\dfrac{1}{2} \pm \dfrac{\sqrt{5}}{2}$
29. $\dfrac{5}{2} \pm \dfrac{\sqrt{53}}{2}$ **31.** $-\dfrac{3}{4} \pm \dfrac{\sqrt{57}}{4}$ **33.** $\dfrac{9}{4} \pm \dfrac{\sqrt{105}}{4}$
35. $2, -8$ **37.** $-11 \pm \sqrt{19}$ **39.** $5 \pm \sqrt{29}$
41. (a) $-\dfrac{7}{2} \pm \dfrac{\sqrt{57}}{2}$; **(b)** $\left(-\dfrac{7}{2} - \dfrac{\sqrt{57}}{2}, 0\right), \left(-\dfrac{7}{2} + \dfrac{\sqrt{57}}{2}, 0\right)$
43. (a) $\dfrac{5}{4} \pm \dfrac{\sqrt{39}}{4}i$; **(b)** no x-intercepts
45. $\dfrac{3}{4} \pm \dfrac{\sqrt{17}}{4}$ **47.** $\dfrac{3}{4} \pm \dfrac{\sqrt{145}}{4}$ **49.** $\dfrac{2}{3} \pm \dfrac{\sqrt{7}}{3}$
51. $-\dfrac{1}{2} \pm \dfrac{\sqrt{7}}{2}i$ **53.** $2 \pm 3i$ **55.** About 0.890 sec
57. About 5.3 sec   **59.** About 4.9 sec   **61.** About 5.9 sec
63. (a) $R(t) = -\frac{2}{15}t + 128$, where t is the number of years after 1981; **(b)** about 122.8 min; **(c)** 2026

64. 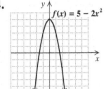 **65.**

66. 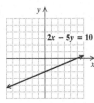 **67.**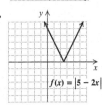

68. $2\sqrt{22}$ **69.** $2x$ **70.** $\dfrac{t^2 + 2t + 4}{t - 3}$ **71.** $\dfrac{2(2x - 5)}{3(x - 1)}$ **72.** 6

73. $\dfrac{t^2}{(t + 1)^2(t - 1)}$ **75.** Left to the student

77. $16, -16$ **79.** $0, \tfrac{7}{2}, -8, -\tfrac{10}{3}$

Calculator Corner, p. 813

1. $-3, 0.8$ **2.** $-1.5, 5$ **3.** $3, 8$ **4.** $2, 4$

Calculator Corner, p. 814

1.–3. Left to the student

Exercise Set 11.2, p. 816

RC1. True **RC2.** False **RC3.** False **RC4.** True
CC1. 3 **CC2.** -1 **CC3.** $5x^2 + x - 9$ **CC4.** 0
1. $-4 \pm \sqrt{14}$ **3.** $\dfrac{-4 \pm \sqrt{13}}{3}$ **5.** $\dfrac{1}{2} \pm \dfrac{\sqrt{3}}{2}i$
7. $2 \pm 3i$ **9.** $\dfrac{-3 \pm \sqrt{41}}{2}$ **11.** $-1 \pm 2i$
13. (a) $0, -1$; (b) $(0, 0), (-1, 0)$
15. (a) $\dfrac{3 \pm \sqrt{229}}{22}$; (b) $\left(\dfrac{3 + \sqrt{229}}{22}, 0\right), \left(\dfrac{3 - \sqrt{229}}{22}, 0\right)$
17. (a) $\tfrac{2}{5}$; (b) $\left(\tfrac{2}{5}, 0\right)$ **19.** $-1, -2$ **21.** $5, 10$
23. $\dfrac{17 \pm \sqrt{249}}{10}$ **25.** $2 \pm i$ **27.** $\tfrac{2}{3}, \tfrac{3}{2}$ **29.** $2 \pm \sqrt{10}$
31. $\tfrac{3}{4}, -2$ **33.** $\tfrac{1}{2} \pm \tfrac{3}{2}i$ **35.** $1, -\dfrac{1}{2} \pm \dfrac{\sqrt{3}}{2}i$
37. $-3 \pm \sqrt{5}$; $-0.764, -5.236$ **39.** $3 \pm \sqrt{5}$; $5.236, 0.764$
41. $\dfrac{3 \pm \sqrt{65}}{4}$; $2.766, -1.266$ **43.** $\dfrac{4 \pm \sqrt{31}}{5}$; $1.914, -0.314$
45. 2 **46.** 8 **47.** No solution **48.** $\tfrac{15}{4}$ **49.** $-1, \tfrac{3}{2}$ **50.** $\tfrac{1}{10}$
51. -2 **52.** $-\tfrac{10}{13}$ **53.** Left to the student; $-0.797, 0.570$
55. $\dfrac{1 \pm \sqrt{1 + 8\sqrt{5}}}{4}$ **57.** $\dfrac{-i \pm i\sqrt{1 + 4i}}{2}$
59. $\dfrac{-1 \pm 3\sqrt{5}}{6}$ **61.** $3 \pm \sqrt{13}$

Translating for Success, p. 825

1. B **2.** G **3.** F **4.** L **5.** N
6. C **7.** J **8.** E **9.** K **10.** A

Exercise Set 11.3, p. 826

RC1. (f) **RC2.** (e) **RC3.** (b)
RC4. (a) **RC5.** (d) **RC6.** (c)
1. Length: 9 ft; width: 2 ft **3.** Length: 18 yd; width: 9 yd
5. Base: 15 mi; height: 8 mi

7. Length: $\dfrac{51 + \sqrt{122{,}399}}{2}$ ft; width: $\dfrac{\sqrt{122{,}399} - 51}{2}$ ft
9. 2 in. **11.** 6 ft, 8 ft **13.** 23 and 24
15. Length: $(2 + \sqrt{14})$ ft ≈ 5.742 ft;
width: $(\sqrt{14} - 2)$ ft ≈ 1.742 ft
17. $\dfrac{17 - \sqrt{109}}{2}$ in. ≈ 3.280 in.
19. $(7 + \sqrt{239})$ ft ≈ 22.460 ft; $(\sqrt{239} - 7)$ ft ≈ 8.460 ft
21. First part: 60 mph; second part: 50 mph
23. 6 km/h **25.** Cessna: 150 mph; Beechcraft: 200 mph;
or Cessna: 200 mph; Beechcraft: 250 mph
27. To Hillsboro: 10 mph; return trip: 4 mph
29. About 11 mph **31.** $s = \sqrt{\dfrac{A}{6}}$ **33.** $r = \sqrt{\dfrac{Gm_1m_2}{F}}$
35. $c = \sqrt{\dfrac{E}{m}}$ **37.** $b = \sqrt{c^2 - a^2}$ **39.** $k = \dfrac{3 + \sqrt{9 + 8N}}{2}$
41. $r = \dfrac{-\pi h + \sqrt{\pi^2 h^2 + 2\pi A}}{2\pi}$ **43.** $g = \dfrac{4\pi^2 L}{T^2}$
45. $H = \sqrt{\dfrac{703W}{I}}$ **47.** $v = \dfrac{c\sqrt{m^2 - (m_0)^2}}{m}$ **49.** $\dfrac{1}{x - 2}$
50. $\dfrac{(x + 1)(x^2 + 2)}{(x - 1)(x^2 + x + 1)}$ **51.** $\dfrac{-x}{(x + 3)(x - 1)}$ **52.** $3x^2\sqrt{x}$
53. $2\sqrt{5}i$ **55.** $\pm\sqrt{2}$ **57.** $A(S) = \dfrac{\pi S}{6}$ **59.** $l = \dfrac{w + w\sqrt{5}}{2}$

Exercise Set 11.4, p. 836

RC1. (b) **RC2.** (a) **RC3.** (c)
RC4. (b) **RC5.** (b) **RC6.** (c)
CC1. $\sqrt{x}$ **CC2.** p^2 **CC3.** x^{-1} **CC4.** $2y + 7$
1. One real **3.** Two nonreal **5.** Two real **7.** One real
9. Two nonreal **11.** Two real **13.** Two real **15.** One real
17. $x^2 - 16 = 0$ **19.** $x^2 + 16 = 0$ **21.** $x^2 - 16x + 64 = 0$
23. $25x^2 - 20x - 12 = 0$ **25.** $12x^2 - (4k + 3m)x + km = 0$
27. $x^2 - \sqrt{3}x - 6 = 0$ **29.** $x^2 + 36 = 0$ **31.** $\pm\sqrt{3}$
33. $1, 81$ **35.** $-1, 1, 5, 7$ **37.** $-\tfrac{1}{4}, \tfrac{1}{9}$ **39.** 1 **41.** $-1, 1, 4, 6$
43. $\pm 2, \pm 5$ **45.** $-1, 2$ **47.** $\pm\dfrac{\sqrt{15}}{3}, \pm\dfrac{\sqrt{6}}{2}$
49. $-1, 125$ **51.** $-\tfrac{11}{6}, -\tfrac{1}{6}$ **53.** $-\tfrac{3}{2}$ **55.** $\dfrac{9 \pm \sqrt{89}}{2}, -1 \pm \sqrt{3}$
57. $\left(\tfrac{4}{25}, 0\right)$ **59.** $(4, 0), (-1, 0), \left(\dfrac{3 + \sqrt{33}}{2}, 0\right), \left(\dfrac{3 - \sqrt{33}}{2}, 0\right)$
61. $(-8, 0), (1, 0)$ **63.** Kenyan: 30 lb; Peruvian: 20 lb
64. Solution A: 4 L; solution B: 8 L **65.** $4x$ **66.** $3x^2$
67. $3a\sqrt[4]{2a}$ **68.** 4
69. 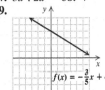 **70.**

71.   **72.**

73. Left to the student **75.** (a) $-\tfrac{3}{5}$; (b) $-\tfrac{1}{3}$
77. $x^2 - \sqrt{3}x + 8 = 0$ **79.** $a = 1, b = 2, c = -3$
81. $\tfrac{100}{99}$ **83.** 259 **85.** $1, 3, -\dfrac{1}{2} \pm \dfrac{\sqrt{3}}{2}i, -\dfrac{3}{2} \pm \dfrac{3\sqrt{3}}{2}i$

Mid-Chapter Review: Chapter 11, p. 840

1. False 2. True 3. True 4. False

5.
$$5x^2 + 3x = 4$$
$$\frac{1}{5}(5x^2 + 3x) = \frac{1}{5} \cdot 4$$
$$x^2 + \frac{3}{5}x = \frac{4}{5}$$
$$x^2 + \frac{3}{5}x + \frac{9}{100} = \frac{4}{5} + \frac{9}{100}$$
$$\left(x + \frac{3}{10}\right)^2 = \frac{89}{100}$$
$$x + \frac{3}{10} = \sqrt{\frac{89}{100}} \quad \text{or} \quad x + \frac{3}{10} = -\sqrt{\frac{89}{100}}$$
$$x + \frac{3}{10} = \frac{\sqrt{89}}{10} \quad \text{or} \quad x + \frac{3}{10} = -\frac{\sqrt{89}}{10}$$
$$x = -\frac{3}{10} + \frac{\sqrt{89}}{10} \quad \text{or} \quad x = -\frac{3}{10} - \frac{\sqrt{89}}{10}$$
The solutions are $-\frac{3}{10} \pm \frac{\sqrt{89}}{10}$.

6.
$$5x^2 + 3x = 4$$
$$5x^2 + 3x - 4 = 0$$
$$a = 5, \ b = 3, \ c = -4$$
$$x = \frac{-b \pm \sqrt{b^2 - 4ac}}{2a}$$
$$x = \frac{-3 \pm \sqrt{3^2 - 4 \cdot 5 \cdot (-4)}}{2 \cdot 5}$$
$$x = \frac{-3 \pm \sqrt{9 + 80}}{10}$$
$$x = \frac{-3 \pm \sqrt{89}}{10}$$
$$x = -\frac{3}{10} \pm \frac{\sqrt{89}}{10}$$

7. $-2 \pm \sqrt{3}$ 8. $-3, \frac{1}{2}$ 9. $-5 \pm \sqrt{31}$ 10. $\frac{1}{2} \pm \frac{\sqrt{21}}{2}$

11. One real solution; one x-intercept
12. Two real solutions; two x-intercepts
13. Two nonreal solutions; no x-intercepts
14. Two nonreal solutions; no x-intercepts
15. Two real solutions; two x-intercepts
16. Two real solutions; two x-intercepts
17. $x^2 - 9x - 10 = 0$ 18. $x^2 - 169 = 0$
19. $x^2 - 2\sqrt{5}x - 15 = 0$ 20. $x^2 + 16 = 0$
21. $x^2 + 12x + 36 = 0$ 22. $21x^2 + 22x - 8 = 0$
23. 60 mph 24. $s = \sqrt{\frac{R}{a}}$ 25. $-\frac{4}{3}, 1$ 26. $\pm\sqrt{3}, \pm\sqrt{5}$
27. $\frac{15 \pm \sqrt{145}}{8}$ 28. $-1, -\frac{2}{7}$ 29. $-1, 0$ 30. $-11, 5$
31. $\pm\frac{4}{7}i$ 32. $\pm\sqrt{6}, \pm 2i$ 33. $\frac{-5 \pm \sqrt{73}}{2}$ 34. $-6 \pm i$
35. $\frac{5}{2} \pm \frac{\sqrt{11}}{2}$ 36. $\frac{7 \pm \sqrt{13}}{6}$ 37. $\frac{1 \pm \sqrt{2}}{2}$ 38. $-1 \pm 4i$
39. $8 \pm \sqrt{3}$ 40. $3 \pm \sqrt{10}i$ 41. $4 \pm \sqrt{26}$ 42. 9

43. Given the solutions of a quadratic equation, it is possible to find an equation equivalent to the original equation but not necessarily expressed in the same form as the original equation. For example, we can find a quadratic equation with solutions -2 and 4:
$$[x - (-2)](x - 4) = 0$$
$$(x + 2)(x - 4) = 0$$
$$x^2 - 2x - 8 = 0.$$
Now $x^2 - 2x - 8 = 0$ has solutions -2 and 4. However, the original equation might have been in another form, such as $2x(x - 3) - x(x - 4) = 8$.

44. Given the quadratic equation $ax^2 + bx + c = 0$, we find
$$x = \frac{-b + \sqrt{b^2 - 4ac}}{2a} \quad \text{or} \quad x = \frac{-b - \sqrt{b^2 - 4ac}}{2a} \text{ using the}$$
quadratic formula.
Then we have $ax^2 + bx + c =$
$$\left(x - \frac{-b + \sqrt{b^2 - 4ac}}{2a}\right)\left(x - \frac{-b - \sqrt{b^2 - 4ac}}{2a}\right).$$
Consider $5x^2 + 8x - 3$. First, we use the quadratic formula to solve $5x^2 + 8x - 3 = 0$:
$$x = \frac{-8 \pm \sqrt{8^2 - 4 \cdot 5 \cdot (-3)}}{2 \cdot 5}$$
$$x = \frac{-8 \pm \sqrt{124}}{10} = \frac{-8 \pm 2\sqrt{31}}{10}$$
$$x = \frac{-4 \pm \sqrt{31}}{5}.$$
Then $5x^2 + 8x - 3 = \left(x - \frac{-4 - \sqrt{31}}{5}\right)\left(x - \frac{-4 + \sqrt{31}}{5}\right)$.

45. Set the product
$(x - 1)(x - 2)(x - 3)(x - 4)(x - 5)(x - 6)(x - 7)$
equal to 0. 46. Write an equation of the form
$a(3x^2 + 1)^2 + b(3x^2 + 1) + c = 0$, where $a \neq 0$. To ensure that this equation has real-number solutions, select a, b, and c so that $b^2 - 4ac \geq 0$ and $3x^2 + 1 \geq 0$.

Exercise Set 11.5, p. 847

RC1. False **RC2.** True **RC3.** True **RC4.** False
CC1. Down **CC2.** Up **CC3.** Up **CC4.** Down

1.
x	$f(x)$
0	0
1	4
2	16
-1	4
-2	16

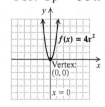

3.
x	$f(x)$
0	0
1	$\frac{1}{3}$
2	$\frac{4}{3}$
-1	$\frac{1}{3}$
-2	$\frac{4}{3}$

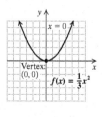

5.
x	$f(x)$
-3	0
-2	1
-1	4
-4	1
-5	4

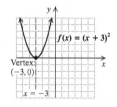

7.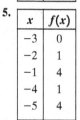

9.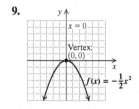

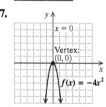

A-36   Answers

11.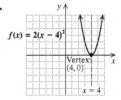

13.

x	f(x)
−2	0
−3	−2
−1	−2
−4	−8
0	−8

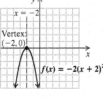

15. 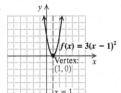 **17.**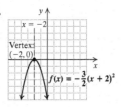

19. 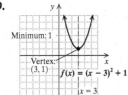 **21.**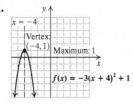

23. 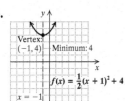 **25.**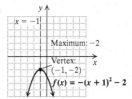

27. $-\frac{2}{5}$ **28.** 0

Visualizing for Success, p. 856
1. F **2.** H **3.** A **4.** I **5.** C **6.** J **7.** G **8.** B
9. E **10.** D

Exercise Set 11.6, p. 857
RC1. upward **RC2.** minimum **RC3.** vertex
RC4. y-intercept
1. (a) $(1, -4)$; **(b)** $x = 1$; **(c)** minimum: -4;
(d)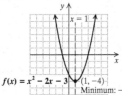

3. (a) $(-2, 2)$; **(b)** $x = -2$; **(c)** maximum: 2;
(d)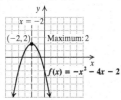

5. (a) $(4, 2)$; **(b)** $x = 4$; **(c)** minimum: 2;
(d)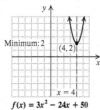

7. (a) $\left(-\frac{1}{2}, \frac{7}{2}\right)$; **(b)** $x = -\frac{1}{2}$; **(c)** maximum: $\frac{7}{2}$;
(d)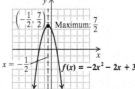

9. (a) $(0, 5)$; **(b)** $x = 0$; **(c)** maximum: 5;
(d)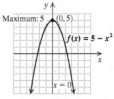

11. (a) $\left(-\frac{5}{4}, -\frac{41}{8}\right)$; **(b)** $x = -\frac{5}{4}$; **(c)** minimum: $-\frac{41}{8}$;
(d)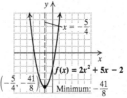

13. y-intercept: $(0, 1)$; x-intercepts: $(3 + 2\sqrt{2}, 0), (3 - 2\sqrt{2}, 0)$
15. y-intercept: $(0, 20)$; x-intercepts: $(5, 0), (-4, 0)$
17. y-intercept: $(0, 9)$; x-intercept: $\left(-\frac{3}{2}, 0\right)$
19. y-intercept: $(0, 8)$; x-intercepts: none
21. $D = 15w$ **22.** $C = \frac{89}{6}t$ **23.** 250; $y = \frac{250}{x}$
24. 250; $y = \frac{250}{x}$ **25.** $\frac{125}{2}$; $y = \frac{125}{2}x$ **26.** $\frac{2}{125}$; $y = \frac{2}{125}x$
27. (a) Minimum: -6.954; **(b)** maximum: 7.014
29. 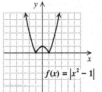 **31.**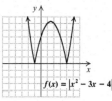

33. $f(x) = \frac{5}{16}x^2 - \frac{15}{8}x - \frac{35}{16}$, or $f(x) = \frac{5}{16}(x - 3)^2 - 5$
35.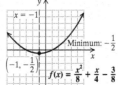

Calculator Corner, p. 862
1. Minimum: 1 **2.** Minimum: 4.875
3. Maximum: 6 **4.** Maximum: 0.5625

Exercise Set 11.7, p. 866

RC1. True **RC2.** False **RC3.** True **RC4.** False
CC1. $4 = 4a + 2b + c$ **CC2.** $0 = 25a + 5b + c$
CC3. $10 = 36a - 6b + c$
1. 180 ft by 180 ft **3.** 3.5 in. **5.** 3.5 hundred, or 350 bicycles
7. 200 ft^2; 10 ft by 20 ft **9.** 11 days after the concert was announced; about 62 tickets **11.** $P(x) = -x^2 + 980x - 3000$; $237,100 at $x = 490$ **13.** 121; 11 and 11 **15.** -4; 2 and -2
17. 36; -6 and -6 **19.** $f(x) = ax^2 + bx + c, a < 0$
21. $f(x) = ax^2 + bx + c, a > 0$ **23.** Polynomial, neither quadratic nor linear **25.** $f(x) = mx + b$
27. $f(x) = 2x^2 + 3x - 1$ **29.** $f(x) = -\frac{1}{4}x^2 + 3x - 5$
31. (a) $A(s) = \frac{3}{16}s^2 - \frac{135}{4}s + 1750$; (b) about 531 per 200,000,000 kilometers driven **33.** $D(x) = -0.008x^2 + 0.8x$; 15 ft **35.** $H(x) = -\frac{1}{32}x^2 + \frac{5}{8}x + 65$, where x is the number of years after 1995; 68.125% **37.** $-2x^2 + 2x - 9$ **38.** $-4m^2n$
39. $c^4d^2 - 4y^2$ **40.** $(10t + 9)(10t - 9)$ **41.** $3x(2x - 5)^2$
42. $(3y - 4)(2y + 3)$

Exercise Set 11.8, p. 876

RC1. negative **RC2.** positive
RC3. positive **RC4.** negative
CC1. Yes **CC2.** No **CC3.** No **CC4.** Yes
1. $\{x \mid x < -2 \text{ or } x > 6\}$, or $(-\infty, -2) \cup (6, \infty)$
3. $\{x \mid -2 \leq x \leq 2\}$, or $[-2, 2]$ **5.** $\{x \mid -1 \leq x \leq 4\}$, or $[-1, 4]$
7. $\{x \mid -1 < x < 2\}$, or $(-1, 2)$ **9.** All real numbers, or $(-\infty, \infty)$ **11.** $\{x \mid 2 < x < 4\}$, or $(2, 4)$
13. $\{x \mid x < -2 \text{ or } 0 < x < 2\}$, or $(-\infty, -2) \cup (0, 2)$
15. $\{x \mid -9 < x < -1 \text{ or } x > 4\}$, or $(-9, -1) \cup (4, \infty)$
17. $\{x \mid x < -3 \text{ or } -2 < x < 1\}$, or $(-\infty, -3) \cup (-2, 1)$
19. $\{x \mid x < 6\}$, or $(-\infty, 6)$ **21.** $\{x \mid x < -1 \text{ or } x > 3\}$, or $(-\infty, -1) \cup (3, \infty)$ **23.** $\{x \mid -\frac{2}{3} \leq x < 3\}$, or $[-\frac{2}{3}, 3)$
25. $\{x \mid 2 < x < \frac{5}{2}\}$, or $(2, \frac{5}{2})$ **27.** $\{x \mid x < -1 \text{ or } 2 < x < 5\}$, or $(-\infty, -1) \cup (2, 5)$ **29.** $\{x \mid -3 \leq x < 0\}$, or $[-3, 0)$
31. $\{x \mid 1 < x < 2\}$, or $(1, 2)$ **33.** $\{x \mid x < -4 \text{ or } 1 < x < 3\}$, or $(-\infty, -4) \cup (1, 3)$ **35.** $\{x \mid 0 < x < \frac{1}{3}\}$, or $(0, \frac{1}{3})$
37. $\{x \mid x < -3 \text{ or } -2 < x < 1 \text{ or } x > 4\}$, or $(-\infty, -3) \cup (-2, 1) \cup (4, \infty)$ **39.** $\frac{5}{3}$ **40.** $\frac{5}{2a}$ **41.** $\frac{4a}{b^2}\sqrt{a}$
42. $\frac{3c}{7d}\sqrt[3]{c^2}$ **43.** $\sqrt{2}$ **44.** $17\sqrt{5}$ **45.** $(10a + 7)\sqrt[3]{2a}$
46. $3\sqrt{10} - 4\sqrt{5}$ **47.** Left to the student
49. $\{x \mid 1 - \sqrt{3} \leq x \leq 1 + \sqrt{3}\}$, or $[1 - \sqrt{3}, 1 + \sqrt{3}]$
51. All real numbers except 0, or $(-\infty, 0) \cup (0, \infty)$
53. $\{x \mid x < \frac{1}{4} \text{ or } x > \frac{5}{2}\}$, or $(-\infty, \frac{1}{4}) \cup (\frac{5}{2}, \infty)$
55. (a) $\{t \mid 0 < t < 2\}$, or $(0, 2)$; (b) $\{t \mid t > 10\}$, or $(10, \infty)$

Summary and Review: Chapter 11, p. 879

Vocabulary Reinforcement
1. quadratic **2.** rational **3.** complete **4.** discriminant
5. quadratic **6.** parabola **7.** symmetry **8.** vertex

Concept Reinforcement
1. False **2.** True **3.** False

Study Guide
1. $2 \pm 3i$ **2.** $6 \pm \sqrt{5}$ **3.** $5 \pm \sqrt{2}$, or 6.414 and 3.586
4. (a) Two real solutions; (b) two nonreal solutions

5. $5x^2 - 13x - 6 = 0$ **6.** $\pm\sqrt{2}, \pm 3$ **7.** Vertex: $(-1, -2)$; line of symmetry: $x = -1$; maximum: -2;

8. y-intercept: $(0, 4)$; x-intercepts: $(3 - \sqrt{5}, 0), (3 + \sqrt{5}, 0)$
9. $\{x \mid x < 4 \text{ or } x > 10\}$, or $(-\infty, 4) \cup (10, \infty)$
10. $\{x \mid 5 < x \leq 11\}$, or $(5, 11]$

Review Exercises
1. (a) $\pm\frac{\sqrt{14}}{2}$; (b) $\left(-\frac{\sqrt{14}}{2}, 0\right), \left(\frac{\sqrt{14}}{2}, 0\right)$ **2.** $0, -\frac{5}{14}$
3. 3, 9 **4.** $\frac{7}{2} \pm \frac{\sqrt{3}}{2}i$ **5.** $\frac{-3 \pm \sqrt{13}}{4}$ **6.** 3, 5
7. $-2 \pm \sqrt{3}$; $-0.268, -3.732$ **8.** 4, -2 **9.** $4 \pm 4\sqrt{2}$
10. $\frac{1 \pm \sqrt{481}}{15}$ **11.** $-3 \pm \sqrt{7}$ **12.** 0.901 sec
13. Length: 14 cm; width: 9 cm **14.** 1 in. **15.** First part: 50 mph; second part: 40 mph **16.** Two real **17.** Two nonreal
18. $25x^2 + 10x - 3 = 0$ **19.** $x^2 + 8x + 16 = 0$
20. $p = \frac{9\pi^2}{N^2}$ **21.** $T = \sqrt{\frac{3B}{2A}}$ **22.** $2, -2, 3, -3$
23. 3, -5 **24.** $\pm\sqrt{7}, \pm\sqrt{2}$ **25.** 81, 16
26. (a) $(1, 3)$; (b) $x = 1$; (c) maximum: 3;
(d)

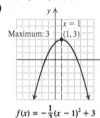

27. (a) $\left(\frac{1}{2}, \frac{23}{4}\right)$; (b) $x = \frac{1}{2}$; (c) minimum: $\frac{23}{4}$;
(d)

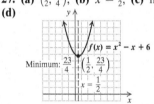

28. (a) $(-2, 4)$; (b) $x = -2$; (c) maximum: 4;
(d)

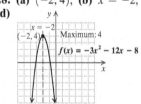

29. y-intercept: $(0, 14)$; x-intercepts: $(2, 0), (7, 0)$
30. y-intercept: $(0, -3)$; x-intercepts: $(2 - \sqrt{7}, 0)$ and $(2 + \sqrt{7}, 0)$ **31.** -121; 11 and -11
32. $f(x) = -x^2 + 6x - 2$ **33.** $\{x \mid -2 < x < 1 \text{ or } x > 2\}$, or $(-2, 1) \cup (2, \infty)$ **34.** $\{x \mid x < -4 \text{ or } -2 < x < 1\}$, or $(-\infty, -4) \cup (-2, 1)$ **35.** (a) $f(x) = \frac{37}{12}x^2 + \frac{121}{6}x + 170$; (b) $462.25 million **36.** B **37.** D
38. $b = 19$ cm, $h = 19$ cm; $A = 180.5$ cm^2 **39.** 18 and 324

Understanding Through Discussion and Writing

1. Yes; for any quadratic function $f(x) = ax^2 + bx + c$, $f(0) = c$, so the graph of every quadratic function has a y-intercept, $(0, c)$.   **2.** If the leading coefficient is positive, the graph of the function opens up and hence has a minimum value. If the leading coefficient is negative, the graph of the function opens down and hence has a maximum value.   **3.** When an input of $y = (x + 3)^2$ is 3 less than (or 3 units to the left of) an input of $y = x^2$, the outputs are the same. In addition, for any input, the output of $f(x) = (x + 3)^2 - 4$ is 4 less than (or 4 units down from) the output of $f(x) = (x + 3)^2$. Thus the graph of $f(x) = (x + 3)^2 - 4$ looks like the graph of $f(x) = x^2$ translated 3 units to the left and 4 units down.   **4.** Find a quadratic function $f(x)$ whose graph lies entirely above the x-axis or a quadratic function $g(x)$ whose graph lies entirely below the x-axis. Then write $f(x) < 0, f(x) \le 0, g(x) > 0$, or $g(x) \ge 0$. For example, the quadratic inequalities $x^2 + 1 < 0$ and $-x^2 - 5 \ge 0$ have no solution.   **5.** No; if the vertex is off the x-axis, then due to symmetry, the graph has either no x-intercept or two x-intercepts.   **6.** The x-coordinate of the vertex lies halfway between the x-coordinates of the x-intercepts. The function must be evaluated for this value of x in order to determine the y-coordinate of the vertex.

Test: Chapter 11, p. 885

1. [11.1a] (a) $\pm \frac{2\sqrt{3}}{3}$; (b) $\left(\frac{2\sqrt{3}}{3}, 0\right), \left(-\frac{2\sqrt{3}}{3}, 0\right)$
2. [11.2a] $-\frac{1}{2} \pm \frac{\sqrt{3}}{2}i$ **3.** [11.4c] 49, 1   **4.** [11.2a] 9, 2
5. [11.4c] $\pm \frac{\sqrt{5}}{2}, \pm \sqrt{3}$ **6.** [11.2a] $-2 \pm \sqrt{6}$; 0.449, -4.449
7. [11.2a] 0, 2   **8.** [11.1b] $2 \pm \sqrt{3}$ **9.** [11.1c] About 6.7 sec
10. [11.3a] About 2.89 mph   **11.** [11.7a] 7 cm by 7 cm
12. [11.1c] About 0.946 sec   **13.** [11.4a] Two nonreal
14. [11.4b] $x^2 - 4\sqrt{3}x + 9 = 0$ **15.** [11.3b] $T = \sqrt{\frac{V}{48}}$, or $\frac{\sqrt{3V}}{12}$
16. [11.6a] (a) $(-1, 1)$; (b) $x = -1$; (c) maximum: 1;
(d)
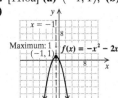
17. [11.6a] (a) $(3, 5)$; (b) $x = 3$; (c) minimum: 5;
(d)
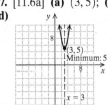
18. [11.6b] y-intercept: $(0, -1)$; x-intercepts: $(2 - \sqrt{3}, 0)$, $(2 + \sqrt{3}, 0)$ **19.** [11.7a] -16; 4 and -4
20. [11.7b] $f(x) = \frac{1}{5}x^2 - \frac{3}{5}x$
21. [11.7b] (a) $P(d) = \frac{1}{64}d^2 + \frac{5}{16}d + \frac{13}{2}$; (b) $13.94
22. [11.8a] $\{x | -1 < x < 7\}$, or $(-1, 7)$
23. [11.8b] $\{x | -3 < x < 5\}$, or $(-3, 5)$
24. [11.8b] $\{x | -3 < x < 1 \text{ or } x \ge 2\}$, or $(-3, 1) \cup [2, \infty)$
25. [11.4b] C
26. [11.6a, b] $f(x) = -\frac{4}{7}x^2 + \frac{20}{7}x + 8$; maximum: $\frac{81}{7}$
27. [11.2a] $\frac{1}{2}$

Cumulative Review: Chapters 1–11, p. 887

1. [10.7a] About 422 yd   **2.** [4.4c] $10x^2 - 8x + 6$
3. [4.5d] $2x^3 - 9x^2 + 7x - 12$ **4.** [6.1d] $\frac{2(a - 4)}{5}$
5. [6.2b] $\frac{1}{y(y + 6)}$ **6.** [6.5b] $\frac{-(m - 3)(m - 2)}{(m + 1)(m - 5)}$
7. [4.8b] $9x^2 - 13x + 26 + \frac{-50}{x + 2}$ **8.** [6.6a] $\frac{y - x}{xy(x + y)}$
9. [10.1b] 0.6   **10.** [10.1b] $3(x - 2)$
11. [10.4a] $12\sqrt{5}$ **12.** [10.5b] $\frac{\sqrt{6} + 9\sqrt{2} - 12\sqrt{3} - 4}{-26}$
13. [10.2d] 256   **14.** [10.8c] $17 + 7i$ **15.** [10.8e] $-\frac{2}{3} - 2i$
16. [5.3a], [5.4a] $(2t + 5)(t - 6)$ **17.** [5.2a] $(a + 9)(a - 6)$
18. [5.1b] $-3a^2(a - 4)$ **19.** [5.5d] $(8a + 3b)(8a - 3b)$
20. [5.5b] $3(a - 6)^2$ **21.** [5.6a] $(\frac{1}{3}a - 1)(\frac{1}{9}a^2 + \frac{1}{3}a + 1)$
22. [5.1c] $(4a + 3)(6a^2 - 5)$ **23.** [5.1b] $(x + 1)(2x + 1)$
24. [2.3c] $\frac{11}{13}$ **25.** [2.4b] $r = \frac{mv^2}{F}$
26. [9.1c]$\{x | x \ge \frac{5}{14}\}$, or $[\frac{5}{14}, \infty)$ **27.** [9.2b] $\{x | x < -\frac{4}{3} \text{ or } x > 6\}$, or $(-\infty, -\frac{4}{3}) \cup (6, \infty)$ **28.** [9.3e] $\{x | -\frac{13}{4} \le x \le \frac{15}{4}\}$, or $[-\frac{13}{4}, \frac{15}{4}]$
29. [8.3a] $(-4, 1)$ **30.** [8.5a] $(\frac{1}{2}, 3, -5)$ **31.** [5.8b] $\frac{1}{5}, -3$
32. [6.7a] $-\frac{5}{3}$ **33.** [6.7a] 3   **34.** [2.4b], [6.7a] $m = \frac{aA}{h - A}$
35. [10.6a] $\frac{37}{2}$ **36.** [10.6b] 11   **37.** [5.8b] 4
38. [11.2a] $\frac{3}{2} \pm \frac{\sqrt{55}}{2}i$ **39.** [11.2a] $\frac{17 \pm \sqrt{145}}{2}$
40. [11.3b] $a = \sqrt{P^2 + b^2}$
41. [11.8b]$\{x | -3 < x < -2 \text{ or } -1 < x < 1\}$, or $(-3, -2) \cup (-1, 1)$
42. [11.8a]$\{x | x < -\frac{5}{2} \text{ or } x > \frac{5}{2}\}$, or $(-\infty, -\frac{5}{2}) \cup (\frac{5}{2}, \infty)$
43. [3.2a]
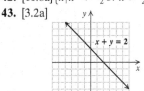
44. [9.4b]
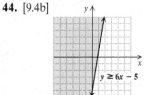
45. [9.4b]
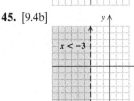
46. [9.4c]
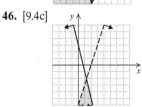
47. [11.6a]
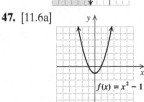
48. [11.6a]
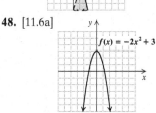
49. [7.5b] $y = \frac{1}{2}x + 4$ **50.** [7.5d] $y = -3x + 1$
51. [11.3a] 16 km/h   **52.** [11.7a] 14 ft by 14 ft; 196 ft^2
53. [8.2b], [8.3b] 36   **54.** [6.8a] 2 hr   **55.** [11.1b] A
56. [11.4c] B   **57.** [11.4c] $\frac{2}{51 + 7\sqrt{61}}$, or $\frac{-51 + 7\sqrt{61}}{194}$
58. [5.6a] $\left(\frac{a}{2} + \frac{2b}{9}\right)\left(\frac{a^2}{4} - \frac{ab}{9} + \frac{4b^2}{81}\right)$

CHAPTER 12

Calculator Corner, p. 894
1. Left to the student

Exercise Set 12.1, p. 898
RC1. True **RC2.** False **RC3.** False **RC4.** True
CC1. $a > 1$ **CC2.** $0 < a < 1$ **CC3.** $0 < a < 1$
CC4. $a > 1$

1.

x	$f(x)$
0	1
1	2
2	4
3	8
-1	$\frac{1}{2}$
-2	$\frac{1}{4}$
-3	$\frac{1}{8}$

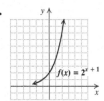

3.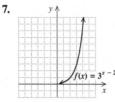

5.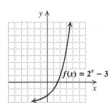

7.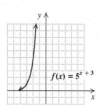

9.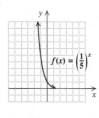

11.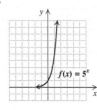

13.

x	$f(x)$
0	1
1	$\frac{1}{2}$
2	$\frac{1}{4}$
3	$\frac{1}{8}$
-1	2
-2	4
-3	8

15.

17.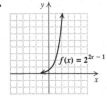

19.

21.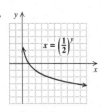

23.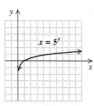

25.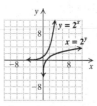

27. (a) $A(t) = \$50{,}000(1.02)^t$; **(b)** $50,000; $51,000; $52,020; $54,121.61; $58,582.97; $60,949.72; $74,297.37;

(c)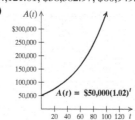

29. $2161.16 **31.** $5287.54

33. (a) 374 nene; about 1812 nene; about 3065 nene;

(b)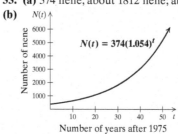

35. (a) $6553; about $2454; about $919;

(b)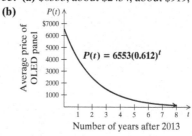

37. (a) 335,000 cans; 150,382 cans; 30,304 cans;

(b)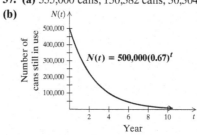

39. $\frac{1}{x^2}$ **40.** $\frac{1}{x^{12}}$ **41.** 1 **42.** 1 **43.** $\frac{2}{3}$ **44.** 2.7

45. $\frac{1}{x^7}$ **46.** $\frac{1}{x^{10}}$ **47.** x **48.** x **49.** 5^4, or 625

51. **53.**

55. 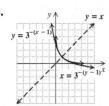   **57.** Left to the student

47. $f^{-1}(x) = 2x + 6$

x	$f(x)$		x	$f^{-1}(x)$
-4	-5		-5	-4
0	-3		-3	0
2	-2		-2	2
4	-1		-1	4

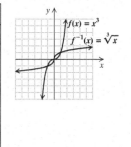

49. $f^{-1}(x) = \sqrt[3]{x}$

x	$f(x)$		x	$f^{-1}(x)$
0	0		0	0
1	1		1	1
2	8		8	2
3	27		27	3
-1	-1		-1	-1
-2	-8		-8	-2
-3	-27		-27	-3

Calculator Corner, p. 911

1. 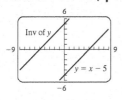   **2.**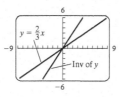

3. 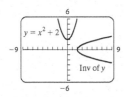   **4.**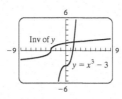

Exercise Set 12.2, p. 915

RC1. relation   **RC2.** inverse   **RC3.** horizontal
RC4. one-to-one   **RC5.** inverse   **RC6.** composition
CC1. 3   **CC2.** 24   **CC3.** 1   **CC4.** -2 **CC5.** 3
CC6. Not defined
1. $-8x + 9; -8x + 18$ **3.** $12x^2 - 12x + 5; 6x^2 + 3$
5. $\dfrac{16}{x^2} - 1; \dfrac{2}{4x^2 - 1}$ **7.** $x^4 - 10x^2 + 30; x^4 + 10x^2 + 20$
9. $f(x) = x^2; g(x) = 5 - 3x$ **11.** $f(x) = \sqrt{x}; g(x) = 5x + 2$
13. $f(x) = \dfrac{1}{x}; g(x) = x - 1$ **15.** $f(x) = \dfrac{1}{\sqrt{x}}; g(x) = 7x + 2$
17. $f(x) = x^4; g(x) = \sqrt{x} + 5$
19. Inverse: $\{(2, 1), (-3, 6), (-5, -3)\}$

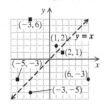

21. Inverse: $x = 2y + 6$

x	y
4	-1
6	0
8	1
10	2
12	3

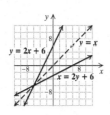

23. Yes   **25.** No   **27.** No   **29.** Yes   **31.** $f^{-1}(x) = \dfrac{x + 2}{5}$
33. $f^{-1}(x) = \dfrac{-2}{x}$ **35.** $f^{-1}(x) = \frac{3}{4}(x - 7)$ **37.** $f^{-1}(x) = \dfrac{2}{x} - 5$
39. Not one-to-one   **41.** $f^{-1}(x) = \dfrac{1 - 3x}{5x - 2}$
43. $f^{-1}(x) = \sqrt[3]{x + 1}$ **45.** $f^{-1}(x) = x^3$

51. $(f^{-1} \circ f)(x) = f^{-1}(f(x)) = f^{-1}(\frac{4}{5}x) = \frac{5}{4}(\frac{4}{5}x) = x;$
$(f \circ f^{-1})(x) = f(f^{-1}(x)) = f(\frac{5}{4}x) = \frac{4}{5}(\frac{5}{4}x) = x$
53. $(f^{-1} \circ f)(x) = f^{-1}(f(x)) = f^{-1}\left(\dfrac{x + 7}{2}\right)$
$= 2\left(\dfrac{x + 7}{2}\right) - 7 = x + 7 - 7 = x;$
$(f \circ f^{-1})(x) = f(f^{-1}(x)) = f(2x - 7)$
$= \dfrac{2x - 7 + 7}{2} = \dfrac{2x}{2} = x$
55. $(f^{-1} \circ f)(x) = f^{-1}(f(x)) = f^{-1}\left(\dfrac{1 - x}{x}\right)$
$= \dfrac{1}{\dfrac{1 - x}{x} + 1} = \dfrac{1}{\dfrac{1}{x}} = x;$

$(f \circ f^{-1})(x) = f(f^{-1}(x)) = f\left(\dfrac{1}{x + 1}\right)$
$= \dfrac{1 - \dfrac{1}{x + 1}}{\dfrac{1}{x + 1}} = \dfrac{\dfrac{x}{x + 1}}{\dfrac{1}{x + 1}} = x$

57. $f^{-1}(x) = \frac{1}{3}x$ **59.** $f^{-1}(x) = -x$ **61.** $f^{-1}(x) = x^3 + 5$
63. (a) 40, 42, 46, 50; (b) $f^{-1}(x) = x - 32$; (c) 8, 10, 14, 18
65. $\sqrt[3]{a}$ **66.** $\sqrt{3}$ **67.** $\sqrt{2xy}$ **68.** $3a^2b^2$ **69.** $-i$
70. $\dfrac{y^{4/3}}{25\, x^2 z^4}$ **71.** $20\sqrt{6}$ **72.** $x^3 + x^2 - 7x - 7$
73. $4y^2 - 20y + 25$ **74.** $49a^4 - c^2$ **75.** No   **77.** Yes
79. (1) C; (2) A; (3) B; (4) D
81.   **83.** $f(x) = \frac{1}{2}x + 3;$
$g(x) = 2x - 6;$ yes

Exercise Set 12.3, p. 928

RC1. False   **RC2.** True   **RC3.** True   **RC4.** True
RC5. False   **RC6.** True   **RC7.** True   **RC8.** True
CC1. 2   **CC2.** 2   **CC3.** 3   **CC4.** 10

1. $x = 2^y$

x, or 2^y	y
1	0
2	1
4	2
8	3
$\frac{1}{2}$	-1
$\frac{1}{4}$	-2
$\frac{1}{8}$	-3

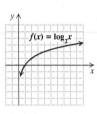

3.   **5.**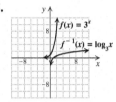

7. $3 = \log_{10} 1000$ **9.** $-3 = \log_5 \frac{1}{125}$ **11.** $\frac{1}{3} = \log_8 2$
13. $0.3010 = \log_{10} 2$ **15.** $2 = \log_e t$ **17.** $t = \log_Q x$
19. $2 = \log_e 7.3891$ **21.** $-2 = \log_e 0.1353$ **23.** $4^w = 10$
25. $6^2 = 36$ **27.** $10^{-2} = 0.01$ **29.** $10^{0.9031} = 8$
31. $e^{4.6052} = 100$ **33.** $t^k = Q$ **35.** 9   **37.** 4   **39.** 4
41. 3   **43.** 25   **45.** 1   **47.** $\frac{1}{2}$ **49.** 2   **51.** 2   **53.** -1
55. 0   **57.** 4   **59.** 2   **61.** 3   **63.** -2 **65.** 0   **67.** 1
69. $\frac{2}{3}$ **71.** 4.8970   **73.** -0.1739 **75.** Does not exist as a real number   **77.** 0.9464   **79.** $6 = 10^{0.7782}$; $84 = 10^{1.9243}$; $987{,}606 = 10^{5.9946}$; $0.00987606 = 10^{-2.0054}$; $98{,}760.6 = 10^{4.9946}$; $70{,}000{,}000 = 10^{7.8451}$; $7000 = 10^{3.8451}$ **81.** $\frac{t-3}{t-2}$ **82.** $c + p$
83. $\dfrac{5}{(x-2)(2x+1)}$ **84.** $(10x+3)(10x-3)$
85. $3x(x-4)^2$ **86.** $(4a-y)(16a^2 + 4ay + y^2)$
87. $(2x-3)(x^2+1)$ **88.** $(5t^3 - 3)(2t^3 + 5)$
89. 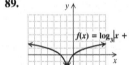   **91.** 25   **93.** 32   **95.** $-\frac{7}{16}$
97. 3   **99.** 0   **101.** -2

Calculator Corner, p. 933
1. Not correct   **2.** Correct   **3.** Not correct   **4.** Correct
5. Not correct   **6.** Correct   **7.** Not correct   **8.** Not correct

Exercise Set 12.4, p. 936
RC1. (b)   **RC2.** (c)   **RC3.** (d)   **RC4.** (a)   **RC5.** (e)
RC6. (f)   **CC1.** (f)   **CC2.** (d)   **CC3.** (b)   **CC4.** (e)
1. $\log_2 32 + \log_2 8$ **3.** $\log_4 64 + \log_4 16$ **5.** $\log_a Q + \log_a x$
7. $\log_b 252$ **9.** $\log_c Ky$ **11.** $4 \log_c y$ **13.** $6 \log_b t$
15. $-3 \log_b C$ **17.** $\log_a 67 - \log_a 5$ **19.** $\log_b 2 - \log_b 5$
21. $\log_c \frac{22}{3}$ **23.** $2 \log_a x + 3 \log_a y + \log_a z$
25. $\log_b x + 2 \log_b y - 3 \log_b z$ **27.** $\frac{4}{3} \log_c x - \log_c y - \frac{2}{3} \log_c z$
29. $2 \log_a m + 3 \log_a n - \frac{3}{4} - \frac{5}{4} \log_a b$ **31.** $\log_a \dfrac{x^{2/3}}{y^{1/2}}$, or $\log_a \dfrac{\sqrt[3]{x^2}}{\sqrt{y}}$
33. $\log_a \dfrac{2x^4}{y^3}$ **35.** $\log_a \dfrac{\sqrt{a}}{x}$ **37.** 2.708   **39.** 0.51   **41.** -1.609
43. $\frac{1}{2}$ **45.** 2.609   **47.** Cannot be found using the properties of logarithms   **49.** t **51.** 5   **53.** 7   **55.** -7 **57.** i
58. -1 **59.** 5   **60.** $\frac{3}{5} + \frac{4}{5}i$ **61.** $23 - 18i$ **62.** $10i$
63. $-34 - 31i$ **64.** $3 - 4i$ **65.** False   **67.** True   **69.** False

Mid-Chapter Review: Chapter 12, p. 938
1. False   **2.** True   **3.** False   **4.** True
5. $\log_5 x = 3$
   $5^3 = x$
   $125 = x$
6. $\log_a 18 = \log_a(2 \cdot 9) = \log_a 2 + \log_a 9$
   $= 0.648 + 2.046 = 2.694$

7. 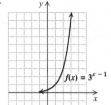   **8.**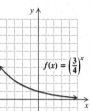

9. 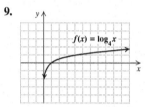   **10.**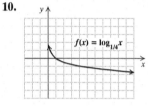

11. (a) $A(t) = \$500(1.04)^t$; (b) $500; $584.93; $740.12
12. $1580.49   **13.** $f^{-1}(x) = \dfrac{x-1}{3}$ **14.** $f^{-1}(x) = \sqrt[3]{x-2}$
15. $1 - 2x$; $8 - 2x$ **16.** $9x^2 - 6x + 2$; $3x^2 + 2$
17. $f(x) = \dfrac{3}{x}$; $g(x) = x + 4$ **18.** $f(x) = \sqrt{x}$; $g(x) = 6x - 7$
19. $(f^{-1} \circ f)(x) = f^{-1}(f(x)) = f^{-1}\left(\dfrac{x}{3}\right) = 3\left(\dfrac{x}{3}\right) = x$;
   $(f \circ f^{-1})(x) = f(f^{-1}(x)) = f(3x) = \dfrac{3x}{3} = x$
20. $(f^{-1} \circ f)(x) = f^{-1}(f(x)) = f^{-1}(\sqrt[3]{x+4})$
   $= (\sqrt[3]{x+4})^3 - 4 = x + 4 - 4 = x$;
   $(f \circ f^{-1})(x) = f(f^{-1}(x)) = f(x^3 - 4)$
   $= \sqrt[3]{x^3 - 4 + 4} = \sqrt[3]{x^3} = x$
21. $3 = \log_7 343$ **22.** $-4 = \log_3 \dfrac{1}{81}$ **23.** $6^t = 12$ **24.** $n^m = T$
25. 3   **26.** 2   **27.** 2   **28.** 5   **29.** 2.3869   **30.** -0.6383
31. $\log_b 2 + \log_b x + 2 \log_b y - 3 \log_b z$
32. $\frac{2}{3} \log_a x + \frac{5}{3} \log_a y - \frac{4}{3} \log_a z$ **33.** $\log_a \dfrac{x\sqrt{z}}{y^2}$
34. $\log_m(b - 4)$ **35.** 0   **36.** 1   **37.** -3 **38.** 5
39. $V^{-1}(t)$ could be used to predict when the value of the stamp will be t, where $V^{-1}(t)$ is the number of years after 2010.
40. $\log_a b$ is the number to which a is raised to get b. Since $\log_a b = c$, then $a^c = b$.   **41.** Express $\dfrac{x}{5}$ as $x \cdot 5^{-1}$ and then use the product rule and the power rule to get
$\log_a\left(\dfrac{x}{5}\right) = \log_a(x \cdot 5^{-1}) = \log_a x + \log_a 5^{-1} =$
$\log_a x + (-1) \log_a 5 = \log_a x - \log_a 5$.
42. The student didn't subtract the logarithm of the entire denominator after using the quotient rule. The correct procedure is as follows:
$\log_b \dfrac{1}{x} = \log_b \dfrac{x}{xx}$
$= \log_b x - \log_b xx$
$= \log_b x - (\log_b x + \log_b x)$
$= \log_b x - \log_b x - \log_b x$
$= -\log_b x$.
(Note that $-\log_b x$ is equivalent to $\log_b 1 - \log_b x$.)

Calculator Corner, p. 944

1. $y = \log_2 x$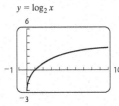

2. $y = \log_3 x$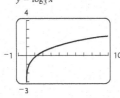

3. $y = \log_{1/2} x$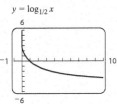

4. $y = \log_{2/3} x$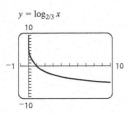

Visualizing for Success, p. 945
1. J **2.** B **3.** O **4.** G **5.** N **6.** F **7.** A **8.** H
9. I **10.** K

Exercise Set 12.5, p. 946
RC1. Base 10 **RC2.** Base e **RC3.** Base 10 **RC4.** Base e
CC1. (b) **CC2.** (a) **CC3.** (b) **CC4.** (a)
1. 0.6931 **3.** 4.1271 **5.** 8.3814 **7.** -5.0832 **9.** -1.6094
11. Does not exist **13.** -1.7455 **15.** 1 **17.** 15.0293
19. 0.0305 **21.** 109.9472 **23.** 5 **25.** 2.5702 **27.** 6.6439
29. 2.1452 **31.** -2.3219 **33.** -2.3219 **35.** 4.6284

37.

x	$f(x)$
0	1
1	2.7
2	7.4
-1	0.4
-2	0.1

39.

41.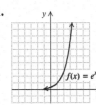

43.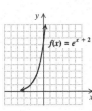

45.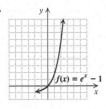

47.

x	$f(x)$
0	0.7
1	1.1
2	1.4
3	1.6
-0.5	0.4
-1	0
-1.5	-0.7

49.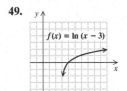

51.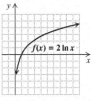

53.

55.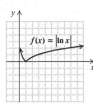

57. 16, 256 **58.** $\frac{1}{4}, 9$ **59.** 49, 121 **60.** $\pm 3, \pm 4$
61. Domain: $(-\infty, \infty)$; range: $[0, \infty)$ **63.** Domain: $(-\infty, \infty)$;
range: $(-\infty, 100)$ **65.** $\left(\frac{5}{2}, \infty\right)$

Exercise Set 12.6, p. 953
RC1. True **RC2.** True **RC3.** True **RC4.** False
CC1. False **CC2.** True **CC3.** False **CC4.** True
CC5. True **CC6.** True
1. 3 **3.** 4 **5.** $\frac{5}{2}$ **7.** $\frac{3}{5}$ **9.** 3.4594 **11.** 5.4263 **13.** $\frac{5}{2}$
15. $-3, -1$ **17.** $\frac{3}{2}$ **19.** 4.6052 **21.** 2.3026 **23.** 140.6705
25. 2.7095 **27.** 3.2220 **29.** 256 **31.** $\frac{1}{32}$ **33.** 10 **35.** $\frac{1}{100}$
37. $e^2 \approx 7.3891$ **39.** $\dfrac{1}{e} \approx 0.3679$ **41.** 121 **43.** 10
45. $\frac{1}{3}$ **47.** 3 **49.** $\frac{2}{5}$ **51.** 5 **53.** No solution
55. $\{x \mid 9 \leq x < 16\}$, or $[9, 16)$
56. $\{x \mid x < 1 \text{ or } x > 4\}$, or $(-\infty, 1) \cup (4, \infty)$ **57.** $-3, 4$
58. $\{x \mid -3 \leq x \leq 4\}$, or $[-3, 4]$ **59.** 65 **60.** 7 **61.** 1
63. (a) 0.3770; (b) -1.9617; (c) 0.9036; (d) -1.5318
65. 3, 4 **67.** -4 **69.** 2 **71.** $\pm\sqrt{34}$ **73.** $10^{100,000}$
75. 1, 100 **77.** 3, -7 **79.** $1, \dfrac{\log 5}{\log 3} \approx 1.465$

Translating for Success, p. 964
1. D **2.** M **3.** I **4.** A **5.** E **6.** H **7.** C **8.** G
9. N **10.** B

Exercise Set 12.7, p. 965
RC1. (b) **RC2.** (d) **RC3.** (c) **RC4.** (a) **CC1.** 550
CC2. 0.13 **CC3.** 2 **CC4.** 713
1. 188 dB **3.** $10^{-7.5}$ W/m²; or about 3.2×10^{-8} W/m²;
10^{-6} W/m² **5.** About 6.8 **7.** 1.58×10^{-8} moles per liter
9. 2.25 ft/sec **11.** 3.40 ft/sec **13.** (a) About 1752 sites;
(b) 2002; (c) 9.7 years **15.** (a) About 1811 deaths;
(b) 2016; (c) 4.0 years **17.** (a) $P(t) = P_0 e^{0.03t}$;
(b) \$5152.27; \$5309.18; \$6749.29; (c) in 23.1 years
19. (a) $P(t) = 7.5e^{0.0106t}$; (b) 8.2 billion; (c) 2082; (d) 65.4 years
21. (a) $k \approx 0.340$; $P(t) = 56e^{0.340t}$; (b) 605 thousand employees;
(c) about 8 years after 2011, or in 2019 **23.** About 2103 years
25. About 7.2 days **27.** 69.3% per year **29.** (a) $k \approx 0.149$;
$A(t) = 451e^{-0.149t}$; (b) about 65 million albums; (c) 2010
31. (a) $k \approx 0.002$; $P(t) = 396{,}000e^{-0.002t}$; (b) about 387,000;
(c) 2031 **33.** (a) $k \approx 0.263$; $V(t) = 19{,}000e^{0.263t}$;
(b) about \$246 million; (c) 2.6 years; (d) 41.3 years **35.** $\left(-1, \frac{1}{2}\right)$
36. $(2, -1, 1)$ **37.** $-1 \pm \sqrt{2}i$ **38.** $-\frac{6}{5}, 4$ **39.** $\frac{9}{2}$
40. $\pm\sqrt{3}i$ **41.** $-0.937, 1.078, 58.770$ **43.** $-0.767, 2, 4$
45. About 80,922 years

Summary and Review: Chapter 12, p. 970

Vocabulary Reinforcement
1. exponential   2. inverse   3. compound   4. common
5. exponent   6. half-life

Concept Reinforcement
1. True   2. False   3. True   4. True   5. False   6. True

Study Guide
1. 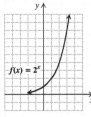   2. $8x + 2; 8x + 1$

3. $f(x) = \dfrac{1}{x}, g(x) = 3x + 2$; answers may vary
4. Yes   5. $g^{-1}(x) = 4 - x$
6.   7.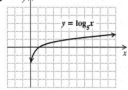

8. $\tfrac{3}{5} \log_a x - \tfrac{2}{5} \log_a y$ 9. $\log_a \dfrac{\sqrt{x}}{y^3}$, or $\log_a \dfrac{x^{1/2}}{y^3}$

10. 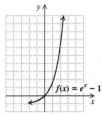   11.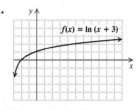

12. $\tfrac{4}{3}$ 13. 3

Review Exercises
1. $\{(2, -4), (-7, 5), (-2, -1), (11, 10)\}$ 2. Not one-to-one
3. $g^{-1}(x) = \dfrac{7x + 3}{2}$ 4. $f^{-1}(x) = \tfrac{1}{2}\sqrt[3]{x}$ 5. $f^{-1}(x) = \dfrac{3x - 4}{2x}$
6.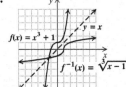

7.
x	$f(x)$
0	$\tfrac{1}{3}$
1	1
2	3
3	9
-1	$\tfrac{1}{9}$
-2	$\tfrac{1}{27}$
-3	$\tfrac{1}{81}$

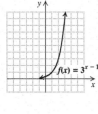

8. 3^y
x, or $3y$	y
1	0
3	1
9	2
27	3
$\tfrac{1}{3}$	-1
$\tfrac{1}{9}$	-2
$\tfrac{1}{27}$	-3

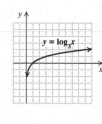

9.
x	$f(x)$
0	2.7
1	7.4
2	20.1
3	54.6
-1	1
-2	0.4
-3	0.1

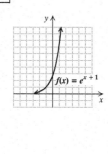

10.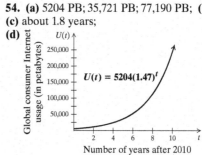

11. $(f \circ g)(x) = 9x^2 - 30x + 25; (g \circ f)(x) = 3x^2 - 5$
12. $f(x) = \sqrt{x}, g(x) = 4 - 7x$; answers may vary
13. $4 = \log 10{,}000$ 14. $\tfrac{1}{2} = \log_{25} 5$ 15. $4^x = 16$
16. $\left(\tfrac{1}{2}\right)^{-3} = 8$ 17. 2   18. -1 19. 1   20. 0
21. -2.7425 22. Does not exist as a real number
23. $4 \log_a x + 2 \log_a y + 3 \log_a z$ 24. $\tfrac{1}{2} \log z - \tfrac{3}{4} \log x - \tfrac{1}{4} \log y$
25. $\log_a 120$ 26. $\log \dfrac{a^{1/2}}{bc^2}$, or $\log \dfrac{\sqrt{a}}{bc^2}$ 27. 17   28. -7
29. 8.7601   30. 3.2698   31. 2.54995   32. -3.6602
33. -2.6921 34. 0.3753   35. 18.3568   36. 0   37. Does not exist   38. 1   39. 0.4307   40. 1.7097   41. $\tfrac{1}{9}$ 42. 2
43. $\tfrac{1}{10{,}000}$ 44. $e^{-2} \approx 0.1353$ 45. $\tfrac{7}{2}$ 46. 1, -5
47. $\dfrac{\log 8.3}{\log 4} \approx 1.5266$ 48. $\dfrac{\ln 0.03}{-0.1} \approx 35.0656$
49. 2   50. 8   51. $\tfrac{17}{5}$ 52. $\sqrt{43}$ 53. 137 dB
54. (a) 5204 PB; 35,721 PB; 77,190 PB;   (b) 2020;
(c) about 1.8 years;
(d)

55. (a) $k \approx 0.094; V(t) = 40{,}000e^{0.094t}$;   (b) $102,399;   (c) 2023
56. $k \approx 0.231$ 57. About 20.4 years   58. About 3463 years
59. C   60. D   61. e^{e^3} 62. $\left(\tfrac{8}{3}, -\tfrac{2}{3}\right)$

Understanding Through Discussion and Writing

1. Reflect the graph of $f(x) = e^x$ across the line $y = x$ and then translate it up one unit.   **2.** Christina mistakenly thinks that, because negative numbers do not have logarithms, negative numbers cannot be solutions of logarithmic equations.

3. $C(x) = \dfrac{100 + 5x}{x}$

$y = \dfrac{100 + 5x}{x}$ Replace $C(x)$ with y.

$x = \dfrac{100 + 5y}{y}$ Interchange variables.

$y = \dfrac{100}{x - 5}$;   Solve for y.

$C^{-1}(x) = \dfrac{100}{x - 5}$ Replace y with $C^{-1}(x)$

$C^{-1}(x)$ gives the number of people in the group, where x is the cost per person, in dollars.
4. To solve $\ln x = 3$, graph $f(x) = \ln x$ and $g(x) = 3$ on the same set of axes. The solution is the first coordinate of the point of intersection of the two graphs.   **5.** You cannot take the logarithm of a negative number because logarithm bases are positive and there is no real-number power to which a positive number can be raised to yield a negative number.   **6.** Answers will vary.

Test: Chapter 12, p. 978

1. [12.1a]

2. [12.3a]
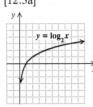

3. [12.5c]
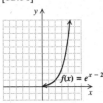

4. [12.5c]

5. [12.2a] $(f \circ g)(x) = 25x^2 - 15x + 2$, $(g \circ f)(x) = 5x^2 + 5x - 2$
6. [12.2b] $\{(3, -4), (-8, 5), (-3, -1), (12, 10)\}$
7. [12.2c, d] $f^{-1}(x) = \dfrac{x + 3}{4}$ **8.** [12.2c, d] $f^{-1}(x) = \sqrt[3]{x} - 1$
9. [12.2c] Not one-to-one   **10.** [12.3b] $\log_{256} 16 = \tfrac{1}{2}$
11. [12.3b] $7^m = 49$ **12.** [12.3c] 3   **13.** [12.4e] 23
14. [12.3c] 0   **15.** [12.3d] -1.9101 **16.** [12.3d] Does not exist as a real number   **17.** [12.4d] $3 \log a + \tfrac{1}{2} \log b - 2 \log c$
18. [12.4d] $\log_a \dfrac{x^{1/3} z^2}{y^3}$ **19.** [12.4d] -0.544 **20.** [12.4d] 1.079
21. [12.5a] 6.6938   **22.** [12.5a] 107.7701   **23.** [12.5a] 0
24. [12.5b] 1.1881   **25.** [12.6b] 5   **26.** [12.6b] 2
27. [12.6b] 10,000   **28.** [12.6b] $e^{1/4} \approx 1.2840$
29. [12.6a] $\dfrac{\log 1.2}{\log 7} \approx 0.0937$
30. [12.6b] 9   **31.** [12.6b] 1   **32.** [12.7a] 4.2
33. [12.7b] **(a)** 4860 people; **(b)** 3.6 days; **(c)** 0.63 day
34. [12.7b] **(a)** $k \approx 0.028$, or 2.8%; $P(t) = 1000e^{0.028t}$; **(b)** \$1251.07; **(c)** after 13 years; **(d)** about 24.8 years
35. [12.7b] About 3%   **36.** [12.7b] About 4684 years
37. [12.6b] B   **38.** [12.6b] 44, -37 **39.** [12.4d] 2

Cumulative Review: Chapters 1–12, p. 981

1. [9.1c] $\{x \mid x \geq -1\}$, or $[-1, \infty)$
2. [9.3e] $\{x \mid x < -6.4 \text{ or } x > 6.4\}$, or $(-\infty, -6.4) \cup (6.4, \infty)$
3. [9.3c] $\{-11, 19\}$ **4.** [8.2a], [8.3a] $\left(-\tfrac{1}{3}, 5\right)$
5. [11.4c] $-3, -2, 2, 3$ **6.** [5.8b] $-1, \tfrac{3}{2}$ **7.** [6.7a] $-\tfrac{2}{3}, 3$
8. [9.3e] $\{x \mid -19 \leq x \leq 7\}$, or $[-19, 7]$ **9.** [5.8b] $-7, -3$
10. [11.2a] $-\tfrac{1}{4} \pm i\dfrac{\sqrt{7}}{4}$ **11.** [2.3c] -2
12. [11.8b] $\{x \mid x < -1 \text{ or } x > 2\}$, or $(-\infty, -1) \cup (2, \infty)$
13. [12.6b] 9   **14.** [11.8a] $\{x \mid x \leq -1 \text{ or } x \geq 1\}$, or $(-\infty, -1] \cup [1, \infty)$ **15.** [12.6b] 1   **16.** [10.6a] 4
17. [9.3d] $\left\{-\tfrac{3}{2}, \tfrac{1}{4}\right\}$ **18.** [12.6a] 1.748   **19.** [8.5a] $(-1, 2, 3)$
20. [9.2a] $\{x \mid -1 \leq x < 6\}$, or $[-1, 6)$ **21.** [6.7a] -16
22. [2.4b] $N = \dfrac{4P - 3M}{6}$ **23.** [12.1c], [12.7b] **(a)** About 104 million barrels per day; about 123 million barrels per day; **(b)** about 37 years; **(c)**

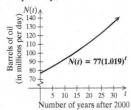

24. [12.7b] **(a)** $A(t) = \$50,000(1.04)^t$; **(b)** \$50,000; \$58,492.93; \$68,428.45; \$74,012.21;
(c)

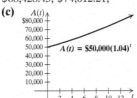

25. [4.5d] $2x^3 - x^2 - 8x - 3$ **26.** [4.4c] $-x^3 + 3x^2 - x - 6$
27. [6.1d] $\dfrac{2m - 1}{m + 1}$ **28.** [6.5b] $\dfrac{x + 2}{x + 1}$ **29.** [6.6a] $\dfrac{1}{x + 1}$
30. [4.8b] $x^3 + 2x^2 - 2x + 1 + \dfrac{3}{x + 1}$ **31.** [10.3b] $5x^2\sqrt{y}$
32. [10.4a] $11\sqrt{2}$ **33.** [10.2d] 8   **34.** [10.8c] $16 + i\sqrt{2}$
35. [10.8e] $\tfrac{3}{10} + \tfrac{11}{10}i$ **36.** [1.2e] $\tfrac{2}{15}$ **37.** [4.1e, f] $-9x^6 y^2$
38. [1.8d] 62.5   **39.** [1.8c] $-x + 30$ **40.** [3.3a] x-intercept: $(2, 0)$; y-intercept: $\left(0, -\tfrac{1}{2}\right)$ **41.** **(a)** [3.4a] $-\tfrac{16}{11}$; **(b)** [3.4a] not defined
42. [7.5c] $y = 2x + 2$ **43.** [7.5d] $y = -\tfrac{1}{2}x + \tfrac{5}{2}$
44. [9.4b]   **45.** [3.2a]   **46.** [3.3a], [7.4a]

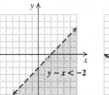 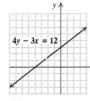

47. [9.4b]   **48.** [7.1c]   **49.** [9.4c]

50. [3.3b], [7.4c]   **51.** [11.6a] 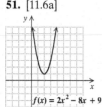 **52.** [12.5c]

53. [12.3a]   **54.** [5.1c] $(x - 6)(2x^3 + 1)$

55. [5.1b], [5.2a] $3(a - 9b)(a + 5b)$
56. [5.2a] $(x - 8)(x - 9)$
57. [5.5d], [5.7a] $(9m^2 + n^2)(3m + n)(3m - n)$
58. [5.5b] $4(2x - 1)^2$ **59.** [5.6a] $3(3a - 2)(9a^2 + 6a + 4)$
60. [5.3a], [5.4a] $2(5x - 2)(x + 7)$
61. [5.1b], [5.3a] $3x(2x - 1)(x + 5)$
62. [12.2d] $f^{-1}(x) = \frac{1}{2}(x + 3)$ **63.** [6.9e] $\frac{4}{5}$ **64.** [7.1b] -10
65. [7.1b] -30 **66.** [7.3c], [7.5e] **(a)** 2018: 74.75 million; 2021: 77.9 million; 2025: 82.1 million;
(b) 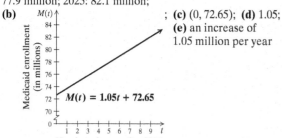 ; **(c)** $(0, 72.65)$; **(d)** 1.05; **(e)** an increase of 1.05 million per year

67. [10.5b] $\dfrac{15 + 8\sqrt{a} + a}{9 - a}$ **68.** [7.2a] All real numbers
69. [7.2a] $\{x \mid x \text{ is a real number and } x \neq -2\}$, or $(-\infty, -2) \cup (-2, \infty)$ **70.** [9.1d] More than 4
71. [6.8a] 612 mi **72.** [2.6a] $11\frac{3}{7}$
73. [8.4a] 24 L of A; 56 L of B **74.** [6.8a] 350 mph
75. [6.8a] $8\frac{2}{5}$ min **76.** [6.9f] 20 **77.** [11.7a] 1250 ft^2
78. [12.7b] 2397 years **79.** [6.9d] 3360 kg
80. [7.5c] $f(x) = -\frac{1}{3}x - \frac{7}{3}$ **81.** [11.7b] $f(x) = -\frac{17}{18}x^2 - \frac{59}{18}x + \frac{11}{9}$
82. [12.3b] $\log r = 6$ **83.** [12.3b] $3^x = Q$
84. [12.4d] $\log_b \left(\dfrac{x^7}{yz^8}\right)^{1/5}$, or $\log_b \dfrac{x^{7/5}}{y^{1/5}z^{8/5}}$
85. [12.4d] $-6\log_b x - 30\log_b y + 6\log_b z$ **86.** [11.7a] 169
87. [12.2c] No **88.** [7.2a] **(a)** -5; **(b)** all real numbers, or $(-\infty, \infty)$; **(c)** $-2, -1, 1, 2$; **(d)** $\{x \mid x \geq -7\}$, or $[-7, \infty)$
89. [12.7b] **(a)** $P(t) = 1{,}998{,}257e^{0.025t}$; **(b)** 3,294,569; **(c)** about 2028 **90.** [6.7a] All real numbers except 0 and -12
91. [12.6b] 81 **92.** [5.8b] $0, \dfrac{b}{a}$
93. [5.6a] $\left(\dfrac{a}{2} + \dfrac{2b}{9}\right)\left(\dfrac{a^2}{4} - \dfrac{ab}{9} + \dfrac{4b^2}{81}\right)$ **94.** [1.8d], [4.1f] $0, \dfrac{328}{3}$

APPENDIXES

Exercise Set B, p. 992
1. Mean: 21; median: 18.5; mode: 29
3. Mean: 5.38; median: 5.7; no mode exists
5. Mean: 239.5; median: 234; mode: 234
7. Mean: $69.8\overline{3}$; median: 13.5; modes: 1, 3
9. Mean: \$4.19; median: \$3.99; mode: \$3.99

Exercise Set C, p. 995
1. $x^2 - x + 1$, R -4; or $x^2 - x + 1 + \dfrac{-4}{x - 1}$
3. $a + 7$, R -47; or $a + 7 + \dfrac{-47}{a + 4}$
5. $x^2 - 5x - 23$, R -43; or $x^2 - 5x - 23 + \dfrac{-43}{x - 2}$
7. $3x^2 - 2x + 2$, R -3; or $3x^2 - 2x + 2 + \dfrac{-3}{x + 3}$
9. $y^2 + 2y + 1$, R 12; or $y^2 + 2y + 1 + \dfrac{12}{y - 2}$
11. $3x^3 + 9x^2 + 2x + 6$ **13.** $x^2 + 2x + 4$
15. $y^3 + 2y^2 + 4y + 8$
17. $y^7 - y^6 + y^5 - y^4 + y^3 - y^2 + y - 1$

Exercise Set D, p. 1000
1. 10 **3.** 0 **5.** -48 **7.** 0 **9.** -10 **11.** -3 **13.** 5
15. 0 **17.** $(2, 0)$ **19.** $\left(-\frac{25}{2}, -\frac{11}{2}\right)$ **21.** $\left(\frac{3}{2}, \frac{5}{2}\right)$ **23.** $(-4, 3)$
25. $(2, -1, 4)$ **27.** $(1, 2, 3)$ **29.** $\left(\frac{3}{2}, -4, 3\right)$ **31.** $(2, -2, 1)$

Exercise Set E, p. 1004
1. $\left(\frac{3}{2}, \frac{5}{2}\right)$ **3.** $(-4, 3)$ **5.** $\left(\frac{1}{2}, \frac{3}{2}\right)$ **7.** $(10, -10)$ **9.** $\left(\frac{3}{2}, -4, 3\right)$
11. $(2, -2, 1)$ **13.** $(0, 2, 1)$ **15.** $\left(4, \frac{1}{2}, -\frac{1}{2}\right)$

Exercise Set F, p. 1007
1. 1 **3.** -41 **5.** 12 **7.** $\frac{13}{18}$ **9.** 5 **11.** 2 **13.** $x^2 - x + 1$
15. 21 **17.** 5 **19.** $-x^3 + 4x^2 + 3x - 12$ **21.** 42 **23.** $-\frac{3}{4}$
25. $\frac{1}{6}$ **27.** $x^2 + 3x - 4; x^2 - 3x + 4; 3x^3 - 4x^2; \dfrac{x^2}{3x - 4}$
29. $\dfrac{1}{x - 2} + 4x^3; \dfrac{1}{x - 2} - 4x^3; \dfrac{4x^3}{x - 2}; \dfrac{1}{4x^3(x - 2)}$
31. $\dfrac{3}{x - 2} + \dfrac{5}{4 - x}; \dfrac{3}{x - 2} - \dfrac{5}{4 - x}; \dfrac{15}{(x - 2)(4 - x)}; \dfrac{3(4 - x)}{5(x - 2)}$

Exercise Set G, p. 1012
1. 5 **3.** $\sqrt{29} \approx 5.385$ **5.** $\sqrt{648} \approx 25.456$ **7.** 7.1
9. $\dfrac{\sqrt{41}}{7} \approx 0.915$ **11.** $\sqrt{6970} \approx 83.487$ **13.** $\sqrt{a^2 + b^2}$
15. $\sqrt{17 + 2\sqrt{14} + 2\sqrt{15}} \approx 5.677$ **17.** $\sqrt{9{,}672{,}400} \approx 3110.048$
19. $\left(\frac{3}{2}, \frac{7}{2}\right)$ **21.** $\left(0, \frac{11}{2}\right)$ **23.** $\left(-1, -\frac{17}{2}\right)$ **25.** $(-0.25, -0.3)$
27. $\left(-\frac{1}{12}, \frac{1}{24}\right)$ **29.** $\left(\dfrac{\sqrt{2} + \sqrt{3}}{2}, \dfrac{3}{2}\right)$
31. 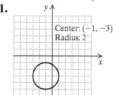 **33.**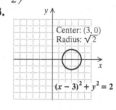

35. 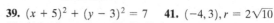 **37.** $x^2 + y^2 = 49$

39. $(x + 5)^2 + (y - 3)^2 = 7$ **41.** $(-4, 3), r = 2\sqrt{10}$
43. $(4, -1), r = 2$ **45.** $(2, 0), r = 2$

Guided Solutions

CHAPTER 1

Section 1.1
5. $A = lw$
$A = (24\,\text{ft})(8\,\text{ft})$
$= (24)(8)(\text{ft})(\text{ft})$
$= 192\,\text{ft}^2$, or
192 square feet

Section 1.3
20. $-\dfrac{1}{5} + \left(-\dfrac{3}{4}\right)$
$= -\dfrac{4}{20} + \left(-\dfrac{15}{20}\right)$
$= -\dfrac{19}{20}$

32. $-x = -(-1.6) = 1.6;$
$-(-x) = -(-(-1.6))$
$= -(1.6) = -1.6$

Section 1.4
11. $2 - 8 = 2 + (-8) = -6$
19. $-12 - (-9) = -12 + 9 = -3$

Section 1.6
21. $\dfrac{4}{7} \div \left(-\dfrac{3}{5}\right) = \dfrac{4}{7} \cdot \left(-\dfrac{5}{3}\right) = -\dfrac{20}{21}$

25. $\dfrac{-5}{6} = \dfrac{5}{-6} = -\dfrac{5}{6}$

Section 1.7
3. $\dfrac{3}{4} = \dfrac{3}{4} \cdot 1 = \dfrac{3}{4} \cdot \dfrac{2}{2} = \dfrac{6}{8}$

4. $\dfrac{3}{4} = \dfrac{3}{4} \cdot 1 = \dfrac{3}{4} \cdot \dfrac{t}{t} = \dfrac{3t}{4t}$

8. $\dfrac{18p}{24pq} = \dfrac{6p \cdot 3}{6p \cdot 4q}$
$= \dfrac{6p}{6p} \cdot \dfrac{3}{4q}$
$= 1 \cdot \dfrac{3}{4q} = \dfrac{3}{4q}$

31. $-2(x - 3)$
$= -2 \cdot x - (-2) \cdot 3$
$= -2x - (-6)$
$= -2x + 6$

44. $16a - 36b + 42$
$= 2 \cdot 8a - 2 \cdot 18b + 2 \cdot 21$
$= 2(8a - 18b + 21)$

52. $3x - 7x - 11 + 8y + 4 - 13y$
$= (3 - 7)x + (8 - 13)y + (-11 + 4)$
$= -4x + (-5)y + (-7)$
$= -4x - 5y - 7$

Section 1.8
13. $5a - 3(7a - 6)$
$= 5a - 21a + 18$
$= -16a + 18$

18. $9 - [10 - (13 + 6)]$
$= 9 - [10 - (19)]$
$= 9 - [-9]$
$= 9 + 9$
$= 18$

25. $-4^3 + 52 \cdot 5 + 5^3 - (4^2 - 48 \div 4)$
$= -64 + 52 \cdot 5 + 125 - (16 - 48 \div 4)$
$= -64 + 52 \cdot 5 + 125 - (16 - 12)$
$= -64 + 52 \cdot 5 + 125 - 4$
$= -64 + 260 + 125 - 4$
$= 196 + 125 - 4$
$= 321 - 4$
$= 317$

CHAPTER 2

Section 2.1
8. $x + 2 = 11$
$x + 2 + (-2) = 11 + (-2)$
$x + 0 = 9$
$x = 9$

Section 2.2
1. $6x = 90$
$\dfrac{1}{6} \cdot 6x = \dfrac{1}{6} \cdot 90$
$1 \cdot x = 15$
$x = 15$
Check: $\dfrac{6x = 90}{6 \cdot 15 \;?\; 90}$
$90 \;|\;$ TRUE

2. $4x = -7$
$\dfrac{4x}{4} = \dfrac{-7}{4}$
$1 \cdot x = -\dfrac{7}{4}$
$x = -\dfrac{7}{4}$

6. $\frac{2}{3} = -\frac{5}{6}y$

$-\frac{6}{5} \cdot \frac{2}{3} = -\frac{6}{5} \cdot \left(-\frac{5}{6}y\right)$

$-\frac{12}{15} = 1 \cdot y$

$-\frac{4}{5} = y$

Section 2.3

4. $-18 - m = -57$
$18 - 18 - m = 18 - 57$
$-m = -39$
$-1(-m) = -1(-39)$
$m = 39$

11. $7x - 17 + 2x = 2 - 8x + 15$
$9 \cdot x - 17 = 17 - 8x$
$8x + 9x - 17 = 17 - 8x + 8x$
$17 \cdot x - 17 = 17$
$17x - 17 + 17 = 17 + 17$
$17x = 34$
$\frac{17x}{17} = \frac{34}{17}$
$x = 2$

13. $\frac{7}{8}x - \frac{1}{4} + \frac{1}{2}x = \frac{3}{4} + x$

$8 \cdot \left(\frac{7}{8}x - \frac{1}{4} + \frac{1}{2}x\right) = 8 \cdot \left(\frac{3}{4} + x\right)$

$8 \cdot \frac{7}{8}x - 8 \cdot \frac{1}{4} + 8 \cdot \frac{1}{2}x = 8 \cdot \frac{3}{4} + 8 \cdot x$

$7x - 2 + 4x = 6 + 8x$
$11x - 2 = 6 + 8x$
$11x - 2 - 8x = 6 + 8x - 8x$
$3x - 2 = 6$
$3x - 2 + 2 = 6 + 2$
$3x = 8$
$\frac{3x}{3} = \frac{8}{3}$
$x = \frac{8}{3}$

Section 2.4

12. $y = mx + b$
$y - b = mx + b - b$
$y - b = mx$
$\frac{y - b}{m} = \frac{mx}{m}$
$\frac{y - b}{m} = x$

Section 2.5

4. 110% of what number is 30?
$110\% \cdot b = 30$

8. 25.3 is 22% of what number?
$25.3 = 22\% \cdot x$
$25.3 = 0.22 \cdot x$
$\frac{25.3}{0.22} = \frac{0.22x}{0.22}$
$115 = x$

Section 2.6

3. Let $x =$ the first marker and $x + 1 =$ the second marker.
Translate and *Solve*:

First marker number + Second marker number = 627

$x + (x + 1) = 627$
$2x + 1 = 627$
$2x + 1 - 1 = 627 - 1$
$2x = 626$
$\frac{2x}{2} = \frac{626}{2}$
$x = 313$

If $x = 313$, then $x + 1 = 314$. The mile markers are 313 and 314.

8. Let $x =$ the principal. Then the interest earned is 5%x.
Translate and *Solve*:

Principal + Interest = Amount

$x + 5\%x = 2520$
$x + 0.05x = 2520$
$(1 + 0.05)x = 2520$
$1.05x = 2520$
$\frac{1.05x}{1.05} = \frac{2520}{1.05}$
$x = 2400$

Section 2.7

10. $5y + 2 \leq -1 + 4y$
$5y + 2 - 4y \leq -1 + 4y - 4y$
$y + 2 \leq -1$
$y + 2 - 2 \leq -1 - 2$
$y \leq -3$

The solution set is $\{y | y \leq -3\}$.

18. $3(7 + 2x) \leq 30 + 7(x - 1)$
$21 + 6x \leq 30 + 7x - 7$
$21 + 6x \leq 23 + 7x$
$21 + 6x - 6x \leq 23 + 7x - 6x$
$21 \leq 23 + x$
$21 - 23 \leq 23 + x - 23$
$-2 \leq x$, or
$x \geq -2$

The solution set is $\{x | x \geq -2\}$.

Section 2.8

9. *Translate* and *Solve:*
$$F < 88$$
$$\frac{9}{5}C + 32 < 88$$
$$\frac{9}{5}C + 32 - 32 < 88 - 32$$
$$\frac{9}{5}C < 56$$
$$\frac{5}{9} \cdot \frac{9}{5}C < \frac{5}{9} \cdot 56$$
$$C < \frac{280}{9}$$
$$C < 31\frac{1}{9}$$

Butter stays solid at Celsius temperatures less than $31\frac{1}{9}°$—that is, $\{C \,|\, C < 31\frac{1}{9}°\}$.

CHAPTER 3

Section 3.1

18. Determine whether $(2, -4)$ is a solution of $4q - 3p = 22$.

$$\begin{array}{c|c} 4q - 3p = 22 \\ \hline 4 \cdot (-4) - 3 \cdot 2 \;?\; 22 \\ -16 - 6 \\ -22 \;\Big|\; \text{FALSE} \end{array}$$

Thus, $(2, -4)$ is not a solution.

Section 3.2

1. Complete the table and graph $y = -2x$.

x	y	(x, y)
-3	6	$(-3, 6)$
-1	2	$(-1, 2)$
0	0	$(0, 0)$
1	-2	$(1, -2)$
3	-6	$(3, -6)$

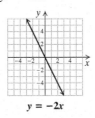

$y = -2x$

9. Graph $5y - 3x = -10$ and identify the y-intercept.

x	y
0	-2	← y-intercept: $(0, -2)$
5	1
-5	-5

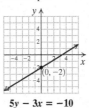

$5y - 3x = -10$

Section 3.3

2. For $2x + 3y = 6$, find the intercepts. Then graph the equation using the intercepts.

x	y
3	0	← x-intercept: $(3, 0)$
0	2	← y-intercept: $(0, 2)$
-3	4	← Check point: $(-3, 4)$

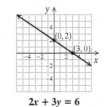

$2x + 3y = 6$

6. Graph: $x = 5$.

x	y
5	-4
5	0
5	3

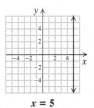

$x = 5$

7. Graph: $y = -2$.

x	y
-1	-2
0	-2
2	-2

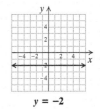

$y = -2$

Section 3.4

1. Graph the line that contains $(-2, 3)$ and $(3, 5)$ and find the slope in two different ways.

$$\frac{5-3}{3-(-2)} = \frac{2}{5}, \text{ or }$$
$$\frac{3-5}{-2-3} = \frac{-2}{-5} = \frac{2}{5}$$

8. Find the slope of the line $5x - 4y = 8$.
$$5x = 4y + 8$$
$$5x - 8 = 4y$$
$$\frac{5x - 8}{4} = \frac{4y}{4}$$
$$\frac{5}{4} \cdot x - 2 = y, \text{ or }$$
$$y = \frac{5}{4} \cdot x - 2$$
Slope is $\frac{5}{4}$.

CHAPTER 4

Section 4.1

18. a) $(4t)^2 = [4 \cdot (-3)]^2$
$= [-12]^2$
$= 144$
b) $4t^2 = 4 \cdot (-3)^2$
$= 4 \cdot (9)$
$= 36$
c) Since $144 \neq 36$, the expressions are not equivalent.

33. $4p^{-3} = 4\left(\frac{1}{p^3}\right) = \frac{4}{p^3}$

Section 4.2

10. $(-2x^4)^{-2} = (-2)^{-2}(x^4)^{-2}$
$= \frac{1}{(-2)^2} \cdot x^{-8}$
$= \frac{1}{4} \cdot \frac{1}{x^8}$
$= \frac{1}{4x^8}$

15. $\left(\dfrac{x^4}{3}\right)^{-2} = \dfrac{(x^4)^{-2}}{3^{-2}} = \dfrac{x^{-8}}{3^{-2}}$

$= \dfrac{\frac{1}{x^8}}{\frac{1}{3^2}} = \dfrac{1}{x^8} \div \dfrac{1}{3^2}$

$= \dfrac{1}{x^8} \cdot \dfrac{3^2}{1} = \dfrac{9}{x^8}$

This can be done a second way.

$\left(\dfrac{x^4}{3}\right)^{-2} = \left(\dfrac{3}{x^4}\right)^2$

$= \dfrac{3^2}{(x^4)^2} = \dfrac{9}{x^8}$

Section 4.3
5. $2x^2 + 5x - 4$
   $= 2(-5)^2 + 5(-5) - 4$
   $= 2(25) + (-25) - 4$
   $= 50 - 25 - 4$
   $= 21$
29. $-2x^4 + 16 + 2x^4 + 9 - 3x^5$
    $= -3x^5 + (-2 + 2)x^4 + (16 + 9)$
    $= -3x^5 + 0x^4 + 25$
    $= -3x^5 + 25$

Section 4.4
14. $(-6x^4 + 3x^2 + 6) - (2x^4 + 5x^3 - 5x^2 + 7)$
    $= -6x^4 + 3x^2 + 6 - 2x^4 - 5x^3 + 5x^2 - 7$
    $= -8x^4 - 5x^3 + 8x^2 - 1$

Section 4.5
12. **a)** $(y + 2)(y + 7)$
    $= y \cdot (y + 7) + 2 \cdot (y + 7)$
    $= y \cdot y + y \cdot 7 + 2 \cdot y + 2 \cdot 7$
    $= y^2 + 7y + 2y + 14$
    $= y^2 + 9y + 14$
    b) The area is $(y + 7)(y + 2)$, or, from part (a), $y^2 + 9y + 14$.
18. $(3y^2 - 7)(2y^3 - 2y + 5)$
    $= 3y^2(2y^3 - 2y + 5) - 7(2y^3 - 2y + 5)$
    $= 6y^5 - 6y^3 + 15y^2 - 14y^3 + 14y - 35$
    $= 6y^5 - 20y^3 + 15y^2 + 14y - 35$

Section 4.6
17. $(6 - 4y)(6 + 4y)$
    $= (6)^2 - (4y)^2$
    $= 36 - 16y^2$
27. $(3x^2 - 5)(3x^2 - 5)$
    $= (3x^2)^2 - 2(3x^2)(5) + 5^2$
    $= 9x^4 - 30x^2 + 25$

Section 4.7
7. The like terms are $-3pt$ and $8pt$, $-5ptr^3$ and $5ptr^3$, and -12 and 4. Collecting like terms, we have
   $(-3 + 8)pt + (-5 + 5)ptr^3 + (-12 + 4)$
   $= 5pt - 8$.

22. $(2a + 5b + c)(2a - 5b - c)$
    $= [2a + (5b + c)][2a - (5b + c)]$
    $= (2a)^2 - (5b + c)^2$
    $= 4a^2 - (25b^2 + 10bc + c^2)$
    $= 4a^2 - 25b^2 - 10bc - c^2$

Section 4.8
5. $(28x^7 + 32x^5) \div (4x^3)$
   $= \dfrac{28x^7 + 32x^5}{4x^3} = \dfrac{28x^7}{4x^3} + \dfrac{32x^5}{4x^3}$
   $= \dfrac{28}{4}x^{7-3} + \dfrac{32}{4}x^{5-3}$
   $= 7x^4 + 8x^2$
10. $(x^2 + x - 6) \div (x + 3)$

$\quad\quad x - 2$
$x + 3 \overline{\smash{\big)}\, x^2 + x - 6}$
$ \underline{x^2 + 3x}$
$ -2x - 6$
$ \underline{-2x - 6}$
$ 0$

CHAPTER 5

Section 5.1
7. Find the GCF of $-24m^5n^6$, $12mn^3$, $-16m^2n^2$, and $8m^4n^4$.
   The coefficients are $-24, 12, -16$, and 8.
   The greatest positive common factor of the coefficients is 4.
   The smallest exponent of the variable m is 1.
   The smallest exponent of the variable n is 2.
   The GCF $= 4mn^2$.
19. $x^3 + 7x^2 + 3x + 21$
    $= x^2(x + 7) + 3(x + 7) = (x + 7)(x^2 + 3)$

Section 5.2
1. Factor: $x^2 + 7x + 12$.
   Complete the following table.

Pairs of Factors	Sums of Factors
1, 12	13
-1, -12	-13
2, 6	8
-2, -6	-8
3, 4	7
-3, -4	-7

   Because both 7 and 12 are positive, we need consider only the positive factors in the table above.
   $x^2 + 7x + 12 = (x + 3)(x + 4)$
8. Factor $a^2 - 40 + 3a$.
   First, rewrite in descending order:
   $a^2 + 3a - 40$.

Pairs of Factors	Sums of Factors
-1, 40	39
-2, 20	18
-4, 10	6
-5, 8	3

   The factorization is $(a - 5)(a + 8)$.

Section 5.4
2. Factor $12x^2 - 17x - 5$.
  1) There is no common factor.
  2) Multiply the leading coefficient and the constant:
     $12(-5) = -60$.
  3) Look for a pair of factors of -60 whose sum is -17. Those factors are 3 and -20.
  4) Split the middle term: $-17x = 3x - 20x$.
  5) Factor by grouping:
     $12x^2 + 3x - 20x - 5$
     $= 3x(4x + 1) - 5(4x + 1)$
     $= (4x + 1)(3x - 5)$.
  6) Check: $(4x + 1)(3x - 5) = 12x^2 - 17x - 5$.

Section 5.5
13. Factor $49 - 56y + 16y^2$.
  Write in descending order:
  $16y^2 - 56y + 49$.
  Factor as a trinomial square:
  $(4y)^2 - 2 \cdot 4y \cdot 7 + (7)^2$
  $= (4y - 7)^2$.

27. $a^2 - 25b^2$
  $= a^2 - (5b)^2$
  $= (a + 5b)(a - 5b)$.

Section 5.6
4. $8y^3 + z^3$
  $= (2y)^3 + z^3$
  $= (2y + z)(4y^2 - 2yz + z^2)$.

Section 5.7
5. Factor $8x^3 - 200x$.
  a) Factor out the largest common factor:
     $8x^3 - 200x = 8x(x^2 - 25)$.
  b) There are two terms inside the parentheses. Factor the difference of squares:
     $8x(x^2 - 25) = 8x(x + 5)(x - 5)$.
  c) We have factored completely.
  d) Check: $8x(x + 5)(x - 5) = 8x(x^2 - 25) = 8x^3 - 200x$.

10. Factor $x^4 + 2x^2y^2 + y^4$.
  a) There is no common factor.
  b) There are three terms. Factor the trinomial square:
     $x^4 + 2x^2y^2 + y^4 = (x^2 + y^2)^2$.
  c) We have factored completely.
  d) Check: $(x^2 + y^2)^2 = (x^2)^2 + 2(x^2)(y^2) + (y^2)^2$
     $= x^4 + 2x^2y^2 + y^4$.

Section 5.8
5. $x^2 - x - 6 = 0$
  $(x + 2)(x - 3) = 0$
  $x + 2 = 0 \quad \text{or} \quad x - 3 = 0$
  $x = -2 \quad \text{or} \quad x = 3$
  Both numbers check.
  The solutions are -2 and 3.

8. $x^2 - 4x = 0$
  $x(x - 4) = 0$
  $x = 0 \quad \text{or} \quad x - 4 = 0$
  $x = 0 \quad \text{or} \quad x = 4$
  Both numbers check.
  The solutions are 0 and 4.

CHAPTER 6
Section 6.1
2. $\dfrac{2x - 7}{x^2 + 5x - 24}$
  $x^2 + 5x - 24 = 0$
  $(x + 8)(x - 3) = 0$
  $x + 8 = 0 \quad \text{or} \quad x - 3 = 0$
  $x = -8 \quad \text{or} \quad x = 3$
  The rational expression is not defined for replacements -8 and 3.

13. $\dfrac{x - 8}{8 - x}$
  $= \dfrac{x - 8}{-(x - 8)}$
  $= \dfrac{1(x - 8)}{-1(x - 8)} = \dfrac{1}{-1} \cdot \dfrac{x - 8}{x - 8}$
  $= -1 \cdot 1 = -1$

16. $\dfrac{a^2 - 4a + 4}{a^2 - 9} \cdot \dfrac{a + 3}{a - 2}$
  $= \dfrac{(a^2 - 4a + 4)(a + 3)}{(a^2 - 9)(a - 2)}$
  $= \dfrac{(a - 2)(a - 2)(a + 3)}{(a + 3)(a - 3)(a - 2)}$
  $= \dfrac{\cancel{(a - 2)}(a - 2)\cancel{(a + 3)}}{\cancel{(a + 3)}(a - 3)\cancel{(a - 2)}}$
  $= \dfrac{a - 2}{a - 3}$

Section 6.2
6. $\dfrac{x}{8} \div \dfrac{x}{5} = \dfrac{x}{8} \cdot \dfrac{5}{x}$
  $= \dfrac{x \cdot 5}{8 \cdot x}$
  $= \dfrac{\cancel{x} \cdot 5}{8 \cdot \cancel{x}}$
  $= \dfrac{5}{8}$

11. $\dfrac{y^2 - 1}{y + 1} \div \dfrac{y^2 - 2y + 1}{y + 1}$
  $= \dfrac{y^2 - 1}{y + 1} \cdot \dfrac{y + 1}{y^2 - 2y + 1}$
  $= \dfrac{(y^2 - 1)(y + 1)}{(y + 1)(y^2 - 2y + 1)}$
  $= \dfrac{(y + 1)(y - 1)(y + 1)}{(y + 1)(y - 1)(y - 1)}$
  $= \dfrac{\cancel{(y + 1)}\cancel{(y - 1)}(y + 1)}{\cancel{(y + 1)}\cancel{(y - 1)}(y - 1)}$
  $= \dfrac{y + 1}{y - 1}$

Section 6.3
1. $16 = 2 \cdot 2 \cdot 2 \cdot 2$
   $18 = 2 \cdot 3 \cdot 3$
   LCM $= 2 \cdot 2 \cdot 2 \cdot 2 \cdot 3 \cdot 3$,
   or 144

5. $\dfrac{3}{16} + \dfrac{1}{18}$
   $= \dfrac{3}{2 \cdot 2 \cdot 2 \cdot 2} + \dfrac{1}{2 \cdot 3 \cdot 3}$
   $= \dfrac{3}{2 \cdot 2 \cdot 2 \cdot 2} \cdot \dfrac{3 \cdot 3}{3 \cdot 3} + \dfrac{1}{2 \cdot 3 \cdot 3} \cdot \dfrac{2 \cdot 2 \cdot 2}{2 \cdot 2 \cdot 2}$
   $= \dfrac{27 + 8}{2 \cdot 2 \cdot 2 \cdot 2 \cdot 3 \cdot 3}$
   $= \dfrac{35}{144}$

Section 6.4
5. $\dfrac{3}{16x} + \dfrac{5}{24x^2}$
   $16x = 2 \cdot 2 \cdot 2 \cdot 2 \cdot x$
   $24x^2 = 2 \cdot 2 \cdot 2 \cdot 3 \cdot x \cdot x$
   LCD $= 2 \cdot 2 \cdot 2 \cdot 2 \cdot 3 \cdot x \cdot x$,
   or $48x^2$
   $\dfrac{3}{16x} \cdot \dfrac{3x}{3x} + \dfrac{5}{24x^2} \cdot \dfrac{2}{2}$
   $= \dfrac{9x}{48x^2} + \dfrac{10}{48x^2}$
   $= \dfrac{9x + 10}{48x^2}$

10. $\dfrac{2x + 1}{x - 3} + \dfrac{x + 2}{3 - x}$
    $= \dfrac{2x + 1}{x - 3} + \dfrac{x + 2}{3 - x} \cdot \dfrac{-1}{-1}$
    $= \dfrac{2x + 1}{x - 3} + \dfrac{-x - 2}{x - 3}$
    $= \dfrac{(2x + 1) + (-x - 2)}{x - 3}$
    $= \dfrac{x - 1}{x - 3}$

Section 6.5
4. $\dfrac{x - 2}{3x} - \dfrac{2x - 1}{5x}$
   LCD $= 3 \cdot x \cdot 5 = 15x$
   $= \dfrac{x - 2}{3x} \cdot \dfrac{5}{5} - \dfrac{2x - 1}{5x} \cdot \dfrac{3}{3}$
   $= \dfrac{5x - 10}{15x} - \dfrac{6x - 3}{15x}$
   $= \dfrac{5x - 10 - (6x - 3)}{15x}$
   $= \dfrac{5x - 10 - 6x + 3}{15x}$
   $= \dfrac{-x - 7}{15x}$

8. $\dfrac{y}{16 - y^2} - \dfrac{7}{y - 4}$
   $= \dfrac{y}{16 - y^2} \cdot \dfrac{-1}{-1} - \dfrac{7}{y - 4}$
   $= \dfrac{-y}{y^2 - 16} - \dfrac{7}{y - 4}$
   $= \dfrac{-y}{(y + 4)(y - 4)} - \dfrac{7}{y - 4} \cdot \dfrac{y + 4}{y + 4}$
   $= \dfrac{-y}{(y + 4)(y - 4)} - \dfrac{7y + 28}{(y + 4)(y - 4)}$
   $= \dfrac{-y - (7y + 28)}{(y + 4)(y - 4)}$
   $= \dfrac{-y - 7y - 28}{(y + 4)(y - 4)}$
   $= \dfrac{-8y - 28}{(y + 4)(y - 4)} = \dfrac{-4(2y + 7)}{(y + 4)(y - 4)}$

Section 6.6

2. $\dfrac{\dfrac{x}{2} + \dfrac{2x}{3}}{\dfrac{1}{x} - \dfrac{x}{2}}$

   LCM of denominators $= 6x$

   $= \dfrac{\dfrac{x}{2} + \dfrac{2x}{3}}{\dfrac{1}{x} - \dfrac{x}{2}} \cdot \dfrac{6x}{6x}$

   $= \dfrac{\left(\dfrac{x}{2} + \dfrac{2x}{3}\right) \cdot 6x}{\left(\dfrac{1}{x} - \dfrac{x}{2}\right) \cdot 6x}$

   $= \dfrac{3x^2 + 4x^2}{6 - 3x^2} = \dfrac{7x^2}{3(2 - x^2)}$

5. $\dfrac{\dfrac{x}{2} + \dfrac{2x}{3}}{\dfrac{1}{x} - \dfrac{x}{2}}$ ← LCD $= 6$
   ← LCD $= 2x$

   $= \dfrac{\dfrac{x}{2} \cdot \dfrac{3}{3} + \dfrac{2x}{3} \cdot \dfrac{2}{2}}{\dfrac{1}{x} \cdot \dfrac{2}{2} - \dfrac{x}{2} \cdot \dfrac{x}{x}}$

   $= \dfrac{\dfrac{3x}{6} + \dfrac{4x}{6}}{\dfrac{2}{2x} - \dfrac{x^2}{2x}} = \dfrac{\dfrac{3x + 4x}{6}}{\dfrac{2 - x^2}{2x}}$

   $= \dfrac{7x}{6} \cdot \dfrac{2x}{2 - x^2} = \dfrac{7 \cdot 2 \cdot x \cdot x}{2 \cdot 3(2 - x^2)}$

   $= \dfrac{7x^2}{3(2 - x^2)}$

Section 6.7

4. $\dfrac{1}{2x} + \dfrac{1}{x} = -12$

LCM $= 2x$

$$2x\left(\dfrac{1}{2x} + \dfrac{1}{x}\right) = 2x(-12)$$

$$2x \cdot \dfrac{1}{2x} + 2x \cdot \dfrac{1}{x} = 2x(-12)$$

$$1 + 2 = -24 \cdot x$$

$$3 = -24x$$

$$\dfrac{3}{-24} = x$$

$$-\dfrac{1}{8} = x$$

7. $\dfrac{4}{x-2} + \dfrac{1}{x+2} = \dfrac{26}{x^2 - 4}$

LCM $= (x-2)(x+2)$

$$(x-2)(x+2)\left(\dfrac{4}{x-2} + \dfrac{1}{x+2}\right) = (x-2)(x+2) \cdot \dfrac{26}{x^2-4}$$

$$4(x+2) + 1(x-2) = 26$$

$$4x + 8 + x - 2 = 26$$

$$5x + 6 = 26$$

$$5x = 20$$

$$x = 4$$

Section 6.9

1. $y = kx$

$8 = k \cdot 20$

$\dfrac{8}{20} = k$

$\dfrac{2}{5} = k$

The variation constant is $\dfrac{2}{5}$. The equation of variation is $y = \dfrac{2}{5} \cdot x$.

5. $y = \dfrac{k}{x}$

$0.012 = \dfrac{k}{50}$

$0.012 \cdot 50 = k$

$0.6 = k$

The variation constant is 0.6. The equation of variation is $y = \dfrac{0.6}{x}$.

7. $y = kx^2$

$175 = k \cdot 5^2$

$175 = k \cdot 25$

$7 = k$

The equation of variation is $y = 7x^2$.

9. $y = kxz$

$65 = k \cdot 10 \cdot 13$

$65 = k \cdot 130$

$\dfrac{65}{130} = k$

$\dfrac{1}{2} = k$

The equation of variation is $y = \dfrac{1}{2}xz$.

CHAPTER 7

Section 7.1

8. a) $f(x) = 2x^2 + 3x - 4$

$f(8) = 2 \cdot 8^2 + 3 \cdot 8 - 4$

$= 2 \cdot 64 + 3 \cdot 8 - 4$

$= 128 + 24 - 4$

$= 152 - 4$

$= 148$

Section 7.2

6. Find the domain of $f(x) = \dfrac{4}{3x+2}$.

Set the denominator equal to 0 and solve for x:

$3x + 2 = 0$

$3x = -2$

$x = -\dfrac{2}{3}$.

Thus, $-\dfrac{2}{3}$ is not in the domain of $f(x)$; all other real numbers are. Domain $= \{x \,|\, x$ is a real number $and\ x \neq -\dfrac{2}{3}\}$.

Section 7.3

6. Graph the line through $(-1, -1)$ and $(2, -4)$ and find its slope.

$m = \dfrac{-4 - (-1)}{2 - (-1)}$

$= \dfrac{-3}{3}$

$= -1$

10. Find the slope and the y-intercept of $5x - 10y = 25$.

First solve for y:

$5x - 10y = 25$

$-10y = -5x + 25$

$y = \dfrac{-5x + 25}{-10}$

$y = \dfrac{1}{2}x - \dfrac{5}{2}$.

Slope is $\dfrac{1}{2}$; y-intercept is $\left(0, -\dfrac{5}{2}\right)$.

Section 7.4

1. Find the intercepts of $4y - 12 = -6x$ and then graph the line.

To find the y-intercept, set $x = 0$ and solve for y:

$4y - 12 = -6 \cdot 0$

$4y - 12 = 0$

$4y = 12$

$y = 3$.

The y-intercept is $(0, 3)$.

To find the x-intercept, set $y = 0$ and solve for x:

$4 \cdot 0 - 12 = -6x$

$-12 = -6x$

$2 = x$.

The x-intercept is $(2, 0)$.

$4y - 12 = -6x$

11. Determine whether the graphs of the lines $x + 4 = y$ and $y - x = -3$ are parallel.

Write each equation in the form $y = mx + b$:

$x + 4 = y \rightarrow y = x + 4$;
$y - x = -3 \rightarrow y = x - 3$.

The slope of each line is 1, and the y-intercepts, $(0, 4)$ and $(0, -3)$, are different. Thus the lines ___are___ parallel.

14. Determine whether the graphs of the lines $2y - x = 2$ and $y + 2x = 4$ are perpendicular.

Write each equation in the form $y = mx + b$:

$2y - x = 2 \rightarrow y = \frac{1}{2}x + 1$;
$y + 2x = 4 \rightarrow y = -2x + 4$.

The slopes of these lines are $\frac{1}{2}$ and -2. The product of the slopes is $\frac{1}{2} \cdot (-2) = -1$. Thus the lines ___are___ perpendicular.

Section 7.5

6. Find an equation of the line containing the points $(4, -3)$ and $(1, 2)$.

First, find the slope:

$$m = \frac{2 - (-3)}{1 - 4} = \frac{5}{-3} = -\frac{5}{3}.$$

Using the point–slope equation,
$y - y_1 = m(x - x_1)$,
substitute 4 for x_1, -3 for y_1, and $-\frac{5}{3}$ for m:

$y - (-3) = -\frac{5}{3}(x - 4)$
$y + 3 = -\frac{5}{3}x + \frac{20}{3}$
$y = -\frac{5}{3}x + \frac{20}{3} - 3$
$y = -\frac{5}{3}x + \frac{20}{3} - \frac{9}{3}$
$y = -\frac{5}{3}x + \frac{11}{3}$.

8. Find an equation of the line containing the point $(2, -1)$ and parallel to the line $9x - 3y = 5$.

Find the slope of the given line:

$9x - 3y = 5$
$-3y = -9x + 5$
$y = 3x - \frac{5}{3}$.

The slope is 3.
The line parallel to $9x - 3y = 5$ must have slope 3.
Using the slope–intercept equation,
$y = mx + b$,
substitute 3 for m, 2 for x, and -1 for y, and solve for b:

$-1 = 3 \cdot 2 + b$
$-1 = 6 + b$
$-7 = b$.

Substitute 3 for m and -7 for b in $y = mx + b$:

$y = 3x + (-7)$
$y = 3x - 7$.

9. Find an equation of the line containing the point $(5, 4)$ and perpendicular to the line $2x - 4y = 9$.

Find the slope of the given line:

$2x - 4y = 9$
$-4y = -2x + 9$
$y = \frac{1}{2}x - \frac{9}{4}$.

The slope is $\frac{1}{2}$.
The slope of a line perpendicular to $2x - 4y = 9$ is the opposite of the reciprocal of $\frac{1}{2}$, or -2.
Using the point–slope equation,
$y - y_1 = m(x - x_1)$,
substitute -2 for m, 5 for x_1, and 4 for y_1:

$y - 4 = -2(x - 5)$
$y - 4 = -2x + 10$
$y = -2x + 14$.

CHAPTER 8

Section 8.1

4. Classify each of the systems in Margin Exercises 1–3 as consistent or inconsistent.

The system in Margin Exercise 1 has a solution, so it is consistent.

The system in Margin Exercise 2 has a solution, so it is consistent.

The system in Margin Exercise 3 does not have a solution, so it is inconsistent.

6. Classify the equations in Margin Exercises 1, 2, 3, and 5 as dependent or independent.

In Margin Exercise 1, the graphs are different, so the equations are independent.

In Margin Exercise 2, the graphs are different, so the equations are independent.

In Margin Exercise 3, the graphs are different, so the equations are independent.

In Margin Exercise 5, the graphs are the same, so the equations are dependent.

Section 8.2

4. $8x - 5y = 12$,   **(1)**
$x - y = 3$ **(2)**

Solve for x in equation (2):

$x - y = 3$
$x = y + 3$.   **(3)**

Substitute $y + 3$ for x in equation (1) and solve for y:

$8x - 5y = 12$
$8(y + 3) - 5y = 12$
$8y + 24 - 5y = 12$
$3y + 24 = 12$
$3y = -12$
$y = -4$.

Substitute -4 for y in equation (3) and solve for x:

$x = y + 3$
$= -4 + 3$
$= -1$.

The ordered pair checks in both equations. The solution is $(-1, -4)$.

Section 8.3

3. Solve by the elimination method:

$$2y + 3x = 12, \quad (1)$$
$$-4y + 5x = -2. \quad (2)$$

Multiply by 2 on both sides of equation (1) and add:

$$\begin{array}{r} 4y + 6x = 24 \\ -4y + 5x = -2 \\ \hline 0 + 11x = 22 \\ 11x = 22 \\ x = 2. \end{array}$$

Substitute 2 for x in equation (1) and solve for y:

$$2y + 3x = 12$$
$$2y + 3(2) = 12$$
$$2y + 6 = 12$$
$$2y = 6$$
$$y = 3.$$

The ordered pair checks in both equations, so the solution is $(2, 3)$.

8. Solve by the elimination method:

$$y + 2x = 3,$$
$$y + 2x = -1.$$

Multiply the second equation by -1 and add:

$$\begin{array}{r} y + 2x = 3 \\ -y - 2x = 1 \\ \hline 0 = 4. \end{array}$$

The equation is <u>false</u> so the system has no solution.

Section 8.4

1.

	White	Red	Total	
	w	r	30	→ $w + r = 30$
	$18.95	$19.50		
	$18.95w$	$19.50r$	572.90	→ $18.95w + 19.50r = 572.90$

3.

	First Investment	Second Investment	Total	
	x	y	$3700	→ $x + y = 3700$
	7%	9%		
	1 year	1 year		
	$0.07x$	$0.09y$	$297	→ $0.07x + 0.09y = 297$

4.

	Attack	Blast	Mixture	
	a	b	60	→ $a + b = 60$
	2%	6%	5%	
	$0.02a$	$0.06b$	0.05×60, or 3	→ $0.02a + 0.06b = 3$

5.

	Distance	Rate	Time	
	d	35 km/h	t	→ $d = 35t$
	d	40 km/h	$t - 1$	→ $d = 40(t - 1)$

6.

	Distance	Rate	Time	
	d	$r + 20$	4 hr	→ $d = (r + 20)4$
	d	$r - 20$	5 hr	→ $d = (r - 20)5$

Section 8.5

3. Solve. Don't forget to check.

$$x + y + z = 100, \quad (1)$$
$$x - y = -10, \quad (2)$$
$$x - z = -30 \quad (3)$$

Add equations (1) and (3):

$$\begin{array}{r} x + y + z = 100 \quad (1) \\ x \quad\quad - z = -30 \quad (3) \\ \hline 2x + y \quad\quad = 70. \quad (4) \end{array}$$

Add equations (2) and (4) and solve for x:

$$\begin{array}{r} x - y = -10 \quad (2) \\ 2x + y = 70 \quad (4) \\ \hline 3x = 60 \\ x = 20. \end{array}$$

Substitute 20 for x in equation (4) and solve for y:

$$2(20) + y = 70$$
$$y = 30.$$

Substitute 20 for x and 30 for y in equation (1) and solve for z:

$$20 + 30 + z = 100$$
$$z = 50.$$

The numbers check. The solution is $(20, 30, 50)$.

Section 8.6

2.

	First Investment	Second Investment	Third Investment	Total
Principal, P	x	y	z	$25,000
Rate of interest, r	3%	4%	5%	
Time, t	1 year	1 year	1 year	
Interest, I	$0.03x$	$0.04y$	$0.05z$	$1120

CHAPTER 9

Section 9.1

11.
$$2x - 3 \geq 3x - 1$$
$$2x - 3 - 2x \geq 3x - 1 - 2x$$
$$-3 \geq x - 1$$
$$-3 + 1 \geq x - 1 + 1$$
$$-2 \geq x, \text{ or }$$
$$x \leq -2$$

The solution set is $\{x \mid x \leq -2\}$, or $(-\infty, -2]$.

15.
$$6 - 5y \geq 7$$
$$6 - 5y - 6 \geq 7 - 6$$
$$-5y \geq 1$$
$$\frac{-5y}{-5} \leq \frac{1}{-5}$$
$$y \leq -\frac{1}{5}$$

The solution set is $\left\{y \mid y \leq -\frac{1}{5}\right\}$, or $\left(-\infty, -\frac{1}{5}\right]$.

Section 9.2

7. $-4 \leq 8 - 2x \leq 4$
$-4 - 8 \leq 8 - 2x - 8 \leq 4 - 8$
$-12 \leq -2x \leq -4$
$\dfrac{-12}{-2} \geq \dfrac{-2x}{-2} \geq \dfrac{-4}{-2}$
$6 \geq x \geq 2$, or
$2 \leq x \leq 6$

The solution set is $\{x | 2 \leq x \leq 6\}$, or $[2, 6]$.

13. $-3x - 7 < -1$ *or* $x + 4 < -1$
$-3x < 6$ *or* $x < -5$
$\dfrac{-3x}{-3} > \dfrac{6}{-3}$ *or* $x < -5$
$x > -2$ *or* $x < -5$

The solution set is $\{x | x < -5 \text{ or } x > -2\}$, or, in interval notation, $(-\infty, -5) \cup (-2, \infty)$.

Section 9.3

6. $|-6 - (-35)| = |-6 + 35|$
$= |29| = 29$

16. $|3x - 4| = 17$
$3x - 4 = -17$ *or* $3x - 4 = 17$
$3x = -13$ *or* $3x = 21$
$x = -\dfrac{13}{3}$ *or* $x = 7$

The solution set is $\left\{-\dfrac{13}{3}, 7\right\}$.

24. $|7 - 3x| \leq 4$
$-4 \leq 7 - 3x \leq 4$
$-11 \leq -3x \leq -3$
$\dfrac{-11}{-3} \geq \dfrac{-3x}{-3} \geq \dfrac{-3}{-3}$
$\dfrac{11}{3} \geq x \geq 1$

The solution set is $\left\{x \Big| 1 \leq x \leq \dfrac{11}{3}\right\}$, or $\left[1, \dfrac{11}{3}\right]$.

25. $|3x + 2| \geq 5$
$3x + 2 \leq -5$ *or* $3x + 2 \geq 5$
$3x \leq -7$ *or* $3x \geq 3$
$x \leq -\dfrac{7}{3}$ *or* $x \geq 1$

The solution set is $\left\{x \Big| x \leq -\dfrac{7}{3} \text{ or } x \geq 1\right\}$, or $\left(-\infty, -\dfrac{7}{3}\right] \cup [1, \infty)$.

Section 9.4

4. $4x + 3y \geq 12$

1. Graph the related equation $4x + 3y = 12$.
2. Since the inequality symbol is $\geq$, draw a solid line.
3. Try the test point $(0, 0)$.

$\dfrac{4x + 3y \geq 12}{4(0) + 3(0) \; ? \; 12}$
$0 \;|\; $ FALSE

$(0, 0)$ is not a solution, so we shade the opposite half-plane.

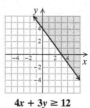

$4x + 3y \geq 12$

CHAPTER 10

Section 10.1

20. $g(x) = \sqrt{6x + 4}$; $g(0), g(3),$ and $g(-5)$
$g(0) = \sqrt{6 \cdot 0 + 4}$
$= \sqrt{0 + 4}$
$= \sqrt{4} = 2$
$g(3) = \sqrt{6 \cdot 3 + 4}$
$= \sqrt{18 + 4}$
$= \sqrt{22}$
$g(-5) = \sqrt{6(-5) + 4}$
$= \sqrt{-30 + 4}$
$= \sqrt{-26}$

-26 is a negative radicand. No real-number function value exists.

33. $\sqrt{x^2 - 6x + 9} = \sqrt{(x - 3)^2}$
$= |x - 3|$

Section 10.2

17. $81^{-3/4} = \dfrac{1}{81^{3/4}} = \dfrac{1}{\left(\sqrt[4]{81}\right)^3}$
$= \dfrac{1}{3^3} = \dfrac{1}{27}$

32. $\dfrac{a^{1/2}b^{3/8}}{a^{1/4}b^{1/8}} = a^{1/2-1/4} \cdot b^{3/8-1/8}$
$= a^{2/4-1/4} \cdot b^{2/8}$
$= a^{1/4} \cdot b^{1/4}$
$= \sqrt[4]{ab}$

Section 10.3

13. $\sqrt{12ab^3c^2}$
$= \sqrt{4 \cdot 3 \cdot a \cdot b^2 \cdot b \cdot c^2}$
$= \sqrt{4} \cdot \sqrt{b^2} \cdot \sqrt{c^2} \cdot \sqrt{3ab}$
$= 2 \cdot b \cdot c \sqrt{3ab}$

17. $\sqrt[3]{3x^2y} \; \sqrt[3]{36x}$
$= \sqrt[3]{3x^2y \cdot 36x}$
$= \sqrt[3]{3 \cdot x \cdot x \cdot y \cdot 2 \cdot 2 \cdot 3 \cdot 3 \cdot x}$
$= 3 \cdot x \sqrt[3]{4y}$

Section 10.4

5. $\sqrt{25x - 25} - \sqrt{9x - 9}$
$= \sqrt{25(x - 1)} - \sqrt{9(x - 1)}$
$= 5\sqrt{x - 1} - 3\sqrt{x - 1}$
$= 2\sqrt{x - 1}$

10. $\left(\sqrt{2} + \sqrt{5}\right)\left(\sqrt{2} - \sqrt{5}\right)$
$= \left(\sqrt{2}\right)^2 - \left(\sqrt{5}\right)^2$
$= 2 - 5$
$= -3$

Section 10.5

3. $\sqrt{\dfrac{4a}{3b}} = \dfrac{\sqrt{4a}}{\sqrt{3b}} \cdot \dfrac{\sqrt{3b}}{\sqrt{3b}}$

$= \dfrac{\sqrt{12ab}}{\sqrt{3^2 \cdot b^2}}$

$= \dfrac{2\sqrt{3ab}}{3b}$

11. $\dfrac{14}{3 + \sqrt{2}}$

$= \dfrac{14}{3 + \sqrt{2}} \cdot \dfrac{3 - \sqrt{2}}{3 - \sqrt{2}}$

$= \dfrac{14(3 - \sqrt{2})}{9 - 2}$

$= \dfrac{14(3 - \sqrt{2})}{7}$

$= 2(3 - \sqrt{2})$

$= 6 - 2\sqrt{2}$

Section 10.6

3. $x + 2 = \sqrt{2x + 7}$

$(x + 2)^2 = (\sqrt{2x + 7})^2$

$x^2 + 4x + 4 = 2x + 7$

$x^2 + 2x - 3 = 0$

$(x + 3)(x - 1) = 0$

$x + 3 = 0 \quad \text{or} \quad x - 1 = 0$

$x = -3 \quad \text{or} \quad x = 1$

The number -3 does not check, but the number 1 does check. The solution is 1.

6. $\sqrt[4]{x - 1} - 2 = 0$

$\sqrt[4]{x - 1} = 2$

$(\sqrt[4]{x - 1})^4 = 2^4$

$x - 1 = 16$

$x = 17$

Section 10.8

5. $\sqrt{-54} = \sqrt{-1 \cdot 54}$

$= \sqrt{-1} \cdot \sqrt{54} = i\sqrt{9 \cdot 6}$

$= i \cdot 3\sqrt{6}, \text{ or } 3\sqrt{6}i$

15. $(1 + 3i)(1 + 5i)$

$= 1 + 5i + 3i + 15i^2$

$= 1 + 8i + 15(-1)$

$= 1 + 8i - 15$

$= -14 + 8i$

25. $6i^{11} + 7i^{14}$

$= 6 \cdot i^{10} \cdot i + 7 \cdot i^{14}$

$= 6(i^2)^5 \cdot i + 7(i^2)^7$

$= 6(-1)^5 \cdot i + 7(-1)^7$

$= 6(-1)i + 7(-1)$

$= -6i - 7$

$= -7 - 6i$

CHAPTER 11

Section 11.1

5. $5x^2 = 15$

$x^2 = 3$

$x = \sqrt{3} \quad \text{or} \quad x = -\sqrt{3}$

The solutions can also be written $\pm \sqrt{3}$. If we round to three decimal places, the solutions are ± 1.732.

9. $x^2 + 16x + 64 = 11$

$(x + 8)^2 = 11$

$x + 8 = \sqrt{11} \quad \text{or} \quad x + 8 = -\sqrt{11}$

$x = -8 + \sqrt{11} \quad \text{or} \quad x = -8 - \sqrt{11}$

The solutions can also be written $-8 \pm \sqrt{11}$.

11. $x^2 - 8x - 20 = 0$

$x^2 - 8x = 20$

$x^2 - 8x + 16 = 20 + 16$

$(x - 4)^2 = 36$

$x - 4 = 6 \quad \text{or} \quad x - 4 = -6$

$x = 10 \quad \text{or} \quad x = -2$

Section 11.2

2. Write the equation in standard form.

$3x^2 + 2x - 7 = 0$

$a = 3, \quad b = 2, \quad c = -7$

$x = \dfrac{-2 \pm \sqrt{2^2 - 4 \cdot 3 \cdot (-7)}}{2 \cdot 3}$

$= \dfrac{-2 \pm \sqrt{88}}{6}$

$= \dfrac{-2 \pm 2\sqrt{22}}{6}$

$= \dfrac{2(-1 \pm \sqrt{22})}{2 \cdot 3}$

$= \dfrac{-1 \pm \sqrt{22}}{3}$

Approximate the solutions and round to three decimal places.

$\dfrac{-1 + \sqrt{22}}{3} \approx 1.230$

$\dfrac{-1 - \sqrt{22}}{3} \approx -1.897$

Section 11.3

4.

	Distance	Speed	Time
Faster ship	3000	$r + 10$	$t - 50$
Slower ship	3000	r	t

6. $V = \pi r^2 h$

$\dfrac{V}{\pi h} = r^2$

$\sqrt{\dfrac{V}{\pi h}} = r$

Section 11.4

3. Determine the nature of the solutions of $3x^2 - 2x + 1 = 0$.

$a = 3, \quad b = -2, \quad c = 1$

$b^2 - 4ac = (-2)^2 - 4 \cdot 3 \cdot 1 = -8$

Since the discriminant is negative, there are two nonreal solutions.

9. Solve: $x^4 - 10x^2 + 9 = 0$.

Let $u = x^2$. Then $u^2 = x^4$.

$x^4 - 10x^2 + 9 = 0$

$u^2 - 10u + 9 = 0$

$(u - 1)(u - 9) = 0$

$u - 1 = 0 \quad$ or $\quad u - 9 = 0$

$u = 1 \quad$ or $\quad u = 9$

$x^2 = 1 \quad$ or $\quad x^2 = 9$

$x = \pm 1 \quad$ or $\quad x = \pm 3$

All four numbers check. The solutions are $-1, 1, -3$, and 3.

Section 11.5

4. Compute $g(x)$ for each value of x shown.

x	$g(x) = (x + 2)^2$
-2	0
-1	1
0	4
1	9
-3	1
-4	4
-5	9

Plot the points and draw the curve.

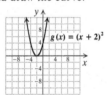

7. Compute $g(x)$ for each value of x shown.

x	$g(x) = (x - 1)^2 - 3$
1	-3
2	-2
3	1
4	6
0	-2
-1	1
-2	6

The vertex is $(1, -3)$.

The line of symmetry is $x = 1$, and the minimum function value is -3.

Plot the points and draw the curve.

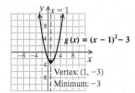

Section 11.6

6. Find the vertex of the parabola given by $f(x) = -4x^2 + 12x - 5$.

The x-coordinate of the vertex is $-\dfrac{b}{2a} = -\dfrac{12}{2(-4)} = \dfrac{3}{2}$.

The second coordinate of the vertex is $f\left(\dfrac{3}{2}\right)$:

$f\left(\dfrac{3}{2}\right) = -4\left(\dfrac{3}{2}\right)^2 + 12\left(\dfrac{3}{2}\right) - 5$

$= -4\left(\dfrac{9}{4}\right) + 18 - 5$

$= -9 + 18 - 5$

$= 4$.

The vertex is $\left(\dfrac{3}{2}, 4\right)$.

7. Find the intercepts of $f(x) = x^2 + 2x - 3$.

The y-intercept is $(0, f(0))$.

$f(0) = 0^2 + 2 \cdot 0 - 3 = -3$

The y-intercept is $(0, -3)$. To find the x-intercepts, we solve $0 = x^2 + 2x - 3$:

$0 = x^2 + 2x - 3$

$0 = (x - 1)(x + 3)$

$x - 1 = 0 \quad$ or $\quad x + 3 = 0$

$x = 1 \quad$ or $\quad x = -3$.

The x-intercepts are $(1, 0)$ and $(-3, 0)$.

Section 11.8

6. Solve: $6x(x + 1)(x - 1) < 0$.

The solutions of $6x(x + 1)(x - 1) = 0$ are $0, -1$, and 1. Divide the number line into four intervals and test values of $f(x) = 6x(x + 1)(x - 1)$.

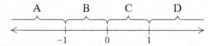

A: Test -2.

$f(-2) = 6(-2)(-2 + 1)(-2 - 1)$

$= -36$

B: Test $-\dfrac{1}{2}$.

$f\left(-\dfrac{1}{2}\right) = 6\left(-\dfrac{1}{2}\right)\left(-\dfrac{1}{2} + 1\right)\left(-\dfrac{1}{2} - 1\right)$

$= \dfrac{9}{4}$

C: Test $\dfrac{1}{2}$.

$f\left(\dfrac{1}{2}\right) = 6\left(\dfrac{1}{2}\right)\left(\dfrac{1}{2} + 1\right)\left(\dfrac{1}{2} - 1\right)$

$= -\dfrac{9}{4}$

D: Test 2.

$f(2) = 6(2)(2 + 1)(2 - 1)$

$= 36$

The expression is negative for values of x in intervals A and C.

The solution set is $\{x \mid x < -1 \text{ or } 0 < x < 1\}$, or $(-\infty, -1) \cup (0, 1)$.

CHAPTER 12

Section 12.1
8. Amount after 1 year:
$40{,}000 + \$40{,}000 \times 0.02$
$= \$40{,}000(1.02)$
$= \$40{,}800$
Amount after 2 years:
$\$40{,}800 + \$40{,}800 \times 0.02$
$= \$40{,}800(1.02)$
$= \$41{,}616$

10. $A = P \cdot \left(1 + \dfrac{r}{n}\right)^{n \cdot t}$

$= 7000 \cdot \left(1 + \dfrac{3.4\%}{4}\right)^{4 \cdot \frac{11}{2}}$

$= 7000 \cdot \left(1 + \dfrac{0.034}{4}\right)^{22}$

$= 7000(1.0085)^{22}$

$\approx \$8432.72$

Section 12.2
2. $(f \circ g)(x) = f(g(x))$
$= f(\sqrt[3]{x})$
$= 4(\sqrt[3]{x}) + 5$
$(g \circ f)(x) = g(f(x))$
$= g(4x + 5)$
$= \sqrt[3]{4x + 5}$

Section 12.3
11. $\log_{10} x = 4$
$10^4 = x$
$10{,}000 = x$

14. Let $\log_{10} 10{,}000 = x$.
$10^x = 10{,}000$
$10^x = 10^4$
$x = 4$

Section 12.4
11. $\log_a \dfrac{x^3 y^4}{z^5 w^9} = \log_a x^3 y^4 - \log_a z^5 w^9$

$= \left(\log_a x^3 + \log_a y^4\right) - \left(\log_a z^5 + \log_a w^9\right)$
$= \log_a x^3 + \log_a y^4 - \log_a z^5 - \log_a w^9$
$= 3\log_a x + 4\log_a y - 5\log_a z - 9\log_a w$

22. $\log_a 16 = \log_a 2^4$
$= 4 \log_a 2$
$= 4(0.301)$
$= 1.204$

Section 12.5
8. (a) $\log_6 7 = \dfrac{\log 7}{\log 6}$;
≈ 1.0860
(b) $\log_6 7 = \dfrac{\ln 7}{\ln 6}$
≈ 1.0860

Section 12.6
3. $7^x = 20$
$\log 7^x = \log 20$
$x \cdot \log 7 = \log 20$
$x = \dfrac{\log 20}{\log 7}$
$x \approx 1.5395$

8. $\log_3 (2x - 1) - \log_3 (x - 4) = 2$
$\log_3 \dfrac{2x - 1}{x - 4} = 2$
$\dfrac{2x - 1}{x - 4} = 3^2$
$\dfrac{2x - 1}{x - 4} = 9$
$2x - 1 = 9(x - 4)$
$2x - 1 = 9x - 36$
$35 = 7x$
$5 = x$

Section 12.7
1. $L = 10 \cdot \log \dfrac{I}{I_0}$
$= 10 \cdot \log \dfrac{3.2 \times 10^{-6}}{10^{-12}}$
$= 10 \cdot \log (3.2 \times 10^6)$
$= 10 \cdot (\log 3.2 + \log 10^6)$
$\approx 10 \cdot (0.5051 + 6)$
≈ 65
The sound level is about 65 decibels.

4. $\text{pH} = -\log [\text{H}^+]$
$7 = -\log [\text{H}^+]$
$-7 = \log [\text{H}^+]$
$10^{-7} = [\text{H}^+]$

Glossary

A

Abscissa The first coordinate in an ordered pair of numbers

Absolute value The distance that a number is from 0 on the number line

ac-method A method for factoring trinomials of the type $ax^2 + bx + c, a \neq 1$, involving the product, ac, of the leading coefficient a and the last term c

Additive identity The number 0

Additive inverse A number's opposite; two numbers are additive inverses of each other if their sum is 0

Algebraic expression An expression consisting of variables, constants, numerals, operation signs, and/or grouping symbols

Area The number of square units that fill a plane region

Arithmetic numbers The whole numbers and the positive fractions; all these numbers can be named with fraction notation $\frac{a}{b}$, where a and b are whole numbers and $b \neq 0$

Ascending order When a polynomial is written with the terms arranged according to degree from least to greatest, it is said to be in ascending order.

Associative law of addition The statement that when three numbers are added, regrouping the addends gives the same sum

Associative law of multiplication The statement that when three numbers are multiplied, regrouping the factors gives the same product

Average A center point of a set of numbers found by adding the numbers and dividing by the number of items of data; also called the *mean*

Axes Two perpendicular number lines used to identify points in a plane

Axis of symmetry A line that can be drawn through a graph such that the part of the graph on one side of the line is an exact reflection of the part on the opposite side; also called *line of symmetry*

B

Base In exponential notation, the number being raised to a power

Binomial A polynomial composed of two terms

C

Circle The set of all points in a plane that are a fixed distance r, called the radius, from a fixed point (h, k), called the center

Circumference The distance around a circle

Coefficient The numerical multiplier of a variable

Common logarithm A logarithm with base 10

Commutative law of addition The statement that when two numbers are added, changing the order in which the numbers are added does not affect the sum

Commutative law of multiplication The statement that when two numbers are multiplied, changing the order in which the numbers are multiplied does not affect the product

Complementary angles Angles whose sum is 90°

Completing the square Adding a particular constant to an expression so that the resulting sum is a perfect square

Complex fraction expression A rational expression that has one or more rational expressions within its numerator and/or denominator

Complex number Any number that can be named as $a + bi$, where a and b are any real numbers

Complex number i The square root of -1; that is, $i = \sqrt{-1}$ and $i^2 = -1$

Complex rational expression A rational expression that has one or more rational expressions within its numerator and/or denominator

Complex-number system A number system that contains the real-number system and is designed so that negative numbers have defined square roots

Composite function A function in which a quantity depends on a variable that, in turn, depends on another variable

Composite number A natural number, other than 1, that is not prime

Compound inequality A statement in which two or more inequalities are joined by the word *and* or the word *or*

Compound interest Interest computed on the sum of an original principal and the interest previously accrued by that principal

Conjugate of a complex number The conjugate of a complex number $a + bi$ is $a - bi$ and the conjugate of $a - bi$ is $a + bi$.

Conjugates of radical terms Pairs of radical terms, like $\sqrt{a} + \sqrt{b}$ and $\sqrt{a} - \sqrt{b}$ or $c + \sqrt{d}$ and $c - \sqrt{d}$, whose product does not have a radical term

Conjunction A statement in which two or more sentences are joined by the word *and*

Consecutive even integers Even integers that are two units apart

Consecutive integers Integers that are one unit apart

Consecutive odd integers Odd integers that are two units apart

Consistent system of equations A system of equations that has at least one solution

Constant A known number

Constant function A function given by an equation of the form $y = b$, or $f(x) = b$, where b is a real number

Constant of proportionality The constant in an equation of direct variation or inverse variation

Coordinates The numbers in an ordered pair

Cube root The number c is the cube root of a, written $\sqrt[3]{a}$, if the third power of c is a.

D

Decay rate The variable k in the exponential decay model $P(t) = P_0 e^{-kt}$

Degree of a polynomial The degree of the term of highest degree in a polynomial

Degree of a term The sum of the exponents of the variables

Denominator The bottom number in a fraction

Dependent equations The equations in a system are dependent if one equation can be removed without changing the solution set.

Descending order When a polynomial is written with the terms arranged according to degree from greatest to least, it is said to be in descending order.

Diameter A segment that passes through the center of a circle and has its endpoints on the circle

Difference of cubes Any expression that can be written in the form $A^3 - B^3$

Difference of squares Any expression that can be written in the form $A^2 - B^2$

Direct variation A situation that gives rise to a linear function $f(x) = kx$, or $y = kx$, where k is a positive constant

Discriminant The expression $b^2 - 4ac$, from the quadratic formula

Disjoint sets Two sets with an empty intersection

Disjunction A statement in which two or more sentences are joined by the word *or*

Distributive law of multiplication over addition The statement that multiplying a factor by the sum of two numbers gives the same result as multiplying the factor by each of the two numbers and then adding

Distributive law of multiplication over subtraction The statement that multiplying a factor by the difference of two numbers gives the same result as multiplying the factor by each of the two numbers and then subtracting

Domain The set of all first coordinates of the ordered pairs in a function

Doubling time The time necessary for a population to double in size

E

Elimination method An algebraic method that uses the addition principle to solve a system of equations

Empty set The set without members

Equation A number sentence that says that the expressions on either side of the equals sign, =, represent the same number

Equation of direct variation An equation described by $y = kx$, with k a positive constant, used to represent direct variation

Equation of inverse variation An equation described by $y = k/x$, with k a positive constant, used to represent inverse variation

Equivalent equations Equations with the same solutions

Equivalent expressions Expressions that have the same value for all allowable replacements

Equivalent inequalities Inequalities that have the same solution set

Evaluate To substitute a value for each occurrence of a variable in an expression and carry out the operations

Even root When the number k in $\sqrt[k]{\ }$ is an even number, we say that we are taking an even root.

Exponent In expressions of the form a^n, the number n is an exponent. For n a natural number, a^n represents n factors of a.

Exponential decay model A decrease in quantity over time that can be modeled by an exponential function of the form $P(t) = P_0 e^{-kt}, k > 0$

Exponential equation An equation in which a variable appears as an exponent

Exponential function The function $f(x) = a^x$, where a is a positive constant different from 1

Exponential growth model An increase in quantity over time that can be modeled by an exponential function of the form $P(t) = P_0 e^{kt}, k > 0$

Exponential growth rate The variable k in the exponential growth model $P(t) = P_0 e^{kt}$

Exponential notation A representation of a number using a base raised to a power

F

Factor *Verb*: To write an equivalent expression that is a product. *Noun*: A multiplier

Factorization of a polynomial An expression that names the polynomial as a product

FOIL To multiply two binomials by multiplying the First terms, the Outside terms, the Inside terms, and then the Last terms

Formula An equation that uses numbers or letters to represent a relationship between two or more quantities

Fraction equation An equation containing one or more rational expressions; also called a *rational equation*

Fraction expression A quotient, or ratio, of polynomials; also called a *rational expression*

Fraction notation A number written using a numerator and a denominator

Function A correspondence that assigns to each member of a set called the domain *exactly one* member of a set called the range

G

Grade The measure of a road's steepness

Graph A picture or diagram of the data in a table; a line, a curve, or a collection of points that represents all the solutions of an equation or an inequality

Greatest common factor (GCF) The common factor of a polynomial with the largest possible coefficient and the largest possible exponent(s)

H

Half-life The amount of time necessary for half of a quantity to decay

Hypotenuse In a right triangle, the side opposite the 90° angle

I

Identity property of 1 The statement that the product of a number and 1 is always the original number

Identity property of 0 The statement that the sum of a number and 0 is always the original number

Imaginary number A number that can be named bi, where b is some real number and $b \neq 0$

Inconsistent system of equations A system of equations for which there is no solution

Independent equations Equations that are not dependent

Index In the expression $\sqrt[k]{a}$, the number k is called the index.

Inequality A mathematical sentence using $<, >, \leq, \geq$, or $\neq$

Input A member of the domain of a function

Integers The whole numbers and their opposites

Intercept The point at which a graph intersects the x- or the y-axis

Intersection of sets A and B The set of all members that are common to A and B

Interval notation The use of a pair of numbers inside parentheses and brackets to represent the set of all numbers between those two numbers

Inverse relation The relation formed by interchanging the coordinates of the ordered pairs in a relation

Inverse variation A situation that gives rise to a function $f(x) = k/x$, or $y = k/x$, where k is a positive constant

Irrational number A real number that cannot be named as a ratio of two integers

J

Joint variation A situation that gives rise to an equation of the form $y = kxz$, where k is a positive constant

L

Leading coefficient The coefficient of the term of highest degree in a polynomial

Leading term The term of highest degree in a polynomial

Least common denominator (LCD) The least common multiple of the denominators

Least common multiple (LCM) The smallest number that is a multiple of two or more numbers

Legs In a right triangle, the two sides that form the right angle

Like radicals Radicals having the same index and radicand

Like terms Terms that have exactly the same variable factors

Line of symmetry A line that can be drawn through a graph such that the part of the graph on one side of the line is an exact reflection of the part on the opposite side; also called *axis of symmetry*

Linear equation in two variables Any equation that can be written in the form $y = mx + b$ or $Ax + By = C$, where x and y are variables

Linear equation in three variables An equation equivalent to one of the type $Ax + By + Cz = D$

Linear function A function that can be described by an equation of the form $y = mx + b$, where x and y are variables

Linear inequality An inequality whose related equation is a linear equation

Logarithmic equation An equation containing a logarithmic expression

Logarithmic function, base a The inverse of an exponential function $f(x) = a^x$

M

Mathematical model A model in which the essential parts of a problem are described in mathematical language

Matrix A rectangular array of numbers

Maximum The largest function value (output) achieved by a function

Mean A center point of a set of numbers found by adding the numbers and dividing by the number of items of data; also called the *average*

Median In a set of data listed in order from smallest to largest, the middle number if there is an odd number of data items, or the average of the two middle numbers if there is an even number of data items

Minimum The smallest function value (output) achieved by a function

Mode The number or numbers that occur most often in a set of data

Monomial An expression of the type ax^n, where a is a real number constant and n is a nonnegative integer

Motion problem A problem that deals with distance, speed (or rate), and time

Multiple A product of a number and some natural number

Multiplication property of 0 The statement that the product of 0 and any real number is 0

Multiplicative identity The number 1

Multiplicative inverses Reciprocals; two numbers whose product is 1

N

Natural logarithm A logarithm with base e

Natural numbers The counting numbers: 1, 2, 3, 4, 5, . . .

Negative integers The integers to the left of zero on the number line

Nonnegative rational numbers The whole numbers and the positive fractions; all these numbers can be named with fraction notation $\frac{a}{b}$, where a and b are whole numbers and $b \neq 0$

Numerator The top number in a fraction

O

Odd root When the number k in $\sqrt[k]{}$ is an odd number, we say we are taking an odd root.

One-to-one function A function for which different inputs have different outputs

Opposite The opposite, or additive inverse, of a number a is denoted $-a$. Opposites are the same distance from 0 on the number line but on different sides of 0.

Opposite of a polynomial To find the opposite of a polynomial, replace each term with its opposite—that is, change the sign of every term.

Ordered pair A pair of numbers of the form (h, k) for which the order in which the numbers are listed is important

Ordinate The second coordinate in an ordered pair of numbers

Origin The point on a graph where the two axes intersect

Output A member of the range of a function

P

Parabola A graph of a quadratic function

Parallel lines Lines in the same plane that never intersect. Two nonvertical lines are parallel if they have the same slope and different y-intercepts.

Parallelogram A four-sided polygon with two pairs of parallel sides

Percent notation A representation of a number as parts per 100

Perfect square A rational number p for which there exists a number a for which $a^2 = p$

Perfect-square trinomial A trinomial that is the square of a binomial

Perimeter The sum of the lengths of the sides of a polygon

Perpendicular lines Two lines are perpendicular if the product of their slopes is -1 or if one line is vertical and the other is horizontal. Lines that form a right angle

Pi (π) The number that results when the circumference of a circle is divided by its diameter; $p \approx 3.14$ or $22/7$

Point–slope equation An equation of the form $y - y_1 = m(x - x_1)$, where m is the slope and (x_1, y_1) is a point on the line

Polynomial A monomial or a combination of sums and/or differences of monomials

Polynomial equation An equation in which two polynomials are set equal to each other

Positive integers The natural numbers, or the integers to the right of zero on the number line

Prime factorization A factorization of a composite number as a product of prime numbers

Prime number A natural number that has *exactly two different factors:* itself and 1

Prime polynomial A polynomial that cannot be factored using only integer coefficients

Principal square root The nonnegative square root of a number

Principle of zero products The statement that an equation $ab = 0$ is true if and only if $a = 0$ is true or $b = 0$ is true or both are true

Proportion An equation stating that two ratios are equal

Proportional numbers Two pairs of numbers having the same ratio

Pythagorean theorem In any right triangle, if a and b are the lengths of the legs and c is the length of the hypotenuse, then $a^2 + b^2 = c^2$.

Q

Quadrants The four regions into which the axes divide a plane

Quadratic equation An equation equivalent to an equation of the type $ax^2 + bx + c = 0$, where $a \neq 0$

Quadratic formula The solutions of $ax^2 + bx + c = 0$ are given by the equation $x = \dfrac{-b \pm \sqrt{b^2 - 4ac}}{2a}$.

Quadratic inequality A second-degree polynomial inequality in one variable

R

Radical The symbol $\sqrt{}$

Radical equation An equation in which a variable appears in one or more radicands

Radical expression An algebraic expression written with a radical

Radicand The expression written under the radical

Radius A segment with one endpoint on the center of a circle and the other endpoint on the circle

Range The set of all second coordinates of the ordered pairs in a function

Rate The ratio of two different kinds of measure

Ratio The quotient of two quantities

Rational equation An equation containing one or more rational expressions; also called *fraction equation*

Rational expression A quotient, or ratio, of polynomials; also called *fraction expression*

Rational inequality An inequality containing a rational expression

Rational number A number that can be written in the form a/b, where a and b are integers and $b \neq 0$

Rationalizing the denominator A procedure for finding an equivalent expression without a radical in the denominator

Real numbers All rational and irrational numbers; the set of all numbers corresponding to points on the number line

Reciprocal A multiplicative inverse; two numbers are reciprocals if their product is 1

Rectangle A four-sided polygon with four right angles

Relation A correspondence between a first set, the domain, and a second set, the range, such that each member of the domain corresponds to *at least one* member of the range

Repeating decimal A decimal in which a number pattern repeats indefinitely

Right triangle A triangle that includes a 90° angle

Rise The change in the second coordinate between two points on a line

Roster notation A way of naming sets by listing all the elements in the set

Rounding Approximating the value of a number; used when estimating

Run The change in the first coordinate between two points on a line

S

Scientific notation A representation of a number of the form $M \times 10^n$, where n is an integer, $1 \leq M < 10$, and M is expressed in decimal notation

Second-order determinant The determinant of a two-by-two matrix $\begin{bmatrix} a_1 & b_1 \\ a_2 & b_2 \end{bmatrix}$ is denoted $\begin{vmatrix} a_1 & b_1 \\ a_2 & b_2 \end{vmatrix}$ and is defined to mean $a_1 b_2 - a_2 b_1$.

Set A collection of objects

Set-builder notation The naming of a set by describing basic characteristics of the elements in the set

Similar triangles Triangles in which corresponding angles have the same measure and the lengths of corresponding sides are proportional

Simplest fraction notation A fraction written with the smallest numerator and denominator

Simplify To rewrite an expression in an equivalent, abbreviated form

Slope The ratio of the rise to the run for any two points on a line

Slope–intercept equation An equation of the form $y = mx + b$, where x and y are variables; the slope is m and the y-intercept is $(0, b)$

Solution A replacement for the variable that makes an equation or inequality true

Solution set The set of all solutions of an equation, an inequality, or a system of equations or inequalities

Solution of a system of linear inequalities An ordered pair (x, y) that is a solution of *both* inequalities

Solution of a system of three equations An ordered triple (x, y, z) that makes *all three* equations true

Solution of a system of two equations An ordered pair (x, y) that makes *both* equations true

Solve To find all solutions of an equation, an inequality, or a system of equations or inequalities; to find the solution(s) of a problem

Speed The ratio of distance traveled to the time required to travel that distance

Square A four-sided polygon with four right angles and all sides of equal length

Square of a number A number multiplied by itself

Square root The number c is a square root of a if $c^2 = a$.

Standard form of a quadratic equation A quadratic equation in the form $ax^2 + bx + c = 0$, where $a \neq 0$

Subsets Sets that are a part of other sets

Substitute To replace a variable with a number

Substitution method A nongraphical method for solving systems of equations

Sum of cubes An expression that can be written in the form $A^3 + B^3$

Sum of squares An expression that can be written in the form $A^2 + B^2$

Supplementary angles Angles whose sum is 180°

Synthetic division A simplified process for dividing a polynomial by a binomial of the type $x - a$

System of equations A set of two or more equations that are to be solved simultaneously

System of linear inequalities A set of two or more inequalities that are to be solved simultaneously

T

Term A number, a variable, or a product or a quotient of numbers and/or variables

Terminating decimal A decimal that can be written using a finite number of decimal places

Third-order determinant The determinant of a three-by-three matrix $\begin{bmatrix} a_1 & b_1 & c_1 \\ a_2 & b_2 & c_2 \\ a_3 & b_3 & c_3 \end{bmatrix}$ is denoted $\begin{vmatrix} a_1 & b_1 & c_1 \\ a_2 & b_2 & c_2 \\ a_3 & b_3 & c_3 \end{vmatrix}$ and is defined to mean $a_1 \begin{vmatrix} b_2 & c_2 \\ b_3 & c_3 \end{vmatrix} - a_2 \begin{vmatrix} b_1 & c_1 \\ b_3 & c_3 \end{vmatrix} + a_3 \begin{vmatrix} b_1 & c_1 \\ b_2 & c_2 \end{vmatrix}$.

Trinomial A polynomial that is composed of three terms

Trinomial square The square of a binomial expressed as three terms

U

Union of sets A and B The set of all elements belonging to A and/or B

V

Value The numerical result after a number has been substituted into an expression

Variable A letter that represents an unknown number

Variation constant The constant in an equation of direct or inverse variation

Vertex The point at which the graph of a quadratic equation crosses its axis of symmetry

Vertical-line test If it is possible for a vertical line to cross a graph more than once, then the graph is *not* the graph of a function.

W

Whole numbers The natural numbers and 0: 0, 1, 2, 3, …

X

x-intercept The point at which a graph crosses the x-axis

Y

y-intercept The point at which a graph crosses the y-axis

Photo Credits

Just-in-Time Review: 1, Ruslan Gusov/Shutterstock; dizanna/123RF. **Chapter 1: 27,** CE Photography/Shutterstock; Wayne Lynch/Shutterstock. **30,** Veniamin Kraskov/Shutterstock. **32,** (left) Zsolt Biczo/Shutterstock; (right) Carlos Santa Maria/Fotolia. **36,** Ed Metz/Shutterstock. **43,** (left) chuyu/123RF; (right) Carlos Villoch/MagicSea.com/Alamy Stock Photo. **50,** Comstock/Getty Images. **56,** Mellowbox/Fotolia. **66,** Ivan Alvarado/Alamy Stock Photos. **69,** Richard Whitcombe/123RF. **105,** Deposit Photos/Glow Images. **Chapter 2: 109,** Angelo Cavalli/AGF Srl/Alamy; Design Pics Inc/Alamy. **133,** Leonid Tit/Fotolia. **134,** Artit Fongfung/123RF. **137,** SoCalBatGal/Fotolia. **138,** (left) Echel, Jochen/Sueddeutsche Zeitung Photo/Alamy; (right) Iakov Filimonov/Shutterstock. **146,** Luca Bertolli/123RF. **149,** (left) iofoto/Shutterstock; (right) John Kavouris/Alamy. **150,** (left) gasparij/123RF; (right) Anja Schaefer/Alamy. **159,** Stan Honda/Getty Images. **160,** Stephen VanHorn/Shutterstock. **164,** (top left) Rodney Todt/Alamy Stock Photo; (top right) Jasminko Ibrakovic/Shutterstock; (bottom left) courtesy of Indianapolis Motor Speedway; (bottom right) Lars Lindblad/Shutterstock. **165,** (left) courtesy of Barbara Johnson; (right) Studio 8. Pearson Education Ltd. **181,** RosaIreneBetancourt 3/Alamy Stock Photo. **184,** Andrey N Bannov/Shutterstock. **187,** (left) Reggie Lavoie/Shutterstock; (right) Monkey Business/Fotolia. **191,** pearlguy/Fotolia. **192,** Don Mammoser/Shutterstock. **Chapter 3: 197,** Africa Studio/Shutterstock; tethysimagingllc/123RF; zimmytws/Shutterstock. **209,** Andrey Popov/Shutterstock. **233,** David Pearson/Alamy. **238,** (left) marchcattle/123RF; (right) Evan Meyer/Shutterstock. **239,** (left) Kostyantine Pankin/123RF; (right) Wavebreak Media Ltd/123RF. **245,** arinahabich/Fotolia. **250,** Wei Ming/Shutterstock. **Chapter 4: 251,** Trevor R A Dingle/Alamy Stock Photo; Universal Images Group North America LLC/DeAgostini/Alamy. **253,** NASA. **264,** (left) Universal Images Group North America LLC/DeAgostini/Alamy; (right) jezper/123RF. **267,** Lorraine Swanson/Fotolia. **268,** Darts/123RF. **270,** (left) luchschen/123RF; (right) Joanne Weston/123RF. **271,** (top) Engine Images/Fotolia; (bottom) NASA. **272,** (left) SeanPavonePhoto/Fotolia; (right) dreamerb/123RF. **282,** (left) Brian Buckland; (right) Johan Swanepoel/123RF. **315,** Johan Swanepoel/123RF. **319,** Mark Harvey/Alamy. **Chapter 5: 341,** Galina Peshkova/123RF; georgejmclittle/123RF; Rawpixel/Shutterstock. **359,** Zoonar/Chris Putnam/Age Fotostock. **373,** Free Spirit Spheres. **409,** georgejmclittle/123RF. **415,** Stephen Barnes/Alamy Stock Photo. **423,** Pattie Steib/Shutterstock. **428,** Vlada Photo/Shutterstock. **Chapter 6: 429,** MediaWorldImages/Alamy Stock Photo; modfos/123RF; Oleksandr Galata/123RF. **448,** Igor Normann/Shutterstock. **483,** Ingrid Balabanova/123RF. **486,** Ortodox/Shutterstock. **491,** (left) Elenathewise/Fotolia; (right) simon johnsen/Shutterstock. **493,** (left) Chris Dorney/123RF; (right) Rolf Nussbaumer Photography/Alamy Stock Photo. **494,** Vicki Jauron, Babylon and Beyond Photography/Getty Images. **495,** Lisa F. Young/Fotolia. **497,** Ikonoklast_hh/Fotolia. **501,** 578foot/123RF. **502,** (left and right) S_oleg/Shutterstock. **503,** Barry Blackburn/123RF. **511,** (left) Dmitry Kalinovsky/Shutterstock; (right) James McConnachie © Rough Guides/Pearson Asset Library. **Chapter 7: 517,** miroslav110/123RF; Softulka/Shutterstock; arcady31/123rf. **519,** Mesut Dogan/123RF. **524,** Geri Lavrov/Getty Images. **527,** (left) WENN Ltd/Alamy Stock Photo; (right) Graham Hardy USA/Alamy Stock Photo. **528,** pstedrak/123RF. **567,** Andrey Cherkasov/123RF. **568,** Thampapon/Shutterstock. **572,** (left) Cathleen A. Clapper/Shutterstock; (right) Mali lucky/Shutterstock. **586,** Steve Peeple/Shutterstock. **Chapter 8: 587,** dotshock/Shutterstock; red mango/Shutterstock; Iculig/123RF. **588,** Cathy Yeulet/123RF. **601,** StephanScherhag/Shutterstock. **608,** John David Mercer/Mobile Press-Register. **611,** (left) Michael Dwyer/Alamy Stock Photo; (right) V. J. Matthew/Shutterstock. **612,** Tyler Olson/123RF. **613,** Vladimir Wrangel/Shutterstock. **621,** (left) REDPIXEL.PL/Shutterstock; (right) dizanna/123RF; **635,** ShutterOK/Shutterstock. **637,** (left) smolaw/Shutterstock; (right) Anton Estrada/123RF. **638,** (left) Alena Ozerova/123RF; (right) Bram Reusen/Shutterstock. **Chapter 9: 651,** mrmohock/Shutterstock; zlomari/123RF. **659,** Vitaliy Vodolazsky/Shutterstock. **660,** gwimages/Fotolia. **666,** Bob Orsillo/Shutterstock. **667,** Icatnews/Shutterstock. **668,** John Kwan/Shutterstock. **675,** Monkey Business/Shutterstock. **678,** Andrey N. Bannov/Shutterstock. **706,** Leksele/Shutterstock. **716,** (left) Sergey Nivens/123RF; (right) Guas/Shutterstock. **Chapter 10: 717,** Blue Jean Images/Alamy Stock Photo; Testing/Shutterstock. **726,** ESB Professional/Shutterstock. **764,** NurPhoto/Contributor/Getty Images. **767,** jivan child/Fotolia. **768,** (left) Akhararat Wathanasing/123RF; (right) Kara Grubis/Shutterstock. **773,** (left) Mo Peerbacus/Alamy Stock Photo; (right) Lasse Kristensen/Fotolia. **791,** Hildegard Williams/123RF. **796,** Jochen Tack/Alamy Stock Photo. **Chapter 11: 797,** Hanna Kuprevich/Shutterstock; Dirk Ercken/123RF. **810,** (left) Ingo Oeland/Alamy Stock Photo; (right) Xinhua/Alamy Stock Photo. **823,** ARENA Creative/Shutterstock. **826,** Bill Bachman/Alamy Stock Photo. **828,** (left) Rob Kints/Shutterstock; (right) Stefan Schurr/Shutterstock. **865,** Bob Cheung/Fotolia. **Chapter 12: 889,** Vaclav Volrab/123RF; Jason Winter/Shutterstock. **902,** (left) Artush/Shutterstock; (right) senrakawa/Shutterstock. **903,** (left) Jenny Thompson/Fotolia; (right) Deco Images II/Alamy Stock Photo. **904,** Alex Van Wyhe and Shane Kimberlin of Tea Mug Collective. **956,** Aaron Amat/Shutterstock. **959,** Roman Fedin/123RF. **962,** Rancz Andrei/Shutterstock. **963,** Reprinted with permission of the Baltimore Sun Media Group. All rights reserved. **965,** (left) Michael Rosskothen/123RF; (right) Fotolia RAW/Fotolia. **966,** (left) Nicky Rhodes/Shutterstock; (right) rawpixel/123RF. **968,** (left) Trial/Shutterstock; (right) pmstephens/Fotolia. **969,** (left) ukartpics/Alamy Stock Photo; (right) David Chedgy/Alamy Stock Photo. **976,** Victor Korchenko/Alamy Stock Photo. **977,** rawpixel/123RF. **981,** xiaoliangge/fotolia. **Appendixes: 990,** National Geographic Creative/Alamy Stock Photo. **992,** (left) Brent Hofacker/123RF; (right) Kvitka Fabian/Shutterstock.

Index

A
Abscissa, 198
Absolute value, 42
   and distance, 42, 683
   on a graphing calculator, 42
   inequalities with, 686, 687
   properties of, 682
   and radical expressions, 722, 724
Absolute-value function, 863
Absolute-value principle, 684
ac-method (grouping method), 368
Addition
   associative law of, 82
   commutative law of, 81
   of complex numbers, 777
   with decimal notation, 15
   of exponents, 254, 257, 268, 731
   of fractions, 12
   of functions, 1005
   of logarithms, 932
   on the number line, 46
   of polynomials, 287, 317
   of radical expressions, 745
   of rational expressions, 449, 451, 452
   of real numbers, 47
Addition method for systems of equations, 603
Addition principle
   for equations, 112
   for inequalities, 170, 655
Additive identity, 7, 79
Additive inverse, 48, 50, 288. *See also* Opposite(s).
Advance of a pipe, 774
Algebra of functions, 1005
Algebraic equation, 29
Algebraic expression(s), 28, 29
   evaluating, 29, 253
   least common multiple, 446
   translating to, 30
   value of, 29
Algebraic-Graphical Connection, 210, 275, 276, 403, 474, 592, 760, 798, 813, 949
"and," 670
Angles
   complementary, 602, 610
   in similar triangles, 486, 487
   supplementary, 602, 610
Annually compounded interest, 896, 958, 970
Approximately equal to ($\approx$), 18
Approximating roots, 39, 719

Approximating solutions of quadratic equations, 814, 815
Area
   of a parallelogram, 33
   of a rectangle, 29, 819
   surface, of a cube, 137
   surface, of a right circular cylinder, 320
   of a triangle, 33
Arithmetic numbers, 7
Ascending order, 280
Associative laws, 81, 82
Asymptotes of an exponential function, 892
At least, 180, 659
At most, 180, 659
Auto mode on graphing calculator, 205
Average, 990
Axes, 198
Axis of symmetry, 843, 851, 854, 879

B
Back-substitution, 1003
Bar graph, 198
Base, 23, 252
   of an exponential function, 891
   of a logarithmic function, 922
   changing, 942
Base-10 logarithms, 926
Binomials, 277
   differences of squares, 378
   as divisor, 325
   product of, 298
      FOIL method, 304
      squares of, 307
      sum and difference of terms, 306
Braces, 94
Brackets, 94

C
Calculator, *see* Graphing calculator
Calculator Corner, 12, 15, 23, 24, 39, 40, 42, 74, 96, 125, 205, 211, 218, 232, 266, 276, 300, 361, 400, 476, 521, 523, 540, 543, 551, 593, 731, 760, 781, 802, 813, 814, 862, 894, 911, 933, 942, 944
Canceling, 9, 439
Carbon dating, 963
Center of a circle, 1010
Center point of data, 990
Central tendency, measure of, 990
Change, rate of, 234, 546. *See also* Slope.
Change-of-base formula, 942

Changing the sign, 50, 74
Check in problem-solving process, 151
Checking
   division of polynomials, 325, 326
   factorizations, 345, 347, 361
   multiplication of polynomials, 300
   solutions
      of applied problems, 151
      of equations, 112, 125, 399, 475, 589, 758, 834, 952
      of inequalities, 655
      of radical equations, 758
      of systems of equations, 589
Circle, 1010
Clearing decimals, 126
Clearing fractions, 125, 473
Coefficients(s), 277, 316
   leading, 350
Collecting like (or similar) terms, 87, 277, 316
   in equation solving, 123
Columns of a matrix, 1001
Combining like (or similar) terms, *see* Collecting like (or similar) terms
Common denominator, 12, 445. *See also* Least common denominator.
Common factor, 4, 86, 344
   greatest, 4, 342, 344
Common multiple, least, 5. *See also* Least common multiple.
Common logarithms, 926
Commutative laws, 80, 81
Complementary angles, 602, 610
Completing the square
   and graphing quadratic functions, 851
   in solving equations, 803–806
Complex conjugate, 779
Complex fraction expression, 467
Complex number(s), 776, 777
   addition, 777
   conjugates, 779
   division, 780
   graphing calculator and, 781
   multiplication, 778
   as solutions of equations, 781
   subtraction, 777
Complex-number system, 777. *See also* Complex number(s).
Complex rational expression, 467
   simplifying, 467, 469
Composite function, 904, 905
   and inverses, 915
Composite number, 3
Composition of functions, 904, 905

Compound inequalities, 668
Compound interest, 896–898, 959, 970
Conjugate
  of a complex number, 779
  of a radical expression, 755
Conjunction of inequalities, 669
Consecutive integers, 154
Consistent system of equations, 590, 607
Constant, 28
  of proportionality, 495, 497
  variation, 495, 497
Constant function, 521, 554
Continuously compounded interest, 959, 970
Converting
  decimal notation to fraction notation, 14
  decimal notation to percent notation, 20
  decimal notation to scientific notation, 265
  between exponential and logarithmic equations, 923
  fraction notation to decimal notation, 17, 40
  fraction notation to percent notation, 21
  a fraction to a mixed numeral, 8
  a mixed numeral to a fraction, 8
  percent notation to decimal notation, 20
  percent notation to fraction notation, 21
  scientific notation to decimal notation, 265
  temperatures, 666
Coordinates, 198
  finding, 200
Correspondence, 518, 519
Cramer's rule, 997–999
Cross products, 485
Cube, surface area, 137
Cube root, 722
Cube-root function, 723
Cubes, factoring sum and difference of, 386–388
Cubic function, 863
Cylinder, right circular, surface area, 320

D
Data analysis, 990, 991
Decay model, exponential, 962, 970
Decay rate, 962
Decimal notation, 14
  addition with, 15
  on a calculator, 15
  converting to/from fraction notation, 14, 17
  converting to/from percent notation, 20
  converting to/from scientific notation, 265

division with, 16
for irrational numbers, 39
multiplication with, 16
for rational numbers, 38, 39
repeating, 38
rounding, 18
subtraction with, 15
terminating, 38
Decimals, clearing, 126
Degree of term and polynomial, 278, 316
Denominator, 7
  least common, 12, 445
  rationalizing, 753–755
Dependent equations, 591, 607, 999
Descending order, 280
Determinant, 996
  solving systems using, 997
Difference, 54. *See also* Subtraction.
  of cubes, factoring, 386–388
  of functions, 1005
  of logarithms, 933
  of squares, 378
    factoring, 379
Direct variation, 495
Directly proportional, 495
Discriminant, 831, 879
Disjoint sets, 671
Disjunction of inequalities, 672
Distance
  and absolute value, 42, 683
  on the number line, 683
  between two points, 1008
Distance traveled, 481
Distributive laws, 83, 84
Dividend(s), 324
  of zero, 71
Division
  checking, 325, 326
  of complex numbers, 780
  with decimal notation, 16
  and dividend of zero, 71
  with exponential notation, 255, 257, 268, 731
  of fractions, 11
  of functions, 1005
  of integers, 70
  of polynomials, 324–327
  of radical expressions, 739
  of rational expressions, 440
  of real numbers, 70, 73
  and reciprocals, 10, 73, 440
  with scientific notation, 266
  synthetic, 993
  by zero, 70
Divisor, 326
  binomial, 325
  monomial, 324
Domain
  of a function, 518, 519, 532, 720, 914
  of a relation, 519
Doubling time, 958, 959

Drawing feature on a graphing calculator, 911

E
e, 940
Elimination method, 603, 627
  using matrices, 1001
Empty set, 671
Endpoints of an interval, 653
Equation(s), 110. *See also* Formula(s).
  with absolute value, 683–685, 687
  algebraic, 29
  checking solutions, 112, 125, 399, 589, 758, 834, 952
  of a circle, 1010
  dependent, 591, 607, 999
  of direct variation, 495
  equivalent, 111
  exponential, 923, 949
  false, 110
  fraction, 473
  graph of, 201. *See also* Graph(s).
  independent, 591, 607
  with infinitely many solutions, 127, 128
  of inverse variation, 497
  linear, 195, 205
  logarithmic, 923, 951
  with no solution, 128
  parentheses in, 126
  point-slope, 562, 563
  polynomial, 275
  quadratic, 399, 799
  quadratic in form, 833
  radical, 758
  rational, 473
  reducible to quadratic, 833
  related, 695
  reversing, 113
  finding slope from, 231
  slope-intercept, 545, 562
  solutions, 110, 200
  solving, *see* Solving equations
  systems of, 589, 627
  translating to, 143, 151
  true, 110
  of variation, 495, 497
Equivalent equations, 111
Equivalent expressions, 7, 79, 431
  for one, 7, 431
Equivalent inequalities, 169, 655
Evaluating
  algebraic expressions, 29, 253
  determinants, 996
  exponential expressions, 23, 253
  formulas, 133
  functions, 520, 521
  polynomials, 274, 275, 276, 315
Even integers, consecutive, 154
Even root, 724
Exponent(s), 23, 252
  adding, 254, 257, 268, 731
  dividing using, 255, 257, 268, 731

I-2 Index

evaluating expressions containing, 253
irrational, 890
laws of, 257, 268, 731
and logarithms, 922
multiplying, 262, 268, 731
multiplying using, 254, 257, 268
negative, 256, 257, 268
one as, 253, 257, 268
power rule, 262, 268, 731
product rule, 263, 268, 731
quotient rule, 262, 268, 731
raising a power to a power, 262, 268, 731
raising a product to a power, 263, 268, 731
raising a quotient to a power, 263, 268, 731
rational, 729–731
rules for, 257, 268, 731
subtracting, 255, 257, 268, 731
zero as, 253, 257, 268
Exponential decay model, 962, 970
Exponential decay rate, 962
Exponential equality, principle of, 949
Exponential equations, 923, 949. *See also* Exponential functions.
converting to logarithmic equations, 923
solving, 949
Exponential expressions, 23, 253
Exponential functions, 890, 891
graphs, 891–895, 943
asymptotes, 892
y-intercept, 893
inverse, 922
Exponential growth model, 959, 970
Exponential growth rate, 959
Exponential notation, 23, 252. *See also* Exponent(s).
Expressions. *See also* Radical expressions; Rational expressions.
algebraic, 28, 29, 253
equivalent, 7, 79, 431
evaluating, 29, 253
exponential, 253
factoring, 85
fraction, 430
radical, 720
rational, 430
simplifying, *see* Simplifying
terms of, 84
value, 29
Extraneous solution, 759

F

Factor(s), 2, 86, 342, 343. *See also* Factoring; Factorization.
common, 4, 86, 344
greatest common, 4, 342, 344
Factor tree, 3
Factoring, 2, 85, 342, 347, 381
ac-method, 368
checking, 345, 347

common factor, 4, 86, 344
completely, 380
difference of cubes, 386–388
difference of squares, 379
and finding LCM, 5
FOIL method, 360
by grouping, 346
grouping method, 368
numbers, 2
polynomials, 344–347, 350–355, 360–364, 368, 369, 377
radical expressions, 737
and solving equations, 401
strategy, 391
sum of cubes, 386–388
sum of squares, 380, 386
trinomial squares, 377
trinomials, 350–355, 360–364, 368, 369, 377
Factorization, 2, 342, 343. *See also* Factoring.
checking, 345, 347
prime, 3
False equation, 110
False inequality, 168, 652
Familiarize in problem-solving process, 151
First coordinate, 198
Fitting functions to data, 862
Five-step problem-solving process, 151
FOIL, 304
and factoring, 360
Formula(s), 133
change-of-base, 942
compound interest, 898, 970
distance between points, 1008
distance traveled, 481
evaluating, 133
games in a league, 415
hang time, 806
horizon, sighting to, 767
for inverse of a function, 910
loudness of sound, 956
midpoint, 1009
motion, 481
period of a pendulum, 744, 768, 822
pH, 957
quadratic, 813, 879
road pavement messages, 770
skidding car, speed of, 767
solving, 134–136, 822, 824
sound, speed of, 764
temperature, 666
water flow, 768
wind chill temperature, 775
Fraction(s). *See also* Fraction notation.
adding, 12
clearing, 125, 473
dividing, 11
multiplying, 10
and sign change, 74

simplifying, 9, 80
subtracting, 12
Fraction bar
as division symbol, 29
as grouping symbol, 24
Fraction equations, 473, 670
Fraction exponents. *See* Rational exponents.
Fraction expressions, 430. *See also* Rational expressions.
complex, 467
Fraction notation, 7, 38. *See also* Fraction(s).
converting to/from decimal notation, 14, 17, 40
converting to/from a mixed numeral, 8
converting to/from percent notation, 21
sign changes in, 74
simplest, 9
simplifying, 9
Function(s), 518, 519, 532
absolute-value, 863
addition of, 1005
algebra of, 1005
composite, 904, 905
constant, 521, 554
cube-root, 723
cubic, 863
difference of, 1005
division of, 1005
domain, 518, 519, 532
evaluating, 520, 521
exponential, 890, 891
fitting to data, 862
graphs, 522, 523
horizontal-line test, 909
inputs, 520
inverse of, 908
library of, 863
linear, 539, 863
logarithmic, 922
multiplication of, 1005
notation, 520
one-to-one, 909
outputs, 520
product of, 1005
quadratic, 842, 863
quartic, 863
quotient of, 1005
range, 518, 519, 532
square-root, 720
subtraction, 1005
sum of, 1005
value(s), 520
vertical-line test, 523

G

Games in a league, 415
GCF, *see* Greatest common factor
Geometry and polynomials, 289
Golden rectangle, 830

Index I-3

Grade, 233, 545
Graph(s), 201. *See also* Graphing.
   of absolute-value function, 863
   bar, 198
   of a circle, 1010
   of cube-root function, 723
   of cubic function, 863
   of direct-variation equations, 496
   of equations, 201, 211, 555
   of exponential functions, 890–895, 943
   of functions, 522, 523
     and their inverses, 911, 913
     vertical-line test, 523
   on a graphing calculator, 523
   of horizontal lines, 219, 220, 553
   of inequalities, 168, 653, 654, 693
   of inverse-variation equations, 498
   of inverses of functions, 911, 912
   of linear equations, 198, 205, 550, 552, 555
   of linear functions, 539, 863
   of logarithmic functions, 921–923, 944
   nonlinear, 211
   of numbers, 37
   of parabolas, 842–847, 851–855, 863
   of quadratic equations, 842–847, 851–855, 863
   of quartic functions, 863
   of radical functions, 721
   reflection of, 895, 913
   slope, 229, 542, 546
   and solving equations, 592, 813
   of square-root functions, 720
   of systems of equations, 589, 591, 593
   of systems of inequalities, 698
   translations of, 540, 844, 894
   of vertical lines, 219, 220, 554
   x-intercept, 216, 403, 550, 855
   y-intercept, 207, 216, 541, 550, 855, 893
Graphing. *See also* Graph(s).
   using intercepts, 216, 550
   linear equations, 205, 220
   numbers, 37
   points on a plane, 198
   using slope and y-intercept, 552
Graphing calculator
   and absolute value, 42
   and approximating solutions of quadratic solutions, 814
   and approximating square roots, 39
   auto mode, 205
   and change-of-base formula, 942
   and checking a factorization, 361
   and checking multiplication of polynomials, 300
   and checking solutions of equations, 125
   and complex numbers, 781
   and converting fraction notation to decimal notation, 40
   and decimal notation, 15
   and drawing feature, 911
   entering equations, 205, 211
   and evaluating polynomials, 276
   and exponential functions, 894
   and exponents, 23
   and fractions, 12
   and function values, 521
   graphing equations, 211
   graphing functions, 523
   graphing logarithmic functions, 944
   and grouping symbols, 96
   and imaginary solutions of quadratic equations, 802
   and intercepts, 218, 551
   intersect feature, 593, 760
   and inverses of functions, 911
   and logarithms, 926, 941
   and maximum and minimum values, 862
   and negative numbers, 40
   and operations with decimal notation, 15
   and operations on fractions, 12
   and operations on real numbers, 74
   and order of operations, 24, 96
   pi key, 39
   and powers, 23
   and properties of logarithms, 933
   and quadratic equations, 400, 813, 814
   and radical equations, 760
   and rational exponents, 911
   real numbers, operations on, 74
   and scientific notation, 266
   slope, visualizing, 232
   and solutions of equations, 205
   solving equations, 400, 760, 802, 813
   solving systems of equations, 593
   and square roots, 39, 719
   table feature, 205
   value feature, 521
   viewing window, 218
   zero feature, 400, 802
Greater than ($>$), 40
Greater than or equal to ($\geq$), 42
Greatest common factor, 4, 342, 344
Grouping
   in addition, 81, 82
   factoring by, 346
   in multiplication, 80, 81
   symbols, 24, 94, 96
Grouping method for factoring (*ac*-method), 368
Growth model, exponential, 959, 970
Growth rate, exponential, 959

H
Half-life, 962
Half-plane, 694

Hang time, 806
Height of a projectile, 878
Horizon, sighting to, 767
Horizontal line, 220, 553, 554
   slope, 554
Horizontal-line test, 909
Hypotenuse, 410

I
i, 776
   powers of, 779
Identity
   additive, 7, 79
   multiplicative, 7, 79
Identity property of one, 7, 79
Identity property of zero, 7, 47, 79
Imaginary number, 776
Inconsistent system of equations, 590, 607, 999
Independent equations, 591, 607
Index of a radical expression, 723
Inequalities, 40, 168, 652. *See also* Solving inequalities.
   with absolute value, 686, 687
   addition principle for, 170, 655
   with "and," 670
   checking solutions, 655
   compound, 668
   conjunction of. 669
   disjunction of, 672
   equivalent, 169, 655
   false, 168, 652
   graphs of, 168, 653, 654, 693
   linear, 693
   multiplication principle for, 172, 656
   with "or," 673
   polynomial, 871
   quadratic, 871
   rational, 874
   solution of, 652
   solution set, 168, 652, 693
   solving, see Solving inequalities
   systems of, 697
   translating to, 180, 728, 729
   true, 168, 652
Infinitely many solutions, 127, 128
Infinity (∞), 654
Inputs, 520
Integers, 35, 36
   consecutive, 154
   division of, 70
   negative, 36
   positive, 36
Intercepts, 216
   of exponential functions, 893
   graphing using, 216, 550
   of a parabola, 855
   of a quadratic function, 855
   x-, 216, 403, 550, 855
   y-, 207, 216, 541, 550, 855, 893
Interest, compound, 896–898, 958, 959, 970

Intersect feature, graphing calculator, 593, 760
Intersection of sets, 668, 670
Interval notation, 653, 654
Introductory Algebra Review, 988
Inverse(s)
  additive, 48, 50, 288. *See also* Opposites.
  of functions, 908
    and composition, 915
    domain, 914
    exponential, 922
    formula for, 910
    logarithmic, 927, 941
    range, 914
  multiplicative, 10, 71. *See also* Reciprocals.
  of relations, 906, 907
Inverse relation, 906, 907. *See also* Inverse(s), of functions.
Inverse variation, 497
Inversely proportional, 497
Irrational numbers, 39
  as exponents, 890

J
Joint variation, 500

K
kth roots, 723, 724
  of quotients, 740
  simplifying, 723, 724, 737

L
Largest common factor, 86
Law(s)
  associative, 81, 82
  commutative, 80, 81
  distributive, 83, 84
  of exponents, 257, 268, 731
LCD, *see* Least common denominator
LCM, *see* Least common multiple
Leading coefficient, 350
Least common denominator, 12, 445
Least common multiple, 5, 445, 446
  and clearing fractions, 125
Legs of a right triangle, 410
Less than ($<$), 40
Less than or equal to ($\leq$), 42
Library of functions, 863
Like radicals, 745
Like terms, 87, 279, 316
Line(s)
  equation of, *see* Linear equations
  horizontal, 219, 220, 553, 554
  parallel, 555
  perpendicular, 556
  slope, 229, 542, 544, 554, 555, 556
  vertical, 219, 220, 554
Line of symmetry, 843, 851, 854, 879
Linear equations, 198, 205
  graphing, 205, 220, 550, 552, 555

graphs of, 198, 205
point-slope form, 562, 563
slope-intercept form, 545, 562
in three variables, 627
Linear functions, 539, 863. *See also* Linear equations.
Linear inequality, 693. *See also* Inequalities.
ln, 940
log, 926
Logarithm(s), 921, 922. *See also* Logarithmic functions.
  of a, base a, 926
  base a, 922
  base a of 1, 925
  base e, 940
  base 10, 926
  of the base to a power, 935
  on a calculator, 926, 941
  change-of-base formula, 942
  common, 926
  difference of, 933
  and exponents, 922
  Naperian, 940
  natural, 940
  of one base a, 925
  of powers, 932
  of products, 932
  properties of, 932, 933, 935
  of quotients, 933
  sum of, 932
Logarithm functions, *see* Logarithmic functions
Logarithmic equality, principle of, 950
Logarithmic equations, 923, 951
  converting to exponential equations, 923
  solving, 924, 951
Logarithmic functions, 922
  graphs of, 921–923, 944
  inverses of, 927, 941
Loudness of sound, 956

M
Mathematical model, 567, 860
Matrix (matrices), 1001
  row-equivalent, 1004
Maximum value, quadratic function, 853
Mean, 990
Measure of central tendency, 990
Median, 990, 991
Midpoint formula, 1009
Minimum value, quadratic function, 851
Missing terms, 280
Mixed numeral, 8
  converting to/from fraction notation, 8
Mixture problems, 613
Mode, 991

Model
  exponential growth and decay, 959, 962, 970
  mathematical, 567, 860
Monomials, 274, 277
  as divisors, 324
  multiplying, 297
Motion formula, 481
Motion problems, 481
Multiples, 5
  least common, 5, 445, 446
Multiplication
  of algebraic expressions, 84
  associative law, 81, 82
  commutative law, 80, 81
  of complex numbers, 778
  with decimal notation, 16
  distributive laws, 83, 84
  with exponential notation, 254, 257, 268, 731
  of exponents, 262, 268, 731
  of fractions, 10
  of functions, 1005
  with negative numbers, 63, 64, 65
  by negative one, 92
  by one, 79, 431
  of polynomials, 297–300, 304–309, 317
  of radical expressions, 736, 746
  of rational expressions, 431
  of real numbers, 63–65, 70
  using exponents, 254, 257, 268
  using scientific notation, 265, 266
  by zero, 64
Multiplication principle
  for equations, 116
  for inequalities, 172, 656
Multiplication property of zero, 64
Multiplicative identity, 7, 79
Multiplicative inverses, 10, 71. *See also* Reciprocals.

N
nth power, 252
Naperian logarithms, 940
Natural logarithms, 940
Natural numbers, 2, 35
Nature of solutions, quadratic equations, 831
Negative exponents, 256, 257, 268
Negative numbers, 41
  on a graphing calculator, 40
  integers, 36
  multiplication with, 63, 64, 65
  square roots of, 718
Negative one, property of, 92
Negative rational exponents, 730
Negative square root, 719
No solutions, equations with, 128
Nonlinear graphs, 211
Nonnegative rational numbers, 7

Index I-5

Notation
  composite function, 904
  decimal, 14
  exponential, 23, 252
  fraction, 7
  function, 520
    composite, 904
    inverse, 910
  interval, 653, 654
  inverse function, 910
  percent, 19
  radical, 720
  for rational numbers, 37, 38
  roster, 35
  scientific, 264, 265, 268
  set, 35, 171, 653, 654
  set-builder, 171, 653
Number line, 37
  addition on, 46
  distance on, 683
  graphing inequalities on, 653, 654
  graphing numbers on, 37
  order on, 40
Numbers
  arithmetic, 7
  complex, 776, 777
  composite, 3
  factoring, 2
  factors of, 2
  graphing on number line, 37
  imaginary, 776
  integers, 35, 36
  irrational, 39
  natural, 2, 35
  negative, 36, 40, 41
  nonnegative rational, 7
  opposites, 36
  order of, 40
  positive, 36, 41
  prime, 3
  rational, 37
    nonnegative, 7
  real, 40
  signs of, 50
  whole, 7, 35
Numeral, mixed, 8
Numerator, 7

O
Odd integers, consecutive, 154
Odd root, 723, 724
Offset of a pipe, 774
Ohm's law, 496
One
  equivalent expressions for, 7
  as exponent, 253, 257, 268
  identity property of, 7, 79
  logarithm of, base a, 925
  multiplying by, 7, 79, 431
  removing a factor of, 9, 80, 432
One-to-one function, 909

Operations
  order of, 24, 95, 96
  row-equivalent, 1004
Opposite(s), 36
  and changing the sign, 50, 74
  in denominators, 451, 452, 458, 459
  and multiplying by negative one, 92
  of numbers, 48, 50
  of an opposite, 49
  of a polynomial, 288
  in rational expressions, 434, 451, 452
  of real numbers, 48, 50
  and subtraction, 55
  of a sum, 92
  sum of, 50
  symbolizing, 49
"or," 673
Order
  ascending, 280
  descending, 280
  on the number line, 40
Order of operations, 24, 95, 96
Ordered pair, 198
Ordinate, 198
Origin, 198
Outputs, 520

P
Pair, ordered, 198
Parabola, 843
  axis of symmetry, 843, 851, 854, 879
  graph of, 842–847, 851–855
  intercepts, 855
  line of symmetry, 843, 851, 879
  vertex, 843, 851, 854, 879
Parallel lines, 555
Parallelogram, area, 33
Parentheses
  in equations, 126
  within parentheses, 94
  removing, 93
Pendulum, period of, 744, 768, 822
Percent notation, 19
  converting to/from decimal notation, 20
  converting to/from fraction notation, 21
Percent problems, 143–146
Perfect square, 721
Perfect-square trinomial, 376
  factoring, 377
Period of a pendulum, 744, 768, 822
Perpendicular lines, 556
pH, 957
pi (π), 39
Place-value chart, 14
Plotting points, 198
Point-slope equation, 562, 563
Points, coordinates, 198
Polynomial(s), 274. See also Binomials; Monomials; Trinomials.
  addition of, 287, 317

  additive inverse of, 288
  ascending order, 280
  binomials, 277
  coefficients, 277, 316
  collecting like terms (or combining similar terms), 279, 316
  degree, 278, 316
  degree of a term, 278, 316
  descending order, 280
  division of, 324–327
  evaluating, 274, 275, 276, 315
  factoring, 344–347, 350–355, 360–364, 368, 369, 377
  and geometry, 289
  leading coefficients, 350
  like terms, 279, 316
  missing terms, 280
  monomials, 274, 277
  multiplication of, 297–300, 304–308, 317
  opposite of, 288
  perfect-square trinomial, 376
  prime, 354
  in several variables, 315
  subtraction of, 288, 317
  terms, 277
  trinomial squares, 376
  trinomials, 277
  value of, 274
Polynomial equation, 275
Polynomial inequality, 871
Population decay, 962, 970
Population growth, 959, 970
Positive integers, 36
Positive numbers, 41
Power(s), 23
  of i, 779
  logarithm of, 932
  principle of, 758
  raising to, 262, 268, 731
Power rule
  for exponents, 262, 268, 731
  for logarithms, 932
Prime factorization, 3
Prime number, 3
Prime polynomial, 354
Principal square root, 719, 721
Principle
  absolute-value, 684
  addition
    for equations, 112
    for inequalities, 170, 655
  of exponential equality, 949
  of logarithmic equality, 950
  multiplication
    for equations, 116
    for inequalities, 172, 656
  of powers, 758
  of square roots, 800, 879
  work, 484
  of zero products, 400
Problem solving, five-step process, 151

Procedure for solving equations, 127
Product(s). *See also* Multiplication.
   cross, 485
   of functions, 1005
   logarithm of, 932
   raising to a power, 263, 268, 731
   of two binomials, 298, 304, 306
Product, raising to a power, 263, 268, 731
Product rule
   for exponents, 254, 257, 268, 731
   for logarithms, 932
   for radicals, 736
Properties
   of absolute value, 682
   of exponents, 257, 268, 731
   of logarithms, 932, 933, 935
   of reciprocals, 72
   of square roots, 718
Property of negative one, 92
Proportion, 484
Proportional, 484
   directly, 495
   inversely, 497
Proportionality, constant of, 495, 497
Pythagorean theorem, 410, 769

Q
Quadrants, 199
Quadratic equations, 399. *See also* Quadratic functions.
   approximating solutions, 814, 815
   discriminant, 831, 879
   graphs of, 842–847, 851–855, 863
   reducible to quadratic, 833
   solutions, nature of, 831
   solving, 814
      by completing the square, 803–806
      by factoring, 401
      graphically, 400, 813
      using principle of square roots, 800
      using principle of zero products, 399, 400
      using quadratic formula, 813
   standard form, 799
   writing from solutions, 832
   x-intercepts, 403
Quadratic in form equations, 833
Quadratic formula, 813, 879
Quadratic functions
   fitting to data, 862
   graphs of, 842–847, 851–855, 863
   intercepts, 855
   maximum value, 853
   minimum value, 851
Quadratic inequalities, 871
Quartic function, 863
Quotient
   of functions, 1005
   of integers, 70
   kth roots of, 740
   logarithm of, 933
   of polynomials, 326

   of radical expressions, 739
   raising to a power, 263, 268, 731
   roots of, 740
Quotient rule
   for exponents, 255, 257, 268, 731
   for logarithms, 933
   for radicals, 739

R
Radical, 720
Radical equations, 758
Radical expressions, 720
   and absolute value, 722, 724
   addition of, 745
   conjugates, 755
   dividing, 739
   factoring, 737
   index, 723
   like, 745
   multiplying, 736, 746
   product rule, 736
   quotient rule, 739
   radicand, 720
   rationalizing the denominator, 753–755
   simplifying, 732
   subtraction of, 745
Radicand, 720
Radius, 1010
Raising a power to a power, 262, 268, 731
Raising a product to a power, 263, 268, 731
Raising a quotient to a power, 263, 268, 731
Range
   of a function, 518, 519, 532, 914
   of a relation, 519
Rate, 484
   of change, 234, 546
   of exponential decay, 962
   of exponential growth, 959
Ratio, 484
Rational equations, 473
Rational exponents, 729–731
Rational expressions, 430. *See also* Rational numbers.
   addition of, 449, 451, 452
   complex, 467
   division of, 440
   multiplying, 431
      by one, 431
   not defined, 430
   opposites in, 434, 451, 452
   reciprocals of, 440
   simplifying, 432, 467, 469
   subtraction of, 457, 458, 459
Rational inequalities, 874
Rational numbers, 37. *See also* Rational expressions.
   decimal notation for, 38
   as exponents, 729–731
   nonnegative, 7

Rationalizing the denominator, 753–755
Real-number system, 39, 40. *See also* Real numbers.
Real number(s), 40
   addition of, 46, 47
   division of, 70, 73
   multiplication of, 63–65, 70
   operations on, on a graphing calculator, 74
   order, 40–42
   subsets, 35, 777
   subtraction of, 54, 55
Reciprocals, 10, 71
   and division, 10, 73
   and exponential notation, 730
   properties, 72
   of rational expressions, 440
   sign of, 72
Rectangle
   area, 29, 819
   golden, 830
Reducible to quadratic equation, 833
Reflection, 895, 913
Related equation, 695
Relation, 519, 532
   inverse, 906, 907
Remainder, 326
Removing a factor of one, 9, 80, 432
Removing parentheses, 93
Repeating decimal notation, 38
Reversing equations, 113
Review of Introductory Algebra, 988
Right circular cylinder, surface area, 320
Right triangle, 410
Rise, 229, 542, 543
Road grade, 545
Roots
   approximating, 39, 719
   cube, 722
   even, 724
   kth, 723, 724, 740
   odd, 723, 724
   square, 376, 718
Roster notation, 35
Rounding decimal notation, 18
Row-equivalent matrices, 1004
Row-equivalent operations, 1004
Rows of a matrix, 1001
Rules for exponents, 257, 258, 731
Run, 229, 542, 543
Run differential, 60

S
Scientific notation, 264, 265, 268
   on a calculator, 266
   converting from/to decimal notation, 265
   dividing with, 266
   multiplying with, 265, 266
Second coordinate, 198
Second-order determinant, 996
Semiannually compounded interest, 897

Set(s), 35
  disjoint, 671
  empty, 671
  intersection, 668, 670
  notation, 35, 171, 653, 654
  of numbers, 777
  solution, 168, 652
  subset, 35
  union, 672, 673
Set-builder, 171, 653
Several variables, polynomial in, 315
Sighting to the horizon, 767
Sign(s), 50
  change of, 50, 74
  in fraction notation, 74
  of a number, 50
  of a reciprocal, 72
Similar terms, collecting, 87, 279, 316
Similar triangles, 486, 487
Simplest fraction notation, 9
Simplifying
  complex rational expressions, 467, 469
  fraction notation, 9, 80
  kth roots, 723, 724, 737
  radical expressions, 732
  rational expressions, 432, 467, 469
  removing parentheses, 109
Skidding car, speed of, 767
Slope, 229, 542, 544
  applications of, 233
  from an equation, 231, 544
  graphing using, 552
  of a horizontal line, 232, 554
  negative, 230
  not defined, 232, 554
  of parallel lines, 555
  of perpendicular lines, 556
  positive, 230
  as rate of change, 234, 546
  of a vertical line, 232, 554
  zero, 232, 553, 554
Slope-intercept equation, 545, 562
Solution(s)
  checking, 112, 125, 399, 475, 589, 655, 758, 834, 952
  complex numbers as, 781
  of equations, 110, 200
  extraneous, 759
  of inequalities, 168, 652, 693
  of quadratic equations, nature of, 831
  of systems of equations, 589, 627
  of systems of inequalities, 697
  writing equations from, 832
Solution sets of inequalities, 168, 652, 693
Solve in problem-solving process, 151
Solving equations, 110, 127. *See also* Solving formulas.
  with absolute value, 683–685, 687
  checking the solution, 112, 125, 399, 475, 589, 758, 834, 952
  clearing decimals, 126

clearing fractions, 126
exponential, 949
by factoring, 401
fraction, 473
graphically, 400, 592, 813
logarithmic, 924, 951
with parentheses, 126
quadratic, 814
  by completing the square, 803–806
  by factoring, 401
  graphically, 400, 813
  using the principle of square roots, 800
  using the principle of zero products, 399
  using the quadratic formula, 813
quadratic in form, 833
radical, 758, 762
rational, 473
reducible to quadratic, 833
systems of, *see* Systems of equations
using the addition principle, 112, 125, 475
using the multiplication principle, 116
Solving formulas, 134–136, 692, 822, 824
Solving inequalities, 168, 655
  with absolute value, 686, 687
  addition principle for, 170, 655
  multiplication principle for, 172, 656
  polynomial, 871
  quadratic, 871
  rational, 874
  solution set, 168
Solving percent problems, 143–146
Solving systems of equations, *see* Systems of equations
Sound
  loudness of, 956
  speed of, 764
Special products of polynomials
  squaring binomials, 307
  sum and difference of two expressions, 306
  two binomials (FOIL), 304
Speed, 484
  of a skidding car, 767
  of sound, 764
  walking, 966
Square(s)
  of binomials, 307
  completing, 803–806, 851
  difference of, 378
    factoring, 379
  perfect, 721
  sum of, 380, 386
  trinomial, 376
Square root(s), 376, 718. *See also* Radical expressions.
  approximating, 39, 719
  negative, 719

of negative numbers, 718
principal, 719, 721
principle of, 800, 879
properties of, 718
Square-root function, 720
Squaring a number, 718
Standard form
  equation of a circle, 1010
  quadratic equation, 799
State the answer in problem-solving process, 151
Studying for Success, 28, 63, 110, 143, 198, 229, 252, 297, 342, 376, 430, 467, 518, 539, 588, 627, 652, 682, 718, 753, 798, 842, 890, 940
Subset, 35
Substituting, 29. *See also* Evaluating algebraic expressions.
Substitution method, 597
Subtraction, 54
  by adding the opposite, 55
  of complex numbers, 777
  with decimal notation, 15
  of exponents, 255, 257, 268, 731
  of fractions, 12
  of functions, 1005
  of logarithms, 933
  of polynomials, 288, 317
  of radical expressions, 745
  of rational expressions, 457, 458, 459
  of real numbers, 54, 55
Sum(s)
  of cubes, factoring, 386–388
  of functions, 1005
  of logarithms, 932
  of opposites, 50
  opposites of, 92
  of squares, 380, 386
Sum and difference of two terms, product of, 306
Supplementary angles, 602, 610
Surface area
  of a cube, 137
  of a right circular cylinder, 320
Symmetry, axis of, 843, 851, 854, 879
Symmetry, line of, 843, 851, 854, 879
Synthetic division, 993
Systems of equations in three variables, 627
Systems of equations in two variables, 589
  consistent, 590, 607
  with dependent equations, 591, 607, 999
  graphs of, 589, 591, 593
  inconsistent, 590, 607, 999
  with independent equations, 591, 607
  with infinitely many solutions, 590, 599, 607
  with no solution, 590, 599, 606, 607
  solution of, 589

I-8 Index

solving
    addition method, 603, 1001
    comparing methods, 607
    Cramer's rule, 997–999
    elimination method, 603, 1001
    graphically, 589, 593
    substitution method, 597
    using matrices, 1001
Systems of inequalities, 697

T

Table feature on a graphing calculator, 205
Table of primes, 3
Temperature, wind chill, 775
Temperatures, converting, 666
Terminating decimal, 38
Terms, 84
  coefficients of, 277
  collecting (or combining) like, 87, 279, 316
  degree of, 278, 316
  like, 87, 279, 316
  missing, 280
  of a polynomial, 277
Theorem, Pythagorean, 410, 769
Third-order determinant, 996
Total-value problems, 612
Translate in problem-solving process, 151
Translating
  to algebraic expressions, 30
  to equations, 143, 151
  to inequalities, 180, 658, 659
  in problem solving, 151
Translating for Success, 162, 413, 488, 620, 661, 771, 825, 964
Translations of graphs, 540, 844, 894

Travel of a pipe, 774
Tree, factor, 3
Trial-and-error factoring, 350
Triangle(s)
  area, 33
  right, 410
  similar, 486, 487
Trinomial(s), 277
  factoring, 350–355, 360–364, 368, 369, 377
Trinomial square, 376
  factoring, 377
True equation, 110
True inequality, 168, 652

U

Undefined rational expression, 430
Undefined slope, 232, 554
Union of sets, 672, 673

V

Value
  absolute, 42
  of a determinant, 996
  of an expression, 29
  of a function, 520, 521
  of a polynomial, 274
Value feature on a graphing calculator, 521
Variable, 28
  substituting for, 29
Variation
  constant of, 495, 497
  direct, 495
  inverse, 497
  joint, 500
  other kinds, 498–500

Variation constant, 495, 497
Vertex of a parabola, 843, 851, 854, 879
Vertical line, 219, 220, 559
  slope, 232, 554
Vertical-line test, 523
Viewing window on a graphing calculator, 218
Visualizing for Success, 221, 310, 557, 702, 856, 945

W

Walking speed, 966
Water flow, 768
Whole numbers, 7, 35
Wind chill temperature, 775
Work principle, 484
Work problems, 483

X

x-intercept, 216, 403, 550, 855

Y

y-intercept, 207, 216, 541, 550, 855, 893

Z

Zero
  degree of, 278
  dividends of, 71
  division by, 70
  as an exponent, 253, 257, 268
  identity property of, 7, 47, 79
  multiplication property of, 64
  as a polynomial, 278
  slope, 232, 553, 554
Zero feature on a graphing calculator, 400, 802
Zero products, principle of, 400

Geometric Formulas

PLANE GEOMETRY

Rectangle
Area: $A = l \cdot w$
Perimeter: $P = 2 \cdot l + 2 \cdot w$

Square
Area: $A = s^2$
Perimeter: $P = 4 \cdot s$

Triangle
Area: $A = \frac{1}{2} \cdot b \cdot h$

Sum of Angle Measures
$A + B + C = 180°$

Right Triangle
Pythagorean Theorem:
$a^2 + b^2 = c^2$

Parallelogram
Area: $A = b \cdot h$

Trapezoid
Area: $A = \frac{1}{2} \cdot h \cdot (a + b)$

Circle
Area: $A = \pi \cdot r^2$
Circumference:
$C = \pi \cdot d = 2 \cdot \pi \cdot r$ ($\frac{22}{7}$ and 3.14 are different approximations for π)

SOLID GEOMETRY

Rectangular Solid
Volume: $V = l \cdot w \cdot h$

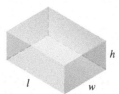

Cube
Volume: $V = s^3$

Right Circular Cylinder
Volume: $V = \pi \cdot r^2 \cdot h$
Surface Area:
$S = 2 \cdot \pi \cdot r \cdot h + 2 \cdot \pi \cdot r^2$

Right Circular Cone
Volume: $V = \frac{1}{3} \cdot \pi \cdot r^2 \cdot h$
Surface Area:
$S = \pi \cdot r^2 + \pi \cdot r \cdot s$

Sphere
Volume: $V = \frac{4}{3} \cdot \pi \cdot r^3$
Surface Area: $S = 4 \cdot \pi \cdot r^2$

Fraction, Decimal, and Percent Equivalents

Fraction Notation	$\frac{1}{10}$	$\frac{1}{8}$	$\frac{1}{6}$	$\frac{1}{5}$	$\frac{1}{4}$	$\frac{3}{10}$	$\frac{1}{3}$	$\frac{3}{8}$	$\frac{2}{5}$	$\frac{1}{2}$	$\frac{3}{5}$	$\frac{5}{8}$	$\frac{2}{3}$	$\frac{7}{10}$	$\frac{3}{4}$	$\frac{4}{5}$	$\frac{5}{6}$	$\frac{7}{8}$	$\frac{9}{10}$	$\frac{1}{1}$
Decimal Notation	0.1	0.125	$0.16\bar{6}$	0.2	0.25	0.3	$0.33\bar{3}$	0.375	0.4	0.5	0.6	0.625	$0.66\bar{6}$	0.7	0.75	0.8	$0.83\bar{3}$	0.875	0.9	1
Percent Notation	10%	12.5% or $12\frac{1}{2}$%	$16.\bar{6}$% or $16\frac{2}{3}$%	20%	25%	30%	$33.\bar{3}$% or $33\frac{1}{3}$%	37.5% or $37\frac{1}{2}$%	40%	50%	60%	62.5% or $62\frac{1}{2}$%	$66.\bar{6}$% or $66\frac{2}{3}$%	70%	75%	80%	$83.\bar{3}$% or $83\frac{1}{3}$%	87.5% or $87\frac{1}{2}$%	90%	100%